Lecture Notes in Computer Science 2400

Edited by G. Goos, J. Hartmanis, and J. van Leeuwen

Springer
Berlin
Heidelberg
New York
Barcelona
Hong Kong
London
Milan
Paris
Tokyo

Burkhard Monien
Rainer Feldmann (Eds.)

Euro-Par 2002 Parallel Processing

8th International Euro-Par Conference
Paderborn, Germany, August 27-30, 2002
Proceedings

Springer

Series Editors

Gerhard Goos, Karlsruhe University, Germany
Juris Hartmanis, Cornell University, NY, USA
Jan van Leeuwen, Utrecht University, The Netherlands

Volume Editors

Burkhard Monien
Rainer Feldmann
Universität Paderborn
Fachbereich 17, Mathematik und Informatik
Fürstenallee 11, 33102 Paderborn
E-mail: {bm/obelix}@upb.de

Cataloging-in-Publication Data applied for

Die Deutsche Bibliothek - CIP-Einheitsaufnahme

Parallel processing : proceedings / Euro-Par 2002, 8th International Euro-Par Conference, Paderborn, Germany, August 27 - 30, 2002. Burkhard Monien ; Rainer Feldmann (ed.). - Berlin ; Heidelberg ; New York ; Barcelona ; Hong Kong ; London ; Milan ; Paris ; Tokyo : Springer, 2002
(Lecture notes in computer science ; Vol. 2400)
ISBN 3-540-44049-6

CR Subject Classification (1998): C.1-4, D.1-4, F.1-3, G.1-2, H2

ISSN 0302-9743
ISBN 3-540-44049-6 Springer-Verlag Berlin Heidelberg New York

Springer-Verlag Berlin Heidelberg New York
a member of BertelsmannSpringer Science+Business Media GmbH

http://www.springer.de

Printed in Germany

Typesetting: Camera-ready by author, data conversion by Steingräber Satztechnik GmbH, Heidelberg
Printed on acid-free paper SPIN: 10873609 06/3142 5 4 3 2 1 0

Preface

Euro-Par – the European Conference on Parallel Computing – is an international conference series dedicated to the promotion and advancement of all aspects of parallel computing. The major themes can be divided into the broad categories of hardware, software, algorithms, and applications for parallel computing. The objective of Euro-Par is to provide a forum within which to promote the development of parallel computing both as an industrial technique and an academic discipline, extending the frontiers of both the state of the art and the state of the practice. This is particularly important at a time when parallel computing is undergoing strong and sustained development and experiencing real industrial take-up. The main audience for and participants in Euro-Par are researchers in academic departments, government laboratories, and industrial organizations. Euro-Par aims to become the primary choice of such professionals for the presentation of new results in their specific areas. Euro-Par is also interested in applications that demonstrate the effectiveness of the main Euro-Par themes.

Euro-Par has its own Internet domain with a permanent website where the history of the conference series is described: `http://www.euro-par.org`. The Euro-Par conference series is sponsored by the Association of Computer Machinery and the International Federation of Information Processing.

Euro-Par 2002 at Paderborn, Germany

Euro-Par 2002 was organized by the Paderborn Center for Parallel Computing (PC^2) and was held at the Heinz Nixdorf MuseumsForum (HNF). PC^2 was founded due to a longlasting concentration on parallel computing at the Department of Computer Science of Paderborn University. It acts as a central research and service center at the university, where research on parallelism is interdisciplinary: groups from the departments of Mathematics and Computer Science, Electrical Engineering, Mechanical Engineering, and Economics are working together on various aspects of parallel computing. The interdisciplinarity is especially visible in SFB 376 (Massively Parallel Computing: Algorithms, Design, Methods, Applications), a large research grant from the German Science Foundation.

HNF includes the largest computer museum in the world, but is also an important conference center. *HNF* unites the classic, historical dimension of a museum with the current functions and future-oriented functions and topics of a forum. Euro-Par 2002 was sponsored by the ACM, IFIP and DFG.

Euro-Par 2002 Statistics

The format of Euro-Par 2002 followed that of the previous conferences and consisted of a number of topics each individually monitored by a committee of four members. There were 16 topics for this year's conference, two of which were included for the first time: Discrete Optimization (Topic 15) and Mobile Computing, Mobile Networks (Topic 16). The call for papers attracted 265 submissions of which 122 were accepted; 67 were presented as regular papers and 55 as research notes. It is worth mentioning that two of the accepted papers were considered to be distinguished papers by the program committee. In total, 990 reports were collected, an average of 3.73 per paper.

Submissions were received from 34 countries (based on the corresponding author's countries), 25 of which were represented at the conference. The principal contributors by country were the USA (19 accepted papers), Spain (16 accepted papers), and then France, Germany, and the UK with 14 accepted papers each.

Acknowledgements

The organization of a large major conference like Euro-Par 2002 is a difficult and time-consuming task for the conference chair and the organizing committee. We are especially grateful to Christian Lengauer, the chair of the Euro-Par steering committee, who gave us the benefit of his experience during the 18 months leading up to the conference. The program committee consisted of 16 topic committees, altogether more than 60 members. They all did a great job and, with the help of more than 400 referees, compiled an excellent academic program.

We owe special thanks to many people in Paderborn: Michael Laska managed the financial aspects of the conference with care. Bernard Bauer, the head of the local organizing team, spent considerable effort to make the conference a success. Jan Hungershöfer was responsible for the webpages of Euro-Par 2002 and the database containing the submissions and accepted papers. He patiently answered thousands of questions and replied to hundreds of emails. Andreas Krawinkel and Holger Nitsche provided us with their technical knowhow. Marion Rohloff and Birgit Farr did a lot of the secretarial work, and Stefan Schamberger carefully checked the final papers for the proceedings. Cornelius Grimm, Oliver Marquardt, Julia Pelster, Achim Streit, Jens-Michael Wierum, and Dorit Wortmann from the Paderborn Center for Parallel Computing spent numerous hours in organizing a professional event.

Last but not least we would like to thank the Heinz-Nixdorf MuseumsForum (HNF) for providing us with a professional environment and hosting most of the Euro-Par 2002 sessions.

June 2002

Burkhard Monien
Rainer Feldmann

Euro-Par Steering Committee

Chair	
Christian Lengauer	University of Passau, Germany
Vice Chair	
Luc Bougé	ENS Cachan, France
European Representatives	
Marco Danelutto	University of Pisa, Italy
Michel Daydé	INP Toulouse, France
Péter Kacsuk	MTA SZTAKI, Hungary
Paul Kelly	Imperial College, UK
Thomas Ludwig	University of Heidelberg, Germany
Luc Moreau	University of Southampton, UK
Rizos Sakellariou	University of Manchester, UK
Henk Sips	Technical University Delft, The Netherlands
Mateo Valero	University Polytechnic of Catalonia, Spain
Non-European Representatives	
Jack Dongarra	University of Tennessee at Knoxville, USA
Shinji Tomita	Kyoto University, Japan
Honorary Members	
Ron Perrott	Queen's University Belfast, UK
Karl Dieter Reinartz	University of Erlangen-Nuremberg, Germany

Euro-Par 2002 Local Organization

Euro-Par 2002 was jointly organized by the Paderborn Center for Parallel Computing and the University of Paderborn.

Conference Chair

Burkhard Monien

Committee

Bernard Bauer	Birgit Farr	Rainer Feldmann
Cornelius Grimm	Jan Hungershöfer	Andreas Krawinkel
Michael Laska	Oliver Marquardt	Holger Nitsche
Julia Pelster	Marion Rohloff	Stefan Schamberger
Achim Streit	Jens-Michael Wierum	Dorit Wortmann

Euro-Par 2002 Program Committee

Topic 1: Support Tools and Environments

Global Chair

Marian Bubak	Institute of Computer Science, AGH and Academic Computer Center CYFRONET Krakow, Poland

Local Chair

Thomas Ludwig	Ruprecht-Karls-Universität, Heidelberg, Germany

Vice Chairs

Peter Sloot	University of Amsterdam, The Netherlands
Rüdiger Esser	Research Center Jülich, Germany

Topic 2: Performance Evaluation, Analysis and Optimization

Global Chair

Barton P. Miller	University of Wisconsin, Madison, USA

Local Chair

Jens Simon	Paderborn Center for Parallel Computing, Germany

Vice Chairs

Jesus Labarta	CEPBA, Barcelona, Spain
Florian Schintke	Konrad-Zuse-Zentrum für Informationstechnik, Berlin, Germany

Topic 3: Scheduling and Load Balancing

Global Chair

Larry Rudolph	Massachusetts Institute of Technology, Cambridge, USA

Local Chair

Denis Trystram	Laboratoire Informatique et Distribution, Montbonnot Saint Martin, France

Vice Chairs

Maciej Drozdowski	Poznan University of Technology, Poland
Ioannis Milis	National Technical University of Athens, Greece

Topic 4: Compilers for High Performance (Compilation and Parallelization Techniques)

Global Chair

Alain Darte	Ecole Normale Supérieure de Lyon, France

Local Chair

Martin Griebl	Universität Passau, Germany

Vice Chairs

Jeanne Ferrante	The University of California, San Diego, USA
Eduard Ayguade	Universitat Politècnica de Catalunya, Barcelona, Spain

Topic 5: Parallel and Distributed Databases, Data Mining and Knowledge Discovery

Global Chair

Lionel Brunie	Institut National de Sciences Appliquées de Lyon, France

Local Chair

Harald Kosch	Universität Klagenfurt, Austria

Vice Chairs

David Skillicorn	Queen's University, Kingston, Canada
Domenico Talia	University of Calabria, Rende, Italy

Topic 6: Complexity Theory and Algorithms

Global Chair

Ernst Mayr	TU München, Germany

Local Chair

Rolf Wanka	Universität Paderborn, Germany

Vice Chairs

Juraj Hromkovic	RWTH Aachen, Germany
Maria Serna	Universitat Politècnica de Catalunya, Barcelona, Spain

Topic 7: Applications of High-Performance Computers

Global Chair

Vipin Kumar	University of Minnesota, USA

Local Chair

Franz-Josef Pfreundt	Institut für Techno- und Wirtschaftsmathematik, Kaiserslautern, Germany

Vice Chairs

Hans Burkhardt	Albert-Ludwigs-Universität, Freiburg, Germany
Jose Laginha Palma	Universidade do Porto, Portugal

Topic 8: Parallel Computer Architecture and Instruction-Level Parallelism

Global Chair

Jean-Luc Gaudiot	University of California, Irvine, USA

Local Chair

Theo Ungerer	Universität Augsburg, Germany

Vice Chairs

Nader Bagherzadeh	University of California, Irvine, USA
Josep L. Larriba-Pey	Universitat Politècnica de Catalunya, Barcelona, Spain

Topic 9: Distributed Systems and Algorithms

Global Chair

Andre Schiper	Ecole Polytechnique Fédérale de Lausanne, Switzerland

Local Chair

Marios Mavronicolas	University of Cyprus, Nicosia, Cyprus

Vice Chairs

Lorenzo Alvisi	University of Texas at Austin, USA
Costas Busch	Rensselaer Polytechnic Institute, Troy, USA

Topic 10: Parallel Programming, Models, Methods and Programming Languages

Global Chair
Kevin Hammond — University of St. Andrews, UK

Local Chair
Michael Philippsen — Universität Karlsruhe, Germany

Vice Chairs
Farhad Arbab — Centrum voor Wiskunde en Informatica (CWI), Amsterdam, The Netherlands
Susanna Pelagatti — University of Pisa, Italy

Topic 11: Numerical Algorithms

Global Chair
Iain Duff — Rutherford Appleton Laboratory, Chilton, UK

Local Chair
Wolfgang Borchers — Universität Erlangen-Nürnberg, Erlangen, Germany

Vice Chairs
Luc Giraud — CERFACS, Toulouse, France
Henk van der Vorst — Utrecht University, The Netherlands

Topic 12: Routing and Communication in Interconnection Networks

Global Chair
Bruce Maggs — Carnegie Mellon University, Pittsburgh, USA

Local Chair
Berthold Vöcking — Max-Planck-Institut für Informatik, Saarbrücken, Germany

Vice Chairs
Michele Flammini — Università di L'Aquila, Italy
Jop Sibeyn — Umeå University, Sweden

Topic 13: Architectures and Algorithms for Multimedia Applications

Global Chair

Andreas Uhl	Universität Salzburg, Austria

Local Chair

Reinhard Lüling	Paderborn, Germany

Vice Chairs

Suchendra M. Bhandarkar	University of Georgia, Athens, USA
Michael Bove	Massachusetts Institute of Technology, Cambridge, USA

Topic 14: Meta- and Grid Computing

Global Chair

Michel Cosnard	INRIA Sophia Antipolis, Sophia Antipolis Cedex, France

Local Chair

Andre Merzky	Konrad-Zuse-Zentrum für Informationstechnik Berlin

Vice Chairs

Ludek Matyska	Masaryk University Brno, Czech Republic
Ronald H. Perrott	Queen's University, Belfast, UK

Topic 15: Discrete Optimization

Global Chair

Catherine Roucairol	Université de Versailles, France

Local Chair

Rainer Feldmann	Universität Paderborn, Germany

Vice Chairs

Laxmikant Kale	University of Urbana-Champaign, USA

Topic 16: Mobile Computing, Mobile Networks

Global Chair

Paul Spirakis	Patras University, Greece

Local Chair

Friedhelm Meyer auf der Heide	Universität Paderborn, Germany

Vice Chairs

Mohan Kumar	University of Texas at Arlington, USA
Sotiris Nikoletseas	Patras University, Greece

Euro-Par 2002 Referees

(not including members of the programme or organization committees)

Alice, Bonhomme
Aluru, Dr. Srinivas
Amestoy, Patrick
Andronikos, Theodore
Angalo, Cosimo
Angel, Eric
Anido, Manuel
Arioli, Mario
Arnold, Dorian
Assmann, Uwe
Atnafu, Solomon
Bagci, Faruk
Baldoni, Roberto
Bal, Henri
Barbuti, Roberto
Beaumont, Olivier
Beauquier, Bruno
Beauquier, Joffroy
Becchetti, Luca
Becker, Jürgen
Benkner, Siegfried
Benkrid, Khaled
Berrendorf, Rudolf
Berthome, Pascal
Bettini, Lorenzo
Bhatia, Karan
Bischof, Holger
Bishop, Benjamin
Blaar, Holger
Blazy, Stephan
Boeres, Cristina
Boufflet, Jean-Paul
Bouras, Christos
Brim, Michael
Brinkschulte, Uwe
Brzezinski, Jerzy
Buck, Bryan
Bull, Mark
Calamoneri, Tiziana
Calder, Brad
Calvin, Christophe
Cannataro, Mario
Cappello, Franck
Casanova, Henri
Cavin, Xavier
Chakravarty, Manuel M.T.
Champagneux, Steeve
Chandra, Surendar
Chaterjee, Mainak
Chatterjee, Siddhartha
Chatzigiannakis, Ioannis
Chaumette, Serge
Chbeir, Richard
Chen, Baoquan
Chin, Kwan-Wu
Choi, Wook
Chrysanthou, Yiorgos
Cicerone, Serafino
Cisternino, Antonio
Clint, Maurice
Codina, Josep M.
Cohen, Albert
Cole, Murray
Coppola, Massimo
Corbal, Jesus
Cortes, Ana
Counilh, Marie-Christine
Crago, Steve
Crainic, Theodor
Cung, Van-Dat
Da Costa, Georges
Danelutto, Marco
Daoudi, El Mostafa
Dasu, Aravind
Datta, Ajoy
Dayde, Michel
Dearle, Al
De Bosschere, Koen
Decker, Thomas
Defago, Xavier
Derby, Dr. Jeffrey
De Sande, Francisco

Desprez, Frederic
de Supinski, Bronis
Deutsch, Andreas
Dhillon, Inderjit
Diaz Bruguera, Javier
Diaz, Luiz
Di Ianni, Miriam
Ding, Yonghua
Di Stefano, Gabriele
Döller, Mario
du Bois, Andre
Ducourthial, Bertrand
Duesterwald, Evelyn
Du, Haitao
Dupont de Dinechin, Florent
Dutot, Pierre
Ecker, Klaus
Egan, Colin
Eilertson, Eric
El-Naffar, Said
Ercal, Dr. Fikret
Eyraud, Lionel
Faber, Peter
Fahle, Torsten
Falcon, Ayose
Farrens, Matthew
Feig, Ephraim
Felber, Pascal
Feldbusch, Fridtjof
Feldmann, Anja
Feo, John
Fernández, Agustín
Ferrari, GianLuigi
Fink, Steve
Fischer, Matthias
Flocchini, Paola
Ford, Rupert
Fraguela, Basilio
Fraigniaud, Pierre
Franke, Hubertus
Franke, Klaus
Frommer, Andreas
Furfaro, Filippo
Furnari, Mario
Galambos, Gabor
Garofalakis, John
Gavoille, Cyril
Gawiejnowicz, Stanislaw
Gendron, Bernard
Gerndt, Michael
Getov, Vladimir
Gibert, Enric
Gimbel, Matthias
Glendinning, Ian
Gorlatch, Sergei
Gratton, Serge
Grothklags, Sven
Guerrini, Stefano
Guillen Scholten, Juan
Guinand, Frederic
Gupta, Sandeep
Hains, Gaétan
Hanen, Claire
Harmer, Terry
Hasan, Anwar
Haumacher, Bernhard
Hegland, Markus
Hellwagner, Hermann
Herzner, Wolfgang
Hladka, Eva
Hogstedt, Karin
Holder, Lawrence
Huard, Guillaume
Hunt, James
Hu, Zhenjiang
Ikonomou, Giorgos
Irigoin, Francois
Jackson, Yin
Jacobs, Josh
Jacquet, Jean-Marie
Jain, Prabhat
Jarraya, Mohamed
Jeannot, Emmanuel
Jeudy, Baptiste
Jiménez, Daniel
Jung, Eunjin
Kaeli, David
Kalyanaraman, Anantharaman
Kanapady, Ramdev
Kang, Jung-Yup

Karl, Wolfgang
Kavi, Krishna
Keller, Jörg
Kelly, Paul
Kielmann, Thilo
Kistler, Mike
Klasing, Ralf
Klein, Peter
Kliewer, Georg
Kluthe, Ralf
Kofler, Andrea
Kokku, Ravindranath
Kothari, Suresh
Kraemer, Eileen
Krzhizhanovskaya, Valeria
Kshemkalyani, Ajay
Kubota, Toshiro
Kuchen, Herbert
Kurc, Wieslaw
Kwok, Ricky Y. K.
Kyas, Marcel
Laforenza, Domenico
Lanteri, Stephane
Laszlo, Boeszoermenyi
Lavenier, Dominique
Le cun, Bertrand
Lee, Jack Y. B.
Lee, Pei-Zong
Lee, Ruby
Lee, Seong-Won
Lee, Walter
Legrand, Arnaud
Lengauer, Christian
Leonardi, Stefano
L'Excellent, Jean-Yves
Libsie, Mulugeta
Lilja, David
Litow, Bruce
Li, Xiang-Yang
Li, X. Sherry
Loechner, Vincent
Loidl, Hans-Wolfgang
Lojewski, Carsten
Loogen, Rita
Lo Presti, Francesco
Loriot, Mark
Lowekamp, Bruce
Lowenthal, David
Löwe, Welf
Maamir, Allaoua
Machowiak, Maciej
Mahjoub, Zaher
Mahmoud, Qusay H.
Maier, Robert
Manco, Giuseppe
Mangione-Smith, Bill
Marcuello, Pedro
Marin, Mauricio
Marlow, Simon
Martin, Jean-Philippe
Martin, Patrick
Martorell, Xavier
Mastroianni, Carlo
Matsuo, Yataka
Mc Cracken, Michael
McQuesten, Paul
Melideo, Giovanna
Michaelson, Greg
Mirgorodskii, Alexandre
Mohr, Bernd
Monfroy, Eric
Monteil, Thierry
Montresor, Alberto
Morajko, Ania
Morin, Christine
Mounie, Gregory
Muller, Jens-Dominik
Müller, Matthias
Müller-Schloer, Christian
Nagel, Wolfgang E.
Nandy, Sagnik
Napper, Jeff
Naroska, Edwin
Naylor, Bruce
Nickel, Stefan
Niktash, Afshin
Nishimura, Satoshi
Noelle, Michael
Noguera, Juanjo
Nölle, Michael

O'Boyle, Mike
O'Donnell, John
Olaru, Vlad
Oliveira, Rui
Ortega, Daniel
Paar, Alex
Padua, David
Pan, Chengzhi
Papadopoulos, George
Papadopoulos, George
Parcerisa, Joan Manuel
Parizi, Hooman
Parmentier, Gilles
Pawlak, Grzegorz
Perego, Raffaele
Perez, Christian
Peserico, Enoch
Petitet, Antoine
Petkovic, Dejan
Petzold, Jan
Pfeffer, Matthias
Picouleau, Christophe
Pierik, Cees
Pietracaprina, Andrea
Pinotti, Cristina
Pinotti, Maria Cristina
Pitoura, Evaggelia
Pizzuti, Clara
Plaks, Toomas
Portante, Peter
Pottenger, Bill
Prasanna, Viktor
Preis, Robert
Pucci, Geppino
Quinson, Martin
Quison, Martin
Rabhi, Fethi
Raffin, Bruno
Rajopadhye, Sanjay
Ramirez, Alex
Rana, Omer
Rauchwerger, Lawrence
Rauhut, Markus
Rehm, Wolfgang
Reinman, Glenn
Rescigno, Adele
Retalis, Symeon
Reuter, Jürgen
Richard, Olivier
Riveill, Michel
Robert, Yves
Robic, Borut
Röblitz, Thomas
Roesch, Ronald
Romagnoli, Emmanuel
Roth, Philip
Ro, Wonwoo
Rus, Silvius
Sanchez, Jesus
Sanders, Peter
Schaeffer, Jonathan
Schiller, Jochen
Schmidt, Bertil
Schmidt, Heiko
Scholtyssik, Karsten
Schroeder, Ulf-Peter
Schulz, Martin
Schütt, Thorsten
Scott, Stan
Sellmann, Meinolf
Senar, Miquel
Sendag, Resit
Seznec, André
Shan, Hongzhang
Shankland, Carron
Shao, Gary
Siebert, Fridtjof
Siemers, Christian
Silc, Jurij
Singhal, Mukesh
Sips, Henk
Smith, James
Snaveley, Allan
Soffa, Mary Lou
Spezzano, Giandomenico
Stenström, Per
Sterna, Malgorzata
Stewart, Alan
Stoyanov, Dimiter
Stricker, Thomas

Striegnitz, Joerg
Strout, Michelle
Suh, Edward
Sung, Byung
Surapaneni, Srikanth
Tabrizi, Nozar
Taillard, Eric
Tantau, Till
Theobald, Kevin
Thiele, Lothar
Torrellas, Josep
Torrellas, Josep
Torres, Jordi
Triantafilloy, Peter
Trichina, Elena
Trinder, Phil
Tseng, Chau-Wen
Tubella, Jordi
Tullsen, Dean
Tuma, Miroslav
Tuminaro, Ray
Turgut, Damla
Uhrig, Sascha
Unger, Andreas
Unger, Walter
Utard, Gil
Valero, Mateo
Vandierendonck, Hans
van Reeuwijk, Kees
Varvarigos, Manos
Venkataramani, Arun
Verdoscia, Lorenzo
Vintan, Lucian
Vivien, Frederic
Vocca, Paola
Vömel, Christof
Walkowiak, Rafal
Walshaw, Chris
Walter, Andy
Watson, Paul
Wolf, Felix
Wolf, Wayne
Wolniewicz, Pawel
Wonnacott, David
Wood, Alan
Worsch, Thomas
Xi, Jing
Xue, Jingling
Yalagandula, Praveen
Yi, Joshua
Zaki, Mohammed
Zaks, Shmuel
Zalamea, Javier
Zandy, Victor
Zehendner, Eberhard
Zhou, Xiaobo
Zhu, Qiang
Zimmermann, Wolf
Zissimopoulos, Vassilios
Zoeteweij, Peter

Table of Contents

Invited Talks

Topic 1

Topic 2

Topic 3

Topic 4

Topic 5

Topic 6

Topic 7

Topic 8

Topic 9

Topic 10

Topic 11

Topic 12

Topic 13

Topic 14

Topic 15

Topic 16

Orchestrating Computations on the World-Wide Web

Young-ri Choi, Amit Garg, Siddhartha Rai, Jayadev Misra, and Harrick Vin

Department of Computer Science
The University of Texas at Austin
Austin, Texas 78712
{yrchoi, amitji, sid, misra,vin}@cs.utexas.edu

Abstract. Word processing software, email, and spreadsheet have revolutionized office activities. There are many other office tasks that are amenable to automation, such as: scheduling a visit by an external visitor, arranging a meeting, and handling student application and admission to a university. Many business applications —protocol for filling an order from a customer, for instance— have similar structure. These seemingly trivial examples embody the computational patterns that are inherent in a large number of applications, of coordinating tasks at different machines. Each of these applications typically includes invoking remote objects, calculating with the values obtained, and communicating the results to other applications. This domain is far less understood than building a function library for spreadsheet applications, because of the inherent concurrency.

We address the task coordination problem by (1) limiting the model of computation to *tree structured concurrency*, and (2) assuming that there is an environment that supports access to remote objects. The environment consists of distributed objects and it provides facilities for remote method invocation, persistent storage, and computation using standard function library. Then the task coordination problem may be viewed as orchestrating a computation by invoking the appropriate methods in proper sequence. Tree structured concurrency permits only restricted communications among the processes: a process may spawn children processes and all communications are between parents and their children. Such structured communications, though less powerful than interactions in process networks, are sufficient to solve many problems of interest, and they avoid many of the problems associated with general concurrency.

1 Introduction

1.1 Motivation

Word processing software, email, and spreadsheet have revolutionized home and office computing. Spreadsheets, in particular, have made effective programmers in a limited domain out of non-programmers. There are many other office tasks that are amenable to automation. Simple examples include scheduling a visit

B. Monien and R. Feldmann (Eds.): Euro-Par 2002, LNCS 2400, pp. 1–20.

by an external visitor, arranging a meeting, and handling student application and admission to a university. Many business applications —protocol for filling an order from a customer, for instance— have similar structure. In fact, these seemingly trivial examples embody the computational patterns that are inherent in a large number of applications. Each of these applications typically includes invoking remote objects, applying certain calculations to the values obtained, and communicating the results to other applications. Today, most of these tasks are done manually by using proprietary software, or a general-purpose software package; the last option allows little room for customization to accommodate the specific needs of an organization.

The reason why spreadsheets have succeeded and general task coordination software have not has to do with the problem domains they address. The former is limited to choosing a set of functions from a library and displaying the results in a pleasing form. The latter requires invocations of remote objects and coordinations of concurrent tasks, which are far less understood than building a function library. Only now are software packages being made available for smooth access to remote objects. Concurrency is still a hard problem; it introduces a number of subtle issues that are beyond the capabilities of most programmers.

1.2 Current Approaches

The computational structure underlying typical distributed applications is *process network*. Here, each process resides at some node of a network, and it communicates with other processes through messages. A computation typically starts at one process, which may spawn new processes at different sites (which, in turn, may spawn other processes). Processes are allowed to communicate in unconstrained manner with each other, usually through asynchronous message passing.

The process network model is the design paradigm for most operating systems and network-based services. This structure maps nicely to the underlying hardware structure, of LANs, WANs, and, even, single processors on which the processes are executed on the basis of time slices. In short, the process network model is powerful.

We contend that the process network model is too powerful, because many applications tend to be far more constrained in their communication patterns. Such applications rarely exploit the facility of communicating with arbitrary processes. Therefore, when these applications are designed under the general model of process networks, they have to pay the price of power: since a process network is inherently concurrent, many subtle aspects of concurrency —synchronization, coherence of data, and avoidance of deadlock and livelock— have to be incorporated into the solution. Additionally, hardware and software failure and recovery are major considerations in such designs.

There have been several theoretical models that distill the essence of process network style of computing. In particular, the models in CSP [9], CCS [15] and π-calculus [16] encode process network computations using a small number of structuring operators. The operators that are chosen have counterparts in the real-world applications, and also pleasing algebraic properties. In spite of the simplicities of the operators the task of ensuring that a program is deadlock-free, for instance, still falls on the programmer; interactions among the components in a process network have to be considered explicitly.

Transaction processing is one of the most successful forms of distributed computing. There is an elaborate theory —see Gray and Reuter [8]— and issues in transaction processing have led to major developments in distributed computing. For instance, locking, commit and recovery protocols are now central to distributed computing. However, coding of transactions remains a difficult task. Any transaction can be coded using remote procedure call (or RMI in Java). But the complexity is beyond the capabilities of most ordinary programmers, for the reasons cited above.

1.3 Our Proposal

We see three major components in the design of distributed applications: (1) persistent storage management, (2) computational logic and execution environment, and (3) methods for orchestrating computations. Recent developments in industry and academia have addressed the points (1) and (2), persistent storage management and distributed execution of computational tasks (see the last paragraph of this subsection). This project builds on these efforts. We address the point (3) by viewing the task coordination problem as orchestration of multiple computational tasks, possibly at different sites. We design a programming model in which the orchestration of the tasks can be specified. The orchestration script specifies *what* computations to perform and *when*, but provides no information on *how* to perform the computations.

We limit the model of computation for the task coordination problem to *tree structured concurrency*. For many applications, the structure of the computation can be depicted as a tree, where each process spawns a number of processes, sends them certain queries, and then receives their responses. These steps are repeated until a process has acquired all needed information to compute the desired result. Each spawned process behaves in exactly the same fashion, and it sends the computed result as a response only to its parent, but it *does not accept unsolicited messages* during its execution. Tree structured concurrency permits only restricted communications, between parents and their children. We exploit this simplicity, and develop a programming model that avoids many of the problems of general distributed applications. We expect that the simplicity of the model will make it possible to develop tools which non-experts can use to specify their scripts.

There has been much work lately in developing solutions for expressing application logic, see, for instance, the .NET infrastructure[13], IBM's WebSphere Application Server [10], and CORBA [6], which provide platforms that distributed applications can exploit. Further, such a platform can be integrated with persistent store managers, such as SQL server [14]. The XML standard [7] will greatly simplify parameter passing by using standardized interfaces. The specification of sequential computation is a well-understood activity (though, by no means, completely solved). An imperative or functional style of programming can express the computational logic. Thus, much of distributed application design reduces to the task coordination problem, the subject matter of this paper.

2 A Motivating Example

To illustrate some aspects of our programming model, we consider a very small, though realistic, example. The problem is for an office assistant in a university department to contact a potential visitor; the visitor responds by sending the date of her visit. Upon hearing from the visitor, the assistant books an airline ticket and contacts two hotels for reservation. After hearing from the airline and any one of the hotels he informs the visitor about the airline and the hotel. The visitor sends a confirmation which the assistant notes. The office assistant's job can be mostly automated. In fact, since the office assistant is a domain expert, he should be able to program this application quite easily given the proper tools.

This example involves a tree-structured computation; the root initiates the computation by sending an email to the visitor, and each process initiates a tree-structured computation that terminates only when it sends a response to its parent. This example also illustrates three major components in the design of distributed applications: (1) persistent storage management, as in the databases maintained by the airline and the hotels, (2) specification of sequential computational logic, which will be needed if the department has to compute the sum of the air fare and hotel charges (in order to approve the expenditure), and (3) methods for orchestrating the computations, as in, the visitor can be contacted for a second time only *after* hearing from the airline and one of the hotels. We show a solution below.

---

task $visit(message :: m, name :: v)\ confirmation$

$;true \quad \rightarrow \quad \boldsymbol{\alpha} : email(m, v)$

$\boldsymbol{\alpha}(date) \quad \rightarrow \quad \boldsymbol{\beta} : airline(date);\ \boldsymbol{\gamma 1} : hotel1(date);\ \boldsymbol{\gamma 2} : hotel2(date)$

$\boldsymbol{\beta}(c1) \wedge (\boldsymbol{\gamma 1}(c2) \vee \boldsymbol{\gamma 2}(c2)) \quad \rightarrow \quad \boldsymbol{\epsilon} : email(\langle c1, c2 \rangle, v)$

$\boldsymbol{\epsilon}(x) \quad \rightarrow \quad x$

end

A *task* is the unit of an orchestration script. It resembles a procedure in that it has input and output parameters. The task *visit* has two parameters, a message m and the name of the visitor, v. It returns a value of type *confirmation*.

On being called, a task executes its constituent actions (which are written as guarded commands) in a manner prescribed in section 3. For the moment, note that an action is executed only when its guard holds, actions are chosen non-deterministically for execution, and no action is executed more than once.

In this example, *visit* has four actions, only the first of which can be executed when the task is called (the guard of the first action evaluates to *true*). The effect of execution of that action is to call another task, *email*, with message m and name v as parameters; the call is identified with a *tag*, α (the tags are shown in bold font in this program). The second action becomes ready to be executed only after a response is received corresponding to the call with tag α. The response carries a parameter called $date$, and the action invokes an $airline$ task and two tasks corresponding to reservations in two different hotels. The next action can be executed only after receiving a response from the airline and response from at least one hotel (response parameters from both hotels are labeled $c2$). Then, an email is sent to v with parameters $c1$ and $c2$. In the last action, her confirmation is returned to the caller of $visit$, and the task execution then terminates.

The task shown here is quite primitive; it assumes perfect responses in all cases. If certain responses, say, from the airline are never received, the execution of the task will never terminate. We discuss issues such as time-out in this paper; we are currently incorporating interrupt (human intervention) into the programming model.

A task, thus, can initiate a computation by calling other tasks (and objects) which may reside at different sites, and transferring parameters among them. A task has no computational ability beyond applying a few standard functions on the parameters. All it can do is sequence the calls on a set of tasks, transfer parameters among them, and then return a result.

3 Programming Model

The main construct of our programming model is a *task*. A task consists of a set of *actions*. Each action has a *guard* and a *command* part. The guard specifies the condition under which the action can be executed, and the command part specifies the requests to be sent to other tasks and/or the response to be sent to the parent. A guard names the specific children from whom the responses have to be received, the structure of each response —an integer, tuple or list, for instance— and any condition that the responses must satisfy, e.g., the hotel's rate must be below \$150 a night. The command part may use the parameters named in the guard. The syntax for tasks is defined in section 3.1.

Each action is executed at most once. A task terminates when it sends a response to its parent.

A guard has three possible values: $\perp$, *true* or *false*. An important property of a guard is that its value is monotonic; the value does not change once it is *true* or *false*. The structure of the guard and its evaluation are of central importance in our work. Therefore, we treat this topic in some detail in section 3.2.

Recursion and the list data structure have proved to be essential in writing many applications. We discuss these constructs in section 3.3.

3.1 Task

A task has two parts, a *header* and a *body*. The header names the task, its formal parameters and their types, and the type of the response. For example,

$$\textbf{task}\ \ visit(message :: m, name :: v)\ confirmation$$

describes a task with name *visit* that has two arguments, of type *message* and *name*, and that responds with a value of type *confirmation*.

The body of a task consists of a set of *actions*. Each action has two parts, a *guard* and a *command*, which are separated by the symbol $\rightarrow$.

When a task is called it is instantiated. Its actions are then executed in arbitrary order according to the following rules: (1) an action is executed only if its guard is *true*, (2) an action is executed at most once, and (3) the task terminates (i.e., its actions are no longer executed) once it sends a response to its caller. A response sent to a terminated task —a *dangling response*— is discarded.

Example (Non-determinism): Send message m to both e and f. After a response is received from any one of them, send the name of the responder to the caller of this task.

---

task $choose(message :: m, name :: e, name :: f)\ name$
  $;true \quad \rightarrow \quad \alpha : email(m, e);\ \beta : email(m, f)$
  $\alpha(x) \quad \rightarrow \quad x$
  $\beta(x) \quad \rightarrow \quad x$
end

A slightly simpler solution is to replace the last two actions with

$$\alpha(x) \vee \beta(x) \rightarrow x$$

Command. The command portion of an action consists of zero or more *requests* followed by an optional *response*. There is no order among the requests.

A request is of the form

$$tag : name(arguments)$$

where *tag* is a unique identifier, *name* is a task name and *arguments* is a list of actual parameters, which are expressions over the variables appearing in the guard (see section 3.2).

A response in the command part is differentiated from a request by not having an associated tag. A response is either an expression or a call on another task. In the first case, the value of the expression is returned to the caller. In the second case, the call appears without a tag, and the response from the called task, if any, is returned to the caller.

An example of a command part that has two requests and a response x is,

$$\alpha : send(e);\ \beta : send(f);\ x$$

Tag. A tag is a variable that is used to label a request and it stores the response, if any, received from the corresponding task. A tag is used in a guard to bind the values received in a response to certain variables, which can then be tested (in the predicate part of the guard) or used as parameters in task calls in the command part. For instance, if tag α appears as follows in a guard

$$\alpha(-, \langle x, y \rangle, b : bs)$$

it denotes that α is a triple, its second component is a tuple where the tuple components are bound to x and y, and the last component of α is a list whose head is bound to b and tail to bs.

Guard. A guard has two parts, *response* and *predicate*. Each part is optional.

guard ::= [response] ; [predicate]
response ::= conjunctive-response
conjunctive-response ::= disjunctive-response {$\wedge$ (disjunctive-response)}
disjunctive-response ::= simple-response {$\vee$ (simple-response)}
simple-response ::= positive-response | negative-response
positive-response ::= [qualifier] tag [(parameters)]
negative-response ::= $\neg$[qualifier] tag(timeout-value)
qualifier ::= *full.* | *nonempty.*
parameters ::= parameter {, parameter}
parameter ::= variable | constant

Response. A response is in conjunctive normal form: it is a conjunction of disjunctive-responses. A disjunctive-response is a disjunction of simple-responses, each of which is either a tag, optionally with parameters, or negation of a tag with a timeout-value. The qualifier construct is discussed in page 11. Shown below are several possible responses.

$\alpha(x)$
$\alpha(x) \ \wedge \ \beta(y)$
$\alpha(x) \ \vee \ \beta(x)$
$\neg\alpha(10\text{ms})$
$\neg\beta(5\text{ms}) \ \wedge \ (\gamma(y) \ \vee \ \delta(y))$

The following restrictions apply to the parameters in a response: (1) all simple responses within a disjunctive-response have the same set of variable parameters, and (2) variable parameters in different disjunctive-responses are disjoint. A consequence of requirement (1) is that a disjunctive-response defines a set of parameters which can be assigned values if any disjunct (simple-response) is true. If a negative-response appears within a disjunctive-response then there is no variable parameter in that disjunctive-response. This is illustrated below; in the last example *Nack* is a constant.

$\neg\alpha(10\text{ms}) \ \vee \ \neg\beta(5\text{ms})$
$\alpha \ \vee \ \neg\beta(5\text{ms})$
$\neg\alpha(10\text{ms}) \ \vee \ \alpha(Nack)$

Predicate. A predicate is a boolean expression over parameters from the response part, and, possibly, constants. Here are some examples of guards which include both responses and predicates.

$\alpha(x);\ 0 \leq x \leq 10$
$\alpha(x)\ \wedge\ \neg\beta(5\text{ms})\ \wedge\ (\gamma(y)\ \vee\ \delta(y));\ x > y$

If a guard has no response part, it has no parameters. So the predicate can only be a constant; the only meaningful constant is *true*. Such a guard can be used to guarantee eventual execution of its command part.

We conclude this subsection with an example to schedule a meeting among A, B and C. Each of A, B and C is an object which has a calendar. Method *lock* in each object locks the corresponding calendar and returns the calendar as its response. *Meet* is a function, defined elsewhere, that computes the meeting time from the given calendars. Method *set* in each object updates its calendar by reserving at the given time; it then unlocks the calendar. The meeting time is returned as the response of *schedule*.

---

task $schedule(object :: A, object :: B, object :: C)\ Time$
$;true\ \rightarrow\ \ \alpha 1 : A.lock;\ \beta 1 : B.lock;\ \gamma 1 : C.lock$
$\alpha 1(Acal)\ \wedge\ \beta 1(Bcal)\ \wedge\ \gamma 1(Ccal)\ \rightarrow$
$\alpha 2 : A.set(t);\ \beta 2 : B.set(t);\ \gamma 2 : C.set(t);\ t$
where $t = Meet(Acal, Bcal, Ccal)$

end

What happens in this example if some process never responds? Other processes then will have permanently locked calendars. So, they must use time-outs. The task has to employ something like a 3-phase commit protocol [8] to overcome these problems.

3.2 Evaluation of Guard

A guard has three possible values, $\perp$, *true* or *false*. It is evaluated by first evaluating its response part, which could be $\perp$, *true* or *false*. The guard is $\perp$ if the response part is $\perp$ and *false* if the response is *false*. If the response is *true* then the variable parameters in the response part are bound to values in the standard way, and the predicate part —which is a boolean expression over variable parameters— is evaluated. The value of the guard is then the value of the predicate part.

An empty response part is taken to be *true*. The evaluation of a response follows the standard rules. A disjunctive-response is *true* if any constituent simple-response is *true*; in that case its variable parameters are bound to the values of any constituent simple-response that is *true*. A disjunctive-response is *false* if all constituent simple-responses are *false*, and it is $\perp$ if all constituent simple-responses are either *false* or $\perp$ and at least one is $\perp$. A conjunctive response is evaluated in a dual manner.

The only point that needs some explanation is evaluation of a negative-response, $\neg\beta(t)$, corresponding to a time-out waiting for the response from β. The response $\neg\beta(t)$ is (1) *false* if the request with tag β has responded within t units of the request, (2) *true* if the request with tag β has not responded within t units of the request, and (3) $\perp$ otherwise (i.e., t units have not elapsed since the request was made and no response has been received yet).

Monotonicity of Guards. A guard is *monotonic* if its value does not change once it is *true* or *false*; i.e., the only possible change of value of a monotonic guard is from $\perp$ to *true* or $\perp$ to *false*. In the programming model described so far, all guards are monotonic. This is an important property that is exploited in the implementation, in terminating a task even before it sends a response, as follows. If the guard values in a task are either *true* or *false* (i.e., no guard evaluates to $\perp$), and all actions with *true* guards have been executed, then the task can be terminated. This is because no action can be executed in the future since all *false* guards will remain *false*, from monotonicity.

3.3 Recursion and Lists

Recursion. The rule of task execution permits each action to be executed at most once. While this rule simplifies program design and reasoning about programs, it implies that the number of steps in a task's execution is bounded by the number of actions. This is a severe limitation which we overcome using recursion. A small example is shown below.

It is required to send messages to e at 10s intervals until it responds. The exact response from e and the response to be sent to the caller of *bombard* are of no importance; we use () for both.

---

task $bombard(message :: m, name :: e)\ ()$
  $;true \rightarrow \alpha : email(m, e)$
  $\alpha \rightarrow ()$
  $\neg\alpha(10\text{s}) \rightarrow bombard(m, e)$
end

In this example, each invocation of *bombard* creates a new instance of the task, and the response from the last instance is sent to the original invoker of *bombard*.

List Data Structure. To store the results of unbounded computations, we introduce *list* as a data structure, and we show next how lists are integrated into our programming model.

Lists can be passed as parameters and their components can be bound to variables by using pattern matching, as shown in the following example. It is required to send requests to the names in a list, f, sequentially, then wait for a day to receive a response before sending a request to the next name in the list. Respond with the name of the first responder; respond with *Nack* if there is no responder.

---

task $hire([name] :: f)\ (Nack \mid Ack\ name)$

$$
\begin{array}{ll}
f([]) & \rightarrow Nack \\
f(x:-) & \rightarrow \alpha : send(x) \\
\alpha(y) & \rightarrow Ack(y) \\
\neg\alpha(\text{1day}) \wedge f(-:xs) & \rightarrow hire(xs)
\end{array}
$$

end

Evolving Tags. Let tsk be a task that has a formal parameter of type t,

task $tsk(t ::\ x)$

We adopt the convention that tsk may be called with a *list* of actual parameters of type t; then tsk is invoked independently for each element of the list. For example,

$$\alpha : tsk(xs)$$

where xs is a list of elements of type t creates and invokes as many instances of tsk as there are elements in xs; if xs is empty, no instances are created and the request is treated as a *skip*.

Tag α is called an *evolving tag* in the example above. An evolving tag's value is the list of responses received, ordered in the same sequence as the list of requests. Unlike a regular tag, an evolving tag always has a value, possibly an empty list. Immediately following the request, an evolving tag value is an empty list. For the request $\alpha : tsk([1,2,3])$ if response r_1 for $tsk(1)$ and r_3 for $tsk(3)$ have been received then $\alpha = [r_1, r_3]$. Given the request $\alpha : tsk(xs)$, where xs is an empty list, α remains the empty list forever.

If a task has several parameters each of them may be replaced by a list in an invocation. For instance, let **task** $tsk(t ::\ x,\ s ::\ y)$ have two parameters. Given

$$\alpha : tsk(xs, ys)$$

where xs and ys are both lists of elements, tsk is invoked for each pair of elements from the cartesian product of xs and ys. Thus, if

$$xs = [1,2,3] \quad ys = [A,B]$$

the following calls to tsk will be made:

$$tsk(1,A)\ \ tsk(1,B)\ \ tsk(2,A)\ \ tsk(2,B)\ \ tsk(3,A)\ \ tsk(3,B)$$

We allow only one level of coercion; tsk cannot be called with a list of lists.

Qualifier for Evolving Tag. For an evolving tag α, $full.\alpha$ denotes that corresponding to the request of which α is the tag all responses have been received, and $nonempty.\alpha$ denotes that some response has been received. If the request corresponding to α is empty then $full.\alpha$ holds immediately and $nonempty.\alpha$ remains *false* forever. An evolving tag has to be preceded by a qualifier, *full* or *nonempty*, when it appears in the response part of a guard.

Examples of Evolving Tags. Suppose we are given a list of names, *namelist*, to which messages have to be sent, and the name of *any* respondent is to be returned as the response.

---

task $choose(message :: m, [name] :: namelist)\ name$
  $;true \rightarrow \alpha : send(m, namelist)$
  $nonempty.\alpha(x : -) \rightarrow x$
end

A variation of this problem is to respond with the list of respondents after receiving a majority of responses, as would be useful in arranging a meeting. In the second action, below, $|\alpha|$ denotes the (current) length of α.

---

task $rsvpMajority([name] :: namelist)\ [name]$
  $;true \rightarrow \alpha : email(namelist)$
  $;2 \times |\alpha| \geq |namelist| \rightarrow \alpha$
end

A much harder problem is to compute the *transitive closure*. Suppose that each person in a group has a list of friends. Given a (sorted) list of names, it is required to compute the transitively-closed list of friends. The following program queries each name and receives a list of names (that includes the queried name). Function *merge*, defined elsewhere, accepts a list of name lists and creates a single sorted list by taking their union.

---

task $tc([name] :: f)\ [name]$
  $;true \rightarrow \alpha : send(f)$
  $full.\alpha;\ f = \beta \rightarrow f$, where $\beta = merge(\alpha)$
  $full.\alpha;\ f \neq \beta \rightarrow tc(\beta)$, where $\beta = merge(\alpha)$
end

Note that the solution is correct for $f = []$.

Evaluation of Guards with Evolving Tags. An evolving tag appears with a qualifier, *full* or *nonempty*, in the response part of a guard. We have already described how a tag with a qualifier is evaluated. We describe next how time-outs with an evolving tag are evaluated. Receiving *some* response within t units of the request makes $\neg nonempty.\alpha(t)$ *false*, receiving no response within t units of the request makes it *true*, and it is $\perp$ otherwise. Receiving *all* responses within t units of the request makes $\neg full.\alpha(t)$ *false*, not receiving any one response within t units of the request makes it *true*, and it is $\perp$ otherwise.

Monotonicity of Guards with Evolving Tags. A guard with evolving tag may not be monotonic. For instance, if its predicate part is of the form $|\alpha| < 5$ where α is an evolving tag. It is the programmer's responsibility to ensure that every guard is monotonic.

3.4 An Example

We consider a more realistic example in this section, of managing the visit of a faculty candidate to a university department. A portion of the workflow is shown schematically in Figure 1. In what follows, we describe the workflow and model it using Orc.

Here is the problem: An office assistant in a university department must manage the logistics of a candidate's visit. She emails the candidate and asks for the following information: dates of visit, desired mode of transportation and research interest. If the candidate prefers to travel by air, the assistant purchases an appropriate airline ticket. She also books a hotel room for the duration of the stay, makes arrangements for lunch and reserves an auditorium for the candidate's talk. She informs the students and faculty about the talk, and reminds them again on the day of the talk. She also arranges a meeting between the candidate and the faculty members who share research interests. After all these steps have been taken, the final schedule is communicated to the candidate and the faculty members.

The following orchestration script formalizes the workflow described above. It is incomplete in that not all actions are shown.

---

task $FacultyCandidateRecruit(String :: candidate, [String] :: faculty,$
$[String] :: student, [String] :: dates,$
$[String] :: transportation,$
$[String] :: interests)\ String$

$;true \quad \rightarrow \quad A : AskUserData(candidate, dates);$
$B : AskUserData(candidate, transportation);$
$C : AskUserData(candidate, interests)$

/* If the candidate prefers to fly, then reserve a seat./
$B(x) \wedge A(y);\ x = \text{``plane''} \rightarrow D : ReserveSeat(y, candidate)$

/* Reserve a hotel room, a lunch table and an auditorium. */
$A(x) \rightarrow E : ReserveHotelRoom(x);$
$F : ReserveAuditorium(x);$
$G : ReserveLunchTable(x)$

/* Arrange a meeting with faculty. */
$C(x) \rightarrow H : [AskUserInterest(l, x) \mid l \leftarrow faculty]$
/* The notation above is for list comprehension */
$H(x) \wedge A(y) \rightarrow I : FindAvailableTime(x, y)$

/* If the auditorum is reserved successfully */
$F(x);\ x \neq \text{``''} \rightarrow J : Inform(x, \text{``}Talk\ Schedule\text{''}, faculty);$
$K : Inform(x, \text{``}Talk\ Schedule\text{''}, student)$
$F(x) \wedge J(y) \rightarrow L : Reminder(x, \text{``}Talk\ Schedule\text{''}, faculty)$
$F(x) \wedge K(y) \rightarrow M : Reminder(x, \text{``}Talk\ Schedule\text{''}, student)$

/* Notify faculty and students about the schedule. */
$H(x) \wedge I(y) \rightarrow N : [Notify(l, y) \mid l \leftarrow x]$
$D(x);\ x \neq \text{``''} \rightarrow O : Notify(candidate, x)$
$F(y) \wedge I(z);\ y \neq \text{``''} \rightarrow P : NotifySchedule(candidate, y, z)$
$L(x) \wedge M(y) \rightarrow \text{``}Done\text{''}$

$D(x);\ x = \text{``''} \rightarrow ErrorMsg(\text{``}assistant@cs\text{''}, \text{``}No\ available\ flight\text{''})$
$F(x);\ x = \text{``''} \rightarrow$
$ErrorMsg(\text{``}assistant@cs\text{''}, \text{``}Auditorium\ reservation\ failed\text{''})$
$\neg E(86400) \rightarrow$
$ErrorMsg(\text{``}assistant@cs\text{''}, \text{``}Hotel\ reservation\ failed\text{''})$

end

3.5 Remarks on the Programming Model

What a Task Is Not. A task resembles a function in not having a state. However, a task is not a function because of non-determinism. A task resembles a transaction, though it is simpler than a transaction in not having a state or imperative control structures. A task resembles a procedure in the sense that it is called with certain parameters, and it may respond by returning values. The main difference is that a task call is asynchronous (non-blocking). Therefore, the caller of a task is not suspended, nor that a response is assured. Since the calling task is not suspended, it may issue multiple calls simultaneously, to different or even the same task, as we have done in this example in issuing two calls to *email*, in the first and the last action. Consequently, our programming model supports concurrency, because different tasks invoked by the same caller may be executed concurrently, and non-determinism, because the responses from the calls may arrive in arbitrary order.

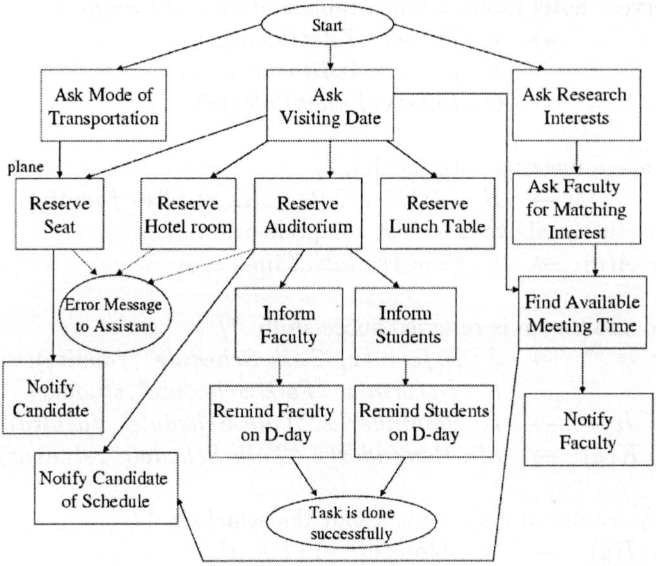

Fig. 1. Faculty candidate recruiting workflow.

A task is not a process. It is instantiated when it is called, and it terminates when its job is done, by responding. A task accepts no unsolicited calls; no one can communicate with a running task except by sending responses to the requests that the task had initiated earlier.

We advocate an asynchronous (non-blocking) model of communication — rather than a synchronous model, as in CCS [15] and CSP [9]— because we anticipate communications with human beings who may respond after long and unpredictable delays. It is not realistic for a task to wait to complete such calls. We intend for each invocation of a task to have finite lifetime. However, this cannot be guaranteed by our theory; it is a proof obligation of the programmer.

Why Not Use a General Programming Language? The *visit* task we have shown can be coded directly in an imperative language, like C++ or Java, which supports creations of threads and where threads may signal occurrences of certain events. Then, each call on a task is spawned off as a thread and receipt of a response to the call triggers a signal by that thread. Each action is a code fragment. After execution of the initial action —which, typically, calls certain tasks/methods— the main program simply waits to receive a signal from some thread it has spawned. On receiving a signal, it evaluates every guard corresponding to the actions that have not yet been executed, and selects an action, if any, whose guard has become true, for execution.

Our proposed model is not meant to compete with a traditional programming language. It lacks almost all features of traditional languages, the only available constructs being task/method calls and non-deterministic selections of actions for executions. In this sense, our model is closer in spirit to CCS [15], CSP [9], or the more recent developments such as π-calculus [16] or Ambient calculus [3]. The notion of action is inspired by similar constructs in UNITY [4], TLA+ [12] and Seuss [17].

One of our goals is to study how little is required conceptually to express the logic of an application, stripping it of data management and computational aspects. Even though the model is minimal, it seems to include all that is needed for computation orchestration. Further, we believe that it will be quite effective in coding real applications because it hides the details of threads, signaling, parameter marshaling and sequencing of the computation.

Programming by Non-experts. The extraordinary success of spreadsheets shows that non-experts can be taught to program provided the number of rules (what they have to remember) is extremely small and the rules are coherent. Mapping a given problem from a limited domain —budget preparation, for instance— to this notation is relatively straightforward. Also, the structure of spreadsheets makes it easy for the users to experiment, with the results of experiments being available immediately.

A spreadsheet provides a simple interface for choosing pre-defined functions from a library, applying them to arguments and displaying the results in a pleasing manner. They are not expected to be powerful enough to specify all functions —elliptic integrals, for instance— nor do they allow arbitrary data structures to be defined by a programmer. By limiting the interface to a small but coherent set, they have helped relative novices to become effective programmers in a limited domain.

In a similar vein, we intend to build a graphical wizard for a subset of this model which will allow non-experts to define tasks. It is easy to depict a task structure in graphical terms: calls on children will be shown by boxes. The parameter received from a response may be bound to the input parameter of a task, not by assigning the same name to them —as would be done traditionally in a programming language— but by merely joining them graphically. The dependency among the tasks is easily understood by a novice, and such dependencies can be depicted implicitly by dataflow: task A can be invoked only with a parameter received from task B; therefore B has to precede A.

One of the interesting features is to exploit spreadsheets for simple calculations. For instance, in order to to compute the sum of the air fare and hotel charges, the user simply identifies certain cells in a spreadsheet with the parameters of the tasks.

4 Implementation

The programming model outlined in this paper has been implemented in a system that we have christened *Orc*. Henceforth, we write "Orc" to denote the programming model as well as its implementation.

The tasks in our model exhibit the following characteristics: (1) tasks can invoke remote methods, (2) tasks can invoke other tasks and themselves, and (3) tasks are inherently non-deterministic. The first two characteristics and the fact that the methods and tasks may run on different machines, require implementation of sophisticated communication protocols. To this end, we take advantage of the Web Service model that we outline below. Non-determinism of tasks, the last characteristic, requires the use of a scheduler that executes the actions appropriately.

Web Services. A web service is a method that may be called remotely. The current standards require web services to use the SOAP[2] protocol for communication and WSDL[5] markup language to publish their signatures. Web services are platform and language independent, thus admitting arbitrary communications among themselves. Therefore, it is fruitful to regard a task as a web service because it allows us to treat remote methods and tasks within the same framework.

The reader should consult the appropriate references for SOAP and WSDL for details. For our needs, SOAP can be used for communication between two parties using the XML markup language. The attractive feature of SOAP is that it is language independent, platform independent and network independent. The WSDL description of a web service provides both a signature and a network location for the underlying method.

4.1 Architecture

Local Server. In order to implement each task as a web service, we host it as an Axis[1] servlet inside a local Tomcat[18] server. A servlet can be thought of as a server-side applet, and the Axis framework makes it possible to expose any servlet as a web service to the outside world.

Translator. The Orc translator is implemented in C and converts an orchestration script into Java. As shown in figure 2, it begins by parsing the input script. In the next step, it creates local java stubs for remote tasks and services. To this end, the URL of the callee task's WSDL description and its name are explicitly described in the Orc script. Thus the translator downloads the WSDL file for each task and uses the WSDL2Java tool, provided by the Axis framework, to create the local stub. Java reflection (described in the next paragraph) is then used to infer the type signature of each task. Finally, Java code is generated based on certain pre-defined templates for Orc primitives like action, evolving tag and timeouts. These templates are briefly described in the following subsection.

Java reflection API [11] allows Java code to discover information about a class and its members in the Java Virtual Machine. Java reflection can be used for applications that require run-time retrieval of class information from a class file. The translator can discover a return type and parameter type by means of Java reflection API.

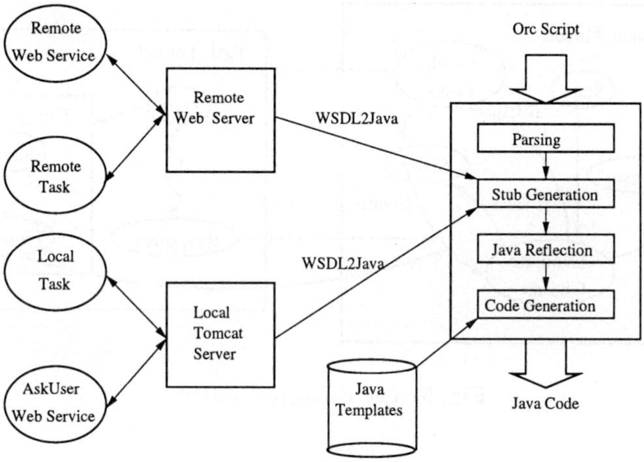

Fig. 2. Components of Orc.

AskUser Web Service. The ability to ask a user a question in an arbitrary stylized format and receive a parsed response is basic to any interactive application. In Orc, this function is captured by the *AskUser* web service. Given a user's email address and an HTML form string, askUser launches an HTTP server to serve the form and receive the reply. It then sends the user an email containing the server's address.

It is interesting to note that the AskUser web service can also be used to implement user interrupts. In order to create a task A that user B can interrupt, we add these two actions to task A:

$$
\begin{array}{lll}
;true & \rightarrow & \alpha : AskUser(B, \text{“}Interrupt\ task?\text{”}) \\
\alpha(\_) & \rightarrow & \beta : \text{ Perform interrupt handling and Return}
\end{array}
$$

The request with tag α asks user B if she wants to interrupt the task, and if a response is received from B, the request with tag β invokes the interrupt procedure and ends the task.

4.2 Java Templates for Orc Primitives

The Manager Class. The Orc translator takes an Orc script as input and emits Java code. The most interesting aspect of the implementation was to build non-determinism into an essentially imperative world. The action system that an Orc script describes is converted into a single thread as shown in figure 3. We call this the Manager thread. All other tasks are invoked by the Manager thread. Every distinct task in the Orc model is implemented as a separate thread class. The manager evaluates the guards of each action in the Orc script and invokes the tasks whose guards are true. When no guard is true it waits for the tasks it has

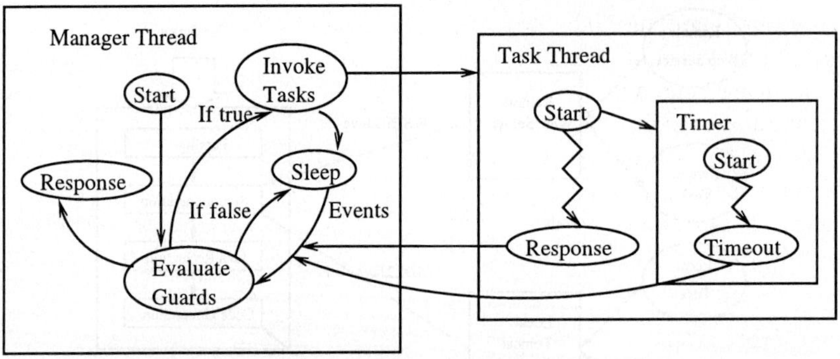

Fig. 3. The Runtime System.

already started to complete, and then checks the guards again. Orc follows the once only semantics. This means that a task in an Orc program may be invoked at most once. Each task follows a particular interface for communicating with the manager. Tasks in Orc may be written directly in Java, or might have been generated from web services. Note that though a web service is essentially a task, once it is invoked it performs some computation and returns a result, the WSDL2Java tool does not translate the tasks in the particular format as required by the manager. We generate a wrapper around the class that the WSDL2Java tool generates, to adhere to the task interface which the manager requires.

Timeouts. Every task in this implementation of Orc includes a timer, as shown in figure 3. The timer is started when the manager invokes a task. A task's timer signals the manager thread if the task does not complete before its designated timeout value.

Evolving Tags. Orc allows the same task to be invoked on a list of input instances. Since the invocations on different input instances may complete at different times, the result list starts out empty and grows as each instance returns a result. Such lists are called evolving tags in our model. The interface used for tasks that return evolving tags is a subclass of the interface used for regular tasks. It adds methods that check if an evolving tag is empty or full, and makes it possible to iterate over the result list.

The templates that we have described here allow a task written in Orc to utilize the already existing web services and extend their capabilities using timeout and evolving tags. The implementation of remaining Orc features is straightforward and not described here.

5 Concluding Remarks

We have identified task coordination as the remaining major problem in distributed application design; the other issues, persistent store management and computational logic, have effective solutions which are widely available. We have suggested a programming model to specify task coordination. The specification uses a scripting language, Orc, that has very few features, yet is capable of specifying complex coordinations. Our preliminary experiments show that the Orc scripts could be two orders of magnitude shorter than coding a problem in a traditional programming language. Our translator, still under development, has been used to coordinate a variety of web services coded by other parties with Orc tasks.

Acknowledgement

This work is partially supported by the NSF grant CCR–9803842.

References

1. Apache axis project. http://xml.apache.org/axis.
2. Don Box, David EhneBuske, Gopal Kakivaya, Andrew Layman, Noah Mendelsohn, Henrik Frystyk Nielson, Satish Thatte, and Dave Winer. Simple object access protocol 1.1. http://www.w3.org/TR/SOAP.
3. Luca Cardelli. Mobility and Security. In Friedrich L. Bauer and Ralf Steinbrüggen, editors, *Proceedings of the NATO Advanced Study Institute on Foundations of Secure Computation*, NATO Science Series, pages 3–37. IOS Press, 2000.
4. K. Mani Chandy and Jayadev Misra. *Parallel Program Design: A Foundation*. Addison-Wesley, 1988.
5. Erik Christensen, Francisco Curbera, Greg Meredith, and Sanjiva Weerawarana. Web services description language 1.1. http://www.w3.org/TR/wsdl.
6. The home page for Corba. `http://www.corba.org`, 2001.
7. Main page for World Wide Web Consortium (W3C) XML activity and information. `http://www.w3.org/XML/`, 2001.
8. Jim Gray and Andreas Reuter. *Transaction Processing: Concepts and Techniques*. Morgan Kaufmann, 1993.
9. C.A.R. Hoare. *Communicating Sequential Processes*. Prentice Hall International, 1984.
10. The home page for IBM's webSphere application server. `http://www-4.ibm.com/software/webservers/appserv`, 2001.
11. Java reflection (API). `http://java.sun.com`, 2001.
12. Leslie Lamport. Specifying concurrent systems with TLA+. In Manfred Broy and Ralf Steinbrüggen, editors, *Calculational System Design*, pages 183–247. IOS Press, 1999.
13. A list of references on Microsoft. Net initiative. `http://directory.google.com/Top/Computers/Programming/Component_Frameworks/NET/`, 2001.
14. The home page for Microsoft SQL server. `http://www.microsoft.com/sql/default.asp`, 2001.
15. R. Milner. *Communication and Concurrency*. International Series in Computer Science, C.A.R. Hoare, series editor. Prentice-Hall International, 1989.

16. Robin Milner. *Communicating and Mobile Systems: the π-Calculus.* Cambridge University Press, May 1999.
17. Jayadev Misra. *A Discipline of Multiprogramming.* Monographs in Computer Science. Springer-Verlag New York Inc., New York, 2001. The first chapter is available at `http://www.cs.utexas.edu/users/psp/discipline.ps.gz`.
18. Jakarta project. http://jakarta.apache.org/tomcat/.

Realistic Rendering in Real-Time

Alan Chalmers and Kirsten Cater

Department of Computer Science
University of Bristol
Bristol, UK
Alan.Chalmers@bris.ac.uk
Cater@cs.bris.ac.uk

Abstract. The computer graphics industry, and in particular those involved with films, games and virtual reality, continue to demand more realistic computer generated images. Despite the ready availability of modern high performance graphics cards, the complexity of the scenes being modeled and the high fidelity required of the images means that rendering such images is still simply not possible in a reasonable, let alone real-time on a single computer. Two approaches may be considered in order to achieve such realism in real-time: Parallel Processing and Visual Perception. Parallel Processing has a number of computers working together to render a single image, which appears to offer almost unlimited performance, however, enabling many processors to work efficiently together is a significant challenge. Visual Perception, on the other hand, takes into account that it is the human who will ultimately be looking at the resultant images, and while the human eye is good, it is not perfect. Exploiting knowledge of the human visual system can save significant rendering time by simply not computing those parts of a scene that the human will fail to notice. A combination of these two approaches may indeed enable us to achieve realistic rendering in real-time.

Keywords: Parallel processing, task scheduling, demand driven, visual perception, inattentional blindness.

1 Introduction

A major goal in virtual reality environments is to achieve very realistic image synthesis at interactive rates. However, the computation time required is significant, currently precluding such realism in real time. The challenge is thus to achieve higher fidelity graphics for dynamic scenes without simultaneously increasing the computational time required to render the scenes. One approach to address this problem is to use parallel processing [2, 8, 11]. However, such parallel approaches have their own inherent difficulties, such as the efficient management of data across multiple processors and the issues of task scheduling to ensure load balancing, which still inhibits their wide-spread use for large complex environments [2].

The perception of a virtual environment depends on the user and the task that he/she is currently performing in that environment. Visual attention is the process by which we humans select a portion of the available visual information for localisation,

B. Monien and R. Feldmann (Eds.): Euro-Par 2002, LNCS 2400, pp. 21–28.

identification and understanding of objects in an environment. It allows our visual system to process visual input preferentially by shifting attention about an image, giving more attention to salient locations and less attention to unimportant regions. When attention is not focused onto items in a scene they can literally go unnoticed. *Inattentional blindness* is the failure of the human to see unattended items in a scene [4]. It is this inattentional blindness that we may exploit to help produce perceptually high-quality images in reasonable times.

2 Realistic Rendering

The concept of realistic image synthesis centers on generating scenes with an authentic visual appearance. The modeled scene should not only be physically correct but also perceptually equivalent to the real scene it portrays [7].

One of the most popular rendering techniques is ray tracing [4, 10, 14]. In this approach, one or more primary rays are traced, for each pixel of the image, into the scene. If a primary ray hits an object, the light intensity of that object is assigned to the corresponding pixel. Shadows, specular reflections and transparency can be simulated by spawning new rays from the intersection point of the ray and the object, as shown in figure 1. These shadow, reflection and transparency rays are treated in exactly the same way as primary rays, making ray tracing a recursive algorithm.

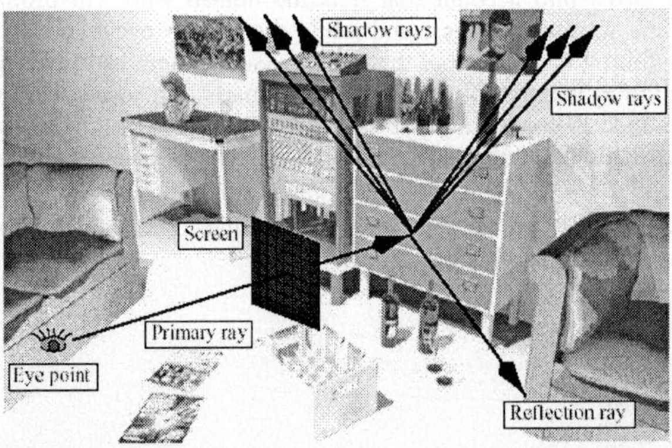

Fig. 1. The ray tracing algorithm, showing shadow and reflection rays, after Reinhard [2].

While most ray tracing algorithms approximate the diffuse lighting component with a constant ambient term, other more advanced systems, in particular the Radiance lighting simulation package [12, 13], accurately computes the diffuse interreflections by shooting a large number of undirected rays into the scene, distributed over a hemisphere placed over the intersection point of the ray with the object. Tracing these diffuse rays is also performed recursively. The recursive ray tracing process has to be carried out for each individual pixel separately. A typical image therefore takes at least a million primary rays and a significant multiple of that for shadow, reflection, transparency and diffuse rays. In addition, often more than one ray is traced per pixel (super-sampling) to help overcome aliasing artifacts.

Despite the enormous amount of computation that is required for ray tracing a single image, this rendering technique is actually well suited to parallel processing as the computation of one pixel is completely independent of any other pixel. Furthermore, as the scene data used during the computation is read, but not modified, there is no need for consistency checking and thus the scene data could be duplicated over every available processor. As such parallel ray tracing has often been referred to as an *embarrassingly parallel problem*.

However, in reality, the scenes we wish to model for our virtual environments are far too complex to enable the data to be duplicated at each processor. This is especially true if, rather than computing a single image of a scene, we wish to navigate through the entire environment. It should be noted, however, that if a shared memory machine is available, the scene does not have to be distributed over a number of processors, nor does data have to be duplicated. As such, parallel ray tracing on shared memory architectures is most certainly a viable approach and has led to implementations that may render complex scenery at interactive rates [8]. However, such shared memory architectures are not easily scalable and thus here we shall consider realistic rendering on the more scalable distributed memory parallel systems.

3 Parallel Processing

The goal of parallel processing remains the solution of a given complex problem more rapidly, or to enable the solution of a problem that would otherwise be impracticable by a single processor [1]. The efficient solution of a problem on a parallel system requires the computational ability of the processors to be fully utilized. Any processor that is not busy performing useful computation is degrading the overall system performance. Careful task scheduling is essential to ensure that all processors are kept busy while there is still work to be done.

The demand driven computational model of parallel processing has been shown to be very effective for parallel rendering [2, 9]. In the demand driven approach for parallel ray tracing, work is allocated to processors *dynamically* as they become idle, with processors no longer bound to any particular portion of pixels. Having produced the result for one pixel, the processors demand the next pixel to compute from some work supplier process.

This approach facilitates dynamic load balancing when there is no prior knowledge as to the complexity of the different parts of the problem domain. Optimum load balancing is still dependent on all the processors completing the last of the work at the same time. An unbalanced solution may still result if a processor is allocated a complex part of the domain towards the end of the solution. This processor may then still be busy well after all the other processors have completed computation on the remainder of the pixels and are now idle as there is no further work to do. To reduce the likelihood of this situation it is important that the computationally complex portions of the domain, the so called *hot spots*, are allocated to processors early on in the solution process. Although there is no *a priori* knowledge as to the exact computational effort associated with any pixel, nevertheless, any insight as to possible hot spot areas, such as knowledge of the computational effort for computing previous pixels, should be exploited. The order in which tasks are supplied to the processors can thus have a significant influence on the overall system performance.

4 Visual Perception

Advances in image synthesis techniques allow us to simulate the distribution of light energy in a scene with great precision. Unfortunately, this does not ensure that the displayed image will have a high fidelity visual appearance. Reasons for this include the limited dynamic range of displays, any residual shortcomings of the rendering process, and the restricted time for processing. Conversely, the human visual system has strong limitations, and ignoring these leads to an over specification of accuracy beyond what can be seen on a given display system [1]. The human eye is "good", but not "that good". By exploiting inherent properties of the human visual system we may be able to avoid significant computational expense *without* affecting the perceptual quality of the resultant image or animation.

4.1 Inattentional Blindness

In 1967, Yarbus [15] showed that the choice of task that the user is performing when looking at an image is important in helping us predict the eye-gaze pattern of the viewer. It is precisely this knowledge of the expected eye-gaze pattern that will allow us to reduce the rendered quality of objects outside the area of interest without affecting the viewer's overall perception of the quality of the rendering.

In human vision, two general processes, called bottom-up and top-down, determine where humans locate their visual attention [4]. The bottom-up process is purely stimulus driven, for example a candle burning in a dark room; a red ball amongst a large number of blue balls; or the lips and eyes of a human face as they are the most mobile and expressive elements of the face. In all these cases, the visual stimulus captures attention automatically without volitional control. The top-down process, on the other hand, is directed by a voluntary control process that focusses attention on one or more objects, which are relevant to the observer's goal when studying the scene. In this case, the attention normally drawn due to conspicuous aspects in a scene may be deliberately ignored by the visual system because of irrelevance to the goal at hand. This is "inattentional blindness" which we may exploit to significantly reduce the computational effort required to render the virtual environment.

4.2 Experiment

The effectiveness of inattentional blindness in reducing overall computational complexity was illustrated by asking a group of users were asked to perform a specific task: to watch two animations and in each of the animations, count the number of pencils that appeared in a mug on a table in a room as he/she moved on a fixed path through four such rooms. In order to count the pencils, the users needed to perform a smooth pursuit eye movement tracking the mug in one room until they have successfully counted the number of pencils in that mug and then perform an eye saccade to the mug in the next room. The task was further complicated and thus retain the viewer's attention, by each mug also containing a number of spurious paintbrushes. The study involved three rendered animations of an identical fly through of four rooms. The only difference being the quality to which the individual animations had been rendered. The three qualities of animation were:

- High Quality(HQ): Entire animation rendered at the highest quality.
- Low Quality(LQ): Entire animation rendered at a low quality with no anti-aliasing.
- Circle Quality(CQ): Low Quality Picture with high quality rendering in the visual angle of the fovea (2 degrees) centered around the pencils, shown by the inner green circle in figure 2. The high quality is blended to the low quality at 4.1 degrees visual angle (the outer red circle in figure 2) [6].

Fig. 2: Visual angle covered by the fovea for mugs in the first two rooms at 2 degrees (smaller circles) and 4.1 degrees (large circles).

Each frame for the high quality animation took on average 18 minutes 53 seconds to render on a Intel Pentium 4 1GHz Processor, while the frames for the low quality animation were each rendered on average in only 3 minute 21 seconds.

A total of 160 subjects were studied which each subject seeing two animations of 30 seconds each displayed at 15 frames per second. Fifty percent of the subjects were asked to count the pencils in the mug while the remaining 50% were simply asked to watch the animations. To minimise experimental bias the choice of condition to be run was randomised and for each, 8 were run in the morning and 8 in the afternoon. Subjects had a variety of experience with computer graphics and all exhibited at least average corrected vision in testing. A count down was shown to prepare the viewers that the animation was about to start followed immediately by a black image with a white mug giving the location of the first mug. This ensured that the viewers focused their attention immediately on the first mug and thus did not have to look around the scene to find it.

On completion of the experiment, each participant was asked to fill in a detailed questionnaire. This questionnaire asked for some personal details, including age, occupation, sex and level of computer graphics knowledge. The participants were then asked detailed questions about the objects in the rooms, their colour, location and quality of rendering. These objects were selected so that questions were asked about objects both near the foveal visual angle (located about the mug with pencils) and in the periphery. They were specifically asked not to guess, but rather state "don't remember" when they had failed to notice some details.

4.3 Results

Figure 3 shows the overall results of the experiment. Obviously the participants did not notice any difference in the rendering quality between the two HQ animations (they were the same). Of interest is the fact that, in the CQ + HQ experiment, 95% of the viewers performing the task consistently failed to notice any difference between the high quality rendered animation and the low quality animations where the area around the mug was rendered to a high quality. Surprisingly 25% of the viewers in the HQ+LQ condition and 18% in the LQ+HQ case were so engaged in the task that they completely failed to notice any difference in the quality between these very different qualities of animation.

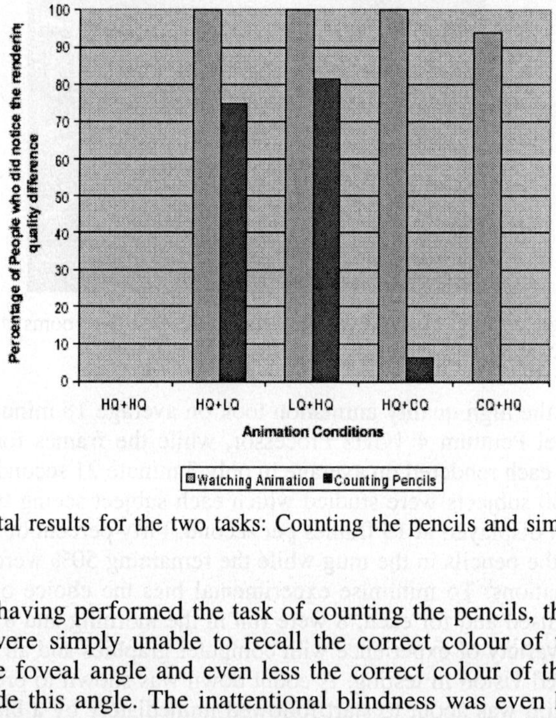

Fig. 3. Experimental results for the two tasks: Counting the pencils and simply watching the animations.

Furthermore, having performed the task of counting the pencils, the vast majority of participants were simply unable to recall the correct colour of the mug (90%) which was in the foveal angle and even less the correct colour of the carpet (95%) which was outside this angle. The inattentional blindness was even higher for "less obvious" objects, especially those outside the foveal angle. Overall the participants who simply watched the animations were able to recall far more detail of the scenes, although the generic nature of the task given to them precluded a number from recalling such details as the colour of specfic objects, for example 47.5% could not recall the correct colour of the mug and 53.8% the correct colour of the carpet.

5 Conclusions

The results presented demonstrate that inattentional blindness may in fact be exploited to significantly reduce the rendered quality of a large portion of a scene without having any affect on the viewer's perception of the scene. This knowledge will enable us to prioritize the order, and the quality level of the tasks that are assigned to the processors in our parallel system. Those few pixels in the visual angle of the fovea (2 degrees) centered around the pencils, shown by the green inner circle in figure 2 should be rendered first and to a high quality, the quality can then be blended to the low quality at 4.1 degrees visual angle (the red outer circle in figure 2).

Perhaps we were too cautious in our study of inattentional blindness. Future work will consider whether in fact we even need to ray trace some of the pixels outside the foveal angle. It could be that the user's focus on the task is such that he/she may fail to notice the colour of many of the pixels outside this angle and that these could simply be assigned an arbitrary neutral colour, or interpolated from a few computed sample pixels.

Visual perception, and in particular inattentional blindness does depend on knowledge of the task being performed. For many applications, for example games and simulators, such knowledge exists offering the real potential of combining parallel processing and visual perception approaches to achieve "perceptually realistic" rendering in real-time.

References

1. Cater K., Chalmers AG. and Dalton C. 2001 Change blindess with varying rendering fidelity: looking but not seeing, Sketch SIGGRAPH 2001, Conference Abstracts and Applications.
2. Chalmers A., Davis T. and Reinhard E. Practical Parallel Rendering, AKPeters, to appear 2002.
3. Glassner A.S. , editor. *An Introduction to Ray Tracing*. Academic Press, San Diego, 1989.
4. James W. 1890 Principles of Psychology, New York: Holt.
5. Mack A. and Rock I. 1998 Inattentional Blindness, Massachusetts Institute of Technology Press.
6. McConkie GW. and Loschky LC. 1997 Human Performance with a Gaze-Linked Multi-Resolutional Display". ARL Federated Laboratory Advanced Displays and Interactive Displays Consortium, Advanced Displays and Interactive Displays First Annual Symposium, 25-34.
7. McNamara, A., Chalmers, A., Troscianko, T. and Reinhard, E., "Fidelity of Graphics Reconstructions: A Psychophysical Investigation". *Proceedings of the 9th Eurographics Workshop on Rendering* (June 1998) Springer Verlag, pp. 237 - 246.
8. Parker S., Martin W., Sloan P.-P., Shirley P., Smits B., and Hansen C. Interactive ray tracing. In *Symposium on Interactive 3D Computer Graphics*, April 1999.
9. Reinhard E., Chalmers A., and Jansen FW. Overview of parallel photo-realistic graphics. In *Eurographics STAR – State of the Art Report*, pages 1–25, August-September 1998.
10. Shirley P. *Realistic Ray Tracing*. A K Peters, Natick, Massachusetts, 2000.
11. I. Wald, P. Slusallek, C. Benthin, and M. Wagner. Interactive rendering with coherent ray tracing. *Computer Graphics Forum*, 20(3):153–164, 2001.
12. Ward GJ, Rubinstein FM., and Clear RD. A ray tracing solution for diffuse interreflection. *ACM Computer Graphics*, 22(4):85–92, August 1988.

13. Ward Larson GJ. and Shakespeare RA. *Rendering with Radiance*. Morgan Kaufmann Publishers, 1998.
14. Whitted T. An improved illumination model for shaded display. *Communications of the ACM*, 23(6):343–349, June 1980.
15. Yarbus AL.1967 Eye movements during perception of complex objects. In L. A. Riggs, Ed., Eye Movements and Vision, Plenum Press, New York, chapter VII, pp. 171-196.

Non-massive, Non-high Performance, Distributed Computing: Selected Issues*

Albert Benveniste

Irisa/Inria, Campus de Beaulieu, 35042 Rennes cedex, France
Albert.Benveniste@irisa.fr
http://www.irisa.fr/sigma2/benveniste/

Abstract. There are important distributed computing systems which are neither massive nor high performance. Examples are: telecommunications systems, transportation or power networks, embedded control systems (such as embedded electronics in automobiles), or Systems on a Chip. Many of them are *embedded systems,* i.e., not directly visible to the user. For these systems, performance is not a primary issue, major issues are reviewed in this paper. Then, we focus on a particular but important point, namely the correct implementation of specifications on distributed architectures.

1 Beware

This is a special and slightly provocative section, just to insist, for the Euro-Par community, that:

> ***there are important distributed computing systems which are neither massive nor high performance.***

Here is a list, to mention just a few:

(a) Telecommunications or web systems.
(b) Transportation or power networks (train, air-traffic management, electricity supply, military command and control, etc.).
(c) Industrial plants (power, chemical, etc.).
(d) Manufacturing systems.
(e) Embedded control systems (automobiles, aircrafts, etc.).
(f) System on Chip (SoC) such as encountered in consumer electronics, and Intellectual Property (IP)-based hardware.

Examples (a,b) are distributed, so to say, by tautology: they are distributed because they are networked. Examples (c,d,e) are distributed by requirement from the physics: the underlying physical system is made of components, each component is computerized, and the components concur at the overall behaviour of the

* This work is or has been supported in part by the following projects: Esprit R&D SAFEAIR, and Esprit NoE ARTIST.

B. Monien and R. Feldmann (Eds.): Euro-Par 2002, LNCS 2400, pp. 29–48.

system. Finally, example (f) is distributed by requirement from the electrons: billion-transistor SoC cannot be globally synchronous.

Now, (almost) all the above examples have one fundamental feature: *they are open systems, which interact continuously with some unspecified environment having its own dynamics.* Furthermore, some of these open systems interact with their environment in a tight way, e.g. (c,d,e) and possibly also (f). These we call

reactive systems,

which will be the focus of this paper. For many reactive systems, computing performance is not the main issue. The extreme case is avionics system, in which the computing system is largely oversized in performance. Major requirements, instead, are [20]:

Correctness: *the system should behave the way it is supposed to.*

Since the computer system interacts with some physical system, we are interested in the resulting closed-loop behaviour, i.e., the joint behaviour of the physical plant and its computer control system. Thus, specifying the signal/data processing and control functionalities to be implemented is a first difficulty, and sometimes even a challenge (think of a flight control system for a modern flight-by-wire aircraft). Extensive virtual prototyping using tools from scientific and control engineering is performed to this end, by using typically Matlab/Simulink with its toolboxes.

Another difficulty is that such reactive systems involve many *modes* of operation (a mode of operation is the combination of a subset of the available functionalities). For example, consider a modern car equipped with computer assisted emergency breaking. If the driver suddendly strongly brakes, then the resulting strong increase in the brake pedal pressure is detected. This causes the fuel injection mode to stop, ABS mode to start, and the maximal braking force is computed on-line and applied automatically, in combination with ABS. Thus mode changes are driven by the pilot, they can also be driven automatically, being indirect consequences of human requests, or due to protection actions.

There are many such modes, some of them can run concurrently, and their combination can yield thousands to million of discrete states. This discrete part of the system interfers with the "continuous" functionalities in a bidirectional way: the monitoring of continuous measurements triggers protection actions, which results in mode changes; symmetrically, continuous functionalities are typically attached to modes. The overall system is called *hybrid,* since it tightly combines both continuous and discrete aspects. This discrete part, and its interaction with the continuous part, is extremely error prone, its correctness is a major concern for the designer.

For some of these systems, real-time is one important aspect. It can be *soft* real-time, where requested time-bounds and throughput are loose, or *hard* real-time, where they are strict and critical. This is different from requesting high performance in terms of average throughput.

As correctness is a major component of safety, it is also critical that the actual *distributed implementation*—also called **distributed deployment** in the sequel—of the specified functionalities and mode changes shall be performed in a correct way. After all, the implementation matters, not the spec! But the implementation adds a lot of nondeterminism: RTOS (real-time operating system), buses, and sometimes even analog-to-digital and digital-to-analog conversions. Thus a careless deployment can impair an otherwise correct design, even if the computer equipment is oversized.

Robustness: *the system should resist to (some amount of) uncertainty or error.* No real physical system can be exactly modeled. Models of different accuracies and complexities are used, for the different phases of the scientific engineering part of the systems design. Accurate models are used for mechanics, aerodynamics, chemical dynamics, etc., when virtual simulation models are developed. Control design uses simple models reflecting only some facets of the systems dynamics. The design of the discrete part for mode switching usually oversimplifies the physics. Therefore, the design of all functionalities, both continuous and discrete, must be robust against uncertainties and approximations in the physics. This is routine for the continuous control engineer, but still requires modern control design techniques. Performing this for the discrete part, however, is still an open challenge today.

Fault-tolerance is another component of robustness of the overall system. Faults can occur, due to failures of physical components. They can be due to the on-board computer and communication hardware. They can also originate from residual faults in the embedded software. *Distributed architectures are a key counter-measure against possible faults*: separation of computers helps mastering the propagation of errors. Now, special principles should be followed when designing the corresponding distributed architecture, so as to limit the propagation of errors, not to increase its risk! For example, rendez-vous communication may be dangerous: a component failing to communicate will block the overall system.

Scope of this paper: Addressing all the above challenges is certainly beyond a single paper, and even more beyond my own capacity. I shall restrict myself to examples (e,f), and to a lesser extend (c,d). There, I shall mainly focus on the issue of correctness, and only express some considerations related to robustness. Moreover, since the correctness issue is very large, I shall focus on the *correctness of the distributed deployment, for so-called embedded systems.*

2 Correct Deployment of Distributed Embedded Applications

As a motivating application example, the reader should think of safety critical embedded systems such as flight control systems in flight-by-wire avionics, or anti-skidding and anti-collision equipment in automobiles. Such systems can be characterized as *moderately distributed,* meaning that:

- The considered system has a "limited scope", in contrast with large distributed systems such as telecommunication or web systems.
- All its (main) components *interact,* as they concur at the overall correct behaviour of the system. Therefore, unlike for large distributed systems, the aim is not that different services or components should not interact, but rather that they should interact in a correct way.
- *Correctness,* of the components and of their interactions with each other and with the physical plant, is critical. This requires tight control of synchronization and timing.
- The design of such systems involves methods and tools from the underlying technical engineering area, e.g., mechanics and mechatronics, control, signal processing, etc. *Concurrency* is a natural paradigm for the systems engineer, not something difficult to be afraid of. The different functionalities run by the computer system operate concurrently, and they are concurrent with the physical plant.
- For systems architecture reasons, not performance reasons, deployment is performed on distributed architectures. The system is distributed, and even some components themselves can be distributed—they can involve intelligent sensors & actuators, and have part of their supervision functionalities embedded in some centralized computer.

Methods and tools used, and corresponding communication paradigms: The methods and tools used are discussed in Fig. 1. In this figure, we show on the left the different tool-sets used throughout the systems design. This diagram is mirrored on the right hand side of the same figure, where the corresponding communication paradigms are shown.

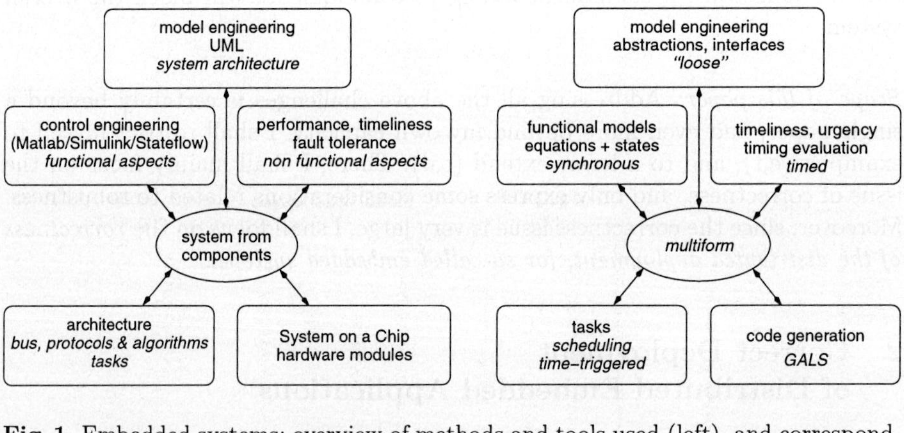

Fig. 1. Embedded systems: overview of methods and tools used (left), and corresponding communication paradigms (right). The top row ("model engineering") refers to the high level system specification, the second row ("control engineering") refers to the the detailed specification of the different components (e.g., anti-skidding control subsystem). And the bottom row refers to the the (distributed) implementation.

Let us focus on the functional aspects first. This is a phase of the design in which scientific engineering tools (such as the Matlab family) are mainly used, for functionalities definition and prototyping. In this framework, there is a natural global time available. Physical continuous time triggers the models developed at the functionalities prototyping phase, in which controllers interact with a physical model of the plant. The digital controllers themselves are discrete time, and refer to some unique global discrete time. Sharing a global discrete time means using a *perfectly synchronous* communication paradigm, this is indicated in the diagram sitting on the right.

Now, some parts of the system are (hard or soft) real-time, meaning that the data handled are needed and are valid only within some specified window of time: buffering an unbounded amount of data, or buffering data for unbounded time, is not possible.

For these first two aspects, tight logical or timed synchronization is essential.

However, when dealing with higher level, global, systems architecture aspects, it may sometimes happen that no precise model for the the components interaction is considered. In this case the communication paradigm is left mostly unspecified. This is a typical situation within the UML (Universal Modeling Language) [19] community of systems engineering.

Focus now on the bottom part of this figure, in which deployment is considered. Of course, there is no such thing like a "loose" communication paradigm, but still different paradigms are mixed. Tasks can be run concurrently or can be scheduled, and scheduling may or may not be based on physical time. Hybrid paradigms are also encountered within Systems on a Chip (SoC), which typically follow a Globally Asynchronous Locally Synchronous (GALS) paradigm.

Fig. 2 shows a different view of the same landscape, by emphasizing the different scheduling paradigms. In this figure, we show a typical control structure of a functional specification (left) with its multi-threaded logical control structure. The horizontal bars figure synchronization points, the (dashed) thick lines figure (terminated) threads, and the diamonds indicate fork/joins. This functional specification can be compiled into non-threaded sequential code by generating

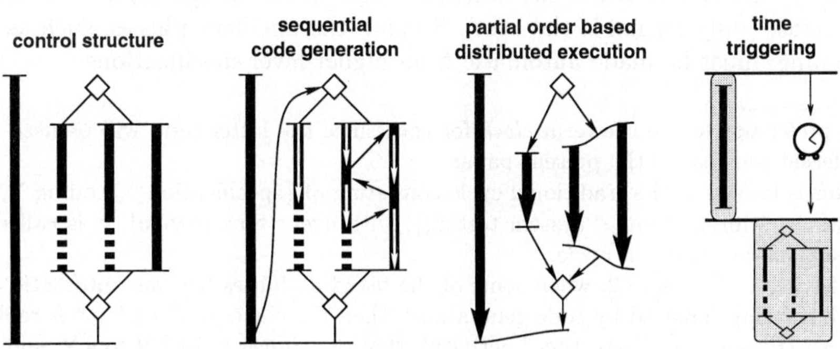

Fig. 2. Embedded systems: scheduling models for execution.

a total order for the threads (mid-left), this has the advantage of producing deterministic executable code for embedding.

But a concurrent, and possibly distributed, execution is also possible (mid-right). For instance, task scheduling is subcontracted to some underlying RTOS, or tasks can be physically distributed.

Finally, task and even component scheduling can be entirely triggered by physical time, by using a distributed infrastructure which provides physically synchronized timers[1], this is usually referred to as "time-triggered architecture" [17].

Objective of this paper. As can be expected from the above discussion, mixed communication paradigms are in use throughout the design process, and are even combined both at early phases of the design and at deployment phase.

This was not so much an issue in the traditional design flow, in which most work was performed manually. In this traditional approach: the physics engineer provides models; the control engineer massages them for his own use and designs the control; then he forwards this as a document in textual/graphical format to the software engineer, who performs programming (in C or assembly language). This holds for each component. Then unit testing follows, and then integration and system testing [2]. Bugs discovered at this last stage are the nightmare of the systems designer! Where and how to find the cause? How to fix them? On the other hand, for this traditional design flow, each engineer has his own skills and underlying scientific background, but there is no need for an overall coherent mathematical foundation for the whole. So the design flow is simple. It uses different skills in a (nearly) independent way. This is why this is mainly the current practice.

However, due to the above indicated drawback, this design flow does not scale up. In very complex systems, many components would mutually interact in an intricate way. There are about 70 ECU's (Electronic Computing Units) in a modern BMW Series 7 car, each of these implements one or more functionalities. Moreover, some of them interact together, and the number of embedded functionalities rapidly increases. Therefore, there is a double need. First, specifications transferred between the different stages of the design must be as formal as possible (fully formal is the best). Second, the ancillary phases, such as programming, must be made automatic from higher level specifications [3].

---

1 we prefer not to use the term *clock* for this, since the latter term will be used for a different purpose in the present paper.

2 This is known as the traditional cycle consisting of {specification $\searrow$ coding $\searrow$ unit testing $\nearrow$ integration $\nearrow$ system testing}, with everything manual. It is called the V-shaped development cycle.

3 Referring to Footnote 2, when some of the listed activities become automatic (e.g., coding being replaced by code generation), then the corresponding $\searrow\nearrow$ is replaced by a $\updownarrow$ (to refer to a "zero-time" activity), thus one moves from a V to a Y, and then further to a T, by relying on extensive virtual prototyping, an approach promoted by the Ptolemy tool [8].

This can only be achieved if we have a full understanding of how the different communication paradigms, attached to the different stages of the design flow, can be combined, and of how migration from a paradigm to the next one can be performed in a provably correct way. A study involving all the above mentioned paradigms is beyond the current state of the research. The purpose of this paper is to focus on the pair consisting of the {*synchronous, asynchronous*} paradigms.

But, before doing so, it is worth discussing in more depth the synchronous programming paradigm and its associated family of tools, as this paradigm is certainly not familiar to the High Performance Computing community. Although many visual or textual formalisms follow this paradigm, it is the contribution of the three "synchronous languages" Esterel, Lustre, and Signal [1] [7] [13] [18] [14] [6] [2], to have provided a firm basis for this concept.

3 Synchronous Programming and Synchronous Languages

The three synchronous languages Esterel, Lustre, and Signal, are built on a common mathematical framework that combines synchrony (i.e., time progresses in lockstep with one or more clocks) with deterministic concurrency.

Fundamentals of synchrony. Requirements from the applications, as resulting from the discussion of Section 2, are the following:

- *Concurrency.* The languages must support functional concurrency, and they must rely on notations that express concurrency in a user-friendly manner. Therefore, depending on the targeted application area, the languages should offer as a notation: block diagrams (also called dataflow diagrams), or hierachical automata, or some imperative type of syntax, familiar to the targeted engineering communities.
- *Simplicity.* The languages must have the simplest formal model possible to make formal reasoning tractable. In particular, the semantics for the parallel composition of two processes must be the cleanest possible.
- *Synchrony.* The languages must support the simple and frequently-used implementation models in Fig. 3, where all mentioned actions are assumed to take finite memory and time.

Combining synchrony and concurrency while maintaining a simple mathematical model is not so straightforward. Here, we discuss the approach taken by the synchronous languages.

Synchrony divides time into discrete instants: a synchronous program progresses according to successive *atomic reactions,* in which the program communicates with its environment and performs computations, see Fig. 3. We write this for convenience using the "pseudo-mathematical" statement $P =_{\text{def}} R^{\omega}$, where R denotes the set of all possible reactions and the superscript ω indicates non-terminating iterations.

```
Initialize Memory
for each input event do
    Compute Outputs
    Update Memory
end
```

```
Initialize Memory
for each clock tick do
    Read Inputs
    Compute Outputs
    Update Memory
end
```

Fig. 3. Two common synchronous execution schemes: event driven (left) and sample driven (right). The bodies of the two loops are examples of *reactions*.

For example, in the block (or dataflow) diagrams of control engineering, the nth reaction of the whole system is the combination of the individual nth reactions for each constitutive component. For component i,

$$\begin{aligned} X_n^i &= f(X_{n-1}^i, U_n^i) \\ Y_n^i &= g(X_{n-1}^i, U_n^i) \end{aligned} \tag{1}$$

where U, X, Y are the (vector) input, state, and output, and combination means that some input or output of component i is connected to some input of component j, say

$$U_n^j(k) = U_n^i(l) \text{ or } Y_n^i(l), \tag{2}$$

where $Y_n^i(l)$ denotes the l-th coordinate of vector output of component i at instant n. Hence the whole reaction is simply the conjunction of the reactions (1) for each component, and the connections (2) between components.

Connecting two finite-state machines (FSM) in hardware is similar. Fig. 4a shows how a finite-state system is typically implemented in synchronous digital logic: a block of acyclic (and hence functional) logic computes outputs and the

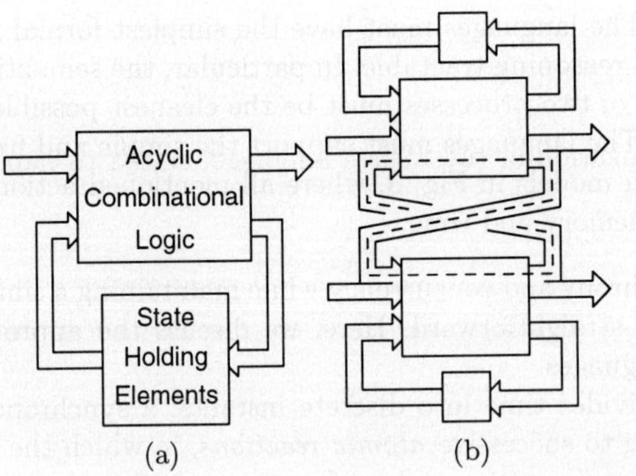

Fig. 4. (a) The usual structure of an FSM implemented in hardware. (b) Connecting two FSMs. The dashed line shows a path with instantaneous feedback that arises from connecting these two otherwise functional FSMs.

next state as a function of inputs and the current state. Fig. 4b shows the most natural way to run two such FSMs concurrently and have them communicate, i.e., by connecting some of the outputs of one FSM to the inputs of the other and vice versa.

Therefore, the following natural definition for parallel composition in synchronous languages was chosen, namely: $P_1 \parallel P_2 =_{\text{def}} (R_1 \wedge R_2)^\omega$, where $\wedge$ denotes conjunction. Note that this definition for parallel composition also fits several variants of the synchronous product of automata. Hence the model of synchrony can be summarized by the following two pseudo-equations:

$$P =_{\text{def}} R^\omega, \tag{3}$$

$$P_1 \parallel P_2 =_{\text{def}} (R_1 \wedge R_2)^\omega. \tag{4}$$

A flavour of the different styles of synchronous languages. Here is an example of a Lustre program, which describes a typical fragment of digital logic hardware. The program:

```
edge = false -> (c and not pre(c));
nat = 0 -> pre(nat) + 1;
edgecount = 0 -> if edge then pre(edgecount) + 1
                 else pre(edgecount);
```

defines edge to be true whenever the Boolean flow c has a rising edge, nat to be the step counter ($\mathsf{nat}_n = n$), and edgecount to count the number of rising edges in c. Its meaning can be expressed in the form of a finite difference equation, with obvious shorthand notations:

$$\begin{cases} e_0 = \mathit{false} \\ N_0 = 0 \end{cases}, \forall n > 0 : \begin{cases} e_n = c_n \text{ and not } c_{n-1} \\ N_n = N_{n-1} + 1 \\ ec_n = \begin{cases} \text{if } e_n = \mathit{true} \text{ then } ec_{n-1} + 1 \\ \qquad\qquad \text{else } ec_{n-1} \end{cases} \end{cases}$$

This style of programming is amenable of graphical formalisms of block-diagram type. It is suited for *computation-dominated* programs. The Signal language is sort of a generalization of the Lustre language, suited to handle open systems, we discuss this point later on.

But reactive systems can also be *control-dominated.* To illustrate how Esterel can be used to describe control behavior, consider the program fragment in Fig. 5 describing the user interface of a portable CD player. It has input signals for play and stop and a lock signal that causes these signals to be ignored until an unlock signal is received, to prevent the player from accidentally starting while stuffed in a bag. Note how the first process ignores the Play signal when it is already playing, and how the suspend statement is used to ignore Stop and Play signals.

The nice thing about synchronous language is that, despite the very different styles of Esterel, Lustre, and Signal, they can be cleanly combined, since they share fully common mathematical semantics.

```
loop
  suspend
    await Play; emit Change
  when Locked;
  abort
    run CodeForPlay
  when Change
end
||
loop
  suspend
    await Stop; emit Change
  when Locked;
  abort
    run CodeForStop
  when Change
end
||
every Lock do
  abort
    sustain Locked
  when Unlock
end
```

emit S	Make signal S present immediately
pause	Stop this thread of control until the next reaction
p ; q	Run p then q
loop p end	Run p; restart when it terminates
await S	Pause until the next reaction in which S is present
p || q	Start p and q together; terminate when both have terminated
abort p when S	Run p up to, but not including, a reaction in which S is present
suspend p when S	Run p except when S is present
sustain S	Means loop emit S; pause end
run M	Expands to code for module M

Fig. 5. An Esterel program fragment describing the user interface of a portable CD player. Play and Stop inputs represent the usual pushbutton controls. The presence of the Lock input causes these commands to be ignored.

Besides the three so-called "synchronous languages", other formalisms or notations share the same type of mathematical semantics, without saying so explicitly. We only mention two major ones. The most widespread formalism is the discrete time part of the Simulink [4] graphical modeling tool for Matlab, it is a dataflow graphical formalism. David Harel's Statecharts [15][16] as for instance implemented in the Statemate tool by Ilogix [5], is a visual formalism to specify concurrent and hierarchical state machines. These formalisms are much more widely used than the previously described synchronous languages. However they do not fully exploit the underlying mathematical theory.

4 Desynchronization

As can be seen from Fig. 1, functionalities are naturally specified using the paradigm of synchrony. In contrast, by looking at the bottom part of the diagrams in the same figure, one can notice that, for larger systems, deployment uses infrastructures that do not comply with the model of synchrony. This problem can be addressed in two different ways.

[4] http://www.mathworks.com/products/

[5] http://www.ilogix.com/frame_html.cfm

1. If the objective is to combine, in the considered system, functionalities that are only loosely coupled, then a direct integration without any special care taken to the nondeterminism of the distributed, asynchronous, infrastructure, will do the job. As an example, think of integrating an air bag system with an anti-skidding system in an automobile. In fact, integrating different functionalities in the overall system, is mostly performed this way in the current practice [11].
2. However, when different functionalities have to be combined, which involve a significant discrete part, and interact together in a tight way, then brute force deployment on a nondeterministic infrastructure can create unexpected combinations of discrete states, a source of risk. As an example to contrast with the previous one, think of combining an air bag system with an automatic door locking control (which decides upon locking/unlocking the doors depending on the driving condition).

For this second case, having a precise understanding of how to perform, in a provably correct way, asynchronous distributed deployment of synchronous systems, is a key issue. In this section, we summarize our theory on the interaction between the two {synchronous, asynchronous} paradigms [5].

4.1 The Models Used

In all the models discussed below, we assume some given underlying finite set V of variables—with no loss of generality, we will assume that each system possesses the same V as its set of variables. Interaction between systems occurs via common variables. The difference between these models lies in the way this interaction occurs, from strictly synchronous to asynchronous. We consider the following three different models:

- *Strictly synchronous*: Think of an intelligent sensor, it possesses a unique clock which triggers the reading of its input values, the processing it performs, and the delivery of its processed values to the bus. The same model can be used for human/machine interfaces, in which the internal clock triggers the scanning of the possible input events: only a subset of these are present at a given tick of the overall clock.
- *Synchronous*: The previous model becomes inadequate when open systems are considered. Think of a generic protection subsystem, it must perform reconfiguration actions on the reception of some alarm event—thus, "some alarm event" is the clock which triggers this protection subsystem, when being designed. But, clearly, this protection subsystem is for subsequent use in combination with some sensoring system which will generate the possible alarm events. Thus, if we wish to consider the protection system separately, we must regard it as an open system, which will be combined with some other, yet unspecified, subsystems. And these additional components may very well be active when the considered open system is silent, cf. the example of the protection subsystem. Thus, the model of a global clock triggering

the whole system becomes inadequate for open systems, and we must go for a view in which *several* clocks trigger different components or subsystems, which would in turn interact at some synchronization points. This is an extension of the strictly synchronous model, we call it synchronous. The Esterel and Lustre languages follow the strictly synchronous paradigm, whereas Signal also encompasses the synchronous one.

- *Asynchronous*: In the synchronous model, interacting components or subsystems share some clocks for their mutual synchronization, this requires some kind of broadcast synchronization protocol. Unfortunately, most distributed architectures are asynchronous and do not offer such a service. Instead, they would typically offer asynchronous communication services satisfying the following conditions: 1/ no data shall be lost, and 2/ the ordering of the successive values, for a given variable, shall be preserved (but the global interleaving of the different variables is not). This corresponds to a network of reliable, point to point channels, with otherwise no synchronization service being provided. This type of infrastructure is typically offered by RTOS or buses in embedded distributed architectures, we refer to it as an asynchronous infrastructure in the sequel.

We formalize these three models as follows.

Strictly synchronous. According to this model, a *state* x assigns an effective value to each variable $v \in V$. A *strictly synchronous behaviour* is a sequence $\sigma = x_1, x_2, \ldots$ of states. A *strictly synchronous process* is a set of strictly synchronous behaviours. A *strictly synchronous signal* is the sequence of values $\sigma_v = v(x_1), v(x_2), \ldots$, for $v \in V$ given. Hence all signals are indexed by the same totally ordered set of integers $\mathbf{N} = \{1, 2, \ldots\}$ (or some finite prefix of it). Hence all behaviours are *synchronous* and are tagged by the same clock, this is why I use the term "strictly" synchronous. In practice, strictly synchronous processes are specified using a set of legal *strictly synchronous reactions* R, where R is some transition relation. Therefore, strictly synchronous processes take the form

$$P = R^{\omega} ,$$

where superscript "$.^{\omega}$" denotes unbounded iterations[6]. Composition is defined as the intersection of the set of behaviours, it is performed by taking the conjunction of reactions:

$$P \,\|\, P' := P \cap P' = (R \wedge R')^{\omega} . \tag{5}$$

This is the classical mathematical framework used in (discrete time) models in scientific engineering, where systems of difference equations and finite state machines are usually considered. But it is also used in synchronous hardware modeling.

---

[6] Now, it is clear why we can assume that all processes possess identical sets of variables: just enlarge the actual set of variables with additional ones, by setting no constraint on the values taken by the states for these additional variables.

Synchronous. Here the model is the same as in the previous case, but every domain of data is enlarged with some *non-informative* value, denoted by the special symbol $\perp$ [3][4][5]. A $\perp$ value is to be interpreted as the considered variable being *absent* in the considered reaction And the process can use the absence of these variables as a viable information for its control. Besides this, things are as before: a *state* x assigns an informative or non-informative value to each state variable $v \in V$. A *synchronous behaviour* is a sequence of states: $\sigma = x_0, x_1, x_2, \ldots$. A *synchronous process* is a set of synchronous behaviours. A *synchronous signal* is the sequence of informative or non-informative values $\sigma_v = v(x_1), v(x_2), \ldots$, for $v \in V$ given. And composition is performed as in (5). Hence, strictly synchronous processes are just synchronous processes involving only informative (or "present") values.

A reaction is called *silent* if all variables are absent in the considered reaction. Now, if $P = P_1 \parallel P_2 \parallel \ldots \parallel P_K$ is a system composed of a set of components, each P_k has its own activation clock, consisting of the sequence of its non-silent reactions. Thus the activation clock of P_k is local to it, and activation clocks provide the adequate notion of local time reference for larger systems. For instance, if P_1 and P_2 do not interact at all (they share no variable), then there is no purpose that they should share some time reference. According to the synchronous model, *non interacting components simply possess independent, non synchronized, activation clocks.*

Thus, our synchronous model can mimic asynchrony. As soon as two processes can synchronize on some common clock, they can also exercise control on the basis of the absence of some variables at a given instant of this shared clock. Of course, sharing a clock needs broadcasting this clock among the different involved processes, this may require some protocol if the considered components are distributed.

Asynchronous. Reactions cannot be observed any more, no clock exists. Instead a *behaviour* is a tuple of *signals,* and each individual signal is a totally ordered sequence of (informative) values: $s_v = v(1), v(2), \ldots$ A process P is a set of behaviours. "Absence" cannot be sensed, and has therefore no meaning. Composition occurs by means of unifying each individual signal shared between two processes:

$$P_1 \parallel_a P_2 := P_1 \cap P_2$$

Hence, in this model, a network of reliable and order-preserving, point-to-point channels is assumed (since each individual signal must be preserved by the medium), but no synchronization between the different channels is required. This models in particular the communications via asynchronous unbounded FIFOs.

4.2 The Fundamental Problems

Many embedded systems use the Globally Asynchronous Locally Synchronous (GALS) architecture, which consists of a network of synchronous processes, in-

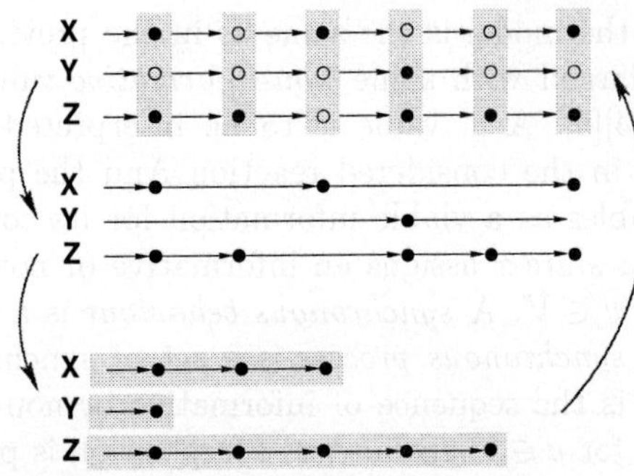

Fig. 6. Desynchronization / resynchronization. Unless desynchronization (shown by the downgoing arrows), resynchronization (shown by the upgoing arrows) is generally non determinate.

terconnected by asynchronous communications (as defined above). The central issue considered in this paper is:

> *what do we preserve when deploying a synchronous specification on a* GALS *architecture?*

The issue is best illustrated in Fig. 6. In this figure, we show a how desynchronization modifies a given run of a synchronous program. The synchronous run is shown on the top, it involves three variables, X, Y, Z. That this is a synchronous run is manifested by the presence of the successive rectangular patches, indicating the successive reactions. A black circle indicates that the considered variable is present in the considered reaction, and a white circle indicates that it is absent; for example, X is present in reactions 1, 3, 6. Desynchronizing this run amounts to 1/ removing the global synchronization clock indicating the successive reactions, and 2/ erasing the absent occurrences, for each variable individually, since absence has no meaning when no more synchronization clock is available. The result is shown in the middle. And there is no difference between the mid and bottom drawings, since time is only logical, not metric. Of course, the downgoing arrows define a proper desynchronization map, we formalize it below. In contrast, desynchronization is clearly not revertible in general, since there are many different possible ways of inserting absent occurrences, for each variable.

Problem 1: What if a synchronous program receives its data from an asynchronous environment? Focus on a synchronous program within a GALS architecture, it receives its inputs as a tuple of (non synchronized) signals. Since some variables can be absent in a given state, it can be the case that some signals will not be involved in a given reaction. But since the environment is asynchronous, this information is not provided by the environment. In other words, the environment does not offer to the synchronous program the correct model for its input stimuli. In general this will drastically affect the semantics of the program. However, some particular synchronous programs are robust against this type of difficulty. How to formalize this?

Let P be such a program, we recall some notations for subsequent use. Symbol $\sigma = x_0, x_1, x_2, \ldots$ denotes a behaviour of P, i.e., a sequence of states compliant with the reactions of P. V is the (finite) set of state variables of P. Each state x is a valuation for all $v \in V$, the valuation for v at state x is written $v(x)$. Hence we can write equivalently

$$\begin{aligned} \sigma &= (v(x_0))_{v\in V}, (v(x_1))_{v\in V}, (v(x_2))_{v\in V}, \ldots \\ &= (v(x_0), v(x_1), v(x_2), \ldots)_{v\in V} \\ &=_{\text{def}} (\sigma_v)_{v\in V} \end{aligned}$$

The valuation $v(x)$ is either an informative value belonging to some domain (e.g., boolean, integer), or it can be the possible special status *absent,* which is denoted by the special symbol $\perp$ in [3][4][5]. Now, for each separate v, remove the $\perp$ from the sequence $\sigma_v = v(x_0), v(x_1), v(x_2), \ldots$, this yields a (strict) *signal*

$$s_v =_{\text{def}} s_v(0), s_v(1), s_v(2), \ldots$$

where $s_v(0)$ is the first non $\perp$ term in σ_v and so on. Finally we set

$$\sigma^a =_{\text{def}} (s_v)_{v\in V}$$

The so-defined map $\sigma \mapsto \sigma^a$ takes a synchronous behaviour, and returns a uniquely defined asynchronous one. This results in a map

$$P \longmapsto P^a$$

defining the *desynchronization* P^a, of P. Clearly, the map $\sigma \mapsto \sigma^a$ is not one-to-one, and thus it is not invertible. However, we have shown in [3][4][5] the first fundamental result that

$$\begin{array}{l} \text{if } P \text{ satisfies a special condition called } \textit{endochrony,} \text{ then} \\ \forall \sigma^a \in P^a \text{ there exists a unique } \sigma \in P \text{ such that } \sigma \mapsto \sigma^a \text{ holds.} \end{array} \quad (6)$$

This means that, by knowing the formula defining reaction R such that $P = R^\omega$, we can uniquely reconstruct a synchronous behaviour, from observing its desynchronized version. In addition, it is shown in [3][4][5] that this reconstruction can be perfomed *on-line* meaning that each continuation of a prefix of σ^a yields a corresponding continuation for the corresponding prefix of σ.

Examples/counterexamples. Referring to Fig. 3, the program shown on the left is *not* endochronous. The environment tells the program which input event is present in the considered reaction, thus the environment provides the structuration of the run into its successive reactions. An asynchronous environment would not provide this service.

In contrast, the program on the right is endochronous. In its simplest form, all inputs are present at each clock tick. In a more complex form, some inputs can be absent, but this the presence/absence, for each input, is *explicitly* indicated by some corresponding always present boolean input. In other words, clocks are encoded using always present booleans; reading the value of these booleans tells the program which input is present in the considered reaction. Thus no extra synchronization role is played by the environment, the synchronization is entirely carried by the program itself (hence the name).

Clearly, if, for the considered program, it is known that the absence of some variable X implies the absence of some other variable Y, then there is no need to read the boolean clock of Y when X is absent. Endochrony introduced in [3][4][5] generalizes this informal analysis. ◇

The important point about result (6) is that endochrony can be model-checked [7] on the reaction R defining the synchronous process P. Also,

$$\text{any } P \text{ can be given a } \textit{wrapper } W \text{ making } P \| W \text{ endochronous.} \quad (7)$$

How can we use (6) to solve Problem 1 ? Let E be the model of the environment. It is an asynchronous process according to our above definition. Hence we need to formalize what it means having "P interacting with E" since they do not belong to the same world. The only possible formal meaning is

$$P^a \parallel_a E$$

Hence having P^a interacting with E results in an asynchronous behaviour $\sigma^a \in P^a$, but using (6) we can reconstruct uniquely its synchronous counterpart $\sigma \in P$. So, this solves Problem 1.

However, considering Problem 1 is not enough, since it only deals with a single synchronous program interacting with its asynchronous environment. It remains to consider the problem of mapping a synchronous network of synchronous programs onto a GALS architecture.

Problem 2 : What if we deploy a synchronous network of synchronous programs onto a GALS architecture ? Consider the simple case of a network of two programs P and Q. Since our communication media behave like a set of FIFOs, one per signal sent from one program to the other, we already know what the desynchronized behaviours of our deployed system will be, namely:

$$P^a \parallel_a Q^a .$$

There is not need for inserting any particular explicit model for the communication medium, since by definition $\|_a$-communication preserves each individual asynchronous signal (but not their global synchronization). In fact, Q^a will be the asynchronous environment for P^a and vice-versa.

[7] *Model checking* consists in exhaustively exploring the state space of a finite state model, for checking whether some given property is satisfied or not by this model. See [12].

Now, if P is endochronous, then, having solved Problem 1 we can uniquely recover a synchronous behaviour σ for P, from observing an asynchronous behaviour σ^a for P^a as produced by $P^a \parallel_a Q^a$.

Yet, we are not happy: it may be the case that there exists some asynchronous behaviour σ^a for P^a produced by $P^a \parallel_a Q^a$, which *cannot be obtained* by desynchronizing the synchronous behaviours of $P \parallel Q$. In fact we only know in general that

$$(P \parallel Q)^a \subseteq (P^a \parallel_a Q^a). \tag{8}$$

However, we have shown in [3][4][5] the second fundamental result that

if (P, Q) satisfies a special condition called *isochrony*, then equality in (8) indeed holds. (9)

The nice thing about isochrony is that it is *compositional*: if P_1, P_2, P_3 are pairwise isochronous, then $((P_1 \| P_2), P_3)$ is an isochronous pair, so we can refer to an *isochronous network* of synchronous processes—also, isochrony enjoys additional useful compositionality properties listed in [3][4][5]. Again, the condition of isochrony can be model-checked on the pair of reactions associated to the pair (P, Q), and

any pair (P, Q) can be given *wrappers* (W_P, W_Q) making $(P\|W_P, Q\|W_Q)$ an isochronous pair. (10)

Examples. A pair (P, Q) of programs having a single clocked communication (all shared variables possess the same clock), is isochronous. More generally, if the restriction of $P \parallel Q$, to the subset of shared variables, is endochronous, the the pair (P, Q) is isochronous: an isochronous pair does not need extra syncrhronization help from the environment, in order to communicate. ◇

Just a few additional words about the condition of isochrony, since isochrony is of interest per se. Synchronous composition $P\|Q$ is achieved by considering the conjunction

$$R_P \wedge R_Q$$

of corresponding reactions of P and Q. In taking this conjunction of relations, we ask in particular that common variables have identical status present/absent in both components, in the considered reaction. Assume we relax this latter requirement by simply requiring that the two reactions should only agree on effective values of common variables, *when they are both present.* This means that a given variable can be freely present in one component but absent in the other. This defines a "weakly synchronous" conjunction of reactions, we denote it by

$$R_P \wedge_a R_Q$$

In general, $R_P \wedge_a R_Q$ has more legal reactions than $R_P \wedge R_Q$. It turns out that the isochrony condition for the pair (P, Q) writes:

$$(R_P \wedge R_Q) \equiv (R_P \wedge_a R_Q).$$

4.3 A Sketch of the Resulting Methodology

How can we use (6) and (9) for a correct deployment on a GALS architecture? Well, consider a synchronous network of synchronous processes

$$P_1 \parallel P_2 \parallel \ldots \parallel P_K,$$

such that

(GALS$_1$) : *Each P_k is endochronous, and*
(GALS$_2$) : *The $P_k, k = 1, \ldots, K$ form an isochronous network.*

Using condition (GALS$_2$), we get

$$P_1^a \parallel_a (P_2^a \parallel_a \cdots \parallel_a P_K^a) \;=\; (P_1 \parallel P_2 \parallel \cdots \parallel P_K)^a.$$

Hence every asynchronous behaviour σ_1^a of P_1^a produced by its interaction with the rest of the asynchronous network $(P_2^a \parallel_a \cdots \parallel_a P_K^a)$ is a desynchronized version of a synchronous behaviour of P_1 produced by its interaction with the rest of the synchronous network. Hence the asynchronous communication does not add spurious asynchronous behaviour. Next, by (GALS$_1$), we can reconstruct on-line this unique synchronous behaviour σ_1, from σ_1^a. Hence,

Theorem 1. *For $P_1 \parallel P_2 \parallel \ldots \parallel P_K$ a synchronous network, assume the deployment is simply performed by using an asynchronous mode of communication between the different programs. If the network satisfies conditions* (GALS$_1$) *and* (GALS$_2$)*, then the original synchronous semantics of each individual program of the deployed* GALS *architecture is preserved (of course the global synchronous semantics is not preserved).*

To summarize, *a synchronous network satisfying conditions* (GALS$_1$) *and* (GALS$_2$) *is the right model for a* GALS*–targetable design,* and we have a correct-by-construction deployment technique for GALS architectures. The method consists in preparing the design to satisfy (GALS$_1$) and (GALS$_2$) by adding the proper wrappers, and then performing bruteforce desynchronization as stated in Theorem 1.

5 Conclusion

There are important distributed computing systems which are neither massive nor high performance, systems of that kind are in fact numerous—they are estimated to constitute more than 80% of the computer systems. Still, their design can be extremely complex, and it raises several difficult problems of interest for computer scientists. These are mainly related to tracking the correctness of the implementation throughout the different design phases. Synchronous languages have emerged as an efficient vehicle for this, but the distributed implementation of synchronous programs raises some fundamental difficulties, which we have briefly reviewed.

Still, this issue is not closed, since not every distributed architecture in use in actual embedded systems complies with our model of "reliable" asynchrony [17]. In fact, the bus architecture used at Airbus does not satisfy our assumptions, and there are excellent reasons for this. Many additional studies are underway to address actual architectures in use in important safety critical systems [10][11].

Acknowledgement

The author is gratefully indebted to Luc Bougé for his help in selecting the focus and style of this paper, and to Joel Daniels for correcting a draft version of it.

References

1. A. Benveniste and G. Berry, The synchronous approach to reactive real-time systems. *Proceedings of the IEEE*, 79, 1270–1282, Sept. 1991.
2. A. Benveniste, P. Caspi, S.A. Edwards, N. Halbwachs, P. Le Guernic, and R. de Simone. The synchronous languages twelve years later. To appear in *Proceedings of the IEEE,* special issue on Embedded Systems, Sastry and Sztipanovits Eds., 2002.
3. A. Benveniste, B. Caillaud, and P. Le Guernic. Compositionality in dataflow synchronous languages: specification & distributed code generation. *Information and Computation*, 163, 125-171, 2000.
4. A. Benveniste, B. Caillaud, and P. Le Guernic. From synchrony to asynchrony. In J.C.M. Baeten and S. Mauw, editors, CONCUR'99, Concurrency Theory, 10th International Conference, *Lecture Notes in Computer Science*, vol. 1664, 162–177, Springer Verlag, 1999.
5. A. Benveniste. Some synchronization issuess when designing embedded systems. In Proc. of the first int. workshop on Embedded Software, EMSOFT'2001, T.A. Henzinger and C.M. Kirsch Eds., *Lecture Notes in Computer Science*, vol 2211, 32–49, Springer Verlag, 2001.
6. G. Berry, *Proof, Language and Interaction: Essays in Honour of Robin Milner*, ch. The Foundations of Esterel. MIT Press, 2000.
7. F. Boussinot and R. de Simone, "The Esterel language," *Proceedings of the IEEE*, vol. 79, 1293–1304, Sept. 1991.
8. J. Buck, S. Ha, E. Lee, and D. Messerschmitt, "Ptolemy: A framework for simulating and prototyping heterogeneous systems," *International Journal of computer Simulation*, special issue on Simulation Software Development, 1994.
9. L.P. Carloni, K.L. McMillan, and A.L. Sangiovanni-Vincentelli. The theory of latency insensitive design. *IEEE Transactions on Computer-Aided Design of Integrated Circuits and Systems,* 20(9), Sept. 2001.
10. P. Caspi and R. Salem. Threshold and Bounded-Delay Voting in Critical Control Systems. Proceedings of Formal Techniques in Real-Time and Fault-Tolerant Systems, Joseph Mathai Ed., *Lecture Notes in Computer Science,* vol. 1926, 68–81, Springer Verlag, Sept. 2000.
11. P. Caspi. Embedded control: from asynchrony to synchrony and back. In Proc. of the first int. workshop on Embedded Software, EMSOFT'2001, T.A. Henzinger and C.M. Kirsch Eds., *Lecture Notes in Computer Science,* vol 2211, 80–96, Springer Verlag, 2001.

12. E.M. Clarke, E.A. Emerson, and A.P. Sistla. Automatic verification of finite-state concurrent systems using temporal logic specifications. *ACM Trans. on Programming Languages and Systems,* 8(2), 244–263, April 1986.
13. N. Halbwachs, P. Caspi, P. Raymond, and D. Pilaud, "The synchronous data flow programming language LUSTRE," *Proceedings of the IEEE*, vol. 79, 1305–1320, Sept. 1991.
14. N. Halbwachs. *Synchronous programming of reactive systems.* Kluwer, 1993.
15. D. Harel, "Statecharts: A visual formalism for complex systems," *Science of Computer Programming,* vol. 8, 231–274, June 1987.
16. D. Harel and M. Politi. *Modeling Reactive Systems with Statecharts.* McGraw-Hill, 1998.
17. H. Kopetz, *Real-time systems, design principles for distributed embedded applications, 3rd edition.* London: Kluwer academic publishers, 1997.
18. P. Le Guernic, T. Gautier, M. Le Borgne, and C. Le Maire, "Programming real-time applications with SIGNAL," *Proceedings of the IEEE*, vol. 79, 1321–1336, Sept. 1991.
19. J. Rumbaugh, I. Jacobson, and G. Booch, *Tne Unified Modeling Language reference manual.* Object technologies series, Addison-Wesley, 1999.
20. J. Sztipanovits and G. Karsai. Embedded software: challenges and opportunities. In Proc. of the first int. workshop on Embedded Software, EMSOFT'2001, T.A. Henzinger and C.M. Kirsch Eds., *Lecture Notes in Computer Science,* vol 2211, 403–415, Springer Verlag, 2001.

The Forgotten Factor: Facts on Performance Evaluation and Its Dependence on Workloads

Dror G. Feitelson

School of Computer Science and Engineering
The Hebrew University, 91904 Jerusalem, Israel
feit@cs.huji.ac.il
http://www.cs.huji.ac.il/~feit

Abstract. The performance of a computer system depends not only on its design and implementation, but also on the workloads it has to handle. Indeed, in some cases the workload can sway performance evaluation results. It is therefore crucially important that representative workloads be used for performance evaluation. This can be done by analyzing and modeling existing workloads. However, as more sophisticated workload models become necessary, there is an increasing need for the collection of more detailed data about workloads. This has to be done with an eye for those features that are really important.

1 Introduction

The scientific method is based on the ability to reproduce and verify research results. But in practice, the research literature contains many conflicting accounts and contradictions — especially multiple conflicting claims to be better than the competition. This can often be traced to differences in the methodology or the conditions used in the evaluation. In this paper we focus on one important aspect of such differences, namely differences in the workloads being used. In particular, we will look into the characterization and modeling of workloads used for the evaluation of parallel systems.

The goal of performance evaluation is typically not to obtain absolute numbers, but rather to differentiate between alternatives. This can be done in the context of system design, where the better design is sought, or as part of a procurement decision, where the goal is to find the option that provides the best value for a given investment. In any case, an implicit assumption is that differences in the evaluation results reflect real differences in the systems under study. But this is not always the case. Evaluation results depend not only on the systems, but also on the metrics being used and on the workloads to which the systems are subjected.

To complicate matters further, there may be various interactions between the system, workload, and metric. Some of these interactions lead to problems, as described below. But some are perfectly benign. For example, an interaction

B. Monien and R. Feldmann (Eds.): Euro-Par 2002, LNCS 2400, pp. 49–60.

between the system and a metric may actually be a good thing. If systems are designed with different objectives in mind, metrics that measure these objectives should indeed rank them differently. In fact, such metrics are exactly what we need if we know which objective function we wish to emphasize. An interaction between the workload and the metric is also possible, and may be meaningless. For example, if one workload contains longer jobs than another, its average response time will also be higher. On the other hand, interactions between a system and a workload may be very important, as they may help identify system vulnerabilities.

But when the effects leading to performance evaluation results are unknown and not understood, this is a problem. Conflicting results cast a shadow of doubt on our confidence in all the results. A solid scientific and experimental methodology is required in order to prevent such situations.

2 Examples of the Importance of Workloads

To support the claim that workloads make a difference, this section presents three specific cases in some detail. These are all related to the scheduling of parallel jobs.

A simple model of parallel jobs considers them as rectangles in processors ×time space: each job needs a certain number of processors for a certain interval of time. Scheduling is then the packing of these job-rectangles into a larger rectangle that represents the available resources. In an on-line setting, the time dimension may not be known in advance. Dealing with this using preemption means that the job rectangle is cut into several slices, representing the work done during each time slice.

2.1 Effect of Job-Size Distribution

The packing of jobs obviously depends on the distribution of job sizes. A good example is provided by the DHC scheme [12], in which a buddy system is used for processor allocation: each request is extended to the next power of two, and allocations are always done is power-of-two blocks of processors. This scheme was evaluated with three different distributions: a uniform distribution in which all sizes are equally likely, a harmonic distribution in which the probability of size s is proportional to $1/s$, and a uniform distribution on powers of two. Both analysis and simulations showed significant differences between the utilizations that could be obtained for the three distributions [12]. This corresponds to different degrees of fragmentation that are inherent to packing with these distributions. For example, with a uniform distribution, rounding each request size up to the next power of two leads to 25% loss to fragmentation — the average between no loss (if the request is an exact power of two) to nearly 50% loss (if the request is just above a power of two, and we round up to the next one). The DHC scheme recovers part of this lost space, so the figure is actually only 20% loss, as shown in Figure 1.

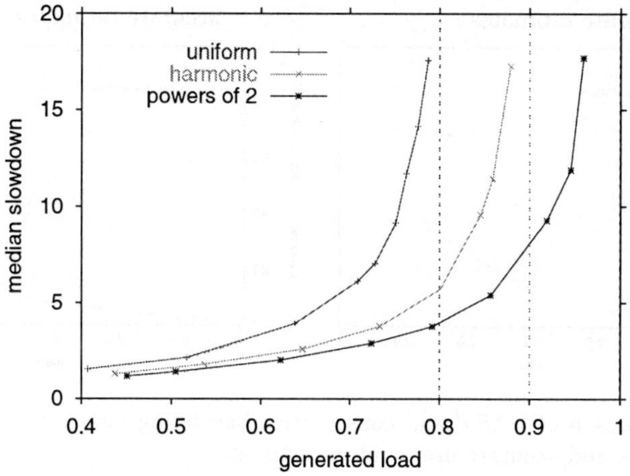

Fig. 1. Simulation results showing normalized response time (slowdown) as a function of load for processor allocation using DHC, from [12]. The three curves are for exactly the same system — the only difference is in the statistics of the workload. The dashed lines are proven bounds on the achievable utilization for the three workloads.

Note that this analysis tells us what to expect in terms of performance, provided we know the distribution of job sizes. But what is a typical distribution encountered in real systems in production use? Without such knowledge, the evaluation cannot provide a definitive answer.

2.2 Effect of Job Scaling Pattern

It is well-known that average response time is reduced by scheduling short jobs first. The problem is that the runtime is typically not known in advance. But in parallel systems scheduling according to job *size* may unintentionally also lead to scheduling by *duration*, if there is some statistical correlation between these two job attributes.

As it turns out, the question of whether such a correlation exists is not easy to settle. Three application scaling models have been proposed in the literature [30, 23]:

– *Fixed work.* This assumes that the work done by a job is fixed, and parallelism is used to solve the same problems faster. Therefore the runtime is assumed to be inversely proportional to the degree of parallelism (negative correlation). This model is the basis for Amdahl's law.
– *Fixed time.* Here it is assumed that parallelism is used to solve increasingly larger problems, under the constraint that the total runtime stays fixed. In this case, the runtime distribution is independent of the degree of parallelism (no correlation).
– *Memory bound.* If the problem size is increased to fill the available memory on the larger machine, the amount of productive work typically grows at

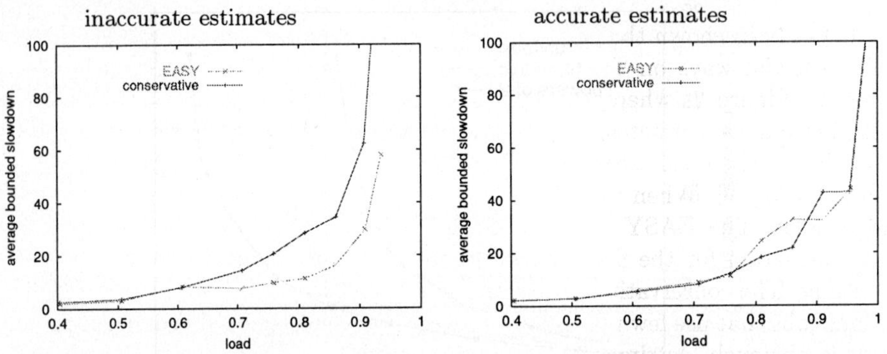

Fig. 2. Comparison of EASY and conservative backfilling using the CTC workload, with inaccurate and accurate user runtime estimates.

least linearly with the parallelism. The overheads associated with parallelism always grow superlinearly. Thus the total execution time actually increases with added parallelism (a positive correlation).

Evaluating job scheduling schemes with workloads that conform to the different models leads to drastically different results. Consider a workload that is composed of jobs the use power-of-two processors. In this case a reasonable scheduling algorithm is to cycle through the different sizes, because the jobs of each size pack well together [16]. This works well for negatively correlated and even uncorrelated workloads, but is bad for positively correlated workloads [16, 17]. The reason is that under a positive correlation the largest jobs dominate the machine for a long time, blocking out all others. As a result, the average response time of all other jobs grows considerably.

But which model actually reflects reality? Again, evaluation results depend on the selected model of scaling; without knowing which model is more realistic, we cannot use the performance evaluation results.

2.3 Effect of User Runtime Estimates

Returning to the 2D packing metaphor, a simple optimization is to allow the insertion of small jobs into holes left in the schedule. This is called backfilling, because new jobs from the back of the queue are used to fill current idle resources. The two common variants of backfilling are conservative backfilling, which makes strict reservations for all queued jobs, and EASY backfilling, which only makes a reservation for the first queued job [19]. Both rely on users to provide estimates of how long each job will run — otherwise it is impossible to know whether a backfill job may conflict with an earlier reservation. Users are expected to be highly motivated to provide accurate estimates, as low estimates improve the chance for backfilling and significantly reduced waiting time, but underestimates will cause the job to be killed by the system.

It has been shown that in some cases performance evaluation results depend in non-trivial ways on the accuracy of the runtime estimates. An example is given in Figure 2, where EASY backfilling is found to have lower slowdown with inaccurate estimates, whereas conservative backfilling is better at least for some loads when the estimates are accurate. This contradiction is the result of the following [8]. When using accurate estimates, the schedule does not contain large holes. The EASY scheduler is not affected too much, as it only heeds the reservation for the first queued job; other jobs do not figure in backfilling decisions. The conservative scheduler, on the other hand, achieves less backfilling of long jobs that use few processors, because it takes all queued jobs into account. This is obviously detrimental to the performance of these long jobs, but turns out to be beneficial for short jobs that don't get delayed by these long jobs. As the slowdown metric is dominated by short jobs, it shows the conservative backfiller to be better when accurate estimates are used, but not when inaccurate estimates are used.

Once again, performance evaluation has characterized the situation but not provided an answer to the basic question: which is better, EASY or conservative backfilling? This depends on the workload, and specifically, on whether user runtime estimates are indeed accurate as we expect them to be.

3 Workload Analysis and Modeling

As shown above, workloads can have a big impact on performance evaluation results. And the mechanisms leading to such effects can be intricate and hard to understand. Thus it is crucially important that representative workloads be used, which are as close as possible to the real workloads that may be expected when the system is actually deployed. In particular, unbased assumptions about the workload are very dangerous, and should be avoided.

3.1 Data-Less Modeling

But how does one know what workload to expect? In some cases, when truly innovative systems are designed, it is indeed impossible to predict what workloads will evolve. The only recourse is then to try and predict the space of possible workloads, and thoroughly sample this space. In making such predictions, one should employ recurring patterns from known workloads as guidelines. For example, workloads are often bursty and self-similar, process or task runtimes are often heavy tailed, and object popularity is often captured by a Zipf distribution [4].

3.2 Data-Based Modeling

The more common case, however, is that new systems are an improvement or evolution of existing ones. In such cases, studying the workload on existing systems can provide significant data regarding what may be expected in the future.

The case of job scheduling on parallel systems is especially fortunate, because data is available in the form of accounting logs [22]. Such logs contain the details of all jobs run on the system, including their arrival, start, and end times, the number of processors they used, the amount of memory used, the user who ran the job, the executable file name, etc. By analyzing this data, a statistical model of the workload can be created [7, 9]. This should focus on recurrent features that appear in logs derived from different installations. At the same time, features that are inconsistent at different installations should also be identified, so that their importance can be verified.

A good example is the first such analysis, published in 1995, based on a log of three months of activity on the 128-node NASA Ames iPSC/860 hypercube supercomputer. This analysis provided the following data [11]:

- The distribution of job sizes (in number of nodes) for system jobs, and for user jobs classified according to when they ran: during the day, at night, or on the weekend.
- The distribution of total resource consumption (node seconds), for the same job classifications.
- The same two distributions, but classifying jobs according to their type: those that were submitted directly, batch jobs, and Unix utilities.
- The changes in system utilization throughout the day, for weekdays and weekends.
- The distribution of multiprogramming level seen during the day, at night, and on weekends. This also included the measured down time (a special case of 0 multiprogramming).
- The distribution of runtimes for system jobs, sequential jobs, and parallel jobs, and for jobs with different degrees of parallelism. This includes a connection between common runtimes and the queue time limits of the batch scheduling system.
- The correlation between resource usage and job size, for jobs that ran during the day, at night, and over the weekend.
- The arrival pattern of jobs during the day, on weekdays and weekends, and the distribution of interarrival times.
- The correlation between the time a job is submitted and its resource consumption.
- The activity of different users, in terms of number of jobs submitted, and how many of them were different.
- Profiles of application usage, including repeated runs by the same user and by different users, on the same or on different numbers of nodes.
- The dispersion of runtimes when the same application is executed many times.

Practically all of this empirical data was unprecedented at the time. Since then, several other datasets have been studied, typically emphasizing job sizes and runtimes [27, 14, 15, 6, 2, 1, 18]. However, some new attributes have also been considered, such as speedup characteristics, memory usage, user estimates of runtime, and the probability that a job be cancelled [20, 10, 19, 2].

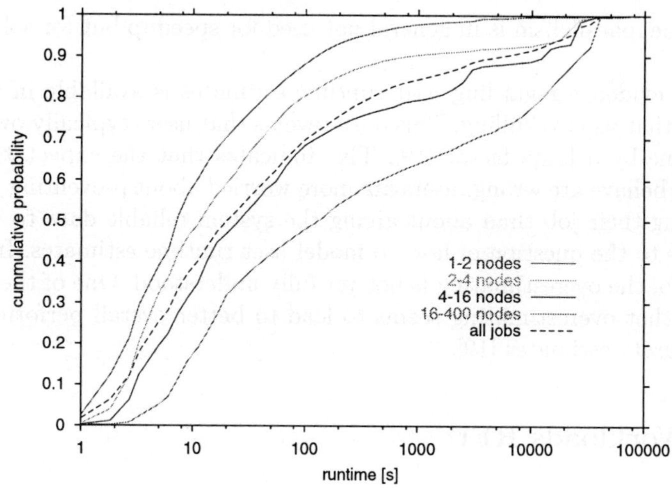

Fig. 3. The cumulative distribution functions of runtimes of jobs with different sizes, from the SDSC Paragon.

3.3 Some Answers and More Questions

Based on such analyses, we can give answers to the questions raised in the previous section. All three are rather surprising.

The distribution of job sizes has often been assumed to be bimodal: small jobs that are used for debugging, and large jobs that use the full power of the parallel machine for production runs. In fact, there are very many small jobs and rather few large jobs, and large systems often do not have *any* jobs that use the full machine. especially surprising is the high fraction of serial jobs, which is typically in the range of 20–30%. Another prominent feature is the emphasis on power-of-two job sizes, which typically account for over 80% of the jobs. This has been claimed to be an artifact of the use of such size limits in the queues of batch scheduling system, or the result of inertia in system where such limits were removed; the claim is supported by direct user data [3]. Nevertheless, the fact remains that users continue to prefer powers of two. The question for workload modeling is then whether to use the "real" distribution or the empirical distribution in models.

It is hard to obtain direct evidence regarding application scaling from accounting logs, because they typically do not contain runs of the same applications using different numbers of nodes, and even if they did, we do not know whether these runs were aimed at solving the same problem. However, we can compare the runtime statistics of jobs that use different numbers of nodes. the result is that there is little if any correlation in the statistical sense. However, the distributions of runtimes for small and large jobs do tend to be different, with large jobs often having longer runtimes [7] (Figure 3). This favors the memory bound or fixed time scaling models, and contradicts the fixed work model. There is also some evidence that larger jobs use more memory [10]. Thus, within a single machine, parallelism is in general not used for speedup but for solving larger problems.

Direct evidence regarding user runtime estimates is available in the logs of machines that use backfilling. This data reveals that users typically overestimate job runtime by a large factor [19]. This indicates that the expectations about how users behave are wrong: users are more worried about preventing the system from killing their job than about giving the system reliable data to work with. This leads to the question of how to model user runtime estimates. In addition, the effect of the overestimating is not yet fully understood. One of the surprising results is that overestimating seems to lead to better overall performance than using accurate estimates [19].

4 A Workloads RFI[1]

There is only so much data that can be obtained from accounting logs that are collected anyway. To get a more detailed picture, active data collection is required. When studying the performance of parallel systems, we need high-resolution data about the behavior of applications, as this affects the way they interact with each other and with the system, and influences the eventual performance measures.

4.1 Internal Structure of Applications

Workload models based on job accounting logs tend to regard parallel jobs as rigid: they require a certain number of processors for a given time. But runtime may depend on the system. For example, runs of the ESP system-level benchmark revealed that executions of the same set of jobs on two different architectures led to completely different job durations [28]. The reason is that different applications make different use of the system in terms of memory, communication, and I/O. Thus an application that requires a lot of fine-grain communication may be relatively slow on a system that does not provide adequate support, but relatively fast on a system with an overpowered communication network.

In order to evaluate advanced schedulers that take multiple resources into account we therefore need more detailed workload models. It is not enough to model a job as a rectangle in processors×time space. We need to know about its internal structure, and model that as well. Such a model can then form the basis for an estimation of the speedup a job will display on a given system, when provided with a certain set of resources.

A simple proposal was given in [13]. The idea is to model a parallel application as a set of tasks, which are either independent of each other, or need to synchronize repeatedly using barriers. The number of tasks, number of barriers, and granularity are all parameters of the model. While this is a step in

[1] Request for Information

the right direction, the modeling of communication is minimal, and interactions with other system resources are still missing. Moreover, representative values for the model parameters are unknown.

There has been some work on characterizing the communication behavior of parallel applications [5, 25]. This has confirmed the use of barrier-like collective communications, but also identified the use of synchronization-avoiding non-blocking communication. The granularity issue has remained open: both very small and very big intervals between communication events have been measured, but the small ones are probably due to multiple messages being sent one after the other in the same communication phase. The granularity of computation phases that come between communication phases is unclear. Moreover, the analysis was done for a small set of applications in isolation; what we really want to know is the distribution of granularities in a complete workload.

More detailed work was done on I/O behavior [21, 24]. Like communication, I/O is repetitive and bursty. But again, the granularity at which it occurs (or rather, the distribution of granularities in a workload) is unknown. An interesting point is that interleaved access from multiple processes to the same file may lead to synchronization that is required in order to use the disks efficiently, even if the application semantics do not dictate any strict synchronization.

Very little work has been done on the memory behavior of parallel applications. The conventional wisdom is that large-scale scientific applications require a lot of memory, and use all of it all the time without any significant locality. Still, it would be nice to root this in actually observations, especially since it is at odds with reports of the different working set sizes of SPLASH applications [29]. Somewhat disturbing also is a single paper that investigated the paging patterns of different processes in the same job, and unexpectedly found them to be very dissimilar [26]. More work is required to verify or refute the generality of this result.

4.2 User Behavior

Workload models typically treat job arrivals as coming from some independent external source. Their statistics are therefore independent of the system behavior. While this makes the evaluation easier, it is unrealistic. In reality, the user population is finite and often quite small; when the users perceive the system as not responsive, they tend to reduce their use (Figure 4). This form of negative feedback actually fosters system stability and may prevent overload conditions.

Another important aspect of user behavior is that users tend to submit the same job over and over again. Thus the workload a system has to handle may be rather homogeneous and predictable. This is very different from a random sampling from a statistical distribution. In fact, it can be called "localized sampling": while over large stretches of time, e.g. a whole year, the whole distribution is sampled, in any given week only a small part of it is sampled.

In terms of performance evaluation, two important research issues may be identified in this regard. One is how to perform such localized sampling, or in other words, how to characterize, model, and mimic the short-range locality

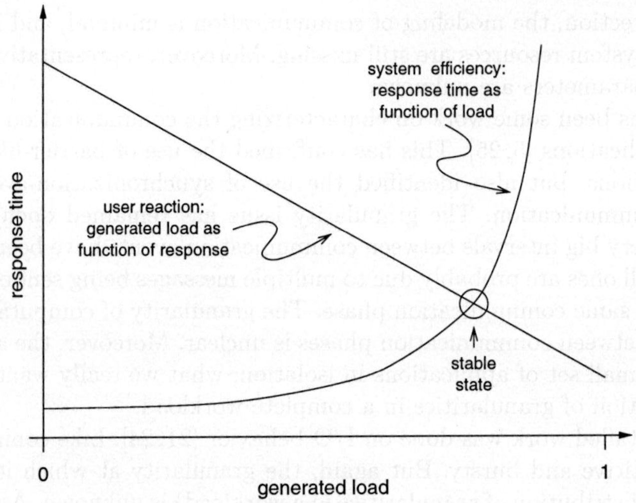

Fig. 4. The workload placed on a system may be affected by the system performance, due to a feedback loop through the users.

of real workloads. the other is to figure out what effect this has on system performance, and under what conditions.

5 The Rocky Road Ahead

Basing performance evaluation on facts rather than on assumptions is important. But it shouldn't turn into an end in itself. As Henri Poincaré said,

> Science is built up with facts, as a house is with stones. But a collection of facts is no more a science than a heap of stones is a house.

The systems we now build are complex enough to require scientific methodology to study their behavior. This must be based on observation and measurement. But knowing what to measure, and how to connect the dots, is not easy.

Realistic and detailed workload models carry with them two dangers. One is clutter and obfuscation — with more details, more parameters, and more options, there are more variations to check and measure. Many of these are probably unimportant, and serve only to hide the important ones. The other danger is the substitution of numbers for understanding. With more detailed models, it becomes harder to really understand the fundamental effects that are taking place, as opposed to merely describing them. This is important if we want to learn anything that will be useful for other problems except the one at hand. These two dangers lead to a quest for Einstein's equilibrium:

> Everything should be made as simple as possible, but not simpler.

The challenge is to identify the important issues, focus on them, and get them right. Unbased assumptions are not good, but excessive detail and clutter is probably not better.

Acknowledgement

This research was supported by the Israel Science Foundation (grant no. 219/99).

References

1. S-H. Chiang and M. K. Vernon, "*Characteristics of a large shared memory production workload*". In *Job Scheduling Strategies for Parallel Processing*, D. G. Feitelson and L. Rudolph (eds.), pp. 159–187, Springer Verlag, 2001. Lect. Notes Comput. Sci. vol. 2221.
2. W. Cirne and F. Berman, "*A comprehensive model of the supercomputer workload*". In 4th *Workshop on Workload Characterization*, Dec 2001.
3. W. Cirne and F. Berman, "*A model for moldable supercomputer jobs*". In 15th *Intl. Parallel & Distributed Processing Symp.*, Apr 2001.
4. M. E. Crovella, "*Performance evaluation with heavy tailed distributions*". In *Job Scheduling Strategies for Parallel Processing*, D. G. Feitelson and L. Rudolph (eds.), pp. 1–10, Springer Verlag, 2001. Lect. Notes Comput. Sci. vol. 2221.
5. R. Cypher, A. Ho, S. Konstantinidou, and P. Messina, "*A quantitative study of parallel scientific applications with explicit communication*". *J. Supercomput.* **10(1)**, pp. 5–24, 1996.
6. A. B. Downey, "*A parallel workload model and its implications for processor allocation*". In 6th *Intl. Symp. High Performance Distributed Comput.*, Aug 1997.
7. A. B. Downey and D. G. Feitelson, "*The elusive goal of workload characterization*". *Performance Evaluation Rev.* **26(4)**, pp. 14–29, Mar 1999.
8. D. G. Feitelson, *Analyzing the Root Causes of Performance Evaluation Results.* Technical Report 2002–4, School of Computer Science and Engineering, Hebrew University, Mar 2002.
9. D. G. Feitelson, "*The effect of workloads on performance evaluation*". In *Performance Evaluation of Complex Systems: Techniques and Tools*, M. Calzarossa (ed.), Springer-Verlag, Sep 2002. Lect. Notes Comput. Sci. Tutorials.
10. D. G. Feitelson, "*Memory usage in the LANL CM-5 workload*". In *Job Scheduling Strategies for Parallel Processing*, D. G. Feitelson and L. Rudolph (eds.), pp. 78–94, Springer Verlag, 1997. Lect. Notes Comput. Sci. vol. 1291.
11. D. G. Feitelson and B. Nitzberg, "*Job characteristics of a production parallel scientific workload on the NASA Ames iPSC/860*". In *Job Scheduling Strategies for Parallel Processing*, D. G. Feitelson and L. Rudolph (eds.), pp. 337–360, Springer-Verlag, 1995. Lect. Notes Comput. Sci. vol. 949.
12. D. G. Feitelson and L. Rudolph, "*Evaluation of design choices for gang scheduling using distributed hierarchical control*". *J. Parallel & Distributed Comput.* **35(1)**, pp. 18–34, May 1996.
13. D. G. Feitelson and L. Rudolph, "*Metrics and benchmarking for parallel job scheduling*". In *Job Scheduling Strategies for Parallel Processing*, D. G. Feitelson and L. Rudolph (eds.), pp. 1–24, Springer-Verlag, 1998. Lect. Notes Comput. Sci. vol. 1459.

14. S. Hotovy, "*Workload evolution on the Cornell Theory Center IBM SP2*". In *Job Scheduling Strategies for Parallel Processing,* D. G. Feitelson and L. Rudolph (eds.), pp. 27–40, Springer-Verlag, 1996. Lect. Notes Comput. Sci. vol. 1162.
15. J. Jann, P. Pattnaik, H. Franke, F. Wang, J. Skovira, and J. Riodan, "*Modeling of workload in MPPs*". In *Job Scheduling Strategies for Parallel Processing,* D. G. Feitelson and L. Rudolph (eds.), pp. 95–116, Springer Verlag, 1997. Lect. Notes Comput. Sci. vol. 1291.
16. P. Krueger, T-H. Lai, and V. A. Dixit-Radiya, "*Job scheduling is more important than processor allocation for hypercube computers*". *IEEE Trans. Parallel & Distributed Syst.* **5(5)**, pp. 488–497, May 1994.
17. V. Lo, J. Mache, and K. Windisch, "*A comparative study of real workload traces and synthetic workload models for parallel job scheduling*". In *Job Scheduling Strategies for Parallel Processing,* D. G. Feitelson and L. Rudolph (eds.), pp. 25–46, Springer Verlag, 1998. Lect. Notes Comput. Sci. vol. 1459.
18. U. Lublin and D. G. Feitelson, *The Workload on Parallel Supercomputers: Modeling the Characteristics of Rigid Jobs.* Technical Report 2001-12, Hebrew University, Oct 2001.
19. A. W. Mu'alem and D. G. Feitelson, "*Utilization, predictability, workloads, and user runtime estimates in scheduling the IBM SP2 with backfilling*". *IEEE Trans. Parallel & Distributed Syst.* **12(6)**, pp. 529–543, Jun 2001.
20. T. D. Nguyen, R. Vaswani, and J. Zahorjan, "*Parallel application characterization for multiprocessor scheduling policy design*". In *Job Scheduling Strategies for Parallel Processing,* D. G. Feitelson and L. Rudolph (eds.), pp. 175–199, Springer-Verlag, 1996. Lect. Notes Comput. Sci. vol. 1162.
21. N. Nieuwejaar, D. Kotz, A. Purakayastha, C. S. Ellis, and M. L. Best, "*File-access characteristics of parallel scientific workloads*". *IEEE Trans. Parallel & Distributed Syst.* **7(10)**, pp. 1075–1089, Oct 1996.
22. *Parallel workloads archive.* URL http://www.cs.huji.ac.il/labs/parallel/workload/.
23. J. P. Singh, J. L. Hennessy, and A. Gupta, "*Scaling parallel programs for multiprocessors: methodology and examples*". *Computer* **26(7)**, pp. 42–50, Jul 1993.
24. E. Smirni and D. A. Reed, "*Workload characterization of input/output intensive parallel applications*". In 9th *Intl. Conf. Comput. Performance Evaluation,* pp. 169–180, Springer-Verlag, Jun 1997. Lect. Notes Comput. Sci. vol. 1245.
25. J. S. Vetter and F. Mueller, "*Communication characteristics of large-scale scientific applications for contemporary cluster architectures*". In 16th *Intl. Parallel & Distributed Processing Symp.,* May 2002.
26. K. Y. Wang and D. C. Marinescu, "*Correlation of the paging activity of individual node programs in the SPMD execution model*". In 28th *Hawaii Intl. Conf. System Sciences,* vol. I, pp. 61–71, Jan 1995.
27. K. Windisch, V. Lo, R. Moore, D. Feitelson, and B. Nitzberg, "*A comparison of workload traces from two production parallel machines*". In 6th *Symp. Frontiers Massively Parallel Comput.,* pp. 319–326, Oct 1996.
28. A. Wong, L. Oliker, W. Kramer, T. Kaltz, and D. Bailey, "*System utilization benchmark on the Cray T3E and IBM SP2*". In *Job Scheduling Strategies for Parallel Processing,* D. G. Feitelson and L. Rudolph (eds.), pp. 56–67, Springer Verlag, 2000. Lect. Notes Comput. Sci. vol. 1911.
29. S. C. Woo, M. Ohara, E. Torrie, J. P. Singh, and A. Gupta, "*The SPLASH-2 programs: characterization and methodological considerations*". In 22nd *Ann. Intl. Symp. Computer Architecture Conf. Proc.,* pp. 24–36, Jun 1995.
30. P. H. Worley, "*The effect of time constraints on scaled speedup*". *SIAM J. Sci. Statist. Comput.* **11(5)**, pp. 838–858, Sep 1990.

Sensor Networks – Promise and Challenges

Pradeep K. Khosla

Electrical and Computer Engineering,
Carnegie Mellon,
Pittsburgh, PA 15213, USA
pkk@ece.cmu.edu

Abstract. Imagine a world in which there exist hundreds of thousands sensors. These sensors monitor a range of parameters – from the mundane such as temperature to more complex such as video imagery. These sensors may be either static or could be mounted on mobile bases. And further, these sensors could be deployed inside or outside and in small or very large numbers. It is anticipated that some of these sensors will not work either due to hardware or software failures. However, it is expected that the system that comprises of these sensors will work all the time – it will be perpetually available. When some of the sensors or their components have to be replaced, this would have to be done in the "hot" mode. And in the ideal situation, once deployed a system such as the one described above will never have to be rebooted.

The world that you have imagined above is entirely within the realm of possibility. However, it is not without significant challenges – both technical and societal – that we will be able to build, deploy, and utilize such a system of sensor networks. A system like the above will be a consequence of the convergence of many technologies and many areas. For the above system to be realized, the areas of networking (wired and wireless), distributed computing, distributed sensing and decision making, distributed robotics, software systems, and signal processing, for example, will have to converge.

In this talk we will describe a vision for a system of sensor networks, we will identify the challenges, and we will show some simple examples of working systems such as the Millibot project at Carnegie Mellon – examples that give hope but are very far from the above described system.

B. Monien and R. Feldmann (Eds.): Euro-Par 2002, LNCS 2400, p. 61.

Concepts and Technologies for a Worldwide Grid Infrastructure

Alexander Reinefeld and Florian Schintke

Zuse Institute Berlin (ZIB)
{ar,schintke}@zib.de

Abstract. Grid computing got much attention lately—not only from the academic world, but also from industry and business. But what remains when the dust of the many press articles has settled? We try to answer this question by investigating the concepts and techniques grids are based on. We distinguish three kinds of grids: the HTML-based *Information Grid*, the contemporary *Resource Grid*, and the newly evolving *Service Grid*.
We show that grid computing is not just another hype, but has the potential to open new perspectives for the co-operative use of distributed resources. Grid computing is on the right way to solve a key problem in our distributed computing world: the discovery and coordinated use of distributed services that may be implemented by volatile, dynamic local resources.

1 Three Kinds of Grids

Grids have been established as a new paradigm for delivering information, resources and services to users. Current grid implementations cover several application domains in industry and academia. In our increasingly networked world, location transparency of services is a key concept. In this paper, we investigate the concepts and techniques grids are based on [7, 8, 10, 13, 17]. We distinguish three categories:

- Information Grid,
- Resource Grid,
- Service Grid.

Figure 1 illustrates the relationship and interdependencies of these three grids with respect to the access, use and publication of meta information.

With the invention of the world wide web in 1990, Tim Berners-Lee and Robert Calliau took the first and most important step towards a global grid infrastructure. In just a few years the exponential growth of the Web created a publicly available network infrastructure for computers—an omnipresent **Information Grid** that delivers information on any kind of topic to any place in the world. Information can be retrieved by connecting a computer to the public telephone network via a modem, which is just as easy as plugging into the electrical power grid.

B. Monien and R. Feldmann (Eds.): Euro-Par 2002, LNCS 2400, pp. 62–71.

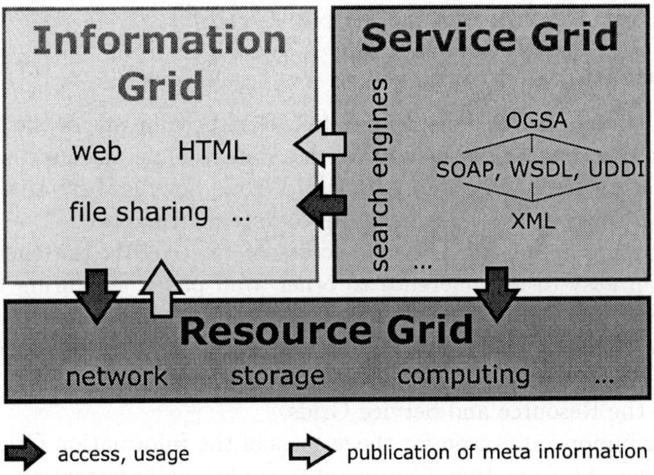

Fig. 1. The three Grids and their relation to each other.

File sharing services like *Gnutella, Morpheus* or *E-Donkey* are also part of today's Information Grid. In contrast to the Web, the shared data is not hosted by an organization or Web site owner. Rather, the file sharing service is set up by individuals who want to exchange files of mp3 audio tracks, video films or software. The bartering service is kept alive by the participants themselves; there is no central broker instance involved. Data is simply referenced by the filename, independent of the current location. This is a distributed, dynamic, and highly flexible environment, which is similar to the *Archie* service that was used in the early years of the Internet to locate files on ftp servers for downloading.

The **Resource Grid** provides mechanisms for the coordinated use of resources like computers, data archives, application services, and special laboratory instruments. The popular *Globus* toolkit [6], for example, gives access to participating computers without the need to bother which computer in the network is actually being used. In contrast to the data supplied by the Information Grid, the facilities of the Resource Grid cannot be given away free of charge and anonymously but are supplied for authorized users only. The core idea behind the Resource Grid is to provide easy, efficient and transparent access to any available resource, irrespective of its location. Resources may be computing power, data storage, network bandwidth, or special purpose hardware.

The third kind of grid, the **Service Grid**, delivers services and applications independent of their location, implementation, and hardware platform. The services are built on the concrete resources available in the Resource Grid. A major point of distinction between the last two grids lies in their different abstraction level: The Service Grid provides abstract, location-independent services, while the Resource Grid gives access to the concrete resources offered at a computer site.

2 Current Status of the Three Grids

2.1 Information Grid

Since its invention in 1990, the Information Grid became one of the biggest success stories in computer technology. With a data volume and user base steadily increasing at an extremely high pace, the Web is now used by a large fraction of the world population for accessing up-to-date information.

One reason for the tremendous success of the Web is the concept of the *hyperlink*, an easy-to-use reference to other Web pages. Following the path of marked hyperlinks is often the fastest way to find related information without typing. Due to the hyperlink the Web quickly dominated ftp and other networks that existed long before. We will show later how this beneficial concept can be adapted to the Resource and Service Grids.

Another important reason for the success of the Information Grid lies in the easy updating of information. Compared to traditional information distribution methods (mail of printed media), it is much easier and more cost-effective for vendors to reach a wide range of customers through the web with up-to-date information.

2.2 Resource Grid

The Internet, providing the means for data transfer bandwidth, is a good example of a Resource Grid. Wide area networks are complex systems where users only pay for the access endpoint, proportional to the subscribed bandwidth and the actual data throughput. The complex relationship between the numerous network providers whose services are used for transmitting the data within the regional networks are hidden from the user. Note that the Internet and other wide area networks were necessary for and pushed by the development of the Web.

Other Resource Grids are more difficult to implement and deploy because resources are costly and hence cannot be given away free of charge. *Computational Grids* give access to distributed supercomputers for time-consuming jobs. Most of them are based on the Globus toolset [6], which became a de-facto standard in this field. Today, there exist prototypes of application-specific grids for CFD, pharmaceutical research, chemistry, astrophysics, video rendering, post production, etc. Some of them use Web portals, others hide the grid access inside the application.

Data Grids provide mechanisms for secure, redundant data storage at geographically distributed sites. In view of the challenges of storing and processing several Petabytes of data at different locations, for example in the EU Datagrid project [4] or in satellite observation projects, this is an increasingly demanding subject. Issues like replication, staging, caching, and data co-scheduling must be solved. On the one hand, the quickly growing capacity of disk drives may tempt users to store data locally, but on the other hand, there are grand challenge projects that require distributed storage for redundancy reasons or simply because the same data sets are accessed by thousands of users at different sites [4].

For individuals *Sourceforge* provides data storage space for open source software projects and *IBP (Internet Backplane Protocol)* [1] provides logistic data management facilities like remote caching and permanent storage space.

Embarrassingly parallel applications like SETI@home, Folding@HOME, fightcancer@home, or distributed.net are easily mapped for execution on distributed PCs. For this application class, no general grid middleware has been developed yet. Instead, the middleware is integrated in the application which also steers the execution of remote slave jobs and the collection of the results. One interesting aspect of such applications is the implicit mutual trust on both sides: the PC owner trusts in the integrity of the software without individual checking authentication and authorization, and the grid software trusts that the results have not been faked by the PC owner.

Access Grids also fall into the category of Resource Grids. They build the technical basis for remote collaborations by providing interactive video conferences and blackboard facilities.

2.3 Service Grid

The Service Grid comprises services available in the Internet like search engines, portals, active server pages and other dynamic content. Email and authorization services (GMX, Hotmail, MS Passport) also fall into this category. They are mostly free of charge due to sponsoring or advertising.

The mentioned services are separate from each other without any calling interface in between. With web services and the *Open Grid Service Architecture OGSA* [8], this state of affairs is currently being changed. Both are designed to provide interoperability between loosely coupled services, independent of their implementation, geographic location or execution platform.

3 Representation Schemes Used in the Three Grids

Because of the different characteristics the representation schemes in the three grids have different capabilities and expressiveness.

In the **Information Grid** the hypertext markup language HTML is used to store information in a structured way. Due to its simple, user-readable format HTML was quickly adopted by Web page designers. However, over the time the original goal of separating the contents from its representation has been more and more compromised. Many Web pages use non-standard language constructs which can not be interpreted by all browsers. The massive growth of data in the Web and the demand to process it automatically revealed a major weakness of HTML, its inability to represent typed data. As an example, it is not possible to clearly identify a number in an HTML document as the product price or as the package quantity. This is due to the missing typing concept in HTML.

An alternative to HTML would have been the *Standard Generalized Markup Language SGML*[1]. However, SGML parsers were found to be too complex and

[1] SGML is a generic descriptive representation method. Used as a meta-language, SGML can be used to specify other languages like XML or HTML.

time-consuming to be integrated into browsers. Later XML [2] started to fill the gap between HTML and SGML and is now used as a common data representation, especially in e-business, where it is now replacing older standards like Edifact and ASN.1.

Although bad practice, XML is often transformed to HTML for presenting data in the Information Grid. Only when the original XML content is available can users process the contents with their own tools and integrating it into their work flow. Meta information conforming to the *Dublin Core*[2] are sometimes included into documents, but mostly hidden from the user, which still restricts their usefulness.

For **Resource Grids** several specification languages have been proposed and are used in different contexts. Globus, for example uses the Resource Specification Language RSL for specifying resource requests and the Grid Resource Information Service GRIS for listing available Globus services. This asymmetric approach (with different schemes for specifying resource offer and request) might be criticised for its lack of orthogonality but it was proven to work efficiently in practice. Condor, as another example, builds on so-called classified advertisements for matching requests with offers. ClassAds use a flexible, semi-structured data model, where not all attributes must be specified. Only matching attributes are checked. A more detailed discussion of resource specification methods can be found in [15].

In the area of **Service Grids** it is difficult to establish suitable representation schemes because there exists a wealth of different services and a lack of generally agreed methods that allow future extension. Hence Service Grids have to restrict to well-defined basic services like file copy, sorting, searching, data conversion, mathematical libraries etc., or distributed software packages like Netsolve. Work is under way to define generic specification schemes for Service Grids.

In cases where remote services are accessed via customized graphical user interfaces, tools like *GuiGen* [16] may be helpful. GuiGen conforms to the Service Grid concept by offering location transparent services, no matter at which site or system they are provided. Data exchange between the user and the remote service provider is based on XML. The user interacts with the application only via the graphical editor—the remote service execution is completely transparent to him.

XML is the most important representation scheme used in grids. Several other schemes build on it. The *Resource Description Framework RDF* is used in the Semantic Web as a higher-level variant. Also the *Web Service Description Language WSDL* [20] for specifying web services [11] has been derived from XML. For accessing remote services, the *Simple Object Access Protocol SOAP* [3] has been devised. Again, it is based on XML.

[2] The Dublin Core Metadata Initiative (DCMI) is an organization dedicated to promoting the widespread adoption of interoperable metadata standards and developing specialized metadata vocabularies for describing resources that enable more intelligent information discovery systems.

4 Organizing Grids

Locating entities is a problem common to all three Grids. Search engines provide a brute force solution which works fine in practice but has several drawbacks. First, the URL of the search engine must be known beforehand, and second, the quality of the search hits is often influenced by web designers and/or payments of advertisement customers. Moreover, certain meta information (like keywords, creation date, latest update) should be disclosed to allow users to formulate more precise queries.

For the proper functioning of Resource Grids, the possibility to specify implicit Web page attributes is even more important, because the resource search mechanisms depend on the Information Grid. The Globus GRIS (grid resource information service), for example, consists of a linked subset of LDAP servers.

4.1 Hyperlinks Specify Relations

As discussed earlier, hyperlinks provide a simple means to find related information in the Web. What is the corresponding concept of hyperlinks in Resource Grids or Service Grids? In Service Grids there is an attempt to modularise jobs to simpler tasks and to link the tasks (sub-services) by hyperlinks. These should not be hidden in the application. Rather, they should be browsable so that users can find them and use them for other purposes, thereby supporting a workflow-style of programming. Figure 2 illustrates an example where boxes represent single or compound services and the links represent calls and data flow between services. With common Web technologies it is possible to zoom into compound boxes to display more details on lower-level services. Note that this approach emulates the Unix toolbox concept on the Grid level. Here applets can be used to compose customized services with visual programming.

The *Universal Description Discovery & Integration UDDI* and the *Web Service Inspection Language WSIL* are current attempts for discovering Web services together with the *Web Service Description Language WSDL*. UDDI is a central repository for Web services and WSIL is a language that can be used between services to exchange information about other services.

UDDI will help to find application services in future grids. When the first Web Services are made publicly available, additional human readable Web pages should be generated from the WSDL documents so that Web search engines can index them just like normal Web pages and people can find them with established mechanisms.

5 Open Grid Service Architecture

For Service Grids the *Open Grid Service Architecture (OGSA)* was proposed [8]. In essence, OGSA marries Web services to grid protocols, thereby making progress in defining interfaces for grid services. It builds extensively on the open

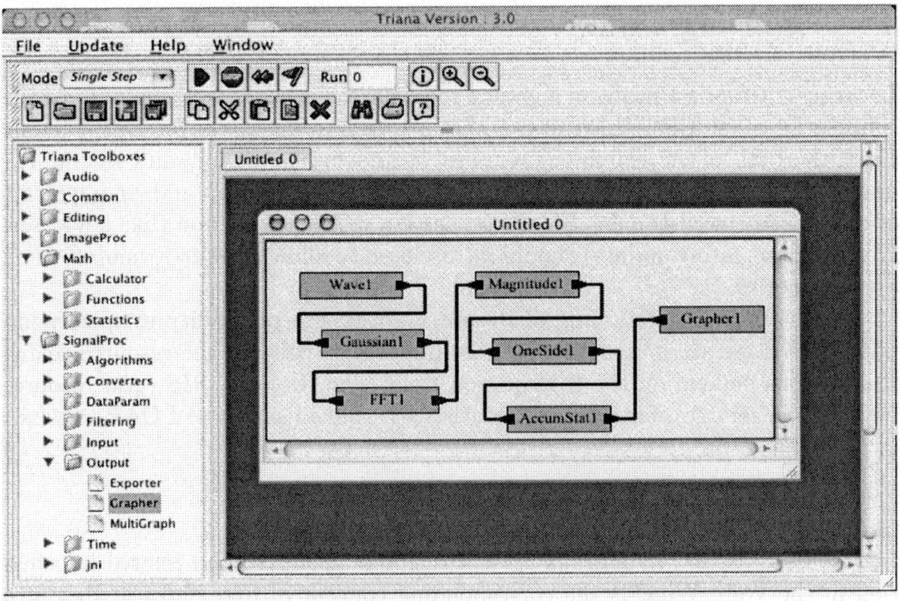

Fig. 2. Representing workflow in Triana [18].

standards SOAP, WSDL and UDDI. By this means, OGSA specifies a standardized behavior of Web services such as the uniform creation/instantiation of services, lifetime management, retrieval of metadata, etc.

As an example, the following functions illustrate the benefits of using OGSA. All of them can be easily implemented in an OGSA conformant domain:

Resilience: When a service request is sent to a service that has just recently crashed, a "service factory" autonomously starts a new instantiation of the service.

Service Directory: With common, uniform metadata on available services, browsable and searchable services can be built.

Substitution: Services can be easily substituted or upgraded to new implementations. The new service implementation just has to conform to the previous WSDL specification and external semantics.

Work-Load Distribution: Service requests may be broadcasted to different service endpoints having the same WSDL specification.

Note that there is a correspondence between the interaction concept of Web services and the object oriented design patterns [9]. The mentioned service factory, for example, corresponds to the Factory pattern. Transforming other design patterns to Web services scheme could be also beneficial, e.g. structural patterns (Adapter, Bridge, Composite, Decorator, Facade, Proxy), but also behavioral patterns like (Command, Interpreter, Iterator, Mediator, Memento, Observer, State, Strategy, Visitor). These patterns will be used in some implementations of services or in the interaction between services. This makes the development and the communication with grid services easier because complex design choices can be easily referred by the names of the corresponding patterns.

Another aspect that makes grid programming easier is virtualizing core services by having a single access method to several different implementations. Figure 3 depicts the concept of a capability layer, that selects the best suited core service and triggers it via adapters.

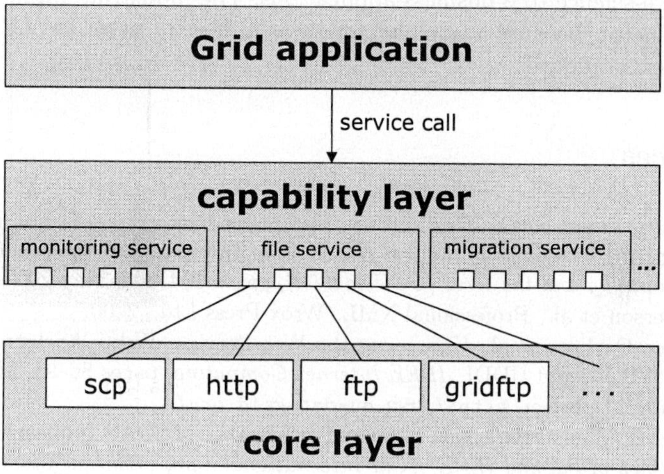

Fig. 3. Virtualizing core services makes grid programming easier.

5.1 OGSA versus CORBA and Legion

Both, CORBA and Legion [14, 10] have been designed for the execution of object oriented applications in distributed environments. Being based on object oriented programming languages, they clearly outperform the slower XML web services. Typical latencies for calling SOAP methods in current implementations range from 15 to 42 ms for a do-nothing call with client and server on the same host. This is an order of magnitude higher than the 1.2 ms latency of a JavaRMI call [5]. In distributed environments this gap will be even more pronounced.

In essence, the OGSA model assumes a more loosely organized grid structure, while CORBA and Legion environments are more coherent and tightly coupled. As a result, remote calls in the latter should be expected to be more efficient than in the former model.

6 Outlook

Grid environments provide an added value by the efficient sharing of resources in dynamic, multi-institutional virtual organizations. Grids have been adopted by academia for e-science applications. For the coming years, we expect the uptake of grids in industry for the broader e-business market as well. Eventually, grid technology may become an integral part of the evolving utility network, that shall bring services to the end user in the not so far future.

"Our immodest goal is to become the 'Linux of distributed computing"' says Ian Foster [12], co-creator of the Globus software toolkit. In order to do so, open standards are needed which are flexible enough to cover the whole range from distributed e-science to e-business applications. The industrial uptake is also an important factor, because academia alone was in history never strong enough to establish new standards.

References

1. Alessandro Bassi et al. The Internet Backplane Protocol: A Study in Resource Sharing. In *Proceedings of the 2nd IEEE/ACM International Symposium on Cluster Computing and the Grid (CCGrid2002)*, pages 194–201, May 2002.
2. R. Anderson et al., Professional XML. Wrox Press Ltd., 2000.
3. Francisco Curbera et al. Unraveling the Web Services Web - An Introduction to SOAP, WSDL, and UDDI. *IEEE Internet Computing*, pages 86–93, March 2002.
4. EU Datagrid project. `http://www.eu-datagrid.org/`.
5. Dan Davis, Manish Parashar. Latency Performance of SOAP Implementations. In *Proceedings of the 2nd IEEE/ACM International Symposium on Cluster Computing and the Grid (CCGrid2002)*, pages 407–412, May 2002.
6. I. Foster, C. Kesselman. Globus: A Metacomputing Infrastructure Toolkit. *International Journal of Supercomputer Applications*, Vol. 11, No. 2, pages 115–128, 1997.
7. Ian Foster, Carl Kesselman, Steven Tuecke. The Anatomy of the Grid - Enabling Scalable Virtual Organizations. *J. Supercomputer Applications*, 2001.
8. Ian Foster, Carl Kesselman, Jeffrey M. Nick, Steven Tuecke. The Physiology of the Grid - An Open Grid Services Architecture for Distributed Systems Integration. draft paper, 2002.
9. Erich Gamma, Richard Helm, Ralph Johnson, John Vlissides. Design Patterns - Elements of a Reusable Object Oriented Software. Addison Wesley, 1995.
10. Andrew S. Grimshaw, William A. Wulf, the Legion team, The Legion Vision of a Worldwide Virtual Computer. *Communications of the ACM*, Vol. 40, No. 1, pages 39–45, January 1997.
11. K. Gottschalk, S. Graham, H. Kreger, J. Snell. Introduction to Web services architecture. *IBM Systems Journal*, Vol. 41, No. 2, pages 170–177, 2002.
12. hpcWire, `http://www.tgc.com/hpcwire.html`, 03.02.2002.
13. Michael J. Litzkow, Miron Livny, Matt W. Mutka. Condor - A Hunter of Idle Workstations. In *Proceedings of the 8th International Conference on Distributed Computing Systems*, pages 104–111, IEEE Computer Society, June 1988.
14. Object Management Group. Corba. `http://www.omg.org/technology/documents/corba_spec_catalog.htm`.
15. A. Reinefeld, H. Stüben, T. Steinke, W. Baumann. Models for Specifying Distributed Computer Resources in UNICORE. *1st EGrid Workshop, ISThmus Conference Proceedings*, pages 313-320, Poznan 2000.

16. A. Reinefeld, H. Stüben, F. Schintke, G. Din. GuiGen: A Toolset for Creating Customized Interfaces for Grid User Communities, To appear in *Future Generation Computing Systems*, 2002.
17. D.B. Skillicorn. Motivating Computational Grids In *Proceedings of the 2nd IEEE/ACM International Symposium on Cluster Computing and the Grid (CC-Grid2002)*, pages 401–406, May 2002.
18. Ian Taylor, Bernd Schutz. Triana - A quicklook data analysis system for gravitational wave detectors. In *Proceedings of the Second Workshop on Gravitational Wave Data Analysis*, Editions Frontières, pages 229–237, Paris, 1998.
19. Steven Tuecke, Karl Czajkowski, Ian Foster, Jeffrey Frey, Steve Graham, Carl Kesselman. Grid Service Specification, 2002.
20. Web Services Description Language (WSDL) 1.1. *W3C Note*, `http://www.w3.org/TR/wsdl/`, 15 March 2001.

Topic 1
Support Tools and Environments

Marian Bubak[1] and Thomas Ludwig[2]

[1] Institute of Computer Science and ACC CYFRONET, AGH Kraków, Poland
[2] Ruprecht-Karls-Universität Heidelberg, Germany

At present days parallel applications are becoming large, heterogeneous, with more and more complex structure of their components and complicated topology of communication channels. Very often they are designed for execution on high-performance clusters and recently we observe the explosion of interest in parallel computing on the grid.

Efficient development of this kind of applications requires supporting tools and environments. In the first stage, support for verification of correctness of communication structure is required. This may be followed by an automatic performance analysis. Next, these tools should allow to observe and manipulate the behavior of an application during run time what is necessary for debugging, performance measurement, visualisation and analysis. The most important problems are measurement of utilisation of system resources and inter-process communication aimed at finding potential bottlenecks to improve the overall performance of an application. Important issues are portability and interoperability of tools.

For these reasons elaboration of supporting tools and environments remains a challenging research problem.

The goal of this Topis was to bring together tool designers, developers, and users and help them in sharing ideas, concepts, and products in this field.

This year our Topic attracted a total of 12 submissions. In fact this is a very low number but we do not want to draw the conclusion that there is no more work necessary with support tools and environments.

From the total of 12 papers with accepted 3 as regular papers and 4 as short papers. The acceptance rate is thus 58%. The papers will be presented in two sessions. Session one focuses on performance analysis. For session two there is no specific focus, instead we find various topics here.

The session on performance analysis presents three full papers of well known research groups.

Hong-Linh Truong and Thomas Fahringer present the tool SCALEA which is a versatile performance analysis tool. The paper gives an overview over its architecture and the various features that guide the programmer through the process of performance tuning. Remarkable is SCALEA's ability to support multi-experiment performance analysis.

Also the paper by Philip C. Roth and Barton P. Miller has its focus on program tuning. They present DeepStart, a new concept for automatic performance diagnosis that uses stack sampling to detect functions that are possible

B. Monien and R. Feldmann (Eds.): Euro-Par 2002, LNCS 2400, pp. 73–74.

bottlenecks. DeepStart leads to a considerable improvement with respect to how quickly bottlenecks can be detected.

The issue of performance collection is also covered in a paper on the scalability of tracing mechanisms. Felix Freitag, Jordi Caubert, and Jesus Labarta present an approach for OpenMP programs where the trace contains only non-iterative data. It is thus much more compact and reveals performance problems faster.

In our second session we find papers that deal with various aspects of tools and environments.

The paper by A.J.G. Hey, J. Papay, A.J. Keane, and S.J. Cox presents a component based problem solving environment (PSE). Based on modern technologies like CORBA, Java, and XML the project supports rapid prototyping of application specific PSEs. Its applicability is shown in an environment for the simulation of photonic crystals.

The paper by Jozsef Kovacs, Gabor Kusper, Robert Lovas, and Wolfgang Schreiner covers the complex topic of parallel debugging. They present their work on the integration of temporal assertions into a debugger. Concepts from model checking and temporal formulas are incorporated and provide means for the programmer to specify and check the temporal behaviour of the program.

Jorji Nonaka, Gerson H. Pfitscher, Katsumi Onisi, and Hideo Nakano discuss time synchronization in PC clusters. They developed low-cost hardware support for clock synchronisation.

Antonio J. Nebro, Enrique Alba, Francisco Luna, and José M. Troya have studied how to adopt JACO to .NET. JACO is a Java-based runtime system for implementing concurrent objects in distributed systems.

We would like to thank the authors who submitted a contribution, as well as the Euro-Par Organizing Committee, and the scores of referees, whose efforts have made the conference and this specific topic possible.

SCALEA: A Performance Analysis Tool for Distributed and Parallel Programs *

Hong-Linh Truong and Thomas Fahringer

Institute for Software Science, University of Vienna
Liechtensteinstr. 22, A-1090 Vienna, Austria
{truong,tf}@par.univie.ac.at

Abstract. In this paper we present SCALEA, which is a performance instrumentation, measurement, analysis, and visualization tool for parallel and distributed programs that supports post-mortem and online performance analysis. SCALEA currently focuses on performance analysis for OpenMP, MPI, HPF, and mixed parallel/distributed programs. It computes a variety of performance metrics based on a novel overhead classification. SCALEA also supports multiple experiment performance analysis that allows to compare and to evaluate the performance outcome of several experiments. A highly flexible instrumentation and measurement system is provided which can be controlled by command-line options and program directives. SCALEA can be interfaced by external tools through the provision of a full Fortran90 OpenMP/MPI/HPF frontend that allows to instrument an abstract syntax tree at a very high-level with C-function calls and to generate source code. A graphical user interface is provided to view a large variety of performance metrics at the level of arbitrary code regions, threads, processes, and computational nodes for single- and multi-experiments.

Keywords: performance analysis, instrumentation, performance overheads

1 Introduction

The evolution of distributed/parallel architectures and programming paradigms for performance-oriented program development challenge the state of technology for performance tools. Coupling different programming paradigms such as message passing and shared memory programming for hybrid cluster computing (e.g. SMP clusters) is one example for high demands on performance analysis that is capable to observe performance problems at all levels of a system while relating low-level behavior to the application program.

In this paper we describe SCALEA, a performance instrumentation, measurement, and analysis system for distributed and parallel architectures that currently focuses on OpenMP, MPI, HPF programs, and mixed programming

* This research is supported by the Austrian Science Fund as part of the Aurora Project under contract SFBF1104.

B. Monien and R. Feldmann (Eds.): Euro-Par 2002, LNCS 2400, pp. 75–85.

paradigms such as OpenMP/MPI. SCALEA seeks to explain the performance behavior of each program by computing a variety of performance metrics based on a novel classification of performance overheads for shared and distributed memory parallel programs which includes data movement, synchronization, control of parallelism, additional computation, loss of parallelism, and unidentified overheads. In order to determine overheads, SCALEA divides the program sources into code regions (ranging from the entire program to single statement) and locates whether performance problems occur in those regions or not. A highly flexible instrumentation and measurement system is provided which can be precisely controlled by program directives and command-line options. In the center of SCALEA's performance analysis is a novel dynamic code region call graph (DRG - [9]) which reflects the dynamic relationship between code regions and their subregions and enables a detailed overhead analysis for every code region. Moreover, SCALEA supports a high-level interface to traverse an abstract syntax tree (AST), to locate arbitrary code regions, and to mark them for instrumentation. The SCALEA overhead analysis engine can be used by external tools as well.

A data repository is employed in order to store performance data and information about performance experiments which alleviates the association of performance information with experiments and the source code. SCALEA also supports multi-experiment performance analysis that allows to examine and compare the performance outcome of different program executions. A sophisticated visualization engine is provided to view the performance of programs at the level of arbitrary code regions, threads, processes, and computational nodes (e.g. single-processor systems, Symetric Multiple Processor (SMP) nodes sharing a common memory, etc.) for single- and multi-experiments.

The rest of this paper is organized as follows: Section 2 presents an overview of SCALEA. In Section 3 we present a classification of performance overheads. The next section outlines the various instrumentation mechanisms offered by SCALEA. The performance data repository is described in the following section. Experiments are shown in Section 6. Related work is outlined in Section 7, followed by conclusions in Section 8.

2 SCALEA Overview

SCALEA is a performance instrumentation, measurement, and analysis system for distributed memory, shared memory, and mixed parallel programs. Figure 1 shows the architecture of SCALEA which consists of several components: SCALEA Instrumentation System (SIS), SCALEA Runtime System, SCALEA Performance Data Repository, and SCALEA Performance Analysis & Visualization System. All components provide open interfaces thus they can be used by external tools as well.

SIS uses the front-end and unparser of the VFC compiler [1]. SIS supports automatic instrumentation of MPI, OpenMP, HPF, and mixed OpenMP/MPI programs. The user can select (by directives or command-line options) code re-

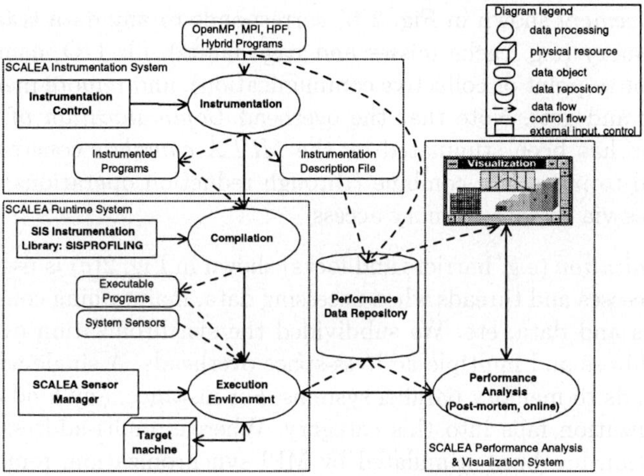

Fig. 1. Architecture of SCALEA

gions and performance metrics of interest. Moreover, SIS offers an interface for other tools to traverse and annotate the AST at a high level in order to specify code regions for which performance metrics should be obtained. SIS also generates an *instrumentation description file* [9] to relate all gathered performance data back to the input program.

The SCALEA runtime system supports profiling and tracing for parallel and distributed programs, and sensors and sensor managers for capturing and managing performance data of individual computing nodes of parallel and distributed machines. The SCALEA profiling and tracing library collects timing, event, and counter information, as well as hardware parameters. Hardware parameters are determined through an interface with the PAPI library [2].

The SCALEA performance analysis and visualization module analyzes the raw performance data which is collected post-mortem or online and stored in the performance data repository. It computes all user-requested performance metrics, and visualizes them together with the input program. Besides single-experiment analysis, SCALEA also supports multi-experiment performance analysis. The visualization engine provides a rich set of displays for various metrics in isolation or together with the source code.

The SCALEA performance data repository holds relevant information about the experiments conducted.

In the following we provide a more detailed overview of SCALEA.

3 Classification of Temporal Overheads

In previous work [9], we presented a preliminary and very coarse grain classification of performance overheads which has been stimulated by [3]. Figure 2 shows our novel and substantially refined overhead classification which includes:

– *Data movement* shown in Fig. 2(b) corresponds to any data transfer within local memory (e.g. cache misses and page faults), file I/O, communication (e.g. point to point or collective communication), and remote memory access (e.g. put and get). Note that the overhead *Communication of Accumulate Operation* has been stimulated by the *MPI_Accumulate* construct which is employed to move and combine (through reduction operations) data at remote sites via remote memory access.

– *Synchronization* (e.g. barriers and locks) shown in Fig. 2(c) is used to coordinate processes and threads when accessing data, maintaining consistent computations and data, etc. We subdivided the synchronization overhead into single address and multiple address-space overheads. A single address space corresponds to memory parallel systems. For instance, any kind of OpenMP synchronization falls into this category. Whereas multi-address space synchronization has been stimulated by MPI synchronization, remote memory locks, barriers, etc.

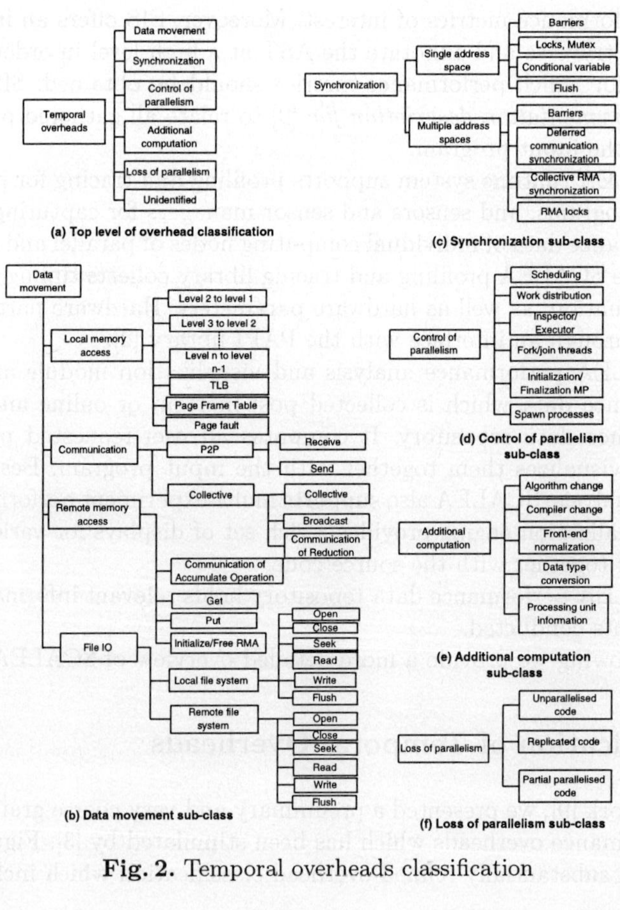

Fig. 2. Temporal overheads classification

– *Control of parallelism* (e.g. fork/join operations and loop scheduling) shown in Fig. 2(d) is used to control and to manage the parallelism of a program which is commonly caused by code inserted by the compiler (e.g. runtime library) or by the programmer (e.g. to implement data redistribution).
– *Additional computation* (see Fig. 2(e)) reflects any change of the original sequential program including algorithmic or compiler changes to increase parallelism (e.g. by eliminating data dependences) or data locality (e.g. through changing data access patterns). Moreover, requests for processing unit identifications, or the number of threads that execute a code region may also imply additional computation overhead.
– *Loss of parallelism* (see Fig. 2(f)) is due to imperfect parallelization of a program which can be further classified: unparallelized code (executed by only one processor), replicated code (executed by all processors), and partially parallelized code (executed by more than one but not all processors).
– *Unidentified overhead* corresponds to the overhead that is not covered by the above categories.

4 SCALEA Instrumentation System (SIS)

SIS provides the user with three alternatives to control instrumentation which includes command-line options, SIS directives, and an instrumentation library combined with an OpenMP/HPF/MPI frontend and unparser. All of these alternatives allow the specification of performance metrics and code regions of interest for which SCALEA automatically generates instrumentation code and determines the desired performance values during or after program execution. In the remainder of this paper we assume that a code region refers to a single-entry single-exit code region. A large variety of predefined mnemonics are provided by SIS for instrumentation purposes. The current implementation of SCALEA supports 49 code region and 29 performance metric mnemonics:

– code region mnemonics: arbitrary code regions, loops, outermost loops, procedures, I/O statements, HPF INDEPENDENT loops, HPF redistribution, OpenMP parallel loops, OpenMP sections, OpenMP critical, MPI send, receive, and barrier statements, etc.
– performance metric mnemonics: wall clock time, cpu time, communication overhead, cache misses, barrier time, synchronization, scheduling, compiler overhead, unparallelized code overhead, HW-parameters, etc. See also Fig. 2 for a classification of performance overheads considered by SCALEA.

The user can specify arbitrary code regions ranging from the entire program unit to single statements and name (to associate performance data with code regions) these regions which is shown in the following:

```
!SIS$ CR region_name BEGIN
        code region
!SIS$ END CR
```

In order to specify a set of code regions $R = \{r_1, ..., r_n\}$ in an enclosing region r and performance metrics which should be computed for every region in R, SIS offers the following directive:

!SIS\$ CR *region_name* [,*cr_mnem-list*] [**PMETRIC** *perf_mnem-list*] **BEGIN**
code region r that includes all regions in R
!SIS\$ END CR

The code region r defines the *scope* of the directive. Note that every (code) region in R is a sub-region of r but r may contain sub-regions that are not in R.

The code region (*cr_mnem-list*) and performance metric (*perf_mnem-list*) mnemonics are indicated as a list of mnemonics separated by commas. One of the code region mnemonics (CR_A) refers to arbitrary code regions. Note that the above specified directive allows to indicate either only code region mnemonics or performance metric mnemonics, or a combination of both. If in a SIS directive d only code region mnemonics are indicated, then SIS is instrumenting all code regions that correspond to these mnemonics inside of the scope of d. The instrumentation is done for a set of default performance metrics which can be overwritten by command-line options.

If only performance metric mnemonics are indicated in a directive d then SIS is instrumenting those code regions that have an impact on the specified metrics. This option is useful if a user is interested in specific performance metrics but doesn't know which code regions may cause these overheads. If both code region and performance metrics are defined in a directive d, then SIS is instrumenting these code regions for the indicated performance metrics in the scope of d. Feasibility checks are conducted by SIS, for instance, to determine whether the programmer is asking for OpenMP overheads in HPF code regions. For these cases, SIS outputs appropriate warnings.

All previous directives are called local directives as the scope of these directives is restricted to part of a program unit (main program, subroutines or functions). The scope of a directive can be extended a full program unit by using the following syntax:

!SIS\$ CR [*cr_mnem-list*] [**PMETRIC** *perf_mnem-list*]

A global directive d collects performance metrics – indicated in the `PMETRIC` part of d – for all code regions – specified in the `CR` part of d – in the program unit which contains d. A local directive implies the request for performance information restricted to the scope of d. There can be nested directives with arbitrary combinations of global and local directives. If different performance metrics are requested for a specific code region by several nested directives, then the union of these metrics is determined. SIS supports command-line options to instrument specific code regions for well-defined performance metrics in an entire application (across all program units).

Moreover, SIS provides specific directives in order to control tracing/profiling. The directives MEASURE ENABLE and MEASURE DISABLE allow the programmer to turn on and off tracing/profiling of a specific code region.

!SIS\$ MEASURE DISABLE
code region
!SIS\$ MEASURE ENABLE

SCALEA also provides an interface that can be used by other tools to exploit SCALEA's instrumentation, analysis and visualization features. We have developed a C-library to traverse the AST and to mark arbitrary code regions for instrumentation. For each code region, the user can specify the performance metrics of interest. Based on the annotated AST, SIS automatically generates an instrumented source code.

In the following example we demonstrate some of the directives as mentioned above by showing a fraction of the application code of Section 6.

```
d1:        !SIS$ CR PMETRIC ODATA_SEND, ODATA_RECV, ODATA_COL
           call MPI_BCAST(nx, 1,MPI_INTEGER, mpi_master,MPI_COMM_WORLD,mpi_err)
           ...
d2:        !SIS$ CR comp_main, CR_A, CR_S PMETRIC WTIME, L2_TCM BEGIN
               ...
d3:            !SIS$ CR init_comp BEGIN
                    dj=real(nx,b8)/real(nodes_row,b8)
                    ...
d4:            !SIS$ END CR
                    ...
d5:            !SIS$ MEASURE DISABLE
                    call bc(psi,i1,i2,j1,j2)
d6:            !SIS$ MEASURE ENABLE
                    ...
                    call do_force(i1,i2,j1,j2)
                    ...
d7:        !SIS$ END CR
```

Directive d_1 is a global directive which instructs SIS to instrument all send, receive and collective communication statements in this program unit. Directives d_2 (begin) and d_7 (end) define a specific code region with the name *comp_main*. Within this code region *comp_main*, SCALEA will determine wall clock times (*WTIME*) and the total number of L2 cache misses (*L2_TCM*) for all arbitrary code regions (based on mnemonic *CR_A*) and subroutine calls (mnemonic *CR_S*) as specified in d_2. Directives d_3 and d_4 specify an arbitrary code region with the name *init_comp*. No instrumentation as well as measurement is done for the code region between directives d_5 and d_6.

5 Performance Data Repository

A key concept of SCALEA is to store the most important information about performance experiments including application, source code, machine information, and performance results in a data repository. Figure 3 shows the structure of the data stored in SCALEA's performance data repository. An *experiment* refers to a sequential or parallel execution of a program on a given target architecture. Every experiment is described by *experiment-related data*, which includes information about the application code, the part of a machine on which the code has been executed, and performance information. An application (program) may have a number of implementations (code versions), each of them consists of a set of source files and is associated with one or several experiments. Every source file has one or several static code regions (ranging from the entire program unit to single statement), uniquely specified by *startPos* and *endPos*

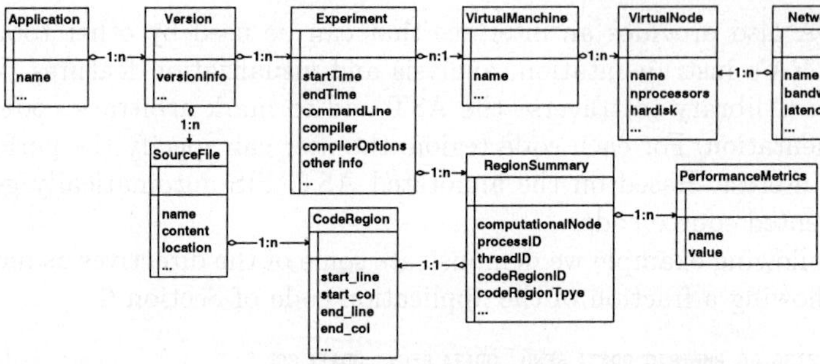

Fig. 3. SCALEA Performance Data Repository

(position – start/end line and column – where the region begins and ends in the source file). Experiments are associated with the virtual machines on which they have been taken. The virtual machine is part of a physical machine available to the experiment; it is described as a set of computational nodes (e.g. single-processor systems, Symetric Multiple Processor (SMP) nodes sharing a common memory, etc.) connected by a specific network. A region summary refers to the performance information collected for a given code region and processing unit (process or thread) on a specific virtual node used by the experiment. The region summaries are associated with performance metrics that comprise performance overheads, timing information, and hardware parameters. Moreover, most data can be exported into XML format which further facilitates accessing performance information by other tools (e.g. compilers or runtime systems) and applications.

6 Experiments

SCALEA as shown in Fig. 1 has been fully implemented. Our analysis and visualization system is implemented in Java which greatly improves their portability. The performance data repository uses PostgreSQL and the interface between SCALEA and the data repository is realized by Java and JDBC. Due to space limits we restrict the experiments shown in this section to a few selected features for post-mortem performance analysis.Our experimental code is a mixed OpenMP/MPI Fortran program that is used for ocean simulation. The experiments have been conducted on an SMP cluster with 16 SMP nodes (connected by Myrinet) each of which comprises 4 Intel Pentium III 700 MHz CPUs.

6.1 Overhead Analysis for a Single Experiment

SCALEA supports the user in the effort to examine the performance overheads for a single experiment of a given program. Two modes are provided for this analysis. Firstly, the *Region-to-Overhead* mode (see the "Region-to-Overhead" window in Fig. 4) allows the programmer to select any code region instance in the DRG for which all detected performance overheads are displayed. Secondly,

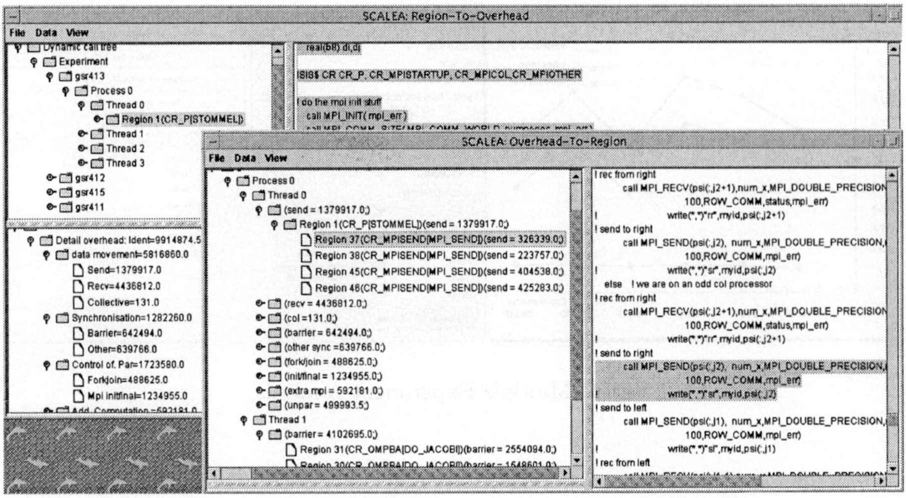

Fig. 4. Region-To-Overhead and Overhead-To-Region DRG View

the *Overhead-to-Region* mode (see the "Overhead-to-Region" window in Fig. 4) enables the programmer to select the performance overhead of interest, based on which SCALEA displays the corresponding code region(s) in which this overhead occurs. This selection can be limited to a specific code region instance, thread or process. For both modi the source code of a region is shown if the code region instance is selected in the DRG by a mouse click.

6.2 Multiple Experiments Analysis

Most performance tools investigate the performance for individual experiments one at a time. SCALEA goes beyond this limitation by supporting also performance analysis for multiple experiments. The user can select several experiments and performance metrics of interest whose associated data are stored in the data repository. The outcome of every selected metric is then analyzed and visualized for all experiments. For instance, in Fig. 5 we have selected 6 experiments (see x-axis in the left-most window) and examine the wall clock, user, and system times for each of them. We believe that this feature is very useful for scalability analysis of individual metrics for changing problem and machine sizes.

7 Related Work

Significant work has been done by Paradyn [6], TAU [5], VAMPIR [7], Pablo toolkit [8], and EXPERT [10]. SCALEA differs from these approaches by providing a more flexible mechanism to control instrumentation for code regions and performance metrics of interest. Although Paradyn enables dynamic insertion of probes into a running code, Paradyn is currently limited to instrumentation of

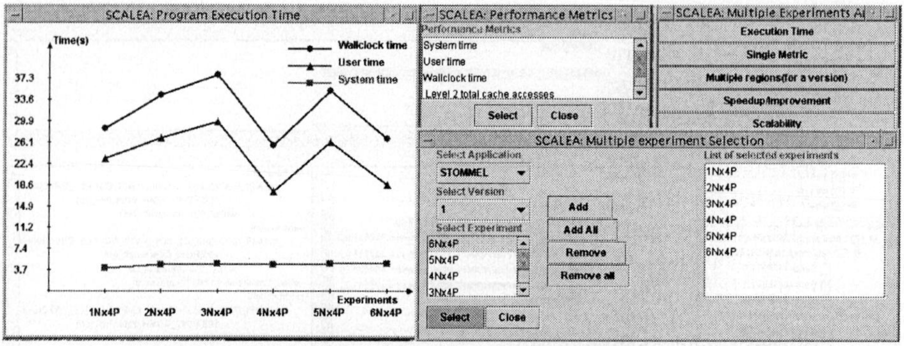

Fig. 5. Multiple Experiment Analysis

subroutines and functions, whereas SCALEA can instrument - at compile-time only - arbitrary code regions including single statements. Moreover, SCALEA differs by storing experiment-related data to a data repository, by providing multiple instrumentation options (directives, command-line options, and high-level AST instrumentation), and by supporting also multi-experiment performance analysis.

8 Conclusions and Future Work

In this paper, we described SCALEA, which is a performance analysis tool for OpenMP/MPI/HPF and mixed parallel programs. The main contributions of this paper are centered around a novel design of the SCALEA architecture, new instrumentation directives, a substantially improved overhead classification, a performance data repository, a visualization engine, and the capability to support both single- and multi-experiment performance analysis.

Currently, SCALEA is extended for online monitoring for Grid applications and infrastructures. SCALEA is part of the ASKALON programming environment and tool set for cluster and Grid architectures [4]. SCALEA is used by various other tools in ASKALON to support automatic bottleneck analysis, performance experiment and parameter studies, and performance prediction.

References

1. S. Benkner. VFC: The Vienna Fortran Compiler. *Scientific Programming, IOS Press, The Netherlands*, 7(1):67–81, 1999.
2. S. Browne, J. Dongarra, N. Garner, K. London, and P. Mucci. A scalable cross-platform infrastructure for application performance tuning using hardware counters. In *Proceeding SC'2000*, November 2000.
3. J.M. Bull. A hierarchical classification of overheads in parallel programs. In *Proc. 1st International Workshop on Software Engineering for Parallel and Distributed Systems*, pages 208–219. Chapman Hall, March 1996.

4. T. Fahringer, A. Jugravu, S. Pllana, R. Prodan, C. Seragiotto, and H.-L. Truong. ASKALON - A Programming Environment and Tool Set for Cluster and Grid Computing. www.par.univie.ac.at/project/askalon, Institute for Software Science, University of Vienna.
5. Allen Malony and Sameer Shende. Performance technology for complex parallel and distributed systems. In *3rd Intl. Austrian/Hungarian Workshop on Distributed and Parallel Systems*, pages 37–46. Kluwer Academic Publishers, Sept. 2000.
6. B. Miller, M. Callaghan, J. Cargille, J. Hollingsworth, R. Irvin, K. Karavanic, K. Kunchithapadam, and T. Newhall. The paradyn parallel performance measurement tool. *IEEE Computer*, 28(11):37–46, November 1995.
7. W. E. Nagel, A. Arnold, M. Weber, H.-C. Hoppe, and K. Solchenbach. VAMPIR: Visualization and analysis of MPI resources. *Supercomputer*, 12(1):69–80, Jan. 1996.
8. D. A. Reed, R. A. Aydt, R. J. Noe, P. C. Roth, K. A. Shields, B. W. Schwartz, and L. F. Tavera. Scalable Performance Analysis: The Pablo Performance Analysis Environment. In *Proc. Scalable Parallel Libraries Conf.*, pages 104–113. IEEE Computer Society, 1993.
9. Hong-Linh Truong, Thomas Fahringer, Georg Madsen, Allen D. Malony, Hans Moritsch, and Sameer Shende. On Using SCALEA for Performance Analysis of Distributed and Parallel Programs. In *Proceeding SC'2001*, Denver, USA, November 2001. IEEE/ACM.
10. Felix Wolf and Bernd Mohr. Automatic Performance Analysis of MPI Applications Based on Event Traces. *Lecture Notes in Computer Science*, 1900:123–??, 2001.

Deep Start: A Hybrid Strategy for Automated Performance Problem Searches[1]

Philip C. Roth and Barton P. Miller

Computer Sciences Department
University of Wisconsin–Madison
1210 W. Dayton Street
Madison, WI 53706–1685 USA
{pcroth,bart}@cs.wisc.edu

Abstract. We present Deep Start, a new algorithm for automated performance diagnosis that uses stack sampling to augment our search-based automated performance diagnosis strategy. Our hybrid approach locates performance problems more quickly and finds problems hidden from a more straightforward search strategy. Deep Start uses stack samples collected as a by-product of normal search instrumentation to find *deep starters*, functions that are likely to be application bottlenecks. Deep starters are examined early during a search to improve the likelihood of finding performance problems quickly.We implemented the Deep Start algorithm in the Performance Consultant, Paradyn's automated bottleneck detection component. Deep Start found half of our test applications' known bottlenecks 32% to 59% faster than the Performance Consultant's current call graphbased search strategy, and finished finding bottlenecks 10% to 61% faster. In addition to improving search time, Deep Start often found more bottlenecks than the call graph search strategy.

1 Introduction

Automated search is an effective strategy for finding application performance problems [7,10,13,14]. With an automated search tool, the user need not be a performance analysis expert to find application performance problems because the expertise is embodied in the tool. Automated search tools benefit from the use of structural information about the application under study such as its call graph [4] and by pruning and prioritizing the search space based on the application's behavior during previous runs [12]. To attack the problem of scalability with respect to application code size, we have developed *Deep Start*, a new algorithm that uses sampling [1,2,8] to augment automated search. Our hybrid approach substantially improves search effectiveness by locating performance problems more quickly and by locating performance problems hidden from a more straightforward search strategy.

[1] This work is supported in part by Department of Energy Grant DE-FG02-93ER25176, Lawrence Livermore National Lab grant B504964, and NSF grants CDA-9623632 and EIA-9870684. The U.S. Government is authorized to reproduce and distribute reprints for Governmental purposes notwithstanding any copyright notation thereon.

B. Monien and R. Feldmann (Eds.): Euro-Par 2002, LNCS 2400, pp. 86–96.

We have implemented the Deep Start algorithm in the Performance Consultant, the automated bottleneck detection component of the Paradyn performance tool [13].

To search for application performance problems, the Performance Consultant (hereafter called the PC) performs experiments that test the application's behavior. Each experiment is based on a *hypothesis* about a potential performance problem. For example, an experiment might use a hypothesis like "the application is spending too much time on blocking message passing operations." Each experiment also reflects a *focus*. A focus names a set of application resources such as a collection of functions, processes, or semaphores. For each of its experiments, the PC uses dynamic instrumentation [11] to collect the performance data it needs to evaluate whether the experiment's hypothesis is true for its focus. The PC compares the performance data it collects against user-configurable thresholds to decide whether an experiment's hypothesis is true. At the start of its search, the PC creates experiments that test the application's overall behavior. If an experiment is true (i.e., its hypothesis is true at its focus), the PC *refines* its search by creating one or more new experiments that are more specific than the original experiment. The new experiments may have a more specific hypothesis or a more specific focus than the original experiment. The PC monitors the cost of the instrumentation generated by its experiments, and respects a user-configurable cost threshold to avoid excessive intrusion on the application. Thus, as the PC refines its search, it puts new experiments on a queue of pending experiments. It *activates* (inserts instrumentation for) as many experiments from the queue as it can without exceeding the cost threshold. Also, each experiment is assigned a priority that influences the order which experiments are removed from the queue.

A *search path* is a sequence of experiments related by refinement. The PC prunes a search path when it cannot refine the newest experiment on the path, either because the experiment was false or because the PC cannot create a more specific hypothesis or focus. The PC uses a Search History Graph display (see Fig. 1) to record the cumulative refinements of a search. This display is dynamic—nodes are added as the PC refines its search. The display provides a mechanism for users to obtain information about the state of each experiment such as its hypothesis and focus, whether it is currently active (i.e., the PC is collecting data for the experiment), and whether the experiment's data has proven the experiment's hypothesis to be true, false, or not yet known.

The Deep Start search algorithm augments the PC's current call-graph-based search strategy with stack sampling. The PC's current search strategy [4] uses the application's call graph to guide refinement. For example, if it has found that an MPI application is spending too much time sending messages, the PC starts at the main function and tries to refine its search to form experiments that test the functions that main calls. If a function's experiment tests true, the search continues with its callees. Deep Start augments this strategy with stack samples collected as a by-product of normal search instrumentation. Deep Start uses its stack samples to guide the search quickly to performance problems. When Deep Start refines its search to examine individual functions, it directs the search to focus on functions that appear frequently in its stack samples. Because these functions are long-running or are called frequently, they are likely to be the application's performance bottlenecks.

Deep Start is more efficient than the current PC search strategy. Using stack samples, Deep Start can "skip ahead" through the search space early in the search. This ability allows Deep Start to detect performance problems more quickly than the

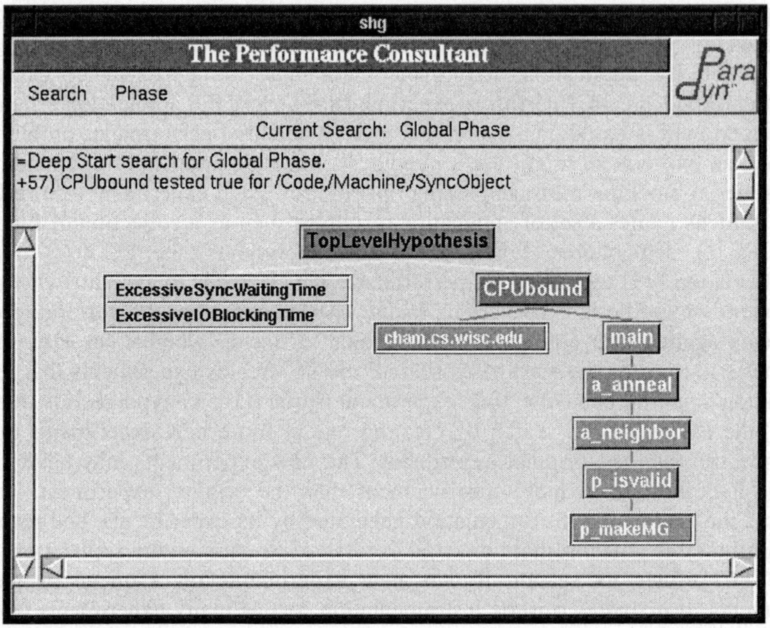

Fig. 1. The Performance Consultant's Search History Graph display

current call graph-based strategy. Due to the statistical nature of sampling and because some types of performance problems such as excessive blocking for synchronization are not necessarily indicated by functions frequently on the call stack, Deep Start also incorporates a call-graph based search as a background task.

Deep Start is able to find performance problems hidden from the current strategy. For example, consider the portion of an application's call graph shown in Fig. 2. If A is a bottleneck but B, C, and D are not, the call graph strategy will not consider E even though E may be a significant bottleneck. Although the statistical nature of sampling does not guarantee that E will be considered by the Deep Start algorithm, if it occurs frequently in the stack samples Deep Start will examine it regardless of the behavior of B, C, and D.

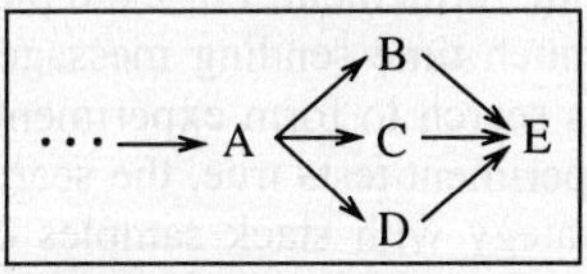

Fig. 2. A part of an application's call graph. Under the Performance Consultant's call graphbased search, if B, C, and D are not bottlenecks, E will not be examined. In contrast, the Deep Start algorithm will examine E if it appears frequently in the collected stack samples C E A B D

2 The Deep Start Search Algorithm

Deep Start uses stack samples collected as a by-product of dynamic instrumentation to guide its search. Paradyn daemons perform a stack walk when they insert instrumentation; this stack walk checks whether it is safe to insert instrumentation code into the ap- plication's processes. Under the Deep Start algorithm, the PC collects these stack samples and analyzes them to find *deep starters*—functions that appear frequently in the samples and thus are likely to be application bottlenecks. It creates experiments to examine the deep starters with high priority so that they will be given preference when the PC activates new experiments.

2.1 Selecting Deep Starters

If an experiment tests true and was examining a Code resource (i.e., an application or library function), the PC triggers its deep starter selection algorithm. The PC collects stack samples from each of the Paradyn daemons and uses the samples to update its *function count graph.* A function count graph records the number of times each function appears in the stack samples. It also reflects the call relationships between functions as indicated in the stack samples. Nodes in the graph represent functions of the application and edges represent a call relationship between two functions. Each node holds a count of the number of times the function was observed in the stack samples. For instance, assume the PC collects the stack samples shown in Fig. 3 (a) (where *x*→*y* denotes that function *x* called function *y*). Fig. 3 (b) shows the function count graph resulting from these samples. In the figure, node labels indicate the function and its count. Once the PC has updated the function count graph with the latest stack sample information, it traverses the graph to find functions whose frequency is higher than a user-configurable *deep starter threshold.* This threshold is expressed as a percentage of the total number of stack samples collected.

In reality, the PC's function count graph is slightly more complicated than the graph shown in Fig. 3. One of the strengths of the PC is its ability to examine application behavior at per-host and per-process granularity. Deep Start keeps global, per-host, and per-process function counts to enable more fine-grained deep starter selections. For example, if the PC has refined the experiments in a search path to examine process 1342 on host `cham.cs.wisc.edu`, Deep Start will only use function counts from that process' stack samples when selecting deep starters to add to the search path. To enable fine-grained deep starter selections, each function count

A→B→C→D
A→E→C→D
A→F→D
A→F→G

(a)

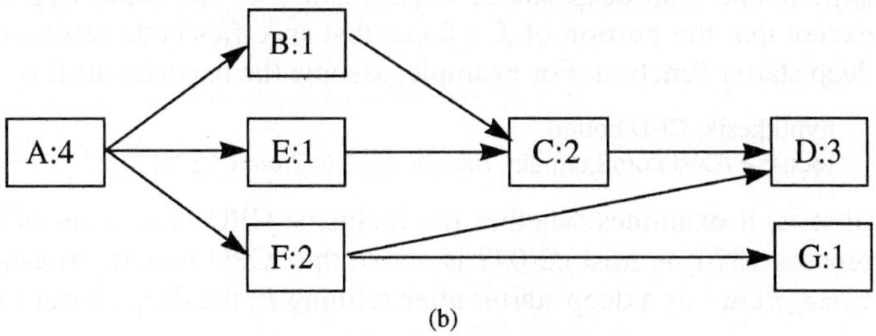

(b)

Fig. 3. A set of stack samples (a) and the resulting function count graph (b) graph node maintains a tree of counts as shown in Fig. 4. The root of each node's *count-tree* indicates the number of times the node's function was seen in all stack samples. Subsequent levels of the counttree indicate the number of times the function was observed in stack samples for specific hosts and specific processes. With count-trees in the function count graph, Deep Start can make per-host and per-process deep starter selections

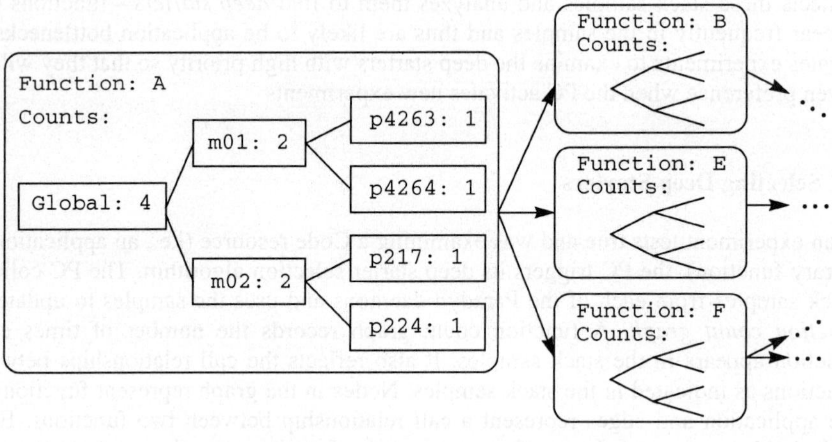

Fig. 4. A function count graph node with count-tree

As Deep Start traverses the function count graph, it may find connected subgraphs whose nodes' function counts are all above the deep starter threshold. In this case, Deep Start selects the function for the deepest node in the subgraph (i.e., the node furthest from the function count graph root) as the deep starter. Given the PC's call-graph-based refinement scheme when examining application code, the deepest node's function in an above-threshold subgraph is the most specific potential bottleneck for the subgraph and is thus the best choice as a deep starter.

2.2 Adding Deep Starters

Once a deep starter function is selected, the PC creates an experiment for the deep starter and adds it to its search. The experiment *E* whose refinement triggered the deep starter selection algorithm determines the nature of the deep starter's experiment. The deep starter experiment uses the same hypothesis and focus as *E*, except that the portion of *E*'s focus that specifies code resources is replaced with the deep starter function. For example, assume the experiment *E* is

hypothesis: CPU bound
focus: < /Code/om3.c/main, /Machine/c2-047/om3{1374} >

(that is, it examines whether the inclusive CPU utilization of the function `main` in process 1374 on host c2-047 is above the "CPU bound" threshold). If the PC selects `time_step` as a deep starter after refining *E*, the deep starter experiment will be

hypothesis: CPU bound
focus: < /Code/om3.c/time_step, /Machine/c2-047/om3{1374} >.

Also, Deep Start assigns a high priority to deep starter experiments so that they are given precedence when the PC activates experiments from its pending queue.

With the PC's current call-graph-based search strategy, the PC's search history graph reflects the application's call graph when the PC is searching through the application's code. Deep Start retains this behavior by creating as many *connecting experiments* as necessary to connect the deep starter experiment to some other experiment already in the search history graph. For example, in the search history graph in Fig. 1 the PC chose `p_makeMG` as a deep starter and added connecting experiments for functions `a_anneal`, `a_neighbor`, and `p_isvalid`. Deep Start uses its function count graph to identify connecting experiments for a deep starter experiment. Deep Start gives medium priority to the connecting experiments so that they have preference over the background call-graph search but not over the deep starter experiments.

3 Evaluation

To evaluate the Deep Start search algorithm, we modified the PC to search using either the Deep Start or the current call graph-based search strategy. We investigated the sensitivity of Deep Start to the deep starter threshold, and chose a threshold for use in our remaining experiments. We then compared the behavior of both strategies while searching for performance problems in several scientific applications. Our results show that Deep Start finds bottlenecks more quickly and often finds more bottlenecks than the call-graph-based strategy.

During our experimentation, we wanted to determine whether one search strategy performed "better" than the other. To do this, we borrow the concept of *utility* from consumer choice theory in microeconomics [15] to reflect a user's preferences. We chose a utility function where t is the elapsed time since the beginning of a search. This function captures the idea that users prefer to obtain results earlier in a search. For a given search, we weight each bottleneck found by U and sum the weighted values to obtain a single value that quantifies the search. When comparing two searches with this utility function, the one with the smallest absolute value is better.

3.1 Experimental Environment

We performed our experiments on two sequential and two MPI-based scientific applications (see Table 1). The MPI applications were built using MPICH [9], version 1.2.2. Our PC modifications were made to Paradyn version 3.2. For all experiments, we ran the Paradyn front-end process on a lightly-loaded Sun Microsystems Ultra 10 system with a 440 MHz Ultra SPARC IIi processor and 256 MB RAM. We ran the sequential applications on another Sun Ultra 10 system on the same LAN. We ran the MPI applications as eight processes on four nodes of an Intel x86 cluster running Linux, kernel version 2.2.19. Each node contains two 933 MHz Pentium III processors and 1 GB RAM. The cluster nodes are connected by a 100 Mb Ethernet switch.

Table 1. Characteristics of the applications used to evaluate Deep Start

Name	Version	Type	Language	Domain	Size
DRACO	6.0	Sequential	Fortran 90	Hydrodynamic simulation	68981 lines 18632 KB 398 functions
ALARA	2.4.4	Sequential	C++	Induced radioactivity analysis	19576 lines 2911 KB 720 functions
om3	1.5	Parallel (MPI)	C	Global ocean climate simulation	2674 lines 385 KB 36 functions
su3_rmd	6	Parallel (MPI)	C	Quantum chromodynamics pure gauge lattice theory simulation	35845 lines 511 KB 189 functions

3.2 Deep Start Threshold Sensitivity

We began by investigating the sensitivity of Deep Start to changes in the deep starter threshold (see Sect. 2.1). For one sequential (ALARA) and one parallel application (om3), we observed the PC's behavior during searches with thresholds 0.2, 0.4, 0.6, and 0.8. We performed five searches per threshold with both applications. We observed that smaller thresholds gave better results for the parallel application. Although the 0.4 threshold gave slightly better results for the sequential application, the difference between Deep Start's behavior with thresholds of 0.2 and 0.4 was small. Therefore, we decided to use 0.2 as the deep starter threshold for our experiments comparing the Deep Start and the call graph search strategy.

3.3 Comparison of Deep Start and Call Graph Strategy

Once we found a suitable deep starter threshold, we performed experiments to compare Deep Start with the PC's existing call graph search strategy. For each of our test applications, we observed the behavior of ten PC searches, five using Deep Start and five using the call graph strategy. Fig. 5 shows search profiles for both Deep Start and call graph search strategies for each of our test applications. These charts relate the bottlenecks found by a search strategy with the time they were found. The charts show the cumulative number of bottlenecks found as a percentage of the total number of known bottlenecks for the application. Each curve in the figure shows the average time over five runs to find a specific percentage of an application's known bottlenecks. Range bars are used to indicate the minimum and maximum time each search strategy needed to find a specific percentage across all five runs. In this type of chart, a steeper curve is better because it indicates that bottlenecks were found earlier and more rapidly in a search. Table 2 summarizes the results of these experiments for each of our test applications, showing the average number of experiments attempted, bottlenecks found, and weighted sum for comparison between the two search strategies.

For each application, Deep Start found bottlenecks more quickly than the current call graph search strategy as evidenced by the average weighted sums in Table 2 and the relative slopes of the curves in Fig. 5. Across all applications, Deep Start found half of the total known bottlenecks an average of 32% to 59% faster than the call graph startegy. Deep Start found all bottlenecks in its search an average of 10% to 61% faster than the call graph strategy. Although Table 2 shows that Deep Start tended to perform more experiments than the call graph search strategy, Deep Start found more bottlenecks when the call graph strategy found fewer than 100% of the known bottlenecks. Our results show that Deep Start finds bottlenecks more quickly and may find more bottlenecks than the current call graph search strategy.

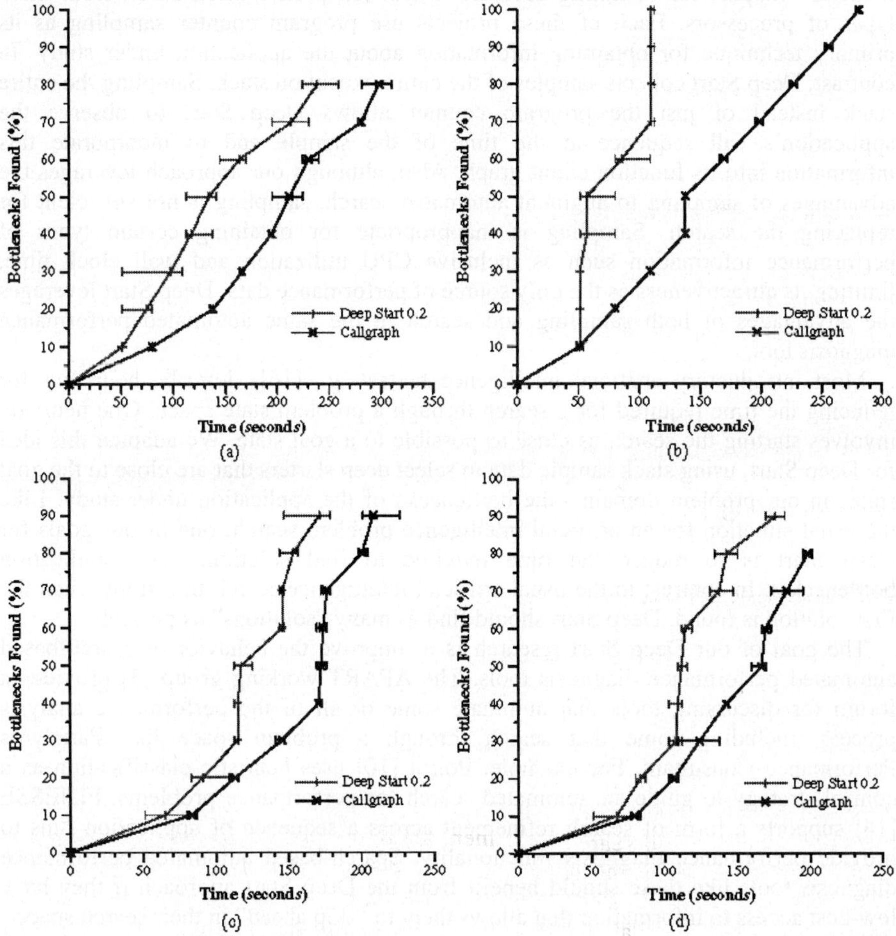

Fig. 5. Profiles for Deep Start and call graph searches on (a) ALARA, (b) DRACO, (c) om3, and (d) su3_rmd. Each curve represents the average time taken over five runs to find a specific percentage of the application's total known bottlenecks. The range bars indicate the best and worst time taken to find each percentage across the five runs

4 Related Work

Whereas Deep Start uses stack sampling to enhance its normal search behavior, several tools use sampling as their primary source of application performance data. Most UNIX distributions include the prof and gprof [8] profiling tools for performing flat and call graph-based profiles, respectively. Quartz [2] addressed the shortcomings of prof and gprof for parallel applications running on shared memory multiprocessors. ProfileMe [1] uses program counter sampling in DCPI to obtain low-level information about instructions executing on in-order Alpha [5] processors. Recognizing the limitations of the DCPI approach for out-of-order processors, Dean et al. [6] designed hardware support for obtaining accurate instruction profile information from these types of processors. Each of these projects use program counter sampling as its primary technique for obtaining information about the application under study. In contrast, Deep Start collects samples of the entire execution stack. Sampling the entire stack instead of just the program counter allows Deep Start to observe the application's call sequence at the time of the sample and to incorporate this information into its function count graph. Also, although our approach leverages the advantages of sampling to augment automated search, sampling is not sufficient for replacing the search. Sampling is inappropriate for obtaining certain types of performance information such as inclusive CPU utilization and wall clock time, limiting its attractiveness as the only source of performance data. Deep Start leverages the advantages of both sampling and search in the same automated performance diagnosis tool.

Most introductory artificial intelligence texts (e.g., [16]) describe heuristics for reducing the time required for a search through a problem state space. One heuristic involves starting the search as close as possible to a goal state. We adapted this idea for Deep Start, using stack sample data to select deep starters that are close to the goal states in our problem domain—the bottlenecks of the application under study. Like the usual situation for an artificial intelligence problem search, one of our goals for Deep Start is to reduce the time required to find solutions (i.e., application bottlenecks). In contrast to the usual artificial intelligence search that stops when the first solution is found, Deep Start should find as many "solutions" as possible.

The goal of our Deep Start research is to improve the behavior of search-based automated performance diagnosis tools. The APART working group [3] provides a forum for discussing tools that automate some or all of the performance analysis process, including some that search through a problem space like Paradyn's Performance Consultant. For example, Poirot [10] uses heuristic classification as a control strategy to guide an automated search for performance problems. FINESSE [14] supports a form of search refinement across a sequence of application runs to provide performance diagnosis functionality. Search-based automated performance diagnosis tools like these should benefit from the Deep Start approach if they have low-cost access to information that allows them to "skip ahead" in their search space.

Table 2. Summary of Deep Start/Call Graph comparison experiments. "Total Known Bottlenecks" is the number of unique bottlenecks observed during any search on the application, regardless of search type and deep starter threshold

Application	Total Known Bottlenecks	Search Type	Average Experiments Attempted	Average Bottlenecks Found	Average Weighted Sum
ALARA	46	Call Graph	174.0	39.4 (86%)	-191.9
		Deep Start	173.4	39.8 (87%)	-134.1
DRACO	18	Call Graph	105.0	18.0 (100%)	-152.7
		Deep Start	105.0	18.0 (100%)	-75.2
om3	145	Call Graph	261.6	141.8 (98%)	-158.1
		Deep Start	269.0	142.2 (98%)	-124.5
su3_rmd	85	Call Graph	260.0	75.6 (89%)	-141.8
		Deep Start	261.4	82.2 (97%)	-114.3

Acknowledgements

This paper benefits from the hard work of Paradyn research group members past and present. We especially wish to thank Victor Zandy and Brian Wylie for fruitful discussions on our topic, and Victor Zandy and Erik Paulson for their help in collecting our MPI application results. We also wish to thank the anonymous reviewers for their helpful comments.

References

[1] Anderson, J.M., Berc, L.M., Dean, J., Ghemawat, S., Henzinger, M.R., Leung, S.-T.A., Sites, R.L, Vandevoorde, M.T., Waldspurger, C.A., Weihl, W.E.: Continuous Profiling: Where Have All the Cycles Gone? *ACM Transactions on Computer Systems* **15**(4) Nov. 1997.

[2] Anderson, T.E., Lazowska, E.D.: Quartz: A Tool For Tuning Parallel Program Performance. *1990 ACM Conf. on Measurement and Modeling of Computer Systems*, Boulder, Colorado, May 1990. Appears in *Performance Evaluation Review* **18**(1) May 1990.

[3] The APART Working Group on Automatic Performance Analysis: Resources and Tools. http://www.gz-juelich.de/apart.

[4] Cain, H.W., Miller, B.P., Wylie, B.J.N.: A Callgraph-Based Search Strategy for Automated Performance Diagnosis. *6th Intl. Euro-Par Conf.*, Munich, Germany, Aug.–Sept. 2000. Appears in *Lecture Notes in Computer Science* **1900**, A. Bode, T. Ludwig, W. Karl, and R. Wismüller (Eds.), Springer, Berlin Heidelberg New York, Aug. 2000.

[5] Compaq Corporation: 21264/EV68A Microprocessor Hardware Reference Manual. Part Number DS-0038A-TE, 2000.

[6] Dean, J., Hicks, J.E., Waldspurger, C.A., Weihl, W.E., Chrysos, G.: ProfileMe: Hardware Support for Instruction-Level Profiling on Out-of-Order Processors. *30th Annual IEEE/ACM Intl. Symp. On Microarchitecture*, Research Triangle Park, North Carolina, Dec. 1997. [7] Gerndt, H.M., Krumme, A.:ARule-Based Approach for Automatic Bottleneck Detection in Programs on Shared Virtual Memory Systems. *2nd Intl. Workshop on High-Level Programming Models and Supportive Environments*, Geneva, Switzerland, Apr. 1997.

[8] Graham, S., Kessler, P., McKusick, M.: An Execution Profiler for Modular Programs. *Software—Practice & Experience* **13**(8) Aug. 1983.

[9] Gropp, W., Lusk, E., Doss, N., Skjellum, A.: A High-Performance, Portable Implementation of the MPI Message Passing Interface Standard. *Parallel Computing* **22**(6) Sept. 1996. [10] Helm, B.R., Malony, A.D., Fickas, S.F.: Capturing and Automating Performance Diagnosis: the Poirot Approach. *1995 Intl. Parallel Processing Symposium*, Santa Barbara, California, Apr. 1995.

[11] Hollingsworth, J.K., Miller, B.P., Cargille, J.: Dynamic Program Instrumentation for Scalable Performance Tools. *1994 Scalable High Perf. Computing Conf.*, Knoxville, Tennessee, May 1994.

[12] Karavanic, K.L., Miller, B.P.: Improving Online Performance Diagnosis by the Use of Historical Performance Data. *SC'99*, Portland, Oregon, Nov. 1999.

[13] Miller, B.P., Callaghan, M.D., Cargille, J.M., Hollingsworth, J.K., Irvin, R.B., Karavanic, K.L., Kunchithapadam, K., Newhall, T.: The Paradyn Parallel Performance Measurement Tool. *IEEE Computer* **28**(11) Nov. 1995.

[14] Mukerjee,N., Riley, G.D., Gurd, J.R.: FINESSE: A Prototype Feedback-Guided Performance Enhancement System. *8th Euromicro Workshop on Parallel and Distributed Processing*, Rhodes, Greece, Jan. 2000.

[15] Pindyck, R.S., Rubinfeld, D.L.: **Microeconomics**. Prentice Hall, Upper Saddle River, New Jersey, 2000.

[16] Rich, E., Knight, K.: **Artificial Intelligence**. McGraw-Hill, New York, 1991.

On the Scalability of Tracing Mechanisms[1]

Felix Freitag, Jordi Caubet, and Jesus Labarta

Departament d'Arquitectura de Computadors (DAC)
European Center for Parallelism of Barcelona (CEPBA)
Universitat Politècnica de Catalunya (UPC)
{felix,jordics,jesus}@ac.upc.es

Abstract. Performance analysis tools are an important component of the parallel program development and tuning cycle. To obtain the raw performance data, an instrumented application is run with probes that take measures of specific events or performance indicators. Tracing parallel programs can easily lead to huge trace files of hundreds of Megabytes. Several problems arise in this context: The storage requirement of the high number of traces from executions under slightly changed conditions; visualization packages have difficulties in showing large traces efficiently leading to slow response time; large trace files often contain huge amounts of redundant information. In this paper we propose and evaluate a dynamic scalable tracing mechanism for OpenMP based parallel applications. Our results show: With scaled tracing the size of the trace files becomes significantly reduced. The scaled traces contain only the non-iterative data. The scaled trace reveals important performance information faster to the performance analyst and identifies the application structure.

1 Introduction

Performance analysis tools are an important component of the parallel program development and tuning cycle. A good performance analysis tool should be able to present the activity of parallel processes and associated performance indices in a way that easily conveys to the analyst the main factors characterizing the application behavior. In some cases, the information is presented by way of summary statistics of some performance index such as profiles of execution time or cache misses per routine. In other cases the evolution of process activities or performance indices along time is presented in a graphical way.

To obtain the raw performance data, an instrumented application is run with probes that take measures of specific events or performance indicators (i.e. hardware counters). In our approach every point of control in the application is instrumented. At the granularity level we are interested in, subroutine and parallel loops are the control points where tracing instrumentation is inserted. The information accumulated in the hardware counters with which modern processors and systems are equipped is read at these points.

[1] This work has been supported by the Spanish Ministry of Science and Technology and by the European Union (FEDER) under TIC2001-0995-C02-01.

B. Monien and R. Feldmann (Eds.): Euro-Par 2002, LNCS 2400, pp. 97–104.

Our approach to the scalability problem of tracing is to limit the traced time to intervals that are sufficient to capture the application behavior. We claim it is possible to dynamically acquire the understanding of the structure of iterative applications and automatically determine the relevant intervals. With the proposed trace scaling mechanism it is possible to dynamically detect and trace only one or several iterations of the repetitive pattern found in scientific applications. The analysis of such a reduced trace can be used to tune the main iterative body of the application.

The rest of the paper is structured as follows: In section 2 we describe scalability problems of tracing mechanisms. Section 3 shows the implementation of the scalable tracing mechanism. Section 4 evaluates our approach. Section 5 describes solutions of other tracing frameworks to the trace scalability. In section 6 we conclude the paper.

2 Scalability Issues of Tracing Mechanisms

Tracing parallel programs can easily lead to huge trace files of hundreds of Megabytes. Several problems arise in this context. The storage requirement of traces can quickly become a limiting factor in the performance analysis cycle. Often several executions of the instrumented application need to be carried out to observe the application behavior under slightly changed conditions.

Visualization packages have difficulties in showing large traces effectively. Large traces make the navigation (zooming, forward/backward animation) through them very slow and require the machine where the visualization package is run to have a large physical memory in order to avoid an important amount of I/O.

Large trace files often contain huge amounts of redundant trace information, since the behavior of many scientific applications is highly iterative. When visualizing such large traces, the search for relevant details becomes an inefficient task for the program analyst. Zooming down to see the application behavior in detail is time-consuming if no hints are given about the application structure.

3 Dynamic Scalable Tracing Mechanism

3.1 OpenMP Based Application Structure and Tracing Tool

The structure of OpenMP based applications usually iterates over several parallel regions, which are marked by directives as code to be executed by the different threads. For each parallel directive the master thread invokes a runtime library passing as argument the address of the outlined routine. The tracing tool intercepts the call and it obtains a stream of parallel function identifiers. This stream contains all executed parallel functions of the application, both in periodic and non-periodic parallel regions.

We have implemented the trace scaling mechanism in the OMPItrace tool [2]. OMPItrace is a dynamic tracing tool to monitor OpenMP and/or MPI applications available for the SGI Origin 2000 and IBM SP platforms. The trace files that OMPItrace generates consist of events (hardware counter values, parallel regions entry/exit, user functions entry/exit) and thread states (computing, idle, fork/join). The traces can be visualized with Paraver [5].

3.2 Pattern Detection

We implemented the periodicity detector (DPD) [3] in the tracing mechanism in order to perform the automatic detection of iterative structures in the trace. The stream of parallel function identifiers is the input to the periodicity detector. The DPD provides an indication whether periodicity exists in the data stream, informs the tracing mechanism on the period length, and segments the data stream into periodic patterns. The periodicity detector is implemented as a library, whose input is a data stream of values from the instrumented parameters.

The algorithm used by the periodicity detector is based on the distance metric given in equation (1).

$$d(m) = sign \sum_{i=0}^{N-1} | x(i) - x(i-m) | \qquad (1)$$

In equation (1), N is the size of the data window, m is the delay (0<m<M), M<=N, x[i] is the current value of the data stream, and d(m) is the value computed to detect the periodicity. It can be seen that equation (1) compares the data sequence with the data sequence shifted m samples. Equation (1) computes the distance between two vectors of size N by summing the magnitudes of the L1-metric distance of N vector elements. The sign function is used to set the values d(m) to 1 if the distance is not zero. The value d(m) becomes zero if the data window contains an identical periodic pattern with periodicity m.

4 Evaluation

We evaluate the following aspects of the scalable tracing mechanism: 1) Trace size reduction; 2) Improvements of the ease of visualization and application structure identification; 3) Tracing overhead; and 4) Trace completeness.

All experiments are carried out on a Silicon Graphics Origin 2000 with 64 processors running with the Irix 6.5.8 operating system. The OpenMP applications are executed in a dedicated environment with 8 CPUs. We configure the scaling mechanism such that after having detected 10 iterative parallel regions it stops writing trace data to the file until it observes a new program behavior. The parameters contained in the trace file are the thread states and OpenMP events, which include two hardware counters.

We use the applications given in Table 1: Four applications from the NAS benchmark suite: Bt (class A), Lu (class A), Cg (class A) and Sp (class A); and five applications of the SPECFp95 suite: Swim, Hydro2d, Apsi, Tomcatv, and Turb3d, all with ref data set. The applications Ep, Ft, Mg from the NAS benchmarks are not used, since their trace files are small and/or they have very few periodic patterns. Column 2 of Table 1, periodicity length, indicates that the NAS benchmarks and Apsi, Swim, and Tomcatv have only one periodicity, while Hydro2d and Turb3d have nested iterative parallel structures (N). The periodicity length is the size of the periodic pattern measured in terms of parallel regions. The number of times each periodic pattern repeats is given in column 3. For instance, the function stream of the Bt application exhibits the same pattern 201 times. In column 4 the data stream length is shown. The data stream length is approximately the product of the periodicity length by the number of iterations, due to the iterative structure of the applications. Additionally, the data stream length includes the number of functions, which do not belong to a periodic pattern.

Table 1. Evaluated benchmarks.

Application	Periodicity length	Number of iterations	Data stream length
NAS Bt	9	201	1827
NAS Cg	4, 106	72,13	1702
NAS Lu	8	251	2021
NAS Sp	14	401	5636
Apsi	6	961	5762
Hydro2D	1(N), 24(N), 269	16, 8, 196	53814
Swim	6	900	5402
Tomcatv	5	751	3750
Turb3d	1(N), 12(N), 142	24, 17, 10	1580

4.1 Reduction of the Trace Size

We study how much the trace file size reduces when using the scalable dynamic mechanism. Fig. 1 shows the size of the trace files obtained with and without scaling mechanism. It can be seen that with scalable tracing the trace files reduced significantly. The NAS Lu trace file, for instance, reduces from 173 Mb to 8 Mb, which is a reduction of 95%.

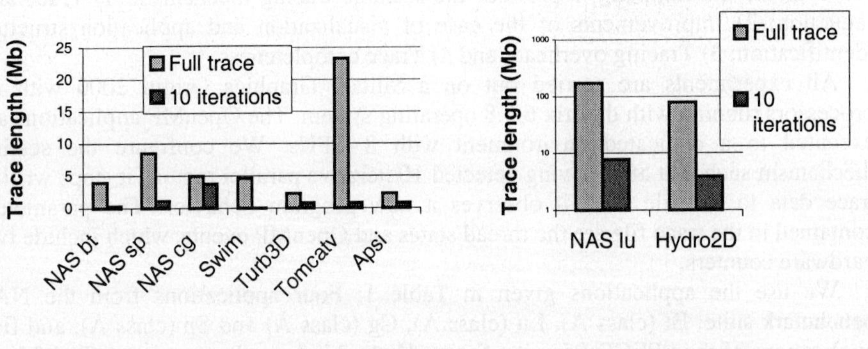

Fig. 1. Comparison of the trace file size with full and scalable tracing.

4.2 Improvement of the Ease of Visualization and Application Structure Identification

We want to show that scaled traces are more visualization-friendly than complete traces. In addition, by selecting the Hydro2d benchmark, we want to illustrate how the scalable tracing mechanism restarts tracing once it observes a change in the application behavior. In Figure 2 we show the visualization of the scaled trace of the Hydro2d application with Paraver. The visualization of the thread states is shown. The activity of the threads is encoded in dark color for actively computing, gray color if the thread is in the idle loop and bright color if it is in a fork/join activity.

In Figure 2 we can easily identify that there is a periodic pattern (period boundaries tagged with flags). It can be observed in the middle part that after a certain number of repetitions this pattern changes and that a new periodic pattern is then repeated. The flags in Figure 2 identify the period, so with the scaled trace it is immediate to zoom to an adequate level to see the actual pattern of behavior. In the visualization of the scaled trace the data describing iterative application behavior is not shown (black area), since the tracing mechanism did not write it to the trace file.

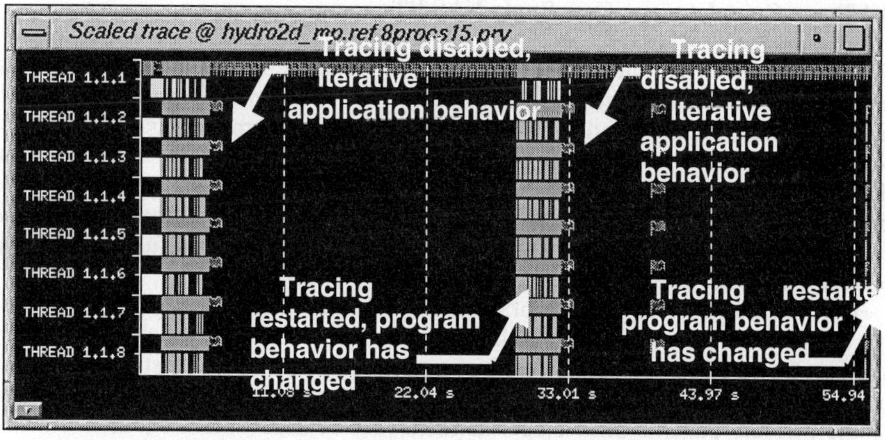

Fig. 2. Visualization of the Hydro2D execution with scalable tracing.

4.3 Overhead of the Scalable Tracing Mechanism

We evaluate the overhead introduced by the scaled tracing mechanism. We compare the execution time of the different benchmarks with and without tracing (Figure 3). The first bar (light gray) shows the execution time of the application without tracing, the second bar (dark gray) shows the execution time when the applications are traced with the original tracing mechanism, and the third bar (white) shows the execution time when the application is traced with the scalable tracing mechanism. The original tracing mechanism adds about 1%-3% to the execution time. With the scalable tracing mechanism, the overhead is about 3%-6%. It can be observed that the overhead introduced by the tracing tool is small in terms of execution time.

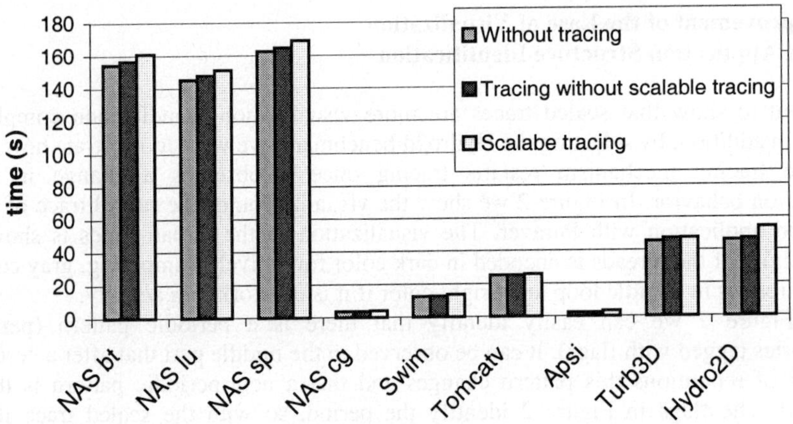

Fig. 3. Overhead evaluation.

4.4 Trace Completeness: Case Study NAS BT

We want to show that the same quantitative performance results are obtained when analysing the full and scaled trace. With Paraver we obtain performance indices such as global load balance, duration of parallel loops, data cache misses, and TLB misses. We wish to demonstrate that using the scaled trace leads to the same conclusions on the application performance as the full trace. In a case study we compare performance indices computed from the scaled trace containing 10 iterations and from the full trace of the NAS BT application. In [1] the comparison of the scaled and full traces of the other applications given in Table 1 can be found.

We compute the load balance of the BT application from the scaled and the full trace. The values obtained from both traces for % running state per thread are very similar. On average we obtain for this application 91 % running state from both traces. Next we compare the TLB and L2 data cache misses in the parallel functions of the periodic pattern. On average the difference in TLB misses between the two traces is less than 1 %, and in L2 data cache misses it is 4%. In Fig. 4 we show the execution time of the parallel functions computed from the two traces. This is the average value for all the executions of each routine during each trace. We show the 95% confidence interval. It can be seen that the function mpdo_z_solve_1 has the longest execution time (0.22 seconds approximately), followed by mpdo_x_solve_1 and mpdo_y_solve_1 (0.155 seconds approximately). This information useful to the analyst can be clearly identified from both traces.

5 Related Work

The most frequent approach to restrict the size of the trace in current practice is to insert calls to start and stop the tracing into the source code. Systems such as VampirTrace [7], VGV [4] rely on this mechanism. OMPItrace [2] also provides this mechanism to the programmer. This approach requires the modification of the source code, which may not always be available to the performance analyst. The totally automatic way we propose in this paper is certainly more convenient as the analyst gets the structure of the application from the trace and can start identifying problems without any need of the source code.

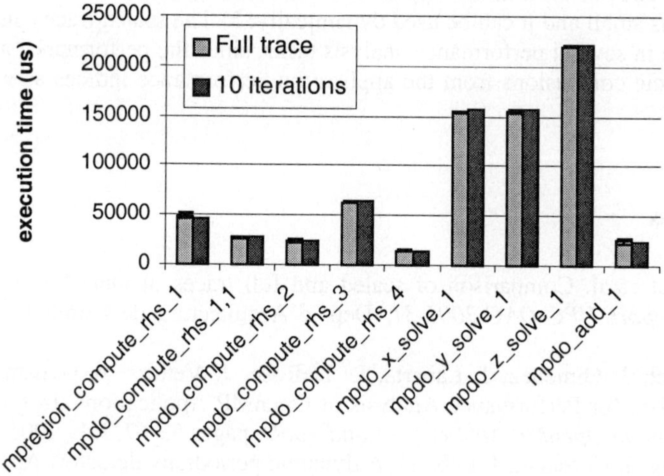

Fig. 4. BT application from the NAS benchmarks. Comparison of the execution time of the parallel functions reported by Paraver from the full and the scaled trace.

In IBM UTE [8] an intermediate approach is followed to partially tackle the problem that very large traces pose to the analysis tool. The tracing facility can generate huge traces of events. Then, some filters are used to extract a trace that focuses on a specific application. To properly handle the fast access to specific regions of a large trace file the SLOG format (scalable logfile format) has been adopted. Using a frame index the Jumpshot visualization tool [9], for instance, improves the access time to trace data.

The Paradyn project [6] developed an instrumentation technology (Dyninst) through which it is possible to dynamically insert and take out probes in a running program. Although no effort is made to automatically detect periods, the methodology behind this approach also relies on the iterative behavior of applications. The automatic periodicity detection idea we present in this paper could also be useful inside the dynamic analysis tool to present to the user the actual structure of the application.

6 Conclusions

We have presented a number of scalability problem of tracing encountered in current performance analysis tools. We described the reasons for large traces and why this is a problem. We showed different approaches currently used to reduce the trace file size. In our approach we implemented a dynamic scalable tracing mechanism, which records only the non-iterative trace data during the application execution and stops writing the redundant data to the trace file. Our results show: 1) The size of the trace file becomes significantly reduced; 2) In order to achieve a reduced trace, our approach does not limit the granularity of tracing, nor the number of read parameters, nor the problem size; 3) The scaled trace lets the analyst faster observe relevant application behavior such as the application structure; 4) The overhead of the scaled tracing tool is small and it can be used dynamically; 5) The scaled trace can substitute the full trace in several performance analysis tasks, since the performance analyst can reach the same conclusions from the application performance indices as when using the full trace.

References

[1] J. Caubet et al. Comparison of scaled and full traces of OpenMP applications. *Tech. Report UPC-DAC-2001-31,* Dep. d'Arquitectura de Computadors, UPC, 2001.

[2] J. Caubet, J. Gimenez, J. Labarta, L. DeRose, J. Vetter. A Dynamic Tracing Mechanism for Performance Analysis of OpenMP Applications. In *International Workshop on OpenMP Applications and Tools*, pages 53-67, July 2001.

[3] F. Freitag, J. Corbalan, J. Labarta. A dynamic periodicity detector: Application to Speedup Computation. In *Proceedings of International Parallel and Distributed Processing Symposium (IPDPS 2001)*, April 2001.

[4] J. Hoeflinger et al. An Integrated Performance Visualizer for MPI/OpemnMP Programs. In *In. Workshop on OpenMP Applications and Tools*, pages 40-52, July 2001.

[5] J. Labarta, S. Girona, V. Pillet, T. Cortes and L. Gregoris. DiP: A Parallel Program Development Environment. In *Proceedings of 2nd International EuroPar Conference (EuroPar 96)*, August 1996.

[6] B. P. Miller et. al. The Paradyn Parallel Performance Measurement Tools. *IEEE Computer* 28(11):37-46, November 1995.

[7] Pallas: *Vampirtrace. Installation and User's Guide.* http://www.pallas.de

[8] C. E. Wu et al. From Trace Generation to Visualization: A Performance Framework for Distributed Parallel Systems. In *Proceedings of SuperComputing (SC2000),* November 2000.

[9] O. Zaki, E. Lusk, W. Gropp, and D. Swider. Toward scalable performance visualization with Jumpshot. In *International Journal of High Performance Computing Applications,* 13(2): pages 277-288, 1999.

Component Based Problem Solving Environment

A.J.G. Hey[1], J. Papay[2], A.J. Keane[3], and S.J. Cox[2]

[1] EPSRC, Polaris House, North Star Avenue, Swindon, SN2 1ET, UK
Tony.Hey@epsrc.ac.uk
http://www.epsrc.ac.uk

[2] Department of Electronics and Computer Science, University of Southampton, Southampton, SO17 1BJ, UK
{jp, sc}@ecs.soton.ac.uk

[3] Computational Engineering and Design Centre, University of Southampton, Southampton, SO17 1BJ, UK
ajk@soton.ac.uk

Abstract. The aim of the project described in this paper was to use modern software component technologies such as CORBA, Java and XML for the development of key modules which can be used for the rapid prototyping of application specific Problem Solving Environments (PSE). The software components developed in this project were a user interface, scheduler, monitor, various components for handling interrupts, synchronisation, task execution and software for photonic crystal simulations. The key requirements for the PSE were to provide support for distributed computation in a heterogeneous environment, a user friendly interface for graphical programming, intelligent resource management, object oriented design, a high level of portability and software maintainability, reuse of legacy code and application of middleware technologies in software design.

1 Introduction

A Problem Solving Environment (PSE) is an application-specific environment that provides the user with support for all stages of the problem solving process - from program design and development to compilation and performance optimization [1]. Such an environment also provides access to libraries and integrated toolsets, as well as support for visualization and collaboration. The main reasons for the development of PSEs are to simplify the usage of existing software modules, simplify problem specification, solution and to maximize the utilisation of distributed computing resources.

In this project we investigated in detail the design, implementation and use of several PSEs applied to large-scale computational problems. These PSEs were Promenvir [2], Toolshed [3], GasTurbnLab [4], Autobench [5] and SCIRUN [6]. The reason for analyzing these projects was to draw lessons and to evaluate the pros and cons of various design options. There were numerous lessons drawn which were related to portability of software components, Object Oriented Design, application of middleware, fault tolerance and performance modelling. Here we summarise the main ideas influencing the software design.

B. Monien and R. Feldmann (Eds.): Euro-Par 2002, LNCS 2400, pp. 105–112.

All software except the FEM solvers used for photonic crystal simulations were developed using the Java [7] programming language. This provided platform independence, portability, high level reliability and security. Moreover, Java proved to be more elegant and more suitable for rigorous object oriented design than the more complex C++ language.

During the project considerable time was spent on the evaluation of emerging middleware technologies and products. Several alternatives for the software architecture were evaluated by taking into account the complexity of the application, capability of handling complex data structures, capability of handling applications written in different languages and running on different platforms. After considering and testing various component technologies, we opted for CORBA based middleware [8]. The use of sockets and other mechanisms was also considered, but these options were later dropped because of the increase in complexity of design. CORBA provides a standard mechanism for defining the interfaces between components, a compiler enabling the use of interfaces in other programming languages, services such as the naming service, implementation repository, event service etc., a communication mechanism enabling objects to interact with each other, programming language and platform independence. CORBA also provides a portable and reliable platform for transparent remote object access and communication and clearly demonstrated the suitability of this technology for the design of large scale distributed systems.

Learning from the lessons of previous projects we decided to use a database rather than implementing complicated data structures such as linked lists, tables, stacks, etc., for storing and manipulating monitoring and scheduling information. The use of the database simplified the software architecture and significantly improved the flexibility of the design and its capacity for evolution.

During the PSE development, issues of fault tolerance and error recovery were given high priority. In this respect the PSE provides availability checking and lifetime control of remote objects. The robustness of the system was tested by simulating various scenarios during which individual servers were switched off.

Performance Engineering (PE) played an important role during the scheduler design. PE is concerned with the reliable prediction and estimation of resource requirements and performance of applications. Several performance models were developed, these models were used by the scheduler and performance estimator for execution time prediction and achieving better utilisation of the available computing resources.

The XML technology was extensively used in the project. The main advantage of XML is that it provides platform independent data exchange and a framework for the specification of data structures. XML was used for the design of user interfaces and as a configuration language for describing the interconnection of tasks, their resource requirements and performance models.

The selection of an appropriate middleware product is of key importance for the software design. Various CORBA based middleware products were tested for their performance, reliability, user support and suitability for distributed PSE development. These products were the Sun ORB, Orbix [9] and Orbacus [10]. Initially the Sun ORB contained in the Java Development Kit (JDK) was used, however, at the time of evaluation it had limited functionality and its speed and reliability were unsatisfactory. Both Orbix and Orbacus offered the required functionality and complied with the CORBA 2.3 specification. Moreover Orbacus is a shareware product which can be an advantage for exploratory academic research, considering the risks associated with a fast moving middleware market. Orbacus includes C++ and Java implementations and provides several services: Naming, Event, Property, Time, Trading, Notification, and Logging. During the lifetime of the project several versions of JDK and Orbacus were released. As a result, components already developed had to be modified in order to maintain compatibility and to take advantage of the improved functionality of new releases. The final version of the PSE prototype was build by using JDK version 1.3 and Orbacus 4.0.5.

2 PSE Architecture

The PSE developed in this project is a distributed component based architecture (see Fig. 1). The reasons for opting for this design were the isolation of components, scalability of the system, flexibility of software deployment and updates, and resistance against the side effects of software modifications.

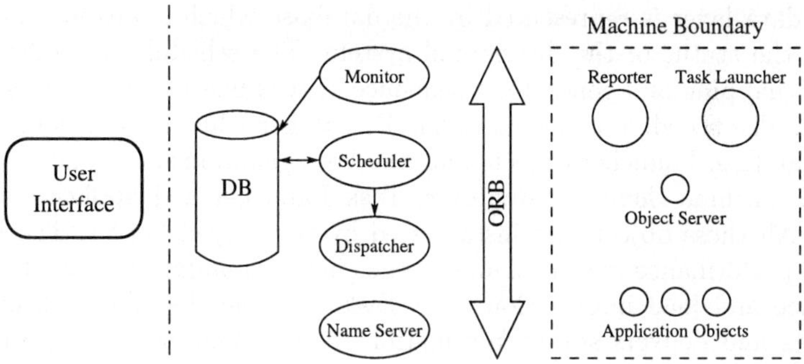

Fig. 1. Architecture of the prototype PSE

The user interface is a graphical Java front-end which provides numerous functions such as task-graph composition, parameter setup, lifetime control of distributed objects, monitoring the status of machines and tasks, computation steering and visualization. The user interface was developed in cooperation with Cardiff University [11]. The output of the graphical composer is a task-graph which describes the execution order, data dependencies and parallelism. This form of problem specification corresponds with the data flow model of computation in which individual tasks consist of programs and parameters which specify their function, characterise their resource requirements and performance. The resource requirements are expressed in terms of memory size, communication and I/O traffic, and disk volume used by the application. An example of a task-graph representing a parallel version of photonic crystal simulation is given in Fig. 2. The user interface also contains a separate control panel which provides life time control of objects, computational steering, task and machine monitoring.

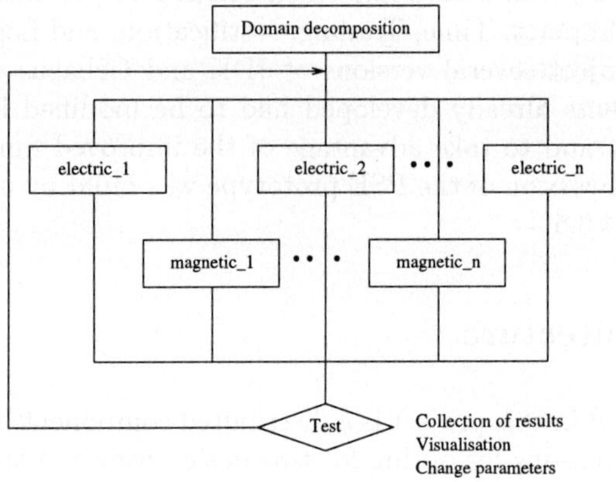

Fig. 2. Task-graph of parallel photonic-crystal simulation

The data layer is represented by the database which stores information related to the status of the distributed system. The scheduler provides task-to-machine mapping and generates a sequence of task executions represented by a schedule. The schedule is passed to the Dispatcher which forwards the tasks to individual Task Launcher objects and handles synchronization and sequencing. On each machine there are Reporter, Task Launcher and application specific objects. All these objects are instantiated by the Object Server. The Reporter provides performance information, i.e., the current amount of available memory, disk space and load information. The Task Launcher handles task execution, interrupts and delivers status information to the Monitor. An important part of the PSE architecture is the Name Server which contains a hierarchy of name and Internet Object Reference (IOR) tuples. This service enables components to be accessed by their name rather than address, this makes application development easier and more transparent. The Application objects represent application specific components described by their IDL specification.

The database is a critical part of the PSE architecture, since it stores all information related to the PSE environment. This information is constantly updated as the computation proceeds. For the database programming the JDBC 2.0 was used which allows database access programmatically, i.e., from the Java code rather than via embedding SQL commands in the code. This provides better portability because it bridges the variations of SQL dialects typical for different DBMS products. The key tables in the database are the Machine and Task tables. The Machine table contains the following fields: machine name, IP-address, number of processors, clock-speed, memory size, disk size, type of operating system, processor's flop rate, I/O rate, communication speed and current load. The Task table reflects the current status of tasks in the system and also contains information on their predicted resource requirements and measured resource usage in terms of flop-count, memory and disk sizes, I/O and communication volume and flop count.

The information collected by the Monitor is stored in a database and used by the Scheduler for task allocation and load balancing. The interactions of the Monitor with the other components of the system are represented in Fig. 3. On each computer there is an Object Server deployed which instantiates all local objects. The Reporter registers with the Name Server, which maintains a list of remote object addresses. The Monitor queries the Reporter objects at regular intervals and updates the Machine and Task tables in the database.

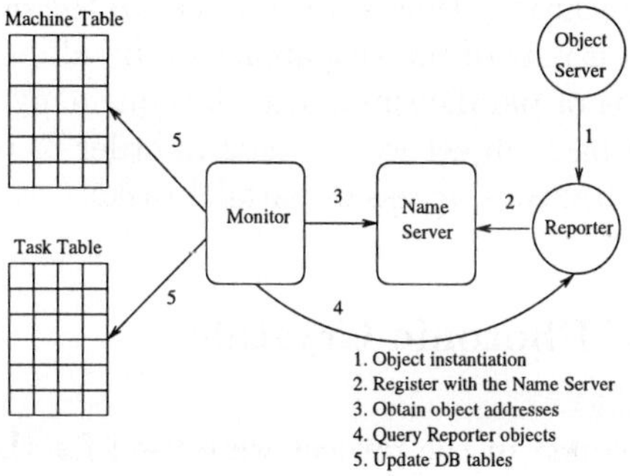

Fig. 3. Monitor's interaction with other modules

The task-to-machine mapping is performed by the scheduler which allows components of the task-graph to run on remote machines in a transparent manner. This operation is based on matching tasks resource requirements with computer parameters. The resource requirements are described by performance models. These were developed by using a characterization technique which involves statistical processing of measurements, identifying the key parameters governing the application's resource requirements and performance, and developing mathematical models. Fig. 4 gives an illustration of performance models of photonic crystal simulations in terms of flop-count, memory and disk usage.

These models are machine independent in that they characterize the resource requirements specific to the given application. The scheduling algorithm is based on the Cartesian product of Machine and Task tables. This operation generates all possible task-to-machine combinations and computes the predicted execution time for each of them. The scheduling algorithm performs the following operations for each task of the task-graph: checks the available memory, disk and load of individual machines, generates a table containing all possible task-to-machine

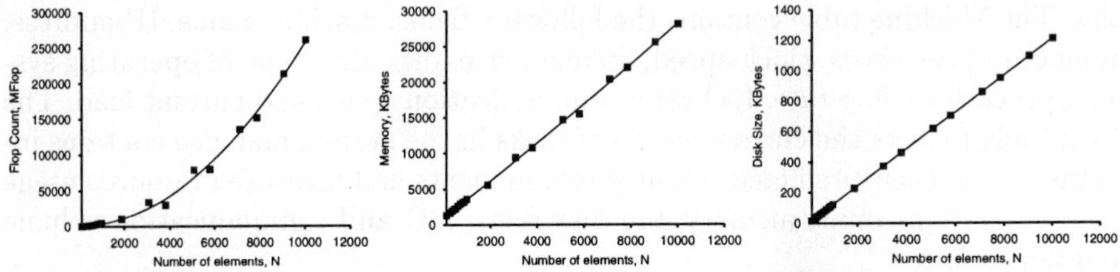

Fig. 4. Resource models for flop-count, memory and disk usage of photonic crystal simulations

mappings,computes the estimated execution time for each combination,generates a schedule by selecting those combinations which minimise the predicted execution time for the whole task-graph.

In the distributed PSE environment where a large number of independent threads are present synchronization is a key issue. In the design we developed a call-back object which is part of the Dispatcher for handling the synchronization between different levels of parallelism and task sequencing. The critical regions of the code in the call-back object were locked in order to prevent corruption of shared data by a simultaneous access of parallel tasks.

3 Simulation of Photonic Crystals

The components developed in the project were used for the construction of an application specific PSE protype. This prototype was used as a parallel computational test-bed for performing various numerical experiments and optimization studies on the example of photonic crystal simulations. The parallel version of the photonic crystal simulation task-graph is presented in Fig. 2. The PSE prototype proved to be an excellent test-bed for performing parametric studies involving changing the material properties and geometry parameters of photonic crystals. Photonic crystals with a band-gap properties have numerous applications for optical computing and for the development of efficient narrow band lasers [12]. These simulations allow optimization of the positioning of air rods in the crystal to achieve band gaps at various frequencies. The measurements of the simulations showed that the complexity of the computations (i.e. flop count) could be approximated by a second order polynomial (see Fig. 4). Substantial performance improvement was achieved by using various approximation methods which significantly increased the speed of computation while preserving the accuracy of results. The prototype PSE was run on a cluster of 13 workstations running the Windows and Linux operating systems. For visualization the Matlab package was used. Fig. 5 gives an illustration of the results by presenting a unit cell of a crystal with an air rod and the density of states for the transverse electric and magnetic components of radiation. Gaps in the density of states indicate frequencies of photons which cannot propagate through the medium.

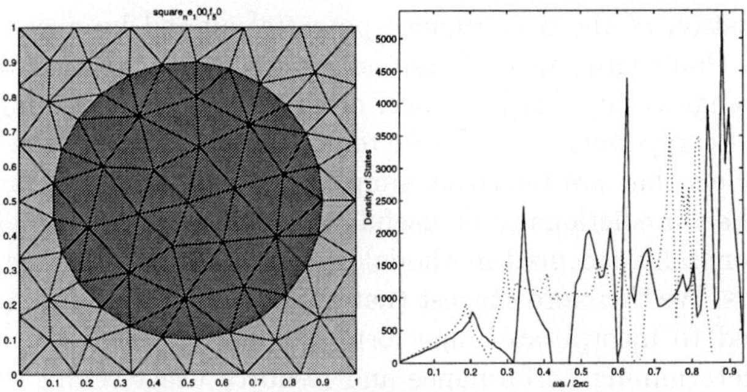

Fig. 5. Unit cell of computation and the distribution of electric and magnetic components of radiation in a photonic crystal

The benefits of the presented PSE can be summarized as follows: this environment integrates all stages of the problem solving process beginning from the task specification to the visualization of results, provides the user with a flexible graphical programming interface which enables to construct various task-graphs, the components developed in the project can be used for the construction of application specific PSEs as it was demonstrated on the example of photonic crystal simulations, the scheduler contains a performance prediction component which provides performance estimation and enables to optimize load balancing on the computing cluster.

4 Conclusions

In this paper the results achieved in a PSE project have been described. The aim of the project was to develop general purpose components such as a user interface, scheduler, monitor, and other components which can be used for rapid prototyping of application specific Problem Solving Environments. The key requirements for this PSE were to provide support for distributed computation in a heterogeneous environment, a user friendly interface for graphical programming, intelligent resource management, object oriented software design, a high level of portability and software maintainability. We conclude the paper with the following remarks:

Although the use of component based middleware technology promises simple distributed application development the reality is that it is not simple to program such systems. The main reason is that the CORBA based middleware technology is still in evolution and numerous features are not mature enough to provide a stable platform for the development of distributed systems. It must be said that the learning curve for CORBA is considerable and a large amount of new knowledge has to be absorbed and utilised in practice. Although programming using CORBA is not a trivial task, nevertheless, this technology is much more suitable for the development of distributed systems than the traditional techniques based on sockets and Remote Procedure Calls.

The utilization of the performance potential offered by a distributed environment is a challenging task. These issues are closely related to scheduling, performance engineering, load balance, ordering of events, dealing with a possible failure of components, etc. These requirements highlight the need for performance models that are relatively simple and easy to use yet are sufficiently accurate in their predictions to be useful as input to a scheduler [3]. At present there is no generally accepted methodology for characterising the performance of applications. We therefore suggest that applications, in addition to specifying interfaces need to incorporate some form of information about the governing parameters determining performance and resource usage. Only with the availability of such performance information will the construction of truly intelligent schedulers become possible. Although the computer science community has been researching performance for a long time, we believe that such research needs to become more systematic and scientific. A common approach to representing performance data together with a methodology that allows independent verification and validation of performance results would be a good start in this direction.

Acknowledgements

This work was sponsored by an EPSRC grant No. GR/M17259/01. We would like to thank to Jacek Generowicz, Ben Hiett, David Lancaster and Tony Scurr for the helpful discussions.

References

1. E.Gallopoulos, E.Houstis and J.R.Rice. Problem Solving Environments for Computational Science, IEEE Comput. Sci. Eng., 1, pp.11-23, 1994.
2. Jacek Marczyk, Principles of Simulation-Based Computer-Aided Engineering FIM Publications, Barcelona, 1999, p.174.
3. N. Floros, K.Meacham, J. Papay, M. Surridge. Predictive Resource Management for Unitary Meta-Applications, Future Generation Computer Systems, 15, 1999, pp.723-734.
4. http://www.cs.purdue.edu/research/cse/gasturbn
5. http://wwwvis.informatik.uni-stuttgart.de/eng/research/proj/autobench/
6. C.Johnson, S.Parker and D.Weinstein. Large Scale Computational Science Applications Using the SCIRun Problem Solving Environment, in Supercomputer 2000.
7. Pat Niemeyer, Jonathan Knudsen, Learning Java 3d revised edition, pp.728, O'Reilly UK, ISBN: 1565927184.
8. CORBA specification by the Object Management Group, http://www.omg.org
9. http://www.iona.com
10. http://www.ooc.com
11. M. S. Shields, O. F. Rana, David W. Walker, Li Maozhen, David Golby. A Java/CORBA based Visual Program Composition Environment for PSEs, Concurrency: Practice and Experience. Volume 12, Issue 8, 2000, pp.687-704.
12. J.M.Generowicz, B.P.Hiett, D.H.Beckett, G.J.Parker, S.J.Cox. Modelling 3 Dimensional Photonic Crystals using Vector Finite Elements. Photonics 2000 (EPSRC, UMIST, Manchester, 4-5 July 2000).

Integrating Temporal Assertions into a Parallel Debugger*

Jozsef Kovacs[1], Gabor Kusper[2], Robert Lovas[1], and Wolfgang Schreiner[2]

[1] Computer and Automation Research Institute (MTA SZTAKI)
Hungarian Academy of Sciences, Budapest, Hungary
{smith,rlovas}@sztaki.hu
http://www.lpds.sztaki.hu

[2] Research Institute for Symbolic Computation (RISC-Linz)
Johannes Kepler University, Linz, Austria
{Gabor.Kusper,Wolfgang.Schreiner}@risc.uni-linz.ac.at
http://www.risc.uni-linz.ac.at

Abstract. We describe the use of temporal logic formulas as runtime assertions in a parallel debugging environment. The user asserts in a message passing program the expected system behavior by one or several such formulas. The debugger allows by "macro-stepping" to interactively elaborate the execution tree (i.e., the set of possible execution paths) which arises from the use of non-deterministic communication operations. In each macro-step, a temporal logic checker verifies that the once asserted temporal formulas are not violated by the current program state. Our approach thus introduces powerful runtime assertions into parallel and distributed debugging by incorporating ideas from the model checking of temporal formulas.

1 Introduction

We report on a system which applies ideas from the model checking of temporal formulas to the area of parallel debugging; its goal is to support the development of correct and reliable parallel programs by runtime assertions that are derived from temporal formulas which describe the expected program behavior.

The behavior of sequential programs can be described with classical logic by a predicate (the output condition) that must hold after the execution of the program. Furthermore, the output condition can be translated (by the technique of weakest preconditions) into conditions that must hold at every step of the program. Such a condition can thus be considered as an *assertion* that must hold at a particular program step; if we restrict our attention to a particular subclass of formulas, such an assertion can be *checked* at runtime. Annotating a program by runtime assertions is a simple but very effective way of increasing the code's reliability and thus the user's confidence in a program's correct behavior.

* Supported by the ÖAD-WTZ Project A-32/2000 "Integrating Temporal Specifications as Runtime Assertions into Parallel Debugging Tools".

B. Monien and R. Feldmann (Eds.): Euro-Par 2002, LNCS 2400, pp. 113–120.

Assertions play an important role in the development of sequential programs, but their role in parallel programming is currently far less dominant. One reason is that (due to non-determinism) a program may exhibit for the same input different executions; furthermore, interesting properties talk about the state of the complete system (and not about the state of a single process). Another reason is that the scopes of properties are usually not defined by specific code locations but by temporal relations to other properties. These problems are difficult to overcome in production runs of parallel programs and with classical logic. We therefore turn our attention to program runs controlled by parallel debuggers and to assertions expressed in the language of temporal logic.

Since debugging parallel programs is an important and difficult task, many projects have been developing tools to support the user in this area; for a survey, see [8]. A particular challenge is the mastering of non-determinism which arises in message passing programs from the *wildcard receive* operation, i.e., a receive operation that non-deterministically accepts messages from different communication partners. The NOPE (Non-deterministic Program Evaluator) deals with this problem by generating in a record phase partial traces which contain ordering information of critical events [9, 8]. During replay these data are used to enforce the same event ordering as occurred in the recording phase. The DI-WIDE debugger [7, 6] applies the technique of macro-stepping which allows to test all branches of an application in a concurrent manner; we will describe this technique in more detail in the remainder of this paper.

Temporal logic has proved as an adequate framework for describing the dynamic behavior of a system (program) consisting of multiple asynchronously executing components (processes) [10]. A temporal logic formula can be considered as the *specification* of a parallel program; in linear time temporal logic, a program is correct if every possible execution satisfies the formula. If a program itself is described in a formal framework, the technique of model checking can be applied to decide about the correctness of the temporal specification, provided that the program only exhibits a finite number of states [1]. There exist tools for the validation of concurrent system designs based on temporal logic [3] and for the generation of test cases from temporal specifications [4]. In the system presented in this paper, we use actual program runs controlled by a debugger as the universe in which a temporal formula is checked. Thus our work combines ideas from parallel debugging and from model checking.

The approach closest to our ideas is that of *pattern-oriented* parallel debugging pioneered by the program analysis tool Belvedere [5]. This approach was included and extended by the post-mortem event-based debugger Ariadne [2], and later applied in the TAU program analysis environment [11]. Ariadne matches user-specified models of intended program behavior against actual program behavior captured in event traces. Ariadne's modeling language for describing program behavior is based on communication patterns with a notation derived from regular expressions. This language is quite simple; the language of temporal logic used in our system is considerably more expressive and allows to describe the intended program behavior in much more detail.

2 Macrostep Debugging in DIWIDE

DIWIDE is a distributed debugger which is part of the visual parallel programming environment P-GRADE. This debugger implements the *macrostep* method which gives the user the ability to execute the application from communication point to communication point [7, 6].

A *macrostep* is the set of executed code regions between two consecutive collective breakpoints. A *collective breakpoint* is a set of local breakpoints, one for each process, that are all placed directly after communication instructions such that a macrostep contains communication instructions only as the last instructions of its regions. In the macro-step execution mode, DIWIDE generates from the current collective breakpoint the next collective breakpoint and then runs the program until the new collective breakpoint is hit. At replay, the progress of the processes is controlled by the stored collective breakpoints; the program is executed again macrostep by macrostep as in the execution phase.

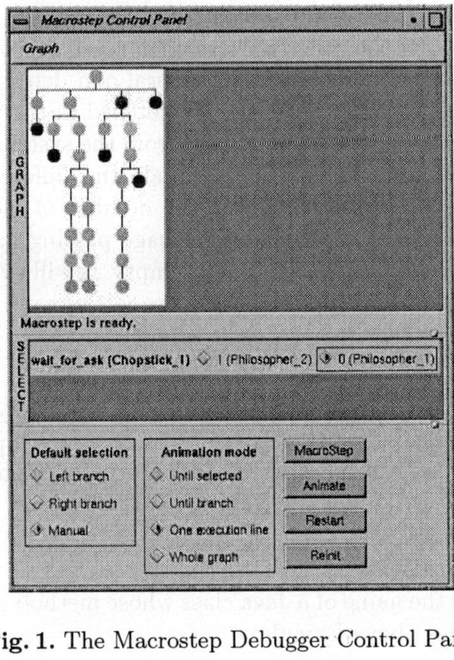

Fig. 1. The Macrostep Debugger Control Panel

When a communication operation in a collective breakpoint is a wildcard receive operation, this breakpoint splits macro-step execution into multiple execution paths. Each path represents one possible selection of a sender/receiver pair for all wildcard receive operations in the originating collective breakpoint.

The set of all possible execution paths can be represents by a tree whose nodes represent collective breakpoints and whose arcs represent macrosteps. The macrostep control panel of the DIWIDE debugger visualizes this tree as far as it has been already constructed and allows the user to control its further elaboration (see Figure 1). The user may select particular branches in the tree or let the system automatically traverse the tree according to some strategy. He may also set a *meta-breakpoint* in some node and let the system replay execution along the corresponding branch until the selected node is hit. The system therefore gives the user very powerful means to control the non-deterministic behavior of a parallel program in the debugging process.

3 Macrostep Debugging with Temporal Assertions

Before we go into technical details, we will illustrate the use of temporal formulas as runtime assertions by a simple example. Take a parallel program which consists of three processes: a *producer* process which generates a finite number of values and sends them to a *buffer* process which receives values from the producer and eventually forwards them to a *consumer* process which receives the values from the buffer and processes them. The buffer has a finite capacity; depending on its fill state (full, empty, not full and not empty) it waits for requests from one or from both of the other processes (to receive or to send a value) and answers them. Its behavior is therefore in general non-deterministic and may be investigated by the macro-step debugger as sketched in the previous section.

A fundamental property which we expect from the system is that the number of messages stored in the buffer always equals the difference of the number of messages sent by the producer and of the number of messages received by the consumer (we assume a synchronous message passing handshake). Another property is the fact that if the buffer is non-empty, it will eventually get empty. In the notation of temporal logic [10], these properties can be written as

$$\Box \text{NoLostMessage} \wedge \Box(\neg \text{BufferEmpty} \Rightarrow \Diamond \text{BufferEmpty})$$

where "NoLostMessage" expresses the core of the first property and "BufferEmpty" the core of the second property. The temporal operator $\Box$ reads as "always" and the temporal operator $\Diamond$ as "eventually". This property can be asserted at the beginning of our program by a C statement

```
assert("BufferSpec");
```

where `BufferSpec` is the name of a Java class whose method `getFormula` returns an object that encodes above formula:

```
class BufferSpec extends Specification {
  public Formula getFormula() {
    return new Conjunction(
      new Always(new Atomic("NoLostMessage", null)),
      new Always(new Implication(
        new Negation(new Atomic("BufferEmpty", null)),
        new Eventually(new Atomic("BufferEmpty", null)))));
  }
}
```

When the debugger encounters the `assert` statement, it instructs the temporal logic checker (TLC) which is implemented in Java to dynamically load this class. TLC is called by the debugger after every subsequent macro-step to verify whether the state of the current collective breakpoint violates the asserted formula or not. The user can follow the checking process in a window that displays the status of the formula in every collective breakpoint: "false" means that the stated assertion has been violated by the current execution, "true" means that the assertion cannot be violated any more, "unknown" means that the assertion may be still violated in the future.

The formula `BufferSpec` refers to two atomic predicates `NoLostMessage` and `BufferEmpty` which are the names of C-functions which are located in a separate library that is dynamically loaded by the debugger. Whenever the TLC asks the debugger for the value of an atomic formula, the debugger executes the corresponding function which returns the truth value of the predicate in the current system state:

```
int NoLostMessage() {
  long number, countP, countC;
  number = getVarLongInt("number", getProcessIndex("Buffer"));
  countP = getVarLongInt("count", getProcessIndex("Producer"));
  countC = getVarLongInt("count", getProcessIndex("Consumer"));
  return number == countP-countC;
}

int BufferEmpty() {
  long number = getVarLongInt("number", getProcessIndex("Buffer"));
  return number == 0;
}
```

The atomic predicate functions can inspect the system state via an interface to the debugger. For instance, the function `getVarLongInt(`*`var`*`, `*`proc`*`)` returns the value of the program variable *var* in process *proc* as a value of type `long`. In this way, the predicate function `NoLostMessage` checks the number of messages in the buffer process with respect of the values of two counter variables in the producer process and in the consumer process.

Summarizing, for using in our system temporal formulas as runtime assertions, the programmer needs to

1. annotate the program to be debugged by the assertions[1],
2. provide a Java encoding of the temporal formulas,
3. provide C functions for the atomic predicates used in the temporal formulas.

This only reflects the current state of the system; in later versions, we plan to develop a meta-language where the Java encoding and the C functions are automatically generated from a high-level specification language.

[1] If the atomic predicates in an assertion refer to program labels, the program must be also annotated by such labels.

4 Temporal Assertions

We are now going to sketch the formal basis of using temporal formulas as runtime assertions. Any system can be described by a tuple $\langle is, ns\rangle$ where is is the set of initial states of the system and ns is the next state relation of the system. A temporal formula F is valid for such a system, written as $\mathbf{T}[[F]] is\, ns$, if for every (finite or infinite) state sequence s induced by $\langle is, ns\rangle$, F holds at position 0 of s. Thus it suffices to define the truth value of a temporal formula F at position i of s, written as $\mathbf{T}[[F]]s\ i$:

$$\mathbf{T}[[\Box F]]s\ i = \text{true iff } \mathbf{T}[[F]]s\ j = \text{true for all } j \text{ with } i \le j < |s|$$
$$\mathbf{T}[[\Diamond F]]s\ i = \text{true iff } \mathbf{T}[[F]]s\ j = \text{true for some } j \text{ with } i \le j < |s|$$

Now let us introduce a "next step" formula $\circ_v F$

$$\mathbf{T}[[\circ_v F]]s\ i = \text{ if } i+1 = |s| \text{ then } v \text{ else } \mathbf{T}[[F]]s\ (i+1)$$

which is true, if F holds in the next step, and if no such step exists, takes the truth value v. We then define a semantics-preserving formula translation $\mathbf{G}[[F]]$

$$\mathbf{G}[[\Box F]] = \mathbf{G}[[F]] \wedge \circ_{\text{true}} \Box F$$
$$\mathbf{G}[[\Diamond F]] = \mathbf{G}[[F]] \vee \circ_{\text{false}} \Box F$$

such that in the result $G := \mathbf{G}[[F]]$ the operators $\Box$ and $\Diamond$ are always guarded by the $\circ_v$ operator. We can therefore reduce the validity of a temporal formula F in a state sequence s at position i to the validity of atomic formulas in state $s(i)$ and to the validity of temporal formulas in s at $i+1$.

Above definition is based on state sequences, but in assertion checking we only have access to the "current" state of the system. We therefore introduce a set of state trees $T(is, ns)$ induced by $\langle is, ns\rangle$. Each node in such a tree t holds a state t_{state}, has a link t_{prev} to its predecessor node, and a set of successor nodes t_{next}. The roots r of these trees (the nodes with $r_{\text{state}} \in is$) have $r_{\text{prev}} = \top$; the leaves l of these trees (for which no state s exists such that $ns(l_{\text{state}}, s)$) have $l_{\text{next}} = \{\top\}$. We can now define the semantics $\mathbf{T}[[G]]t$ of a guarded formula G with respect to such a tree t such that the relationship to the original semantics is preserved, i.e., $\mathbf{T}[[F]] is\ ns = \mathbf{T}[[G]]\, T(is, ns)$.

However, during assertion checking, we only have access to a part of the tree referenced by a "current" state node whose children (the nodes of the successor states) may not yet be completely (or not at all) evaluated. We represent such "partial trees" by trees that contain "unknown subtrees" denoted by $\bot$ and extend the semantics $\mathbf{T}$ on complete trees to a semantics $\mathbf{T}_3$ on partial trees using a 3-valued logic with an additional logical value $\bot$ ("unknown").

The new semantics is compatible with the original one and monotonic with respect to a partial ordering $\sqsubseteq$ of trees according to their information content: $s \sqsubseteq t \Rightarrow \mathbf{T}_3[[G]]s \sqsubseteq \mathbf{T}_3[[G]]t$. We therefore have defined the semantics of a guarded temporal formula G on the set of partial state trees induced by a system. While above explanation only describes temporal operators $\Box$ and $\Diamond$ which refer to the "future" of a state, our framework also supports corresponding operators which talk about the "past". The temporal logic checker TLC described in the following section implements $\mathbf{T}_3$ to determine the validity of an assertion G.

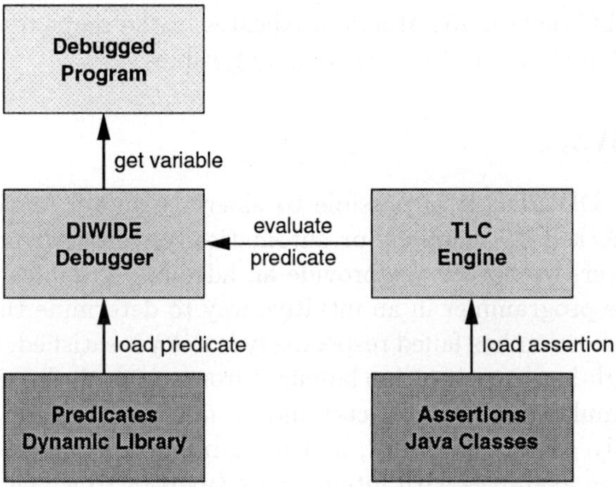

Fig. 2. Integrating TLC with DIWIDE

5 Checking Temporal Assertions with TLC in DIWIDE

The temporal logic checker TLC interacts with the DIWIDE debugger by a specified protocol. This protocol operates in a sequence of rounds that correspond to the states of an execution sequence. In each round,

1. TLC may receive from DIWIDE a new (additional) temporal formula whose validity is to be checked in the subsequent state,
2. TLC may ask DIWIDE questions about the truth values of atomic formulas in the current state (see Figure 2),
3. TLC announces its knowledge about the truth of the set of temporal formulas it has received up to now (true, false, unknown).

When a round has ended, DIWIDE informs TLC about the beginning of a new round (when a new state in the current execution sequence is available).

TLC evaluates in each round the truth of all formulas with respect to the round in which the corresponding formula has been submitted by the external partner. If a formula refers to the future, the result will be frequently "unknown". However, if more and more rounds are performed, the added knowledge may let the knowledge about such a formula change to "true" (the formula cannot be falsified any more after the current round) respectively "false" (the formula has been falsified by the current round). If a formula has been falsified, the corresponding assertion has been violated.

TLC does not repeatedly evaluate (sub)formulas whose final value ("true" or "false") is already known. Such results are cached such that only those formulas are re-evaluated whose values are not yet known but are required to determine the value of the overall formula. To support temporal "past operators", TLC prefetches in each round the values of all atomic predicates whose results may be required in the future to evaluate a "past formula". Whenever such a formula is submitted, TLC records the atomic predicates in the scope of such operators in order to start prefetching the corresponding values.

6 Future Work

With TLC and DIWIDE it is possible to assert temporal formulas and have their validity checked in (manually or automatically) selected runs of a parallel program. However, we do not yet provide an adequate graphical user interface which allows the programmer in an intuitive way to determine the fundamental reason why an assertion has failed respectively is not yet satisfied. Second, we are still lacking a high-level specification language from which the low level encodings of temporal formulas (as Java objects) and atomic predicates (as C functions) are automatically generated. Third, and most important, we need to evaluate by larger program examples with interesting properties to which extent the use of temporal assertions actually helps to improve the understanding of program behaviors and detect errors in them. In any case, the presented system will serve as a good starting point for these investigations on the usefulness of extending a parallel debugger with model checking capabilities.

References

1. E. M. Clarke, Jr., O. Grumberg, and D. A. Peled. *Model Checking.* MIT Press, Cambridge, MA, 1999.
2. J. Cuny et al. The Ariadne Debugger: Scalable Application of Event-Based Abstraction. *SIGPLAN Notices*, 28(12):85–95, December 1993.
3. D. Drusinsky. The Temporal Rover and the ATG Rover. In *SPIN Model Checking and Software Verification, 7th International SPIN Workshop*, volume 1885 of *LNCS*, pages 323–330, Stanford, CA, August 30 - September 1, 2000. Springer.
4. J. Hakansson. Automated Generation of Test Scripts from Temporal Logic Specifications. Master's thesis, Uppsala University, Sweden, 2000.
5. A. Hough and J. Cuny. Initial Experiences with a Pattern-Oriented Parallel Debugger. *SIGPLAN Notices*, 24(1):195–205, January 1988.
6. P. Kacsuk. Systematic Macrostep-by Macrostep Debugging of Message Passing Parallel Programs. *Future Generation Computer Systems*, 16(6):609–624, 2000.
7. P. Kacsuk, R. Lovas, and J. Kovács. Systematic Debugging of Parallel Programs in DIWIDE Based on Collective Breakpoints and Macrosteps. In P. Amestoy et al., editors, *5th Euro-Par Conference*, volume 1685 of *Lecture Notes in Computer Science*, pages 90–97, Toulouse, France, August 31 – September 3, 1999. Springer.
8. D. Kranzlmüller. *Event Graph Analysis for Debugging Massively Parallel Programs.* PhD thesis, Johannes Kepler University, September 2000.
9. D. Kranzlmüller and J. Volkert. NOPE: A Nondeterministic Program Evaluator. In *Parallel Computation, 4th International ACPC Conference*, volume 1557 of *LNCS*, pages 490–499, Salzburg, Austria, February 16–18, 1999. Springer.
10. Z. Manna and A. Pnueli. *The Temporal Logic of Reactive and Concurrent Systems — Specification.* Springer, Berlin, 1992.
11. S. Shende et al. Event- and State-based Debugging in TAU: A Prototype. In *ACM SIGMETRICS Symposium on Parallel and Distributed Tools*, pages 21–30, Philadelphia, PA, May 1996.

Low-Cost Hybrid Internal Clock Synchronization Mechanism for COTS PC Cluster

Jorji Nonaka[1,*], Gerson H. Pfitscher[1], Katsumi Onisi[2], and Hideo Nakano[2]

[1] Department of Computer Science, University of Brasilia, Brasília-DF, Brazil
jorji@ais.sys.i.kyoto-u.ac.jp, gerson@unb.br

[2] Media Centre, Osaka-City University, Osaka, Japan
{onisi,nakano}@media.osaka-cu.ac.jp

Abstract. NTP is a well-known and widely used clock synchronization mechanism for PC cluster environments. However, like other software-based clock synchronization algorithms, its precision depends on the estimated accuracy of the network latency for synchronization messages. This paper presents a low-cost internal clock synchronization mechanism that uses a simple TTL signal distributor to support remote clock reading to obtain the time drift necessary to execute local clock adjustments. The objective is to improve precision, thus eliminating the need to estimate the network latency as in usual methods.

1 Introduction

Synchronized clocks are useful in distributed systems for performance measurements, auditing, event ordering and task scheduling, among other things. Typically, COTS (commodity off-the-shelf) PC clusters do not have access to a global clock or possess dedicated clock synchronization support. The goal of the internal clock synchronization is to minimize the maximum difference between any two clocks. Most internal clock synchronization algorithms, such as the *Network Time Protocol* (NTP)[1], are software-based. They are flexible and economical, but the performance is limited by the synchronization message transit delay. This paper presents a low-cost hybrid internal clock synchronization mechanism and its implementation on a COTS PC cluster running Linux. A simple signal distributor hardware and the parallel printer ports of the machines are used to support remote clock reading to improve the precision, thus eliminating the need to estimate the network latency.

Clock synchronization has been extensively studied for the last two decades and *hardware*, *software* and *hybrid* approaches have been proposed[2, 3]. *Hardware* approaches[2–4] achieve μs level precision through the use of dedicated hardware at each node and a separate network solely for the propagation of clock signals. Cost, however, has been a limiting factor. *Software* approaches do not need any extra hardware but the performance is limited by the synchronization message transit delay and provides a precision in the ms range[5, 2].

* Currently with Graduate School of Informatics, Kyoto University, Japan.

B. Monien and R. Feldmann (Eds.): Euro-Par 2002, LNCS 2400, pp. 121–124.

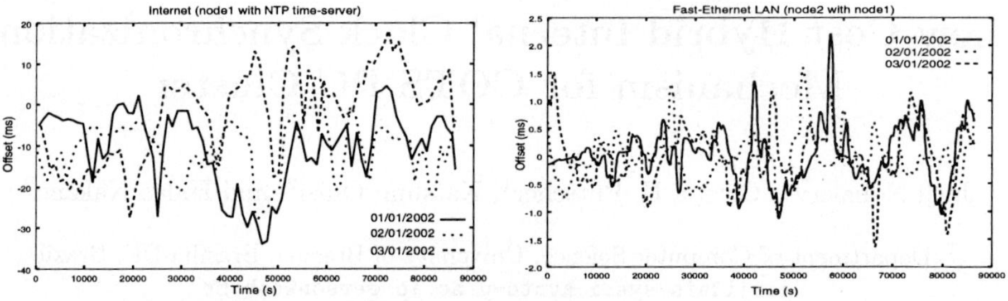

Fig. 1. External and Internal clock synchronization via NTP.

Figure 1 shows the interference of network latency on the clock synchronization precision obtained via NTP in two different network environments: Internet and Fast-Ethernet LAN.

Due to that limitation, a *hybrid* approach[3, 5] has been proposed. This approach requires minimum extra hardware and uses the available network infrastructure for synchronization message exchanges. Although the extra hardware is considered to be minimum, these solutions usually require expensive apparatus such as a precise quartz crystal oscillator[3, 5], a GPS signal receiver[5, 6] or a custom LSI[4, 5]. Even the most recent version of NTP, based on a *Nanokernel*[6]algorithm, needs a precise pulse per second (PPS) signal produced by an external device to obtain better performance. We refrained from using expensive hardware, such as those used in other methods, to match with the Beowulf-class PC cluster philosophy.

2 Clock Synchronization Mechanism

The on-going method uses the usually inactive PCs' I/O ports, such as serial or parallel ports, which can generate hardware interruptions. The TTL level signal distributor hardware is used to simultaneously signal the I/O ports to start a reading process of each local clock. After that, we obtain the instantaneous image of all clocks' value involved in the synchronization. The available network infrastructure is used to transmit the reference clock's value to the other nodes, to calculate the time offset in order to adjust each local clock in an appropriate way. This way, although the clock value is transmitted through the available network there is no need to calculate the message transmission latency as would be the case in a purely software based algorithm. Figure 2 shows a simplified scheme of the synchronization mechanism and the implemented hardware.

3 Implementation Issue and Results

The PC cluster on which we have implemented our model consists of 8 Pentium II 350 MHz with 64MB RAM interconnected by a Fast-Ethernet network. We used kernel version 2.2.16 of the linux operating system and NTP version

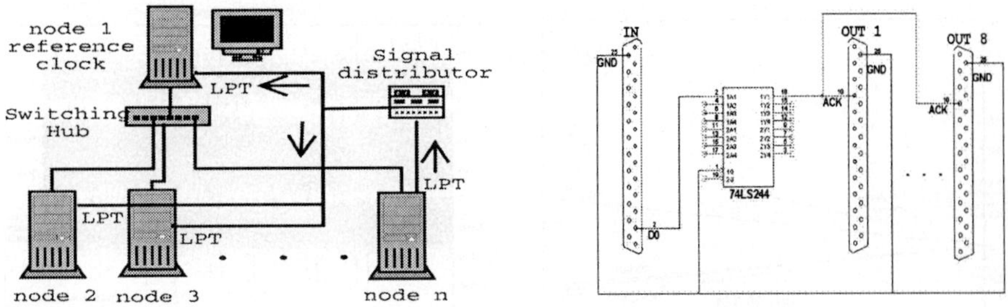

Fig. 2. Hybrid internal clock synchronization mechanism.

4.0.99k. The NTP was used for external clock synchronization of the reference node's clock. We have elaborated a simple synchronization signal distributor hardware, a character oriented device driver for parallel ports, and the master and slave processes. The choice of the parallel printer port was only due to its easy manipulation and programming.

The synchronization signal distributor hardware distributes a TTL level signal to the *Acknowledge* (ACK) parallel port pin to generate hardware interruption *IRQ7* which causes the device driver to read the local clock through the kernel function *do_gettimeofday* with 1 μs resolution and makes its value available in the *"/dev/sincro"* device. Then it wakes up the slave processes to begin the resynchronization by requesting the master process for the reference clock value to adjust each local clock. This is done through the system call *adjtimex* derived from the NTP clock discipline algorithm[1, 6], which has been used by Linux since 1992.

During the experiments, the processes running on the cluster were reduced to a minimum level and we observed the clock drift behavior of all node clocks (*node2* to *7*) in relation to the reference clock (*node1*)without any local clock adjustments. The result can be viewed in figure 3. We used several resynchronization interval periods to observe the clock synchronization behavior and obtained hundreds, tens and few microseconds precision using, respectively, 30s, 3s and 1s as the pulse intervals. Figure 3 for 3s pulse interval, unlike figure 1 shows only the clock offset before each resynchronization process instead of the real-time behavior. We did not work on the clock adjustment algorithm and only used available function *adjtimex* to cancel out the time-offset. Before each resynchronization process, each clock drifts according to its own drifting rate.

In comparison to the NTP we have obtained better and more homogeneous results in the time-offset variation as time passes even using 30s as the resynchronization interval. Using smaller intervals the processing and communication overheads increase. It is therefore necessary to choose an optimum interval matching the application's requirements. The computational cost of the slave process that consumes most processing time is less than 0.03% of the CPU time. During the experiments, we observed some interferences that appear in the graph as spikes. One of them is the *reading error* that produces data inconsistency and the other

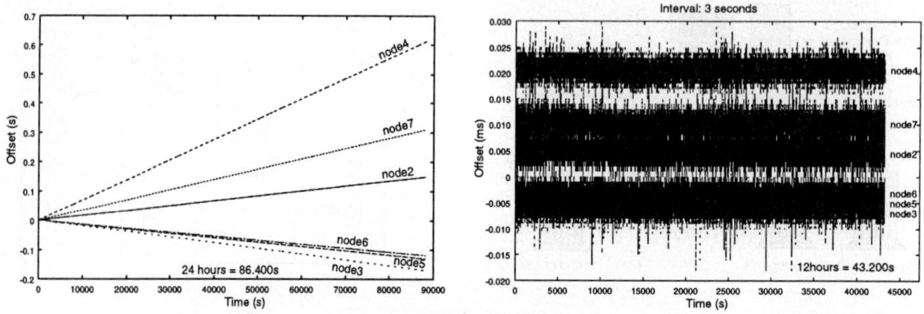

Fig. 3. Clock drift and internal synchronization using an interval of 3s.

is the noise generated by external devices, such as monitors and transformers. These values are ignored by the data filtering subroutine.

4 Conclusions

This paper has described an overview of a low-cost hybrid internal clock synchronization mechanism. The performance evaluation in a small size COTS PC cluster running Linux has shown that this method can be a good alternative to improve internal clock synchronization without using any expensive extra hardware. However, the implemented model still requires improvements to be suitable for the real world. This model has great scalability in systems with the condition that the nodes are placed physically close. A next step might be to analyze the dependence between the precision and the signal generation interval, making it possible to construct an adaptive clock synchronization mechanism.

References

1. Mills, D. L.: Internet Time Synchronization: The Network Time Protocol. *IEEE Micro Trans. Communications*, 39(1):1482-1493, January 1991.
2. Anceaume, E., Puaut, I.: Performance Evaluation of Clock Synchronization Algorithms. *Technical report 3526*, INRIA, France, October 1998.
3. Ramanathan, P., Kandlur, Dilip D., Shin, Kang G.: Hardware-Assisted Software Clock Synchronization for Homogeneous Distributed Systems. *IEEE Transactions on Computers*, 39(4):514-524, April 1990.
4. Kopetz, H., Ochsenreiter W.: Clock Synchronization in Distributed Real-Time Computer Systems. *IEEE Transactions on Computers*, C-36(8):933-940, August 1987.
5. Horauer, M.: Hardware Support for Clock Synchronization in Distributed Systems. *Proceedings of the International Conference on Dependable Systems and Networks (DSN'01)*, Gōteborg, Sweden, 2001.
6. Mills, D.L.,Kamp, P.: The Nanokernel. *Proc. Precision Time and Time Interval (PTTI) Applications and Planning Meeting*, Reston VA, USA, 2000.

.NET as a Platform for Implementing Concurrent Objects

Antonio J. Nebro, Enrique Alba, Francisco Luna, and José M. Troya

Departamento de Lenguajes y Ciencias de la Computación
Universidad de Málaga (Spain) a
{antonio,eat,troya}@lcc.uma.es

Abstract. *JACO* is a Java-based runtime system designed to study techniques for implementing concurrent objects in distributed systems. The use of Java has allowed us to build a system that permits to combine heterogeneous networks of workstations and multiprocessors as a unique metacomputing system. An alternative to Java is Microsoft's .NET platform, that offers a software layer to execute programs written in different languages, including Java and C#, a new language specifically designed to exploit the full advantages of .NET. In this paper, we present our experiences in porting *JACO* to .NET. Our goal is to analyze how Java parallel code can be re-used in .NET. We study two alternatives. The first one is to use J#, the implementation of Java offered by .NET. The second one is to rewrite the Java code in C#, using the native .NET services. We conclude that porting JACO from Java to C# is not difficult, and that our sequential programs run faster in .NET than in Java, while internode communications have a higher cost in .NET.

1 Introduction

Concurrent object-oriented languages are characterized by combining concurrent programming and object-oriented programming. However, there is not a unique way to combine these two paradigms [1]. An alternative is to consider programs as collections of concurrent objects that communicate and synchronize by invoking the operations they define in their interfaces.

In the past, we have investigated implementing concurrent objects in parallel and distributed systems [2][3] in an efficient way. As a result, we have developed *JACO*, a runtime implemented in Java. With *JACO* (JAva based Concurrent Object System), we can use Java to write programs according to a concurrent object model. *JACO* offers services to concurrent object creation, object communication, and object replication plus migration.

The choice of Java to implement *JACO* is justified by the suitability of some Java features, such as object-orientation, multithreading support, socket based communication, heterogeneity, reflexion, and XML support. These features can be also found in the new language C# [4], which is implemented on top of Microsoft's .NET platform [5]. The resulting runtime system allows to combine heterogeneous networks of workstations and multiprocessors as a unique metacomputing system. Both together, .NET and C#, appear as an alternative to

B. Monien and R. Feldmann (Eds.): Euro-Par 2002, LNCS 2400, pp. 125–129.

Java. One of the most interesting features of .NET is that it is a multi-language platform. Whereas the Java Virtual Machine is bounded to only one language, the .NET runtime allows to execute programs written in C++, Visual Basic, C#, and even Java. The Java implementation on .NET is called J#, although it only provides the functionality of certain JDK classes, while some features, such as applets or RMI, are not supported. A drawback of .NET is that it is bound only to the Windows family of operating systems, although there are currently some initiatives to port .NET to other platforms [6]. Given that C# is similar to Java in many aspects, it is interesting to study whether parallels programs written in Java can be ported easily to this new environment. Furthermore, the availability of J# should allow us to execute Java programs directly on .NET.

In this paper, we present our experiences in porting *JACO* to .NET. We have developed a C# version of *JACO*, which is named $JACO^{C\#}$. The original Java version will be referred as $JACO^{J}$. After recompiling $JACO^{J}$ using J#, we have got a version named $JACO^{J\#}$.

The paper is organized as follows. In Section 2, we give an overview of the *JACO* runtime system. In Section 3, we compare the implementation of *JACO* in Java, J# and C#. In Section 4, we present results from preliminary experiments. Finally, we provide some conclusions in Section 5.

2 The *JACO* Runtime System

The object model assumed by *JACO* considers a concurrent object as an active entity, with an internal state and a public interface. Objects can also have synchronization constraints, which disable some operations when they are not allowed. There are two kinds of operations: *commands*, which are asynchronous operations, and *queries*, which are read-only synchronous operations. To ensure that a number of operations are executed in mutual exclusion, objects need to be acquired in shared or exclusive mode before being accessed.

Internally, *JACO* runs a process by node, containing the runtime system and the concurrent objects. The main components of the system are the object table, the object scheduler, and a communication agent for controlling internode communication. The *JACO* scheduler manages a pool of threads, and its mission is to assign an object ready to run to a thread. The object table contains references to object handlers, which are proxies used to access *JACO* objects.

3 Java versus .NET Implementations

In this section, we compare the three implementations of *JACO*. We begin with the description of $JACO^{J}$, and later we discuss $JACO^{J\#}$ and $JACO^{C\#}$.

As discussed in last section, $JACO^{J}$ uses threads to execute objets. Internode communication is carried out using TCP sockets. Each concurrent object in $JACO^{J}$ has an object identifier, which is a data structure containing an identifier of the Java class of the object, among other information. Thus, given an object identifier, the runtime can create an instance of the objcct handler using the Java's reflexion mechanism. $JACO^J$ uses configuration files which contain network information. These are XML files, which are processed using the JAXP API of Java. Apart from these features, $JACO^J$ programs are pure Java programs. We do not use graphics, applets, nor RMI.

The simplest way to port $JACO^J$ to .NET is to use J#. In theory, we only have to recompile the Java code with the J# compiler. However, J# does not implement the JAXP API for XML processing, because this functionality is offered by the underlying .NET platform. So, a solution is to use the .NET XML services from J#. For this work, we took the simple approach of removing the XML code and including the information contained in the configuration files as constants objects in header files. After removing the XML code, $JACO^{J\#}$ compiled without problems. When running some *JACO* applications (see Section 4), we found some problems. For example, the Java random objects (`java.util.Random`) did not work well in J#, but they can be due to the fact that we have used a beta version of J# (Visual J# .NET Beta 1). Anyway, the problem was solved easily by invoking the equivalent service (`System.Random`) offered by .NET.

The implementation of *JACO* in C#, as well as the applications developed on top of it, required to rewrite all the Java code in C#. Syntactically, the two languages are similar: their object-orientation model is basically the same, and the .NET services to manage threads, sockets, XML, and object serialization are almost equivalent in C# and Java. Furthermore, we were able of maintaining the same package structure of the original Java code, by simply replacing packages by namespaces, so the translation was not a complicated task.

4 Performance Comparison

In this section, we present a performance comparison of the three *JACO* implementations. The experiments we have carried out must be considered as preliminary ones, because we have used a beta version of .NET. Nevertheless, the results obtained show the current differences between Java and .NET, and they can give us an insight of what we can expect from the upcoming releases of .NET.

To measure performance, we have computed the cost of invoking object operations and analyzed two distributed applications, a branch and bound algorithm and a genetic algorithm [7]. The experiments were executed on a network of 6 PCs, each one having an Intel Pentium III 550MHz processor, 128MB of real memory, and a Fast Ethernet 100 Mbps adapter. They run Windows 2000 (SP2). We have used JDK 1.3.1-b24, and the Java programs were compiled with the `-O` optimization flag. The J# programs were compiled using Visual J# .NET Beta 1, on top of Visual Studio .NET Beta 2. We run the release version of J# programs on top of the CLR V1.0.2914. Finally, we compiled and run the C# programs using the Framework SDK .NET, CLR V1.0.3705. C# programs were compiled using the `/o` optimization flag.

Let us begin by measuring basic communication costs in *JACO*. In Table 1 we include the cost of invoking a command and a query operation on a concurrent

Table 1. Costs of local and remote object communication (in ms)

JACO version	Java		J#		C#	
	Local	Remote	Local	Remote	Local	Remote
Command	0.007	3.6	0.005	40.3	0.005	13.3
Query	0.209	11.7	0.076	89.2	0.256	29.2

Table 2. Times (in sec) and speed-ups obtained with the branch and bound program

JACO version	Java	J#	C#
Sequential	756	722	737
Distributed (6 nodes)	134	384	202
Speed-up	5.6	1.8	3.6

`double` object. In the case of local communication, we observe that commands take a similar time, while queries in Java perform slightly better than in C#, but roughly tree times worst than in J#. However, remote communications in Java are significantly better.

In Table 2 we report the times and speed-ups obtained when running the distributed branch and bound algorithm to solve a 100-city instance of the Traveling Saleman Problem. This algorithm is characterized by a high degree of communication, needed to enhance the load balancing. The execution of the sequential program shows that the three versions yield comparable times, being faster the J# version. The speed-ups obtained in the parallel executions using 6 nodes are the consequence of the high cost of remote communication in J# and C#, which work out 1.8 and 3.6, respectively, while the speed-up in Java is 5.6.

The second application is a distributed genetic algorithm (DGA) that tries to optimize the following ONEMAX function: $f_{ONEMAX}(x) = \sum_{i=1}^{n} x_i$. This algorithm is the *JACO* version of the one used in [8], which was implemented in Java using sockets and threads. The DGA program is characterized by a low computation/communication ratio, and its results are strongly dependent of the random number generator, because it uses stochastic operators. In Table 3 we report the results of running each version of the program on our 6 node network. Here, the execution time does not provide a complete measure of performance, because we observed that the optimum was found rapidly by the Java version, while it was hard to find with the J# and C# versions. Considering that the DGA is exactly the same in all the tests, the explanation has to do with the random number generator of .NET. However, an insight of performance can be

Table 3. Results obtained with the distributed genetic algorithm

JACO version	Java		J#		C#	
	Local	Distributed	Local	Distributed	Local	Distributed
Time (in ms)	74.02	13.27	158.90	66.91	153.00	19.46
Evaluations	2,451	14,557	3,066	11,075	5,421	29,942

obtained if we analyze the mean number of evaluations per second of the three programs: the C# version is roughly twice faster than the Java and J# versions.

5 Conclusions

In this paper we have presented our experiences in porting *JACO*, a Java-based runtime system for implementing concurrent objects, to .NET. We have used two .NET languages, J# and C#. We can conclude that Java parallel programs that use standard mechanisms, such as threads and sockets, can compile and run on .NET using J# with few problems, and that the similarities between Java and C# allow to rewrite Java programs in C# with little effort.

Preliminary performance results show that our .NET-based programs perform slightly better than the Java versions of the same programs in sequential execution, while there is a significant advantage for Java concerning remote communications. However, our tests using C# reveal a reduction in the communication time when comparing with the same tests using J#. Since C# programs use a more recent version of the Framework SDK .NET than J# programs (V.1.0.3705 versus V.1.0.2914), we can conclude that this issue is being improved by Microsoft. Despite C# and J# run over .NET, there are differences in the execution time of the programs. This can be due to the fact that the compilers are different, and probably they do not generate the same code.

.NET is recent, while Java JDK has been continuously improved for many years, so we can expect that future releases of .NET will allow distributed programs to run efficiently. A more exhaustive evaluation of *JACO*, including more applications on top of Java and .NET is a matter of future work.

References

1. Philippsen, M.: A survey of concurrent object-oriented languages. Concurrency: Practice and Experience **12** (2000) 917–980
2. Nebro, A.J., Pimentel, E., Troya, J.M.: Distributed objects: An approach based on replication and migration. The Journal of Object-Oriented Programming (JOOP) **12** (1999) 22–27
3. Nebro, A.J., Pimentel, E., Troya, J.M.: Integrating an entry consistency memory model and concurrent object-oriented programming. In: Third International Euro-Par Conference. (1997) Passau, Alemania.
4. Liberty, J.: Programming C#. O'Reilly (2001)
5. Platt, D.S., Ballinger, K.: Introducing Microsoft .NET. Microsoft Press (2001)
6. de Icaza, M.: The Mono Project: An Overview (2001) HTTP://WWW.XIMIAN.COM/DEVZONE/TECH/MONO.HTML.
7. UEA Calma Group: Parallelism in combinatorial optimisation. Technical report, School of Information Systems, University of East Anglia, Norwich, UK (1995)
8. Alba, E., Nebro, A.J., Troya, J.M.: Heterogeneous computing and parallel genetic algorithms. Accepted for publication in the Journal of Parallel and Distributed Computing (2002)

Topic 2
Performance Evaluation, Analysis and Optimization

Barton P. Miller, Jesus Labarta, Florian Schintke, and Jens Simon

Topic Chairs

Performance is the key to parallel computing. Seymore Cray went so far as to say that it was more important to get the result fast than to get it correct, and David Dewitt has commented that "...first you do performance debugging, then you do correctness tuning."

There are several crucial aspects to achieve high-performance. These aspects include algorithm design, analytic modeling, simulation, measurement, and tuning. While there has been significant previous research in each of these areas, there continues to be a strong need for new techniques and tools that increase our understanding of performance limitations and opportunities and simplify the task of the programmer. Experience has shown us that when a team pays serious attention to any of these aspects, programs run faster or more efficiently.

Out of 18 submitted papers to this topic, 4 have been accepted as regular papers (22%), 3 as short papers (17%), and 11 contributions have been rejected (61%). In total, 79 reviews have been received, in average more than 4 reviews per paper.

The papers presented this year cover the spectrum of approaches to high performance in parallel computing and present valuable insights and experiences. The papers, and the resulting discussions, should continue the important progress in this area

B. Monien and R. Feldmann (Eds.): Euro-Par 2002, LNCS 2400, p. 131.

Performance of MP3D on the SB-PRAM Prototype

Roman Dementiev, Michael Klein, and Wolfgang J. Paul

Saarland University
Computer Science Department
D-66123 Saarbrücken, Germany
{rd,ogrim,wjp}@cs.uni-sb.de

Abstract. The SB-PRAM is a shared memory machine which hides latency by simple interleaved context switching and which can be expected to behave almost exactly like a PRAM if all threads can be kept busy. We report measured run times of various versions of the MP3D benchmark on the completed hardware of a 64 processor SB-PRAM. The main findings of these experiments are: 1) parallel efficiency is 79% for 32 processors and 56% for 64 processors. 2) Parallel efficiency is limited by the number of available threads.

1 Introduction and Previous Work

The SPLASH benchmark suite was designed in 1991 to test the performance of shared memory parallel machines[9]. MP3D is a program from this benchmark suite simulating the laminar flow of 40 000 particles in a wind tunnel around a space vehicle. As particles only interact by collisions a parallelization by spatial decomposition would be very efficient even on distributed memory machines. But MP3D partitions work among processors quite differently: *every particle is always processed by the same processor.*

On their way through the wind tunnel particles mix in an irregular way leading to non local communication patterns between processors. Most cache based machines react to this situation with thrashing. The DASH machine [8] is typical for this behaviour. Parallel efficiency is 23% for 16 processors and 16% for 32 processors.[1]. However, using a cache protocol optimised for 'migratory sharing' of data the MIT Alewife machine [4] achieves for optimized versions of MP3D an efficiency of up to 84% for 16 processors and 68% for 32 processors.

Frequent irregular accesses to shared memory do not inherently slow down parallel machines provided network congestion and memory module congestion can be avoided (e.g by address hashing and combining) and the latency of the network can be hidden (e.g. by multithreading). Various combinations of address hashing, combining and multithreading are used in machines like the NYU ultracomputer [7, 11], the Fluent Machine [12], Cray T3D [1], TERA [5] and the SB-PRAM [2]. The latter machine is a reengineered version of the Fluent machine using address hashing, combining at the memory modules and in the network nodes as well as simple interleaved contex switching. With p processors it uses a certain number of threads $th(p)$ per processor in order to hide latency. Simulations suggest that for $th(p) \geq 3 \cdot \log p$ such a machine with cycle time τ

[1] MP3D is not present in the SPLASH 2 benchmark suite [16]

B. Monien and R. Feldmann (Eds.): Euro-Par 2002, LNCS 2400, pp. 132–136.

will look to the user almost exactly like a priority CRCW-PRAM with $p \cdot th(p)$ processors and cycle time $th(p) \cdot \tau$ processing one instruction per cycle *provided all threads can be kept busy.* [3]. The results reported here will use $p \leq 64$ processors, each with $th(p) = 32$ threads. Simulations of a 16 processor SB-PRAM predict for an optimized version of MP3D an efficiency of 72% [6]. In this note we present *measured* run times of optimized versions of MP3D on the completed prototype of a 64 processor SB-PRAM. On such a machine one tries to keep up to $32 \cdot 64 = 2048$ threads busy. The instruction set of the SB-PRAM includes multiprefix instructions. For details about the prototype see [10].

In section 2 we show that for such large numbers of threads the handling of collisions between particles in MP3D becomes a bottleneck. With 32 processors the efficiency drops to 59% and with 64 processors even to 27%, i.e. the machine is even slightly slower than with 32 processors. Parallelizing the handling of collisions brings the efficiency back up to 79% for 32 processors and 56% for 64 processors. Increasing the number of particles to 80 000 increases the efficiency to 71% for 64 processors. In section 3 we develop and validate a simple model of the run time based on threads and available tasks. The model shows that the less than perfect speedup figures are due to a shortage of tasks. We draw some conclusions in section 4.

2 Optimizing MP3D on the SB-PRAM

2.1 Parallel Prefix Operations and Parallel Task Arrays

If locality of memory access is not an issue and multiprefix instructions are supported by a combining network, then the method of choice for the parallelization and load balancing of many scientific codes is to make use of fine grained parallel task queues [15, 14] or even more simply by two-dimensional *task arrays* where each row contains the data for one task. Parallel reading from or writing to different rows of a task array is easily realized with the help of multprefix adds.

Each time step of MP3D can essentially be divided into 5 phases i, each with a number of tasks N_i in the following way :

1. 'Initialize data structures'. One task initializes for one cell the data structure handling collissions of particles in the cell. The number N_1 of tasks equals the number of cells, i.e. $N_1 = 840$.
2. 'move and collide particle'. One task updates for 1 particle the position. In case of collisions it also updates direction and velocity. The number of tasks N_2 equals the number of particles, i.e. $N_2 = 40000$. This phase consumes sequentially 95% of the time.
3. 'add to free stream'. One task is to insert a new particle from the reservoirs into the tunnel. The average number of tasks is $N_3 = 30$.
4. 'move reservoir particle'. One task moves one particle in the reservoir. The number of tasks equals the number N_4 of particles in the reservoir. The reservoir contains an extra 2% of particles, i.e. $N_4 = N_2 \cdot 2/100$.
5. 'collide particles in reservoir'. A single task collides a pair of particles in the reservoir. The number of tasks N_5 equals $N_4/2$.

Note that N_4 and N_5 depend on N_2.

Partitioning and parallelizing work in this way [13] leads to the simulated figures reported in [6] for up to 16 processors. Running the program on the real hardware instead of the simulator reproduced the predicted run times and speedups to within 1 %. Running the same program on 32 or 64 processors leads to the speed ups shown in the left column of Table 1. One sees that doubling the number of processors from 16 to 32 only increases the speed up by only 46%. Doubling the number of processors again from 32 to 64 processors even leads to slightly deteriorated performance.

Table 1. Measured speedups of MP3D implementations for SB-PRAM

procs.	old (40000 part.)	new (40000 part.)	new (80000 part.)
1	1.00	1.00	1.00
2	1.90	1.94	1.91
4	3.72	3.84	3.81
8	7.19	7.53	7.56
16	12.88	14.40	14.78
32	19.16	25.28	27.84
64	18.69	35.97	45.69

2.2 Parallelizing the Handling of Collisions

The deterioration of performance comes from the fact that in MP3D the processing of collisions of particles serializes for each cell. Collisions are handled in the following way: A processor processing particle u in cell c checks if there is a particle waiting in the cell. If no particle is waiting, the processor leaves particle u waiting in the cell and quits. Otherwise there is exactly one particle v waiting in the cell. The processor collides the pair of particles u and v. None of them is waiting any more. Locks are used in order to guarantee that at most 1 processor at a time is accessing the cell. As long as the number of cells is large compared with the number of threads this is not a problem, but for $p = 32$ processors the number $32 \cdot p = 1028$ of threads already exceeds the number of cells. Adding more processors does not speed up this part of the computation which happens to dominate the run time.

Fortunately the handling of collisions during one time step can be parallelized for each cell. The problem to pair the particles in the cell is easily solved: Each cell gets a task array with enough lines for all particles which can be in the cell simultaneously (200 lines suffice) and a pointer q indicating that already q processors have accessed the cell. Lines in the array are numbered 0,1,2,.... Pointer q is initialized with 0. Suppose processors $i_1, \ldots, i_k$ are simultaneously accessing the cell with particles $u_1, \ldots, u_k$. With a multiprefix add each processor i_j determines $q(j) = q + j - 1$. If $q(j)$ is even processor i_j leaves particle u_j in row $q(j)/2$ waiting and quits. If $q(j)$ is odd processor i_j waits until a particle y is waiting in row $\lfloor q(j)/2 \rfloor$ and then collides particles u_j and y.

With the handling of collisions parallelized in this way one gets the speed ups in the middle column of Table 1. Parallel efficiency is back up to 78% for 32 processors and 56% for 64 processors. For 64 and 32 processors this is the best speed up for MP3D known to us. Increasing the number of particles to 80 000 improves things further as can be seen in the right column of Table 1. Efficiency is now 87% for 32 processors and 71% for 64 processors.

3 Speedup as a Function of the Number of Threads

The less than perfect speed up figures can be explained with a simple run time model. Let $t_i(N_i, th)$ be the run time of phase i with N_i tasks and th threads. Measurements show that with the constants C_i and T_i from Table 2 the formula $t_i(N_i, th) = C_i + T_i \cdot \lceil N_i/th \rceil$ models the run times of the phases with an error of at most 4% for the computation in for phases 2 to 5 and at most 7.6% for phase 1.

Table 2. Model parameters for different stages

i	phase	C_i (cycles)	T_i (cycles)	$N_i(40000)$	$N_i(80000)$
1	reset	296237	159203	840	840
2	move	1457627	306214	40000	80000
3	add to free stream	511190	69054	30	30
4	reservoir move	304000	28800	800	1600
5	reservoir collide	236406	701946	400	800

In all phases $i \neq 2$, i.e. except phase 'move', we have $\lceil N_i(40000)/1024 \rceil = \lceil N_i(40000)/2048 \rceil = 1$. Hence these phases do not become faster as we double the number of processors from 32 to 64. For N_2 particles and th threads the model predicts a run time of $t(N_2, th) = \sum_{i=1}^{5} t_i(N_i(N_2), th)$. In particular $t(N_2, 1)$ gives the run time of a single thread running on a single processor Therefore with respect to threads the parallel efficiency is $eff(N_1, th) = t(N_1, 1)/(th \cdot t(N_1, th))$. The efficiency for p processors is obtained by $eff(N_1, 32 \cdot p)$ which matches the measured values up to 7.3%. [2]

4 Conclusions

MP3D is a benchmark with highly irregular and non local access patterns to the memory. We have optimized this benchmark for the SB-PRAM, an architecture designed to deal effortlessly with irregular access patterns. We have measured parallel efficiencies of 79% for 32 processors and 56 % for 64 processors. We have shown that on the SB-PRAM architecture the efficiency is bounded by the number of available threads. Increasing the number of particles from 40 000 to 80 000 increases the efficiency to 71% for 64 processors.

[2] The 5 phases do not cover the entire work done in a computation step

References

1. Cray Research, Inc. Cray T3D System Architecture Overview, March 1993.
2. F. Abolhassan, R. Drefenstedt, J. Keller, W. J. Paul, and D. Scheerer. On the physical design of PRAMs. *The Computer Journal*, 36(8):756–762, 1993.
3. F. Abolhassan, J. Keller, and W. J. Paul. On the cost-effectiveness of PRAMs. *Acta Informatica*, 36(6):463–487, 1999.
4. A. Agarwal, R. Bianchini, D. Chaiken, K. Johnson, D. Kranz, J. Kubiatowicz, B.-H. Lim, K. Mackenzie, and D. Yeung. The MIT alewife machine: Architecture and performance. In *Proc. of the 22nd Annual Int'l Symp. on Computer Architecture (ISCA'95)*, pages 2–13, 1995.
5. R. Alverson, D. Callahan, D. Cummings, B. Koblenz, A. Porterfield, and B. Smith. The Tera computer system. In *Proc. of the 1990 International Conference on Supercomputing*, pages 1–6, 1990.
6. A. Formella, T. Grun, J. Keller, W. Paul, T. Rauber, and G. Runger. Scientific applications on the SB-PRAM. In *Proc. of International Conference on Multi-Scale Phenomena and Their Simulation, pages 272–281. World Scientific, Singapore*, 1997.
7. A. Gottlieb, R. Grishman, C. P. Kruskal, K. P. McAuliffe, L. Rudolph, and M. Snir. The NYU ultracomputer - designing an MIMD shared memory parallel computer. *IEEE Transactions on Computers*, 32(2):175–189, 1983.
8. D. Lenoski, J. Laudon, T. Joe, D. Nakahira, L. Stevens, A. Gupta, and J. Hennessy. The DASH prototype: Logic overhead and performance. In *IEEE Transactions on Parallel and Distributed Systems, 4(1):41–61*, 1993.
9. J. Pal Singh, W. Weber, and A. Gupta. SPLASH: Stanford Parallel Applications for Shared-Memory. Technical Report CSL-TR-91-469, Stanford University, 1991.
10. W. J. Paul, P. Bach, M. Bosch, J. Fischer, C. Lichtenau, and J. Roehrig. Real PRAM Programming. In *Proc. of the Europar'02*, 2002.
11. G. Pfister, W. Brantley, D. George, S. Harvey, W. Kleinfelder, K. McAuliffe, E. Melton, V. Norton, and J. Weiss. The IBM research parallel processor prototype. In *Proc. Int.l Conf. on Parallel Processing 764–771*, 1985.
12. A. G. Ranade. The Fluent Abstract Machine. Technical Report TR-12 BA87-3, 1987.
13. T. Rauber, G. Runger, and C. Scholtes. Shared-memory implementation of an irregular particle simulation method. In *Proc. of the EuroPar'96, number 1123 in Springer LNCS.*, pages 822–827, 1996.
14. J. Roehrig. Implementierung der P4-Laufzeitbibliothek auf der SB-PRAM. Master's thesis, Universitaet des Saarlandes, Saarbruecken, 1996.
15. J. Wilson. *Operating System Data Structures for Shared-Memory MIMD Machines with Fetch-and-Add.* PhD thesis, 1988.
16. S. Woo, M. Ohara, E. Torrie, J. Singh, and A. Gupta. The SPLASH-2 programs: Characterization and methodological considerations. In *Proc. of the 22nd International Symposium on Computer Architecture, pages 24–38*, 1995.

Multi-periodic Process Networks: Prototyping and Verifying Stream-Processing Systems

Albert Cohen[1], Daniela Genius[1,2], Abdesselem Kortebi[2], Zbigniew Chamski[2], Marc Duranton[2], and Paul Feautrier[1]

[1] INRIA Rocquencourt, A3 Project
[2] Philips Research

Abstract. This paper aims at modeling video stream applications with structured data and multiple clocks. Multi-Periodic Process Networks (MPPN) are real-time process networks with an adaptable degree of synchronous behavior and a hierarchical structure. MPPN help to describe stream-processing applications and deduce resource requirements such as parallel functional units, throughput and buffer sizes.

1 Context and Goals

The need arises for hardware units to handle new kinds of video applications, combining multiple streams, graphics and MPEG movies, leading to increased system complexity. When beginning the design of a video system, the engineer is primarily interested in quickly determining the hardware requirements to run an application under specific real-time constraints.

Multi-Periodic Process Networks (MPPN) model heterogeneous video-stream applications and help resource allocation. They describe an application's structure and temporal behavior, not precise functionality of processes, and they may interact with a high-level language from which scheduling and resource allocation are determined. However, MPPN are *not* intended to model *reactive* systems with unpredictable input events [2] or *dynamic* process creation. On the opposite, our model provides precise information regarding the *steady state* of a *deterministic* application mapped to a parallel architecture. We believe MPPN are well suited to help the mapping of a video filter or 3D graphics pipeline to explicitly parallel micro-architectures, e.g. clustered VLIW embedded processors.

2 Related Work

Three theoretical models have influenced MPPN: Petri nets, data-flow graphs and Kahn Process Networks (KPN). Petri nets are inherently asynchronous and handle time constraints [3, 13, 8]. MPPN may be simulated by timed Petri nets but this does not bring precise schedule information. The sub-class of discrete event systems [1, 4] enables scheduling and performance analysis but does not model token assembling/splitting. Data-flow graphs are a well-established means

B. Monien and R. Feldmann (Eds.): Euro-Par 2002, LNCS 2400, pp. 137–146.

to describe asynchronous processing: various properties can be verified, such as bounded memory [11, 5]. Both models capture repetitive actions through cyclic paths whose production rate compel performance, whereas stream-processing applications benefit from alternative descriptions such as lazy streams. Indeed, KPN [9] are closer to our approach: they provide (unbounded) FIFO buffers with blocking reads and non-blocking writes while enforcing deterministic control. But real-time is not considered and processes have no observable semantics. Synchronous approaches [6] are based on clock calculi and enable synchronous code generation, but static steady-state properties are not available. In our deterministic stream-processing context, properties such as degree of parallelism, buffer size and bandwidth are out of reach of these popular models. Other approaches are complementary to MPPN. Alpha [12] is a high-level language for semi-automatic scheduling and mapping of numerical applications to VHDL. Within the Ptolemy project [5], Compaan targets automatic KPN generation from MatLab loop nests [10]; it is also a powerful simulation tool. KPN modeling, though frequently used in co-design, is insufficient for streams of structured data. As an introductory example consider *downscaling* of a video image is decomposed into a sequence of horizontal and vertical filtering. The former operates on pixels and the latter operates on lines — see Figure 1 for a simplified KPN model. A certain number of pixels/lines is used to determine the new, smaller number of pixels/lines. We assume a horizontal downscaling of 8:3 and a vertical downscaling of 9:4 (High Definition to Single Definition). Figure 1 describes the "data reordering" occurring within stripes, between the horizontal and the vertical filters. First of all, the hierarchy captures non-FIFO communication without resorting to an explicit reorder process. More importantly, each passage through a hierarchy boundary corresponds to an explicit synchronization, where larger messages are considered, consisting of a fixed number of smaller messages. This hierarchical synchronization of events is called *multi-periodic*: it will be characterized through multiple, hierarchically layered, periodic schemes.

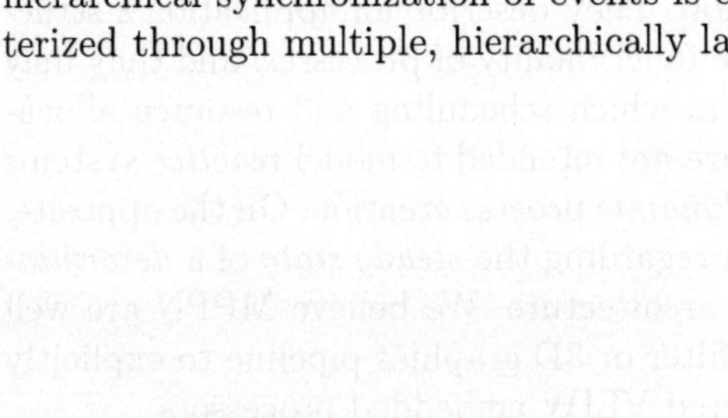

Fig. 1. Example: downscaler

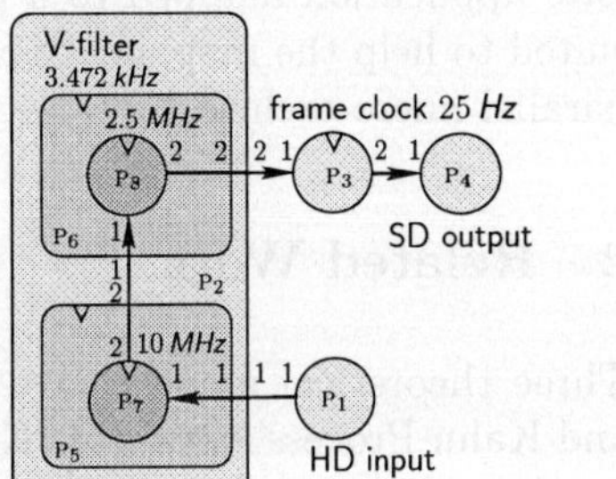

Fig. 2. Simple model of the downscaler

3 Network Structure

A Multi-Periodic Process Network (MPPN) is a 5-tuple $(\mathbf{P}, \sqsubseteq, \mathbf{C}, \text{IN}, \text{OUT})$, where $\mathbf{P}$ is a set of *processes*, $\sqsubseteq$ is a *hierarchical ordering* on $\mathbf{P}$ (its Hasse diagram is a forest), $\mathbf{C}$ is a set of *channels*, process $\text{P}_i \in \mathbf{P}$ is associated with *input ports* in $\text{IN}(\text{P}_i)$ and *output ports* in $\text{OUT}(\text{P}_i)$. P_i^j denotes port j of process P_i. Ordering $\sqsubseteq$ describes the hierarchy among processes: P_i is a *sub-process enclosed* by P_k if and only if $\text{P}_i \sqsubset \text{P}_k$. Moreover, a process P_i is *immediately* enclosed by P_k if and only if $\text{P}_i \sqsubset \text{P}_k$ and there is no other process enclosing P_i and enclosed by P_k. Processes which do not enclose other processes are called *atomic*; conversely, *compound* processes enclose of one or more sub-processes.

A channel connects process ports: $\text{P}_i^j \text{P}_k^l$ represents a channel whose *source* is port P_i^j and whose *sink* is port P_k^l. Any port must belong to exactly one channel. Channels are defined inductively:

– if P_i and P_k are *immediately* enclosed by the same process or at the upper level, $j \in \text{OUT}(\text{P}_i)$, $l \in \text{IN}(\text{P}_k)$, then $\text{P}_i^j \text{P}_k^l$ is a *flat atomic channel*;
– if P_i is *immediately* enclosed by P_k, $j \in \text{IN}(\text{P}_i)$, $l \in \text{IN}(\text{P}_k)$ (resp. $\text{OUT}(\text{P}_i)$, $\text{OUT}(\text{P}_k)$), then $\text{P}_k^l \text{P}_i^j$ is a *downward atomic channel* (resp. $\text{P}_i^j \text{P}_k^l$ is an *upward atomic channel*);
– if $\text{P}_i^j \text{P}_k^l \in \mathbf{C}$ and $\text{P}_k^l \text{P}_{k'}^{l'} \in \mathbf{C}$, then $\text{P}_i^j \text{P}_{k'}^{l'}$ is also a channel (not an atomic one); $\text{P}_i^j \text{P}_k^l$ and $\text{P}_k^l \text{P}_m^n$ are called *sub-channels* of $\text{P}_i^j \text{P}_m^n$.

$\mathbf{C}_{\text{FLAT}}$, $\mathbf{C}_{\text{DOWN}}$ and $\mathbf{C}_{\text{UP}}$ are the sets of flat, downward and upward atomic channels, respectively. The MPPN for the downscaler in Figure 2 illustrates these definitions. It is built from three compound processes, P_2, P_5, P_7, and five atomic ones. Digits accross process boundaries are port numbers: $\text{P}_5^2 \text{P}_6^1$ and $\text{P}_2^2 \text{P}_3^1$ are flat channels, $\text{P}_5^1 \text{P}_6^1$ is a downward channel, $\text{P}_6^2 \text{P}_2^2$ is an upward channel, etc.

A *path* π over the network is a list $\text{P}_{i_1}^{j_1} \text{P}_{i_2}^{j_2} \cdots \text{P}_{i_n}^{j_n}$ such that: $\text{P}_{i_m}^{j_m} \text{P}_{i_{m+1}}^{j_{m+1}}$ is either an *atomic* channel or a pair of input/output ports of the *same* process. E.g., any channel is a path, and $\text{P}_2^1 \text{P}_5^1 \text{P}_7^1 \text{P}_7^2 \text{P}_5^2 \text{P}_6^1 \text{P}_8^1 \text{P}_8^2 \text{P}_6^2 \text{P}_2^2$ is a path in Figure 2. A port or process is *reachable* from another port or process if there exists a path from the latter to the former. Eventually, any output port of a compound process P_i must be reachable from an input port of P_i. For the sake of clarity, we only consider *acyclic* networks with *periodic* input streams (see Section 7 for extensions).

4 Network Semantics

We now enrich the network structure with data-flow activation and message semantics to model the execution of a stream-processing application.

During the course of execution, processes exchange messages and activate in response to receiving such messages. For each process P_i (resp. port P_i^j), the activation *count* is defined as the number of activations of P_i (resp. the number of messages hitting port P_i^j) *since the last activation of the enclosing process* — or since the beginning of the execution if P_i is at the highest level of the hierarchy. Activation and message *dates* are defined likewise: date 0 corresponds to the last activation of the enclosing process — or the beginning of the execution of an outermost process — and all activation/message dates of sub-processes and ports are relative to this activation.

The model is designed such that local event dates only depend on the local event count, i.e., previous activations of the enclosing process have no "memory" effect. This locality property is one of the keys to compositionality — see Section 4.3. It also enables the following definition: an *execution* of a MPPN is a pair of non-decreasing functions (ACT, MSG), such that ACT : $\mathbf{P} \times \mathbb{N} \to \mathbb{R}$ maps processes and activation counts to activation dates, MSG : $\{P_i^j \mid P_i \in \mathbf{P} \wedge j \in \mathrm{IN}(P_i) \cup \mathrm{OUT}(P_i)\} \times \mathbb{N} \to \mathbb{R}$ maps process ports and message counts to message dates; $\mathrm{ACT}(P_i, n)$ is the date of activation n of process P_i, $\mathrm{MSG}(P_i^j, n)$ is the date of message n hitting port P_i^j.

4.1 Propagation in Atomic Channels

When a big message is sent through a *flat* channel and decomposed into smaller ones, the first small message is received right after the big one is sent (pending some communication latency). Conversely, building a big message out of smaller ones takes additional time: many small messages must be sent before a big one is received. In both cases, n messages of size Q_k^l hitting port l of P_k through $P_i^j P_k^l$ correspond to $\lceil nQ_k^l/Q_i^j \rceil$ messages sent by P_i.

In addition, we assume a constant communication latency $c_{i,k}^{j,l}$ for an elementary message sent through an *atomic* channel $P_i^j P_k^l$. Recalling that $\mathrm{MSG}(P_k^l, n)$ is the date of the $(n+1)$-th message hitting port l of process P_k,

$$\forall P_i^j P_k^l \in \mathbf{C}_{\mathrm{FLAT}} : \mathrm{MSG}(P_k^l, n) = \mathrm{MSG}\big(P_i^j, \lceil (n+1) Q_k^l / Q_i^j \rceil - 1\big) + c_{i,k}^{j,l}. \quad (1)$$

Considering an upward channel, the propagation equation sums the activation date of the enclosing process and the local (relative) date of the last small message assembled to build an output message at port P_k^l:

$$\forall P_i^j P_k^l \in \mathbf{C}_{\mathrm{UP}} : \mathrm{MSG}(P_k^l, n) = \mathrm{ACT}(P_k, n) + \mathrm{MSG}\big(P_i^j, \lceil Q_k^l / Q_i^j \rceil - 1\big). \quad (2)$$

Considering a downward channel, hierarchical composition enforces that *no message enters a compound process before it activates*. More precisely, when a message reaches an input port of a compound process P_i, this message is *not* propagated further on the channel (and possibly decomposed) before P_i activates on this very message. Activation of P_i coincides with the reception of a message at port P_k^l, since $Q_k^l \le Q_i^j$ (decomposition into smaller messages). Therefore,

$$\forall P_i^j P_k^l \in \mathbf{C}_{\mathrm{DOWN}} : \mathrm{MSG}(P_k^l, n) = 0. \quad (3)$$

4.2 Activation Model

We consider a data-flow scheme: process activation starts as soon as there is at least one message on each input port. Let Q_i^j be the size of messages sent or received on port j of P_i. A process P_i enters activation n as soon as the following data-flow condition is met: every input port $j \in \mathrm{IN}(\mathrm{P}_i)$ has been hit by message n of size Q_i^j (except for special *clocked* processes, see Sections 4.4). The data-flow activation scheme is formalized as follows:

$$\mathrm{ACT}(\mathrm{P}_i, n) = \max_{j \in \mathrm{IN}(\mathrm{P}_i)} \mathrm{MSG}(\mathrm{P}_i^j, n). \tag{4}$$

This definition allows multiple overlapping activations of a process.

Considering an *atomic* process P_i, we call ℓ_i^j the latency of P_i for sending a message through output port j. It is defined as the elapsed time between an activation of P_i and the corresponding output of a message through port j, supposed *constant* for all executions of the process:

$$\mathrm{MSG}(\mathrm{P}_i^j, n) = \mathrm{ACT}(\mathrm{P}_i, n) + \ell_i^j. \tag{5}$$

This constant latency will be extended to *compound* processes in Section 4.3.

In the following, details and proofs that had to be left out can be found in [7]. While any size change of messages is possible when traversing a process or channel, the same is not true when traversing a path. As a consequence of the previous equations, in order to ensure compositionality, message sizes must obey a strict *scaling rule* when traversing hierarchy boundaries. Let us consider a *compound* process P_i, an input port $j \in \mathrm{IN}(\mathrm{P}_i)$ and an output port $l \in \mathrm{OUT}(\mathrm{P}_i)$. If these ports are connected through a path π of channels and sub-processes of P_i, one single message hitting P_i^j may traverse several assembling/splitting stages through π, but it must yield one single message hitting P_i^l. This is intrinsic to the hierarchical model; the following path constraints enforce the scaling rule:[1]

$$\prod\nolimits_{\mathrm{P}_i^j \mathrm{P}_k^l \in \pi} (Q_i^j / Q_k^l) = 1 \text{ and } \forall \pi' \text{ prefix of } \pi : \prod\nolimits_{\mathrm{P}_i^j \mathrm{P}_k^l \in \pi'} (Q_i^j / Q_k^l) \geq 1. \tag{6}$$

Activation of a *source* (input-less) process P_i is not constrained by any data-flow scheme. We assume a *periodic behavior* instead: considering two real numbers $\mathrm{ACT}(\mathrm{P}_i, 0)$ — the *reference date* — and $\mathrm{PER}(\mathrm{P}_i)$ — the *period*,

$$\mathrm{ACT}(\mathrm{P}_i, n) = \mathrm{ACT}(\mathrm{P}_i, 0) + n\mathrm{PER}(\mathrm{P}_i). \tag{7}$$

4.3 Latencies of Compound Processes

Consider an output port j of a compound process P_i. We can prove that every activation of P_i sends one message through P_i^j after a *constant* latency. This result lies in the data-flow equations (message propagation and activation rules are time invariant) and in the scale factor constraint (6) which ensures that any single activation of P_i sends exactly one message through P_i^j. We may thus extend (5) to *compound* processes: $\mathrm{MSG}(\mathrm{P}_i^j, n) - \mathrm{ACT}(\mathrm{P}_i, n)$ is a constant, ℓ_i^j can be computed for $n = 0$ based on information at the lower level:

$$\forall \mathrm{P}_i \in \mathbf{P}, j \in \mathrm{OUT}(\mathrm{P}_i) : \ell_i^j = \mathrm{MSG}(\mathrm{P}_i^j, 0) - \mathrm{ACT}(\mathrm{P}_i, 0). \tag{8}$$

Thus, MPPN exhibit compositional semantics. Latencies of compound processes do not depend on the surrounding network: they are computed once and for all.

[1] This equation applies when messages are multiples of one another.

4.4 Clocked Processes

We provide an extended kind of process to synchronize streams over a fixed clock period: *clocked* processes. These processes can be either atomic or compound, and their activation rule is generalized from ordinary processes. Considering a clocked process P_i, an *internal clock* starts at the reference date $\mathrm{ACT}(\mathrm{P}_i, 0)$, and subsequent activations may only occur *one at a time* when all input messages are present and when an *internal clock tick* occurs. To put it simple, a clocked process has a double role of (local) sampling and delay. A clocked process P_i is characterized through a *clock frequency* f_i such that $f_i \geq 1/\mathrm{PER}(\mathrm{P}_i)$; equality enforces periodicity (e.g., to enable stream resynchronization for video output).

When enclosed in compound processes, MPPN clocks behave differently from hardware clocks (whose semantics is exclusive): two independent activations of a compound process may trigger overlapping streams of events on clocked sub-processes. This is required to preserve compositionality. In practice, the designer may want the enclosing process to be clocked itself so that executions of the clocked sub-process are sequential, e.g., in dividing the frequency of the clocked sub-process by the hierarchical scale factor.

5 High Level Properties

From the abovementioned MPPN semantics, one may deduce resource requirements of the application. In this paper, we focus on conservative estimates for the number of functional units, bandwidth and buffers. We derive global (absolute) properties from the product of local evaluations.

5.1 Asymptotic Periodic Execution

We have proven that all processes follow a steady-state scheme which extends and relaxes the periodic constraint (7) on source processes; starting from (7), this follows inductively from (1), (4) and (5). For each process P_i, there exist an *average period* $\mathrm{PER}(\mathrm{P}_i)$ and a *burstiness* $\mathrm{ADV}(\mathrm{P}_i)$ such that

$$\forall n \geq 0 : (n - \mathrm{ADV}(\mathrm{P}_i))\mathrm{PER}(\mathrm{P}_i) \leq \mathrm{ACT}(\mathrm{P}_i, n) - \mathrm{ACT}(\mathrm{P}_i, 0) \leq n\mathrm{PER}(\mathrm{P}_i) \qquad (9)$$

$$\forall n \geq 0 : (n - \mathrm{ADV}(\mathrm{P}_i^j))\mathrm{PER}(\mathrm{P}_i) \leq \mathrm{MSG}(\mathrm{P}_i^j, n) - \mathrm{MSG}(\mathrm{P}_i^j, 0) \leq n\mathrm{PER}(\mathrm{P}_i). \qquad (10)$$

The burstiness is the maximal number of *advance* activations of P_i, i.e., activations ahead of the periodic execution scheme. This parameter encompasses both *deterministic bursts* of early messages and *jittering streams* with earliest/latest bounds (and, possibly, periodic resynchronization). Even under the worst-case conditions, better evaluations of $\mathrm{ACT}(\mathrm{P}_i, n)$ can be hoped for (possibly exact ones): deterministic event bursts can be characterized effectively within the relaxed periodic scheme.

Considering an output port j of a process P_i, messages hit j after a constant delay, hence output message burstiness is equal to activation burstiness:

$$\forall j \in \mathrm{OUT}(\mathrm{P}_i) : \mathrm{ADV}(\mathrm{P}_i^j) = \mathrm{ADV}(\mathrm{P}_i). \qquad (11)$$

This equation stands for both *upward* channels (in compound processes) and atomic processes.

Sending one message and $\text{ADV}(\text{P}_i^j)$ advance messages at port P_i^j corresponds to sending $Q_i^j(1+\text{ADV}(\text{P}_i^j))$ bytes of data. These data are received as one message plus $\text{ADV}(\text{P}_k^l)$ advance messages at port P_k^l, i.e., $Q_k^l(1+\text{ADV}(\text{P}_k^l))$ bytes. For *flat* channels, the result is the following:

$$\forall \text{P}_i^j \text{P}_k^l \in \mathbf{C}_{\text{FLAT}} : Q_i^j(1+\text{ADV}(\text{P}_i^j)) = Q_k^l(1+\text{ADV}(\text{P}_k^l)); \tag{12}$$

for *downward* channels, a single activation of the enclosing process is considered:

$$\forall \text{P}_i^j \text{P}_k^l \in \mathbf{C}_{\text{DOWN}} : Q_i^j = Q_k^l(1+\text{ADV}(\text{P}_k^l)). \tag{13}$$

Notice that (12) and (13) may require burstinesses to be non-integer. In addition, communication may be implemented through *bounded buffers* as long as asymptotic data throughput is the same at both ends of a channel:

$$\forall \text{P}_i^j \text{P}_k^l \in \mathbf{C} : Q_i^j/\text{PER}(\text{P}_i) = Q_k^l/\text{PER}(\text{P}_k). \tag{14}$$

Activation burstiness can be deduced from the data-flow scheme, replacing $\text{ACT}(\text{P}_i, n)$ and $\text{MSG}(\text{P}_i^j, n)$ by their lower bounds in (4). The result is that processes tend to "smooth" message bursts and initiation delays:

$$\text{ADV}(\text{P}_i) = \min_{j \in \text{IN}(\text{P}_i)} \left\{ \text{ADV}(\text{P}_i^j) + \frac{\text{ACT}(\text{P}_i,0) - \text{MSG}(\text{P}_i^j,0)}{\text{PER}(\text{P}_i)} \right\}. \tag{15}$$

This is a two-phase computation: on a given stream, sum up the burstiness and the number of messages that precede activation 0, then minimize these adjusted burstinesses. If the process is clocked, the former result is multiplied by $1 - 1/(f_i \text{PER}(\text{P}_i))$; one expectedly get $\text{ADV}(\text{P}_i) = 0$ when $\text{PER}(\text{P}_i) = 1/f_i$.

5.2 Global Properties

We call ℓ_i the latency for P_i to complete an execution, i.e., the maximum of ℓ_i^j at output ports j of P_i: $\ell_i = \max_{j \in \text{OUT}(\text{P}_i)} \ell_i^j$. Let $\mathsf{overlap}(\text{P}_i, d)$ denote the maximum number of executions of P_i during a given period of time d, and triggered by a single activation of the enclosing process (if P_i is a sub-process):

$$\mathsf{overlap}(\text{P}_i, d) = \min\left(\lceil f_i d \rceil, \lceil d/\text{PER}(\text{P}_i) + \text{ADV}(\text{P}_i) \rceil\right).$$

We proved that the (absolute) maximal number of parallel executions of a process P_i is bounded by the product of the local maximal number of activations of P_i and all its enclosing processes during the same duration ℓ_i:

$$\mathsf{maxpll}(\text{P}_i) = \prod_{\text{P}_i \sqsubseteq \text{P}_k} \mathsf{overlap}(\text{P}_k, \ell_i).$$

Depending on the architecture and the resource allocation strategy, ports associated with physical input/output may be identified. On this subset, it is legitimate to ask for an estimate of the average and maximal bandwidths. Such estimates can be built from the periods, burstinesses and overlapping factors, see [7] for details. Our current model assumes that actual loads/stores are distributed evenly over the whole access period ℓ_i^j. This hypothesis is optimistic, but finer evaluations can be crafted following the same reasoning. Port bandwidth is of critical interest when implementing process communications through shared-memory buffers, whereas channel bandwidth provides some insight about network contention when focusing on distributed architectures.

We describe a method to bound buffer size for any *atomic* channel, not considering architecture-specific buffer requirements. The minimal size of a buffer for channel $\mathrm{P}_i^j \mathrm{P}_k^l$ is the maximum amount of temporary data that must be stored during message propagation through this channel; it is denoted by $\mathsf{maxbuf}(\mathrm{P}_i^j \mathrm{P}_k^l)$. Such a buffer must hold all the messages sent to the channel by P_i and not yet received by P_k, i.e., the difference between data sent and received. An upper bound of this difference is evaluated from the liveness of a message and the product of overlapping factors.

6 Network Analysis

The analysis of a multi-periodic process network consists in solving the above equations. Let us sketch an algorithm for MPPN analysis and verification.

Input. A multi-periodic process network, Q_i^j for each port, $c_{i,k}^{j,l}$ for each atomic flat channel, ℓ_i^j for each atomic process, reference date $\mathrm{ACT}(\mathrm{P}_i, 0)$ for each source process (e.g., 0), period $\mathrm{PER}(\mathrm{P}_i)$ for one process per weakly-connected component of the network, burstiness $\mathrm{ADV}(\mathrm{P}_i)$ for each source process. Optional values for other parameters, e.g., burstinesses and periods at sink processes.

Output. Values for all parameters or contradiction.

Resolution. The algorithm is decomposed into four phases.

1. Perform a topological sort of the network. Check the scaling rule of all compound processes, using (6).
2. Compute $\mathrm{PER}(\mathrm{P}_i)$ traversing the network incrementally, starting from processes with known periods using (14) and checking for consistency.
3. Traverse the hierarchical structure bottom-up, applying the following steps:
   - choose a compound process P_i whose sub-processes have known latencies;
   - compute (relative) reference dates for sub-processes, using (1), (4) and (5);
   - deduce latency ℓ_i^j for every output port j, using (8).
4. Compute $\mathrm{ADV}(\mathrm{P}_i)$ and $\mathrm{ADV}(\mathrm{P}_i^j)$ through a top-down traversal, following the topological ordering at each hierarchical level, using (11), (12) and (15).

From the output of this algorithm, one may deduce the degree of parallelism, bandwidths and buffer sizes for all processes, ports and channels. We may now show the results on the introductory example.

Input. We consider a pixel unit for all message sizes, $Q_1^1 = Q_2^1 = 1920 \times 1080$, $Q_2^2 = Q_3^1 = Q_3^2 = Q_4^1 = 720 \times 480$, $Q_7^1 = 8$, $Q_5^1 = 1920$, $Q_7^2 = 3$, $Q_5^2 = 720$, $Q_8^1 = 9$, $Q_6^1 = 720 \times 9$, $Q_8^2 = 4$, $Q_6^2 = 720 \times 4$. Some latencies, activation dates, burstinesses, and periods are already given: $\mathrm{ACT}(\mathrm{P}_1, 0) = 0$, $\mathrm{ADV}(\mathrm{P}_1) = 0$, $\mathrm{PER}(\mathrm{P}_4) = 40\,ms$ ($25\,Hz$),

$\text{ADV}(\text{P}_4) = 0$, $\ell_1^1 = 0$ (source process), $\ell_3^2 = 1\,ms$, $\ell_7^2 = 200\,ns$, $\ell_8^2 = 1\,\mu s$, $c_{1,2}^{1,1} = c_{2,3}^{2,1} = c_{3,4}^{2,1} = 1\,\mu s$ (communication), $c_{5,6}^{2,1} = 100\,ns$ (local buffer). Clocks must be such that $f_i \geq 1/\text{PER}(\text{P}_i)$: we choose $f_7 = 10\,MHz$ and $f_8 = 2.5\,MHz$, and we deduce clocks for P_5 and P_6 from the hierarchical scale factors: $f_5 = f_7 \times (3/720) = 41.67\,kHz$, $f_6 = f_8 \times (4/(4 \times 720)) = 3.47\,kHz$.

Output. Compound process latencies: $\ell_5^2 = 24.1\,\mu s$, $\ell_6^2 = 288.6\,\mu s$ and $\ell_2^2 = 34.752\,ms$; periods: $\text{PER}(\text{P}_5) = 37.04\,\mu s$, $\text{PER}(\text{P}_6) = 333.33\,\mu s$, $\text{PER}(\text{P}_7) = 154.3\,ns$, $\text{PER}(\text{P}_8) = 462.96\,ns$. Burstinesses are large, $\text{ADV}(\text{P}_7) = 154.9$, $\text{ADV}(\text{P}_8) = 621.2$, $\text{ADV}(\text{P}_5) = 699.2$ and $\text{ADV}(\text{P}_6) = 102.8$, but this only reports a high variability around the average period. The clock's effect on resource usage is more significant: the parallelism degree is $\mathsf{maxpll}(\text{P}_7) = 2$ and $\mathsf{maxpll}(\text{P}_8) = 3$. This demonstrates how MPPN achieve a precise description of the burst rate within an average periodic scheme. Finally, we get the bandwidth and buffer results: $\mathsf{maxbw}(\text{P}_7^1) = 80\,Mpixel/s$, $\mathsf{maxbw}(\text{P}_7^2) = 30\,Mpixel/s$, $\mathsf{maxbw}(\text{P}_8^1) = 22.5\,Mpixel/s$, $\mathsf{maxbw}(\text{P}_8^2) = 10\,Mpixel/s$ and $\mathsf{maxbuf}(\text{P}_5^2\text{P}_6^1) = 15.12\,kpixel$, which expectedly corresponds to a single stripe buffer between the two filters plus 116.7% overhead, making conservative assumptions on message liveness.

7 Extensions, Conclusion and Future Work

Applicability of the previous model is vastly improved when considering three simple extensions. First of all, we provide two special kinds of atomic processes for multiplexing and demultiplexing streams. For example, such processes refine the modeling of a picture-in-picture application, where a rectangle of a larger frame is replaced by a downscaled frame from another video stream. Activation of a *splitter* process sends one message alternatively through each of its output ports, whereas activation of a *selector* process receives one message alternatively from each of its input ports. We proved that a periodic alternation of process ports preserves the periodic nature of message and activations events.

Moreover, it is quite natural to relax the periodic constraint on input streams, and only require an average periodicity. This is easily achieved in adding a burstiness parameter to source processes. We proved that conservative reference dates can be deduced from the "latest" execution scheme associated with the "latest" valid schedule of source processes (a periodic one).

Eventually, we consider cyclic networks whose semantics differs from iteration modeling in Petri nets: in stream-processing frameworks, cycles model feedback or data reuse. The latency for a message to traverse a given circuit must be less than or equal to the average period of the initiating process. In other words, a cyclic path is legal as long as it has *no* influence on the global throughput; it can be statically checked through a path constraint (analogue to the scaling rule) and an additional *bootstrap* constraint. Dynamic noise reduction is a typical example: a noise threshold is updated to provide dynamic control over the filtering stage.

Multi-periodic process networks are an expressive model and a powerful tool for statically manipulating real-time properties, concurrency, and resource requirements of regular stream-processing applications. Primarily influenced by synchronous extensions to Kahn process networks, they exploit the application's regularity and hierarchical structure to extend the range of possible analyses and transformations. Six major properties can be expressed:

abstraction: nested processes allow for different levels of specification, both for messages (hierarchical nature of data structures) and activation events;

composition: the same property set describes all processes (latency, period...); nested processes are analyzed only once and reused through MPPN libraries;

synchronization: nesting of processes and hierarchical data-flow activation provides an elegant and efficient tool for modeling synchronization;

jitter: event dates — whether messages or process activations — are bounded within "earliest" and "latest" deterministic functions;

bursts: deterministic bursts of events can be captured explicitly within periodic event schemes, using hierarchical layers of periodic characterizations;

sequencing: communication uses First-In First-Out (FIFO) channels, but implicit reordering is allowed when assembling/splitting messages.

A prototype of the verifier was implemented in Java; it uses XML representations of MPPN for easy integration into application design environments.

Acknowledgments: We received many contributions from the further members of the SANDRA team, Christine Eisenbeis, Laurent Pasquier, Valerie Rivierre-Vier, Francois Thomasset and Qin Zhao. We thank them for their support and in-depth reviews.

References

1. F. Baccelli, G. Cohen, G. J. Olsder, and J.-P. Quadrat. *Synchronization and Linearity.* Wiley, 1992.
2. G. Berry, P. Couronné, and G. Gonthier. The synchronous approach to reactive and real-time systems. *Proc. IEEE*, 79(9):1270–1282, Sept. 1991.
3. B. Berthomieu and M. Diaz. Modelling and Verification of Time-Dependent Systems Using Time Petri Nets. *IEEE Trans. on Software Eng.*, 17, Mar. 1991.
4. J.-Y. L. Boudec and P. Thiran. *Network Calculus.* Springer LNCS 2050, Jan. 2002.
5. J. T. Buck, S. Ha, E. A. Lee, and D. G. Messerschmitt. Ptolemy: A framework for simulating and prototyping heterogeneous systems. *J. Comp. Simulation*, 4, 1992.
6. P. Caspi and M. Pouzet. Synchronous Kahn Networks. In *ACM SIGPLAN Int. Conference on Functional Programming (ICFP)*, Philadelphia, May 1996. ACM.
7. A. Cohen and D. Genius. Multi-periodic process networks: Technical report. http://www-rocq.inria.fr/~acohen/publications/mppn, 2002.
8. G. Cohen, S. Gaubert, and J.-P. Quadrat. Algebraic system analysis of timed petri nets. *Idempotency, Cambridge University Press*, 1997.
9. G. Kahn. The Semantics of a Simple Language for Parallel Programming. In *IFIP 74 Congress*, Amsterdam, 1974. North-Holland.
10. B. Kienhuis, E. Rijpkema, and E. Deprettere. Compaan: Deriving process networks from matlab for embedded signal processing architectures. In *Proc. 8th workshop CODES*, pages 13–17, NY, May 3–5 2000. ACM.
11. E. A. Lee and J. C. Bier. Architectures for statically scheduled dataflow. *J. Parallel and Distributed Computing*, 10(4):333–348, Dec. 1990.
12. H. Leverge, C. Mauras, and P. Quinton. The ALPHA language and its use for the design of systolic arrays. *J. of VLSI Signal Processing*, 3:173–182, 1991.
13. P. Sénac and M. Diaz. Time Streams Petri Nets, A Model for Timed Multimedia Informations. In *16th Int. Conf. Appl. and Theory of Petri Nets*, Turin, June 1995.

Symbolic Cost Estimation of Parallel Applications

Arjan J.C. van Gemund

Faculty of Information Technology and Systems
Delft University of Technology
P.O. Box 5031, NL-2600 GA Delft, The Netherlands
a.j.c.vangemund@its.tudelft.nl

Abstract. Symbolic cost models are an important performance engineering tool because of their diagnostic value and their very low solution cost when the computation features regularity. However, especially for parallel applications their derivation, including the symbolic simplifications essential for low solution cost, is an effort-intensive and error-prone process. We present a tool that automatically compiles process-oriented performance simulation models into symbolic cost models that are symbolically simplified to achieve extremely low solution cost. As the simulation models are intuitively close to the parallel program and machine under study, derivation effort is significantly reduced. Apart from its use as a stand-alone tool, the compiler is also used within a symbolic cost estimator for data-parallel programs. With minimal program annotation by the user, symbolic cost models are automatically generated in a matter of seconds, while the evaluation time of the models ranges in the milliseconds. Experimental results on four data-parallel programs show that the average prediction error is less than 15 %. Apart from providing program scalability assessment, the models correctly predict the best design alternative in all cases.

1 Introduction

Symbolic cost modeling is a performance modeling technique where parallel programs are mapped into explicit, algebraic performance expressions that are parameterized in terms of, e.g., the number of processors, problem size, and machine computation and communication parameters. Symbolic models are analytical, providing diagnostic insight in the complex interplay between program and machine parameters. An important benefit of symbolic cost models is their potentially low solution cost, compared to, e.g., simulation [13], Petri nets [2], queuing networks [1, 11], and process algebras [10]. As most parallel programs and machines employ some form of (sequential or parallel) replication, cost models typically have a regular structure. This regularity enables symbolic simplification that dramatically reduces solution complexity by many orders of magnitude. As interactive parameter experimentation during parallel program development requires cost models to evaluate in (milli)seconds rather than minutes or hours, this feature is extremely valuable.

Although attractive in terms of solution complexity, the derivation of low-complexity cost models from parallel programs is an effort-intensive and error-prone process, that is akin to manual complexity analysis of parallel algorithms [4, 14]. Aimed to provide tool support in this derivation process, a performance simulation formalism called PAMELA

B. Monien and R. Feldmann (Eds.): Euro-Par 2002, LNCS 2400, pp. 147–156.

(PerformAnce ModEling LAnguage) has been presented [7]. Both the parallel program and machine are modeled in terms of PAMELA. Instead of simulation, the PAMELA model is mechanically mapped into a symbolic cost model that has the lowest possible solution complexity, while offering a prediction accuracy that is sufficient to discriminate between various program design alternatives. As the simulation language is intuitively close to the parallel program (and machine), cost model derivation efficiency is significantly improved. While previous publications presented the methodology and calculus, in this paper we present a language implementation, featuring a compiler that automatically compiles PAMELA models into symbolic cost models.

An interesting feature of the PAMELA methodology is that for data-parallel programs the mapping of programs to PAMELA models can also be mechanized. As part of a data-parallel compiler project, a so-called Modeling Engine has been built that automatically generates PAMELA models from data-parallel programs that have been annotated with missing information on, e.g., loop bounds and branch probabilities [9]. Combined with the PAMELA compiler, a data-parallel program can be subsequently *compiled* into a symbolic cost model within seconds, while the cost model predicts performance within milliseconds. In this paper we also describe the results of this automatic cost estimator. The accuracy of the generated cost models is more than sufficient to allow a correct ranking of various coding and/or data partitioning strategies.

The paper is organized as follows. In Section 2 we present our tool implementation based on the PAMELA methodology. In Section 3 we briefly describe how PAMELA models are generated from data-parallel programs. In Section 4 we demonstrate the utility of the symbolic cost estimation process in the design of four well-known numeric applications. In Section 5 we summarize our contributions.

2 Symbolic Cost Estimation

In this section we briefly summarize the PAMELA modeling language and the compiler. As the language is intuitively close to the description of algorithms, the use of PAMELA is more cost effective than manual oriented approaches which do not offer tool support. Apart from its use as a stand-alone tool, the PAMELA compiler is also used in conjunction with the Modeling Engine that generates PAMELA models from data-parallel programs.

2.1 Language

PAMELA is a process-algebraic language that allows a parallel program to be modeled in terms of a sequential, conditional, and parallel composition of processes that model workload. Work is described by the `use` *process*. The construct `use(r,t)` exclusively acquires service from *resource* `r` for `t` units time (excluding possible queuing delay). The scheduling policy that is currently supported is First-Come-First-Served (FCFS) with non-deterministic conflict arbitration. A resource `r` has a multiplicity that may be larger than unity. As in queuing networks, it is convenient to define an infinite-server resource called `rho` that has infinite multiplicity. Instead of writing `use(rho,t)` we will simply write `delay(t)`.

PAMELA features the following process composition operators: `;` for binary sequential composition, `seq (<index> = <lb>, <ub>)` for n-ary sequential composition, `||` for binary parallel composition, `par (<index> = <lb>, <ub>)` for n-ary parallel composition, `if (<cond>) [else]` for conditional composition.

PAMELA is a strongly typed language. Variables can be of three types: `process`, `resource`, or `numeric`. The latter type is used for time expressions, parameters, indices, and loop bounds. Each lhs variable can have a formal parameter list. The scope of these parameters is limited to the rhs expression.

The following PAMELA equations model the well-known Machine Repair Model (MRM) in which `P` clients either spend a mean time `t_l` on local processing, or request service from a single server `s` with service time `t_s` with a total cycle count of `N` iterations (unlike steady-state analysis, in our approach we require models to terminate; yet `N` may be symbolic).

```
numeric parameter P                 % # clients
numeric parameter N                 % # iterations
numeric t_l = 10                    % local think time
numeric t_s = 0.1                   % service time

resource parameter fcfs(i)          % predefined FCFS array
resource s = fcfs(0,1)              % s is FCFS type resource
                                    % args: index, multiplicity

process main = par (p = 1, P)
                    seq (i = 1, N) {
                        delay(t_l) ;
                        use(s,t_s)
                    }
```

The example illustrates the high, "problem" level at which an application is modeled. In the PAMELA top-down modeling approach, problem parallelism is modeled explicitly in terms of `par` (or `;`) operators, to be constrained by mutual exclusion (`use`) when the processes are mapped onto a limited number of resources, such as software locks, file servers (cf. above example), (co)processors, communication links, memories, I/O disk handlers, etc. This high-level modeling approach allows virtually all parallel algorithms to be modeled in terms of *series-parallel* (SP) process expressions. This SP synchronization structure is essential to allow mechanization of the symbolic timing analysis. A lower, "implementation" level approach would require a message-passing paradigm which cannot be statically analyzed [8]. Also note, that due to the ultra-low solution complexity of the performance model high parameter values can be specified as the model typically evaluates in milliseconds.

As in ordinary mathematics, the semantics of a PAMELA model is based on expression substitution. Although for readability a model may be coded in terms of many equations, internally each expression is evaluated by recursively substituting every global variable by its corresponding rhs expression. Consequently, the above MRM model is internally rewritten to the equivalent normal-form model below (relevant equations shown only).

```
numeric parameter P
numeric parameter N
process main = par (p = 1, P)
                    seq (i = 1, N) {
                        delay(10) ;
                        use(fcfs(0,1),0.1)
                    }
```

where `s`, `t_l`, and `t_s` have been substituted. Note that the optional `parameter` modifier blocks this substitution process. In the above example this will cause `P` and `N` to appear as parameters in the eventual cost model.

Apart from the process operators mentioned above, PAMELA includes the usual numeric operators such as +, *, `mod`, `div`, ==, <, `max`, etc., as well as the reductions `sum (<index> = <lb>, <ub>)` and `max (<index> = <lb>, <ub>)`. Conditional numeric expressions are described using `if-then` just like conditional process expressions. Furthermore, as parts of the analysis result is expressed in terms of vectors, the `numeric` abstract data-type includes vectors as well as scalars, thus overloading all operators. A vector is denoted `[<scalar>, ..., <scalar>]` Hence, the expression `[1,2,3] * 4` is legal and, incidentally, will be compiled to `[4,8,12]` as a result of the compiler's internal numeric optimization engine. In order to generate unbounded, symbolic vectors PAMELA features the `unitvec` operator which returns a unit vector (base 0) in the dimension given by its argument. For instance, the expression `10 * unitvec(3)` will be compiled to `[0,0,0,10]`.

2.2 Compilation

A PAMELA model is translated to a time-domain performance model by substituting every `process` equation by a `numeric` equation that models the execution time associated with the original process. The lhs is derived from the original lhs by prefixing `T_`. Thus the cost model of a process expression `main` is denoted `T_main`. The result is a PAMELA model that only comprises `numeric` equations as the original `process` and `resource` equations are no longer relevant. The fact that the cost model is again a PAMELA model is for reasons of convenience as explained later on. In the following we briefly describe the translation process. A more detailed background can be found in [8].

The analytic approach underlying the translation process is based on critical path analysis of the delays due to condition synchronization [5, 12, 16] ("task synchronization"), combined with a lower bound approximation of the delays due to mutual exclusion synchronization ("queuing delay") as a result of resource contention [8]. In the following we assume a PAMELA model in which all expressions have already been substituted as the result of the normalization pass described earlier.

Per `process` equation four `numeric` equations are generated, whose lhs identifiers are derived from the original process variable by prefixing specific strings. Let `L` denote the lhs of a process equation. The first equation generated is `phi_L` which computes the condition synchronization delay by recursively applying the following transformation rules:

```
L = a ; b                -> phi_L = phi_a + phi_b
```

```
L = a || b               -> phi_L = max(phi_a,phi_b)
L = use(fcfs(a,b),t) -> phi_L = t / b
```

The second equation generated is `delta_L` which computes the mutual exclusion synchronization delay by

```
L = a ; b                -> delta_L = delta_a + delta_b
L = a || b               -> delta_L = max(delta_a,delta_b)
L = use(fcfs(a,b),t) -> delta_L = unitvec(a) * (t / b)
```

The delta vectors represent the aggregate workload per resource (index). The effective mutual exclusion delay is computed by the third equation, which is generated by the following transformation rule:

```
L = ...                  -> omega_L = max(delta_L)
```

Finally, the execution time `T_L` is generated by the following transformation rules:

```
L = a ; b                -> T_L = T_a + T_b
L = a || b               -> T_L = max(max(T_a,T_b),omega_L)
L = use(fcfs(a,b),t) -> T_L = phi_L
```

The above `max(max(T_a,T_b),omega_L)` computation shows how each of the delays due to condition synchronization and mutual exclusion are combined in one execution time estimate that effectively constitutes a lower bound on the actual execution time. The recursive manner in which both delays are combined guarantees a bound that is the sharpest possible for an automatic symbolic estimation technique (discussed later on). Conditional composition is simply transferred from the process domain to the time domain, according to the transformation

```
L = if (c) a else b  -> X_L = if (c) X_a else X_b
```

where `X` stands for the `phi`, `delta`, `omega`, and `T` prefixes. The numeric condition, representing an average truth probability when embedded within a sequential loop, is subsequently reduced, based on the numeric (average truth) value of `c` according to

```
if (c) X_a else X_b -> c * X_a + (1 - c) * X_b
```

An underlying probabilistic calculus is described in [6].

Returning to the MRM example, based on the above translation process the PAMELA model of the MRM is internally compiled to the following time domain model (`T_main` shown only):

```
numeric parameter P
numeric parameter N
numeric T_main = max(max (p = 1, P) {
                          sum (i = 1, N) {
                            10.1
                          }
                       },
                       max(sum (p = 1, P) {
                             sum (i = 1, N) {
                               [ 0.1 ]
                             }
                          }))
```

Although this result is a symbolic cost model, evaluation of this model would be similar to simulation. Due to the regularity of the original (MRM) computation, however, this model is amenable to simplification, a crucial feature of our symbolic cost estimation approach. The simplification engine within the PAMELA compiler automatically yields the following cost model:

```
numeric parameter P
numeric parameter N
numeric T_main = max((N * 10.1),(P * (N * 0.1)))
```

which agrees with the result of bounding analysis in queueing theory (the steady-state solution is obtained by symbolically dividing by `N`). This result can be subsequently evaluated for different values of `P` and `N`, possibly using mathematical tools other than the PAMELA compiler. In PAMELA further evaluation is conveniently achieved by simply recompiling the above model after removing `parameter` modifiers while providing a numeric rhs expression. For example, the following instance

```
numeric P = 1000
numeric N = 1000000
numeric T_main = max((N * 10.1),(P * (N * 0.1)))
```

is compiled (i.e., evaluated) to

```
numeric T_main = 100000000
```

While the prediction error of the symbolic model compared to, e.g., simulation is zero for $P = 0$ and $P \to \infty$, near to the saturation point ($P = 100$) the error is around 8%. It is shown that for very large PAMELA models (involving $\mathcal{O}(1000+)$ resources) the worst case average error is limited to 50% [8]. However, these situations seldom occur as typically systems are either dominated by condition synchronization or mutual exclusion, in which case the approximation error is in the percent range [8].

Given the ultra-low solution complexity, the accuracy provided by the compiler is quite acceptable in scenarios where a user conducts, e.g., application scalability studies as a function of various machine parameters, to obtain an initial assessment of the parameter sensitivities of the application. This is shown by the results of Section 4. In particular, note that on a Pentium II 350 MHz the symbolic performance model of the MRM only requires 120 μs per evaluation (*irrespective* of N and P), while the evaluation of the original model, in constrast, would take approximately 112 Ks. The $\mathcal{O}(10^9)$ time reduction provides a compelling case for symbolic cost estimation.

3 Automatic Cost Estimation

In this section we describe an application of the PAMELA compiler within an automatic symbolic cost estimator for data-parallel programs. The tool has been developed as part of the JOSES project, a European Commission funded research project aimed at developing high-performance Java compilation technology for embedded (multi)processor systems [9]. The cost estimator is integrated as part of the Timber compiler [15], which compiles parallel programs written in Spar/Java (a Java dialect with data-parallel features similar to HPF) to distributed-memory systems. The cost estimator is based on a combination of a so-called Modeling Engine and the PAMELA compiler. The Modeling Engine is a Timber compiler engine that generates a PAMELA model from a Spar/Java program. The PAMELA compiler subsequently compiles the PAMELA model to a symbolic cost model. While symbolic model compilation is automatic, PAMELA model generation by the Timber compiler cannot always be fully automatic, due to the undecidability problems inherent to static program analysis. This problem is solved by using simple compiler pragmas which enables the programmer to portably annotate the source program, supplying the compiler with the information required (e.g., branch probabilities, loop bounds). Experiments with a number of data-parallel programs show that only minimal user annotation is required in practice.

For all basic (virtual) machine operations such as +, ..., *, (computation), and = (local and global communication) specific PAMELA process calls are generated. During PAMELA compilation, each call is substituted by a corresponding PAMELA machine model that is part of a separate PAMELA source file that models the target machine. All parallel, sequential, and conditional control flow constructs are modeled in terms of similar PAMELA constructs, except unstructured statements such as `goto`, `break`, which cannot be modeled in PAMELA. In order to enable automatic PAMELA model generation, the following program annotations are supported: the `lower` and `upper` bound pragmas (when loop bounds cannot be symbolically determined at compile-time), the `cond` pragma (for data-dependent branch conditions), and the `cost` pragma (for assigning an entire, symbolic cost model for, e.g., some complicated sequential subsection).

A particular feature of the automatic cost estimator is the approach taken to modeling program parallelism. Instead of modeling the generated SPMD message-passing code, the modeling is based on the source code which is still expressed in terms of the original data-parallel programming model. Despite the fact that a number of low-level compiler-generated code features are therefore beyond the modeling scope, this high-level approach to modeling is essential to modeling correctness [9]. As a simple modeling example, let the vector `V` be cyclically partitioned over P processors. A (pseudo code) statement

```
forall (i = 1 .. N) V[i] = .. * ..;
```

will generate (if the compiler would use a simple owner-computes rule)

```
par (i = 1, N) { ... ; ... ; mult(i mod P) ; ... }
```

The PAMELA machine model includes a model for `mult` according to

```
resource cpu(p) = fcfs(p,1)
...
mult(p) = use(cpu(p),t_mult)
...
```

which models multiplication workload being charged to processor (index) `p`.

4 Experimental Results

In the following we apply the automatic cost estimator to four test codes, i.e., MATMUL (Matrix Multiplication), ADI (Alternate Implicit Integration), GAUSS (Gaussian

Elimination), and PSRS (Parallel Sorting by Regular Sampling). The actual application performance is measured on a 64 nodes partition of the DAS distributed-memory machine [3], of which a PAMELA machine model has been derived, based on simple computation and communication microbenchmarks [9]. In these microbenchmarks we measure local and global vector load and store operations at the Spar/Java level, while varying the access stride to account for cache effects. The first three regular applications did not require any annotation effort, while PSRS required 6 annotations.

The MATMUL experiment demonstrates the consistency of the prediction model for various N and P. MATMUL computes the product of $N \times N$ matrices A and B, yielding C. A is block-partitioned on the i axis, while B and C are block-partitioned on the j-axis. In order to minimize communication, the row of A involved in the computation of the row of C is assigned to a replicated vector (i.e., broadcast). The results for $N =$ 256, 512, and 1,024 are shown in Figure 1. The prediction error is 5 % on average with a maximum of 7 %.

The ADI (horizontal phase) speedup prediction for a $1,024 \times 1,024$ matrix, shown in Figure 2, clearly distinguishes between the block partitioning on the j-axis (vertical) and the i-axis (horizontal). The prediction error of the vertical version for large P is caused by the fact that the PAMELA model generated by the compiler does not account for the loop overhead caused by the SPMD level processor ownership tests. The maximum prediction error is therefore 77 % but must be attributed to the current Modeling Engine, rather than the PAMELA method. The average prediction error is 15 %.

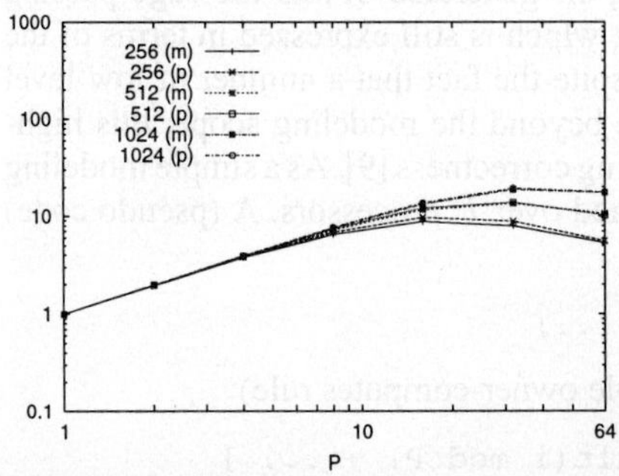

Fig. 1. MATMUL execution time [s] ($N = 256$, 512, and 1,024)

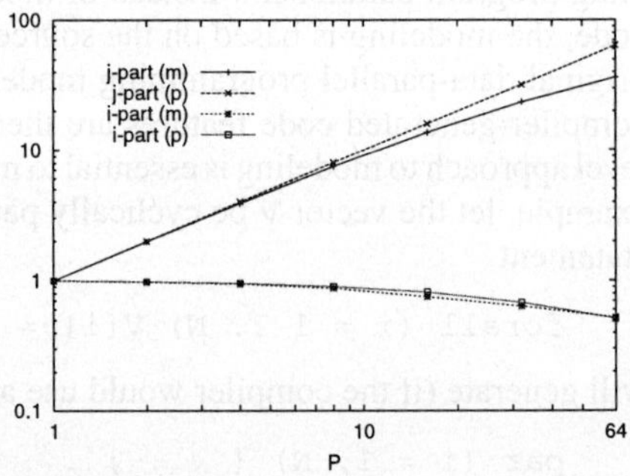

Fig. 2. ADI speedup (j and i-axis data partitioning)

The GAUSS application illustrates the use of the PAMELA model in predicting the difference between cyclic and block partitioning. The 512×512 matrix is partitioned on the j-axis. The submatrix update is coded in terms of a j loop, nested within an i loop, minimizing cache misses by keeping the matrix access stride as small as possible. The speedup predictions in Figure 3 clearly confirm the superior performance of block partitioning. For cyclic partitioning the access stride increases with P which causes delayed speedup due to increasing cache misses. The prediction error for large P is caused

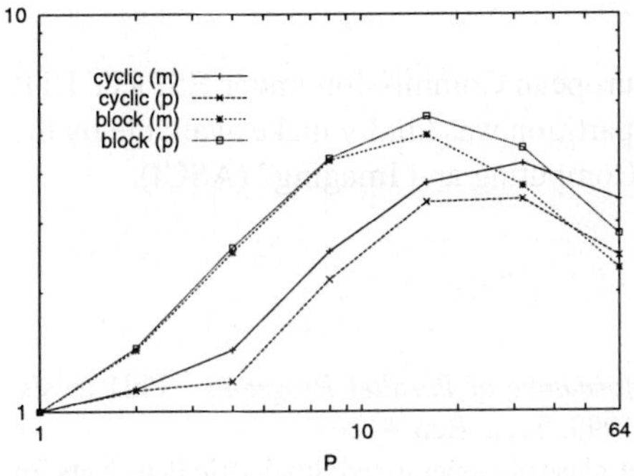

Fig. 3. GAUSS speedup (cyclic and block mapping)

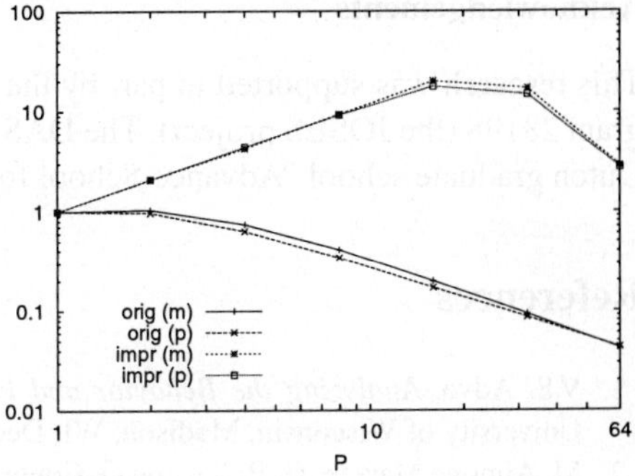

Fig. 4. PSRS speedup (original and improved data mapping)

by the fact that individual broadcasts partially overlap due to the use of asynchronous communication, which is not modeled by our PAMELA machine model. The prediction error is 13 % on average with a maximum of 35 %.

The PSRS application sorts a vector X of N elements into a result vector Y. The vectors X and Y are block-partitioned. Each X partition is sorted in parallel. Using a global set of pivots X is repartitioned into Y, after which each Y partition is sorted in parallel. Figure 4 shows the prediction results for $N = 819,200$ for two different data mapping strategies. Due to the dynamic, data-dependent nature of the PSRS algorithm, six simple loop and branching annotations were necessary. Most notably, the Quicksort procedure that is executed on each processor in parallel, required a few sequential profiling runs in order to enable modeling by the Modeling Engine. In the original program all arrays except X and Y are replicated (i.e., pivot vector and various index vectors). This causes a severe $\mathcal{O}(NP)$ communication bottleneck. In the improved program version this problem is solved by introducing a new index vector that is also partitioned. The prediction error is 12 % on average with a maximum of 26 %.

5 Conclusion

In this paper we present a tool that automatically compiles process-oriented performance simulation models (PAMELA models) into symbolic cost models that are symbolically simplified to achieve extremely low evaluation cost. As the simulation models are intuitively close to the parallel program and machine under study, the complex and error-prone effort of deriving symbolic cost models is significantly reduced. The PAMELA compiler is also used within a symbolic cost estimator for data-parallel programs. With minimal program annotation by the user, symbolic cost models are automatically generated in a matter of seconds, while the evaluation time of the models ranges in the milliseconds. For instance, the 300 s execution time of the initial PSRS code for 64 processors on the real parallel machine is predicted in less than 2 ms, whereas simulation would have taken over 32,000 s. Experimental results on four data-parallel programs show that the average error of the cost models is less than 15 %. Apart from providing a good scalability assessment, the best design choice is correctly predicted in all cases.

Acknowledgements

This research was supported in part by the European Commission under ESPRIT LTR grant 28198 (the JOSES project). The DAS I partition was kindly make available by the Dutch graduate school "Advance School for Computing and Imaging" (ASCI).

References

1. V.S. Adve, *Analyzing the Behavior and Performance of Parallel Programs.* PhD thesis, University of Wisconsin, Madison, WI, Dec. 1993. Tech. Rep. #1201.
2. M. Ajmone Marsan, G. Balbo and G. Conte, "A class of Generalized Stochastic Petri Nets for the performance analysis of multiprocessor systems," *ACM TrCS*, vol. 2, 1984, pp. 93–122.
3. H. Bal *et al.*, "The distributed ASCI supercomputer project," *Operating Systems Review*, vol. 34, Oct. 2000, pp. 76–96.
4. D. Culler, R. Karp, D. Patterson, A. Sahay, K.E. Schauser, E. Santos, R. Subramonian and T. von Eicken, "LogP: Towards a realistic model of parallel computation," in *Proc. 4th ACM SIGPLAN Symposium on PPoPP*, May 1993, pp. 1–12.
5. T. Fahringer, "Estimating and optimizing performance for parallel programs," *IEEE Computer*, Nov. 1995, pp. 47–56.
6. H. Gautama and A.J.C. van Gemund, "Static performance prediction of data-dependent programs," in *ACM Proc. on The Second International Workshop on Software and Performance (WOSP'00)*, Ottawa, ACM, Sept. 2000, pp. 216–226.
7. A.J.C. van Gemund, "Performance prediction of parallel processing systems: The PAMELA methodology," in *Proc. 7th ACM Int'l Conf. on Supercomputing*, Tokyo, 1993, pp. 318–327.
8. A.J.C. van Gemund, *Performance Modeling of Parallel Systems.* PhD thesis, Delft University of Technology, The Netherlands, Apr. 1996.
9. A.J.C. van Gemund, "Automatic cost estimation of data parallel programs," Tech. Rep. 1-68340-44(2001)09, Delft University of Technology, The Netherlands, Oct. 2001.
10. N. Götz, U. Herzog and M. Rettelbach, "Multiprocessor and distributed system design: The integration of functional specification and performance analysis using stochastic process algebras," in *Proc. SIGMETRICS/PERFORMANCE'93*, LNCS 729, Springer, 1993.
11. H. Jonkers, A.J.C. van Gemund and G.L. Reijns, "A probabilistic approach to parallel system performance modelling," in *Proc. 28th HICSS, Vol. II*, IEEE, Jan. 1995, pp. 412–421.
12. C.L. Mendes and D.A. Reed, "Integrated compilation and scalability analysis for parallel systems," in *Proc. PACT '98*, Paris, Oct. 1998, pp. 385–392.
13. H. Schwetman, "Object-oriented simulation modeling with C++/CSIM17," in *Proc. 1995 Winter Simulation Conference*, 1995.
14. L. Valiant, "A bridging model for parallel computation," *CACM*, vol. 33, 1990, pp. 103–111.
15. C. van Reeuwijk, A.J.C. van Gemund and H.J. Sips, "Spar: A programming language for semi-automatic compilation of parallel programs," *Concurrency: Practice and Experience*, vol. 9, Nov. 1997, pp. 1193–1205.
16. K-Y. Wang, "Precise compile-time performance prediction for superscalar-based computers," in *Proc. ACM SIGPLAN PLDI'94*, Orlando, June 1994, pp. 73–84.

Performance Modeling and Interpretive Simulation of PIM Architectures and Applications

Zachary K. Baker and Viktor K. Prasanna

University of Southern California, Los Angeles, CA USA
zbaker@halcyon.usc.edu, prasanna@ganges.usc.edu
http://advisor.usc.edu

Abstract. Processing-in-Memory systems that combine processing power and system memory chips present unique algorithmic challenges in the search for optimal system efficiency. This paper presents a tool which allows algorithm designers to quickly understand the performance of their application on a parameterized, highly configurable PIM system model. This tool is not a cycle-accurate simulator, which can take days to run, but a fast and flexible performance estimation tool. Some of the results from our performance analysis of 2-D FFT and biConjugate gradient are shown, and possible ways of using the tool to improve the effectiveness of PIM applications and architectures are given.

1 Introduction

The von Neumann bottleneck is a central problem in computer architecture today. Instructions and data must enter the processing core before execution can proceed, but memory and data bus speeds are many times slower than the data requirements of the processor. Processing-In-Memory (PIM) systems propose to solve this problem by achieving tremendous memory-processor bandwidth by combining processors and memory together on the same chip substrate. Notre Dame, USC ISI, Berkeley, IBM, and others are developing PIM systems and have presented papers demonstrating the performance and optimization of several benchmarks on their architectures. While excellent for design verification, the proprietary nature and the time required to run their simulators are the biggest detractors of their tools for application optimization. A cycle-accurate, architecture-specific simulator, requiring several hours to run, is not suitable for iterative development or experiments on novel ideas. We provide a simulator which will allow faster development cycles and a better understanding of how an application will port to other PIM architectures [4, 7]. For more details and further results, see [2].

[1] Supported by the US DARPA Data Intensive Systems Program under contract F33615-99-1-1483 monitored by Wright Patterson Airforce Base and in part by an equipment grant from Intel Corporation. The PIM Simulator is available for download at http://advisor.usc.edu

B. Monien and R. Feldmann (Eds.): Euro-Par 2002, LNCS 2400, pp. 157–161.

2 The Simulator

The simulator is a wrapper around a set of models. It is written in Perl, because the language's powerful run-time interpreter allows us to easily define complex models. The simulator is modular; external libraries, visualization routines, or other simulators can be added as needed. The simulator is composed of various interacting components. The most important component is the data flow model, which keeps track of the application data as it flows through the host and the PIM nodes. We assume a host with a separate, large memory. Note that as the PIM nodes make up the main memory of the host system in some PIM implementations. The host can send and receive data in a unicast or multicast fashion, either over a bus or a non-contending, high-bandwidth, switched network. The bus is modeled as a single datapath with parameterized bus width, startup time and per element transmission time. Transmissions over the network are assumed to be scheduled by the application to handle potential collisions. The switched network is also modeled with the same parameters but with collisions defined as whenever any given node attempts to communicate with more than one other node(or host), except where multicast is allowed. Again, the application is responsible for managing the scheduling of data transmission. Communication can be modeled as a stream or as packets.

Computation time can be modeled at an algorithmic level, e.g. $n \lg(n)$ based on application parameters, or in terms of basic arithmetic operations. The accuracy of the computation time is dependent entirely on the application model used. We assume that the simulator will be commonly used to model kernel operations such as benchmarks and stressmarks, where the computation is well understood, and can be distilled into a few expressions. This assumption allows us to avoid the more complex issues of the PIM processor design and focus more on the interactions of the system as a whole.

3 Performance Results

3.1 Conjugate Gradient Results

Figure 1 shows the overall speedup of the biConjugate Gradient stressmark with respect to the number of active PIM elements. It compares results produced by our tool using a DIVA parameterized architecture to the cycle-accurate simulation results in [4]. Time is normalized to a simulator standard. The label of our results, "Overlap 0.8", denotes that 80% of the data transfer time is hidden underneath the computation time, via prefetching or other latency hiding techniques. The concept of overlap is discussed later in this paper.

BiConjugate Gradient is a DARPA DIS stressmark [1]. It is used in matrix arithmetic to find the solution of $y = Ax$, given y and A. The complex matrices in question tend to be sparse, which makes the representation and manipulation of data significantly different than in regular data layout of FFT. The application model uses a compressed sparse row matrix representation of A, and load balances based on the number of elements filling a row. This assumes that the number of rows is significantly higher than the number of processors. All PIM nodes are sent the vector y and can thus execute on their sparse elements independently of the other PIM nodes.

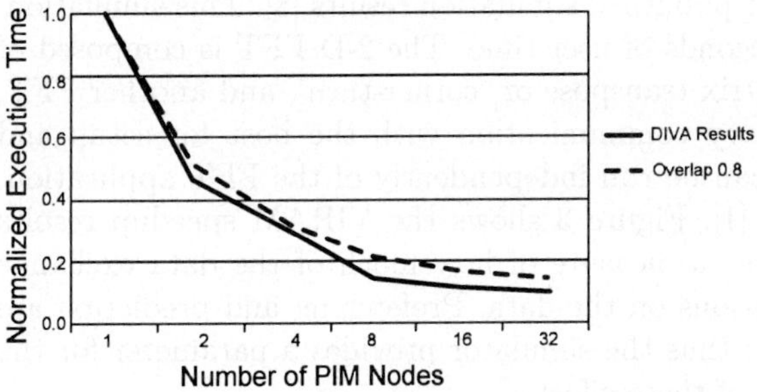

Fig. 1. Speedup from one processor to n processors with DIVA model

Figure 2 is a graph of the simulator output for a BiCG application with parameters similar to that of the DIVA architecture with a parallel, non-contending network model, application parameters of n(row/column size of the matrix)=14000 and nz(non zero ele- ments)=14 elements/row. Figure 2(left) shows the PIM-to-PIM transfer cost, Host-to-PIM transfer costs, computation time, and total execution time(total) as the number of PIM nodes increases under a DIVA model. The complete simulation required 0.21 seconds of user time on a Sun Ultra250 with 1024 MB of memory.

The graph shows that the computation time decreases linearly with the number of PIM nodes, and the data transfer time increases non-linearly. We see in the graph that PIM-to-PIM transfer time is constant– this is because the number of PIM nodes in the system does not dramatically affect the amount of data (a vector of size n in each iteration) sent by the BiCG model. Host-to-PIM communication increases logarithmically with number of PIM; the model is dependent mostly on initial setup of the matrices and final collection of the solution vectors. The Host-to-PIM communication increases toward the end as the communications setup time for each PIM becomes non-negligible compared to the total data transferred. Figure 2(right) shows a rescaled version of the total execution time for the same parameters. Here, the optimal number of PIM under the BiCG model and architectural parameters is clear– this particular application seems suited to a machine of 64 to 128 PIM nodes most optimally in this architecture model.

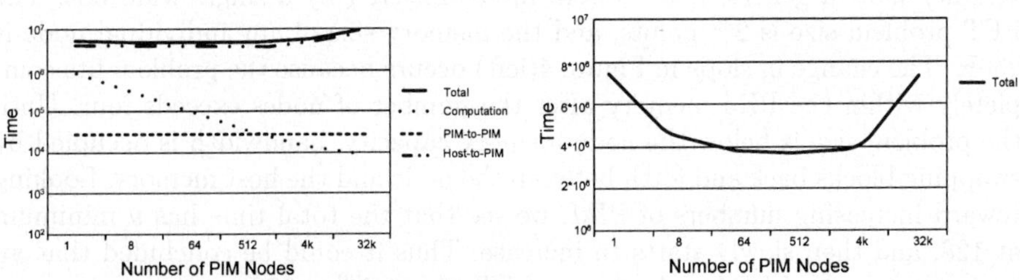

Fig. 2. BiConjugate Gradient Results; unit-less timings for various amounts of PIM nodes. (left: all results, right: total execution time only)

3.2 FFT

Another stressmark modeled is the 2-D FFT. Figure 3 shows execution time versus the number of FFT points for the Berkeley VIRAM architecture, comparing our results against their published simulation results [8]. This simulation, for all points, required 0.22 seconds of user time. The 2-D FFT is composed of a one dimensional FFT, a matrix transpose or 'corner-turn', and another FFT, preceded and followed by heavy communication with the host for setup and cleanup. Corner turn, which can be run independently of the FFT application, is a DARPA DIS stressmark [1]. Figure 3 shows the VIRAM speedup results against various overlap factors– a measure of how much of the data exchange can overlap with actual operations on the data. Prefetching and prediction are highly architecture dependent; thus the simulator provides a parameter for the user to specify the magnitude of these effects.

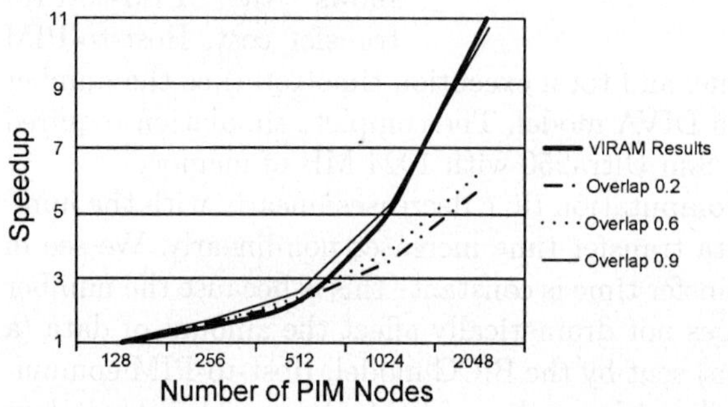

Fig. 3. Speedup versus number of FFT Points for various fetch overlaps, normalized to 128 points.

In the graph we see that the VIRAM results match most closely with an overlap of 0.9; that is, virtually all of the data transfer is hidden by overlapping with the computation time. This 'overlap' method is similar to the 'clock multiplier factor N' used by Rsim in that it depends on the application and the system and cannot to determined without experimentation [5].

Inspecting the VIRAM architecture documentation, we see that it includes a vector pipeline explicitly to hide the DRAM latency [6]. Thus our simulation results suggest the objective of the design has been achieved.

The simulator can be used to understand the performance of a PIM system under varying application parameters, and the architecture's effect on optimizing those parameters. A graph of the simulator output in Figure 4(left) and 4(right) show a generic PIM system interconnected by a single wide bus. The FFT problem size is 2^{20} points, and the memory size of any individual node is 256K. The change in slope in Figure 4(left) occurs because the problem fits completely within the PIM memory after the number of nodes exceeds four. Until the problem size is below the node memory capacity, bandwidth is occupied by swapping blocks back and forth between the node and the host memory. Looking toward increasing numbers of PIM, we see that the total time has a minimum at 128, and then slowly starts to increase. Thus it could be concluded that an optimal amount of PIM nodes for an FFT of size 2^{20} is 128.

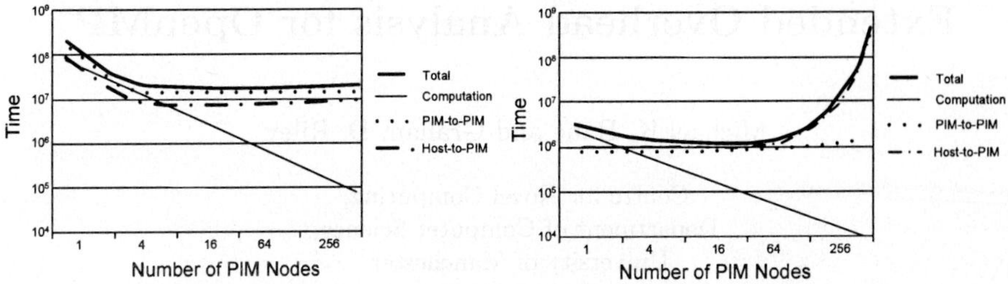

Fig. 4. 2-D FFT Results (left: Small memory size, right: Small problem size)

4 Conclusions

In this paper we have presented a tool for high-level modeling of Processing-In-Memory systems and its uses in optimization and evaluation of algorithms and architectures. We have focused on the use of the tool for algorithm optimization, and in the process have given validation of the simulator's models of DIVA and VIRAM. We have given a sketch of the hardware abstraction, and some of the modeling choices made to provide an easier-to-use system. We have shown some of the application space we have modeled, and presented validation for those models against simulation data from real systems, namely DIVA from USC ISI and VIRAM from Berkeley.

This work is part of the Algorithms for Data IntensiVe Applications on Intelligent and Smart MemORies (ADVISOR) Project at USC [3]. In this project we focus on developing algorithmic design techniques for mapping applications to architectures. Through this we understand and create a framework for application developers to exploit features of advanced architectures to achieve high performance.

References

1. Titan Systems Corporation Atlantic Aerospace Division. DIS Stressmark Suite. http://www.aaec.com/projectweb/dis/, 2000.
2. Z. Baker and V.K. Prasanna. Technical report: Performance Modeling and Interpretive Simulation of PIM Architectures and Applications. In preparation.
3. V.K. Prasanna et al. ADVISOR project website. http://advisor.usc.edu.
4. M. Hall, P. Kogge, J. Koller, P. Diniz, J. Chame, J. Draper, J. LaCoss, J. Granacki, A. Srivastava, W. Athas, J. Brockman, V. Freeh, J. Park, and J. Shin. Mapping Irregular Applications to DIVA, a PIM-based Data-Intensive Architecture. In *SC99*.
5. C.J. Hughes, V.S. Pai, P. Ranganathan, and S.V. Adve. Rsim: Simulating Shared-Memory Muliprocessors with ILP Processors, Feb 2002.
6. Christoforos Kozyrakis. A Media-Enhanced Vector Architecture for Embedded Memory Systems Technical Report UCB//CSD-99- 1059, July 1999.
7. D. Patterson, T. Anderson, N. Cardwell, R. Fromm, K. Keeton, C. Kozyrakis, R. Thomas, and K. Yelick. A Case for Intelligent RAM: IRAM, 1997.
8. Randi Thomas. An Architectural Performance Study of the Fast Fourier Transform on Vector IRAM. Master's thesis, University of California, Berkeley, 2000.

Extended Overhead Analysis for OpenMP

Michael K. Bane and Graham D. Riley

Centre for Novel Computing,
Department of Computer Science,
University of Manchester,
Oxford Road,
Manchester, UK
{bane, griley}@cs.man.ac.uk

Abstract. In this paper we extend current models of overhead analysis to include complex OpenMP structures, leading to clearer and more appropriate definitions.

1 Introduction

Overhead analysis is a methodology used to compare achieved parallel performance to the ideal parallel performance of a reference (usually sequential) code. It can be considered as an extended view of Amdahl's Law [1]:

$$T_p = \frac{T_s}{p} + \frac{p-1}{p}(1-\alpha)\,T_s \tag{1}$$

where T_s and T_p are the times spent by a serial and parallel implementation of a given algorithm on p threads, and α is a measure of the fraction of parallelized code. The first term is the time for an ideal parallel implementation. The second term can be considered as an overhead due to unparallelized code, degrading the performance. However, other factors affect performance, such as the implementation of the parallel code and the effect of different data access patterns. We therefore consider (1) to be a specific form of

$$T_p = \frac{T_s}{p} + \sum_i O_i \tag{2}$$

where each O_i is an overhead. Much work has been done on the classification and practical use of overheads of parallel programs eg ([2], [3], [4], [5]).

A hierarchical breakdown of temporal overheads is given in [3]. The top level overheads are information movement, critical path, parallelism management, and additional computation. The critical path overheads are due to imperfect parallelization. Typical components will be load imbalance, replicated work and insufficient parallelism such as unparallelized or partially parallelized code. We extend the breakdown of overheads with an "unidentified overheads" category that includes those overheads that have not yet been, or cannot be, determined during the analysis of a particular experiment. It is possible for an overhead

B. Monien and R. Feldmann (Eds.): Euro-Par 2002, LNCS 2400, pp. 162–166.

to be negative and thus relate to an improvement in the parallel performance. For example, for a parallel implementation the data may fit into a processor's memory cache whereas it does not for the serial implementation. In such a case, the overhead due to data accesses would be negative.

The practical process of quantifying overheads is typically a refinement process. The main point is not to obtain high accuracy for all categories of overheads, but to optimize the parallel implementation.

Overhead analysis may be applied to the whole code or to a particular region of interest.

2 Overhead Analysis Applied to OpenMP

This paper argues that the current formalization of overhead analysis as applied to OpenMP [6] is overly simplistic, and suggests an improved scheme.

Consider two simple examples to illustrate the definition and measurement of an overhead. A simple `OMP PARALLEL DO` loop may lead to load imbalance overhead, defined as the difference between the time taken by the slowest thread and the average thread time. The definition of the Load Imbalance overhead in [3] is given as

Load imbalance: time spent waiting at a synchronization point, because, although there are sufficient parallel tasks, they are asymmetric in the time taken to execute them.

We now turn our attention to the simplest case of unparallelized code overhead, where only one thread executes code in a given parallel region – for example, an `OMP PARALLEL` construct consisting solely of an `OMP SINGLE` construct. From [3] we have the following definitions:

Insufficient parallelism: processors are idle because there is an insufficient number of parallel tasks available for execution at that stage of the program; with subdivisions:

Unparallelized code: time spent waiting in sections of code where there is a single task, run on a single processor;

Partially parallelized code: time spent waiting in sections of code where there is more than one task, but not enough to keep all processors active.

For the above examples we have a synchronization point at the start and end of the region of interest, and only one construct within the region of interest. However, analysis of such examples is of limited use. OpenMP allows the creation of a parallel region in which there can be a variety of OpenMP constructs as well as replicated code that is executed by the team of threads [1]. The number of threads executing may also depend on program flow; in particular when control is determined by reference to the value of the function `OMP_GET_THREAD_NUM`,

[1] OpenMP allows for differing numbers of threads for different parallel regions, either determined by the system or explicitly by the user. In this paper, we assume that there are p threads running for each and every parallel region. Cases where there a different number of threads for a parallel region is beyond the scope of this introductory paper.

which returns the thread number. Various OpenMP constructs can also *not* have an implicit barrier at the exit point (for example, `OMP END DO NOWAIT`). Thus a given OpenMP parallel region can be quite sophisticated leading to several different overheads within a region which may interfere constructively or destructively. The remainder of this paper discusses appropriate overhead analysis for non-trivial OpenMP programs.

Let us now consider an OpenMP parallel region consisting of a `SINGLE` region followed by a distributed `DO` loop:

```
C$OMP PARALLEL PRIVATE(I)
C$OMP SINGLE
      CALL SINGLE_WORK()
C$OMP END SINGLE NOWAIT
C$OMP DO SCHEDULE(DYNAMIC)
      DO I=1, N
         CALL DO_WORK()
      END DO
C$OMP END DO
C$OMP END PARALLEL
```

Since the `SINGLE` region does not have a barrier at the exit point, those threads not executing `SINGLE_WORK()` will start `DO_WORK()` immediately. We could therefore have a situation shown in Figure 1, where the double line represents the time spent in `SINGLE_WORK()`, the single line the time spent in `DO_WORK` and the dashed line being thread idle time. One interpretation of the above

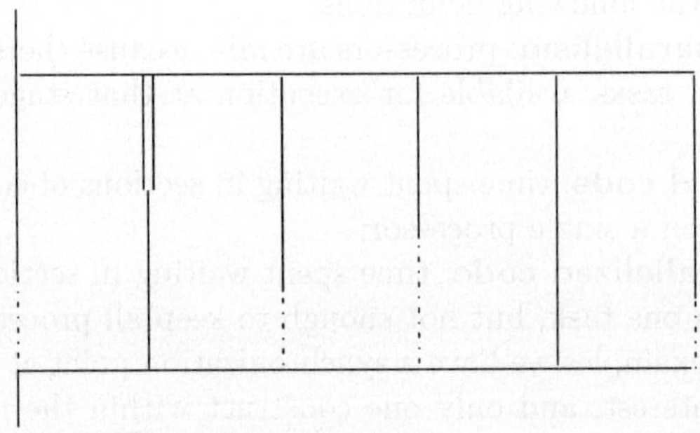

Fig. 1. Time Graph for Complex Example #1

definitions would be that this example has an element of unparallelized code overhead.

Depending upon the amount of time it takes to perform `SINGLE_WORK()` it is possible to achieve ideal speed up for such an example, despite a proportion of code being executed on only one thread, which would normally imply unparallelized code overhead.

Assume the time spent on one thread is t_{sing} for `SINGLE_WORK()`and t_{do} for `DO_WORK()` then for this region the sequential time $T_s = t_{sing} + t_{do}$ and the ideal time on p threads is thus $T_{ideal} = \frac{T_s}{p} = \frac{t_{sing}+t_{do}}{p}$. During the time that one thread has spent in the `SINGLE` region a total of $(p-1)\,t_{sing}$ seconds have been allocated to `DO_WORK()`. There is therefore $t_{do} - (p-1)\,t_{sing}$ seconds worth of work left to do, now over p threads. So, the actual time taken is

$$T_p = t_{sing} + \max\left(0, \frac{t_{do} - (p-1)\,t_{sing}}{p}\right) \tag{3}$$

Thus either the work in the `SINGLE` region dominates (all the other threads finish first), or there is sufficient work for those threads executing `DO_WORK()` compared to `SINGLE_WORK()` in which case (3) reduces to $T_p = T_{ideal}$. That is, we may achieve a perfect parallel implementation despite the presence of a `SINGLE` region; perfection is not guaranteed, depending on the size of the work quanta in `DO_WORK`.

Therefore, we can see that the determination of overheads needs to take into account interactions between OpenMP constructs in the region in question.

Consider a slight variation to the above case, where an OpenMP parallel region contains just an `OMP SINGLE` construct and an `OMP DO` loop without an exit barrier (ie `OMP END DO NOWAIT` is present). As long as the work is independent, we can write such a code in two different orders, one with the `SINGLE` construct followed by the `DO` loop and the other in the opposite order.

At first glance, one might be tempted to define the overheads in terms of that OpenMP construct which leads to lost cycles immediately before the final synchronization point. Thus overhead in the first case would be mainly load imbalance with an unparallelized overhead contribution, and in the second case, mainly unparallelized overhead with a load imbalance overhead contribution.

Given such "commutability" of overheads, together with the previous examples, it is obvious we need a clearer definition of overheads.

3 An Improved Schema

We now give a new, extended schema for defining overheads for real life OpenMP programs where we assume that the run time environment allocates the requested number of threads, p, for each and every parallel region.

1. Overheads can be defined only between two synchronization points. Overheads for a larger region will be the sum of overheads between each consecutive pair of synchronization points in that region.
2. Overheads exist only if the time taken between two synchronization points by the parallel implementation on p threads, T_p, is greater than the ideal time, T_{ideal}.
3. Unparallelized overhead is the time spent between two consecutive synchronization points of the parallel implementation when only one thread is executing.

4. Partially parallelized overhead is the time spent between two synchronization points when the number of threads being used throughout this region, p', is given by $1 < p' < p$. This would occur, for example, in an `OMP PARALLEL SECTIONS` construct where there are less `SECTION`s than threads.
5. Replicated work overhead occurs between two synchronization points when members of the thread team are executing the same instructions on the same data in the same order.
6. Load imbalance overhead is the time spent waiting at the exit synchronization point when the same number of threads, $p'' > 1$, execute code between the synchronization points, irrespective of the cause(s) of the imbalance. In the case $p'' < p$, we can compute load imbalance overhead with respect to p'' threads and partially parallelized overhead with respect to $p - p''$ threads.

In computing overheads for a synchronization region, point (2) should be considered first. That is, if there is ideal speed up, there is no need to compute other overheads – ideal speed up being the "goal". There may, of course, by some negative overheads which balance the positive overheads but this situation is tolerated because the speed up is acceptable.

4 Conclusions and Future Work

In this paper we have outlined an extension to the current analysis of overheads, as applied to OpenMP. Our future work will involve expanding the prototype Ovaltine [5] tool to include these extensions, and an in-depth consideration of cases where different parallel regions have different numbers of threads, either as a result of dynamic scheduling or at the request of the programmer.

References

1. G.M. Amdahl, *Validity of the Single Processor Approach to Achieving Large Scale Computing Capabilities*, AFIPS Conference Proceedings, vol. 30, AFIPS Press, pp. 483-485, 1967.
2. M.E. Crovella and T.J. LeBlanc, Parallel Performance Prediction Using Lost Cycles Analysis, Proceedings of Supercomputing '94, IEEE Computer Society, pp. 600-609, November 1994.
3. J.M. Bull. *A Hierarchical Classification of Overheads in Parallel Programs*, Proceedings of First IFIP TC10 International Workshop on Software Engineering for Parallel and Distributed Systems, I. Jelly, I. Gorton and P. Croll (Ed.s), Chapman Hall, pp. 208-219, March 1996.
4. G.D. Riley, J.M. Bull and J.R. Gurd, *Performance Improvement Through Overhead Analysis: A Case Study in Molecular Dynamics*, Proc. 11th ACM International Conference on Supercomputing, ACM Press, pp. 36-43, July 1997.
5. M.K. Bane and G.D. Riley, *Automatic Overheads Profiler for OpenMP Codes*, Proceedings of the Second European Workshop on OpenMP (EWOMP2000), September 2000.
6. http://www.openmp.org/specs/

CATCH – A Call-Graph Based Automatic Tool for Capture of Hardware Performance Metrics for MPI and OpenMP Applications

Luiz DeRose[1] and Felix Wolf[2,*]

[1] Advanced Computing Technology Center
IBM Research
Yorktown Heights, NY 10598 USA laderose@us.ibm.com
[2] Research Centre Juelich
ZAM
Juelich, Germany
f.wolf@fz-juelich.de

Abstract. CATCH is a profiler for parallel applications that collects hardware performance counters information for each function called in the program, based on the path that led to the function invocation. It automatically instruments the binary of the target application independently of the programming language. It supports MPI, OpenMP, and hybrid applications and integrates the performance data collected for different processes and threads. Functions representing the bodies of OpenMP constructs are also monitored and mapped back to the source code. Performance data is generated in XML for visualization with a graphical user interface that displays the data simultaneously with the source code sections they refer to.

1 Introduction

Developing applications that achieve high performance on current parallel and distributed systems requires multiple iterations of performance analysis and program refinements. Traditional performance tools, such as SvPablo [7], TAU [11], Medea [3], and AIMS [14], rely on experimental performance analysis, where the application is instrumented for data capture, and the collected data is analyzed after the program execution. In each cycle developers instrument application and system software, in order to identify the key program components responsible for the bulk of the program's execution time. Then, they analyze the captured performance data and modify the program to improve its performance. This optimization model requires developers and performance analysts to engage in a laborious cycle of instrumentation, program execution, and code modification, which can be very frustrating, particularly when the number of possible

* This work was performed while Felix Wolf was visiting the Advanced Computing Technology Center at IBM Research.

B. Monien and R. Feldmann (Eds.): Euro-Par 2002, LNCS 2400, pp. 167–176.

optimization points is large. In addition, static instrumentation can inhibit compiler optimizations, and when inserted manually, could require an unreasonable amount of the developer's time.

Moreover, most users do not have the time or desire to learn how to use complex tools. Therefore, a performance analysis tool should be able to provide the data and insights needed to tune and optimize applications with a simple to use interface, which does not create additional burden to the developers. For example, a simple tool like the GNU gprof [9] can provide information on how much time a serial program spent in which function. This "flat profile" is refined with a call-graph profiler, which tells the time separately for each caller and also the fraction of the execution time that was spent in each of the callees. This call-graph information is very valuable, because it not only indicates the functions that consume most of the executions time, but also identifies in which context it happened. However, a high execution time does not necessarily indicate inefficient behavior, since even an efficient computation can take a long time. Moreover, as computer architectures become more complex, with clustered symmetric multiprocessors (SMPs), deep-memory hierarchies managed by distributed cache coherence protocols, and speculative execution, application developers face new and more complicate performance tuning and optimization problems. In order to understand the execution behavior of application code in such complex environments, users need performance tools that are able to support the main parallel programming paradigms, as well as, access hardware performance counters and map the resulting data to the parallel source code constructs. However, the most common instrumentation approach that provides access to hardware performance counters also augments source code with calls to specific instrumentation libraries (e.g., PAPI [1], PCL [13], SvPablo [7] and the HPM Toolkit [5]). This static instrumentation approach lacks flexibility, since it requires re-instrumentation and recompilation, whenever a new set of instrumentation is required.

In this paper we present CATCH (Call-graph-based Automatic Tool for Capture of Hardware-performance-metrics), a profiler for MPI and OpenMP applications that provides hardware performance counters information related to each path used to reach a node in the application's call graph. CATCH automatically instruments the binary of the target application, allowing it to track the current call-graph node at run time with only constant overhead, independently of the actual call-graph size. The advantage of this approach lies in its ability to map a variety of expressive performance metrics provided by hardware counters not only to the source code but also to the execution context represented by the complete call path. In addition, since it relies only on the binary, CATCH is programming language independent.

CATCH is built on top of DPCL [6], an object-based C++ class library and run-time infrastructure, developed by IBM, which is based on the Paradyn [10] dynamic instrumentation technology, from the University of Wisconsin. DPCL flexibly supports the generation of arbitrary instrumentation, without requiring access to the source code. We refer to [6] for a more detailed description of DPCL

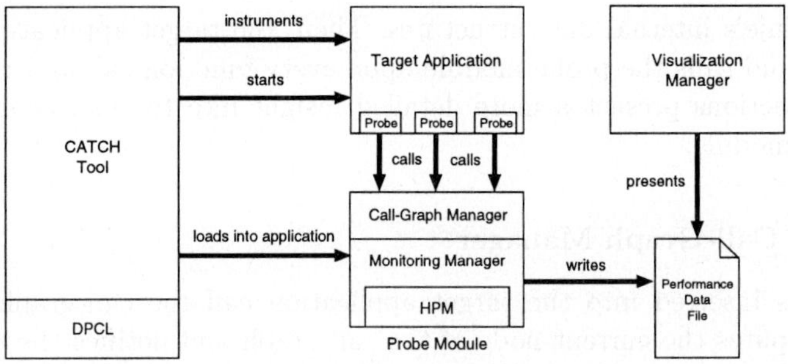

Fig. 1. Overall architecture of CATCH.

and its functionality. CATCH profiles the execution of MPI, OpenMP, and hybrid application and integrates the performance data collected for different processes and threads. In addition, based on the information provided by the native AIX compiler, CATCH is able to identify the functions the compiler generates from OpenMP constructs and to link performance data collected from these constructs back to the source code. To demonstrate the portability of our approach, we additionally implemented a Linux version, which is built on top of Dyninst [2].

The remainder of this article is organized as follows: Section 2 contains a description of the different components of CATCH and how they are related to each other. Section 3 presents a detailed explanation of how CATCH tracks a call-graph node at run time. Section 4 discusses related work. Finally, Section 5 presents our conclusions and plans for future work.

2 Overall Architecture

As illustrated in Figure 1, CATCH is composed of the CATCH *tool*, which instruments the target application and controls its execution, and the CATCH *probe module*, which is loaded into the target application by CATCH to perform the actual profiling task. The probe module itself consists of the *call-graph manager* and the *monitoring manager*. The former is responsible for calculating the current call-graph position, while the latter is responsible for monitoring the hardware performance counters. After the target application finishes its execution, the monitoring manager writes the collected data into an XML file, whose contents can be displayed using the *visualization manager*, a component of the HPM Toolkit, presented in [5].

When CATCH is invoked, it first creates one or more processes of the target application in suspended state. Next, it computes the static call graph and performs the necessary instrumentation by inserting calls to probe-module functions into the memory image of the target application. Finally, CATCH writes the call graph into a temporary file and starts the target application.

Before entering the main function, the instrumented target application first initializes the probe-module, which reads in the call-graph file and builds up the probe module's internal data structures. Then, the target application resumes execution and calls the probe module upon every function call and return. The following sections present a more detailed insight into the two components of the probe module.

2.1 The Call-Graph Manager

The probes inserted into the target application call the call-graph manager, which computes the current node of the call graph and notifies the monitoring manager of the occurrence of the following events:

Initialization. The application will start. The call graph and the number of threads are provided as parameters. The call graph contains all necessary source-code information on modules, functions, and function-call sites.

Termination. The application terminated.

Function Call. The application will execute a function call. The current call-graph node and the thread identifier are provided as parameters.

Function Return. The application returned from a function call. The current call-graph node and the thread identifier are provided as parameters.

OpenMP Fork. The application will fork into multi-threaded execution.

OpenMP Join. Multi-threaded execution finished.

MPI Init. MPI will be initialized. The number of MPI processes and the process identifier are provided as parameters. When receiving this event, the monitoring manager knows that it can execute MPI statements. This event is useful, for example, to synchronize clocks for event tracing.

MPI Finalize. MPI will be finalized. It denotes the last point in time, where the monitoring manager is able to execute an MPI statement, for example, to collect the data gathered by different MPI processes.

Note that the parameterized events listed above define a very general profiling interface, which is not limited to profiling, but is also suitable for a multitude of alternative performance-analysis tasks (e.g., event tracing). The method of tracking the current node in the call graph is described in Section 3.

2.2 The Monitoring Manager

The monitoring manager is an extension of the HPM data collection system, presented in [5]. The manager uses the probes described above to activate the HPM library. Each node in the call graph corresponds to an application section that could be instrumented. During the execution of the program, the HPM library accumulates the performance information for each node, using tables with unique identifiers for fast access to the data structure that stores the information during run time. Thus, the unique identification of each node in the call graph, as described in Section 3, is crucial for the low overhead of the data collection system. The HPM library supports nested instrumentation and multiple calls to any node. When the program execution terminates, the HPM library reads and traverses the call graph to compute exclusive counts and durations for each node.

In addition, it computes a rich set of derived metrics, such as cache hit ratios and MFLOP/sec rates, that can be used by performance analysts to correlate the behavior of the application to one or more of the hardware components. Finally, it generates a set of *performance files*, one for each parallel task.

3 Call-Graph Based Profiling with Constant Overhead

In this section we describe CATCH's way of instrumenting an application, which provides the ability to calculate the current node in the call graph at run time by introducing only constant overhead independently of the actual call-graph size. Our goal is to be able to collect statistics for each function called in the program, based on the path that led to the function invocation. For simplicity, we first discuss serial non-recursive applications and later explain how we treat recursive and parallel ones.

3.1 Building a Static Call Graph

The basic idea behind our approach is to compute a static call graph of the target application in advance before executing it. This is accomplished by traversing the code structure using DPCL. We start from the notion that an application can be represented by a multigraph with functions represented as nodes and call sites represented as edges. If, for example, a function f calls function g from k different call sites, the correspondent transitions are represented with k arcs from node f to node g in the multigraph. A sequence of edges in the multigraph corresponds to a path.

The multigraph of non-recursive programs is acyclic. From the application's acyclic multigraph, we build a static call tree, which is a variation of the call graph, where each node is a simple path that start at the root of the multigraph. For a path $\pi = \sigma e$, where σ is a path and e is an edge in the multigraph, σ is the parent of π in the tree. We consider the root of the multigraph to be the function that calls the application's main function. This *start* function is assumed to have an empty path to itself, which is the root of the call tree.

3.2 Instrumenting the Application

The probe module holds a reference to the call tree, where each node contains an array of all of its children. Since the call sites within a function can be enumerated and the children of a node correspond to the call sites within the function that can be reached by following the path represented by that node, we arrange the children in a way that child i corresponds to call site i. Thus, child i of node n in the tree can be accessed directly by looking up the ith element of the array in node n.

In addition, the probe module maintains a pointer to the current node n_c, which is moved to the next node n_n upon every function call and return. For a function call made from a call site i, we assign:

$$n_n := child_i(n_c)$$

That is, the ith call site of the function currently being executed causes the application to enter the ith child node of the current node. For this reason, the probe module provides a function `call(int i)`, which causes the pointer to the current node to be moved to child i. In case of a function return, we assign:

$$n_n := parent(n_c)$$

That is, every return just causes the application to re-enter the parent node of the current node, which can be reached via a reference maintained by CATCH. For this reason, the probe module provides a function `return()`, which causes the pointer to the current node to be moved to its parent.

Since DPCL provides the ability to insert calls to functions of the probe module before and after a function-call site and to provide arguments to these calls, we only need for each function f to insert `call(i)` before a function call at call site i and to insert `return()` after it.

Because `call(int i)` needs only to look up the ith element of the children array, and `return()` needs only to follow the reference to the parent, calling these two functions introduces only constant execution-time overhead independently of the application's call-tree size.

3.3 Recursive Applications

Trying to build a call tree for recursive applications would result in a tree of infinite size. Hence, to be able to support recursive applications, CATCH builds a call graph that may contain loops instead. Every node in this call graph can be described by a path π that contains not more than one edge representing the same call site.

Suppose we have a path $\pi = \sigma d \rho d$ that contains two edges representing the same call site, which is typical for recursive applications. CATCH builds up its graph structure in a way, such that $\sigma d = \sigma d \rho d$, that is, both paths are considered to be the same node. That means, we now have a node that can be reached using different paths. Note that each path has still a unique parent, which can be obtained by collapsing potential loops in the path.

However, in case of loops in the call graph we can no longer assume that a node was entered from its parent. Instead, CATCH pushes every new node it enters upon a function call onto a stack and retrieves it from there upon a function return:

$$push(n_n) \quad \text{(call)}$$
$$n_n := pop() \quad \text{(return)}$$

Since the stack operations again introduce not more than constant overhead in execution time, the costs are still independent of the call-graph size.

3.4 Parallel Applications

OpenMP*:* OpenMP applications follow a fork-join model. They start as a single thread, fork into a team of multiple threads at some point, and join together after the parallel execution has been finished. CATCH maintains for each thread a separate stack and a separate pointer to the current node, since each thread may call different functions at different points in time. When forking, each slave thread inherits the current node of the master.

The application developer marks code regions that should be executed in parallel by enclosing them with compiler directives or pragmas. The native AIX compiler creates functions for each of these regions. These functions are indirectly called by another function of the OpenMP run-time library (i.e., by passing a pointer to this function as an argument to the library function). Unfortunately, DPCL is not able to identify indirect call sites, so we cannot build the entire call graph only relying on the information provided by DPCL. However, the scheme applied by the native AIX compiler to name the functions representing OpenMP constructs enables CATCH to locate these indirect call sites and to build the complete call graph in spite of their indirect nature.

MPI*:* CATCH maintains for each MPI process a separate call graph, which is stored in a separate instance of the probe module. Since there is no interference between these call graphs, there is nothing extra that we need to pay specific attention to.

3.5 Profiling Subsets of the Call-Graph

If the user is only interested in analyzing a subset of the application, it would be reasonable to restrict instrumentation to the corresponding part of the program in order to minimize intrusion and the number of instrumentation points. Hence, CATCH offers two complementary mechanisms to identify an interesting subset of the call graph. The first one allows users to identify subtrees of interest, while the second is used to filter out subtrees that are not of interest.

- *Selecting* allows the user to select subtrees associated with the execution of certain functions and profile these functions only. The user supplies a list of functions as an argument, which results in profiling being switched off as soon as a subtree of the call graph is entered that neither contains call sites to one of the functions in the list nor has been called from one of the functions in the list.
- *Filtering* allows the user to exclude subtrees associated with the execution of certain functions from profiling. The user specifies these subtrees by supplying a list of functions as an argument, which results in profiling being switched off as soon as one of the function in the list is called.

Both mechanisms have in common that they require switching off profiling when entering and switching it on again when leaving certain subtrees of the call graph. Since the number of call sites that can be instrumented by DPCL may be limited, CATCH recognizes when a call no longer needs to be instrumented due to a subtree being switched off and does not insert any probes there. By default, CATCH instruments only function-call sites to user functions and OpenMP and MPI library functions.

3.6 Limitations

The main limitations of CATCH result from the limitations of the underlying instrumentation libraries. Since DPCL identifies a function called from a function-call site only by name, CATCH is not able to cope with applications defining a function name twice for different functions. In addition, the Linux version, which is based on Dyninst, does not support MPI or OpenMP applications. Support for parallel applications on Linux will be available when the DPCL port to Linux is completed.

CATCH is not able to statically identify indirect calls made via a function pointer passed at run-time. Hence, CATCH cannot profile applications making use of those calls, which limits its usability in particular for C++ applications. However, CATCH still provides full support for the indirect calls made by the OpenMP run-time system of the native AIX compiler as described in Section 3.4.

4 Related Work

The most common instrumentation approach augments source code with calls to specific instrumentation libraries. Examples of these static instrumentation systems include the Pablo performance environment toolkit [12] and the Automated Instrumentation Monitoring System (AIMS) [14]. The main drawbacks of static instrumentation systems are the possible inhibition of compiler optimization and the lack of flexibility, since it requires application re-instrumentation, recompilation, and a new execution, whenever new instrumentation is needed. CATCH, on the other hand, is based on binary instrumentation, which does not require recompilation of programs and does not affect optimization. Binary instrumentation can be considered as a subset of the dynamic instrumentation technology, which uses binary instrumentation to install and remove probes during execution, allowing users to interactively change instrumentation points during run time, focusing measurements on code regions where performance problems have been detected. Paradyn [10] is the exemplar of such dynamic instrumentation systems. Since Paradyn uses probes for code instrumentation, any probe built for CATCH could be easily ported to Paradyn. However, the main contributions of CATCH, which are not yet provided in Paradyn, are the OpenMP support, the precise distinction between different call paths leading to the same program location when assessing performance behavior, the flexibility of allowing users to select different sets of performance counters, and the presentation of a rich set of derived metrics for program analysis.

OMPTrace [4] is a DPCL based tool that combines traditional tracing with binary instrumentation and access to hardware performance counters for the performance analysis and optimization of OpenMP applications. Performance data collected with OMPTrace is used as input to the Paraver visualization tool [8] for detailed analysis of the parallel behavior of the application. Both OMPTrace and CATCH use a similar approach to exploit the information provided by the native AIX compiler to identify and instrument functions the compiler generates from OpenMP constructs. However, OMPTrace and CATCH differ completely in their data collection techniques, since the former collects traces, while CATCH is a profiler.

GNU gprof [9] creates execution-time profiles for serial applications. In contrast to our approach, gprof uses sampling to determine the time fraction spent in different functions of the program. Besides plain execution times gprof estimates the execution time of a function when called from a distinct caller only. However, since the estimation is based on the number of calls from this caller it can introduce significant inaccuracies in cases where the execution time highly depends on the caller. In contrast, CATCH creates a profile for the full call graph based on measurement instead of estimation.

Finally, PAPI [1] and PCL [13] are application programming interfaces that provide a common set of interfaces to access hardware performance counters across different platforms. Their main contribution is in providing a portable interface. However, as opposed to CATCH, they still require static instrumentation and do not provide a visualization tool for presentation.

5 Conclusion

CATCH is a profiler for parallel applications that collects hardware performance counters information for each function called in the program, based on the path that led to the function invocation. It supports MPI, OpenMP, and hybrid applications and integrates the performance data collected for different processes and threads. Functions representing the bodies of OpenMP constructs, which have been generated by the compiler, are also monitored and mapped back to the source code. The user can view the data using a GUI that displays the performance data simultaneously with the source code sections they refer to.

The information provided by hardware performance counters provide more expressive performance metrics than mere execution times and thus enable more precise statements about the performance behavior of the applications being investigated. In conjunction with CATCH's ability not only to map these data back to the source code but also to the full call path, CATCH provides valuable assistance in locating hidden performance problems in both the source code and the control flow. Since CATCH works on the unmodified binary, its usage is very easy and independent of the programming language.

In the future, we plan to use the very general design of CATCH's profiling interface to develop a performance-controlled event tracing system that tries to identify interesting subtrees at run time using profiling techniques and to record the performance behavior at those places using event tracing, because tracing allows a more detailed insight into the performance behavior. Since now individual event records can carry the corresponding call-graph node in one of their data fields, they are aware of the execution state of the program even when event tracing starts in the middle of the program. Thus, we are still able to map the observed performance behavior to the full call path. The benefit of selective tracing would be a reduced trace-file size and less program perturbation by trace-record generation and storage in the main memory.

References

1. S. Browne, J. Dongarra, N. Garner, K. London, and P. Mucci. A Scalable Cross-Platform Infrastructure for Application Performance Tuning Using Hardware Counters. In *Proceedings of Supercomputing'00*, November 2000.
2. B. R. Buck and J. K. Hollingsworth. An API for Runtime Code Patching. *Journal of High Performance Computing Applications*, 14(4):317–329, Winter 2000.
3. Maria Calzarossa, Luisa Massari, Alessandro Merlo, Mario Pantano, and Daniele Tessera. Medea: A Tool for Workload Characterization of Parallel Systems. *IEEE Parallel and Distributed Technology*, 3(4):72–80, November 1995.
4. Jordi Caubet, Judit Gimenez Jesus Labarta, Luiz DeRose, and Jeffrey Vetter. A Dynamic Tracing Mechanism for Performance Analysis of OpenMP Applications. In *Proceedings of the Workshop on OpenMP Applications and Tools - WOMPAT 2001*, pages 53 – 67, July 2001.
5. Luiz DeRose. The Hardware Performance Monitor Toolkit. In *Proceedings of Euro-Par*, pages 122–131, August 2001.
6. Luiz DeRose, Ted Hoover Jr., and Jeffrey K. Hollingsworth. The Dynamic Probe Class Library - An Infrastructure for Developing Instrumentation for Performance Tools. In *Proceedings of the International Parallel and Distributed Processing Symposium*, April 2001.
7. Luiz DeRose and Daniel Reed. SvPablo: A Multi-Language Architecture-Independent Performance Analysis System. In *Proceedings of the International Conference on Parallel Processing*, pages 311–318, August 1999.
8. European Center for Parallelism of Barcelona (CEPBA). *Paraver - Parallel Program Visualization and Analysis Tool - Reference Manual*, November 2000. http://www.cepba.upc.es/paraver.
9. J. Fenlason and R. Stallman. *GNU prof - The GNU Profiler*. Free Software Foundation, Inc., 1997. http://www.gnu.org/manual/gprof-2.9.1/gprof.html.
10. Barton P. Miller, Mark D. Callaghan, Jonathan M. Cargille, Jeffrey K. Hollingsworth, R. Bruce Irvin, Karen L. Karavanic, Krishna Kunchithapadam, and Tia Newhall. The Paradyn Parallel Performance Measurement Tools. *IEEE Computer*, 28(11):37–46, November 1995.
11. Bernd Mohr, Allen Malony, and Janice Cuny. TAU Tuning and Analysis Utilities for Portable Parallel Programming. In G. Wilson, editor, *Parallel Programming using C++*. M.I.T. Press, 1996.
12. Daniel A. Reed, Ruth A. Aydt, Roger J. Noe, Phillip C. Roth, Keith A. Shields, Bradley Schwartz, and Luis F. Tavera. Scalable Performance Analysis: The Pablo Performance Analysis Environment. In Anthony Skjellum, editor, *Proceedings of the Scalable Parallel Libraries Conference*. IEEE Computer Society, 1993.
13. Research Centre Juelich GmbH. *PCL - The Performance Counter Library: A Common Interface to Access Hardware Performance Counters on Microprocessors.*
14. J. C. Yan, S. R. Sarukkai, and P. Mehra. Performance Measurement, Visualization and Modeling of Parallel and Distributed Programs Using the AIMS Toolkit. *Software Practice & Experience*, 25(4):429–461, April 1995.

SIP: Performance Tuning through Source Code Interdependence

Erik Berg and Erik Hagersten

Uppsala University, Information Technology, Deparment of Computer Systems
P.O. Box 337, SE-751 05 Uppsala, Sweden
{erikberg,eh}@docs.uu.se

Abstract. The gap between CPU peak performance and achieved application performance widens as CPU complexity, as well as the gap between CPU cycle time and DRAM access time, increases. While advanced compilers can perform many optimizations to better utilize the cache system, the application programmer is still required to do some of the optimizations needed for efficient execution. Therefore, profiling should be performed on optimized binary code and performance problems reported to the programmer in an intuitive way. Existing performance tools do not have adequate functionality to address these needs. Here we introduce *source interdependence profiling*, SIP, as a paradigm to collect and present performance data to the programmer. SIP identifies the performance problems that remain after the compiler optimization and gives intuitive hints at the source-code level as to how they can be avoided. Instead of just collecting information about the events directly caused by each source-code statement, SIP also presents data about events from some interdependent statements of source code.
A first SIP prototype tool has been implemented. It supports both C and Fortran programs. We describe how the tool was used to improve the performance of the SPEC CPU2000 183.equake application by 59 percent.

1 Introduction

The peak performance of modern microprocessors is increasing rapidly. Modern processors are able to execute two or more operations per cycle at a high rate. Unfortunately, many other system properties, such as DRAM access times and cache sizes, have not kept pace. Cache misses are becoming more and more expensive. Fortunately, compilers are getting more advanced and are today capable of doing many of the optimizations required by the programmer some years ago, such as blocking. Meanwhile, the software technology has matured, and good programming practices have been developed. Today, a programmer will most likely aim at, first, getting the correct functionality and good maintainability; then, profile to find out where in the code the time is spent; and, finally optimizing that fraction of the code. Still, many applications spend much of their execution time waiting for slow DRAMs.

B. Monien and R. Feldmann (Eds.): Euro-Par 2002, LNCS 2400, pp. 177–186.

Although compilers evolve, they sometimes fail to produce efficient code. Performance tuning and debugging are needed in order to identify where an application can be further optimized as well as how it should be done. Most existing profiling tools do not provide the information the programmer needs in a straightforward way. Often the programmer must have deep insights into the cache system and spend a lot of time interpreting the output to identify and solve possible problems.

Profiling tools are needed to explain the low-level effects of an application's cache behavior in the context of the high level language. This paper describes a new paradigm that gives straightforward aid to identify and remove performance bottlenecks. A prototype tool, implementing a subset of the paradigm, has proven itself useful to understand home-brewed applications at the Department of Scientific Computing at Uppsala University. In this paper we have chosen the SPEC CPU2000 183.equake benchmark as an example.

The paper is outlined as follows. Section 2 discusses the ideas behind the tool and general design considerations. Section 3 gives the application writer's view of our first prototype SIP implementation; Section 4 demonstrates how it is used for tuning of equake. Section 5 describes the tool implementation in more detail, section 6 compares SIP to other related tools, before the final conclusion.

2 SIP Design Considerations

The semantic gap between hardware and source code is a problem of application tuning. Code-centric profilers, which for example present the cache miss rate per source-code statement, reduces this gap, but the result can be difficult to interpret. We have no hints as to why the misses occurred. High cache miss ratios are often not due to one single source code statement, but depend on the way different statements interact, and how well they take advantage of the particular data layout used. Data-centric profilers instead collect information about the cache utilization for different data structures in the program. This can be useful to identify a poorly laid out, or misused, data structure. However, it provides little guidance to exactly where the code should be changed.

We propose using a profiler paradigm that presents data based on the interdependence between source code statements: *Source Interdependence Profiler*, SIP. SIP is both code-centric, in that statistics are mapped back on the source code, and data-centric, in that the collected statistics can be subdivided for each data structure accessed by a statement. The interdependence information for individual data structures accessed by a statement tells the programmer which data structures that may be restructured or accessed in a different way to improve performance.

The interdependence between different memory accesses can be either positive or negative. *Positive cache interdependence*, i.e., a previously executed statement has touched the same cache line, can cause a cache hit; *negative cache interdependence*, i.e., a more recent executed statement has touched a different cache line indexing to the same cache set and causing it to be replaced, may cause a cache miss. A statement may be interdependent with itself because of loop constructs or because it contains more than one access to the same memory location.

To further help the programmer, the positive cache interdependence collected during a cache line's tenure in the cache is subdivided into spatial and temporal locality. The spatial locality tells how large fraction of the cache line was used before eviction, while the temporal locality tells how many times each piece of data was used on average.

3 SIP Prototype Overview

The prototype implementation works in two phases. In the first phase, the studied application is run on the Simics [10] simulator. A cache simulator and a statistics collector is connected to the simulator. During the execution of the studied application, cache events are recorded and associated with load and store instructions in the binary executable. In the second phase, an analyzer summarizes the gathered information and correlates it to the studied source code. The output from the analyzer consists of a set of HTML files viewable by a standard browser. They contain the source code and the associated cache utilization.

Figure 1 shows a sample output from the tool. The browser shows three panes. To the left is an index pane where the source file names and the data structures are presented, the upper right pane shows the source code, and the lower right contains the results of the source-interdependence analysis. A click on a source file name in the index pane will show the content of the original source file with line numbers in the source pane. It will also show estimated relative execution costs to the left of the last line of every statement in the file. Source statements with high miss rates or execution times are colored and boldfaced. The source-interdependence analysis results for a statement can be viewed by clicking on the line number of the last line of the statement. It will show in the left lower pane in the following three tables:

- **Summary**
  A summary of the complete statement. It shows the estimated relative cost of the statement as a fraction of the total execution time of the application, the fraction of load/store cost caused by floating point and integer accesses, and miss rates for first- and second-level caches.
- **Spatial and Temporal Use**
  Spatial and temporal use is presented for integer and floating point loads and stores. The Spatial and Temporal use measures are chosen to be independent from each other to simplify the interpretation.
  - **Spatial use**
    Indicates how large fraction of the data brought into cache that is ever used. It is the percentage, on average, of the number of bytes allocated into cache by this statement that are ever used before evicted. This includes used by this same statement again, e.g. in next iteration of a loop, or used by another statement elsewhere in the program.

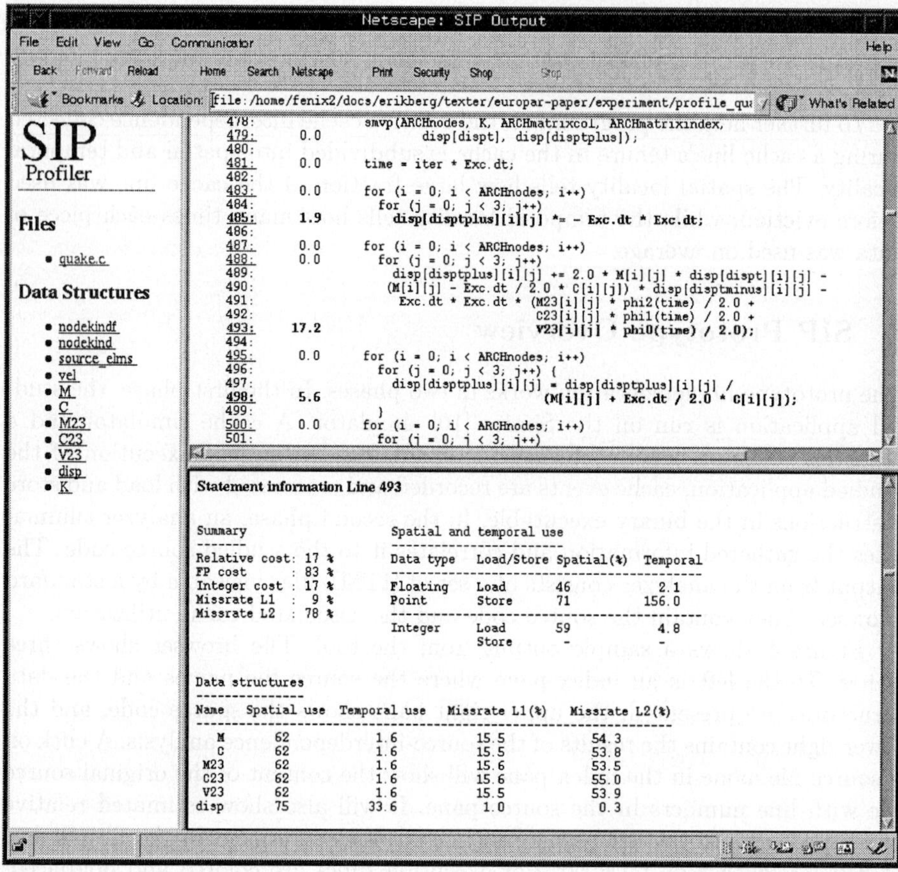

Fig. 1. A screen dump from experiments with SPEC CPU2000 183.equake, 32 bit binary on UltraSPARCII. It shows the index pane (left), source pane (right) and profile information pane (bottom). It shows that the application exhibits poor spatial locality (46 percent) and temporal locality (2.1 times) for floating point loads.

- **Temporal use**
  The average number of times data is reused during its tenure in the cache. First touch is not counted, i.e. a temporal use equal to zero indicates that none of the data is not touched more than once before it is evicted. Data that is never touched is disregarded, and therefore this measure does not depend on the spatial use.

– **Data Structures:**
Miss ratios, spatial use and temporal use are presented for the individual data structures, or arrays, accessed by the statement.

This prototype SIP implementation does not implement the explicit pointers to other statements where data is reused, but only the implicit interdependence in spatial and temporal use. We anticipate that future enhancements of the tool will include the explicit interdependencies.

4 Case Study: SPEC 183.equake

A case study shows how SIP can be used to identify and help understanding of performance problems. We have chosen the 183.equake benchmark from the SPEC [15] CPU2000 suite. It is an earthquake-simulator written in C. First, SIP was used to identify the performance bottlenecks in the original[1] application and examine their characteristics. Figure 1 shows a screen dump of the result. The statement on lines 489-493 accounts for slightly more than 17 percent of the total execution time. Click on "493", and the browser will show the statement information in the lower pane as in the figure. As can be seen under Summary, the cost of floating-point loads and stores is large. Miss rates are also large, especially in the Level 2 cache.

4.1 Identifying Spatial Locality Problems

The spatial use shows poor utilization of cached data. Floating-point loads show the worst behavior. As can be seen in the lower right pane under "Spatial and temporal use", not more than 46 percent of the floating-point data fetched into cache by loads in this statement are ever used. Floating-point store and integer loads behave better, 71 and 59 percent respectively. The information about individual data structures, in bottom table of the same pane, points in the same direction. All but one, the array disp, have only 62 percent spatial use. When examining the code, the inner-most loop, beginning on line 488, corresponds to the last index of the data accesses on lines 489 - 492. This should result in good spatial behavior and contradicts the poor spatial percentage reported by the tool.

These results caused us to take a closer look at the memory layout. We found a problem in the memory allocation function. The data structure in the original code is a tree, where the leafs are vectors containing three doubles each. The memory allocation function does not allocate these vectors adjacent to each other, but leaves small gaps between them. Therefore not all of the data brought into the cache are ever used, causing the poor cache utilization.

A simple modification of the original memory-allocation function substantially increases performance. The new function allocates all leaf vectors adjacent to each other and the SIP tool shows that the spatial use of data improves. The speedups caused by the memory-allocation optimization are 43 percent on a 64-bit (execution time reduced from 1446s to 1008s) and 10 percent on a 32-bit executable. The probable reason of the much higher speedup on the 64-bit

[1] In the prototype, the main function must be instrumented with a start call to tell SIP that the application has started. Recognizable data structures must also be instrumented. For heap-allocated data structures, this can be done automatically.

binary is that the larger pointers cause larger gaps between the leafs in the original memory allocation. The SIP tool also revealed other code spots that benefit from this optimization. Therefore the speedup of the application is larger than the 17 percent execution cost of the statement on lines 489-493. A matrix-vector multiplication especially benefits by the above optimization. All speedup measurements were conducted with a Sun Forte version 6.1 C compiler and a Sun E450 server with 16KB level 1 data cache, 4MB unified level 2 cache and 4GB of memory, running SunOS 5.7. Both 64- and 32-bit executables were created with the *-fast* optimization flag. All speed gains were measured on real hardware.

4.2 Identifying Temporal Problems

The temporal use of data is also poor. For example, Figure 1 shows that floating-point data fetched into the cache from the statement are only reused 2.1 times on average. The code contains four other loop nests that access almost the same data structures as the loop nest on lines 487-493. They are all executed repeatedly in a sequence. Because the data have not been reused more, the working sets of the loops are too large to be contained in the cache. Code inspection reveals that loop merging is possible. Profiling an optimized version of the program with the loops merged shows that the data reuse is much improved. The total speedups with both this and the previous memory allocation optimizations are 59 percent on a 64-bit and 25 percent on a 32-bit executable.

5 Implementation Details

The prototype implementation of SIP is based on the Simics full-system simulator. Simics[10] simulates the hardware in enough detail to run an unmodified operating system and, on top of that, the application to be studied. This enables SIP to collect data non-intrusively and to take operating-system effects, such as memory-allocation and virtual memory system policies, into account. SIP is built as a module of the simulator, so large trace files are not needed. The tool can profile both Fortran and C code compiled with Sun Forte compilers and can handle highly optimized code. As described earlier, the tool works in two phases, the collecting phase and the analyzing phase.

5.1 SIP Collecting Phase

During the collecting phase, the studied application is run on Simics to collect cache behavior data. A memory-hierarchy simulator is connected to Simics. It simulates a multilevel data-cache hierarchy. The memory-hierarchy simulator can be configured for different cache parameters to reflect the characteristics of the computer, for which the studied application is to be optimized. The parameters are cache sizes, cache line sizes, access times, etc. The slowdown of the prototype tool's analyzing phase is around 450 times, mostly caused by the simulator, Simics.

The memory hierarchy reports every cache miss and evicted data to a statistics collector. Whenever some data is brought to a higher level of the cache hierarchy, the collector starts to record the studied application's use of it. When data are evicted from a cache, the recorded information is associated with the instruction that originally caused the data to be allocated into the cache. All except execution count and symbol reference are kept per cache level. The information stored for each load or store machine instruction includes the following:

- **Execution count**
  The total number of times the instruction is executed.
- **Cache misses**
  The total number of cache misses caused by the instruction.
- **Reuse count**
  The reuse count of one cache-line-sized piece of data is the number of times it is touched from the time it is allocated in the cache until it is evicted. Reuse count is the sum of the reuse counts of all cache-line-sized pieces of data allocated in the cache.
- **Total spatial use**
  The sum of the spatial use of all cache-line-sized pieces of data allocated in cache. The spatial use of one cache line-sized-piece of data is the number of different bytes that have been touched from the time it is allocated in cache until it is evicted.
- **Symbol reference**
  Each time a load or store instruction accesses memory, the address is compared to the address ranges of known data structures. The addresses of the data structures comes from instrumenting the source code. If a memory-access address matches any known data structure, a reference to that data structure is associated with the instruction PC. This enables the tool to relate caching information with specific data structures.

5.2 SIP Analyzing Phase

The analyzer uses the information from the statistics collector and produces the output. First, a mapping from machine instructions to source statements is built. This is done for every source file of the application. Second, for each source code statement, every machine instruction that is related to it is identified. Then, the detailed cache behavior information can be calculated for every source statement; and finally, the result is output as HTML files.

SIP uses compiler information to relate the profiling data to the original source code. To map each machine instruction to a source-code statement, the analyzer reads the debugging information [16] from the executable file and builds a translation table between machine-instruction addresses and source-code line numbers. The machine instructions are then grouped together per source statement. This is necessary since the compiler reorganizes many instructions from different source statements during optimization and the tool must know which load and store instructions that belongs to any source statement. The accurate machine-to-source-code mapping generated by Sun Forte C and F90 compilers makes this grouping possible. It can often be a problem to map optimized machine code to source code, but in this case it turned out to work quite well.

Derived measures are calculated at source-statement level. The information collected for individual machine instructions are summarized over their respective source-code statements, i.e. total spatial use for one statement is the sum of the total spatial uses of every load and store instruction that belongs to that statement. Reuse count is summarized analogous. To calculate the information that is presented in the table "Spatial and temporal use" in Figure 1, instructions are further subdivided into integer load, integer store, floating point load and floating point store for each source statement. For example, the total spatial use for floating-point load of one statement is the sum of the total spatial uses of every floating-point load instruction that belongs to that statement. The spatial use for a statement is calculated as:

$$Spatial\ use(\%) = 100 \cdot \frac{total\ spatial\ use\ of\ the\ statement}{\#cache\ misses\ of\ the\ statement \cdot cache\ line\ size}$$

Temporal use is calculated as:

$$Temporal\ use = \frac{reuse\ count\ of\ the\ statement}{total\ spatial\ use\ of\ the\ statement} - 1$$

The output is generated automatically in HTML format. It is easy to use and it does not need any specialized viewer. SIP creates two output files for each source file, one that contains the source code with line numbers, and one that contains the detailed cache information. It also produces a main file that sets up frames and links to the other files.

6 Related Work

Source-code interdependence can be investigated at different levels. Tools that simply map cache event counts to the source code do not give enough insights in how different parts of the code interact. Though useful, they fail to fully explain some performance problems. Cacheprof [14] is a tool that annotates source-code statements with the number of cache misses and the hit-and-miss ratios. It is based on assembly code instrumentation of all memory access instructions. For every memory access, a call to a cache simulator is inserted.

MemSpy [11] is based on the tango [6] simulator. For every reference to dynamically allocated data, the address is fed to a cache simulator. It can be used for both sequential and parallel applications. The result is presented at the procedure and data-structure level and indicates whether the misses were caused by communication or not. The FlashPoint tool [12] gathers similar information using the programmable cache-coherence controllers in the FLASH multiprocessor computer. CPROF [8] uses a binary executable editor to insert calls to a cache simulator for every load and store instruction. It annotates source code with cache-miss ratios divided into the categories of compulsory, conflict and capacity. It also gives similar information for data-structures. It does not investigate how different source statements relate to each other through data use, except for the implicit information given by the division into conflict and capacity. The full system simulator SimOS[9] has also been used to collect similar data and to optimize code. MTOOL[5] is a tool that compares estimated cycles due to pipeline stalls with measurements of actual performance. The difference is assumed to be due to cache miss stalls. Buck and Hollingsworth [2] present two methods for finding memory bottlenecks; counter overflow and n-way search based on the number of cache misses to different memory regions.

DCPI [1] is a method to get systemwide profiles. It collects information about such things as cache misses and pipeline stalls and maps this information to machine or source code. It uses the ProfileMe[3] hardware mechanism in the Alpha processor to accurately annotate machine instructions with different event counters, such as cache misses and pipeline stalls. The elaborate hardware support and sampling of nearby machine instructions can find dependencies between different machine instructions, but the emphasis is on detailed pipeline dependencies rather than memory-system interaction. SvPablo [4] is a graphical viewer for profiling information. Data can be collected from different hardware counters and mapped to source code. The information is collected by instrumenting source-code with calls to functions that read hardware counters and records there values. Summaries are produced for procedures and loop constructs.

MHSIM[7] is the tool that is most similar to SIP. It is based on source-code instrumentation of Fortran programs. A call to a memory-hierarchy simulator is inserted for every data access in the code. It gives spatial and temporal information at loop, statement and array-reference levels. It also gives conflict information between different arrays. The major difference is that it operates at source-code level and therefore gives no information as to whether the compiler managed to remove any performance problems. The temporal measure in MHSIM is also less elaborate. For each array reference, it counts the fraction of accesses that hit previously used data.

7 Conclusions and Future Work

We have found that source-code interdependence profiling is useful to optimize software. In a case study we have shown how the information collected by SIP, Source code Interdependence Profiling, can be used to substantially improve an application's performance. The mechanism to detect code interdependencies increases the understanding of an application's cache behavior. The comprehensive measures of spatial and temporal use presented in the paper also proved useful. It shows that further investigation should prove profitable.

Future work includes adding support to relate different pieces of code to each other through their use of data. Further, we intend to reduce the tool overhead by collecting the information by assembly code instrumentation and analysis.

We also plan to incorporate this tool into DSZOOM [13], a software distributed shared memory system.

References

1. J. Anderson, L. Berc, J. Dean, S. Ghemawat, M. Henzinger, S. Leung, D. Sites, M. Vandevoorde, C. Waldspurger, and W. Weihl. Continuous profiling: Where have all the cycles gone? *ACM Transactions on Computer Systems*, 1997.
2. B. Buck and J. Hollingsworth. Using hardware performance monitors to isolate memory bottlenecks. In *Proceedings of Supercomputing*, 2000.
3. J. Dean, J. Hicks, C. Waldspurger, W. Weihl, and G. Chrysos. ProfileMe: Hardware support for instruction-level profiling on out-of-order processors. In *Proceedings of the 30th Annual International Symposium on Microarchitecture*, 1997.
4. L. DeRose and D. Reed. Svpablo: A multi-language architecture-independent performance analysis system. In *10th International Conference on Performance Tools*, pages 352–355, 1999.
5. A. Goldberg and J. Hennessy. MTOOL: A method for isolating memory bottlenecks in shared memory multiprocessor programs. In *Proceedings of the International Conference on Parallel Processing*, pages 251–257, 1991.
6. S. Goldschmidt H. Davis and J. Hennessy. Tango: A multiprocessor simulation and tracing system. In *Proceedings of the International Conference on Parallel Processing*, 1991.
7. R. Fowler J. Mellor-Crummey and D. Whalley. Tools for application-oriented performance tuning. In *Proceedings of the 2001 ACM International Conference on Supercomputing*, 2001.
8. Alvin R. Lebeck and David A. Wood. Cache profiling and the SPEC benchmarks: A case study. *IEEE Computer*, 27(10):15–26, 1994.
9. S. Devine M. Rosenblum, E. Bugnion and S. Herrod. Using the simos machine simulator to study complex systems. *ACM Transactions on Modelling and Computer Simulation*, 7:78–103, 1997.
10. P. Magnusson, F. Larsson, A. Moestedt, B. Werner, F. Dahlgren, M. Karlsson, F. Lundholm, J. Nilsson, P. Stenström, and H. Grahn. SimICS/sun4m: A virtual workstation. In *Proceedings of the Usenix Annual Technical Conference*, pages 119–130, 1998.
11. M. Martonosi, A. Gupta, and T. Anderson. Memspy: Analyzing memory system bottlenecks in programs. In *ACM SIGMETRICS International Conference on Modeling of Computer Systems*, pages 1–12, 1992.
12. M. Martonosi, D. Ofelt, and M. Heinrich. Integrating performance monitoring and communication in parallel computers. In *Measurement and Modeling of Computer Systems*, pages 138–147, 1996.
13. Z. Radovic and E. Hagersten. Removing the overhead from software-based shared memory. In *Proceedings of Supercomputing 2001*, November 2001.
14. J. Seward. The cacheprof home page `http://www.cacheprof.org/`.
15. SPEC. Standard performance evaluation corporation `http://www.spec.org/`.
16. Sun. *Stabs Interface Manual, ver.4.0.* Sun Microsystems, Inc, Palo Alto, California, U.S.A., 1999.

Topic 3
Scheduling and Load Balancing

Maciej Drozdowski, Ioannis Milis, Larry Rudolph, and Denis Trystram

Topic Chairpersons

Despite the large number of papers that have been published, scheduling and load balancing continue to be an active area of research. The topic covers all aspects related to scheduling and load balancing including application and system level techniques, theoretical foundations and practical tools. New aspects of parallel and distributed systems, such as clusters, grids, and global computing require new solutions in scheduling and load balancing.

There were 27 papers submitted to Topic 3 track of Euro-Par 2001. As the result of each submission being reviewed by at least three referees, a total of 10 papers were chosen to be included in the conference program; 5 as regular papers and 5 as research notes.

Four papers present new theoretical results for selected scheduling problems. S.Fujita in *A Semi-Dynamic multiprocessor scheduling algorithm with an asymptotically optimal performance ratio* considers the on-line version of the classical problem of scheduling independent tasks on identical processors and proposes a new clustering algorithm which beats the competitive ratio of the known ones. E.Angel et al. in *Non-approximability results for the hierarchical communication problem with a bounded number of clusters* explore the complexity and approximability frontiers between several variants of the problem of scheduling precedence constrained tasks in the presence of hierarchical communications. For the same problem, but in the case of bulk synchronous processing, N.Fujimoto and K.Hagihara in *Non-approximability of the bulk synchronous task scheduling problem* show the first known approximation threshold. W.Loewe and W.Zimmermann in *On Scheduling Task-Graphs to LogP-Machines with Disturbance* propose a probabilistic model for the prediction of the expected makespan of executing task graphs to the realistic model of LogP-machines, when computation andcommunication may be delayed.

Another four papers propose scheduling and load balancing algorithms which are tested experimentally and exhibit substantially improved performance.

D.T.Altilar and Y.Paker in *Optimal scheduling algorithms for communication constrained parallel processing* consider video processing applications and propose periodic real-time scheduling algorithms based on optimal data partition and I/O utilization. F.Gine et al. in *Adjusting time slices to apply coscheduling techniques in a non-dedicated NOW* present an algorithm for adjusting dynamically the time slice length to the needs of the distributed tasks while keeping good response time for local processes. E.Krevat et al. in *Job Scheduling for the BlueGene/L System* measure the impact of migration and backfilling, as enhancements to the pure FCFS scheduler, to the performance parameters of BlueGene/L system developed for protein folding analysis. D.Kulkarni and

B. Monien and R. Feldmann (Eds.): Euro-Par 2002, LNCS 2400, pp. 187–188.

M.Sosonkina in *Workload Balancing in Distributed Linear System Solution: a Network-Oriented Approach* propose a dynamic adaptation of the application workload based on a network information collection and call-back notification mechanism.

Finally, two papers propose practical tools and ideas for automatic mapping and scheduler selection. X.Yuan et al. in *AMEEDA: A General-Purpose Mapping Tool for Parallel Applications on Dedicated Clusters* combine formalisms, services and a GUI into an integrated tool for automatic mapping tasks on PVM platform. M.Solar and M.Inostroza in *Automatic Selection of Scheduling Algorithms* propose a static layering decision model for selecting an adequate algorithm from a set of schedulingalgorithms which carry out the best assignment for an application.

We would like to express our thanks to the numerous experts in the field for their assistance in the reviewing process. They all worked very hard and helped to make this is a coherent and thought provoking track.

Larry Rudolph - general chair
Denis Trystram - local chair
Maciej Drozdowski, Ioannis Milis - vice chairs

On Scheduling Task-Graphs to LogP-Machines with Disturbances

Welf Löwe[1] and Wolf Zimmermann[2]

[1] Växjö University, School of Mathematics and Systems Engineering, Software Tech. Group, S-351 95 Växjö, Sweden, `Welf.Lowe@msi.vxu.se`
[2] Martin-Luther-Universität Halle-Wittenberg, Institut für Informatik, D-06099 Halle/Saale, Germany, `zimmer@informatik.uni-halle.de`

Abstract. We consider the problem of scheduling task-graphs to LogP-machines when the execution of the schedule may be delayed. If each time step in the schedule is delayed with a certain probability, we show that under LogP the expected execution time for a schedule s is at most $\mathcal{O}(TIME(s))$ where $TIME(s)$ is the makespan of the schedule s.

1 Introduction

Schedules computed by scheduling algorithms usually assume that the execution time of each task is precisely known and eventually that communication parameters such as latencies are also known precisely. Almost all scheduling algorithms base on this assumption. On modern parallel computers, however, the processors asynchronously execute their programs. These programs might be further delayed due to operating system actions etc. Thus, it is impossible to know the precise execution time of the tasks and the exact values of the communication parameters.

Many scheduling algorithms only assume computation times and latencies to compute schedules. The processor are supposed to be able to send or receive an arbitrary number of messages within time 0 (e.g [17, 6]). In practice however, this assumption is unrealistic. We assume LogP-machines [3] capturing the above model as a special case and, in general, considering other properties such as network bandwidth and communication costs on processors. Under LogP, a processor can send or receive only one message for each time step, i.e. sending and receiving a message requires processor time.

The LogP model has been confirmed for quite a large number of parallel machines including the CM-5 [3], the IBM SP1 machine [4], a network of workstations and a powerXplorer [5], and the IBM RS/6000 SP [10]. Theoretic predictions on execution times of programs showed them to be adequate in practice even under the assumption of deterministic computation and communication times. However, to get adequate predictions, computation and communication times ought to be measured in experiments rather than derived analytically from hardware parameters. Our contribution in this paper explains this observation.

We assume that each step on each processor and on each message transmission is transmitted with a fixed probability q, $0 < q < 1$. If a schedule s has

B. Monien and R. Feldmann (Eds.): Euro-Par 2002, LNCS 2400, pp. 189–196.

makespan $T(s)$, we show that the expected execution time with disturbances under the above probability model is at most $c \cdot T(s)$ for a constant c. We distinguish two cases: First, we derive such a constant c under the assumption that the network has infinite bandwidth. In this case the constant c is independent of the communication parameters. Second, we extend the result to the case of finite bandwidth.

We propose the following strategy to scheduling problem: schedule the task-graph under LogP with optimistic assumptions, i.e., the execution times of the tasks and the communication parameters are known exactly by analyzing the program and the hardware. Here any scheduling algorithm can be used (e.g.[5, 9, 11, 18]). Account for the expected delay by considering the probability q using our main result.

Section 2 introduces the LogP-model, LogP-schedules, and the probability model. Section 3 discusses the case of infinite bandwidth and Section 4 discusses the case of finite bandwidth. Section 5 compares our work with related works.

2 Basic Definitions

We assume a HPF-like programming with data parallel synchronous program but without any data distribution. For simplicity, we further assume that the programs operate on a single composite data structure which is an array a. The size of an input a, denoted by $|a|$, is the length of the input array a.

We can model the execution of programs on an input x by a family task-graphs $G_x = (V_x, E_x, \tau_x)$. The tasks $v \in V_x$ model local computations without access to the shared memory, $\tau(v)$ is the execution time of task v on the target machine, and there is a directed edge from v to w iff v writes a value into the shared memory that is read later by task w. Therefore, task-graphs are always acyclic. G_x does not always depend on the actual input x. In many cases of practical relevance it only depends on the problem size n. We call these program *oblivious* and denote its task graphs by G_n. In the following, we consider oblivious programs and write G instead of G_n if n is arbitrary but fixed. The *height* of a task v, denoted by $h(v)$, is the length of the longest path from a task with in-degree 0 to v.

Machines are modelled by LogP [3]: in addition to the computation costs τ, it models communication costs with parameters *L*atency, *o*verhead, and *g*ap (which is actually the inverse of the bandwidth per processor). In addition to L, o, and g, parameter P describes the number of processors. Moreover, there is a capacity constraint: at most $\lceil L/g \rceil$ messages are in transmission in the network from any processor to any processor at any time. A send operation that exceeds this constraint stalls.

A LogP-schedule is a schedule that obeys the precedence constraints given by the task-graph and the constraints imposed by the LogP-machine, i.e., sending and receiving a message takes time o, between two consequential send or receive operations must be at least time g, between the end of a send task and the beginning of the corresponding receive task must be a least time L, and the capacity

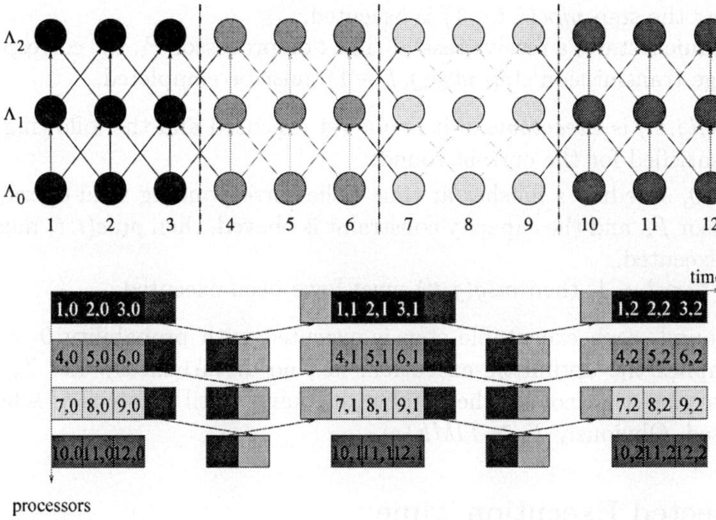

Fig. 1. Partitioning the task graph according to block-wise data distribution (left) and the corresponding LogP-schedule (right) with parameters $L = 2$, $o = 1$, $g = 2$.

constraint must be obeyed. For simplicity, we only consider LogP-schedules that use all processors and no processor sends a message to itself. A *LogP-schedule* is a set of sequences of computations, send, and receive operations and their starting times corresponding to the tasks and edges of the task-graph. For each task, its predecessors must be computed either on the same processor or their outputs must be received from other processors. The schedules must guarantee the following constraints: (i) sending and receiving a message of size k takes time $o(k)$, (ii) between two sends or two receives on one processor, there must be at least time $g(k)$, (iii) a receive must correspond to a send at least $L(k)+o(k)$ time units earlier in order to avoid waiting times, (iv) computing a task v takes time $\tau(v)$, and (v) a correct LogP-schedule of a task-graph G must compute all tasks at least once. $TIME(s)$ denotes the execution time of schedule s, i.e., the time when the last task finishes. Figure 1 shows a task graph, sketches a scheduling algorithms according to block-wise distribution of the underlying data array and gives the resulting schedule.

Finally, we introduce the probability model. Suppose s is a LogP-schedule. For the probability model, we enumerate the processors from 0 to $P-1$ and the message transmissions from 0 to $M-1$ (if there are M message transmissions in the schedule) in any order. This leads to two kinds of steps: $proc(i,t)$ denotes the t-th time step on the i-th processor and $msg(j,t')$ is the t'-th time step of message transmission j. Observe that $0 \leq t' < L$. In the following, these pairs are uniformly denoted by *steps*.

The execution of s proceeds in rounds. At each round, there are steps that are executable. A step $proc(i,t)$ is *executable* iff it is not yet executed and the following conditions are satisfied for the current round:

1. $t = 0$ or the step $proc(i, t-1)$ is executed.
2. If schedule s starts a receive task at time t on processor P_i, the corresponding message transmission step $msg(j, L-1)$ must be completed.

A step $msg(j, t')$ is *executable* iff it is not yet executed and the following conditions are satisfied for the current round:

1. If $t' = 0$, schedule s finishes at time t the corresponding send-operation on processor P_i, and the capacity constraint is obeyed, then $proc(i, t)$ must have been executed.
2. If $0 < t' < L-1$, then $msg(j, t')$ must have been executed.

At each round, each executable step is executed with probability $0 < q < 1$ ($q = 1$ implies the optimistic execution, i.e., no disturbances). Let T_s be the random variable that counts the number of rounds until all steps of schedule s are executed. Obviously $T_s \geq TIME(s)$.

3 Expected Execution Time for Networks with Infinite Bandwidth

For a first try, we we assume $g = 0$, i.e., the network has infinite capacity. Therefore the capacity constraint can be ignored, i.e. messages never stall. In particular, a step $msg(j, 0)$ of a LogP-schedule s is executable iff it is not executed, schedule s finishes at time t the corresponding send-operation on processor P_i, and $proc(i, t)$ is executed.

The proof for analyzing the random variable T_s uses the following lemma, firstly proved in [12]. Another proof can be found in [14].

Lemma 1 (Random Circuit Lemma). *Let $G = (V, E)$ be a directed acyclic graph with depth h and with n distinct (but not necessarily disjoint) paths from input vertices (in-degree 0) to output vertices (out-degree 0). If, in each round, any vertex which has all its predecessors marked is itself marked with probability at least $q > 0$ in this step, then the expected number of rounds to mark all output vertices is at most $(6/q)(h + \log n)$ and, for any constant $c > 0$, the probability that more than $(5c/q)(h + \log n)$ steps are used is less than $1/n^c$.*

The graph G_s can be directly obtained from a LogP-schedule s: the vertices are the steps, and there are the following edges:

1. $proc(i, t) \to proc(i, t+1)$
2. $msg(j, t') \to msg(j, t'+1)$
3. $msg(j, L-1) \to proc(i, t)$ if schedule s starts at time t on processor P_i the corresponding receive task.
4. $proc(i, t) \to msg(j, 0)$ if schedule s finishes at time t on processor P_i the corresponding send task.

Fig. 2 shows the graph G_s for the schedule of Fig. 1. With the graph G_s, the probability model described in the random circuit lemma corresponds exactly to our probability model, when a a vertex in G_s is marked iff the corresponding step is executed.

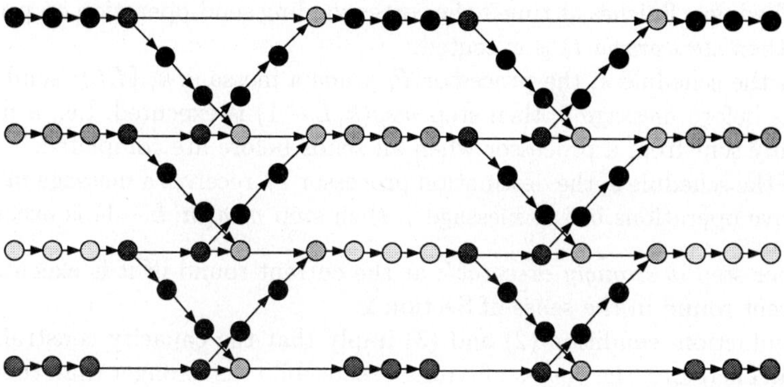

Fig. 2. Graph G_s for the Schedule of Figure 1.

Corollary 1. *Let s be a LogP-schedule. If $g = 0$, then under the probability model of Section 2, it holds*

i) $\Pr[T_s > \frac{5c}{q}(2 \cdot TIME(s) + \log P)] \leq P^{-c}$ *for any constant $c > 0$, and*

ii) $\mathrm{E}[T_s] \leq \frac{6}{q}(2 \cdot TIME(s) + \log P)$

Proof. We apply the Random Circuit Lemma to the graph G_s. The depth of G_s is by construction $h = TIME(s)$. Since at each time, a processor can send at most one message, the out-degree of each vertex of G_s is at most 2. Furthermore, there are P vertices with in-degree 0. Hence, for the number n of paths, it holds: $P \leq n \leq P \cdot 2^{TIME(s)}$. The claims directly follow from these bounds.

Hence, if the execution of schedule s is disturbed according to the probability model and $g = 0$, the expected delay of the execution is at most a constant factor (approximately $12/q$).

4 The General Case

We now generalize the result of Section 3 if the network has finite bandwidth, i.e. $g > 0$. In this case, sending a message might be blocked because there are too many messages in the network. Thus, the construction of the graph G_s of Section 4 cannot be directly applied because a vertex $msg(j, 0)$ might be marked although more than $\lceil L/g \rceil$ messages are in transit from the source processor or to the target processor, respectively.

The idea to tackle the problem is to define a stronger notion of executability. Let s be a schedule and T'_s the number of rounds required to execute all steps with the stronger notion of executability. Then, it holds

$$\mathrm{E}[T_s] \leq \mathrm{E}[T'_s] \quad \text{and} \quad \Pr[T_s > t] \leq \Pr[T'_s > t] \tag{1}$$

For a schedule s, a step $msg(j, 0)$ is *strongly executable* at the current round iff it is not yet executed and the following conditions are satisfied.

1. If schedule s finishes at time t the corresponding send-operation on processor P_i, then step $proc(i,t)$ is executed.
2. If in the schedule s, the processor P_i sends a message k, $\lceil L/g \rceil$ send operations before message j, then step $msg(k, L-1)$ is executed. I.e., a message is only sent from a processor when all sends before are completed.
3. If in the schedule s, the destination processor P_h receives a message m, $\lceil L/g \rceil$ receive operations before message j, then step $msg(m, L-1)$ is executed.

Any other step is *strongly executable* at the current round iff it is executable at the current round in the sense of Section 2.

By induction, condition (2) and (3) imply that the capacity constraints are satisfied. Therefore, the notion of strong executability is stronger than the notion of executability. If at each round each strongly executable step is executed with probability q and T'_s denotes the random variable counting the number rounds required to execute all steps, then (1) is satisfied.

Theorem 1. *Let s be a LogP-schedule. Then under the probability model of Section 2, it holds*

i) $\Pr[T_s > \frac{5c}{q}((1+\log P)\cdot \mathit{TIME}(s)+\log P)] \leq P^{-c}$ *for any constant $c>0$, and*

ii) $\mathrm{E}[T_s] \leq \frac{6}{q}((1+\log P)\cdot \mathit{TIME}(s)+\log P)$

Proof. By (1) and the above remarks, it is sufficient to show

$$\Pr[T'_s > \frac{5c}{q}((1+\log P)\cdot \mathit{TIME}(s)+\log P)] \leq P^{-c} \qquad \text{for any constant } c>0$$

$$\mathrm{E}[T'_s] \leq \frac{6}{q}((1+\log P)\cdot \mathit{TIME}(s)+\log P)$$

For proving these propositions, we extend the graph G_s defined in Section 3 by edges reflecting conditions (2) and (3), i.e., we have edge $msg(k, L-1) \to msg(j, 0)$ if a processor sends message k just before message j by schedule s or a processor receives message k just before message j by schedule s, respectively. These additional edges ensure that the capacity constraint is satisfied. Furthermore these additional edges do not change the order of sending and receiving messages.

With this new graph, the probability model described in the random circuit lemma corresponds exactly to the stronger probability model as defined above.

Since s obeys the capacity constraints, the new edges do not increase the depth. Thus, the depth of the extended G_s is $\mathit{TIME}(s)$. Furthermore, if there are two edges $msg(k, L-1) \to msg(j, 0)$ and $msg(k, L-1) \to msg(m, 0)$, then messages j and m are sent from different processors to the same processor. Since a processor never sends a message to itself, the source of messages j and m must be different from the destination of message k. Therefore, the out-degree of these steps is at most P – the number n of paths is $P \leq n \leq \mathit{TIME}(s) \cdot P^{\mathit{TIME}(s)}$. With these bounds, we obtain the claim using Lemma 1.

5 Related Work

Related works consider disturbances in the scheduling algorithms themselves. The approach of [15] statically allocates the tasks and schedules them on-line. The performance analysis is experimental. [8] presents another approach using a similar two-phase approach as [15]. This work includes a theoretical analysis. Both approaches are based on the communication delay model. [15] only considers disturbances in the communications.

Our work differs from [15, 8] in two aspects: First, our machine-model is the LogP-machine. Second, we analyze – under a probability model – the expected makespan of schedules produced by static scheduling algorithms. The approach follows the spirit of analyzing performance parameters of asynchronous machine models [1, 2, 7, 13, 14, 16]. [1, 7, 13, 14, 16] introduce work-optimal asynchronous parallel algorithms. Time-optimal parallel algorithms are discussed only in [2]. These works consider asynchronous variants of the PRAM.

6 Conclusions

The present paper accounts for the asynchronous processing paradigm on today's parallel architectures. With a simple probabilistic model, we proved that the expectation of the execution time of a parallel program under this asynchrony assumption is delayed by a constant factor compared to the execution time in an idealistic synchronous environment. Our main contribution shows this for schedules for the general LogP model.

This asynchronous interpretation of the LogP model could explain our previous practical results comparing estimations in the synchronous setting with practical measurements: if the basic LogP parameters and the computation times for the single tasks are obtained in preceding experiments then estimations and measurements nicely match. If, in contrast, the LogP parameters and the computation times are derived analytically, measurements did not confirm our estimations. In the former experiments, the probability for the delay q (disturbance) is implicitly regarded, in the latter they are not.

Future work should support this assumption by the following experiment: we derive the disturbance q by comparing execution time estimations of an example program based on analytic parameters with those based on measured parameters. Then q should be generally applicable for other examples. Thereby it could turn out that we have to assume different disturbances for computation and for network parameters, which would require also an extension of our theory.

References

1. R. Cole and O. Zajicek. The aPRAM: Incorporating asynchrony into the PRAM model. In *1st ACM Symp. on Parallel Algorithms and Architectures*, pp 169 – 178, 1989.
2. R. Cole and O. Zajicek. The expected advantage of asynchrony. In *2nd ACM Symp. on Parallel Algorithms and Architectures*, pp 85 – 94, 1990.

3. D. Culler, R. Karp, D. Patterson, A. Sahay, K. E. Schauser, E. Santos, R. Subramonian, and T. von Eicken. LogP: Towards a realistic model of parallel computation. In *4th ACM SIGPLAN Symposium on Principles and Practice of Parallel Programming (PPOPP 93)*, pp 1–12, 1993. published in: SIGPLAN Notices (28) 7. also published in: *Communications of the ACM*, 39(11):78–85, 1996.
4. B. Di Martino and G. Ianello. Parallelization of non-simultaneous iterative methods for systems of linear equations. In *Parallel Processing: CONPAR 94 – VAPP VI*, volume 854 of *LNCS*, pp 253–264. Springer, 1994.
5. J. Eisenbiegler, W. Löwe, and W. Zimmermann. Optimizing parallel programs on machines with expensive communication. In *Europar' 96 Parallel Processing Vol. 2*, volume 1124 of *LNCS*, pp 602–610. Springer, 1996.
6. A. Gerasoulis and T. Yang. On the granularity and clustering of directed acyclic task graphs. *IEEE Trans. Parallel and Distributed Systems*, 4:686–701, Jun. 1993.
7. P. Gibbons. A more practical PRAM model. In *1st ACM Symp. on Parallel Algorithms and Architectures*, pp 158 – 168, 1989.
8. A. Gupta, G. Parmentier, and D. Trystram. Scheduling precedence task graphs with disturbances. *RAIRO Operational Research Journal*, 2002. accepted.
9. W. Löwe and W. Zimmermann. Upper time bounds for executing pram-programs on the logp-machine. In M. Wolfe, editor, *9th ACM International Conference on Supercomputing*, pp 41–50. ACM, 1995.
10. W. Löwe, W. Zimmermann, S. Dickert, and J. Eisenbiegler. Source code and task graphs in program optimization. In *HPCN'01: High Performance Computing and Networking*, LNCS, 2110, pp 273ff. Springer, 2001.
11. W. Löwe, W. Zimmermann, and J. Eisenbiegler. On linear schedules for task graphs for generalized logp-machines. In *Europar'97: Parallel Processing*, LNCS, 1300, pp 895–904. Springer, 1997.
12. M. Luby. On the parallel complexity of symmetric connection networks. Technical Report 214/88, University of Toronto, Departement of Computer Science, 1988.
13. C. Martel, A. Park, and R. Subramonian. Asynchronous PRAMs are (almost) as good as synchronous PRAMs. In *31st Symp. on Foundations of Computer Science*, pp 590–599, 1990.
14. C. Martel, A. Park, and R. Subramonian. Work-optimal asynchronous algorithms for shared memory parallel computers. *SIAM J. on Computing*, 21(6):1070–1099, Dec 1992.
15. A. Moukrim, E. Sanlaville, and F. Guinand. Scheduling with communication delays and on-line disturbances. In P. Amestoy et. al., editor, *Europar'99: Parallel Processing*, number 1685 in LNCS, pp 350–357. Springer-Verlag, 1999.
16. M. Nishimura. Asynchronous shared memory parallel computation. In *2nd ACM Symp. on Parallel Algorithms and Architectures*, pp 76 – 84, 1990.
17. C.H. Papadimitriou and M. Yannakakis. Towards an architecture-independent analysis of parallel algorithms. *SIAM J. on Computing*, 19(2):322 – 328, 1990.
18. W. Zimmermann and W. Löwe. An approach to machine-independent parallel programming. In *Parallel Processing: CONPAR 94 – VAPP VI*, volume 854 of *LNCS*, pp 277–288. Springer, 1994.

Optimal Scheduling Algorithms for Communication Constrained Parallel Processing

D. Turgay Altılar and Yakup Paker

Dept. of Computer Science, Queen Mary, University of London
Mile End Road, E1 4NS, London, United Kingdom
{altilar, paker}@dcs.qmul.ac.uk

Abstract. With the advent of digital TV and interactive multimedia over broadband networks, the need for high performance computing for broadcasting is stronger than ever. Processing a digital video sequence requires considerable computing. One of the ways to cope with the demands of video processing in real-time, we believe, is parallel processing. Scheduling plays an important role in parallel processing especially for video processing applications which are usually bounded by the data bandwidth of the transmission medium. Although periodic real-time scheduling algorithms have been under research for more than a decade, scheduling for continuous data streams and impact of scheduling on communication performancc are still unexplored. In this paper we examine periodic real-time scheduling assuming that the application is communication constrained where input and output data sizes are not equal.

1 Introduction

The parallel video processing scheduling system studied here assumes a real-time processing with substantial amount of periodic data input and output. Input data for such a real-time system consists of a number of video sequences that naturally possess continuity and periodicity features. Continuity and periodicity of the input leads one to define predictable and periodic scheduling schemes for data independent algorithms. Performance of a scheduling scheme relies upon both the system architecture and the application. Architectural and algorithmic properties enable to define relations among the number of processors, required I/O time, and processing time. I/O bandwidth, processor power, and data transmission time, could be considered as architectural properties. Properties of the algorithm indicate the requirements of an application such as the need of consecutive frames for some computation. In this paper, two scheduling and data partitioning schemes for parallel video processing system are defined by optimising the utilisation of first I/O channels and then processors. Although it is stated that the goal of high performance computing is to minimise the response time rather than utilising processors or increasing throughput [1], we have concentrated both on utilisation side and response time. In the literature, there are a number of cost models such as the ones defined in [1],[2],[3],[4],[5] and [6]. We defined scheduling and data partitioning schemes that can work together. The

B. Monien and R. Feldmann (Eds.): Euro-Par 2002, LNCS 2400, pp. 197–206.

parameters for defined schemes reflect the features of the chosen parallel system architecture and algorithm class. The defined schemes could be used for finding the optimal number of processors and partitions to work on for each scheduling model. In an other way around, system requirements could also be computed for a specific application, which enables us to build the parallel processing system.

The target parallel architecture is a client-server based system having a point-to-point communication between the server and client processors, which are required to implement Single Program Multiple Data (SPMD) type of programming. A typical hardware configuration comprises a server processor, a frame buffer and a number of client processors connected via a high speed I/O bus and signal bus. Video data transfer occurs over the high speed I/O bus between clients and the frame buffer. The frame buffer is a specially developed memory to save video streams. Since the frame buffer can provide only one connection at a time, any access to the frame buffer should be under the control of an authority, the server, to provide the mutual exclusion. The server is responsible to initialise clients, to partition data, to sent data addresses to clients to read and write, and to act as arbiter of the high speed I/O bus. No communication or data transfer exists between client processors.

Digital video processing algorithms can be classified under two groups considering their dependency on the processing of the previous frames. If an algorithm runs over consecutive frames independently we call it *stream based processing* [7] which is not considered in this paper. If an algorithm requires the output from the previous frame of a stream, the computation of a frame can proceed when the previous frame is processed. We call this mode *frame by frame processing.* In order to run a frame by frame computation in parallel, a frame can be split into tiles to be distributed to client processors. These tiles are processed and then collected by the server to re-compose the single processed frame.

Parallel Recursive (PR) and Parallel Interlaced (PI) scheduling algorithms are suggested in this paper for parallel video processing applications that require the output from the preceding frame to start with a new one. Video input/output is periodic. A new frame appears for every 40 ms for a PAL sequence. Input and output size are unequal for many of the video processing algorithms . Such as in mixing two sequences outputs size is roughly one third of the input.

The rest of the paper is organised as follows: Section 2 introduces the mathematical modelling and relevant definitions that are used in analysis of scheduling models. Equal data partitioning scenarios are discussed and analysed in Section 3. Scheduling for unequal input and output are investigated and new algorithms are proposed and analysed in Section 4. Section 5 compares all the introduced methods via a case study. Paper ends with conclusions and further research.

2 Mathematical Modeling and Definitions

Read and write times can be best defined as a linear function of input data size and bus characteristics. The linear functions include a constant value, p for read and s for write, which identifies the cost of overhead. These constant costs are considered as initialisation costs due to the system (latency) and/or due to the algorithm (data structure initialisations). Data transfer cost is proportional to another constant q for read and t for write. Computation time is accepted as proportional to the data input size. r is computational cost per unit data. It is important to note that r is not a complexity term. d_i indicates the partition of the data in percentage to sent i^{th} processor. Throughout the following derivations only input data size is taken as a variable. Consistant with the existing literature and cost models referred in the introduction, the developed a cost model includes first degree equations for cost analysis although numeric solutions always exist for higher degree equations. For the ith processor read R_i, compute C_i and write W_i times can be expressed as follows where the sum of all d_i is 1;

$$R_i = p + qd_i,\ C_i = rd_i,\ W_i = w + td_i \tag{1}$$

Sending data from frame buffer to client processors, processing in parallel and receiving processed data from all of the available client processors constitutes a cycle. Processing of a single frame finishes by the end of a cycle. Since our intention is to overlap compute time of a processor with I/O times of the others, starting point of the analysis is always an equation between read, compute and write times. In order to make a comparative analysis the response time, T_{cycle}, is essential. Also note that T_{cycle} provides a means to compute speed up.

3 Equal Data Partitioning

Partitioning data in equal sizes is the simplest and standard way of data partitioning to provide load balancing. Data is partitioned into N equal sizes to be dispatched to N processors. While processors compute their part of data, they obviously leave the I/O bus free. Utilisation of the I/O Bus depends on these idle durations. Whenever a processor starts computation another one starts a read. This continues in the same manner for other processors until the very first one finishes processing its data and becomes ready to write its output via the I/O bus. One could envisage a scenario that the computation time of the first processor is equal to the sum of the read time of others so that no I/O wait time is lost for the first processor. Therefore, the maximum number of processors is determined by the number of read time slots available for other processors within computation time of the first processor. In order to ensure the bus becomes free when the first processor completes computation, the compute time for the first processor must be equal to or greater than the sum of all reads. Similarly for the second processor's computation time can be defined as the sum of read times of the successor processors and the write time of the first one. If one continues for the subsequent processors, it is easy to see that compute time for i^{th} processor must be greater than or equal to the sum of read times of the successor processors and the sum of write times of the predecessor processors. Assuming that N is the number of processors, to achieve the full utilisation of data bus computation time should equal to sum of communication times:

$$C_i = \sum_{k=1}^{i-1} W_k + \sum_{j=i+1}^{N} R_j \tag{2}$$

By substituting definitions of R, W and C given in Eq.1 in Eq.2 and solving the produced quadratic function the positive root can be found as follows:

$$N = \left((p-q) + \sqrt{(p-q)^2 + 4pr} \right) / 2p \tag{3}$$

The lower bound of N is the optimal value for N for the utilisation of I/O bus. Moreover, the cycle time (T_{cycle}), posing another constraint to be met in real time video processing can be computed as the sum of all writes and reads, i.e. $T_{\text{cycle}} = 2\left(\lfloor N \rfloor p + q\right)$.

3.1 Equal Data Partitioning with Unequal I/O

However, when input and output data sizes (or cost factors) become different equal partitioning can not provide the best solution. There can two cases of unequal input and output data transfer: input data takes longer to transfer than output or vice-versa.

Write time greater than read time. The first case is for a generic class of the algorithms with larger output data size than input such as rendering a 3D scene. Rendering synthetic images, the data size of 3D modelling parameters (polygons) to construct the image is less than the rendered scene (pixels). If processors receive equal amount of data they all produce output after a computation time which is almost the same for each of them. As writing output data takes longer than reading input data, the successor processor waits the predecessor to finish its writing back. Although the load balancing is perfect, i.e. each processor spends the same amount of time for computation, I/O channel is not fully utilised. In Fig.1a, L_2, L_3, and L_4 indicate the time that processors spend while waiting to write back to the frame buffer. We keep the same approach as we analyse the equal input and output case: computation time should overlap data transfer time (either read or write) of the other processors. It can be seen in Fig.1a that computation time of the first processor can be made equal to the read time of the rest of the processors. For the second processor however, the derivation introduces a new period called L for the idle duration of the processor as W_1 R_2 (Note that all read times are equal as well as write times). Therefore the difference between read and write time produces an idle duration for the successor processor. The latency for the second processor is $L2 = W1 - R2$. The sum of all idle durations for all client processors is $L_{\text{total}} = \left(N^2 - N\right) L_2/2$ As shown in Fig.1a, although I/O channel is fully utilised client processors are not. Moreover, the cycle time is extended by the idle time of the last client processor taken part in the computation. The overall parallel computation cycle time is: $T_{cycle} = N(R + W)$.

Read time greater than write time. The second generic case (Fig.1b) occurs when writing takes less time than reading data. Consider motion estimation of MPEG video compression which reads a core block (called "macro block" in MPEG terminology) of a by a pixels from the current frame to be matched with neighbouring blocks of previous frame within a domain of $(2b+1)(2b+1)$ pixels centred on the macro block where b could be up to $16a$ [8]. However the output is only a motion vector determining the direction of the macro block. The second step of the derivation introduces a new duration called I for the idle duration of the I/O bus. The difference between read and write time produces an idle duration for the I/O bus which can be given as I=R-W. As a processor finishes writing earlier than the start of writing of its successor there is no queuing effect. The sum of idle durations of the I/O bus, it I_T, is proportional to the number processors, $I_T = (N-1)I$, and T_{cycle} becomes: $T_{cycle} = (2N-1)R + W$.

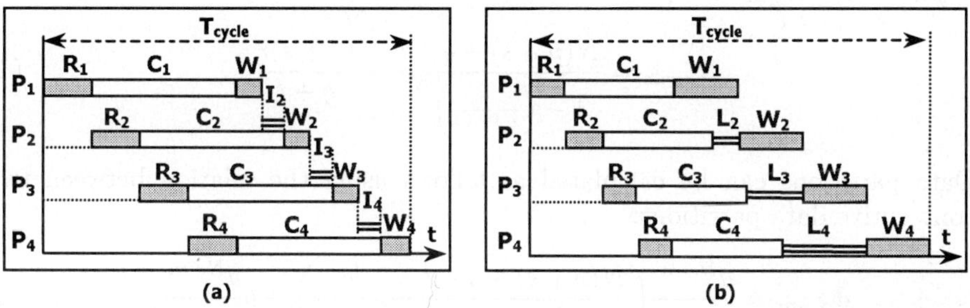

Fig. 1. Equal data partitioning with (a) write time greater than read time and (b) read time greater than write time

4 Scheduling for Unequal I/O

We have shown in Section 3 that equal data partitioning for equal load distribution does not always utilise the I/O channel and/or the processors fully. The following figures (Fig.2a, Fig.2b and Fig.2c) show the three possible solutions based on two new partitioning approaches. The main objective is to maximise I/O Bus utilisation since we assume applications are bounded by data transfer. We also assume that the algorithm is data independent and data can be partitioned and distributed in arbitrary sizes.

4.1 PR Scheduling and Data Partitioning

Parallel Recursive (PR) data partitioning and scheduling method exploits the computation duration of a processor for its successor to proceed with its I/O. As the successor processor starts computation, the next one can start its I/O. This basic approach can be recursively applied until the compute time becomes not long enough for read-compute-write sequence of the successor processor. Although utilisation of the I/O channel would be high, and cycle time would be better than the equal data partitioning (PE) method, it suffers from the under utilisation of processors. Recursive structure of the scheduling and partitioning provides a repetitive pattern for all the processors. Since subsequent processors exploit duration between read and write times of the first processor, cycle time is determined by the first processor. The computation time of the first processor which leaves I/O bus idle is used by the second one. The same relationship exists between the second processor and the third one and so on. Although read time is greater than write time in Fig.2a, the following equations are also valid for the other two scenarios in which (i) write time is greater than read time and (ii) write time and read time are equal. The first processor dominates the cycle time. One can define compute time considering Fig 6 for N processors: $C_i = R_{i+1} + C_{i+1} + W_{i+1}$ Since the sum of all reads and writes is equal to $T_{cycle} - C_N$. T_{cycle} can be derived as follows in terms of system constants:

$$T_{cycle} = \frac{N(p+s)+q+t}{1-\left(\frac{r}{q+r+t}\right)^N} - \frac{(p+s)r}{q+t} \tag{4}$$

Data partitions can be calculated as follows using the relation between two consecutive data partitions:

$$d_{N-m} = \frac{a^{N-m}}{r}\left(\frac{N(p+s)+q+t}{1-a^N} - \frac{br}{a-1}\right) + b\frac{a^{N-m}-1}{a-1} \tag{5}$$

where $a = r/(q+r+t)$ and $b = -(p+s)/(q+r+t)$

The number of processors to maximise the utilisation of I/O channel is also a question worth considering. The recursive structure of the model leaves smaller task for a processor than its predecessor. After a number of successive iterative steps the compute time of a processor will not be sufficient for its successor for read and write as the overall working time becomes smaller for the successor processors. This constraint poses a limit for the number of processors. In the case of computing data partition size for an insufficient slot the computed data size would be negative. N can be computed numerically via the following inequality:

$$\frac{Na^N(p+s)+q+t}{1-a^N} \rangle \frac{br}{a-1} + q + t \tag{6}$$

4.2 PI Scheduling and Data Partitioning

Parallel Interlaced (PI) scheduling and data partitioning method is another proposed method to maximise the utilisation of the I/O bus. Unlike PR the basic approach is for each processor to complete its read-compute-write cycle after its predecessor but before its successor. This is the same approach that we use to analyse equal input and output. The two other possible scenarios is analysed in this section. Fig.2a and Fig.2b show the possible solutions for unequal

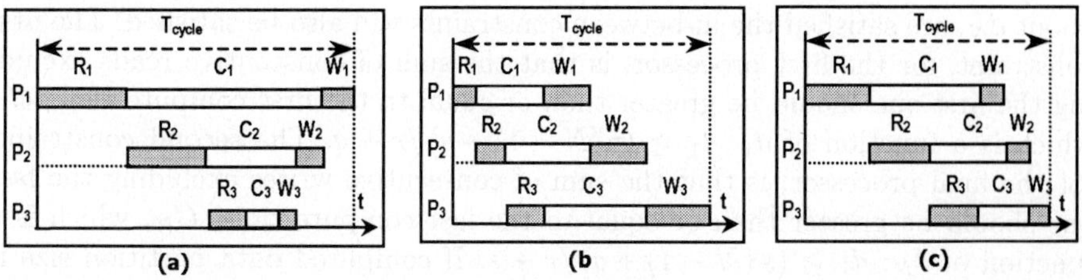

Fig. 2. Optimal data partitioning with (a) write time greater than read time and (b) read time greater than write time

input/output. For the first case given in Fig.2b, since writing requires more time than reading, computation time should increase with the processor number in order to accommodate longer writing times. Since read, compute, and write times are proportional to data size, from Fig.2b we can say that ascending read, compute and write times increases with the increasing index of processors provides full utilisation of the I/O channel for an application with longer write time than read. The second case is shown in Fig.2c where reading requires more time than writing. Thus, computation time should decrease with the increase of processor number in order to accommodate shorter writing times. A close look at Fig.2c shows that with increased processor numbers the read compute and write times are also increased. So long as the read time is longer than the write time, the difference reduces the time for the successor processor to read and compute. Although the difference between write time and read time provides an additional time for the successor processor in one case (Fig.2b), and reduces the time for the other case (Fig.2c) the compute time and response time satisfy the following equations for both of the cases:

$$T_{cycle} = C_i + \sum_{k=1}^{i} R_k + \sum_{j=i}^{N} W_j \tag{7}$$

$$C_i + W_i = R_{i+1} + C_{i+1} \tag{8}$$

One can solve these equations for d_n as follows

$$d_n = (r+t)^{n-1}(r+q)^{N-n}\left(\frac{t-q+N(s-p)}{(r+t)^N-(r+q)^N}\right) - \frac{s-p}{r-q} \tag{9}$$

Thus for a given number of processors N and systems and algorithmic constants, data partitions can be computed. We dealt with a relation between two consecutive data partitions and which allows us to derive recursively all the others. However, since the aim is high utilisation of the I/O channel, data partitions should also fulfil the constraints. These constraints derived from the relations between compute time of one processors with read and write times of the others. We are going to deal with two constraints, which could be considered as upper and lower bounds. If these two constraints, one is about d1 and the other is about d_N, are satisfied the in-between constraints will also be satisfied. The first constraint, for the first processor, is that the sum of consecutive reads excluding the first one should be greater than or equal to the first compute time, C_1, which is a function of d_1: $d_1 \geq (p\,(N-1)+q)\,/r + q$. The second constraint, for the final processor, is that the sum of consecutive writes excluding the last one should be greater than or equal to the last compute time, C_N, which is a function of d_N: $d_1 \geq (s\,(N-1)+q)\,/r + s$. If computed data partition size is less than any of these two limits values, data transmission time will be less than the compute time which yields poor utilisation of the I/O bus and increase in cycle time.

5 Comparison of Data Partitioning Schemes

In order to compare the given three methods, PE, PR, and PI, data partitions and cycle times for a single frame (T_{cycle}) have to be computed. This comparison will indicate the shortest cycle time which is crucial for real-time video processing. On the other hand, there are constraints to be satisfied in order to utilise the I/O channel. Fig.3 and Fig.4 give a brief information about the data partitions with constraints and cycle times. Since we are dealing with a colour PAL video sequences of 576*720 pixels, 24 bit colour and 25 frames per second a single PAL frame is approximately 1.2 Mbytes and has to be processed within 40 ms.

The algorithm considered in this example is mixing which requires three video streams: two streams to mix and one alpha frame to define the layering. Initialisation for reading which includes both algorithmic and systems delay is assumed to be 3.00 ms. Initialisation duration for writing is assumed to be less than reading and is 1.20 s, i.e., p=3.00 ms and s=1.20 ms. Assuming that the bus is rated 1GBytes/sec and since three streams are required for input and one is produced for output for mixing overall read and write times are $R_{overall}$=3.6ms and $W_{overall}$=1.2ms. Therefore q=3.60ms and t=1.20 ms. Given a CPU with a clock rate of 300MHz, assume that the algorithm requires 30 cycles per pixel - which can be found either by rehearsal runs on a single CPU or analysing the machine code of the program - to compute ends with a total processing time $W_{overall}$ 120ms i.e, r=120 ms. Fig.3 and Fig.4 are produced for p=3.00 ms, q=3.60 ms, r=120 ms, s=1.20 ms, and t=1.20 ms with regard to the given analysis and derived equations. Partition percentages and cycle times per frame for equal partitioning (PE) method is given in Table.1. The first row of the table indicates cycle times for different numbers of processors. Obviously the best result is 43.00 ms for 6 processors. The last two lines of constraints for data partitions are also satisfied for 6 processors. Therefore the best overall process cycle can be declared as of 6 processors. Data partitions would be equal for processors and each processors would receive approximately 17% of the input to process. However overall processing time of 43ms does not satisfy the real-time constraint of video processing for a PAL sequence of 25 frames per second. The number of processors can be computed by Eq.3 as 6.3195. The lower bound of N is equal to 6. Therefore 6 processors give the best solution for highly utilisation of

(PE)

		PROCESSORS						
		1	2	3	4	5	6	7
T_{cycle}		129.0	71.40	54.20	47.10	44.04	43.00	43.11
PARTITIONS	1	1.00	0.50	0.33	0.25	0.20	0.17	0.14
	2		0.50	0.33	0.25	0.20	0.17	0.14
	3			0.33	0.25	0.20	0.17	0.14
	4				0.25	0.20	0.17	0.14
	5					0.20	0.17	0.14
	6						0.17	0.14
	7							0.14
$d_1 \geq$ ...		-	0.05	0.08	0.10	0.13	0.15	0.17
$d_N \geq$ ...		-	0.02	0.03	0.04	0.05	0.06	0.07

(PR)

		PROCESSORS						
		1	2	3	4	5	6	7
T_{cycle}		129.0	69.96	51.75	43.76	39.88	38.07	37.45
PARTITIONS	1	1.00	0.53	0.38	0.32	0.29	0.27	0.27
	2		0.47	0.33	0.27	0.24	0.23	0.22
	3			0.29	0.23	0.20	0.18	0.18
	4				0.18	0.16	0.14	0.14
	5					0.12	0.10	0.10
	6						0.07	0.06
	7							0.03

Fig. 3. Data partitions and cycle times for PE and PR

I/O channel. Rounding the number to its lower bound yields a deviation from the optimal solution. Fig.3 shows the results for recursive partitioning (PR) method. Best cycle time is found for 7 processors, i.e., 37.45 ms. As PR method is recursive there is no constraint due to the data size except for the fact that partitions should be positive percentages. For eight processors, the size of data partition for the eight processor is computed to be less than zero. Therefore the maximum number of processors for this case is 7. The results for interlaced partitioning is shown in Fig.4. The best overall cycle time is 36.82ms for 8 processors. However partitions given for 8 processors do not satisfy the constraints given in last two rows of the table. The first column fulfilling the constraints is for 7 processors. The overall cycle time is 36.85ms which also satisfies 40 ms maximum processing time constraint. The cycle time values for the three methods are drawn in Fig.4.

Obviously PI has the best performance, where PR comes the second and PE the third. One can see the change of slopes of the curves at different values. The point on which slope is zero indicates optimum number of processors to provide the shortest cycle time if this value satisfies the constraints as well.

6 Conclusion and Further Research

In this paper, we proposed two optimal data partitioning and scheduling algorithms, Parallel Recursive (PR) and Parallel Interlaced (PI), for real-time fram by frame processing. We also provide analysis and simulation results to compare these two with the conventional Parallel Equal (PE) method. We aimed at highly utilisation of I/O bus or I/O channel under the assumptions of being dealt with data bandwidth bounded applications having different input and output data sizes. The proposed algorithms are developed considering some parallel digital video processing applications representing a wide range of applications. These algorithms apply on any data independent algorithm requiring substantial amount of data to process where arbitrary data partitioning is available. In the systems side, an optimal value for the number of processors can be computed for given characteristics of both application and systems which is modeled with

(PI)		PROCESSORS							
		1	2	3	4	5	6	7	8
T_{cycle}		129.0	69.91	51.63	43.56	39.57	37.63	36.85	36.82
PARTITIONS	1	1.00	0.51	0.35	0.28	0.24	0.21	0.20	0.19
	2		0.49	0.33	0.26	0.22	0.19	0.18	0.17
	3			0.31	0.24	0.20	0.18	0.16	0.15
	4				0.22	0.18	0.16	0.14	0.13
	5					0.16	0.14	0.12	0.12
	6						0.12	0.11	0.10
	7							0.09	0.08
	8								0.07
$d_1 \geq$...		-	0.05	0.08	0.10	0.13	0.15	0.17	0.20
$d_N \geq$...		-	0.02	0.03	0.04	0.05	0.06	0.07	0.08

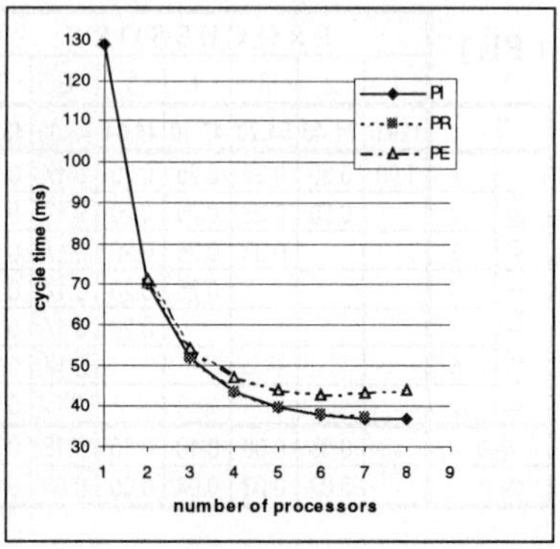

Fig. 4. Data partitions and cycle times for PI and comparison of cycle times

five parameters. Suggested algorithms were evaluated only on a bus based architecture with video based applications in this paper. Hierarchical structures such as tree architectures, mathematical applications such as domain decomposition are yet to be investigated using the same cost model and analysis method.

References

[1] Crandall P. E., Quinn M. J., A Partitioning Advisory System for Networked Data-parallel Processing, Concurrency: Practice and Experience, 479-495, August 1995.

[2] Agrawal R, Jagadish H V, Partitioning Techniques for Large-Grained Parallelism, IEEE Transactions on Computers, Vol.37, No.12, December,1988.

[3] Culler D, Karp R, Patterson D, Sahay A, Schauser K, Santos E, Subramonian R and Eicken T, LogP: Towards a realistic mode of parallel computation, Proceedings of 4th ACM SIGPLAN Symposium on Principles and Practices of Parallel Programming, Vol.28, May 1993.

[4] Lee C., Hamdi M., Parallel Image Processing Applications on a Network of Workstations, Parallel Computing, 21 (1995), 137-160.

[5] Moritz C A, Frank M, LoGPC: Modeling Network Contention in Message-Passing Programs, ACM Joint International Conference on Measurement and Modeling of Computer Systems, ACM Sigmetrics/Performance 98, Wisconsin, June 1998.

[6] Weissman J.B., Grimshaw A. S., A Framework for Partitioning Parallel Computations in Heterogeneous Environments, Concurrency: Practice and Experience, Vol.7(5),455-478,August 1995.

[7] Altilar D T, Paker Y, An Optimal Scheduling Algorithm for Parallel Video Processing, Proceedings of International Conference on Multimedia Computing and Systems'98, Austin Texas USA, 245-258, July 1998.

[8] ISO/IEC, MPEG 4 Video Verification Model Ver7.0, N1642, Bristol, April 1997.

Job Scheduling for the BlueGene/L System

Elie Krevat[1], José G. Castaños[2], and José E. Moreira[2]

[1] Massachusetts Institute of Technology, Cambridge, MA 02139-4307
krevat@mit.edu

[2] IBM T. J. Watson Research Center, Yorktown Heights, NY 10598-0218
{castanos,jmoreira}@us.ibm.com

Abstract. Cellular architectures with a toroidal interconnect are effective at producing highly scalable computing systems, but typically require job partitions to be both rectangular and contiguous. These restrictions introduce fragmentation issues which reduce system utilization while increasing job wait time and slowdown. We propose to solve these problems for the BlueGene/L system through scheduling algorithms that augment a baseline first come first serve (FCFS) scheduler. Our analysis of simulation results shows that migration and backfilling techniques lead to better system performance.

1 Introduction

BlueGene/L (BG/L) is a massively parallel cellular architecture system. 65,536 self-contained computing nodes, or *cells*, are interconnected in a three-dimensional toroidal pattern [7]. While toroidal interconnects are simple, modular, and scalable, we cannot view the system as a flat, fully-connected network of nodes that are equidistant to each other. In most toroidal systems, job partitions must be both rectangular (in a multidimensional sense) and contiguous. It has been shown in the literature [3] that, because of these restrictions, significant machine fragmentation occurs in a toroidal system. The fragmentation results in low system utilization and high wait time for queued jobs.

In this paper, we analyze a set of scheduling techniques to improve system utilization and reduce wait time of jobs for the BG/L system. We analyze two techniques previously discussed in the literature, backfilling [4–6] and migration [1, 8], in the context of a toroidal-interconnected system. Backfilling is a technique that moves lower priority jobs ahead of other higher priority jobs, as long as execution of the higher priority jobs is not delayed. Migration moves jobs around the toroidal machine, performing on-the-fly defragmentation to create larger contiguous free space for waiting jobs.

We conduct a simulation-based study of the impact of those techniques on the system performance of BG/L. We find that migration can improve maximum system utilization, while enforcing a strict FCFS policy. We also find that backfilling, which bypasses the FCFS order, can lead to even higher utilization and lower wait times. Finally, we show that there is a small benefit from combining backfilling and migration.

2 Scheduling Algorithms

This section describes four job scheduling algorithms that we evaluate in the context of BG/L. In all algorithms, arriving jobs are first placed in a queue of waiting jobs,

B. Monien and R. Feldmann (Eds.): Euro-Par 2002, LNCS 2400, pp. 207–211.

prioritized according to the order of arrival. The scheduler is invoked for every job arrival and job termination event, and attempts to schedule new jobs for execution.

First Come First Serve (FCFS). For FCFS, we adopt the heuristic of traversing the waiting queue in order and scheduling each job in a way that maximizes the largest free rectangular partition left in the torus. If we cannot fit a job of size p in the system, we artificially increase its size and retry. We stop when we find the first job in the queue that cannot be scheduled.

FCFS With Backfilling. Backfilling allows a lower priority job j to be scheduled before a higher priority job i as long as this reschedule does not delay the estimated start time of job i. Backfilling increases system utilization without job starvation [4, 9]. It requires an estimation of job execution time. Backfilling is invoked when FCFS stops because a job does not fit in the torus and there are additional jobs in the waiting queue. A reservation time for the highest-priority job is then calculated, based on the worst case execution time of jobs currently running. If there are additional jobs in the waiting queue, a job is scheduled out of order as long as it does not prevent the first job in the queue from being scheduled at the reservation time.

FCFS With Migration. The migration algorithm rearranges the running jobs in the torus in order to increase the size of the maximal contiguous rectangular free partition, counteracting the effects of fragmentation. The migration process is undertaken immediately after the FCFS phase fails to schedule a job in the waiting queue. Running jobs are organized in a queue of migrating jobs sorted by size, from largest to smallest. Each job is then reassigned a new partition, using the same algorithm as FCFS and starting with an empty torus. After migration, FCFS is performed again in an attempt to start more jobs in the rearranged torus.

FCFS with Backfilling and Migration. Since backfilling and migration are independent scheduling concepts, an FCFS scheduler may implement both of these functions. First, we schedule as many jobs as possible via FCFS. Next, we rearrange the torus through migration to minimize fragmentation, and then repeat FCFS. Finally, the backfilling algorithm from Scheduler 2 is performed.

3 Experiments

We used an event-driven simulator to process actual job logs of supercomputing centers. The results of simulations for all four schedulers were then studied to determine the impact of their respective algorithms. The BG/L system is organized as a $32 \times 32 \times 64$ three-dimensional torus of nodes (cells). The unit of allocation for job execution in BG/L is a 512-node ensemble organized in an $8 \times 8 \times 8$ configuration. Therefore, BG/L behaves as a $4 \times 4 \times 8$ torus of these *supernodes*. We use this supernode abstraction when performing job scheduling for BG/L. That is, we treat BG/L as a machine with 128 (super)nodes.

A job log contains information on the arrival time, execution time, and size of all jobs. Given a torus of size N, and for each job j the arrival time t_j^a, execution time t_j^e and size s_j, the simulation produces values for the start time t_j^s and finish time t_j^f of each job. These results were analyzed to determine the following parameters for each job: (1)

wait time $t_j^w = t_j^s - t_j^a$, (2) response time $t_j^r = t_j^f - t_j^a$, and (3) bounded slowdown $t_j^{bs} = \frac{\max(t_j^r, \Gamma)}{max(t_j^e, \Gamma)}$ for $\Gamma = 10$ s. The Γ term appears according to recommendations in [4], because jobs with very short execution time may distort the slowdown.

Global system statistics are also determined. Let the simulation time span be $T = \max_{\forall j}(t_j^f) - \min_{\forall k}(t_k^a)$. We then define system utilization (also called *capacity utilized*) as $w_{\text{util}} = \sum_{\forall j} \frac{s_j t_j^e}{TN}$. Similarly, let $f(t)$ denote the number of free nodes in the torus at time t and $q(t)$ denote the total number of nodes requested by jobs in the waiting queue at time t. Then, the total amount of unused capacity in the system, w_{unused}, is defined as $w_{\text{unused}} = \int_{\min(t_j^a)}^{\max(t_j^f)} \max(0, f(t) - q(t))dt$. This parameter is a measure of the work unused by the system because there is a lack of jobs requesting free nodes. The balance of the system capacity is lost despite the presence of jobs that could have used it. The lost capacity in the system is then derived as $w_{\text{lost}} = 1 - w_{\text{util}} - w_{\text{unused}}$.

We performed experiments on 10,000-job segments of two job logs obtained from the *Parallel Workloads Archive* [2]. The first log is from NASA Ames's 128-node iPSC/860 machine (from the year 1993). The second log is from the San Diego Supercomputer Center's (SDSC) 128-node IBM RS/6000 SP (from the years 1998-2000). In the NASA log, job sizes are always powers of 2. In the SDSC log, job sizes are arbitrary. Using these two logs as a basis, we generated logs of varying workloads by multiplying the execution time of each job by a constant coefficient.

Figure 1 presents a plot of average job bounded slowdown (t_j^{bs}) $\times$ system utilization (w_{util}) for each of the four schedulers considered and each of the two job logs. (B+M is the backfilling and migration scheduler.) We also include results from the simulation of a fully-connected (*flat*) network. This allows us to assess how effective the schedulers are in overcoming the difficulties imposed by a toroidal interconnect. The overall shapes of the curves for wait time are similar to those for bounded slowdown. The most significant performance improvement is attained through backfilling, for both the NASA and SDSC logs. Also, for both logs, there is a certain benefit from migration, whether combined with backfilling or not.

With the NASA log, all four schedulers provide similar average job bounded slowdown for utilizations up to 65%. The FCFS and Migration schedulers saturate at about 77% and 80% utilization respectively. Backfilling (with or without migration) allows utilizations above 80% with a bounded slowdown of less than a hundred. We note that migration provides only a small improvement in bounded slowdown for most of the utilization range. In the NASA log, all jobs are of sizes that are powers of 2, which results in a good packing of the torus. Therefore, the benefits of migration are limited.

With the SDSC log, the FCFS scheduler saturates at 63%, while the stand-alone Migration scheduler saturates at 73%. In this log, with jobs of more varied sizes, fragmentation occurs more frequently. Therefore, migration has a much bigger impact on FCFS, significantly improving the range of utilizations at which the system can operate. However, we note that when backfilling is used there is again only a small benefit from migration, more noticeable for utilizations between 75 and 85%.

Migration by itself cannot make the results for a toroidal machine as good as those for a flat machine. For the SDSC log, in particular, a flat machine can achieve better than 80% utilization with just the FCFS scheduler. However, the backfilling results are closer

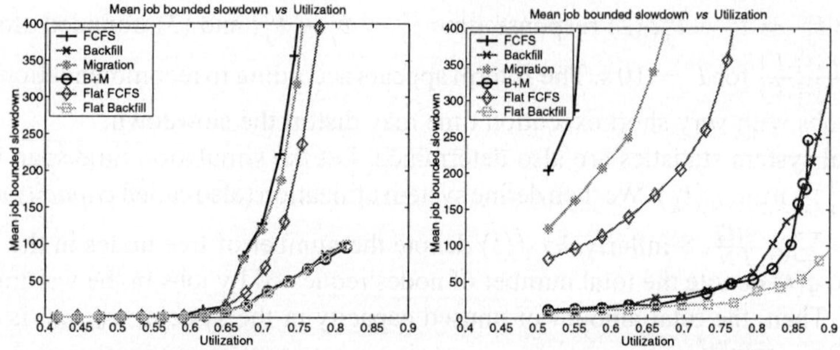

(a) NASA iPSC/860 (b) SDSC RS/6000 SP

Fig. 1. Mean job bounded slowdown *vs* utilization for the NASA and SDSC logs, comparing toroidal and flat machines.

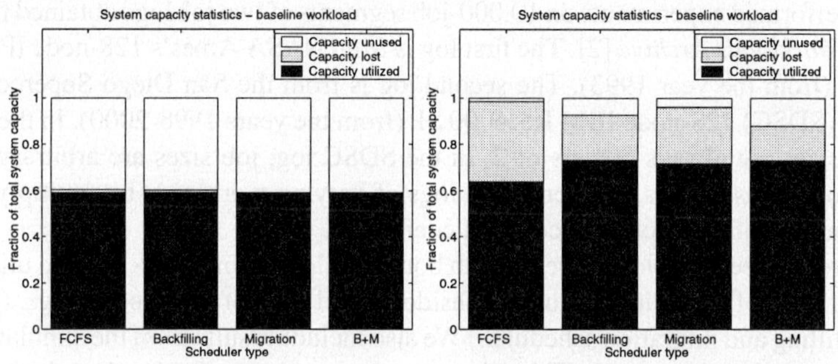

(a) NASA iPSC/860 (b) SDSC RS/6000 SP

Fig. 2. Capacity utilized, lost, and unused as a fraction of the total system capacity.

to each other. For the NASA log, results for backfilling with migration in the toroidal machine are just as good as the backfilling results in the flat machine. For the SDSC log, backfilling on a flat machine does provide significantly better results for utilizations above 85%.

The results of system capacity utilized, unused capacity, and lost capacity for each scheduler type and both job logs (scaling coefficient of 1.0) are plotted in Figure 2. The utilization improvements for the NASA log are barely noticeable – again, because its jobs fill the torus more compactly. The SDSC log, however, shows the greatest improvement when using B+M over FCFS, with a 15% increase in capacity utilized and a 54% decrease in the amount of capacity lost. By themselves, the Backfill and Migration schedulers each increase capacity utilization by 15% and 13%, respectively, while decreasing capacity loss by 44% and 32%, respectively. These results show that B+M is significantly more effective at transforming lost capacity into unused capacity.

4 Related and Future Work

The topics of our work have been the subject of extensive previous research. In particular, [4–6] have shown that backfilling on a flat machine like the IBM RS/6000 SP is an effective means of improving quality of service. The benefits of combining migration and gang-scheduling have been demonstrated both for fully connected machines [10] and toroidal machines like the Cray T3D [3]. This paper applies a combination of backfilling and migration algorithms, exclusively through space-sharing techniques, to improve system performance on a toroidal-interconnected system. As future work, we plan to study the impact of different FCFS scheduling heuristics for a torus. We also want to investigate time-sharing features enabled by preemption.

5 Conclusions

We have investigated the behavior of various scheduling algorithms to determine their ability to increase processor utilization and decrease job wait time in the BG/L system. We have shown that a scheduler which uses only a backfilling algorithm performs better than a scheduler which uses only a migration algorithm, and that migration is particularly effective under a workload which produces a large amount of fragmentation. We show that FCFS scheduling with backfilling and migration shows a slight performance improvement over just FCFS and backfilling. Backfilling combined with migration converts significantly more lost capacity into unused capacity than just backfilling.

References

1. D. H. J. Epema, M. Livny, R. van Dantzig, X. Evers, and J. Pruyne. **A worldwide flock of Condors: Load sharing among workstation clusters**. *Future Generation Computer Systems*, 12(1):53–65, May 1996.
2. D. G. Feitelson. **Parallel Workloads Archive**. `http://www.cs.huji.ac.il/labs/parallel/workload/index.html`.
3. D. G. Feitelson and M. A. Jette. **Improved Utilization and Responsiveness with Gang Scheduling**. In *IPPS'97 Workshop on Job Scheduling Strategies for Parallel Processing*, volume 1291 of *Lecture Notes in Computer Science*, pages 238–261. Springer-Verlag, 1997.
4. D. G. Feitelson and A. M. Weil. **Utilization and predictability in scheduling the IBM SP2 with backfilling**. In *12th International Parallel Processing Symposium*, April 1998.
5. D. Lifka. **The ANL/IBM SP scheduling system**. In *IPPS'95 Workshop on Job Scheduling Strategies for Parallel Processing*, volume 949 of *Lecture Notes in Computer Science*, pages 295–303. Springer-Verlag, April 1995.
6. J. Skovira, W. Chan, H. Zhou, and D. Lifka. **The EASY-LoadLeveler API project**. In *IPPS'96 Workshop on Job Scheduling Strategies for Parallel Processing*, volume 1162 of *Lecture Notes in Computer Science*, pages 41–47. Springer-Verlag, April 1996.
7. H. S. Stone. **High-Performance Computer Architecture**. Addison-Wesley, 1993.
8. C. Z. Xu and F. C. M. Lau. **Load Balancing in Parallel Computers: Theory and Practice**. Kluwer Academic Publishers, Boston, MA, 1996.
9. Y. Zhang, H. Franke, J. E. Moreira, and A. Sivasubramaniam. **Improving Parallel Job Scheduling by Combining Gang Scheduling and Backfilling Techniques**. In *Proceedings of IPDPS 2000*, Cancun, Mexico, May 2000.
10. Y. Zhang, H. Franke, J. E. Moreira, and A. Sivasubramaniam. **The Impact of Migration on Parallel Job Scheduling for Distributed Systems**. In *Proceedings of the 6th International Euro-Par Conference*, pages 242–251, August 29 - September 1 2000.

An Automatic Scheduler for Parallel Machines

Mauricio Solar and Mario Inostroza

Universidad de Santiago de Chile, Departamento de Ingenieria Informatica,
Av. Ecuador 3659, Santiago, Chile
{msolar, minostro}@diinf.usach.cl

Abstract. This paper presents a static scheduler to carry out the best assignment of a Directed Acyclic Graph (DAG) representing an application program. Some characteristics of the DAG, a decision model, and the evaluation parameters for choosing the best solution provided by the selected scheduling algorithms are defined. The selection of the scheduling algorithms is based on five decision levels. At each level, a subset of scheduling algorithms is selected. When the scheduler was tested with a series of DAGs having different characteristics, the scheduler's decision was right 100% of the time in those cases in which the number of available processors is known. [1]

1 Introduction

This paper is included in the framework of a research project aimed at creating a parallel compiler [1] for applications written in C programming language, in which the scheduling algorithms for generating an efficient parallel code to be carried out on a parallel machine are automatically selected. The input program in C is represented by a task Directed Acyclic Graph (DAG) which is assigned by means of scheduling algorithms, depending on the DAGs characteristics. The stage of this project which is presented in this paper is the implementation of the scheduler in charge of automatically selecting the scheduling algorithms which make the best assignment of the DAG, depending on the latter's characteristics.

The paper introduces the theoretical framework (some definitions). Section 3 describes the scheduler design and the scheduler's decision model. Section 4 shows the results obtained. Finally, the conclusions of the work are given.

2 Theoretical Framework

The applications that it is desired to parallelize may be represented by a task graph in the DAG form, which is a graph that has the characteristic of being acyclic and directed, and can be regarded as a tuple $D = (V, E, C, T)$, where V is the set of DAG tasks; $v = |V|$ is the number of DAG tasks; v_i is the ith DAG task; E is the set of DAG edges, made of e_{ij} elements, e_{ij} is the edges from task v_i to task v_j; $e = |E|$ is the number of edges; C is the set of DAG

[1] This project was partially funded by FONDECYT 1000074.

B. Monien and R. Feldmann (Eds.): Euro-Par 2002, LNCS 2400, pp. 212–216.

communication costs c_{ij}; T is the set of execution time t_i of the DAG tasks; t_i is the execution time of v_i; t_m is the average value of the execution time of the tasks, $\sum t_i/v$; c_m is the average value of the communication costs, $\sum c_{ij}/e$; G is the granularity, which is the t_m/c_m ratio in the DAG; L is the total number of DAG levels; R_{vn} is the level to task ratio, $1-(L-1)/(v-1)$; $blevel(v_x)$ is the length of the longest path between v_x (included) and an output task; $tlevel(v_x)$ is the length of the longest path between v_x (not included) and an input task; PT is the total parallel time for executing the assignment; and p is the number of processors available for carrying out the assignment.

3 Scheduler Design

The scheduler uses the DAG and its characteristics to choose the assignment heuristics which best assign the DAG on the target parallel machine. The scheduler uses both the DAG characteristics as well as those of the scheduling algorithms to carry out the selection of the latter for assigning the DAG. A Gantt chart with the DAG's planning is given. Two types of scheduling algorithms are considered: List and Clustering [2]. Table 1 shows a summary of the main characteristics of the algorithms that are considered. The 2nd column shows an order of time complexity of each algorithm. The 3rd and 4th columns indicate whether the algorithm considers some special restriction in terms of t_i and/or c_{ij}, respectively. The 5th and 6th column show if the priority function considers the calculation of *blevel* and *tlevel*, respectively. The last column shows if the algorithm serves for some special case of G.

The Scheduler model input corresponds to the DAG and its characteristics, and the output is the best DAG planning found by the scheduler. The design is made of six blocks: *Block 1 (DAG and its characteristics)* represents the input to the system; *Block 2 (Scheduler decision)* makes the decision of which scheduling algorithms to use, depending on the specific characteristics of the analyzed DAG. The scheduler's decision model has five stages as shown in Fig. 1; *Block 3 (Scheduling algorithms)* has a set of algorithms for planning the execution of the DAG; *Block 4 (Gantt chart proposals)* delivers as output a Gantt chart with the planning of the input DAG; *Block 5 (Analysis of Gantt charts)* selects the best planning delivered by the selected scheduling algorithms by comparing a set of evaluation parameters; *Block 6 (Final Gantt chart)* corresponds to the planning that gave the best yield according to the evaluation parameters, which are: PT, p, and total real communication time.

When stage 2 (Analysis of Characteristic k in Fig. 1) of the implemented scheduler's decision model is applied, five decision levels ($k = 5$) are obtained (shown in Table 2).

Level 1: Sarkar's algorithm sorts C according to their c_{ij}, giving higher priority to those which have a greater cost, with the purpose of minimizing the PT when assigning the higher c_{ij} to the same cluster. So, the unitary C does not consider Sarkar's algorithm. If c_{ij} is arbitrary, the LT algorithm does not have a good behavior. So, the arbitrary C does not consider LT algorithm.

Table 1. Summary of the scheduling algorithms considered

Algorithm	O()	t_i	c_{ij}	*blevel*	*tlevel*	G
LT [3]	v^2	Unitary	Unitary	Yes	Yes	Fine
MCP [2]	$v^2 \log v$	Arbitrary	Arbitrary	No	Yes	—
ISH [4]	v^2	Arbitrary	Arbitrary	Yes	No	Fine
KBL [5]	$v(v+e)$	Arbitrary	Arbitrary	No	No	—
SARKAR [6]	$e(v+e)$	Arbitrary	Arbitrary	No	No	—
DSC [7]	$(v+e) \log v$	Arbitrary	Arbitrary	Yes	Yes	—
RC [8]	$v(v+e)$	Arbitrary	Arbitrary	Yes	No	—

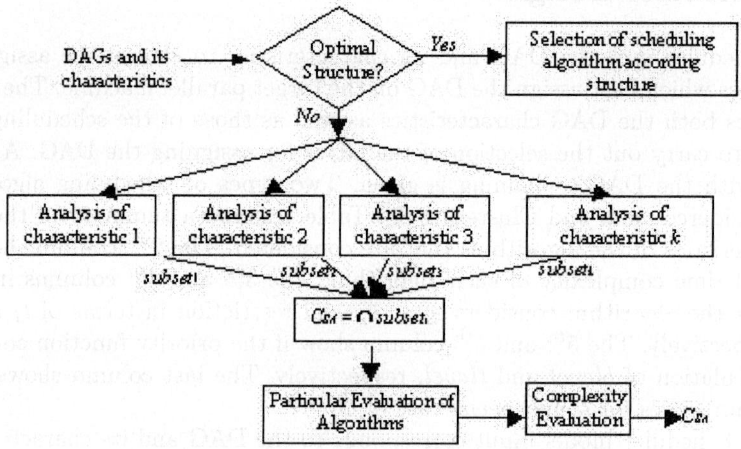

Fig. 1. The Scheduler's decision model (Block 2)

Table 2. Decision Levels of the Scheduler

Level	Characteristic	Subsets
1	Communication Cost, c_{ij}	Unitary: LT, MCP, ISH, KBL, DSC, RC
		Arbitrary: MCP, ISH, KBL, SARKAR, DSC, RC
2	Execution Time, t_i	Unitary: LT, MCP, ISH, KBL, SARKAR, DSC, RC
		Arbitrary: MCP, ISH, KBL, SARKAR, DSC, RC
3	Level to Task Ratio, R_{vn}	$R_{vn} \geq 0.7$: LT, ISH, DSC, RC
		$R_{vn} \leq 0.5$: LT, MCP, DSC
		Other: LT, MCP, ISH, KBL, SARKAR, DSC, RC
4	Granularity, G	$G \leq 3$: LT, MCP, ISH, KBL, SARKAR, DSC, RC
		Other: LT, MCP, KBL, SARKAR, DSC, RC
5	Number of Processors, p	Bounded: LT, MCP, ISH, RC
		Unbounded LT, MCP, ISH, KBL, SARKAR, DSC

Level 2: If the tasks have arbitrary cost, the LT algorithm is not selected.

Level 3: First, R_{vn} is obtained which provides the relation between DAG tasks and levels, giving an idea of the DAG's degree of parallelism. For $v > 1$, this index takes values in the range of [0..1] (expressed in equation 1).

$$R_{vn} = \{1 \Rightarrow parallel; 0 \Rightarrow sequential\}. \quad (1)$$

In general [2], assignment in the order of decreasing *blevel* tends to assign first the critical path tasks, while assignment in the order of increasing *tlevel* tends to assign the DAG in topological order. Those scheduling algorithms which consider the *blevel* within their priority function are more adequate for assigning DAGs with a high degree of parallelism ($R_{vn} \geq 0,7$), and those scheduling algorithms which consider the *tlevel* within their priority function are more adequate for DAGs with a low degree of parallelism, i.e. with greater sequentiality ($R_{vn} \leq 0,5$). In case the DAG does not show a marked tendency in the degree of parallelism, it is assumed that any scheduling algorithm can give good results.

Level 4: The ISH algorithm is the only one of the algorithms considered which shows the characteristic of working with fine grain DAGs. The particular characteristic of ISH is the possibility of inserting tasks in the slots produced as a result of communication between tasks. If the DAG has coarse grain, the communication slots are smaller than t_i, so it is not possible to make the insertion.

Level 5: The LT, ISH, MCP and RC algorithms carry out an assignment on a limited p. The remaining algorithms are unable to make an assignment on a bounded p, but rather these algorithms determine p required for the assignment that they make.

4 Tests and Analysis of Results

The model and the scheduling algorithms considered were implemented in C programming language under the Linux operating system. The model was tested with a set of 100 different DAGs (regular and irregular graphs). For each of the test DAGs, three different assignments were made on different p [3]. First, considering an architecture with $p = 2$ and $p = 4$, and then an architecture with an unbounded p.

Table 3 shows the percentage of effectiveness in both the choosing and the nonchoosing of an algorithm by the scheduler. In the case of the choosing, 100% means that of all the times that the algorithm was chosen, the best solution was always found with this chosen algorithm. On the contrary, 0% means that the times that the algorithm was chosen, the best solution was never found. In other words, a better solution was found by other algorithm.

For the case of nonchoosing, 100% means that of all the times that the algorithm was not selected, it did not find the best solution, and 0% means that of all the times that the algorithm was not selected, it found the best solution.

Table 3. Performance of the scheduler for each algorithm when choosing it or not

Algorithm	% choice		effectiveness	% no choice		effectiveness
	$p = 2$	$p = 4$	unbounded	$p = 2$	$p = 4$	unbounded
LT	100%	100%	57.1%	100%	100%	100%
MCP	85.7%	57.1%	42.9%	0%	0%	0%
ISH	100%	85.7%	28.6%	0%	20%	80%
KBL	–	–	0%	–	–	77.7%
SARKAR	–	–	50%	–	–	100%
DSC	–	–	25%	–	–	100%
RC	0%	0%	–	50%	50%	–

5 Conclusions

The implemented scheduler gave good overall results. The 100% success in its main objective shows that the design and decision levels that were created are right. It is noteworthy that this design is based only on the assignment characteristics of the scheduling algorithms.

One of the main problems found in this design appears when the architecture has an unbounded p. For the time being it is not possible to estimate a priori p that an algorithm will use when there is a limited number of them, but in practical terms it is always a known parameter.

References

1. Lewis, T., El-Rewini, H.: Parallax: A Tool for Parallel Program Scheduling. IEEE Parallel and Distributed Technology, Vol. 1. **2** (1993)
2. Kwok, Y., Ahmad, I.: Benchmarking and Comparison of the Task Graph Scheduling Algorithms. J. of Parallel and Distributed Processing. Vol. 59. **3** (1999) 381-422
3. Solar, M., Inostroza, M.: A Parallel Compiler Scheduler. XXI Int. Conf. of the Chilean Computer Science Society, IEEE CS Press, (2001) 256-263
4. Kruatrachue, B., Lewis, T.: Duplication Scheduling Heuristics: A New Precedence Task Scheduler for Parallel Processor Systems. Oregon State University. (1987)
5. Kim, S., Browne, J.: A General Approach to Mapping of Parallel Computation upon Multiprocessor Architecture. Int. Conf. on Parallel Processing, Vol. 3. (1988)
6. Sarkar, V.: Partitioning and Scheduling Parallel Programs for Multiprocessors. MIT Press, Cambridge, MA, (1989)
7. Yang, T., Gerasoulis, A.: DSC: Scheduling Parallel Tasks on an Unbounded Number of Processors. IEEE Trans. Parallel and Distributed Systems. Vol. 5. **9** (1994)
8. Zhou, H.: Scheduling DAGs on a Bounded Number of Processors. Int. Conf. on Parallel & Distributed Processing. Sunnyvale (1996)

Non-approximability Results for the Hierarchical Communication Problem with a Bounded Number of Clusters*
(Extended Abstract)

Eric Angel, Evripidis Bampis, and Rodolphe Giroudeau

LaMI, CNRS-UMR 8042, Université d'Evry Val d'Essonne
523, Place des Terrasses
F–91000 Evry France
{angel, bampis, giroudea}@lami.univ-evry.fr

Abstract. We study the hierarchical multiprocessor scheduling problem with a constant number of clusters. We show that the problem of deciding whether there is a schedule of length three for the hierarchical multiprocessor scheduling problem is $\mathcal{NP}$-complete even for bipartite graphs i.e. for precedence graphs of depth one. This result implies that there is no polynomial time approximation algorithm with performance guarantee smaller than 4/3 (unless $\mathcal{P} = \mathcal{NP}$). On the positive side, we provide a polynomial time algorithm for the decision problem when the schedule length is equal to two, the number of clusters is constant and the number of processors per cluster is arbitrary.

1 Introduction

For many years, the standard communication model for scheduling the tasks of a parallel program has been the *homogeneous communication model* (also known as the *delay model*) introduced by Rayward-Smith [12] for unit-execution-times, unit-communication times (UET-UCT) precedence graphs. In this model, we are given a set of identical processors that are able to communicate in a uniform way. We wish to use these processors in order to process a set of tasks that are subject to precedence constraints. Each task has a processing time, and if two adjacent tasks of the precedence graph are processed by two different processors (resp. the same processor) then a communication delay has to be taken into account explicitly (resp. the communication time is neglected). The problem is to find a trade-off between the two extreme solutions, namely, execute all the tasks sequentially without communications, or try to use all the potential parallelism but in the cost of an increased communication overhead. This model has been extensively studied these last years both from the complexity and the (non)-approximability point of views [7].

* This work has been partially supported by the APPOL II (IST-2001-32007) thematic network of the European Union and the GRID^2 project of the French Ministry of Research.

B. Monien and R. Feldmann (Eds.): Euro-Par 2002, LNCS 2400, pp. 217–224.

In this paper, we adopt the *hierarchical communication model* [1,3] in which we assume that the communication delays are not homogeneous anymore; the processors are connected in *clusters* and the communications inside the same cluster are much faster than those between processors belonging to different clusters. This model captures the hierarchical nature of the communications in todays parallel computers, composed by many networks of PCs or workstations (NOWs). The use of networks (clusters) of workstations as a parallel computer has renewed the interest of the users in the domain of parallelism, but also created new challenging problems concerning the exploitation of the potential computation power offered by such a system. Most of the attempts to model these systems were in the form of programming systems rather than abstract models [4,5,13,14]. Only recently, some attempts concerning this issue appeared in the literature [1,6]. The one that we adopt here is the *hierarchical communication model* which is devoted to one of the major problems appearing in the attempt of efficiently using such architectures, the *task scheduling problem.*

The proposed model includes one of the basic architectural features of NOWs: the hierarchical communication assumption i.e. a level-based hierarchy of the communication delays with successively higher latencies.

The hierarchical model. In the *precedence constrained multiprocessor scheduling problem with hierarchical communication delays*, we are given a set of multiprocessor machines (or clusters) that are used to process n precedence constrained tasks. Each machine (cluster) comprises several identical parallel processors. A couple (c_{ij}, ϵ_{ij}) of communication delays is associated to each arc (i,j) of the precedence graph. In what follows, c_{ij} (resp. ϵ_{ij}) is called intercluster (resp. interprocessor) communication, and we consider that $c_{ij} \geq \epsilon_{ij}$. If tasks i and j are executed on different machines, then j must be processed at least c_{ij} time units after the completion of i. Similarly, if i and j are executed on the same machine but on different processors then the processing of j can only start ϵ_{ij} units of time after the completion of i. However, if i and j are executed on the same processor then j can start immediately after the end of i. The communication overhead (intercluster or interprocessor delay) does not interfere with the availability of the processors and all processors may execute other tasks.

Known results and our contribution. In [2], it has been proved that there is no hope (unless $\mathcal{P} = \mathcal{NP}$) to find a ρ-approximation algorithm with ρ strictly less than 5/4, even for the simple UET-UCT ($p_i = 1; (c_{ij}, \epsilon_{ij}) = (1,0)$)case where an unbounded number of bi-processor machines, denoted in what follows by $\bar{P}(P2)$ is considered, $(\bar{P}(P2)|prec; (c_{ij}, \epsilon_{ij}) = (1,0); p_i = 1|C_{\max})$. For the case where each machine contains m processors, where m is a fixed constant (i.e. for $\bar{P}(Pm)|prec;\ (c_{ij}, \epsilon_{ij}) = (1,0); p_i = 1|C_{\max}$), a $\frac{4m}{2m+1}$-approximation algorithm has been proposed in [1]. However, no results are known for arbitrary processing times and/or communication delays. The small communication times (SCT) assumption where the intercluster communication delays are smaller than or equal to the processing times of the tasks, *i.e.* $\Phi = \frac{\min p_i,\ i \in V}{\max c_{kj},\ (k,j) \in E} \geq 1$, have been adopted in [3], where, as in [1], the interprocessor communication delays have been considered as negligible. The authors presented a $\frac{12(\Phi+1)}{12\Phi+1}$-approximation algorithm, which is based on linear programming and rounding. Notice that for the case where $c_{ij} = \epsilon_{ij}$, *i.e.* in the classical model with communication delays, Hanen and Munier [10] proposed a $\frac{2(1+\Phi)}{2\Phi+1}$-approximation algorithm for the problem with an unbounded number of machines.

In this paper, we consider for the first time the case where the number of clusters is bounded and more precisely we examine the non-approximability of the problem with two clusters composed by a set of identical processors ($P2(P)|prec; (c_{ij}, \epsilon_{ij}) = (1,0); p_i = 1|C_{max}$). In Section 2, we prove that the problem of deciding whether there is a schedule of length three is $\mathcal{NP}$-complete even for bipartite graphs i.e. for precedence graphs of depth one. This result implies that there is no polynomial time approximation algorithm with performance guarantee smaller than 4/3 (unless $\mathcal{P} = \mathcal{NP}$). In Section 3, we provide a polynomial time algorithm for the decision problem when the schedule length is equal to two, the number of clusters is constant and the number of processors per cluster is arbitrary.

2 The Non-approximability Result

In this section, we show that the problem of deciding whether an instance of $P2(P)|bipartite; (c_{ij}, \epsilon_{ij}) = (1,0); p_i = 1|C_{max}$ has a schedule of length at most three is $\mathcal{NP}$-complete. We use a polynomial time reduction from the $\mathcal{NP}$-complete problem balanced independent set (BBIS) problem [15].

Definition 1. Instance of BBIS: *An undirected balanced bipartite graph* $B = (X \bigcup Y, E)$ *with* $|X| = |Y| = n$*, and an integer* k*.*
Question: *Is there in* B*, an independent set with* k *vertices in* X *and* k *vertices in* Y*?*

If such an independent set exists, we call it *balanced independent set of order* k. Notice that, the problem remains $\mathcal{NP}$-complete even if $k = \frac{n}{2}$, n is even (see [15]). In what follows, we consider BBIS with $k = \frac{n}{2}$ as the source problem.

Theorem 1. *The problem of deciding whether an instance of* $P2(P)|bipartite; (c_{ij}, \epsilon_{ij}) = (1,0); p_i = 1|C_{max}$ *has a schedule of length at most three is* $\mathcal{NP}$*-complete.*

Proof. It is easy to see that the problem $P2(P)|bipartite; (c_{ij}, \epsilon_{ij}) = (1,0); p_i = 1|C_{max} \in \mathcal{NP}$. The rest of the proof is based on a reduction from $BBIS$.

Given an instance of $BBIS$, i.e. a balanced bipartite graph $B = (X \cup Y, E)$, we construct an instance of the scheduling problem $P2(P)|bipartite; (c_{ij}, \epsilon_{ij}) = (1,0); p_i = 1|C_{max} = 3$, in the following way:

– We orient all the edges of B from the tasks of X to the tasks of Y.

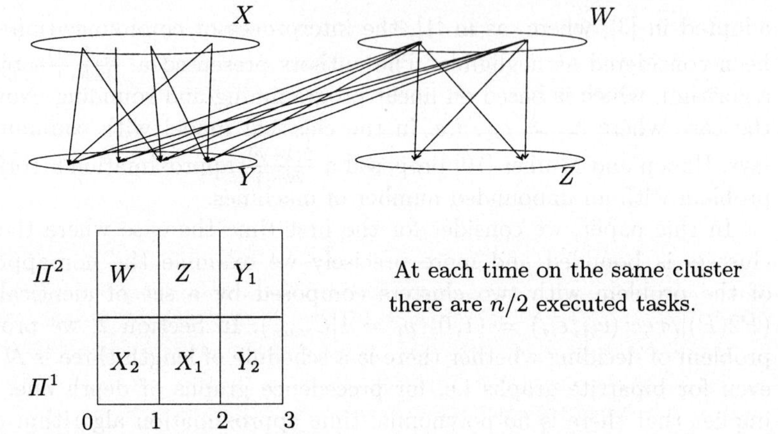

Fig. 1. The precedence graph and an associated schedule corresponding to the polynomial reduction $BBIS \propto P2(P)|bipartite;(c_{ij}, \epsilon_{ij}) = (1,0); p_i = 1|C_{max}$.

- We add two sets of tasks: $W = \{w_1, w_2, \ldots, w_{n/2}\}$ and $Z = \{z_1, z_2, \ldots, z_{n/2}\}$. The precedence constraints among these tasks are the following:

$$w_i \rightarrow z_j,\ \forall i \in \{1, 2, \ldots, n/2\},\ \forall j \in \{1, 2, \ldots, n/2\}.$$

- We also add the precedence constraints:

$$w_i \rightarrow y_j,\ \forall i \in \{1, 2, \ldots, n/2\},\ \forall j \in \{1, 2, \ldots, n\}.$$

- We suppose that the number of processors per cluster is equal to $m = n/2$, and that all the tasks have unit execution times.

The construction is illustrated in the first part of Figure 1. The proposed reduction can be computed in polynomial time.

Notation: The first (resp. second) cluster is denoted by Π^1 (resp. Π^2).

- Let us first consider that B contains a balanced independent set of order $\frac{n}{2}$, call it (X_1, Y_1) where $X_1 \subset X$, $Y_1 \subset Y$, and $|X_1| = |Y_1| = n/2$. Let us show now that there exists a feasible schedule in three units of time. The schedule is as follows.
  - At $t = 0$, we execute on the processors of cluster Π^1 the $n/2$ tasks of $X - X_1 = X_2$, and on the cluster Π^2 the $n/2$ tasks of W.
  - At $t = 1$, we execute on Π^1 the $n/2$ tasks of X_1 and on Π^2 the $n/2$ tasks of Z.
  - We execute at $t = 2$ on the cluster Π^2 the $n/2$ tasks of Y_1 and on the cluster Π^1 the $n/2$ tasks of $Y - Y_1 = Y_2$.

  The above way of scheduling the tasks preserves the precedence constraints and the communication delays and gives a schedule of length three, whenever there exists in B a balanced independent set of order $\frac{n}{2}$.

- Conversely, we suppose that there is a schedule of length three. We will prove that any schedule of length three implies the existence of a balanced independent set (X_1', Y_1'), in the graph B, where $X_1' \subset X$, $Y_1' \subset Y$ and $|X_1'| = |Y_1'| = n/2$.
  We make four essential observations. In every feasible schedule of length at most three:
  1. Since the number of tasks is $3n$ there is no idle time.
  2. All the tasks of W must be executed at $t = 0$, since every such task precedes $\frac{3n}{2}$ tasks, and there is only $\frac{n}{2}$ processors per cluster (n in total). Moreover, all the tasks of W must be executed on the same cluster. Indeed, if two tasks of W are scheduled at $t = 0$ on different clusters, then no task of Z or Y can be executed at $t = 1$. Thus, the length of the schedule is greater than 3 because $|Z \bigcup Y| = \frac{3n}{2}$.
     Assume w.l.o.g. that the tasks of W are executed on Π^1.
  3. No task of Y or Z can be executed at $t = 0$.
     Let X_2' be the subset of X executed on the processors of cluster Π^2 at $t = 0$. It is clear that $|X_2'| = \frac{n}{2}$, because of point 1.
  4. No task of Y or Z can be executed at $t = 1$ on Π^2. Hence, at $t = 1$, the only tasks that can be executed on Π^2, are tasks of X, and more precisely the tasks of $X - X_2' = X_1'$.

  Let Y_1' be the subset of tasks of Y which have a starting time at $t = 1$ or at $t = 2$ on the cluster Π^1. This set has at least $\frac{n}{2}$ elements and together with the $\frac{n}{2}$ elements of X_1', they have to form a balanced independent set in order the schedule to be feasible.

Corollary 1. *The problem of deciding whether an instance of* $P2(P)|bipartite; (c_{ij}, \epsilon_{ij}) = (1,0); p_i = 1; dup|C_{max}$ *has a schedule of length at most three is* $\mathcal{NP}$*-complete.*

Proof. The proof comes directly from the one of Theorem 1. In fact, no task can be duplicated since otherwise the number of tasks would be greater than $3n$, and thus the schedule length would be greater than three.

Corollary 2. *There is no polynomial-time algorithm for the problem* $P2(P)|bipartite; (c_{ij}, \epsilon_{ij}) = (1,0); p_i = 1|C_{max}$ *with performance bound smaller than* $\frac{4}{3}$ *unless* $\mathcal{P} = \mathcal{NP}$.

Proof. The proof is an immediate consequence of the Impossibility Theorem (see [9, 8]).

3 A Polynomial Time Algorithm for $C_{max} = 2$

In this section, we prove that the problem of deciding whether an instance of $Pk(P)|prec; (c_{ij}, \epsilon_{ij}) = (1,0); p_i = 1|C_{max}$ has a schedule of length at most two is polynomial by using dynamic programming. In order to prove this result, we show that this problem is equivalent to a generalization of the well known problem $P2||C_{max}$.

Theorem 2. *The problem of deciding whether an instance of $Pk(P)|prec;(c_{ij},\epsilon_{ij})=(1,0);p_i=1|C_{max}$ has a schedule of length at most two is polynomial.*

Proof. We assume that we have $k=2$ clusters. The generalization for a fixed $k>2$ is straightforward. Let π be an instance of the problem $Pk(P)|prec;(c_{ij},\epsilon_{ij})=(1,0);p_i=1|C_{max}=2$. We denote by G the oriented precedence graph, and by G^* the resulting non oriented graph when the orientation on each arc is removed. In the sequel we consider that G has a depth of at most two, since otherwise the instance does not admit a schedule of length at most two. It means that $G=(X\bigcup Y,A)$ is a bipartite graph. The tasks belonging to X (resp. Y), i.e. tasks without predecessors (resp. without successors), will be called source (resp. sink) tasks. In the sequel we assume that G does not contain any tasks without successors and predecessors, i.e. isolated tasks. We shall explain how to deal with these tasks later.

Let denote W_j the j-th connected component of graph G^*. The set of tasks which belong to a connected component W_j will be called a *group of tasks* in the sequel.

Each *group of tasks* constitutes a set of tasks that have to be executed by the same cluster in order to yield a schedule within two time units. Consequently the following condition holds: there is no feasible schedule within two time units, if there exists a *group of tasks* W_j such that $|W_j\bigcap X|\geq m+1$, or $|W_j\bigcap Y|\geq m+1$. Recall that m denotes the number of processors per cluster.

The problem of finding such a schedule can be converted to a variant of the well known $P2||C_{max}$ problem. We consider a set of n jobs $\{1,2,\dots n\}$. Each job j has a couple of processing times $p_j=(p_j^1,p_j^2)$. We assume that $\sum_{j=1}^n p_j^1\leq 2m$ and $\sum_{j=1}^n p_j^2\leq 2m$. The goal is to find a partition $(S,\overline{S})$ of the jobs such that the makespan is at most m if we consider either the first or second processing times, i.e. determine $S\subset\{1,2,\dots n\}$ such that $\sum_{j\in S}p_j^1\leq m$, $\sum_{j\in S}p_j^2\leq m$, $\sum_{j\in\overline{S}}p_j^1\leq m$ and $\sum_{j\in\overline{S}}p_j^2\leq m$.

Now, to each *group of tasks* W_j we can associate a job with processing times $p_j^1=|W_j\bigcap X|$ and $p_j^2=|W_j\bigcap Y|$. The Figure 2 presents the transformation between the problem $P2(P)|prec;(c_{ij},\epsilon_{ij})=(1,0);p_i=1|C_{max}$ and the variant of $P2||C_{max}$.

The problem $P2||C_{max}$ can be solved by a pseudo-polynomial time dynamic programming algorithm [11]. In the sequel we show that there exists a polynomial time algorithm for the problem we consider.

Let us define $I(j,z_1,z_2)=1$, with $1\leq j\leq n$, $0\leq z_1,z_2\leq m$, if there exists a subset of jobs, $S(j,z_1,z_2)\subseteq\{1,2,\dots,j-1,j\}$, for which the sum of processing times on the first (resp. second) coordinate is exactly z_1 (resp. z_2). Otherwise $I(j,z_1,z_2)=0$.

The procedure basically fills the $0-1$ entries of a n by $(m+1)^2$ matrix row by row, from left to right. The rows (resp. columns) of the matrix are indexed by j (resp. (z_1,z_2)). Initially we have $I(1,p_1^1,p_1^2)=1$, $S(1,p_1^1,p_1^2)=\{1\}$, and $I(1,z_1,z_2)=0$ if $(z_1,z_2)\neq(p_1^1,p_1^2)$. The following relations are used to fill the matrix:

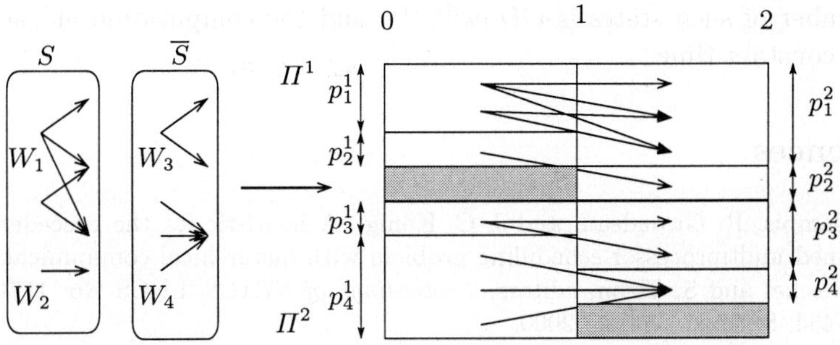

Fig. 2. llustration of the transformation with $m = 4$ (idle time is in grey).

- If $I(j, z_1, z_2) = 1$ then $I(j + 1, z_1, z_2) = 1$. Moreover $S(j + 1, z_1, z_2) = S(j, z_1, z_2)$.
- If $I(j, z_1, z_2) = 1$ then $I(j + 1, z_1 + p^1_{j+1}, z_2 + p^2_{j+1}) = 1$. Moreover $S(j + 1, z_1 + p^1_{j+1}, z_2 + p^2_{j+1}) = S(j, z_1, z_2) \bigcup \{j + 1\}$.

Now, we examine the last row of the matrix, and look for a state (n, m_1, m'_1) such that $I(n, m_1, m'_1) = 1$, with $|X| - m \leq m_1 \leq m$ and $|Y| - m \leq m'_1 \leq m$. It is easy to see that the instance π admits a schedule within two time units if and only if there exists such a state. From such a state (n, m_1, m'_1) we can find a schedule of length at most two in the following way. Let W (resp. $\overline{W}$) the set of *group of tasks* associated with jobs in $S(n, m_1, m'_1)$ (resp. $\overline{S(n, m_1, m'_1)}$). The $m_1 \leq m$ source (resp. $|X| - m_1 \leq m$ sink) tasks of W are scheduled on the first cluster, during the first (resp. second) unit of time. The $m'_1 \leq m$ source (resp. $|Y| - m'_1 \leq m$ sink) tasks of $\overline{W}$ are scheduled on the second cluster, during the first (resp. second) unit of time.

In the case where the graph G contains a set of isolated tasks, we remove those tasks from set X, compute the previous matrix, and look for the same state as before. The instance π admits a schedule within two time units if and only we can fill the gaps of the previous schedule with the isolated tasks.

For $k > 2$ clusters we consider the $Pk||C_{max}$ scheduling problem in which each job has a couple of processing times. The goal is to find a partition $(S_1, \ldots, S_{k-1}, \overline{S_1 \cup \ldots \cup S_{k-1}})$ of the jobs such that the makespan is at most m if we consider either the first or second processing times. As before this problem can be solved by a pseudo-polynomial time dynamic programming algorithm using the states $(j, z_1, z_2, \ldots z_{2(k-1)})$, with $1 \leq j \leq n$ and $1 \leq z_i \leq m$, $i = 1, \ldots, 2(k - 1)$. We have $I(j, z_1, z_2, \ldots z_{2(k-1)}) = 1$ if there exists a partition $(S_1, \ldots, S_{k-1})$ of jobs such that $\sum_{j \in S_{2l+1}} p^1_j = z_{2l+1}$ and $\sum_{j \in S_{2l+2}} p^2_j = z_{2l+2}$ for $0 \leq l \leq k - 2$.

Let us now evaluate the running time of the overall algorithm for a problem instance with m processors per cluster (m is part of the input of the instance).

Lemma 1. *The complexity of the algorithm is equal to $\mathcal{O}(nm^{2(k-1)})$.*

Proof. Each state of the dynamic programming algorithm is a tuple $(j, z_1, z_2, \ldots z_{2(k-1)})$, with $1 \leq j \leq n$ and $1 \leq z_i \leq m$, $i = 1, \ldots, 2(k - 1)$.

The number of such states is $\mathcal{O}(nm^{2(k-1)})$ and the computation at each state needs a constant time.

References

1. E. Bampis, R. Giroudeau, and J.-C. König. A heuristic for the precedence constrained multiprocessor scheduling problem with hierarchical communications. In H. Reichel and S. Tison, editors, *Proceedings of STACS*, LNCS No. 1770, pages 443–454. Springer-Verlag, 2000.
2. E. Bampis, R. Giroudeau, and J.C. König. On the hardness of approximating the precedence constrained multiprocessor scheduling problem with hierarchical communications. Technical Report 34, LaMI, Université d'Évry Val d'Essonne, to appear in RAIRO Operations Research, 2001.
3. E. Bampis, R. Giroudeau, and A. Kononov. Scheduling tasks with small communication delays for clusters of processors. In *SPAA*, pages 314–315. ACM, 2001.
4. S.N. Bhatt, F.R.K. Chung, F.T. Leighton, and A.L. Rosenberg. On optimal strategies for cycle-stealing in networks of workstations. *IEEE Trans. Comp.*, 46:545–557, 1997.
5. R. Blumafe and D.S. Park. Scheduling on networks of workstations. In *3d Inter Symp. of High Performance Distr. Computing*, pages 96–105, 1994.
6. F. Cappello, P. Fraignaud, B. Mans, and A. L. Rosenberg. HiHCoHP-Towards a Realistic Communication Model for Hierarchical HyperClusters of Heterogeneous Processors, 2000. to appear in the Proceedings of IPDPS'01.
7. B. Chen, C.N. Potts, and G.J. Woeginger. A review of machine scheduling: complexity, algorithms and approximability. Technical Report Woe-29, TU Graz, 1998.
8. P. Chrétienne and C. Picouleau. Scheduling with communication delays: a survey. In P. Chrétienne, E.J. Coffman Jr, J.K. Lenstra, and Z. Liu, editors, *Scheduling Theory and its Applications*, pages 65–90. Wiley, 1995.
9. M.R. Garey and D.S. Johnson. *Computers and Intractability, a Guide to the Theory of NP-Completeness.* Freeman, 1979.
10. A. Munier and C. Hanen. An approximation algorithm for scheduling dependent tasks on m processors with small communication delays. In *IEEE Symposium on Emerging Technologies and Factory Automation*, Paris, 1995.
11. M. Pinedo. *Scheduling : theory, Algorithms, and Systems.* Prentice Hall, 1995.
12. V.J. Rayward-Smith. UET scheduling with unit interprocessor communication delays. *Discr. App. Math.*, 18:55–71, 1987.
13. A.L. Rosenberg. Guidelines for data-parallel cycle-stealing in networks of workstations I: on maximizing expected output. *Journal of Parallel Distributing Computing*, pages 31–53, 1999.
14. A.L. Rosenberg. Guidelines for data-parallel cycle-stealing in networks of workstations II: on maximizing guarantee output. *Intl. J. Foundations of Comp. Science*, 11:183–204, 2000.
15. R. Saad. Scheduling with communication delays. *JCMCC*, 18:214–224, 1995.

Non-approximability of the Bulk Synchronous Task Scheduling Problem

Noriyuki Fujimoto and Kenichi Hagihara

Graduate School of Information Science and Technology,
Osaka University
1-3, Machikaneyama, Toyonaka, Osaka, 560-8531, Japan
{fujimoto, hagihara}@ist.osaka-u.ac.jp

Abstract. The mainstream architecture of a parallel machine with more than tens of processors is a distributed-memory machine. The bulk synchronous task scheduling problem (BSSP, for short) is an task scheduling problem for distributed-memory machines. This paper shows that there does not exist a ρ-approximation algorithm to solve the optimization counterpart of BSSP for any $\rho < \frac{6}{5}$ unless $\mathcal{P} = \mathcal{NP}$.

1 Introduction

Existing researches on the task scheduling problem for a distributed-memory machine (**DMM** for short) simply model DMMs as the parallel machines with large communication delays [2,12,13]. In contrast to this, in the papers [4,5,7], one noticed the following things by both analysis of architectural properties of DMMs and experiments to execute parallel programs which corresponds to schedules generated by existing task scheduling algorithms:

- It is essential to task scheduling for a DMM to consider the software overhead in communication, even if a DMM is equipped with a dedicated communication co-processor per processor.
- Existing task scheduling algorithms would ignore the software overhead.
- For the above reasons, it is hard for existing algorithms to generate schedules which become fast parallel programs on a DMM.

To remedy this situation, in the papers [4–7], one proposed an optimization problem named the bulk synchronous task scheduling problem (**BSSPO** for short), i.e., the problem of finding a bulk synchronous schedule with small makespan. Formally, BSSPO is an optimization problem which restricts output, rather than input, of the general task scheduling problem with communication delays. A bulk synchronous schedule is a restricted schedule which has the following features:

- The well-known parallel programming technique to reduce the software overhead significantly , called message aggregation [1], can be applied to the parallel program which corresponds to the schedule.
- Makespan of the schedule approximates well the execution time of the parallel program applied message aggregation.

B. Monien and R. Feldmann (Eds.): Euro-Par 2002, LNCS 2400, pp. 225–233.

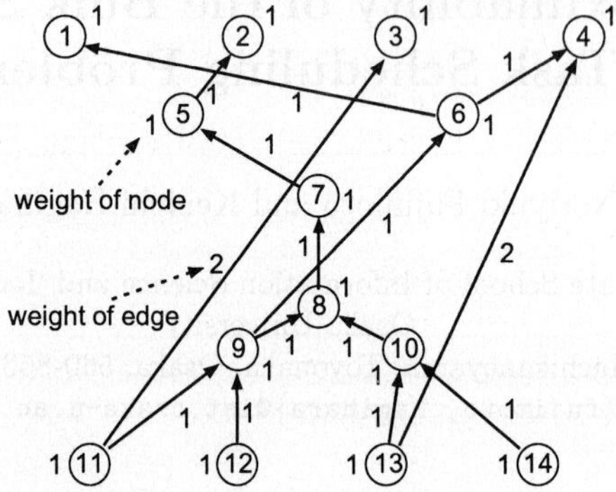

Fig. 1. An example of a task graph

Hence a good BSSPO algorithm generates a schedule which becomes a fast parallel program on a DMM.

In this paper, we consider non-approximability of BSSPO. The decision counterpart of BSSPO (**BSSP**, for short) is known to be $\mathcal{NP}$-complete even in the case of unit time tasks and positive integer constant communication delays [6]. For BSSPO, two heuristic algorithms [4, 5, 7] for the general case and several approximation algorithms [6] for restricted cases are known. However, no results are known on non-approximability of BSSPO. This paper shows that there does not exist a ρ-approximation algorithm to solve BSSPO for any $\rho < \frac{6}{5}$ unless $\mathcal{P} = \mathcal{NP}$.

The remainder of this paper is organized as follows. First, we give some definitions in Section 2. Next, we review a bulk synchronous schedule in Section 3. Then, we prove non-approximability of BSSPO in Section 4. Last, in Section 5, we summarize and conclude the paper.

2 Preliminaries and Notation

A parallel computation is modeled as a task graph [3]. A **task graph** is represented by a weighted directed acyclic graph $G = (V, E, \lambda, \tau)$, where V is a set of nodes, E is a set of directed edges, λ is a function from a node to the weight of the node, and τ is a function from a directed edge to the weight of the edge. We write a directed edge from a node u to a node v as (u, v). A node in a task graph represents a task in the parallel computation. We write a task represented by a node u as T_u. The value $\lambda(u)$ means that the execution time of T_u is $\lambda(u)$ unit times. An edge (u, v) means that the computation of T_v needs the result of the computation of T_u. The value $\tau(u, v)$ means that interprocessor communication delay from the processor p which computes T_u to the processor q which computes T_v is at most $\tau(u, v)$ unit times if p is not equal to q. If p and q are identical, no interprocessor communication delay is needed.

Thurimella gave the definition of a schedule in the case that $\lambda(v)$ is equal to a unit time for any v and $\tau(u,v)$ is a constant independent of u and v [14]. For general task graphs, we define a schedule as extension of Thurimella's definition as follows. For a given number p of available processors, a **schedule** S of a task graph $G = (V, E, \tau, \lambda)$ for p is a finite set of triples $\langle v, q, t\rangle$, where $v \in V$, $q (1 \le q \le p)$ is the index of a processor, and t is the **starting time** of task T_v. A triple $\langle v, q, t\rangle \in S$ means that the processor q computes the task T_v between time t and time $t + \lambda(v)$. We call $t + \lambda(v)$ the **completion time** of the task T_v. A schedule which satisfies the following three conditions R1 to R3 is called **feasible** (In the following of this paper, we abbreviate a feasible schedule as a schedule.):

R1 For each $v \in V$, there is at least one triple $\langle v, q, t\rangle \in S$.
R2 There are no two triples $\langle v, q, t\rangle, \langle v', q, t'\rangle \in S$ with $t \le t' \le t + \lambda(v)$.
R3 If $(u, v) \in E$ and $\langle v, q, t\rangle \in S$, then there exists a triple $\langle u, q', t'\rangle \in S$ either with $t' \le t - \lambda(u)$ and $q = q'$, or with $t' \le t - \lambda(u) - \tau(u, v)$ and $q \ne q'$.

Informally, the above rules can be stated as follows. The rule $R1$ enforces each task T_v to be executed at least once. The rule $R2$ says that a processor can execute at most one task at any given time. The rule $R3$ states that any task must receive the required data (if exist) before its starting time. The **makespan** of S is $\max\{t + \lambda(v) | \langle v, q, t\rangle \in S\}$. An **optimal schedule** is a schedule with the smallest makespan among all the schedules.

A schedule within a factor of α of optimal is called an **α-optimal schedule**. A **ρ-approximation algorithm** is a polynomial-time algorithm that always finds a ρ-optimal schedule.

3 Review of a Bulk Synchronous Schedule

As shown in Fig. 2, a bulk synchronous schedule is a schedule such that no-communication phases and communication phases appear alternately (In a general case, no-communication phases and communication phases appear repeatedly). Informally, a no-communication phase is a set of task instances in a time interval such that the corresponding program executes computations only. A communication phase is a time interval such that the corresponding program executes communications only. A bulk synchronous schedule is similar to BSP (Bulk Synchronous Parallel) computation proposed by Valiant [15] in that local computations are separated from global communications. A no-communication phase corresponds to a super step of BSP computation. In the following, we first define a no-communication phase and a communication phase. Then, we define a bulk synchronous schedule using them.

Let S be a schedule of a task graph $G = (V, E, \lambda, \tau)$ for a number p of available processors. We define the following notation: For S, t_1, and t_2 with $t_1 < t_2$,

$$S[t_1, t_2] = \{\langle v, q, t\rangle \in S | t_1 \le t \le t_2 - \lambda(v)\}$$

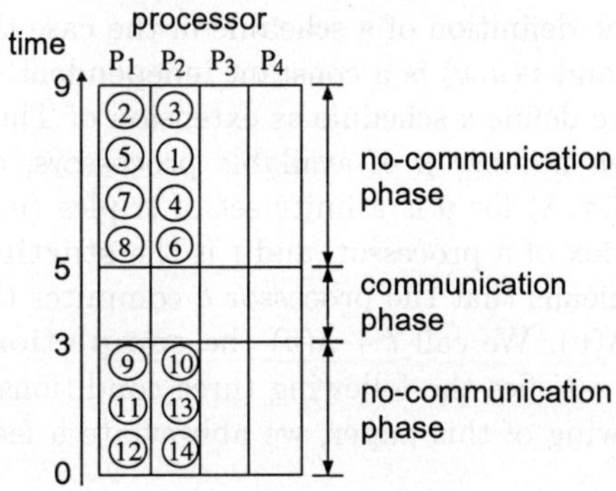

Fig. 2. An example of a bulk synchronous schedule

Notation $S[t_1, t_2]$ represents the set of all the triples such that both the starting time and the completion time of the task in a triple are between t_1 and t_2. A set $S[t_1, t_2] \subseteq S$ of triples is called a **no-communication phase** of S iff the following condition holds.

C1 If $(u, v) \in E$ and $\langle v, q, t \rangle \in S[t_1, t_2]$, then there exists a triple $\langle u, q', t' \rangle \in S$ either with $t' \leq t - \lambda(u)$ and $q = q'$, or with $t' \leq t_1 - \lambda(u) - \tau(u, v)$ and $q \neq q'$.

The condition C1 means that each processor needs no interprocessor communication between task instances in $S[t_1, t_2]$ since all the needed results of tasks are either computed by itself or received from some processor before t_1.

Let $S[t_1, t_2]$ be a no-communication phase. Let t_3 be $\min\{t | \langle u, q, t \rangle \in (S - S[0, t_2])\}$. Assume that a no-communication phase $S[t_3, t_4]$ exists for some t_4. We say that $S[t_1, t_2]$ and $S[t_3, t_4]$ are **consecutive no-communication phases**. We intend that in the execution of the corresponding program each processor sends the results, which are computed in $S[t_1, t_2]$ and are required in $S[t_3, t_4]$, as packaged messages at t_2 and receives all the needed results in $S[t_3, t_4]$ as packaged messages at t_3. A **communication phase** between consecutive no-communication phases is the time interval where each processor executes communications only. To reflect such program's behavior in the time interval on the model between consecutive no-communication phases, we assume that the result of $\langle u, q, t \rangle \in S[t_1, t_2]$ is sent at t_2 even in case of $t + \lambda(u) < t_2$ although the model assumes that the result is always sent at $t + \lambda(u)$. Let $Comm(S, t_1, t_2, t_3, t_4)$ be $\{(u, v) | (u, v) \in E, \langle u, q, t \rangle \in S[t_1, t_2], \langle v, q', t' \rangle \in S[t_3, t_4], \langle u, q', t'' \rangle \notin S, q \neq q', t'' \leq t' - \lambda(u)\}$. A set $Comm(S, t_1, t_2, t_3, t_4)$ of edges corresponds to the set of all the interprocessor communications between task instances in $S[t_1, t_2]$ and task instances in $S[t_3, t_4]$. Note that task duplication [8] is considered in the definition of $Comm(S, t_1, t_2, t_3, t_4)$. We define the following notation: For $C \subseteq E$,

$$\tau_{suff}(C) = \begin{cases} 0 & if\ C = \emptyset \\ \max\{\tau(u, v) | (u, v) \in C\} & otherwise \end{cases}$$

Consider simultaneous sendings of all the results in C. The value $\tau_{suff}(C)$ represents the elapsed time on the model till all the results are available to any processor. So, the value $\tau_{suff}(Comm(S, t_1, t_2, t_3, t_4))$ represents the minimum communication delay on the model between the two no-communication phases.

We say S is a **bulk synchronous schedule** iff S can be partitioned into a sequence of no-communication phases $\langle S[st_1, ct_1], S[st_2, ct_2], \cdots, S[st_m, ct_m]\rangle$ $(m \geq 1)$ which satisfies the following condition C2.

C2 For any i, j $(1 \leq i < j \leq m)$, $ct_i + \tau_{suff}(Comm(S, st_i, ct_i, st_j, ct_j)) \leq st_j$

Note that C2 considers communications between not only consecutive no-communication phases but also non consecutive ones. Fig. 2 shows an example of a bulk synchronous schedule $\langle S[0,3], S[5,9]\rangle$ of the task graph in Fig. 1 for four processors. The set $Comm(S, 0, 3, 5, 9)$ of edges is $\{(9,6), (10,8), (11,3)\}$. The edge with maximum weight of all the edges in $Comm(S, 0, 3, 5, 9)$ is $(11, 3)$. So, the weight of the edge $(11, 3)$ decides that $\tau_{suff}(Comm(S, 0, 3, 5, 9))$ is two.

4 A Proof of Non-approximability of BSSP

4.1 An Overview of Our Proof

In this section, we prove that a ρ-approximation algorithm for BSSPO does not exist for any $\rho < \frac{6}{5}$ unless $\mathcal{P} = \mathcal{NP}$. For this purpose, we use the following lemma [11].

Lemma 1. *Consider a combinatorial minimization problem for which all feasible solutions have non-negative integer objective function value. Let k be a fixed positive integer. Suppose that the problem of deciding if there exists a feasible solution of value at most k is $\mathcal{NP}$-complete. Then, for any $\rho < (k+1)/k$, there does not exist a ρ-approximation algorithm unless $\mathcal{P} = \mathcal{NP}$.*

To extract our non-approximability result using Lemma 1, we prove $\mathcal{NP}$-completeness of BSSP in the case of a given fixed constant communication delay c and makespan at most $3 + 2c$ (3BSSP(c), for short) by reducing to 3BSSP(c) the unit time precedence constrained scheduling problem in the case of makespan at most 3 [10] (3SP, for short). These problems are defined as follows:

- 3BSSP(c) where c is a constant communication delay (positive integer).
  Instance: A task graph G such that all the weights of nodes are unit and all the weights of edges are the same as c, a number p of available processors
  Question: Is there a bulk synchronous schedule S_{BSP} whose makespan is at most $3 + 2c$?
- 3SP
  Instance: A task graph G such that all the weights of nodes are unit and all the weights of edges are the same as zero, a number p of available processors
  Question: Is there a schedule S whose makespan is at most 3 ?

$\mathcal{NP}$-completeness of 3SP was proved by Lenstra and Rinnooy Kan [10]. In the following, we denote an instance of 3BSSP(c) (3SP, resp.) as (G, p, c) ((G, p), resp.).

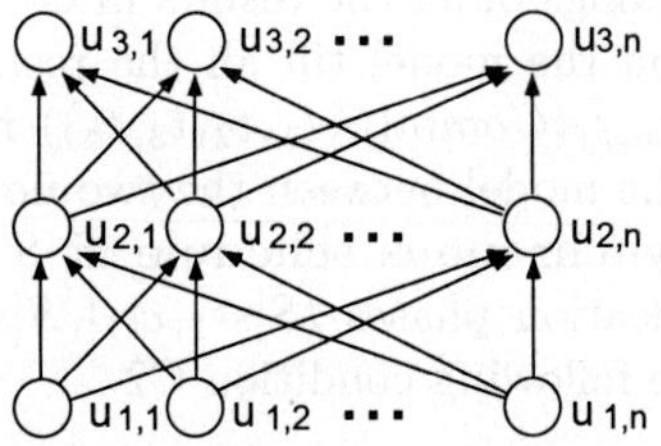

The weight of each node is unit.
The weight of each edge is c units.

Fig. 3. A ladder graph $LG(n, c)$

4.2 A Ladder Graph and Its Bulk Synchronous Schedule

A **ladder graph** $LG(n, c)$ is a task graph such that $V = \{u_{i,j} | 1 \leq i \leq 3, 1 \leq j \leq n\}$, $E = \{(u_{i,j}, u_{i+1,k}) | 1 \leq i < 3, 1 \leq j \leq n, 1 \leq k \leq n\}$, $\lambda(v) = 1$ for any $v \in V$, and $\tau(e) = c$ for any $e \in E$. Fig. 3 shows a ladder graph $LG(n, c)$. Then, the following lemma follows.

Lemma 2. *For any positive integer c, any bulk synchronous schedule for a ladder graph $LG(3 + 2c, c)$ onto at least $(3 + 2c)$ processors within deadline $(3 + 2c)$ consists of three no-communication phases with one unit length.*

Proof. Let D be $3 + 2c$. Let S_{BSP} be a bulk synchronous schedule for a ladder graph $LG(D, c)$ onto p $(\geq D)$ processors within deadline D. Any $u_{i+1,j}(1 \leq i < 3, 1 \leq j \leq D)$ cannot construct a no-communication phase with all of $\{u_{i,k} | 1 \leq k \leq D\}$ because the computation time $(D + 1)$ of these nodes on one processor is greater than the given deadline D. That is, any $u_{i+1,j}(1 \leq i < 3, 1 \leq j \leq D)$ must communicate with at least one of $\{u_{i,k} | 1 \leq k \leq D\}$. Hence, there exists a sequence $\{u_{1,k_1}, u_{2,k_2}, u_{3,k_3}\}$ of nodes such that $u_{i+1,k_{i+1}}$ communicates with u_{i,k_i} for any $i(1 \leq i < 3)$. This means that S_{BSP} includes at least two communication phases. On the other hand, S_{BSP} cannot include more than two communication phases because the deadline D is broken. Therefore, S_{BSP} includes just two communication phases. Consequently, S_{BSP} must consist of just three no-communication phases with one unit length. One of schedules possible as S_{BSP} is $\{\langle u_{i,j}, j, (i - 1)(c + 1)\rangle | 1 \leq i \leq 3, 1 \leq j \leq D\}$ (See Fig. 4). □

4.3 A Polynomial-Time Reduction

Now, we show the reduction from 3SP to 3BSSP(c). Let $(G = (V_G, E_G, \lambda_G, \tau_G), p)$ be an instance of 3SP. Let c be any positive integer. Let $LG(3 + 2c, c) = (V_{LG}, E_{LG}, \lambda_{LG}, \tau_{LG})$ be a ladder graph. Let G' be a task graph $(V_G \cup V_{LG}, E_G \cup E_{LG}, \lambda_{G'}, \tau_{G'})$ where $\lambda_{G'}(v) = 1$ for any $v \in V_G \cup V_{LG}$, and $\tau_{G'}(e) = c$ for any $e \in E_G \cup E_{LG}$.

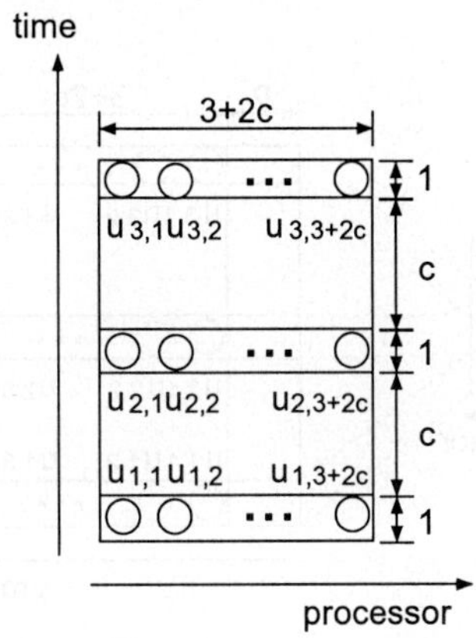

Fig. 4. A bulk synchronous schedule of a ladder graph $LG(3+2c, c)$ onto at least $(3+2c)$ processors within deadline $(3+2c)$

Lemma 3. *The transformation from an instance (G, p) of 3SP to an instance $(G', p+3+2c, c)$ of 3BSSP(c) is a polynomial transformation such that (G, p) is a yes instance iff $(G', p+3+2c, c)$ is a yes instance.*

Proof. If (G, p) is a "yes" instance of 3SP, then let S be a schedule for (G, p). A set $\{\langle v, q, t(c+1)\rangle | \langle v, q, t\rangle \in S\} \cup \{\langle u_{i,j}, p+j, (i-1)(c+1)\rangle | 1 \le i \le 3, 1 \le j \le 3+2c\}$ of triples is a bulk synchronous schedule for $(G', p+3+2c)$ with three no-communication phases and two communication phases (See Fig. 5).

Conversely, if $(G', p+3+2c, c)$ is a "yes" instance of 3BSSP(c), then let S'_{BSP} be a schedule for $(G', p+3+2c, c)$. From Lemma 2, $LG(3+2c, c)$ must be scheduled into a bulk synchronous schedule which consists of just three no-communication phases with one unit length. Therefore, whole S'_{BSP} must consists of just three no-communication phases with one unit length. Hence, S'_{BSP} must become a schedule as shown in Fig. 5. A subset $\{\langle v, q, t\rangle | \langle v, q, t(c+1)\rangle \in S'_{BSP}, 1 \le q \le p\}$ of S'_{BSP} is a schedule for (G, p). □

Theorem 1. *For any positive integer c, 3BSSP(c) is $\mathcal{NP}$-complete.*

Proof. Since BSSP(c) is $\mathcal{NP}$-complete [6], it is obvious that 3BSSP(c) is in $\mathcal{NP}$. Hence, from Lemma 3, the theorem follows. □

Let BSSPO(c) be the optimization counterpart of BSSP(c).

Theorem 2. *Let c be any positive integer. Then, a ρ-approximation algorithm for BSSPO(c) does not exist for any $\rho < \frac{4+2c}{3+2c}$ unless $\mathcal{P} = \mathcal{NP}$.*

Proof. From Theorem 1 and Lemma 1, the theorem follows. □

Theorem 3. *A ρ-approximation algorithm for BSSPO does not exist for any $\rho < \frac{6}{5}$ unless $\mathcal{P} = \mathcal{NP}$.*

Proof. From Theorem 2, a ρ'-approximation algorithm for BSSPO(1) does not exist for any $\rho' < \frac{6}{5}$ unless $\mathcal{P} = \mathcal{NP}$. If a ρ-approximation algorithm A for

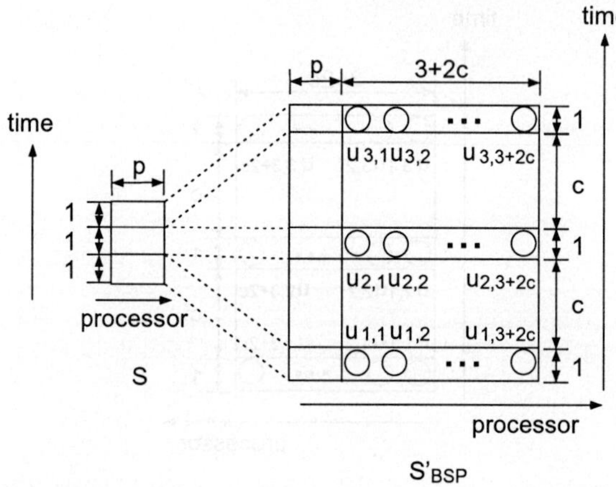

Fig. 5. A yes instance to a yes instance correspondence

BSSPO exists for some $\rho < \frac{6}{5}$, A can be used as a ρ-approximation algorithm for BSSPO(1). Hence, the theorem follows. □

5 Conclusion and Future Work

For the bulk synchronous task scheduling problem, we have proved that there does not exist a ρ-approximation algorithm for any $\rho < \frac{6}{5}$ unless $\mathcal{P} = \mathcal{NP}$.

In order to prove that, we have showed that generating a bulk synchronous schedule of length at most 5 is $\mathcal{NP}$-hard. However, the complexity of the problem for a schedule of length at most 4 is unknown. The $\mathcal{NP}$-hardness means non-approximability stronger than our result. So, one of the future work is to clear the complexity like Hoogeveen et al.'s work [9] for the conventional (i.e., not bulk synchronous) task scheduling problem.

Acknowledgement

This research was supported in part by the Kayamori Foundation of Informational Science Advancement.

References

1. Bacon, D.F. and Graham, S.L. and Sharp, O.J.: Compiler Transformations for High-Performance Computing, ACM computing surveys, Vol.26, No.4 (1994) 345-420
2. Darbha, S. and Agrawal, D. P.: Optimal Scheduling Algorithm for Distributed-Memory Machines, IEEE Trans. on Parallel and Distributed Systems, Vol.9, No.1 (1998) 87-95

3. El-Rewini, H. and Lewis, T.G. and Ali, H.H.: TASK SCHEDULING in PARALLEL and DISTRIBUTED SYSTEMS, PTR Prentice Hall (1994)
4. Fujimoto, N. and Baba, T. and Hashimoto, T. and Hagihara, K.: A Task Scheduling Algorithm to Package Messages on Distributed Memory Parallel Machines, Proc. of 1999 Int. Symposium on Parallel Architectures, Algorithms, and Networks (1999) 236-241
5. Fujimoto, N. and Hashimoto, T. and Mori, M. and Hagihara, K.: On the Performance Gap between a Task Schedule and Its Corresponding Parallel Program, Proc. of 1999 Int. Workshop on Parallel and Distributed Computing for Symbolic and Irregular Applications, World Scientific (2000) 271-287
6. Fujimoto, N. and Hagihara, K.: NP-Completeness of the Bulk Synchronous Task Scheduling Problem and Its Approximation Algorithm, Proc. of 2000 Int. Symposium on Parallel Architectures, Algorithms, and Networks (2000) 127-132
7. Fujimoto, N. and Baba, T. and Hashimoto, T. and Hagihara, K.: On Message Packaging in Task Scheduling for Distributed Memory Parallel Machines, The International Journal of Foundations of Computer Science, Vol.12, No.3 (2001) 285-306
8. Kruatrachue, B., "Static task scheduling and packing in parallel processing systems", Ph.D. diss., Department of Electrical and Computer Engineering, Oregon State University, Corvallis, 1987
9. Hoogeveen, J. A., Lenstra, J. K., and Veltman, B.: "Three, Four, Five, Six or the Complexity of Scheduling with Communication Delays", Oper. Res. Lett. Vol.16 (1994) 129-137
10. Lenstra, J. K. and Rinnooy Kan, A. H. G.: Complexity of Scheduling under Precedence Constraints, Operations Research, Vol.26 (1978) 22-35
11. Lenstra, J.K. and Shmoys, D. B.: Computing Near-Optimal Schedules, Scheduling Theory and its Applications, John Wiley & Sons (1995) 1-14
12. Palis, M. A. and Liou, J. and Wei, D. S. L.: Task Clustering and Scheduling for Distributed Memory Parallel Architectures, IEEE Trans. on Parallel and Distributed Systems, Vol.7, No.1 (1996) 46-55
13. Papadimitriou, C. H. and Yannakakis, M.: Towards An Architecture-Independent Analysis of Parallel Algorithms, SIAM J. Comput., Vol.19, No.2 (1990) 322-328
14. Thurimella, R. and Yesha, Y.: A scheduling principle for precedence graphs with communication delay, Int. Conf. on Parallel Processing, Vol.3 (1992) 229-236
15. Valiant, L.G.: A Bridging Model for Parallel Computation, Communications of the ACM, Vol.33, No.8 (1990) 103-111

Adjusting Time Slices to Apply Coscheduling Techniques in a Non-dedicated NOW*

Francesc Giné[1], Francesc Solsona[1], Porfidio Hernández[2], and Emilio Luque[2]

[1] Departamento de Informática e Ingeniería Industrial, Universitat de Lleida, Spain.
{sisco,francesc}@eup.udl.es

[2] Departamento de Informática, Universitat Autónoma de Barcelona, Spain.
{p.hernandez,e.luque}@cc.uab.es

Abstract. Our research is focussed on keeping both local and parallel jobs together in a time-sharing NOW and efficiently scheduling them by means of coscheduling mechanisms. In such systems, the proper length of the time slice still remains an open question. In this paper, an algorithm is presented to adjust the length of the quantum dynamically to the necessity of the distributed tasks while keeping good response time for interactive processes. It is implemented and evaluated in a Linux cluster.

1 Introduction

The challenge of exploiting underloaded workstations in a NOW for hosting parallel computation has led researchers to develop techniques to adapt the traditional uniprocessor time-shared scheduler to the new situation of mixing local and parallel workloads. An important issue in managing parallel jobs in a non-dedicated cluster is how to coschedule the processes of each running job across all the nodes. Such simultaneous execution can be achieved by means of identifying the coscheduling need during execution [3, 4] from local implicit runtime information, basically communication events. Our efforts are addressed towards developing coscheduling techniques over a non-dedicated cluster.

In such a system, parallel jobs performance is very sensitive to the quantum [1, 6]. The quantum length is a compromise; according to the local user necessity, it should not be too long in order not to degrade the responsive time of interactive applications, whereas from the point of view of the parallel performance [1] shorter time slices can degrade the cache performance, since each process should reload the evicted data every time it restarts the execution. However, an excessively long quantum could degrade the performance of coscheduling techniques [6]. A new technique is presented in this paper to adjust dynamically the quantum of every local scheduler in a non-dedicated NOW according to local user interactivity, memory behavior of each parallel job and coscheduling decisions. This technique is implemented in a Linux NOW and compared with other alternatives.

* This work was supported by the MCyT under contract TIC 2001-2592 and partially supported by the Generalitat de Catalunya -Grup de Recerca Consolidat 2001SGR-00218.

B. Monien and R. Feldmann (Eds.): Euro-Par 2002, LNCS 2400, pp. 234–239.

2 DYNAMICQ: An Algorithm to Adjust the Quantum

Our framework is a non-dedicated cluster, where every node has a time sharing scheduler with process preemption based on ranking processes according to their priority. The scheduler works by dividing the CPU time into *epochs*. In a single epoch, each process (*task*) is assigned a specified quantum ($task_i.q_n$:time slice of task i for the nth epoch), which it is allowed to run. When the running process has expired its quantum or is blocked waiting for an event, another process is selected to run from the Ready Queue (RQ). The epoch ends when all the processes in the RQ have exhausted their quantum. The next epoch begins when the scheduler assigns a fresh quantum to all processes.

It is assumed that every node has a two level cache memory (L1 and L2), which is not flushed at a context switch. In this kind of environment, the proper length of time slices should be set according to process locality in order to amortize the context switch overhead associated with processes with large memory requirements [1, 5]. For this reason, we propose to determine the proper length of the next time slice ($task.q_{n+1}$) according to the L2 cache miss-rate ($mr_n = \frac{L2\_cache\_misses_n}{L1\_cache\_misses_n}$),where $Li\_cache\_misses_n$ is the number of misses of Li cache occurred during the nth epoch. It can be obtained from the hardware counters provided by current microprocessors [2].

It is assumed that every local scheduler applies a coscheduling technique, named *predictive coscheduling*, which consists of giving more scheduling priority to tasks with higher receive-send communication rates. This technique has been chosen because of the good performance achieved in a non-dedicated NOW [4].

Algorithm 1 shows the steps for calculating the quantum. This algorithm, named *DYNAMICQ*, will be computed by every local scheduler every time that a new epoch begins and will be applied to all active processes (*line 1*).

In order to preserve the performance of local users, the algorithm, first of all, checks if there is an interactive user in such a node. If there were any, the predicted quantum ($task_p.q_{n+1}$) would be set to a constant value, denoted as *DEFAULT_QUANTUM*[1] *(line 3).* When there is no interactivity user, the quantum is computed according to the cache miss-rate (mr_n) and the length of the previous quantum ($task_p.q_n$). Although some authors assume that the miss-rate decreases as the quantum increases, the studies carried out in [1] reveal that when a time slice is long enough to pollute the memory but not enough to compensate for the misses caused by context switches, the miss-rate may increase in some cases since more data, from previous processes, are evicted as the length of time slice increases. For this reason, whenever the miss-rate is higher than a threshold, named MAX_MISS, or if it has been increased with respect to the preceding epoch ($mr_{n-1} < mr_n$), the quantum will be doubled (line 6).

When applying techniques, such as the *predictive coscheduling technique* [4], an excessively long quantum could decrease the performance of parallel tasks. Since there is no global control, which could schedule all the processes of a parallel job concurrently, a situation could occur quite frequently in which scheduled

[1] Considering the base time quantum of Linux o.s., it is set to 200ms.

Algorithm 1 DYNAMICQ Algorithm.

```
1 for_each_active_task(p)
2   if (INTERACTIVE_USER)
3     task_p.q_{n+1} =DEFAULT_QUANTUM;
4   else
5     if((mr_n¿MAX_MISS) —— (mr_{n-1} < mr_n)) && (task_p.q_n¡=MAX_SLICE)
6       task_p.q_{n+1} = task_p.q_n * 2;
7     else if(task_p.q_n >MAX_SLICE)
8       task_p.q_{n+1} = task_p.q_n/2;
9     else
10      task_p.q_{n+1} = task_p.q_n;
11    endelse;
12  endelse;
13 endfor;
```

processes that constitute different parallel jobs contended for scheduling their respective correspondents. Thus, if the quantum was too long, the context switch request through sent/received messages could be discarded and hence the parallel job would eventually be stalled until a new context-switch was initiated by the scheduler. In order to avoid this situation, a maximum quantum *(MAX_SLICE)* was established. Therefore, if the quantum exceeds this threshold, it will be reduced to half (*line* 8). Otherwise, the next quantum will be fixed according to the last quantum computed (*line 10*).

3 Experimentation

DYNAMICQ was implemented in the Linux Kernel v.2.2.15 and tested in a cluster of eight Pentium III processors with 256MB of main memory and a L2 four-way set associative cache of 512KB. They were all connected through a Fast Ethernet network. *DYNAMICQ* was evaluated by running four PVM NAS parallel benchmarks [5] with class A: *IS, MG, SP and BT*. Table 1 shows the time ratio corresponding to each benchmarks's computation and communication cost. The local workload was carried out by means of running one synthetic benchmark, called *local.* This allows the CPU activity to alternate with interactive activity. The CPU is loaded by performing floating point operations over an array with a size and during a time interval set by the user (in terms of time rate). Interactivity was simulated by means of running several system calls with an exponential distribution frequency (mean=500ms by default) and different data transferred to memory with a size chosen randomly by means of a uniform distribution in the range [1MB,...,10MB]. At the end of its execution, the benchmark returns the system call latency and wall-clock execution time.

Four different workloads (table 1) were chosen in these trials. All the workloads fit in the main memory. Three environments were compared, the plain Linux scheduler (*LINUX*), predictive coscheduling with a static quantum (*STATICQ*) and predictive coscheduling applying the DYNAMICQ algorithm.

Table 1. local(z) means that one instance of local task is executed in z nodes.

Bench.	%Comp.	%Comm.	Workload (Wrk)	
IS.A	62	38	1	SP+BT+IS
SP.A	78	22	2	BT+SP+MG
BT.A	87	13	3	SP+BT+*local(z)*
MG.A	83	17	4	BT+MG+*local(z)*

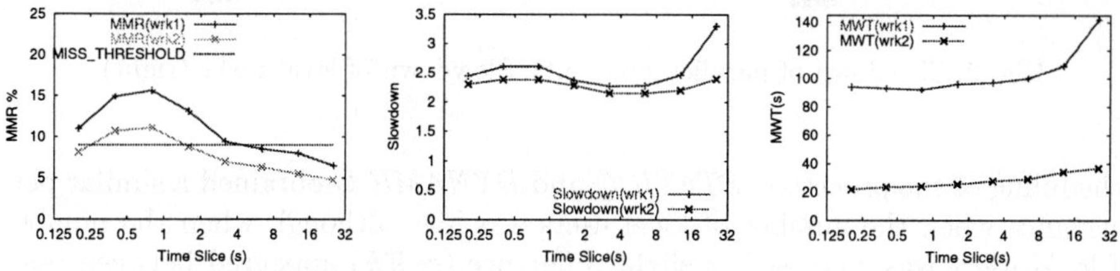

Fig. 1. STATICQ mode. MMR (left), Slowdown (centre) and MWT (right) metrics.

In the *STATICQ* mode, all the tasks in each node are assigned the same quantum, which is set from a system call implemented by us.

Its performance was validated by means of three metrics: *Mean Cache Miss-rate:* ($MMR = \frac{\sum_{k=1}^{8}\sum_{n=1}^{N_k}(\frac{mr_{nk}}{N_k})}{8} x100$) where N_k is the number of epochs passed during execution in node k; *Mean Waiting Time (MWT)*, which is the average time spent by a task waiting on communication; and *Slowdown* averaged over all the jobs of every workload.

3.1 Experimental Results

Fig. 1(left) shows the MMR parameter for Wrk1 and Wrk2 in the STATICQ mode. In both cases, we can see that for a quantum smaller than 0.8s, the cache performance is degraded because the time slice is not long enough to compensate the misses caused by the context switches. In order to avoid this degradation peak, a *MAX_MISS* threshold equal to 9% was chosen for the rest of the trials.

Fig. 1 examines the effect of the time slice length on the slowdown (centre) and MWT (right) metrics. The rise in slowdown for a quantum smaller than 1s reveals the narrow relationship between the cache behavior and the distributed job performance. For a quantum greater than 6.4s, the performance of Wrk1 is hardly affected by the coscheduling policy, as we can see in the analysis of the MWT metric. In order to avoid this coscheduling loss, the *DYNAMICQ* algorithm works by default with a *MAX_SLICE* equal to 6.4s.

Fig. 2 (left) shows the slowdown of parallel jobs for the three environments (*STATICQ* with a quantum= 3.2s) when the number of local users (*local* benchmark was configured to load the CPU about 50%) is increased from 2 to 8. *LINUX* obtained the worst performance due to the effect of uncoordinated

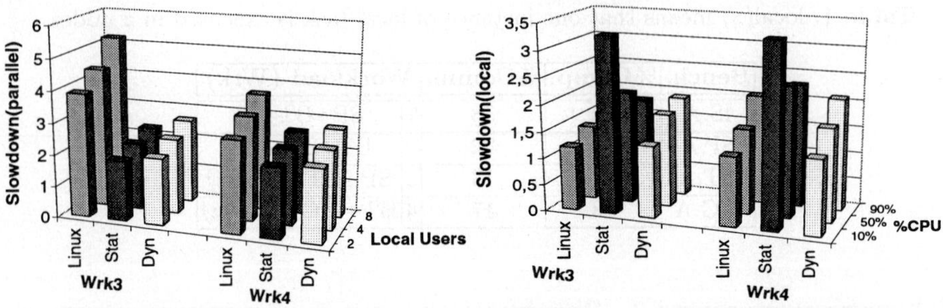

Fig. 2. Slowdown of parallel jobs (left). Slowdown of local tasks (right).

scheduling of the processes. *STATICQ* and *DYNAMICQ* obtained a similar performance when the number of local users was low, although when the number of local users was increased, a slight difference ($\simeq$ 9%) appeared between both modes due to the heterogeneous quantum present in the cluster in DYNAMICQ mode. Fig. 2 (right) shows the overhead introduced into the *local task* (the CPU requirements were decreased from 90% to 10%*)*. It can be seen that the results obtained for *Linux* are slightly better than those for *DYNAMICQ,* whereas *STATICQ* obtains the worst results. This is because the *STATICQ and DYNAMICQ* modes give more execution priority to distributed tasks with high communication rates, thus delaying the scheduling of local tasks until distributed tasks finish their quantum. This priority increase has little effect on local tasks with high CPU requirements but provokes an overhead proportional to the quantum length in the interactive tasks. This is reflected in the high slowdown in *STATICQ* mode when local tasks have low CPU requirements (10%).

4 Conclusions and Future Work

This paper discusses the need to fix the quantum accurately to apply scoscheduling techniques in a non-dedicated NOW. An algorithm is proposed to adjust the proper quantum dynamically according to the cache miss-rate, coscheduling decisions and local user performance. Its good performance is proved experimentally over a Linux cluster. Future work will be directed towards extending our analysis to a wider range of workloads and researching the way to set both thresholds, *MAX_SLICE* and *MAX_MISS* automatically from runtime information.

References

1. G. Edward Suh and L. Rudolph. "Effects of Memory Performance on Parallel Job Scheduling". LNCS, vol.2221, 2001.
2. Performance-Monitoring Counters Driver, *http://www.csd.uu.se/~mikpe/linux/perfctr*
3. P.G. Sobalvarro, S. Pakin, W.E. Weihl and A.A. Chien. "Dynamic Coscheduling on Workstation Clusters". *IPPS'98,* LNCS, vol.1459, 1998.

4. F. Solsona, F. Giné, P. Hernández and E. Luque. "Predictive Coscheduling Implementation in a non-dedicated Linux Cluster". *EuroPar'2001*, LNCS, vol.2150, 2001.
5. F.C. Wong, R.P. Martin, R.H. Arpaci-Dusseau and D.E. Culler "Architectural Requirements and Scalability of the NAS Parallel Benchmarks". *Supercomputing'99.*
6. A. Yoo and M. Jette. "An Efficient and Scalable Coscheduling Technique for Large Symmetric Multiprocessors Clusters". LNCS, vol.2221, 2001.

A Semi-dynamic Multiprocessor Scheduling Algorithm with an Asymptotically Optimal Competitive Ratio

Satoshi Fujita*

Department of Information Engineering
Graduate School of Engineering, Hiroshima University
Higashi-Hiroshima, 739-8527, Japan

Abstract. In this paper, we consider the problem of assigning a set of n independent tasks onto a set of m identical processors in such a way that the overall execution time is minimized provided that the precise task execution times are not known a priori. In the following, we first provide a theoretical analysis of several conventional scheduling policies in terms of the worst case slowdown compared with the outcome of an optimal scheduling policy. It is shown that the best known algorithm in the literature achieves a worst case competitive ratio of $1+1/f(n)$ where $f(n) = O(n^{2/3})$ for any fixed m, that approaches to one by increasing n to the infinity. We then propose a new scheme that achieves a better worst case ratio of $1+1/g(n)$ where $g(n) = \Theta(n/\log n)$ for any fixed m, that approaches to one more quickly than the other schemes.

1 Introduction

In this paper, we consider the problem of assigning a set of n independent tasks onto a set of m identical processors in such a way that the overall execution time of the tasks will be minimized. It is widely accepted that, in the multiprocessor scheduling problem, both dynamic and static scheduling policies have their own advantages and disadvantages; for example, under dynamic policies, each task assignment incurs (non-negligible) overhead that is mainly due to communication, synchronization, and the manipulation of date structures, and under static policies, unpredictable faults and the delay of task executions will significantly degrade the performance of the scheduled parallel programs.

The basic idea of our proposed method is to adopt the notion of clustering in a "balanced" manner in terms of the worst case slowdown compared with the outcome of an optimal scheduling policy; i.e., we first partition the given set of independent tasks into several clusters, and apply static and dynamic schedulings to them in a mixed manner, in such a way that the worst case competitive ratio will be minimized. Note that this method is a generalization of two extremal cases in the sense that the case in which all tasks are contained in a single cluster

* This research was partially supported by the Ministry of Education, Culture, Sports, Science and Technology of Japan (# 13680417).

B. Monien and R. Feldmann (Eds.): Euro-Par 2002, LNCS 2400, pp. 240–247.

corresponds to a static policy and the case in which each cluster contains exactly one task corresponds to a dynamic policy. In the following, we first provide a theoretical analysis of several scheduling policies proposed in the literature; it is shown that the best known algorithm in the literature achieves a worst case competitive ratio of $1 + 1/f(n)$ where $f(n) = O(n^{2/3})$ for any fixed m, that approaches to one by increasing n to the infinity. We then propose a new scheme that achieves a better worst case ratio of $1 + 1/g(n)$ where $g(n) = \Theta(n/\log n)$ for any fixed m, that approaches to one more quickly than the other schemes.

The remainder of this paper is organized as follows. In Section 2, we formally define the problem and the model. A formal definition of the competitive ratio, that is used as the measure of goodness of scheduling policies, will also be given. In Section 3, we derive upper and lower bounds on the competitive ratio for several conventional algorithms. In Section 4, we propose a new scheduling policy that achieves a better competitive ratio than conventional ones.

2 Preliminaries

2.1 Model

Let S be a set of n independent tasks, and $P = \{p_1, p_2, \ldots, p_m\}$ be a set of identical processors connected by a complete network. The execution time of a task u, denoted by $\tau(u)$, is a real satisfying $\alpha_u \le \tau(u) \le \beta_u$ for predetermined boundaries α_u and β_u, where the precise value of $\tau(u)$ can be known only when the execution of the task completes. Let $\alpha \stackrel{\text{def}}{=} \min_{u \in S} \alpha_u$ and $\beta \stackrel{\text{def}}{=} \max_{u \in S} \beta_u$. A **scheduling** of task u is a process that determines: 1) the processor on which the task is executed, and 2) the (immediate) predecessor of the task among those tasks assigned to the same processor[1]. Scheduling of a task can be conducted in either static or dynamic manner. In a static scheduling, each task can start its execution immediately after the completion of the predecessor task, although in a dynamic scheduling, each task execution incurs a scheduling overhead ϵ before starting, the value of which depends on the configuration of the system and the sizes of S and P.

A **scheduling policy** $\mathcal{A}$ for S is a collection of schedulings for all tasks in S. A scheduling policy $\mathcal{A}$ is said to be "static" if all schedulings in $\mathcal{A}$ are static, and is said to be "dynamic" if all schedulings in $\mathcal{A}$ are dynamic. A scheduling policy that is neither static nor dynamic will be referred to as a semi-dynamic policy.

In this paper, we measure the goodness of scheduling policies in terms of the worst case slowdown of the resultant schedule compared with the outcome of an optimal off-line algorithm, where term "off-line" means that it knows precise value of $\tau(u)$'s before conducting a scheduling; i.e., an off-line algorithm can generate an optimal static scheduling with overhead zero, although in order to

[1] Note that in the above definition, a scheduling does not fix the start time of each task; it is because we are considering cases in which the execution time of each task can change dynamically depending on the runtime environment.

obtain an optimal solution, it must solve the set partition problem that is well known to be NP-complete [1].

Let $\mathcal{A}(S, m, \tau)$ denote the length of a schedule generated by scheduling policy $\mathcal{A}$, which assigns tasks in S onto a set of m processors under an on-line selection τ of execution times for all $u \in S$. Let OPT denote an optimal off-line scheduling policy. Then, the (worst case) competitive ratio of $\mathcal{A}$ is defined as

$$r(\mathcal{A}, m, n) \stackrel{\text{def}}{=} \sup_{|S|=n,\tau} \frac{\mathcal{A}(S, m, \tau)}{OPT(S, m, \tau)}.$$

Note that by definition, $r(\mathcal{A}, m, n) \geq 1$ for any $\mathcal{A}$, $n \geq 1$, and $m \geq 2$. In the following, an asymptotic competitive ratio is also used, that is defined as follows:

$$r(\mathcal{A}, m) \stackrel{\text{def}}{=} \sup_{n \geq 1} r(\mathcal{A}, m, n).$$

2.2 Related Work

In the past two decades, several semi-dynamic scheduling algorithms have been proposed in the literature. Their main application is the parallelization of nested loops, and those semi-dynamic algorithms are commonly referred to as "chunk" scheduling schemes.

In the chunk self-scheduling policy (CSS, for short), a collection of tasks is divided into several chunks (clusters) with an equal size K, and those chunks are assigned to processors in a greedy manner [3] (note that an instance with $K = 1$ corresponds to a dynamic scheduling policy). CSS with chunk size K is often referred to as $CSS(K)$, and in [3], the goodness of $CSS(K)$ is theoretically analyzed under the assumption that the execution time of each task (i.e., an iteration of a loop) is an independent and identically distributed (i.i.d) random variable with an exponential distribution.

Polychronopoulos and Kuck proposed a more sophisticated scheduling policy called guided self-scheduling (GSS, for short) [4]. This policy is based on the intuition such that in an early stage of assignment, the size of each cluster can be larger than those used in later stages; i.e., the size of clusters can follow a decreasing sequence such as geometrically decreasing sequences. More concretely, in the ith assignment, GSS assigns a cluster of size R_i/m to an idle processor, where R_i is the number of remaining loops at that time; e.g., R_1 is initialized to n, and R_2 is calculated as $R_1 - R_1/m = n(1 - 1/m)$. That is, under GSS, the cluster size geometrically decreases as

$$n/m,\ n/m(1 - 1/m),\ n/m(1 - 1/m)^2, \ldots.$$

Factoring scheduling proposed in [2] is an extension of GSS and CSS in the sense that a "part" of remaining loops is equally divided among the available processors; Hence, by using a parameter ℓ, that is a function of several parameters such as the mean execution time of a task and its deviation, the decreasing sequence of the cluster size is represented as $\underbrace{(n/m)\ell, \ldots, (n/m)\ell}_{m}, \underbrace{(n/m)\ell^2, \ldots, (n/m)\ell^2}_{m}, \ldots$.

Trapezoid self-scheduling (TSS, for short) proposed in [5] is another extension of GSS; in the scheme, the size of clusters decreases linearly instead of exponentially, and the sizes of maximum and minimum clusters can be specified as a part of the policy. (Note that since the total number of tasks is fixed to n, those two parameters completely define a decreasing sequence.) In [5], it is claimed that TSS is more practical than GSS in the sense that it does not require a complicated calculation for determining the size of the next cluster.

3 Analysis of Conventional Algorithms

This section gives an analysis of conventional algorithms described in the last section in terms of the competitive ratio.

3.1 Elementary Bounds

Recall that $\alpha \stackrel{\text{def}}{=} \min_{u \in S} \alpha_u$ and $\beta \stackrel{\text{def}}{=} \max_{u \in S} \beta_u$. The competitive ratio of any static and dynamic policies is bounded as in the following two lemmas (proofs are omitted in this extended abstract).

Lemma 1 (Static). *For any static policy $\mathcal{A}$ and for any $m \geq 2$, $r(\mathcal{A}, m) \geq 1 + \frac{\beta - \alpha}{\alpha + \beta/(m-1)}$, and the bound is tight in the sense that there is an instance that achieves it.*

Lemma 2 (Dynamic). *1) For any dynamic policy $\mathcal{A}$ and for any $m, n \geq 2$, $r(\mathcal{A}, m, n) \geq 1 + \epsilon/\alpha$, and 2) for any $2 \leq m \leq n$, there is a dynamic policy $\mathcal{A}^*$ such that $r(\mathcal{A}^*, m, n) \leq 1 + \frac{\epsilon}{\alpha} + \left(\frac{\beta+\epsilon}{\alpha}\right)\left(\frac{m}{n}\right)$.*

The goodness of chunk self-scheduling (CSS) in terms of the competitive ratio could be evaluated as follows.

Theorem 1 (CSS). *$r(CSS, m, n)$ is at least $1 + \left(\frac{2\sqrt{\beta\epsilon}}{\alpha}\right)\sqrt{\frac{m}{n}} + \frac{(\beta+\epsilon)m}{\alpha n}$, which is achieved when the cluster size is selected as $K = \left\lceil \sqrt{n\epsilon/m\beta} \right\rceil$.*

Since the largest cluster size in GSS is n/m, by using a similar argument to Lemma 1, we have the following theorem.

Corollary 1 (GSS). *$r(GSS, m, n) \geq 1 + \frac{\beta - \alpha}{\alpha + \frac{\beta}{m-1}}$.*

A similar claim holds for factoring method, since it does not take into account two boundaries α and β to determine parameter ℓ; i.e., for large β such that $\beta(n/m)\ell > \{n - (n/m)\ell\}\alpha$, we cannot give a good competitive ratio that approaches to one.

3.2 Clustering Based on Linearly Decreasing Sequence

Let Δ be a positive integer that is given as a parameter. Consider a sequence of integers $s_1, s_2, \ldots$, defined as follows: $s_i = s_1 - \Delta(i-1)$ for $i = 1, 2, \ldots$. Let k be an integer such that $\sum_{i=1}^{k-1} s_i < n \leq \sum_{i=1}^{k} s_i$. Trapezoid self-scheduling (TSS) is based on a sequence of k clusters $S_1, S_2, \ldots, S_k$, such that the sizes of the first $k-1$ clusters are $s_1, s_2, \ldots, s_{k-1}$, respectively, and that of the last cluster is $n - \sum_{i=1}^{k-1} s_i$. (A discussion for rational Δ's is complicated since it depends on the selection of m and n; hence we leave the analysis for rational Δ's as a future problem.)

In this subsection, we prove the following theorem.

Theorem 2. *$r(TSS, m, n) \geq 1 + 1/f(n)$ where $f(n) = O(n^{2/3})$ for fixed m.*

Proof. If $k \leq m$, then the same bound with Lemma 1 holds since in such cases, $|S_1| > n/m$ must hold. So, we can assume that $k > m$, without loss of generality. Let t be a non-negative integer satisfying the following inequalities:

$$(t+1)m \;<\; k \;\leq\; (t+2)m.$$

In the following, we consider the following three cases separately in this order; i.e., when t is an even greater than or equal to 2 (Case 1), when t is odd (Case 2), and when $t = 0$ (Case 3).

Case 1: For even $t \geq 2$, we may consider the following assignment τ of execution times to each task:

- if $|S_{tm+1}| \geq 2|S_{(t+1)m+1}|$ then the $(tm+1)$st cluster S_{tm+1} consists of tasks with execution time β, and the other clusters consist of tasks with execution time α, and
- if $|S_{tm+1}| < 2|S_{(t+1)m+1}|$ then the $(tm+m+1)$st cluster $S_{(t+1)m+1}$ consists of tasks with execution time β, and the other clusters consist of tasks with execution time α.

Since S contains at most $|S_{tm+1}|$ tasks with execution time β and all of the other tasks have execution time α, the schedule length of an optimal (off-line) algorithm is at most

$$OPT = \frac{n\alpha + (\beta - \alpha)|S_{tm+1}|}{m} + \beta \tag{1}$$

where the first term corresponds to the minimum completion time among m processors and the second term corresponds to the maximum difference of the completion times. On the other hand, for given τ, the length of a schedule generated by TSS is at least

$$TSS = \frac{n\alpha}{m} + t\epsilon + \frac{(\beta - \alpha)|S_{tm+1}|}{2} \tag{2}$$

where the first term corresponds to an optimal execution time of tasks provided that the execution time of each task is α, the second term corresponds to the total overhead (per processor) incurred by the dynamic assignment of tasks, and the third term corresponds to the minimum difference of the completion times between the longest one and the others, provided that the execution time of tasks of one cluster (i.e., S_{tm+1} or S_{tm+m+1}) becomes β from α. Note that under TSS, clusters are assigned to m processors in such a way that all processors complete their $(2i)$th cluster simultanesously for each $1 \leq i \leq t/2$, and either S_{tm+1} or S_{tm+m+1} will be selected as a cluster consisting of longer tasks. Note also that by the rule of selection, at least $|S_{tm+1}|/2$ tasks contribute to the increase of the schedule length, according to the change of execution time from α to β.

Hence the ratio is at least

$$\begin{aligned} r(GSS, m, n) &= \frac{n\alpha/m + t\epsilon + (\beta-\alpha)|S_{tm+1}|/2}{n\alpha/m + (\beta-\alpha)|S_{tm+1}|/m + \beta - \alpha} \\ &= 1 + \frac{t\epsilon m + (\beta-\alpha)(|S_{tm+1}|m/2 - |S_{tm+1}| - m)}{n\alpha + (\beta-\alpha)(|S_{tm+1}| + m)} \\ &\geq 1 + \frac{\epsilon k - (\beta-\alpha+\epsilon)m + (\beta-\alpha)|S_{tm+1}|(m/2-1)}{n\alpha + (\beta-\alpha)(|S_{tm+1}| + m)} \end{aligned}$$

where the last inequality is due to $tm < k - m$.

Now consider the following sequence of clusters $S'_1, S'_2, \ldots, S'_k$ such that $|S'_i| = |S'_1| - \Delta'(i-1)$ for some Δ', $|S'_k| = 1$, and $\sum_{i=1}^{k} |S'_i| = n$. It is obvious that $|S_i| \geq |S'_i|$ for $k/2 \leq i \leq k$, and $tm+1 \geq k/2$ holds since $t \geq 2$; i.e., $|S_{tm+1}| \geq |S'_{tm+1}|$. On the other hand, since $|S'_1| = 2n/k$ and $tm+1-k > m$, we can conclude that $|S_{tm+1}| \geq 2nm/k^2$. By substituing this inequality to the above formula, we have

$$\begin{aligned} r(TSS, m, n) &\geq 1 + \frac{\epsilon k - (\beta-\alpha+\epsilon)m + (\beta-\alpha)\frac{2nm}{k^2}(m/2-1)}{n\alpha + (\beta-\alpha)(|S_{tm+1}| + m)} \\ &= 1 + \frac{\epsilon k - (\beta-\alpha+\epsilon)m + (\beta-\alpha)\frac{2nm}{k^2}(m/2-1)}{\beta n + (\beta-\alpha)m}, \end{aligned}$$

where the right hand side takes a minimum value when $\epsilon k = (\beta-\alpha)\frac{2nm}{k^2}(m/2-1)$, i.e., when $k \simeq \sqrt[3]{\frac{(\beta-\alpha)nm}{\epsilon}}$. Hence by letting $k = \Theta(\sqrt[3]{n})$, we have $r(TSS, m, n) \geq 1 + 1/f(n)$ where $f(n) = O(n^{2/3})$ for any fixed m.

Case 2: For odd $t \geq 1$, we may consider the following assignment τ of execution times to each task:

- if $|S_{tm+1}| \geq 2|S_{(t+1)m+1}|\alpha/(\beta-\alpha)$ then the $(tm+1)$st cluster S_{tm+1} consists of tasks with execution time β, and the other clusters consist of tasks with execution time α.
- if $|S_{tm+1}| < 2|S_{(t+1)m+1}|\alpha/(\beta-\alpha)$ then the $(tm+m+1)$st cluster $S_{(t+1)m+1}$ consists of tasks with execution time β, and the other clusters consist of tasks with execution time α.

For such τ, an upper bound on the schedule length of an optimal (off-line) algorithm is given as in Equation (1), and the length of a schedule generated by

TSS can be represented in a similar form to Equation (2), where the last term should be replaced by

$$(\beta - \alpha)\,|S_{tm+1}| - \alpha|S_{tm+m+1}| \geq \frac{(\beta - \alpha)|S_{tm+1}|}{2}$$

when S_{tm+1} is selected, and by

$$(\beta - \alpha)|S_{tm+m+1}| > \frac{(\beta - \alpha)^2|S_{tm+1}|}{2\alpha}$$

when $S_{(t+1)m+1}$ is selected. Note that in both cases, a similar argument to Case 1 can be applied.

Case 3: When $t = 0$, we may use τ such that either S_1 or S_{m+1} is selected as a cluster with longer tasks as in Case 1, and for such τ, a similar argument to Lemma 1 holds. Q.E.D.

4 Proposed Method

In this section, we propose a new semi-dynamic scheduling policy that exhibits a better worst case performance than the other policies proposed in the literature. Our goal is to prove the following theorem.

Theorem 3. *There exists a semi-dynamic policy $\mathcal{A}$ such that $r(\mathcal{A}, m, n) = 1 + 1/g(n)$ where $g(n) = \Theta(n/\log n)$ for any fixed m.*

In order to clarify the explanation, we first consider the case of $m = 2$. Consider the following (monotonically decreasing) sequence of integers $s_0, s_1, s_2, \ldots$:

$$s_i \stackrel{\text{def}}{=} \begin{cases} n & \text{if } i = 0 \\ \left\lceil \left(\frac{\beta}{\alpha+\beta}\right) s_{i-1} \right\rceil & \text{if } i \geq 1. \end{cases}$$

Let k be the smallest integer satisfying $s_k \leq 2 + \frac{\beta}{\alpha}$. Note that such a k always exists, since $s_i \geq s_{i-1}$ for any $i \geq 1$, and if $s_{k'} = s_{k'-1}$ for some k', then $k' > k$ must hold (i.e., $s_1, s_2, \ldots, s_k$ is a strictly decreasing sequence). In fact, $s_{k'} = s_{k'-1}$ implies $\left(\frac{\beta}{\alpha+\beta}\right) s_{k'-1} > s_{k'-1} - 1$; i.e., $s_{k'-1} < 1 + \frac{\beta}{\alpha}$ $(< 2 + \frac{\beta}{\alpha})$.

By using a (finite) sequence $s_0, s_1, \ldots, s_k$, we define a partition of S, i.e., $\{S_1, S_2, \ldots, S_k, S_{k+1}\}$, as follows:

$$|S_i| \stackrel{\text{def}}{=} \begin{cases} s_{i-1} - s_i & \text{if } i = 1, 2, \ldots, k \text{ and} \\ s_k & \text{if } i = k + 1. \end{cases}$$

By the above definition, we have

$$\sum_{u \in S_i} \tau(u) \leq \sum_{v \in S_{i+1} \cup \ldots \cup S_k} \tau(v)$$

for any i and τ, provided that it holds $\alpha \leq \tau(u) \leq \beta$ for any $u \in S$. Hence, by assigning clusters $S_1, S_2, \ldots, S_{k+1}$ to processors in this order, we can bound the difference of completion times of two processors by at most $\beta|S_k| + \epsilon$; i.e., we can bound the competitive ratio as

$$r(\mathcal{A}, 2) \leq \frac{(X + (k+1)\epsilon)/2 + \beta|S_{k+1}| + \epsilon}{X/2} \leq 1 + \frac{k\epsilon + 2\beta|S_{k+1}| + 3\epsilon}{n\alpha} \quad (3)$$

Since we have known that $|S_{k+1}| \leq 2 + \frac{\beta}{\alpha}$, the proof for $m = 2$ completes by proving the following lemma.

Lemma 3.

$$k \leq \frac{\log_2 n}{\log_2(1 + \alpha/\beta)}.$$

Proof. Let a be a constant smaller than 1. Let $f_a(x) \stackrel{\text{def}}{=} \lceil ax \rceil$, and let us denote $f_a^i(x) \stackrel{\text{def}}{=} f_a(f_a^{i-1}(x))$, for convenience. Then, by a simple calculation, we have

$$f_a^i(x) \leq a^i \times x + a^{i-1} + a^{i-2} + \cdots + 1 \leq a^i \times x + \frac{1}{1-a}.$$

Hence, when $a = \left(\frac{\beta}{\alpha+\beta}\right)$ and $i = \log_{(1+\alpha/\beta)} n$, since $a^i = 1/n$, we have

$$f_a^i(n) \leq \frac{1}{n} \times n + \frac{1}{1 - \left(\frac{\beta}{\alpha+\beta}\right)} = 2 + \frac{\beta}{\alpha}$$

Hence, the lemma follows. Q.E.D.

We can extend the above idea to general m, as follows: Given sequence of clusters $S_1, S_2, \ldots, S_{k+1}$, we can define a sequence of $(k+1)m$ clusters by partitioning each cluster into m (sub)clusters equally (recall that this is a basic idea that is used in the factoring method). By using a similar argument to above, we can complete the proof of Theorem 3.

References

1. M. R. Garey and D. S. Johnson. *Computers and Intractability: A Guide for the Theory of NP-Completeness.* Freeman, San Francisco, CA, 1979.
2. S. F. Hummel, E. Schonberg, and L. E. Flynn. Factoring, a method for scheduling parallel loops. *Communications of the ACM*, 35(8):90–101, August 1992.
3. C. P. Kruscal and A. Weiss. Allocationg independent subtasks on parallel processors. *IEEE Trans. Software Eng.*, SE-11(10):1001–1016, October 1985.
4. C. Polychronopoulos and D. Kuck. Guided self-scheduling: A practical self-scheduling scheme for parallel supercomputers. *IEEE Trans. Comput.*, C-36(12):1425–1439, December 1987.
5. T. H. Tzen and L. M. Ni. Trapezoid self-scheduling: A practical scheduling scheme for parallel compilers. *IEEE Trans. Parallel and Distributed Systems*, 4(1):87–98, January 1993.

AMEEDA: A General-Purpose Mapping Tool for Parallel Applications on Dedicated Clusters*

X. Yuan[1], C. Roig[2], A. Ripoll[1], M.A. Senar[1], F. Guirado[2], and E. Luque[1]

[1] Universitat Autònoma de Barcelona, Dept. of CS
xiaoyuan@aows10.uab.es, a.ripoll@cc.uab.es,
miquelangel.senar@uab.es, e.luque@cc.uab.es

[2] Universitat de Lleida, Dept. of CS
roig@eup.udl.es, fernando@eup.udl.es

Abstract. The mapping of parallel applications constitutes a difficult problem for which very few practical tools are available. AMEEDA has been developed in order to overcome the lack of a general-purpose mapping tool. The automatic services provided in AMEEDA include instrumentation facilities, parameter extraction modules and mapping strategies. With all these services, and a novel graph formalism called TTIG, users can apply different mapping strategies to the corresponding application through an easy-to-use GUI, and run the application on a PVM cluster using the desired mapping.

1 Introduction

Several applications from scientific computing, e.g. from numerical analysis, image processing and multidisciplinary codes, contain different kinds of potential parallelism: task parallelism and data parallelism [1]. Both data and task parallelism can be expressed using parallel libraries such as PVM and MPI. However, these libraries are not particularly efficient in exploiting the potential parallelism of applications. In both cases, the user is required to choose the number of processors before computation begins, and the processor mapping mechanism is based on very simple heuristics that take decisions independently of the relationship exhibited by tasks. However, smart allocations should take these relationships into account in order to guarantee that good value for the running time is achieved.

In general, static mapping strategies make use of synthetic models to represent the application. Two distinct kinds of graph models have been extensively used in the literature [2]. The first is the TPG (Task Precedence Graph), which models parallel programs as a directed acyclic graph with nodes representing tasks and arcs representing dependencies and communication requirements. The second is the TIG (Task Interaction Graph) model, in which the parallel application is modeled as an undirected graph, where vertices represent the tasks and

* This work was supported by the MCyT under contract 2001-2592 and partially sponsored by the Generalitat de Catalunya (G. de Rec. Consolidat 2001SGR-00218).

B. Monien and R. Feldmann (Eds.): Euro-Par 2002, LNCS 2400, pp. 248–252.

edges denote intertask interactions. Additionally, the authors have proposed a new model, TTIG (Temporal Task Interaction Graph) [3], which represents a parallel application as a directed graph, where nodes are tasks and arcs denote the interactions between tasks. The TTIG arcs include a new parameter, called degree of parallelism, which indicates the maximum ability of concurrency of communicating tasks. This means that the TTIG is a generalized model that includes both the TPG and the TI G.

In this work, we present a new tool called AMEEDA (Automatic Mapping for Efficient Execution of Distributed Applications). AMEEDA is an automatic general-purpose mapping tool that provides a unified environment for the efficient execution of parallel applications on dedicated cluster environments. In contrast to the tools existing in the literature [4] [5], AMEEDA is not tied to a particular synthetical graph model.

2 Overview of AMEEDA

The AMEEDA tool provides a user-friendly environment that performs the automatic mapping of tasks to processors in a PVM platform. First, the user supplies AMEEDA with a C+PVM program whose behavior is synthesized by means of a tracing mechanism. This synthesized behavior is used to derive the task graph model corresponding to the program, which will be used later to automatically allocate tasks to processors, in order to subsequently run the application. Figure 1 shows AMEEDA's overall organization and its main modules, together with the utility services that it is connected with, whose functionalities are described below.

2.1 Program Instrumentation

Starting with a C+PVM application, the source code is instrumented using the TapePVM tool *(ftp://ftp.imag.fr/pub/APACHE/TAPE)*. We have adopted this technique, in which instructions or functions that correspond to instrumentation probes are inserted in users' code before compilation, because of its simplicity. Using a representative data set, the instrumented application is executed in the PVM platform, where a program execution trace is obtained with TapePVM and is recorded onto a trace file.

2.2 Synthesized Behaviour

For each task, the trace file is processed to obtain the computation phases where the task performs sequential computation of sets of instructions, and the communication and synchronization events with their adjacent tasks. This information is captured in a synthetic graph called the Temporal Flow Graph (TFG).

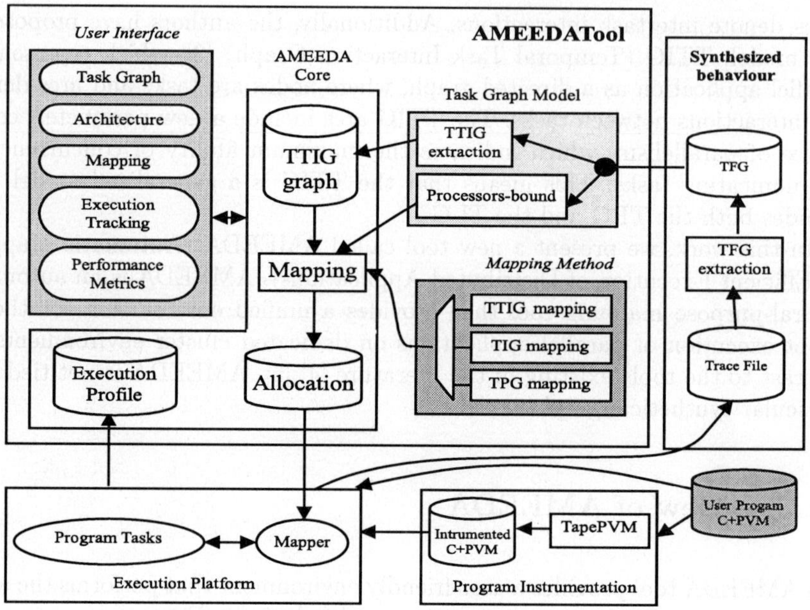

Fig. 1. Block diagram of AMEEDA.

2.3 AMEEDA Tool

With the synthesized behavior captured in the TFG graph, the AMEEDA tool executes the application using a specific task allocation. The necessary steps to physically execute the application tasks using the derived allocation are carried out by the following AMEEDA modules.

1. Task Graph Model
Starting from the TFG graph, the TTIG model corresponding to the application is calculated. Note that, although different traces may be collected if an application is executed with different sets of data, only one TTIG is finally obtained, which captures the application's most representative behavior.

The *Processors-bound* sub-module estimates the minimum number of processors to be used in the execution that allows the potential parallelism of application tasks to be exploited. This is calculated using the methodology proposed in [6] for TPGs, adapted to the temporal information summarized in the TFG graph.

2. Mapping Method
Currently, there are three kinds of mapping policies integrated within AMEEDA that can be applied to the information captured in the TTIG graph of an application.

- *(a) TTIG mapping.* This option contains the MATE (Mapping Algorithm based on Task Dependencies) algorithm, based on the TTIG model [3]. The assignment of tasks to processors is carried out with the main goal of joining the most dependent tasks to the same processor, while the least-dependent tasks are assigned to different processors in order to exploit their ability for concurrency.
- *(b) TIG mapping.* In this case, allocation is carried out through using the CREMA heuristic [7]. This heuristic is based on a two-stage approach that first merges the tasks into as many clusters as number of processors, and then assigns clusters to processors. The merging stage is carried out with the goal of achieving load balancing and minimization of communication cost.
- *(c) TPG mapping.* Allocation is based on the TPG model. In particular, we have integrated the ETF heuristic (Earliest Task First) [8], which assigns tasks to processors with the goal of minimizing the starting time for each task, and has obtained good results at the expense of relatively high computational complexity.

3. User Interface

This module provides several options through a window interface that facilitates the use of the tool. The *Task Graph* sub-module allows the information from the TTIG graph to be visualized. The *Architecture* sub-module shows the current configuration of the PVM virtual machine. The execution of the application, with a specific allocation chosen in the *Mapping* option, can be visualized by using the *Execution tracking* submodule that graphically shows the execution state for the application. The *Mapping* can also be used to plug-in other mapping methods. Finally, the *Performance* option gives the final execution time and speedup of a specific run. It can also show historical data recorded in previous executions in a graphical way, so that performance analysis studies are simplified. Figure 2 corresponds to the AMEEDA window, showing the TTIG graph for a real application in image processing, together with the speedup graphic generated with the Performance sub-module, obtained when this application was executed using the PVM default allocation and the three different mapping strategies under evaluation.

3 Conclusions

We have described the AMEEDA tool, a general-purpose mapping tool that has been implemented with the goal of generating efficient allocations of parallel programs on dedicated clusters. AMEEDA provides a unified environment for computing the mapping of long-running applications with relatively stable computational behavior. The tool is based on a set of automatic services that instrumentalize the application and generate the suitable synthetic information. Subsequently, the application will be executed following the allocation computed by AMEEDA, without any user code re-writing. Its graphical user interface constitutes a flexible environment for analyzing various mapping algorithms and

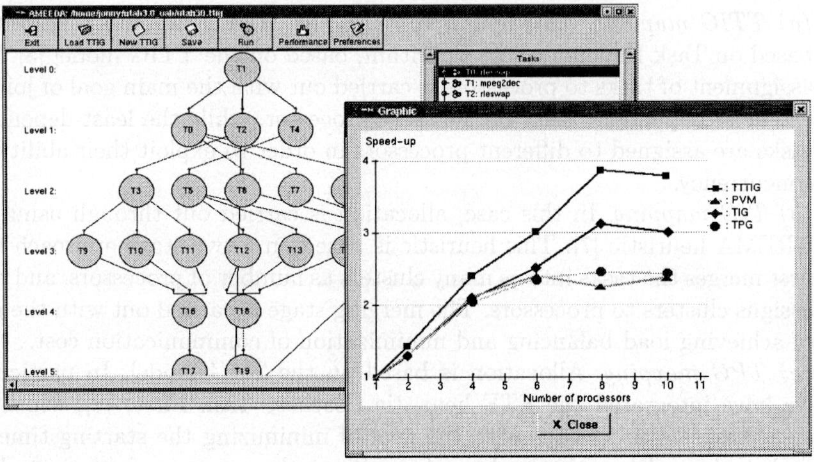

Fig. 2. AMEEDA windows showing the TTIG graph and the speedup for a real application.

performance parameters. In its current state of implementation, the graphical tool includes a small set of representative mapping policies. Further strategies are easy to include, which is also a highly desirable characteristic in its use as a teaching and learning aid for understanding mapping algorithms. As future work, AMEEDA will be enhanced in such a way that the most convenient mapping strategy is automatically chosen, according to the characteristics of the application graph, without user intervention.

References

1. Subhlok J. and Vongran G.: Optimal Use of Mixed Task and Data Parallelism for Pipelined Computations. J. Par. Distr. Computing. vol. 60. pp 297-319. 2000.
2. Norman M.G. and Thanisch P.: Models of Machines and Computation for Mapping in Multicomputers. ACM Computing Surveys, 25(3). pp 263-302. 1993.
3. Roig C., Ripoll A., Senar M.A., Guirado F. and Luque E.: A New Model for Static Mapping of Parallel Applications with Task and Data Parallelism. IEEE Proc. of IPDPS-2002 Conf. ISBN: 0-7695-1573-8. Apr. 2002.
4. Ahmad I. and Kwok Y-K.: CASCH: A Tool for Computer-Aided Scheduling. IEEE Concurrency. pp 21-33. oct-dec. 2000.
5. Decker T. and Diekmann R.: Mapping of Coarse-Grained Applications onto Workstation Clusters. IEEE Proc. of PDP'97. pp 5-12. 1997.
6. Fernandez E.B. and Bussel B.: Bounds on the Number of Processors and Time for Multiprocessor Optimal Schedule. IEEE Tr. on Computers. pp 299-305. Aug. 1973.
7. Senar M. A., Ripoll A., Cortés A. and Luque E.: Clustering and Reassignment-base Mapping Strategy for Message-Passing Architectures. Int. Par. Proc Symp&Sym. On Par. Dist. Proc. (IPPS/SPDP 98) 415-421. IEEE CS Press USA, 1998.
8. Hwang J-J., Chow Y-C., Anger F. and Lee C-Y.: Scheduling Precedence Graphs in Systems with Interprocessor Communication Times. SIAM J. Comput. pp: 244-257, 1989.

Topic 4
Compilers for High Performance (Compilation and Parallelization Techniques)

Martin Griebl

Topic chairperson

Presentation

This topic deals with all issues concerning the automatic parallelization and the compilation of programs for high-performance systems, from general-purpose platforms to specific hardware accelerators. This includes language aspects, program analysis, program transformation and optimization concerning the use of diverse resources (processors, functional units, memory requirements, power consumption, code size, etc.).

Of the 15 submissions, 5 were accepted as regular papers and 3 as research notes.

Organization

The topic is divided into two sessions. The papers in the first session focus on locality.

- "Tiling and memory reuse for sequences of nested loops" by Youcef Bouchebaba and Fabien Coelho combines fusion, tiling, and the use of circular buffers into one transform, in order to improve data locality for regular loop programs.
- "Reuse Distance-Based Cache Hint Selection" by Kristof Beyls and Erik H. D'Hollander exploits the full cache control of the EPIC (IA-64) processor architecture, and shows how this allows to specify the cache level at which the data is likely to be found.
- "Improving Locality in the Parallelization of Doacross Loops" by María J. Martín, David E. Singh, Juan Touriño, and Francisco F. Rivera is an inspector/executor run time approach to improve locality of doacross loops with indirect array accesses on CC-NUMA shared memory computers; the basic concept is to partition a graph of memory accesses.
- "Is Morton array layout competitive for large two-dimensional arrays?" by Jeyarajan Thiyagalingam and Paul Kelly focuses on a specific array layout. It demonstrates experimentally that this layout is a good all-round option when program access structure cannot be guaranteed to follow data structure.

The second session is mainly dedicated to loop parallelization.

B. Monien and R. Feldmann (Eds.): Euro-Par 2002, LNCS 2400, pp. 253–254.

- "Towards Detection of Coarse-Grain Loop-Level Parallelism in Irregular Computations" by Manuel Arenaz, Juan Tourino, and Ramon Doallo presents an enhanced compile-time method for the detection of coarse-grain loop-level parallelism in loop programs with irregular computations.
- "On the Optimality of Feautrier's Scheduling Algorithm" by Frédéric Vivien is a kind of meta paper: it shows that the well known greedy strategy of Feautrier's scheduling algorithm for loop programs is indeed an optimal solution.
- "On the Equivalence of Two Systems of Affine Recurrences Equations" by Denis Barthou, Paul Feautrier, and Xavier Redon goes beyond parallelization of a given program; it presents first results on algorithm recognition for programs that are expressed as systems of affine recurrence equations.
- "Towards High-Level Specification, Synthesis, and Virtualization of Programmable Logic Designs" by Thien Diep, Oliver Diessel, Usama Malik, and Keith So completes the wide range of the topic at the hardware end. It tries to bridge the gap between high-level behavioral specification (using the Circal process algebra) and its implementation in an FPGA.

Comments

In Euro-Par 2002, Topic 04 has a clear focus: five of the eight accepted papers deal with locality improvement or target coarse granularity. These subjects – even if not new – seem to become increasingly important, judging by their growing ratio in the topic over recent years.

Except for one paper, all contributions treat very traditional topics of compilers for high performance systems. This is a bit surprising since the topic call explicitly mentions other optimization goals. It seems that there is enough work left in the central area of high-performance compilation.

Furthermore, it is interesting to see that none of the proposed compilation techniques is specific to some programming language, e.g., Java, HPF, or OpenMP.

Acknowledgements

The local topic chair would like to thank the other three PC members, Alain Darte, Jeanne Ferrante, and Eduard Ayguade for a very harmonious collaboration.

Also, we are very grateful for the excellent work of our referees: every submission (except for two, which have identically been submitted elsewhere, and were directly rejected) received four reviews, and many of the reviewers gave very detailed comments.

Last but not least, we also thank the organization team of Euro-Par 2002 for their immediate, competent, and friendly help on all problems that arose.

Tiling and Memory Reuse for Sequences of Nested Loops

Youcef Bouchebaba and Fabien Coelho

CRI, ENSMP, 35, rue Saint Honoré, 77305 Fontainebleau, France
{boucheba, coelho}@cri.ensmp.fr

Abstract. Our aim is to minimize the electrical energy used during the execution of signal processing applications that are a sequence of loop nests. This energy is mostly used to transfer data among various levels of memory hierarchy. To minimize these transfers, we transform these programs by using simultaneously loop permutation, tiling, loop fusion with shifting and memory reuse. Each input nest uses a stencil of data produced in the previous nest and the references to the same array are equal, up to a shift. All transformations described in this paper have been implemented in *pips*, our optimizing compiler and cache misses reductions have been measured.

1 Introduction

In this paper we are interested in the application of fusion with tiling to a sequence of loop nests and in memory reuse in the merged and tiled nest. Our transformations aim at improving data locality so as to replace costly transfers from main memory to cheaper cache or register memory accesses. Many authors have worked in tiling [9, 16, 14], fusion [5, 4, 11, 17], loop shifting [5, 4] and memory reuse [6, 13, 8]. Here, we combine these techniques to apply them to sequence of loop nests.

We assume that input programs are sequences of loop nests. Each of these nests uses a stencil of data produced in the previous nest and the references to the same array are equal, up to a shift. Consequently, the dependences are uniform. We limit our method to this class of code (chains of jobs), because the problem of loop fusion with shifting in general (graphs of jobs) is *NP_hard* [4].

Our tiling is used as a loop transformation [16] and is represented by two matrices: (1) a matrix $\mathcal{A}$ of hierarchical tiling that gives the various tile coefficients and (2) a permutation matrix $\mathcal{P}$ that allows to exchange several loops and so to specify the organization of tiles and to consider all possible schedules.

After application of fusion with tiling, we have to guarantee that all necessary data for the computation of a given iteration has already been computed by the previous iterations. For this purpose, we shift the computation of each nest by a delay $\boldsymbol{h_k}$. Contrary to the other works, it is always possible to apply our fusion with tiling. To avoid loading several times the same data, we use the notion of live data introduced initially by Gannon et al [8] and applied by Einsenbeis et al [6], to fusion with tiling. Our method replaces the array associated to each nest by a set of buffers that will contain the live data of the corresponding array.

B. Monien and R. Feldmann (Eds.): Euro-Par 2002, LNCS 2400, pp. 255–264.

2 Input Code

The input codes are signal processing applications [10], that are sequences of loop nests of equal but arbitrary depth (see *Figure 1 (a)*). Each of these nests uses a stencil of data produced in the previous nest and represented by a set $V^k = \{\boldsymbol{v}_1^k, \boldsymbol{v}_2^k, \cdots, \boldsymbol{v}_{m_k}^k\}$. The references to the same array are equal, up to a shift. The bounds of these various nests are numerical constants and the various arrays have the same dimension.

```
do i_1 ∈ D_1
 A_1(i_1) = A_0(i_1 + v_1^1) ⊗ .... ⊗ A_0(i_1 + v_{m_1}^1)
enddo
.
do i_k ∈ D_k
 A_k(i_k) = A_{k-1}(i_k + v_1^k) ⊗ .... ⊗ A_{k-1}(i_k + v_{m_k}^k)
enddo
.
do i_n ∈ D_n
 A_n(i_n) = A_{n-1}(i_n + v_1^n) ⊗ .... ⊗ A_{n-1}(i_n + v_{m_n}^n)
enddo
(a) Input code in general form
```

```
do (i = 4, N - 5)
 do (j = 4, N - 5)
  A_1(i, j) = A_0(i - 4, j) + A_0(i, j - 4)
   +A_0(i, j) + A_0(i, j + 4) + A_0(i + 4, j)
 enddo
enddo
do (i = 8, N - 9)
 do (j = 8, N - 9)
  A_2(i, j) = A_1(i - 4, j) + A_1(i, j - 4)
   +A_1(i, j) + A_1(i, j + 4) + A_1(i + 4, j)
 enddo
enddo
(b) Specific example
```

Fig. 1. General input code and a specific example. Where $\otimes$ represents any operation.

Domain D_0 associated with the array A_0 is defined by the user. To avoid illegal accesses to the various arrays, the domains D_k $(1 \leq k \leq n)$ are derived in the following way: $D_k = \{\boldsymbol{i} \mid \forall \boldsymbol{v} \in V^k : \boldsymbol{i} + \boldsymbol{v} \in D_{k-1}\}$. We suppose that vectors of the various stencils are lexicographically ordered: $\forall k : \boldsymbol{v}_1^k \preceq \boldsymbol{v}_2^k \preceq \ldots\ldots \preceq \boldsymbol{v}_{m_k}^k$. In this paper, we limited our study to codes given in *Figure 1 (a)*, $A_k(\boldsymbol{i})$ is computed using elements of array A_{k-1}. Our method is generalizable easily to a code, such as the computation of the element $A_k(\boldsymbol{i})$ in the nest k, it will be according to the arrays $A_0, \cdots A_{k-1}$.

3 Loop Fusion

To merge all the nests into one, we should make sure that all the elements of array A_{k-1} that are necessary for the computation of an element $A_k(\boldsymbol{i}_k)$ at iteration $\boldsymbol{i}_k$ in the merged nest have already been computed by previous iterations. To satisfy this condition, we shift the iteration domain of every nest by a delay $\boldsymbol{h}_k$. Let Time_k be the shifting function associated to nest k and defined in the following way: $\text{Time}_k : D_k \rightarrow Z^n$ so that $\boldsymbol{i}_k \longmapsto \boldsymbol{i} = \boldsymbol{i}_k + \boldsymbol{h}_k$.

The fusion of all nests is legal if and only if each shifting function Time_k meets the following condition:

$$\forall \boldsymbol{i}_k,\ \forall \boldsymbol{i}_{k+1}\ , \forall \boldsymbol{v} \in V^{k+1}\ : \boldsymbol{i}_k = \boldsymbol{i}_{k+1} + \boldsymbol{v} \Rightarrow \text{Time}_k(\boldsymbol{i}_k) \preceq \text{Time}_{k+1}(\boldsymbol{i}_{k+1}) \quad (1)$$

The condition (1) means that if an iteration $\boldsymbol{i}_k$ produces an element that will be consumed by iteration $\boldsymbol{i}_{k+1}$, then the shift of the iteration $\boldsymbol{i}_k$ by Time_k should be lexicographically lower than the shift of the iteration $\boldsymbol{i}_{k+1}$ by Time_{k+1}.

The merged code after shifting of the various iteration domains is given in *Figure 2*. S_k is the instruction label and $D_{iter} = \cup_{k=1}^{n}(D'_k)$, with $D'_k = \{\boldsymbol{i} = \boldsymbol{i}_k + \boldsymbol{h}_k \mid \boldsymbol{i}_k \in D_k\}$ the shift of domain D_k by vector $\boldsymbol{h}_k$. This domain is not necessarily convex. If not, we use its convex hull to generate the code. As instruction S_k might not be executed at each iteration of domain D_{iter}, we guard it by condition $C_k(\boldsymbol{i})$ = if $(\boldsymbol{i} \in D'_k)$, which can be later eliminated [12].

```
do i ∈ D_iter
  S_1 : C_1(i) A_1(i − h_1) = A_0(i − h_1 + v_1^1) ⊗ .... ⊗ A_0(i − h_1 + v_{m_1}^1)
  .
  S_k : C_k(i) A_k(i − h_k) = A_{k−1}(i − h_k + v_1^k) ⊗ .... ⊗ A_{k−1}(i − h_k + v_{m_k}^k)
  .
  S_n : C_n(i) A_n(i − h_n) = A_{n−1}(i − h_n + v_1^n) ⊗ ··· ⊗ A_{n−1}(i − h_n + v_{m_n}^n)
enddo
```

Fig. 2. Merged nest.

As $\boldsymbol{v}_1^k \preceq \boldsymbol{v}_2^k \preceq \ldots\ldots \preceq \boldsymbol{v}_{m_k}^k$, the validity condition of fusion given in (1) will be equivalent to $-\boldsymbol{h}_k \succeq -\boldsymbol{h}_{k+1} + \boldsymbol{v}_{m_{k+1}}^{k+1}$ $(1 \le k \le n-1)$.

3.1 Fusion with Buffer Allocation

To save memory space and to avoid loading several times the same element, we replace the arrays A_1, A_2,..and A_{n-1} by circular buffers B_1, B_2,..and B_{n-1}. Buffer B_i is a one-dimensional array that will contain the live data of array A_i. Each of these buffers will be managed in a circular way and an access function will be associated with it to load and store its elements.

Live data. Let $\boldsymbol{O}_k$ and $\boldsymbol{N}_k + \boldsymbol{O}_k - \mathbf{1}$ respectively the lower and upper bound of domain D'_k : $D'_k = \{\boldsymbol{i} \mid \boldsymbol{O}_k \le \boldsymbol{i} \le \boldsymbol{N}_k + \boldsymbol{O}_k - \mathbf{1}\}$. The memory volume $M_k(\boldsymbol{i})$ corresponding to an iteration $\boldsymbol{i} \in D'_k$ $(2 \le k \le n)$ is the number of elements of the array A_{k-1} that were defined before $\boldsymbol{i}$ and that are not yet fully used: $M_k(\boldsymbol{i}) = |E_k(\boldsymbol{i})|$ with $E_k(\boldsymbol{i}) = \{\boldsymbol{i}_1 \in D'_{k-1} \mid \exists \boldsymbol{v} \in V^k, \exists\, \boldsymbol{i}_2 \in D'_k : \boldsymbol{i}_1 - \boldsymbol{h}_{k-1} = \boldsymbol{i}_2 - \boldsymbol{h}_k + \boldsymbol{v}$ and $\boldsymbol{i}_1 \preceq \boldsymbol{i} \preceq \boldsymbol{i}_2\}$.

At iteration $\boldsymbol{i}$, to compute $A_k(\boldsymbol{i} - \boldsymbol{h}_k)$, we use m_k elements of array A_{k-1} produced respectively by $\boldsymbol{i}_1, \cdots, \boldsymbol{i}_{m_k}$ such that $\boldsymbol{i}_q = \boldsymbol{i} - (\boldsymbol{h}_k - \boldsymbol{h}_{k-1} - \boldsymbol{v}_q^k)$.

The oldest of these productions is $\boldsymbol{i}_1$. Consequently the volume $M_k(\boldsymbol{i})$ is bounded by the number of iterations in D'_{k-1} between $\boldsymbol{i}_1$ and $\boldsymbol{i}$. This upper boundary is given by $\mathrm{Sup}_k = \boldsymbol{C}_k.\ (\boldsymbol{h}_k - \boldsymbol{h}_{k-1} - \boldsymbol{v}_1^k) + 1$ with

$\boldsymbol{C}_k = (\prod_{i=2}^{n} N_{k-1,i}, \prod_{i=3}^{n} N_{k-1,i}, \cdots, \prod_{i=n-1}^{n} N_{k-1,i},\ N_{k-1,n},\ 1)^t$ and $N_{k,i}$ is the ith component of $\boldsymbol{N}_k$.

Code generation. Let B_k $(1 \le k \le n-1)$ be the buffers associated with arrays A_k $(1 \le k \le n-1)$ and *succ*$(\boldsymbol{i})$ the successor of $\boldsymbol{i}$ in the domain D'_k. Sup_{k+1}, given previously, represents an upper bound for the number of live data of array A_k. Consequently the size of buffer B_k can safely be set to Sup_{k+1} and we associate with it the access function $F_k : D'_k \rightarrow N$ such that:

1. $F_k(\boldsymbol{O_k}) = 0$
2. $F_k(succ(\boldsymbol{i})) = \begin{cases} F_k(\boldsymbol{i}) + 1 \text{ if } (F_k(\boldsymbol{i}) \neq \text{Sup}_{k+1} - 1) \\ 0 \qquad\qquad \text{otherwise} \end{cases}$

To satisfy these two conditions, it is sufficient to choose $F_k(\boldsymbol{i}) = (\boldsymbol{C_k} \,.\, (\boldsymbol{i} - \boldsymbol{O_k})) \mod \text{Sup}_{k+1}$.

Let's consider statement S_k of the merged code in *Figure* 2. At iteration $\boldsymbol{i}$, we compute the element $A_k(\boldsymbol{i} - \boldsymbol{h_k})$ as a function of the m_k elements of array A_{k-1} produced respectively by $\boldsymbol{i_1}$, $\boldsymbol{i_2}$,$\cdots$and $\boldsymbol{i_{m_k}}$. The element $A_k(\boldsymbol{i} - \boldsymbol{h_k})$ is stored in the buffer B_k at position $F_k(\boldsymbol{i})$. The elements of array A_{k-1} are already stored in the buffer B_{k-1} at positions $F_{k-1}(\boldsymbol{i_1})$, $F_{k-1}(\boldsymbol{i_2})$,..,$F_{k-1}(\boldsymbol{i_{m_k}})$ $(\boldsymbol{i_q} = \boldsymbol{i} - (\boldsymbol{h_k} - \boldsymbol{h_{k-1}} - \boldsymbol{v_q^k}))$. Thus the statement S_k will be replaced by

$C_k(\boldsymbol{i})\ B_k(F_k(\boldsymbol{i})) = B_{k-1}(F_{k-1}(\boldsymbol{i_1})) \otimes \cdots \otimes B_{k-1}(F_{k-1}(\boldsymbol{i_{m_k}}))$.

4 Tiling with Fusion

A lot of work on tiling has been done but most of it is only dedicated to a single loop nest. In this paper, we present a simple and effective method that simultaneously applies tiling with fusion to a sequence of loop nests. Our tiling is used as a loop transformation [16] and is represented by two matrices: (1) a matrix $\mathcal{A}$ of hierarchical tiling that gives the various coefficients of tiles and (2) a permutation matrix $\mathcal{P}$ that allows to exchange several loops and so to specify the organization of tiles and to consider all possible tilings. As for fusion, the first step before applying tiling with fusion to a code similar to the one in *Figure 1 (a)* is to shift the iteration domain of every nest by a delay $\boldsymbol{h_k}$. We note by $D'_k = \{\boldsymbol{i} = \boldsymbol{i_k} + \boldsymbol{h_k} | \boldsymbol{i_k} \in D_k\}$ the shift of domain D_k by vector $\boldsymbol{h_k}$.

4.1 One-Level Tiling

In this case, we are interested only in data that lives in the cache memory. Thus our tiling is at one level.

Matrix $\mathcal{A}$. Matrix $\mathcal{A}(n,\ 2n)$ defines the various coefficients of tiles and allows us to transform every point $\boldsymbol{i} = (i_1, \cdots, i_n)^t \in \cup_{i=1}^{n}(D'_k)$ into a point $\boldsymbol{i'} = (i'_1, \cdots, i'_{2n})^t \in Z^{2n}$ (*figure 3*). This matrix has the following shape:

$$\mathcal{A} = \begin{pmatrix} a_{1,1} & 1\,0 \cdots 0 & & 0\,0 \cdots 0 & 0 \\ \vdots & \vdots\ \vdots & \vdots & \vdots\ \vdots & \vdots & \vdots \\ 0 & 0\,0 \cdots & a_{i,2i-1} & 1\,0 \cdots 0 & & 0 \\ \vdots & \vdots\ \vdots & \vdots & \vdots\ \vdots & \vdots & \vdots \\ 0 & 0\,0 \cdots 0 & & 0\,0 \cdots & a_{n,2n-1} & 1 \end{pmatrix}$$

All the elements of the ith line of this matrix are equal to zero except: 1) $a_{i,2i-1}$, which represents the size of tiles on the ith axis and 2) $a_{i,2i}$, which is equal to 1.

```
do (i' = ...)
S_1 : C_1(i') A_1(Ai' - h_1) = A_0(Ai' - h_1 + v_1^1) ⊗ .... ⊗ A_0(Ai' - h_1 + v_{m_1}^1)
.
S_k : C_k(i') A_k(Ai' - h_k) = A_{k-1}(Ai' - h_k + v_1^k) ⊗ .... ⊗ A_{k-1}(Ai' - h_k + v_{m_k}^k)
.
S_n : C_n(i') A_n(Ai' - h_n) = A_{n-1}(Ai' - h_n + v_1^n) ⊗ .... ⊗ A_{n-1}(Ai' - h_n + v_{m_n}^n)
enddo
```

Fig. 3. Code after application of $\mathcal{A}$.

The relationship between $\boldsymbol{i}$ and $\boldsymbol{i}'$ is given by:

1. $\boldsymbol{i} = \mathcal{A}\boldsymbol{i}'$
2. $\boldsymbol{i}' = (\lfloor \frac{i_1}{a_{1,1}} \rfloor,\ i_1 \bmod a_{1,1}, \cdots,\ \lfloor \frac{i_m}{a_{m,2m-1}} \rfloor,\ i_m \bmod a_{m,2m-1}, \cdots,$
$\lfloor \frac{i_n}{a_{n,2n-1}} \rfloor,\ i_n \bmod a_{n,2n-1})^t.$

Matrix $\mathcal{P}$. The matrix $\mathcal{A}$ has no impact on the execution order of the initial code. Permutation matrix $\mathcal{P}(2n,\ 2n)$ allows to exchange several loops of code in *Figure 3* and is used to specify the order in which the iterations are executed. This matrix transforms every point $\boldsymbol{i}' = (i'_1,\ i'_2, \cdots,\ i'_{2n})^t \in Z^{2n}$ (*Figure 3*) into a point $\boldsymbol{l} = (l_1,\ l_2, \cdots,\ l_{2n})^t \in Z^{2n}$ such as $\boldsymbol{l} = \mathcal{P}\ \boldsymbol{i}'$. Every line and column of this matrix has one and only one element that is equal to 1.

Tiling modeling. Our tiling is represented by a transformation ω_1:
$\omega_1 : Z^n \rightarrow Z^{2n}$

$$\boldsymbol{i} \longmapsto \boldsymbol{l} = \mathcal{P} \,.\, (\lfloor \frac{i_1}{a_{1,1}} \rfloor,\ i_1 \bmod a_{1,1}, \cdots,\ \lfloor \frac{i_m}{a_{m,2m-1}} \rfloor,\ i_m \bmod a_{m,2m-1}, \cdots,$$
$$\lfloor \frac{i_n}{a_{n,2n-1}} \rfloor,\ i_n \bmod a_{n,2n-1})^t.$$

As mentioned in our previous work [1, 2], the simultaneous application of tiling with fusion to the code in *Figure 1(a)* is valid if and only if:

$$\forall\ k, \forall \boldsymbol{i} \in D'_k, \forall q : \omega_1(\boldsymbol{i} + \boldsymbol{v}_q^{k+1} - \boldsymbol{h}_{k+1} + \boldsymbol{h}_k) \preceq \omega_1(\boldsymbol{i}) \qquad (2)$$

One legal delay of *formula (2)*, is $-\boldsymbol{h}_k = -\boldsymbol{h}_{k+1} + (\max_l\ v_{l,1}^{k+1}, \cdots,\ \max_l\ v_{l,n}^{k+1})^t$ and $\boldsymbol{h}_n = \boldsymbol{0}$. Where $v_{l,i}^{k+1}$ is the ith component of vector $\boldsymbol{v}_l^{k+1}$. The choice of this delay makes the merged nest fully permutable. We know that if a loop nest is fully permutable, we can apply to it any tiling parallel to its axis [15].

Buffer allocation. To maintain in memory the live data and to avoid loading several times the same data, we suggested in our previous work [1, 2] to replace arrays A_1, A_2,...and A_{n-1} by circular buffers B_1, B_2,...and B_{n-1}. A buffer B_i is a one-dimensional array that contains the live data of array A_i. This technique is effective for the fusion without tiling. On the other hand, in the case of fusion with tiling, this technique has two drawbacks: 1) dead data are stored in these buffers to simplify access functions and 2) the size of these buffers increases when the tile size becomes large. For the purpose of eliminating these two problems, we replace every array A_k by $n+1$ buffers.

a) Buffers associated with external loops: One-level tiling allows to transform a nest of depth n into another nest of depth $2n$. The n external loops iterate over tiles, while the n internal loops iterate over iterations inside these tiles. For every external loop m, we associate a buffer $B_{k,m}$ (k corresponds to array A_k) that will contain the live data of array A_k produced in tiles such that $(l_m = b)$ and used in the next tiles such that $(l'_m = b+1)$. To specify the size of these buffers, we use the following notations:

- $E(n,\ n)$, the permutation matrix of external loops: $E_{i,j} = \begin{cases} 1 \text{ if } \mathcal{P}_{i,2j-1} = 1 \\ 0 \quad \text{otherwise} \end{cases}$
- $I(n,\ n)$, the permutation matrix of internal loops: $I_{i,j} = \begin{cases} 1 \text{ if } \mathcal{P}_{i+n,2j} = 1 \\ 0 \quad \text{otherwise} \end{cases}$
- $\boldsymbol{T} = (T_1, \cdots, T_n)^t$, the tile size vector: $T_i = a_{i,2i-1}$;
- $\boldsymbol{N}_k = (N_{k,1}, \cdots, N_{k,n})^t$ where $N_{k,m}$ is the number of iterations of loop i_m in *nest k* of code in *Figure 1(a)*.
- $\boldsymbol{d}_k = (d_{k,1}, \cdots, d_{k,n})^t$, where $d_{k,m}$ is the maximum of the projections of all dependences on the mth axis (dependences connected to array A_k);
- $\boldsymbol{T'} = E\ \boldsymbol{T}$, $\boldsymbol{N}'_k = E\ \boldsymbol{N}_k$ and $\boldsymbol{d}'_k = E\ \boldsymbol{d}_k$.

The memory volume required for buffer $B_{k,m}$ associated with array A_k and the mth external loop is less than $V_{k,m} = \prod_{i=1}^{m-1} T'_i * d'_{k,m} * \prod_{i=m+1}^{n} N'_{k,i}$. Every coefficient in this formula corresponds to a dimension in the buffer $B_{k,m}$. There are $n!$ ways to organize the dimensions of this buffer. In this paper, we will consider the following organization:$[T'_1, .., T'_{m-1},\ d'_{k,m}, N'_{k,m+1}, .., N'_{k,n}]$.

To locate the elements of array A_k in the various buffers associated with it, we define for every buffer $B_{k,m}$ an access function $F_{k,m}$:

$$F_{k,m}(\boldsymbol{i}') = (E_1 \boldsymbol{i}'_{in}, \cdots,\ E_{m-1}\boldsymbol{i}'_{in},\ E_m(\boldsymbol{i}'_{in} - (\boldsymbol{T} - \boldsymbol{d}_k)),\ (E_{m+1}\boldsymbol{T})\ (E_{m+1}\boldsymbol{i}'_E) \\ + E_{m+1}\boldsymbol{i}'_{in}, \cdots,\ (E_n\ \boldsymbol{T})(E_n \boldsymbol{i}'_E) + E_n \boldsymbol{i}'_{in}), where:$$

- E_m represents the mth line of matrix E;
- $\boldsymbol{i}'_E$ is sub vector of $\boldsymbol{i}'$ which iterate over tiles;
- $\boldsymbol{i}'_{in}$ is sub vector of $\boldsymbol{i}'$ which iterate over iterations inside tiles.

b) Buffers associated with internal loops: For all the internal loops, we define a single buffer $B_{k,n+1}$ which contains the live data inside the same tile. The memory volume of this buffer is bounded by $V_{k,n+1} = (I_1\ \boldsymbol{d}_k + 1) * \prod_{k=2}^{n}(I_k\ \boldsymbol{T})$.

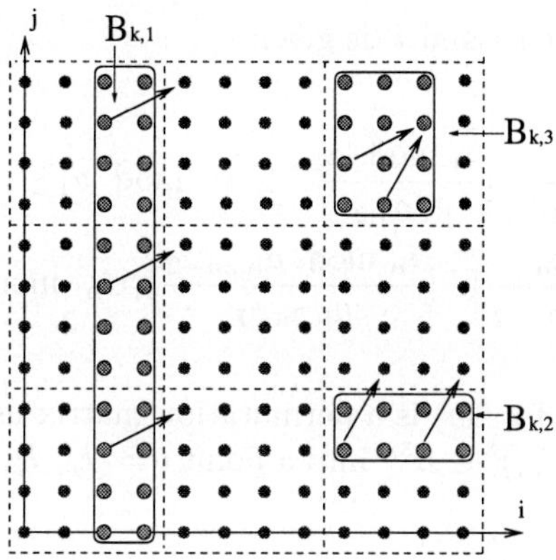

Fig. 4. Example with allocation of three buffers.

As in the previous case, every coefficient in this formula corresponds to a dimension in buffer $B_{k,n+1}$. There are $n!$ ways to organize these dimensions. To obtain the best locality in that case, we choose the following organization: $[I_1\ \boldsymbol{d}_k + 1,\ I_2\ \boldsymbol{T}, \cdots,\ I_n\ \boldsymbol{T}]$. The access function associated with buffer $B_{k,n+1}$ is defined by: $F_{k,n+1}(\boldsymbol{i}') = ((I_1\ \boldsymbol{i}'_{in}) \bmod\ (I_1\ \boldsymbol{d}_k + 1),\ I_2\ \boldsymbol{i}'_{in}, \cdots,\ I_n\ \boldsymbol{i}'_{in})$.

As shown in *figure 4*, if the nest depth is 2 ($n = 2$), every array A_k will be replaced by three buffers : $B_{k,1}$, $B_{k,2}$ and $B_{k,3}$.

4.2 Two-Level Tiling

In this case we are interested in data that lives in the cache and registers. Thus our tiling is at two levels.

Matrix $\mathcal{A}$. Matrix $\mathcal{A}(n,\ 3n)$ allows to transform every point $\boldsymbol{i} = (i_1, \cdots,\ i_n)^t \in Z^n$ into a point $\boldsymbol{i}' = (i'_1, \cdots,\ i'_{3n})^t \in Z^{3n}$ and has the following shape:

$$\mathcal{A} = \begin{pmatrix} a_{1,1} & a_{1,2} & 1\,0 \cdots 0 & 0 & 0\,0 \cdots 0 & 0 & 0 \\ \vdots & \vdots & \vdots\ \vdots\ \ \vdots & \vdots & \vdots\ \vdots\ \ \vdots & \vdots & \vdots \\ 0 & 0 & 0\,0 \cdots a_{i,3i-2} & a_{i,3i-1} & 1\,0 \cdots 0 & 0 & 0 \\ \vdots & \vdots & \vdots\ \vdots\ \ \vdots & \vdots & \vdots\ \vdots\ \ \vdots & \vdots & \vdots \\ 0 & 0 & 0\,0 \cdots 0 & 0 & 0\,0 \cdots a_{n,3n-2} & a_{n,3n-1} & 1 \end{pmatrix}$$

All elements of the ith line of this matrix are equal to zero except:

- $a_{i,3i-2}$, which represents the external tile size on the ith axis.
- $a_{i,3i-1}$, which represents the internal tile size on the ith axis.
- $a_{i,3i}$, which is equal at 1.

The relationship between $\boldsymbol{i}$ and $\boldsymbol{i}'$ is given by:

1. $\boldsymbol{i} = \mathcal{A}\boldsymbol{i}'$
2. $\boldsymbol{i}' = (\lfloor \frac{i_1}{a_{1,1}} \rfloor,\ \lfloor \frac{i_1 \bmod a_{1,1}}{a_{1,2}} \rfloor,\ i_1 \bmod a_{1,2}, \cdots,$
$\lfloor \frac{i_n}{a_{n,3n-2}} \rfloor,\ \lfloor \frac{i_n \bmod a_{n,3n-2}}{a_{n,3n-1}} \rfloor,\ i_n \bmod a_{n,3n-2})^t.$

Matrix $\mathcal{P}$. Matrix $\mathcal{P}(3n,\ 3n)$ is a permutation matrix used to transform every point $\boldsymbol{i}' = (i'_1,\ i'_2, \cdots,\ i'_{3n})^t \in Z^{3n}$ into a point $\boldsymbol{l} = (l_1,\ l_2, \cdots,\ l_{3n})^t \in Z^{3n}$, with $\boldsymbol{l} = \mathcal{P}.\boldsymbol{i}'$.

Tiling modeling. Our tiling is represented by a transformation ω_2 : $\omega_2 : Z^n \rightarrow Z^{3n}$

$$\boldsymbol{i} \longmapsto \boldsymbol{l} = \mathcal{P} \cdot (\lfloor \frac{i_1}{a_{1,1}} \rfloor,\ \lfloor \frac{i_1 \bmod a_{1,1}}{a_{1,2}} \rfloor,\ i_1 \bmod a_{1,2}, \cdots,$$
$$\lfloor \frac{i_n}{a_{n,3n-2}} \rfloor,\ \lfloor \frac{i_n \bmod a_{n,3n-2}}{a_{n,3n-1}} \rfloor,\ i_n \bmod a_{n,3n-2})^t$$

As for one-level tiling, to apply tiling at two levels with fusion to the code in *figure 1(a)*, we have to shift every domain D_k by a delay $\boldsymbol{h}_k$ and these various delays should satisfy the following condition:

$$\forall\ k, \forall \boldsymbol{i} \in D'_k, \forall q : \omega_2(\boldsymbol{i} + \boldsymbol{v}_q^{k+1} - \boldsymbol{h}_{k+1} + \boldsymbol{h}_k) \preceq \omega_2(\boldsymbol{i}) \qquad (3)$$

As before, one possible solution is $-\boldsymbol{h}_k = -\boldsymbol{h}_{k+1} + (\max_l\ v_{l,1}^{k+1}, \cdots,\ \max_l\ v_{l,n}^{k+1})^t$.

5 Implementation and Tests

All transformations described in this paper have been implemented in *Pips* [7]. To measure the external cache misses caused by the various transformations of the example in *Figure 1 (b)*, we used an *UltraSparc10* machine with 512 MB main memory, 2 MB external cache ($L2$) and 16 KB internal cache ($L1$). *Figure 5* gives the experimental results for the external cache misses caused by these various transformations. As one can see from this figure, all the transformations considerably decrease the number of external cache misses when compared to the initial code. Our new method of buffer allocation for tiling with fusion gives the best result and reduces the cache misses by almost a factor of 2 when compared to the initial code.

As often with cache we obtained a few points incompatible with the average behavior. We haven't explained them yet but they have not occurred with the tiled versions. The line of rate $16/L$ (L is size of external cache line) represents the theoretical values for cache misses of the initial code. We do not give the execution times, because we are interested in the energy consumption, which is strongly dependent on cache misses [3].

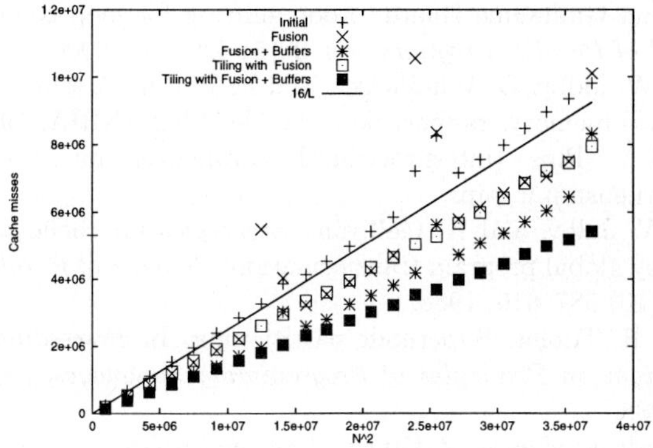

Fig. 5. External cache misses caused by transformations of code in *Figure 1(b)*.

6 Conclusion

There is a lot of work on the application of tiling[9, 16, 14], fusion [5, 4, 11, 17], loop shifting [5, 4] and memory allocations[6, 13, 8]. To our knowledge, the simultaneous application of all these transformations has not been treated. In this paper, we combined all these transformations to apply them to a sequence of loop nests.

We gave a system of inequalities that takes into account the relationships between the added delays, the various stencils, and the two matrices $\mathcal{A}$ and $\mathcal{P}$ defining the tiling. For this system of inequalities, we give a solution for a class of tiling. We have proposed a new method to increase data locality that replaces the array associated with each nest by a set of buffers that contain the live data of the corresponding array. Our tests show that the replacement of the various arrays by buffers considerably decreases the number of external cache misses.

All the transformations described in this paper have been implemented in *pips* [7]. In our future work, we shall study the generalization of our method of buffer allocations in tiling at two levels and we shall look at the issues introduced by combining for buffer and register allocations.

References

1. Youcef Bouchebaba and Fabien Coelho. Buffered tiling for sequences of loops nests. In *Compilers and Operating Systems for Low Power 2001*.
2. Youcef Bouchebaba and Fabien Coelho. Pavage pour une séquence de nids de boucles. *To appear in Technique et science informatiques*, 2000.
3. F. Cathoor and al. *Custom memory management methodology-Exploration of memory organisation for embedded multimedia system design.* Kluwer Academic Publishers, 1998.
4. Alain Darte. On the complexity of loop fusion. *Parallel Computing*, 26(9):1175–1193, 2000.

5. Alain Darte and Guillaume Huard. Loop shifting for loop compaction. *International Journal of Parallel Programming*, 28(5):499–534, 2000.
6. C. Eisenbeis, W. Jalby, D. Windheiser, and F. Bodin. A strategy for array management in local memory. rapport de recherche 1262, INRIA, 1990.
7. Equipe PIPS. Pips (interprocedural parallelizer for scientific programs) http://www.cri.ensmp.fr/pips.
8. D. Gannon, W. Jalby, and K. Gallivan. Strategies for cache and local memory management by global program transformation. *Journal of Parallel and Distibuted Computing*, 5(10):587–616, 1988.
9. F. Irigoin and R. Triolet. Supernode partitioning. In *Proceedings of 15th Annual ACM Symposium on Principles of Programming Languages*, pages 319–329, San Diego, CA, 1988.
10. N. Museux. *Aide au placement d'applications de traitement du signal sur machines parallèles multi-spmd.* Phd thesis, École Nationale Supérieure des Mines de Paris, 2001.
11. W. Pugh and E. Rosser. Iteration space slicing for locality. In *LCPC99*, pages 165–184, San Diego, CA, 1999.
12. F. Quilleré, S. Rajopadhye, and D. Wild. Generation of efficient nested loops from polyhedra. *International journal of parallel programming*, 28(5):496–498, 2000.
13. Fabien Quilleré and Sanjay Rajopadhye. Optimizing memory usage in the polyhedral model. *Transactions on Programming Languages and Systems*, 22(5):773–815, 2000.
14. M. Wolf, D. Maydan, and Ding-Kai-Chen. Combining loop transformations considering caches and scheduling. *International Journal of Parallel Programming*, 26(4):479–503, 1998.
15. M. E. Wolf. *Improving locality and parallelism in nested loops.* Phd thesis, University of stanford, 1992.
16. J. Xue. On tiling as a loop transformation. *Parallel Processing Letters*, 7(4):409–424, 1997.
17. H. P. Zima and B. M. Chapman. *Supercompilers for parallel and vector computers*, volume 1. Addison-Wesley, 1990.

Reuse Distance-Based Cache Hint Selection

Kristof Beyls and Erik H. D'Hollander

Department of Electronics and Information Systems
Ghent University
Sint-Pietersnieuwstraat 41
9000 Ghent, Belgium
{kristof.beyls,erik.dhollander}@elis.rug.ac.be

Abstract. Modern instruction sets extend their load/store-instructions with cache hints, as an additional means to bridge the processor-memory speed gap. Cache hints are used to specify the cache level at which the data is likely to be found, as well as the cache level where the data is stored after accessing it. In order to improve a program's cache behavior, the cache hint is selected based on the data locality of the instruction. We represent the data locality of an instruction by its reuse distance distribution. The reuse distance is the amount of data addressed between two accesses to the same memory location. The distribution allows to efficiently estimate the cache level where the data will be found, and to determine the level where the data should be stored to improve the hit rate. The Open64 EPIC-compiler was extended with cache hint selection and resulted in speedups of up to 36% in numerical and 23% in non-numerical programs on an Itanium multiprocessor.

1 Introduction

The growing speed gap between the memory and the processor push computer architects, compiler writers and algorithm designers to conceive ever more powerful data locality optimizations. However, many programs still stall more than half of their execution time, waiting for data to arrive from a slower level in the memory hierarchy. Therefore, the efforts of reducing memory stall time should be combined on the three different program levels: hardware, compiler and algorithm. In this paper, a combined approach at the compiler and hardware level is described.

Cache hints are emerging in new instruction set architectures. Typically they are specified as attachments to regular memory instructions, and occur in two kinds: source and target hints. The first kind, the *source cache specifier*, indicates at which cache level the accessed data is likely to be found. The second kind, the *target cache specifier*, indicates at which cache level the data is kept after the instruction is executed. An example is given in fig. 1, where the effect of the load instruction LD_C2_C3 is shown. The source cache specifier C2 suggests that at the start of the instruction, the data is expected in the L2 cache. The target cache specifier C3 causes the data to be kept in the L3 cache, instead of keeping

B. Monien and R. Feldmann (Eds.): Euro-Par 2002, LNCS 2400, pp. 265–274.

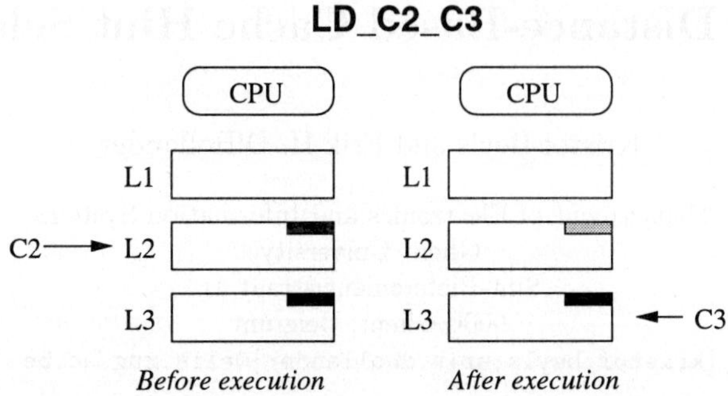

Fig. 1. Example of the effect of the cache hints in the load instruction LD_C2_C3. The source cache specifier C2 in the instruction suggests that the data resides in the L2-cache. The target cache specifier C3 indicates that the data should be stored no closer than the L3-cache. As a consequence, the data is the first candidate for replacement in the L2-cache.

it also in the L1 and L2 caches. After the execution, the data becomes the next candidate for replacement in the L2 cache.

In an Explicitly Parallel Instruction Computing architecture (EPIC), the source and destination cache specifiers are used in different ways.

The *source cache specifiers* are used by the compiler to know the estimated data access latency. Without these specifiers, the compiler assumes that all memory instructions hit in the L1 cache. Using the source cache specifier, the compiler is able determine the true memory latency of instructions. It uses this information to schedule the instructions explicitly in parallel.

The *target cache specifiers* are used by the processor, where they indicate the highest cache level at which the data should be kept. A carefully selected target specifier will maintain the data at a fast cache level, while minimizing the probability that it is replaced by intermediate accesses.

Small and fast caches are efficient when there is a high data locality, while for larger and slower caches lower data locality suffices. To determine the data locality, the reuse distance is measured and used as a discriminating function to determine the most appropriate cache level and associated cache hints.

The reuse distance-based cache hint selection was implemented in an EPIC-compiler and tested on an Itanium multiprocessor. On a benchmark of general purpose and numerical programs, up to 36% speedup is measured, with an average speedup of 7%.

The emerging cache hints in EPIC instruction sets are discussed in sect. 2. The definition of the reuse distance, and some interesting lemmas are stated in sect. 3. The accurate selection of cache hints in an optimizing compiler is discussed in sect. 4. The experiments and results can be found in sect. 5. The related work is discussed in sect. 6. In sect. 7, the conclusion follows.

2 Software Cache Control in EPIC

Cache hints and cache control instructions are emerging in both EPIC[4, 7] and superscalar[6, 10] instruction sets. The most expressive and orthogonal cache hints can be found in the HPL-PD architecture[7]. Therefore, we use them in this work. The HPL-PD architecture defines 2 kinds of cache hints: *source cache specifiers* and *target cache specifiers*. An example of a load instruction can be found in fig. 1.

source cache specifier The source cache specifier indicates the highest cache level where the data is assumed to be found,

target cache specifier The target cache specifier indicates the highest cache level where the data should be stored. If the data is already present at higher cache levels, it becomes the primary candidate for replacement at those levels.

In an EPIC-architecture, the compiler is responsible for instruction scheduling. Therefore, the source cache specifier is used inside the compiler to obtain good estimates of the memory access latencies. Traditional compilers assume L1 cache hit latency for all load instructions. The source cache specifier allows the scheduler to have a better view on the latency of memory instructions. In this way, the scheduler can bridge the cache miss latency with parallel instructions. After scheduling, the source cache specifier is not needed anymore.

The target cache specifier is communicated to the processor, so that it can influence the replacement policy of the cache hierarchy. Since the source cache specifier is not used by the processor, only the target cache specifier needs to be encoded in the instruction. As such, the IA-64 instruction set only defines target cache specifiers. Our experiments are executed on an IA-64 Itanium-processor, since it is the only processor available with this rich set of target cache hints. E.g., in the IA-64 instruction set, the target cache hints `_C1`, `_C2`, `_C3`, `_C4` are indicated by the suffixes `.t1`, `.nt1`, `.nt2`, `.nta`[4]. Further details about the implementation of those cache hints in the Itanium processor can be found in [12].

In order to select the most appropriate cache hints, the locality of references to the same data is measured by the reuse distance.

3 Reuse Distance

The reuse distance is defined within the framework of the following definitions. When data is moved between different levels of the cache hierarchy, a complete cache line is moved. To take this effect into account when measuring the reuse distance, a memory line is considered as the basic unit of data.

Definition 1. *A* memory line[2] *is an aligned cache-line-sized block in the memory. When data is loaded from the memory, a complete memory line is brought into the cache.*

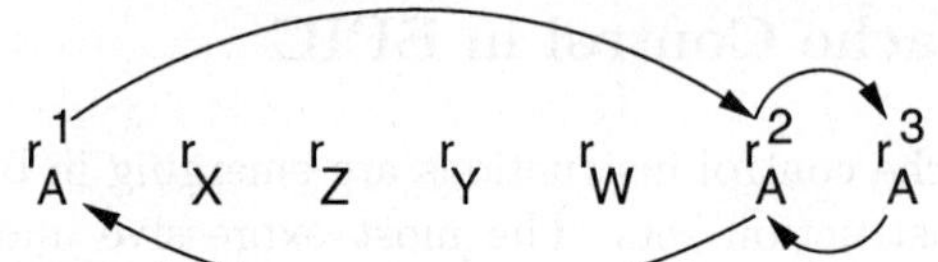

Fig. 2. A short reference stream with indication of the reuses. The subscript of the references indicates which memory line the reference accesses. The references r_X, r_Z, r_Y and r_W are not part of a reuse pair, since memory lines W, X, Y and Z are accessed only once in the stream. Reuse pair $\langle r_A^1, r_A^2 \rangle$ has reuse distance 4, while the reuse pair $\langle r_A^2, r_A^3 \rangle$ has reuse distance 0. The forward reuse distance of r_A^1 is 4, its backward reuse distance is ∞. The forward reuse distance of r_A^2 is 0, its backward reuse distance is 4.

Definition 2. *A* reuse pair $\langle r_1, r_2 \rangle$ *is a pair of references in the memory reference stream, accessing the same memory line, without intermediate references to the same memory line. The set of reuse pairs of a reference stream s is denoted by* $\mathcal{R}_s$*. The* reuse distance *of a reuse pair* $\langle r_1, r_2 \rangle$ *is the number of unique memory lines accessed between references* r_1 *and* r_2*.*

Corollary 1. *Every reference in a reference stream s occurs at most 2 times in* $\mathcal{R}_s$*: once as the first element of a reuse pair, once as the second element of a reuse pair.*

Definition 3. *The* forward reuse distance *of a memory access* x *is the reuse distance of the pair* $\langle x, y \rangle$*. If there is no reuse pair where* x *is the first element, its forward reuse distance is* ∞*. The* backward reuse distance *of* x *is the reuse distance of* $\langle w, x \rangle$*. If there is no such pair, the backward reuse distance is* ∞*.*

Example 1. Figure 2 shows two reuse pairs in a short reference stream.

Lemma 1. *In a fully associative LRU cache with* n *lines, a reference with backward reuse distance* $d < n$ *will hit. A reference with backward reuse distance* $d \geq n$ *will miss.*

Proof. In a fully-associative LRU cache with n cache lines, the n most recently referenced memory lines are retained. When a reference has a backward reuse distance d, exactly d different memory lines were referenced previously. If $d \geq n$, the referenced memory line is not one of the n most recently referenced lines, and consequently will not be found in the cache. □

Lemma 2. *In a fully associative LRU cache with* n *lines, the memory line accessed by a reference with forward reuse distance* $d < n$ *will stay in the cache until the next access of that memory line. A reference with forward reuse distance* $d \geq n$ *will be removed from the cache before the next access.*

Proof. If the forward reuse distance is infinite, the data will not be used in the future, so there is no next access. Consider the forward reuse distance of reference

r_1 and assume that the next access to the data occurs at reference r_2, resulting in a reuse pair $\langle r_1, r_2 \rangle$. By definition, the forward reuse distance d of r_1 equals the backward reuse distance of r_2, i.e. d. Lemma 1 stipulates that the data will be found in the cache at reference r_2, if and only if $d < n$. □

Lemmas 1 and 2 indicate that the reuse distance can be used to precisely indicate the cache behavior of fully-associative caches. However, previous research[1] indicates that also for lower-associative, and even for direct mapped caches, the reuse distance can be used to obtain a good estimation of the cache behavior.

4 Cache Hint Selection

4.1 Reuse Distance-Based Selection

The cache hint selection is based on the forward and backward reuse distances of the accesses. Lemma 1 is used to select the most appropriate *source cache specifier* for a fully associative cache, i.e. the smallest and fastest cache level where data will be found upon reference. This is the smallest cache level with a size larger than the *backward reuse distance.* Similarly, lemma 2 yields the following *target cache specifier* selection: the specifier must indicate the smallest cache where the data will be found upon the next reference, i.e. the cache level with a size larger than the *forward reuse distance.* This mapping from reuse distance to cache hint is graphically shown in fig. 3(a). Notice that a single reuse distance metric allows to handle all the cache levels. Cache hint selection based on a cache hit/miss metric would need a separate cache simulation for all cache levels.

For every memory access, the most appropriate cache hint can be determined. However, a single memory instruction can generate multiple memory accesses during program execution. Those accesses can demand different cache hints. It is not possible to specify different cache hints for them, since the cache hint is specified on the instruction. As a consequence, all accesses originating from the same instruction share the same cache hint. Because of this, it is not possible to assign the most appropriate cache hint to all accesses. In order to select a cache hint which is reasonable for most memory accesses generated by an instruction, we use a threshold value. In our experiments, the cache hint indicates the smallest cache level appropriate for at least 90% of the accesses, as depicted in fig. 3(b).

4.2 Cache Data Dependencies

The source cache specifier makes the compiler aware of the cache behavior. However, adding cache dependencies, in combination with source cache specifiers further refines the compilers view on the latency of memory instructions. Consider fig 4. Two loads access data from the same cache line in a short time period. The first load misses the cache. Since the first load brings the data into the fastest cache level, the second load hits the cache. However, the second load

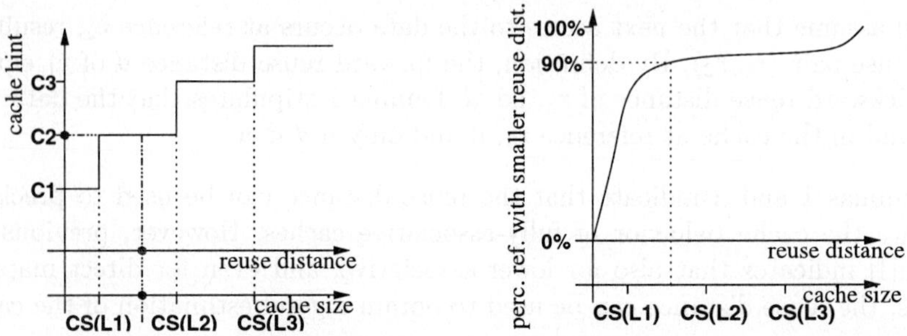

(a) Cache hint in function of the reuse distance of a single access.

(b) Cumulative reuse distance distribution (CDF) of an instruction The 90th percentile determines the cache hint.

Fig. 3. The selection of cache hints, based on the reuse distance. In (a), it is shown how the reuse distance of a single memory access maps to a cache level and an accompanying cache hint. For example, a reuse distance larger than the cache size of L1, but smaller than L2 results in cache hints C2. In (b), a cumulative reuse distance distribution for an instruction is shown and how a threshold value of 90% maps it to cache hint C2.

can only hit the cache if the first load had enough time to bring the data into the cache. Therefore, the second load is cache dependent on the first load. If this dependence is not visible to the scheduler, it could schedule the second load with cache hit latency, before the first load has brought the data into the cache. This can lead to a schedule where the instructions dependent on the second load would be issued before their input data is available, leading to processor stall on an in-order EPIC machine.

One instruction can generate multiple accesses, with the different accesses coming from the same instruction dictating different cache dependencies. A threshold is used to decide if an instruction is cache dependent on another instruction. If a load instruction y accesses a memory line at a certain cache level, and that memory line is brought to that cache level by instruction x in at least 5% of the accesses, a cache dependence from instruction x to instruction y is inserted.

5 Experiments

The Itanium processor, the first implementation of the IA-64 ISA, was chosen to test the cache hint selection scheme described above. The Itanium processor provides cache hints as described in sect. 2.

5.1 Implementation

The above cache hint selection scheme was implemented in the Open64 compiler[8], which is based on SGI's Pro64 compiler. The reuse distance distribution

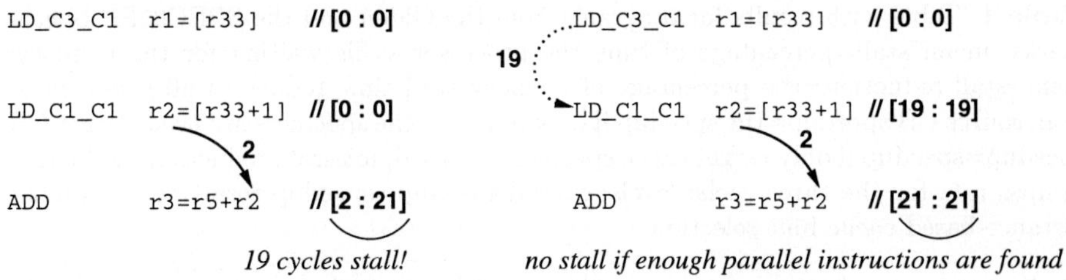

Fig. 4. An example of the effect of cache dependence edges in the instruction scheduler. The two load instructions access the same memory line. The first number between square brackets indicates the schedulers idea of the first cycle in which the instruction can be executed. The second number shows the real cycle in which the instruction can be executed. On the left, there is no cache dependence edge and a stall of up to 19 cycles can occur, while the instruction scheduler is not aware of it. On the right hand, the cache dependence is visible to the compiler, and the scheduler can try to move parallel instruction between the first and the second load instruction to hide the latency.

for the memory instructions, and the necessary information needed to create cache dependencies are obtained by instrumenting and profiling the program. The source and target cache hints are annotated to the memory instruction, based on the profile data. After instruction scheduling, the compiler produces the EPIC assembly code with target cache hints.

All compilations were performed at optimization level -O2, the highest level at which instrumentation and profiling is possible in the Open64 compiler. The existing framework doesn't allow to propagate the feedback information through some optimizations phases at level -O3.

5.2 Measurements

The programs were executed on a HP rx4610 multiprocessor, equipped with 733MHz Itanium processors. The data cache hierarchy consists of a 16KB L1, 96KB L2 and a 2MB L3 cache. The hardware performance counters of the processor were used to obtain detailed micro-architectural information, such as processor stall time because of memory latency and cache miss rates.

The programs were selected from the Olden and the Spec95fp benchmarks. The Olden benchmark contains programs which uses dynamic data structures, such as linked lists, trees and quadtrees. The Spec95fp programs are numerical programs with mostly regular array accesses. For the Spec95fp, the profiling was done using the train input sets, while the speedup measurements were done with the large input sets. For Olden, no separate input sets are available, and the training input was identical to the input for measuring the speedup. The results of the measurements can be found in table 1.

The table shows that the programs run 7% faster on average, with a maximum execution time reduction of 36%. In the worst case, a slight performance

Table 1. Table with results for programs from the Olden and the SPEC95FP benchmarks: mem. stall=percentage of time the processor stalls waiting for the memory; mem. stall reduction=the percentage of memory stall time reduction after optimization; source CH speedup=the speedup if only source cache specifiers are used; target CH speedup=speedup if only target cache specifiers are used; missrate reduction=reduction in miss rate for the three cache levels; overall speedup=speedup resulting from reuse distance-based cache hint selection.

	program	mem. stall	mem. stall reduction	source CH speedup	target CH speedup	missrate reduction L1	L2	L3	overall speedup
Olden	bh	26%	0%	0%	-1%	1%	-20%	-3%	**-1%**
	bisort	32%	0%	0%	0%	0%	6%	-5%	**0%**
	em3d	77%	25%	6%	20%	-28%	-3%	35%	**23%**
	health	80%	19%	2%	16%	0%	-1%	15%	**20%**
	mst	72%	1%	0%	0%	-10%	1%	2%	**1%**
	perimeter	53%	-1%	-1%	-1%	-11%	-56%	-6%	**-2%**
	power	15%	0%	0%	0%	-14%	2%	0%	**0%**
	treeadd	48%	0%	-2%	-1%	-2%	26%	17%	**0%**
	tsp	20%	0%	0%	0%	2%	7%	7%	**0%**
	Olden avg.	47%	5%	0%	4%	-6%	-6%	7%	**5%**
Spec95fp	swim	78%	0%	0%	1%	32%	0%	0%	**0%**
	tomcatv	69%	33%	7%	4%	-11%	-43%	6%	**9%**
	applu	49%	10%	4%	1%	-9%	-1%	-1%	**4%**
	wave5	43%	-9%	4%	15%	-26%	-7%	-5%	**5%**
	mgrid	45%	13%	36%	0%	13%	-24%	25%	**36%**
	Spec95fp avg.	57%	9%	10%	4%	0%	-15%	5%	**10%**
	overall avg.	51%	7%	4%	4%	-5%	-8%	6%	**7%**

degradation of 2% is observed. On average, the Olden benchmarks do not profit from the source cache specifiers. To take advantage of the source cache specifiers, the instruction scheduler must be able to find parallel instructions to fit in between a long latency load and its consuming instructions. In the pointer-based Olden benchmarks, the scheduler finds little parallel instructions, and cannot profit from its better view on the cache behavior. On the other hand, in the floating point programs, on average a 10% speedup is found because of the source cache hints. Here, the loop parallelism allows the compiler to find parallel instructions, mainly because it allows it to software pipeline the loops with long latency loads. In this way, the latency is overlapped with parallel instructions from different loop iterations. Some of the floating point programs didn't speedup a lot when employing source cache specifiers. The scheduler couldn't generate better code since the long latency of the loads demanded too many software pipeline stages to overlap it. Because of the large number of pipeline stages, not enough registers were available to actually create the software pipelining code.

The table also shows that the target cache specifiers improve both kind of programs by the same percentage. This improvement is caused by an average reduction in the L3 cache misses of 6%. The reduction is due to the improved cache replacement decisions made by the hardware, based on the target cache specifiers.

6 Related Work

Much work has been done to eliminate cache misses by loop and data transformations. In our approach, the remaining cache misses after these transformations are further diminished in two orthogonal ways: target cache specifiers and source cache specifiers. In the literature, ideas similar to either the target cache specifier or the source cache specifier are proposed, but not both.

Work strongly related to target cache specifiers is found in [5], [11], [13] and [14]. In [13], it is shown that less than 5% of the load instructions cause over 99% of all cache misses. In order to improve the cache behavior, the authors propose not allocating the data in the cache when the instruction has a low hit ratio. This results in a large decrease of the memory bandwidth requirement, while the hit ratio drops only slightly. In [5], keep and kill instructions are proposed. The keep instruction locks data into the cache, while the kill instruction indicates it as the first candidate to be replaced. Jain et al. also proof under which conditions the keep and kill instructions improve the cache hit rate. In [14], it is proposed to extend each cache line with an EM(Evict Me)-bit. The bit is set by software, based on compiler analysis. If the bit is set, that cache line is the first candidate to be evicted from the cache. In [11], a cache with 3 modules is presented. The modules are optimized respectively for spatial, temporal and spatial-temporal locality. The compiler indicates in which module the data should be cached, based upon compiler analysis or a profiling step. These approaches all suggest interesting modifications to the cache hardware, which allow the compiler to improve the cache replacement policy. However, the proposed modifications are not available in present day architectures. The advantage of our approach is that it uses cache hints available in existing processors. The results show that the presented cache hint selection scheme is able to increase the performance on real hardware.

The source cache specifiers hide the latency of cache misses. Much research has been performed on software prefetching, which also hides cache miss latency. However, prefetching requires extra prefetch instructions to be inserted in the program. In our approach, the latency is hidden without inserting extra instructions. Latency hiding without prefetch instructions is also proposed in [3] and [9]. In [3] the cache behavior of numerical programs is examined using miss traffic analysis. The detected cache miss latencies are hidden by techniques such as loop unrolling and shifting. In comparison, our technique also applies to non-numerical programs and the latencies are compensated by scheduling low level instructions. The same authors also introduce cache dependency, and propose to shift data accesses with cache dependencies to previous iterations. In the present paper, cache dependencies are treated as ordinary data dependencies. In [9], load instructions are classified into normal, list and stride access. List and stride accesses are maximally hidden by the compiler because they cause most cache misses. However the classification of memory accesses in two groups is very coarse. The reuse distance provides a more accurate way to measure the data locality, and as such permits the compiler to generate a more balanced schedule.

Finally, all the approaches mentioned above apply only to a single cache level. In contrast, reuse distance based cache hint selection can easily be applied to multiple cache levels.

7 Conclusion

Cache hints emerge in new processor architectures. This opens the perspective of new optimization schemes aimed at steering the cache behavior from the software level. In order to generate appropriate cache hints, the data locality of the program must be measured. In this paper, the reuse distance is proposed as an effective locality metric. Since it is independent of cache parameters such as cache size or associativity, the reuse distance can be used for optimizations which target multiple cache levels. The properties of this metric allow a straightforward generation of appropriate cache hints. The cache hint selection was implemented in an EPIC compiler for Itanium processors. The automatic selection of source and target cache specifiers resulted in an average speedup of 7% in a number of integer and numerical programs, with a maximum speedup of 36%.

References

1. K. Beyls and E. H. D'Hollander. Reuse distance as a metric for cache behavior. In *Proceedings of PDCS'01*, 2001.
2. S. Ghosh. *Cache Miss Equations: Compiler Analysis Framework for Tuning Memory Behaviour.* PhD thesis, Princeton University, November 1999.
3. P. Grun, N. Dutt, and A. Nicolau. MIST: An algorithm for memory miss traffic management. In *ICCAD*, 2000.
4. *IA-64 Application Developer's Architecture Guide*, May 1999.
5. P. Jain, S. Devadas, D. Engels, and L. Rudolph. Software-assisted replacement mechanisms for embedded systems. In *CCAD'01*, 2001.
6. G. Kane. *PA-RISC 2.0 architecture.* Prentice Hall, 1996.
7. V. Kathail, M. S. Schlansker, and B. R. Rau. HPL_PD architecture specification: Version 1.1. Technical Report HPL-93-80(R.1), Hewlett-Packard, February 2000.
8. Open64 compiler. http://sourceforge.net/projects/open64.
9. T. Ozawa, Y. Kimura, and S. Nishizaki. Cache miss heuristics and preloading techniques for general-purpose programs. In *MICRO'95*.
10. K. R.E. The alpha 21264 microprocessor. *IEEE Micro*, pages 24–36, mar 1999.
11. J. Sanchez and A. Gonzalez. A locality sensitive multi-module cache with explicit management. In *Proceedings of the 1999 Conference on Supercomputing.*
12. H. Sharangpani and K. Arora. Itanium processor microarchitecture. *IEEE Micro*, 20(5):24–43, Sept./Oct. 2000.
13. G. Tyson, M. Farrens, J. Matthews, and A. R. Pleszkun. A modified approach to data cache management. In *MICRO'95.*
14. Z. Wang, K. McKinley, and A. Rosenberg. Improving replacement decisions in set-associative caches. In *Proceedings of MASPLAS'01*, April 2001.

Improving Locality in the Parallelization of Doacross Loops*

María J. Martín[1], David E. Singh[2], Juan Touriño[1], and Francisco F. Rivera[2]

[1] Dep. of Electronics and Systems, University of A Coruña, Spain
{mariam,juan}@udc.es

[2] Dep. of Electronics and Computer Science, University of Santiago, Spain
{david,fran}@dec.usc.es

Abstract. In this work we propose a run-time approach for the efficient parallel execution of doacross loops with indirect array accesses by means of a graph partitioning strategy. Our approach focuses not only on extracting parallelism among iterations of the loop, but also on exploiting data access locality to improve memory hierarchy behavior and thus the overall program speedup. The effectiveness of our algorithm is assessed in an SGI Origin 2000.

1 Introduction

This work addresses the parallelization of doacross loops, that is, loops with loop-carried dependences. These loops can be partially parallelized by inserting synchronization primitives to force the memory access order imposed by these dependences. Unfortunately, it is not always possible to determine the dependences at compile-time as, in many cases, they involve input data that are only known at run-time and/or the access pattern is too complex to be analyzed.

There are in the literature a number of run-time approaches for the parallelization of doacross loops [1–4]. All of them follow an inspector-executor strategy and they differ on the kinds of dependences that are considered and the level of parallelism exploited (iteration-level or operation-level parallelism). A comparison between strategies based on iteration-level and operation-level parallelism is presented in [5]. The work shows experimentally that operation-level methods outperform iteration-level methods.

In this paper we present a new operation-level algorithm based on graph partitioning techniques. Our approach not only maximizes parallelism, but also (and basically) increases data locality to better exploit memory hierarchy in order to improve code performance. The target computer assumed throughout this paper is a CC-NUMA shared memory machine. We intend, on the one hand, to increase cache line reuse in each processor and, on the other hand, to reduce false sharing of cache lines, which is an important factor of performance degradation in CC-NUMA architectures.

* This work has been supported by the Ministry of Science and Technology of Spain and FEDER funds of the European Union (ref. TIC2001-3694-C02)

B. Monien and R. Feldmann (Eds.): Euro-Par 2002, LNCS 2400, pp. 275–279.

2 Run-Time Strategy

Our method follows the inspector-executor strategy. During the inspector stage, memory access and data dependence information is collected. The access information, which determines the iteration partition approach, is stored in a graph structure. Dependence information is stored in a table called *Ticket Table* [1]. So, the inspector phase consists of three parts:

- *Construction of a graph* representing memory accesses. It is a non-directed weighted graph; both nodes and graph edges are weighted. Each node represents m consecutive elements of array A, m being the number of elements of A that fit in a cache line. The weight of each node is the number of iterations that access that node for write. Moreover, a table which contains the indices of those iterations is assigned to each node. The edges join nodes that are accessed in the same iteration. The weight of each edge corresponds to the number of times that the pair of nodes is accessed in an iteration.
- *Graph partitioning.* The graph partitioning will result in a node distribution (and, therefore, an iteration distribution) among processors. Our aim is to partition the graph so that a good node balance is achieved and the number of edges being cut is minimum. Node balance results in load balance and cut minimization involves a decrease in the number of cache invalidations, as well as an increase in cache line reuse. Besides, as each node represents a cache line with consecutive elements of A, false sharing is eliminated. We have used the *pmetis* program [6] from the METIS software package to distribute the nodes among the processors according to the objectives described above.
- *Creation of a Ticket Table* containing data dependence information. The creation of the Ticket Table is independent of the graph construction and partitioning, and thus these stages can be performed in parallel.

The executor phase makes use of the dependence information recorded in the Ticket Table to execute, in each processor, the set of iterations assigned in the inspector stage. An array reference can be performed if and only if the preceding references are finished. All accesses to the target array are performed in parallel except for the dependences specified in the Ticket Table. The iterations with dependences can be partially overlapped because we consider dependences between accesses instead of between iterations. In [7] we propose an inspector that considers an iteration partitioning based on a block-cyclic distribution.

3 Performance Evaluation

In this section, the experimental results obtained for our strategy are evaluated and compared with the classical approach, an algorithm that uses a cyclic distribution of the iterations. The cyclic distribution maximizes load balancing and favors parallelism, without taking into account data access locality.

Although for illustrative purposes a loop with one read and one write per loop iteration will be used as case study, our method is a generic approach that can also be applied to loops with more than one indirect read access per iteration.

3.1 Experimental Conditions

The parallel performance of the irregular doacross loop is mainly characterized by three parameters: loop size, workload cost and memory access pattern. In order to evaluate a set of cases as large as possible, we have used the loop pattern shown in Figure 1, where N represents the problem size, the computational cost of the loop is simulated through the parameter W, and the access pattern is determined by the array $INDEX$ and the size of array A. Examples of this loop pattern can be found in the solution of sparse linear systems (see, for instance, routines *lsol*, *ldsol* and *ldsoll* of the Sparskit library [8]), where the loop size and the access pattern depend on the sparse coefficient matrix. These systems have to be solved in a wide variety of codes, including linear programming applications, process simulation, finite element and finite difference applications, and optimization problems, among others. Therefore, we have used in our experiments as indirection arrays the patterns of sparse matrices from the Harwell-Boeing collection [9] that appear in real codes. The test matrices are characterized in Figure 1, where the size of the indirection array $INDEX$ is $2 \times N$, and M is the size of array A.

```
REAL A(M)
DO i = 1,N
   tmp1 = A(INDEX(i*2-1))
   A(INDEX(i*2)) = tmp2
   DO j = 1,W
     dummy loop simulating useful work
   ENDDO
ENDDO
```

	N	M
gemat1	23684	4929
gemat12	16555	4929
mbeacxc	24960	496
beaflw	26701	507
psmigr_2	270011	3140

Fig. 1. Loop used as experimental workload and benchmark matrices

Our target machine is an SGI Origin 2000 CC-NUMA multiprocessor with R10k at 250 MHz. The R10k utilizes a two level cache hierarchy: L1 instruction and data caches of 32 KB each, and a unified L2 cache of 4MB (cache line size of 128 bytes). All tests were written in Fortran using OpenMP directives. All data structures were cache aligned. In our experiments, the cost per iteration of the outer loop of Figure 1 can be modeled as $T(W)=8.02\times10^{-5}+ 8\times10^{-5}\ W$ ms. The cost per iteration depends on the application. For illustrative purposes, typical values of W range from 5 to 30 using HB matrices for the loop patterns of the aforementioned Sparskit routines that solve sparse linear systems.

3.2 Experimental Results

We have used the R10k event counters to measure L1 and L2 cache misses as well as the number of L2 invalidations. Figure 2 shows the results (normalized

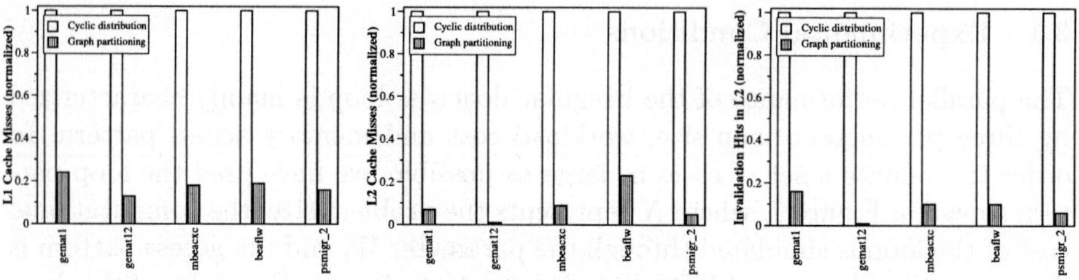

Fig. 2. Cache behavior

with respect to the cyclic distribution) for each test matrix on 8 processors. As can be observed, the reduction in the number of cache misses and invalidations is very significant.

Figure 3 shows the overall speedups (inspector and executor phases) on 8 processors for different workloads. Speedups were calculated with respect to the sequential execution of the code of Figure 1. Our proposal works better for loops with low W because, in this case, memory hierarchy performance has a greater influence on the overall execution time. As W increases, the improvement falls because load balancing and waiting times become critical factors for performance. The increase in the speedups illustrated in Figure 3 is a direct consequence of the improvement in data locality introduced by our approach. The best memory hierarchy optimization achieved by matrix *gemat12* results in the highest increase in speedup.

In many applications, the loop to be parallelized is contained in one or more sequential loops. In this case, if the access pattern to array A does not change across iterations, the inspector can be reused and thus its cost is amortized. An example of such applications are iterative sparse linear system solvers. Figure 4 shows the executor speedups on 8 processors for different workloads. Note that not only speedups increase, but also the improvement with respect to the cyclic iteration distribution strategy.

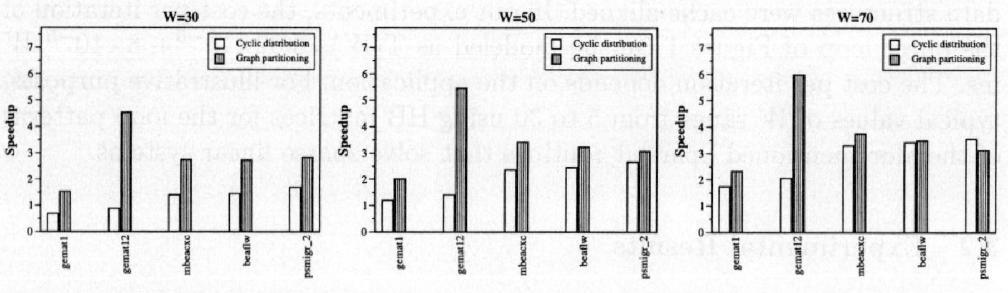

Fig. 3. Overall speedups on 8 processors for different workloads

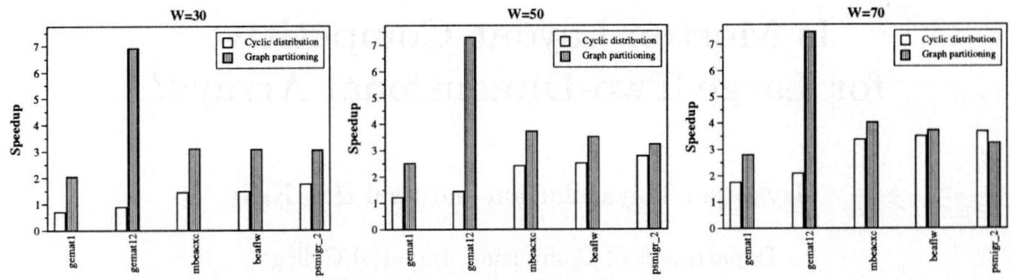

Fig. 4. Executor speedups on 8 processors for different workloads

4 Conclusions

Cache misses are becoming increasingly costly due to the widening gap between processor and memory performance. Therefore, it is a primary goal to increase the performance of each memory hierarchy level. In this work we have presented a proposal to parallelize doacross loops with indirect array accesses using run-time support. It is based on loop restructuring, and achieves important reductions in the number of cache misses and invalidations. It results in a significant increase in the achieved speedups (except for high workloads), and this improvement is even more significant if the inspector can be reused.

References

1. D.-K. Chen, J. Torrellas and P.-C. Yew: An Efficient Algorithm for the Run-Time Parallelization of DOACROSS Loops, Proc. Supercomputing Conf. (1994) 518–527
2. J.H. Saltz, R. Mirchandaney and K. Crowley: Run-Time Parallelization and Scheduling of Loops, IEEE Trans. on Computers **40**(5) (1991) 603–612
3. C.-Z. Xu and V. Chaudhary: Time Stamp Algorithms for Runtime Parallelization of DOACROSS Loops with Dynamic Dependences, IEEE Trans. on Parallel and Distributed Systems **12**(5) (2001) 433–450
4. C.-Q. Zhu and P.-C. Yew: A Scheme to Enforce Data Dependence on Large Multiprocessor Systems, IEEE Trans. on Soft. Eng. **13**(6) (1987) 726–739
5. C. Xu: Effects of Parallelism Degree on Run-Time Parallelization of Loops, Proc. 31st Hawaii Int. Conf. on System Sciences (1998)
6. G. Karypis and V. Kumar: A Fast and High Quality Multilevel Scheme for Partitioning Irregular Graphs, SIAM J. on Scientific Comp. **20**(1) (1999) 359–392
7. M.J. Martín, D.E. Singh, J. Touriño and F.F. Rivera: Exploiting Locality in the Run-time Parallelization of Irregular Loops, Proc. 2002 Int. Conf. on Parallel Processing (2002)
8. Y. Saad: SPARSKIT: a Basic Tool Kit for Sparse Matrix Computations (Version 2), at http://www.cs.umn.edu/Research/darpa/SPARSKIT/sparskit.html (1994)
9. I.S. Duff, R.G. Grimes and J.G.Lewis: User's Guide for the Harwell-Boeing Sparse Matrix Collection, Tech. Report TR-PA-92-96, CERFACS (1992)

Is Morton Layout Competitive for Large Two-Dimensional Arrays?

Jeyarajan Thiyagalingam and Paul H.J. Kelly

Department of Computing, Imperial College
180 Queen's Gate, London SW7 2BZ, U.K.
{jeyan,phjk}@doc.ic.ac.uk

Abstract. Two-dimensional arrays are generally arranged in memory in row-major order or column-major order. Sophisticated programmers, or occasionally sophisticated compilers, match the loop structure to the language's storage layout in order to maximise spatial locality. Unsophisticated programmers do not, and the performance loss is often dramatic — up to a factor of 20. With knowledge of how the array will be used, it is often possible to choose between the two layouts in order to maximise spatial locality. In this paper we study the Morton storage layout, which has substantial spatial locality whether traversed in row-major or column-major order. We present results from a suite of simple application kernels which show that, on the AMD Athlon and Pentium III, for arrays larger than 256×256, Morton array layout, even implemented with a lookup table with no compiler support, is always within 61% of both row-major and column-major — and is sometimes faster.

1 Introduction

Every student learns that multidimensional arrays are stored in "lexicographic" order: row-major (for Pascal etc) or column-major (for Fortran). Modern processors rely heavily on caches and spatial locality, and this works well when the access pattern matches the storage layout. However, accessing a row-major array in column-major order leads to dismal performance (and vice-versa). The Morton layout for arrays (for background and history see [2,7]) offers a compromise, with some spatial locality whether traversed in row-major or column-major order — although in neither case is spatial locality as high as the best case for row-major or column-major. A further disadvantage is the cost of calculating addresses. So, should language implementors consider using Morton layout for all multidimensional arrays? This paper explores this question, and provides some qualified answers.

Perhaps controversially, we confine our attention to "naively" written codes, where a mismatch between access order and layout is reasonably likely. We also assume that the compiler does not help, neither by adjusting storage layout, nor by loop nest restructuring such as loop interchange or tiling. Naturally, we fervently hope that users will be expert and that compilers will successfully

B. Monien and R. Feldmann (Eds.): Euro-Par 2002, LNCS 2400, pp. 280–288.

analyse and optimise the code, but we recognise that very often, neither is the case.

The idea is this: if we know how the array is going to be used, we could choose optimally between the two lexicographic layouts. If we don't know how the array will be used, we can guess. If we guess right, we can expect good performance. If wrong, we may suffer very badly. In this paper, we investigate whether the Morton layout is a suitable compromise for avoiding such worst-case behaviour. We use a small suite of simple application kernels to test this hypothesis and to evaluate the slowdown which occurs when the wrong layout is chosen.

2 Related Work

Compiler techniques. Locality can be enhanced by restructuring loops to traverse the data in an appropriate order [6, 8]. Tiling can suffer disappointing performance due to associativity conflicts, which, in turn, can be avoided by copying the data accessed by the tile into contiguous memory [5]. Copying can be avoided by building the array in this layout. More generally, storage layout can be selected to match execution order [4]. While loop restructuring is limited by what the compiler can infer about the dependence structure of the loops, adjusting the storage layout is always valid. However, each array is generally traversed by more than one loop, which may impose layout constraint conflicts which can be resolved only with foreknowledge of program behaviour.

Blocked and recursively-blocked array layout. Wise *et al.* [7] advocate Morton layout for multidimensional arrays, and present a prototype compiler that implements the dilated arithmetic address calculation scheme which we evaluate in Section 4. They found it hard to overcome the overheads of Morton address calculation, and achieve convincing results only with recursive formulations of the loop nests.

Chatterjee *et al.* [2] study Morton layout and a blocked "4D" layout (explained below). They focus on tiled implementations, for which they find that the 4D layout achieves higher performance than the Morton layout because the address calculation problem is easier, while much or all the spatial locality is still exploited. Their work has similar goals to ours, but all their benchmark applications are tiled (or "shackled") for temporal locality; they show impressive performance, with the further advantage that performance is less sensitive to small changes in tile size and problem size, which can result in cache associativity conflicts with conventional layouts.

In contrast, the goal of our work is to evaluate whether Morton layout can simplify the performance programming model for unsophisticated programmers, without relying on very powerful compiler technology.

3 Background

3.1 Lexicographic Array Storage

For an $M \times N$ two dimensional array A, a mapping $\mathcal{S}(i,j)$ is needed, which gives the memory offset at which array element $A_{i,j}$ will be stored. Conventional solutions are row-major (for e.g. in Pascal) and column-major (as used by Fortran) mappings expressed by

$$\mathcal{S}_{rm}^{(N,M)}(i,j) = N \times i + j \quad \text{and} \quad \mathcal{S}_{cm}^{(N,M)}(i,j) = i + M \times j$$

respectively. We refer to row-major and column-major as lexicographic layouts, i.e. the sort order of the two indices (another term is "canonical"). Historically, array layout has been mandated in the language specification.

3.2 Blocked Array Storage

How can we reduce the number of code variants needed to achieve high performance? An attractive strategy is to choose a storage layout which offers a compromise between row-major and column-major. For example, we could break the $N \times M$ array into small, $P \times Q$ row-major subarrays, arranged as a $N/P \times M/Q$ row-major array. We define the blocked row-major mapping function (this is the 4D layout discussed in [2]) as:

$$\mathcal{S}_{brm}^{(N,M)}(i,j) = (P \times Q) \times \mathcal{S}_{rm}^{(N/P,M/Q)}(i/P, j/P) + \mathcal{S}_{rm}^{(P,Q)}(i\%P, j\%Q)$$

This layout can increase the cache hit rate for larger arrays, since every load of a block will satisfy multiple future requests.

3.3 Bit-Interleaving

Assume for the time being that, for an $N \times M$ array, $N = 2^n$, $M = 2^m$. Write the array indices i and j as

$$\mathcal{B}(i) = i_{n-1}i_{n-2}\cdots i_3 i_2 i_1 i_0 \quad \text{and} \quad \mathcal{B}(j) = j_{n-1}j_{n-2}\cdots j_3 j_2 j_1 j_0$$

respectively. Now the lexicographic mappings can be expressed as bit-concatenation (written "$\|$"):

$$\mathcal{S}_{rm}^{(N,M)}(i,j) = \mathcal{B}(i)\|\mathcal{B}(j) = i_{n-1}i_{n-2}\dots i_3 i_2 i_1 i_0 j_{n-1}j_{n-2}\cdots j_3 j_2 j_1 j_0$$
$$\mathcal{S}_{cm}^{(N,M)}(i,j) = \mathcal{B}(j)\|\mathcal{B}(i) = j_{n-1}j_{n-2}\cdots j_3 j_2 j_1 j_0 i_{n-1}i_{n-2}\dots i_3 i_2 i_1 i_0$$

If $P = 2^p$ and $Q = 2^q$, the blocked row-major mapping is

$$\mathcal{S}_{brm}^{(N,M)}(i,j) = \mathcal{B}(i)_{(n-1)\dots p}\|\mathcal{B}(j)_{(m-1)\dots q}\|\mathcal{B}(i)_{(p-1)\dots 0}\|\mathcal{B}(j)_{(q-1)\dots 0}.$$

Now, with $N = M$ choose $P = Q = 2$, and apply blocking recursively:

$$\mathcal{S}_{mz}(i,j) = i_{n-1}j_{n-1}i_{n-2}j_{n-2}\dots i_3 j_3 i_2 j_2 i_1 j_1 i_0 j_0$$

This mapping is called the Morton Z-order [2], and is illustrated in Fig. 1.

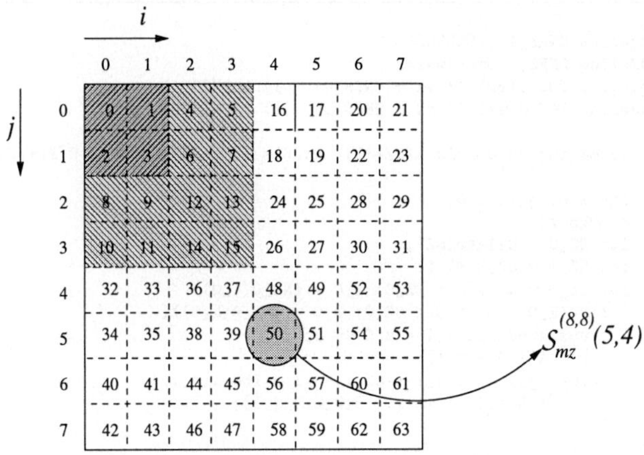

Fig. 1. Morton storage layout for 8×8 array. Location of element $A[4,5]$ is calculated by interleaving "dilated" representations of 4 and 5 bitwise: $\mathcal{D}_0(4) = 010000_2$, $\mathcal{D}_1(5) = 100010_2$. $\mathcal{S}_{mz}(5,4) = \mathcal{D}_0(5) \mid \mathcal{D}_1(4) = 110010_2 = 50_{10}$. A 4-word cache block holds a 2×2 subarray; a 16-word cache block holds a 4×4 subarray. Row-order traversal of the array uses 2 words of each 4-word cache block on each sweep of its inner loop, and 4 words of each 16-word block. Column-order traversal achieves the same hit rate.

3.4 Cache Performance with Morton-Order Layout

Given a cache with any even power-of-two block size, with an array mapped according to the Morton order mapping $\mathcal{S}_{mz}$, the cache hit rate of a row-major traversal is the same as the cache-hit rate of a column-major traversal. In fact, this applies given any cache hierarchy with even power-of-two block size at each level. This is illustrated in Fig. 1. The problem of calculating the actual cache performance with Morton layout is somewhat involved; an interesting analysis for matrix multiply is presented in [3].

4 Morton-Order Address Calculation

4.1 Dilated Arithmetic

Bit-interleaving is too complex to execute at every loop iteration. Wise *et al.* [7] explore an intriguing alternative: represent each loop control variable i as a "dilated" integer, where the i's bits are interleaved with zeroes. Define $\mathcal{D}_0$ and $\mathcal{D}_1$ such that

$$\mathcal{B}(\mathcal{D}_0(i)) = 0i_{n-1}0i_{n-2}0\ldots0i_20i_10i_0 \quad \text{and} \quad \mathcal{B}(\mathcal{D}_1(i)) = i_{n-1}0i_{n-2}0\ldots i_20i_10i_00$$

Now we can express the Morton address mapping as $\mathcal{S}_{mz}(i,j) = \mathcal{D}_0(i) \mid \mathcal{D}_1(j)$, where "|" denotes bitwise-or. At each loop iteration we increment the loop control variable; this is fairly straightforward:

$$\mathcal{D}_0(i+1) = ((\mathcal{D}_0(i) \mid \text{Ones}_0) + 1) \ \& \ \text{Ones}_1$$
$$\mathcal{D}_1(i+1) = ((\mathcal{D}_1(i) \mid \text{Ones}_1) + 1) \ \& \ \text{Ones}_0$$

```
#define ONES_1 0x55555555
#define ONES_0 0xaaaaaaaa
#define INC_1(vx) (((vx + ONES_0) + 1) & ONES_1)
#define INC_0(vx) (((vx + ONES_1) + 1) & ONES_0)

void mm_ikj_da(double A[SZ*SZ], double B[SZ*SZ], double C[SZ*SZ])
{
  int i_0, j_1, k_0;
  double r;
  int SZ_0 = Dilate(SZ);
  int SZ_1 = SZ_0 << 1;
  for (i_0 = 0; i_0 < SZ_0; i_0 = INC_0(i_0))
    for (k_0 = 0; k_0 < SZ_0; k_0 = INC_0(k_0)){
      unsigned int k_1 = k_0 << 1;
      r = A[i_0 + k_1];
      for (j_1 = 0; j_1 < SZ_1; j_1 = INC_1(j_1))
        C[i_0 + j_1] += r * B[k_0 + j_1];
    }
}
```

Fig. 2. Morton-order matrix-multiply implementation using dilated arithmetic for the address calculation. Variables `i_0` and `k_0` are dilated representations of the loop control counter $\mathcal{D}_0$(`i`) and $\mathcal{D}_0$(`k`). Counter `j` is represented by `j_1`= $\mathcal{D}_1$(`j`). The function `Dilate` converts a normal integer into a dilated integer.

where "&" denotes bitwise-and, and

$$\mathcal{B}(\text{Ones}_0) = 01010\ldots10101 \qquad \text{and} \qquad \mathcal{B}(\text{Ones}_1) = 10101\ldots01010$$

This is illustrated in Fig. 2, which shows the ikj variant of matrix multiply.

The dilated arithmetic approach works when the array is accessed using an induction variable which can be incremented using dilated addition. We found that a much simpler scheme often works nearly as well: we simply pre-compute a table for the two mappings $\mathcal{D}_0(i)$ and $\mathcal{D}_1(i)$. We illustrate this for the ikj matrix multiply variant in Fig. 3. Note that the table accesses are very likely cache hits, as their range is small and they have unit stride. One small but important detail: we use addition instead of logical "or". This may improve instruction selection. It also allows the same loop to work on lexicographic layout using suitable tables. If the array is non-square, $2^n \times 2^m$, $n < m$, we construct the table so that the j index is dilated only up to bit n.

Fig. 4 shows the performance of these two variants on a variety of computer systems. In the remainder of the paper, we use the table lookup scheme exclusively. With compiler support, many applications could benefit from the dilated arithmetic approach, leading in many cases to more positive conclusions.

5 Experimental Results

We have argued that Morton layout is a good compromise between row-major and column-major. To test this experimentally, we have collected a suite of simple implementations of standard numerical kernels operating on two-dimensional arrays:

```
void mm_ikj_tb(double A[SZ*SZ], double B[SZ*SZ], double C[SZ*SZ],
               unsigned int MortonTabEven[],
               unsigned int MortonTabOdd[])
{
  int i, j, k;
  double r;
  for (i = 0; i < SZ; i++)
    for (k = 0; k < SZ; k++){
      r = A[MortonTabEven[i] + MortonTabOdd[k]];
      for (j = 0; j < SZ; j++)
        C[MortonTabEven[i] + MortonTabOdd[j]]
          += r * B[MortonTabEven[k] + MortonTabOdd[j]];
    }
}
```

Fig. 3. Morton-order matrix-multiply implementation using table lookup for the address calculation. The compiler detects that `MortonTabEven[i]` and `MortonTabEven[k]` are loop invariant, leaving just one table lookup in the inner loop.

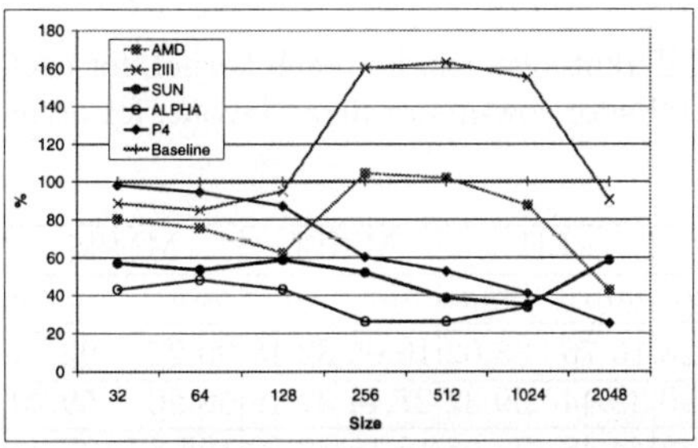

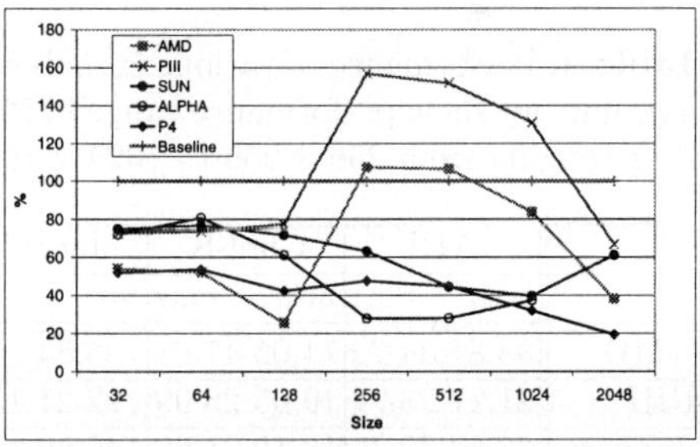

Fig. 4. Matrix multiply (ikj) performance (in MFLOPs) of *(left)* dilated arithmetic Morton address calculation (see Fig. 2) versus *(right)* table-based Morton address calculation (see Fig. 3). The graphs show MFLOPs normalised to the performance achieved by the standard row-major ikj implementation at each problem size on each system. Details of the systems are given in Table 1. The worst slowdown of the table lookup scheme over the dilated-arithmetic scheme is observed on the P4 and is 46%. For problem sizes larger than 256 the worst figure is 24% on PIII. On the SunFire 6800 the lookup table implementation is always faster. In these graphs, larger numbers represent better performance.

MMijk	Matrix multiply, ijk loop nest order (usually poor due to large stride)
MMikj	Matrix multiply, ikj loop nest order (usually best due to unit stride)
LU	LU decomposition with pivoting (based on Numerical Recipes)
Jacobi2D	Two-dimensional four-point stencil smoother
ADI	Alternating-direction implicit kernel, ij,ij order
Cholesky	k variant (usually poor due to large stride)

In each case we run the code on square arrays of various sizes, repeating the calculation if necessary to ensure adequate timing resolution. The system

Table 1. Cache and CPU configurations used in the experiments.

Alpha Compaq AlphaServer ES40	Alpha 21264 (EV6) 500MHz, L1 D-cache: 2-way, 64KB, 64B cache block L2 cache: direct mapped, 4MB. Compiler: Compaq C V6.1-020 "-fast"
Sun SunFire 6800	UltraSparc III (v9) 750MHz L1 D-cache: 4-way, 64KB, 32B cache block L2 cache: direct-mapped, 8MB. Compiler: Sun Workshop 6 "-xO5" (update 1 C 5.2 Patch 109513-07)
PIII	Intel Pentium III Coppermine, 1GHz L1 D-cache: 4-way, 16KB, 32B cache block L2 cache: 8-way 256KB, sectored 32B cache block 512MB SDRAM. Compiler "gcc-2.95 -O3"
P4	Pentium 4, 1.3 GHz L1 D-cache: 8-way, 8KB, sectored 64B cache block L2 cache: 8-way, 256KB, sectored 64B cache block 256MB RDRAM. Compiler "gcc-2.95 -O3"
AMD	AMD Athlon Thunderbird, 1.4GHZ L1 D-Cache: 2-way, 64KB, 64B cache block L2 cache: 8-way, 256KB, 64B cache block 512MB DDR RAM. Compiler "gcc-2.95 -O3"

Table 2. Performance of various kernels on different systems. For each kernel, for each machine, we show performance range in MFLOPs for row-major array layout, for array sizes ranging from 256 × 256 to 1024 × 1024.

	ADI		Chol-K		Jacobi2D		LU		MMijk		MMikj	
	min	*max*	*min*	*max*	*min*	*max*	*min*	*max*	*min*	*max*	*min*	*max*
AMD	33.81	34.72	11.05	47.61	195.84	199.25	16.76	83.02	10.05	32.18	90.27	92.72
PIII	21.17	23.71	16.05	26.99	122.21	128.90	32.44	69.32	27.44	37.19	58.90	59.20
SunFire	37.64	40.35	16.12	21.62	140.69	411.78	44.48	77.08	16.16	69.90	125.57	137.24
Alpha	49.77	63.47	12.02	41.90	120.23	245.53	30.22	112.28	14.41	95.34	148.78	254.13
P4	65.04	67.56	23.05	43.15	410.16	419.32	41.72	73.98	32.35	34.98	293.51	297.92

configurations are detailed in Table 1. Table 2 shows the baseline performance achieved by each machine using standard row-major layout.

Results using Morton layout are summarised in Fig. 5. We have not used non-square arrays in this paper, but the approach handles them reasonably effectively (see Section 4), at the cost of padding each dimension to the next power of two.

Our results show that Morton layout is not effective for arrays smaller than 256 × 256. We therefore confine our attention to larger problem sizes. On the AMD Athlon and Pentium III, we find that Morton layout is often faster than both row-major and column-major, and is never more than 61% slower. Furthermore, the costs of poor layout choice on these machines are particularly acute - in extreme cases a factor of 20. We have only studied up to 2048 × 2048 (32MB), and further investigation is needed for very large problems. On the other machines, the picture is less clear. Kernels with high spatial locality, such as MMikj and Jacobi2D, run close to the machine's peak performance; so bandwidth to L1 cache for table access is probably a major factor.

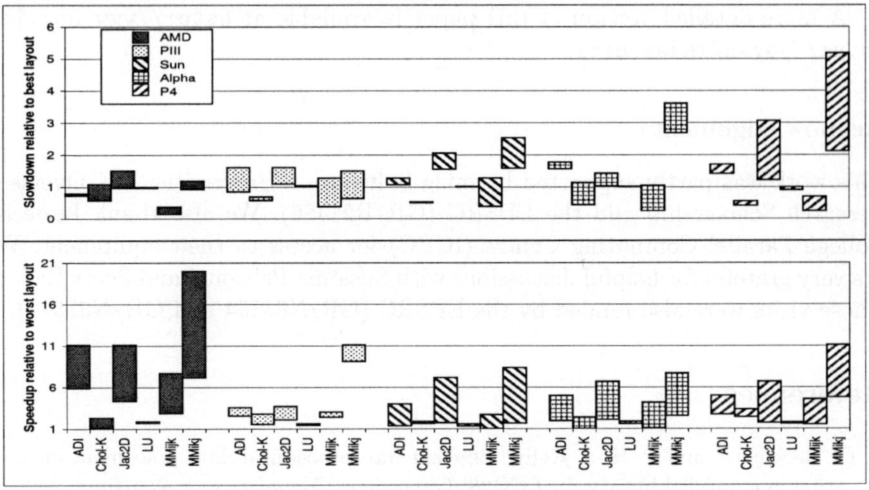

Fig. 5. Performance of table-lookup-based implementation of Morton layout for various common dense kernels. In the upper graph we show how much slower Morton layout can be compared with row-major layout (which for our benchmarks are usually fastest). In each case we show the maximum and minimum slowdown over a range of problem sizes from 256 × 256 to 2048 × 2048. In the lower graph, we show how much faster Morton layout can be compared with column-major layout. In each case we show the maximum and minimum speedup over the same range of problem sizes. A more detailed version of this paper is available at http://www.doc.ic.ac.uk/~jeyan/index.html.

6 Conclusions and Directions for Further Research

Using a small suite of dense kernels working on two-dimensional arrays, we have studied the impact of poor array layout. On some machines, we found that Morton array layout, even implemented with a lookup table with no compiler support, is always within 61% of both row-major and column-major. We also found that using a lookup-table for address calculation allows flexible selection of fine-grain non-linear array layout, while offering attractive performance compared with lexicographic layouts on untiled loops.

The next step is building Morton layout into a compiler, or perhaps a self-optimising BLAS library [1] (which would allow run-time layout selection). It should be possible to achieve better results using competitive redistribution — i.e. instrument memory accesses and copy the array into a more appropriate distribution if indicated.

In our brief analysis of spatial locality using Morton layout (Section 3.4, Fig. 1), we assumed that cache blocks and VM pages are a *square* (even) power of two. This depends on the array's element size, and is often not the case. Then, row-major and column-major traversal of Morton layout lead to differing spatial locality. A more subtle non-linear layout could address this. It seems less likely that Morton layout can offer a competitive compromise for arrays with more than two dimensions.

A more detailed version of this paper is available at `http://www.doc.ic.ac.uk/~jeyan/index.html`.

Acknowledgements

This work was partly supported by mi2g Software, a Universities UK Overseas Research Scholarship and the EPSRC (GR/R21486). We also thank Imperial College Parallel Computing Centre (ICPC) for access to their equipment. We are very grateful for helpful discussions with Susanna Pelagatti and Scott Baden, whose visits were also funded by the EPSRC (GR/N63154 and GR/N35571).

References

1. O. Beckmann and P. H. J. Kelly. Efficient interprocedural data placement optimisation in a parallel library. In *LCR98: Languages, Compilers and Run-time Systems for Scalable Computers*, number 1511 in LNCS, pages 123–138. Springer-Verlag, May 1998.
2. S. Chatterjee, V. V. Jain, A. R. Lebeck, S. Mundhra, and M. Thottethodi. Nonlinear array layouts for hierarchical memory systems. In *Int. Conf. on Supercomputing*, pages 444–453, 1999.
3. P. J. Hanlon, D. Chung, S. Chatterjee, D. Genius, A. R. Lebeck, , and E. Parker. The combinatorics of cache misses during matrix multiplication. *JCSS*, 63, 2001.
4. M. T. Kandemir, A. N. Choudhary, J. Ramanujam, N. Shenoy, and P. Banerjee. Enhancing spatial locality via data layout optimizations. In *Euro-Par*, pages 422–434, 1998.
5. M. S. Lam, E. E. Rothberg, and M. E. Wolf. The cache performance and optimizations of blocked algorithms. *SIGPLAN Notices*, 26(4):63–74, 1991.
6. K. S. McKinley, S. Carr, and C.-W. Tseng. Improving data locality with loop transformations. *ACM Transactions on Programming Languages and Systems*, 18(4):424–453, July 1996.
7. D. S. Wise, J. D. Frens, Y. Gu, , and G. A. Alexander. Language support for Morton-order matrices. In *Proc. 2001 ACM Symp. on Principles and Practice of Parallel Programming, SIGPLAN Not. 36, 7*, 2001.
8. M. E. Wolf and M. S. Lam. A data locality optimizing algorithm. In *Proceedings of ACM SIGPLAN '91 Conference on Programming Language Design and Implementation*, 1991.

Towards Detection of Coarse-Grain Loop-Level Parallelism in Irregular Computations

Manuel Arenaz, Juan Touriño, and Ramón Doallo

Computer Architecture Group
Department of Electronics and Systems
University of A Coruña, Spain
{arenaz,juan,doallo}@udc.es

Abstract. This paper presents a new algorithm for the detection of coarse-grain parallelism in loops with complex irregular computations. Loops are represented as directed acyclic graphs whose nodes are the strongly connected components (SCC) that appear in the Gated Single Assignment (GSA) program form, and whose edges are the use-def chains between pairs of SCCs. Loops that can be executed in parallel using run-time support are recognized at compile-time by performing a demand-driven analysis of the corresponding SCC graphs. A prototype was implemented using the infrastructure provided by the Polaris parallelizing compiler. Encouraging experimental results for a suite of real irregular programs are shown.

1 Introduction

The automatic parallelization of irregular codes is still a challenge for current parallelizing compilers, which are faced with the analysis of complex computations that involve subscripted subscripts. Several approaches to detect loop-level parallelism in irregular codes can be found in the literature. Keßler [6] describes a pattern-driven automatic parallelization technique based on the recognition of syntactical variations of sparse matrix computation kernels, which limits its scope of application. Pottenger and Eigenmann [9] describe a pattern-matching-based method to recognize induction variables and reduction operations. Techniques that recognize patterns directly from the source code have two major drawbacks: dependence on the code quality and difficulty in analyzing complex control constructs. Furthermore, [9] only addresses the detection of recurrence forms whose recurrence variable is not tested in any if–endif construct within the loop body. We call these programming constructs *structural recurrence forms.* Suganuma et al. [11] address the detection of *semantic recurrence forms*, where the recurrence variable is tested within the loop body. However, they use a non-standard program representation, and they limit the scope of application to scalar variables. Semantic recurrences are associated with computational kernels that are widely used in real codes; a representative, yet simple, example is the computation of the minimum/maximum value of an array (either dense or sparse).

B. Monien and R. Feldmann (Eds.): Euro-Par 2002, LNCS 2400, pp. 289–298.

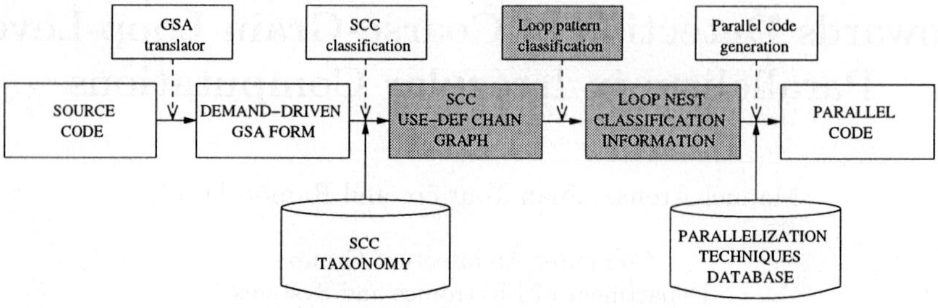

Fig. 1. Compiler framework block diagram.

In this paper, we present a new algorithm to enhance the detection of coarse-grain loop-level parallelism in codes with irregular computations. The method enables the detection of structural and semantic recurrence forms in a unified manner, even in the presence of complex control constructs. The recognition of semantic recurrences can enable the parallelization of a wider set of loops that would not be detected otherwise. We propose a parallelizing compiler framework (Figure 1) that is based on a demand-driven implementation of GSA program form [12]. Recurrence forms are detected in two steps: first, the SCCs that appear within a loop body are classified according to the SCC taxonomy shown in Figure 2, and the loop is represented as a directed acyclic SCC use-def chain graph; second, the loop pattern is determined at compile time by performing a demand-driven analysis of the SCC graph. The SCC classification algorithm is presented in [1]. This paper is focused on the description of the loop pattern classification algorithm. The final stage of the framework, which is out of the scope of this work, is the generation of parallel code using specific methods proposed in the literature; for instance [2, 7, 8, 13, 14].

This paper is organized as follows. In Section 2 the GSA program form is briefly described, as well as the advantages of building the compiler framework on top of GSA rather than on top of the source code. Basic notations and definitions that support our further analysis are introduced, too. In Section 3 the SCC classification algorithm is outlined and the SCC classes that appear in the case study presented in this paper are described. In Section 4 a new algorithm to identify coarse-grain parallelism in irregular loops is presented. Section 5 is devoted to experimental results using the SparsKit-II library [10], which was chosen because it covers a wide range of irregular patterns. We close with the conclusions of the paper in Section 6.

2 Background Concepts

2.1 Gated Single Assignment (GSA)

GSA [12] is a program representation that captures data-flow information for scalar and array variables by inserting *pseudo-functions* at the confluence nodes of the control flow graph where multiple reaching definitions of a variable exist.

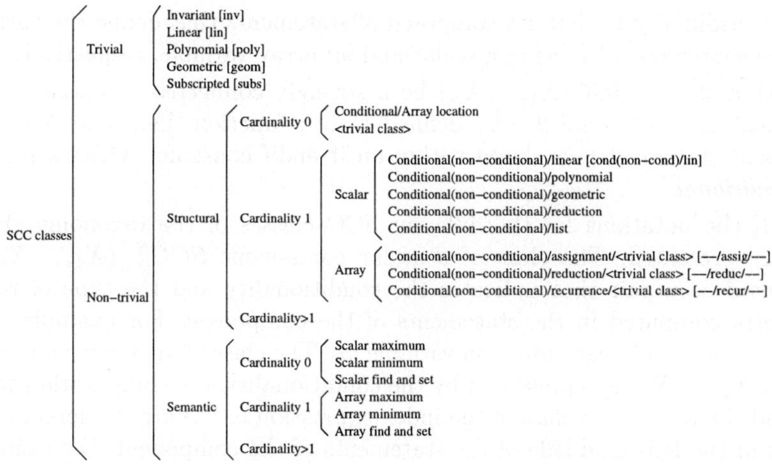

Fig. 2. Taxonomy of strongly connected components in GSA graphs. Abbreviations of SCC classes are written within brackets.

Several types of pseudo-functions are defined in GSA. In this work we use the μ-function, which appears at loop headers and selects the initial and loop-carried values of a variable; the γ-function, which is located at the confluence node associated with a branch and captures the condition for each definition to reach the confluence node; and the α-function, which replaces an array assignment statement.

The idea underlying this kind of representation is to rename the variables in a program according to a specific naming discipline which assures that left-hand sides of assignment statements are pairwise disjoint [4]. As a consequence, each use of a variable is reached by one definition at most. From the point of view of dependence analysis, this property of GSA assures that false dependences are removed from the program, both for scalar and array definitions (not for array element references). As a result, detection techniques based on GSA are only faced with the analysis of true dependences for scalars and with the analysis of the dependences that arise for arrays at the element level.

2.2 Basic Notations and Definitions

Let $SCC(X_1, ..., X_n)$ denote a strongly connected component composed of n nodes of a GSA dependence graph. The nodes are associated with the GSA statements where the variables X_k $(k = 1, ..., n)$ are defined.

Definition 1. Let $X_1, ..., X_n$ be a set of variables defined in the GSA form. The *cardinality* of a SCC is defined as the number of different variables of the source code that are associated with $X_1, ..., X_n$.

In this paper, only SCCs with cardinality zero or one are considered as the percentage of loops that contain SCCs with cardinality greater than one is very low in SparsKit-II. Let $SCC^S_{\#C}(X_1, ..., X_n)$ and $SCC^A_{\#C}(X_1, ..., X_n)$ denote

SCCs of cardinality C that are composed of statements that define the variable X in the source code, X being a scalar and an array variable, respectively.

Definition 2. Let $SCC(X_1, ..., X_n)$ be a strongly connected component. The component is *conditional* if $\exists X_j$ defined in a γ-function, i.e., if at least one assignment statement is enclosed within an if–endif construct. Otherwise, it is *non-conditional.*

In [1] the notations for the different SCC classes of the taxonomy (Figure 2) are presented. The class of a scalar component $SCC^S_{\#C}(X_1, ..., X_n)$ is represented as a pair that indicates the conditionality and the type of recurrence form computed in the statements of the component. For example, *non-cond/lin* denotes a linear induction variable [5]. The class of an array component $SCC^A_{\#C}(X_1, ..., X_n)$ is represented by the conditionality, the computation structure, and the recurrence class of the index expression of the array reference that appears in the left-hand side of the statements of the component. For example, *cond/reduc/subs* denotes an irregular reduction.

Definition 3. Let $SCC(X_1, ..., X_n)$ be a strongly connected component. The component is *trivial* if it consists of exactly one node of the GSA dependence graph ($n = 1$). Otherwise, the component is *non-trivial* ($n > 1$). Trivial components are non-conditional.

Definition 4. Let $SCC(X_1, ..., X_n)$ be a strongly connected component. The component is *wrap-around* if it is only composed of μ–statements. Otherwise, it is *non-wrap-around.*

Definition 5. Let $SCC(X_1, ..., X_n)$ and $SCC(Y_1, ..., Y_m)$ be strongly connected components. A *use-def chain* $SCC(X_1, ..., X_n) \rightarrow SCC(Y_1, ..., Y_m)$ exists if the assignment statements associated with $SCC(X_1, ..., X_n)$ contain at least one occurrence of the variables $Y_1, ..., Y_m$.

During the SCC classification process, some information about use-def chains is compiled. This information is denoted as *pos:exp.* The tag *exp* represents the expression within $SCC(X_1, ..., X_n)$ where the recurrence variable Y defined in $SCC(Y_1, ..., Y_m)$ is referenced. The tag *pos* represents the location of the reference within the corresponding statement of $SCC(X_1, ..., X_n)$. The reference to variable Y may appear in the index expression of an array reference located in the left-hand side (*lhs_index*), or in the right-hand side (*rhs_index*) of an assignment statement; it may also be located in the right-hand side, but not within an index expression (*rhs*).

Definition 6. Let G be the SCC use-def chain graph of a program in GSA form. Let $SCC(X_1, ..., X_n)$ be a non-wrap-around component associated with a source node of G. The *non-wrap-around source node (NWSN) subgraph* of $SCC(X_1, ..., X_n)$ in G is the subgraph composed of the nodes and edges that are accessible from $SCC(X_1, ..., X_n)$.

Definition 7. Let $SCC(X_1, ..., X_n) \rightarrow SCC(Y_1, ..., Y_m)$ be a use-def chain between two SCCs of cardinality zero or one. The use-def chain is *structural* if one of the following conditions is fulfilled: (a) $SCC(X_1, ..., X_n)$ and $SCC(Y_1, ..., Y_m)$ are scalar SCCs associated with the same scalar variable in the source code; (b)

$SCC(X_1, ..., X_n)$ is an array SCC, and the class of $SCC(Y_1, ..., Y_m)$ and that of the index expression in the class of $SCC(X_1, ..., X_n)$ are the same. Otherwise, the use-def chain is *non-structural.*

3 SCC Classification

In [1] we presented a non-deadlocking demand-driven algorithm to classify the SCCs that appear in the GSA program representation according to the taxonomy of Figure 2. The class of a $SCC(X_1, ..., X_n)$ is determined from the number of nodes of the GSA graph that compose the SCC, and from the properties of the operands and the operators that appear in the definition expression of the recurrence. This class provides the compiler with information about the type of recurrence form that is computed in the statements associated with $X_1, ..., X_n$.

In this section we describe the SCC classes that support our further analysis. For illustrative purposes, Figure 3 shows the source code, the GSA form and the SCC use-def chain graph corresponding to an interesting loop nest extracted from the SparsKit-II library. A trivial SCC (oval nodes in Figure 3) is associated with a scalar variable that is not defined in terms of itself in the source code, for example, a scalar temporary variable. Two classes are used in this paper: *subs*, which represents a scalar that is assigned the value of a different array entry in each iteration of a loop (Figure 3, wrap-around $SCC^S_{\#0}(k_1)$); and *lin*, which indicates that the scalar variable follows a linear progression (Figure 3, shaded oval $SCC^S_{\#1}(ii_1)$ associated with the index of the outermost loop).

In contrast, non-trivial SCCs (rectangular nodes in Figure 3) arise from the definition of variables whose recurrence expression depends of the variable itself, for example, reduction operations. In this paper we use: *non-cond/lin*, which represents a *linear induction variable* [5] of the source code (Figure 3, $SCC^S_{\#1}(ko_3, ko_4)$); *non-cond/assig/lin*, which captures the computation of consecutive entries of an array (Figure 3, $SCC^A_{\#1}(jao_1, jao_2, jao_3)$), as the corresponding assignment statements are not enclosed within an if–endif construct; and *cond/assig/lin*, which is distinguished from *non-cond/assig/lin* by the fact that at least one assignment statement is enclosed within an if–endif (Figure 3, $SCC^A_{\#1}(ao_1, ao_2, ao_3, ao_4)$).

4 Loop Classification

Loops are represented in our compiler framework as SCC use-def chain graphs. The class of a $SCC(X_1, ..., X_n)$ provides the compiler with information about the recurrence class that is computed in the statements associated with $X_1, ..., X_n$. However, the recurrence class computed using X in the source code may be different because X may be modified in other statements that are not included in $SCC(X_1, ..., X_n)$. In our framework, these situations are captured as dependences between SCCs that modify the same variable X. The analysis of the SCC use-def chain graph enables the classification of the recurrence computed in the loop body.

```
DO ii = 1, nrow
  ko = iao(perm(ii))
  DO k = ia(ii), ia(ii + 1) − 1
    jao(ko) = ja(k)
    IF (values) THEN
      ao(ko) = a(k)
    END IF
    ko = ko + 1
  END DO
END DO
```

(a) Source code.

```
DO ii1 = 1, nrow, 1
  jao1 = μ(jao0, jao2)
  k1 = μ(k0, k2)
  ko1 = μ(ko0, ko3)
  ao1 = μ(ao0, ao2)
  ko2 = iao(perm(ii1))
  DO k2 = ia(ii1), ia(ii1 + 1) − 1, 1
    jao2 = μ(jao1, jao3)
    ko3 = μ(ko2, ko4)
    ao2 = μ(ao1, ao4)
    jao3(ko3) = α(jao2, ja(k2))
    IF (values1) THEN
      ao3(ko3) = α(ao2, a(k2))
    END IF
    ao4 = γ(values1, ao3, ao2)
    ko4 = ko3 + 1
  END DO
END DO
```

(b) GSA form.

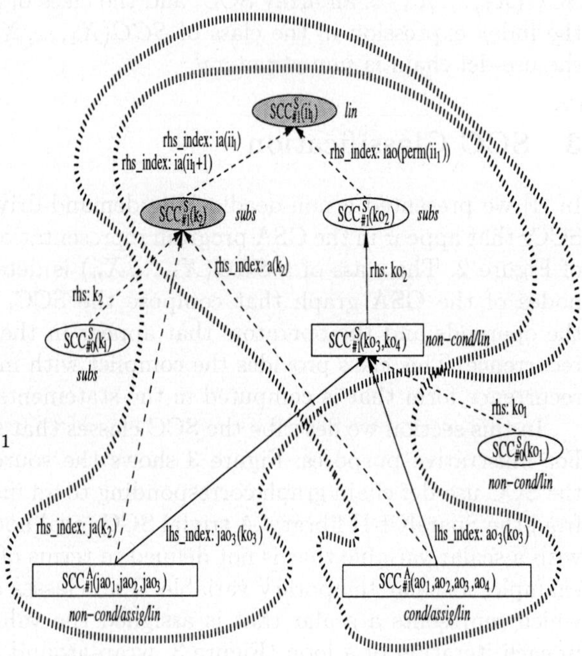

(c) SCC graph.

Fig. 3. Permutation of the rows of a sparse matrix (extracted from module UNARY of SparsKit-II, subroutine *rperm*).

4.1 SCC Use-Def Chain Graph Classification Procedure

The classification process of a loop begins with the partitioning of the SCC use-def chain graph into a set of connected subgraphs. For each connected subgraph, a recurrence class is derived for every NWSN subgraph (see Def. 6). The loop class is a combination of the classes of all the NWSN subgraphs. The core of the loop classification stage is the algorithm for classifying NWSN subgraphs (nodes and edges inside curves in Figure 3). A post-order traversal starts from the NWSN. When a node $SCC(X_1, ..., X_n)$ is visited, structural use-def chains (see Def. 7 and solid edges in Figure 3) are analyzed, as they supply all the information for determining the type of recurrence form computed using X in the source code. The analysis of non-structural use-def chains (dashed edges in Figure 3) provides further information that is useful, for example, in the parallel code generation stage, which is out of the scope of this paper. If $SCC(X_1, ..., X_n)$ was not successfully classified, the classification process stops, the loop is classified as unknown, and the classification process of inner loops starts. Otherwise, the algorithm derives the class of the NWSN subgraph, which belongs to the same class as the NWSN.

During this process, the class of some SCCs may be modified in order to represent more complex recurrence forms than those presented in the SCC taxonomy. In this work we refer to two of such classes. The first one consists of a linear induction variable that is reinitialized to a loop-variant value in each iteration of an outer loop (Figure 3, $SCC^S_{\#1}(ko_3, ko_4) \rightarrow SCC^S_{\#1}(ko_2)$). It is denoted as *non-cond/lin r/subs*. The second one represents consecutive write operations on an array in consecutive loop iterations, using an induction variable. This kind of computation was reported as a *consecutively written array* in [8] (Figure 3, $SCC^A_{\#1}(ao_1, ao_2, ao_3, ao_4) \rightarrow SCC^S_{\#1}(ko_3, ko_4)$).

4.2 Case Study

The example code presented in Figure 3 performs a permutation of the rows of a sparse matrix. Inner loop *do_k* contains an induction variable *ko* that is referenced in two consecutively written arrays *jao* and *ao* (note that the condition *values*, which is used to determine at run-time if the entries *ao* of the sparse matrix are computed, is loop invariant). Loop *do_k* can be executed in parallel, for example, by computing the closed form of *ko*. However, coarser-grain parallelism can be extracted from the outer loop *do_ii*. A new initial value of *ko* is computed in each *do_ii* iteration. Thus, a set of consecutive entries of arrays *jao* and *ao* is written in each *do_ii* iteration. As a result, *do_ii* can be executed in parallel if those sets do not overlap. As arrays *iao*, *perm* and *ia* are invariant with respect to *do_ii*, a simple run-time test would determine whether *do_ii* is parallel or serial. In the parallel code generation stage, this test can be inserted by the compiler just before *do_ii* in the control flow graph of the program.

In our compiler framework, *do_ii* is represented as one connected subgraph composed of two NWSN subgraphs that are associated with the source nodes $SCC^A_{\#1}(jao_1, jao_2, jao_3)$ and $SCC^A_{\#1}(ao_1, ao_2, ao_3, ao_4)$. Let us focus on the NWSN subgraph of $SCC^A_{\#1}(jao_1, jao_2, jao_3)$. During the post-order traversal of this subgraph, structural use-def chains are processed in the following order. The class of $SCC^S_{\#0}(ko_3, ko_4)$ is adjusted from *non-cond/lin* to *non-cond/lin r/subs* because there exists only one structural use-def chain $SCC^S_{\#1}(ko_3, ko_4) \rightarrow SCC^S_{\#1}(ko_2)$ where:

1. $SCC^S_{\#1}(ko_3, ko_4)$ and $SCC^S_{\#1}(ko_2)$ belong to classes *non-cond/lin* and *subs*, respectively.
2. The loop *do_k* contains the statements of $SCC^S_{\#1}(ko_3, ko_4)$, and the statement of $SCC^S_{\#1}(ko_2)$ belongs to the outer loop *do_ii* and precedes *do_k* in the control flow graph of the loop body.

The following step of the NWSN subgraph classification algorithm is faced with a structural use-def chain $SCC^A_{\#1}(jao_1, jao_2, jao_3) \rightarrow SCC^S_{\#1}(ko_3, ko_4)$ where:

1. $SCC^A_{\#1}(jao_1, jao_2, jao_3)$ belongs to class *non-cond/assig/lin*.
2. $SCC^S_{\#1}(ko_3, ko_4)$ is *non-cond/lin r/subs*, and all the operations on *ko* are increments (or decrements) of a constant value.
3. The use-def chain is labeled as *lhs_index* (see Section 2.2).

As these properties are fulfilled, array *jao* is called a candidate consecutively written array. Next, consecutively written arrays are detected using an algorithm proposed in [8], which basically consists of traversing the control flow graph of the loop body and check that every time an array entry *jao*(*ko*) is written, the corresponding induction variable *ko* is updated. In [8] a heuristic technique to detect these candidate arrays is roughly described. However, note that our framework enables the recognition of candidate arrays in a deterministic manner.

5 Experimental Results

We have developed a prototype of our loop classification algorithm using the infrastructure provided by the Polaris parallelizing compiler [3]. A set of costly operations for the manipulation of sparse matrices was analyzed, in particular: basic linear algebra operations (e.g. matrix-matrix product and sum), non-algebraic operations (e.g. extracting a submatrix from a sparse matrix, filter out elements of a matrix according to their magnitude, or performing mask operations with matrices), and some sparse storage conversion procedures.

Table 1 presents, for each nest level, the number of serial and parallel loops that appear in the modules *matvec*, *blassm*, *unary* and *formats* of the SparsKit-II library [10]. The last two rows summarize the information for all the nest levels, level-1 being the innermost level. The first four columns list in detail the structural and semantic recurrence forms detected in parallel irregular loops. Blank entries mean zero occurrences of the loop class. The last column *#loops* summarizes the total number of occurrences for each loop class.

The statistics were obtained by processing 256 loop nests (382 regular and irregular loops were analyzed in total), where approximately 47% carry out irregular computations. According to Table 1, 48 out of 382 loops were classified as parallel irregular loops. However, we have checked that many irregular loops are currently classified as serial because the loop body contains either jump-like statements (goto, return, exit), or recurrence forms whose SCC is not recognized by the SCC classification method.

In the context of irregular codes, current parallelizing compilers usually parallelize simple recurrence forms that appear in the innermost loop. Experimental results show that our method is able to recognize complex recurrence forms, even in outer loops, as stated in Table 1. In SparsKit-II, the prototype mainly detected irregular reductions (classes *non-cond/reduc/subs* and *cond/reduc/subs*) in level-2 and level-4 loops, and consecutively written arrays in level-1 and level-2 loops. Note that effectiveness decreases as nest level rises because outer loops usually compute more complex recurrence forms.

In SparsKit-II, a small percentage of parallel loops compute semantic recurrence forms only. We have also checked that some loops that contain a combination of structural and semantic recurrences are currently classified as serial.

Table 1. Classification of the loops from modules of SparsKit-II.

	matvec	blassm	unary	formats	#loops
Level-1 loops	22	32	88	107	249
Serial loops	2	11	42	56	111
Parallel loops	20	21	46	51	138
Structural recurrences					
non-conditional/assignment/subscripted		6	2	1	9
non-conditional/reduction/subscripted	4			2	6
conditional/assignment/subscripted		1			1
conditional/reduction/subscripted		1			1
consecutively written array			3	4	7
Semantic recurrences					
scalar maximum			2	1	3
scalar minimum with array location			1		1
Level-2 loops	16	8	34	59	117
Serial loops	11	6	27	50	94
Parallel loops	5	2	7	9	23
Structural recurrences					
non-conditional/assignment/subscripted		2	2	2	6
non-conditional/reduction/subscripted	3		1	2	6
consecutively written array			3	4	7
Level-3 loops	1	1	4	6	12
Serial loops	1	1	4	6	12
Parallel loops					0
Level-4 loops	1		1	2	4
Serial loops			1	2	3
Parallel loops	1				1
Structural recurrences					
non-conditional/reduction/subscripted	1				1
Serial loops	14	18	74	114	220
Parallel loops	26	23	53	60	162

6 Conclusions

Previous works on detection of parallelism in irregular codes addressed the problem of recognizing specific and isolated recurrence forms (usually using pattern-matching to analyze the source code). Unlike these techniques, we have described a new loop-level detection method that enables the recognition of structural and semantic recurrences in a unified manner, even in outer levels of loop nests.

Experimental results are encouraging and show the effectiveness of our method in the detection of coarse-grain parallelism in loops that compute complex structural and semantic recurrence forms. Further research will focus on the improvement of the SCC and the loop classification methods to cover a wider range of irregular computations.

Acknowledgements

This work was supported by the Ministry of Science and Technology of Spain and FEDER funds of the European Union (Project TIC2001-3694-C02-02).

References

1. Arenaz, M., Touriño, J., Doallo, R.: A Compiler Framework to Detect Parallelism in Irregular Codes. In Proceedings of 14th International Workshop on Languages and Compilers for Parallel Computing, LCPC'2001, Cumberland Falls, KY (2001)
2. Arenaz, M., Touriño, J., Doallo, R.: Run-time Support for Parallel Irregular Assignments. In Proceedings of 6th Workshop on Languages, Compilers, and Run-Time Systems for Scalable Computers, LCR'02, Washington D.C. (2002)
3. Blume, W., Doallo, R., Eigenmann, R., Grout, J., Hoeflinger, J., Lawrence, T., Lee, J., Padua, D.A., Paek, Y., Pottenger, W.M., Rauchwerger, L., Tu, P.: Parallel Programming with Polaris. IEEE Computer **29**(12) (1996) 78–82
4. Cytron, R., Ferrante, J., Rosen, B.K., Wegman, M.N., Zadeck, F.K.: Efficiently Computing Static Single Assignment Form and the Control Dependence Graph. ACM Transactions on Programming Languages and Systems **13**(4) (1991) 451–490
5. Gerlek, M.P., Stoltz, E., Wolfe, M.: Beyond Induction Variables: Detecting and Classifying Sequences Using a Demand-Driven SSA Form. ACM Transactions on Programming Languages and Systems **17**(1) (1995) 85–122
6. Keβler, C.W.: Applicability of Automatic Program Comprehension to Sparse Matrix Computations. In Proceedings of 7th International Workshop on Compilers for Parallel Computers, Linköping, Sweden (1998) 218–230
7. Knobe, K., Sarkar, V.: Array SSA Form and Its Use in Parallelization. In Proceedings of 25th ACM SIGACT-SIGPLAN Symposium on the Principles of Programming Languages (1998) 107–120
8. Lin, Y., Padua, D.A.: On the Automatic Parallelization of Sparse and Irregular Fortran Programs. In Proceedings of 4th Workshop on Languages, Compilers, and Run-Time Systems for Scalable Computers, LCR'98, Pittsburgh, PA, Lecture Notes in Computer Science, Vol. 1511 (1998) 41–56
9. Pottenger, W.M., Eigenmann, R.: Idiom Recognition in the Polaris Parallelizing Compiler. In Proceedings of 9th ACM International Conference on Supercomputing, Barcelona, Spain (1995) 444–448
10. Saad, Y.: SPARSKIT: A Basic Tool Kit for Sparse Matrix Computations. http://www.cs.umn.edu/Research/darpa/SPARSKIT/sparskit.html (1994)
11. Suganuma, T., Komatsu, H., Nakatani, T.: Detection and Global Optimization of Reduction Operations for Distributed Parallel Machines. In Proceedings of 10th ACM International Conference on Supercomputing, Philadelphia, PA (1996) 18–25
12. Tu, P., Padua, D.: Gated SSA-Based Demand-Driven Symbolic Analysis for Parallelizing Compilers. In Proceedings of 9th ACM International Conference on Supercomputing, Barcelona, Spain (1995) 414–423
13. Yu, H., Rauchwerger, L.: Adaptive Reduction Parallelization Techniques. In Proceedings of 14th ACM International Conference on Supercomputing, Santa Fe, NM (2000) 66–77
14. Xu, C.-Z., Chaudhary, V.: Time Stamps Algorithms for Runtime Parallelization of DOACROSS Loops with Dynamic Dependences. IEEE Transactions on Parallel And Distributed Systems **12**(5) (2001) 433–450

On the Optimality of Feautrier's Scheduling Algorithm

Frédéric Vivien

ICPS-LSIIT, Université Louis Pasteur, Strasbourg, Pôle Api, F-67400 Illkirch, France.

Abstract. Feautrier's scheduling algorithm is the most powerful existing algorithm for parallelism detection and extraction. But it has always been known to be suboptimal. However, the question whether it may miss some parallelism because of its design was still open. We show that this is not the case. Therefore, to find more parallelism than this algorithm does, one needs to get rid of some of the hypotheses underlying its framework.

1 Introduction

One of the fundamental steps of automatic parallelization is the detection and extraction of parallelism. This extraction can be done in very different ways, from the try and test of *ad hoc* techniques to the use of powerful scheduling algorithms. In the field of dense matrix code parallelization, lots of algorithms have been proposed along the years. Among the main ones, we have the algorithms proposed by Lamport [10], Allen and Kennedy [2], Wolf and Lam [15], Feautrier [7,8], and Darte and Vivien [5]. This collection of algorithm spans a large domain of techniques (loop distribution, unimodular transformations, linear programming, etc.) and a large domain of dependence representations (dependence levels, direction vectors, affine dependences, dependence polyhedra). One may wonder which algorithm to chose from such a collection. Fortunately, we have some theoretical comparative results on these algorithms, as well as some optimality results.

Allen and Kennedy's, Wolf and Lam's, and Darte and Vivien's algorithms are optimal for the representation of the dependences they respectively take as input [4]. This means that each of these algorithms extracts all the parallelism contained in its input (some representation of the code dependences). Wolf and Lam's algorithm is a generalization of Lamport's; Darte and Vivien's algorithm is a generalization of those of Allen and Kennedy, and of Wolf and Lam, and is generalized by Feautrier's [4]. Finally, Feautrier's algorithm can handle any of the dependence representations used by the other algorithms [4].

It appears from these results that Feautrier's algorithm is the most powerful algorithm we have at hand. Although this algorithm has always be known to be suboptimal, its exact efficiency was so far unknown. Hence the questions we address in this paper: What are its weaknesses? Is its suboptimality only due to its framework or also to its design? What can be done to improve this algorithm? How can we build a more powerful algorithm?

B. Monien and R. Feldmann (Eds.): Euro-Par 2002, LNCS 2400, pp. 299–308.

In Section 2 we briefly recall Feautrier's algorithm. Then we discuss its weaknesses in Section 3. In Section 4 we present what seems to be a "better" algorithm. Section 5 presents the major new result of this paper: to find "more" parallelism than Feautrier's algorithm one needs to use far more powerful techniques.

2 The Algorithm

Feautrier uses schedules to detect and extract parallelism. This section gives an overview of his algorithm. The missing details can be found either in [7, 8] or [4].

Framework: Static Control Programs. To enable an exact dependence analysis, the control-flow must be predictable at compile time. The necessary restrictions define the class of the *static control programs.* These are the programs:

- whose only data structures are integers, floats, arrays of integers, and arrays of floats, with no pointers or pointer-like mechanisms;
- whose elementary statements are assignments of scalars or array elements;
- whose only control structure are sequences and do loops with constant steps;
- where the array subscripts and the loop bounds are affine functions of surrounding loop indices and structural parameters.

Static control programs are mainly sets of nested loops. Figure 1 presents an example of such a program. Let S be any statement. The iteration domain of S, denoted $\mathcal{D}_S$, is the set of all possible values of the vector of the indices (the *iteration vector*) of the loops surrounding S: in Example 1, $\mathcal{D}_S = \{(i,j) \mid 1 \leq i \leq N, 1 \leq j \leq i\}$. An iteration domain is always a polyhedron. In other words, there always exist a matrix A and a vector b such that : $\mathcal{D}_S = \{x \mid A.x \leq b\}$.

```
DO i=1, N
 DO j=1, i
S: a(i,i+j+1) = a(i-1,2*i-1) + a(j,2*j)
 ENDDO
ENDDO
```

Fig. 1. Example 1.

e_1: $S(i-1,i-1) \rightarrow S(i,j), h_{e_1}(i,j)=(i-1,i-1)$
$\mathcal{D}_{e_1} = \{(i,j) \mid 2 \leq i \leq N, 1 \leq j \leq i\}$
e_2: $S(j,j-1) \rightarrow S(i,j), h_{e_2}(i,j)=(j,j-1)$
$\mathcal{D}_{e_2} = \{(i,j) \mid 1 \leq i \leq N, 2 \leq j \leq i\}$

Fig. 2. Dependences for Example 1.

Dependence Representation. In the framework of static control programs, an exact dependence analysis is feasible [6] and each exact *dependence relation* e from statement S_e to statement T_e is defined by a polyhedron $\mathcal{D}_e$, the domain of existence of the dependence relation, and a quasi-affine [1] function h_e as follows:

[1] See the original paper [6] for more details.

for any value $j \in \mathcal{D}_e$, operation $T_e(j)$ depends on operation $S_e(h_e(j,N))$:

$$j \in \mathcal{D}_e \quad \Rightarrow \quad S_e(h_e(j,N)) \rightarrow T_e(j)$$

where N is the vector of structural parameters. Obviously, the description of the exact dependences between two statements may involve the union of many such dependence relations. A dependence relation e describes for any value $j \in \mathcal{D}_e$ a dependence between the two operations $S_e(h_e(j,N))$ and $T_e(j)$, what we call an *operation to operation dependence.* In other words, a *dependence relation* is a set of elementary *operation to operation dependences.* Figure 2 presents the dependence relations for Example 1.

Following Feautrier [7], we suppose that all the quasi-affine functions we have to handle are in fact affine functions (at the possible cost of a conservative approximation of the dependences).

Searched Schedules. Feautrier does not look for any type of functions to schedule affine dependences. He only considers nonnegative functions, with rational values, that are affine functions in the iteration vector and in the vector of structural parameters. Therefore he only handles (affine) schedules of the form:

$$\Theta(S,j,N) = X_S.j + Y_S.N + \rho_S \tag{1}$$

where X_S and Y_S are non-parameterized rational vectors and ρ_S is a rational constant. The hypothesis of nonnegativity of the schedules is not restrictive as all schedules must be lower bounded.

Problem Statement. Once chosen the form of the schedules, the scheduling problem *seems* to be simple. For a schedule to be valid, it must (and only has to) satisfy the dependences. For example, if operation $T(j)$ depends on operation $S(i)$, $T(j)$ must be scheduled *after* $S(i)$: $\Theta(T,j,N) > \Theta(S,i,N)$. Therefore, for each statement S, we just have to find a vector X_S, a vector Y_S, and a constant ρ_S such that, for each dependence relation e, the schedule satisfies: [2]

$$j \in \mathcal{D}_e \quad \Rightarrow \quad \Theta(S_e, h_e(j,N), N) + 1 \;\le\; \Theta(T_e, j, N). \tag{2}$$

The set of constraints is linear, and one can imagine using linear system solvers to find a solution. Actually, there are now two difficulties to overcome:

1. Equation (2) must be satisfied for any possible value of the structural parameters. If polyhedron $\mathcal{D}_e$ is parameterized, Equation (2) may correspond to an infinite set of constraints, which cannot be enumerated. There are two means to overcome this problem: the polyhedron vertices (cf. Section 4) and the affine form of Farkas' lemma (see below). Feautrier uses the latter.
2. There does not always exist a solution for such a set of constraints. We will see how the use of multidimensional schedules can overcome this problem.

[2] The transformation of the inequality, from $a > b$ to $a \ge 1+b$, is obvious for schedules with integral values and classical for schedules with rational values [12].

The Affine Form of Farkas' Lemma and Its Use. This lemma [7, 13] predicts the shape of certain affine forms.

Theorem 1 (Affine Form of Farkas' Lemma). *Let $\mathcal{D}$ be a nonempty polyhedron defined by p inequalities: $a_k x + b_k \geq 0$, for any $k \in [1, p]$. An affine form Φ is nonnegative over $\mathcal{D}$ if and only if it is a nonnegative affine combination of the affine forms used to define $\mathcal{D}$:*

$$\Phi(x) \equiv \lambda_0 + \sum_{k=1}^{p} \lambda_k (a_k x + b_k), \text{ with } \lambda_k \geq 0 \text{ for any } k \in [0, p].$$

This theorem is useful as, in static control programs, all the important sets are polyhedra : iteration domains, dependence existence domains [6], etc. Feautrier uses it to predict the shape of the schedules and to simplify the set of constraints.

Schedules. By hypothesis, the schedule $\Theta(S, j, N)$ is a nonnegative affine form defined on a polyhedron $\mathcal{D}_S$: the iteration domain of statement S. Therefore, the affine form of Farkas' lemma states that $\Theta(S, j, N)$ is a nonnegative affine combination of the affine forms used to define $\mathcal{D}_S$. Let $\mathcal{D}_S = \{x \mid \forall i \in [1, p_S], A_{S,i}.x + B_{S,i}.N + c_{S,i} \geq 0\}$ ($\mathcal{D}_S$ is thus defined by p_S inequalities). Then Theorem 1 states that there exist some nonnegative values $\mu_{S,0}$, ..., μ_{S,p_S} such that:

$$\Theta(S, j, N) \equiv \mu_{S,0} + \sum_{i=1}^{p_S} \mu_{S,i}(A_{S,i}.j + B_{S,i}.N + c_{S,i}). \tag{3}$$

Dependence Constraints. Equation (2) can be rewritten as an affine function that is nonnegative over a polyhedron because the schedules and the function h_e are affine functions:

$$j \in \mathcal{D}_e \quad \Rightarrow \quad \Theta(T_e, j, N) - \Theta(S_e, h_e(j, N), N) - 1 \geq 0.$$

Once again we can apply the affine form of Farkas' lemma. Let $\mathcal{D}_e = \{x \mid \forall i \in [1, p_e], A_{e,i}.x + B_{e,i}.N + c_{e,i} \geq 0\}$ ($\mathcal{D}_e$ is thus defined by p_e inequalities). Theorem 1 states that there exist some nonnegative values $\lambda_{e,0}$, ..., λ_{e,p_e} such that:

$$\Theta(T_e, j, N) - \Theta(S_e, h_e(j, N), N) - 1 \equiv \lambda_{e,0} + \sum_{i=1}^{p_e} \lambda_{e,i}(A_{e,i}.j + B_{e,i}.N + c_{e,i}).$$

Using Equation (3), we rewrite the left-hand side of this equation:

$$\begin{aligned} &\left(\mu_{T_e,0} + \sum_{i=1}^{p_{T_e}} \mu_{T_e,i}(A_{T_e,i}.j + B_{T_e,i}.N + c_{T_e,i}) \right) \\ &\quad - \left(\mu_{S_e,0} + \sum_{i=1}^{p_{S_e}} \mu_{S_e,i}(A_{S_e,i}.h_e(j, N) + B_{S_e,i}.N + c_{S_e,i}) \right) - 1 \\ &\qquad \equiv \lambda_{e,0} + \sum_{i=1}^{p_e} \lambda_{e,i}(A_{e,i}.j + B_{e,i}.N + c_{e,i}). \end{aligned} \tag{4}$$

Equation 4 is a formal equality ($\equiv$). Thus, the coefficients of a given component of either of the vectors j and N must be the same on both sides. The constant terms on both sides of this equation must also be equal. This identification process leads to a set of $(n+q+1)$ equations, equivalent to Equation (4), where n is the size of the iteration vector j, and q the size of the parameter vector N.

The way Feautrier uses the affine form of Farkas' lemma enables him to obtain a finite set of linear equations and inequations, equivalent to the original scheduling problem, and that can be solved using any solver of linear systems.

Extension to Multidimensional Scheduling. There exist some static control programs that cannot be scheduled with (monodimensional) affine schedules (e.g. Example 1, cf. Section 4). Hence the need for multidimensional schedules, i.e. schedules whose values are not rationals but rational *vectors* (ordered by lexicographic ordering). The solution proposed by Feautrier is simple and greedy. For the first dimension of the schedules one looks for affine functions that 1) respect all the dependences; 2) satisfy as many dependence relations as possible. The algorithm is then recursively called on the unsatisfied dependence relations. This, plus a strongly connected component distribution[3] that reminds us of Allen and Kennedy's algorithm, defines the algorithm below. G denotes the multigraph defined by the statements and the dependence relations. The multidimensional schedules built satisfy the dependences according to the lexicographic order [4].

FEAUTRIER(G)

1. Compute the strongly connected components of G.
2. **For each** strongly connected component G_i of G **do** in topological order:
   (a) Find, using the method exposed above, an affine function that satisfies

   $$\forall e,\ j \in \mathcal{D}_e \Rightarrow \Theta(S_e, h_e(j,N), N) + z_e \leq \Theta(T_e, j, N) \text{ with } 0 \leq z_e \leq 1 \quad (5)$$

   and which maximizes the sum $\sum_e z_e$.
   (b) Build the subgraph G'_i generated by the unsatisfied dependences. If G'_i is not empty, recursively call FEAUTRIER(G'_i).

3 The Algorithm's Weaknesses

Definitions of Optimality. Depending on the definition one uses, an algorithm extracting parallelism is optimal if it finds all the parallelism: 1) that can be extracted in its framework (only certain program transformations are allowed, etc.); 2) that is contained in the representation of the dependences it handles; 3) that is contained in the program to be parallelized (not taking into account the dependence representation used nor the transformations allowed). For example, Allen, Callahan, and Kennedy uses the first definition [1], Darte and Vivien the second [5], and Feautrier the third [8]. We now recall that FEAUTRIER is not optimal under any of the last two definitions.

[3] This distribution is rather esthetic as the exact same result can be achieved without using it. This distribution is intuitive and ease the computations.

The Classical Counter-Example to Optimality. Feautrier proved in his original article [7] that his algorithm was not optimal for parallelism detection in static control programs. In his counterexample (Example 2, Figure 3) the source of any dependence is in the first half of the iteration domain and the sink in the second half. Cutting the iteration domain "in the middle" enables a trivial parallelization (Figure 4). The only loop in Example 2 contains some dependences. Thus, Feautrier's schedules must be of dimension at least one (hence at least one sequential loop after parallelization), and FEAUTRIER finds no parallelism.

```
DO i=0, 2n
  x(i) = x(2n-i)
ENDDO
```

Fig. 3. Example 2.

```
DOPAR i=0, n
  x(i) = x(2n-i)
ENDDOPAR
DOPAR i=n+1, 2n
  x(i) = x(2n-i)
ENDDOPAR
```

Fig. 4. Parallelized version of Example 2.

Weaknesses. The weaknesses in Feautrier's algorithm are either a consequence of the algorithm framework, or of the algorithm design.

Framework. Given a program, we extract its implicit parallelism and then we rewrite it. The new order of the computations must be rather regular to enable the code generation. Hence the restriction on the schedule shape: affine functions. The parallel version of Example 2 presented Figure 4 can be expressed by a non affine schedule, but not by an affine schedule. The restriction on the schedule shape is thus a cause of inefficiency. Another problem with Example 2 is that Feautrier looks for a transformation conservative in the number of loops. Breaking a loop into several loops, i.e., cutting the iteration domain into several subdomains, can enable to find more parallelism (even with affine schedules). The limitation here comes from the hypothesis that all instances of a statement are scheduled the same way, i.e., with the same affine function. (Note that this hypothesis is almost always made [10, 2, 15, 5], [9] being the exception.)

Some of the weaknesses of FEAUTRIER are thus due to its framework. Before thinking of changing this framework, we must check whether one can design a more powerful algorithm, or even improve FEAUTRIER, in Feautrier's framework.

Algorithm design. FEAUTRIER is a greedy algorithm which builds multidimensional schedules whose first dimension satisfies as many *dependence relations* as possible, and not as many *operation to operation dependences* as possible. We may wonder with Darte [3, p. 80] whether this can be the cause of a loss of parallelism. We illustrate this possible problem with Example 1.

The first dimension of the schedule must satisfy Equation (5) for both dependence relations e_1 and e_2. This gives us respectively Equations (6) and (7):

$$X_S\begin{vmatrix} i-1 \\ i-1 \end{vmatrix} + z_{e_1} \leq X_S \begin{vmatrix} i \\ j \end{vmatrix} \Leftrightarrow z_{e_1} \leq X_S \begin{vmatrix} 1 \\ j-i+1 \end{vmatrix} \Leftrightarrow z_{e_1} \leq \alpha + \beta(j-i+1) \text{ with } \begin{matrix} 2 \leq i \leq N \\ 1 \leq j \leq i \end{matrix} \quad (6)$$

$$X_S\begin{vmatrix} j \\ j-1 \end{vmatrix} + z_{e_2} \leq X_S \begin{vmatrix} i \\ j \end{vmatrix} \Leftrightarrow z_{e_2} \leq X_S \begin{vmatrix} i-j \\ 1 \end{vmatrix} \Leftrightarrow z_{e_2} \leq \alpha(i-j) + \beta \text{ with } \begin{matrix} 1 \leq i \leq N \\ 2 \leq j \leq i \end{matrix} \quad (7)$$

if we note $X_S = (\alpha, \beta)$ [4]. Equation (6) with $i = N$ and $j = 1$ is equivalent to $z_{e_1} \leq \alpha + \beta(2 - N)$. The schedule must be valid for any (nonnegative) value of the structural parameter N, this implies $\beta \leq 0$. Equation (7) with $i = j$ is equivalent to $z_{e_2} \leq \beta$. Hence $z_{e_2} \leq 0$. As z_{e_2} must be nonnegative $z_{e_2} = 0$ (cf. Equation (5)). This means that the first dimension of any affine schedule cannot satisfy the dependence relation e_2.

The dependence relation e_1 can be satisfied, a solution being $X_S = (1, 0)$ ($\alpha = 1$, $\beta = 0$). Therefore, FEAUTRIER **is called recursively on the whole dependence relation** e_2. However, most of the dependences described by e_2 are satisfied by the schedule $\Theta(S, (i, j), N) = i$ (defined by $X_S = (1, 0)$). Indeed, Equation (6) is then satisfied for any value $(i, j) \in \mathcal{D}_{e_2}$ except when $i{=}j$. Thus, **one only needed to call recursively** FEAUTRIER **on the dependence relation** e'_2: $S(j, j{-}1) \to S(i, j)$, $h_{e_2}(i, j) = (j, j{-}1)$, $\mathcal{D}_{e'_2} = \{(i, j) \mid 2{\leq}i{\leq}N, i = j\}$. The search for the schedules in FEAUTRIER is thus overconstrained by design.

We may now wonder whether this overconstraining may lead FEAUTRIER to build some affine schedules of non minimal dimensions and thus to miss some parallelism. We first present an algorithm which gets rid of this potential problem. Later we will show that no parallelism is lost because of this design particularity.

4 A Greedier Algorithm

The Vertex Method. A polyhedron can always be decomposed as the sum of a polytope (i.e. a bounded polyhedron) and a polyhedral cone, called the characteristic cone (see [13] for details). A polytope is defined by its vertices, and any point of the polytope is a nonnegative barycentric combination of the polytope vertices. A polyhedral cone is finitely generated and is defined by its rays and lines. Any point of a polyhedral cone is the sum of a nonnegative combination of its rays and any combination of its lines. Therefore, a polyhedron $\mathcal{D}$ can be equivalently defined by a set of *vertices*, $\{v_1, \ldots, v_\omega\}$, a set of *rays*, $\{r_1, \ldots, r_\rho\}$, and a set of *lines*, $\{l_1, \ldots, l_\lambda\}$. Then $\mathcal{D}$ is the set of all vectors p such that

$$p = \sum_{i=1}^{\omega} \mu_i v_i + \sum_{i=1}^{\rho} \nu_i r_i + \sum_{i=1}^{\lambda} \xi_i l_i \quad (8)$$

[4] Example 1 contains a single statement S. Therefore, the components Y_S and ρ_S of Θ (cf. Equation (1)) have no influence here on Equation (5) which is equivalent to: $(X_S.h_e(j, N) + Y_S.N + \rho_S) + z_e \leq (X_S.j + Y_S.N + \rho_S) \Leftrightarrow X_S.h_e(j, N) + z_e \leq X_S.j$.

with $\mu_i \in \mathbb{Q}^+$, $\nu_i \in \mathbb{Q}^+$, $\xi_i \in \mathbb{Q}$, and $\sum_{i=1}^{\omega} \mu_i = 1$. As we have already stated, all the important sets in static control programs are polyhedra, and any nonempty polyhedron is fully defined by its vertices, rays, and lines, which can be computed even for parameterized polyhedra [11]. The vertex method [12] explains how we can use the vertices, rays, and lines to simplify set of constraints.

Theorem 2 (The Vertex Method). *Let $\mathcal{D}$ be a nonempty polyhedron defined by a set of vertices, $\{v_1, \ldots, v_\omega\}$, a set of rays, $\{r_1, \ldots, r_\rho\}$, and a set of lines, $\{l_1, \ldots, l_\lambda\}$). Let Φ be an affine form of linear part A and constant part b ($\Phi(x) = A.x + b$). Then the affine form Φ is nonnegative over $\mathcal{D}$ if and only if 1) Φ is nonnegative on each of the vertices of $\mathcal{D}$ and 2) the linear part of Φ is nonnegative (respectively null) on the rays (resp. lines) of $\mathcal{D}$. This can be written :*

$$\forall p \in \mathcal{D},\ A.p + b \geq 0 \quad \Leftrightarrow$$
$$\forall i \in [1, \omega],\ A.v_i + b \geq 0,\ \forall i \in [1, \rho],\ A.r_i \geq 0,\ \textit{and}\ \ \forall i \in [1, \lambda],\ A.l_i = 0.$$

The polyhedra produced by the dependence analysis of programs are in fact polytopes. Then, according to Theorem 2, an affine form is nonnegative on a polytope if and only if it is nonnegative on the vertices of this polytope. We use this property to simplify Equation (2) and define a new scheduling algorithm.

The Greediest Algorithm. Feautrier's algorithm is a greedy heuristic which maximizes the number of *dependence relations* satisfied by the first dimension of the schedule. The algorithm below is a greedy heuristic which maximizes the number of *operation to operation dependences* satisfied by the first dimension of the schedule, and then proceeds recursively. To achieve this goal, this algorithm greedily considers the vertices of the existence domain of the dependence relations. Let e_1, ..., e_n be the dependence relations in the studied program. For any $i \in [1, n]$, let $v_{i,1}$, ..., v_{i,m_i} be the vertices of $\mathcal{D}_{e_i}$, and let, for any $j \in [1, m_i]$, $e_{i,j}$ be the operation to operation dependence from $S_{e_i}(h_{e_i}(v_{i,j}), N)$ to $T_{e_i}(v_{i,j})$. G denotes here the multigraph generated by the dependences $e_{i,j}$.

GREEDY(G)

1. Compute the strongly connected components of G.
2. **For each** strongly connected component G_k of G **do** in topological order:
   (a) Find an *integral* affine function Θ that satisfies
   $$\forall e_{i,j},\ \ \Theta(S_{e_i}, h_{e_i}(v_{i,j}, N), N) + z_{i,j} \ \leq\ \Theta(T_{e_i}, v_{i,j}, N)\ \text{ with }\ 0 \leq z_{i,j} \leq 1$$
   and which maximizes the sum $\sum_{e_{i,j}} z_{i,j}$.
   (b) Build the subgraph G'_k generated by the unsatisfied dependences. If G'_k is not empty, recursively call GREEDY(G'_k).

Lemma 1 (Correctness and Maximum Greediness). *The output of algorithm* GREEDY *is a schedule and the first dimension of this schedule satisfies all the operation to operation dependences that can be satisfied by the first dimension of an affine schedule (of the form defined in Section 2).*

5 Schedules of Minimal Dimension

As GREEDY is greedier than FEAUTRIER, one could imagine that the former may sometimes build schedules of smaller dimension than the latter and thus may find more parallelism. The following theorem shows that this never happens.

Theorem 3 (The Dimension of Feautrier's Schedules is Minimal). *Let us consider a loop nest whose dependences are all affine, or are represented by affine functions. If we are only looking for one affine schedule per statement of the loop nest, then the dimension of the schedules built by* FEAUTRIER *is minimal, for each statement of the loop nest.*

Note that this theorem cannot be improved, as the study of Example 2 shows. The proof is direct (not using algorithm GREEDY) and can be found in [14].

Principle of the proof. Let σ be an affine schedule whose dimension is minimal for each statement in the studied loop nest. Let e be a dependence relation, of existence domain $\mathcal{D}_e$. We suppose that e is not fully, but partially, satisfied by the first dimension of σ (otherwise there is no problem with e). The operation to operation dependences in e not satisfied by the first dimension of the schedule σ define a subpolyhedron $\mathcal{D}_e^1$ of $\mathcal{D}_e$: this is the subset of $\mathcal{D}_e$ on which the first dimension of σ induces a null delay. $\mathcal{D}_e^1$ is thus defined by the equations defining $\mathcal{D}_e$ and by the null delay equation involving the first dimension of σ $(\sigma_1(T_e, j, N) - \sigma_1(S_e, h_e(j, N), N) = 0)$. The second dimension of σ must respect the dependences in $\mathcal{D}_e^1$, i.e., must induce a nonnegative delay over $\mathcal{D}_e^1$. Therefore, the second dimension of σ is an affine form nonnegative over a polyhedron. Using the affine form of Farkas' lemma, we obtain that the second dimension of σ is defined from the (null delay equation on the) first dimension of σ and from the equations defining $\mathcal{D}_e$. From the equations obtained using Farkas' lemma, we build a nonnegative linear combination of the first two dimensions of σ which induces a nonnegative delay over $\mathcal{D}_e$ (and not only on $\mathcal{D}_e^1$), and which satisfies all the operation to operation dependences in e satisfied by any of the first two dimensions of σ. This way we build a schedule *a la* Feautrier of same dimension than σ: a whole dependence relation is kept as long as all its operation to operation dependences are not satisfied by the same dimension of the schedule.

Consequences. First, a simple and important corollary of the previous theorem:

Corollary 1. FEAUTRIER *is well-defined: it always outputs a valid schedule when its input is the exact dependences of an existing program.*

The original proof relied on an assumption on the dependence relations that can be easily enforced but which is not always satisfied: all operation to operation dependences in a dependence relation are of the same dependence level. For example, dependence relation e_2 in Example 1 does not satisfy this property.

More important, Theorem 3 shows that Feautrier's algorithm can only miss some (significant amount of) parallelism because of the limitations of its framework, but not because of its design: as the dimension of the schedule is minimal, the magnitude of the schedule's makespan is minimal, for any statement.

6 Conclusion

Feautrier's scheduling algorithm is the most powerful existing algorithm for parallelism detection and extraction. But it has always been known to be suboptimal. We have shown that Feautrier's algorithm do not miss any significant amount of parallelism *because* of its design, even if one can design a greedier algorithm. Therefore, to improve Feautrier's algorithm or to build a more powerful algorithm, one must get rid of some of the restrictive hypotheses underlying its framework: affine schedules — but more general schedules will cause great problems for code generation — and one scheduling function by statement — Feautrier, Griebl, and Lengauer have already begun to get rid of this hypothesis by splitting the iteration domains [9].

What Feautrier historically introduced as a "greedy heuristic" is nothing but the most powerful algorithm in its class!

References

1. J. Allen, D. Callahan, and K. Kennedy. Automatic decomposition of scientific programs for parallel execution. In *Proceedings of the Fourteenth Annual ACM Symposium on Principles of Programming Languages*, pages 63–76, Munich, Germany, Jan. 1987.
2. J. R. Allen and K. Kennedy. PFC: A program to convert Fortran to parallel form. Technical Report MASC-TR82-6, Rice University, Houston, TX, USA, 1982.
3. A. Darte. De l'organisation des calculs dans les codes répétitifs. *Habilitation thesis*, école normale supérieure de Lyon, 1999.
4. A. Darte, Y. Robert, and F. Vivien. *Scheduling and Automatic Parallelization.* Birkhäuser Boston, 2000. ISBN 0-8176-4149-1.
5. A. Darte and F. Vivien. Optimal Fine and Medium Grain Parallelism Detection in Polyhedral Reduced Dependence Graphs. *Int. J. of Parallel Programming*, 1997.
6. P. Feautrier. Dataflow analysis of array and scalar references. *International Journal of Parallel Programming*, 20(1):23–51, 1991.
7. P. Feautrier. Some efficient solutions to the affine scheduling problem, part I: One-dimensional time. *Int. J. Parallel Programming*, 21(5):313–348, Oct. 1992.
8. P. Feautrier. Some efficient solutions to the affine scheduling problem, part II: Multi-dimensional time. *Int. J. Parallel Programming*, 21(6):389–420, Dec. 1992.
9. M. Griebl, P. Feautrier, and C. Lengauer. Index set splitting. *International Journal of Parallel Programming*, 28(6):607–631, 2000.
10. L. Lamport. The parallel execution of DO loops. *Communications of the ACM*, 17(2):83–93, Feb. 1974.
11. V. Loechner and D. K. Wilde. Parameterized polyhedra and their vertices. *International Journal of Parallel Programming*, 25(6), Dec. 1997.
12. P. Quinton. *Automata Networks in Computer Science*, chapter The systematic design of systolic arrays. Manchester University Press, 1987.
13. A. Schrijver. *Theory of Linear and Integer Programming.* John Wiley & Sons, New York, 1986.
14. F. Vivien. On the Optimality of Feautrier's Scheduling Algorithm. Technical Report 02-04, ICPS-LSIIT, ULP-Strasbourg I, France, http://icps.u-strasbg.fr, 2002.
15. M. E. Wolf and M. S. Lam. A data locality optimizing algorithm. In *SIGPLAN Conference PLDI*, pages 30–44. ACM Press, 1991.

On the Equivalence of Two Systems of Affine Recurrence Equations

Denis Barthou[1], Paul Feautrier[2], and Xavier Redon[3]

[1] Université de Versailles Saint-Quentin, Laboratoire PRiSM, F-78035 Versailles, France, Denis.Barthou@prism.uvsq.fr,
[2] INRIA, F-78153 Le Chesnay, France, Paul.Feautrier@inria.fr,
[3] Université de Lille I, École Polytech. Univ. de Lille & Laboratoire LIFL, F-59655 Villeneuve d'Ascq, France, Xavier.Redon@eudil.fr,

Abstract. This paper deals with the problem of deciding whether two Systems of Affine Recurrence Equations are equivalent or not. A solution to this problem would be a step toward algorithm recognition, an important tool in program analysis, optimization and parallelization. We first prove that in the general case, the problem is undecidable. We then show that there nevertheless exists a semi-decision procedure, in which the key ingredient is the computation of transitive closures of affine relations. This is a non-effective process which has been extensively studied. Many partial solutions are known. We then report on a pilot implementation of the algorithm, describe its limitations, and point to unsolved problems.

1 Introduction

1.1 Motivation

Algorithm recognition is an old problem in computer science. Basically, one would like to submit a piece of code to an analyzer, and get answers like "Lines 10 to 23 are an implementation of Gaussian elimination". Such a facility would enable many important techniques: program comprehension and reverse engineering, program verification, program optimization and parallelization, hardware-software codesign among others.

Simple cases of algorithm recognition have already been solved, mostly using pattern matching as the basic technique. An example is reduction recognition, which is included in many parallelizing compilers. A reduction is the application of an associative commutative operator to a data set. See [9] and its references. This approach has been recently extended to more complicated patterns by several researchers (see the recent book by Metzger [8] and its references).

In this paper, we wish to explore another approach. We are given a library of algorithms. Let us try to devise a method for testing whether a part of the source

B. Monien and R. Feldmann (Eds.): Euro-Par 2002, LNCS 2400, pp. 309–313.

program is equivalent to one of the algorithms in the library. The stumbling block is that in the general case, the equivalence of two programs is undecidable. Our aim is therefore to find sub-cases for which the equivalence problem is solvable, and to insure that these cases cover as much ground as possible.

The first step is to normalize the given program as much as possible. One candidate for such a normalization is conversion to a System of Affine Recurrence Equations (SARE)[3]. It has been shown that *static control programs* [4] can be automatically converted to SAREs. The next step is to design an equivalence test for SAREs. This is the main theme of this paper.

1.2 Equivalence of Two SAREs

Suppose we are given two SAREs with their input and output variables. Suppose furthermore that we are given a bijection between the input variables of the two SAREs, and also a bijection between the output variables. In what follows, two corresponding input or output variables are usually denoted by the same letter, one of them being accented.

The two SAREs are equivalent with respect to a pair of output variables, iff the outputs evaluate to the same values provided that the input variables are equal. In order to avoid difficulties with non-terminating computations, we will assume that both SAREs have a schedule.

The equivalence of two SAREs depends clearly on the domain of values used in the computation. In this preliminary work, we will suppose that values belong to the Herbrand universe (or the initial algebra) of the operators occurring in the computation. The Herbrand universe is characterized by the following property:

$$\omega(t_1,\ldots,t_n) = \omega'(t'_1,\ldots,t'_{n'}) \Leftrightarrow \omega = \omega', n = n' \text{ and } t_i = t'_i, i = 1\ldots n. \quad (1)$$

where ω and ω' are operators and $t_1,\ldots,t_n$, $t'_1,\ldots,t'_{n'}$ are arbitrary terms. The general case is left for future work.

It can be proved that, even in the Herbrand universe, the equivalence of two SAREs is undecidable. The proof is rather technical and can be found in [1]. In Sect. 2 we define and prove a semi-decision procedure which may prove or disprove the equivalence of two SAREs, or fails. In Sect. 3 we report on a pilot implementation of the semi-decision procedure. We then conclude and discuss future work.

2 A Semi-decision Procedure

From the above result, we know that any algorithm for testing the equivalence of two SAREs is bound to be incomplete. It may give a positive or negative answer, or fail without reaching a decision. Such a procedure may nevertheless be useful, provided the third case does not occur too often. We are now going to design such a semi-decision procedure. To each pair of SAREs we will associate a memory state automaton (MSA) [2] in such a way that the equivalence of our SAREs can be expressed as problems of reachability in the corresponding MSA. Let us consider the two parametric SAREs (with parameter n):

$$O[i] = 1, \quad i = 0, \qquad = f(I[i]), \; 1 \le i \le n, \tag{2}$$

$$\begin{aligned} O'[i'] &= 1, & i' &= 0, \\ &= f(X'[i', n]), & 1 &\le i' \le n, \\ X'[i', j'] &= I'[i'], & 0 &\le i' \le n, j' = 0, \\ &= X'[i', j' - 1], & 0 &\le i' \le n, 1 \le j' \le n. \end{aligned} \tag{3}$$

The reader familiar with systolic array design may have recognized a much simplified version of a transformation known as pipelining or uniformization, whose aim is to simplify the interconnection pattern of the array.

The equivalence MSA is represented by the following drawing. Basically, MSA are finite state automata, where each state is augmented by an *index vector*. Each edge is labelled by a *firing relation*, which must be satisfied by the index vector for the edge to be traversed.

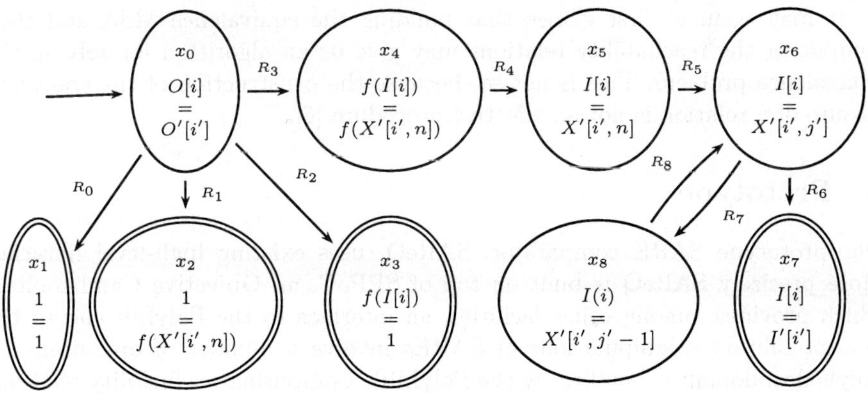

The automaton is constructed on demand from the initial state $O[i] = O'[i']$, expressing the fact that the two SAREs have the same output. Other states are equations between subexpressions of the left and right SARE. The transitions are built according to the following rules: If the lhs of a state is $X[u(i_x)]$, it can be replaced in its successors by $X[i_y]$, provided the firing relation includes the predicate $i_y = u(i_x)$ (R_8). If the lhs is $X[i_x]$ where X is defined by n clauses $X[i] = \omega_k(\ldots Y[u_Y(i)]\ldots), i \in D_k$ then it can be replaced in its n successors by $\omega_k(\ldots Y[u_Y(i_y)]\ldots)$ provided the firing relation includes $\{i_x \in D_k, i_y = i_x\}$ ($R_0, \ldots, R_3$ and R_6, R_7). There are similar rules for the rhs. Note that equations of the successor states are obtained by simultaneous application of rules for lhs and rhs. Moreover, the successors of a state with equation $\omega(\ldots) = \omega(\ldots)$ are states with equations between the parameters of the function ω. The firing relation is in this case the identity relation (R_4). For instance, R_3 and R_8 are:

$$R_3 = \left\{ \begin{bmatrix} i_{x0} \\ i'_{x0} \end{bmatrix} \to \begin{bmatrix} i_{x4} \\ i'_{x4} \end{bmatrix}, \left\{ \begin{array}{l} i_{x4} = i_{x0} \\ i'_{x4} = i'_{x0} \\ 1 \le i_{x0} \le n \\ 1 \le i'_{x0} \le n \end{array} \right\} \right\}, R_8 = \left\{ \begin{bmatrix} i_{x8} \\ i'_{x8} \\ j'_{x8} \end{bmatrix} \to \begin{bmatrix} i_{x6} \\ i'_{x6} \\ j'_{x6} \end{bmatrix}, \left\{ \begin{array}{l} i_{x6} = i_{x8} \\ i'_{x6} = i'_{x8} \\ j'_{x6} = j'_{x8} - 1 \end{array} \right\} \right\}.$$

States with no successors are final states. If the equation of a final state is always true, then this is a success (x_1, x_7), otherwise this is a failure state (x_2, x_3). The access path from the initial state x_0 to the failure state x_2 is $R_{x_2} = R_1$ and to x_7 is $R_{x_7} = R_3.R_4.R_5.(R_7.R_8)^*.R_6$. When actual relations are substituted to letters, the reachability relations of these states are:

$$R_{x_2} = \left\{ \begin{bmatrix} i_{x0} \\ i'_{x0} \end{bmatrix} \to \begin{bmatrix} i_{x2} \\ i'_{x2} \end{bmatrix}, \left\{ \begin{array}{c} i'_{x_2} = i'_{x0} \\ i_{x0} = 0 \\ i_{x_2} = 0 \\ 1 \le i'_{x_2} \le n \end{array} \right\} \right\}, R_{x7} \left\{ \begin{bmatrix} i_{x0} \\ i'_{x0} \end{bmatrix} \to \begin{bmatrix} i_{x7} \\ i'_{x7} \end{bmatrix}, \left\{ \begin{array}{c} i_{x7} = i_{x0} \\ i'_{x7} = i'_{x0} \\ 1 \le i_{x0} \le n \\ 1 \le i'_{x0} \le n \end{array} \right\} \right\}.$$

Theorem 1. *Two SAREs are equivalent for outputs O and O' iff the equivalence MSA with initial state $O[i] = O'[i']$ is such that all failure states are unreachable and the reachability relation of each success state is included in the identity relation.*

In our example, reachability relations of success states are actually included in the main diagonal (obviously true for R_{x_7} since $i_{x_0} = i'_{x0}$ implies $i_{x_7} = i'_{x_7}$) and it can be shown that the relations for the failure states are empty (verified for R_{x_2} since $i_{x_0} = i'_{x0}$ implies $1 \le 0$). Hence, the two SAREs are equivalent.

It may seem at first glance that building the equivalence MSA and then computing the reachability relations may give us an algorithm for solving the equivalence problem. This is not so, because the construction of the transitive closure of a relation is not an effective procedure [6].

3 Prototype

Our prototype SARE comparator, SAReQ, uses existing high-level libraries. More precisely SAReQ is built on top of SPPoC, an Objective Caml toolbox which provides, among other facilities, an interface to the PolyLib and to the Omega Library. Manipulations of SAREs involve a number of operations on polyhedral domains (handled by the PolyLib). Computing reachability relations of final states boils down to operations such as composition, union and transitive closure on relations (handled by the Omega Library).

The SAREs are parsed using the camlp4 preprocessor for OCaml, the syntax used is patterned after the language Alpha [7]. We give below the text of the two SAREs of section 2 as expected by SAReQ:

```
pipe [n] {
  O[i] = { { i=0 } : 1 ;
           { 1<=i<=n } : f(I[i]) }
}
```

```
pipe' [n] {
  X'[i',j'] = { { 0<=i'<=n, j'=0 } : I[i'] ;
                { 0<=i'<=n, 1<=j'<=n } : X'[i', j'-1]
              } ;
  O'[i'] = { { i'=0 } : 1 ;
             { 1<=i'<=n } : f(X'[i',n]) }
}
```

To make the program more friendly a WEB interface is available at the URL `http://sareq.eudil.fr`. This interface gives access to a library of examples, allows the testing of new problems and presents the results in a readable way.

4 Conclusions and Future Work

We believe that our SARE comparator has about the same analytic power as most automatic parallelization tools. It can handle only affine array subscripts and affine loop bounds. Comparison with the work of Metzger et. al. [8] is difficult, since we do not have access to an implementation. We believe our normal form is more powerful than theirs, since we can upgrade an array to arbitrary dimension, while they are limited to scalar expansion. Also, it does not seem that they can deal with most loop modifications (interchange, skewing, index set splitting) and are limited to loop distibution. On the other hand, provided the program give them the necessary clues, they can handle some forms of associativity and commutativity.

We believe that the most important problem with the present tool is the fact that it cannot use semantical information on the underlying operators. We would like to specify a semantics by a set of simplification rules or algorithms. In the present prototype, the possibility of applying simplifications is limited, since computation rules are never combined. One suggestion is to add "forward" substitution rules. However, we still have to find a heuristics for driving the substitution process.

The present tool is just a building block in a complete program comparator. In the first place, we have to connect it to an array dataflow analyzer ([4], [5]). Secondly, we must build a library of reference algorithms, and this will depends on the application domain. Lastly, many source programs are built by composition from several reference algorithms. Our tool can only be applied if we have delineated the several components, and if we have identified inputs and outputs. At the time of writing, we believe that this has to be handled by heuristics, but this can only be verified by experiments.

References

1. D. Barthou, P. Feautrier, and X. Redon. On the equivalence of two systems of affine recurrence equations. Technical Report RR-4285, INRIA, Oct. 2001.
2. B. Boigelot and P. Wolper. Symbolic verification with periodic sets. In *Proceedings of the 6th International Conference on Computer-Aided Verification*, volume 818 of *Lecture Notes in Computer Science*, pages 55–67. Springer-Verlag, 1994.
3. A. Darte, Y. Robert, and F. Vivien. *Scheduling and automatic Parallelization.* Birkhäuser, 2000.
4. P. Feautrier. Dataflow analysis of scalar and array references. *Int. J. of Parallel Programming*, 20(1):23–53, Feb. 1991.
5. M. Griebl and C. Lengauer. The loop parallelizer LooPo – Announcement. In *9th Languages and Compilers for Parallel Computing Workshop.* Springer, LNCS 1239, 1996. http://www.fmi.uni-passau.de/cl/loopo.
6. W. Kelly, W. Pugh, E. Rosser, and T. Shpeisman. Transitive closure of infinite graphs and its applications. *Int. J. of Parallel Programming*, 24(6):579–598, 1996.
7. H. Leverge, C. Mauras, and P. Quinton. The ALPHA language and its use for the design of systolic arrays. *Journal of VLSI Signal Processing*, 3:173–182, 1991.
8. R. Metzger and Z. Wen. *Automatic Algorithm Recognition: A New Approach to Program Optimization.* MIT Press, 2000.
9. X. Redon and P. Feautrier. Detection of scans in the polytope model. *Parallel Algorithms and Applications*, 15:229–263, 2000.

Towards High-Level Specification, Synthesis, and Virtualization of Programmable Logic Designs

Oliver Diessel, Usama Malik, and Keith So

School of Computer Science & Engineering
University of New South Wales
UNSW Sydney NSW 2052
Australia

Abstract. Current FPGA design flows do not readily support high-level, behavioural design or the use of run-time reconfiguration. Designers are thus discouraged from taking a high-level view of their systems and cannot fully exploit the benefits of programmable hardware. This paper reports on our advances towards the development of design technology that supports behavioural specification and compilation of FPGA designs and automatically manages FPGA chip virtualization.

1 Introduction

A significant barrier to the wider adoption of reconfigurable computing is the problem of mapping applications into circuit structures that can easily be implemented on a given FPGA device. Ideally we should be able to describe the desired functionality of hardware or its components, and have a compiler map these specifications into effective logic allocations and reconfiguration schedules. Being oriented towards static hardware configurations, current FPGA design flows do not support such dynamic circuit design and configuration.

Our research aims to identify the key techniques and principles that underlie the design of suitable languages and compilers for the high-level design and implementation of run-time reconfigurable applications. We are currently investigating how to model and express the parallelism inherent in FPGA circuits and how to implicitly control run-time reconfiguration in an abstract and machine independent manner.

For a number of reasons, we have so far focused on the use of a process algebra as the high-level specification language [7]. Process algebras (PAs) are simple yet powerful formalisms in which it is easier to explore fundamental language issues than with hardware description languages and programming languages. PAs are well-suited to the behavioural description of interacting finite state machines, and as such can be used to model control-parallel and systolic FPGA applications. Furthermore, there is the hope that a top-down, hierarchical, and modular focus, as emphasized by a PA such as Circal, will aid logic synthesis because ever more complex structures may be built through assembly, while the effort required to design each module remains relatively constant.

B. Monien and R. Feldmann (Eds.): Euro-Par 2002, LNCS 2400, pp. 314–317.

Our approach differs from other efforts to develop high-level programming languages for reconfigurable computing, which commonly augment sequential languages such as C, C++, and Java with support for data-parallel operations and hardware layouts [1, 6, 2, 9]. While such languages have been sucessfully used to design efficient applications, they emphasize a signal-oriented view of computation, do not allow concurrency to be expressed naturally, and they do not attempt to target the run-time reconfigurable capabilities of the hardware. Our goal is to model the capabilities of the hardware in order to have a well-understood target for compilers that can exploit those capabilities and languages that allow them to be expressed.

2 An FPGA Interpreter for Circal

In [3] we described a compiler that derives and implements a digital logic representation of high-level behavioural descriptions of systems specified using Circal. At the topmost design level, the circuit is clustered into blocks of logic that correspond to the processes (individual finite state machines) of a system. The process logic blocks implement circuits with behaviours corresponding to the component processes of the specification. Below the process level in the hierarchy, the circuits are partitioned into component circuit modules that implement logic functions of minor complexity. The circuit modules are rectangular in shape and are laid out onto abutting regions of the array surface, allowing signals to flow from one module to another via aligned ports.

The current implementation of the compiler targets the Virtex chipset [10], which is substantially more coarse-grained, operates at considerably higher frequencies, and is availble in much greater logic densities than the original Xilinx XC6200 target. Module placement follows a decomposition approach similar to that of the XC6200-based compiler, albeit with an arrangement that utilizes the fast logic cascade chains and the enriched routing fabric available on Virtex, while taking into account that partial reconfiguration is column-oriented.

2.1 Developing Support for Run-Time Deployment of Circal Models

The compiler may produce circuits that cannot be implemented because they are too large for the available FPGA resource. We have therefore developed support for automatically partitioning the circuit and swapping the resulting partitions with those on chip so as to give the effect of having enough FPGA area to implement circuits of any size.

The approach we are exploring is to insert a virtual hardware manager (VHM) between the front-end of the compiler, which derives a hardware independent representation of the circuit in a modularized form, and the back-end, which maps and places these modules onto a particular FPGA. In doing so, we preserve some off-line features of a compiler, and incorporate some of the on-line functions of an interpreter. The VHM stores the circuits off-chip in a state transition graph form. When a new partition is needed, the module parameters for the corresponding sub-graph are passed to the back-end for bitstream generation and loading. The back-end makes use of the JBits API for configuring the device [4].

The FPGA area is partitioned into regions that are reserved for holding the logic for a single porcess. The area provided is large enough to implement the logic for a single state in the worst case. We attempt to provide sufficient space to map a sizable portion of the state transition graph by expanding the area allocated to each process. At initialization, the VHM selects a sub-graph rooted at the initial state for each process. The graph is traversed in a breadth-first manner in order to identify the amount of logic that can be accommodated on chip. To avoid backtracking, the search is guided by circuit size estimates based on the number of transitions each included state adds to the sub-graph. The identified sub-graph is then given to the back-end which maps it to the FPGA. When a state transition leads to a state that lies on the boundary of the implemented sub-graph, an exception is generated and the VHM selects a new sub-graph rooted at the boundary state.

2.2 Example

Figure 1(a) depicts a cruise control system for an automobile. The system consists of a Cruise Controller and a Speed Controller. Inputs are provided by sensors not shown in the diagram. In Figure 1(b) we model the Cruise Controller process. Suppose the area avilable to this process suffices to implement the complete transition logic for 2 states together with any outedges from these leading to a boundary state. In that case, the initial partition includes the *Inactive* and *Active* states, as well as the state flip-flop for the *Cruising* state. When the *Cruising* state is entered, an exception is triggered, and a new sub-graph consisting of the *Cruising* and *Standby* states is configured, together with a state flip-flop for the *Inactive* state.

The Virtex compiler needs 7 columns and 5 rows of configurable logic blocks (CLBs) to implement the complete state graph. The interpreter allows the first partition to be implemented in 4 columns and 4 rows and the second partition to be implemented in 5 columns and 4 rows. Reconfiguring the 5 columns takes

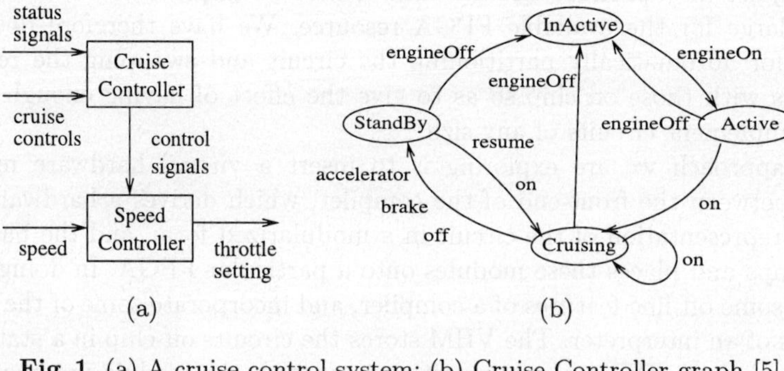

Fig. 1. (a) A cruise control system; (b) Cruise Controller graph [5]

about 1.25ms on an XCV1000 chip, which contains 64 rows and 96 columns of CLBs altogether. The time to generate the partition is likely to be larger.

Reconfiguration delays are expected to be amortized in practice by aligning processes on top of each other. The stack of processes could thus be reconfigured with a single frame update. In order to reduce the cost of generating the partitions, they might be cached once generated in case they are to be used once more.

3 Future Work

In its current form, Circal is suited to the specification of control-flow applications like logic controllers, protocols checkers, and cellular automata. In order to fully exploit the capabilities of modern high density FPGAs, support for datapath designs must be provided by Circal. One approach we intend to pursue is to consider building a domain specific language into which the control features developed with Circal are embedded.

References

1. M. Aubury, I. Page, G. Randall, J. Saul, and R. Watts. Handel-C *Language Reference Guide.* Oxford Univerity Computing Laboratory, Oxford, UK, Aug. 1996.
2. P. Bellows and B. Hutchings. JHDL — An HDL for reconfigurable systems. In Pocek and Arnold [8], pages 175 – 184.
3. O. Diessel and G. Milne. A hardware compiler realizing concurrent processes in reconfigurable logic. *IEE Proceedings — Computers and Digital Techniques*, 148(4):152 – 162, Sept. 2001.
4. S. A. Guccione and D. Levi. XBI: A java-based interface to FPGA hardware. In J. Schewel, editor, *Configurable Computing: Technology and Applications, Proc. SPIE 3526*, pages 97–102. SPIE – The International Society for Optical Engineering, Nov. 1998.
5. J. Magee and J. Kramer. *Concurrency: State Models & Java Programs.* Worldwide Series in Computer Science. John Wiley & Sons, New York, NY, 1999.
6. O. Mencer, M. Morf, and M. J. Flynn. PAM–Blox: High performance FPGA design for adaptive computing. In Pocek and Arnold [8], pages 167 – 174.
7. G. J. Milne. CIRCAL and the representation of communication, concurrency, and time. *ACM Transactions on Programming Languages and Systems*, 7(2):270 – 298, Apr. 1995.
8. K. L. Pocek and J. M. Arnold, editors. *The 6th Annual IEEE Symposium on FPGAs for Custom Computing Machines (FCCM'98)*, Los Alamitos, CA, Apr. 1998. IEEE Computer Society Press.
9. G. Snider, B. Shackleford, and R. J. Carter. Attacking the semantic gap between application programming languages and configurable hardware. In *FPGA'01 Ninth International Symposium on Field Programmable Gate Arrays*, pages 115 – 124, New York, NY, Feb. 2001. ACM Press.
10. Xilinx. *Virtex 2.5V Field Programmable Gate Arrays.* Xilinx, Inc., Oct. 2000. Version 1.3.

Topic 5
Parallel and Distributed Databases, Data Mining and Knowledge Discovery

Harald Kosch, David Skilicorn, and Domenico Talia

Topic Chairpersons

We would like to welcome you to Paderborn and to the Europar 2002 topic on Parallel and Distributed Databases, Data Mining and Knowledge Discovery.

Current research and applications in parallel and distributed database systems are stimulated both by advances in technology (e.g., parallel architectures, high performance networks) and by the requirements of new applications (e.g., multimedia, medicine). These applications are mostly data-hungry and are running on very large databases with the goal of extracting information diamonds. Data mining is one of the key techniques here. However, these intensive data-consuming applications suffer from performance problems and and bottlenecks caused by single sources of data. Introducing data distribution and parallel processing helps to overcome these resource problems and to achieve guaranteed throughput, quality of service, and system scalability.

This year, 13 papers were submitted, more than in previous years. The range and quality of the submitted papers was impressive and reflects advances in technology, as well as describing solutions for application-specific problems. Each paper was reviewed by at least three reviewers and, in the end, we were able to select 6 regular papers and 3 short ones. This shows the high-quality of the submissions. From this, three full sessions have been scheduled. The first one is dedicated to aspects of parallel and distributed database systems, the second one deals with strategies for parallel data mining, and the final third one contains interesting papers on data-grid applications and distribution aspects in database systems.

Parallel databases are currently making a shift from relational query processing and traditional transaction management towards the handling of new data-hungry applications, like multimedia, introducing new operators to query processing. The first paper in the session, "Dynamic Query Scheduling in Parallel Data Warehouses", by H. Märtens, E. Rahm and T. Stöhr introduces a new skew-aware query scheduling strategy for shared-disk parallel systems which considers both disks and processors. The tested queries stem from data warehouse applications. The second paper, "Speeding Up Navigational Requests in a Parallel Object Database System" by J. Smith, P. Watson, S. Sampaio and N. Paton is dedicated to query processing in object-oriented parallel databases. This is, in our opinion, the first framework which completely covers parallelization of so-called path queries including functions. Finally, the third paper, "Retrieval of Multispectral Satellite Imagery on Cluster Architectures", by T. Bretschneider and O. Kao falls into parallel image retrieval. It reflects the need for parallel

B. Monien and R. Feldmann (Eds.): Euro-Par 2002, LNCS 2400, pp. 319–320.

query processing in images, and multimedia databases in general, and parallelizes efficiently similarity-based operators on the image database systems.

The need to analyze large amounts of data is the main motivation for the implementation of parallel data mining algorithms. This trend also benefits from the wider availability of cost-effective parallel machines such as clusters and SMPs. Session 2 includes three papers that discuss parallel data mining algorithms and systems. The first paper "Shared Memory Parallelization of Decision Tree Construction Using a General Data Mining Middleware" by Jin and Agrawal, describes the use of a data-mining framework for developing shared-memory parallel implementations of classifiers based on decision trees. Experiments presented in the paper show that applying a set of techniques results in good performance. The second paper "Characterizing the Scalability of Decision-Support Workloads on Clusters and SMP Systems" by Zhang, Sivasubramaniam, Nagar, and Franke, discusses the scalability of the TPC-H decision support benchmark on a cluster and on an SMP machine. The paper focus is on the impact of various hardware parameters such as CPU, memory, disk and network. The evaluation results show that for both the cluster and the SMP environments, the CPU and memory resources are not major contributors to performance beyond a point, but the I/O parallelism. The last paper "A Parallel Learning Algorithm for Text Classification on PIRUN Beowulf Cluster" by Kruengkrai and Jaruskulch, presents a parallel learning algorithm for text classification based on the combination of the expectation-maximization algorithm and the naive Bayes. The preliminary experimental results discussed in the paper show that the proposed parallel implementation in the mining of up to 10000 documents has reasonable speedup.

The growing size of online data, and the fact that it is often distributed, create new challenges for data-intensive applications. The application papers in Session 3 illustrate this trend. The paper by Orlando *et al.* "Scheduling High Performance Data Mining Tasks on a Data Grid Environment" explores issues of scheduling distributed data mining algorithms using cost information. Unlike conventional grid resource discovery, this paper shows how to use samples to determine the likely behavior of computing resources. The second paper, by Kwok *et al.*, "Parallel Fuzzy c-Means Clustering for Large Data Sets" illustrates the use of parallelism in data mining. Clustering data is a major application. This paper goes beyond the standard k-means approach to clustering to parallelize the fuzzy c-means algorithm using MPI. Finally, the paper by Boukerche and Tuck "A delayed-Initiation Risk-Free Multiversion temporally correct algorithms" present a general algorithm for risk-free multiversion concurrency, extending work presented at Europar in 2001.

In closing, we would like to thank the authors who submitted a contribution, as well as the Europar Organizing Committee, and the referees with there highly useful comments, whose efforts have made this conference, and Topic 05 possible.

Dynamic Query Scheduling in Parallel Data Warehouses

Holger Märtens, Erhard Rahm, and Thomas Stöhr

University of Leipzig, Germany
{maertens|rahm|stoehr}@informatik.uni-leipzig.de

Abstract. Data warehouse queries pose challenging performance problems that often necessitate the use of parallel database systems (PDBS). Although dynamic load balancing is of key importance in PDBS, to our knowledge it has not yet been investigated thoroughly for parallel data warehouses. In this study, we propose a scheduling strategy that simultaneously considers both processors and disks while utilizing the load balancing potential of a Shared Disk architecture. We compare the performance of this new method to several other approaches in a comprehensive simulation study, incorporating skew aspects and typical data warehouse features such as star schemas.

1 Introduction

A successful data warehouse must ensure acceptable response times for complex analytical queries. Along with measures such as new query operators [8], specialized index structures [13, 19], intelligent data allocation [18], and materialized views [3], *parallel database systems (PDBS)* are used to provide high performance [5]. For effective parallelism, good load balancing is a must, and many algorithms have been proposed for general PDBS. But we are not aware of load balancing studies for data warehouses with characteristic features such as star schemas and bitmap indices.

In this paper, we evaluate a new approach to *dynamic load balancing* in parallel data warehouses based on the simultaneous consideration of both CPUs and disks. These are frequent bottlenecks in the voluminous scan/aggregation queries characteristic of data warehouses. A balanced utilization of both resources depends not only on the *location* (on which CPU) but also on the *timing* of load units such as subqueries. We thus propose to perform both decisions in an integrated manner based on the resource requirements of queued subqueries as well as the current system state.

To this end, we exploit the flexibility of the *Shared Disk (SD)* architecture [16] in which each processing node can execute any subquery. For scan workloads, the balance of CPU load does not depend on the data allocation, permitting query scheduling with shared job queues for all nodes. Disk contention is harder to control because the total load per disk is predetermined by the data allocation and cannot be shifted at runtime.

In a detailed simulation study, we compare the new integrated strategy to several simpler methods of dynamic query scheduling. We use a data warehouse setting based on the APB-1 benchmark comprising a star schema with a huge fact table supported by bitmap indices, both declustered across many disks for parallel access. The large

B. Monien and R. Feldmann (Eds.): Euro-Par 2002, LNCS 2400, pp. 321–331.

scan/aggregation queries we regard stress both disks and CPUs, creating a challenging scheduling problem. We particularly consider the often neglected but performance-critical treatment of *skew effects*. As a first step in the field, we focus on single-user mode, but our scheduling approaches can also be applied in multi-user mode.

In Section 2 of this paper, we briefly review some related work. Section 3 outlines our general load balancing paradigm, whereas our specific scheduling heuristics are defined in Section 4. Section 5 describes our simulation system and presents the performance evaluation of the scheduling strategies. We conclude in Section 6. Details omitted due to space constraints can be found in an extended version of this paper [11].

2 Related Work

We are not aware of any load balancing studies for parallel data warehouses. For general PDBS, load balancing problems have been widely researched, for a variety of workloads and architectures [4, 6, 9, 10, 16]. Many of these approaches rely on extensive data redistribution too costly in a large data warehouse. Furthermore, most previous studies have been limited to balancing CPU load, sometimes including main memory [14]. Even so, the need for dynamic scheduling has been emphasized [2, 14]. Conversely, load distribution on disks has largely been considered in isolation from CPU-side processing. Most of these studies have focused either on data partitioning and allocation [7, 15, 17] or on limiting disk contention through reduced parallelism [16]. Integrated load balancing as proposed in this paper has not been addressed.

The Shared Disk architecture has been advocated due to its superior load balancing potential especially for read-only workloads as in data warehouses [9, 12, 16]. It also offers great freedom in data allocation [15]. But the research on how to exploit this potential is still incomplete. SD is also supported by some commercial PDBS from *IBM*, *ORACLE*, and *SYBASE*. These and other data warehouse products (e.g., *INFORMIX*, *RED BRICK*, *MICROSOFT*, and *TERADATA*) support star schemas and (mostly) bitmap indices as well as adequate data fragmentation and parallel processing. But since no documentation is available on disk-sensitive scheduling methods, we believe that dynamic disk load balancing is not yet supported in current products.

3 Dynamic Load Balancing for Parallel Scan Processing

This section presents our basic approach to dynamic load balancing, which is not restricted to data warehouse environments. We presume a horizontal partitioning of relational tables into disjoint *fragments*. If bitmap indices or similar access structures exist, they must be partitioned analogously so that each table fragment with its corresponding bitmap fragments can form an independent unit of processing. We focus on the optimization of scan queries and exploit the flexibility of the Shared Disk architecture.

The two performance-critical types of resources for scans are processing nodes and disks, but their respective load balance depends on different conditions: CPU utilization is largely determined by *how much* data each processor is assigned. A balanced disk load, on the other hand, hinges on *when* the data residing on each device are processed because their location is fixed. We thus aim for an integrated view on both resources.

When a query enters the system, a *coordinator* node that controls its execution partitions the query into subqueries based on the presumed horizontal fragmentation. Each subquery scans either a fragment or a *partition* of the relevant table, where a partition comprises all table fragments residing on one disk. Fragments known to contain no hit rows are excluded. We thus obtain independent subqueries that can be processed on any processing node, yielding great flexibility in the subsequent scheduling step.

Fragments, being smaller than partitions, permit a more even load distribution especially in case of skew. Partition-sized subqueries, however, reduce the scheduling and communication overhead as well as disk contention as no two subqueries will process the same table partition, although some interference may still stem from index access.

Scheduling. Presuming full parallelism for the large queries we examine, we are left with the task of allocating subqueries to processors and timing their execution. We consider this *scheduling* step particularly important as it finalizes the actual load distribution in the system. To this end, the coordinator maintains a list of subqueries that are dispatched following a given ordering policy (cf. Section 4) and processed locally as described below. All processors obtain the same number of subqueries (±1) up to a given limit roughly corresponding to the performance ratio of CPUs to disks; remaining tasks are kept in a central queue. When a processor finishes a subquery and reports the local result to the coordinator, it is assigned new work from the queue until all subqueries are done. Finally, the coordinator returns the overall query result to the user.

This simple, highly dynamic approach already provides a good balance of processor load. A node that has been assigned a long-running subquery will automatically obtain less load as execution progresses, thus nearly equalizing CPU load. Since no two subqueries address the same fragment, we may also achieve low disk contention depending on the order in which subqueries are dispatched. This aspect is elaborated in Section 4.

Local Processing of Subqueries. When a node is assigned a fragment-sized subquery, it processes any required bitmap fragments and the respective table fragment simultaneously, minimizing memory consumption while exploiting prefetching and parallel I/O. For the scan/aggregation queries we assume, the measures contained in the selected tuples are aggregated locally to avoid a shipping of large datasets, and the partial results are returned to the coordinator at subquery termination. For partition-sized subqueries, a node will process its partition sequentially, skipping irrelevant fragments. Multiple subqueries on the same processor coexist without any need for intra-node coordination.

4 Scheduling Order of Subquery Execution

Since we regard the scheduling of subquery execution as the most important aspect of load balancing in our processing model, we now present four scheduling policies based on either static (Section 4.1) or dynamic (Section 4.2) ordering of subqueries. Detailed calculations and some variant strategies can be found in the extended paper [11].

4.1 Statically Ordered Scheduling

Our simpler heuristics employ a *static ordering* of subqueries. Even under these strategies, however, our scheduling scheme as such is still dynamic as the allocation of workload to processing nodes is determined at runtime based on the progress of execution.

Strategy *LOGICAL*. This heuristic – taken from our previous study on star schema allocation [18] as a baseline reference – assigns fragment-sized subqueries in the logical or- der of the fragments they refer to. Since the allocation scheme applied here does not maintain this order (cf. Section 5.1), *LOGICAL* will not yield optimal performance.

Strategy *PARTITION*. Partition-sized subqueries are dispatched in a round-robin fashion with respect to their logical disk numbers. In single-user mode, this means that each table partition is accessed by only one processor at a time. However, bitmap access (if required) can cause each subquery to read from multiple disks, so that access conflicts may not be avoided completely. Still, we expect this policy to minimize disk contention.

Strategy *SIZE*. This method starts fragment-sized subqueries in decreasing order of size, based on the expected number of referenced pages. It implements an LPT (*longest processing time first*) scheme that provides good load balancing for many scheduling problems. It does not consider disk allocation but may be expected to optimize the balance of processor load based on the total amount of data processed per node.

4.2 Dynamically Ordered Scheduling

The static policies above tend to optimize the balance of *either* CPU *or* disk load. For an improved, integrated load balancing we reckon with both criteria based on a *dynamic ordering*. To distribute disk load over time and reduce contention, we try to execute concurrently subqueries with minimum overlap in disk access. To simultaneously balance CPU load, we also consider subquery sizes similar to the previous section. Figure 1 illustrates the following considerations using a 4-disk example with 4 subqueries.

Strategy *INTEGRATED*. We model the disk load characteristics of each subquery in the shape of a *load vector* ① containing the expected number of pages referenced on each disk. This number is calculated from the query's estimated selectivity and

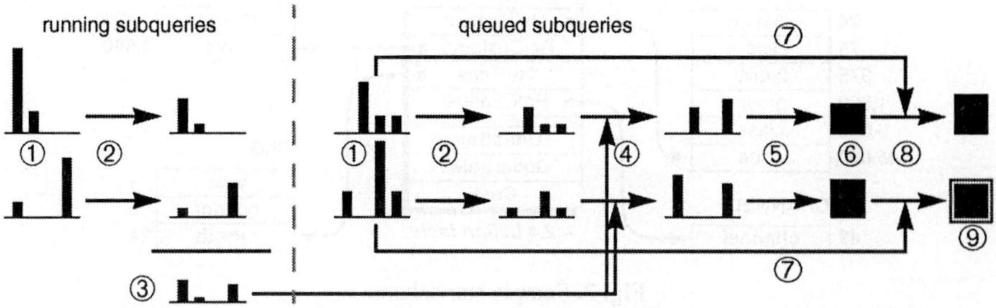

Fig. 1. Sequence of load vector calculation in strategy *Integrated* (graph scaling varies).

includes both table and bitmap fragments. The load vector is normalized ② to represent the relative load distribution across the disks *at a given point in time* rather than its total magnitude.

In addition to the single load vectors for each subquery, we keep a global vector of current disk load, defined as the sum of the load vectors of all subqueries currently running ③. We can then compute an expected rate of access conflict between the current load and any queued subquery by comparing their respective load vectors. Specifically, the products of local intensities per disk ④, added over all disks ⑤, yield a measure of the total access conflict between each candidate and the current load ⑥. To integrate disk conflict estimates with the distribution of CPU load, we divide the expected disk access conflict for each subquery by its total size ⑦⑧, so that long-running tasks may be executed earlier than shorter ones even if they incur a slight increase in disk contention. The subquery that minimizes the resulting ratio ⑨ (thus optimizing the trade-off between both criteria) will be dispatched in the next scheduling step.

5 Simulation Study

We now present our simulation study, first introducing the simulation system used (Section 5.1), then discussing the performance of our scheduling schemes (Section 5.2), and finally testing the scalability of our methods in speed-up experiments (Section 5.3).

5.1 Simulation System and Setup

Our proposed strategies were implemented in a comprehensive simulation system for parallel data warehouses that has been used successfully in previous studies [18]. Simulating a Shared Disk PDBS with 20 processors and 100 disks, it realistically reflects resource contention by modeling both CPUs and disks as servers. CPU overhead is reckoned for all relevant operations, and seek times in the disk modules depend on the location (track number) of the desired data within a disk. Each processor owns a buffer module with separate LRU queues for fact table and bitmap

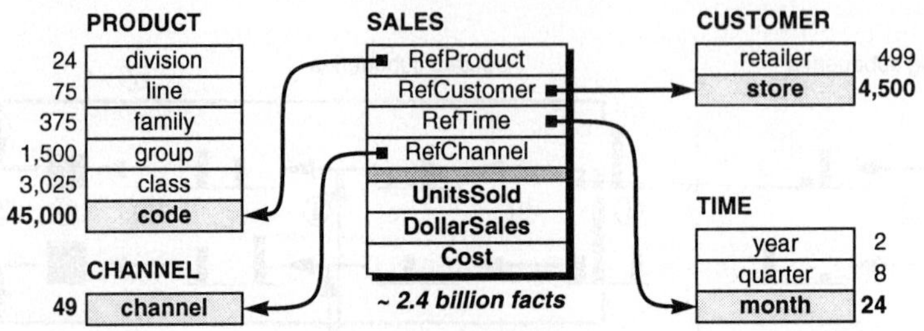

Fig. 2. Sample star schema.

access. The network incurs communication delays proportional to message sizes but models no contention, so as to avoid specific network topologies unduly influencing experimental results.

Our sample data warehouse is modeled as a relational star schema for a sales analysis environment (Figure 2) derived from the *APPLICATION PROCESSING BENCHMARK (APB- 1)* [1]. The denormalized *dimension tables PRODUCT*, *CUSTOMER*, *CHANNEL* and *TIME* each define a *hierarchy* (such as product divisions, lines, families, and so on). The *fact table SALES* comprises several *measure* attributes (turnover, cost etc.) and a foreign key to each dimension. With a *density factor* of 1%, it contains a tuple for 1/100 of all value combinations. We incorporate common *bitmap join indices* [13] to avoid costly full scans of the fact table. We employ *standard bitmaps* for the low-cardinality dimensions *TIME* and *CHANNEL*, but use *hierarchically encoded bitmaps* [19] for the more voluminous dimensions *PRODUCT* and *CUSTOMER* to save disk space and I/O. With these indices, queries can avoid explicit join processing between fact table and dimension table(s) in favor of a simple selection using the respective precomputed bitmap(s).

We follow a horizontal, multi-dimensional fragmentation strategy for star schemas that we proposed and evaluated in [18]. Specifically, we choose a two-dimensional fragmentation based on *TIME.MONTH* and *PRODUCT.FAMILY*. Each resulting fact table fragment thus combines all rows referring to one particular product family within one particular month, creating $375 \cdot 24 = 9000$ fragments. This can significantly reduce work for queries referencing one or both of the fragmentation dimensions; it also supports both processing and I/O parallelism and scales well. As demanded in Section 3, the fragmentation of bitmaps follows that of the fact table.

Since one focus of our study is on skew effects, we explicitly model *attribute value skew* in the fact table, using zipf-like frequency distributions with respect to dimension values. This leads to varying densities and sizes of table fragments, potentially causing severe load imbalance. To help alleviate such *density skew*, we employ a *greedy data allocation algorithm* similar to [17] which allocates fact table fragments in decreasing order of size onto the least occupied disk at each time to keep disk partitions balanced. Corresponding bitmap fragments of each bitmap are stored on adjacent disks to support parallel bitmap access. Note, however, that a smart allocation scheme is merely a complement, not a replacement for intelligent scheduling techniques employed at runtime.

As our study regards single-user mode for the time being, queries are executed strictly sequentially. Focusing on fact table access, we assume simple aggregation queries that do not require joins to the dimension tables. All queries within a single experiment are of the same type (e.g., *QDIVISION*, aggregating data from one product division) but with random parameters (e.g., the specific division selected). However, different simulation runs will use the same set of queries, facilitating a fair comparison of results.

5.2 Scheduling Strategies

Since the performance of our strategies will depend in part on the type of query being processed, we consider both disk-bound and CPU-bound workloads, as well as borderline cases that shift between categories. In our case, the *selectivity* of a query *within* the relevant fact table fragments determines the ratio of CPU to I/O load. Our CPU-intensive queries each have a 100% selectivity within the fragments they access. I/O-bound loads, in contrast, select only some of the tuples in each fragment, causing less CPU work per I/O, and use bitmap indices, which are also cheap to process on the CPU side.

All queries are tested for our four scheduling strategies under varying degrees of skew on the two fragmentation dimensions, *TIME* and *PRODUCT*, using the same degree of skew to both dimensions. Under the zipf-like distributions we employ, the skew parameter may range from 0 (no skew) to values around 1 (heavy skew).

Disk-Bound Queries. In Figure 3, we show simulation results for two disk-bound queries, $Q_{CHANNEL}$ and Q_{STORE}. With our greedy allocation scheme, balanced disk partitions can be processed in constant time regardless of skew under a proper scheduling method. *PARTITION* achieves the best response times as would be expected for disk-bound workloads, keeping disks optimally loaded at nearly 100%. It also minimizes the inevitable disk contention caused by concurrent access to fact table and bitmap fragments. *INTEGRATED* performs equally well as *PARTITION* for the $Q_{CHANNEL}$ query with only 1% deviation; it is only slightly worse on Q_{STORE} with at most 15% response time increase. Apparently, the conflict analysis it performs is similarly effective to avoid disk contention as a strict separation of partitions, despite the additional size criterion.

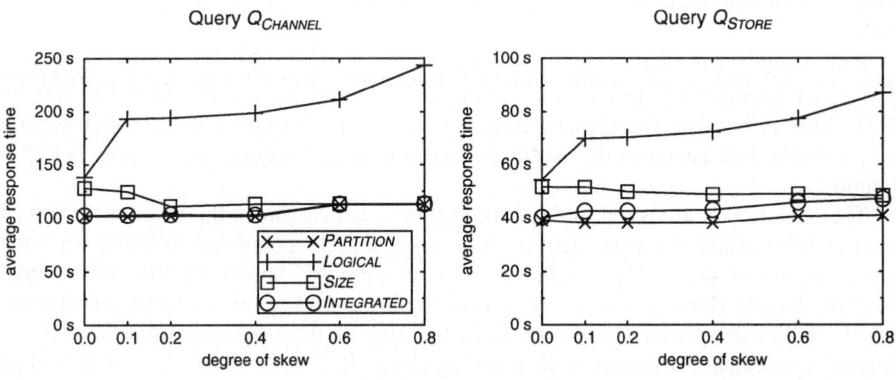

Fig. 3. Disk-bound queries

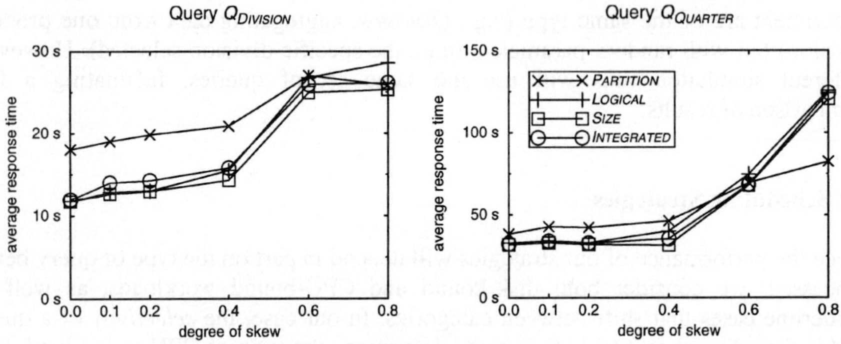

Fig. 4. CPU-bound queries

The other strategies are less successful here as they do not respect disk allocation to the same degree. The worst case is *LOGICAL*, which processes fragments in their logical order that is unrelated to their disk location under the greedy scheme, more than doubling the response time. *SIZE* mimics partitionwise scheduling to some extent because it processes fragments in the same size-based order in which they were allocated. Still, it cannot contend with the near-optimal *PARTITION*, with differences of up to 35%.

CPU-Bound Queries. The CPU-bound queries $Q_{DIVISION}$ and $Q_{QUARTER}$ perform a selection on the skewed fragmentation dimensions *PRODUCT* and *TIME*, respectively, and thus respond markedly to skew effects (Figure 4). Although partition sizes are well balanced for the database as a whole, this is not the case for single product divisions or calendar quarters and the largest fragment *within* such a subset can dominate the query's response time. This can be corrected by data allocation only to a limited extent.

The best results are achieved by *SIZE* as it balances the sheer amount of data processed per node, which is essential for CPU-bound queries. *PARTITION* performs worst (up to 58% for $Q_{DIVISION}$ and 46% for $Q_{QUARTER}$) because it does not permit more than one processor to access the same disk even under low disk utilization. The other two strategies achieve good success; *INTEGRATED* approximates *SIZE* most closely with only 10% deviation, demonstrating good performance for CPU-bound workloads as well.

Increasing skew changes the ranking in favor of *PARTITION*. With $Q_{QUARTER}$, *PARTITION* becomes by far the best strategy for extreme skew, now offsetting *SIZE* by 46%. This is because the skewed fragment sizes turn the query *locally disk-bound*, i.e., a single disk becomes the bottleneck even though the query as a whole is CPU-bound!

This situation is analyzed in detail in Figure 5, which shows the response times of queries referencing the least densely and most densely populated quarters for each given degree of skew. The smaller queries remain CPU-bound for the entire range because density skew is less severe toward the lower end of our zipf-like distribution curve. For large quarters, however, both the size of the respective quarter and the fragment imbalance increase with growing skew. It is only these queries that shift from CPU-bound to locally disk-bound so that *PARTITION* wins out by 43% for high skew.

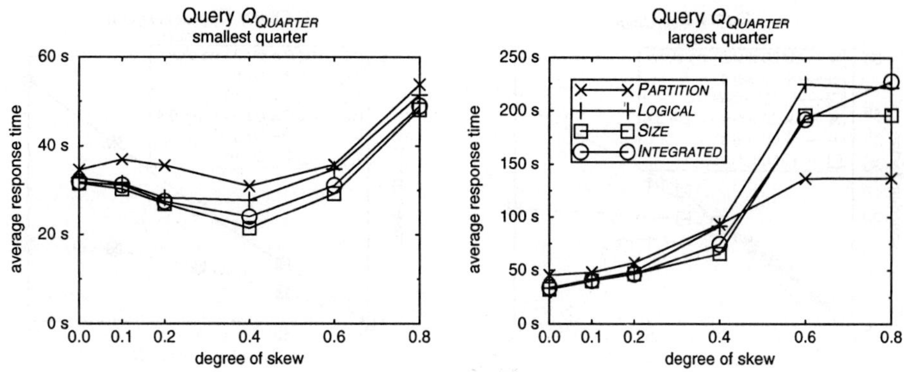

Fig. 5. Shift from CPU-bound to disk-bound

Discussion. The results show that no single scheduling scheme is optimal for all situations. For (globally or locally) disk-bound queries, minimal response times are normally achieved under the *PARTITION* heuristic, whereas CPU-bound workloads are best processed using *SIZE*. The choice of the truly best strategy then depends on the 'boundness' of a query, as determined by its selectivity and index utilization, the degree of skew, and a number of other parameters. A cost-based query optimizer of a PDBS might make a sensible decision by comparing the total (estimated) processing cost on the CPU and disk side, respectively, although locally disk-bound queries may be hard to detect.

On the other hand, our dynamic scheduling scheme based on the *INTEGRATED* heuristic was able to adapt to different types of queries and performed near-optimally in most experiments. Using this strategy thus promises to be more robust for complex workloads and avoids the need to select among different scheduling approaches based on error-prone cost estimates. Especially in a multi-user environment, we expect such an adaptive method to react more gracefully to the inevitable fluctuations in system load. In contrast, the correct selection between *PARTITION* and *SIZE* will be very difficult against a continually changing background load alternating between CPU-bound and disk-bound states. This aspect, however, needs to be investigated in future studies.

5.3 Speed-Up Behavior

In this simulation series, we test the scalability of our query processing and scheduling strategies with varying numbers of disks and processors. For each configuration, we run the queries $Q_{QUARTER}$ and $Q_{RETAILERMONTH}$ under a medium skew degree of 0.4 and against skewless data, respectively. Results are shown in Figure 6.

Since $Q_{QUARTER}$ is CPU-bound, we test its speed-up in relation to the number of processors, using *SIZE* as the scheduling strategy according to the previous results. Against skewless data, $Q_{QUARTER}$ shows linear speed-up until the disks of the system become bottlenecks and speed-up with respect to processors is no longer achievable. With skew (dashed graph), the curves decline earlier because response times are

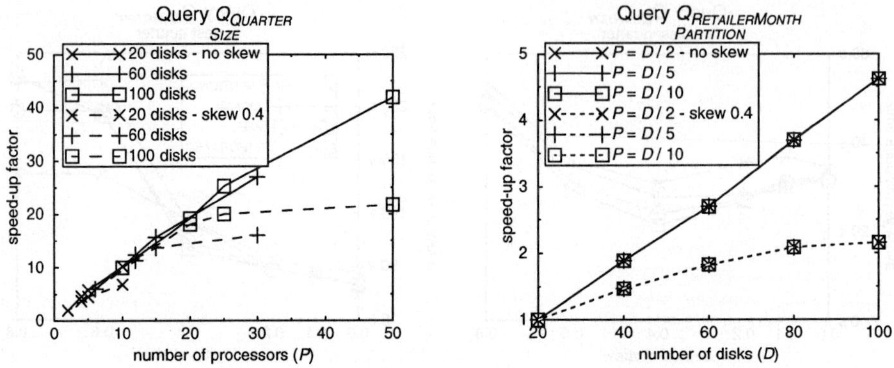

Fig. 6. Speed-up behavior of queries $Q_{QUARTER}$ and $Q_{RETAILERMONTH}$

dominated by the work on the largest fragment, causing locally disk-bound processing.

To test the speed-up for disk-bound queries, we use $Q_{RETAILERMONTH}$ which is more responsive to skew than $Q_{CHANNEL}$ and Q_{STORE} used above. $Q_{RETAILERMONTH}$ is scheduled using *PARTITION*, and speed-up is evaluated in relation to the number of disks.

As in the the previous case, speed-up is near-linear with skewless data but limited by the largest fragment in case of skew. The effect is even stronger this time as skew is more pronounced on lower hierarchy levels (months) than on higher ones (quarters).

For both types of workload, the *INTEGRATED* policy we proposed achieved equivalent results to the above (not shown here). Overall, our load balancing method scales very well for all relevant scheduling policies; limitations due to skewed fragment sizes are not caused by scheduling and must be treated at the time of data allocation.

6 Conclusions

In this paper, we have investigated load balancing strategies for the parallel processing of star schema fact tables with associated bitmap indices. We found that simple scheduling heuristics like *PARTITION* and *SIZE* can be very effective. But the selection of the appropriate method depends on whether a query is disk-bound or CPU-bound, which can be difficult to determine especially under skew conditions. As an alternative, we proposed a more complex, dynamically ordered scheduling approach (*INTEGRATED*) that yields only slightly worse performance but naturally adapts to different query types.

While we assumed a Shared Disk environment, most of the results can be transferred to other architectures, in particular, Shared Everything. Shared Nothing systems are restricted to strategies similar to *PARTITION*, which we found to be non-optimal. This demonstrates the benefits of Shared Disk and justifies our architectural choice. The extension of our findings to multi-user mode is not trivial. As the simple heuristics *PARTITION* and *SIZE* may no longer be sufficient, we expect our integrated strategy to gain importance. Verifying this assumption will be a focus of our future work.

References

1. APB-1 OLAP Benchmark, Release II. OLAP Council, Nov. 1998.
2. R. Avnur, J.M. Hellerstein: Eddies: Continuously Adaptive Query Processing. Proc. ACM SIGMOD Conf., Dallas, 2000.
3. E. Baralis, S. Paraboschi, E. Teniente: Materialized View Selection in a Multidimensional Database. Proc. 23rd VLDB Conf., Athens, 1997.
4. L. Bouganim, D. Florescu, P. Valduriez: Dynamic Load Balancing in Hierarchical Parallel Database Systems. Proc. 22nd VLDB Conf., Bombay, 1996.
5. S. Chaudhuri, U. Dayal: An Overview of Data Warehousing and OLAP Technology. SIGMOD Record 26(1), 1997.
6. D.J. DeWitt, J.F. Naughton, D.A. Schneider, S. Seshadri: Practical Skew Handling in Parallel Joins. Proc. 18th VLDB Conf., Vancouver, 1992.
7. S. Ghandeharizadeh, D.J. DeWitt, W. Qureshi: A Performance Analysis of Alternative Multi- Attribute Declustering Strategies. Proc. ACM SIGMOD Conf., San Diego, 1992.
8. J. Gray et al.: Data Cube: A Relational Aggregation Operator Generalizing Group-By, Cross-Tab, and Sub-Totals. In: U. Fayyad, H. Mannila, G. Piatetsky-Shapiro: Data Mining and Knowledge Discovery 1, 1997.
9. H. Lu, K.-L. Tan: Dynamic and Load-balanced Task-Oriented Database Query Processing in Parallel Systems. Proc. 3rd EDBT Conf., Vienna, 1992.
10. S. Manegold, J.K. Obermaier, F. Waas: Load Balanced Query Evaluation in Shared-Everything Environments. Proc. 3rd Euro-Par Conf., Passau, 1997.
11. H. Märtens, E. Rahm, T. Stöhr: Dynamic Query Scheduling in Parallel Data Warehouses. Techn. report, University of Leipzig, 2001. `http://dol.uni-leipzig.de/pub/2001-38/en`
12. C. Mohan, I. Narang: Recovery and Coherency-Control Protocols for Fast Intersystem Page Transfer and Fine-Granularity Locking in a Shared Disks Transaction Environment. Proc. 17th VLDB Conf., Barcelona, 1991.
13. P. O'Neil, G. Graefe: Multi-Table Joins Through Bitmapped Join Indices. ACM SIGMOD Record 24 (3), 1995.
14. E. Rahm, R. Marek: Dynamic Multi-Resource Load Balancing in Parallel Database Systems. Proc. 21st VLDB Conf., Zurich, 1995.
15. E. Rahm, H. Märtens, T. Stöhr: On Flexible Allocation of Index and Temporary Data in Parallel Database Systems. Proc. 8th HPTS Workshop, Asilomar, 1999.
16. E. Rahm, T. Stöhr: Analysis of Parallel Scan Processing in Parallel Shared Disk Database Systems. Proc. 1st Euro-Par Conf., Stockholm, 1995.
17. P. Scheuermann, G. Weikum, P. Zabback: Data Partitioning and Load Balancing in Parallel Disk Systems, VLDB Journal 7(1), 1998.
18. T. Stöhr, H. Märtens, E. Rahm: Multi-Dimensional Database Allocation for Parallel Data Warehouses. Proc. 26th VLDB Conf., Cairo, 2000.
19. M.-C. Wu, A.P. Buchmann: Encoded Bitmap Indexing for Data Warehouses. Proc. 14th ICDE Conf., Orlando, 1998.

Speeding Up Navigational Requests in a Parallel Object Database System

Jim Smith[1], Paul Watson[1],
Sandra de F. Mendes Sampaio[2], and Norman W. Paton[2]

[1] Department of Computing Science, University of Newcastle upon Tyne, Newcastle upon Tyne, NE1 7RU UK
{Jim.Smith, Paul.Watson}@ncl.ac.uk

[2] Department of Computer Science, University of Manchester, Manchester, M13 9PL UK
{sampaios, norm}@cs.man.ac.uk

Abstract. In data intensive applications, both programming and declarative query languages have attractions, the former in comprehensiveness and the latter for ease of use. Databases sometimes support the calling of side-effect free user defined functions from within declarative queries. As well as enabling more efficient coding of computationally intensive functions, this provision not only moves computation to data in a client-server setting, but also enables speedup through data parallel execution if the server is parallel. There has been little work on the combined use of query and program based database access in the context of parallel servers. We believe Polar is the first parallel object-oriented database which supports this arbitrary navigation both in a client application and in functions (operations) which may be called from within declarative queries. This work introduces Polar's support for navigation and for calling operations from within parallel queries and presents performance results for example navigational requests for the large OO7 benchmark.

1 Introduction

Suppose a large engineering design company has a collection of engine designs to be stored in a database, and that an engine is modelled as a hierarchical collection of component objects, such as a gearbox, cylinders etc, each having 3d shape and orientation. Suppose also a client seeks a particular type of engine that fits into some space which has arbitrary 3d shape, and can tolerate various possible orientations. The overall request is suited to expression in a declarative query language, but the computation which determines whether a particular engine fits into the available space may be too complex, or too CPU intensive, to express in a high level query language. On the other hand the request could be implemented entirely in a programming language. In this case however, not only must the implementor handle the data related optimisation issues which the query compiler would handle for a declarative expression, but in the typical client-server context the cost of transporting the data comprising all the engine

B. Monien and R. Feldmann (Eds.): Euro-Par 2002, LNCS 2400, pp. 332–341.

designs to the client is incurred. If the server is parallel, then there is a greater incentive to execute complex code on the server since the query language part of the request is transparently executed in parallel, offering the potential to gain a speedup in execution of the complex/CPU intensive portion of the request expressed in a programming language, even though that language is sequential, by calling the code on multiple objects in parallel.

The need to combine query and programming languages to meet user requests is widely appreciated, and stored procedures in SQL [9] support the calling of programming language code from within an SQL query. In [12] a four quadrant classification is presented of database applications, with those in the fourth quadrant, that require both query and programatic access, being considered the most challenging. The engine design example postulated above is typical of those problems in this class where, not only the query language part of the request, but also the programming language part needs to access complex stored data. The Object Database Management Group (ODMG) standard [6] defines a marriage of Object Query Language (OQL) and programming languages (e.g. C++) for an OODB that addresses these problems and others in the fourth quadrant. In an OODB, the complex computation of the example described earlier could be expressed as an operation of the *Engine* class, such as *Engine::fitsInto(...)*, implemented in a programming language, such as C++, and called both from a client based navigational program and from an OQL expression.

There is not much work on parallel database support for both navigational and programatic access. Monet is a parallel OODB, optimised for main memory use, but while supporting a library of parallel traversals [1], it doesn't provide for arbitrary navigation in a programming language. Parsets [13] are sets of object identifiers (OIDs) which support global operations on the referenced objects which are distributed over multiple Shore [5] instances. When such an operation is called, the OIDs are distributed to their home processors and the operation performed there. One of these operations, *Apply* allows a method of each object to be called. While Parsets implement certain global operations familiar in query languages, such as *Select* and *Reduce*, execution is planned explicitly by the user rather than by a query compiler based on a full physical algebra and exploiting data related statistics maintained in the database metadata.

Saving previous results e.g. [8, 10], can benefit performance of queries calling user-defined functions where a particular function is called with the same parameters multiple times, but not where a particular function is called for the same object multiple times with different parameters, or where multiple functions are called for the same object. The work described here is concerned with situations where functions must be called and so employs the complementary approach of object state caching.

Polar [11], unlike Monet is disk based and supports programming language bindings in the ODMG style, and unlike the Parsets approach integrates operation calls into a physical algebra supporting OQL. Polar claims to be the first parallel disk based ODMG compliant architecture implemented. While earlier work on Polar [11, 3, 4] emphasised query processing facilities, this paper expands the architectural description to show how navigation is supported and presents measurements for a prototype implementation.

In the following, Section 2 gives an overview of the Polar architecture, while Section 3 describes its support for OQL based operation calls. By way of example, Section 4 describes how selected navigational requests may be parallelised, and Section 5 presents measured results for these examples implemented on a prototype of the Polar architecture. Finally, Section 6 concludes.

2 Architecture

As illustrated in [11], the Polar parallel store comprises multiple store units accessing global metadata, but distributed application data, each comprising an object manager and query execution engine. The object store may be accessed by both an OQL client and a navigational client, the latter executing an application written in a programming language. Either may generate OQL query expressions, which it directs to a compiler unit.

Logically, data is organised by extents which are directly accessible to code, be it a declarative query or a navigational program, executed on either the server or a client machine. Physically however, these extents are partitioned across a number of volumes distributed over a number of store units. Documenting this partitioning in the metadata allows both the query compiler and a navigational operator, using the volume mapping information also stored in the metadata, to identify the store units containing partitions of a particular extent.

An OQL expression undergoes logical, physical and parallel optimisation to yield a data flow graph of parallel algebra operators (query plan), which is distributed between object store and client. Typically one partition of this plan runs in the client and the remainder in a selection of store units, the latter creating and filtering tuples, former collating and presenting tuples. A *tuple* is a complex but generic structure which holds an instance of an intermediate result collection within an executing query plan.

The operators follow the *iterator model* [7] whereby each implements a common interface comprising three main functions: *open* performs setup and causes the operator to begin computing its result collection; *next* returns the next tuple in that result; and *close* is called to clean up, after receipt of an *eof* tuple. As described in [11], parallelism in a query is encapsulated in the exchange operator which implements a partition between two threads of execution, and a configurable data redistribution, the latter implementing a flow control policy.

Figure 1 shows the runtime support in a navigational client and store unit. Both contain a query execution engine, support for language binding and query setup/control together with some basic infrastructure. Control is initiated in the client based application, but following query compilation the resulting query plan is controlled by execution services in the separate machines while the application waits for results to stream through.

In the store unit, programming language code is encapsulated within operation libraries that are loaded on demand under control of such an executing

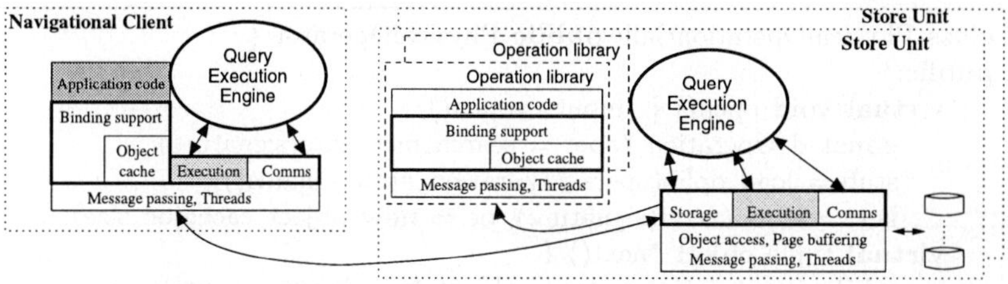

Fig. 1. Runtime Support Services

query. The language binding support, which contains an object cache which transparently faults in objects as required, is also present in a query compiler and an OQL client to support access to the distributed metadata. In general, the execution of declarative queries is planned to access data once only so there is little scope for exploiting an object cache. As described elsewhere [4], where objects are reaccessed many times in a query, the approach adopted in Polar is to access data at a lower level than the object cache in scans and to use specific tuple caching where this is beneficial.

The query execution engine implements the algorithms of the parallel algebra operators, building on a support library. At the lowest level is basic support, for: management of untyped objects, message exchange and multi-threading. On top of this, a storage service supports page based access to objects either by OID directly or through an iterator interface to an extent partition and a communications service implements the flow control underlying exchange. In support of inter store navigation, the storage service can request and relay pages of objects stored at other store units.

3 Operation Calls

Schema compilation generates a description of the application schema, including operations within the metadata. From this, a header file in the appropriate programming language can be derived to be included when compiling operation implementations and application code. Each operation body conforms to its definition in the header file. An operation is called from within a query via a stub file which converts between data representations used within the execution engine and those used within language bindings and constructs the call parameters. Specifically, a different representation of complex structures and collections is used in the query execution engine from those of the language bindings. A stub generator automatically creates this code from the metadata. Currently in Polar, an operation is globally identified by a signature comprised of its fully scoped name and parameter types, for example:

OO7::Document::replaceText(IN(string),IN(string))

The signature is stored in the metadata entry for the corresponding operation.

```
class PhysicalOperationCall: public PhysicalOperator {
public:
    virtual void open() { input->open();
        const d_Operation *oper = search_metadata(signature);
        stub = load_oplib(oper->operation_library_name());
        key = stub->xref(signature); oc = new object_cache(oc_size); }
    virtual tuple_object *next() {
        while (! t->eof()) { tuple_object *t = input->next();
        for (list<expr>::iterator it = expression.begin();
              it != expression.end(); it++)
            arg[i+1] = t->evaluate(*it); // arg[0] is result
        stub->call_operation(key, arg);
        t->insert(arg[0]); if (t->evaluate(predicate)) return t; }
    virtual void close() {delete oc; unload_oplib(stub); input->close();}
private:
    string signature; list<expr*> expression;
    list<genform*> *predicate; int oc_size;
    vector<concrete_object*> arg; int key;
    object_cache *oc; class oplib_stub; oplib_stub *stub; };
```

Fig. 2. Implementation of a simple Operation call operator

Compiled operation implementations can be linked into either a client program or, with an appropriate stub, a shared library. The installation utility records the name of the operation library in the metadata describing each of the operations implemented in that library.

Execution of such operation calls in OQL is realised through a separate physical operator, operation call. This facilitates arbitrary placement of a call within the query plan. Figure 2 outlines a realisation of a simple operation call operator. An object cache is created in *open()*, shared by all operation calls made in *next()*, and finally shut down in *close()*. The appropriate size for a particular object cache may be determined by user hint and/or runtime statistics. The stub code is responsible for actually initiating an operation call, by first creating a reference of the correct type in the object cache from the OID stored in the tuple, and then calling the appropriate method through that reference.

The basic parameters of the simple Operation call operator are:

– the signature of the function to be executed;
– the name by which the result is to be known in the result tuple;
– a list of (possibly constant) expressions, identifying actual parameters;
– optionally, a predicate to filter tuples resulting from the operation call; and
– parameters defining the environment in which the function call is executed, for instance the size of the object cache.

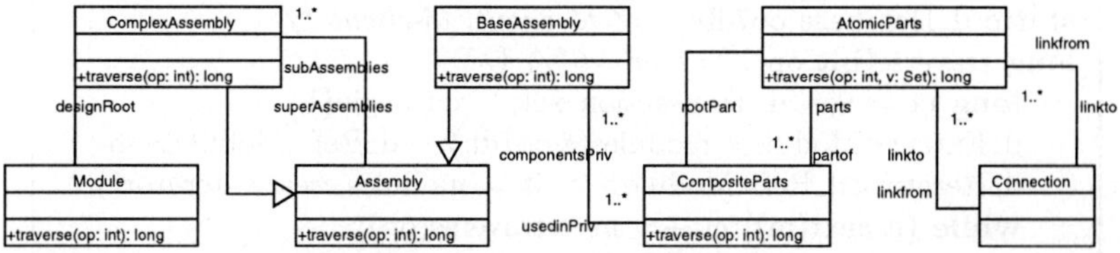

Fig. 3. A possible representation of *t1* and *t6* in the database schema

4 Parallelising Navigational Requests

To illustrate the parallelisation of navigational requests, the following shows how two navigational requests of the OO7 benchmark [2] may be executed first sequentially and then in parallel. The examples chosen are traversals *t1* and *t6* which perform read only traversals of the design hierarchy, in the former case completing a depth first traversal of the atomic parts in each composite part and in the latter case touching only the single root atomic part of each composite part. Figure 3 shows how the two traversals might be reflected in a single operation *traverse(op)* at each level in the design hierarchy, where *op* is an integer value selecting traversal 1 or 6. Figure 4 shows C++-OML code implementing these operations.

Originally OO7 [2] specified the traversal benchmarks with respect to a single module. As in the Parsets work, where a configuration having multiple modules is used, traversals are applied to all those modules. However, no ordering of the sep-

```
long Module::traverse(int op) { return designRoot->traverse(op); }
long ComplexAssembly::traverse(int op) {
    long ct = 0;  d_Ref<Assembly> assm;
    d_Iterator<d_Ref<Assembly> > it = subAssemblies.create_iterator();
    while (it.next(assm)) ct += assm->traverse(op);  return ct; }
long BaseAssembly::traverse(int op) {
    long ct = 0;  d_Ref<CompositePart> cp;
    d_Iterator<d_Ref<CompositePart> > it = componentsPriv.create_iterator();
    while (it.next(cp)) ct += cp->traverse(op);  return ct; }
long CompositePart::traverse(int op) {
    set<long> visited;  return rootPart->traverse(op, visited); }
long AtomicPart::traverse(int op, set<int> &v) {
    if (op == 6) return 1;  // else, for op == 1, do DFS
    long ct = 0;  d_Ref<AtomicPart> ap;  d_Ref<Connection> cn;
    d_Iterator<d_Ref<Connection> > it = linkto.create_iterator();
    while (it.next(cn))
        {ap = cn->linkto; if (!v.find(ap->id)) ct += (ap->traverse(op, v));}
    return ct; }
```

Fig. 4. Implementation of *t1* and *t6* in C++-OML

```
static d_Database oo7db;   /* t1-client, t6-client */
long traverse(int op /*=1 or =6*/) {
    long ct = 0;  d_Transaction xct;   xct.begin();
    d_Extent<Module> modules(&oo7db);   d_Ref<Module> m;
    d_Iterator<d_Ref<Module> > it = modules.create_iterator();
    while (it.next(m)) ct += m->traverse(op);
    xct.commit();   return ct; }
```

Fig. 5. C++-OML client code to call *t1* or *t6*

arate module traversals is specified, though each module traversal is assumed to be depth first. In the original benchmark presentation a navigational client program iterates through the extent of modules and calls the *Module::traverse(...)* operation of each. Figure 5 shows essential parts of such client C++-OML code.

OQL does not currently support a general form of recursion such as that required to traverse the assembly hierarchy purely within OQL. However, for a given database size it is possible to unroll the recursion by hand. In the main three OO7 configurations where there are 7 levels of assembly objects *t6* may be implemented using the OQL expression below.

```
select count(c.rootPart.id)   /* t6-q2*/
from m in modules, a2 in m.designRoot.subAssemblies, a3 in a2.subAssemblies,
   a4 in a3.subAssemblies, a5 in a4.subAssemblies, a6 in a5.subAssemblies,
   a7 in a6.subAssemblies, b in a7.subAssemblies, c in b.componentsPriv;
```

Alternatively, either request may be implemented by appropriately calling *Module::traverse(...)* within an OQL query.

```
select m.traverse(1)  /* t1-q1 */     select m.traverse(6)  /* t6-q1 */
   from m in modules;                    from m in modules;
```

A possible parallel query plan corresponding to the OQL expression *t1-q1* is shown in figure 6. The query is distributed across the parallel store by executing sub-plan 1 in each store processor and sub-plan 2 in the coordinator. The operation call operator receives a stream of Module OIDs and, for each of these, calls

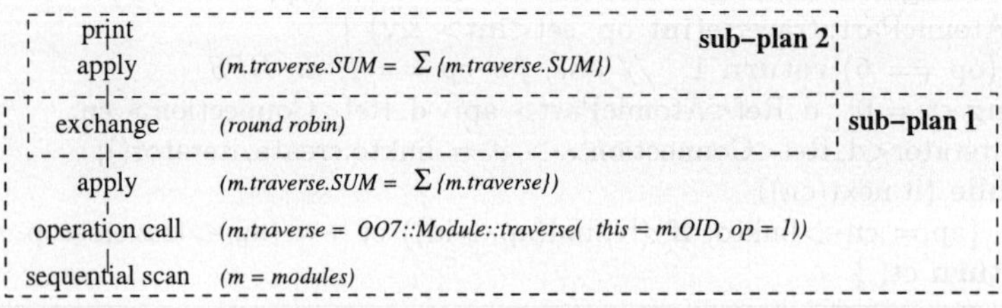

Fig. 6. A possible parallel query plan implementing the OQL expression *t1-q1*

the operation *OO7::Module::traverse(1)*. The call returns are summed locally and these local sums are accumulated in the coordinator.

5 Experimental Evaluation

Initial results are reported here for the implementations of the traversals described in section 4. The experiments are conducted using instantiations of the OO7 database on Polar within a shared nothing environment comprising a cluster of 10 860MHz Pentium III machines running RedHat Linux version 7.2, each with 512MB main memory and local disks, connected via a 100Mbps Fast ethernet hub. For each experiment, data is partitioned over a number of machines but located on one disk per machine, this being a SEAGATE ST39204LC. User requests are submitted, except where otherwise indicated, from an identical but separate machine on the same network. The OO7 configuration used is large with fan-out 3 and occupies a total of about 750M Bytes.

Polar's bulk loading facility supports a number of data distribution policies including the ability to cluster by store unit one object with a related object. In these experiments, modules are distributed *round robin* across the store units and all remaining levels of the OO7 hierarchy, i.e. assemblies, manuals, composite parts, etc, are clustered with the related module. This clustering policy ensures that all inter store traffic is confined to the physical algebra operators, so it is easier to understand the expected performance. However, since there are only 10 modules in large OO7, the performance graphs show obvious granularity effects.

Figure 7 (a) compares the measured performance of the three implementations of *t6*: *t6-client* in the single store configuration; *t6-q1* and *t6-q2* across the range of parallel configurations. The performance of *t6-q1* is better than

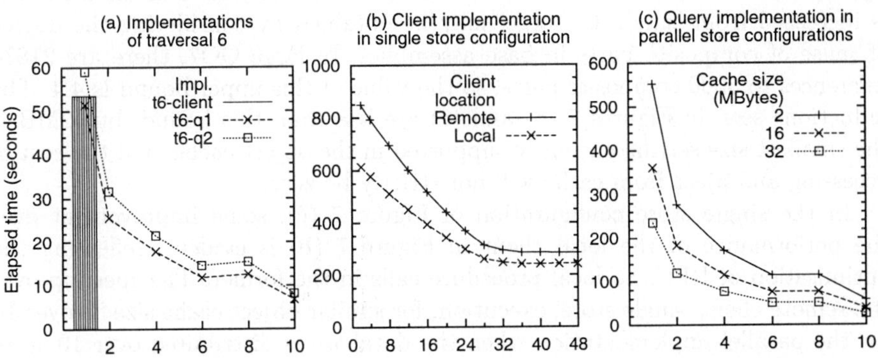

Fig. 7. Measured performance over large OO7 (a) of implementations of *t6* (b) of *t1-client*, running in separate cases remotely from the store unit and locally, and (c) of *t1-q1*. In (a), the client performance, shown as a bar, is only measured over the single store configuration

that of *t6-client* even in the single store configuration. This is to be expected since the operation code in *t6-client* is running remotely and therefore pays some cost for data transfer across the network while in *t6-q1*, the same operation code is running in the same context as the server process.

Logically the two parallel implementations are performing identical traversals, but the implementations are quite different. The following are likely to contribute to the higher cost of the OQL based implementation.

- The object store is built on Shore [5] and the disk format of object states is closer to that of the C++ binding than to that of the generic tuple structure, so the cost of object materialisation is likely to be higher in the latter case.
- The version of Polar used for these experiments did not include the tuple caching facility of [4]. Therefore, while *t6-q1* can re-access cheaply from the object cache composite part and atomic part objects which are touched more than once in the traversal, *t6-q2* must repeat the tuple materialisation multiple times, even if all such objects are retained in file-system buffers or in the storage manager buffer pool

Overall in this experiment highest performance is achieved through encoding the bulk of the user level query in class operations, installed on the server and called via an OQL query. It seems likely that this pattern will recur in other examples, but set against the performance advantage of the installed operations is the administrative overhead and development cost.

Figure 7 (b) shows how the performance of the sequential implementation, *t1-client*, varies with object cache size for the single store onfiguration where the client is located: firstly on a separate machine from the store; and secondly on the same machine as the store. Figure 7 (c) shows the performance of the parallel implementation, *t1-q1*, on different store configurations.

Assuming that traversing an already cached composite part has zero cost, an upper bound on the reduction in the cost of the overall traversal as the cache size is increased from 'small' to 'large' may be obtained by considering the degree of reuse of composite parts in base assemblies. In large OO7, there are 21870 references to 5000 composite parts, so the value of this upper bound is 4.4. The reductions seen in Figures 7 (b) and (c) are less than this bound, but 2MB is the smallest size setting currently supported in the object cache, and the cost of accessing an object from cache will not strictly be zero.

In the single store configuration of Figure 7 (c), some improvement over the performance of the local client in Figure 7 (b) is evident, reflecting the optimisation of RPCs to local procedure calls in the former. The speedup over the remote client, single store, execution, for similar object cache size, is over 10 for the parallel implementation when the database is distributed over 10 store processors. In fact, the speedup is over 9 with respect to the local client for each of the object cache sizes.

6 Conclusions

This paper has described the implementation and evaluation of a mixed query and programming language environment over a parallel OODB. The work has shown how the query and procedural languages present in a parallel OODB can be combined to speedup operations which traverse complex structures of objects. Furthermore, experimental results have shown good absolute speedup in benchmark tests. We believe that Polar is the first implementation of a parallel OODB which supports this combination. In these experiments, a highly favourable data distribution is employed. Furthermore, the load is as balanced as possible, all-be-it at the large granularity of a module. Further work will explore these issues, along with that of implementing more complex user requests, e.g. ones implying multiple operation calls.

References

1. P. A. Boncz, F. Kwakkel, and M. L. Kersten. High performance support for OO traversals in Monet. In *BNCOD*, pages 152–169. Springer, July 1996.
2. M. J. Carey, D. J. DeWitt, and J. F. Naughton. The OO7 benchmark. In *SIGMOD Conference*, pages 12–21. ACM Press, May 1993.
3. S. de F. Mendes Sampaio, J. Smith, N. W. Paton, and P. Watson. An experimental performance evaluation of join algorithms for parallel object databases. In *Euro-Par*. Springer, August 2001.
4. S. de F. Mendes Sampaio, J. Smith, N. W. Paton, and P. Watson. Experimenting with object navigation in parallel object databases. In *Workshop on Parallel and Distributed Databases, with DEXA'01*, Munich, Germany, August 2001.
5. M. J. Carey et al. Shoring up persistent applications. In *SIGMOD Conference*, pages 383–394. ACM Press, 1994.
6. R. G. G. Cattell et al., editor. *The Object Database Standard: ODMG 2.0.* Morgan Kaufmann, San Francisco, 1997.
7. G. Graefe. Query evaluation techniques for large databases. *ACM Computing Surveys*, 25(2):73–170, June 1993.
8. A. Kemper, C. Kilger, and G. Moerkotte. Function materialization in object bases: Design, realization, and evaluation. *TKDE*, 6(4):587–608, August 1994.
9. J. Melton. *Understanding SQL's Persistent Stored Modules.* Morgan Kaufmann, San Francisco, 1997.
10. J. M. Hellerstein J. F. Naughton. Query execution techniques for caching expensive methods. In *SIGMOD Conference*, pages 423–434. ACM Press, June 1996.
11. J. Smith, P. Watson, S. de F. Mendes Sampaio, and N. W. Paton. Polar: An architecture for a parallel ODMG compliant object database. In *CIKM*, pages 352–359. ACM Press, November 2000.
12. M. Stonebraker and P. Brown. *Object Relational DBMSs Tracking The Next Great Wave.* Morgan kauffman, September 1998.
13. D. J. De Witt, J. F. Naughton, J. C. Shafer, and S. Venkataraman. Parallelising OODBMS traversals: A performance evaluation. *VLDB Journal*, 5(1):3–18, January 1996.

Retrieval of Multispectral Satellite Imagery on Cluster Architectures

T. Bretschneider[1] and O. Kao[2]

[1] School of Computer Engineering, Nanyang Technological University, Singapore
[2] Department of Computer Science, Paderborn University, Germany
okao@upb.de

Abstract. The retrieval of images in remote sensing databases is based on world-oriented information like the location of the scene, the utilised scanner, and the date of acquisition. However, these descriptions are not meaningful for many users who have a limited knowledge about remote sensing but nevertheless have to work with satellite imagery. Therefore a content-based dynamic retrieval technique using a cluster architecture to fulfil the resulting computational requirements is proposed. Initially the satellite images are distributed evenly over the available computing nodes and the retrieval operations are performed simultaneously. The dynamic strategy creates the need for a workload balancing before the sub-results are joined in a final ranking.

1 Introduction

The development and application of remote sensing platforms result in the production of huge amounts of image data. The obtained image data has to be systematically collected, registered, organised, and classified. Furthermore, adequate search procedures and methods to formulate queries have to be provided. The retrieval of images usually starts with a query based on world-oriented descriptions for the images, e.g. location, scanner, date. These characteristics allow a remote sensing professional to retrieve the required data with respect to the specific requirements determined by the application. However, the provided mechanisms are not suitable for users with limited expertise in remote sensing and the acquisition systems' characteristics, respectively. A possible scenario with the corresponding query, which is not supported by the conventional retrieval systems for satellite data, is the following ecological application: *A certain region reveals symptoms of increasing salinity and a satellite image of the area was purchased. It is of interest to find other regions which suffered from the same phenomenon and which were successfully recovered or preserved, respectively. Thus the applied strategies in these regions can help to develop effective counteractions for the specific case under investigation.*

This and other related examples [1] require a search by content rather than by related information, i.e. world-oriented features. Therefore a completely new understanding of the demands for remote sensing databases towards powerful retrieval methods is mandatory which led to the development of the *Retrieval System for Remotely Sensed Imagery* or short $(RS)^2I$.

B. Monien and R. Feldmann (Eds.): Euro-Par 2002, LNCS 2400, pp. 342–345.

2 Feature Extraction and Retrieval Strategy

The techniques for content-based retrieval in remote sensing databases are based on spatial content, spectral information, and a combination of both. The published results either require supervision of the feature extraction process or are limited to a specific type of satellite. Although the results might be satisfying for the utilised image sets, the retrieval result in an archive of data from a variety of scanners will lack in quality, i.e. in too many inadequate image answers.

This paper uses a new powerful feature extraction approach solely based on the spectral content by assigning each individual pixel a class membership accordingly to the multispectral radiance values. The successive feature extraction is based on the classes and uses a variety of different criteria for the description, like the class mean, class variance, spatial class neighbours, class compactness in the image etc. – in total the feature vector consists of more than 24 elements. For a detailed description of the proposed method refer to [1].

A satellite image is classificated, i.e. the features are extracted, when it is inserted in the database. Prior sub-division of the whole image is necessary before the processing due to the size of the data and the manifold classes in an entire scene which make computation burdensome. Unfortunately, every a-priori chosen sub-division technique for the images in the database restricts the flexibility of the system. As a conclusion this paper extends the query possibilities by a dynamic component, which has already proven its positive impact on the quality of the retrieval result in general image databases [3]. Instead of calculating the feature vectors solely a-priori they are generated at run-time accordingly to the requirements set by the query image. The combination of static and dynamic features reduces the needed time to obtain the retrieval result. Nevertheless, the dynamic processing is the bottleneck which can be overcome by utilising parallel architectures. Earlier investigations [3] already found that clusters are the most suitable architecture for the considered dynamic image database.

3 Parallel Execution of the Retrieval Operation

Dynamic retrieval of satellite imagery requires the analysis of all *image sections* in the remote sensing database and produces an enormous computational load since the requirements grow exponentially by using arbitrary sections instead of pre-defined tiles. Therefore, the utilisation of parallel architectures is necessary for the solution of the performance problem. The developed prototype of a parallel remote sensing database is based on a Beowulf cluster with:

- Master node controls the cluster, receives the query specification, and broadcasts the algorithms to the computing nodes. Furthermore, it unifies the intermediate results of the compute nodes and produces the final ranking.
- Computing nodes perform the image processing and comparisons. Each of these nodes contains a disjunctive subset of the existing images and executes all operations with the data stored on the local devices. The sub-results are sent to the master node.

For the initial distribution of the sections over the available nodes a content independent partitioning strategy is selected, such that the memory size of the sub-images stored on the local devices is approximately equal for all nodes. The resulting set of partitions $P = \{P_1, P_2, \ldots, P_m\}$ over the image set $\mathcal{B}$ for m nodes has following characteristics:

$$\forall P_i, P_j \subset \mathcal{B} : P_i \cap P_j = \emptyset,\ size(P_i) \approx size(P_j), \quad i, j = 1, \ldots, m,\ i \neq j. \quad (1)$$

The advantages of this strategy are the simple implementation and management. All nodes have uniform processing times, if a query needs to analyse all images in the database and a homogenous, dedicated platform is assumed. However, queries combining world-oriented and dynamically extracted features distort the even image distribution and lead to varying node processing times. Therefore, workload balancing is a major issue in the $(\mathrm{RS})^2\mathrm{I}$.

Let s denote the operation sequence performed with world-oriented features and d operation sequence with dynamic feature extraction. The query $q(\mathcal{B}) = d \circ s(\mathcal{B})$ transforms P_i into $\mathcal{A} = \{A_1, A_2, \ldots, A_m\}$ with $A_i := s(P_i)$. The images from $A_i = \{b_{i1}, b_{i2}, \ldots, b_{in_i}\}$ are analysed with dynamically extracted features in the next step. As they do not fulfil necessarily the size condition defined in (1), the even image distribution over the cluster nodes is distorted. A migration of images within the cluster and thus workload balancing is required in order to equalise the processing times of all nodes. Each feature extraction algorithm p implemented in the image database is assigned a measure describing the processing time t_p as a function of the number of pixels. This is usually calculated experimentally and stored in the database. With this information the processing time per image $t_p(b_{ij})$ as well as the system response time t_r, minimal processing time t_{min} and the optimal processing time t_{opt} can be estimated [5]. After t_{min} at least one node idles and an image re-distribution is necessary in order to avoid idling compute resources. Note that storing images permanently on the new assigned processing node if the requirements of the succeeding are invariable, e.g. always urban related queries are submitted, can reduce the necessity for a re-distribution. However, experiments under read-world conditions did show that in general a partitioning like given by Equation (1) is favourable.

It can be proven that this problem is NP-complete [4], therefore the heuristic LTF (*Largest Task First*) strategy [5] was developed. Large images with long compute times are mainly processed on the local node and thus the transfer effort is reduced. The processing of a query is divided into three stages. After processing the images on the local storage devices from $0, \ldots, t_{min}$, a temporal transfer of images from overloaded nodes to the other nodes according to the computed schedule follows. Finally, the rest of the images is analysed. The LTF strategy has with $O(n \log n)$ a low computational complexity. Disadvantages result mainly from the image communication between all nodes at nearly the same time, as a network overload is caused [5].

The performance measurements are executed using the implemented prototype of the parallel remote sensing database, which consists of six SMP-nodes (Dual Pentium III, 667 MHz, 256 Mbyte memory) and a 100 Mbit FastEthernet. The retrieval method was applied to multispectral scenes of the satellites Land-

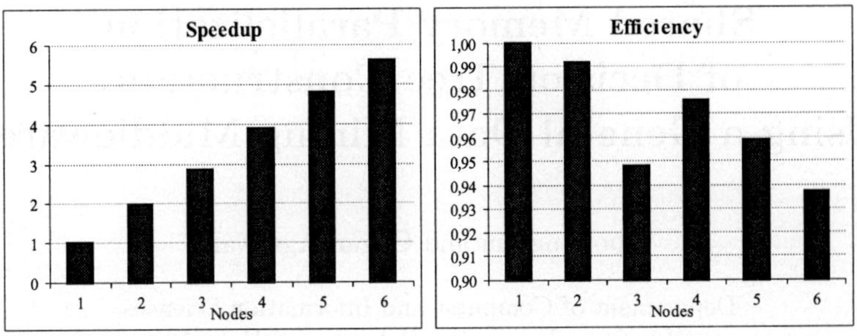

Fig. 1. Speedup and efficiency values for the parallel retrieval using the $(RS)^2I$

sat, SPOT, MOMS-02, and IKONOS which add up to an image archive covering an area of more than 204,150 km^2 in the 1–30m spatial resolution range. The images show a mixture of urban, agriculture, and forestry areas around the world. An in-depth analysis of the retrieval results can be found in [1] and is omitted here due to the limited space. Figure 1 shows, that a linear speedup with the number of nodes is achieved. Due to the minimised communication between the cluster nodes during query processing the efficiency values remain in a narrow interval between 94% and 98%.

4 Conclusions

In this paper an in-sight in the parallel aspects of the $(RS)^2I$ was provided. The system enables dynamic retrieval of remotely sensed data in an image database and shows that the utilisation of a content-based approach enables the retrieval of data even for persons without in-depth remote sensing background knowledge. The developed cluster-based architecture satisfies the computational demands and enables the use of the proposed system in real world applications.

References

1. T. Bretschneider, R. Cavet, and O. Kao, "Retrieval of remotely sensed imagery using spectral information content", Geoscience and Remote Sensing Symposium, to be published 2002.
2. T. Bretschneider, O. Kao, "Indexing strategies for content-based retrieval in satellite image databases", Conference on Imaging Science, to be published 2002.
3. S. Geisler, O. Kao, T. Bretschneider, "Analysis of cluster topologies for workload balancing strategies in image databases", Conference on Parallel and Distributed Processing and Applications, pp. 874–880, 2001.
4. R. Karp, "Reducibility among combinatorial problems", in: Complexity of Computer Computations, Plenum Press, pp. 85–104, 1972.
5. O. Kao, G. Steinert, F. Drews, "Scheduling aspects for image retrieval in cluster-based image databases", IEEE/ACM Symposium on Cluster Computing, pp. 329–336, 2001.

Shared Memory Parallelization of Decision Tree Construction Using a General Data Mining Middleware*

Ruoming Jin and Gagan Agrawal

Department of Computer and Information Sciences
Ohio State University, Columbus, OH 43210
{jinr,agrawal}@cis.ohio-state.edu

1 Introduction

Decision tree construction is a very well studied problem in data mining, machine learning, and statistics communities [3, 2, 7–9]. The input to a decision tree construction algorithm is a database of *training records*. Each record has several *attributes*. An attribute whose underlying domain is totally ordered is called a *numerical* attribute. Other attributes are called *categorical* attributes. One particular attribute is called *class label*, and typically can hold only two values, true and false. All other attributes are referred to as *predictor* attributes.

A number of algorithms for decision tree construction have been proposed. In recent years, particular attention has been given to developing algorithms that can process datasets that do not fit in main memory [3, 6, 10]. Another development in recent years has been the emergence of more scalable shared memory parallel machines. To the best of our knowledge, there is only one effort on shared memory parallelization of decision tree construction on disk-resident datasets, which is by Zaki *et al.* [11].

In our previous work, we have developed a middleware for parallelization of data mining tasks on large SMP machines and clusters of SMPs. This middleware was used for apriori association mining, k-means clustering, and k-nearest neighbor classifiers [4, 5]. In this paper, we demonstrate the use of the same middleware for decision tree construction. We particularly focus on parallelizing the *RainForest* framework for scalable decision tree construction [3].

The rest of the paper is organized as follows. We describe the original RainForest framework in Section 2. We review our middleware and parallelization techniques in Section 3. Our parallel algorithms and implementation are presented in Section 4. Section 5 presents the experimental results. We conclude in Section 6.

* This research was supported by NSF CAREER award ACI-9733520, NSF grant ACR-9982087, and NSF grant ACR-0130437.

B. Monien and R. Feldmann (Eds.): Euro-Par 2002, LNCS 2400, pp. 346–354.

2 Decision Tree Construction Using RainForest Framework

Though a large number of decision tree construction approaches have been used in the past, they are common in an important way. The decision tree is constructed in a top-down, recursive fashion. Initially, all training records are associated with the root of the tree. A criteria for splitting the root is chosen, and two or more children of this node are created. The training records are partitioned (physically or logically) between these children. This procedure is recursively applied, till either all training records associated with a node have the same class label, or the number of training records associated with a node is below a certain threshold. The different approaches for decision tree construction differ in the way criteria for splitting a node is selected, and the data-structures required for supporting the partitioning of the training sets.

RainForest is a general approach for scaling decision tree construction to larger datasets, while also effectively exploiting the available main memory. This is done by isolating an AVC (Attribute-Value, Classlabel) set for a given attribute and a node being processed. The size of the AVC-set for a given node and attribute is proportional to the number of distinct values of the attribute and the number of distinct class labels. For example, in a SPRINT like approach, AVC-set for a categorical attribute will simply be the count of occurrence of each distinct value the attribute can take. Therefore, the AVC-set can be constructed by taking one pass through the training records associated with the node.

A number of algorithms have been proposed within the RainForest framework to split decision tree nodes at lower levels. In this paper, we will mainly focus on parallelizing the RF-read algorithm. In the algorithm RF-read, the dataset is never partitioned. The algorithm progresses level by level. In the first step, AVC-group for the root node is built and a splitting criteria is selected. At any of the lower levels, all nodes at that level are processed in a single pass if the AVC-group for all the nodes fit in main memory. If not, multiple passes over the input dataset are made to split nodes at the same level of the tree. Because the training dataset is not partitioned, this can mean reading each record multiple times for one level of the tree.

3 Middleware and Parallelization Techniques

Our middleware is based upon the observation that a number of popular algorithms for association mining, clustering, and classification have a common structure. Specifically, the core of the computing in these algorithms follows a canonical loop that is shown in Figure 1. In our earlier work, we have argued how various association mining, clustering and classification algorithms have such a structure [5].

We next describe the parallelization techniques we use.

Full Replication: One simple way of avoiding race conditions is to replicate the reduction object and create one copy for every thread. The copy for each

```
{* Outer Sequential Loop *}
While() {
    {* Reduction Loop *}
    Foreach(element e) {
        (i, val)    =  Compute(e) ;
        RObj(i)     =  Reduc(RObj(i),val) ;
    }
}
```

Fig. 1. Canonical Loop Depicting the Structure of Common Data Mining Algorithms

thread needs to be initialized in the beginning. Each thread simply updates its own copy, thus avoiding any race conditions. After the local reduction has been performed using all the data items on a particular node, the updates made in all the copies are *merged.*

We next describe the locking schemes.

Full Locking: One obvious solution to avoiding race conditions is to associate one lock with every element in the reduction object. After processing a data item, a thread needs to acquire the lock associated with the element in the reduction object it needs to update.

This scheme has overheads of three types, doubled memory requirements, increased number of cache misses due to accessing locks and elements, and false sharing.

Optimized Full Locking: Optimized full locking scheme overcomes the the large number of cache misses associated with full locking scheme by allocating a reduction element and the corresponding lock in consecutive memory locations. This results in at most one cache miss when a reduction element and the corresponding lock is accessed.

Cache-Sensitive Locking: The final technique we describe is *cache-sensitive locking.* Consider a 64 byte cache block and a 4 byte reduction element. We use a single lock for all reduction elements in the same cache block. Moreover, this lock is allocated in the same cache block as the elements. So, each cache block will have 1 lock and 15 reduction elements.

Cache-sensitive locking reduces each of three types of overhead associated with full locking. This scheme results in lower memory requirements than the full locking and optimized full locking schemes. Each update operation results in at most one cache miss, as long as there is no contention between the threads. The problem of false sharing is also reduced because there is only one lock per cache block.

4 Parallel RainForest Algorithm and Implementation

In this section, we will present the algorithm and implementation details for parallelizing RF-read using our middleware. The algorithm is presented in Figure 5.

The algorithm takes several passes over the input dataset D. The dataset is organized as a set of chunks. During every pass, there are a number of nodes that are *active* or belong to the set AQ. These are the nodes for which AVC-group is built and splitting criteria is selected.

This processing is performed over three consecutive loops. In the first loop, the chunks in the dataset are read. For each training record or *tuple* in each chunk that is read, we determine the *node* at the current level to which it belongs. Then, we check if the node belongs to the set AQ. If so, we increment the elements in the AVC-group of the node.

The second loop finds the best splitting criteria for each of the active nodes, and creates the children. Before that, however, it must check if a *stop condition* holds for this node, and therefore, it need not be partitioned. For the nodes that are partitioned, no physical rewriting of data needs to be done. Instead, just the tree should be updated, so that future invocations to *classify* point to the appropriate children. The nodes that have been split are removed from the set AQ and the newly created children are added to the set Q.

At the end of the second loop, the set AQ is empty and the set Q contains the nodes that still need to be processed. The third loop determines the set of the nodes that will be processed in the next phase. We iterate over the nodes in the set Q, remove a node from Q and move it to AQ. This is done till either no more memory is available for AVC-groups, or Q is empty.

The last loop contains only a very small part of the overall computing. Therefore, we focus on parallelizing the first and the second loop. Parallelization of the second loop is straight-forward and discussed first.

A simple multi-threaded implementation is used for the second loop. There is one thread per processor. This thread gets a node from the set AQ and processes the corresponding AVC-group to find the best splitting criteria. The computing done for each node is completely independent. The only synchronization required is for getting a node from AQ to process. This is implemented by simple locking.

Next, we focus on the first loop. Note that this loop fits nicely with the structure of the canonical loop we had shown in Figure 1. The set of AVC-groups for all nodes that are currently active is the reduction object. As different consumer threads try to update the same element in a AVC-set, race conditions can arise. The elements of the reduction object that are updated after processing a tuple cannot be determined without processing the tuple.

Therefore, the parallelization techniques we had presented in the last section are applicable to parallelizing the first loop. Both memory overheads and locking costs are important considerations in selecting the parallelization strategy. At lower levels of the tree, the total size of the reduction object can be very large. Therefore, memory overhead of the parallelization technique used is an important consideration. Also, the updates to the elements of the reduction object are fine-grained. After getting a lock associated with an element or a set of elements, the only computing performed is incrementing one value. Therefore, locking overheads can also be significant.

Next, we discuss the application of the techniques we have developed to parallelization of the first loop. Recall that the memory requirements of the three techniques are very different. If R is the size of reduction object, N is the size of consumer threads, and L is the number of elements per cache line, the memory requirement of full replication, optimized full locking and cache sensitive locking are $N \times R$, $2 \times R$, and $\frac{N}{N-1} \times R$, respectively. This has an important implication for our parallel algorithm. Choosing a technique with larger memory requirements means that the set AQ will be smaller. In other words, a larger number of passes over the dataset will be required.

An important property of the reduction object in RF-read is that updates to each AVC-set are independent. Therefore, we can apply different parallelization techniques to nodes at different levels, and for different attributes. Based upon this observation, we developed a number of approaches for applying one or more of the parallelization techniques we have. These approaches are, *pure*, *horizontal*, *vertical*, and *mixed*.

In the pure approach, the same parallelization approach is used for all AVC-sets, i.e., for nodes at different levels and for both categorical and numerical attributes.

The vertical approach is motivated by the fact that the sum of sizes of AVC-groups for all nodes at a level is quite small at upper levels of the tree. Therefore, full replication can be used for these levels without incurring the overhead of additional passes. Moreover, because the total number of elements in the reduction object is quite small at these levels, locking schemes can result in high overhead of waiting for locks and coherence cache misses. Therefore, in the vertical approach, replication is used for the first few levels (typically between 3 to 5) in the tree, and either optimized full locking or cache-sensitive locking is used at lower levels.

In determining the memory overheads, the cost of waiting for locks, and coherence cache misses, one important consideration is the number of distinct values of an attribute. If the number of the distinct values of an attribute is small, the corresponding AVC-set is small. Therefore, the memory overhead in replicating such AVC-sets may not be a significant consideration. At the same time, because the number of elements is small, the cost of waiting for locks and coherence cache misses can be significant. Note that typically, categorical attributes have a small number of distinct values and numerical attributes can have a large number of distinct values in a training set.

Therefore, in the horizontal approach, full replication is used for attributes with small number of distinct values, and one of the locking schemes is used for attributed with a large number of distinct values. For any attribute, the same technique is used at all levels of the tree.

Finally, the mixed strategy combines the two approaches. Here, full replication is used for all attributes at the first few levels, and for attributes with small number of distinct values at the lower levels. One of the locking schemes is used for the attributes with a large number of distinct values at lower levels of the tree.

5 Experimental Results

We used a Sun Fire 6800. Each processor in this machine is a 64 bit, 750 MHz Sun UltraSparc III. Each processor has a 64 KB L1 cache and a 8 MB L2 cache. The total main memory available is 24 GB. The Sun Fireplane interconnect provides a bandwidth of 9.6 GB per second. We experimented with up to 8 consumer threads on this machine.

The dataset we used for our experiments was generated using a tool described by Agrawal *et al.* [1]. The dataset is nearly 1.3 GB, with 32 million records in the training set. Each record has 9 attributes, of which 3 are categorical and other 6 are numerical. Every record belongs to 1 of 2 classes.

In Section 4, we had described *pure*, *vertical*, *horizontal*, and *mixed* approaches for using one or more of the parallelization techniques we support in the middleware. Based upon these, a total of 9 different versions of our parallel implementation were created. Obviously, there are three pure versions, corresponding to the use of full replication (`fr`), optimized full locking (`ofl`) and cache sensitive locking (`csl`). Optimized full locking can be combined with full replication using vertical, horizontal, and mixed approach, resulting in three versions. Similarly, cache sensitive locking can be combined with full replication using vertical, horizontal, and mixed approach, resulting in three additional versions, for a total of 9 versions.

Figure 2 shows the performance of pure versions. With 1 thread, `fr` gives the best performance. However, the parallel speedups are not good. This is because the the use of full replication for AVC-sets at all levels results in very high memory requirements. Locking schemes result in a 20 to 30% overhead on 1 thread, but the relative speedups are better. Using 8 threads, the relative speedups for `ofl` and `csl` are 5.37 and 4.95, respectively.

Figure 3 shows the experimental results from combining `fr` and `ofl`. As stated earlier, the two schemes can be combined in three different ways, horizontal, vertical, and mixed. The performance of these three versions is quite similar.

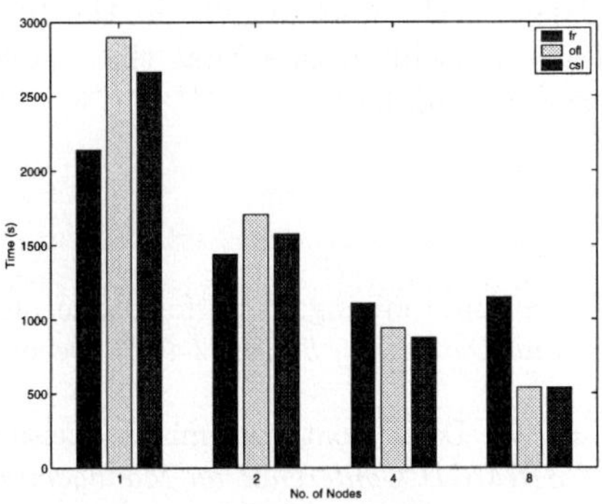

Fig. 2. Performance of *pure* versions

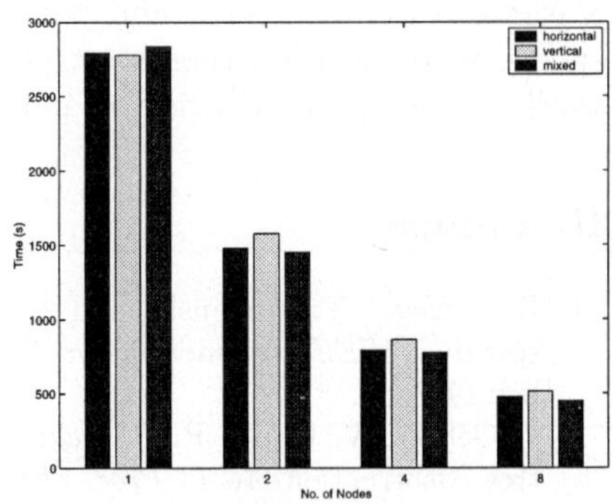

Fig. 3. Combining full replication and full locking, `dataset 2`

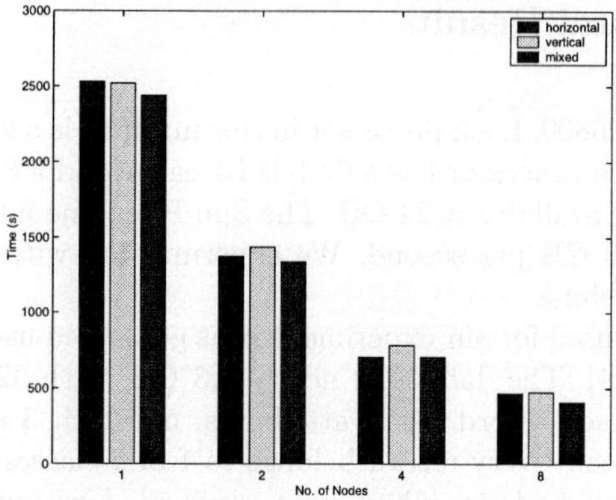

Fig. 4. Combining full replication and cache-sensitive locking

`vertical` is the slowest on 2, 4, and 8 threads, whereas `mixed` is the best on 2, 4, and 8 threads.

Figure 4 presents the experimental results from combining `fr` and `csl`. Again, the `mixed` version is the best among the three versions, for 2, 4, and 8 threads.

6 Conclusions

In this paper, we have presented a shared memory parallelization of a RainForest based decision tree construction algorithm, using a middleware framework we had developed in our earlier work.

Our work has lead to a number of interesting observations. First, we have shown that a RainForest based decision tree construction algorithm can be parallelized in a way which is very similar to the way association mining and clustering algorithms have been parallelized. Therefore, a general middleware framework for decision tree construction can simplify the parallelization of algorithms for a variety of mining tasks. Second, unlike the algorithms for other mining tasks, a combination of locking and replication based techniques results in the best speedups for decision tree construction. Thus, it is important that the framework used supports a variety of parallelization techniques.

References

1. R. Agrawal, T. Imielinski, and A. Swami. Database mining: A performance perspective. *IEEE Transactions on Knowledge and Data Eng., 5(6):914-925,*, December 1993.
2. J. Gehrke, V. Ganti, R. Ramakrishnan, and W. Loh. Boat– optimistic decision tree construction. In *In Proc. of the ACM SIGMOD Conference on Management of Data*, June 1999.
3. J. Gehrke, R. Ramakrishnan, and V. Ganti. Rainforest - a framework for fast decision tree construction of large datasets. In *VLDB*, 1996.

```
RF-Read(dataset D)
 global Tree root, Queue Q ,AQ;
 local Node node;
 Q ← NULL; AQ ← NULL;
 add(root, AQ);
 while not (empty(Q) and empty(AQ))
      { Loop1: build AVC-group for nodes in AQ}
      foreach (chunk C ∈ D)
           foreach (tuple t ∈ C)
                node ← classify(root, t);
                if node ∈ AQ
                     foreach (attribute a ∈ t)
                        reduction(node.avc_group, a, t.class);
      { Loop2: split the nodes in AQ}
      foreach (node ∈ AQ)
           if not satisfy_stop_condition(node)
                find_best_split(node);
                foreach (Node child ∈ create_children(node))
                     add(child, Q);
      { Loop3: build new AQ}
      AQ ← NULL;
      done ← false;
      while not empty(Q) and not done
           get(node, Q);
           if enough_memory(Q)
                remove(node, Q);
                add(node, AQ);
             else done ← true;
```

Fig. 5. Algorithm for Parallelizing RF-read Using Our Middleware

4. Ruoming Jin and Gagan Agrawal. A middleware for developing parallel data mining implementations. In *Proceedings of the first SIAM conference on Data Mining*, April 2001.
5. Ruoming Jin and Gagan Agrawal. Shared Memory Parallelization of Data Mining Algorithms: Techniques, Programming Interface, and Performance. In *Proceedings of the second SIAM conference on Data Mining*, April 2002.
6. M. V. Joshi, G. Karypis, and V.Kumar. Scalparc: A new scalable and efficient parallel classification algorithm for mining large datasets. In *In Proc. of the International Parallel Processing Symposium*, 1998.
7. F. Provost and V. Kolluri. A survey of methods for scaling up inductive algorithms. *Knowledge Discovery and Data Mining*, 3, 1999.
8. J. Ross Quinlan. *C4.5: Programs for Machine Learning.* Morgan Kaufmann, San Mateo, CA, 1993.

9. S. Ruggieri. Efficient c4.5. Technical Report TR-00-01, Department of Information, University of Pisa, February 1999.
10. J. Shafer, R. Agrawal, and M. Mehta. SPRINT: A scalable parallel classifier for data mining. In *Proceedings of the 22nd International Conference on Very Large Databases (VLDB)*, pages 544–555, September 1996.
11. M. J. Zaki, C.-T. Ho, and R. Agrawal. Parallel classification for data mining on shared-memory multiprocessors. *IEEE International Conference on Data Engineering*, pages 198–205, May 1999.

Characterizing the Scalability of Decision-Support Workloads on Clusters and SMP Systems

Yanyong Zhang[1], Anand Sivasubramaniam[1], Jianyong Zhang[1], Shailabh Nagar[2], and Hubertus Franke[2]

[1] Dept. of Computer Science & Engg., The Pennsylvania State University
University Park, PA 16802, USA
{yyzhang,anand,jzhang}@cse.psu.edu
[2] IBM Thomas J. Watson Research Center,
Yorktown Heights NY 10598, USA
{nagar, frankeh}@us.ibm.com

Abstract. Using a public domain version of a commercial clustered database server and TPC-H like[1] decision support queries, this paper studies the performance and scalability issues of a Pentium/Linux cluster and an 8-way Linux SMP. The execution profile demonstrates the dominance of the I/O subsystem in the execution, and the importance of the communication subsystem for cluster scalability. In addition to quantifying their importance, this paper provides further details on how these subsystems are exercised by the database engine.

1 Introduction

Commercial workloads have long been used to benchmark the performance of server systems. These workloads have influenced the design of all aspects of computer systems from hardware to operating systems, middleware and applications. The TPC series of benchmarks is an important set of representative commercial workloads that can be used to test the performance of computer systems.

The goal of this paper is to study the scalability characteristics of the TPC-H decision support benchmark. Specifically, it examines the impact of scaling important system resources such as CPU, memory, disks and network on the performance of a midsized TPC-H benchmark implemented on a standard database engine. The study uses the DB2 database engine from International Business Machine Corp. (IBM) on two commonly used server platforms : a cluster of 2-way symmetric multiprocessors (SMPs) and an 8-way SMP, both running the Linux operating system (OS).

In our study, the DB2 database engine, the Linux operating system and the hardware characteristics, such as the speeds of the processor, memory and disk, are all considered to be part of the environment. As such, we have done

[1] These results have not been audited by the Transaction Processing Performance Council and should be denoted as "TPC-H like" workload.

B. Monien and R. Feldmann (Eds.): Euro-Par 2002, LNCS 2400, pp. 355–364.

only a reasonable amount of optimization of DB2 and the Linux kernel. The focus of the study is on understanding the impact of varying the amount of hardware resources on the performance of the TPC-H benchmark. Our work neither attempts to maximize the performance of the benchmark nor does it try to evaluate the database engine or the operating system. The same reporting standards that serve to make TPC-H an important workload also prevent us from reporting absolute performance numbers. Since the study focusses on scalability rather than performance, this does not limit the value of the reported results.

Studying the impact of varying amounts of hardware resources on TPC-H performance offers two benefits. It aids system administrators in capacity planning and tuning. Further, middleware and OS designers can gain insights into scalability bottlenecks which helps them make design tradeoffs.

The methodology followed by our study is as follows. We first characterize the various queries in the TPC-H benchmark on a cluster with respect to their usage of CPU, memory and I/O bandwidth. We then run the benchmark on a cluster of 2-way SMPs and study the performance impact of adding nodes to the cluster. Each added node increases the CPU, memory and disk resources available for the workload and incurs a potential penalty of increased network costs. We vary the number of CPUs and memory in the cluster to try to identify which of the resources affects performance the most. A similar exercise is done on an 8-way SMP platform. Besides eliminating the effects of networking, the SMP platform allows a greater flexibility in varying the number of CPUs and memory. Finally, we analyze these three sets of results to glean characteristics of the workload.

The results of the experiments broadly show that I/O is by far the most important scalability bottleneck on both the cluster and the SMP platform. Individual queries demonstrate different attributes which can be correlated to their characteristics. The importance of I/O bandwidth suggests that a more aggressive overlap of computation and I/O would be desirable while designing the next generation databases and operating systems. It also suggests that it is better to explicitly increase the I/O parallelism in a cluster rather than rely on an implicit increase through node addition. Similarly in an SMP environment, more than 4 CPUs and 1.5 GB memory does not increase performance for even a 10GB dataset size. It is more useful to add disks and increase the I/O bandwidth.

The second important conclusion of this study is that the networking overheads of a cluster are not a significant scalability inhibitor. The benefits of the added disks, CPU and memory outweigh the additional networking cost when a cluster is scaled by adding a node. Such scaling is seen to be a viable option even with the current levels of clustering support in hardware, database and OS.

2 Experimental Setup

The TPC-H benchmark is best described by TPC's own website as follows: "TPC Benchmark H (TPC-H) is a decision support benchmark. It consists of a suite of oriented ad-hoc queries and data modifications. The queries and the data populating the database have been chosen to have broad industry-wide relevance. This benchmark illustrates decision systems that examine large volumes of data, execute queries with a high degree of complexity, and give answers to critical business questions".

In the interest of space, we are not including all the details about the distribution of TPC-H tables across the cluster nodes or the implementation of the queries. This workload contains a sequence of 22 queries (Q1 to Q22), that are fired one after another to the database engine. Queries 21 and 22 take too long to run on clusters and hence are omitted from the results shown. There are several measures that are used to determine their performance as specified in [2]. We choose query completion time as our performance metric.

TPC-H workloads use standard dataset sizes ranging from 1 to 3000 GB. We have chosen to run TPC-H with a 10GB dataset size, keeping in mind our resource and time constraints. This is a representative dataset size for a small business. TPC-H workload was run on two platforms : a cluster of 2-way SMPs and an 8-way SMP. Henceforth we shall refer to the former as the cluster and the latter as the SMP.

The cluster consists of eight nodes with each node having two Pentium II CPUs, 256 MB RAM and one disk of 9 GB capacity. The nodes are connected by both switched Myrinet [3] and Ethernet. Unless stated otherwise, TCP over Myrinet is used for network communication. Ethernet is used in one set of results to show that the network bandwidth is not a scalability bottleneck. The cluster nodes run Redhat Linux 7.2 with kernel version 2.4.8. This kernel has been instrumented in detail to gather different statistics, and also modified to provide insight on the database engine execution.

During the runs, a client machine (not part of the cluster), sends the TPC-H queries to a database coordinator node on the cluster, which then distributes the work and gives back the results to the client. A freely available version of DB2 Extended Enterprise Edition (EEE) version 7.2 [1] was used. The EEE version of DB2 is specifically written to take advantage of cluster hardware. The database was run in partitioned mode with one partition per cluster node. Hardware resources were scaled mainly through the addition of nodes. In some experiments, individual node configurations were modified to use only one CPU or lesser RAM through the `maxcpus` and `mem` Linux boot parameters.

The SMP experiments were conducted on an 8-way IBM Netfinity 8500R server with PIII processors, 2MB L2 cache and 2.5GB of main memory. The operating system was Red Hat Linux 7.2 running the 2.4.17 kernel. The default OS installation was modified to increase kernel resource limits such as the number of open files and semaphores. A scalable timer patch was also applied to the base kernel to take care of known timer issues. No TPC-H specific tuning was done to the OS. The freely available version of DB2 Enterprise Edition (EE) version 7.2 [1] was used on the SMP. The number of CPU's and physical memory configuration was varied using `maxcpus` and `mem` boot parameters.

Table 1. OS profile for 8-node cluster (statistics are collected from node 1)

query	user CPU (%)	system CPU (%)	page faults per jiffy	blocks read per jiffy	blocks written per jiffy	packets sent per jiffy	packets received per jiffy	CPU utilization during IO (%)
Q1	53.80	17.71	1.50	51.01	23.1151	0.0012	0.0015	26.87
Q2	70.78	15.67	0.39	21.97	1.7718	1.9077	1.9248	53.73
Q3	69.93	17.19	0.99	55.47	4.9591	0.9778	0.9938	55.40
Q4	53.38	17.10	0.00	22.97	1.0478	0.3517	0.3652	68.41
Q5	60.92	17.09	0.05	16.73	1.1369	2.0779	2.0849	42.81
Q6	41.41	25.44	1.73	90.72	0.0020	0.0012	0.0012	33.49
Q7	68.54	14.76	0.00	16.89	1.9467	2.3880	2.3521	31.04
Q8	28.52	23.22	0.01	27.63	4.8078	0.0261	0.0228	12.91
Q9	63.42	15.51	0.00	6.69	1.9276	0.0133	0.0136	23.21
Q10	39.99	20.40	0.17	41.99	1.8880	0.8774	0.8859	18.71
Q11	79.63	13.16	0.49	18.87	0.0020	2.3794	2.4011	66.46
Q12	16.07	21.96	0.25	40.79	0.0197	0.0102	0.0101	4.8
Q13	49.79	22.31	1.62	45.86	0.0034	2.0966	2.0935	32.71
Q14	36.91	23.27	1.01	84.78	0.0025	0.1156	0.1175	29.75
Q15	55.37	18.69	0.60	75.26	0.0033	0.6260	0.7703	51.68
Q16	51.84	15.30	2.32	15.36	0.8454	2.4836	2.4993	46.24
Q17	33.02	20.49	0.00	32.76	5.2532	0.0036	0.0037	13.94
Q18	65.77	20.15	0.04	17.16	0.6261	0.3890	0.3803	35.11
Q19	38.11	22.29	0.29	78.76	0.0032	0.3343	0.3329	23.38
Q20	29.16	19.35	0.00	15.74	0.1601	0.3456	0.2973	47.36

3 Operating System Profile

Before we study how the queries scale when we increase different hardware resources, it is important for us to first characterize the executions of these queries. These statistics will give us indications what resources are the main limiting factors to the workloads. There are four main hardware components which are being exercised by TPC-H, namely, CPU, memory, disks and network (the last is only applicable in the clustered version). In Linux, one can obtain resource usage statistics through the proc file system which are updated every jiffy (10 milliseconds). We sample these statistics roughly every 500 milliseconds to minimize the perturbation due to the sampling. By sampling the numbers given by files (`stat`, `net/dev`, `process/stat`), we obtain a number of interesting statistics for each query. These are shown in Table 1 for the cluster and in Table 2 for the SMP environment.

We observe that the bulk of the query execution time is spent in waiting for disk I/O. High number of disk I/Os result in not only poor CPU utilization, but also high system call overheads. Amongst the I/O operations, reads are much more common than writes. Individual queries have specific characteristics which are referred to later.

4 Scalability Issues in Clustered Database Engine

In the first set of experiments, we study how the average job response time scales with the number of nodes in the cluster. When we increase the number of nodes, the total amount of memory and the number of disks will also increase

Table 2. OS profile for 8-way SMP

id	user CPU (%)	system CPU (%)	page faults per unit time	blocks read per unit time	blocks written per unit time
1	44.70	39.35	0.00	0.11	0.1109
3	27.86	31.26	0.02	0.05	0.1069
4	16.11	29.29	0.00	0.08	0.1886
5	24.39	24.23	0.02	0.15	0.1695
6	14.90	19.30	0.01	0.71	0.5748
7	27.92	13.17	0.01	0.06	0.0822
8	15.97	9.98	0.01	0.02	0.0151
9	17.56	9.40	0.06	0.04	0.1620
10	31.23	35.54	0.63	0.02	0.2277
11	5.01	7.12	0.35	0.01	0.0118
12	4.93	5.49	0.04	0.02	0.0081
13	45.11	9.83	1.47	0.02	0.0729
14	9.39	16.80	1.47	0.21	0.5112
15	19.93	25.27	1.26	0.02	0.0698
16	25.15	13.23	7.08	0.12	0.0107
17	16.00	9.26	4.20	0.15	0.2397
18	41.44	18.89	0.03	0.05	0.1501
19	39.78	41.32	0.05	0.25	0.0955
20	6.07	7.98	0.28	0.19	0.0078

proportionally. Fig.1 shows the results as the nodes in the cluster are increased from 1 to 8. It may be noted that for the one node configuration, a software RAID using two 9 GB disks was utilized to accomodate the 10GB dataset. Six of the bars shown are truncated in the interest of clarity; their values are: 8.85, 2.4, 1.38, 1.47, 57.64 and 4.07 respectively.

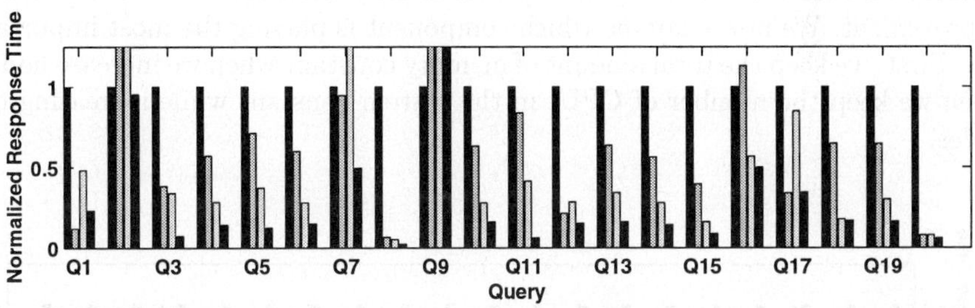

Fig. 1. For each query, we show (left to right) the query response times with a configuration using (respectively): (i) one 2-way SMP node; (ii) two 2-way SMP nodes, myrinet; (iii) four 2-way SMP nodes, myrinet, and (iv) eight 2-way SMP nodes, myrinet. Execution times are normalized with respect to configuration (i).

We have the following observations from the results :

– Increasing the number of nodes can increase the disk parallelism, the total amount of memory which can be used as the bufferpool, and CPU processing power. On the other hand, it will also incur/increase the network overhead. In general, we find that the benefits of having more nodes offset the drawbacks. We can decrease the response time when we have more nodes.

- Queries Q4, Q6, Q10, Q13, Q14, Q15 and Q19 have a more or less linear decrease in response times. This is because these queries either have high number of disk accesses like in Q6, Q13, Q14, and Q19 (blocks read per jiffy column in table 1), large scope to hide IO cost with useful computation as in Q4 (the last column in. table 1), a mix of these two factors as in Q10 and Q15. In these cases, more nodes can bring in disk parallelism, and they can hide disk overhead with computation.
- Queries Q2, Q3, Q5, Q11 have a super-linear decrease in response times when we increase the number of nodes. These queries also have the characteristics of the above queries. Besides, their CPU utilization is very high. As a result, they can benefit not only from more disk parallelism, but also from more CPU parallelism by more nodes.
- Queries Q7, Q8, Q9, Q16, Q18, and Q20 do not seem to scale from 2/4 nodes to 8 nodes. It can be explained as follows. Both of these queries have low disk accesses, so they cannot benefit much from the disk parallelism. Also, they incur quite high network overhead (packets sent per jiffy column in table 1 when we have more nodes in the system.
- Queries Q1, Q12 and Q17 do not benefit much from adding more nodes in general. It can be explained by the fact that all three have low overall CPU utilization which make them not able to benefit much from the parallelism.

From this study, we can see that performance is improved when we increase number of nodes in the system. When the number of nodes increases, CPU, memory and disk all increase linearly, which all contribute to the performance improvement. We now examine which component is playing the most important role. First, we keep the total amount of memory constant when we increase nodes. Then we keep the number of CPUs in the system constant while increasing the nodes.

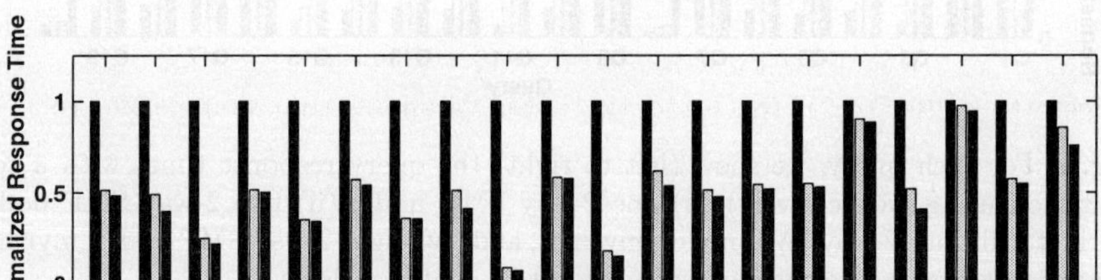

Fig. 2. For each query, we show (left to right) the query response times with a configuration using (respectively): (i) four 2-way SMP nodes, myrinet, with 256M RAM on each node; (ii) eight 2-way SMP nodes, myrinet, with 128M RAM on each node; and (iii) eight 2-way SMP nodes, myrinet, with 256M RAM on each node. Execution times are normalized with respect to configuration (i).

To help us understand the impact of available physical memory on workload performance, we use three system configurations : 4 nodes with 256MB RAM, 8 nodes with 128MB RAM and 8 nodes with 256M RAM, henceforth referred by (4,256), (8,128) and (8,256). In Fig. 2, we see great performance improvement for most of the queries as we move from (4,256) to (8,256). On the other hand, going from (8,128) to (8,256) shows only marginal performance gains. This shows that memory is not a major contributor to the performance, though it can have some impact. In TPC-H, most of the queries involve sequential scans through the data leading to working sets which are much larger than the available memory. Furthermore, the data reuse rate is very low. In general, in the presence of both these characteristics i.e. large working set size and low data locality, adding memory only provides marginal benefits. This is further confirmed by the relatively large performance improvements seen for queries 17 and 20. An examination of their data access patterns shows that they both have relatively smaller working sets.

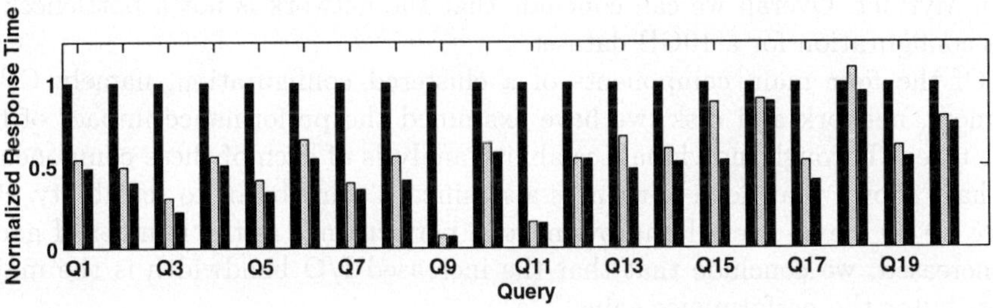

Fig. 3. For each query, we show (left to right) the query response times with a configuration using (respectively): (i) four 2-way SMP nodes ; (ii) eight uniprocessor nodes and (iii) eight 2-way SMP nodes; Execution times are normalized with respect to configuration (i).

Next we want to see if CPU is the limiting factor for queries. We compare the performance of 4 dual cpu nodes, 8 single cpu nodes and 8 dual cpu nodes, henceforth called (4,2), (8,1) and (8,2). Fig. 3 shows that going from (8,1) to (8,2) does not help most of the queries except Q15 (which has a high CPU utilization as well as a good computation overlap with I/O as is seen in Table 1. But we see a substantial gain while going from (4,2) to (8,1). From this we can conclude that the major benefits of increased nodes do not come from the addition of CPUs.

Having eliminated the CPU and memory as overall performance boosters, we next look at the major performance inhibitor in a cluster environment, namely the network. In Fig. 4 we compare the performance under two different network subsystem configurations. In the first setup which is our default network configuration, we run TCP/IP over Myrinet. The Myrinet link bandwidth is 1.06 Gbps which is substantially higher than that of 100 Mbps Ethernet which is the

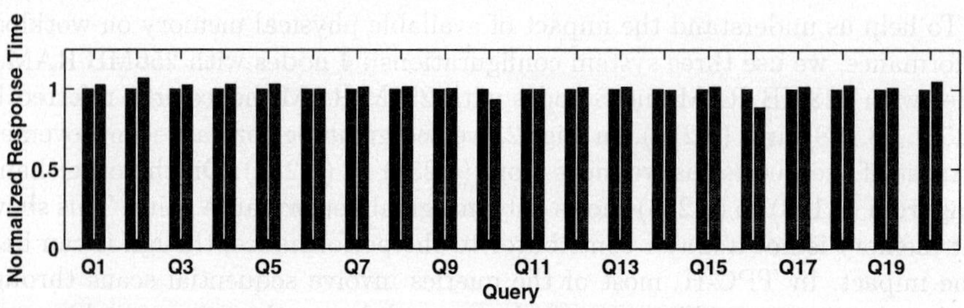

Fig. 4. For each query, we show (left to right) the query response times with a configuration using (respectively): (i) eight 2-way SMP nodes, myrinet, and (ii) eight 2-way SMP nodes, 100Mbps Ethernet. Execution times are normalized with respect to configuration (i).

second setup. While one would expect the faster network hardware to show some benefit, we see less than 5% improvement. Q10 and Q15 even show a degradation with Myrinet. Overall we can conclude that the network is not a bottleneck in this configuration for a 10GB dataset.

Of the four main components of a clustered configuration, namely CPU, memory, network and disk, we have examined the performance impact of the first three. Through individual scalability analysis of each of these components, we have shown that none of them is a significant contributor to scalability. But since we do see an overall improvement in performance as the number of nodes is increased, we conclude that that the increased I/O bandwidth is the major contributor the performance gains.

5 Scalability on SMP

In this section, we analyze the scalability of TPC-H on the 8-way SMP platform. We do not intend to compare the scalability of a cluster with that of an SMP as the underlying hardware and middleware is significantly different. For an SMP server, there are mainly three components which are exercised by the workloads: CPU, memory and disks. We investigate the impact of each of these components individually.

Fig. 5 shows the performance impact when system memory is increased from 1024 MB to 2560 MB. For 8 of the 20 queries shown, we see significant performance gains as memory increases from 1024 to 1536 MB. After that and for all queries, increasing memory barely improves response time. This indicates that memory is not a limiting resource for these queries. Three bars are truncated in the interests of clarity; their values are: 1.54, 1.70 and 1.71 respectively.

Next, we vary the number of available CPUs to investigate its impact on the response time for each query. Though adding CPUs might increase overall parallelism, on an SMP it might also lead to increased contention in the middleware and OS. Fig. 6 shows the response times for each query as the number of CPUs is varied from 1 to 8.

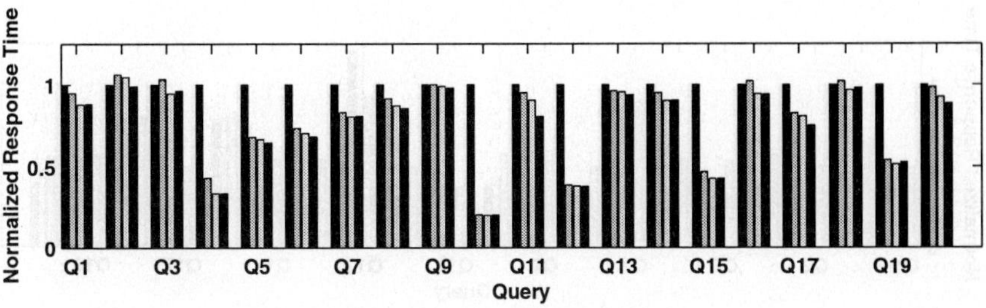

Fig. 5. For each query, we show query response times with a configuration using respectively (left to right): (i) 8-way SMP, 1024M RAM; (ii) 8-way SMP, 1536M RAM; (iii) 8-way SMP, 2048M RAM; and (iv) 8-way SMP, 2560M RAM; Execution times are normalized with respect to configuration (i).

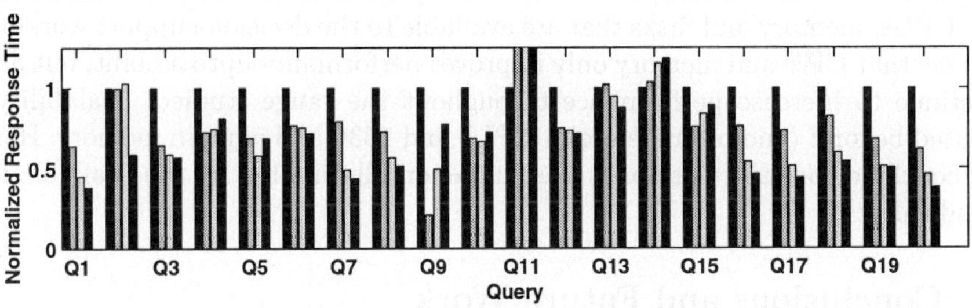

Fig. 6. For each query, we show query response times with a configuration using respectively (left to right): (i) 1-way SMP (ii) 2-way SMP; (iii) 4-way SMP; and (iv) 8-way SMP. Execution times are normalized with respect to configuration (i).

Most queries benefit when we increase number of CPUs. We see consistent response time decrease when we move from 1-way to 8-way SMP for queries Q1, Q3, Q7, Q8, Q12, Q13, Q16, Q17, Q18, Q19, and Q20. The maximum gains from increasing CPUs count are seen when we go from 1 to 4 cpus. The performance gain from 4-way to 8-way are, in general, marginal. The performance for queries Q11 and Q14 consistently degrades with number of CPUs. This is probably due to the aforementioned increase in contention. In some cases like queries Q2, Q4, Q5, Q9, Q10, and Q15, we see an optimal CPU count. Performance improves with cpu count upto the optimal point and degrades thereafter. The most common optimal point is 2 CPUs.

Finally, we vary the number of disks that are used by the database. In our setup all disks are managed by a single controller and share a single I/O bus. Hence, increasing the number of disks potentially increases the contention for these shared elements. Fig. 7 shows the benefits of I/O parallelism as the number of disks increases from 2 to 10. Most queries are seen to benefit significantly, with the exception of Q13 and Q18. More significantly, response time are lowest with the maximum number of disks.

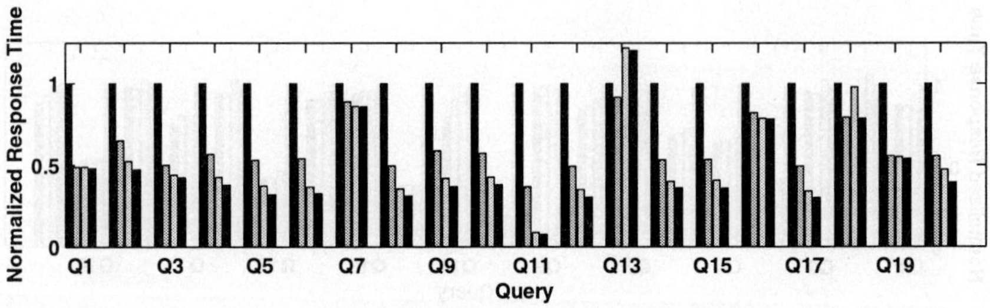

Fig. 7. For each query, we show (left to right) the query response (execution) time with a configuration using respectively: (i) 2 disks (ii) 5 disks; (iii) 8 disks; and (iv) 10 disks. Execution times are normalized with respect to configuration (i).

As in the clustered case, we have selectively examined the impact of increasing the CPUs, memory and disks that are available to the decision-support workload. We see that CPU and memory only improves performance upto a point, but disks continue to increase performance throughout the range studied. Scalability is limited beyond (and often before) 4 CPUs and 1536 MB of main memory. Hence we conclude that the workloads are fundamentally limited by the available I/O bandwidth.

6 Conclusions and Future Work

In this study we set out to examine the scalability of the TPC-H decision support benchmark on cluster and SMP environments. Our main focus was not on performance but on gaining an insight into the impact of various hardware parameters such as CPU, memory, disk and network. Consequently, we did not attempt to optimize the middleware or OS used in this paper.

Our results show that for both the cluster and the SMP environments, the CPU and memory resources are not major contributors to performance beyond a point. For a cluster, adding nodes improves performance even though it increases network overhead. This leads us to conclude that the decision support workload is most sensitive to an increase in I/O parallelism. The hypothesis is confirmed by the SMP results in which query response times consistently improve with an increase in the number of disks.

There is a considerable amount of work that remains to be done as part of this scalability study. We would like to expand the scope of our investigation and look into the Linux operating system to identify potential scalability inhibitors.

References

1. DB2 Universal Database. http://www-3.ibm.com/software/data/db2/udb/.
2. TPC-H Benchmark. `http://www.tpc.org/tpch/default.asp/`.
3. N. J. Boden et al. Myrinet: A Gigabit-per-second Local Area Network. *IEEE Micro*, 15(1):29–36, February 1995.

Parallel Fuzzy c-Means Clustering for Large Data Sets

Terence Kwok[1], Kate Smith[1], Sebastian Lozano[2], and David Taniar[1]

[1] School of Business Systems, Faculty of Information Technology, Monash University, Australia
{terence.kwok, kate.smith, david.taniar}@infotech.monash.edu.au
[2] Escuela Superior de Ingenieros, University of Seville, Spain
slozano@cica.es

Abstract. The parallel fuzzy c-means (PFCM) algorithm for clustering large data sets is proposed in this paper. The proposed algorithm is designed to run on parallel computers of the Single Program Multiple Data (SPMD) model type with the Message Passing Interface (MPI). A comparison is made between PFCM and an existing parallel k-means (PKM) algorithm in terms of their parallelisation capability and scalability. In an implementation of PFCM to cluster a large data set from an insurance company, the proposed algorithm is demonstrated to have almost ideal speedups as well as an excellent scaleup with respect to the size of the data sets.

1 Introduction

Clustering is the process of grouping a data set in such a way that the similarity between data within a cluster is maximised while the similarity between data of different clusters is minimised. It offers unsupervised classification of data in many data mining applications. A number of clustering techniques have been developed, and these can be broadly classified as hierarchical or partitional [1]. Hierarchical methods produce a nested series of partitions of the data, while partitional methods only produce one partitioning of the data. Examples of hierarchical clustering algorithms include the single-link, complete-link, and minimum variance algorithms [1]. Examples of partitional methods include the k-means algorithm [2], fuzzy c-means [3, 4], and graph theoretical methods [5].

For increasingly large data sets in today's data mining applications, scalability of the clustering algorithm is becoming an important issue [6]. It is often impossible for a single processor computer to store the whole data set in the main memory for processing, and extensive disk access amounts to a bottleneck in efficiency. With the recent development of affordable parallel computing platforms, scalable and high performance solutions can be readily achieved by the implementation of parallel clustering algorithms. In particular, a computing cluster in the context of networked workstations is becoming a low cost alternative in high performance computing. Recent research in parallel clustering algorithms has demonstrated their implementations on these machines can yield

B. Monien and R. Feldmann (Eds.): Euro-Par 2002, LNCS 2400, pp. 365–374.

large benefits [7]. A parallel k-means clustering algorithm has been proposed by Dhillon and Modha [8], which was then implemented on an IBM POWERparallel SP2 with a maximum of 16 nodes. Their investigation with test data sets has achieved linear relative speedups and linear scaleup with respect of the size of data sets and the desired number of clusters. Stoffel and Belkoniene [9] have implemented another parallel version of the k-means algorithm over 32 PC's on an Ethernet and shown a near linear speedup for large data sets. Scalability of the parallel k-means algorithm has also been demonstrated by others [10, 11].

For clustering techniques generating *crisp* partitions, each data point belongs to exactly one cluster. However, it is often useful for each data point to admit multiple and non-dichotomous cluster memberships. This requirement has led to the development of fuzzy clustering methods. One of the widely used fuzzy clustering methods is the fuzzy c-means (FCM) algorithm [3, 4]. FCM is a fuzzy partitional clustering approach, and can be seen as an improvement and a generalisation of k-means. Because of FCM's high computational load and similarity to k-means, it is anticipated that a parallel implementation of the FCM algorithm would greatly improve its performance. To the best of our knowledge, such a parallel implementation of the FCM algorithm has not yet been developed.

It is the aim of this paper to propose a parallel version of the FCM clustering algorithm, and to measure its speedup and scaleup capabilities. The algorithm is implemented to cluster a large data set provided by an Australian insurance company, and its performance is compared to a parallel implementation of the k-means algorithm using the same data set. In Sect. 2, a brief outline of the FCM algorithm is presented. We propose the parallel fuzzy c-means (PFCM) clustering algorithm in Sect. 3. A brief description of an existing parallel k-means (PKM) algorithm [8] is given in Sect. 4 for comparison. In Sect. 5, we describe how the PFCM is implemented to cluster an insurance data set, with experimental results comparing the performance of PFCM and PKM. Conclusions are drawn in Sect. 6.

2 The Fuzzy c-Means Algorithm

Suppose that there are n data points $x_1, x_2, \ldots, x_n$ with each data point x_i in $\mathbb{R}^d$, the task of traditional crisp clustering approaches is to assign each data point to exactly one cluster. Assuming that c clusters are to be generated, then c centroids $\{v_1, v_2, \ldots, v_c\}|v_j \in \mathbb{R}^d$ are calculated, with each v_i as a prototype point for each cluster. For fuzzy clustering, c membership values are calculated for each data point x_i, which are denoted by $u_{ji} \in [0, 1], j = 1, \ldots, c;\ i = 1, \ldots, n$. This concept of membership values can also be applied to crisp clustering approaches, for which $u_{ji} \in \{0, 1\}$ and $\sum_{j=1}^{c} u_{ji} = 1\ \forall i$.

The FCM clustering algorithm was proposed by Dunn [3] and generalised by Bezdek [4]. In this method, the clustering is achieved by an iterative optimisation process that minimises the objective function:

$$J = \sum_{i=1}^{n} \sum_{j=1}^{c} (u_{ji})^m \left\| x_i - v_j \right\|^2 \tag{1}$$

subject to

$$\sum_{j=1}^{c} u_{ji} = 1 \ . \tag{2}$$

In (1), $\|\cdot\|$ denotes any inner product norm metric. FCM achieves the optimisation of J by the iterative calculations of v_j and u_{ji} using the following equations:

$$v_j = \frac{\sum_{i=1}^{n}(u_{ji})^m x_i}{\sum_{i=1}^{n}(u_{ji})^m} \tag{3}$$

and

$$u_{ji} = \left(\sum_{k=1}^{c} \left(\frac{d_{ji}}{d_{ki}} \right)^{\frac{2}{m-1}} \right)^{-1} \quad \text{where } d_{ji} = \|x_i - v_j\| \ . \tag{4}$$

The iteration is halted when the condition $\mathrm{Max}\,\{\|u_{ji}^{(t+1)} - u_{ji}^{(t)}\|\} < \epsilon \ \forall j, i$ is met for successive iterations t and $t+1$, where ϵ is a small number. The parameter $m \in [1, \infty)$ is the weighting exponent. As $m \rightarrow 1$, the partitions become increasingly crisp; and as m increases, the memberships become more fuzzy. The value of $m = 2$ is often chosen for computational convenience.

From (3) and (4), it can be seen that the iteration process carries a heavy computational load, especially when both n and c are large. This prompts the development of our proposed PFCM for parallel computers, which is presented in the next section.

3 The Parallel Fuzzy c-Means Algorithm

The proposed PFCM algorithm is designed to run on parallel computers belonging to the Single Program Multiple Data (SPMD) model incorporating message-passing. An example would be a cluster of networked workstations with the Message Passing Interface (MPI) software installed. MPI is a widely accepted system that is both portable and easy-to-use [12]. A typical parallel program can be written in C (or C++ or FORTRAN 77), which is then compiled and linked with the MPI library. The resulted object code is distributed to each processor for parallel execution.

In order to illustrate PFCM in the context of the SPMD model with message-passing, the proposed algorithm is presented as follows in a pseudo-code style with calls to MPI routines:

Algorithm 1: Parallel Fuzzy c-Means (PFCM)

```
1:   P = MPI_Comm_size();
2:   myid = MPI_Comm_rank();
3:   randomise my_uOld[j][i] for each x[i] in fuzzy cluster j;
4:   do {
5:       myLargestErr = 0;
6:       for j = 1 to c
7:           myUsum[j] = 0;
8:           reset vectors my_v[j] to 0;
9:           reset my_u[j][i] to 0;
10:      endfor;
11:      for i = myid * (n / P) + 1 to (myid + 1) * (n / P)
12:          for j = 1 to c
13:              update myUsum[j];
14:              update vectors my_v[j];
15:          endfor;
16:      endfor;
17:      for j = 1 to c
18:          MPI_Allreduce(myUsum[j], Usum[j], MPI_SUM);
19:          MPI_Allreduce(my_v[j], v[j], MPI_SUM);
20:          update centroid vectors:
                          v[j] = v[j] / Usum[j];
21:      endfor;
22:      for i = myid * (n / P) + 1 to (myid + 1) * (n / P)
23:          for j = 1 to c
24:              update my_u[j][i];
25:              myLargestErr = max{|my_u[j][i] - my_uOld[j][i]|};
26:              my_uOld[j][i] = my_u[j][i];
27:          endfor;
28:      endfor;
29:      MPI_Allreduce(myLargestErr, Err, MPI_MAX);
30:  } while (Err >= epsilon)
```

In Algorithm 1, subroutine calls with the MPI prefix are calls to the MPI library [12]. These calls are made whenever *messages* are passed between the processes, meaning that a transfer of data or a computation requires more than one processes. Line 1 means that P processors are allocated to the parallel processing jobs (or *processes*). Each process is assigned an identity number of `myid` $= 0, \ldots, P-1$ (line 2). Following the indexing notation in Sect. 2, each data point is represented by the vector variable `x[i]` where $i = 1, \ldots, n$, and each cluster is indentified by the index j where $j = 1, \ldots, c$. The algorithm requires that the data set be evenly divided into equal number of data points, so that each process computes with its n/P data points loaded into its own local memory. If a computation requires data points stored in other processes, an MPI call is required. Likewise, the fuzzy membership function u_{ji} is divided up among the processes, with the local representation my_u[j][i] storing the membership of the local data only. This divide-and-conquer strategy in parallelising the storage of data and variables allows the heavy computations to be carried out solely in the main memory without the need to access the secondary storage such as the disk. This turns out to enhance performance greatly, when compared to serial algorithms where the data set is often too large to reside solely in the main memory. In line 3, my_uOld[j][i] represents the old value of my_u[j][i], and its values are initialised with random numbers in $[0,1]$ such that (2) is satisfied.

Lines 4–30 of Algorithm 1 are the parallelisations of the iterative computation of (3) and (4). Lines 5–10 reset several quantities to 0: the variable myUsum[j] stores the local summation of (my_uOld[j][i])^m , which corresponds to the denominator of (3); my_v[j] stores the vectorial value of cluster centroid j; and also my_u[j][i], the membership function for the next iteration.

Lines 11–21 compute the cluster centroids v_j. The first half of the computation (lines 11–16) deals with the intermediate calculations carried out within each process, using only the data points local to the process. These calculations are the local summation versions of (3). Since the evaluation of v_j requires putting together all the intermediate results stored locally in each process, two MPI calls are requested (lines 18, 19) for the second half of the computation. The MPI_Allreduce() subroutine performs a parallel computation over P processes using the first argument as its input variable, and the output is stored in its second argument. Here a summation is carried out by using MPI_SUM to yield an output. For example, in line 18 each process would have a different value of myUsum[1] after the calculations in lines 11–16 using local data points; but only one value of Usum[1] would be generated after the summation, which is then stored in the Usum[1] variable in every process.

Lines 22–28 are used to compute the fuzzy membership function my_u[j][i]. Equation (4) is employed for the calculation in line 24, using the most updated values of v[j]. In order to terminate the algorithm upon convergence of the system, we calculate the largest difference between u_{ji}'s of successive iterations (line 25, 29). In line 25, this value is temporarily obtained within each process; then a subsequent value Err covering all processes is obtained by calling MPI_Allreduce() with MPI_MAX (line 29). The algorithm comes to a halt when Err is smaller than the tolerance epsilon (ϵ).

In the next section, an existing parallel k-means (PKM) algorithm is outlined for comparison. The focus is on the similarities and differences between the PFCM and PKM algorithms, which accounts for the respective performances to be presented in Sect. 5.

4 The Parallel k-Means Algorithm

The PKM algorithm described here was proposed by Dhillon and Modha [8] in 1999. It is suitable for parallel computers fitting the SPMD model and with MPI installed. Algorithm 2 below is transcribed from [8] with slightly modified notations for consistency with Algorithm 1 in the last section.

Algorithm 2: Parallel k-Means (PKM)

```
1:  P = MPI_Comm_size();
2:  myid = MPI_Comm_rank();
3:  MSE = LargeNumber;
4:  if (myid = 0)
5:     Select k initial cluster centroids m[j], j = 1...k;
6:  endif;
7:  MPI_Bcast(m[j], 0), j = 1...k;
8:  do {
9:     OldMSE = MSE;
10:    my_MSE = 0;
11:    for j = 1 to k
12:       my_m[j] = 0; my_n[j] = 0;
13:    endfor;
14:    for i = myid * (n / P) + 1 to (myid + 1) * (n / P)
15:       for j = 1 to k
16:          compute squared Euclidean
                      distance d_Sq(x[i], m[j]);
17:       endfor;
18:       find the closest centroid m[r] to x[i];
19:       my_m[r] = my_m[r] + x[i]; my_n[r] = my_n[r] + 1;
20:       my_MSE = my_MSE + d_Sq(x[i], m[r]);
21:    endfor;
22:    for j = 1 to k
23:          MPI_Allreduce(my_n[j], n[j], MPI_SUM);
24:       MPI_Allreduce(my_m[j], m[j], MPI_SUM);
25:       n[j] = max(n[j], 1); m[j] = m[j] / n[j];
26:    endfor;
27:    MPI_Allreduce(my_MSE, MSE, MPI_SUM);
28: } while (MSE < OldMSE)
```

In Algorithm 2, the same strategy of dividing the data set into smaller portions for each process is used. Intermediate results using local data within each process are stored in local variables with the `my_` prefix. The variable `MSE` stores the mean-squared-error for the convergence criterion. Since the k-means algorithm forms crisp partitions, the variable `n[j]` is used to record the number of data points in cluster j. The initialisation of the centroids takes place in lines 4–7. This is performed locally in one process, then the centroids are broadcast to every process using `MPI_Bcast()`. This approach is not adopted for the initialisation of the fuzzy membership function in Algorithm 1 (line 3) because of two reasons: `my_uOld[j][i]` is a local quantity to each process, so the initialisation can be done locally; and there are $c \times n$ membership values in total, which should be computed in parallel across all processes for sharing the load. It is interesting to note that although the PFCM algorithm appears to be more complicated, it only requires three `MPI_Allreduce`'s, which is the same number of calls as in

PKM (Algorithm 2). This is certainly desirable as extra `MPI_Allreduce` calls means extra time spent for the processes to communicate with each other.

In the next section, we illustrate how the PFCM algorithm is implemented to cluster a large data set in business. The computational performance of PFCM is then measured and compared to PKM.

5 Experiments

The proposed PFCM algorithm is implemented on an AlphaServer computing cluster with a total of 128 processors and 64 Gigabytes of main memory. 32 Compaq ES40 workstations form the cluster, each with 4 Alpha EV68 processors (nodes) running at 833 MHz with 2 Gigabytes of local main memory. A Quadrics interconnect provides a very high-bandwidth (approx. 200 Megabytes/sec per ES40) and low-latency (6 msec) interconnect for the processors. We use C for the programming, and MPI is installed on top of the UNIX operating system.

Since it is our aim to study the speedup and scaleup characteristics of the proposed PFCM algorithm, we use `MPI_Wtime()` calls in our codes to measure the running times. We do not include the time taken for reading in the data set from the disk, as that is not part of our algorithm. The data set for our experiments is provided by an Australian motor insurance company, which consists of insurance policies and claim information. It contains 146,326 data points of 80 attributes, totalling to around 24 Megabytes.

First of all, we study the speedup behaviour of the PFCM algorithm. Speedup measures the efficiency of the parallel algorithm when the number of processors used is increased, and is defined as

$$\text{Speedup} = \frac{\text{running time with 1 processor}}{P \times \text{running time with } P \text{ processors}} \,. \tag{5}$$

For an ideal parallelisation, speedup = 1. Fig. 1 shows the speedup measurements for five different sizes of the data set. The five speedup curves correspond to $n = 2^9, 2^{11}, 2^{13}, 2^{15}$, and 2^{17} respectively. These parameters are kept constant for Fig. 1: $m = 2, \epsilon = 0.01$, and $c = 8$. A log_{10} is applied on the running times for better presentation. For a particular n, the dotted line represents the measured running time, while the solid line represents the running time of an ideal speedup. In other words, the closer the two curves are to each other, the better the speedup. It can be seen from Fig. 1 that the speedups are almost ideal for large values of n. But for smaller n's, the speedup deteriorates. The cause for the deterioration is likely to be that when many processors are used and n is relatively small, the running time becomes very short, but the time spent in inter-processors communication cannot be reduced in the same rate, thus worsening the speedup.

For a comparison of speedup behaviours between PFCM (Algorithm 1) and PKM (Algorithm 2), a set of corresponding measurements are plotted in Fig. 2 for the PKM algorithm with the same values of n and $k = 8$. By comparing Fig. 1 and Fig. 2, it can be seen that although the running times for PFCM are longer in general, it has a better speedup than PKM, especially for small n's.

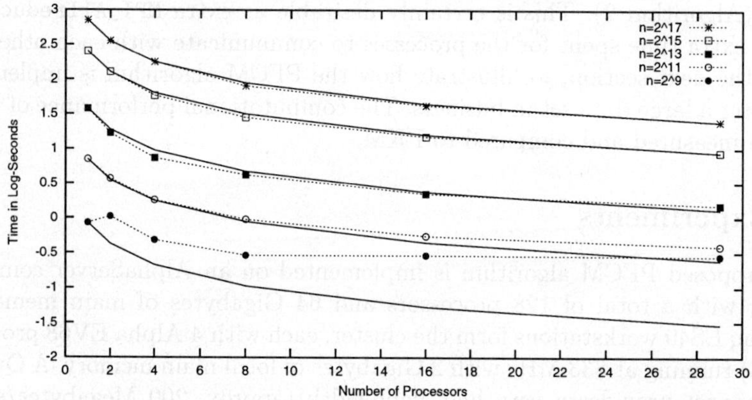

Fig. 1. Speedup curves of PFCM with $m = 2$ and $c = 8$. Time measurements with five data set sizes are shown. Dotted lines represent recorded times and solid lines represent ideal speedup times

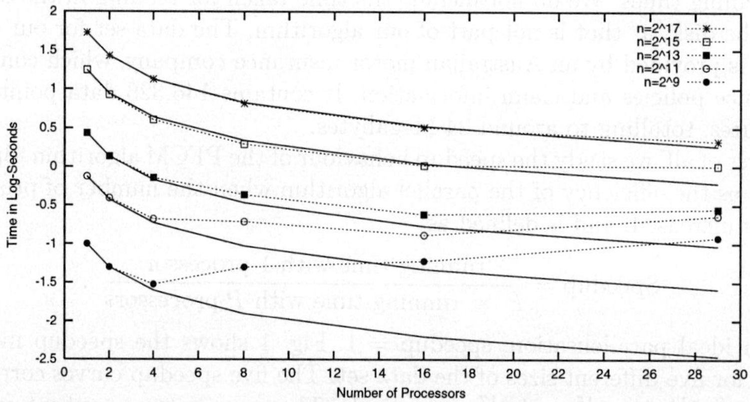

Fig. 2. Speedup curves of PKM with $k = 8$. Time measurements with five data set sizes are shown. Dotted lines represent recorded times and solid lines represent ideal speedup times

In Fig. 3, the size of the data set is fixed to $n = 2^{17}$ while the number of desired clusters is changed, with $c = 2, 4, 8, 16, 32$ respectively for the five speedup curves. This experiment is to measure how well the PFCM algorithm performs in speedup when c increases. Fig. 3 shows that the speedup is almost ideal for small c's, and close to ideal for large c's. Thus we can say that the speedup behaviour of the proposed PFCM algorithm is not sensitive to the number of desired clusters.

For an effective deployment of PFCM to data mining applications involving large data sets, it is important to study its scaleup behaviour. Scaleup measures the effectiveness of the parallel algorithm in dealing with larger data sets when

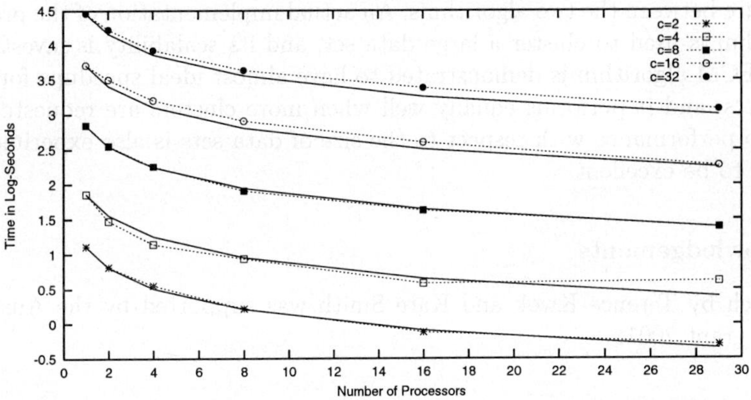

Fig. 3. Speedup curves of PFCM with $m = 2$ and $n = 2^{17}$. Time measurements with five values of c are shown. Dotted lines represent recorded times and solid lines represent ideal speedup times

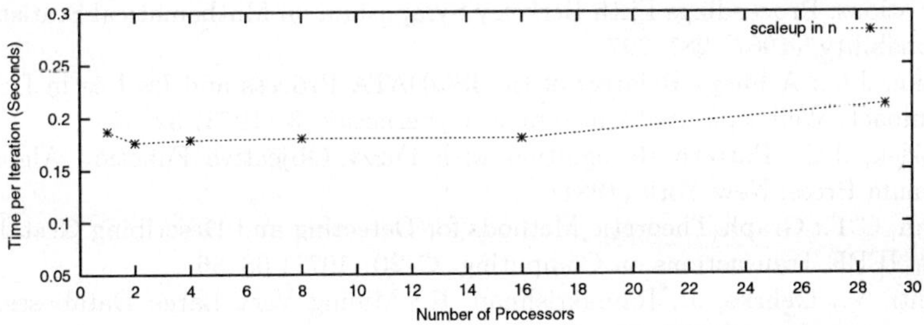

Fig. 4. Scaleup of PFCM with $m = 2, c = 8$, and $n = 2^{12} \times P$. Time per iteration in seconds are measured

more processors are used. In Fig. 4, $n = 2^{12} \times P$, which means the size of the data set is made to increase proportionally with the number of processors used. Other parameters are kept constant: $c = 8, m = 2$, and $\epsilon = 0.01$. It can be observed from Fig. 4 that the running time per iteration varies by 0.025 second only. This means that as far as running time is concerned, an increase in the size of the data set can almost always be balanced by a proportional increase in the number of processors used. We can thus say that the proposed PFCM algorithm has an excellent scaleup behaviour with respect to n.

6 Conclusions

A parallel version of the fuzzy c-means algorithm for clustering large data sets is proposed. The use of the Message Passing Interface (MPI) is also incorporated for ready deployment of the algorithm to SPMD parallel computers. A comparison is made between PFCM and PKM, which reveals a similar parallelisation structure between the two algorithms. An actual implementation of the proposed algorithm is used to cluster a large data set, and its scalability is investigated. The PFCM algorithm is demonstrated to have almost ideal speedups for larger data sets, and it performs equally well when more clusters are requested. The scaleup performance with respect to the size of data sets is also experimentally proved to be excellent.

Acknowledgements

Research by Terence Kwok and Kate Smith was supported by the Australian VPAC grant 2001.

References

1. Jain, A.K., Murty, M.N., Flynn, P.J.: Data Clustering: A Review. ACM Computing Surveys. **31** (1999) 264–323
2. McQueen, J.: Some Methods for Classification and Analysis of Multivariate Observations. Proceedings Fifth Berkeley Symposium on Mathematical Statistics and Probability. (1967) 281–297
3. Dunn, J.C.: A Fuzzy Relative of the ISODATA Process and Its Use in Detecting Compact, Well-separated Clusters. J. Cybernetics. **3** (1973) 32–57
4. Bezdek, J.C.: Pattern Recognition with Fuzzy Objective Function Algorithms. Plenum Press, New York (1981)
5. Zahn, C.T.: Graph-Theoretic Methods for Detecting and Describing Gestalt Clusters. IEEE Transactions on Computing. **C-20** (1971) 68–86
6. Ganti, V., Gehrke, J., Ramakrishnan, R.: Mining Very Large Databases. IEEE Computer. **Aug.** (1999) 38–45
7. Judd, D., McKinley, P., Jain, A.: Large-Scale Parallel Data Clustering. Proceedings of the International Conference on Pattern Recognition. (1996) 488–493
8. Dhillon, I.S., Modha, D.S.: A Data-Clustering Algorithm on Distributed Memory Multiprocessors. In: Zaki, M.J., Ho, C.-T. (eds.): Large-Scale Parallel Data Mining. Lecture Notes in Artificial Intelligence, Vol. 1759. Springer-Verlag, Berlin Heidelberg (2000) 245–260
9. Stoffel, K., Belkoniene, A.: Parallel k-Means Clustering for Large Data Sets. In: Parallel Processing. Lecture Notes in Computer Science, Vol. 1685. Springer-Verlag, Berlin (1999) 1451–1454
10. Nagesh, H., Goil, S., Choudhary, A.: A Scalable Parallel Subspace Clustering Algorithm for Massive Data Sets. Proceedings International Conference on Parallel Processing. IEEE Computer Society. (2000) 477–484
11. Ng, M.K., Zhexue, H.: A Parallel k-Prototypes Algorithm for Clustering Large Data Sets in Data Mining. Intelligent Data Engineering and Learning. **3** (1999) 263–290
12. Gropp, W., Lusk, E., Skjellum, A.: Using MPI: Portable Parallel Programming with the Message Passing Interface. The MIT Press, Cambridge, MA (1996)

Scheduling High Performance Data Mining Tasks on a Data Grid Environment

S. Orlando[1], P. Palmerini[1,2], R. Perego[2], and F. Silvestri[2,3]

[1] Dipartimento di Informatica, Università Ca' Foscari, Venezia, Italy
[2] Istituto CNUCE, Consiglio Nazionale delle Ricerche (CNR), Pisa, Italy
[3] Dipartimento di Informatica, Università di Pisa, Italy

Abstract. Increasingly the datasets used for data mining are becoming huge and physically distributed. Since the distributed knowledge discovery process is both data and computational intensive, the *Grid* is a natural platform for deploying a high performance data mining service. The focus of this paper is on the core services of such a Grid infrastructure. In particular we concentrate our attention on the design and implementation of specialized broker aware of data source locations and resource needs of data mining tasks. Allocation and scheduling decisions are taken on the basis of performance cost metrics and models that exploit knowledge about previous executions, and use sampling to acquire estimate about execution behavior.

1 Introduction

In the last years we observed an explosive growth in the number and size of electronic data repositories. This gave researchers the opportunity to develop effective data mining (DM) techniques for discovering and extracting knowledge from huge amounts of information. Moreover, due to their size and also to social or legal restrictions that may prevent analysts from gathering data in a single site, the datasets are often physically distributed. If we also consider that data mining algorithms are computationally expensive, we can conclude that the *Grid* [4] is a natural platform for deploying a High Performance service for the Parallel and Distributed Knowledge Discovery (PDKD) process. The Grid environment may in fact furnish coordinated resource sharing, collaborative processing, and high performance data mining analysis of the huge amounts of data produced and stored. Since PDKD applications are typically *data intensive*, one of the main requirements of such a PDKD Grid environment is the efficient management of storage and communication resources.

A significative contribution in supporting data intensive applications is currently pursued within the Data Grid effort [1], where a data management architecture based on storage systems and metadata management services is provided. The data considered here are produced by several scientific laboratories geographically distributed among several institutions and countries. Data Grid services are built on top of Globus [3], and simplify the task of managing computations that access distributed and large data sources. The above Data Grid

B. Monien and R. Feldmann (Eds.): Euro-Par 2002, LNCS 2400, pp. 375–384.

framework seems to have a core of common requirements with the realization of a PDKD Grid, where data involved may originate from a larger variety of sources. Even if the Data Grid project is not explicitly concerned with data mining issues, its basic services could be exploited and extended to implement higher level grid services dealing with the process of discovering knowledge from larger and distributed data repositories. Motivated by these considerations, in [8] a specialized grid infrastructure named *Knowledge Grid* (K-Grid) has been proposed. This architecture was designed to be compatible with lower-level grid mechanisms and also with the Data Grid ones.

In this paper we analyze some of the issues encountered in the design and implementation of a specialized K-Grid *broker* for PDKD computations: given a user request for performing a DM analysis, the broker takes allocation and scheduling decisions, and builds the *execution plan*, establishing the sequence of actions that have to be performed in order to prepare execution (e.g., resource allocation, data and code deployment), actually execute the task, and return the results to the user. The execution plan has to satisfy given requirements (such as performance, response time, and mining algorithm) and constraints (such as data locations, available computing power, storage size, memory, network bandwidth and latency) and to maximize or minimize some metrics of interest (e.g. throughput, average service time). In its decision making process, this service has to exploit a composite performance model which consider the actual status of the Grid, the location of data sources, and the task execution behavior. The broker needs quite detailed knowledge about computation and communication costs to evaluate the profitability of alternative mappings and related dataset transfer/partitioning. For example, the broker could evaluate when it is profitable to launch a given expensive mining analysis in parallel. Unfortunately, the performance costs of many DM tools are difficult to predict in advance, because they depend not only on the size of data, but also on the specific mining parameters provided by the user. For example, the complexity of the Association Rule Mining (ARM) analysis also depends on the user-provided support and confidence thresholds, as well as the specific features of the input dataset.

In the absence of knowledge about costs, the K-Grid broker would make blind allocation and scheduling decisions. To overcome this problem, we suggest to exploit *sampling* as a method to acquire preventive knowledge about the rough execution costs of specific, possibly expensive, DM jobs. Sampling has also been suggested as an efficient and effective approach to speed up data mining process, since in some case it may be possible to extract accurate knowledge from a sample of an huge dataset [9]. Unfortunately, the accuracy of the mined knowledge depends on the size of the sample in a non-linear way, and determining how much data has to be used is not possible a priori, thus making the approach impractical. In this paper we investigate an alternative use of sampling: in order to forecast the actual execution cost of a given DM algorithm on the whole dataset. Many DM algorithms demonstrate optimal scalability with respect to the size of the processed dataset, thus making our performance estimate possible and accurate enough. Moreover, even if a wrong estimate is made, this can only affect the optimal use of the Grid and not the results of the final DM analysis to be performed on the whole dataset. Besides execution costs, with sampling we can also estimate the size of the mined results, as well as to predict the amount of I/O and main memory required. These costs will thus feed specific performance models used by the K-Grid broker in order to forecast communication overhead, the effect of resource sharing, and the possible gain deriving from parallelism exploitation.

The paper is organized as follows. Section 2 introduce our K-Grid scheduler and presents the cost model on which it is based on. In Section 3 we discuss the methodology to be used to predict performances by sampling. Section 4 discusses our mapper and the relative simulative framework, and reports some preliminary results. Finally, Section 5 draws some conclusions and outlines future work.

2 Distributed Scheduling on the Knowledge Grid

The main difference among existing Grod brokers regard their organization (e.g., centralized, hierarchical, distributed), and the scheduling policy (e.g., we may optimize system throughput or application completion time). Moreover, scheduling algorithms may consider the state of the system as unchanged in the future, i.e. only depending on the decisions taken by the scheduler, or may try to *predict* possible changes in the system state. In both cases, it is important to know in advance information about task durations under several resource availability constraints and possible resource sharing. The algorithms used to schedule jobs may be classified as dynamic or static [7]. Dynamic scheduling may be on-line, i.e. when a task is assigned to a machine as soon as it arrives, or batch, i.e. when the arrived tasks are collected into a set that is examined for mapping at pre-scheduled times. On the other hand, static approaches, which exploit very expensive mapping strategies, are usually adopted to map long-running applications. Due to the characteristics of DM jobs, which are often interactive, we believe that the best scheduling policy to be used in the design of a K-Grid scheduler should be a dynamic one. In this preliminary study, we thus evaluate feasibility and benefits of adopting a *centralized local on-line scheduler* for a K-Grid organization, which includes several clusters. This local scheduler will be part of a more complex hierarchical superscheduler/broker for the K-Grid.

2.1 Task Scheduling Issues

Before sketching the dynamic scheduling algorithm used for mapping DM jobs, we introduce the issues encountered in scheduling these kind of computations and the simple cost model proposed. Most of the terminology used and the model of performance adopted have been inspired by [5, 7].

Several decisions have to be taken in order to map and schedule the execution of a given (data mining) task on the Grid. First consider that a DM task t_i is completely defined in terms of the DM analysis requested, the dataset $\mathcal{D}_i$ (of size $|\mathcal{D}_i|$) to analyze, and the user parameters u_i that specify and affect the analysis behavior. Let $\alpha_i(\mathcal{D}_i)$ be the knowledge model extracted by t_i, where $|\alpha_i(\mathcal{D}_i)|$

is its size. In general the knowledge model extracted has to to be returned to a given site, where further analysis or visualization must be performed. Before discussing in detail the mapping algorithm and the simulation environment, let us make the following assumptions:

- A centralized local scheduler controls the mapping of DM tasks onto a small Grid organization, which is composed of a set $\mathcal{M} = \{m_1, \ldots, m_{|\mathcal{M}|}\}$ of $|\mathcal{M}|$ machines, where a known performance factor p_j is associated with each machine m_j. This performance factor measures the relative speed of the various machines. Since in this paper we do not consider node multitasking, we do not take into account possible external machine loads that could affect these performance factors. Moreover, the machines are organized as a set of clusters $\mathcal{CL} = \{cl_1, \ldots, cl_{|\mathcal{CL}|}\}$, where each cluster cl_J comprises a disjoint set of machines in $\mathcal{M}$ interconnected by a high-speed network. In particular, $cl_J = \{m_1^J, \ldots, m_{|cl_J|}^J\}$. Each cluster cl_J is thus a candidate for hosting a parallel implementation of a given DM analysis. The performance factors of a cluster cl_J is p_J, which is equal to the factor of the slowest machine in the cluster.
- The code (sequential or parallel) that implements each DM tool is considered to be available at each Grid site. So the mapping issues, i.e. the evaluation of the benefits deriving from the assignment of a task to a given machine, only concern the communication times needed to move input/output data, and also the *ready times* of machines and communication links.
- On the basis of sampling or historical data we assume that it is possible to estimate e_i, defined as the base (normalized) sequential computational cost of task t_i, when executed on dataset $\mathcal{D}_i$ with user parameters u_i. Let $e_{ij} = p_j \cdot e_i$ be the execution time of t_i on machine m_j.
  When an analysis is performed in parallel on a cluster cl_J, we assume that, in the absence of load imbalance, task t_i can be executed in parallel with a quasi perfect speedup. In particular, let e_{iJ} be the execution time of task t_i on a cluster cl_J, defined as $max_{m_t^J \in cl_J}(e_{it}/|cl_J|) + ovh = max_{m_t^J \in cl_J}((p_t \cdot e_i)/|cl_J|) + ovh$. The term ovh models the overhead due to the parallelization and heterogeneity of the cluster. Consider that when a cluster is homogeneous and e_i is large enough, ovh is usually very small.
- A dataset $\mathcal{D}_i$ may be centralized, i.e. stored in a single site, or distributed. In the following we will not consider the inherent distribution of datasets, even if we could easily add such a constraint into our framework. So we only assume that a dataset is moved when it is advantageous for reducing the completion time of a job. In particular, a centralized dataset stored in site h can be moved to another site j, and the cost depends on the average network bandwidth b_{hj} between the two sites. For example, $\mathcal{D}_i$ can be transferred with a cost of $\frac{|\mathcal{D}_i|}{b_{hj}}$.
  Moving datasets between sites has to be carried out by the *replica manager* of the lower Grid services, which is also responsible of the coherence of copies. Future accesses to the a dataset may take advantage of the existence of different copies disseminated on the Grid. So, when a task t_i must be mapped,

we have to consider that, for each machine, we have to choose the most advantageous copy of a dataset to be moved or accessed.

2.2 Cost Model

In the following cost model we assume that each input dataset is initially stored on at least a single machine m_h, while the knowledge model extracted must be moved to a machine m_k. Due to decisions taken by the scheduler, datasets may be replicated onto other machines, or partitioned among the machines composing a cluster.

Sequential execution. Dataset $\mathcal{D}_i$ is stored on a single machine m_h. Task t_i is sequentially executed on machine m_j, and its execution time is e_{ij}. The knowledge model extracted $|\alpha_i(\mathcal{D}_i)|$ must be returned to machine m_k. We have to consider the communications needed to move $\mathcal{D}_i$ from m_h to m_j, and those to move the results to m_k. Of course, the relative communication costs involved in dataset movements are zeroed if either $h = j$ or $j = k$. The total execution time is thus:

$$E_{ij} = \frac{|\mathcal{D}_i|}{b_{hj}} + e_{ij} + \frac{|\alpha_i(\mathcal{D}_i|)}{b_{jk}}$$

Parallel execution. Task t_i is executed in parallel on a cluster cl_J, with an execution time of e_{iJ}. In general, we have also to consider the communications needed to move and partition $\mathcal{D}_i$ from machine m_h to cluster cl_J, and to return the results $|\alpha_i(\mathcal{D}_i)|$ to machine m_k. Of course, the relative communication costs are zeroed if the dataset is already distributed, and is allocated on the machines of cl_J. The total execution time is thus:

$$E_{iJ} = \sum_{m_t^J \in cl_J} \frac{|\mathcal{D}_i|/|cl_J|}{b_{ht}} + e_{iJ} + \sum_{m_t^J \in cl_J} \frac{|\alpha_i(\mathcal{D}_i)|/|cl_J|}{b_{tk}}$$

Finally, consider that the parallel algorithm we are considering requires coallocation and coscheduling of all the machines of the cluster. A different model of performance should be used if we adopted a more asynchronous distributed DM algorithm, where first independent computations are performed on distinct dataset partitions, and then the various results of distributed mining analysis are collected and combined to obtain the final results.

Performance metrics. To optimize scheduling, our batch mapper has to forecast the *completion time* of a task t_i when executed on a machine m_j. To this end, the mapper has also to consider the tasks that were previously scheduled, and that are still queued or running. Let C_{ij} be the wall-clock time at which all communications and sequential computation involved in the execution of t_i on machine m_j complete. Let s_{hj} be the starting time of the communication needed to move $\mathcal{D}_i$ from m_h to m_j, s_j the starting time of the sequential execution of task t_i on m_j, and, finally, s_{jk} the starting time of the communication needed to move $\alpha_i(\mathcal{D}_i)$ from m_j to m_k. All the starting times must be derived on the basis of the ready times of interconnection links and machines. From the above definitions:

$$C_{ij} = (s_{hj} + \frac{|\mathcal{D}_i|}{b_{hj}}) + \delta_1 + e_{ij} + \delta_2 + \frac{|\alpha_i(\mathcal{D}_i)|}{b_{jk}} = s_{hj} + E_{hj} + \delta_1 + \delta_2$$

where $\delta_1 = s_j - (s_{hj} + \frac{|\mathcal{D}_i|}{b_{hj}}) \geq 0$ and $\delta_2 = s_{jk} - (s_j + e_{ij}) \geq 0$.

If $m_{\bar{j}}$ is the specific machine chosen by our scheduling algorithm for executing a task t_i, where T is the set of all the tasks to be scheduled, we can define $C_i = C_{i\bar{j}}$. The *makespan* for the complete scheduling is thus defined as $\max_{t_i \in T}(C_i)$, and its minimization roughly corresponds to the maximization of the system thoughput.

3 Sampling as a Method for Performance Prediction

Before discussing our mapping strategy based on the cost model outlined in the previous Section, we want to discuss the feasibility of *sampling* as a method to predict the performance of a given DM analysis. The rationale of our approach is that, since DM tasks may be very expensive, it may be more profitable to spend a small additional time to sample their execution in order to estimate performances and schedule tasks more accurately, than adopting a blind scheduling strategy. For example, if a task is guessed to be expensive, it may be worth schedule it to a remote machine characterized by an early ready time, or distribute it on a parallel cluster. Differently from [9], we are not interested in the accuracy of the knowledge extracted from a sampled dataset, but only in an approximate performance prediction of the task. To this end, it becomes important to study and analyze memory requirements and completion times of a DM algorithm as a function of the size of the sample exploited, i.e. to study the scalability of the algorithm. From this scalability study we expect to derive, for each algorithm, functions that, given the measures obtained with sampling, return predicted execution time and memory requirement for running the same analysis on the whole dataset.

Suppose that a given task t_i is first executed on a sample $\widehat{\mathcal{D}}_i$ of dataset $\mathcal{D}_i$ on machine m_j. Let $\widehat{e}_{ij}$ be this execution time, and let $\widehat{e}_i = \widehat{e}_{ij}/p_j$ be the normalized execution time on the sample. Sampling is feasible as a method to predict performance of task t_i iff, on the basis of the results of sampling, we can derive a cost function $F()$, such that $e_i = F(|\mathcal{D}_i|)$. In particular, the coefficients of $F()$ must be derived on the basis of the sampled execution, i.e., in terms of $\widehat{e}_i$, $\widehat{\mathcal{D}}_i$, and $|\widehat{\mathcal{D}}_i|$. The simplest case is when the algorithm scales linearly, so that $F()$ is a linear function of the size of the dataset, i.e. $e_i = \gamma \cdot |\mathcal{D}_i|$, where $\gamma = \widehat{e}_i \,/\, |\widehat{\mathcal{D}}_i|$.

We analyzed two DM algorithms: DCP, an ARM algorithm which exploits out-of-core techniques to enhance scalability [6], and *k-means*, the popular clustering algorithm. We ran DCP and *k-means* on synthetic datasets by varying the size of the sample considered. The results of the experiments are promising: both DCP and *k-means* exhibit quasi linear scalability with respect to the size of the sample of a given dataset, when user parameters are fixed. Figure 1.(a) reports the DCP completion times on a dataset of medium size (about 40 MB)

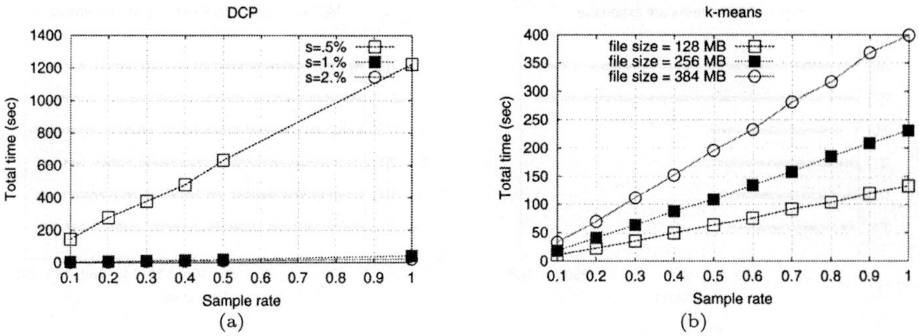

Fig. 1. Execution time of the DCP ARM algorithm (a), and the k-means clustering one (b), as a function of the sample rate of the input dataset.

as a function of the size of the sample, for different user parameters (namely the minimum support $s\%$ of frequent itemsets). Similarly, in Figure 1.(b) the completion time of *k-means* is reported for different datasets, but for identical user parameters (i.e., the number k of clusters to look for). The results obtained for other datasets and other user parameters are similar, and are not reported here for sake of brevity. Note that the slopes of the various linear curves depend on both the specific user parameters and the features of the input dataset $\mathcal{D}_i$. Therefore, given a dataset and the parameters for executing one of these DM algorithms, the slope of each curve can be captured by running the same algorithm on a smaller sampled dataset $\widehat{\mathcal{D}}_i$. For other algorithms, scalability curves may be more complex than a simple linear one. Another problem that may occur in some DM algorithms, is the generation of false patterns for small sampling sizes [9]. Hence, in order to derive an accurate performance model for a given algorithm, it should be important to perform an off-line training of the model, for different dataset characteristics and different parameter sets.

4 On-Line Scheduling of DM Tasks

We analyzed the effectiveness of a centralized on-line mapper based on the MCT (Minimum Completion Time) heuristics [5, 7], which schedules DM tasks on a small organization of a K-Grid. The mapper does not consider node multitasking, is responsible for scheduling both dataset transfers and computations involved in the execution of a given task t_i, and also is informed about their completions. The MCT mapping heuristics adopted is very simple. Each time a task t_i is submitted, the mapper evaluates the *expected ready time* of each machine and communication links. The expected ready time is an estimate of the *ready time*, the earliest time a given resource is ready after the completion of the jobs previously assigned to it. On the basis of the expected ready times, our mapper evaluates all possible assignment of t_i, and chooses the one that reduces the completion time of the task. Note that such estimate is based on both estimated and actual execution times of all the tasks that have been assigned to the resource in

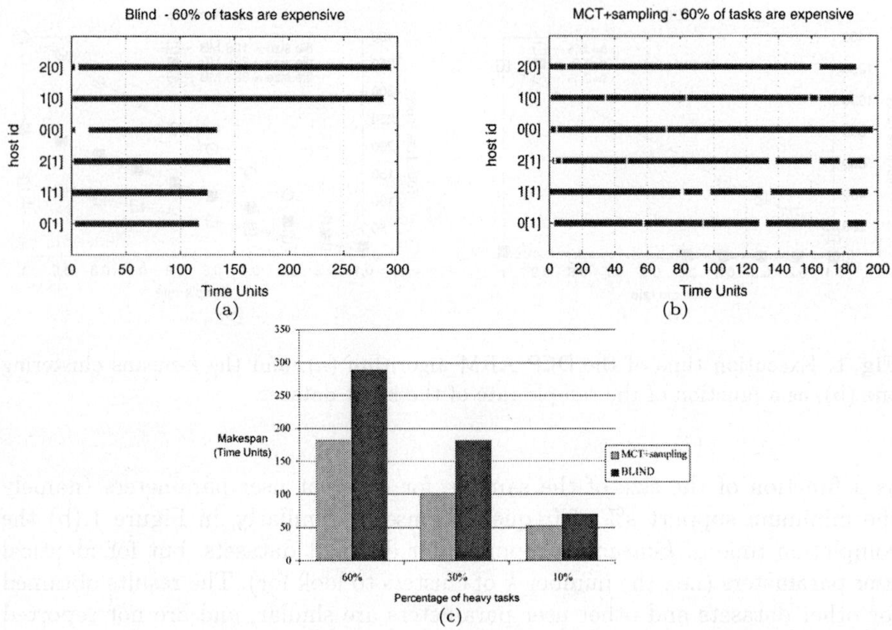

Fig. 2. (a,b): Gannt charts showing the busy times (in time units of 100 sec.) of our six machines when the 60% of the tasks are expensive: (a) blind scheduling heuristics, (b) MCT+sampling scheduling heuristics. (c): Comparison of makespans observed for different percentages of expensive tasks.

the past. To update resource ready times, when data transfers or computations involved in the execution of t_i complete, a report is sent to the mapper.

Note that any MCT mapper can make correct scheduling decisions only if the expected execution time of a task is known. When no performance prediction is available for t_i, our mapper first generates and schedules $\widehat{t_i}$, i.e. the task t_i executed on the sampled dataset $\widehat{\mathcal{D}_i}$. Unfortunately, the expected execution time of sampled task $\widehat{t_i}$ is unknown, so that the mapper has to assume that it is equal to a given small constant. Since our MCT mapper can not be able to optimize the assignment of $\widehat{t_i}$, it simply assigns $\widehat{t_i}$ to the machine that hosts the corresponding input dataset, so that no data transfers are involved in the execution of $\widehat{t_i}$. When $\widehat{t_i}$ completes, the mapper is informed about its execution time. On the basis of this knowledge, it can predict the performance of the actual task t_i, and optimize its subsequent mapping and scheduling.

4.1 Simulation Framework and Some Preliminary Results

We designed a simulation framework to evaluate our MCT on-line scheduler, which exploits sampling as a technique for performance prediction. We thus compared our *MCT+sampling strategy* with a *blind mapping strategy.* Since the blind strategy is unaware of actual execution costs, it can only try to minimize data transfer costs, and thus always maps each task on the machine that holds the corresponding input dataset. Moreover, it can not evaluate the profitability of parallel executions, so that sequential implementations are always preferred.

The simulated environment is similar to an actual Grid environment we have at disposal, and is composed of two clusters of three machines. Each cluster is interconnected by a switched fast Ethernet, while a slow WAN interconnection exists between the two clusters. The two clusters are homogeneous, but the machines of one cluster are *two times faster* than the machines of the other one. To fix simulation parameters, we actually measured average bandwidths b_{WAN} and b_{LAN} of the WAN and LAN interconnections, respectively. Unfortunately, the WAN interconnection is characterized by long latency, so that, due to the TCP default window size, single connections are not able to saturate the actual bandwidth available. This effect is exacerbated by some packet losses, so that retransmissions are necessary and the TCP pipeline can not be filled. Under these hypotheses, we can open a limited number of concurrent sockets, each one characterized by a similar average bandwidth b_{WAN} (100KB/s).

We assumed that DM tasks to be scheduled arrive in a burst, according to an exponential distribution. They have random execution costs, but the $x\%$ of them corresponds to expensive tasks (1000 sec. as mean sequential execution time on the slowest machine), while the $(100 - x)\%$ of them are cheap tasks (50 sec. as mean sequential execution time on the slowest machine). Datasets D_i are all of medium size (50MB), and are randomly located on the machines belonging to the two clusters.

In these first simulation tests, we essentially checked the feasibility of our approach. Our goal was thus to evaluate mapping quality, in terms of *makespan*, of an optimal on-line MCT+sampling technique. This mapper is optimal because it is supposed to also know in advance (through an oracle) the exact costs of the sampled tasks. In this way, we can evaluate the maximal improvement of our technique over the blind scheduling one.

Figures 2.(a,b) illustrate a pair of Gannt charts, which show the busy times of the six machines of our Grid testbed, when either the blind or the MCT+sampling strategy is adopted. The charts are relative to a burst of 100 tasks, where the 60% of them are expensive. Machine i of cluster j is indicated with the label i[j]. Note that when the blind scheduling strategy is adopted, since cluster 0 is slower and no datasets are moved, the makespan on the slower machines results higher. Note that our MCT+sampling strategy sensibly outperforms the blind one, although it introduces higher computational costs due to the sampling process. Finally, Figure 2.(c) shows the improvements in makespans obtained by our technique over the blind one when the percentage of heavy tasks is varied.

5 Conclusions and Future Works

In this paper we have discussed an on-line MCT heuristic strategy for scheduling high performance DM tasks onto a local organization of a Knowledge Grid. Scheduling decisions are taken on the basis of cost metrics and models based on information collected during previous executions, and use sampling to forecast execution costs. We have also reported the results of some preliminary simulations showing the improvements in the makespan (system throughput) of our strategy over a blind one. Our scheduling technique might be adopted by a centralized on-line mapper, which is part of a more complex hierarchical superscheduler/broker, where the higher levels of the superscheduler might be responsible for taking rough schedule-decisions over multiple administrative organizations. The on-line mapper we have discussed does not permit node multitasking, and schedules tasks in batch. In future works we plan to consider also this feature, e.g., the mapper could choose to concurrently execute a compute-bound and an I/O-bound task on the same machine.

A possible drawback of our technique is the additional cost of sampling. Of course, knowledge models extracted by sampling tasks could in some cases be of interest for the users, who might decide on the basis of the sampling results to abort or continue the execution on the whole dataset. On the other hand, since the results obtained with sampling actually represent a partial knowledge model extracted from a *partition* of the dataset, we could avoid to discard these partial results. For example, we might exploit a different DM algorithm, also suitable for distributed environments, where independent DM analysis are performed on different dataset partitions, and then the partial results are merged. According to this approach, the knowledge extracted from the sample $\widehat{\mathcal{D}}_i$ might be retained, and subsequently merged with the one obtained by executing the task on the rest of the input dataset $\mathcal{D}_i \setminus \widehat{\mathcal{D}}_i$.

References

1. A. Chervenak, I. Foster, C. Kesselman, C. Salisbury, and S. Tuecke. The Data Grid: towards an architecture for the distributed management and analysis of large scientific datasets. *J. of Network and Comp. Appl.*, (23):187–200, 2001.
2. S. Fitzgerald, I. Foster, C. Kesselman, G. von Laszewski, W. Smith, and S. Tuecke. A directory service for configuring high-performance distributed computations. In *Proc. 6th IEEE Symp. on High Perf. Distr. Comp.*, pages 365–375. IEEE Computer Society Press, 1997.
3. I. Foster and C. Kesselman. Globus: A metacomputing infrastructure toolkit. *Intl J. of Supercomputer Appl.*, 11(2):115–128, 1997.
4. I. Foster and C. Kesselman, editors. *The Grid: Blueprint for a Future Computing Infrastructure.* 1999.
5. M. Maheswaran, A. Shoukat, H. J. Siegel, D. Hensgen, and R. F. Freund. Dynamic matching and scheduling of a class of independent tasks onto heterogeneous computing systems. In *8th HCW*, 1999.
6. S. Orlando, P. Palmerini, and R. Perego. Enhancing the Apriori Algorithm for Frequent Set Counting. In *Proc. of 3rd DaWaK 01.* LNCS Spinger-Verlag, 2001.
7. H. J. Siegel and Shoukat Ali. Techniques for Mapping Tasks to Machines in Heterogeneous Computing Systems. *Journal of Systems Arch.*, (46):627–639, 2000.
8. D. Talia and M. Cannataro. Knowledge grid: An architecture for distributed knowledge discovery. *Comm. of the ACM*, 2002. to appear.
9. M. J. Zaki, S. Parthasarathy, W. Li, and M. Ogihara. Evaluation of sampling for data mining of association rules. In *7th Int. Work. on Research Issues in Data Eng.*, pages 42–50, 1997.

A Delayed-Initiation Risk-Free Multiversion Temporally Correct Algorithm[1]

Azzedine Boukerche and Terry Tuck

Department of Computer Sciences, University of North Texas
{boukerche, tuck}@cs.unt.edu

Abstract. In this paper, we devise a temporally reordering" mechanism of supporting update transactions that are impacted by delays (e.g., network delays) to the extent that they cannot be executed because of the irreversible progress of other conflicting transactions (i.e., data dependent and temporally dependant) extend this scheme with a delayed-initiation mechanism. This mechanism allows (a) the impacted update transaction to be repositioned to the earliest supportable point in the temporal ordering of transactions, and (b) the associated transaction manager (and in turn, the application or entity that submitted the transaction) to be notified of the new position, thereby providing the opportunity for adjustments or transaction termination. We describe the risk-free MVTC (RF-MVTC) algorithm and its delay-initiation variation (RF-MVTCD). We also present the set of experiments we have carried out to study the performance of MVTC and its variations: RF-MVTC, and RF-MVTCD.

1 Introduction

In our earlier work, we have proposed table-level writeset predeclarations as a method for identifying a priori inter-transactional conflict. This novel method allows transaction concurrency (i.e., speed) to be increased without a corresponding increase in the risk of encountering unidentified conflict. Consequently, a risk-free MVTC concurrency control algorithm as a risk free alternative to Conservative MVTC was proposed in [1,2]. In this paper, we devise a temporally reordering" mechanism of supporting update transactions that are impacted by delays (e.g., network delays) to the extent that they cannot be executed because of the irreversible progress of other conflicting transactions (i.e., data dependent and temporally dependant) extend this scheme with a delayed-initiation mechanism. This mechanism allows (a) the impacted update transaction to be repositioned to the earliest supportable point in the ordering of transactions, and (b) the associated transaction manager (and in turn, the application or entity that submitted the transaction) to be notified of the new position, thereby providing the opportunity for adjustments or transaction termination.

In our database model, the requirement of immediate execution of write operations requires support for multiple versions of each data item. Reducing semantic incorrectness associated with the return of an incorrect data item version suggests a non-

[1] This work was supported by the Texas Advanced Research Program ARP/ATP and UNT Faculty Research Grants.

B. Monien and R. Feldmann (Eds.): Euro-Par 2002, LNCS 2400, pp. 385–390.

aggressive, if not conservative, synchronization technique. The requirement for an execution schedule that is equivalent to a timestamp-ordered serial schedule necessitates the use of timestamp ordering. Finally, the requirement for improved concurrency combined with a non-aggressive synchronization technique requires predeclaration of the data items to be accessed. Combining this with the requirement for ease of use for the application developer requires that predeclaration be done at other than the data-item level.

2 Risk-Free MVTC and Its Delayed-Initiation Variation

The correctness constraint for all schedules produced by the MVTC concurrency control algorithm is that they are conflict serializable and computationally equivalent to a temporally ordered serial schedule for the same set of transactions. As its name implies, the operation of the Risk-free MVTC algorithm is further constrained to be risk-free with respect to the semantic correctness of all responses to *read* messages. In other words, the risk associated with the temporary return of incorrect values followed by an abort is avoided. In our earlier work, we have shown that both conservative and risk-free MVTC exhibit a better performance than previous concurrency control schemes. However, due to unexpected events such as network failures or site failures in a distributed environment, these schemes may fail. Hence, win this paper, we introduce a delay-initiation variant of the Risk-free MVTC algorithm. The key idea of this scheme is that if we delay the initiation of every transaction's first *read* message, we provide a delay for transactions that offsets anomalous[2] network delays incurred by *begin* messages of older transactions. This delaying strategy effectively normalizes the network delays incurred by transactions during the first part of their execution. This scheme is relatively simple in comparison with the original MVTC algorithm, yet is potentially more robust with respect to late-arriving *begin* messages. The tradeoff for this increased robustness is an expected delay in the response to a transaction's first *read* operation.

While Write and Read rules are the same as those for MVTC. The Read rule, when combined with the design constraint of being risk-free (i.e., avoiding semantic incorrectness associated with the return of an incorrect version for the targeted data item), requires that a database delay its response to a *read* request until the correct version is both available and committed. The Delay rule ensures that every *read* request is returned to the correct and committed data-item value. When combined with the Read rule, the Delay rule allows databases to respond to *read* requests with only data values that have been written by committed transactions that immediately precede the reader in the temporal order in terms of data-item-level conflict. When combined with the Reorder rule (below), the second part of Delay rule offers the advantage that all responses are risk-free, and no transaction will ever need to be restarted for updates of a late-arriving writing transaction. The Delay rule is an effective *read-write* synchronization mechanism as long as the writing transaction's *begin* message is received at the database prior to its processing of a conflicting *read* from a younger transaction. However, it is possible for the writer's *begin* message to be affected by network de-

[2] If network delays are identical for all messages there can be no "late" *begins*, as all younger *read* msgs will be equally late.

lays (or similar) to the extent that it is received at the local database after the servicing of a conflicting *read*. In such cases, it is necessary to reposition the writer in the temporal order of transactions to the earliest position that maintains that the writer is younger than all serviced readers of the declared tables.

The concept of delayed initiation is best explained by first detailing the events that motivate its use. With the original Risk-free MVTC algorithm, whenever a transaction's *begin* message arrives at a local database at which some younger reader has already accessed, there is a chance that allowing the older transaction to proceed will lead to a non-serializable schedule. This chance exists when the reader has read from one of the tables included in the table list of the delinquent *begin*, and is a tested condition in the algorithm's processing of *begin* messages. In order to avoid the risk of a non-serializable schedule, the corrective action with the Risk-free MVTC algorithm is to reject the *begin*, and return with the rejection a new timestamp that will reorder the associated transaction to a younger temporal position. Recall that the key to the concept of delayed initiation is the realization that a delay between the receipt and the processing of the younger transaction's *read* message decreases the likelihood of a reorder.

With respect to transaction turnaround times, the delayed initiation approach may appear to be a costly approach for the sake of reducing transaction reorders. Since the issue is one of trading longer turnaround times for fewer transaction reorders, the actual cost is application specific, and calculable only after identifying within the application the impact of reordered transactions. However, the general cost of the approach is limited by several factors that are identified in the following points: (i) Update-only transactions need not be delayed. If a transaction includes no *read* operations, cannot cause another older transaction to be reordered; (ii) At most one delay is needed per transaction. By delaying the processing of a transaction's first *read*, all *reads* within the transaction are effectively delayed; and (iii) Transaction timestamps facilitate delay calculations. Since timestamps reflect the global system time of a transaction's initiation at its associated transaction manager, the duration of the delay between receipt and processing of a particular *read* can be limited to the needed amount.

To be more in-line with the intent of delayed-initiation, it is more desirable to induce a fixed-duration delay between the start of transactions and the processing of their first *read* messages. For example, it might be desirable to ensure that (a) no first *read* message is processed without a delay of, say, twice the expected network delay, and (b) no first *read* message is further delayed if it has already incurred a delay of more than twice the expected network delay. Transaction timestamps provide a simple way to calculate the duration of delays. Since timestamps reflect the global system time of a transaction's initiation at its associated transaction manager, the actual delay incurred by a transaction's first *read* can be calculated: $delay_{actual} = t_{receipt} - timestamp$. Given a specific value for the desired delay before the processing of transactions' first *read* messages, $delay_{total}$, the delayed-initiation delay is the difference of the two: $delay_{di} = delay_{total} - delay_{actual}$, such that $delay_{actual} < delay_{total}$.

3 Simulation Experiments

In our experiments to evaluate the performance of MVTC and its Risk-Free delayed initiated variation scheme, we have used two types of platforms interconnected with a 10 Mbs LAN. In order to reduce the likelihood of conflict between update transactions, the writeset size was fixed at two data items. The data items selected for update were chosen at random, thereby distributing the probability of update uniformly across all data items within the database. In order to increase the likelihood of conflict with the read-only transactions, the ratio of update to read-only transactions was set at 4-to-1. This is accomplished in the experiment runs by restricting each TM to one of the two types of transactions, and allowing the TMs to execute as many transactions as possible within a run. In all runs, 10 TMs were executed simultaneously against a single database. The readsets for the read-only transactions were selected using a sequential pattern in order to produce readsets consisting of adjacent data items. This pattern was chosen in an effort to focus the accesses of each read-only transaction to the fewest tables without reading any data item more than once. Using this pattern, experiment runs were executed with mean readset sizes for the read-only transactions of 5- 100 data items. As a percentage of the total database size, these readset sizes correspond to 1-10, and 20%, respectively.

Let us now turn to our results.

(a) Our results indicate that each of the algorithms executed the most update transactions for the runs with smaller readset sizes. For the proposed algorithms, with only two TMs executing read-only transactions, resource contention at the database is relatively low with the smaller transactions, and the update transactions are consequently able to execute more quickly. With C-MVTO, however, the higher throughput for the runs with smaller read-only transactions is an indirect result of the blocks by the update transactions on read-only transactions being relatively short. Unlike the proposed algorithms, as the size of the read-only transactions is increased, the throughput with C-MVTO decreases substantially, as the duration of blocks is increased by larger readset sizes.

Regarding the relative performance of the different algorithms, our results clearly shows the poor performance of C-MVTO; it is able to execute update transactions at an average throughput of only 37% of that with RFMVTC-D, the poorest performing of the proposed algorithms. This result was expected, since the lack of table predeclarations with C-MVTO means that progress on each TM's transaction must be blocked until no other younger transactions are active. RFMVTC, on the other hand, provided the highest throughput at an average of over 112 offsets the negative impact on turnaround time. The most significant positive effect caused by the delayed initiation is a smaller fraction of update transactions that experience blocks on conflicting transactions. Indeed, we have observed that the fraction transactions per second. MVTC was the middle-performing proposed algorithm. It averaged 84% of the throughput of RFMVTC, and 9% better than RFMVTC-D.

Our results also indicate that the higher performance for throughput in comparison to turnaround time, a positive effect caused by the delayed initiation that of blocks for RFMVTC-D was less than 50% of that for the other two algorithms. By delaying the initiation of a given transaction, it becomes more likely that conflicting transactions will commit prior to the transaction's initiation. Consequently, it becomes less likely that blocks are required for any given transaction.

(b) During the course of our experiments, we have observed that as the size of the read-only transactions is increased, the necessary decrease in throughput occurs with all algorithms. For experiment runs with the smallest readset sizes (i.e., those with readsets fewer than 25 data items), the proposed MVTC algorithm and its delayed initiated risk free variation yield significantly greater throughput than C-MVTO. Without the benefit of table-level writeset predeclarations, C-MVTO must block the execution of every read-only transaction while younger transactions are active, thereby missing the advantage of concurrency experienced with the proposed algorithms. As the size of the readsets is increased, the duration of the blocks decreases relative to the time required for reading more data items. For runs with readset sizes of 50 or more, throughput with C-MVTO surpasses that with MVTC and RFMVTC. This is due to contention at the database, and is covered in the discussion section.

We have also observed that RFMVTC outperforms both MVTC and RFMVTC-D in terms of throughput for runs with the smallest readset sizes. The impact of additional overhead with MVTC and delayed initiation with RFMVTC-D is too great in comparison to the potential benefits, constrained by only two TMs executing relatively small read-only transactions. For runs with readsets of five data items, throughput with MVTC and RFMVTC-D lag that with RFMVTC by 10% and 14%, respectively. For runs with readset sizes of 25 or more data items, RFMVTC-D replaces RFMVTC as the best-performing algorithm. This appears to be an indirect benefit of the delayed initiation mechanism. With eight TMs concurrently executing the smaller update transactions, the fraction of their time spent idle during the delayed initiation is significant. The execution of the read-only transactions benefits from the reduced-contention database resulting from the delays with the update transactions.

(c) Our results indicate that the turnaround time increases as the readsets become larger. Our results indicate that RFMVTC provides the highest performance for smaller readsets, and RFMVTC-D the highest for larger readsets. The explanations for the relative performance differences provided in the throughput discussion hold here.

Read-only transaction performances with the MVTC and RFMVTC algorithms are contrasted. As the readset size increases from 5 data items, the relative performance of MVTC drops. At 25 data items, the advantage for RFMVTC reaches its maximum: the overall read-only transaction performance for MVTC is only 70% of that with RFMVTC. However, at this point the trend reverses, and the relative performance of MVTC increases with larger readset sizes. For runs with the largest readset size, performance with MVTC is within 5% of that with RFMVTC.

4 Conclusion

In this paper, we have proposed to enhance the MVTC scheme by introducing the delayed-initiation variant to the Risk-free MVTC concurrency control algorithm. With the inclusion of the delayed-initiation mechanism, this algorithm maintains the benefits of the original Risk-free MVTC algorithm while addressing its shortcoming: increased temporal reordering. We have also presented a set of experiments to study the performance of MVTC and its risk-free and delay initiation variations. Our results indicate that for less-tolerant applications, the delayed-initiation variant of Risk-free MVTC is the algorithm of choice; our experimental results show that it provides good performance even when long-duration delays are used for the delayed-initiation mechanism.

References

[1] Boukerche, A., S. K. Das, A. Datta and T. LeMaster. "Implementation of a Virtual Time Synchronizer for Distributed Databases." *Proceedings of EuroPar '98, LNCS,* 534-538.

[2] Boukerche A, Tuck T., "Improving Conservative Concurrency Control in Distributed Databases", *Proc. EuroPar 2001, LNCS 2150*, Springer Verlag, pp. 301-309, 2001.

[3] Georgakopoulos, D., M. Rusinkiewicz and W. Litwin. "Chronological Scheduling of Transactions with Temporal Dependencies." *VLDB Journal* 3:1 (January 1994): 1-28.

[4] Jefferson, D., and A. Motro. "The Time Warp Mechanism for database concurrency control." *Proc. of the 2nd Int'l Conference on Data Engineering*, (1986): 474-481.

[5] Nicol, D.M. and X. Liu. "The Dark Side of Risk (What your mother never told you about Time Warp)." *Proc. of the 11th Workshop on Parallel and Distributed Simulation*, (1997): 188-195.

[6] Özsu, M.T. and P.Valduriez. *Principles of Distributed Database Systems.* Prentice Hall, 1999.

Topic 6
Complexity Theory and Algorithms

Ernst W. Mayr

Global Chair

The goal of algorithm design and complexity theory in parallel/distributed computing is to study efficient algorithms for (and limitations on the complexity of) problems, taking into account such parallel complexity measures as the number of processing nodes or the amount of communication, in addition to classical measures like time and space. Research areas such as development of efficient parallel algorithms on realistic parallel computation models, communication complexity, parallel complexity classes, and lower bounds for specific problems have received a lot of attention in recent years, but many important problems remain open.

For this topic, three papers have been selected from the submissions. They deal with new methods for the design of parallel algorithms for several classical problems. Tiskin's paper presents, within the model of bulk-synchronous parallel computation (BSP), an optimal deterministic parallel algorithm for computing the convex hull in 3D. Niculescu explains how to design, using several fundamental data structures as tools for the functional description of parallel programs, several variants of FFT. The third paper, by Han et al., describes a parallel algorithm based on branch and bound for the capacitated minimum spanning tree problem, supporting its efficiency by extensive experimental data.

We would like to express our sincere appreciation to the other members of the Program Committee, Prof. Maria Serna, Prof. Juraj Hromkovic, and Dr. Rolf Wanka, for their invaluable help in the entire selection process. We would also like to thank all the other reviewers for their time and effort.

B. Monien and R. Feldmann (Eds.): Euro-Par 2002, LNCS 2400, p. 391.

Parallel Convex Hull Computation by Generalised Regular Sampling

Alexandre Tiskin*

Department of Computer Science
University of Warwick, Coventry CV4 7AL, UK
tiskin@dcs.warwick.ac.uk

Abstract. The model of bulk-synchronous parallel (BSP) computation is an emerging paradigm of general-purpose parallel computing. We propose the first optimal deterministic BSP algorithm for computing the convex hull of a set of points in three-dimensional Euclidean space. Our algorithm is based on known fundamental results from combinatorial geometry, concerning small-sized, efficiently constructible *ϵ-nets* and *ϵ-approximations* of a given point set. The algorithm generalises the technique of *regular sampling*, used previously for sorting and two-dimensional convex hull computation. The cost of the simple algorithm is optimal only for extremely large inputs; we show how to reduce the required input size by applying regular sampling in a multi-level fashion.

1 Introduction

The model of *bulk-synchronous parallel (BSP) computation* (see [21, 13]) provides a simple and practical framework for general-purpose parallel computing. A slightly restricted version of the BSP model is known in literature as *CGM (coarse grained multicomputer)*.

In this work we propose a new deterministic BSP algorithm for computing the convex hull of a set of points in $\mathbb{R}^3$. Computation of convex hulls in $\mathbb{R}^2$ and $\mathbb{R}^3$ is a classic problem of computational geometry, extensively studied in the context of both sequential and parallel computation (see e.g. [18, 7]).

Consider computing the convex hull of n points in $\mathbb{R}^2$ or $\mathbb{R}^3$. The lower bound on sequential computation time is $\Omega(n \log n)$ in both cases. This lower bound is provided by reduction of the sorting problem, and is attained by many existing $O(n \log n)$-time algorithms (see [18] for references). A parallel (EREW PRAM), processor-optimal, deterministic 2D convex hull algorithm with time complexity $O(\log n)$ was given in [14]. For 3D convex hulls, a CREW PRAM, processor-optimal randomized $O(\log n)$ algorithm was proposed in [17], and an EREW PRAM, processor-optimal deterministic $O((\log n)^2)$ algorithm in [1]. These results were further refined in papers [2, 8], which describe respectively a deterministic $O(\alpha^{-1} \log n)$-time CREW PRAM algorithm using $O(n^{1+\alpha})$ processors, and a randomised $O(\log n)$-time processor-optimal EREW PRAM algorithm.

* Partially supported by the Future and Emerging Technologies programme of the EU under contract number IST-1999-14186 (ALCOM-FT).

B. Monien and R. Feldmann (Eds.): Euro-Par 2002, LNCS 2400, pp. 392–399.

For the BSP/CGM model, optimal deterministic 2D convex hull algorithms are given in [8, 4, 5]. Assuming that the input size n is sufficiently large with respect to the number of processors p, these algorithms perform $O(n \log n/p)$ computation work, $O(n/p)$ communication work and $O(1)$ synchronisation steps. For 3D convex hulls, papers [8, 3] describe randomised BSP/CGM algorithms, achieving the same optimal cost values with high probability. To the author's knowledge, no optimal deterministic BSP/CGM algorithms for 3D convex hull computation have been known previously.

This work attempts to fill this gap by presenting an optimal deterministic BSP algorithm for 3D convex hull computation. The algorithm is based on known fundamental results from combinatorial geometry, concerning small-sized, efficiently constructible *ϵ-nets* and *ϵ-approximations* of a given point set. The algorithm generalises the technique of *regular sampling*, used in [19] for sorting, and in [4] for two-dimensional convex hull computation.

The rest of this work is organised as follows. Section 2 gives a brief introduction to the BSP computation model; this section may be skipped by a reader already familiar with the model. Section 3 describes previously known regular sampling algorithms for sorting and 2D convex hull computation. Section 4 introduces the necessary concepts from combinatorial geometry, describes the proposed generalisation of the regular sampling technique, and presents the new BSP convex hull algorithm. Section 5 draws conclusions and identifies directions for future research.

2 The BSP Model

A *BSP computer*, introduced in [21], consists of *p processors* connected by a communication network. Each processor has a fast *local memory*. The processors may follow different threads of computation. A BSP computation is a sequence of *supersteps*. The processors are synchronised between supersteps. The computation within a superstep is asynchronous.

Let *cost unit* be the cost of performing a basic arithmetic operation or a local memory access. If, for a particular superstep, w is the maximum number of local operations performed by each processor, h' (respectively, h'') is the maximum number of data units received (respectively, sent) by each processor, and $h = h' + h''$ (another possible definition is $h = \max(h', h'')$), then the cost of the superstep is defined as $w + h \cdot g + l$. Here g and l are the BSP parameters of the computer. If a computation consists of S supersteps with costs $w_s + h_s \cdot g + l$, $1 \leq s \leq S$, then its total cost is $W + H \cdot g + S \cdot l$, where $W = \sum_{s=1}^{S} w_s$ is the local computation cost, $H = \sum_{s=1}^{S} h_s$ is the communication cost, and S is the synchronisation cost. The values of W, H and S typically depend on the number of processors p and on the problem size.

Many BSP algorithms are only defined for input sizes that are sufficiently large with respect to the number of processors. This requirement is loosely referred to as *slackness*. In this work, we present a simple 3D convex high algorithm requiring very high slackness, and then show how this requirement can be reduced to a reasonable amount of slackness.

For the sake of simplicity, we ignore small irregularities that arise from imperfect matching of integer parameters. For example, when we say that an array of size n is divided equally across p processors, the value n may not be an exact multiple of p, and therefore the shares may differ in size by ± 1.

3 Regular Sampling

Sorting is a classical problem of both sequential and parallel computing. Probably the simplest parallel sorting algorithm is *parallel sorting by regular sampling (PSRS)*, proposed in [19] (see also [16, 20]). The method of regular sampling relies on the principle of representing the processor's local data by a set of regular samples, the number of which does not depend on the size of the input. From the union of all samples we can select a subset of regular pivots, which represent the global data structure, and can be used for efficient data repartitioning intended to reflect this global structure.

We consider comparison-based sorting of an array of size n. Without loss of generality, we may assume that all elements of the array are distinct (otherwise, we distinguish equal elements by attaching to each a unique tag). The algorithm proceeds as follows.

Algorithm 1 (Sorting by regular sampling).
Input: array of size n, with n/p elements in each processor.
Output: the same array rearranged in increasing order.
Description. First, the processor subarrays are sorted independently by an optimal sequential algorithm. The problem now consists in merging the p sorted subarrays.

In the first stage of merging, $p+1$ regularly spaced *samples* are selected from each subarray (the first and the last elements of a subarray are included among the samples). The samples divide each subarray into p blocks of size at most n/p^2. Then, $p \cdot (p+1)$ samples are collected in one processor and merged by the optimal sequential algorithm. After that, we select $p+1$ regularly spaced *pivots* from the sorted array of samples (the first and the last samples are again included in the pivots). The pivots partition the set of all elements into p *buckets*, each bucket being distributed across the processors. It is easy to show that the size of a bucket is at most $2n/p$. Now it remains to broadcast the pivots, to assign each bucket to a particular processor, and then to sort the elements of the buckets by an optimal sequential algorithm.
Cost analysis. The computation cost of the algorithm is $O(n \log n/p)$. In the worst case, a processor sends and receives at most $2n/p$ values, therefore the communication cost is $O(n/p)$. The algorithm performs $O(1)$ global synchronisations. The total BSP cost of the algorithm is

$$W = O(n \log n/p) \qquad H = O(n/p) \qquad S = O(1)$$

For the above resource bounds to hold, it is necessary that the parallel part of the algorithm dominate the sequential part. In particular, we need $n/p \geq p^2$, therefore $n \geq p^3$. □

We can reduce the slackness requirement by introducing additional levels of regular sampling.

Let us introduce a constant parameter $\delta > 0$. Instead of taking $p+1$ regular samples, initially each processor selects p^δ regular samples from its local subarray. All $O(p^{1+\delta})$ samples are collected in one processor, which selects from them $O(p^\delta)$ "first-level pivots". These are used to partition the array into p^δ "first-level buckets". We redistribute the global array, so that each first-level bucket is assigned to a group of $p^{1-\delta}$ processors.

Next, each group computes in the same manner p^δ "second-level pivots". The original point set is now partitioned into $p^{2\delta}$ "second-level buckets"; we assign each to a group of $p^{1-2\delta}$ processors. The process continues until p buckets are obtained.

The total BSP cost of the resulting BSP algorithm is still

$$W = O(n \log n/p) \qquad H = O(n/p) \qquad S = O(1)$$

However, now the constant factors in each of the cost components depend on δ. For the above resource bounds to hold, we need $n/p \geq p^{1+\delta}$, therefore it sufficient to take $n \geq p^{2+\delta}$ for any constant $\delta > 0$.

A BSP/CGM algorithm for 2D convex hull computation, based on an idea similar to PSRS, is described in [4]. In the first stage in the algorithm, the array of points is sorted by x-coordinate with an optimal BSP algorithm, such as PSRS. This initial pre-sorting of points is a crucial part of the algorithm. Therefore, it is hard to modify it for computing convex hulls in higher dimensions, where pre-sorting does not help. In the next section we develop an alternative, more general approach to geometric regular sampling.

4 Generalised Regular Sampling

This section presents a simple, optimal BSP algorithm for the 3D convex hull problem. To apply the regular sampling technique, we need to find a way of representing an arbitrary, unsorted set of points by a small-sized sample. Such a representation is provided by the concept of *ϵ-approximation* (see e.g. [15, 10, 11]).

Let X be a finite set of points in $\mathbb{R}^d$. Take any ϵ, $0 \leq \epsilon \leq 1$. A subset $A \subseteq X$ is *an ϵ-approximation to* X with respect to halfspaces, if for any halfspace H, we have

$$\left| \frac{|A \cap H|}{|A|} - \frac{|X \cap H|}{|X|} \right| \leq \epsilon,$$

that is, the relative number of points of A in H approximates the relative number of points of X in H with accuracy ϵ. In other words, any hyperplane that cuts off $\alpha|A|$ points of A must cut off $(\alpha \pm \epsilon)|X|$ points of X. The above definition is a special case of a general definition from VC-dimension theory. Since we only consider this special type of ϵ-approximation, from now on we will omit the words "with respect to halfspaces".

The following theorem is a corollary of a more general result on ϵ-approximation with respect to simplices, proved in [9].

Theorem 1. *Let X be a set of n points in $\mathbb{R}^d$. For any r, $1 \le r \le n$, a $1/r$-approximation to X of size $O\big(r^d(\log r)^{O(1)}\big)$ can be computed deterministically in time $O(n \log r)$.*

The general VC-dimension theory guarantees the existence of smaller ϵ-approximations, but they are harder to compute.

In our BSP convex hull algorithm, a $1/p$-approximation to a local point set will play the role of the sample set. Note that for a 2D array of points of size n in convex position, a regular sample containing every n/p-th point is an $1/p$-approximation of size p. However, we will have to use larger approximations provided by Theorem 1, in order to deal with non-convex, three-dimensional sets.

The relatively large size of an approximation guaranteed by Theorem 1 prevents us from using an ϵ-approximation as the pivot set, since such a set would not provide an efficient problem partitioning. Fortunately, the approximation properties required from pivots are weaker compared to those of samples. Therefore, we can trade approximation accuracy for set size. A suitable concept here is that of *ϵ-net* (see e.g. [15, 10, 11]). As before, let X be a finite set of points in $\mathbb{R}^d$, and ϵ an arbitrary real number, $0 \le \epsilon \le 1$. A subset $A \subseteq X$ is *an ϵ-net for* X, if for any halfspace H, such that $|X \cap H|/|X| > \epsilon$, we have $|A \cap H| > 0$, i.e. H contains at least one point from A. In other words, any hyperplane that does not cut off any points of A can cut off at most $\epsilon|X|$ points of X. Note that by definition, every ϵ-approximation is an ϵ-net, but not vice versa.

The following simple, but important facts can be proved by straightforward application of the definitions (cf. [10]).

Lemma 1. *Let $X = X_1 \cup \cdots \cup X_m$, where $X_1, \ldots, X_m$ are disjoint subsets of equal cardinality. For each i, let $A_i \subseteq X_i$ be an ϵ-approximation to X_i. Then $A_1 \cup \cdots \cup A_m$ is an ϵ-approximation to X.*

Lemma 2. *Let A be an ϵ-approximation to X. Let B be a δ-approximation to (respectively, a δ-net for) A. Then B is an $(\epsilon+\delta)$-approximation to (respectively, an $(\epsilon+\delta)$-net for) X.*

The crucial property of geometric 3D ϵ-nets is proved in [12].

Theorem 2. *Let X be a set of n points in $\mathbb{R}^3$. For any r, $1 \le r \le n$, a $1/r$-net for X of size $O(r)$ can be computed deterministically in time $O(n^3)$.*

The analogue of Theorem 2 holds for $\mathbb{R}^2$, but is not known for dimensions higher than 3. The general VC-dimension theory only guarantees the existence of $1/r$-nets of size $O(r \log r)$, where the constant factor is a fast-growing function of the dimension (see e.g. [15, 10, 11]).

We are now ready to describe the optimal BSP algorithm for computing the convex hull of n points in 3D. Without loss of generality, we may assume that the points are in general position (otherwise, we achieve this by a slight perturbation of the points). The algorithm proceeds as follows.

Algorithm 2 (3D convex hull by generalised regular sampling).
Input: array of points in $\mathbb{R}^3$ of size n, with n/p points in each processor.
Output: edges of the convex hull of the point.
Description. In the first stage, each processor computes, by Theorem 1, a $1/p$-approximation of size $O\big(p^3(\log p)^{O(1)}\big)$ to its local point set, in time $O(n/p{\cdot}\log p)$. The points of this approximation are the *samples.* (In practice, it might be useful to compute the local convex hulls before determining the samples; we view the local convex hull computation as optional, since it does not affect the asymptotic complexity of the algorithm.)

Then, all $p \cdot O\big(p^3(\log p)^{O(1)}\big) = O\big(p^4(\log p)^{O(1)}\big)$ samples are collected together in one processor. By Lemma 1, these samples form a $1/p$-approximation to the global point set. By Theorem 2, we compute sequentially a $1/p$-net of size $O(p)$ for the set of all samples, in time $O\big(p^{12}(\log p)^{O(1)}\big)$. The points of this net are the *pivots.* By Lemma 2, these pivots form a $2/p$-net for the global point set.

We now compute sequentially the convex hull of the pivot set. This pivot hull defines a convex polytope with at most p vertices, and therefore with $O(p)$ edges and $O(p)$ faces. By definition of a $2/p$-net, every tangent plane of the polytope cuts off at most $2n/p$ points from the global point set. Each edge of the polytope defines a *bucket*, which is the set of all points visible (assuming the polytope is opaque) from the relative interior of the defining edge. A bucket is the union of points cut off by the supporting planes of two adjacent polytope faces, therefore its size is at most $4n/p$. After broadcasting the pivots, we assign each bucket to a separate processor, and compute the convex hull of the bucket by an efficient sequential algorithm. Every edge of the global convex hull must be completely visible from the relative interior of at least one edge of the pivot polytope, therefore all global convex hull edges are included among the edges of the locally computed convex hulls.

The total number of edges in the local hulls is $O(n)$, with $O(n/p)$ edges per processor. We now apply the technique of [3] to detect and remove all edges that are not in the global convex hull. This can be achieved in a constant number of supersteps, with local computation cost $O(n \log n/p)$ and communication cost $O(n/p)$. See [3] for details.
Cost analysis. The computation cost of the algorithm is $O(n \log n/p)$. In the worst case, $4n/p$ points are collected in a single processor, therefore the communication cost is $O(n/p)$. The algorithm performs $O(1)$ global synchronisations. The total BSP cost of the algorithm is

$$W = O(n \log n/p) \qquad H = O(n/p) \qquad S = O(1)$$

For the above resource bounds to hold, it is necessary that the parallel part of the algorithm dominate the sequential part. In particular, we need $n \log n/p \geq p^{12}(\log p)^{O(1)}$, therefore $n \log n \geq p^{13}(\log p)^{O(1)}$. □

We have obtained an optimal deterministic BSP algorithm for 3D convex hull; however, it reqires an unrealistically high amount of slackness. We can reduce the slackness requirement by introducing additional levels of regular sampling.

Let us introduce a constant parameter $\delta > 0$. Instead of computing a $1/p$-approximation, initially each processor computes a $1/p^\delta$-approximation of size $O(p^{3\delta}(\log p)^{O(1)})$ to its local point set, in time $O(n/p \cdot \log p)$. All $O(p^{1+3\delta}(\log p)^{O(1)})$ samples are collected in one processor, which computes a $2/p^\delta$-net for the global point set in time $O(p^{3+9\delta}(\log p)^{O(1)})$. The points of this net are used as "first-level pivots", which partition the point set into p^δ "first-level buckets". We redistribute the global set of points, so that each first-level bucket is assigned to a group of $p^{1-\delta}$ processors.

Next, each group computes in the same manner p^δ "second-level pivots". The original point set is now partitioned into $p^{2\delta}$ "second-level buckets"; we assign each to a group of $p^{1-2\delta}$ processors. The process continues until p buckets are obtained.

The total BSP cost of the resulting BSP algorithm is still

$$W = O(n \log n/p) \qquad H = O(n/p) \qquad S = O(1)$$

However, now the constant factors in each of the cost components depend on δ. For the above resource bounds to hold, we need $n \log n/p \geq p^{3+9\delta}(\log p)^{O(1)}$, therefore it sufficient to take $n \log n \geq p^{4+\gamma}$ for any constant $\gamma > 0$.

5 Conclusion

We have presented the first optimal deterministic BSP algorithm for computing the convex hull of a finite set of points in $\mathbb{R}^3$. Our algorithm is based on known fundamental results from combinatorial geometry, concerning small-sized, efficiently constructible *ϵ-nets* and *ϵ-approximations* of a given point set. The algorithm generalises the technique of *regular sampling*, used previously for sorting and two-dimensional convex hull computation. The BSP cost of the simple algorithm is optimal only for extremely large inputs; however, it is possible to reduce the input size to reasonable values by applying the regular sampling technique in a multi-level fashion.

More sophisticated regular sampling techniques and future advances in methods for constructing geometric ϵ-nets may further improve the required lower bound on the input size. The question of existence of a fully-scalable deterministic BSP algorithm for 3D convex hull remains open.

Despite the results of this and other recent papers on optimal deterministic sampling, it appears that random sampling still has better constant-factor efficiency and scalability, and therefore remains the method of choice for the practice.

Acknowledgement

The author thanks the anonymous referees of several previous versions of this paper for their valuable comments.

References

1. N. M. Amato, M. T. Goodrich, and E. A. Ramos. Parallel algorithms for higher-dimensional convex hulls. In *Proc. of the 35th IEEE FOCS*, pages 683–694, 1994.
2. N. M. Amato and F. P. Preparata. A time-optimal parallel algorithm for three-dimensional convex hulls. *Algorithmica*, 14:169–182, 1995.
3. F. Dehne, Xiaotie Deng, P. Dymond, A. Fabri, and A. A. Khokhar. A randomized parallel 3D convex hull algorithm for coarse grained multicomputers. *Theory of Computing Systems*, 30(6):547–558, 1997.
4. M. Diallo, A. Ferreira, A. Rau-Chaplin, and S. Ubéda. Scalable 2D convex hull and triangulation algorithms for coarse grained multicomputers. *Journal of Parallel and Distributed Computing*, 56:47–70, 1999.
5. P. Dymond, Jieliang Zhou, and Xiaotie Deng. A 2D parallel convex hull algorithm with optimal communication phases. *Parallel Computing*, 27:243–255, 2001.
6. J. E. Goodman and J. O'Rourke, editors. *Handbook of Discrete and Computational Geometry.* The CRC Press Series on Discrete Mathematics and Its Applications. CRC Press, 1997.
7. M. T. Goodrich. Parallel algorithms in geometry. In Goodman and O'Rourke [6], chapter 36, pages 669–681.
8. M. T. Goodrich. Randomized fully-scalable BSP techniques for multi-searching and convex hull construction. In *Proc. of the 8th ACM-SIAM SODA*, 1997.
9. J. Matoušek. Reporting points in halfspaces. *Computational Geometry: Theory and Applications*, 2/3:169–186, 1992.
10. J. Matoušek. Approximations and optimal geometric divide-and-conquer. *Journal of Computer and System Sciences*, 50:203–208, 1995.
11. J. Matoušek. Derandomization in computational geometry. In J.-R. Sack and J. Urrutia, editors, *Handbook of Computational Geometry*, pages 559–596. 2000.
12. J. Matoušek, R. Seidel, and E. Welzl. How to net a lot with little: Small ϵ-nets for disks and halfspaces. In *Proc. of the 6th ACM Symposium on Computational Geometry*, pages 16–22, 1990. An updated version available from `http://kam.mff.cuni.cz/~matousek/oldpaps.html`.
13. W. F. McColl. Universal computing. In L. Bougé et al., editors, *Proc. of Euro-Par (Part I)*, volume 1123 of *Lecture Notes in Computer Science*, pages 25–36. Springer-Verlag, 1996.
14. R. Miller and Q. F. Stout. Efficient parallel convex hull algorithms. *IEEE Transactions on Computers*, C–37(12):1605–1618, 1988.
15. K. Mulmuley and O. Schwarzkopf. Randomized algorithms. In Goodman and O'Rourke [6], chapter 34, pages 633–652.
16. M. J. Quinn. *Parallel Computing: Theory and Practice.* McGraw-Hill Series in Computer Science. McGraw-Hill, second edition, 1994.
17. J. H. Reif and S. Sen. Optimal parallel randomized algorithms for three-dimensional convex hulls and related problems. *SIAM Journal of Computing*, 21(3):466–485, 1992.
18. R. Seidel. Convex hull computations. In Goodman and O'Rourke [6], chapter 19, pages 361–375.
19. H. Shi and J. Schaeffer. Parallel sorting by regular sampling. *Journal of Parallel and Distributed Computing*, 14(4):361–372, 1992.
20. A. Tiskin. The bulk-synchronous parallel random access machine. *Theoretical Computer Science*, 196(1–2):109–130, April 1998.
21. L. G. Valiant. A bridging model for parallel computation. *Communications of the ACM*, 33(8):103–111, August 1990.

Parallel Algorithms for Fast Fourier Transformation Using *PowerList*, *ParList* and *PList* Theories

Virginia Niculescu

Department of Computer Science
Babeş-Bolyai University
Cluj-Napoca, Romania
gina@cs.ubbcluj.ro

Abstract. *PowerList*, *ParList* and *PList* data structures are efficient tools for functional descriptions of parallel programs that are *divide & conquer* in nature.
The goal of this work is to develop three parallel variants for Fast Fourier Transformation using these theories. The variants are implied by the degree of the polynomial, which can be a power of two, a prime number, or a product of prime factors. The last variant includes the first two, and represents a general and efficient parallel algorithm for Fast Fourier Transformation. This general algorithm has a very good time complexity, and can be mapped on a recursive interconnection network.

1 Introduction

PowerList, *ParList* and *PList* are data structures that can be successfully used for simple functional descriptions of parallel programs that are *divide&conquer* in nature. To assure methods for verification of the parallel programs correctness, algebras and induction principles are defined on these data structures[1].

A *PowerList* is a linear data structure whose elements are all of the same type. The length of a *PowerList* data structure is a power of two. A *PowerList* with 2^n elements of type X is specified by $PowerList.X.n$. Two similar *PowerLists* can be combined into a *PowerList* data structure with double length in two different ways: using *tie* operator $(p \mid q)$ when the result contains elements from p, followed by elements from q, and using *zip* operator $(p \natural q)$ when the result contains elements from p and q, alternatively taken.

Example 1. (*Polynomial Value*)

A polynomial with coefficients $(a_i, 0 \leq i < 2^n)$, where $n \geq 0$, may be represented by a *PowerList* – p, whose ith element is a_i. The following function vp evaluates a polynomial p; vp accepts an arbitrary *PowerList*, which contains the points, as its second argument.

$$vp : PowerList.X.n \times PowerList.X.m \rightarrow PowerList.X.m$$

$$\begin{aligned} vp.[a].[z] &= [a] \\ vp.p.(u|v) &= vp.p.u \mid vp.p.v \\ vp.(p\natural q).w &= vp.p.w^2 + (w \cdot (vp.q.w^2)) \end{aligned} \quad (1)$$

B. Monien and R. Feldmann (Eds.): Euro-Par 2002, LNCS 2400, pp. 400–403.

The length of a *ParList* data structure is not always a power of two. A *ParList* with n elements of type X is specified by $ParList.X.n$. It is necessary to use two other operators: *cons*($\triangleright$) and *snoc*($\triangleleft$); they allow the adding of an element to a *ParList*, at the beginning or at the end of the *ParList*.

Example 2. (*Polynomial Value*)

$$\begin{array}{ll} vp : ParList.X.n \times ParList.X.m \rightarrow ParList.X.m & \\ vp.[a].[z] \quad = a & \\ vp.(p \natural q).w \ = vp.p.w^2 + w \cdot vp.q.w^2 & vp.p.(u \mid v) = vp.p.u \mid vp.p.v \\ vp.(a \triangleright p).w = a + w \cdot vp.p.w & vp.p.(z \triangleright w) = vp.p.[z] \triangleright vp.p.w \end{array} \tag{2}$$

PLists are constructed with the n-way $\mid$ and $\natural$ operators; for a positive integer n, the n-way $\mid$ takes n similar *PLists* and returns their concatenation, and the n-way $\natural$ returns their interleaving. Functions over *PList* are defined using two arguments. The first argument is a list of arities of type *PosList* (*PosList* is the type of linear lists with positive integers), and the second is the *PList* argument. Functions over *PList* are only defined for certain pairs of these input values.

Example 3. (*Polynomial Value*)

We extend the strategy used for *PowerList* case by using different radices, and we define the function *vp* on *PLists* (we use the notation $\overline{n} = \{0, \ldots, n-1\}$):

$$\begin{array}{l} vp : PosList \times PList.X.n \times PosList \times PList.X.m \rightarrow PList.X.m \\ defined.vp.lx.p.ly.w \equiv prod.lx = length.p \wedge prod.ly = length.w \\ vp.[].[a].[].[z] \qquad\qquad\quad = a \\ vp.lx.p.(y \triangleright ly).[\| i : i \in \overline{y} : w.i] \ = [\| i : i \in \overline{y} : vp.lx.p.ly.(w.i)] \\ vp.(x \triangleright lx).[\natural i : i \in \overline{x} : p.i].ly.w = (+i : i \in \overline{x} : w^i \cdot vp.lx.(p.i).ly.w^x) \end{array} \tag{3}$$

2 Fast Fourier Transformation

We consider a polynomial pf with coefficients $(a_i, 0 \leq i \leq n)$. A scalar function will be used in all the cases: function $root : Nat \rightarrow Com$ applied to n returns the principal nth order unity root (*Com* is the type of complex numbers). Three functions named $powers : Com \times PXxxList.Com.n \rightarrow PXxxList.Com.n$ will be used; they each return a *PXxxList* of the same length as the input list(p) containing the powers of the first argument from 0 up to the length of p (*Xxx* could be *Power*, *Par*, or *P*).

2.1 The Case $n = 2^k$

In this case, for the parallel program specification, *PowerList* data structures can be used. The function $fft : PowerList.Com.n \rightarrow PowerList.Com.n$ can be defined as:

$$\begin{array}{l} fft.[a] \quad\ = [a] \\ fft.(p \natural q) = (r + u \cdot s) \mid (r - u \cdot s) \\ where \\ \quad \begin{array}{l|l} r = fft.p & u = powers.z.p \\ s = fft.q & z = root.(length.(p \natural q)) \end{array} \end{array} \tag{4}$$

The values r, s, and u are independent and can be computed in parallel. So, the time complexity is $O(\log_2 n)$, where n is the length of p.

2.2 The Case n Prime

In this case, it is necessary to compute directly the polynomial values:

$$\begin{aligned} &fft : ParList.Com.n \to ParList.Com.n \\ &fft.p = vp.p.(powers.z.p) \end{aligned} \tag{5}$$

The maximum complexity for the computation of a polynomial value at one point is obtained when $n = 2^k - 1$ ($n = length.p$), when there are $2k-2$ parallel steps, and the minimum complexity is achieved when $n = 2^k$. If we can compute the n polynomial values in parallel, the time complexity is still $O(\log_2 n)$, even if the constant is larger.

2.3 The Case $n = r_1 \cdot \ldots \cdot r_k$

If n is not a power of two, but a product of two numbers r_1 and r_2, the formula for computing the polynomial value can be generalized in the following way:

$$pf.w_j = \sum_{k=0}^{r_1-1} \{ \sum_{t=0}^{r_2-1} a_{tr_1+k} e^{\frac{2\pi ijt}{r_2}} \} e^{\frac{2\pi ijk}{n}}, \quad 0 \leq j < n \tag{6}$$

Theorem 1. *The best factorization $n = r_1 \cdot r_2$ for* FFT *(from the complexity point of view) is to choose r_1 from the prime factors of n [4].*

Therefore, for the specification of the parallel algorithm, we consider the decomposition in prime factors $n = r_1 \cdot \ldots \cdot r_k$. The *PList* data structures will be used. The *PosList* is formed by the prime factors of $n : [r_1, r_2, \cdots, r_p]$.

We start the derivation from the classic definition: $fft.l.p = vp.l.p.l.w$, where $w = powers.z.l.p$. Function vp is the one defined on $PLists$ (*Example 3*).

We use the notation $W.z.l = powers.z.l.p$. The following properties are true, due to the properties of the unity roots:

$$\begin{aligned} W.z.(x \rhd l) &= [|i : i \in \overline{x} :< z^{\frac{n}{x}i} \cdot > .(W.z.l)], \text{where } n = x \cdot prod.l \\ (W.z.(x \rhd l))^x &= [|i : i \in \overline{x} : W.z^x.l] \\ (W.z.l)^i &= W.(z^i).l \end{aligned} \tag{7}$$

We derive a new expression for fft, based on the induction principle:
Base case:

$$\begin{aligned} & fft.[x].[\natural i : i \in \overline{x} : [a.i]] \\ = & \{ \text{ definition of } fft, \text{ calculus} \} \\ & [|j : j \in \overline{x} : (+i : i \in \overline{x} : a.i \cdot z^{(i \cdot j)})] \end{aligned} \tag{8}$$

Inductive Step:

$$\begin{aligned} & fft.(x \rhd l).[\natural i : i \in \overline{x} : p.i] \\ = & \{\text{definitions of } fft \text{ and } vp, \text{ properties of function } W.z.l, \text{ calculus}\} \\ \ldots & \\ & (+i : i \in \overline{x} : [|j : j \in \overline{x} :< z^{\frac{n}{x}ij} \cdot > .(W.(z^i).l) \cdot fft.l.(p.i)] \end{aligned} \tag{9}$$

So, the definition of fft is now:

$$
\begin{array}{l}
fft : PosList \times PList.Com.n \rightarrow PList.Com.n \\
defined.fft.l.p \equiv (prod.l = length.p) \\
fft.[x][\natural i : i \in \overline{x} : [a.i]] \quad = [|j : j \in \overline{x} : (+i : i \in \overline{x} : a.i \cdot z^{(i \cdot j)})] \\
where \quad z = root.x \\
fft.(x \triangleright l).[\natural i : i \in \overline{x} : p.i] = [|j : j \in \overline{x} : (+i : i \in \overline{x} : r.i \cdot u.i.j)] \\
where \\
\begin{array}{l|l}
r.i = fft.l.(p.i) & z = root.n \\
u.i.j =< z^{(ij \cdot \frac{n}{x})} \cdot > .W.(z^i).l & n = length.[\natural i : i \in \overline{x} : p.i]
\end{array}
\end{array}
\tag{10}
$$

(A special case of the high order function *map* was used: $< z \cdot > .p = map.(z \cdot).p.$)

For the base case, the algorithm presented when n is prime can be used, which is more efficient. If the list of arities contains just values equal to 2, the algorithm becomes the one specified in the first case.

The algorithm for Fast Fourier Transformation can be done simultaneously with the decomposition of n in prime factors. If the prime factors become too large, then we can stop and apply the algorithm used when n is prime.

The time complexity of the algorithm depends on the prime factors of n and on their number m. If we consider that all the prime factors are less than a number M, then the time complexity is $O(m)$, with a constant that depends on M. If, for example, $n = 3^k$, then the time complexity is $O(\log_3 n)$.

We can implement this algorithm using a *recursive* interconnection network[2], which has the same arity list of the nodes like the arity list used for the calculation of fft. The implementation has two stages: a descendent stage and an ascendent stage.

3 Conclusions

The last algorithm for Fast Fourier Transformation is a general parallel algorithm that does not depend on the degree of the polynomial. It was formally derived, and so its correctness was proved.

The time complexity that can be obtained with this algorithm is better than the complexity of the other two (and so better than the classic one); and also the algorithm can be mapped on a classic interconnection network.

References

1. Kornerup, J.: Data Structures for Parallel Recursion. PhD thesis, University of Texas at Austin (1997)
2. Kornerup, J.: PLists: Taking PowerLists Beyond Base Two. In: Gorlatch, S. (ed.): First International Workshop on Constructive Methods for Parallel Programming, MIP-9805 (1998) 102-116
3. Misra, J. : PowerList: A structure for parallel recursion. ACM Transactions on Programming Languages and Systems, Vol. 16 No.6 (1994) 1737-1767
4. Wilf, H.S.: Algorithms and Complexity. Mason & Prentice Hall (1985)

A Branch and Bound Algorithm for Capacitated Minimum Spanning Tree Problem

Jun Han, Graham McMahon, and Stephen Sugden

School of Information Technology, Bond University
jhan@bond.edu.au, gmcmahon@bond.edu.au, ssugden@bond.edu.au

Abstract. This paper studies the capacitated minimum spanning tree problem (CMST), which is one of the most fundamental and significant problems in the optimal design of communication networks. CMST has a great variety of applications, such as in the design of local access networks, the design of minimum cost teleprocessing networks, the vehicle routing and so on. A solution method using branch and bound technique is introduced. Computational experiences demonstrate the algorithm's effectiveness.

1 Introduction

In the capacitated minimum spanning tree problem the objective is to find a minimum cost tree spanning a given set of nodes such that some capacity constraints are observed. We consider a connected graph $G=(V,A,b,c)$ with node set $V=\{0,1,...,n\}$ and arc set A. Each node i in V has a unit node weight $b_i = 1$ with $b_0 = 0$. The node weights may be interpreted as flow requirements whereas an arc weight c_{ij} represents the cost of using arc (i,j) in A. Node 0 is called center node and will be the root of the tree. We define a rooted sub-tree of a tree spanning V as its maximal sub-graph that is connected to the center by arc $(0,i)$. To satisfy the capacity constraint the flow requirement of each rooted sub-tree must not exceed a given capacity K. By means of these definitions the CMST problem is the problem finding a minimum cost tree spanning node set V where all rooted sub-trees satisfy the capacity constraint. Assume $b_i = 1$ for all $i = 1,...,n$ and $b_i = 0$ for $i = 0$. Define $x_{ij} = 1$, if arc (i,j) is included in the solution, and $x_{ij} = 0$, otherwise. Let y_{ij} denote the flow on arc (i,j). The following formulation gives a minimum cost directed capacitated spanning tree with center node 0 being the root:

$$Minimize \sum_{i=0}^{n}\sum_{j=1}^{n} c_{ij}x_{ij} \quad (1)$$

$S.T.$

B. Monien and R. Feldmann (Eds.): Euro-Par 2002, LNCS 2400, pp. 404–407.

$$\sum_{i=0}^{n} x_{ij} = 1 \qquad i = 1,\dots,n \quad j = 1,\dots,n \tag{2}$$

$$\sum_{i=0}^{n} y_{ij} - \sum_{i=1}^{n} y_{ji} = 1 \qquad j = 1,\dots,n \tag{3}$$

$$x_{ij} \le y_{ij} \le (k - b_i) \cdot x_{ij} \qquad i = 1,\dots,n \quad j = 1,\dots,n \tag{4}$$

$$x_{ij} \in \{0,1\} \quad y_{ij} \ge 0 \qquad i = 1,\dots,n \quad j = 1,\dots,n \tag{5}$$

The capacitated minimum spanning tree problem is NP-hard when $2<k<n/2$.

A significant amount of research has been done in designing heuristics for solving CMST problem [1],[2],[6],[7],[8],[9],[11],[13],[17],[18]. Existing exact algorithms include [4],[5],[11],[14],[15], and computation results with 20-node problems to 40-node problems were reported.

2 The Branch and Bound Algorithm

We consider the solution as a collection of n-1 arcs (for n-node problem) chosen from all candidate arcs, so we will introduce an arc-oriented branch and bound method. We construct a binary searching tree in which each stage corresponds to the decision of either adding a certain arc (ordered in ascendant cost) to the solution arc set or not.

At each stage, we construct a minimum spanning tree under the presupposition that some arcs must be included and some arcs must not be included, and we use the cost of this tree as a bound. An efficient feasibility check procedure is essential for pruning the searching tree. For capacity constrain check, instead of direct calculating of traffic on the arcs, we maintain some sets of nodes, with each set containing the nodes that have been connected by selected arcs. Thus the total number of nodes in each node set reflects the traffic on each central arc. This will foretell whether a sub-tree will overflow even if it hasn't yet been connected to the center node. Detailed description of the searching process is omitted here due to limited space.

3 Computational Experiences

We used a number of sets of instances generated randomly to test our branch and bound algorithm, and the results (time used in seconds) of one set of 50-Node problems and one set of 100-node problems are shown in Table 1. The computations were performed on a machine with Pentium II 400MHz CPU and 128M RAM.

It has to be pointed out that comparability is not guaranteed in all cases, since different authors have proposed various ways of conducting experiments. To the CMST problem, instances with larger value of capacity constraint K will be easier to solve. Moreover, if we locate all nodes in the two-dimensional Euclidean plane and

Moreover, if we locate all nodes in the two-dimensional Euclidean plane and use the Euclidean distance between two nodes as the cost of the arc between them, the location of the root has considerable impact on the performance of most algorithms.

OR-Library (http://mscmga.ms.ic.ac.uk/info.html) is a collection of test data sets for a variety of operations research problems. Algorithms that have been reported to have found the optimal solution of some of problem instances maintained in OR-Library include Hall [12] and Gouviea [10]. Table 2 is a comparison of our test results with Hall and Gouviea's work when n=41 and k=10.Regarding Hall and Gouviea's results, "Gap" refers to (UB-LB)/LB where UB and LB refer to the upper and lower bounds generated respectively. "CPU'" is the Number of CPU seconds.

From Table 2, we can see that our branch and bound algorithm found the optimal solutions of more instances in center 41-node problem, but it is much slower. All methods couldn't find the optimal solution of corner 41-node problem. Actually, Hall and Gouviea's algorithms are called "cutting plane algorithm", which is not an exact algorithm but makes effort on finding better lower bounds of a problem and may verify the optimality of a solution if there is no gap between the lower bound and the feasible solution found by other heuristics. Our branch and bound algorithm is an exact algorithm and can guarantee finding the optimal solution, thus it is slower and is more sensitive with the "difficulty" of sample problems.

No exact algorithm has been reported to have solved the sample CMST problems provided by OR-Library or random generated problem with more than 50 nodes.

All the test problems and solutions found by our algorithm can be accessed at: http://www.geocities.com/ilhll/cmst.html.

Table 1. Results of Random Instances

50-Node	P1	P2	P3	P4	P5	P6	P7	P8	P9	P10
K=20	0.00	0.9	0.00	0.22	0.5	0.03	0.06	0.08	0.1	0.4
K=15	0.02	3.4	0.5	0.03	84	9.4	0.08	30	0.1	13
K=10	0.2	1.2k	329	6	39k	351	2	178	532	1k
K=5	33k		59k	82k			38k			
100-Node	PP1	PP2	PP3	PP4	PP5	PP6	PP7	PP8	PP9	PP10
K=30	156	27k		6	4k			28k		

Table 2. Results of OR-Library Problems

Prob...	Root	B&BCPU	Hall Gap	Hall CPU	Gou Gap	Gou CPU
Tc41-1	Center	379	min 0.00	min 1.21	min 0.00	min 177
Tc41-2	Center	3617				
Tc41-3	Center	12173	avg 0.08	avg 2.60	avg 1.603	avg 1620
Tc41-4	Center	56330				
Tc41-5	Center	92199	max 0.39	max 6.78	max 2.76	max 3877
Te41-1	Corner	Unsolved	min 0.70	min 8.59	min 0.71	min 1332
Te41-2	Corner	Unsolved				
Te41-3	Corner	Unsolved	avg 2.00	avg 15.24	avg 2.092	avg 3473
Te41-4	Corner	Unsolved				
Te41-5	Corner	Unsolved	max 3.69	max 28.32	max 4.08	max 9360

4 Conclusion

We have proposed a branch and bound algorithm that can solve up to 100-node problem instances and have presented the comparison with other algorithms. Our experiences also support the argument that the difficulty in solving CMST problem is highly sensitive to the location of the root and to the magnitude of the constrained capacity.

References

1. Ahuja,R.K., J. B. Orlin and D. Sharma. 2000. Very large-scale neighborhood Search. Operational Research, 7: 301-317.
2. Altinkemer, K. and B. Gavish. 1988. Heuristics with constant error guarantees for the design of tree networks. Management Science, 32: 331 - 341.
3. Amberg, A., W. Domscheke and S.Braunschweig. 1996. Capacitated Minimum Spanning Trees: Algorithms using intelligent search. Combinatorial optimization: Theory and Practice, 1:9-39.
4. Chandy, K.M. and R.A. Russell. 1972. The design of multipoint linkages in a teleprocessing tree network. IEEE Transactions on Computers, 21: 1062-1066.
5. Chandy, K.M. and T. Lo. 1973. The capacitated minimum spanning tree. Networks, 3: 173-181.
6. Elias, D. and M.J. Ferguson. 1974. Topological design of multipoint teleprocessing networks. IEEE Transactions on Communications, 22: 1753 - 1762.
7. Frank, H., I.T. Frisch, R. Van Slyke and W.S. Chou. 1971. Optimal design of centralized computer design network. Networks, 1: 43-57.
8. Gavish, B. 1991. Topological design of telecommunication networks-local access design methods. Annals of Operations Research, 33: 17-71.
9. Gavish, B. and K. Altinkemer. 1986. Parallel savings heuristics for the topological design of local access tree networks. Proc. IEEE INFOCOM 86 Conference, 130-139.
10. Gouveia, L. and P.Martins.1995. An extended flow based formulation for the capacitated minimum spanning tree, presented at the third ORSA Telecommunications Conference, Boca Raton, FL.
11. Gouveia, L. and J. Paixao. 1991. Dynamic programming based heuristics for the topological design of local access networks. Annals of Operations Research, 33: 305-327.
12. Hall,L. 1996. Experience with a cutting plane algorithm for the capacitated spanning tree problem. INFORMS Journal on Computing, 8(3): 219-234.
13. Karnaugh, M. 1976. A new class of algorithms for multipoint network optimization. IEEE Transactions on Communications, 24: 500 - 505.
14. Kershenbaum, A. and P.R. Boorstyn. 1983. Centralized teleprocessing network design. Networks, 13: 279-293.
15. Malik, K. and G. Yu 1993. A branch and bound algorithm for the capacitated minimum spanning tree problem. Networks, 23: 525-532.
16. Papadimitriou, C.H. 1978. The Complexity of the capacitated tree problem. Networks, 8: 217-230.
17. Schneider, G.M. and M. N. Zastrow. 1982. An algorithm for the design of multilevel concentrator networks. Computer Networks, 6: 1-11.
18. Thangiah, S.R., I.H. Osman and T. Sun. 1994. Hybrid genetic algorithms, simulated annealing and tabu search methods for vehicle routing problems with time windows. Working Paper, Univ. of Kent, Canterbury.

Topic 7
Applications on High Performance Computers

Vipin Kumar, Franz-Josef Pfreundt, Hans Burkhard, and Jose Laghina Palma

Topic Chairpersons

High Performance computers are used now in industry on a day-to-day basis. Companies that need parallel computers spread now over all branches of industry. With the availability of cost effective PC Clusters parallel computing is today even a topic for start-up companies or for more traditional companies like foundries. From this spread of parallel technology a broad and interesting spectrum of parallel applications was expected for this years conference. The topic description for this section asks especially for papers related to the problems connected with data management and scalability on Clusters, for papers about multidisciplinary applications and has a special focus on application from image analysis, multimedia and visualization.

This year 16 papers were submitted. The majority of the papers are from the areas related to image analysis, Physics and Biology, resp. Bioinformatics. Other areas covered are forestry, hydrology and optimisation.

"Stochastic simulation of a marine host-parasite system using a hybrid MPI / Open MP programming". The authors describe a stochastic simulation model of a marine host parasite system. The presented numerical approach uses a multilevel parallel software architecture to solve the demanding computing problem resulting from the mathematical model, which characterises the individual behavior of each species. The hardware used to carry out the simulation has the typical structure of today's parallel machines consisting of coupled SMP systems.

"Demand Driven Parallel Ray Tracing", a paper dealing with a new variant of a ray tracing algorithm based on image space division , where image parts assigned to individual processes is optimal with respect to load balancing. Two load balancing approaches are presented.

"Parallel Controlled Conspiracy Number Search" is about parallelizing a best-first game-tree search algorithm. The very nature of best-first search means that this is a hard problem for achieving good parallel performance. Starvation for work is a serious issue. As well, the difficulty of sharing information between processors causes inefficiencies. This application is an instance of the most challenging types of problems for parallel computing.

As short papers the topics *texture analysis, numerical methods for the Boltzmann Equation and Fire Propagation Simulation* are presented.

In summary, unlike many submitters of papers those in applications areas typically have to satisfy two widely different sets of goals, in furthering both the science in the application area and developing methodologies that constitute high-performance computing research. We thank all of the authors who submitted contributions and succeeded in both goals, and the many referees who helped the committee in spanning the many applications fields represented.

B. Monien and R. Feldmann (Eds.): Euro-Par 2002, LNCS 2400, p. 409.

Perfect Load Balancing for Demand-Driven Parallel Ray Tracing*

Tomas Plachetka

University of Paderborn, Department of Computer Science, Fürstenallee 11, D-33095 Paderborn, Germany

Abstract. A demand-driven parallelization of the ray tracing algorithm is presented. Correctness and optimality of a perfect load balancing algorithm for image space subdivision are proved and its exact message complexity is given. An integration of antialiasing into the load balancing algorithm is proposed. A distributed object database allows rendering of complex scenes which cannot be stored in the memory of a single processor. Each processor maintains a permanent subset of the object database as well as a cache for a temporary storage of other objects. A use of object bounding boxes and bounding hierarchy in combination with the LRU (Last Recently Used) caching policy reduces the number of requests for missing data to a necessary minimum. The proposed parallelization is simple and robust. It should be easy to implement with any sequential ray tracer and any message-passing system. Our implementation is based on POV-Ray and PVM.

Key words: parallel ray tracing, process farming, load balancing, message complexity, antialiasing, distributed database, cache

1 Introduction

Ray tracing [16] computes an image of a 3D scene by recursively tracing rays from the eye through the pixels of a virtual screen into the scene, summing the light path contributions to pixels' colors. In spite of various optimization techniques [3], typical sequential computation times range from minutes to hours.

Parallel ray tracing algorithms (assuming a message passing communication model) can be roughly divided into two classes [4]:

Image space subdivision (or screen space subdivision) algorithms [5], [4], [1], [6], [9], [11] exploit the fact that the primary rays sent from the eye through the pixels of the virtual screen are independent of each other. Tracing of primary rays can run in parallel without a communication between processors. The problem of an unequal workload in processors must be considered. Another problem arises by rendering large scenes—a straightforward parallelization requires a copy of the whole 3D scene to be stored in memories of all processors. On the other hand, these algorithms are usually easy to implement.

* Partially supported by the Future and Emerging Technologies programme of the EU under contract number IST-1999-14186 (ALCOM-FT) and VEGA 1/7155/20.

B. Monien and R. Feldmann (Eds.): Euro-Par 2002, LNCS 2400, pp. 410–419.

Object space subdivision algorithms [2], [10], [1], [6], [15] geometrically divide the 3D scene into disjunct regions which are distributed in processors' memories. The computation begins with shooting of primary rays into processors storing the regions through which the primary rays pass first. The rays are then recursively traced in the processors. If a ray leaves a processor's region, it is passed to the processor which stores the adjacent region (or discarded if there is no adjacent region in the ray's direction). An advantage of object space subdivision algorithms is that the maximum size of the rendered 3D scene is theoretically unlimited because it depends only of the total memory of all processors. Potential problems are an unequal workload and a heavy communication between processors. Moreover, an implementation of these algorithms may be laborious.

Some parallelizations (*hybrid algorithms*) [7], [8], [13] extend the data-driven approach of object space subdivision algorithms with additional demand-driven tasks running in processors in order to achieve a better load balance.

A lot of work has been done concerning ray tracing parallelization. However, the majority of the research papers is interested only in the performance of the algorithms. Software engineering issues are seldom mentioned. Among these belong questions like: Can the parallel algorithm be easily integrated into an existing sequential code? Will it be possible to continue the development of the sequential code without the need to reimplement the parallel version?

This paper presents a parallelization based on the algorithm of Green and Paddon [5]. The basic idea is to combine the image space subdivision with a distributed object database. Our implementation named POV||Ray (read "parallel POV-Ray") builds upon the (sequential) freeware ray tracer POV-Ray version 3.1. Many practical aspects addressed in the implementation apply generally.

Screen space subdivision using a process farming is explained in section 2. Correctness and optimality of a perfect load balancing algorithm as well as its exact message complexity are proved. It is also shown how antialiasing can be integrated into the load balancing algorithm. Section 3 describes the management of the distributed object database. Results of experiments are presented in section 4, followed by conclusions in section 5.

2 Image Space Subdivision

The computed image consists of a finite number of pixels. Computations on pixels are independent of each other. However, computation times on different pixels differ and they are not known beforehand. A load balancing is therefore needed for an efficient parallelization. There are two principal approaches to achieving a balanced load in processors: work stealing [1] and process farming. A problem of both is tuning their parameters. We focus on process farming and analyze a perfect load balancing algorithm which is tuned using two intuitive parameters.

2.1 Process Farming

In process farming a central process (*load balancer*) distributes image parts to *worker processes* (*workers*) on demand. When a worker becomes idle, it sends a *work request* to the load balancer. Once a worker gets a work (an image part), it must compute it in a whole. Workers do not communicate with each other.

The problem is to determine an appropriate granularity of the assigned parts. The choice of granularity influences the load balance and the number of exchanged messages. There are two extreme cases: 1. The assigned parts are minimal (pixels). In this case the load is balanced perfectly but the number of work requests is large (equal to the number of pixels which is usually much greater than the number of workers); 2. The assigned parts are maximal (the whole image is partitioned into as many parts as the number of workers). In this case the number of work requests is low but the load imbalance may be great.

The following algorithm is a compromise between these two extremes (a version of this algorithm is given in [9] and [11]). The algorithm is perfect in the sense that it guarantees a perfect load balancing, and at the same time it minimizes the number of work requests if the ratio of computation times on any two parts of the same size can be bounded by a constant T ($T \geq 1.0$). W denotes the total number of atomic parts (e.g. image pixels or image columns), N denotes the number of workers (running on a homogeneous parallel machine).

Claim. The algorithm in Fig. 1 always assigns as much work as possible to idle workers, while still ensuring the best possible load balance.

Proof. The algorithm works in rounds, one round being one execution of the while-loop. In the first round the algorithm assigns image parts of size

$$s_{max} = max\,(1, \lfloor \mathrm{W}/(1 + T \cdot (\mathrm{N} - 1)) \rfloor)$$

(measured in the number of atomic parts). In each of the following rounds the parts are smaller than in the previous round. Obviously, the greatest imbalance is obtained when a processor p_{max} computes a part of the size s_{max} from the first round as long as possible (whereby the case of $s_{max} = 1$ is trivial and will not be considered here) and all the remaining $N-1$ processors compute the rest of the image as quickly as possible (in other words, the load of the remaining $N-1$

```
loadbalancer(float T, int W, int N)
  int part_size;
  int work = W;
  while (work > 0)
    part_size = max (1, ⌊work/(1 + T · (N − 1))⌋);
    for (counter = 0; counter < N; counter++)
      wait for a work request from an idle worker;
      if (work > 0)
        send work of size part_size to the worker;
        work = work − part_size;
  collect work requests from all workers;
  send termination messages to all workers;
```

Fig. 1. The perfect load balancing algorithm

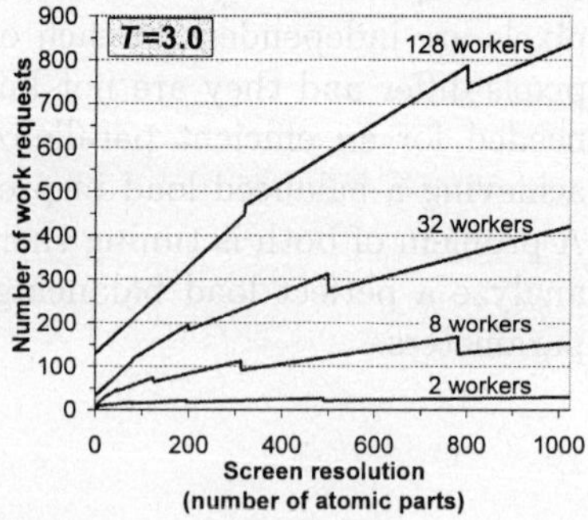

Fig. 2. Number of work requests

processors is perfectly balanced). The number of parts computed in parallel by all processors except of p_{max} is $W - s_{max}$. The ratio of the total workload (in terms of the number of processed atomic parts) of one of the $N-1$ processors (let p_{other} denote the processor and let s_{other} denote its total workload) and s_{max} is then

$$\frac{s_{other}}{s_{max}} = \frac{\frac{W-s_{max}}{N-1}}{s_{max}} = \frac{\frac{W-\lfloor W/(1+T\cdot(N-1))\rfloor}{N-1}}{\lfloor W/(1+T\cdot(N-1))\rfloor}$$

This ratio is greater or equal to T. This means that the processor p_{other} does at least T times more work than the processor p_{max} in this scenario. From this and from our assumptions about T and about the homogeneity of processors follows that the processor p_{max} must finish computing its part from the first round at the latest when p_{other} finishes its part from the last round. Thence, a perfect load balance is achieved even in the worst case scenario.

It follows directly from the previous reasoning that the part sizes s_{max} assigned in the first round cannot be increased without affecting the perfect load balance. (For part sizes assigned in the following rounds a similar reasoning can be used, with a reduced image size.) This proves the optimality of the above algorithm. □

Claim. The number of work requests (including final work requests that are not going to be fulfilled) in the algorithm in Fig. 1 is equal to

$$N\cdot(r+1) + \left\lceil W\cdot\left(1-\frac{N}{1+T\cdot(N-1)}\right)^{r}\right\rceil$$

where

$$r = \max\left(0, \left\lfloor \log_{1+\frac{N}{1+T\cdot(N-1)}} (W/N)\right\rfloor\right)$$

Proof. It is easy to observe that

$$\frac{N\cdot W}{1+T\cdot(N-1)}\cdot\left(1-\frac{N}{1+T\cdot(N-1)}\right)^{i-1}$$

atomic parts get assigned to workers during the ith execution of the while-loop and that

$$W\cdot\left(1-\frac{N}{1+T\cdot(N-1)}\right)^{i}$$

atomic parts remain unassigned after the ith execution of the while-loop.

r is the total number of executions of the while-loop minus 1. The round r is the last round on the beginning of which the number of yet unassigned atomic parts is greater than the number of workers N. r can be determined from the fact that the number of yet unassigned atomic parts after r executions of the while-loop is at most N:

$$W\cdot\left(1-\frac{N}{1+T\cdot(N-1)}\right)^{r} \leq N$$

which yields (r is an integer greater than or equal to 0)

$$r = \max\left(0, \left\lfloor \log_{1+\frac{N}{1+T\cdot(N-1)}}(W/N)\right\rfloor\right)$$

There are N work requests received during each of the r executions of the while-loop, yielding a total of $N \cdot r$ work requests. These do not include the work requests received during the last execution of the while-loop. The number of work requests received during the last execution of the while-loop is equal to

$$\left\lceil W \cdot \left(1 - \frac{N}{1+T\cdot(N-1)}\right)^r \right\rceil$$

Finally, each of the workers sends one work request which cannot be satisfied. Summed up,

$$N \cdot (r+1) + \left\lceil W \cdot \left(1 - \frac{N}{1+T\cdot(N-1)}\right)^r \right\rceil$$

is the total number of work requests. □

Two parameters must be tuned in the algorithm. The first parameter is the constant T. The second parameter is the size of an atomic work part in pixels (it makes sense to pack more pixels into a single message because a computation of several pixels usually costs much less than sending several messages instead of one). We experimentally found that a combination of T between 2.5 and 4.0 and an atomic part size of a single image column yields the best performance for most scenes (if no antialiasing is applied). Fig. 2 shows the total number of work requests as a function of image resolution for a varying number of workers.

We use a process farm consisting of three process types in our implementation. The *master* process is responsible for interaction with the user, for initiation of the ray tracing computation, and for assembling the image from the computed image parts. The *load balancer* process runs the perfect load balancing algorithm, accepting and replying work requests from workers. *Worker* processes receive parts of the image from the load balancer, perform (sequential) ray tracing computations on those parts and send the computed results to the master process. The idea of this organization is to separate the process of image partitioning (which must respond as quickly as possible to work requests in order to minimize the idle periods in workers) from collecting and processing of results.

Problems arising by parsing of large scenes and computing camera animations are discussed in [11]. A solution to another important problem—an efficient implementation of antialiasing—is given in the following section.

2.2 Antialiasing

It is useful to restrict the work parts to rectangular areas because they are easy to encode (a rectangular work part is identified by coordinates of its upper-left and bottom-right corners). If we further restrict the work parts to *whole neighboring columns* or *whole neighboring rows*, then columns should be used by landscape-shaped images and rows otherwise. This leads to a finer granularity of image parts in the perfect load balancing algorithm (because of the integer arithmetic used in the algorithm).

A good reason for using *whole* image columns or rows as atomic parts is an integration of antialiasing in the load balancing algorithm. A usual (sequential) antialiasing technique computes the image using one ray per pixel and then for each pixel compares its color to the colors of the left and above neighbors. If a significant difference is reported, say, in the colors of the pixel and its left neighbor, both pixels are resampled using more primary rays than one (whereby the already computed sample may be reused). Each pixel is resampled at most once—if a color difference in two pixels is found, whereby one of them has already been resampled, then only the other one is going to be resampled. In a parallel algorithm using an image subdivision the pixels on the border of two image parts must either be computed twice or an additional communication must be used. The additional work is minimized when whole image columns or rows are used as atomic job parts.

We use an additional communication. No work is duplicated in workers. If antialiasing is required by the user, the workers apply antialiasing to their image parts except of *critical parts.* (A critical part is a part requiring information that is not available on that processor in order to apply antialiasing.) Critical parts are single image columns (from now on we shall assume that whole image columns are used as atomic image parts). A worker marks pixels of a critical part which have already been resampled in an *antialiasing map* (a binary map). Once a worker finishes its computation of an image part, it packs the aliasing map and the corresponding part of the computed image to the job request and sends the message to the load balancer. The load balancer runs a slightly modified version of the perfect load balancing algorithm from section 2.1. Instead of sending the termination messages at the end of the original algorithm, it replies idle workers' requests with *antialiasing jobs.* An antialiasing job consists of a critical part and the part of which the critical part depends on. The computed colors and antialiasing maps of these two parts (two columns) are comprised in the antialiasing job. Only after all critical parts have been computed, the load balancer answers all pending work requests with termination messages. A worker, upon a completion of an antialiasing job, sends as before another job request to the load balancer and the computed image part to the master process. The master process updates the image in its frame buffer.

The computation of antialiasing jobs is interleaved with the computation of "regular" jobs. There is no delay caused by adding the antialiasing phase to the original load balancing algorithm. The antialiasing phase may eventually *compensate* a load imbalance caused by an underestimation of the constant T in the load balancer. Note that the above reasoning about using whole columns or rows as atomic image parts does not apply to antialiasing jobs. The image parts of antialiasing jobs can be horizontally (in landscape-shaped images) or vertically (in portrait-shaped images) cut into rectangular pieces to provide an appropriate granularity for load balancing.

3 Distributed Object Database

A disadvantage of image subdivision is that the parallel algorithm is not data-scalable, that means, the entire scene must fit into memory of each processor. This problem may be overcome by using a distributed object database. We followed the main ideas described for instance in [5], [4], [13], [14]. In the following we talk about our implementation.

Before coming to a design of a distributed object database, we make some observations. A 3D scene consists of objects. Ray tracers support a variety of object types most of which are very small in memory. Polygon meshes are one of a few exceptions. They require a lot of memory and occur very frequently in 3D models. If a scene does not fit into memory of a single processor, it is usually because it consists of several objects modeled of large polygon meshes. However, any single object (or any few objects) can usually be stored in memory. Moreover, it should also be said that a vast majority of scenes *does* fit into memory of a single processor.

For the above reasons we decided to distribute polygon meshes in processors' memories. All other data structures are replicated—in particular object bounding hierarchy including the bounding boxes of polygon meshes. We assume one worker process running on each processor. At the beginning of the computation data of each polygon mesh reside in exactly one worker (we shall refer to the worker as to the *mesh's owner*). Each worker must be able to store—besides the meshes it owns and besides its own memory structures—at least the largest polygon mesh of the remaining ones.

The initial distribution of objects on processors is quasi-random. The master process is the first process which parses the scene. When it recognizes a polygon mesh, it makes a decision which of the worker processes will be the object's owner. The master process keeps a track of objects' ownership and it always selects the worker with the minimum current memory load for the object's owner. Then it broadcasts the mesh together with the owner's ID to all workers. Only the selected owner keeps the *mesh data* (vertices, normals, etc.) in memory. The remaining workers preprocess the mesh (update their bounding object hierarchies) and then release the mesh's data. However, they do not release the *object envelope* of the mesh. In case of meshes, this object envelope contains the mesh's bounding box, an internal bounding hierarchy tree, the size of the missing data, the owner's ID, a flag whether the mesh data are currently present in the memory, etc.

A mesh's owner acts as a server for all other workers who need those mesh's data. When it receives a mesh request from some other worker (meshes have their automatically generated internal IDs known to all workers), it responds with the mesh's data. Two things are done at the owner's side in order to increase efficiency. First, the mesh's data are prepacked (to a ready-to-send message form) even though it costs additional memory. Second, the owner runs a thread which reacts to mesh requests by sending the missing data without interrupting the main thread performing ray tracing computations. (This increases efficiency on parallel computers consisting of multiprocessor machines connected in a network, such as Fujitsu-Siemens hpcLine used for our experiments.)

The mesh data are needed at two places in the (sequential) ray tracing algorithm: by intersection computations and by shading. The problem of missing data is solved by inserting an additional code at these two places. This code first checks whether the mesh's data are in memory (this information is stored in the mesh's envelope). If the data are not present, the worker checks whether there is enough memory to receive and unpack the data. If there is not enough memory, some objects are released from the *object cache memory* (we use Last Recently Used replacement policy which always releases objects that have not been used for the longest time from the cache to make space for the new object) and then sends a data request to the owner. Upon the arrival of the response, the data are unpacked from the message into the cache memory. After that the original computation is resumed.

Efficiency can be slightly improved also on the receiver's side. Note that in the ray tracing algorithm shading of a point P is always preceded by an intersection test resulting in the point P. However, there can be a large number of intersection and shading operations between these two. A bad situation happens when the mesh data needed for shading of the point P have meanwhile been released from the cache. In order to avoid this, we precompute all necessary information needed for shading of point P in the intersection code and store it in the mesh's object envelope. The shading routine is modified so that it looks only into the mesh's envelope instead of into mesh's data. By doing so we avoid an eventual expensive communication for the price of a much less expensive unnecessary computation (not all intersection points are going to be shaded).

There is a communication overhead at the beginning of the computation also when the cache memory is large enough to store the entire scene. It must be assumed during the parsing phase that the entire scene will not fit into workers' cache memories, and therefore mesh data are released in all workers but one after they have been preprocessed. Cache memories of all workers are thus empty after the parsing phase. This is called a *cold start.* A cold start can be avoided by letting the user decide whether to delete meshes from workers' cache memories during parsing. If an interaction with the user is not desirable and if computed scenes may exceed a single worker's memory, then a cold start cannot be avoided.

Note that the above mechanisms can be applied to any object types, not only to polygon meshes. Certain scenes may include large textures which consume much memory. It makes then sense to distribute textures over workers' memories as well. The caching policy (LRU, for instance) can be easily extended to handle any object types.

4 Experiments

The following results were measured on the hpcLine parallel computer by Fujitsu-Siemens. hpcLine consists of double-processor nodes (Pentium III, 850 MHz) with 512 MB of memory per node. The underlying PVM 3.4 message-passing library uses Fast Ethernet network for communication.

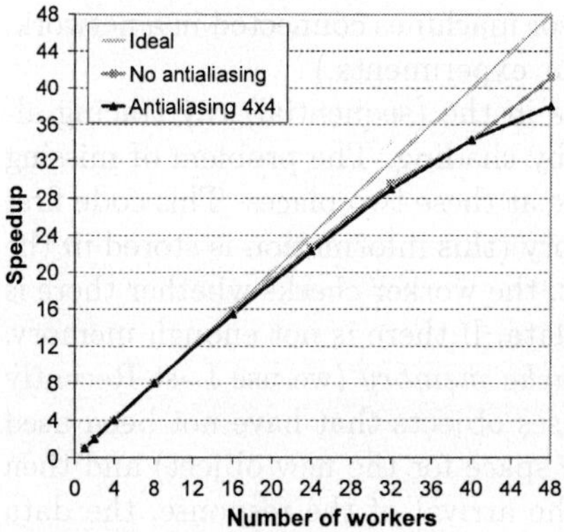

Fig. 3. No distributed database

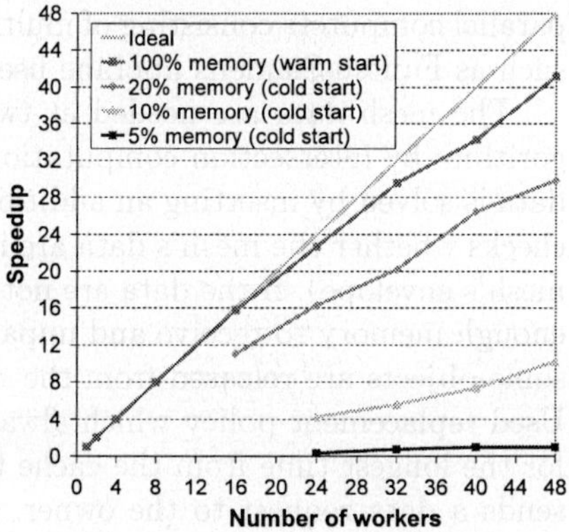

Fig. 4. Distributed database

We used a model of a family house for the measurements. The model consists of about 614 objects (approximately 75,000 triangles) and 22 light sources. The resolution of the rendered images was 720x576 pixels (PAL).

The use of the distributed database was switched off in the speedup measurements of Fig. 3. $T = 4.0$ (see section 2.1) was used for all measurements without antialiasing. In the measurements with antialiasing, 16 samples per pixel were used when the colour difference of neighboring pixels exceeded 10%, and the constant T was set to 16.

The memory limit was simulated in the software by the experiments with the distributed object database (the whole scene can be stored in the memory of one node of hpcLine). The limit says how much memory may a worker use at most for storing the objects it owns. The *rest* of memory below the limit is used for the object cache. The memory limit is relative to the size of the distributed data (e.g., "100% memory" means that each worker is allowed to store the whole scene in its memory). The speedups in Fig. 4 are relative to the computational time of a single worker with a warm start (no antialiasing). The cache hit ratio was above 99% and the total number of object requests about 3,000 in the 20% memory case (only about 20% of all objects are relevant for rendering the image). The cache hit ratio dropped only by a promile in the 10% memory case—however, the total number of object requests grew to about 500,000. In the 5% memory case, the cache hit ratio was about 85% and the total number of object requests about 7,000,000. Some data are missing in the graph in Fig. 4. In these cases the scene could not be stored in the distributed memory of worker processes due to the (simulated) memory limits.

The LRU caching policy performs very well. The falloff in efficiency in Fig. 4 is caused by the idle periods in the worker process beginning when the worker sends an object request and ending when the requested object arrives. (This latency depends mainly of the thread switching overhead in the object's owner—the computation thread gets interrupted by the thread which handles the request.)

5 Conclusions

A simple and robust parallelization of the ray tracing algorithm was presented which allows rendering of large scenes that do not fit into memory of a single processor. An analysis of a load balancing strategy for image space partitioning was given. A parallel ray tracer implementing the described ideas was integrated as a component of a remote rendering system during the project HiQoS [12]. The proposed design of the distributed object database may serve as a basis for developing other parallel global illumination algorithms.

References

1. D. Badouel, K. Bouatouch, and T. Priol. Distributing data and control for ray tracing in parallel. *Comp. Graphics and Applications*, 14(4):69–77, 1994.
2. J. G. Cleary, B. Wyvill, G. M. Birtwistle, and R. Vatti. Multi-processor ray tracing. *Comp. Graphics Forum*, 5(1):3–12, 1986.
3. A. S. Glassner, editor. *Introduction to ray tracing.* Academic Press, Inc., 1989.
4. S. Green. *Parallel Processing for Comp. Graphics.* Research Monographs in Parallel and Distributed Computing. The MIT Press, 1991.
5. S. Green and D. J. Paddon. Exploiting coherence for multiprocessor ray tracing. *Comp. Graphics and Applications*, 9(6):12–26, 1989.
6. M. J. Keates and R. J. Hubbold. Interactive ray tracing on a virtual shared-memory parallel computer. *Comp. Graphics Forum*, 14(4):189–202, 1995.
7. W. Lefer. An efficient parallel ray tracing scheme for distributed memory parallel computers. In *Proc. of the Parallel Rendering Symposium*, pages 77–80, 1988.
8. M. L. Netto and B. Lange. Exploiting multiple partitioning strategies for an evolutionary ray tracer supported by DOMAIN. In *First Eurographics Workshop on Parallel Graphics and Visualisation*, 1996.
9. I. S. Pandzic, N. Magnenat-Thalmann, and M. Roethlisberger. Parallel raytracing on the IBM SP2 and T3D. *EPFL Supercomputing Review (Proc. of First European T3D Workshop in Lausanne)*, (7), 1995.
10. P. Pitot. The Voxar project. *Comp. Graphics and Applications*, pages 27–33, 1993.
11. T. Plachetka. POV||Ray: Persistence of vision parallel ray tracer. In L. Szirmay-Kalos, editor, *Proc. of Spring Conference on Comp. Graphics (SCCG 1998)*, pages 123–129. Comenius University, Bratislava, 1998.
12. T. Plachetka, O. Schmidt, and F. Albracht. The HiQoS rendering system. In L. Pacholski and P. Ruzicka, editors, *Proc. of the 28th Annual Conference on Current Trends in Theory and Practice of Informatics (SOFSEM 2001)*, Lecture Notes in Comp. Science, pages 304–315. Springer-Verlag, 2001.
13. E. Reinhard and A. Chalmers. Message handling in parallel radiance. In *Recent Advances in Parallel Virtual Machine and Message Passing Interface*, pages 486–493. Springer-Verlag, 1997.
14. E. Reinhard, A. Chalmers, and F. W. Jansen. Hybrid scheduling for parallel rendering using coherent ray tasks. In *1999 IEEE Parallel Visualization and Graphics Symposium*, pages 21–28. ACM SIGGRAPH, 1999.
15. E. Reinhard, A. J. F. Kok, and A. Chalmers. Cost distribution prediction for parallel ray tracing. In *Second Eurographics Workshop on Parallel Graphics and Visualisation*, pages 77–90, 1998.
16. T. Whitted. An improved illumination model for shaded display. *Communications of the ACM*, 23(6):343–349, 1980.

Parallel Controlled Conspiracy Number Search*,**

U. Lorenz

Department of Mathematics and Computer Science
Paderborn University
Germany

Abstract. Tree search algorithms play an important role in many applications in the field of artificial intelligence. When playing board games like chess etc., computers use game tree search algorithms to evaluate a position. In this paper, we present a procedure that we call Parallel Controlled Conspiracy Number Search (Parallel CCNS). Briefly, we describe the principles of the sequential CCNS algorithm, which bases its approximation results on irregular subtrees of the entire game tree. We have parallelized CCNS and implemented it in our chess program P.ConNerS, which now is the first in the world that could win a highly ranked Grandmaster chess-tournament. We add experiments that show a speedup of about 50 on 159 processors running on an SCI workstation cluster.

1 Introduction

When a game tree is so large that it is not possible to find a correct move, there are two standard approaches for computers to play games. In the first approach, the algorithms work in two phases. Initially, a subtree of the game tree is chosen for examination. This subtree may be a full width, fixed depth tree, or any other subtree rooted at the starting position. Thereafter, a search algorithm heuristically assigns evaluations to the leaves and propagates these numbers up the tree according to the minimax principle. Usually the chosen subtree is examined by the help of the $\alpha\beta$-algorithm [6][5][4] or one of its variants. As far as the error frequency is concerned, it does not make any difference whether the chosen game tree is examined by the $\alpha\beta$-algorithm or by a pure minimax algorithm. In both cases, the result is the same. Only the effort to get the result differs drastically.

The heuristic minimax value of such a static procedure already leads to high quality approximations of the root value. However, there are several improvements that more specifically form the selected subtree. These lead us to a second class of algorithms which work in only one phase and which form the tree shape dynamically at execution time. Some of the techniques are domain independent such as Nullmoves [2], Fail High Reductions [3], Singular Extensions [1], min/max approximation [12], or 'Conspiracy Number Search'. Conspiracy Number Search was introduced by D. McAllister [11]. J. Schaeffer [14] interpreted the idea and developed a search algorithm that behaves well

* Supported by the German Science Foundation (DFG) project *Efficient Algorithms For Discrete Problems And Their Applications*

** A short version of this paper was presented at the SPAA'01 revue [9].

B. Monien and R. Feldmann (Eds.): Euro-Par 2002, LNCS 2400, pp. 420–430.

on tactical chess positions. The startup point of Conspiracy Number Search (CNS) is the observation that, in a certain sense, the $\alpha\beta$-algorithm computes decisions with low security. The changing of the value of one single leaf (e.g. because of a fault of the heuristic evaluation function) can change the decision at the root. Thus, the $\alpha\beta$-algorithm takes decisions with security (i.e. conspiracy number) 1. The aim of CNS is to distribute the available resources in a way that decisions are guaranteed to be made with a certain conspiracy number $c > 1$. Such decisions are stable against up to $c - 1$ changes of leaf-values.
We introduced the Controlled Conspiracy Number Search, an efficient flexible search algorithm which can as well deal with Conspiracy Numbers[7]. In this paper, we describe the parallel version of our search algorithm. It is implemented in the chess program 'P.ConNerS', which was the first one that could win an official FIDE Grandmaster Tournament [8]. The success was widely recognized in the chess community.
In section 2, some basic definitions and notations are presented. Section 3 briefly describes the principles of the Ccns-algorithm, and the parallel algorithm in more detail. Section 4 deals with experimental results from the domain of chess.

2 Definitions and Notations

A *game tree* G is a tuple (T, h), where $T = (V, K)$ is a tree (V a set of nodes, $K \subset V \times V$ the set of edges) and $h : V \to \mathbb{Z}$ is a function. $L(G)$ is the set of leaves of T. $\Gamma(v)$ denotes the set of successors of a node v. We identify the nodes of a game tree G with positions of the underlying game and the edges of T with moves from one position to the next. Moreover, there are two *players*: MAX and MIN. MAX moves on even and MIN on odd levels. The so called minimax values of tree nodes are inductively defined by *minimax(v)* := $h(v)$ if $v \in L(G)$, $\max\{$*minimax(v')* $\mid (v, v') \in K\}$, if $v \notin L(G)$ and MAX to move, and $\min\{$*minmax(v')* $\mid (v, v') \in K\}$ if $v \notin L(G)$ and MIN to move.
Remark: Let $\mathcal{A}$ be a game tree search algorithm. We distinguish between the *universe*, an *envelope* and a (current) *search tree*. We call the total game tree of a specific game the *universe*. A subtree E of a game tree G (G being a universe) is called an *envelope* if, and only if, the root of E is the root of G and each node v of E either contains all or none of the successors of v in G, and E is finite. Last but not least, a search tree is a subtree of an envelope. E.g., the minimax-algorithm and the $\alpha\beta$-algorithm may examine the same envelope, but they usually examine different search trees.

A *MIN-strategy (MAX-strategy)* is a subtree of a game tree G, which contains the root of G and all successors of each MAX-node (MIN-node) and exactly one successor of each MIN-node (MAX-node). A MIN-strategy proves an upper bound of the minimax value of G, and a MAX-strategy a lower one. A *strategy* is either a MIN- or a MAX-strategy. The figure on the right shows how two leaf-disjoint strategies prove the lower bound 6 at the root.

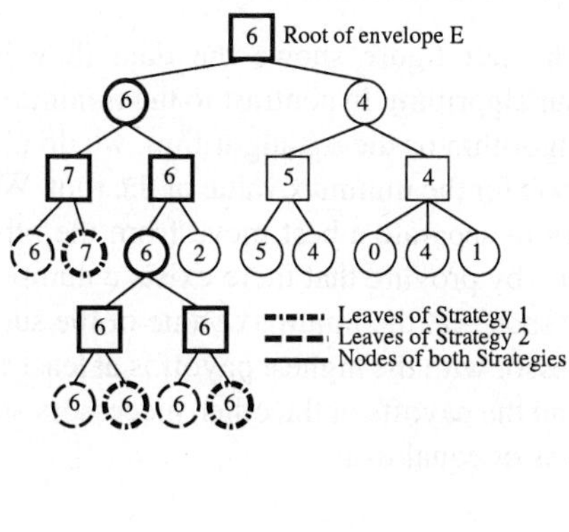

Definition 1. *(Best Move) Let $G = (T, h) = ((V, K), h)$ be a game tree. A best move is a move from the root to a successor which has the same minimax-value as the root has. Let $m = (v, v')$ be such a move. We say m is* secure *with conspiracy number C and depth d if there exists an $x \in \mathbb{Z}$ so that a) there are at least C leaf-disjoint strategies, with leaves at least in depth d, showing the minimax value of v' being greater than or equal to x, and b) for all other successors of the root there are at least C leaf disjoint strategies, with leaves at least in depth d, showing the minimax value of them being less than or equal to x. C is a lower bound for the number of terminal nodes of G that must change their values in order to change the best move at the root of G. The aim of a CCNS is to base all results on envelopes that contain secure decisions.*

Remark: An error analysis in game trees [10] has lead us to the assumption that 'leaf-disjoint strategies' are one of THE key-terms in the approximation of game tree values.

Definition 2. *(Value) Let $G = ((V, K), h)$ be a game tree. A* value *is a tuple $w = (a, z) \in \{'\leq', '\geq', '\#'\} \times \mathbb{Z}$. a is called the* attribute *of w, and z the* number *of w. $W = \{'\geq', '\leq', '\#'\} \times \mathbb{Z}$ is the* set of values. *We denote $w_v = (a_v, z_v)$ the value of the node v, with $v \in V$.*

Remark: *Let v be a node. Then $w_v = ('\leq', x)$ will express that there is a subtree below v, the minimax-value of which is $\leq x$. $w_v = ('\geq', x)$ is analogously used. $w_v = ('\#', x)$ implies that there exists a subtree below v the minimax-value of which is $\leq x$, and there is a subtree below v whose minimax-value is $\geq x$. The two subtrees need not be identical. A value w_1 can be 'in contradiction' to a value w_2 (e.g. $w_1 = ('\leq', 5), w_2 = ('\geq', 6)$), 'supporting' (e.g. $w_1 = ('\leq', 5), w_2 = ('\leq', 6)$), or 'unsettled' (e.g. $w_1 = ('\geq', 5), w_2 = ('\leq', 6)$).*

Definition 3. *Target:* A target *is a tuple $t = (\omega, \delta, \gamma)$ with ω being a value and $\delta, \gamma \in \mathbb{N}_0$. Remark: Let $t_v = (\omega, \delta, \gamma) = ((a, z), \delta, \gamma)$ be a target which is associated with a node v. δ expresses the demanded distance from the current node to the leaves of the final envelope. γ is the* conspiracy number *of t_v. It informs a node of how many leaf-disjoint strategies its result must base. If the demand, expressed by a target, is fulfilled, we say that the target t_v is* fulfilled.

3 Description of CCNS

3.1 Search Framework

The left figure shows the data flow in our algorithm. In contrast to the minimax-algorithm or the $\alpha\beta$-algorithm, we do not look for the minimax value of the root. We try to separate a best move from the others, by proving that there exists a number x such that the minimax value of the successor with the highest payoff is at least x, and the payoffs of the other successors are less or equal to x.

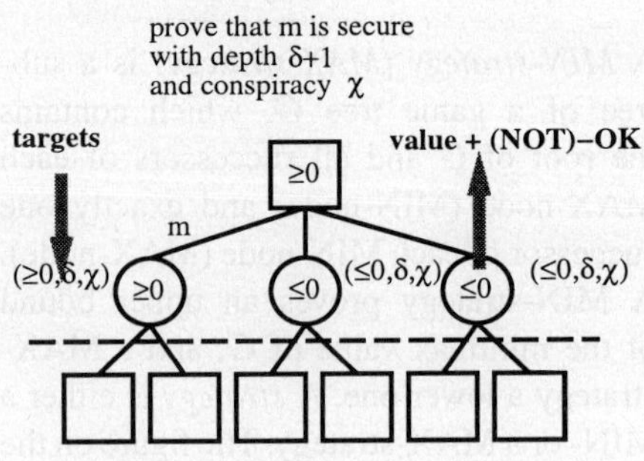

Fig. 1. Principle behavior of CCNS

As we work with heuristic values which are available for all nodes, the searched tree offers such an x and a best move m, at any point of time. As long as m is not *secure* enough, we take x and m as a hypothesis only, and we commission the successors of the root either to show that new estimates make the hypothesis fail, or to verify it. The terms of 'failing' and 'verifying' are used in a weak sense: they are related to the best possible knowledge at a specific point of time, not to the absolute truth. New findings can cancel former 'verifications'. The verification is handled by the help of the targets, which are split and spread over the search tree in a top down fashion. A target t expresses a demand to a node. Each successor of a node, which is supplied by a target, takes its own value as an expected outcome of a search below itself, and commissions its successors to examine some sub-hypotheses, etc. A target t will be fulfilled when t demands a leaf, or when the targets of all of v's successors are fulfilled. When a target is fulfilled at a node v, the result 'OK' is given to the father of v. If the value of v changes in a way that it contradicts the value component of t then the result will be 'NOT-OK'.

3.2 Skeleton of the Sequential Search Algorithm

In the following, we assume that there is an evaluation-procedure which can either return the heuristic value $h(v)$ of a given node v, or which can answer whether $h(v)$ is smaller or greater than a given number y.

We call the starting routine (no figure) at the root DetermineMove(root r, d, $c = 2$). d stands for the remaining depth and c for the conspiracy number which the user wants to achieve. If the successors of r have not been generated yet, then DetermineMove will generate them and assign heuristic values of the form ('#' , ...) to all successors. It picks up the successor which has the highest value x, and assigns a lower bound target of the form $(('\geq' , x), d, c)$ to the best successor and targets of the form $(('\leq' , x), d, c)$ to all other successors. Then it starts the procedure Ccns on all successors. DetermineMove repeats the previous steps, until Ccns returns with OK from all of r's successors.

```
bool Ccns(node v, target t_v = (α_v, β_v, δ_v, γ_v))
  if (δ_v = 0 and γ_v ≤ 1) or
     |Γ(v)| = 0 return OK;
  r := NOT_OK;
  while r = NOT_OK do {
    PartialExpansion(v, t_v);
    if not OnTarget(v, t_v) return NOT_OK;
    Split(v, t, v_1 ... v_|Γ'(v)|);
    /* assigns targets to sons */
    for i := 1 to |Γ(v)'| do {
      r := Ccns(v_i, t_i);
      w_v := UpdateValue(v);
      if not OnTarget(v, t_v) return NOT_OK;
      if r = NOT_OK break ;
      /* Leave the loop, goto l.3 */
    }
  } /* while ... */;
  return OK;
```

Fig. 2. Recursive Search Procedure

Let $\Gamma'(v)$ be the set of successors of v, as far as the current search tree is concerned. Let t_v be the target for v and w_v the value of v. Let $v_1 \ldots v_{|\Gamma'(v)|}$ be the successors of v concerning the current search tree. Let $t_1 \ldots t_{|\Gamma'(v)|}$ be the targets of the nodes $v_1 \ldots v_{|\Gamma'(v)|}$, and let $w_1 \ldots w_{|\Gamma'(v)|}$ be their values. We say that a node is OnTarget(v, t_v) when the value of v is not in contradiction with the value component of t_v. This will express that Ccns still is on the right way. When Ccns (figure 2) enters a node v, it is guaranteed that v is on target and that the value of v supports t_v (either by DetermineMove, or

because of figure 2, ll. 5-6). Firstly, Ccns checks whether v is a leaf, i.e. whether t_v is trivially fulfilled (l.1). This is the case when the remaining search depth of the target is zero ($\delta_v = 0$) and the demanded conspiracy number (i.e. the number of leaf-disjoint bound-proving strategies) is 1 ($\gamma_v \leq 1$). If v is not a leaf, the procedure PartialExpansion (no figure) will try to find successors of v which are well suited for a splitting operation. Therefore, it starts the evaluation of successors which either have not yet been evaluated, or which have an unsettled value in relation to the target $t_v = (\ldots, x)$. If a successor s has been evaluated once before and is examined again it will get a *point value* of the form ('#' $, y)$. For a not yet examined successor s the evaluation function is inquired whether the value of s supports or contradicts the value component of t_v. s gets a value of the form ('$\geq$' $, x)$ or ('$\leq$' $, x)$.
If v is an allnode and a partial expansion changes the value of v in a way that it contradicts the target t, Ccns will immediately stop and leave v by line 11. If v is a cutnode, PartialExpansion will evaluate successors which have not yet been examined or which are unsettled in relation to t, until it will have found γ_v-many successors which support the value of v.
After that, the target of the node v is 'split', i.e. sub-targets are worked out for the successors of v. The resulting sub-targets are given to the successors of v, and Ccns examines the sons of v, until either all sons of v will have fulfilled their targets (some successors may get so called null-targets, i.e. a target that is always fulfilled), or v itself is not 'on target' any longer, which means that the value of v contradicts the current target of v. When a call of Ccns returns with the result OK at a node $v.i$ (line 8), the node $v.i$ could fulfill its subtarget. When Ccns returns with NOT-OK, some values below $v.i$ have changed in a way that it seems impossible that the target of $v.i$ can be fulfilled any more. In this case, Ccns must decide, whether to report a NOT-OK to its father (line 10), or to rearrange new sub-targets to its sons (ll. 11 and 3).
For all further details of the sequential algorithm, as well as for correctness- and termination-proofs etc. we refer to [7].

3.3 The Distributed Algorithm

In the following, let tuples (v, t_v) represent subproblems, where v is a root of a game tree $G = ((V, E), f)$ and t_v a target, belonging to v. The task is to find a subtree of G, rooted at v, which fulfills all demands that are described by t_v.
Our parallelization of the CCNS-algorithm is based on a dynamic decomposition of the game tree to be searched, and on parallel evaluation of the resulting subproblems. Although the sequential CCNS-algorithm is a best-first search algorithm, which is able to jump irregularly in the searched tree, the algorithm prefers to work in a depth-first manner. To do so, a current variation is additionally stored in memory. On the left and on the right of this current variation, there exist nodes which have got targets. These nodes, together with their targets describe subproblems. All the nodes to the left of the current variation have been visited, and all nodes to the right of this variation have not yet been examined. The idea of our parallelization of such a tree search is to make as many of the not yet examined subproblems available for parallel evaluation as possible. By this, several processors may start a tree search on a

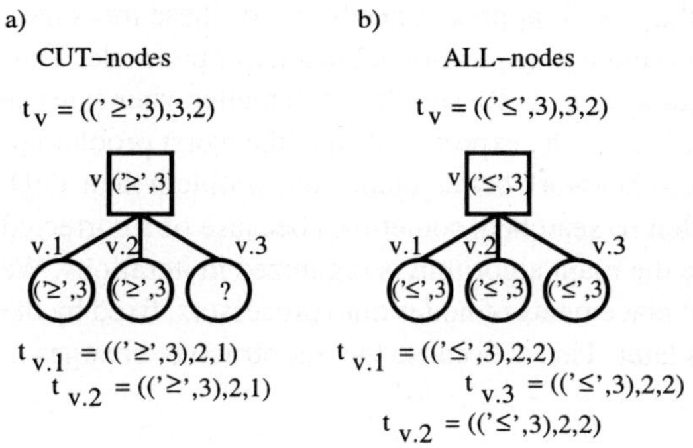

Fig. 3. Parallelism

subtree of the whole game tree. These processors build up current variations by themselves.

Selection of Subproblems. In order to fulfill target t_v in figure 3b) the targets $t_{v.1} \ldots t_{v.3}$ must be fulfilled. The resulting subproblems are enabled for parallel computations.
CUT-nodes (figure 3a)) provide parallelism, as well, when two or more successors get non-trivial targets.
The Ccns-algorithm is a kind of speculative divide and conquer algorithm. Thus, if all targets could be fulfilled during a computation, we would be able to evaluate many subproblems simultaneously and the sequential and the parallel versions would search exactly the same tree. This is not reality, but usually, most of the targets are fulfilled.

Start of an Employer-Worker Relationship. Initially, all processors are idle. The host processor reads the root position and sends it to an specially marked processor. In the following this processor behaves like all other processors in the network.
A processor, which receives a subproblem, is responsible for the evaluation of this subproblem. It starts the search of this subproblem as in the sequential algorithm, i.e. it builds up a current variation of subproblems. Other subproblems are generated, which are stored for later evaluation. Some of these subproblems are sent to other processors by the following rule:
We suppose that there is a hash function *hash* from nodes into the set of processors. If v is a node of a subproblem p on processor P and if $hash(v) \neq P$, the processor P will self-responsibly send the subproblem p to the destination $Q = hash(v)$. It does so by the help of a WORK-message. If P already knows the address of the node v on processor Q, the WORK-message will also contain this address. An employer-worker relationship has been established. The worker Q starts a search below the root of its new subproblem. It maps the root v of its new subproblem into its memory. If v has not yet been examined on processor Q before, Q sends an ADDRESS-message to processor P, which contains the address of the node v on processor Q. By this, processor Q later can find the node v again, for further computations.
A note on work-stealing: The work-stealing technique is mostly used for the pupose of dynamic load balancing. Nevertheless, we have never been able to implement a work-stealing mechanism as efficient, as our transpositiontable driven (TPD) approach [1]. When we tested work-stealing, we had a further performance loss of at least 10-15%. We observed several pecularities which may serve as an explanation: **a)** Roughly spoken, the transposition table driven approach needs three messages per node (i.e. WORK,

[1] A very similar approach has simultaniously been developed for the IDA* algorithm in the setting of 1-person games [13].

ADDRESS, ANSWER). A workstealing (WS) approach needs less of these messages, but it needs a transposition table query and a transposition table answer per node. Thus the total number of messages, sent in the system, is not significantls smaller when you use the work stealing approach. **b)** We will see in the experiments that the worst problem of parallel CCNS is the low load. WS does not work better against this problem than TPD. **c)** In the course of time, nodes are often re-searched, sometimes because of a corrected value-estimation, but mostly, because the main algorithm is organized in iterations. We could, in some examples, observe that placements of nodes onto processors, fixed by the WS, became misstructured mappings later. This is because the tree structure changes in the course of time.

End of an Employer-Worker Relationship. There are two ways how an employer-worker relationship may be resolved:
A processor Q, which has solved a subproblem by itself or with the help of other processors, sends an ANSWER-message to its employer P. This message contains a heuristic value computed for the root of the concerned subproblem, as well as OK or NOT-OK, depending on whether the target of that subproblem could be fulfilled. The incoming result is used by P as if P had computed it by itself.
An employer P can also resolve an employer-worker relationship. Let v be the root of a subproblem p, sent to Q. Let v' be the predecessor of v. If processor P must reorganize its search at the node v' (either because of figure 2, lines 10-11, or because P has received a STOP-message for the subproblem p) it sends a STOP-message to Q, and Q cancels the subproblem which belongs to the node v.
Altogether, we have got half-dynamic load balancing mechanism. On the one hand, the hash function which maps nodes into the network contains a certain accidentalness because we do not know the resulting search tree in advance. On the other hand, a node that has been generated on a processor Q once, in future can be accessed over processor Q only.

3.4 Local Behaviour of the Distributed Algorithm

The overall distributed algorithm has a simple structure. It consists of a loop with alternating calls to a search step of an iterative version of the sequential CCNS-algorithm, a process handling incoming and outgoing messages, and a selection of a task. This is necessary because the work is distributed sender-initiated so that a processor may get several worck packeges at a time. The overall process structure is shown in figure 4. The process 'Communication' is described in figure 5.
The parallel CCNS-algorithm usually does not exchange tasks. It does so only, when the active task cannot be further examined, because it only waits for answers from other processors. In the latter case, the first non-waiting task in the list of tasks becomes active.

3.5 Search Overhead and Loss of Work Load

The efficiency of the algorithm mainly depends on the following aspects:
1. When the parallel algorithm visits more nodes than the sequential one we say that *search overhead* arises. Here, it can arise by the fact that some targets are not fulfilled in

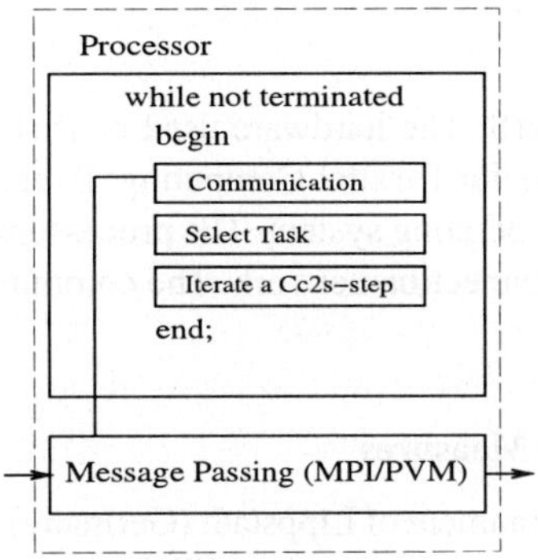

Fig. 4. Process Structure

```
process Communication;
  if problem received then
    initialize subproblem rooted at v;
    if employer does not know the address of v then
        send address
  if address received then save address;
  if answer received then incorporate answer;
  if stop received then
    find task T which belongs to stop-message;
    send stop-messages to all processors which work
      on subproblems of the current variation of T;
    terminate T;
```

Fig. 5. Communication

the course of computations. As the algorithm is driven by heuristic values of inner nodes of the game tree, it may occur that the sequential algorithm quickly comes to a result, while the parallel version examines the complete game tree. In that case there will not even be a termination in acceptable time. Nevertheless, experiments in the domain of chess show that the heuristic values of inner nodes, which guide the search, are stable enough so that the search overhead stays in an acceptable range.
Remark: Such large differences in the computing time between the sequential and a parallel version are not possible in the known variants of parallel $\alpha\beta$-game tree search. Nevertheless, in its worst case a parallel $\alpha\beta$-algorithm examines all nodes up to a fixed, predefined depth, whilst the sequential version only examines the nodes of a minimal solution tree, which must be examined by any minimax-based algorithm. Then, the parallel $\alpha\beta$-algorithm does not stop in acceptable time either.
2. Availability of subproblems: If there are not enough subproblems available at any point of time, the network's load will decrease. This is the hardest problem for the distributed CCNS-algorithm.
3. Speed of communication: The time which a message needs to reach a remote processor will play a decisive role. If the communication fails to supply the processors with information similar to the sequential search algorithm, then the algorithm will search many nodes on canceled and stopped subproblems. Thus, the time which a STOP-message needs to reach its destination will increase the search overhead.
The time which a WORK-message needs in order to reach its destination will decrease the work load.

Moreover, messages which are exchanged in the network, must be processed, and the management of several quasi-simultaneous tasks costs time, as well. Nevertheless, as these periods seem to be negligible in our distributed chess program we have not further considered the costs.

4 Experimental Results

All results are taken with our chess program P.ConNerS. The hardware used consists of the PSC-workstation cluster at the Paderborn Center for Parallel Computing. Every processor is a Pentium II/450 MHz running the Linux operating system. The processors are connected as a 2D-Torus by a Scali/Dolphin interconnection network. The communication is implemented on the basis of MPI.

4.1 Grandmaster Tournament and Other Quality Measures

In July 2000 P.ConNerS won the 10th Grandmaster tournament of Lippstadt (Germany). Thus, for the very first time, a chess program has won a strong, categorized FIDE tournament (category 11). In a field of 11 human Grandmasters P.ConNerS had only to resign against the experienced Grandmaster J. Speelman and the former youth-worldchampion Slobodjan. P. ConNerS won the tournament with 6 victories, 3 draws, and 2 losses. The opponents had an average ELO-strength of 2522 points and P.ConNerS played a performance of 2660 ELO-points. (The ELO-system is a statistical measure for the strength of chess players. An International Master has about 2450 ELO, a Grandmaster about 2550, the top-ten chess players have about 2700 ELO on the average, and the human World Champion about 2800 ELO.)
Although, in former years, chess programs could win against Grandmasters here and there, the games mostly were either blitz- or rapid games. In Lippstadt the human players competed under optimal conditions for human beings [8].

Another possibility of getting an impression of the strength of a program is to compare its strength by means of a set of selected positions, with other programs and human players. On the widely accepted BT2630 test suite, P.ConNerS achieves 2589 points, which is a result that has not yet been reported by any other program.

4.2 Speedup, Search Overhead, and Work Load

The tables below presents us with the data from the parallel evaluations (averaged over the 30 instances of the BT2630 test). As can be seen, the overall speedup (SPE) is about 50 on 159 processors. The search overhead (SO) is kept in an acceptable range (we experienced that keeping the search overhead small is a good first-order heuristic), so that the

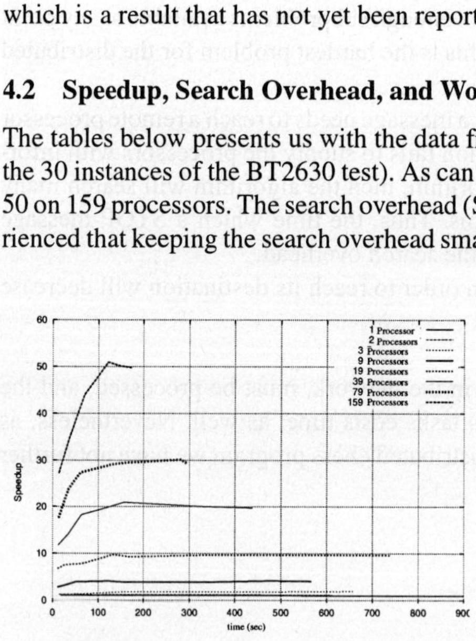

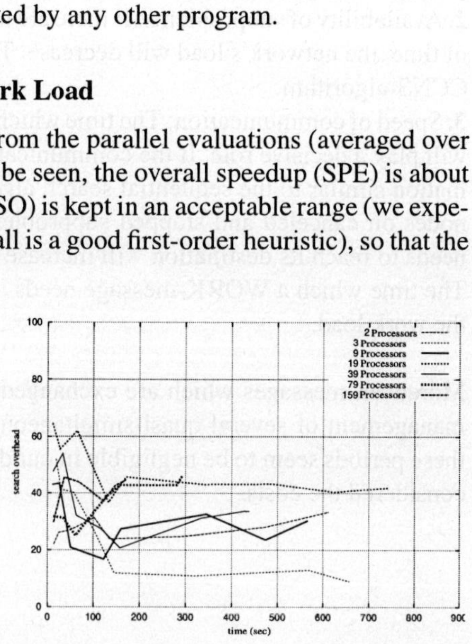

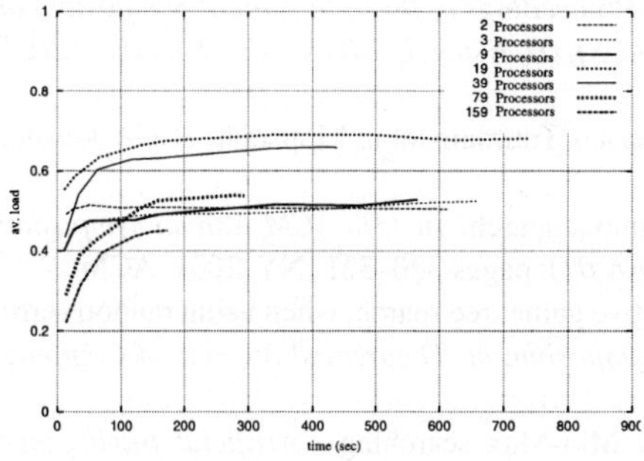

main disabler for even better speedups is the limited average-load (LO). There are two reasons for this effect: One is the limited granularity of work-packages.
As the search tree is irregular, there are periods of time, when the amount of distributable work is too small for a good load-balancing. We used a depth-2 alphabeta search together with quiescence search as 'static' evaluation procedure. A more fine-grained parallelization showed a remarkably higher effort for the management of those small subproblems. Moreover, the quality of the sequential/parallel search became remarkably worse on tactical test positions. When we tried more coarse-grained work packeges, spreading the work over the network took too much time.
The other reason is that subproblems are placed more or less randomly onto the network. The average length of a tree edge is half of the diameter of the processor network. Thus, the speed of the communication network directly limits the performance. Additinal experiments with fast-ethernet on the same machines (using the MPICH library, no figures here) showed the effect: We observed speedups no better than 12.

5 Conclusion

The Parallel CCNS algorithm, as described here, dynamically embeds its search tree into a processor network. We achieved efficiencies of 30% on an SCI workstation cluster, using 160 processors. In consideration of the fact that the results are mesured on a workstaion cluster (and not on a classic parallel computer), and under consideration that not only the work load but also the space must be distributed (which makes the application to an instance of the most challenging types of problems for parallel computing), the results are remarkably nice. They are comparable to the best known results of the parallel alphabeta algorithm on workstation clusters.

References

1. T.S. Anantharaman. Extension heuristics. *ICCA Journal*, 14(2):47–63, 1991.
2. C. Donninger. Null move and deep search. *ICCA Journal*, 16(3):137–143, 1993.
3. R. Feldmann. Fail high reductions. *Advances in Computer Chess 8 (ed. J. van den Herik)*, 1996.
4. R. Feldmann, M. Mysliwietz, and B. Monien. Studying overheads in massively parallel min/max-tree evaluation. In *6th ACM Annual symposium on parallel algorithms and architectures (SPAA'94)*, pages 94–104, New York, NY, 1994. ACM.
5. R.M. Karp and Y. Zhang. On parallel evaluation of game trees. In *First ACM Annual symposium on parallel algorithms and architectures (SPAA'89)*, pages 409–420, New York, NY, 1989. ACM.
6. D.E. Knuth and R.W. Moore. An analysis of alpha-beta pruning. *Artificial Intelligence*, 6(4):293–326, 1975.

7. U. Lorenz. Controlled Conspiracy-2 Search. *Proceedings of the 17th Annual Symposium on Theoretical Aspects of Computer Science (STACS), (H. Reichel, S.Tison eds), Springer LNCS*, pages 466–478, 2000.
8. U. Lorenz. P.ConNers wins the 10th Grandmaster Tournament in Lippstadt. *ICGA Journal*, 23(3), 2000.
9. U. Lorenz. Parallel controlled conspiracy number search. In *13th ACM Annual symposium on parallel algorithms and architectures (SPAA'01)*, pages 320–321, NY, 2001. ACM.
10. U. Lorenz and B. Monien. The secret of selective game tree search, when using random-error evaluations. *Accepted for the 19th Annual Symposium on Theoretical Aspects of Computer Science (STACS) 2002*, to appear.
11. D.A. McAllester. Conspiracy Numbers for Min-Max searching. *Artificial Intelligence*, 35(1):287–310, 1988.
12. R.L. Rivest. Game tree searching by min/max approximation. *Artificial Intelligence*, 34(1):77–96, 1987.
13. J.W. Romein, A. Plaat, H.E. Bal, and J. Schaeffer. Transposition Table Driven Work Scheduling in Distributed Search. In *Proceedings of the Sixteenth National Conference on Artificial Intelligence (AAAI-99)*, pages 725–731, 1999.
14. J. Schaeffer. Conspiracy numbers. *Artificial Intelligence*, 43(1):67–84, 1990.

A Parallel Solution in Texture Analysis Employing a Massively Parallel Processor

Andreas I. Svolos, Charalambos Konstantopoulos, and Christos Kaklamanis

Computer Technology Institute and Computer Engineering & Informatics Dept.,
Univ. of Patras, GR 265 00 Patras, Greece,
svolos@cti.gr

Abstract. Texture is a fundamental feature for image analysis, classification, and segmentation. Therefore, the reduction of the time needed for its description in a real application environment is an important objective. In this paper, a texture description algorithm running over a hypercube massively parallel processor, is presented and evaluated through its application in real texture analysis. It is also shown that its hardware requirements can be tolerated by modern VLSI technology.

Key words: texture analysis, co-occurrence matrix, hypercube, massively parallel processor

1 Introduction

Texture is an essential feature that can be employed in the analysis of images in several ways, e.g. in the classification of medical images into normal and abnormal tissue, in the segmentation of scenes into distinct objects and regions, and in the estimation of the three-dimensional orientation of a surface. Two major texture analysis methods exist: statistical and syntactic or structural. Statistical methods employ scalar measurements (features) computed from the image data that characterize the analyzed texture. One of the most significant statistical texture analysis methods is the Spatial Gray Level Dependence Method (SGLDM). SGLDM is based on the assumption that texture information is contained in the overall spatial relationship that the gray levels have to one another. Actually, this method characterizes the texture in an image region by means of features derived from the spatial distribution of pairs of gray levels (second-order distribution) having certain inter-pixel distances (separations) and orientations [1].

Many comparison studies have shown SGLDM to be one of the most significant texture analysis methods [2]. The importance of this method has been shown through its many applications, e.g. in medical image processing [3]. However, the co-occurrence matrix [1], which is used for storing the textural information extracted from the analyzed image, is inefficient in terms of the time needed for its computation. This disadvantage limits its applicability in real-time applications and prevents the extraction of all the texture information that can be captured

B. Monien and R. Feldmann (Eds.): Euro-Par 2002, LNCS 2400, pp. 431–435.

by the SGLDM. The parallel computation of the co-occurrence matrix is a potential solution to the computational time inefficiency of this data structure. The first attempt to parallelize this computation was made in [4]. However, to the best of our knowledge, the only previous research effort of parallelization using a massively parallel processor was made in [5]. The reason is that until recently, full parallelization was possible only on very expensive machines. The cost of the parallel computation was prohibitive in most practical cases. For this reason, even in [5], there was a compromise between hardware cost and computational speed. The parallel co-occurrence matrix computation ran over a Batcher network topology. However, this parallel scheme had two significant drawbacks. First, the Batcher network requires a very large number of processing elements and interconnection links limiting its usefulness to the analysis of very small image regions. Second, the parallel algorithm proposed in [5] assumes an off-line pre-computation of the pairs of pixels that satisfy a given displacement vector in each analyzed region. This pre-computation has to be performed by another machine, since the Batcher network does not have this capability.

The rapid evolution of CMOS VLSI technology allows a large number of processing elements to be put on a single chip surface [6], dramatically reducing the hardware cost of the parallel implementation. Employing a regular parallel architecture also helps towards achieving a larger scale of integration. In this paper, a parallel algorithm for the computation of the co-occurrence matrix running on a hypercube massively parallel processor, is presented and evaluated through its application to the analysis of real textures.

2 The Parallel Algorithm for the Co-occurrence Matrix Computation

The processing elements of the massively parallel processor employed in this paper are interconnected via a hypercube network. The hypercube is a general-purpose network proven to be efficient in a large number of applications, especially in image processing (2D-FFT, Binary Morphology) [7]. It has the ability to efficiently compute the gray level pairs in an analyzed image region for any displacement vector. Moreover, its large regularity makes feasible the VLSI implementation of this parallel architecture.

In this paper, a modified odd-even-merge sort algorithm is employed for the parallel computation of the co-occurrence matrix. In the proposed algorithm, each element is associated with a counter and a mark bit. The counter gives the number of times an element has been compared with an equal element up to the current point of execution. The mark bit shows whether this element is active, i.e. it participates in the parallel computation (bit = 0), or is inactive (bit = 1). Each time two equal elements are compared, the associated counter of one of these two elements increases by the number stored in the counter of the other element. Also, the mark bit of the other element becomes 1, that is, the element becomes inactive. Inactive elements are considered to be larger than the largest element in the list. In the case that the compared elements are not equal,

```
for i := 1 to m do
    for j := 1 to i − 1 do                /* transposition sub-steps */
        parbegin
            P¹ = P_m P_{m−1} ... P_{i+1} 1 P_{i−1} ... P_{j+1} 0 P_{j−1} ... P_1;
            P² = P_m P_{m−1} ... P_{i+1} 1 P_{i−1} ... P_{j+1} 1 P_{j−1} ... P_1;
            P¹ ↔ P²;
        parend
    od
    for j := i to 1 do                    /* comparison sub-steps */
        parbegin
            P¹ = P_m P_{m−1} ... P_{j+1} 0 P_{j−1} ... P_1;
            P² = P_m P_{m−1} ... P_{j+1} 1 P_{j−1} ... P_1;
            P² → P¹;              /* the content of element P² is transferred to element P¹ */
            if P¹.M == 0 and P².M == 0 and P¹.(A_1, B_1) == P².(A_2, B_2) then
                P¹.C := P¹.C + P².C;
                P².M := 1;
                P¹ → P²;          /* the updated content of P² is sent back to P² */
            else if P¹.M == 1 or P¹.(A_1, B_1) > P².(A_2, B_2) then
                P¹ → P²;          /* P² gets the content of P¹ */
                P¹ := P²;         /* P¹ gets the content sent from P² */
            else
                nop;
            endif
        parend
    od
od
```

Fig. 1. The pseudocode of the proposed parallel algorithm for the co-occurrence matrix computation

the classical odd-even-merge sort algorithm is applied. At the end, the modified algorithm gives for each active element its times of repetition in the initial list. If each element in the list is a pair of gray levels in the analyzed region that satisfies a given displacement vector, it is straightforward to see that the above algorithm eventually computes the corresponding co-occurrence matrix.

The pseudocode of the algorithm is shown in Fig. 1. In this figure, the language construct **parbegin...parend** encloses the instructions, which are executed by all processing elements, concurrently. The " = " operator declares equivalence of notations. Actually, the right operand is the binary representation of processing element P in the hypercube. The " $\leftrightarrow$ " operator performs a transposition of the contents of its operands (processing elements) through the hypercube network. The "$\rightarrow$" operator transfers data from its left operand to its right operand over the hypercube network. Finally, $P.(A, B)$ is the pair of gray levels stored in processing element P, $P.C$ is the counter associated with gray level pair (A, B) and $P.M$ is the corresponding mark bit.

3 Results and Discussion

In order to show the time performance of the proposed parallel algorithm in a practical case, a large number of samples from natural textures were analyzed employing the SGLDM (fur, water, weave, asphalt, and grass) [8]. The co-occurrence matrices were computed using the proposed parallel algorithm running on the hypercube, the algorithm running on the Batcher network and the fastest serial algorithm. Each image had a dynamic range of 8 bits (256 gray levels). From each image, data sets of 64 non-overlapping sub-images of size 64 × 64, 256 non-overlapping sub-images of size 32 × 32, and 1024 non-overlapping sub-images of size 16 × 16 were extracted. 8 displacement vectors were employed in the texture analysis of all five categories of samples, namely (1,0), (0,1), (1,1), (1,-1), (2,0), (0,2), (2,2), and (2,-2). In this experiment, both parallel architectures (hypercube and Batcher network) were assumed to be consisted of all processing elements required to fully take advantage of the parallelism inherent in the co-occurrence matrix computation for a specific image size. The compared architectures were simulated on the Parallaxis simulator [9]. The total computational time from the analysis of all images in each of the 15 data sets was estimated. Then, an averaging of the computational time over all data sets corresponding to the same image size was performed. The estimated average times were employed in the computation of the speedups. Fig. 2 a) shows the speedup of the hypercube over the serial processor whereas Fig. 2 b) shows the speedup of the hypercube over the Batcher network.

The hypercube attains a greater speedup in all compared cases (see Fig. 2). From Fig. 2 a), it is clear that the speedup increases as the size of the analyzed images increases. It becomes about 2183 for the analyzed sets of the 64 × 64 images. The reason for this increase is that the proposed algorithm running on the hypercube can fully utilize the inherent parallelism in co-occurrence matrix computation. As we increased the number of processing elements in the performed experiment to handle the larger image size the proposed parallel algorithm became much faster than the serial one. This phenomenon also appears in Fig. 2 b), where the speedup rises from about 6, in the case of the 16 × 16 images, to about 30, in the case of the 64 × 64 images. From this figure, it is obvious that in all analyzed cases the hypercube network was superior to the Batcher network. However, in this performance comparison the achieved speedup was mainly due to the efficient way of deriving the gray level pairs for a given displacement vector employing the proposed architecture.

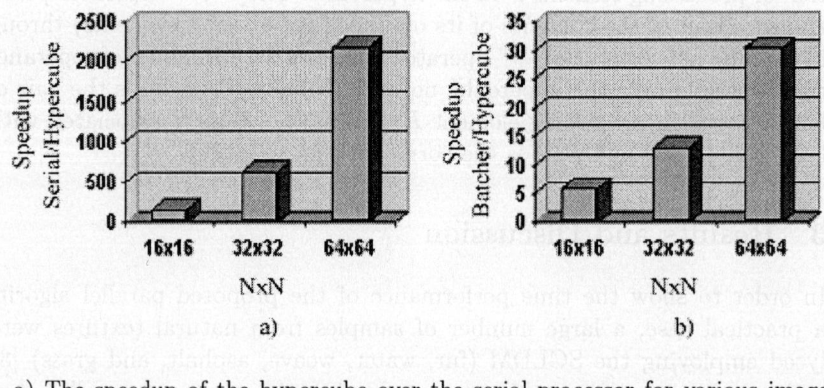

Fig. 2. a) The speedup of the hypercube over the serial processor for various image sizes. b) The speedup of the hypercube over the Batcher network for various image sizes

Even though the degree of the hypercube increases logarithmically with the number of nodes, which is actually its biggest disadvantage, the rapid evolution of the VLSI technology and the large regularity of this type of architecture made possible the manufacturing of large hypercubes. With the current submicron CMOS technology [6], hundreds of simple processing elements can be put on a single chip allowing the implementation of a massively parallel system on a single printed circuit board for the simultaneous processing of the pixels of a 64×64 gray level image with a dynamic range of 8 bits (256 gray levels). Moreover, from the pseudocode in Fig. 1, it is clear that the structure of each processing element in the proposed parallel architecture can be very simple.

4 Conclusions

The parallel algorithm for the SGLDM proposed in this paper was shown to be superior in all compared cases, in terms of computational time. The analysis of real textures showed that the algorithm has the ability to fully exploit the parallelism inherent in this computation. Furthermore, the employed parallel architecture needs much less hardware than the previously proposed massively parallel processors, which can be tolerated by modern VLSI technology.

Acknowledgements

This work was supported in part by the European Union under IST FET Project ALCOM-FT and Improving RTN Project ARACNE.

References

1. Haralick, R., Shanmugam, K., Dinstein, I.: Textural features for image classification. IEEE Trans. Syst. Man. Cybern. **SMC-3** (1973) 610–621
2. Ohanian, P., Dubes, R.: Performance evaluation for four classes of textural features. Patt. Rec. **25** (1992) 819–833
3. Kovalev, V., Kruggel, F., et al.: Three-dimensional texture analysis of MRI brain datasets. IEEE Trans. Med. Imag. **MI-20** (2001) 424–433
4. Kushner, T., Wu, A., Rosenfeld, A.: Image processing on ZMOB. IEEE Trans. on Computers **C-31** (1982) 943–951
5. Khalaf, S., El-Gabali, M., Abdelguerfi, M.: A parallel architecture for co-occurrence matrix computation. In Proc. 36th Midwest Symposium on Circuits and Systems (1993) 945–948
6. Ikenaga, T., Ogura, T.: CAM^2: A highly-parallel two-dimensional cellular automaton architecture. IEEE Trans. on Computers **C-47** (1998) 788–801
7. Svolos, A., Konstantopoulos, C., Kaklamanis, C.: Efficient binary morphological algorithms on a massively parallel processor. In IEEE Proc. 14th Int. PDPS. Cancun, Mexico (2000) 281–286
8. Brodatz, P.: Textures: a Photographic Album for Artists and Designers. Dover Publ. (1966)
9. http://www.informatik.uni-stuttgart.de/ipvr/bv/p3

Stochastic Simulation of a Marine Host-Parasite System Using a Hybrid MPI/OpenMP Programming

Michel Langlais[1,*], Guillaume Latu[2,*], Jean Roman[2,*], and Patrick Silan[3]

[1] MAB, UMR CNRS 5466, Université Bordeaux 2,
146 Léo Saignat, 33076 Bordeaux Cedex, France
Michel.Langlais@sm.u-bordeaux2.fr

[2] LaBRI, UMR CNRS 5800, Université Bordeaux 1 & ENSEIRB
351, cours de la Libération, 33405 Talence, France
{latu|roman}@labri.fr

[3] UMR CNRS 5000, Université Montpellier II, Station Méditerranéenne de l'Environnement Littoral, 1 Quai de la Daurade, 34200 Sète, France
silan@univ-montp2.fr

Abstract. We are interested in a host-parasite system occuring in fish farms, *i.e.* the sea bass - *Diplectanum aequans* system. A discrete mathematical model is used to describe the dynamics of both populations. A deterministic numerical simulator and, lately, a stochastic simulator were developed to study this biological system. Parallelization is required because execution times are too long. The Monte Carlo algorithm of the stochastic simulator and its three levels of parallelism are described. Analysis and performances, up to 256 processors, of a hybrid MPI/OpenMP code are then presented for a cluster of SMP nodes. Qualitative results are given for the host-parasite system.

1 Introduction

Host-parasite systems can present very complex behaviors and can be difficult to analyse from a purely mathematical point of view [12]. Ecological and epidemiologic interests are motivating the study of their population dynamics. A deterministic mathematical model (using some stochastic elements) for the sea bass–*Diplectanum aequans* system was introduced in [3, 6]. It concerns a pathological problem in fish farming. Numerical simulations and subsequent quantitative analysis of the results can be done, and a validation of the underlying model is expected. Our first goal in this work is to discover the hierarchy of various mechanisms involved in this host-parasite system. A second one is to understand the sensitivity of the model with respect to the initial conditions. In our model, many factors are taken into account to accurately simulate the model, *e.g.* spatial and temporal heterogeneities. Therefore, the realistic deterministic

* Research action ScAlApplix supported by INRIA.

B. Monien and R. Feldmann (Eds.): Euro-Par 2002, LNCS 2400, pp. 436–446.

simulator has a significant computation cost. Parallelization is required because execution times of the simulations are too long [7].

Individual-Based Models (IBM) are becoming more and more useful to describe biological systems. Interactions between individuals are simple and local, yet can lead to complex patterns at a global scale. The principle is to replicate several times the simulation program to obtain statistically meaningful results. In fact, a single simulation run driven by a sequence of pseudo-random numbers is not representative for a set of input parameters. Then, outputs are averaged for all theses *simulation runs* (or *replicates*). The Individual-Based Model approach contrasts with a more aggregate population modeling approach, and provides a mechanistic rather than a descriptive approach to modeling. Stochastic simulations reproduce elementary processes and often lead to prohibitive computations. Hence, parallel machines were used to model complex systems [1, 8, 9].

In this work, a description of the biological background and of performances of the deterministic simulator is briefly given. Next, we present the main issues concerning the parallel stochastic simulator. We point out the complexity of computations and, then, we develop our parallel algorithmic solution and investigate its performances. Hybrid MPI and OpenMP programming is used to achieve nested parallelization. Finally, we present some of the biological results obtained for an effective implementation on a SP3 IBM machine. This work received a grant from ACI bio-informatique. This project is a collaborative effort in an interdisciplinary approach: population dynamics with CNRS, mathematics with Université Bordeaux 2, computer science with Université Bordeaux 1.

2 Description of the Biological Background

In previous works [3, 6, 12], the mathematical model of the host-parasite system was presented; a summary is given now. The numerical simulation is mainly intended to describe the evolution of two populations, hosts and parasites, over one year in a fish farm. After a few time steps, any parasite egg surviving natural death becomes a larva. A time step $\Delta t = 2$ days corresponds to the average life span of a larva. The larva population is supplied by eggs hatching and by an external supply (larvae coming from open sea by pipes). An amount of $L(t)$ larvae is recruited by hosts, while others die. Highly parasitized hosts tend to recruit more parasites than others do. This means that the parasite population is overdispersed or aggregated with the host population. Most parasites are located on a few hosts. A detailed age structure of the parasites on a given host is required because only adult parasites lay eggs, while both juvenile and adult parasites have a negative impact on the survival rate of hosts. The population of parasites is divided into $K = 10$ age classes, with 9 classes of juvenile parasites and one large class for adult parasites. We consider that only a surfeit of parasites can lead to the death of a host. Environmental and biological conditions are actually used in the simulations, *e.g.* water temperature $T(t)$, death rate of parasites $\mu(T(t))$. The final goal is to obtain values of state variables at each time step.

3 Deterministic Numerical Simulation

The elementary events of one time step are quantified into probabilistic functions describing interactions between eggs, larvae, parasites and hosts. The frequency distribution of parasite numbers per host is updated with deterministic equations (without random number generation). Let $C(K, S)$ be the complexity of one *time step*. The S variable is limited to the minimum number of parasites that is lethal for a fish (currently $S \leq 800$); K is the number of age classes used ($K = 10$). A previous study [5] led to a reduced update cost of $C(K, S) = K\,S^4$ for one *time step* Δt, and one has $C(10, 800) = 950$ GFLOP. This large cost comes from the fine distribution of parasites within the host population, taking care of the age structure of parasites. A matrix formulation of the algorithm allows us to use BLAS 3 subroutines intensively, and leads to large speedups. Different mappings of data and computations have been investigated. A complete and costly simulation of 100 TFLOP lasts only 28 minutes on 128 processors (IBM SP3 / 16-way NH2 SMP nodes of the CINES[1]) and 9 minutes on 448 processors. The performance analysis has established the efficiency and the scalability of the parallel algorithm [7]. Relative efficiency of a 100 TFLOP simulation reached 83% using 128 processors and 75% using 448 processors.

4 Stochastic Model of Host-Parasite Interactions

For an Individual-Based Model, basic interactions are usually described between the actors of the system. Hosts and settled parasites are represented individually in the system, while eggs and larvae are considered globally. This allows to compare the deterministic and stochastic simulators, because only the inter-relationship between host and parasite populations are modeled differently. The deterministic simulator produces one output for a set of input parameters, whereas the stochastic simulator needs the synthesis of multiple different simulation runs to give a representative result. The number of replications R will depend on the desired accuracy of the outputs. We now describe how to manage host-parasite interactions. Let $\mathbb{H}_i$ be the host object indexed by i. Let $\mathbb{H}_i^p$ be the amount of parasites on the host $\mathbb{H}_i$.

- The probability for the host $\mathbb{H}_i$ to die, between time t and $t + \Delta t$, is given by $\pi(\mathbb{H}_i^p)$. A random number x is uniformly generated on $[0, 1]$ at each time step and for each living host $\mathbb{H}_i$. If $x \leq \pi(\mathbb{H}_i^p)$, then $\mathbb{H}_i$ dies.
- Consider that $\mathbb{P}_i(q)$ is the amount of parasites of age $q\Delta t$ settled on the host $\mathbb{H}_i$. Assume that the water temperature is $T(t)$, the death rate of parasites is $\mu(T(t))$. A binomial distribution $\mathcal{B}(\mathbb{P}_i(q); \mu(T(t)))$ is used to compute how many parasites among $\mathbb{P}_i(q)$ are dying during the time step t. All surviving parasites are moved to $\mathbb{P}_i(q + 1)$ (for $0 < q < K$), see figure 1. In short, for each host and each age class of parasites, a random number is generated using a binomial distribution to perform the aging process of parasites.

[1] Centre Informatique National de l'enseignement supérieur - Montpellier, France.

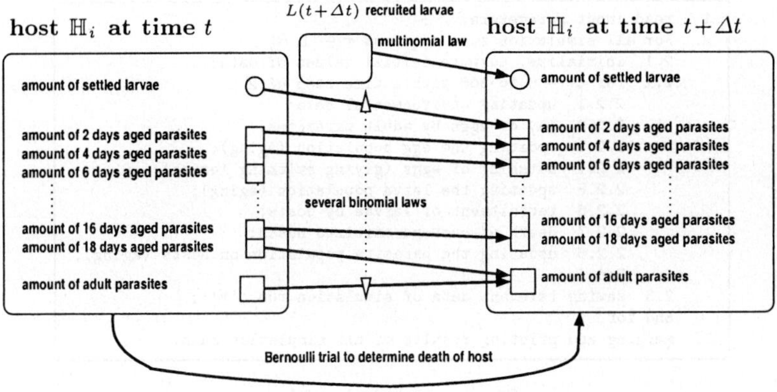

Fig. 1. Update of all living hosts and parasites at time t

- A function $f(p,t)$ gives the average percentage of larvae that are going to settle on a host having p parasites. Let $L(t)$ be the number of recruited larvae, one has:

$$\sum_{i/with\ \mathbb{H}_i\ living\ at\ time\ t} f(\mathbb{H}_i^p, t)\, L(t) = L(t)\,. \qquad (1)$$

The recruitment of $L(t)$ larvae on $H(t)$ hosts must be managed. Each host $\mathbb{H}_i$ recruits a larva with mean $f(\mathbb{H}_i^p,t)$. Let R_i be the variable giving the number of larvae recruited by $\mathbb{H}_i$ at time $t+\Delta t$. Let $i_1, i_2.., i_{H(t)}$ be the indices of living hosts at time t. To model this process, a multinomial distribution is used: $(R_{i_1}, R_{i_2}, ..R_{i_{H(t)}})$ follows the multinomial distribution $\mathcal{B}(L(t);\, f(\mathbb{H}_{i_1}^p,t),\, f(\mathbb{H}_{i_2}^p,t)..,\, f(\mathbb{H}_{i_{H(t)}}^p,t))$. One has the property that $R_{i_1}+R_{i_2}+..+R_{i_{H(t)}}=L(t)$.

5 Stochastic Algorithm

The algorithm used in the stochastic model is detailed in figure 2. Parts related to direct interactions between hosts and parasites (*i.e.* 2.2.6, 2.2.7 and 2.2.8) represent the costly part of the algorithm.

On a set of benchmarks, these correspond to at least 89% of execution time for all simulation runs. For simulations with long execution times, parasites and hosts appear in large numbers. For this kind of simulations, the epizooty develops for six months. One can observe more than 4×10^3 hosts and 10^6 parasites at a single time step. The most time consuming part of this problem is the calculation of the distribution of larvae among the host population (2.2.6 part). With the elementary method to reproduce a multinomial law, it means a random trial per recruited larva; the complexity is then $\Theta(L(t))$. In the 2.2.7 part, the number of Bernoulli trials to establish the death of hosts corresponds to a complexity $\Theta(H(t))$. In the 2.2.8 part, each age class $q\Delta t$ of parasites of each

```
1.  read input parameters;
2.  For all simulation runs required r ∈ [1, R] ;
    2.1  initialize, compute initial values of data;
    2.2  for t := 0 to 366 with a time step of 2
         2.2.1  updating environmental data;
         2.2.2  lay of eggs by adult parasites;
         2.2.3  updating the egg population (aging);
         2.2.4  hatching of eggs (giving swimming larvae);
         2.2.5  updating the larva population (aging);
         2.2.6  recruitment of larvae by hosts;
         2.2.7  death of over-parasitized hosts;
         2.2.8  updating the parasite population on hosts (aging);
         End for
    2.3  saving relevant data of simulation run ''r'';
    End for
3.  merging and printing results of all simulation runs.
```

Fig. 2. Global algorithm

host i is considered to determine the death of parasites. For each and every one, one binomial trial $\mathcal{B}(\mathbb{P}_i(q); \mu(T(t)))$ is done, giving a $\Theta(K \times H(t))$ complexity. So, one time step of one simulation run grows as $\Theta(H(t)+L(t))$. For a long simulation, the 2.2.6 part can take up to 90 % of the global simulation execution time, and after a few time steps, one has $H(t) \ll L(t)$. In that case, the overall complexity of the simulation is $\Theta(\sum_{t\in[0,366]} L(t))$. The sum of recruited larvae over one year reaches 2×10^8 in some simulations. Considering R replications, the complexity is then $R\,\Theta(\sum_{t\in[0,366]} L(t))$.

The main data used in the stochastic simulator are hosts and age classes of parasites. The memory space taken for these structures is relatively small in our simulations: $\Theta(K\,H(t))$. Nevertheless, to keep information about each time step, state variables are saved to do statistics. For J saved variables and 183 steps, the space required for this record is $\Theta(183\,J\,R)$, for all simulation runs.

6 Multilevel Parallelism for Stochastic Simulations

Several strategies of parallelization are found in the literature for stochastic simulations. First, all available processors could be used to compute one simulation run; simulation runs are then performed one after the other. Generally, a spatial decomposition is carried out. In multi-agent systems, the space domain of agent interactions is distributed over processors [8, 9]. For a cellular automaton based algorithm, the lattice is split among processors [1]. Nevertheless, this partitioning technique is available only if the granularity of computation is large enough, depending on the target parallel machine.

A more general approach for a stochastic simulation consists in mapping replicates onto different processors. Then, totally independent sequences of instructions are executed. At the end of all simulation runs, outputs are merged to generate a synthesis, *i.e.* means and standard deviations of state variables for each time step. However, this approach shows limitations. If simulation runs have not equal execution times, it leads to load imbalance. This potential penalty could be

partly solved with dynamic load balancing, if simulation runs could be mapped onto idle processors, whenever possible. The required number of simulation runs is a limitation too, because one has for P processors, $P \leq R$. Finally, the overhead of the step used to generate final outputs must be significantly lower than the cost of simulation runs. This second approach is often described [2], because it leads to massive parallelization. The problem remains of generating uncorrelated and reproducible sequences of random numbers on processors.
Finally, the validation of simulation models may require a sensitivity analysis. Sensitivity analysis consists in assessing how the variation in the output of a model can be apportioned, qualitatively or quantitatively, to different sources of variation in the input. It provides an understanding of how the output variables respond to changes in the input variables, and how to calibrate the data used. Exploration of input space may require a considerable amount of time, and may be difficult to perform in practice. Aggregation and structuring of results consume time and disk space. Now, a sequence of simulations using different input sets could be automated and parallelized. The synthesis of final outputs need the cooperation of all processors. This third level of parallelism is described in [4], however often unreachable for costly simulations. As far as we know, no example of combining these different levels of parallelism appears in the literature.

7 Parallel Algorithm

Most recent parallel architectures contain a large number of SMP nodes connected by a fast network. The hybrid programming paradigm combines two layers of parallelism: implementing OpenMP [11] shared-memory codes within each SMP node, while using MPI between them. This mixed programming method allows codes to potentially benefit from loop-level parallelism and from coarse-grained parallelism. Hybrid codes may also benefit from applications that are well-suited to take advantage of shared-memory algorithms. We shall evaluate the three levels of parallelism described above within the framework of such SMP clusters. Our parallel algorithm is presented in figure 3.

At the first level of parallelism, the process of larvae recruitment can be distributed (2.2.6 part of the algorithm). A sequence of random numbers is generated, then the loop considering each larva is split among the processors. Each OpenMP thread performs an independent computation on a set of larvae. This fine-grain parallelism is well suited for a shared-memory execution, avoiding data redundancy and communication latencies.

Suppose we do not use the first level of parallelism; the second level of parallelism means to map simulation runs onto the parallel machine. Typically, each processor gets several simulation runs, and potentially there is a problem of load imbalance. However, benchmarks have established that execution times of simulation runs do not have large variability for a given set of input parameters of a costly simulation. So, if each processor has the same number of replicates to carry out, the load is balanced. MPI is used to perform communications. In fact, the use of OpenMP is not a valuable choice here, because it prevents the

```
For all simulations a ∈ [1,A] of the sensitivity analysis do in // {
.  read input parameters;
.  For all simulations runs r ∈ [1,R] do in //
.  .  compute initial values of state variables;
.  .  For t:=0 to 366 with a time step of 2 do
.  .  .  update of steps 2.2.1, 2.2.2, 2.2.3, 2.2.4, 2.2.5;
.  .  .  parallel update of step 2.2.6; (* OpenMP threads *)
.  .  .  update of steps 2.2.7, 2.2.8;
.  .  }
.  }
.  gather outputs of simulation a (MPI collective communication);
}
print outputs;
```

Fig. 3. Parallel algorithm

execution on several SMP nodes. When all simulation runs are finished, a gather step is performed with a MPI global communication routine.

When performing a sensitivity analysis (third level of parallelism), the external loop (a variable) is distributed among sp sets of processors. Each set has m processors, so the total number of processors is $sp \times m = P$. The values of the a incices are assigned to the sp sets in a cyclic manner to balance the load. Next, the values of the r indices are mapped onto m processors. To get a high-quality load-balancing at the second level, we assume that m divides R. A new potential load imbalance exists at the third level of parallelism. If we suppose the cost of one simulation to be a constant, the load will be well balanced only if A divides sp.

A pseudo-random sequence generator is a procedure that starts with a specified random number seed and generates random numbers. We currently use the library PRNGlib [10], which provides several pseudo-random number generators through a common interface on parallel architecture. Common routines are specified to initialize the generators with appropriate seeds on each processor, and to generate in particular uniform distributed random vectors. The proposed generators are successful in most empirical and theoretical tests and have a long period. They can be quickly computed in parallel, and generate the same random sequence independently of the number of processors. This library is used to generate $A \times R$ independent random sequences. It is necessary to make an adequate number of simulation runs so that the mean and standard deviation of the wanted statistics fall within the prescribed error at the specified tolerance. For $R = 32$ and a confidence interval of 95%, the average number of hosts and parasites is known with a relative error of 2%. This is sufficient for a single simulation (without the third level of parallelism) and for the sensitivity analysis on most cases. On the other hand, if a spectral analysis is wanted, $R = 512$ simulation runs are usually performed. The frequency distribution around the mean is then obtained, and constitutes a significant result of the system dynamic.

8 Hybrid OpenMP/MPI Parallelization

Simulations have been performed on an IBM SP3. The machine has 28 NH2 nodes (16-way Power 3, 375 Mhz) with 16 GBytes of memory per node; a Colony switch manages the interconnection of nodes. The code has been developed in FORTRAN 90 with the XL Fortran compiler and using the MPI message-passing library (IBM proprietary version). For performance evaluation and analysis, a representative set of input parameters of a costly simulation were chosen.

First, we evaluate the performances of a single simulation. Let m be the number of MPI processes (parallelization of the r loop in figure 3), nt be the number of OpenMP threads within a MPI process, and $P = m \times nt$ the number of processors ($sp{=}1$). If $R{=}32$, the fine-grain parallelism allows us to use more processors than the number of replicates. In our experiments, between one and four OpenMP threads were allocated to compute simulation runs. Figure 4 shows that the execution times decrease for a given number P of processors and an increasing number nt of OpenMP threads (*e.g.* for 32 processors the sequence $m \times nt = 32 \times 1,\ 16 \times 2,\ 8 \times 4$). For these representative results, performances of the MPI-only code always exceed those of the hybrid code. But, we can use 128 processors for $R{=}32$ with the hybrid code. That means execution times of 81 s on 64 processors and 59,7 s on 128 processors.

Number of threads (nt)	Number of MPI processes (m)				
	1	4	8	16	32
1	3669,9s	935,3 s	471,3s	238,8s	123,8s
2	2385,5s	609,1s	307,5s	155,9s	81,0s
3	1963,5s	500,3s	252,4s	127,8s	67,3s
4	1745,9s	469,4s	228,1s	119,1s	59,7s

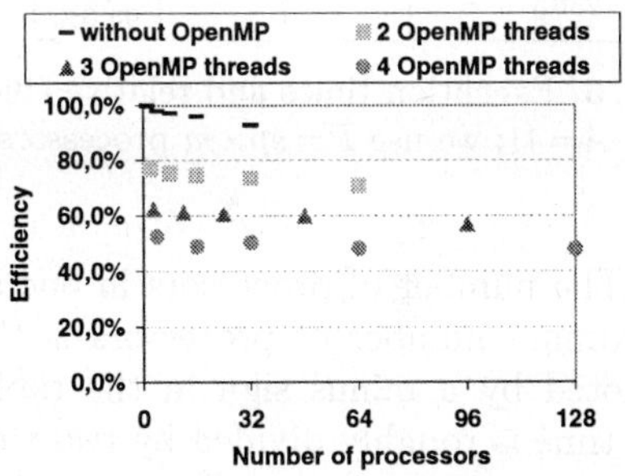

Fig. 4. Execution times and relative efficiency of a simulation for $R{=}32$; with m MPI process, nt threads in each MPI process, using $m \times nt$ processors

The OpenMP directives add a loop-level parallelism to the simulator. The first level of parallelism consists in the parallelization of a loop (step 2.2.6) with usually many iterations (*e.g.* 2×10^8). The arrays used inside that loop can be shared on the node with hybrid programming. But if we consider an MPI version of this loop-level parallelism, it would imply an overhead due to the communication of these arrays between processors. Precisely, these communication costs would be the main overhead of a MPI implementation.

Furthermore, the computation time spent in that loop represents in average 81 % of the sequential execution time t_{seq}. Let $T_p = 0,81\ t_{seq}$ be the portion of computation time that may be reduced by way of parallelization, and $T_s = 0,19\ t_{seq}$ be the time for the purely sequential part of the program. The Amdhal's law says, that for n processors the computation time is $T(n)=T_s+\frac{T_p}{n}$. Therefore, the parallel efficiency should be equal theoretically to 84 % for 2 processors (the effective performance is shown on figure 4, $m=1$, $nt=2$) and to 64 % for 4 processors ($m=1$, $nt=4$). These efficiencies are, in fact, upper limits. They induce a quickly decreasing efficiency for one to several OpenMP threads (nt). A version of our code using POSIX threads was tested and gave the same performances as OpenMP did. In our case, for one parallel loop, there is no overhead between the OpenMP version compared to the POSIX version.

The combination of the first two levels of parallelism were described. In the following, we will focus on the use of the second and third levels, excluding the first one. Each set of processors is not carrying out the same number of simulations. In figure 5, performances of two sensitivity analysis are presented with $A=15$, $A=41$.

A=15	Number of processor sets (sp)			
	1	2	4	16
$P=32$	1677s 100,0%	1725s 97,2%	1754s 95,6%	1697s 98,8%
$P=64$	–	897s 93,5%	861s 97,4%	861s 97,4%
$P=128$	–	–	445s 94,2%	438s 95,7%
$P=256$	–	–	–	223s 94,0%

A=41	Number of processor sets (sp)			
	1	2	4	16
$P=32$	4444s 100,0%	4607s 96,5%	4531s 98,1%	5438s 81,7%
$P=64$	–	2319s 95,8%	2298s 96,7%	2566s 86,6%
$P=128$	–	–	1197 92,8%	1344s 82,6%
$P=256$	–	–	–	682s 81,4%

Fig. 5. Execution times and relative efficiency of two sensitivity analysis with $A=15$ and $A=41$; we use $P=sp\times m$ processors with $R=32$

The number of processors in one set is at most $R=32$; we deduce that the maximum number of processors is then $sp\times R$ (impossible configurations are denoted by a minus sign in the tables). For a sensitivity analysis, note that the time is roughly divided by two when the number of processors doubles. For up to 256 processors, really costly simulations can be run with a good parallel efficiency; we can conclude that our implementation is scalable. Nevertheless, efficiency seems lower for $A=41$ and $sp=16$. Assume run-times of simulation runs are close to rts. The sequential complexity comes to $A\times R\times rts$. With the cyclic distribution at the third level, the parallel cost is given by $P\times\lceil A/sp\rceil\times(R\times rts/m)$. This implies an efficiency lower than $A/(sp\times\lceil A/sp\rceil)$. For $A=41$, $sp=16$, the parallel efficiency is then theoretically limited up to 85%. The assumption of equal execution times is approximate, but it explains why performances for $A=41$, $sp=16$ are not so good. However, an expensive sensitivity analysis ($A=41$) spends less than 12 minutes on 256 processors.

9 Biological Results

The results given by the new stochastic and the deterministic simulators come from two distinct computation methods. Both are based on a single bio-mathe-

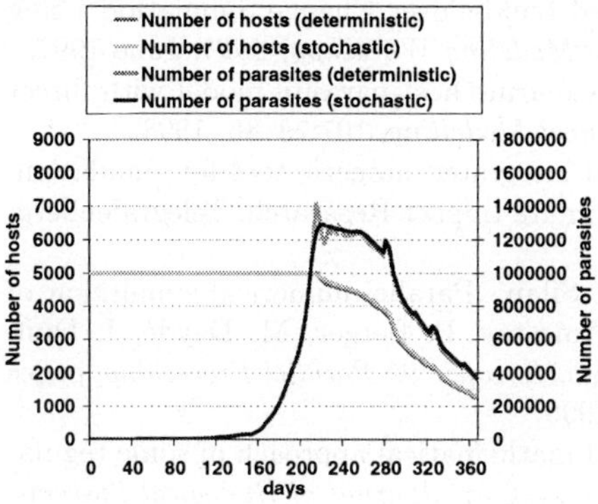

Fig. 6. Experiment with temporary endemic state at the end of simulation

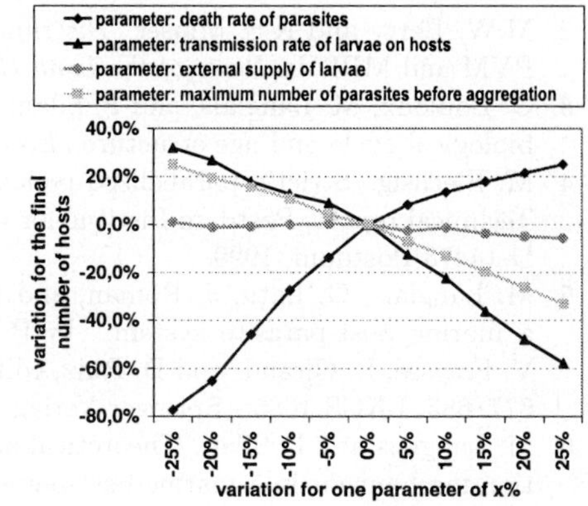

Fig. 7. Sensitivity analysis for 41 sets of input parameters

matical model and then outputs should not be very different. In fact, similarities are clearly observed, the number of hosts and parasites are given for one experiment in figure 6. For the stochastic simulation, the mean of $R=400$ simulation runs is given. For some parameter values, variations are observed between the two simulators. We already know that some interactions in the host-parasite system cannot be reproduced in the deterministic simulator (without modeling at a finer scale). Figure 7 corresponds to one result of the sensitivity analysis introduced in figure 5 ($A=41$); the intersection point (0%,0%) corresponds to a reference simulation. It shows the variation in percentage of the final number of hosts depending on the variation of four distinct input parameters. We conclude that the system is very sensitive to the death rate of parasites.

10 Conclusion

For similar outputs, a complete and costly stochastic simulation of the host-parasite system lasts only 1 minute on 128 processors versus 28 minutes for the deterministic simulation. A performance analysis has established the efficiency and the scalability of the stochastic algorithm using three levels of parallelism. The hybrid implementation allows us to use more processors than the number of simulation runs. The stochastic simulation gives the frequency distribution around the mean for outputs, providing new insights into the system dynamics. The sensitivity analysis, requiring several series of simulations is now accessible. An expensive sensitivity analysis spends less than 12 minutes on 256 processors.

References

1. M. Bernaschi, F. Castiglione, and S. Succi. A parallel algorithm for the simulation of the immune response. In *WAE'97 Proceedings: Workshop on Algorithm Engineering, Venice Italy*, September 1997.

2. M.W. Berry and K.S. Minser. Distributed Land-Cover Change Simulation Using PVM and MPI. In *Proc. of the Land Use Modeling Workshop, 1997*, June 1997.
3. C. Bouloux, M. Langlais, and P. Silan. A marine host-parasite model with direct biological cycle and age structure. *Ecological Modelling*, 107:73–86, 1998.
4. M. Flechsig. Strictly parallelized regional integrated numeric tool for simulation. Technical report, Postdam Institue for Climate Impact Reasearch, Telegrafenberg, D-14473 Postdam, 1999.
5. M. Langlais, G. Latu, J. Roman, and P. Silan. Parallel numerical simulation of a marine host-parasite system. In P. Amestoy, P. Berger, M. Daydé, I. Duff, V. Frayssé, L. Giraud, and D. Ruiz, editors, *Europar'99 Parallel Processing*, pages 677–685. LNCS 1685 - Springer Verlag, 1999.
6. M. Langlais and P. Silan. Theoretical and mathematical approach of some regulation mechanisms in a marine host-parasite system. *Journal of Biological Systems*, 3(2):559–568, 1995.
7. G. Latu. Solution parallèle pour un problème de dynamique de population. *Technique et Science Informatiques*, 19:767–790, June 2000.
8. H. Lorek and M. Sonnenschein. Using parallel computers to simulate individual-oriented models in ecology: a case study. In *Proceedings: ESM'95 European Simulation Multiconference, Prag*, June 1995.
9. B. Maniatty, B. Szymanski, and T. Caraco. High-performance computing tools for modeling evolution in epidemics. In *Proc. of the 32nd Hawaii International Conference on System Sciences*, 1999.
10. N. Masuda and F. Zimmermann. *PRNGlib : A Parallel Random Number Generator Library*, 1996. TR-96-08, ftp://ftp.cscs.ch/pub/CSCS/libraries/PRNGlib/.
11. OpenMP. A Proposed Industry Standard API for Shared Memory Programming. October 1997, OpenMP Forum, http://www.openmp.org/.
12. P. Silan, M. Langlais, and C. Bouloux. Dynamique des populations et modélisation : Application aux systèmes hôtes-macroparasites et à l'épidémiologie en environnement marin. In C.N.R.S eds, editor, *Tendances nouvelles en modélisation pour l'environnement*. Elsevier, 1997.

Optimization of Fire Propagation Model Inputs: A Grand Challenge Application on Metacomputers*

Baker Abdalhaq, Ana Cortés, Tomás Margalef, and Emilio Luque

Departament d'Informàtica, E.T.S.E, Universitat Autònoma de Barcelona,
08193-Bellaterra (Barcelona) Spain
baker@aows10.uab.es
{ana.cortes,tomas.margalef,emilio.luque}@uab.es

Abstract. Forest fire propagation modeling has typically been included within the category of grand challenging problems due to its complexity and to the range of disciplines that it involves. The high degree of uncertainty in the input parameters required by the fire models/simulators can be approached by applying optimization techniques, which, typically involve a large number of simulation executions, all of which usually require considerable time. Distributed computing systems (or metacomputers) suggest themselves as a perfect platform to addressing this problem. We focus on the tuning process for the *ISStest* fire simulator input parameters on a distributed computer environment managed by Condor.

1 Introduction

Grand Challenge Applications (GCA) address fundamental computation-intensive problems in science and engineering that normally involves several disciplines. Forest fire propagation modeling/simulation is a relevant example of GCA; it involves several features from different disciplines such as meteorology, biology, physics, chemistry or ecology. However, due to a lack of knowledge in most of the phases of the modeling process, as well as the high degree of uncertainty in the input parameters, in most cases the results provided by the simulators do not match real fire propagation and, consequently, the simulators are not useful since their predictions are not reliable. One way of overcoming these problems is that of using a method external to the model that allows us to rectify these deficiencies, such as, for instance, optimization techniques. In this paper, we address the challenge of calibrating the input values of a forest fire propagation simulator on a distributed computing environment managed by Condor [1] (a software system that runs on a cluster of workstations in order

* This work has been supported by MCyT-Spain under contract TIC2001-2592, by the EU under contract EVG1-CT-2001-00043 and partially supported by the Generalitat de Catalunya- Grup de Recerca Consolidat 2001SGR-00218. This research is made in the frame of the EU Project SPREAD - Forest Fire Spread Prevention and Mitigation.

B. Monien and R. Feldmann (Eds.): Euro-Par 2002, LNCS 2400, pp. 447–451.

to harness wasted CPU cycles from a group of machines called a Condor pool). A Genetic Algorithm (GA) scheme has been used as optimization strategy. In order to evaluate the improvement provided by this optimization strategy, its results have been compared against a Pure Random Search.

The rest of this paper is organized as follows. In section 2, the main features of forest fire propagation models are reported. Section 3 summarizes the experimental results obtained and, finally, section 4 presents the main conclusions.

2 Forest Fire Propagation Model

Classically, there are two ways of approaching the modeling of forest fire spread. These two alternatives essentially differ from one other in their degree of scaling. On one hand, we refer to local models when one small unit (points, sections, arcs, cells, ...) is considered as the propagation entity. These local models take into account the particular conditions (vegetation, wind, moisture, ...) of each entity and also of its neighborhood in order to calculate its evolution. On the other hand, as a propagation entity, global models consider the fire line view as a whole unit (geometrical unit) that evolves in time and space.

The basic cycle of a forest fire simulator involves the execution of both local and global models. On the basis of an initial fire front and simulating the path for a certain time interval, the result expected from the simulator is the new situation of the real fire line, once the said time has passed. Many factors influence the translation of the fire line. Basically, these factors can be grouped into three primary groups of inputs: vegetation features, meteorological and topographical aspects. The parameter that possibly provides the most variable influence on fire behavior is the wind [2]. The unpredictable nature of wind caused by the large number of its distinct classes and from its ability to change both horizontal and vertical direction, transforms it into one of the key points in the area of fire simulation. In this work, we focus on overcoming wind uncertainty regardless of the model itself and of the rest of the input parameters, which are assumed to be correct. The *ISStest* forest fire simulator [3], which incorporates the Rothermel model [4] as a local model and the global model defined by André and Viegas in [5], has been used as a working package for forest fire simulation.

3 Experimental Study

The experiments reported in this section were executed on a Linux cluster composed of 21 PC´s connected to a Fast Ether Net 100 Mb. All the machines were configured to use NFS (Network File System) and the Condor system; additionally, PVM were installed on every machine. The *ISStest* forest fire simulator assumes that the wind remains fixed during the fire-spread simulation process; consequently, it only considers two parameters in quantifying this element: wind speed (w_s) and wind direction

(w_d). We refer to the two-component vector represented by $\theta=(w_s \;\; w_d)$ as a *static* wind vector. However, in order to be more realistic, we have also considered a different scenario where in which the wind vector changes over time. The new wind vector approach will be referred to as a *dynamic* wind vector and is represented as follows:

$$\theta = (w_{s0} \quad w_{d0} \quad w_{s1} \quad w_{d1} \quad w_{s2} \quad w_{d2} \quad \ldots \;\ldots \quad w_{s(t-1)} \quad w_{d(t-1)}) \qquad (1)$$

where *t* corresponds to the number of wind changes considered. In order to tune these values as closely as possible to their optimum values, a Genetic Algorithm (GA) [6] as optimization technique has been applied. We also conducted the same set of experiments using a Pure Random approach to optimize the *wind* vector parameters in order to have a reference point for measuring the improvement provided by GA. The real fire line, which was used as a reference during the optimization process, was obtained in a synthetic manner for both the *static* and *dynamic* scenarios. Furthermore, we used the Hausdorff distance [7], which measures the degree of mismatch between two sets of points, in our case the real and simulated fire line, to measure the quality of the results.

For optimization purposes, the Black-Box Optimization Framework (BBOF) [8] was used. BBOF was implemented in a plug&play fashion, where both the optimized function and optimization technique can easily be changed. This optimization framework works in an iterative fashion, moving step-by-step from an initial set of guesses about the vector θ to a final value that is expected to be closer to the optimal vector of parameters than were the initial guesses. This goal is achieved because, at each iteration (or evaluation) of this process, the preset optimization technique (GA or Pure Random) is applied to generate a new set of guesses that should be better than the previous set.

We will now outline some preliminary results obtained on both the *static* and *dynamic* wind vector scenarios.

3.1 Static Wind Vector

As is well known, GA's need to be tuned in order to ensure maximum exploitation. Therefore, previous to the fire simulation experimental study, we conducted a tuning process on the GA, taking into account the particular characteristics of our problem. Since the initial set of guesses used as inputs by the optimization framework (BBOF) were obtained in a random way, we conducted 5 different experiments and the corresponding results were averaged. Table 1 shows the Hausdorff distance, on average, obtained for both strategies (GA and Random). As can be observed, GA provides considerable improvement in results compared to the case in which no optimization strategy has been applied.

In the following section, we will outline some preliminary results obtained on the *dynamic* wind vector scenario.

Table 1. Final Haussdorf distance (m.) obtained by GA and a Pure Random scheme under the static wind vector scenario.

Algorithm	Genetic	Random
Hausdorff dist. (m)	11	147,25
Evaluations	200	200

3.2 Dynamic Wind Vector

Two different experiments were carried out in order to analyze the *dynamic* wind vector scenario. In the first study, the wind changes were supposed to occur twice, the first change after 15 minutes with the second change coming 30 minutes later. Therefore, the vector to be optimized will include 4 parameters and is represented by: $\theta = (w_{s1} \;\; w_{d1} \;\; w_{s2} \;\; w_{d2})$. In the second case, three change instants have been considered, each separated from the next by 15 minutes. Consequently, the vector to be optimized will be: $\theta = (w_{s1} \;\; w_{d1} \;\; w_{s2} \;\; w_{d2} \;\; w_{s3} \;\; w_{d3})$. In both cases, the optimization process was run 5 times with different initial sets of guesses and, for each one, 20000 evaluations had been executed. Table 2 shows the Hausdorff distance, on average, for GA and Random strategies and for both dimensions setting for the *dynamic* wind vector. We observe that the results obtained when the vector dimension is 6 are worse than those obtained for dimension 4. Although the number of evaluations has been increased by two orders of magnitude with respect to the experiment performed when the wind vector was considered as *static*, the results are considerably poorer in the case of the *dynamic* wind vector.

As can be observed in table 2, GA provides a final Hausdorff distance, on average, which, in the case of a tuned vector composed of 4 components, is five times better than that provided by the Random approach, which represents the case in which no external technique is applied. In the other tested case (6 components), we also observed improvements in the results. Therefore, and for this particular set of experiments, we have determined that GA is a good optimization technique in overcoming the uncertainty input problem presented by forest fire simulators. Since the improvement shown by this approach is based on the execution of a large number of simulations, the use of a distributed platform to carry out the experiments was crucial.

Table 2. Final Haussdorf distance (m.) obtained by GA and a Pure Random scheme under the dynamic wind vector scenario for 4 and 6 vector dimensions and after 20000 objective function evaluations

Parameters	4	6
Random	97.5	103.5
Genetic	18.8	84.75

4 Conclusions

Forest fire propagation is evidently a challenging problem in the area of simulation. Uncertainties in the input variables needed by the fire propagation models (temperature, wind, moisture, vegetational features, topographical aspects...) can play a substantial role in producing erroneous results, and must be considered. For this reason, we have provided optimization methodologies to adjust the set of input parameters for a given model, in order to obtain results that are as close as possible to real values.

In general, it has been observed that better results are obtained by the application of some form of optimization technique in order to rectify deficiency in wind fields, or in their data, than by not applying any method at all. The method applied in our experimental study was that of GA. In the study undertaken, we would draw particular attention to that fact that, in order to emulate the real behavior of wind once a fire has started, and in order to attain results that can be extrapolated to possible future emergencies, a great number of simulations need to be carried out. Since these simulations do not have any response-time requirements, these applications are perfectly suited to distributed environments (metacomputers), in which it is possible to have access to considerable computing power over long periods of time

References

1. M. Livny and R. Raman. High-throughput resource management. In Ian Foster and Carl Kesselman, editors, The Grid: Blueprint for a New Computing Infrastructure. Morgan Kauffmann, (1999)
2. Lopes, A.,: Modelaçao numérica e experimental do escoamento turbulento tridimensional em topografia complexa: aplicaçao ao caso de um desfiladeiro, PhD Dissertation, Universidade de Coimbra, Portugal, (1993)
3. Jorba J., Margalef T., Luque E., J. Campos da Silva Andre, D. X Viegas: Parallel Approah to the Simulation Of Forest Fire Propagation. Proc. 13 Internationales Symposium "Informatik fur den Umweltshutz" der Gesellshaft Fur Informatik (GI). Magdeburg (1999)
4. Rothermel, R. C., "A mathematical model for predicting fire spread in wildland fuels", USDA-FS, Ogden TU, Res. Pap. INT-115, 1972.
5. André, J.C.S. and Viegas, D.X.,: An Unifying theory on the propagation of the fire front of surface forest fires, Proc. of the 3nd International Conference on Forest Fire Research. Coimbra, Portugal, (1998).
6. Baeck T., Hammel U., and Schwefel H.P.: Evolutionary Computation: Comments on the History and Current State. IEEE Transactions on Evolutionary Computation, Vol. 1, num.1 (April 1997) 3–17
7. Reiher E., Said F., Li Y. and Suen C.Y.: Map Symbol Recognition Using Directed Hausdorff Distance and a Neural Network Classifier. Proceedings of International Congress of Photogrammetry and Remote Sensing, Vol. XXXI, Part B3, Vienna, (July 1996) 680–685
8. Abdalhaq B., Cortés A., Margalef T. and Luque E.: Evolutionary Optimization Techniques on Computational Grids, In Proceeding of the 2002 International Conference on Computational Science LNCS 2329, 513-522

Parallel Numerical Solution of the Boltzmann Equation for Atomic Layer Deposition

Samuel G. Webster[1], Matthias K. Gobbert[1],
Jean-François Remacle[2], and Timothy S. Cale[3]

[1] Department of Mathematics and Statistics,
University of Maryland, Baltimore County,
1000 Hilltop Circle, Baltimore, MD 21250, U.S.A.

[2] Scientific Computing Research Center, Rensselaer Polytechnic Institute,
110 8th Street, Troy, NY 12180-3590, U.S.A.

[3] Focus Center — New York, Rensselaer: Interconnections for Gigascale Integration,
Rensselaer Polytechnic Institute, CII 6015,
110 8th Street, Troy, NY 12180-3590, U.S.A.

Abstract. Atomic Layer Deposition is one step in the industrial manufacturing of semiconductor chips. It is mathematically modeled by the Boltzmann equation of gas dynamics. Using an expansion in velocity space, the Boltzmann equation is converted to a system of linear hyperbolic equations. The discontinuous Galerkin method is used to solve this system. The speedup becomes near-perfect for the most complex two-dimensional cases. This demonstrates that the code allows for efficient parallel computation of long-time studies, in particular for the three-dimensional model.

1 Introduction

Atomic Layer Deposition (ALD) provides excellent film thickness uniformity in high aspect ratio features found in modern integrated circuit fabrication. In an ideal ALD process, the deposition of solid material on the substrate is accomplished one atomic or monolayer at a time, in a self-limiting fashion which allows for complete control of film thickness. The ALD process is appropriately modeled by a fully transient, Boltzmann equation based transport and reaction model [1, 4, 6].

The flow of the reactive gases inside an individual feature of typical size less than 1 μm on the feature scale is described by the Boltzmann equation [1], stated here in dimensionless form as

$$\frac{\partial f}{\partial t} + v \cdot \nabla_x f = \frac{1}{\mathrm{Kn}} Q(f, f). \tag{1}$$

The unknown variable is the density distribution function $f(x, v, t)$, that gives the scaled probability density that a molecule is at position $x = (x_1, x_2) \in \Omega \subset \mathbb{R}^2$ with velocity $v = (v_1, v_2) \in \mathbb{R}^2$ at time $t \geq 0$. The velocity integral of $f(x, v, t)$ gives the dimensionless number density of the reactive species

B. Monien and R. Feldmann (Eds.): Euro-Par 2002, LNCS 2400, pp. 452–456.

$c(x,t) = \int f(x,v,t)\,dv$. The left-hand side of (1) describes the convective transport of the gaseous species while the right-hand side of the Boltzmann equation models the effect of collisions among molecules. For feature scale models the Knudsen number Kn is large and hence the transport is free molecular flow. Mathematically, this corresponds to a special case of (1) with zero right-hand side. The stated model is two-dimensional, a generalization to three dimensions is straightforward and first results are presented in [9].

Initial coarse and fine meshes of the domain Ω for the feature scale model are shown in Fig. 1. The fine mesh contains approximately twice as many elements as the coarse mesh. The meshes are slightly graded from top to bottom with a higher resolution near the wafer surface.

We model the inflow at the top of the domain ($x_2 = 0.25$) by prescribing a Maxwellian velocity distribution. We assume specular reflection on the sides of the domain ($x_1 = -0.25$ and $x_1 = +0.25$). Along the remainder of the boundary, which represents the wafer surface of the feature, a reaction model is used to describe the adsorption of molecules to the surface and diffusive emission describes the re-emission of molecules from the surface [1, 4, 6]. Initially, no molecules of the reactive species are present in the domain.

2 The Numerical Method

To numerically solve (1) with the given boundary conditions and initial condition, the unknown f for the reactive species is expanded in velocity space $f(x,v,t) = \sum_{k=0}^{K-1} f_k(x,t)\varphi_k(v)$, where the $\varphi_k(v)$, $k = 0, 1, \ldots, K-1$, form an orthogonal set of basis functions in velocity space with respect to some inner product $\langle \cdot, \cdot \rangle_C$. Using a Galerkin ansatz and choosing the basis functions judiciously, the linear Boltzmann equation (1) is converted to a system of linear hyperbolic equations

$$\frac{\partial F}{\partial t} + A^{(1)} \frac{\partial F}{\partial x_1} + A^{(2)} \frac{\partial F}{\partial x_2} = 0, \tag{2}$$

where $F(x,t) = (f_0(x,t), ..., f_{K-1}(x,t))^T$ is the vector of coefficient functions. $A^{(1)}$ and $A^{(2)}$ are $K \times K$ diagonal matrices with components $A^{(\ell)} = \operatorname{diag}(A^{(\ell)}_{kk})$ ($\ell = 1, 2$) [5]. Therefore, each equation for component function $f_k(x,t)$

$$\frac{\partial f_k}{\partial t} + a_k \cdot \nabla_x f_k = 0 \tag{3}$$

is a hyperbolic equation with constant velocity vector $a_k = (A^{(1)}_{kk}, A^{(2)}_{kk})^T$ given by the diagonal elements of $A^{(1)}$ and $A^{(2)}$. Note that the equations remain coupled through the reaction boundary condition at the wafer surface [6].

This system is then solved using the discontinuous Galerkin method (DGM) [2]. In the implementation in the code DG [7], we choose to use a discontinuous L^2-orthogonal basis in space and an explicit time-discretization (Euler's method). This leads to a diagonal mass matrix so that no system of equations has to be solved. The degrees of freedom are the values of the K solution components $f_k(x,t)$ on all three vertices of each of the N_e triangles. Hence, the complexity of the computational problem is given by $3KN_e$; it is proportional both to the system size K and to the number of elements N_e.

The domain is partitioned in a pre-processing step, and the disjoint subdomains are distributed to separate parallel processors. The code uses local mesh refinement and coarsening and dynamic load-balancing using the Zoltan library as load balancer [3] and the graph partitioning software ParMETIS [8].

3 Results

Numerical studies were conducted for three different velocity discretizations. The demonstration results presented below were computed using four and eight discrete velocities in each spatial direction, respectively; hence, there are $K = 16$ and $K = 64$ equations, respectively. In each case, simulations were run for the two different initial meshes of Fig. 1. The solutions are presented in [4, 6].

The studies were performed on a 8-processor cluster of four dual Linux PCs with 1000 MHz Pentium III processors with 256 KB L1 cache and 1 GB of RAM per node. The nodes are connected by 100 Mbit commodity cables on a dedicated network, forming a Beowulf cluster. Files are served centrally from one of the nodes using a SCSI harddrive.

Figure 2 shows observed speedup for up to eight processes for the various numerical studies conducted; the speedup measures the improvement in wall-

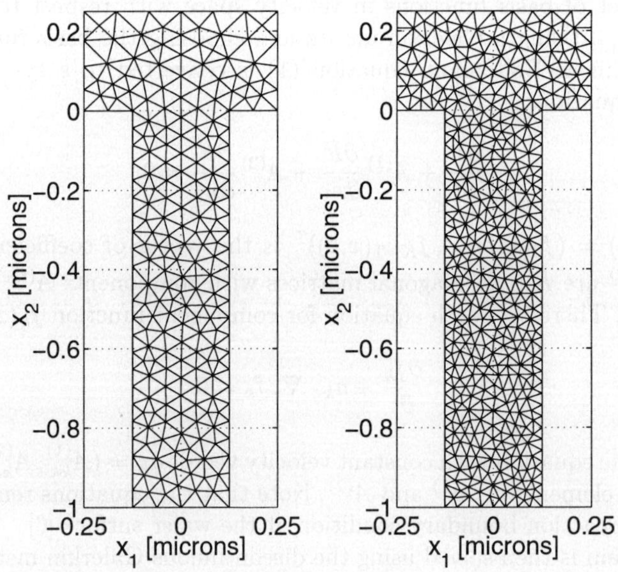

Fig. 1. (a) Coarse initial mesh, (b) fine initial mesh.

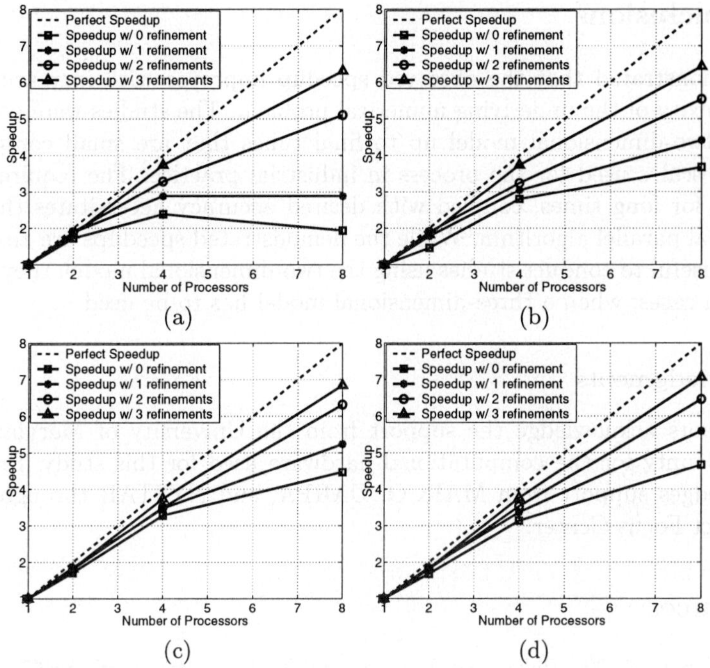

Fig. 2. Observed speedup for (a) coarse mesh / $K = 16$, (b) fine mesh / $K = 16$, (c) coarse mesh / $K = 64$, (d) fine mesh / $K = 64$.

clock time of the parallel code using p processes over the serial version of the code. The first row of plots in the figure corresponds to four discrete velocities ($K = 16$), and the second row corresponds to eight discrete velocities ($K = 64$). The left-hand column and right-hand column of Fig. 2 correspond to the coarse initial mesh and fine initial mesh of Figs. 1(a) and (b), respectively. Figure 2(a) compares the speedup for different levels of refinement of the initial coarse mesh with $K = 16$. Observe the decay in speedup without refinement due to the small number of degrees of freedom per process. Thus, as the maximum allowable refinement level increases and, consequently, the number of degrees of freedom increases, the speedup improves. Figures 2(a) and (b) demonstrate speedup for $K = 16$ for the two initial meshes. The finer mesh contains approximately twice as many elements as the coarse mesh; hence, the number of degrees of freedom increases by a factor of two. A comparison of the respective mesh refinement levels between the two plots shows that speedup improves because the degrees of freedom for the finer mesh is larger than for the coarse mesh. Figures 2(a) and (c) display speedup for the coarse mesh for the two studies $K = 16$ and $K = 64$. The finer velocity discretization in Fig. 2(c) introduces additional degrees of freedom which again improves speedup. Figure 2(d) combines the effect of the fine initial mesh and the finer velocity discretization. Observe that this is the most complex numerical study and thus possesses the best speedup.

4 Conclusions

It is demonstrated that the observed speedup improves with increasing levels of complexity of the underlying numerical problem. The studies were conducted using a two-dimensional model up to final times that are small compared to the time scales used for the process in industrial practice. The requirement to compute for long times, coupled with desired accuracy necessitates the use of an optimal parallel algorithm. While the demonstrated speedups are already extremely useful to conduct studies using the two-dimensional model, they become crucial in cases, when a three-dimensional model has to be used.

Acknowledgments

The authors acknowledge the support from the University of Maryland, Baltimore County for the computational hardware used for this study. Prof. Cale acknowledges support from MARCO, DARPA, and NYSTAR through the Interconnect Focus Center.

References

1. C. Cercignani. *The Boltzmann Equation and Its Applications*, volume 67 of *Applied Mathematical Sciences.* Springer-Verlag, 1988.
2. B. Cockburn, G. E. Karniadakis, and C.-W. Shu, editors. *Discontinuous Galerkin Methods: Theory, Computation and Applications*, volume 11 of *Lecture Notes in Computational Science and Engineering.* Springer-Verlag, 2000.
3. K. Devine, B. Hendrickson, E. Boman, M. St.John, and C. Vaughan. *Zoltan: A Dynamic Load-Balancing Library for Parallel Applications; User's Guide.* Technical report, Sandia National Laboratories Tech. Rep. SAND99-1377, 1999.
4. M. K. Gobbert and T. S. Cale. A feature scale transport and reaction model for atomic layer deposition. In M. T. Swihart, M. D. Allendorf, and M. Meyyappan, editors, *Fundamental Gas-Phase and Surface Chemistry of Vapor-Phase Deposition II*, volume 2001-13, pages 316–323. The Electrochemical Society Proceedings Series, 2001.
5. M. K. Gobbert, J.-F. Remacle, and T. S. Cale. A spectral Galerkin ansatz for the deterministic solution of the Boltzmann equation on irregular domains. In preparation.
6. M. K. Gobbert, S. G. Webster, and T. S. Cale. Transient adsorption and desorption in micron scale features. *J. Electrochem. Soc.*, in press.
7. J.-F. Remacle, J. Flaherty, and M. Shephard. An adaptive discontinuous Galerkin technique with an orthogonal basis applied to Rayleigh-Taylor flow instabilities. *SIAM J. Sci. Comput.*, accepted.
8. K. Schloegel, G. Karypis, and V. Kumar. Multilevel diffusion algorithms for repartitioning of adaptive meshes. *Journal of Parallel and Distributed Computing*, 47:109–124, 1997.
9. S. G. Webster, M. K. Gobbert, and T. S. Cale. Transient 3-D/3-D transport and reactant-wafer interactions: Adsorption and desorption. *The Electrochemical Society Proceedings Series*, accepted.

Topic 8
Parallel Computer Architecture and Instruction-Level Parallelism

Jean-Luc Gaudiot

Global Chair
University of California, Irvine

Welcome to this topic of the Euro-Par conference held this year in picturesque Paderborn, Germany. I was extremely honored to serve as the global chair for these sessions on Parallel Computer Architecture and Instruction-Level Parallelism and I look forward to meeting all practitioners of the field, researchers, and students at the conference.

Today, Instruction-Level Parallelism is present in all contemporary microprocessors. Thread-level parallelism will be harnessed in next generation of high-performance microprocessors. The scope of this topic includes parallel computer architectures, processor architecture (architecture and microarchitecture as well as compilation), the impact of emerging microprocessor architectures on parallel computer architectures, innovative memory designs to hide and reduce the access latency, multi-threading, as well as the influence of emerging applications on parallel computer architecture design.

A total of 28 papers were submitted to this topic. The overall quality of the submissions rendered our task quite difficult and caused quite a flurry of messages back and forth between the topic organizers. Most papers were refereed by at least four experts in the field and some received five reports. In the end, we settled on 6 regular papers and 6 short papers spread across three sessions: Instruction-level Parallelism 1 and 2, Multiprocessors and Reconfigurable Architectures.

I would like to thank the other members of this topic organizing committee: Professor Theo Ungerer (Local Chair), Professor Nader Bagherzadeh, and Professor Josep Larriba-Pey (Vice-Chairs) who each painstakingly provided reviews for each of the submissions and participated with insight in our electronic "Program Committee meetings." A special note of thanks goes to Professor Ungerer for his representing us at the Euro-Par Program Committee meeting. Of course, all this was made possible by the referees who lent us their time and expertise with their high quality reviews.

B. Monien and R. Feldmann (Eds.): Euro-Par 2002, LNCS 2400, p. 457.

Independent Hashing as Confidence Mechanism for Value Predictors in Microprocessors

Veerle Desmet, Bart Goeman, and Koen De Bosschere

Department of Electronics and Information Systems, Ghent University
Sint-Pietersnieuwstraat 41, B-9000 Gent, Belgium
{vdesmet,bgoeman,kdb}@elis.rug.ac.be

Abstract. Value prediction is used for overcoming the performance barrier of instruction-level parallelism imposed by data dependencies. Correct predictions allow dependent instructions to be executed earlier. On the other hand mispredictions affect the performance due to a penalty for undoing the speculation meanwhile consuming processor resources that can be used better by non-speculative instructions. A confidence mechanism performs speculation control by limiting the predictions to those that are likely to be correct.

When designing a value predictor, hashing functions are useful for compactly representing prediction information but suffer from collisions or *hash-aliasing*. This hash-aliasing turns out to account for many mispredictions. Our new confidence mechanism has its origin in detecting these aliasing cases through a second, independent, hashing function. Several mispredictions can be avoided by not using predictions suffering from hash-aliasing.

Using simulations we show a significant improvement in confidence estimation over known confidence mechanisms, whereas no additional hardware is needed. The combination of independent hashing with saturating counters performs better than pattern recognition, the best confidence mechanism in literature, and it does not need profiling.

1 Introduction

Nowadays computer architects are using every opportunity to increase the IPC, the average number of Instructions executed Per Cycle. The upper bound on achievable IPC is generally imposed by data dependencies. To overcome these data dependencies, the outcome of instructions is predicted such that dependent instructions can be executed in parallel using this prediction.

As correct program behaviour has to be guaranteed, mispredictions require recovery techniques to undo the speculation by restarting the execution from a previous processor state. This recovery takes some cycles and therefore every predictor design tries to avoid mispredictions. To further prevent mispredictions, applying selective prediction [3] or predicting only for a subset of instructions is recommended as over 40% of predictions made may not be useful in enhancing performance [8]. The selection of appropriate predictions can be done by using

B. Monien and R. Feldmann (Eds.): Euro-Par 2002, LNCS 2400, pp. 458–467.

a confidence mechanism based upon history information. Common techniques include saturating counters and pattern recognition [2].

We propose a new confidence mechanism for prediction schemes using a hashing function. Specifically for these predictors many mispredictions occur due to hashing: collisions or hash-aliasing occurs when different unhashed elements are mapped on a same hashing value. Our confidence mechanism tries detecting this interaction by using a second hashing function, independent on the original one. If hash-aliasing is detected the corresponding prediction will be ignored resulting in higher prediction accuracies. We evaluate it for a Sazeides predictor [10], the most accurate non-hybrid [11] value predictor known today.

This paper starts with an introduction to value prediction and the problem of aliasing. In section 3 we discuss the need for a confidence mechanism, explain previously proposed confidence mechanisms and describe metrics for comparing different confidence mechanisms. The use of an independent hashing function for detecting hash-aliasing is introduced in section 4. In section 5 we evaluate our independent hashing mechanism. Section 6 summarises the main conclusions.

2 Value Prediction

Most instructions need the outcome of preceding instructions and therefore have to wait until the latter are finished before their execution can be started. These so-called data dependencies can be eliminated by predicting the outcome of instructions so that dependent instructions can be executed earlier using the predicted value. This predicted value is simply referred as *prediction* to distinguish it from the *computed value* which verifies the prediction. The prediction is made during fetching while the computed value is available when the execution is completed. In case of a misprediction the speculation has to be undone according to a recovery policy: re-fetching or selective recovery. *Re-fetching* is used in branch prediction and involves all instructions following the misprediction to be re-fetched. This is a very costly operation and makes high prediction accuracies necessary. It is however easy to implement since the branch recovery hardware can be reused. *Selective recovery* only re-executes those instructions that depend on the misprediction resulting in lower misprediction penalties but requires additional hardware to keep track of dependency chains.

2.1 FCM Predictor

The finite context method (FCM) is a context-based prediction scheme [7] using recently computed values, called *history*, to determine the next prediction. The number of computed values forming the history is the *order* of the prediction scheme. One of the most accurate FCM predictors was introduced by Sazeides [10] and the actions taken during prediction are illustrated in Figure 1(a). Two prediction tables are managed. The first one is the *history table* and is indexed by the program counter. It contains the history, which is hashed in order to reduce the total number of bits to store. The hashed history is then used as an

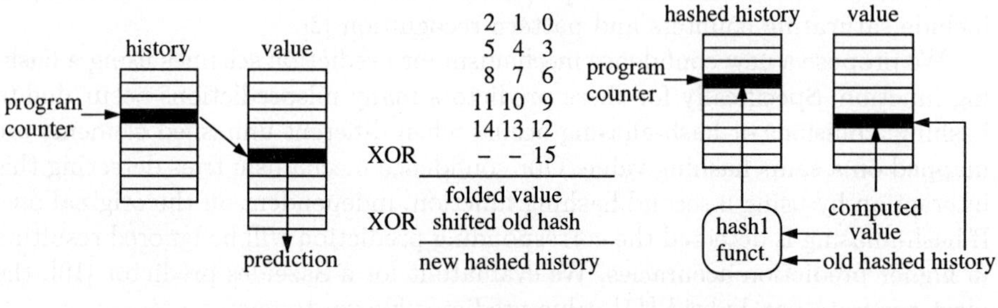

(a) Predicting (b) Hashing (c) Updating

Fig. 1. FCM predictor: Sazeides

index in the *value table*, where the prediction is found. An accuracy up to 78% (order 3) is reached by the Sazeides predictor when using infinite tables [10].

Throughout this paper, the history is hashed according to Sazeides' FS R-5 hashing function because it provides high prediction accuracy for a wide range of predictor configurations [9]. This hashing function incrementally calculates a new hashed history, by only using the old hashed history and the computed value to add to it. For a value table of 2^b entries we need a hashing function that maps the history consisting of *order* values to b bits. The construction is illustrated in Figure 1(b) for a 16-bit computed value and $b = 3$. The computed value is folded by splitting into sequences of b consecutive bits and combining these sequences by XORing. According to the definition of the order, we shift the old hashed history over $\lceil \frac{b}{order} \rceil$ bits. By XORing the shifted old hashed history and the folded value we obtain the new hashed history.

All actions that have to be taken during update, i.e. when the computed value is known, are shown in Figure 1(c). They include storing the computed value in the entry pointed by the old hashed history, calculating the new hashed history and storing it in the history table.

2.2 Instruction Aliasing

The discussed prediction scheme uses the program counter to index the history table. Using infinite tables every instruction has its own table entry. For finite tables however, only part of the program counter is used as an index resulting in many instructions sharing the same entry, which is called *instruction aliasing.* Although the interaction between instructions could be constructive, it is mostly destructive [4].

3 Confidence Mechanism

Basically, a value predictor is capable to make a prediction for each instruction. However, sometimes the prediction tables do not contain the necessary information to make a correct prediction. In such a case, it is better not to use the prediction because mispredictions incur a penalty for undoing the speculation, whereas making no prediction does not. From this point, we can influence the prediction accuracy of value predictors by selectively ignoring some predictions. To perform this selection we associate to each prediction a degree of reliability or *confidence.* Along this gradation a confidence mechanism assigns *high confidence* or *low confidence* such that assigning high confidence goes together with little chance of making a misprediction. High-confident predictions will be used whereas for low-confident ones it behaves as if no value predictor is available. Confidence mechanisms are based on *confidence information*, stored in the prediction tables together with the prediction. We first describe saturating counters and patterns as types of confidence information and we then explain how different confidence mechanisms can be compared.

3.1 Saturating Counters

A *saturating counter* directly represents the confidence of the corresponding prediction [6]. If the counter value is lower than a certain *threshold* the prediction is assigned low confidence, otherwise high confidence. The higher the threshold, the stronger the confidence mechanism. Regardless of the assigned confidence, the counter is updated at the moment the computed value is known. For this update we increment the counter (e.g. by one) for a correctly predictable value saturating at the maximum counter value and decrement (e.g. by one) the counter down to zero if the value was not correctly predictable. This way a saturating counter is a metric for the prediction accuracy in the recent past.

3.2 Patterns

Pattern recognition as proposed in [2] is based on *prediction outcome histories* keeping track of the outcome of the last value predictions. To identify the predictable history patterns, a wide range of programs are profiled (i.e. looking at their behaviour). Patterns precisely represent the recent history of prediction outcomes and do not suffer from saturating effects. Typically, patterns require more confidence bits than saturating counters and perform slightly better.

3.3 Metrics for Comparing Confidence Mechanisms

For a given value predictor the predictions are divided into two classes: predictions that are correct and those that are not. Confidence assignment does not change this classification. The improvement of adding a confidence mechanism is thus limited by the power of the underlying value predictor. For each prediction, the confidence mechanism distinguishes high-confident predictions from low-confident ones. Bringing together the previous considerations we categorise each prediction in one of the quadrants shown in Figure 2 [5]. A perfect confidence mechanism only puts predictions into classes HCcorr and LCntcorr. In a

	Correctly predictable	Not correctly predictable
High Confidence	HCcorr	HCntcorr
Low Confidence	LCcorr	LCntcorr

Fig. 2. Classification of predictions

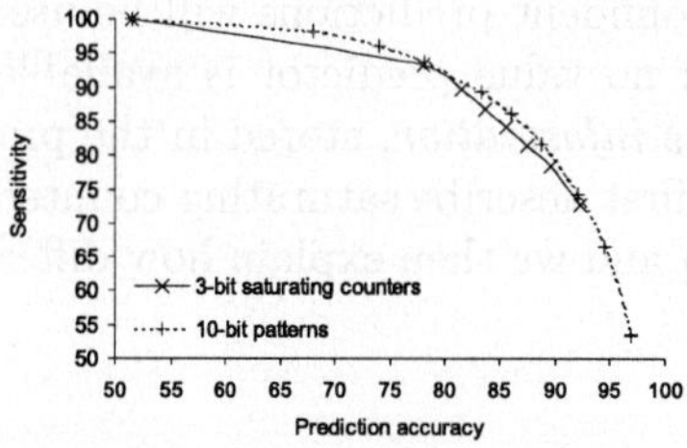

Fig. 3. Counters and patterns

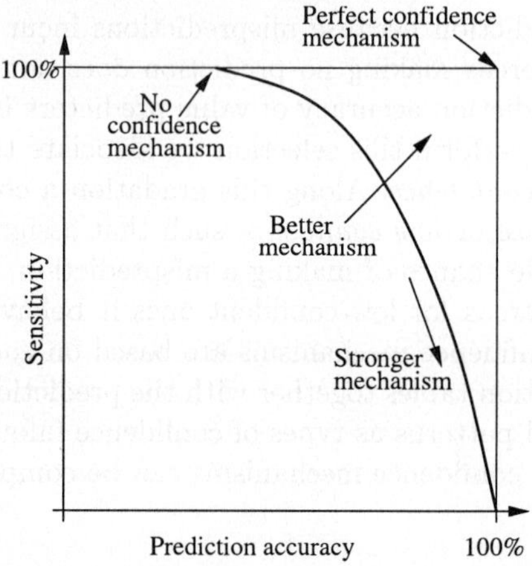

Fig. 4. Sensitivity versus prediction accuracy

realistic situation all quadrants are populated even the classes LCcorr and HCntcorr. We note that these 'bad' classes are not equivalent because the impact of a misprediction is usually different from that of missing a correct prediction.

We now describe a way for comparing different confidence strategies against the same value predictor without fixing the architecture as proposed in [5] for comparing confidence mechanisms in branch predictors. We will use the following independent metrics which are both "higher-is-better": *prediction accuracy* representing the probability that a high confidence prediction is correct and *sensitivity* being the fraction of correct predictions identified as high confidence.

$$\textbf{Prediction accuracy} = Prob[\text{correct prediction}|HC] = \frac{HCcorr}{HCcorr+HCntcorr}$$

$$\textbf{Sensitivity} = Prob[HC|\text{correctly predictable}] = \frac{HCcorr}{HCcorr+LCcorr}$$

We will plot figures of sensitivity versus prediction accuracy as sketched in Figure 4. Values closer to the upper right corner are better as perfect confidence assignment reaches 100% sensitivity and 100% prediction accuracy. A value predictor without confidence mechanism uses all predictions and achieves the highest possible sensitivity in exchange for lower prediction accuracy. A *stronger* confidence mechanism ignores more predictions by assigning low confidence to them and necessarily reaches lower sensitivities because the number of predictions in class HCcorr decreases (stronger mechanism) and the number of correctly predictable predictions is constant (fixed value predictor). The same reasoning in terms of the prediction accuracy is impossible, but a stronger mechanism should avoid more mispredictions than it looses correct predictions so that the prediction accuracy increases. In the limit when the sensitivity decreases down to 0% by using none of the predictions, prediction accuracy is strictly undefined, but we assume it approaches 100%. Figure 3 shows the sensitivity versus prediction accuracy for confidence mechanisms with 3-bit saturating counters and 10-bit patterns (threshold is varied along the curve).

4 Independent Hashing

Using a hashing function in the Sazeides predictor causes different unhashed histories to be mapped on the same level-2 entry. This interaction is called *hash-aliasing* and occurs in 34% of all predictions, for a predictor of order 3 and both tables with 2^{12} entries. Only in 4% this results in a correct prediction whereas the other 30% end up in a misprediction [4]. In order to avoid these mispredictions we propose detecting hash-aliasing, assigning low confidence to the corresponding predictions and so eliminating predictions suffering from hash-aliasing.

First, the detection can be done perfectly by storing the complete unhashed history in both prediction tables. This requires a hardware budget that exceeds many times that of the value predictor itself and is not acceptable, but it gives an upper limit for sensitivity and prediction accuracy. Figure 5 shows a sensitivity of 96% and a prediction accuracy of more than 90%, a considerable improvement over counters and patterns. Note that only hash-aliasing is detected and that this technique does not try to estimate predictability.

Secondly, we perform the detection by a *second hashing function*, independent on the one used in the value predictor. This second hashing function maps the history on a *second hashing value*. The actions taken to locate the prediction and to compute the corresponding confidence are illustrated in Figure 7(a). The history table contains two independent hashing values based on the same complete history, while *val_hash2* corresponds to the history on which the value stored in the value field follows. High confidence is assigned when the second hashing values match, otherwise the prediction is of low confidence. Confidence information is spread over both prediction tables.

The second hashing function has to satisfy the following requirements:

1. If the Sazeides hashing function computes the same hashing value for two different unhashed histories, the second hashing function should map these histories to different values with a good chance. In other words, the hashing functions have to be *independent* meaning that none of the hashing bits can be derived by XORing any other combination of hashing bits.
2. All history bits should be used.
3. A new hashing value must be computable from the old hashing value and the computed value.

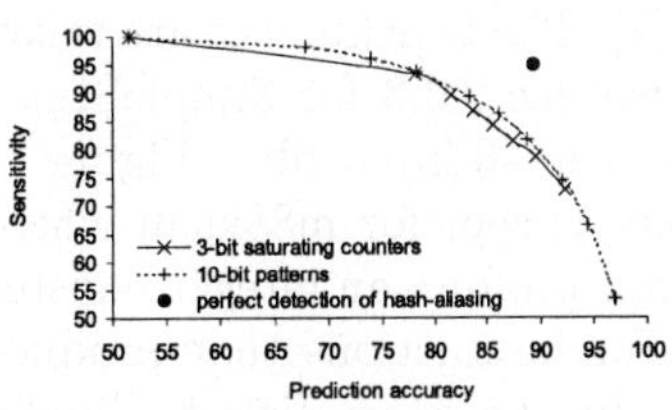

Fig. 5. Perfect detection of hash-aliasing compared to counters and patterns

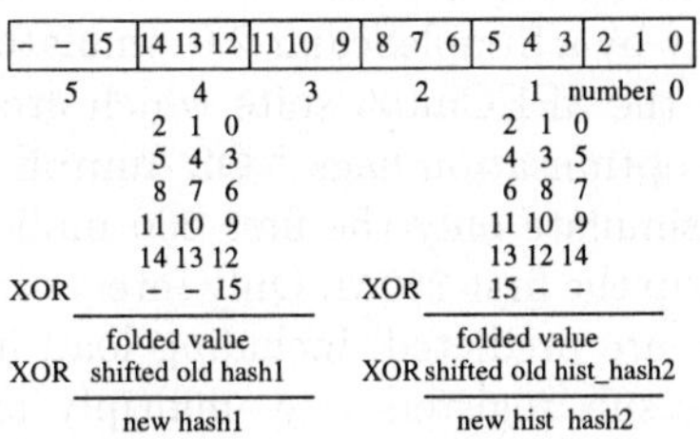

Fig. 6. Second, independent, hashing function on a 16-bit computed value and $b = 3$

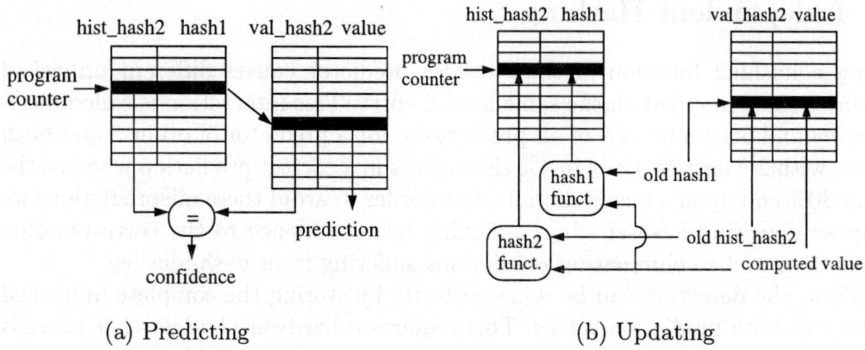

Fig. 7. Independent hashing

By analogy with the hashing function from Sazeides we propose a second hashing function based on the fold-and-shift principle. Again we assume a hashing function mapping the history on a value of b bit illustrated in Figure 6 for $b = 3$. After splitting the computed value into sequences of b consecutive bits, the second hashing function first rotates the sequences to the left before XORing. If we number these sequences starting by zero, the rotation of each sequence is done over (*number* MOD b) bits. Once the folded value is computed, the calculation of both hashing values is similar.

The above-described second hashing function is easy to compute (shifting, rotating and XORing) and uses all history bits. We also examined the independence of the second hashing function upon the original one. Therefore we use a matrix that represents the hashing functions such that after multiplication with a column representing the unhashed history, both hashing values are computed. By verifying the independence of rows in this matrix we prove the independence of both hashing functions.

When the computed value is known the content in the prediction tables is updated. This update phase is shown in Figure 7(b).

5 Evaluation

For each configuration, we use trace-based simulations using traces generated on-the-fly by a SimpleScalar 2.0 simulator (sim-safe) [1]. The benchmarks are taken from the SPECint95 suite which are compiled with gcc 2.6.3 for SimpleScalar with optimisation flags "-O2 -funroll_loops". We use small input files (Figure 8) and simulate only the first 200 million instructions, except for m88ksim where we skip the first 250M. Only integer instructions that produce an integer register value are predicted, including load instructions. For instructions that produce two result registers (e.g. multiply and divide) only one is predicted. Finally value prediction was not performed for branch and jump instructions and the presented results show the weighted average over all SPECint benchmarks. When not explicitly mentioned we consider a FCM-based Sazeides value predictor of

Program	options, input	predictions	Program	options, input	predictions
cc1	cccp.SS.i	140M	li	7queens.lsp	123M
compress	test.in	133M	m88ksim	-c ctl.raw.lit	139M
go	30 8	157M	perl	scrabbl.pl	126M
ijpeg	-image_file	155M		scrabbl7_train.in	
	vigo_ref.ppm -GO		vortex	vortex.ref.lit	122M

Fig. 8. Description of the benchmarks

order 4 with 2^{12} entries in both tables. The original hashing function in the value predictor then folds each history into 12 bits. First we evaluate adding the confidence mechanism to the value predictor and not its embedding in an actual processor. Afterwards we check if the higher accuracy and higher sensitivity translate in an actual speedup.

5.1 Independent Hashing

In this section we evaluate our second hashing function as confidence mechanism and we compare it to saturating counters and patterns, placed at the history table since this provides the best results. We found that using 4 bits in the second hashing value is a good choice as it assigns in 90% of the predictions the same confidence as perfect detection of hash-aliasing. Using more bits in the second hashing value do slightly better but require more hardware. The result of a 4-bit second hashing function is shown in Figure 9.

Our independent hashing function performs well in the sense that interaction between histories is detected and assigned low confidence. Nevertheless this technique does not account for predictability itself as high confidence is assigned every time no interaction occurs or can be detected. To correct this we propose combining detection of hash-aliasing with other confidence mechanisms. In a combined mechanism high confidence is assigned if both confidence mechanisms indicate high confidence. We can put the additional confidence in either of the two tables. If we add a simple 2-bit saturating counter with varying threshold, we get Figure 10. We also show the combination with a perfect detection system as well as the two possibilities for placing the saturating counter. The second hashing function approaches perfect detection when both are combined with a saturating counter. It gets even closer for higher thresholds. In the situation where we placed the counters at the value table only the highest threshold could be a meaningful configuration.

For a fair comparison in terms of hardware requirement we should compare 10-bit pattern recognition against the combination of a 4-bit second hashing function with 2-bit saturating counters. The difference is significant and moreover patterns need profiling, while our technique does not.

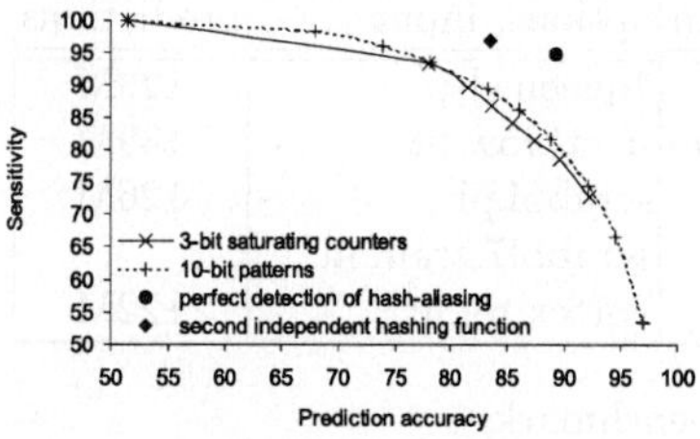

Fig. 9. 4-bit independent hashing

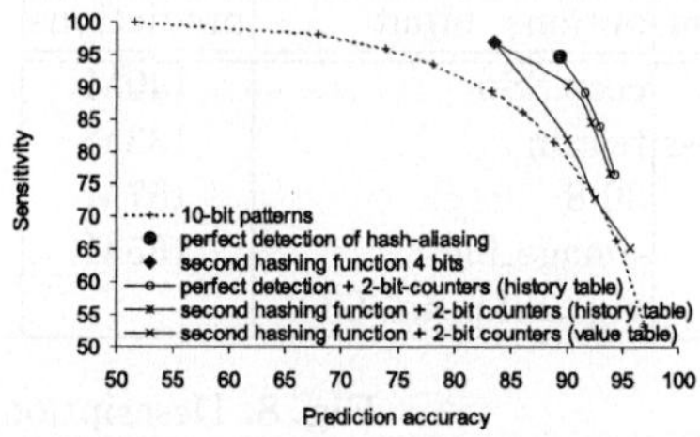

Fig. 10. 4-bit independent hashing combined with 2-bit saturating counters

Fetch, decode, issue, commit: 8
RUU/LSQ queue: 64/16
Functional units: 4
Branch predictor: perfect
L1 Icache: 128KB
L1/L2 latency: 3/12
L1 Dcache: 128
L2 cache (shared): 2MB
Recovery policy: selective

Fig. 11. Out-of-order architecture

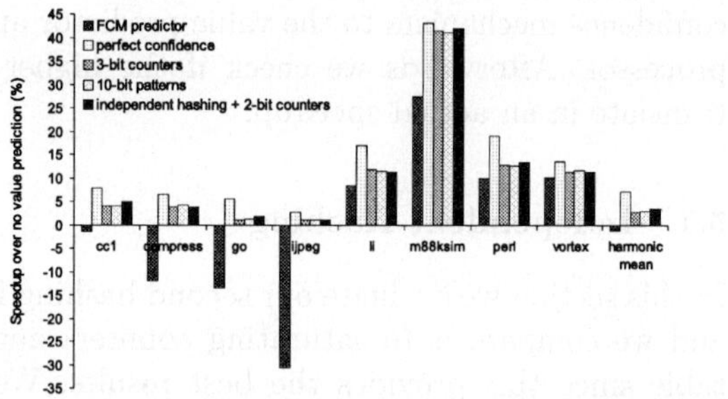

Fig. 12. Speedup over no value prediction

5.2 IPC

In this section we test if the higher prediction accuracy and higher sensitivity reached by independent hashing translate in actual speedup. Simulations are done by an out-of-order architecture (sim-outorder) as shown in Figure 11. In Figure 12 speedup over using no value prediction is plotted in 4 different cases: value prediction without confidence mechanism, perfect confidence mechanism, 3-bit saturating counter, 10-bit patterns and finally the combination of independent hashing with saturating counters. Independent hashing reaches a speedup that is only a slight improvement over patterns. An important aspect to increase performance by value prediction is *criticality* [3, 8]. Only correct predictions on the critical path can increase the performance while mispredictions are not dramatic when not on the critical path. None of the described confidence mechanisms consider about criticality of instructions and hence it is not evident that using more correct predictions do augment the IPC.

6 Conclusion

This paper studies confidence mechanisms for a context based Sazeides value predictor. We explain that many mispredictions are a result of using a hashing function and that detecting hash-aliasing can avoid a lot of mispredictions. Detection of hash-aliasing is done through a second, independent hashing function as confidence mechanism. In case of detecting hash-aliasing the confidence mechanism assigned low confidence forcing the processor not to use the prediction. We evaluate our confidence mechanism and show a significant improvement according to saturating counters and patterns. Especially the combination of our technique with saturating counters translates in a slight speedup, needs the same storage as patterns and eliminates the use of profiling.

References

1. D. Burger, T. M. Austin, and S. Bennett. Evaluating future microprocessors: The SimpleScalar Tool Set. Technical report, Computer Sciences Department, University of Wisconsin-Madison, July 1996.
2. M. Burtscher and B. G. Zorn. Prediction outcome history-based confidence estimation for load value prediction. *Journal of Instruction-Level Parallelism*, 1, May 1999.
3. B. Calder, G. Reinman, and D. M. Tullsen. Selective value prediction. In *Proceedings of the 26th Annual International Symposium on Computer Architecture*, pages 64–74, May 1999.
4. B. Goeman, H. Vandierendonck, and K. D. Bosschere. Differential FCM: Increasing value prediction accuracy by improving table usage efficiency. In *Proceedings of the 7th International Symposium on High Performance Computer Architecture*, pages 207–216, Jan. 2001.
5. D. Grunwald, A. Klauser, S. Manne, and A. Pleszkun. Confidence estimation for speculation control. In *Proceedings of the 25th Annual International Symposium on Computer Architecture*, pages 122–131, 1998.
6. M. H. Lipasti and J. P. Shen. Exceeding the dataflow limit via value prediction. In *Proceedings of the 29th Annual International Symposium on Microarchitecture*, Dec. 1996.
7. T. N. Mudge, I.-C. K. Chen, and J. T. Coffey. Limits to branch prediction. Technical Report CSE-TR-282-96, The University of Michigan, Ann Arbor, Michigan, 48109-2122, 1996.
8. B. Rychlik, J. Faistl, B. Krug, and J. P. Shen. Efficacy and performance impact of value prediction. In *Parallel Architectures and Compilation Techniques (PACT)*, Oct. 1998.
9. Y. Sazeides and J. E. Smith. Implementations of context based value predictors. Technical Report ECE97-8, Department of Electrical and Computer Engineering, University of Wisconsin-Madison, Dec. 1997.
10. Y. Sazeides and J. E. Smith. The predictability of data values. In *Proceedings of the 30th Annual International Symposium on Microarchitecture*, Dec. 1997.
11. K. Wang and M. Franklin. Highly accurate data value prediction using hybrid predictors. In *Proceedings of the 30th Annual International Symposium on Microarchitecture*, pages 281–290, Dec. 1997.

Exploiting the Prefetching Effect Provided by Executing Mispredicted Load Instructions

Resit Sendag[1], David J. Lilja[1], and Steven R. Kunkel[2]

[1] Department of Electrical and Computer Engineering
Minnesota Supercomputing Institute
University of Minnesota
200 Union St. S.E., Minneapolis, MN 55455, USA
{rsgt, lilja}@ece.umn.edu
[2] IBM, Rochester, MN, USA
srkunkel@us.ibm.com

Abstract. As the degree of instruction-level parallelism in superscalar architectures increases, the gap between processor and memory performance continues to grow requiring more aggressive techniques to increase the performance of the memory system. We propose a new technique, which is based on the wrong-path execution of loads far beyond instruction fetch-limiting conditional branches, to exploit more instruction-level parallelism by reducing the impact of memory delays. We examine the effects of the execution of loads down the wrong branch path on the performance of an aggressive issue processor. We find that, by continuing to execute the loads issued in the mispredicted path, even after the branch is resolved, we can actually reduce the cache misses observed on the correctly executed path. This wrong-path execution of loads can result in a speedup of up to 5% due to an indirect prefetching effect that brings data or instruction blocks into the cache for instructions subsequently issued on the correctly predicted path. However, it also can increase the amount of memory traffic and can pollute the cache. We propose the Wrong Path Cache (WPC) to eliminate the cache pollution caused by the execution of loads down mispredicted branch paths. For the configurations tested, fetching the results of wrong path loads into a fully associative 8-entry WPC can result in a 12% to 39% reduction in L1 data cache misses and in a speedup of up to 37%, with an average speedup of 9%, over the baseline processor.

1 Introduction

Several methods have been proposed to exploit more instruction-level parallelism in superscalar processors and to hide the latency of the main memory accesses, including speculative execution [1-7] and data prefetching [8-21]. To achieve high issue rates, instructions must be fetched beyond the basic block-ending conditional branches. This

B. Monien and R. Feldmann (Eds.): Euro-Par 2002, LNCS 2400, pp. 468–480.

can be done by speculatively executing instructions beyond branches until the branches are resolved. This speculative execution will allow many memory references to be issued that turn out to be unnecessary since they are issued from the mispredicted branch path. However, these incorrectly issued memory references may produce an indirect prefetching effect by bringing data or instruction lines into the cache that are needed later by instructions that are subsequently issued along correct execution path. On the other hand, these incorrectly issued memory references will increase the amount of memory traffic and can potentially pollute the cache with unneeded cache blocks [2].

Existing processors with deep pipelines and wide issue units do allow memory references to be issued speculatively down wrongly-predicted branch paths. In this study, however, we go one step further and examine the effects of continuing to execute the loads down the mispredicted branch path *even after the branch is resolved*. That is, we allow all speculatively issued loads to access the memory system if there is an available memory port. These instructions are marked as being from the mispredicted branch path when they are issued so they can be squashed in the write-back stage of the processor pipeline to prevent them from altering the target register after they access the memory system. In this manner, the processor is allowed to continue accessing memory with loads that are known to be from the wrong branch path. No store instructions are allowed to alter the memory system, however, since they are known to be invalid.

While this technique very aggressively issues load instructions to produce a significant impact on cache behavior, it has very little impact on the implementation of the processor's pipeline and control logic. The execution of wrong-path loads can make a significant performance improvement with very low overhead when there exists a large disparity between the processor cycle time and the memory speed. However, executing these loads can reduce performance in systems with small data caches and low associativities due to cache pollution. This cache pollution occurs when the wrong-path loads move blocks into the data cache that are never needed by the correct execution path. It also is possible for the cache blocks fetched by the wrong-path loads to evict blocks that still are required by the correct path.

In order to eliminate the cache pollution caused by the execution of the wrong-path loads, we propose the *Wrong Path Cache (WPC)*. This small fully-associative cache is accessed in parallel with the L1 cache. It buffers the values fetched by the wrong-path loads plus the blocks evicted from the data cache. Our simulations show that the WPC can be very effective in eliminating the pollution misses caused by the execution of wrong path loads while simultaneously reducing the conflict misses that occur in the L1 data cache.

The remainder of the paper is organized as follows -- Section 2 describes the proposed wrong path cache. In Section 3, we present the details of the simulation environment with the simulation results given in Section 4. Section 5 discusses some related work with the conclusions given in Section 6.

2 Wrong Path Cache (WPC)

For small low-associativity data caches, the execution of loads down the incorrectly-predicted branch path can reduce performance since the cache pollution caused by these wrong-path loads might offset the benefits of their indirect prefetching effect. To eliminate the pollution caused by the indirect prefetching effect of the wrong-path loads, we propose the Wrong Path Cache (WPC). The idea is simply to use a small fully associative cache that is separate from the data cache to store the values returned by loads that are executed down the incorrectly-predicted branch path. Note that the WPC handles the loads that are *known* to be issued from the wrong path, that is, after the branch result is known. The loads that are executed before the branch is resolved are speculatively put in the L1 data cache.

If a wrong-path load causes a miss in the data cache, the required cache block is brought into the WPC instead of the data cache. The WPC is queried in parallel with the data cache. The block is transferred simultaneously to the processor and the data cache when it is not in the data cache but it is in the WPC. When the address requested by a wrong-path load is in neither the data cache nor the WPC, the next cache level in the memory hierarchy is accessed. The required cache block is then placed into the WPC only to eliminate the pollution in the data cache that could otherwise be caused by the wrong-path loads. Note that misses due to loads on the correct execution path, and misses due to the loads issued from the wrong path before the branch is resolved, move the data into the data cache but not into the WPC.

The WPC also caches copies of blocks recently evicted by cache misses. That is, if the data cache must evict a block to make room for a newly referenced block, the evicted block is transferred to the WPC, as is done in the victim cache [9].

3 Experimental Setup

3.1 Microarchitecture

Our microarchitectural simulator is built on top of the SimpleScalar toolset [22], version 3.0. The simulator is modified to compare the processor configurations described in Section 3.2. The processor/memory model used in this study is an aggressively pipelined processor capable of issuing 8 instructions per cycle with out-of-order execution. It has a 128-entry reorder buffer with a 64-entry load/store buffer. The store forwarding latency is increased to 3 cycles in order to compensate for the added complexity of disambiguating loads and stores in a large execution window. There is a 6-cycle branch misprediction penalty. The processor has 8 integer ALU units, 2-integer MULT/DIV units, 4 load/store units, 6-FP Adders and 2-FP MULT/DIV units. The latencies are: ALU=1 cycle, MULT=3 cycles, integer DIV=12 cycles, FP Adder=2 cycles, FP MULT=4 cycles, and FP DIV=12 cycles. All the functional units, except the divide units, are fully pipelined to allow a new instruction to initiate execution each cycle.

The processor has a first-level 32 KB, 2-way set associative instruction cache. Various sizes of the L1 data cache (4KB, 8KB, 16KB, 32KB) with various associativities (direct-mapped, 2-way, 4-way) are examined in the following simulations. The first-level data cache is non-blocking with 4 ports. Both caches have block sizes of 32 bytes and 1-cycle hit latency. Since the memory footprints of the benchmark programs used in this paper are somewhat small, a relatively small 256K 4-way associative unified L2 cache is used for all of the experiments in order to produce significant L2 cache activity. The L2 cache has 64-byte blocks and a hit latency of 12 cycles. The round-trip main memory access latency is 200 cycles for all of the experiments, unless otherwise specified. We model the bus latency to main memory with a 10 cycle bus occupancy per request. Results are shown for bus bandwidth of 8 bytes/cycle. The effect on the WPC performance of varying the cache block size is examined in the simulations. There is a 64-entry 4-way set associative instruction TLB and 128-entry 4-way set associative data TLB, each with a 30-cycle miss penalty. For this study, we used the GAp branch predictor [24, 25]. The predictor has a 4K-entry Pattern History Table (PHT) with 2-bit saturating counters.

3.2 Processor Configurations Tested

The following superscalar processor configurations are simulated to determine the performance impact of executing wrong-path loads, and the performance contributions of the Wrong Path Cache. The configurations, *all, vc,* and *wpc,* are modifications of the SimpleScalar [22] baseline processor described above.

orig: This configuration is the SimpleScalar baseline processor. It is an 8-issue processor with out-of-order execution and support for speculative execution of instructions issued from a predicted branch path. Note that this processor can execute loads from a mispredicted branch path. These loads can potentially change the contents of the cache, although they cannot change the contents of any registers. These wrong-path loads are allowed to access the cache memory system until the branch result is known. After the branch is resolved, they are immediately squashed and the processor state is restored to the state prior to the predicted branch. The execution then is restarted down the correct path.

all: In this configuration, the processor allows as many fetched loads as possible to access the memory system regardless of the predicted direction of conditional branches. This configuration is a good test of how the execution of the loads down the wrong branch path affects the memory system. Note that, in contrast to the *orig* configuration, the loads down the mispredicted branch direction are allowed to continue execution even after the branch is resolved. Wrong-path loads that are not ready to be issued before the branch is resolved, either because they are waiting for the effective address calculation or for an available memory port, are issued to the memory system if they become ready after the branch is resolved, even though they are known to be from the wrong path. Instead of being squashed after the branch is resolved as in the *orig* configuration, they are allowed to access the memory. However, they are squashed before being allowed to write to the destination register. Note that a wrong-path load that is dependent upon another instruction that gets flushed after the branch is resolved also is flushed in the same cycle. Wrong-path stores are not allowed to execute and are squashed as soon as the branch result is known.

orig_vc: This configuration is the *orig* configuration (the baseline processor) with the addition of an 8-entry victim cache.

all_vc: This configuration is the *all* configuration with the addition of an 8-entry victim cache. It is used to compare against the performance improvement made possible by caching of the wrong-path loads in the WPC.

wpc: This configuration adds an 8-entry Wrong Path Cache (WPC) to the *all* configuration.

3.3 Benchmark Programs

The test suite used in this study consists of the combination of SPEC95 and SPEC2000 benchmark programs. All benchmarks were compiled using gcc 2.6.3 at optimization level O3 and each benchmark ran to completion. The SPEC2000 benchmarks are run with the MinneSPEC input data sets to limit their total simulation time while maintaining the fundamental characteristics of the programs' overall behaviors [23].

4 Results

The simulation results are presented as follows. First, the performances of the different configurations are compared using the speedups relative to the baseline (*orig*) processor. Next, several important memory system parameters are varied to determine the sensitivity of the WPC to these parameters. The impact of executing wrong-path loads both with and without the WPC also is analyzed.

Having used small or reduced input sets to limit the simulation time, most of the results are given for a relatively small L1 data cache to mimic more realistic workloads with higher miss rates. The effect of different cache sizes is investigated in Section 4.2. In this paper, our focus is on improving the performance of on-chip direct-mapped data caches. Therefore, most of the comparisons for the WPC are made against a victim cache [9]. We do investigate the impact of varying the L1 associativity in Section 4.2, however.

4.1 Performance Comparisons

4.1.1 Speedup Due to the WPC

Figure 1 shows the speedups obtained relative to the *orig* configuration when executing each benchmark on the different configurations described in Section 3.2. The WPC and the victim cache each have eight entries in those configurations that include these structures.

Of all of the configurations, *wpc*, which executes loads down the wrong branch path with an 8-entry WPC, gives the greatest speedup. From Figure 1, we can see that, for small caches, the *all* configuration actually produces a slowdown due to the large number of wrong-path loads polluting the L1 cache. However, by adding the WPC, the new configuration, *wpc*, produces the best speedup compared to the other configurations. In particular, *wpc* outperforms the *orig_vc* and *all_vc* configurations, which use a simple victim cache to improve the performance of the baseline processor. While both the WPC and the victim cache reduce the impact of conflict misses in the data cache by storing recent evictions near the processor, the WPC goes further by acting like a prefetch buffer and thus preventing pollution misses due to the indirect prefetches caused by executing the wrong-path loads in the *all* configuration.

While we will study the effect of different cache parameters in later sections, Figure 2 shows the speedup results for an 8KB L1 data cache with 4-way associativity. When increasing the associativity of the L1 cache, the speedup obtained by the *orig_vc* seen in Figure 1 disappears. However, the *wpc* still provides

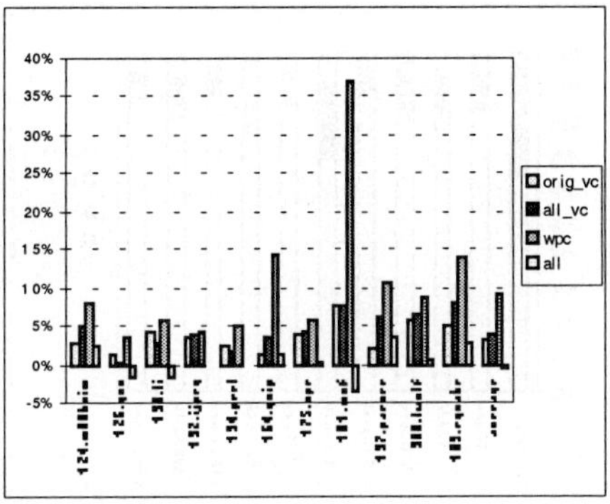

Fig. 1. The Wrong Path Cache (*wpc*) produces consistently higher speedups than the victim cache (*vc*) or the *all* configuration, which does not have a WPC but does execute all ready wrong–path loads if there is a free port to the memory system. The data cache is 8KB direct-mapped and has 32-byte blocks. All speedups are relative to the baseline (*orig*)processor.

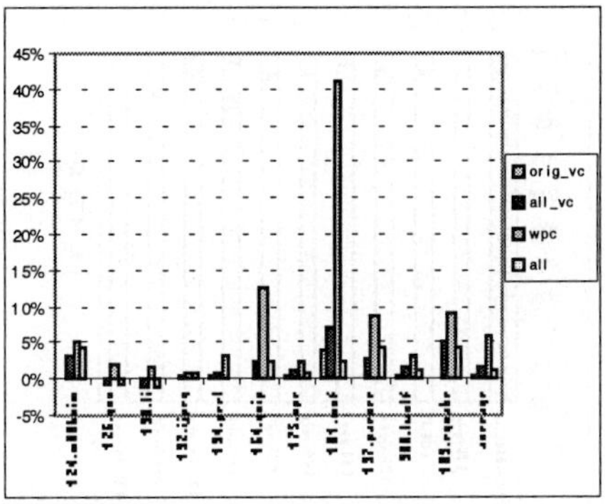

Fig. 2. With a data cache of 8KB with 4-way associativity, the speedup obtained by *orig_vc* disappears. However, *wpc* continues to provide significant speedup and substantially outperforms the *all_vc* configuration. The *all* configuration also shows significant speedup for some benchmarks. The data cache has 32-byte blocks. All speedups are relative to the baseline (*orig*) processor.

significant speedup as the associativity increases and it substantially outperforms the *all_vc* configuration. The *mcf* program shows generally poor cache behavior and increasing the L1 associativity does not reduce its miss rate significantly. Therefore, we see that the speedup produced by the *wpc* for *mcf* remains the same in Figures 1 and 2. As expected, a better cache with lower miss rates reduces the benefit of the *wpc*. From Figure 2, we also see that the *all* configuration can produce some speedup. There is still some slowdown for a few of the benchmarks due to pollution from the wrong path execution of loads. However, the slowdown for the *all* configuration is less than in Figure 1, where the cache is direct-mapped.

4.1.2 A Closer look at the WPC Speedups

The speedup results shown in Figures 1 and 2 can be explained at least partially by examining what levels of the memory hierarchy service the memory accesses. Figure 3 shows that the great majority of all memory accesses in the benchmark programs are serviced by the L1 cache, as is to be expected. While a relatively small fraction of the memory accesses cause misses, these misses add a disproportionately large amount of time to the memory access time. The values for memory accesses that miss in the L1 cache must be obtained from one of three possible sources, the wrong-path cache (WPC), the L2 cache, or the memory. Figure 3 shows that a substantial fraction of the misses in these benchmark programs are serviced by the WPC. For example, 4% of all memory accesses issued by *twolf* are serviced by the WPC. However, this fraction corresponds to 32% of the L1 misses generated by this program. Similarly, 3.3% of *mcf*'s memory accesses, and 1.9% of *equake*'s, are serviced by the WPC, which corresponds to 21% and 29% of their L1 misses, respectively. Since the WPC is accessed in parallel with the L1 cache, misses serviced by the WPC are serviced in the same amount of time as a hit in the L1 cache, while accesses serviced by the L2 cache require 12 cycles and accesses that must go all the way to memory require 200

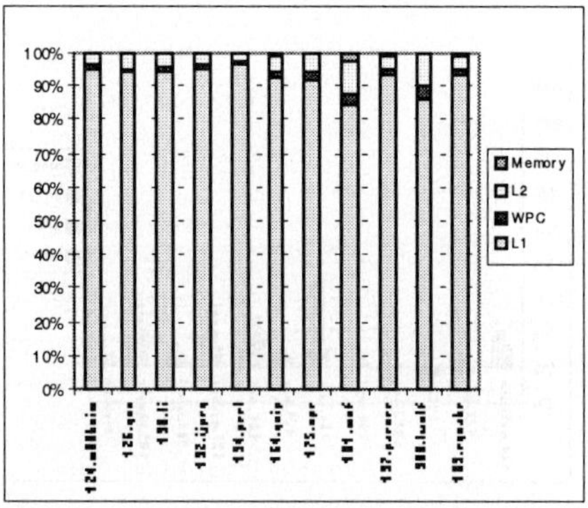

Fig.3. The fraction of memory references on the *correct* execution path that are serviced by the L1 cache, the WPC, the L2 cache, and memory. The L1 data cache is 8KB direct-mapped and has 32-byte blocks.

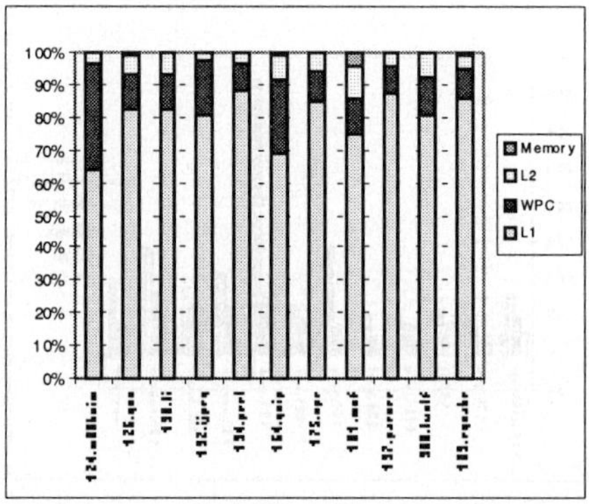

Fig 4. The fraction of memory references on the *wrong* execution path that are serviced by the L1 cache, the WPC, the L2 cache, and memory. The L1 data cache is 8KB direct-mapped and has 32-byte blocks.

cycles. For most of these programs, we see that the WPC converts approximately 20-35% of misses that would have been serviced by the L2 cache or the memory into accesses that are equivalent to an L1 hit.

While the above discussion explains some of the speedups seen in Figures 1 and 2, it does not completely explain the results. For instance, *twolf* has the largest fraction of memory accesses serviced by the WPC in Figure 3. However, *mcf*, *gzip*, and *equake* show better overall speedups. This difference in speedup is explained in Figure 4. This figure shows which levels of the memory hierarchy service the speculative loads issued on what is subsequently determined to be the wrong branch path. Speculative loads that miss in both the L1 cache and the WPC are serviced either by the L2 cache or by the memory. These values are placed in the WPC in the hope that the values will be subsequently referenced by a load issued on the correct branch path.

In Figure 4, we see that 30 percent of the wrong path accesses that miss in both the L1 and the WPC are serviced by memory, which means that this percentage of the blocks in the WPC are loaded from memory. So, from Figure 3 we can say that 30 percent of the correct path accesses that hit in the WPC for *mcf* would have been serviced by the memory in a system without the WPC. That is, the WPC effectively converts a large fraction of this program's L1 misses into the equivalent of an L1 hit. In *twolf*, on the other hand, most of the hits to the WPC would have been hits in the L2 cache in the absence of the WPC. We see in Figure 4 that less than 1% of the wrong path accesses for *twolf* that miss both in the L1 and the WPC are serviced by memory, while 99% of these misses are serviced by the L2 cache. That is, almost all the data in the WPC comes from the L2 cache for *twolf*. Thus, the WPC does a better job of hiding miss delays for *mcf* than for *twolf*, which explains why *mcf* obtains a higher overall speedup with the WPC than does *twolf*. A similar argument explains the speedup results observed in the remainder of the programs, as well.

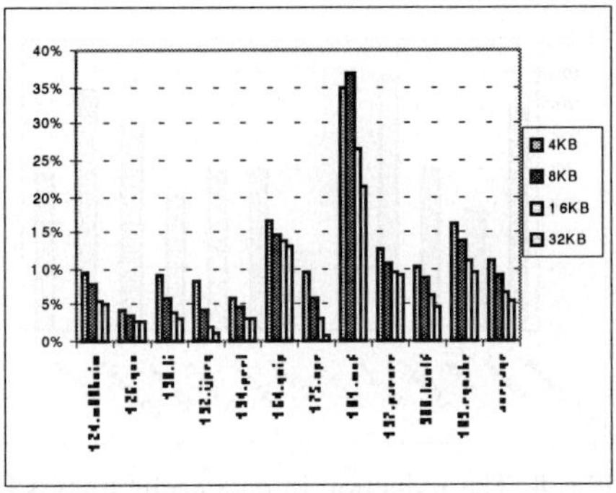

Fig. 5. Speedup obtained with the *wpc* configuration as the L1 cache size is varied. The L1 data cache is direct-mapped with 32-byte blocks. All speedups are relative to the baseline (*orig*)processor.

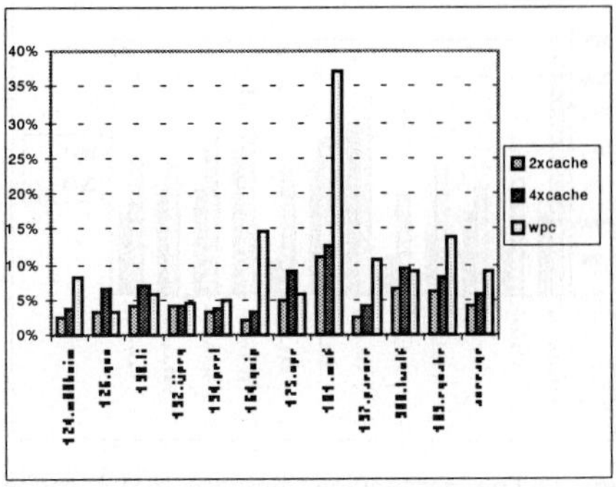

Fig. 6. The speedup obtained with the WPC compared to configurations with larger L1 caches but without a WPC. The base cache size is 8KB and is direct-mapped with 32- byte blocks.

4.2 Sensitivity to Cache Parameters

There are several parameters that affect the performance of a cache memory system. In this study, we examine the effects of the cache size, the associativity, and the cache block size on the cache performance when allowing the execution of wrong-path loads both with and without the WPC. Due to lack of space, the effects of memory latency and the size of WPC are not given in this paper. See [26] for information on the effects of these parameters.

Figure 5 shows that the relative benefit of the *wpc* decreases as the L1 cache size increases. However, the WPC size is kept constant in these simulations so that the relative size of the WPC to the data cache is reduced. With a smaller cache, wrong-path loads cause more misses compared to configurations with larger caches. These additional misses tend to prefetch data that is put into the WPC for use by subsequently executed correct branch paths. The WPC eliminates the pollution in the L1 data cache for the *all* configuration that would otherwise have occurred without the WPC, which then makes these indirect prefetches useful for the correct branch path execution.

While the WPC is a relatively small hardware structure, it does consume some chip area. Figure 6 shows the performance obtained with an 8-entry WPC used in conjunction with an 8KB L1 cache compared to the performance obtained with the original processor configuration using a 16KB L1 cache or a 32KB L1 cache but without a WPC. We find that, for all of the test programs, the small WPC with the 8KB cache exceeds the performance of the processor when the cache size is doubled, but without the WPC. Furthermore, the WPC configuration exceeds the performance obtained when the size of the L1 cache is quadrupled for all of the test programs except *gcc*, *li*, *vpr*, and *twolf.* We conclude that this small WPC is an excellent use of the chip area compared to simply increasing the L1 cache size.

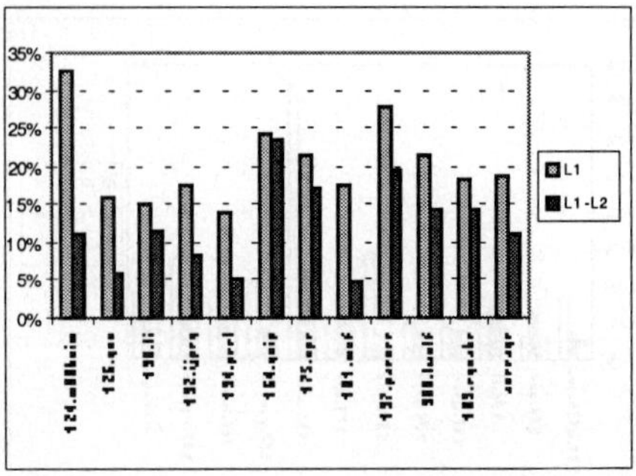

Fig. 7. The percentage increase in L1 cache accesses and traffic between the L1 cache and the L2 cache for the *wpc* configuration compared to the *orig* configuration. The L1 cache is 8 KB, direct-mapped and has 32-byte blocks.

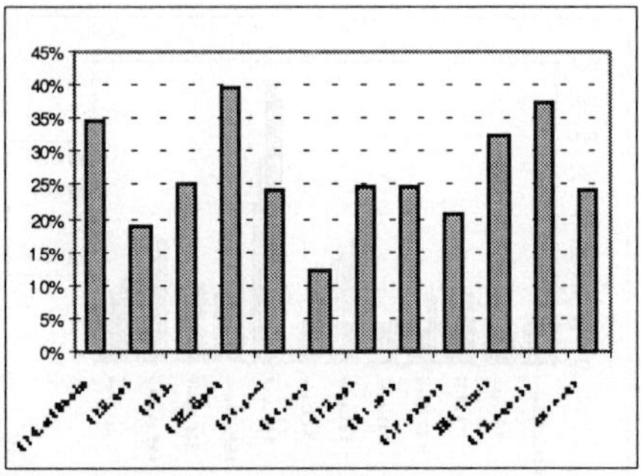

Fig. 8. The reduction in data cache misses for the *wpc* configuration compared to the *orig* configuration. The L1 cache is 8 KB, direct-mapped and has 32-byte blocks.

Figure 7 shows that executing the loads that are known to be down the wrong path typically increases the number of L1 data cache references by about 15-25% for most of the test programs. Furthermore, this figure shows that executing these wrong-path loads increases the bus traffic (measured in bytes) between the L1 cache and the L2 cache by 5-23%, with an average increase of 11%. However, the WPC reduces the total data cache miss ratio for loads on the correct path by up to 39%, as shown in Figure 8.

Increasing the L1 cache associativity typically tends to reduce the number of L1 misses on both the correct path [8] and the wrong path. This reduction in misses reduces the number of indirect prefetches issued from the wrong path, which then reduces the impact of the WPC, as shown in Figure 9. The *mcf* program is the exception since its overall cache behavior is less sensitive to the L1 associativity than the other test programs.

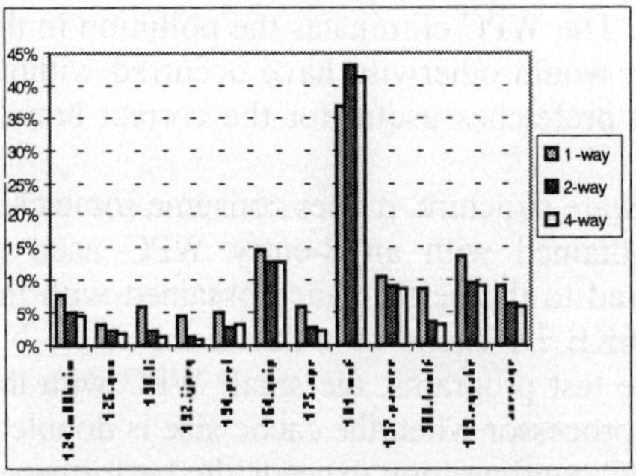

Fig. 9. The effect of the L1 cache associativity on the speedup of the *wpc* configuration compared to the *orig* configuration. The L1 cache size is 8 KB with 32-byte blocks.

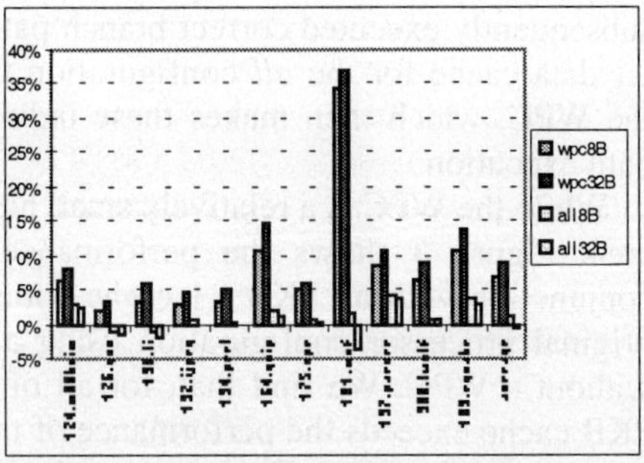

Fig. 10. The effect of the cache block size on the speedup of the *all* and *wpc* configurations compared to the *orig* configuration. The L1 cache is direct-mapped and 8 KB. The WPC is 256B, i.e, 8–entries with 32-byte blocks (wpc32B), or 32-entries with 8-byte blocks (wpc8B).

As the block size of the data cache increases, the number of conflict misses also tends to increase [8, 27]. Figure 10 shows that smaller cache blocks produce better speedups for configurations without a WPC when wrong-path loads are allowed to execute since larger blocks more often displace useful data in the L1 cache. However, for the systems with a WPC, the increasing conflict misses in the data cache due to the larger blocks increases the number of misses that hit in the WPC because of the victim-caching behavior of the WPC. In addition, the indirect prefetches provide a greater benefit for large blocks since the WPC eliminates their polluting effects. We conclude that larger cache blocks work well with the WPC since the strengths and weaknesses of larger blocks and the WPC are complementary.

5 Related Work

There have been several studies examining how speculation affects multiple issue processors [1-7]. Farkas *et al* [1], for example, looked at the relative memory system performance improvement available from techniques such as non-blocking loads, hardware prefetching, and speculative execution, used both individually and in combination. The effect of deep speculative execution on cache performance has been studied by Pierce and Mudge [2]. Several other authors [3-7] examined speculation and pre-execution in their studies Wallace et al. [4] introduced instruction recycling, where previously executed wrong path instructions are injected back into the rename stage instead of being discarded. This technique increases the supply of instructions to the execution pipeline and decreases fetch latency.

Prefetching, which overlaps processor computations with data accesses, has been shown to be one of several effective approaches that can be used to tolerate large memory latencies. Prefetching can be hardware-based, software-directed, or a combination of both [21]. Software prefetching relies on the compiler to perform static program analysis and to selectively insert prefetch instructions into the executable code [16-19]. Hardware-based prefetching, on the other hand, requires no compiler support, but it does require some additional hardware connected to the cache [8-15]. This type of prefetching is designed to be transparent to the processor.

Jouppi [9] proposed *victim caching* to tolerate conflict misses. While several other prefetching schemes have been proposed, such as adaptive sequential prefetching [11], prefetching with arbitrary strides [11, 14], fetch directed prefetching [13], and selective prefetching [15], Pierce and Mudge [20] have proposed a scheme called wrong path instruction prefetching. This mechanism combines next-line prefetching with the prefetching of all instructions that are the targets of branch instructions regardless of the predicted direction of conditional branches.

Most of the previous prefetching schemes require a significant amount of hardware to implement. For instance, they require a *prefetcher* that prefetches the contents of the missed address into the data cache or into an on-chip prefetch buffer. Furthermore, a *prefetch scheduler* is needed to determine the right time to prefetch. On the other hand, this work has shown that executing loads down the wrongly-predicted branch paths can provide a form of indirect prefetching, at the potential expense of some cache pollution. Our proposed Wrong Path Cache (WPC) is essentially a combination of a very small prefetch buffer and a victim cache [9] to eliminate this pollution effect.

6 Conclusions

This study examined the performance effects of executing the load instructions that are issued along the incorrectly predicted path of a conditional branch instruction. While executing these wrong-path loads increases the total number of memory references, we find that allowing these loads to continue executing, even after the branch is resolved, can reduce the number of misses observed on the correct branch path. Executing these wrong-path loads thus provides an indirect prefetching effect. For small caches, however, this prefetching can pollute the cache causing an overall slowdown in performance.

We proposed the Wrong Path Cache (WPC), which is a combination of a small prefetch buffer and a victim cache, to eliminate the pollution caused by the execution of the wrong-path loads. Simulation results show that, when using an 8 KB L1 data cache, the execution of wrong-path loads without the WPC can result in a speedup of up to 5%. Adding a fully-associative eight-entry WPC to an 8 KB direct-mapped L1 data cache, though, allows the execution of wrong path loads to produce speedups of 4% to 37% with an average speedup of 9%. The WPC also shows substantially higher speedups compared to the baseline processor equipped with a victim cache of the same size.

This study has shown that the execution of loads that are known to be from a mispredicted branch path has significant potential for improving the performance of aggressive processor designs. This effect is even more important as the disparity between the processor cycle time and the memory speed continues to increase. The Wrong Path Cache proposed in this paper is one possible structure for exploiting the potential benefits of executing wrong-path load instructions.

Acknowledgement

This work was supported in by National Science Foundation grants EIA-9971666 and CCR-9900605, of the IBM Corporation, of Compaq's Alpha Development Group, and of the Minnesota Supercomputing Institute.

References

[1] K. I. Farkas, N. P. Jouppi, and P. Chow, "How Useful Are Non-Blocking Loads, Stream Buffers, and Speculative Execution in Multiple Issue Processors?" *Technical Report WRL RR 94/8*, Western Research Laboratory – Compaq, Palo Alto, CA, August 1994.

[2] J. Pierce and T. Mudge, "The effect of speculative execution on cache performance," *IPPS 94, Int. Parallel Processing Symp.*, Cancun Mexico, pp. 172-179, Apr. 1994.

[3] G. Reinman, T. Austin, and B. Calder, "A Scalable Front-End Architecture for Fast Instruction Delivery," *26th International Symposium on Computer Architecture*, pages 234-245, May 1999.

[4] S. Wallace, D. Tullsen, and B. Calder, "Instruction Recycling on a Multiple-Path Processor," *5th International Symposium On High Performance Computer Architecture*, pages 44-53, January 1999.

[5] G. Reinman and B. Calder, "Predictive Techniques for Aggressive Load Speculation," *31st International Symposium on Microarchitecture*, pages 127-137, December 1998.

[6] J. D. Collins, H. Wang, D. M. Tullsen, C. Hughes, Y.-F. Lee, D. Lavery, J. P. Shen, "Speculative Precomputation: Long-range Prefetching of Delinquent Loads," In *28th International Symposium on Computer Architecture*, July, 2001.
[7] J. Dundas and T. Mudge, "Improving data cache performance by pre-executing instructions under a cache miss," *Proc. 1997 ACM Int. Conf. on Supercomputing*, July 1997, pp. 68-75.
[8] A.J. Smith, "Cache Memories," *Computing Surveys*, Vol. 14, No. 3, Sept. 1982, pp. 473-530.
[9] N.P. Jouppi, "Improving Direct-Mapped Cache Performance by the Addition of a Small Fully-associative Cache and Prefetch Buffers," *Proc. 17th Annual International Symposium on Computer Architecture*, Seattle, WA, May 1990, pp. 364-373.
[10] F. Dahlgren, M. Dubois and P. Stenstrom, "Fixed and Adaptive Sequential Prefetching in Shared-memory Multiprocessors," *Proc. First IEEE Symposium on High Performance Computer Architecture*, Raleigh, NC, Jan. 1995, pp. 68-77.
[11] T.F. Chen and J.L Baer, "Effective Hardware-Based Data Prefetching for High Performance Processors," *IEEE Transactions on Computers*, Vol. 44, No.5, May 1995, pp. 609-623.
[12] D. Joseph and D. Grunwald, "Prefetching using markov predictors," *IEEE Transactions on Computers,* Vol. 48, No 2, 1999, pp 121-133.
[13] G. Reinman, B. Calder, and T. Austin, "Fetch Directed Instruction Prefetching," *In proceedings of the 32nd International Symposium on Microarchitecture*, November 1999.
[14] T.F. Chen and J.L Baer, "A Performance Study of Software and Hardware Data Prefetching Schemes," *Proc. of the 21st Annual International Symposium on Computer Architecture*, Chicago, Il, April 1994, pp. 223-234.
[15] R. Pendse and H. Katta, "Selective Prefetching: Prefetching when only required," *Proc. of the 42nd IEEE Midwest Symposium on Circuits and Systems*, volume 2, 2000, pp. 866-869.
[16] C-K. Luk and T. C. Mowry. "Compiler-based prefetching for recursive data structures," *In Seventh International Conference on Architectural Support for Programming Languages and Operating Systems*, pages 222--233, Oct. 1996.
[17] Bernstein, D., C. Doron and A. Freund, "Compiler Techniques for Data Prefetching on the PowerPC," *Proc. International Conf. on Parallel Architectures and Compilation Techniques*, June 1995, pp. 19-26.
[18] E.H. Gornish, E.D. Granston and A.V. Veidenbaum, "Compiler-directed Data Prefetching in Multiprocessors with Memory Hierarchies," *Proc. 1990 International Conference on Supercomputing*, Amsterdam, Netherlands, June 1990, pp. 354-368.
[19] M.H. Lipasti, W.J. Schmidt, S.R. Kunkel and R.R. Roediger, "SPAID: Software Prefetching in Pointer and Call-Intensive Environments," *Proc. 28th Annual International Symposium on Microarchitecture*, Ann Arbor, MI, November 1995, pp. 231-236.
[20] J. Pierce and T. Mudge, "Wrong-Path Instruction Prefetching," *Proc. of 29th Annual IEEE/ACM Symp. Microarchitecture (MICRO-29)*, Dec. 1996, pp. 165-175.
[21] S. P. VanderWiel and D. J. Lilja, "Data Prefetch Mechanisms," ACM Computing Surveys, Vol. 32, Issue 2, June 2000, pp. 174-199.
[22] D.C. Burger, T.M. Austin, and S. Bennett, "Evaluating future Microprocessors: The SimpleScalar Tool Set," *Technical Report CS-TR-96-1308*, University of Wisconsin-Madison, July 1996.
[23] AJ KleinOsowski, J. Flynn, N. Meares, and D. J. Lilja, "Adapting the SPEC 2000 Benchmark Suite for Simulation-Based Computer Architecture Research," *Workload Characterization of Emerging Computer Applications,* L. Kurian John and A. M. Grizzaffi Maynard (eds.), Kluwer Academic Publishers, pp 83-100, (2001).
[24] S-T Pan, K. So, and J.T. Rahmeh, "Improving the Accuracy of Dynamic Branch Prediction Using Branch Correlation," *Proc. of the 5th International Conference on Architectural Support for Programming Languages and Operating Systems*, 1992, pp. 76-84.

[25] T.Y. Yeh and Y. N. Patt, "A Comparison of Dynamic Branch Predictors that Use Two Levels of Branch History," *Proc. of the International Symposium on Computer Architecture*, 1993, pp. 257--267.

[26] R. Sendag, D. J. Lilja, and S. R. Kunkel, "Exploiting the Prefetching Effect provided by Executing Misprediced Load Instructions," *Laboratory for Advanced Research in Computing Technology and Compilers, Technical Report No. ARCTIC 02-05, May 2002.*

[27] D. A. Patterson and J. L. Hennessy: *Computer Architecture: A Quantitative Approach*, 2nd edition, Morgan Kaufmann press, 1995, pp. 393-395.

Increasing Instruction-Level Parallelism with Instruction Precomputation

Joshua J. Yi, Resit Sendag, and David J. Lilja

Department of Electrical and Computer Engineering
Minnesota Supercomputing Institute
University of Minnesota - Twin Cities
Minneapolis, MN 55455
{jjyi, rsgt, lilja}@ece.umn.edu

Abstract. Value reuse improves a processor's performance by dynamically caching the results of previous instructions and reusing those results to bypass the execution of future instructions that have the same opcode and input operands. However, continually replacing the least recently used entries could eventually fill the value reuse table with instructions that are not frequently executed. Furthermore, the complex hardware that replaces entries and updates the table may necessitate an increase in the clock period. We propose *instruction precomputation* to address these issues by profiling programs to determine the opcodes and input operands that have the highest frequencies of execution. These instructions then are loaded into the precomputation table before the program executes. During program execution, the precomputation table is used in the same way as the value reuse table is, with the exception that the precomputation table does not dynamically replace any entries. For a 2K-entry precomputation table implemented on a 4-way issue machine, this approach produced an average speedup of 11.0%. By comparison, a 2K-entry value reuse table produced an average speedup of 6.7%. Instruction precomputation outperforms value reuse, especially for smaller tables, with the same number of table entries while using less area and having a lower access time.

1 Introduction

A program may repeatedly perform the same computations during the course of its execution. For example, in a nested pair of FOR loops, an add instruction in the inner loop will repeatedly initialize and increment a loop induction variable. For each iteration of the outer loop, the computations performed by that add instruction are exactly identical. An optimizing compiler typically cannot remove these operations since the induction variable's initial value may change for each iteration.

Value reuse [3, 4] exploits this program characteristic by dynamically caching an instruction's opcode, input operands, and result into a value reuse table (VRT). For each instruction, the processor checks if its opcode and input operands match an entry in the VRT. If a match is found, then the processor can use the result stored in the VRT instead of re-executing the instruction.

B. Monien and R. Feldmann (Eds.): Euro-Par 2002, LNCS 2400, pp. 481–485.

Since the processor constantly updates the VRT, a redundant computation could be stored in the VRT, evicted, re-executed, and re-stored. As a result, the VRT could hold redundant computations that have a very low frequency of execution, thus decreasing the effectiveness of this mechanism.

To address this frequency of execution issue, instruction precomputation uses profiling to determine the redundant computations with the highest frequencies of execution. The opcodes and input operands for these redundant computations are loaded into the precomputation table (PT) before the program executes. During program execution, the PT functions like a VRT, but with two key differences: 1) The PT stores only the highest frequency redundant computations, and 2) the PT does not replace or update any entries. As a result, this approach selectively targets those redundant computations that have the largest impact on the program's performance.

This paper makes the following contributions:

1. It shows that a large percentage of a program is spent repeatedly executing a handful of redundant computations.
2. It describes a novel approach of using profiling to improve the performance and decrease the cost (area, cycle time, and ports) of value reuse.

2 Instruction Precomputation

Instruction precomputation consists of two main steps: profiling and execution. The profiling step determines the redundant computations with the highest frequencies of execution. An instruction is a redundant computation if its opcode and input operands match a previously executed instruction's opcode and input operands.

After determining the highest frequency redundant computations, those redundant computations are loaded into the PT before the program executes. At run-time, the PT is checked to see if there is a match between a PT entry and the instruction's opcode and input operands. If a match is found, then the instruction's output is simply the value in the output field of the matching entry. As a result, that instruction can bypass the execute stage. If a match is not found, then the instruction continues through the pipeline as normal.

For instruction precomputation to be effective, the high frequency redundant computations have to account for a significant percentage of the program's instructions. To determine if this is the situation in typical programs, we profiled selected benchmarks from the SPEC 95 and SPEC 2000 benchmark suites using two different input sets ("A" and "B") [2]. For this paper, all benchmarks were compiled using the gcc compiler, version 2.6.3 at optimization level O3 and were run to completion.

To determine the amount of redundant computation, we stored each instruction's opcode and input operands (hereafter referred to as a "unique computation"). Any unique computation that has a frequency of execution greater than one is a redundant computation. After profiling each benchmark, the unique computations were sorted by their frequency of execution. Figure 1 shows the percentage of the total dynamic instructions that were accounted for by the top 2048 unique computations. (Only arithmetic instructions are shown here because they are the only instructions that we allowed into the PT.) As can be seen in Figure 1, the top 2048 arithmetic unique computations account for 14.7% to 44.5% (Input Set A) and 13.9% to 48.4% (B) of the total instructions executed by the program.

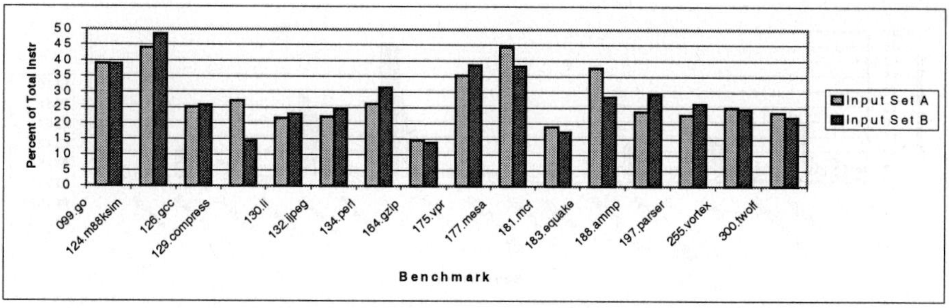

Fig 1. Percentage of the Total Dynamic Instructions Due to the Top 2048 Arithmetic Unique Computations

3 Results and Analysis

To determine the performance of instruction precomputation, we modified sim-outorder from the Simplescalar tool suite [1] to include a precomputation table. The PT can be accessed in both the dispatch and issue stages. In these two stages, the current instruction's opcode and input operands are compared against the opcode and input operands that are stored in the PT. If a match is found in the dispatch stage, the instruction obtains its result from the PT and is removed from the pipeline (i.e. it waits only for in-order commit to complete its execution). If a match is found in the issue stage, the instruction obtains its result from the PT and is removed from the pipeline only if a free functional unit *cannot* be found. Otherwise, the instruction executes as normal.

The base machine was a 4-way issue processor with 2 integer and 2 floating-point ALUs; 1 integer and 1 floating-point multiply/divide unit; a 64 entry RUU; a 32 entry LSQ; and 2 memory ports. The L1 D and I caches were set to 32KB, 32B blocks, 2-way associativity, and a 1 hit cycle latency. The L2 cache was set to 256KB, 64B blocks, 4-way associativity, and a 12 cycle hit latency. The memory latency of the first block was 60 cycles while each following block took 5 cycles. The branch predictor was a combined predictor with 8K entries.

To reiterate one key point, the profiling step is used only to determine the highest frequency unique computations. Since it is extremely unlikely that the same input set that is used for profiling also will be used during execution, we simulate a combination of input sets, that is, we profile the benchmark using one input set, but run the benchmark with another input set (i.e. Profile A, Run B or Profile B, Run A).

Figure 2 shows the speedup of instruction precomputation as compared to the base machine for Profile B, Run A. We see that instruction precomputation improves the performance of all benchmarks by an average of 4.1%–11.0% (16 to 2048 entries). Similar results also occur for the Profile A, Run B combination. *These results show that the highest frequency unique computations are common across benchmarks and are not a function of the input set.*

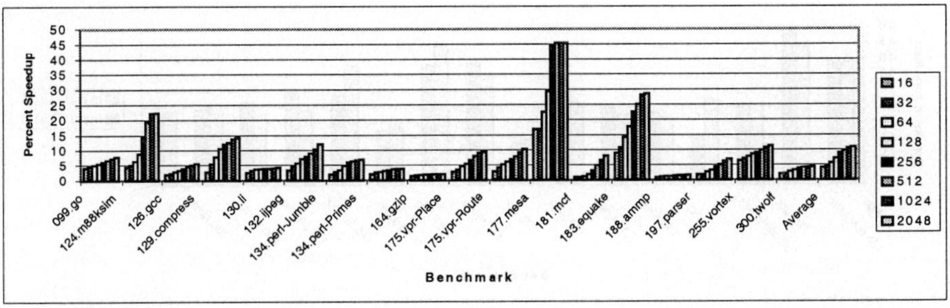

Fig 2. Percent Speedup Due To Instruction Precomputation for Various Table Sizes; Profile Input Set B, Run Input Set A

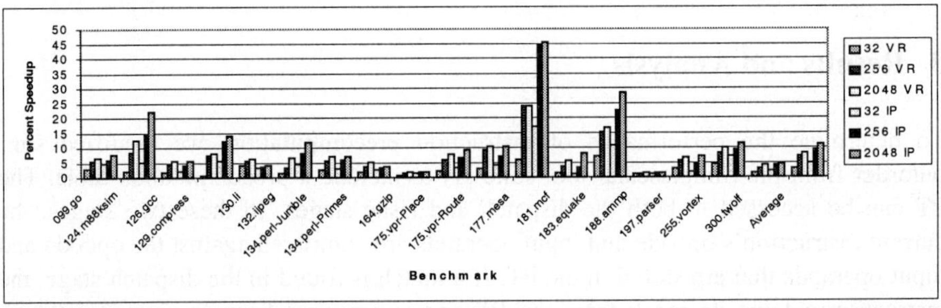

Fig 3: Speedup Comparison Between Value Reuse (VR) and Instruction Precomputation (IP) for Various Table Sizes; Profile Input Set A, Run Input Set B

In addition to having a lower area and access time, instruction precomputation also outperforms value reuse for tables of similar size. Figure 3 shows the speedup of instruction precomputation and value reuse, as compared to the base machine, for three different table sizes. For almost all table sizes and benchmarks, instruction precomputation yields a higher speedup than value reuse does. A more detailed comparison of instruction precomputation and value reuse can be found in [5].

4 Related Work

Sodani and Sohi [4] found speedups of 6% to 43% for a 1024 entry dynamic value reuse mechanism. While their speedups are comparable to those presented here, our approach has a smaller area footprint and a lower access time.

Molina *et. al.* [3] implemented a dynamic value reuse mechanism that exploited value reuse at the both the global (PC-independent) and local levels (PC-dependent). However, their approach is very area-intensive and their speedups are tied to the area used. For instance, for a realistic 36KB table size, the average speedup was 7%.

5 Conclusion

This paper presents a novel approach to value reuse that we call instruction precomputation. This approach uses profiling to determine the unique computations with the highest frequencies of execution. These unique computations are preloaded into the PT before the program begins execution. During execution, for each instruction, the opcode and input operands are compared to the opcodes and input operands in the PT. If there is a match, then the instruction is removed from the pipeline. For a 2048 entry PT, this approach produced an average speedup of 11.0%. Furthermore, the speedup for instruction precomputation is greater than the speedup for value reuse for almost all benchmarks and table sizes. Instruction precomputation also consumes less area and has a lower table access time as compared to value reuse.

Acknowledgements

This work was supported in by National Science Foundation grants EIA-9971666 and CCR-9900605, by IBM, and by the Minnesota Supercomputing Institute.

References

1. D. Burger and T. Austin; “The Simplescalar Tool Set, Version 2.0”; University of Wisconsin Computer Sciences Department Technical Report 1342.
2. A. KleinOsowski, J. Flynn, N. Meares, and D. Lilja; "Adapting the SPEC 2000 Benchmark Suite for Simulation-Based Computer Architecture Research"; Workload Characterization of Emerging Computer Applications, L. Kurian John and A. M. Grizzaffi Maynard (eds.),Kluwer Academic Publishers, (2001) 83-100
3. C. Molina, A. Gonzalez, and J. Tubella; "Dynamic Removal of Redundant Computations"; International Conference on Supercomputing, (1999)
4. A. Sodani and G. Sohi; "Dynamic Instruction Reuse"; International Symposium on Computer Architecture, (1997)
5. J. Yi, R. Sendag, and D. Lilja; " Increasing Instruction-Level Parallelism with Instruction Precomputation "; University of Minnesota Technical Report: ARCTiC 02-01

Runtime Association of Software Prefetch Control to Memory Access Instructions

Chi-Hung Chi and JunLi Yuan

School of Computing, National University of Singapore
Lower Kent Ridge Road, Singapore 119260

Abstract. In this paper, we introduce a new concept of run-time collaboration between hardware and software prefetching mechanisms. An association bit is added to a memory access instruction (MAI) to indicate if any software PREFETCH instruction corresponding to the MAI has been inserted into the program. This bit is set by the compiler. Default hardware prefetching might be triggered for a MAI only if a "1" is detected in this bit. Simulation on SPEC95 shows that this association concept is very useful in HW/SW hybrid prefetching; its performance improvement in floating point applications ranges from a few percent to about 60%, with an average of 28.63%. This concept is important because its requirements for hardware and compiler support are very minimal. Furthermore, most existing architectures actually have unused encoding space that can be used to hold the association information.

1 Challenges to Hybrid HW/SW Prefetching

Research in data prefetching often focuses on two main issues: accuracy and coverage [2]. The accuracy of a prefetch scheme refers to the probability that a prefetched data is actually referenced in cache. The coverage of a prefetch scheme refers to the portion of the memory data in a program whose reference pattern might potentially be predicted by the scheme prior to the actual execution. A prefetch scheme is said to be efficient if it has large coverage and high accuracy. However, it is not easy for a prefetch scheme to have good qualities on these two factors at the same time. As a result, the concept of hybrid prefetching arises. With multiple predictors supported by a prefetch unit, each predictor can fine-tune to just one selected group of data references. While hybrid prefetching schemes with hardware-only predictors or software-only predictors are proposed [4,5], the more promising hybrid prefetching with a mix of hardware and software predictors is a challenge to computer architects. This is due to the lack of association between the memory access instruction (MAI) for hardware prefetching and the PREFETCH instruction for software prefetching.

To get a deeper understanding why the association between a MAI and its PREFETCH instruction is so difficult to obtain at run-time, let us go back to their basic instruction definition. Under the current ISA of most microprocessors, PREFETCH instructions are defined just like LOAD instructions except that they do

B. Monien and R. Feldmann (Eds.): Euro-Par 2002, LNCS 2400, pp. 486–489.

not have destination registers [1]. For each `PREFETCH` instruction inserted into a program by the compiler, there should be a corresponding MAI involved. However, due to the lack of architectural support, this association information is not recorded in the program. As a result, after the compilation of a program, the association information will be lost. And it will be extremely difficult (if possible) for the hardware to recover this relationship in real-time, during the program execution. Compiler optimization and program transformation make the real-time recovery process even more difficult. The inserted `PREFETCH` instructions might be moved to any place of the program by the compiler.

The lack of run-time association between a `PREFETCH` instruction and its corresponding MAI results in an embarrassing position when software prefetching tries to incorporate a default hardware oriented prefetch scheme into it. It is because the prefetch hardware does not have any knowledge on when the "default" cases should occur. The default case refers to the situation where a MAI does not have without any associated `PREFETCH` instruction inserted in a program.

2 Runtime Association between PREFETCH and MAI

We argue that collaboration among all possible prefetch requests of a MAI is very important to obtain good performance for SW/HW hybrid prefetching. This collaboration should have at least three main properties. The first one is the exclusive triggering of prefetch request. Since only one prefetch request can possibly be correct, triggering multiple prefetch requests for the execution of a MAI is likely to result in cache pollution. The second one is the selection of prefetch request for action. Obviously, given multiple possible prefetch requests for a MAI, the one with the highest accuracy should be chosen. The third one is related to the order of defining the prefetch actions. Once a `PREFETCH` instruction for a MAI is inserted into a program, no hardware prefetching for the MAI should be triggered. It is because there is no mechanism to remove `PREFETCH` instructions from a program dynamically; the exclusive triggering rule also needs to be observed. However, this should not be a problem as the accuracy of software prefetching is usually at least as good as the hardware ones and the run-time overhead problem should have been considered before `PREFETCH` instructions are inserted into a program.

To achieve the goal of collaborated SW/HW hybrid prefetching, we propose to extend the definition of MAI in its ISA. There is a bit, called the association bit, in each MAI. This bit will determine if any hardware prefetch mechanism can be triggered for a MAI. If the bit is "0", it means that the MAI does not have any associated `PREFETCH` instruction inserted in the program. Hence, the hardware is free to trigger its own prefetch action. On the other hand, if the bit is "1", all hardware oriented prefetch mechanisms for a given MAI should be suppressed; no hardware prefetch requests should be triggered in this case. Compiler support to set this association bit for MAIs in a program is trivial; it can be done in the same procedure of adding `PREFTCH` instructions into the program code.

Incorporating the association bit into the MAI definition of the processor's ISA is quite simple. The benefit of this association bit in cache performance is already large enough to justify its existence. Furthermore, there is often unused encoding space in existing processor's architecture that can be used to hold this association information. For example, in HP's PA architecture [3], there is a 2-bit cache control field, *cc*, defined in the MAIs. These bits are mainly used to provide hints about the spatial locality and the block copy of memory references. More importantly, the *cc* bits are set by the compiler and the bit pattern "11" is still unused. As a result, this is the ideal place for the association information to be stored. Similarly, in SPARC architecture, bit 5-11 of each MAI can also be used to encode the association information.

There are other alternative solutions to encode the association information in a program for existing architectures. For example, the compiler can insist that any insertion of a `PREFETCH` instruction must occur immediately after (or before) its MAI. In this way, the hardware can look for this during the program execution and can take the appropriate action based on the finding. The advantage of this solution is that there is absolutely no change to the architecture. However, it puts too much constraint to the compiler optimization. Consequently, we do not recommend this solution.

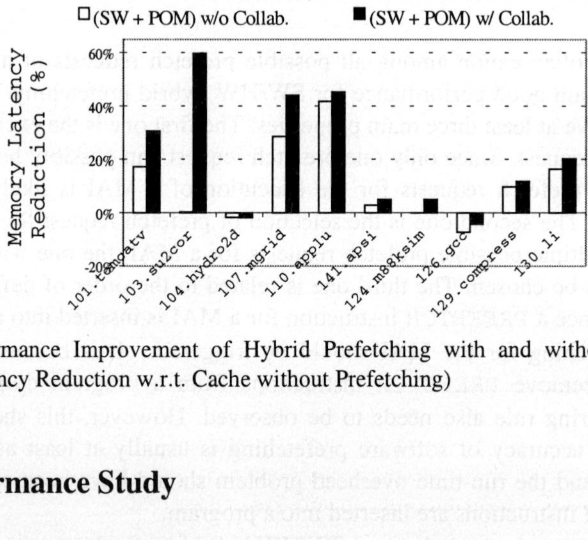

Fig. 1. Performance Improvement of Hybrid Prefetching with and without Collaboration (Memory Latency Reduction w.r.t. Cache without Prefetching)

3 Performance Study

To study the effect of our proposed association concept, Figure 1 shows the performance improvement of SW/HW hybrid prefetching with and without collaboration. Here, we assume the "default" hardware prefetch scheme is the "prefetch-on-miss (POM)" scheme and the software prefetch scheme focuses on linear stride access [1]. The benchmark suite is the SPEC95 and the simulated architecture is a superscalar, Ultra-SPARC ISA compatible processor with the 1st level separated 32 Kbytes instruction cache and 32 Kbytes cache and a 2nd level 256 Kbytes unified cache, all being direct-mapped.

For floating point benchmark programs, the improvement is very significant; it ranges from a few percents to about 60%, with an average of 29.86%. This is compared to the average performance improvement of 14.87% in the non-collaborated case, almost double in cache performance with collaboration. In the integer benchmark programs, the performance gain by the hybrid prefetching is lesser, only in the range of a few percent. This is expected because the chance for a compiler to insert `PREFETCH` instructions into an integer program for linear array accesses in loops is much less. Hence, the pollution effect of the wrong default hardware prefetching becomes smaller.

4 Conclusion

In this paper, we argue that while the concept of default prefetching can improve cache performance by increasing the coverage, it cannot be applied to software prefetching. This is mainly due to the lack of association information between a MAI and its corresponding `PREFETCH` instruction. Detailed analysis on the behavior of software prefetching with "always default" hardware prefetching shows that there are rooms for cache performance improvement because over two-third of the triggered hardware prefetch requests are actually either redundant or inaccurate. To make up for this situation, we propose a novel concept of run-time association between MAIs and their corresponding software prefetch controls. With thc help of a one-bit field per MAI to hold the association information, we see that significant improvement in cache performance can be obtained. This concept is very attractive to processor design because most ISAs have unused encoding space in their MAI instructions that can be used to hold the association information.

References

1. Callahan, D., Kennedy, K., Porterfield, A., "Software Prefetching," *Proceedings of the Four International Conference on Architectural Support for Programming Languages and Operating Systems*, April 1991, pp. 40-52.
2. Chi, C.H., Cheung, C.M., "Hardware-Driven Prefetching for Pointer Data References," *Proceedings of the 1997 ACM International Conference on Supercomputing,* July 1998.
3. Kane, G., PA-RISC 2.0 Architecture, Prentice-Hall Press, 1996.
4. Manku, G.S., Prasad, M.R., Patterson, D.A., "A New Voting Based Hardware Data Prefetch Scheme," *Proceedings of 4th International Conference on High Performance Computing,* Dec. 1997, pp. 100-105.
5. Wang, K., Franklin, M., "Highly Accurate Data Value Prediction using Hybrid Predictors," *Proceedings of the MICRO-30,* 1997, pp. 281-290.

Realizing High IPC Using Time-Tagged Resource-Flow Computing

Augustus Uht[1], Alireza Khalafi[2], David Morano[2], Marcos de Alba[2], and David Kaeli[2]

[1] University of Rhode Island, Kingston, RI, USA,
uht@ele.uri.edu

[2] Northeastern University, Boston, MA, USA,
{akhalafi,dmorano,mdealba,kaeli}@ece.neu.edu

Abstract. In this paper we present a novel approach to exploiting ILP through the use of *resource-flow computing*. This model begins by executing instructions independent of data flow and control flow dependencies in a program. The rest of the execution time is spent applying *programmatic* data flow and control flow constraints to end up with a programmatically-correct execution. We present the design of a machine that uses *time tags* and *Active Stations*, realizing a registerless data path.

In this contribution we focus our discussion on the Execution Window elements of our machine, present Instruction Per Cycle (IPC) speedups for SPECint95 and SPECint2000 programs, and discuss the scalability of our design to hundreds of processing elements.

1 Introduction

A number of ILP studies have concluded that there exists a significant amount of parallelism in common applications [9, 15, 17]. So why haven't we been able to obtain these theoretical speedups? Part of the reason is that we have not been aggressive enough with our execution model.

Lam and Wilson showed us that if a machine could follow multiple flows of control while utilizing a simple branch predictor and limited control dependencies (i.e., instructions after a forward branch's target are independent of the branch), a speedup of 40 could be obtained on average [9]. If an Oracle (i.e., perfect) branch predictor was used, speedups averaged 158. Research has already been reported that overcomes many control flow issues using limited multi-path execution [2, 15].

To support rampant speculation while maintaining scalable hardware, we introduce a statically ordered machine that utilizes instruction *time tags* and *Active Stations* in the Execution Window. We call our machine *Levo* [16]. Next we will briefly describe our machine model.

B. Monien and R. Feldmann (Eds.): Euro-Par 2002, LNCS 2400, pp. 490–499.

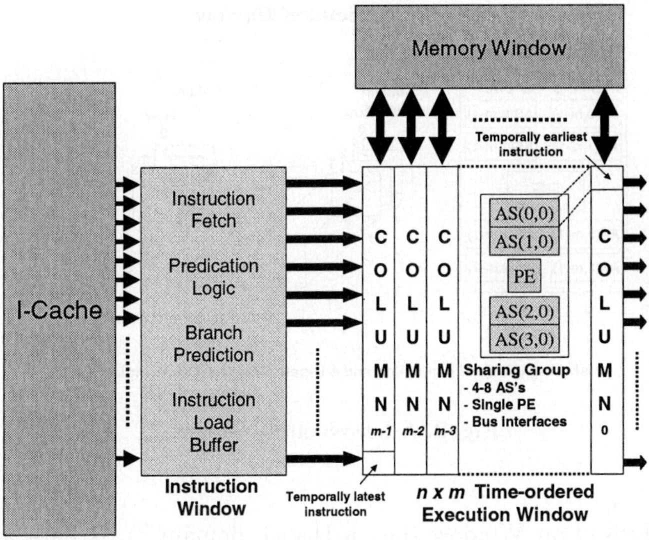

Fig. 1. The Levo machine model.

2 The Levo Machine Model

Figure 1 presents the overall model of Levo, which consists of 3 main components: 1) the Instruction Window, 2) the Execution Window, and 3) the Memory Window.

The Instruction Window fetches instructions from an instruction memory, performs dynamic branch prediction, and generates predicates. Instructions are fetched in the *static order* in which they appear in the binary image (similar to assuming all conditional branches are not taken). By fetching down the not-taken path, we will capture the taken and not taken paths of most branch hammocks [3, 8]. We exploit this opportunity and spawn execution paths to cover both paths (taken and not taken) for hard-to-predict hammocks.

Some exceptions to our static fetch policy are:

1. unconditional jump paths are followed,
2. loops are unrolled dynamically [14] in the Execution Window, and
3. in the case of conditional branches with *far targets,* [1] if the branch is strongly predicted taken in the branch predictor, begin static fetching from its target.

We utilize a conventional two-level gshare predictor [11] to guide both instruction fetch (as in case 3 above), as well as to steer instruction issue. Levo utilizes full run-time generated predicates, such that every branch that executes

[1] Far implies that the branch target is farther than two-thirds the size of the Execution window size. For a machine with 512 ASs, this distance is equal to 341 instructions.

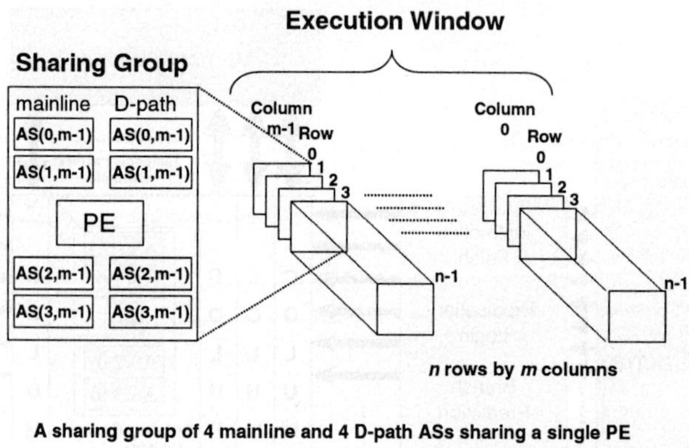

Fig. 2. A Levo Sharing Group.

within the Execution Window (i.e., a branch domain [2]), is data and control independent of all other branches.

Levo is an in-order issue, in-order completion machine, though it supports a high degree of speculative *resource-flow-order* execution. The Execution Window is organized as a grid; columns of *processing elements* (PEs) are arranged in a number of *Sharing Groups* (SGs) per column. A SG shares a common PE (see Figure 2).

Levo assigns PEs to the highest priority instruction in a SG that has not been executed, independent of whether the instruction's inputs or operands are known to be correct (data flow independent), and regardless of whether this instruction is known to be on the actual (versus mispredicted) control path (control flow independent). The rest of the execution time is spent applying programmatic data flow (re-executions) and control flow constraints (squashes), so as to end up with a programmatically-correct execution of the program. Instructions are retired in order, when all instructions in the column have completed execution.

Each sharing group contains a number of *Active Stations* (ASs); instructions are issued in static order to ASs in a column. Each issued instruction is assigned a *time tag*, based on its location in the column. Time tags play a critical role in the simplicity of Levo by labeling each instruction and operand in our Execution Window. This label is used during the maintenance/enforcement of program order in our highly speculative machine.

Our ASs are designed after Tomasulo's reservation stations [13]. There is one instruction per Active Station. Levo ASs are able to snoop and snarf [3] data from buses with the help of the time tags. ASs are also used to evaluate predicates, and to squash redundant operand updates (again using time tags).

---

[2] A branch domain includes the static instructions starting from the branch to its target, exclusive of the target and the branch itself [15].

[3] Snarfing entails snooping address/data buses, and when the desired address value is detected, the associated data value is read.

ASs within a Sharing Group compete for the resources of the group, including the single pipelined PE and the broadcast bus outputs. Each spanning bus is connected to adjacent Sharing Groups. The spanning bus length is constant and does not change with the size of the Execution Window; this addresses scalability of this busing structure.

A *column* in the Execution Window is completely filled with the sequence of instructions as they appear in the Instruction Window. During execution, hardware runtime predication is used for all forward branches with targets within the Execution Window. Backward branches are handled via dynamic loop unrolling [14] and runtime conversion to forward branches.

2.1 Levo Execution Window Datapath

Levo's *spanning buses* play a similar role as Tomasulo's reservation stations' *Common Data Bus.* Spanning buses are comprised of both forwarding and backwarding buses. Forwarding buses are used to broadcast register, memory and predicate values. If an AS needs an input value, it sends the request to earlier ASs via a backwarding bus and the requested data is returned on a forwarding bus.

An AS connects to the spanning buses corresponding to the position of the AS in the column. Each AS performs simple comparison operations on the time tags and addresses broadcast on the spanning buses to determine whether or not to snarf data or predicates. Figure 3 shows the structure for this function of an AS.

2.2 Scalability

So far we have described a machine with ASs all connected together with some small number of spanning buses. In effect, so far there is little difference between a Levo spanning bus and Tomasulo's Common Data Bus. This microarchitecture may reduce the number of cycles needed to execute a program via resource flow, but having the buses go everywhere will increase the cycle time unacceptably.

The Multiscalar project demonstrated that register lifetimes are short, typically spanning only one or two basic blocks (32 instructions at the high end) [1, 5]. Based on this important observation, we partition each bus into short segments, limiting the number of ASs connected to any segment; this has been set to the number of ASs in a column for the results presented in this paper.

We interconnect broadcast buses with buffer registers; when a value is transmitted on the bus from the preceding bus segment the sourcing AS needs to compete with other ASs for the bus segment. Local buffer space is provided. Thus, there can be a one or more cycle delay for sending values across bus segments.

In Levo there is no centralized register file, there are no central renaming buffers nor reorder buffer. Levo uses locally-consistent register values distributed throughout the Execution Window and among the PEs. A register's contents are likely to be globally inconsistent, but locally usable. A register's contents will eventually become consistent at instruction commit time. In Levo, PEs broadcast their results directly to only a small subset of the instructions in the Execution Window, which includes the instructions within the same Sharing Group.

2.3 Time Tags and Renaming

A time tag indicates the position of an instruction in the original sequential program order (i.e., in the order that instructions are issued). ASs are labeled with time tags starting from zero and incrementing up to one minus the total number of ASs in the microarchitecture. A time tag is a small integer that uniquely identifies a particular AS.

Similar to a conventional reservation station, operand results are broadcast forward for use by waiting instructions. With ASs, all operands that are forwarded after the execution of an instruction are also tagged with the time tag value of the AS that generated the updated operand. This tag will be used by subsequent ASs to determine if the operand should be snarfed as an input operand that will trigger the execution of its loaded instruction. Essentially all values within the Execution Window are tagged with time tags. Since our microarchitecture can also allow for the concurrent execution of disjoint paths, we also introduce a path ID.

The microarchitecture that we have devised requires the forwarding of three types of operands. These are register operands, memory operands, and instruction predicate operands. These operands are tagged with time tags and path IDs that are associated with the ASs that produced them. The information broadcast from an AS to subsequent ASs in future program ordered time is referred to as a *transaction*, and consists of :

- a path ID
- the time tag of the originating AS
- the identifier of the architected operand
- the actual data value for this operand

Figure 3 shows the registers inside an active station for one of its input operands. The *time-tag*, *address*, and *value* registers are reloaded with new values on each snarf, while the *path* and *AS time-tag* (column indices) are only loaded when the AS is issued an instruction, with the path register only being reloaded upon a taken disjoint path execution (disjoint execution will be discussed later).

This scheme effectively eliminates the need for rename registers or other speculative registers as part of the reorder buffer. The whole of the microarchitecture thus provides for the full renaming of all operands, thus avoiding all false dependencies. There is no need to limit instruction issue or to limit speculative instruction execution due to a limit on the number of non-architected registers for holding those temporary results. True flow dependencies are enforced through continuous snooping by each AS.

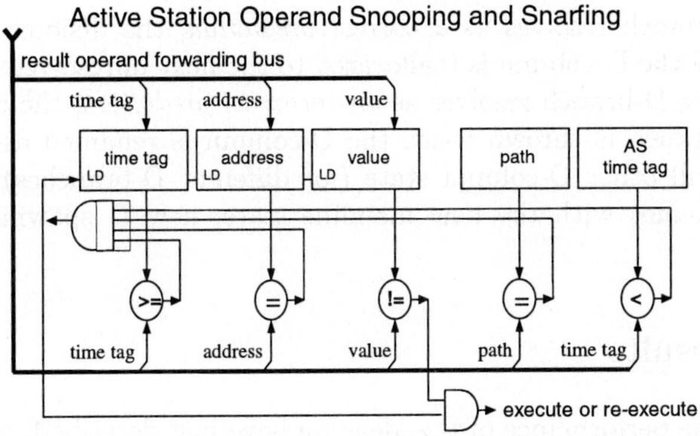

Fig. 3. *AS Operand Snooping and Snarfing.* The registers and snooping operation of one of several possible source operands is shown. Time tags, addresses and values are compared against each active stations corresponding values. Just one bus is shown being snooped, though several operand forwarding buses are snooped simultaneously.

2.4 Disjoint Execution

Our resource flow Execution Window can only produce high IPC if it contains the stream of instructions that will be committed next. In an effort to insure that we can handle the ill-effects of branch mispredictions, we have utilized *disjoint execution* to handle the cases where branch prediction is wrong.

In Figure 2 we showed a Sharing Group containing both a mainline and disjoint (D-path) set of ASs. The D-path ASs will share the common PE with the mainline execution, though will receive a lower priority when attempting to execute an instruction. The disjoint path is used to hide potential latencies associated with branch mispredictions. The disjoint path is copied from a mainline path in a cycle when the instruction loading buses are free. The disjoint path will use a copy of the mainline path, though will start execution from a point after a branch instruction (if the branch was predicted taken), or at a branch target (if the branch was predicted as not taken) in the mainline execution. For branches that exhibit a chaotic behavior (changing from taken to not-taken often), spawning disjoint paths should be highly beneficial. For more predictable branches (e.g., a loop-ending branch), we can even reap some benefit by executing down the loop exit path.

In [7], we discuss how to select the best path to spawn disjoint paths. In this work, we always start spawning from the column prior to the current column being loaded, and spawn up to 5 paths (3 for 4 column configurations). If we look at one of the D-columns, the code and state above the D-branch (the point at which we spawned a disjoint path) is the same as in the mainline path. The code is the same for the entire column (static order). The sign of the predicate of the D-branch is set to the not-predicted direction of the original branch. All other branch predications in the column follow those of the same branches in the mainline column.

If the D-branch resolves as a *correct prediction*, the disjoint path state is discarded, and the D-column is reallocated to the next unresolved branch in the mainline. If the D-branch resolves as an *incorrect prediction*, the mainline state after the D-branch is thrown away, the D-column is renamed as the mainline column, and all other D-column state (for different D-branches) is discarded. Execution resumes with the new mainline state; new D spawning can begin again.

3 ILP Results

To evaluate the performance of the ideas we have just described, we have developed a trace-driven simulation model of the Levo machine. The simulator takes as input a trace containing instructions and their associated operand values. We include results for 5 programs taken from the SPECint2000 and SPECint95 suites, using the reference inputs. Our current environment models a MIPS-1 ISA with some MIPS-2 and MIPS-3 instructions included which are used by the SGI compiler or are in SGI system libraries. While we use a specific ISA in this work, Levo is not directly ISA dependent.

For our baseline system (BL), we assume a machine that is bound by true dependencies in the program, and does no forwarding or backwarding of values. The machine follows a single path of execution (no disjoint paths are spawned). We compare the baseline to a variety of Levo systems that implement resource flow (RF). We also show results for a machine that uses D-path spawning (D). We study the effects of different memory systems, assuming both a conventional hierarchical memory system (CM) and a perfect cache memory (PM). All speedup results are relative to a baseline system that uses a conventional memory (BL-CM).

Table 1 summarizes many of the machine parameters we use in the set of results presented. The table includes the parameters for the conventional data memory system. Table 2 shows the 5 different machine configurations studied and presents our baseline IPC numbers which we will use to compare against.

Table 1. Common model simulation parameters.

Feature	Size	Comment
Fetch width	1-column each cycle	
L1 I-Cache		100% hit
Branch predictor	2-level gshare 1024 PAg 4096 GPHT	multi-ported
L1 D-Cache	32KB 2-way 32B line	4-way interleaved
L1 D-hit time	1 cycle	
L1 D-miss penalty	10 cycles	
L2 and Memory		100% hit
Forwarding/Backwarding unit delay	1 cycle	
Bus delay	1 cycle	

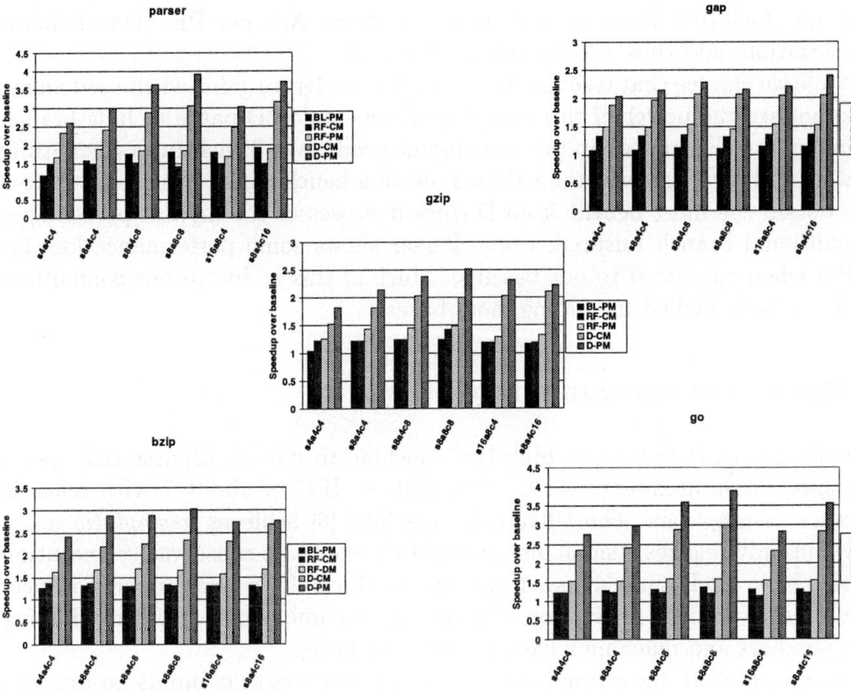

Fig. 4. IPC comparison for baseline perfect memory (BL-PM), resource flow conventional memory (RF-CM), resource flow perfect memory (RF-PM), D-paths conventional memory (D-CM) and D-paths perfect memory (D-PM). All speedup factors are versus our baseline assuming conventional memory (BL-CM).

Table 2. Levo machine configurations and BL-CM IPC values for the 5 benchmarks. s = SGs per column, a = ASs per SG and c = Columns.

Machine Config	SGs per Column	ASs per SG	Columns	gzip BL-CM IPC	gap BL-CM IPC	parser BL-CM IPC	bzip BL-CM IPC	go BL-CM IPC
s4a4c4	4	4	4	2.3	2.4	1.8	1.9	1.7
s8a4c4	8	4	4	2.8	3.5	2.4	2.5	2.4
s8a4c8	8	4	8	2.9	3.9	2.5	2.7	2.5
s8a8c8	8	8	8	4.1	4.4	2.7	2.9	2.7
s16a8c4	16	8	4	3.1	3.9	2.5	2.6	2.5
s8a4c16	8	4	16	3.1	4.2	2.5	2.5	2.5

Figure 4 shows the relative speedup in IPC for our five benchmarks, for the six machine configurations described. All results are relative to our Baseline system with a conventional data cache memory hierarchy, as described in Table 2. The s8a8c8 configuration provides the highest IPC. The first 3 configurations have fewer hardware resources; s16a8c4 does not have enough columns to hide latency and while s8a4c16 has enough columns, with fewer ASs per PE, there is lower PE utilization, and thus a lower IPC is obtained.

While we can see that resource flow provides moderate gains when used alone, we do not see the power of this model until we employ D-paths to hide branch mispredictions. For parser and go, we obtain speedups of 3 to 4 times when using D-paths, with IPCs greater than 10 in 3 out of 5 benchmarks. We should expect go to obtain the most benefit from D since it possesses the highest percentages of conditional branch mispredictions. Parser shows some performance loss for RF-PM when compared to our baseline. Much of this is due to bus contention, which can be remedied by adding more buses.

4 Discussion and Summary

Probably the most successful high-IPC machine to date is Lipasti and Shen's Superspeculative architecture [10], achieving an IPC of about 7 with realistic hardware assumptions. The Ultrascalar machine [6] achieves *asymptotic* scalability, but only realizes a small amount of IPC, due to its conservative execution model. The Warp Engine [4] uses time tags, like Levo, for a large amount of speculation; however their realization of time tags is cumbersome, utilizing floating point numbers and machine wide parameter updating.

Nagarajan et al. have proposed a *Grid Architecture* that builds an array of ALUs, each with limited control, connected by a operand network [12]. Their system achieves an IPC of 11 on SPEC2000 and Mediabench benchmarks. While this architecture presents many novel ideas in attempt to reap high IPC, it differs greatly in its interconnect strategy and register design. They also rely on a compiler to obtain this level of IPC, whereas Levo does not.

In this paper we have described the Levo machine model. We have illustrated the power of resource flow and especially D-path execution. We have been successful in obtaining IPCs above 10.

We still believe that there remains substantial ILP to be obtained. In Table 3 we select the configuration that obtained the best D-path result (8,8,8) and show IPC speedup (relative to our D-path result in Figure 4) using an *Oracle predictor* [9]. As we can see, there still remains a lot of IPC that can be obtained

Table 3. IPC Speedup for Oracle branch prediction for an 8,8,8 configuration with D-Cache and Perfect Memory. Speedup is relative to the D-CM and D-PM results in Figure 4.

Benchmark	IPC Speedup Factor D-Cache Memory	IPC Speedup Factor Perfect Memory
go	1.6	1.8
bzip	2.1	2.4
gzip	1.5	1.6
parser	1.4	1.4
gap	1.0	1.1

through improved control flow speculation. We plan to look at spawning dynamic paths (versus the static path approach described in this work).

References

1. Austin T.M and Sohi G.S. Dynamic Dependency Analysis of Ordinary Programs. In *Proceedings of ISCA-19*, pages 342–351, May 1992.
2. Chen T.F. Supporting Highly Speculative Execution via Adapative Branch Trees. In *Proceedings of the 4th Annual International Symposium on High Performance Computer Architecture*, pages 185–194, January 1998.
3. Cher C.-Y. and Vijaykumar T.N. Skipper: A Microarchitecture For Exploiting Control-Flow Indepdendence. In *Proceedings of MICRO-34*, December 2001.
4. Cleary J.G, Pearson M.W and Kinawi H. The Architecture of an Optimistic CPU: The Warp Engine. In *Proceedings of HICSS*, pages 163–172, January 1995.
5. Franklin M. and Sohi G.S. Register Traffic Analysis for Streamlining Inter-Operation Communication in Fine-Grain Parallel Processors. In *Proceedings of MICRO-25*, pages 236–247, Dec 1992.
6. Henry D.S and Kuszmaul B.C. and Loh G.H. and Sami R. Circuits for Wide-Window Superscalar Processors. In *Proceedings of ISCA-27*, pages 236–247. ACM, June 2000.
7. Khalafi A., Morano D., Uht A. and Kaeli D. Multipath Execution on a Large-Scale Distributed Microarchitecture. Technical Report 022002-001, University of Rhode Island, Department of Electrical and Computer Engineering, Feb 2002.
8. Klauser A., Austin T., Grunwald D. and Calder B. Dynamic Hammock Predication for Non-Predicated Instruction Set Architectures. In *Proceedings of PACT*, pages 278–285, 1998.
9. Lam M.S. and Wilson R.P. Limits of Control Flow on Parallelism. In *Proceedings of ISCA-19*, pages 46–57. ACM, May 1992.
10. Lipasti M.H and Shen J.P. Superarchitecture Microarchitecture for Beyond AD 200. *IEEE Computer Magazine*, 30(9), September 1997.
11. McFarling S. Combining Branch Predictors. Technical Report DEC WRL TN-36, Digital Equipment Western Research Laboratory, June 1993.
12. Nagarajan R. and Sankaralingam K. and Burger D. and Keckler S. A Design Space Evaluation of Grid Processor Architectures. In *Proceedings of MICRO-30*, December 2001.
13. Tomasulo R.M. An Efficient Algorithm for Exploiting Multiple Arithmetic Units. *IBM Journal of Research and Development*, 11(1):25–33, Jan 1967.
14. Tubella J. and Gonzalez A. Control speculation in multithreaded processors through dynamic loop detection. In *Proceedings of HPCA-4*, pages 14–23, January 1998.
15. Uht, A. K. and Sindagi, V. Disjoint Eager Execution: An Optimal Form of Speculative Execution. In *Proceedings of MICRO-28*, pages 313–325. ACM-IEEE, November/December 1995.
16. Uht A., Morano D., Khalafi A., Wenisch T., Ashouei M. and Kaeli D. IPC in the 10's via Resource Flow Computing with Levo. Technical Report 092001-001, Unversity of Rhode Island, Department of Electrical and Computer Engineering, Sept 2001.
17. Wall D.W. Limits of Instruction-Level Parallelism. In *Proceedings of ASPLOS-4*, pages 176–188. ACM, April 1991.

A Register File Architecture and Compilation Scheme for Clustered ILP Processors*

Krishnan Kailas[1], Manoj Franklin[2], and Kemal Ebcioğlu[1]

[1] IBM T. J. Watson Research Center, Yorktown Heights, NY, U.S.A.,
{krish, kemal}@watson.ibm.com

[2] Department of ECE, University of Maryland, College Park, MD, U.S.A.,
manoj@eng.umd.edu

Abstract. In Clustered Instruction-level Parallel (ILP) processors, the function units are partitioned and resources such as register file and cache are either partitioned or replicated and then grouped together into on-chip clusters. We present a novel partitioned register file architecture for clustered ILP processors which exploits the temporal locality of references to remote registers in a cluster and combines multiple inter-cluster communication operations into a single broadcast operation using a new `sendb` instruction. Our scheme makes use of a small *Caching Register Buffer* (CRB) attached to the traditional partitioned local register file, which is used to store copies of remote registers. We present an efficient code generation algorithm to schedule `sendb` operations on-the-fly. Detailed experimental results show that a windowed CRB with just 4 entries provides the same performance as that of a partitioned register file with infinite non-architected register space for keeping remote registers.

1 Introduction

Registers are the primary means of inter-operation communication and inter-operation dependency specification in an ILP processor. In contemporary processors, as the issue width increase, additional register file ports are required to cater to the large number of function units. Further, a large number of registers are required for exploiting many aggressive ILP compilation techniques, as well as for reducing the memory traffic due to spilling. Unfortunately, the area, access delay, and power dissipation of the register file grows rapidly with the number of ALUs [1]. Clearly, huge monolithic register files with large number of ports are either impractical to build or limit the cycle time of the processor. The partitioned and replicated register file structures used in clustered ILP processors [1–8] are two promising approaches to effectively address the issues related to large number of ports, area and power of register files. A typical cluster consists of a set of function units and a local register file, as shown in Fig. 1. A local register file can provide faster access times than its monolithic counter-part, because each local register file has to cater only to a subset of all the function

* Supported in part by U.S. National Science Foundation (NSF) through a CAREER grant (MIP 9702569) and a regular grant (CCR 0073582).

B. Monien and R. Feldmann (Eds.): Euro-Par 2002, LNCS 2400, pp. 500–511.

Fig. 1. Generic Clustered ILP Processor Model

units in the processor. If the register file is *replicated* (each local register file shares the entire architected name space), then copy operations are required to maintain the coherency among the local register files. On the other hand, if the register file is *partitioned* (each local register file has a non-intersecting subset of the architected name space), then copy operations are needed to access registers from remote clusters. The main advantage of a partitioned register file over a replicated register file is that it can reduce the number of write ports as well as the size of the local register file, thereby providing shorter access delays, smaller area and lower power requirements. In this paper, we shall concentrate on partitioned register file architectures for statically scheduled processors.

In general, the partitioned register file schemes used in statically scheduled clustered ILP processors belong to one of the following two categories:

- Registers are copied to remote clusters ahead of their use to hide the inter-cluster communication delay (eg. Multiflow Trace [9], M-machine [10], and Limited Connectivity VLIW [11]). Often several copies of a register are maintained in the local register files of several clusters, increasing the register pressure due to inefficient use of architected name space in this scheme.
- Remote registers are requested either on demand (eg. TI's clustered DSPs[1] [4]) or using send-receive operations (eg. Lx [2]), each time they are referenced. The potential drawback of this approach is that it demands more inter-cluster communication bandwidth, and hence may possibly delay the earliest time a remote register can be used when the interconnect is saturated.

In this paper, we propose a new partitioned register file with a *Caching Register Buffer (CRB)* structure, which tries to overcome the above drawbacks. The CRB is explicitly managed by the compiler using a single new primitive instruction called `sendb`.

The rest of this paper is organized as follows. Section 2 describes our CRB-based partitioned register file scheme, and discuss some related programming and hardware complexity issues. In section 3, we briefly discuss the code generation framework used and present an on-the-fly scheduling algorithm to explicitly manage the CRB using the new `sendb` operations. The experimental evaluation of the proposed CRB-based partitioned register scheme is discussed in section 4. We present related work in section 5, followed by conclusion in section 6.

[1] In addition, TMS320C6x series DSPs also support `rcopy` OPs described in section 4.

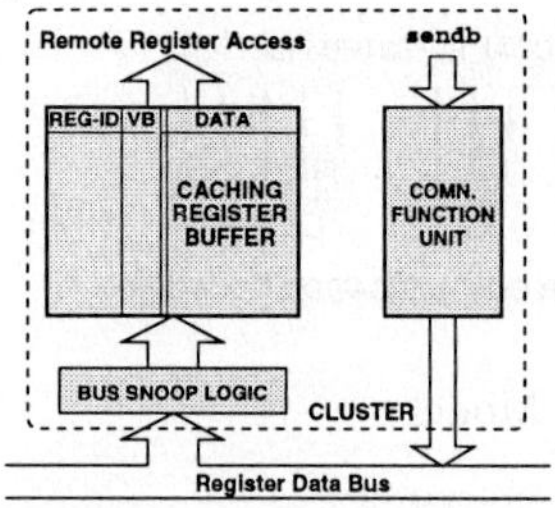

Fig. 2. Caching Register Buffer and Communication Function Unit

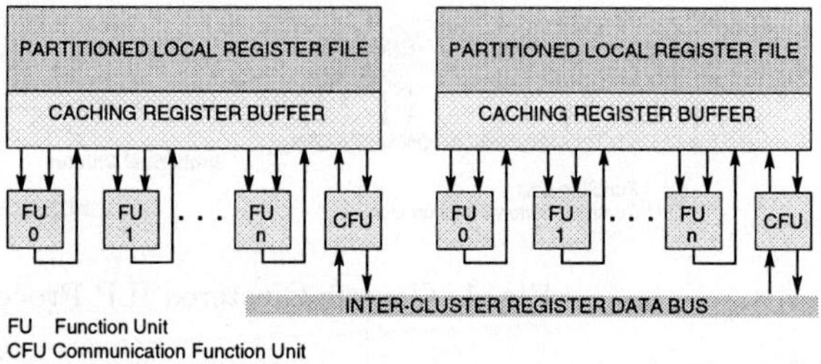

Fig. 3. A 2-cluster processor with CRB-based partitioned register file

2 Partitioned Register File with CRB

The basic idea behind our scheme is to use compile-time techniques to exploit the temporal locality of references to remote registers in a cluster. Instead of allocating and copying a remote register value to a local register (and thereby wasting an architected register name), the remote register contents are stored in a *Caching Register Buffer* (CRB), which is a fast local buffer for caching the frequently accessed remote registers ahead of their use. The CRB is therefore functionally similar to a small fully-associative cache memory, and it may be implemented using an array of registers. There is a tag field (REG-ID) and a one bit flag (VALID-BIT) associated with each CRB data entry as shown in Fig. 2. The tag (REG-ID) contains the unique address (architected name) of the cached remote register, and the VALID-BIT is used for validating the data and tag.

An operation trying to access a remote register first performs an associative search in the REG-ID array. If a valid copy of the remote register being accessed is available in the CRB, then the data value is returned from the appropriate CRB data register (akin to a cache hit). A hardware-initiated *send-stall* bubble is injected into the pipeline automatically if a valid copy of the requested remote register is not found in the CRB. During the send-stall cycle, which is similar to a cache miss, the data is loaded from the remote register file and the execution of the pipe is resumed.

Any local register can be sent to the CRBs of remote clusters via an explicit *send-broadcast* (`sendb`) operation using a *communication function unit* (CFU). The CRBs and the CFUs of all clusters are interconnected via one or more *register data buses* as shown in Fig. 3. All the resources involved in inter-cluster communication — CRBs, CFUs, and register data buses — are reserved on a cycle-by-cycle basis at compile-time. The `sendb` operation (OP) is scheduled on the CFU by the compiler such that the remote register is cached in the CRB of a remote cluster ahead of its use. The `sendb` OP has the following format:

`sendb`	`SrcRegID`	`ClusterIDbitvector`

where `SrcRegID` is the ID of the register that has to be sent to other clusters, and `ClusterIDbitvector` is a bit vector with bit positions indicating the remote cluster IDs to which the register value is to be sent. The width of the `ClusterIDbitvector` is therefore equal to the number of clusters in the processor. The CRB's bus interface unit snoops the register data bus; if the bit position in `ClusterIDbitvector` matches its cluster ID then it updates the appropriate CRB data register, concurrently initializing its tag REG-ID entry and setting its VALID-BIT to 1.

2.1 Precise Exceptions and Calling Conventions

An important issue that needs to be addressed is how to support precise exceptions in the presence of a CRB. Only the contents of the architected registers (the local partitioned register file) need to be saved on an interrupt by the interrupt handler code. In order to make the CRB structure transparent to the user program, the interrupt return (`rfi`) OP will invalidate the contents of all CRBs by resetting all VALID-BITs to 0. Therefore, after returning from the interrupt handler, if a valid copy of the requested remote register is not found in the CRB, it is loaded directly from the register file during send-stall cycles. An alternative approach is to save and restore the entire CRB image on an interrupt. However, this approach may result in large interrupt latencies and slower context switches.

In addition to physical partitioning, there is also a *logical partitioning* of register name space due to calling conventions. The logical partitioning due to calling conventions should co-exist with the physical partitioning in the new partitioned register file with CRB as well. We use a partitioning scheme in which both caller-save and callee-save registers are partitioned and distributed equally among all clusters [8]. However, the logical partitioning of register name space and the calling convention provide only a protocol to preserve the callee-save registers across function calls. This brings up the important issue of dealing with the changes in the cached callee-save registers in the CRBs across function calls. Note that, for the correct execution of the program, the data in any valid CRB register and the architected register specified by its REG-ID tag should be the same. A simple solution to this problem is to invalidate the CRB contents upon return from a function call and load a copy of the register directly into the CRB and function unit. Such send-stall cycles can be completely eliminated by scheduling a `sendb` OP after the `call` OP for each one of the registers that are live across the function call. Our preliminary experiments showed that such schemes can either increase the number of send-stall cycles or increase the code size as well as schedule length, especially for function-call-intensive benchmark programs. This led us to the following hardware-based solution. The basic idea is to use "windowed" CRBs, which allow saving different versions of the CRB contents in different register windows. Like any register windowing scheme, such as SUN PicoJava processor's "dribbling" mechanism, each call nesting level selects a different CRB window automatically. When the nesting level of calls exceeds the maximum number of hardware CRB windows in the processor, the contents of the oldest CRB window are flushed to cache.

The number of ports in a typical CRB structure is less than the number of ports in a partitioned register file [8]. Because we need only a 4-way associative search logic for good performance (based on our simulation results presented in section 4), one can argue that it is feasible to build a windowed CRB structure that has the same access time of the partitioned register file attached to it. However, CRB may need a larger area than a traditional register file.

3 Compilation Scheme

3.1 Overview of Code Generation Framework

We used the CARS code generation framework [12, 8] which combines the cluster assignment, register allocation, and instruction scheduling phases to generate efficient code for clustered ILP processors. The input to the code generator is a dependence flow graph (DFG) [13] with nodes representing OPs and directed edges representing data/control flow. The basic scheduling unit can be a set of OPs created using any "region" formation or basic block grouping heuristics.

During code generation, CARS dynamically partitions the OPs in the scheduling region into a set of mutually exclusive aggregates (groups) — `unscheduled`, `ready list`, and `vliws`. The data ready nodes in the `unscheduled` aggregate are identified and moved into the `ready list`. The nodes in the `ready list` are selected based on some heuristics for combined cluster assignment, register allocation and instruction scheduling (henceforth referred to as CARS*cheduling*). CARS also performs on-the-fly global register allocation using usage counts during CARScheduling. This process is repeated until all the nodes of the DFG are CARScheduled and moved to the appropriate `vliws`.

The CARS algorithm extends the list-scheduling algorithm [14]. In order to find the best cluster to schedule an OP selected from the `ready list`, the algorithm first computes the resource-constrained *schedule_cycle* in which the OP can be scheduled in each cluster, and then the minimum value of *schedule_cycle* as *earliest_cycle*. Based on the *earliest_cycle* computed, the OP is either scheduled in the current cycle on one of the clusters corresponding to *earliest_cycle* or pushed back into the `ready list`. Often inter-cluster copy OPs have to be inserted in the DFG and retroactively scheduled in order to access operands residing in remote clusters. An operation-driven version of the CARS algorithm is used for this purpose. In the list-scheduling mode, only the most recently scheduled `vliws` aggregate's resource availability is searched to make a cluster-scheduling decision. On the other hand, in the operation driven scheduling mode[2], resource availability in a set of `vliws` aggregates are considered to make the cluster-scheduling decision. In order to find the best VLIW to schedule the inter-cluster copy OP, the algorithm searches all the `vliws` aggregates starting from the Def cycle of the operand (or from the cycle in which the join node of the current region is scheduled, if operand is not defined in the current region) to the current cycle. In the next section, we describe how we have extended the CARS algorithm for scheduling `sendb` operations on-the-fly.

[2] Due to the large search space involved with operation-driven scheduling, use of this mode has been restricted to scheduling inter-cluster copy OPs and spill store OPs.

3.2 On-the-fly Scheduling of sendb Operations

Our basic approach is to schedule a `sendb` OP whenever it is beneficial to schedule a node on a cluster in which one or both of its remote source registers are not cached in the CRB and the inter-cluster communication bus is available for broadcasting the register value. We also want to combine multiple inter-cluster copy operations into a single `sendb` OP on-the-fly while CARScheduling nodes in the DFG in topological order, without back-tracking. In order to combine multiple inter-cluster copy OPs, we keep track of `sendb` OPs scheduled for accessing each live register. The information is maintained by the code generator as a list of `sendb` OPs inside each live physical register's resource structure.

Our scheme may be explained using an example of scheduling nodes in a basic block. Assume that a `sendb` OP has already been scheduled for accessing a remote register, say `r10`, from a cluster C_1, and that we want to use the same register `r10` in a different cluster C_2 at some time later while scheduling a different Use-node of the Def mapped to `r10`. Our algorithm first checks if there are any `sendb` OPs scheduled for register `r10`. As there is one such `sendb` OP scheduled already for `r10`, we simply set the bit corresponding to the new destination cluster C_2 in the `ClusterIDbitvector` operand of the `sendb` OP (instead of scheduling a new inter-cluster operation as in traditional schemes). The process is repeated whenever a Use-node of the Def mapped to `r10` needs the register in a different cluster. Therefore, by the end of scheduling all the OPs in DFG, the algorithm would have combined all the inter-cluster copy OPs into the single `sendb` OP by incrementally updating the `ClusterIDbitvector` operand of the `sendb` OP.

The scheme discussed above for scheduling `sendb` OPs in a basic block can be extended, as shown in Algorithm 1, to the general case of scheduling `sendb` OPs for accessing a remote register assigned to a web of Def-Use chains (corresponding to a life that spans multiple basic blocks). The nodes that use such registers that are allocated to webs of Def-Use chains may have multiple reaching definitions. Clearly, there are two options: either schedule a `sendb` OP for each one of the Def nodes that can be reached from the Use node, or schedule one `sendb` OP before the Use node in the current basic block. Our preliminary experiments showed that former is attractive only when there are empty issue slots in already scheduled VLIWs (low ILP applications), because if there are no issue slots then we have to open a new VLIW just for scheduling a `sendb` OP. Moreover, if there are not many remote uses for the register, then the latter option is better than scheduling multiple `sendb` OPs. In view of these observations, our algorithm uses the following scheme to reduce the number of `sendb` OPs. If `sendb` OPs have already been scheduled for all the reaching definitions corresponding to a Use, then we update the `ClusterIDbitvector` of all those `sendb` OPs (lines 9-12). Otherwise, we schedule a `sendb` OP in the current basic block using the operation-driven version of CARS algorithm (line 5) [8].

Algorithm 1 Algorithm for CARScheduling `sendb` OPs

```
SCHEDULE-SENDB(Op, RegId, ClusterID)
/* Op is an OP trying to access remote register≡RegId in cluster≡ClusterID */
find LifeInfo of the life to which register RegId is currently assigned
{LifeInfo is a list of Defs in the web of Def-Use chain corresponding to a life (live
range) in DFG}
find the subset of Defs SrcDefs ∈ LifeInfo that can be reached from Op
for each Def D in SrcDefs do
  if no sendb OP is scheduled for D then
    schedule a new sendb OP for D using OPERATION-CARS in the current block
    return
  end if
end for
for each Def D in SrcDefs do
  find the sendb OP S scheduled for D
  set the bit corresponding to ClusterID in ClusterIDbitvector of S
end for
```

4 Experimental Results

We used the chameleon VLIW research compiler [15] (gcc version) with a new back-end based on the CARS code generation framework [12] for the experimental evaluation of our new partitioned register file scheme. We have developed a cycle-accurate simulator to model the clustered ILP processor with CRB and inter-cluster buses. We use the *compiled simulation* approach: each VLIW instruction is instrumented and translated into PowerPC assembly code that calls the CRB simulator. For each VLIW instruction, the simulator processes all the local and remote register file access requests, updates the CRB contents and keeps track of the number of CRB hits and misses based on whether the requested register was found in the CRB or not. The send-stall cycles are computed based on the assumption that multiple CRB misses are serviced serially subject to the availability of inter-cluster buses. FIFO replacement policy is used in the CRB.

We compared the CRB-`sendb` scheme with the traditional partitioned register file scheme using inter-cluster copy (`rcopy`) OPs. We generated executable compiled simulation binaries for the 14 benchmark programs used (8 from SPEC CINT95, 2 from SPEC CINT2000, and 4 from MediaBench). The binaries are run to completion with the respective input data sets of the benchmark programs. We used the execution time of benchmark programs in cycles as a metric to compare the performance of 4 different clustered ILP processors listed in Table 1. Infinite cache models were used for all configurations and the penalty for flushing/restoring a CRB window is assumed to be 5 cycles. The `rcopy` OPs are semantically equivalent to register `copy` OPs; however their source and destination registers reside in different clusters. For each remote register access, an `rcopy` OP is scheduled on the destination cluster to copy a register from a remote register file to an *infinitely* large non-architected name space in the local

Table 1. Configurations of different 8 ALU processors with partitioned register files studied.[1] Each CRB has 4 windows.

Configuration		Local resources per cluster				Global resources	
		ALUs	GPRs	FPRs	CRB/buf[1]	CCRs	Buses
single cluster (base)		8	64	64	-	16	
`rcopy`	1 bus 4 clusters	2	16	16	∞	16	1
	1 bus 2 clusters	4	32	32	∞	16	1
	2 buses 4 clusters	2	16	16	∞	16	2
	2 buses 2 clusters	4	32	32	∞	16	2
`sendb` + 4/8/16-entry CRB	1 bus 4 clusters	2	16	16	[4/8/16]x4	16	1
	1 bus 2 clusters	4	32	32	[4/8/16]x4	16	1
	2 buses 4 clusters	2	16	16	[4/8/16]x4	16	2
	2 buses 2 clusters	4	32	32	[4/8/16]x4	16	2

register file. The rationale is to compare the performance of CRB-`sendb` scheme with the (unrealistic) upper-bound on performance that may be achieved via traditional partitioned register file scheme using `rcopy` OPs. We studied three partitioned register file configurations with CRB sizes 4, 8 and 16.

The clustered machines are configured such that the issue width and resources of the base machine are evenly divided and assigned to each cluster as shown in Table 1. Condition code registers (CCRs) and buses are treated as shared resources. The function units, which are fully pipelined, have the following latencies: Fix, Branch, and Communication: 1 cycle; Load and FP: 2 cycles; FP divide and sqrt: 9 cycles. Figures 4, 5, 6 and 7 show the speed up (ratio of execution times) with respect to the base single cluster processor for the four different 8 ALU clustered processors for all the benchmarks studied. In 2-cluster configurations, CRB-`sendb` scheme does not provide any significant performance improvement for most of the benchmarks. This is mainly because, in a 2-cluster

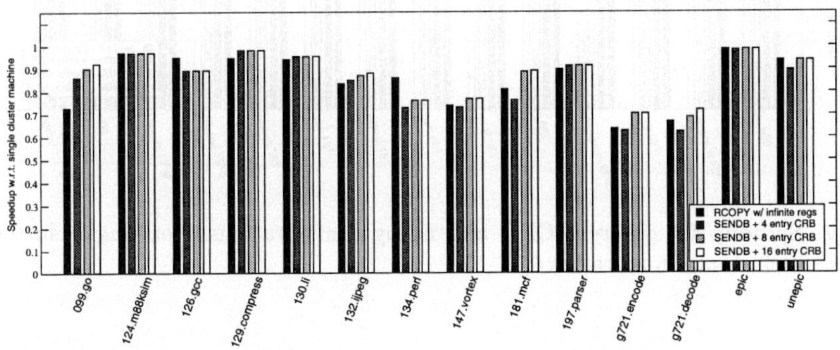

Fig. 4. Speedup for different CRB and `rcopy` configurations over single cluster for 1 bus 2 clusters configuration

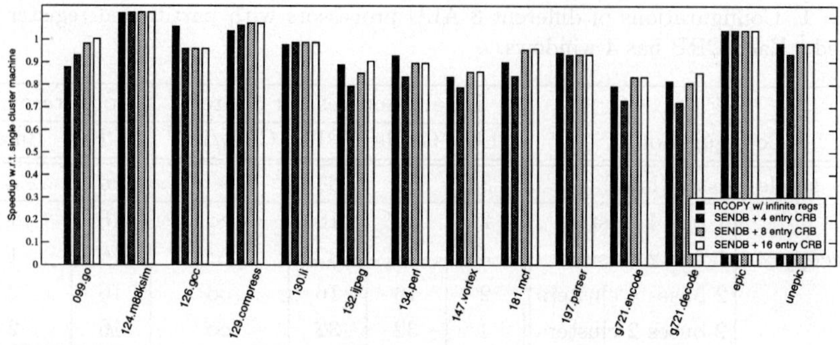

Fig. 5. Speedup for different CRB and `rcopy` configurations over single cluster for 2 buses 2 clusters configuration

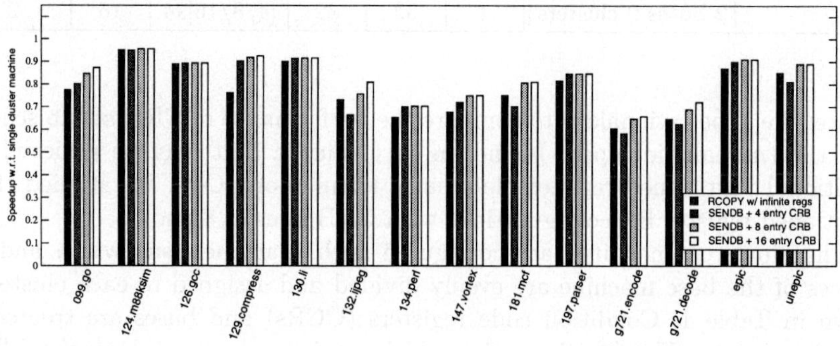

Fig. 6. Speedup for different CRB and `rcopy` configurations over single cluster for 1 bus 4 clusters configuration

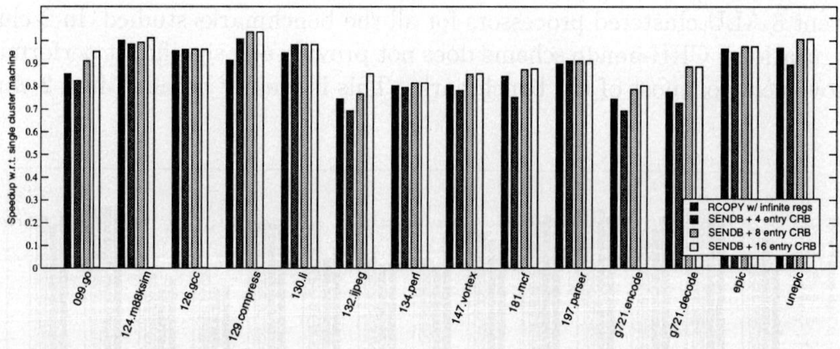

Fig. 7. Speedup for different CRB and `rcopy` configurations over single cluster for 2 buses 4 clusters configuration

machine, the `sendb` OP degenerates into an `rcopy` OP because there is only one target cluster. However, in the 4-cluster configurations, the CRB-`sendb` scheme outperforms the traditional `rcopy` scheme. In 4-cluster configurations, on an average an additional speedup of 2%, 4.9%, and 5.7% are observed over `rcopy` scheme when 4, 8 and 16-entry CRB is used with partitioned register file respectively. Speedup higher than one was observed on clustered machines for some benchmarks (124.m88ksim). This is primarily due to the non-linearity of the cluster-scheduling process and also due to the aggressive peephole optimizations done after CARScheduling; similar observations have been reported for the code generated using other algorithms [12]. For some benchmarks, especially for 126.gcc and 134.perl, the `rcopy` scheme is observed to perform better than `sendb` scheme. This is because of the combined effect of factors such as finite size of CRB, and redundant sendb OPs scheduled for some long live ranges with multiple reaching Defs. It may be possible to eliminate these `sendb` OPs (or replace them with a single local `sendb` OP) by making an additional pass on the scheduled code performing a data flow analysis. A partitioned register file with only a 4-entry CRB is therefore sufficient to achieve a performance identical to a clustered machine with an infinitely large local register space for keeping remote registers. Note that the rcopy experiments provide a conservative comparison: we have not modeled the effects of increase in register pressure, the potential possibility of spills due to the increased register pressure, and the impact of increase in code size in our `rcopy` experiments. These effects would have further lowered the performance of the `rcopy` scheme reported herein.

5 Related Work

The partitioned register file structures used in prior clustered ILP processor architectures [9, 11, 4, 10, 2] are different from our CRB-based scheme as explained in section 1. Caching of processor registers in general is not a new concept and has been used in different contexts such as the Named-State Register file [16] and Register-Use Cache [17] for multi-threaded processors, the pipelined register cache [18], and Register Scoreboard and Cache [19]. Our scheme is different from these as none of them are aimed at keeping a local copy of a remote register.

Fernandes *et al.* [20] proposed a partitioned register file scheme in which individually addressable registers are replaced by queues. In contrast, our scheme permits random access to the cached remote registers in the CRB, and also provides more freedom to the cluster scheduler for scheduling the Use-nodes of remote registers in any arbitrary order limited only by true data dependencies. Llosa *et al.* proposed a dual register file scheme which consists of two replicated, yet not fully consistent register files [21]. Cruz *et al.* proposed a multiple-bank register file [22] for dynamically scheduled processors in which the architected registers are assigned at run time to multiple register banks. Zalamea *et al.* have proposed a two-level hierarchical register file [23] explicitly managed using two new spill instructions. In contrast, CRB is transparent to the register allocator; it is not assigned any architected name-space, which in turn allows it to cache any architected register. Due to space limitations, readers are referred to [8] for a more comprehensive list of related work and comparison of those with our scheme.

6 Conclusion

We presented a new partitioned register file architecture that uses a caching register buffer (CRB) structure to cache the frequently accessed remote registers without using any architected registers in the local register file. The scheme reduces register pressure without increasing the architected name space (which would have necessitated increasing the number of bits used for specifying register operands in instruction encoding). The CRB structure requires only one write port per inter-cluster communication bus, and the partitioned register file with CRB can be realized without affecting register file access time of a clustered ILP processor. The CRB is managed explicitly by the compiler using a new `sendb` operation. The `sendb` OP can update multiple CRBs concurrently via broadcasting the register value over inter-cluster communication bus/network. The `sendb` OP can thus eliminate multiple send-receive OP pairs used in prior schemes. It can also combine several inter-cluster copy OPs used by the `rcopy`-based prior partitioned register file schemes. Experimental results indicate that in a clustered ILP processor with non-trivial number of clusters, a 4-entry windowed CRB structure can provide the same performance as that of a partitioned register file with infinite non-architected register space for keeping remote registers. Because clustered ILP processors are an attractive complexity-effective alternative to contemporary monolithic ILP processors, these results are important in the design of future processors.

References

1. V. Zyuban and P. M. Kogge, "Inherently Lower-Power High-Performance Superscalar Architectures," *IEEE Trans. on Computers*, vol. 50, pp. 268–285, Mar. 2001.
2. P. Faraboschi, J. Fisher, G. Brown, G. Desoli, and F. Homewood, "Lx: A Technology Platform for Customizable VLIW Embedded Processing," in *Proceedings of the 27th International Symposium on Computer Architecture*, June 2000.
3. K. Ebcioğlu, J. Fritts, S. Kosonocky, M. Gschwind, E. Altman, K. Kailas, and T. Bright, "An Eight-Issue Tree-VLIW Processor for Dynamic Binary Translation," in *Proc. of Int. Conf. on Computer Design (ICCD'98)*, pp. 488–495, 1998.
4. Texas Instruments, Inc., *TMS320C62x/C67x Technical Brief*, Apr. 1998.
5. J. Fridman and Z. Greenfield, "The TigerSHARC DSP architecture," *IEEE Micro*, vol. 20, pp. 66–76, Jan./Feb. 2000.
6. R. E. Kessler, "The Alpha 21264 microprocessor," *IEEE Micro*, vol. 19, pp. 24–36, Mar./Apr. 1999.
7. R. Canal, J. M. Parcerisa, and A. Gonzalez, "Dynamic cluster assignment mechanisms," in *Proc. of the 6th Int. Conference on High-Performance Computer Architecture (HPCA-6)*, pp. 133–142, Jan. 2000.
8. K. Kailas, *Microarchitecture and Compilation Support for Clustered ILP Processors.* PhD thesis, Dept. of ECE, University of Maryland, College Park, Mar 2001.
9. R. P. Colwell *et al.*, "A VLIW architecture for a trace scheduling compiler," *IEEE Transactions on Computers*, vol. C-37, pp. 967–979, Aug. 1988.
10. S. Keckler, W. Dally, D. Maskit, N. Carter, A. Chang, and W. Lee, "Exploiting fine-grain thread level parallelism on the MIT Multi-ALU processor," in *Proc. of the 25th Annual Int. Symposium on Computer Architecture*, pp. 306–317, 1998.

11. A. Capitanio, N. Dutt, and A. Nicolau, "Partitioned register files for VLIWs: A preliminary analysis of tradeoffs," in *Proceedings of the 25th Annual International Symposium on Microarchitecture*, pp. 292–300, Dec. 1–4, 1992.
12. K. Kailas, K. Ebcioğlu, and A. Agrawala, "CARS: A New Code Generation Framework for Clustered ILP Processors," in *Proceedings of the 7th International Symposium on High-Performance Computer Architecture (HPCA-7)*, pp. 133–143, 2001.
13. K. Pingali, M. Beck, R. Johnson, M. Moudgill, and P. Stodghill, "Dependence flow graphs: an algebraic approach to program dependencies," in *Proc. of the 18th annual ACM symposium on Principles of programming languages*, pp. 67–78, 1991.
14. R. Sethi, *Algorithms for minimal-length schedules*, ch. 2. Computer and job-shop scheduling theory (E. G. Coffman, ed.), John Wiley & Sons, Inc., New York., 1976.
15. M. Moudgill, "Implementing an Experimental VLIW Compiler," *IEEE Technical Committee on Computer Architecture Newsletter*, pp. 39–40, June 1997.
16. P. R. Nuth, *The Named-State Register File.* PhD thesis, MIT, AI Lab, Aug. 1993.
17. H. H. J. Hum, K. B. Theobald, and G. R. Gao, "Building multithreaded architectures with off-the-shelf microprocessors," in *Proceedings of the 8th International Symposium on Parallel Processing*, pp. 288–297, 1994.
18. E. H. Jensen, "Pipelined register cache." U.S. Patent No. 5,117,493, May 1992.
19. R. Yung and N. C. Wilhelm, "Caching processor general registers," in *International Conference on Computer Design*, pp. 307–312, 1995.
20. M. M. Fernandes, J. Llosa, and N. Topham, "Extending a VLIW Architecture Model," technical report ECS-CSG-34-97, Dept. of CS, Edinburgh University, 1997.
21. J. Llosa, M. Valero, and E. Ayguade, "Non-consistent dual register files to reduce register pressure," in *Proceedings of the First International Symposium on High-Performance Computer Architecture*, pp. 22–31, 1995.
22. J.-L. Cruz, A. González, M. Valero, and N. P. Topham, "Multiple-banked register file architectures," in *Proceedings of the 27th Annual International Symposium on Computer Architecture*, pp. 316–325, 2000.
23. J. Zalamea, J. Llosa, E. Ayguade, and M. Valero, "Two-level hierarchical register file organization for VLIW processors," in *Proceedings of the 33rd Annual International Symposium on Microarchitecture (MICRO-33)*, 2000.

A Comparative Study of Redundancy in Trace Caches

Hans Vandierendonck[1], Alex Ramírez[2],
Koen De Bosschere[1], and Mateo Valero[2]

[1] Dept. of Electronics and Information Systems, Ghent University,
Sint-Pietersnieuwstraat 41, B-9000 Gent, Belgium,
{hvdieren,kdb}@elis.rug.ac.be

[2] Computer Architecture Department, Universitat Politècnica de Catalunya,
c/ Jordi Girona 1-3, Module D6, 08034 Barcelona, Spain,
{aramirez,mateo}@ac.upc.es

Abstract. Trace cache performance is limited by two types of redundancy: duplication and liveness. In this paper, we show that duplication is not strongly correlated to trace cache performance. Generally, the best-performing trace caches also introduce the most duplication. The amount of dead traces is extremely high, ranging from 76% in the smallest trace cache to 35% in the largest trace cache studied. Furthermore, most of these dead traces are never used between storing them and replacing them from the trace cache.

1 Introduction

The performance of wide-issue superscalar processors is limited by the the rate at which the fetch unit can supply useful instructions. The trace cache [1] can fetch across multiple control transfers per clock cycle without increasing the latency of the fetch unit [2]. Fundamental to the operation of the trace cache is that consecutively executed instructions are *copied* into consecutive locations, such that fetching is facilitated.

In this paper, we analyse the relation between redundancy and trace cache performance. We use the term redundancy to mean *those traces and instructions stored in the trace cache that do not contribute to performance.* We consider two types of redundancy. The first type of redundancy detects multiple copies of the same instruction and is called *duplication.* The second type of redundancy is *liveness.* Dead traces are redundant (need not be stored in the trace cache), because they do not contribute to performance. We show in this paper that liveness is a greater concern in small trace caches than duplication. We study various ways to build traces and inspect their effect on redundancy.

2 Metrics

After every trace cache access, we make a break-down of the contents of the trace cache and average this break-down over all accesses. We consider two break-downs: one to quantify duplication and one to quantify liveness.

B. Monien and R. Feldmann (Eds.): Euro-Par 2002, LNCS 2400, pp. 512–516.

We measure 4 categories to characterise duplication. **Unused** traces correspond to trace cache frames that are never filled during the execution of the programs. Fragmentation (**frag**) is caused by traces shorter than the maximum length of 16 instructions. Duplicated intructions (**dup**) are those instructions for which there is another copy of the same static instruction in the trace cache. The first copy of each static instruction is present in the category **uniq**. This part of the trace cache would also be present in a regular instruction cache.

When studying liveness, we consider following types of traces. **Unused** traces are the same as in the previous break-down. **Quick-dead** traces are never used between being built and being replaced. **Dead** traces are stored in the trace cache but will not be used again before being replaced. Other traces are **live**.

3 Methodology

We measure the effect of the trace termination policy on duplication and liveness. The baseline policy base allows traces to grow up to 16 instructions and 3 branches. The whole blocks policy wb requires that a trace ends in a branch instruction, unless a basic block is longer than 16 instructions. The backward branch policy bb terminates traces in a backward branch only. Other variations are the *software trace cache* (STC) [3], *selective trace storage* (STS) [4] and the combination of STS and STC, abbreviated STCS. STS discriminates between blue traces that have all their internal branches not-taken and red traces which contain taken branches. Blue traces can be fetched from the instruction cache in a single cycle and need not be stored in the trace cache. With STC, the program is rewritten to bias branches towards not-taken, resulting in more blue traces.

The analysis is performed on 6 SPECint95 benchmarks running training inputs. At most 1 billion instructions are simulated per benchmark. We use a 32 kB 2-way set-associative instruction cache with 32 byte blocks. It can fetch at most one taken branch and 16 instructions per cycle. We use an MGag multiple branch predictor with 16K entries and a 2K-entry 4-way set-associative BTB and a 256-entry RAS. The trace cache size is varied from 32-4096 traces and is always 4-way set-associative. A trace can hold at most 16 instructions and 3 conditional branches.

4 Analysis

The performance is measured as FIPA: useful number of fetched instructions per fetch unit access (Figure 1). The performance results follow the trends known from the literature [3–5].

4.1 Duplication in the Trace Cache

The break-down of duplication (Figure 2) shows that **unused** trace cache frames only occur often for the biggest trace cache investigated, because for most programs the number of different traces is around 4K or less.

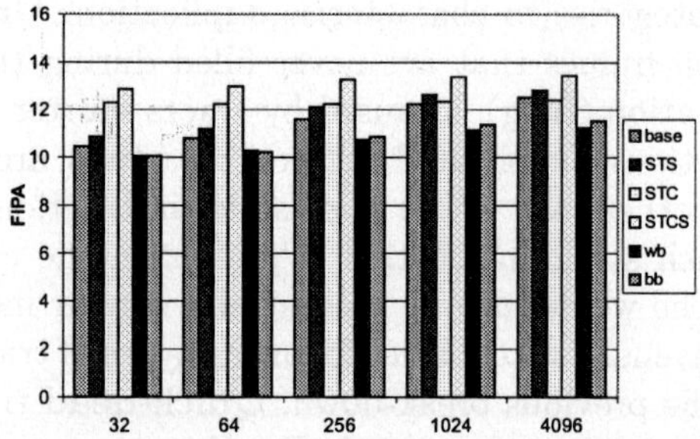

Fig. 1. FIPA for the trace cache. The horizontal axis shows the size of the trace cache expressed in traces.

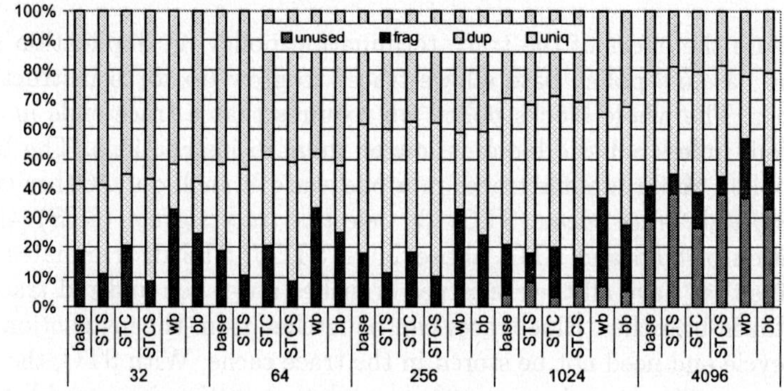

Fig. 2. Breakdown of wasted instruction frames in the trace cache.

At least 10% of the trace cache is wasted on **fragmentation**. The impact of fragmentation is largely independent of the trace cache size. STS and STCS reduce fragmentation because red traces are longer than blue traces [4]. The wb and bb policies increase fragmentation. Because the traces are terminated on all or some branch instructions, they are a lot shorter than with the base policy. For this reason, the performance of the wb and bb policies is lower than with the other policies.

Duplication varies strongly over the different trace termination policies and trace cache sizes. In the base case, 20% to 50% of the trace cache stores duplicates of instructions already present in the trace cache. This number is the largest for the 1K-entry trace cache. For the larger trace cache it is less. Instead, trace cache frames are left empty. This trend occurs for all trace termination policies.

STS increases duplication over the base policy, e.g.: from 43% to 48% in the 256-entry trace. Duplication is increased because the STS policy does not store blue traces but stores red traces instead. The red traces are longer and hence additional copies of instructions are inserted in the trace cache. This way, fragmented space is converted into duplicated instructions. When the trace cache is very large (e.g.: 4K traces) then the fragmented space is turned into unused trace frames.

The wb policy reduces duplication but increased fragmentation removes all benefits. STS is more effective to reduce the number of constructed traces, because it does not impact the fragmentation.

A large part of the trace cache is wasted, e.g. 60% of the 256-entry trace cache. This indicates that the trace cache is very inefficient in storing the program: A trace cache needs to be about 2.5 times as big as an instruction cache to store the same piece of code.

The trace termination policy influences duplication but not in a straightforward way. The best performing policy (STCS in the 1K-entry trace cache) also has the highest duplication. Therefore, when looking for new techniques to improve trace cache performance, one should not be concerned with reducing duplication per se, because the amount of duplication my increase with increasing performance, as is the case when using selective trace storage.

4.2 Liveness of Traces

The liveness analysis (Figure 3) shows that the trace cache contains many dead traces, around 76% for the smallest to 35% for the biggest trace caches. For the small trace caches, the majority of the dead traces is built but never used. This part slowly decreases with increasing trace cache size and for the 1K-entry trace cache, around 50% of the dead traces have never been used since they were built. This trend is largely independent of the trace termination policy.

In short, the contention in the trace cache is very high and most of the traces are dead. It is therefore worthwhile to consider not storing some traces, as these traces will not be used. In fact, STS gains performance this way.

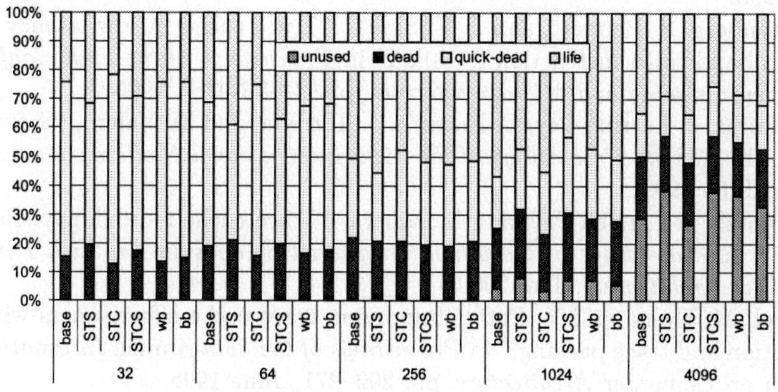

Fig. 3. Breakdown of life and dead traces.

5 Conclusion

This paper studies duplication and liveness under several optimisations. Duplication occurs frequently, e.g.: 43% in the 256-entry trace. However, duplication does not seem to be strongly correlated to trace cache performance, although the best-performing trace caches also introduce the most duplication.

The amount of dead traces is extremely high, ranging from 76% in the smallest trace cache to 35% in the largest trace cache studied. Furthermore, most of these dead traces are never used between storing them and replacing them from the trace cache.

Acknowledgements

We want to thank Joseph L. Larriba Pey for his useful comments on this work. Hans Vandierendonck is supported by the Flemish Institute for the Promotion of Scientific-Technological Research in the Industry (IWT). This work has been supported by the Ministry of Science and Technology of Spain and the European Union (FEDER funds) under contract TIC2001-0995-C02-01 and by a research grant from Intel Corporation. This research is partially supported by the Improving the Human Potential Programme, Access to Research Infrastructures, under contract HPR1-1999-CT-00071 "Access to CESCA and CEPBA Large Scale Facilities" established between The European Community and CESCA-CEPBA.

References

1. A. Peleg and U. Weiser, "Dynamic flow instruction cache memory organized around trace segments independent of virtual address line," *U.S. Patent Number 5.381.533*, Jan. 1995.
2. E. Rotenberg, S. Bennett, and J. E. Smith, "Trace cache: A low latency approach to high bandwidth instruction fetching," in *Proceedings of the 29th Conference on Microprogramming and Microarchitecture*, pp. 24–35, Dec. 1996.
3. A. Ramírez, J. Larriba-Pey, C. Navarro, J. Torrella, and M. Valero, "Software trace cache," in *ICS'99. Proceedings of the 1999 International Conference on Supercomputing*, June 1999.
4. A. Ramírez, J. Larriba-Pey, and M. Valero, "Trace cache redundancy: Red & blue traces," in *Proceedings of the 6th International Symposium on High Performance Computer Architecture*, Jan. 2000.
5. S. Patel, M. Evers, and Y. Patt, "Improving trace cache effectiveness with branch promotion and trace packing," in *Proceedings of the 25th Annual International Symposium on Computer Architecture*, pp. 262–271, June 1998.

Speeding Up Target Address Generation Using a Self-indexed FTB

Juan C. Moure, Dolores I. Rexachs, and Emilio Luque[1]

Computer Architecture and Operating Systems Group,
Universidad Autónoma de Barcelona. 08193 Barcelona (Spain)
{JuanCarlos.Moure, Dolores.Rexachs, Emilio.Luque}@uab.es

Abstract. The fetch target buffer (FTB) holds information on basic blocks to predict taken branches in the fetch stream and also their target addresses. We propose a variation to FTB, the self-indexed FTB, which, through an extra level of indirection, provides the high hit rate of a relatively large, high-associative FTB with the fast access delay of a small, direct-mapped FTB. The critical and most frequent operation –predicting the next FTB entry– is speeded up, whilst less frequent operations –such as recovering from FTB misses– are slightly slowed down. The new design is both analyzed and simulated. Performance increase on a 512-entry FTB is estimated at between 15% and 30%.

1 Introduction

The first stage on a processor's pipeline provides a stream of instruction addresses to the instruction cache (iCache). Since taken branches are detected later in the pipeline, they create holes in the instruction fetch sequence, and limit the processor's performance. The Branch Target Buffer (BTB) represents a hardware solution to the problem, and consists of storing previous history –branch addresses, target addresses, and previous branch behavior– to anticipate the occurrence of taken branches, [6,7]. Two kinds of prediction are carried out: *target address generation* –predicting the occurrence of taken branches and their target address– and *branch prediction* –choosing the direction (taken or not) of conditional branches–. More flexibility is provided by using a dedicated structure for conditional branch prediction, distinct from the BTB [3]. This paper focuses exclusively on target address generation.

Address generation throughput and efficiency is increased by storing the size of each basic block (BB) –the group of sequential instructions starting with a target instruction and ending with a branch– within the target buffer [10]. Reinmann et al. proposed a multi-level, BB-based structure, called the Fetch Target Buffer (FTB), [8], which combines the low access delay of a small, first-level table with the high prediction accuracy of a large second-level table. Their approach is based on technological trends that seem to indicate that, as feature sizes shrink, circuit delays become increasingly limited by the amount of memory in the critical path, [1,4]. The

[1] This work was supported by the MCyT-Spain under contract TIC 2001-2592 and partially supported by the *Generalitat de Catalunya - Grup Recerca Consolidat 2001* SGR-00218

B. Monien and R. Feldmann (Eds.): Euro-Par 2002, LNCS 2400, pp. 517–521.

size and associativity of the L1 FTB determine its access delay, and cannot be overly reduced, or an excessive L1 miss rate will hamper overall FTB performance.

Following this approach, we propose the Self-indexed FTB (SiFTB), which reduces the average Level-1 FTB cycle time without increasing its miss rate. An additional table containing FTB indexes, rather than instruction addresses, provides an extra level of indirection that avoids costly associative searches on consecutive FTB hits. Associative searches are still required on FTB misses, but they should be infrequent. Section 2 provides fuller details on the FTB and SiFTB functionality.

Using CACTI, [9], we estimate lower and upper bounds for the SiFTB's delay advantage, ranging from between 30% to 70%. Using simulation, we then calculate the frequency of hits and misses for each design. Combining delay bounds and simulation results, we found speedup to be very dependent on FTB size. For most useful sizes, speedup ranges from between 15% to 30%. This is presented in section 3. Section 4 outlines the conclusions and sets out future lines of research.

2 The FTB and the Self-indexed FTB

In this section, we describe the FTB and SiFTB schemes and compare their advantages and disadvantages. We then discuss certain related issues.

Figure 1 depicts a simplified FTB. Each FTB's entry is tagged by the starting address of a BB (*BBaddr* field), and contains the type of the ending branch (*type*) as well as the addresses of the taken and not-taken BBs (*tkBB* and *ntBB*), also called target and fall-through BBs. A separate Conditional Branch Predictor (CBP) is used, and a Return Address Stack (RAS) predicts the target of subroutine returns.

A Fetch Target Queue (FTQ) decouples the FTB and iCache stages, allowing each of these to be designed as a multi-cycle, pipelined hierarchy. It allows the FTB to run ahead of the iCache, and hides part of the FTB and iCache penalties, whilst its delay overhead is only exposed on mispredictions.

The current BB address is compared against the *BBaddr* tags every cycle. On a hit, the *type* value and CBP outcome select the next BB address from three options: *tkBB* value (for unconditional jumps or predicted-taken conditional branches), *ntBB* value (for predicted-not-taken conditional branches), and top of RAS (for return jumps). Call instructions also push the *ntBB* value on the RAS. On a miss, a fixed constant *k* is added to the current BB address to generate the next BB address.

When a BB not found in the FTB is identified in the decode stage, it is inserted into the FTB (LRU replacement). After a misprediction, the correct BB address is used on the FTB to resume target prediction. The CBP is updated as branches are solved.

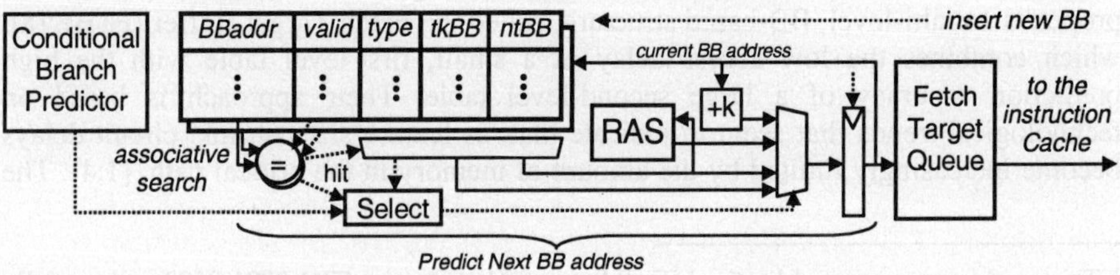

Fig. 1. Block diagram of a Fetch Target Buffer (FTB).

The SiFTB includes a new self-indexed table (SiTable) containing the previous *type* field and two new index fields, *tkIdx* and *ntIdx*, which are pointers to the SiFTB. A separate FTB array contains the rest of the fields (see Figure 2).

Every cycle, both tables are addressed in parallel using the current index: the SiTable provides the next SiFTB index, and the FTB array provides the next BB address. Predicting the next index is carried out as in the FTB. The *type* value and the CBP outcome select the next BB index from: *tkIdx* value (unconditional or predicted-taken conditional), *ntIdx* value (predicted-not-taken conditional), and top of RAS (returns). It should be noticed that the RAS also contains indexes (calls push the ntIdx value on the RAS). The BB address corresponding to a SiFTB index is obtained one cycle before the index is obtained. At this time, target misses are checked by comparing the value of the *BBaddr* field to the previously predicted BB address.

BBs are inserted in the SiFTB when the index for the following BB is known, so that one of the index fields is correctly set, while the other index field initially points to a random entry within the set. If an index field points to an incorrect entry and the expected BB is stored in the other entry, the SiFTB may incur a *false* miss. This may also occur due to entry replacements.

On hits, the BB generation rate is speeded up because the SiTable is small (few and short fields) and direct-mapped (no associative comparison). SiFTB misses, however, do need associative search logic on the *BBaddr* fields to re-enter the chain of BB pointers. The associative search is avoided on conditional branch mispredictions by storing the index for the alternative path along with the predicted branch.

The SiFTB scheme is similar to a predictive set-associative cache (PSAC), [5,3,11]. The PSAC embeds a BTB into the iCache by using an index field to predict the next cache block set and way. It may incur way misses that are similar to those we call false misses. However, separating the FTB and the iCache, as in our proposal, allows for traversing full BBs instead of fixed-size cache blocks, thus maximizing instruction bandwidth per branch prediction, and permits the design of multi-cycle, pipelined iCaches and FTBs.

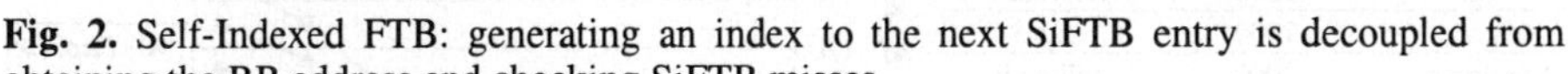

Fig. 2. Self-Indexed FTB: generating an index to the next SiFTB entry is decoupled from obtaining the BB address and checking SiFTB misses.

3 Performance Evaluation

We compare the performance of both designs by first estimating lower and upper bounds for the circuit delay differences on all possible execution cases; by using simulation, we then evaluate the frequency of these cases on several benchmarks.

The access time of a table for a varying number of entries (64 to 4K), entry sizes (14 bits to 70 bits), and associative degrees (direct-mapped and 4-way associative) was obtained using the CACTI cache simulator, [9]. The result is that a set-associative FTB is between 30% and 70% slower than a direct-mapped SiTable (with entries from 14 to 26 bits), and between 20% and 60% slower than a direct-mapped FTB array (with larger entries). On consecutive hits, then, the SiTable generates BBs that are between 30% to 70% faster than the FTB, which suffers the associative logic overhead. Storing the index for the alternative path on a conditional branch prediction avoids the overhead of the associative logic on a misprediction.

Recovering from a BB target miss always requires an associative search on both designs. Since the SiFTB is optimized for direct access, we assume that the delay for the associative access should be larger than on the FTB (between 5% and 25%). False misses on the SiFTB will therefore involve an extra penalty equal to the difference between a direct and an associative SiFTB access. We assume that the insertion of BBs into the FTB or SiFTB do not increase the miss penalty.

In our limit analysis, we assume the FTB/SiFTB is the only system bottleneck. Therefore, we only simulate hit/miss FTB/SiFTB behavior, ignoring time issues and conditional and indirect branch prediction, and assuming that tables are immediately updated. Because we assume that sequential instructions are injected by default in the case of misses, the penalty for recovering from a miss is only suffered after a taken branch, and many misses between two consecutive taken branches are only penalized as a single miss.

We have used release 3.0 of the *SimpleScalar-Alpha* simulation tools, [2], and the SPECint95 and SPECint00 benchmarks that impose more problems on the FTB/SiFTB (Figure 3.a). For each benchmark, we simulate the first thousand million instructions, and count the number of misses and false misses for a number of entries, ***n***, varying from 32 to 2048. The proportion of false misses compared to true misses is small, but increases with the number of entries (from 2% to 10% of the total misses).

Figure 3.b shows the result of combining miss rates and cycle time speedup (*Csp* from 1.2 to 1.6, considering values from the worst to the best case). When the number of entries is large enough to reduce misses bellow 1%, overall performance speedup comes close to cycle time speedup, as expected. An excessive miss rate hampers the cycle time advantage of the SiFTB, as misses have a higher penalty on the SiFTB than on the FTB. Our selected benchmarks have large instruction working sets and require more than 512 entries to allow the SiFTB to achieve a total performance increase of 60% of cycle time speedup. With *Csp* between 1.25 and 1.6, the average performance increase of the SiFTB versus the FTB will be 15%-30%.

benchmark	SPEC version	input data
vpr	2000	reference
gcc	2000	expr.i
perlbmk	2000	splitmail
crafty	2000	reference
eon	2000	cook
twolf	2000	reference
vortex	95	reference
li	95	reference
go	95	reference

a) Benchmarks: DEC C compiler (-O4)

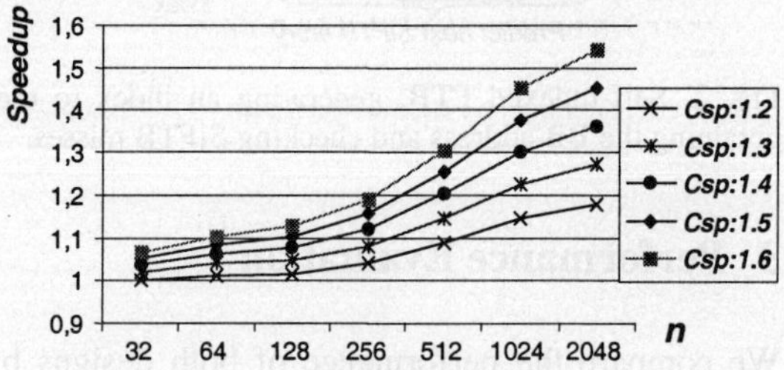

b) Performance Speedup of SiFTB versus FTB

Fig. 3. *Csp* is cycle time speedup and *n* is the number of FTB/SiFTB entries. Results are averaged for all the benchmarks. Placement is 4-way associative and replacement is LRU.

4 Conclusions and Future Work

We have proposed a modification to the FTB, the Self-indexed FTB, which uses a small and direct-mapped table of SiFTB pointers to increase BB prediction rate on hits. Misses, however, have a slightly larger penalty and new false misses may occur. Using simulation, we have calculated the frequency of misses generated by the FTB and the SiFTB and found a small proportion of false misses. Nevertheless, the SiFTB's delay advantage is only exploited when the number of total misses is small, i.e., for tables with a relatively large number of entries.

In order to exploit all the proposal's possibilities, other bottlenecks should be removed. For example, conditional branch mispredictions will waste a substantial part of the SiFTB advantage. We have also ignored the fact that a very fast SiFTB requires a very fast CBP. Reinmann et al. embedded the CBP into the FTB, [8], and there are techniques, such as the lookahead prediction, [10], and the overridden prediction, [4] to avoid branch prediction from becoming a cycle-time bottleneck..

There are many FTB variations that we have not analyzed in this paper. Reinmann et al. considered dynamic BBs that may embed several not-taken conditional branches, [8], and analyze a two-level FTB hierarchy. We will consider these issues in the future. Since a single SiFTB suffers if the number of entries is reduced, adding an extra L2 SiFTB may reduce the penalty of L1 misses and extend the delay advantage on hits to very small L1 SiFTBs.

References

1. V.Agarwal, M.S.Hrishikesh, S.W.Keckler, D.Burger: Clock Rate versus IPC: The End of the Road for Conventional Microarchitectures. Proc. ISCA-27, (2000) 248–259
2. D.Burger, T.M.Austin: The SimpleScalar Tool Set. *Univ. Wisconsin-Madison Computer Science Department*, Technical Report TR-1342, 1997
3. B.Calder, D.Grunwald: Next Cache Line and Set Prediction. Proc. ISCA-22, (1995) 287–296
4. D.A. Jimenez, S.W. Keckler, C. Lin: The Impact of Delay on the Design of Branch Predictors. Proc. MICRO-33, (2000) 67–76
5. M.Johnson: Superscalar Microprocessor Design. *Innovative Technology. Prentice-Hall* Inc., Englewood Cliffs, NJ, (1991)
6. J.K.F.Lee, A.J.Smith: Branch Prediction Strategies and Branch Target Buffer Design. *IEEE Computer* (1984) 17(2): 6-22
7. C.H.Perleberg, A.J.Smith: Branch Target Buffer Design and Optimization. *IEEE Trans. on Computers* (1993) 42(4): 396-412
8. G.Reinman, T.Austin, B.Calder: A Scalable Front-End Architecture for Fast Instruction Delivery. Proc. ISCA-26, (1999) 234-245
9. G.Reinman, N.Jouppi: An Integrated Cache Timing and Power Model. *COMPAQ Western Research Lab,* http://www.research.digital.com/wrl/people/jouppi/CACTI.html (1999)
10. T.-Y.Yeh, Y.N.Patt: A Comprehensive Instruction Fetch Mechanism for a Processor Supporting Speculative Execution. Proc. MICRO-25, (1992) 129-139
11. R. Yung: Design Decisions Influencing the Ultrasparc's Instruction Fetch Architecture. Proc. MICRO-29, (1996) 178-190

Real PRAM Programming

Wolfgang J. Paul, Peter Bach*, Michael Bosch*, Jörg Fischer**,
Cédric Lichtenau***, and Jochen Röhrig***

The work reported here was done while all the authors were affiliated with
Computer Science Department, Saarland University,
Postfach 151150, 66041 Saarbrücken, Germany† `http://www-wjp.cs.uni-sb.de/`

Abstract. The SB-PRAM is a parallel architecture which uses i) multithreading in order to hide latency, ii) a pipelined combining butterfly network in order to reduce hot spots and iii) address hashing in order to randomize network traffic and to reduce memory module congestion. Previous work suggests that such a machine will efficiently simulate shared memory with constant access time independent of the number of processors (i.e. the theoretical PRAM model) provided enough threads can be kept busy. A prototype of a 64 processor SB–PRAM has been completed. We report some technical data about this prototype as well as performance measurements. On all benchmark programs measured so far the performance of the real machine was at most 1,37 % slower than predicted by simulations which assume perfect shared memory with uniform access time.

1 Introduction

Successful commercial parallel machines have to use standard processors because their performance/price ratio has to grow with that of the standard processors. Standard processors use caches in order to reduce memory latency and to overcome the bandwidth bottleneck imposed by the pins of the processor chip. Due to the caches a local programming style is strongly rewarded by successful commercial machines and techniques to maintain locality are highly developed.

On the other hand asymptotically work optimal emulations of PRAMs, where locality plays no role whatsoever, are well known in the theoretical literature. For a survey see [1]. They are based on interleaved multithreading and for p processors they require networks with cost $O(p \log\ p)$. With respect to a gate level model of hardware a construction with small constants was given in [2]. Based on this construction a prototype with 64 processors was developpped in the SB–PRAM project over the last decade.

* currently ETAS GmbH, Stuttgart, `{peter.bach,michael.bosch}@etas.de`

** after July 2002: Jörg Schmittler, `jofis@graphics.cs.uni-sb.de`

*** currently IBM Deutschland Entwicklung GmbH, Böblingen Lab, `{lichtenau,roehrig}@de.ibm.com`

† This work was supported by the German Science Foundation (DFG) under contract SFB 124, TP D4.

B. Monien and R. Feldmann (Eds.): Euro-Par 2002, LNCS 2400, pp. 522–531.

The foremost goal of the project was to make the PRAM programming style a reality on hardware which would scale say to hundreds of processors. The construction of a research prototype with 64 processors was completed. This paper contains technical data about the hardware as well as measured performance figures from this prototype confirming that the behaviour of the real hardware and the abstract user model of a PRAM differ by less than 1.37 % on all runs observed.

The paper is organized in the following way. In section 2 the architecture of the SB–PRAM and its relation to other architectures is sketched. In sections 3 and 4 the hardware and system software are described. Section 5 contains performance figures. We draw conclusions in section 6.

2 SB–PRAM Architecture

Table 1 lists for various shared memory machines how the 4 basic problems of hot spots, module congestion, network traffic and latency are addressed. It can serve as a rough definition of the SB–PRAM architecture and it shows the relation of the SB–PRAM to other architectures.

1. In the SB–PRAM architecture hot spots are avoided by a combining butterfly network like in the NYU Ultracomputer [3] and the RP3 [4] machine.
2. Module congestion is made unlikely by address hashing with a randomly chosen bijective linear hash function. Nontrivial hash functions are also employed in the RP3 and TERA machines [5].
3. The randomization of network traffic and its positive effects on network delays come for free with address hashing.
4. Like HEP [6] and TERA the SB–PRAM does not use a cache at the processors hereby missing the positive effect caches have on network bandwidth. Instead the latency is hidden by multi-threading. Every processor can support a number of threads which grows with $L(p) = c \cdot log\ p$ for a constant c depending only on the relative speed of processors, network and memory. The intention is to hide almost completely the latency of the network as long as all threads can be kept busy.

Table 1. Techniques used in order to avoid network congestion

	multi threading	address hashing	combining	caches
HEP	√	-	-	-
TERA	√	√	-	-
T3E	√	√	at memory	√
RP3	-	√	√	√
Ultra	-	√	√	√
SB-PRAM	√	√	√	-

For details see [2, 7] and [8].

Let $L(p) \geq 3 \cdot log\ p$ be the number of threads per processor on an SB–PRAM. Simulations [2] show that except in rare situations of network and memory module congestion no time at all is lost during the emulation of $p \cdot L(p)$ threads on a machine with p processors. If t is the cycle time of the processors, then the system exhibits *almost exactly* the behavior of a PRAM with $p \cdot L(p)$ processors and cycle time $t \cdot L(p)$. Note that this is a bulk synchronous computation with latency $L(p)$ and throughput $g = 1$ in the sense of [9, 10].

Moreover the constants for the hardware cost and cycle time are very small, at least in a gate level model of hardware [2]. Although the cost of the network grows asymptotically faster than the cost of the processors, the number of gates in the network of a 64 (resp. 128) processor machine is only 26 % (resp. 28 %) of the total cost of the machine [11] (assuming the cost of memory modules equals the cost of processors).

3 Hardware

The hardware of the SB–PRAM prototype is surveyed in [12]. Theses and papers documenting the design can be found at our web site:

`http://www-wjp.cs.uni-sb.de/projects/sbpram/`

The hardware effort had three major parts: chip design, board design and hardware integration.

3.1 Chips and Cycle Time

The chips designed were

1. The processor [13]. It has a Berkeley RISC instruction set and a 32 bit floating point unit manipulating normal numbers in the IEEE 754 format. The simplest context switching scheme supporting the PRAM programming style was realized: simple interleaved context switching. The processor was designed for machines with up to $p = 128$ processors. This required at least $3 \cdot 7 = 21$ contexts. For machines of different sizes, the processor can be configured to store 8, 16, 24 or 32 contexts. The register files storing up to 32 copies of the processor's register set were realized by external SRAM. The processor was designed in the year 1994 in $0.7 \mu m$ technology and fabricated by Motorola. It has about 80,000 gate equivalents, 230 signal pins and a cycle time of 16 ns [13, 14].
2. The sorting array [15]. One sorting array per processor is used. In each round it sorts the packets entering the network from the processor by destination address. This is part of the hardware support for Ranade's routing scheme [16, 17]. The sorting array also supports multiprefix operations. The sorting array was designed in the year 1995 in 0.7 μm technology and fabricated by Motorola. It has 37,000 gate equivalents, 175 signal pins and a cycle time of 25 ns [15].

3. The network chip [18]. It also supports multiprefix operations for the operators $\wedge$, $\vee$, $+$ and max and implements Ranade's routing scheme. In order to save pins each networkchip contains 4 (interconnected) halves of network nodes [19]. The chip was manufactured in the year 1996 in 0.8 μm technology by Thesis. It has 67,000 gate equivalents and a worst case cycle time of 37 ns.

In the prototype the network chips work at a cycle time of 37 ns, thus the network can be clocked with 27 MHz. The processors are operated at 1/4 of the network frequency, i.e. at 6.75 MHz. The prototype is based on old low cost technology. A redesign based on 1996's technology, where the processors are clocked at 93.6 MHz is sketched in [20]. Running the processors at 1 GHz would require optical links between boards. Moreover the data rates between chips would require very advanced technology e.g. optical I/O at chip boundaries.

3.2 Boards

The boards designed were

1. the processor card [21, 22] with 32 MByte local memory for programs and a PCI-bus interface [23]. It also contains two SCSI-controllers to connect hard disks directly to the processor card. It has an area of 802 cm^2.
2. the memory card. Each memory module [24, 25] contains 64 MByte in 4 banks of memory for each processor. Every memory card contains two memory modules. The card has an area of 743 cm^2.
3. the network card [26] contains 8 network chips. It has an area of 1464 cm^2.
4. various small cards for the global clock and reset distribution, for the interconnection of back-planes and for the connection to the host computer. They form a small portion of the machine.

3.3 Geometry of Wiring

Most interconnections between boards were realized as ribbon cables. Transmission by ECL or optical fibres would have been possible [26] but not affordable. In order not to repeat painful experiences from earlier projects, theorems on the geometrical arrangement of boards, connectors and cables were proven extremely early in the project [18, 27]. These theorems bound the length of wires in order to make timing analysis possible, and they show how to remove single boards without disassembling large portions of the machine.Figure 1 shows a part of the wires arranged as prescribed by the theorems. The whole machine has 1.3 km of ribbon cables with a total of 166 km of wires.

3.4 Relative Cost of Processors and Network

Figure 1 shows the entire prototype with 64 processors with the network boards, memory boards and processor boards marked separately. We show that the network boards occupy roughly 1/3 of the total printed circuit board area of the machine:

Fig. 1. left half of the 64–SB–PRAM and the complete 64–SB–PRAM

Let A_{PM}= 802 cm^2 + 371.5 cm^2 be the area of a processor and half a memory board. Let A_N = 1464 cm^2 be the area of a network board. A network board contains 8 chips and the equivalent of 16 network nodes. A machine with p processors needs $p \cdot log\ p$ network nodes, i.e. $(p/16) \cdot log\ p$ network boards (for appropriate values of p, e.g. 64). Thus a machine with p processors occupies an area of

$$A(p) \approx p \cdot A_{PM} + ((p/16) \cdot log\ p) \cdot A_N.$$
$$= 1173.5 \cdot p + 91.5 \cdot p \cdot log\ p$$

The relative size of the network is $(91.5 \cdot p \cdot log\ p)/A(p)$. For $p = 64$ this evaluates to about 32 %. This is slightly higher than the 26 % estimated by the gate model [11]. The number of processors p for which the network occupies half the total area is determined by

$$1173.5 \cdot p = 91.5 \cdot p \cdot log\ p$$

which is between 4096 and 8192.

3.5 System Integration and Debugging

Debugging the 4 processor machine in 1996 took unreasonable amounts of time because not enough testability was designed into the boards. Therefore all boards were redesigned using JTAG [28] wherever possible. Test procedures were specified and theorems about the error coverage were proven [22, 25, 29]. The error model includes shorts (AND, OR and dominating errors), opens and dynamic fault models for timing violations, ground bounce and cross talk [29].

Technical difficulties in the design of the test procedures came from 3 sources of which the last two are relevant for future designs:

1. The chips designed earlier in the project have no JTAG paths.
2. The clock lines and control lines of the JTAG circuitry themselves have to be tested [25].

3. given the size of the machine the time for running the test procedures became a serious issue. A detailed self test of the whole machine takes 4 hours [29].

4 System Software

The system software has 4 major components: 1) A UNIX-like operating system which heavily uses the multiprefix operations of the SB–PRAM for memory management. 2) A compiler for the Language FORK[8, 30]. 3) A compiler for C (later C++) extended by the concept of shared variables. 4) A library of communication primitives for the C dialect. Roughly speaking it included the P4 macros [31] needed for the direct compilation of the SPLASH benchmarks and the communication primitives from the NYU ultracomputer project [32], most notably the parallel job queue ported to the SB–PRAM. The migration was reasonably straight forward; on the SB–PRAM the primitives have *predictable* run times (which is not surprising because they were inspired by the theoretical literature in the first place).

5 Measured Run Times

Before completion of the hardware numerous applications were implemented on a simulator, which simulated the *idealized user model* of an SB-PRAM on an instruction by instruction basis. Recall that for p processors with cycle time t and $L(p)$ threads per processor the user is intended to see a priority CRCW PRAM with $p \cdot L(p)$ processors and cycle time $t \cdot L(p)$, Berkeley RISC instruction set and multi–prefix operations processing one instruction per cycle. The simulations allowed to make optimistic predictions of run times and speed-ups and to compare the PRAM programming style with the common programming style which exploits locality. For a survey see [33].

With the hardware up and running the predictions can of course be checked against reality. We compare here the predicted and measured run times of 4 programs from the benchmark suites SPLASH and SPLASH 2: Radiosity, Locus Route, PTHOR and MP3D. All speed-ups reported below are about *absolute* speed-ups, i.e. the run time of a parallel program with p processors is compared with the run time of the *fastest known sequential* program running on a single processor. On the SB–PRAM sequential computation is performed by using a single thread on a single processor, but that of course makes only use of $1/L(p) = 1/32$ of the power of a processor. In order to determine speed-up in a fair way, we measure the run time t_1 of 1 thread on 1 procesor, compare with time t_{32p} using 32 threads on each of p processors and scale by 32: $speedup = (1/32) \cdot (t_1/t_{32p})$.

5.1 Measured Speed-Ups

Figure 2 concerns the benchmarks Radiosity and Locus Route. It compares the speed-up of non optimized and optimized PRAM programs with that on the

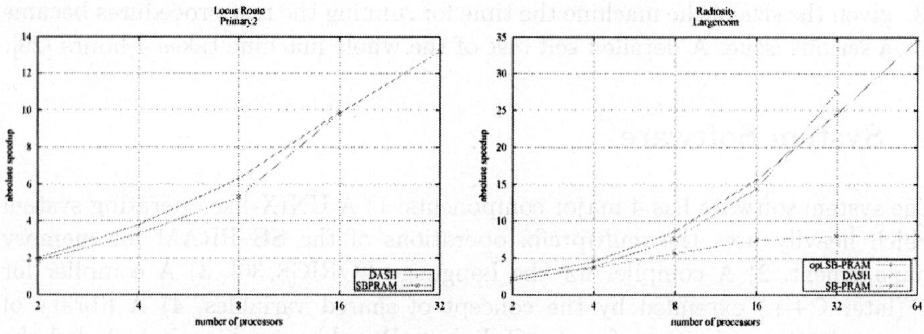

Fig. 2. Results: Locusroute and Radiosity

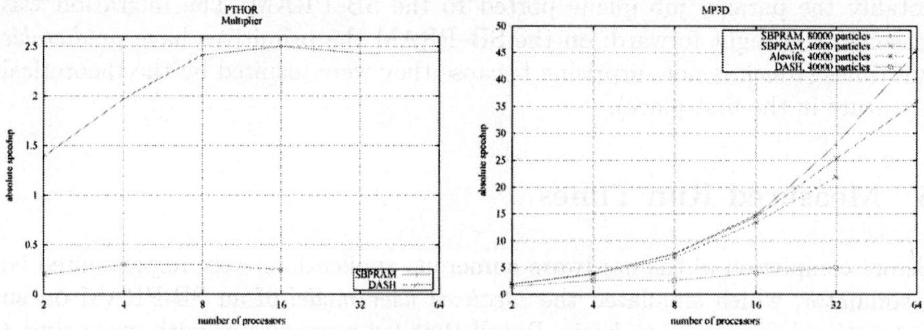

Fig. 3. Results: PTHOR and MP3D

(cache based) DASH machine [34]. Both programs parallelize well on cache based machines. Migrating such programs in a naive way to the SB–PRAM leads to *deteriorated parallel efficiency*: with p PRAM processors one has to keep at least $3p \cdot log\ p$ threads busy, and naive migration tends not to achieve this. If one optimizes the programs and recodes portions of the programs using multiprefix operations and the powerful communication primitives one regains most or all of the lost ground.

Figure 3 shows speed-ups for PTHOR. The only good news here is, that a nontrivial speed-up is reached at all on a real machine. The discrete event simulation PTHOR is difficult to parallelize for 3 reasons: i) access patterns to memory are non local, ii) the overhead for parallelization is enormous. The parallel program for 1 processor is about 5.4 times slower than the fastest sequential program, iii) the number of available tasks is limited. Reasons ii) and iii) are problematic for the SB–PRAM too.

Figure 3 also shows speed-ups for the particle simulation MP3D. In the simulation the particles mix in an irregular fashion. Because every particle is always processed by the same processor the handling of collisions leads to non local memory accesses. Using a cache protocol optimized for migratory sharing the MIT Alewife machine [35] handles this situation remarkably well.

This seems like the perfect benchmark for the SB–PRAM which is designed to handle non local access pattern effortlessly. Yet the parallel efficiency for 64 processors is still far from 100 %, at least with 40 000 particles. The reason is, that the large number of threads of a 64 processor SB–PRAM can only be kept busy for the central tasks of moving and colliding particles. But other portions of the program with only hundreds of parallelizable tasks begin to dominate the run time. For details see [36].

5.2 Overhead Due to Congestion

Consider a run of the simulator with r rounds of computation for each of $L(p)$ user threads per processor. Then with cycle time t the prediction for the wall clock time w is $w' = r \cdot L(p) \cdot t$. This prediction is optimistic and assumes that latency due to congestion can be hidden completely. The quantity overhead $ovh = (w - w')/w' = w/(r \cdot L(p) \cdot t) - 1$ measures the relative overhead incurred due to congestion in the network and at the memory modules which cannot be hidden. We report measurements of this quantity which were performed with various combinations of processor numbers and threads per processor. Runs were performed with 2 hash factors h for address hashing:

1. with the trivial hash factor $h = 1$. In this case the maximal overhead observed was 26% for 16 processors and 223% for 64 processors. Thus switching off the address hashing leads to significant deterioration of the performance due to congestion.
2. with hash factor $h = 0x826586F3$ generated by a random number generator. Here the result is very encouraging: in 771 runs for 16 processors the maximal overhead observed was under 1 %. In 324 runs for 64 processors the maximal overhead observed was 1.37 %. Thus deterioration of performance due to congestion was almost absent for practical purposes.

Note that the optimized application programs make heavy use of communication primitives like a fine grained parallel job queue which put a heavy load on the communication network and the memory modules.

6 Conclusion

The prototype of a 64 processor SB–PRAM has been completed. The cost of the construction grows asymptotically with $p \log p$. But in the particular (by now old) technology used, the network boards would contain less than half the area even for a machine with 4096 processors. Thus the construction is reasonably scalable.

The results of section 5.2 show that the PRAM simulation using the Ranade routing scheme is performing for 64 processors with extremely low overhead. Thus running PRAM algorithms with a run time matching the theoretical run time to within a few percent is a reality. Performance deteriorates if the address hashing is switched off.

The results of section 5.1 show that on scientific application programs the strength of the SB–PRAM in handling non local memory accesses translates at least in some situations like MP3D (and PTHOR) into a measurable gain in efficiency. The large number of threads however, which need to be kept busy, is an issue requiring careful attention.

References

1. Valiant, L.G.: General Purpose Parallel Architectures. In van Leeuwen, J., ed.: Handbook of Theoretical Computer Science, Vol. A. Elsevier Science Publishers and MIT Press (1990) 943–971
2. Abolhassan, F., Keller, J., Paul, W.J.: On the Cost-Effectiveness of PRAMs. In: Proceedings of the 3rd IEEE Symposium on Parallel and Distributed Processing. (1991) 2–9
3. Gottlieb, A., Grishman, R., Kruskal, C.P., McAuliffe, K.P., Rudolph, L., Snir, M.: The NYU Ultracomputer – Designing an MIMD Shared Memory Parallel Computer. In: IEEE Transactions on Computers, C-32(2). (Feb 1983) 175–189
4. Pfister, G.F., Brantley, W.C., George, D.A., Harvey, S.L., Kleinfelder, W.J., McAuliffe, K.P., Melton, E.A., Norton, V.A., Weiss, J.: The IBM Research Parallel Processor Prototype (RP3): Introduction and Architecture. In: International Conference on Parallel Processing, Los Alamitos, Ca., USA, IEEE Computer Society Press (1985) 764–771
5. Alverson, R., Callahan, D., Cummings, D., Koblenz, B., Porterfield, A., Smith, B.: The Tera computer system. In: Proceedings of the1990 International Conference on Supercomputing. (1990) 1–6
6. Smith, B.J.: Architecture and applications of the HEP multiprocessor computer system. SPIE Real-Time Signal Processing IV **298** (1981) 241–248
7. Abolhassan, F., Keller, J., Paul, W.J.: On the Cost-Effectiveness of PRAMs. In: Acta Informatica 36. Springer-Verlag (1999) 463–487
8. Keller, J., Kessler, C.W., Träff, J.L.: Practical PRAM Programming. Wiley Interscience Series on Parallel and Distributed Computing (2000)
9. Valiant, L.G.: Bulk-synchronous parallel computers. In: Parallel Processing and Artificial Intelligence. (1989) 15–22
10. Valiant, L.G.: A Bridging Model for Parallel Computation. In: Communications of the ACM. (August 1990) 103–111
11. Abolhassan, F.: Vergleich von Parallelen Maschinen mit gemeinsamen und verteilten Speichern. PhD thesis, Universität des Saarlandes, Saarbrücken (1994)
12. Abolhassan, F., Drefenstedt, R., Keller, J., Paul, W.J., Scheerer, D.: On the Physical Design of PRAMs. In: Computer Journal 1993 36(8). (1993) 756–762
13. Scheerer, D.: Der Prozessor der SB-PRAM. PhD thesis, Universität des Saarlandes, Saarbrücken (1995)
14. Keller, J., Paul, W.J., Scheerer, D.: Realization of PRAMs: Processor Design. In: Proc. WDAG '94, 8th Int. Workshop on Distributed Algorithms. Springer Lecture Notes in Computer Science No. 857 (1994) 17–27
15. Göler, T.: Der Sortierknoten der SB-PRAM. Master's thesis, Universität des Saarlandes, Saarbrücken (1996)
16. Ranade, A.G.: How to Emulate Shared Memory. In: Journal of Computer and System Sciences 42. (1991) 307–326

17. Ranade, A.G., Bhatt, S.N., Johnson, S.L.: The Fluent Abstract Machine. In: Proceedings of the 5th MIT Conference on Advanced Research in VLSI, Cambridge, MA, MIT Press (1988) 71–93
18. Walle, T.: Das Netzwerk der SB-PRAM. PhD thesis, Universität des Saarlandes, Saarbrücken (1997)
19. Cross, D., Drefenstedt, R., Keller, J.: Reduction of Network Cost and Wiring in Ranade's Butterfly Routing. In: Information Processing Letters, vol. 45 no. 2. (1993) 63–97
20. A. Formella, J. Keller, T.W.: HPP - A High Performance PRAM. Volume 2. (1996) 425–434
21. Bach, P.: Entwurf und Realisierung der Prozessorplatine der SB-PRAM. Master's thesis, Universität des Saarlandes, Saarbrücken (1996)
22. Bach, P.: Schnelle Fertigungsfehlersuche am Beispiel der Prozessorplatine CPU-LIGHT. Dissertation (paul), Universität des Saarlandes, Saarbrücken (2000)
23. Janocha, S.: Design der PCIPro-Karte. Master's thesis, Universität des Saarlandes, Saarbrücken (2000)
24. Lichtenau, C.: Entwurf und Realisierung des Speicherboards der SB-PRAM. Diplomarbeit an der universität des saarlandes fb 14 (paul), Universität des Saarlandes, Saarbrücken (1996)
25. Lichtenau, C.: Entwurf und Realisierung des Aufbaus und der Testumgebung der SB-PRAM. Dissertation, Universität des Saarlandes, Saarbrücken (2000)
26. Fischer, J.: Entwurf und Realisierung der Netzwerkplatinen der SB-PRAM. Diplomarbeit an der Universität des Saarlandes FB 14 (paul), Universität des Saarlandes, Saarbrücken (1998)
27. Keller, J.: Zur Realisierbarkeit des PRAM Modelles. PhD thesis, Universität des Saarlandes, Saarbrücken (1992)
28. 1990, I.S..: Standard test access port and boundary scan architecture. Institute of Electrical and Electronics Engineers (1993)
29. Bosch, M.: Fehlermodelle und Tests für das Netzwerk der SB-PRAM. Dissertation, Universität des Saarlandes, Saarbrücken (2000)
30. Kessler, C., Seidl, H.: Fork95 Language and Compiler for the SB–PRAM. In: Proceedings of the 5th International Workshop on Compilers for Parallel Computers. (1995) 408–420
31. Röhrig, J.: Implementierung der P4-Laufzeitbibliothek auf der SB-PRAM. Master's thesis, Universität des Saarlandes, Saarbrücken (1996)
32. Wilson, J.M.: Operating System Data Structures for Shared-Memory MIMD Machines with Fetch–and–Add. PhD thesis, Courant Institute, New York University (1998)
33. Formella, A., Grün, T., Keller, J., Paul, W., Rauber, T., Rünger, G.: Scientific applications on the sb-pram. In: Proceedings of International Conference on Multi-Scale Phenomena and their Simulation in Parallel, World Scientific (1997)
34. Lenoski, D., Laudon, J., Joe, T., Nakahira, D., Stevens, L., Gupta, A., Hennessy, J.: The DASH prototype: Logic overhead and performance. IEEE Transactions on Parallel and Distributed Systems **4** (1993) 41–61
35. Agarwal, A., Bianchini, R., Chaiken, D., Johnson, K., Kranz, D., Kubiatowicz, J., Lim, B.H., Mackenzie, K., Yeung, D.: The MIT Alewife Machine: Architecture and Performance. In: International Symposium on Computer Architecture 1995. (1995)
36. Dementiev, R., Klein, M., Paul, W.J.: Performance of MP3D on the SB-PRAM prototype. In: Proc. of the Europar'02. (2002)

In-memory Parallelism for Database Workloads

Pedro Trancoso

Department of Computer Science, University of Cyprus,
75 Kallipoleos Str., P.O. Box 20537, CY-1678 Nicosia, Cyprus,
pedro@ucy.ac.cy, http://www.cs.ucy.ac.cy/~pedro

Abstract. In this work we analyze the parallelization of database workloads for an emerging memory technology: Processing-In-Memory (PIM) chips. While most previous studies have used scientific workloads to evaluate PIM architectures, we focus on database applications as they are a dominant class of applications.
For our experiments we built a simple DBMS prototype, which contains modified parallel algorithms, an in-chip data movement algorithm, and a simple query optimizer. Compared to the single processing execution, the average speedup for a PIM with 32 processing elements is 43 times. Other results show that an n-way multiprocessor of similar cost cannot perform as well. Overall, the results obtained indicate that PIM chips are an architecture with large potential for database workloads.

1 Introduction

One of the major bottlenecks in the execution of an application is the memory access. Both processor and memory technology are evolving at a very fast rate but the gap between their speed is increasing. While caches have been extensively used to alleviate this problem, recent developments in the integration of logic and memory on the same chip offer us new solutions. This technology, known as Processing-In-Memory (PIM), Intelligent Memory or Embedded Memory, offers a higher bandwidth and lower latency from the logic to the memory elements.

So far PIM technology has been mostly used for dedicated applications such as disk controllers or media processors. In addition, the target applications are mostly scientific workloads. The focus of this work is to investigate how we can use this new technology for database workloads as they are a dominant class of applications.

We can summarize the contribution of this paper as: (a) Implementation of a DBMS-prototype for PIM architectures (PIM-DB); (b) Detailed evaluation of different database operations on a PIM; and (c) Comparison of a PIM architecture with an n-way multiprocessor for Database applications.

The results of this work show that large speedup may be achieved for systems configured with PIM chips. Nevertheless, the speedup values depend on the algorithm used. We identify three categories of algorithms: (1) *sequential scan*, which achieved a large speedup ; (2) *hash join* or *sort*, which achieved a moderate speedup ; and (3) *index scan*, which achieved no speedup.

B. Monien and R. Feldmann (Eds.): Euro-Par 2002, LNCS 2400, pp. 532–542.

This paper is organized as follows. Section 2 presents an overview of relevant related work. Section 3 describes the general PIM architecture while Section 4 describes the PIM-DB system. In Section 5 the experimental setup is presented and in Section 6 the results of the experiments. Finally the conclusions and future work are discussed in Section 7.

2 Related Work

Recent performance evaluation studies of database workloads [1–3] have identified memory access as one of the factors inhibiting performance improvement. Regarding the reduction of the memory bottleneck, several studies have proposed both hardware and software solutions to this problem. Keeton *et al.* [4] and Acharya *et al.* [5] propose disks equipped with PIM chips in their controllers to filter the data from disk to the system. Several cache miss reduction techniques for database workloads have also been presented. Among others, Shatdal *et al.* [6] proposed the use of cache conscious database algorithms, Trancoso and Torrellas [7] proposed cache-aware optimizations, Rao and Ross [8] proposed cache conscious indexing, Nyberg *et al.* [9] proposed a cache-aware sorting algorithm, and Boncz *et al.* [10] proposed memory access tuning for the radix algorithm. While these studies all focus on the data access overhead, Ramirez *et al.* have addressed the reduction of the instruction access overhead [11].

Another way to improve database workload performance is to exploit its parallelism. Two recent studies address database parallelism: Lo *et al.* [12] studied the use of a simultaneous multithreaded processor and Barroso *et al.* [13] proposed a chip-multiprocessor with high speed memory links to improve database workload performance.

Consequently, as the performance of database workloads depends on the reduction of the memory bottleneck and the increase of parallelism, it seems obvious that this type of application should benefit from the use of PIM chips. A number of research groups has proposed different PIM designs from special purpose chips, such as C-RAM [14] and Imagine [15], to general purpose chips such as IRAM [16], Active Pages [17], DIVA [18], Smart Memories [19] and FlexRAM [20]. The processing elements in these chips range from a single general purpose processor for IRAM, up to thousands of processing elements containing few logic gates for C-RAM. In this work we study a solution that exploits on-chip parallelism but is able to execute general purpose applications.

3 PIM Architecture

In this section we present an overview of the PIM architecture used in our study. This chip, which we call *PIM-n* is composed of n blocks each containing a Memory (PIM-M) and a Logic (PIM-L) portion. A representation of PIM-n is shown in Figure 1.

Although the PIM-n design shows two arrows that represent the communication links between the different blocks, in order to keep the chip design simple

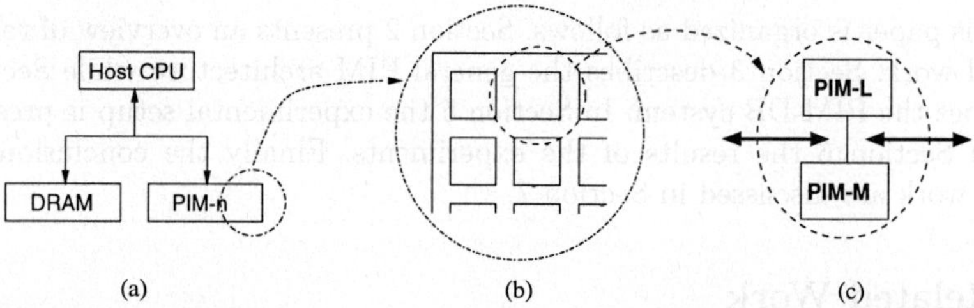

Fig. 1. (a) System configured with one PIM-n chip and regular DRAM; (b) Section of PIM-n chip; (c) Detail of PIM-n block.

these links are limited. While the host CPU may access any memory location, each Logic block accesses only its own Memory block and in addition to this it may access also its two nearest neighbors (left and right). This assumption was proposed by the FlexRAM design [20]. Notice that although two Logic blocks may access the same Memory block, there is no hardware mechanism to assure coherency, therefore this becomes the responsibility of the programmer.

PIM-n can be mapped to many different PIM designs that have been previously proposed. For example, Active Pages [17] may be mapped to a PIM-n chip (1GB DRAM technology) with 128 blocks, each PIM-M has capacity of 512KB and each PIM-L contains 256K transistors of logic.

4 PIM-DB

To exploit the characteristics of the PIM-n architecture we developed a simple prototype of a DBMS called PIM-DB. This system is composed of two major components: Optimizer and Algorithms. The query optimization is performed in two steps [21]. First we use a traditional optimizer to obtain the best sequential query plan and then the system determines which parallel algorithm to use. The system needs to decide between a high degree of parallelism, which may require costly data exchanges, and single processor execution, with no data movement. To evaluate these options we have developed simple mathematical cost models for PIM-DB, which account for the number of instructions and memory accesses.

4.1 Parallel Algorithms

In this section we discuss the techniques we used to map some of the existing parallel algorithms [22] into the PIM-n architecture. Notice that all the algorithms assume that the data is memory resident.

Data Repartition. One characteristic of this architecture is that it does not have an interconnection between all the blocks. Therefore it is not possible to use traditional repartitioning algorithms, where each computation node analyzes its data and ships it directly to the destination node [23]. Instead, we can use the ability to access the neighbors' memory in order to build a *software virtual* *ring* within the memory chip, and the host CPU to distribute the data across chips. The algorithm works as follows. Each block has four bins: *Left*, *Right*, *Local*, and *Remote*. In the first step each PIM-L reads its data and places it in the corresponding bin. In the second step, the host CPU reads all the data in the *Remote* bins and places it on the respective destination. In parallel, each PIM-L reads the left neighbor's *Right* bin and the right neighbor's *Left* bin and places the data in one of the three bins *Left*, *Right*, or *Local* accordingly. This intra-chip data exchange is repeated until all the data arrives to its destination.

Scan. In this work we implemented two scan algorithms: *Sequential* and *Index Scan*. *Sequential Scan*, is fully parallel and no data repartitioning is required. For *Index Scan* to execute in parallel, each block must contain a partial index structure which covers all the tuples in that local memory.

Join. We consider only one join algorithm: *Hash Join*. Its implementation is similar to the Shared-Nothing parallel hash join. The algorithm is executed in two steps: data repartition as described at the beginning of this section (for the data that requires to be moved), and hash join on the partitioned data.

Sort. In this work we use *quicksort*. To execute this algorithm in parallel we need to start by using the data partitioning algorithm previously described to perform a range-partitioning of the data to be sorted. Then we use the quicksort algorithm on each block.

Group. Grouping is a trivial operation when performed after sorting. It requires only to check if consecutive tuples belong to the same group or not.

Aggregate. This operation is performed in a hierarchical way. First each block applies the aggregate function to its local tuples and produces a single value. Then the host CPU collects all the values from each block and produces the result value.

4.2 Validation

To validate PIM-DB we compared its execution with PostgreSQL, a source free system that was developed from the research project Ingres [24]. The queries used were the same throughout this work and are described in Section 5.2. Three sets of different data sizes were used ranging from the original size (650KB), up to twenty times larger (13MB). The execution was performed on a PC equipped with an AMD Athlon 1.1GHz processor, 128MB RAM, running Linux Red Hat 7.1. The measurements of the execution cycles were done using the processor's performance counters.

The results obtained were very satisfactory. Although the absolute execution time is different for the two systems, as the data size is increased, the execution time for both systems scale in the same way. For four out of five queries, the drift in the scaling between the two systems is in average 5%. In absolute values, in average, a query executing on PIM-DB executes 25 times faster than a query executing on PostgreSQL. For the query that uses the index scan algorithm PIM-DB executes it 400 times faster. We attribute this result to the difference between the implementation of the two algorithms as one handles data resident on disk and the other memory-resident data.

5 Experimental Setup

5.1 Hardware

In this work we simulate a system configured with PIM-n chips using a MINT-based [25] execution-driven simulator that is able to model dynamic superscalar processors [26]. The simulation environment is the one developed for the FlexRAM project [20].

The system is configured with a main processor (host CPU), and one or more PIM-n chips. Comparing with other work we believe it is reasonable to assume that the PIM-n chip will be configured with 32 memory blocks of 1MB each and 32 logic blocks, each comparable to an R5000 processor [19]. We call this configuration PIM-32.

The details of the different computation elements are summarized in Table 1. In addition, the memory latency of this system for a row buffer miss is 91 cycles from the CPU and 12 cycles from the PIM-L. For a row buffer hit, the latencies are 85 cycles from the CPU and 6 from the PIM-L. Also, the CPU's model includes a 2-way 32KB L1 Data Cache that is write-through with a 2-cycle hit time, and a 4-way 1MB L2 Data Cache that is write-back with a 10-cycle hit time. The system bus modeled supports split transactions and is 16B wide.

Notice that we present two versions of PIM-L: *Conservative* and *Aggressive*. The first represents the type of technology that can be implemented today using, for example, ASICs [27], while the second represents a solution using a dedicated design chip. Previous work [28] has determined a penalty of 10% in the density in order to achieve such high speeds.

Table 1. Parameters used in the simulation.

	CPU	PIM-L
Freq	1 GHz	500 MHz (Cons)/ 1 GHz (Aggr)
Issue Width	out-of-order 6-issue	in-order 2-issue
Func. Units	4 Int + 4 FP + 2 Ld/St	1 Int + 1 FP + 1 Ld/St
Pending Ld/St	8/16	1/1
Branch Penalty	4 cycles	1 cycles

5.2 Database Workload

The database system used in this work is the PIM-DB prototype as described in Section 4. To obtain the serial query execution plans we use PostgreSQL.

For this work we selected five different queries from the Wisconsin Benchmark [29] (Query 1, Query 3, Query 9, Query 20, and Query 21), based on the different operations that each one includes. Query 1 (*W1*) is a simple selection query with a 1% selectivity factor. The algorithm used to implement this query is the *sequential scan*. Query 3 (*W3*) is the same as Query 1, except that it uses the *index scan* algorithm. Query 9 (*W9*) is a join query of two tables. Before the join operation, one of the tables is scanned with a selectivity factor of 10%. The selectivity factor for the join is 1/100000. For this query we use the *sequential scan* algorithm to implement the scan operation and then we use the *hash join* algorithm to perform the join operation. Query 20 (*W20*) is a simple aggregate query that finds the minimum value for a certain attribute in a table. For this query we use the *hierarchical aggregate* algorithm as described in Section 4.1. Query 21 (*W21*) is the same aggregate query, but in this case the results are grouped on a certain attribute. The algorithms used for this table are: *sort* to create the groups and *hierarchical aggregate* to obtain the result for each group. All input tables have 100000 tuples of 208 Bytes each.

The use of these queries is without loss of generality as it was shown by Ailamaki *et al.* [1] who justified that simple queries such as these are representative of the queries in more complex benchmarks, such as the ones in the TPC-H Decision Support System Benchmark.

In this study we consider the database to be memory-resident. This fact does not seem to be a limitation because of the developments in hiding I/O latency as described by Barroso *et al.* [2]. The mapping of the data tuples is done in a round-robin fashion among the different PIM-M blocks.

6 Experimental Results

6.1 Query Characteristics

First we present an overview of the characteristics of the queries that were studied. In Table 2 we present the execution cycles for single processor execution for each query, broken down into three categories: *Busy* which accounts for the useful instructions, *Memory* which accounts for the waiting cycles of memory requests, and *Hazard* which accounts for stalls in the pipeline due to hazards.

Notice that for all queries the dominant portion of time is the *Memory* time. Also, for most queries the *Busy* time is relatively small. Both *W3* and *W21* have large *Hazard* time, which may be a limiting factor for the performance once the PIM-L processing units have fewer resources than the host CPU.

Table 2. Single processor execution cycles for all queries.

	Busy	Memory	Hazard	Total
Query 1 (W1)	416690 (7.3%)	4933654 (86.7%)	337296 (5.9%)	5687640
Query 3 (W3)	444 (13.8%)	2132 (66.1%)	651 (20.2%)	3227
Query 9 (W9)	2026716 (7.4%)	20878790 (76.2%)	4482948 (16.4%)	27388454
Query 20 (W20)	58362 (2.4%)	2334782 (97.2%)	7998 (0.3%)	2401142
Query 21 (W21)	4885082 (19.9%)	12022034 (49.0%)	7610712 (31.0%)	24517828

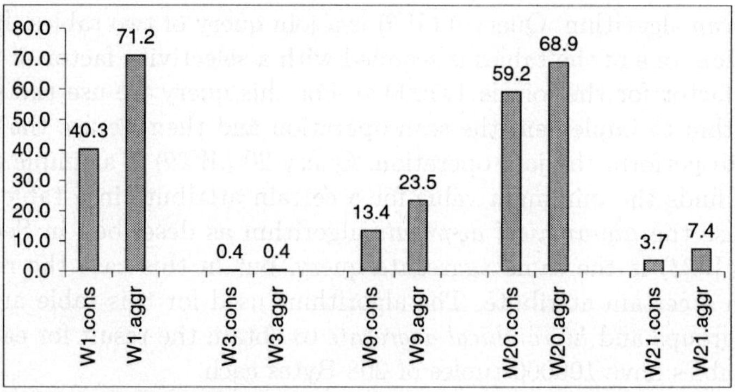

Fig. 2. Speedup results for a system configured with one PIM-32 chip for both the *Conservative* and *Aggressive* setup.

6.2 Query Parallelism

In Figure 2 we present the speedup values for the different queries for a system configured with one PIM-32 chip. For each query we present two bars: *Conservative* and *Aggressive* system configuration. It is important to notice that even though the results for the *Conservative* setup show significant speedup, this speedup increases significantly, for almost all queries, for the *Aggressive* setup.

According to their speedup, we may divide the different queries into three groups. In the first group we include both *W1* and *W20* which achieve high speedup for a single PIM-32 system. The high degree of parallelism exploited by the *sequential scan* algorithm justifies these results.

In the second group we have both *W9* and *W21* which achieve moderate to low speedup. Although the *hash join* algorithm of Query *W9* is fully parallel, in order for this algorithm to perform the data needs to be repartitioned using the *software virtual ring* as described in Section 4.1. For the largest table, this operation consumes approximately 80% of the total query execution time. This penalty is a consequence of the limited connectivity between the blocks. In a simulation where the data was allowed to be shipped directly to the destination memory block, the speedup increased by an order of magnitude. As for *W21*, the reason for its lower speedup is the time consumed with the sorting (approximately 80% of the total execution time). Given that this query is compute-intensive, the penalty observed is a consequence of the small-issue offered by the PIM-L elements compared to the wide-issue offered by the host CPU.

In the last group we include query *W3*. This query achieves no speedup at all. If we analyze the index search algorithm used, which is based on binary search, we conclude that although the size of the index decreases, the difference in the number of elements traversed by the search ($\log n$ in the worst case) is very small. Therefore, the work performed by each processing element is close to the work performed by the host CPU. In addition, the host CPU is able to issue more instructions in the same clock cycle and contains more functional units.

Overall, the average speedup for four out of the five queries is 43x for the *Aggressive* setup. The parallelism offered by the multiple PIM-L computation elements is the main reason for the high speedup observed for some queries. The limited connectivity, the limited instruction issue width of the PIM-L elements and the index algorithm seem to be limiting factors for higher speedup.

6.3 Further Parallelism

In this section we analyze how to exploit further the parallelism by configuring the system with multiple PIM-32 chips. In addition, we compare the results obtained with a traditional n-way system. We simulated systems with 1 to 8 PIM-32 chips (*Aggressive*) and compared these results with 1-way to 8-way multiprocessor systems. For all the queries except *W3* we observed a near-linear speedup increase for the multiple PIM-32 configurations and the n-way configurations. Due to space limitations we only show the results for *W9* in Figure 3. Other results are presented in [30].

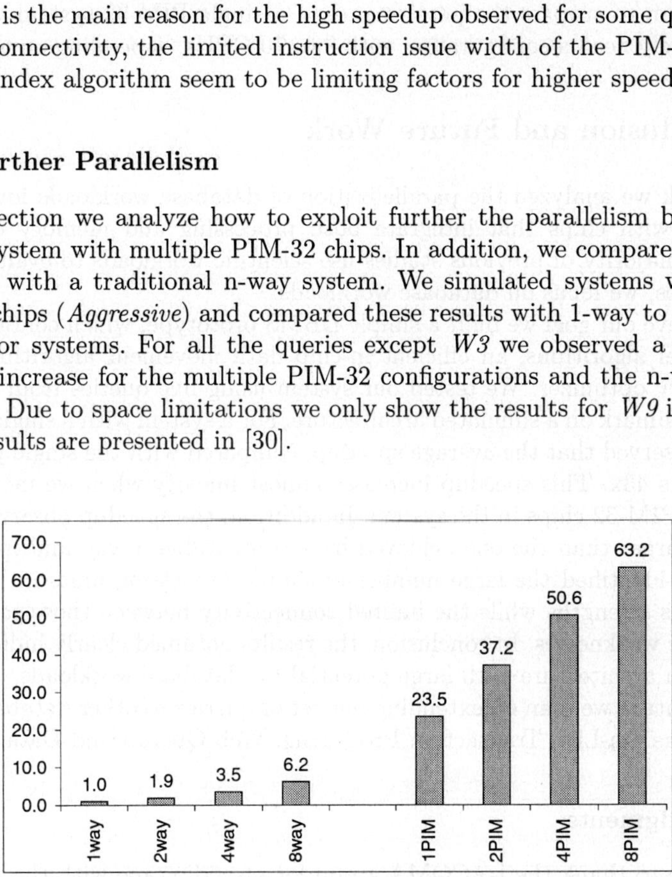

Fig. 3. Speedup results for *W9* for n-way and multiple PIM-n systems.

For the PIM-n systems, the slower increase rate for the speedup is due to the fact that with the larger number of chips the data is more spread and therefore: (1) the overheads such as initialization become significant; and (2) the data repartitioning becomes more costly. For the n-way configurations the reason for the near linear speedup is that although the algorithms scale well, the processor-to-memory interconnection becomes a bottleneck, as it may not support the necessary bandwidth.

Finally we may use these results to compare the PIM-n approach against the traditional n-way approach. Although it is very difficult to compare two such systems of equal cost (as there is no pricing information for the PIM-n chip) we can observe that the PIM-n approach performs significantly better. If we use the die size as an estimate of pricing and we use the values of the FlexRAM chip [20] we can estimate that the 32 processing elements are equivalent to 4 medium range CPUs. Nevertheless, assuming the best case of a linear speedup increase, for the n-way system to surpass the performance of a single PIM-32 system we would approximately need a configuration with 8 to 70 CPUs, depending on the query.

7 Conclusion and Future Work

In this work we analyzed the parallelization of database workloads for systems configured with chips that integrate both processing and memory elements. While the majority of previous studies use scientific workloads to evaluate such architectures, we focus on database workloads.

To achieve our goal we built a simple DBMS prototype, which contains modified parallel algorithms, an efficient in-chip data movement algorithm, and a simple query optimizer. We tested our system using five queries from the Wisconsin benchmark on a simulated architecture. For a system with a single PIM-32 chip, we observed that the average speedup, compared with the single processor execution, is 43x. This speedup increases almost linearly when we increase the number of PIM-32 chips in the system. In addition, the speedup observed is significantly larger than the one achieved by a comparable n-way multiprocessor. Finally, we identified the large number of simple processing units in PIM-n to be one of its strengths while the limited connectivity between those same units as one of its weaknesses. In conclusion, the results obtained clearly indicate that PIM-n is an architecture with large potential for database workloads.

In the future we plan on extending our set of queries to other database workloads such as On-Line Transaction Processing, Web Queries and Data Mining.

Acknowledgments

I would like to thank the I-ACOMA group for providing me with the resources to execute my experiments. I would also like to thank Yiannakis Sazeides and the anonymous reviewers for their comments.

References

1. Ailamaki, A., DeWitt, D., Hill, M., Wood, D.: DBMSs On A Modern Processor: Where Does Time Go? In: Proceedings of the 25th VLDB. (1999)
2. Barroso, L., Gharachorloo, K., Bugnion, E.: Memory System Characterization of Commercial Workloads. In: Proceedings of the 25th ISCA. (1998)
3. Trancoso, P., Larriba-Pey, J.L., Zhang, Z., Torrellas, J.: The Memory Performance of DSS Commercial Workloads in Shared-Memory Multiprocessors. In: Proceedings of the HPCA-3. (1997)
4. Keeton, K., Patterson, D., Hellerstein, J.: A Case for Intelligent Disks (IDISKs). SIGMOD Record (1998)

5. Acharya, M., Uysal, M., Saltz, J.: Active Disks: Programming Model, Algorithms and Evaluation. In: Proceedings of ASPLOS VIII. (1998)
6. Shatdal, A., Kant, C., Naughton, J.: Cache Conscious Algorithms for Relational Query Processing. In: Proceedings of the 20th VLDB. (1994)
7. Trancoso, P., Torrellas, J.: Cache Optimization for Memory-Resident Decision Support Commercial Workloads. In: Proceedings of the 1999 ICCD. (1999)
8. Rao, J., Ross, K.: Cache conscious indexing for decision-support in main memory. In: Proceedings of the VLDB. (1999)
9. Nyberg, C., Barclay, T., Cvetanovic, Z., Gray, J., Lomet, D.: AlphaSort: A RISC Machine Sort. In: Proceedings of the 1994 ACM-SIGMOD International Conference on Management of Data. (1994) 233–242
10. Boncz, P., Manegold, S., Kersten, M.: Database Architecture Optimized for the new Bottleneck: Memory Access. In: Proceedings of the 25th VLDB. (1999)
11. Ramirez, A., Barroso, L., Gharachorloo, K., Cohn, R., Larriba-Pey, J., Lowney, P., Valero, M.: Code Layout Optimizations for Transaction Processing Workloads. In: Proceedings of the Intl. Symposium on Computer Architecture. (2001) 155–164
12. Lo, J., Barroso, L., Eggers, S., Gharachorloo, K., Levy, H., Parekh, S.: An Analysis of Database Workload Performance on Simultaneous Multithreaded Processors. In: Proceedings of the 25th ISCA. (1998)
13. Barroso, L., Gharachorloo, K., McNamara, R., Nowatzyk, A., Qadeer, S., Sano, B., Smith, S., Stets, R., Verghese, B.: Piranha: A Scalable Architecture Based on Single-Chip Multiprocessing. In: Proceedings of the 27th ISCA. (2000)
14. Elliot, D., Snelgrove, W., Stumm, M.: Computational Ram: A Memory-SIMD Hybrid and its Application to DSP. In: Proceedings of the Custom Integrated Circuits Conference. (1992)
15. Rixner, S., Dally, W., Kapasi, U., Khailany, U., Lopez-Lagunas, A., Matterson, P., Owens, J.: A Bandwidth-Efficient Architecture for Media Processing. In: Proceedings of the 31st Micro. (1998)
16. Patterson, D., Anderson, T., Cardwell, N., Fromm, R., Keeton, K., Kozyrakis, C., Tomas, R., Yelick, K.: A Case for Intelligent DRAM. IEEE Micro (1997) 33–44
17. Oskin, M., Chong, F., Sherwood, T.: Active Pages: A Computation Model for Intelligent Memory. In: Proceedings of the 1998 ISCA. (1998)
18. Hall, M., Kogge, P., Koller, J., Diniz, P., Chame, J., Draper, J., LaCoss, J., Granacki, J., Brockman, J., Srivastava, A., Athas, W., Freeh, V., Shin, J., Park, J.: Mapping Irregular Applications to DIVA, a PIM-based Data-Intensive Architecture. In: Proceedings of Supercomputing 1999. (1999)
19. Mai, K., Paaske, T., Jayasena, N., Ho, R., Horowitz, M.: Smart Memories: A Modular Reconfigurable Architecture. In: Proceedings of the 27th ISCA. (2000)
20. Kang, Y., Huang, W., Yoo, S.M., Keen, D., Ge, Z., Lam, V., Pattnaik, P., Torrellas, J.: FlexRAM: Toward an Advanced Intelligent Memory System. In: Proceedings of the 1999 ICCD. (1999)
21. Stonebraker, M., Aoki, P., Seltzer, M.: The Design of XPRS. In: Proceedings of the VLDB. (1988)
22. Graefe, G.: Query Evaluation Techniques for Large Databases. ACM Computing Surveys **25** (1993) 73–170
23. DeWitt, D., Gerber, R., Graefe, G., Heytens, M., Kumar, K., Muralikrishna, M.: GAMMA: A high performance dataflow database machine. In: Proceedings of the VLDB. (1986)
24. Stonebraker, M.: The Design and Implementation of Distributed INGRES. In: The INGRES Papers. Addison-Wesley (1986)

25. Veenstra, J., Fowler, R.: MINT: A Front End for Efficient Simulation of Shared-Memory Multiprocessors. In: Proceedinds of the MASCOTS'94. (1994)
26. Krishnan, V., Torrellas, J.: An Execution-Driven Framework for Fast and Accurate Simulation of Superscalar Processors. In: Proceedings of the PACT. (1998)
27. Microelectronics, I.: Blue Logic SA-27E ASIC. News and Ideas of IBM Microelectronics (1999)
28. Iyer, S., Kalter, H.: Embedded dram technology: opportunities and challenges. IEEE Spectrum (1999)
29. Bitton, D., DeWitt, D., Turbyfill, C.: Benchmarking Database Systems, a Systematic Approach. In: Proceedings of the 9th VLDB. (1983)
30. Trancoso, P.: In-Memory Parallelism for Database Workloads. Technical report, University of Cyprus (In preparation)

Enforcing Cache Coherence at Data Sharing Boundaries without Global Control: A Hardware-Software Approach

H. Sarojadevi[1], S.K. Nandy[1], and S. Balakrishnan[2]

[1] Indian Institute of Science, India,
{saroja,nandy}@serc.iisc.ernet.in
[2] Philips Research Laboratories, The Netherlands,
srinivasan.balakrishnan@philips.com

Abstract. The technology and application trends leading to current day multiprocessor architectures such as chip multiprocessors, embedded architectures, and massively parallel architectures, demand faster, more efficient, and more scalable cache coherence schemes than the existing ones. In this paper we present a new scheme that has a potential to meet such a demand. The software support for our scheme is in the form of program annotations to detect shared accesses as well as release synchronizations that represent data sharing boundaries. A small hardware called Coherence Buffer (CB) with an associated controller, local to each processor forms the control unit to locally enforce cache coherence actions which are off the critical path. Our simulation study shows that a 8 entry 4-way associative CB helps achieve a speedup of 1.07 – 4.31 over full-map 3-hop directory scheme for five of the SPLASH-2 benchmarks (representative of migratory sharing, producer-consumer and write-many workloads), under Release Consistency model.

1 Introduction

Existing mechanisms to maintain cache coherence in Distributed Shared-Memory Multiprocessors (DSM) are hardware directory based, or compiler directed [1]. The directory based schemes have the disadvantage of large storage overhead, and limited scalability. They incur increased memory access latency and network traffic, since the coherence transactions are in the critical path of shared accesses. Optimized directory schemes [3–6] attempting to improve the cost-effectiveness of a directory protocol however inherit these disadvantages. Compiler assisted schemes with simple hardware support have been suggested as a viable alternative since they maintain cache coherence locally without the need for interprocessor communication and expensive hardware. However, their conservative approach results in inaccurate detection of stale data, resulting in unnecessary cache misses.

Building on the idea [7] of auto-invalidating cache lines at the expiry of their life time, we propose a programmer-centric approach to cache coherence in DSM systems, using a small hardware unit called Coherence Buffer (CB), that – (a)

B. Monien and R. Feldmann (Eds.): Euro-Par 2002, LNCS 2400, pp. 543–546.

obviates the need to maintain a directory by ensuring that the memory is consistent at release boundaries [1]; (b) enforces early and local coherence which are not in the critical path of a memory access – early coherence actions attempt to hide the latency by improving overlap of operations; (c) tends to reduce the bandwidth requirement by reducing the processor-memory traffic; (d) improves performance compared to directory based DSMs, is complexity effective, and scalable. The rest of the paper is organized as follows. Section 2 provides the details of our scheme. Section 3 presents the performance. In section 4 we summarize the contributions of our work.

2 The Coherence Buffer Based Scheme

This scheme derives support from the application by way of program annotations that identify shared variables and synchronizations, and from the architecture by way of using CB, which is a small cache like structure. Assuming a release consistent memory model, we start by identifying all shared accesses between two consecutive release boundaries. Status of distinct cache blocks corresponding to the detected accesses are recorded in the CB, using *state bits* – `Clean` (unmodified), `Dirty` (modified), and `Invalid` (free entry). The block address is further split into CB tag and CB index to facilitate CB lookup. A CB controller (CBC) carries out all CB related operations. At a release boundary, clean cache blocks corresponding to the CB entries are invalidated (using special request, AUTOINVAL), and dirty subblocks are written back as well as invalidated (using special request, AUTOINV&WB). On a capacity or conflict miss in the CB, between release boundaries, corresponding cache blocks are replaced, writing back only the dirty subblocks. These are early coherence actions which may seem disadvantageous because of the additional coherence transactions associated with the misses. But in reality, the performance improves due to increased overlap of coherence transactions with other memory operations.

A release boundary is marked by release fence instructions in a program. When a release fence is ready to graduate, the processor issues a special memory request, INVL_WB, to cache controller for flushing the CB. On a INVL_WB request, the CBC selects a valid CB entry, and sends the special coherence requests to the cache controller to enforce coherence. On receiving acknowledgments for all coherence requests, which may reach in an overlapped order, the CBC informs the cache controller, which in turn signals the processor about completion of the INVL_WB request, indicating that all coherence actions intended at the present release boundary are complete.

3 Performance Evaluation

Performance of the CB based scheme is evaluated using `RSIM` [8], an execution driven simulator for DSM systems. A 16 node system is simulated, with each

[1] We call the position of the *release* synchronization instance, indicated by a release fence in the execution trace, as *release boundary.*

node having 450MHz MIPS R10000 like processor, 64KB 2-way L1, 2MB 4-way L2 caches, both with 64 byte line size. The processor-memory interconnect is a 256 bit wide split-transaction bus with 100MHz clock. A 2D mesh, working at 150MHz connects various nodes. Details of the simulation environment is given in [2]. A full-map, 3-hop, three state (MSI) directory based scheme is used as the basis for comparison, for its simplicity and cost-effectiveness compared to enhanced directory schemes. Input includes a subset of SPLASH-2 benchmarks – Mp3d with 50000 particles, Water-Nsquared with 512 molecules, Radix with 512K keys, LU with 512X512 matrix and block size of 16, FFT with 256K points, and Quicksort with 32K integers – all compiled using Sparc V9 gcc compiler with `-O2 -funrollloop` option. LUOPT and FFTOPT are versions of LU and FFT respectively, compiled using ILP-specific optimizations [2] for function inlining and loop interchange. Through a parameter space study of the CB, we have found that a 8 entry, 4-way associative CB performs the best. Results corresponding to this CB configuration are presented in figure 1. The `Cohe` component represents stall time due to bursty coherence actions at a release boundary, which are present only in a CB based system. Miss rate in a CB based system is much higher than that in a directory based system (results tabulated in Fig. 1(b)). Whereas, the miss latency [3] significantly reduces due to reduced network traffic. With reduced average miss latency, coherence operations can find better overlap with data operations, which results in improved latency tolerance. These observations are supported by the fact that the CB scheme shows significant performance gains in all applications except Water-Nsquared which suffers from large synchronization overhead. We also observe that the ratio of the memory

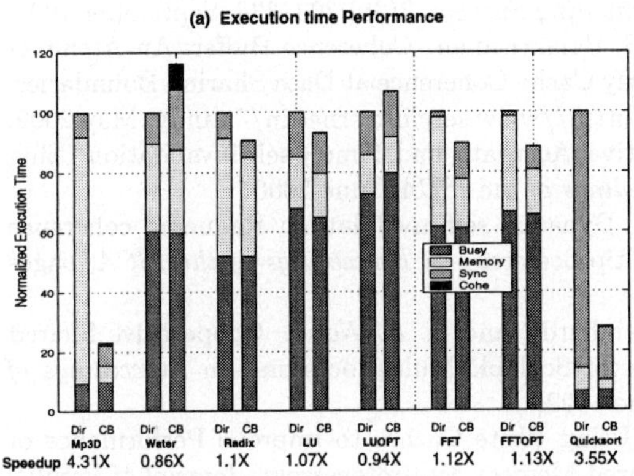

(b) Impact on Cache Performance

Application	Directory based DSM		CB based DSM	
	Miss rate	Average Miss latency	Miss rate	Average Miss latency
Mp3d	3.6330	524.31	4.2967	42.04
Water	1.4220	172.64	3.550	123.48
Radix	3.0169	70.99	29.464	16.25
LU	1.6745	21.05	6.7691	10.45
LUOPT	4.3906	16.75	60.470	14.88
FFT	2.2574	263.94	9.3236	47.66
FFTOPT	2.2874	293.44	9.6360	31.21
Quicksort	0.3546	1120.72	0.2160	150.56

Fig. 1. Performance; `Dir`: directory, `CB`: CB–based DSMs; `Busy`: CPU time, `Memory`: Memory stall time,`Sync`: Synchronization stall time, `Cohe`: Coherence stall time

[2] These optimizations cluster load misses close together to maximize the benefits of overlapping in a processor with support for Instruction Level Parallelism (ILP).

[3] Miss latency is the cycle time measured from an address generation point to the data arrival point of a memory reference.

stall time to the CPU busy time is the maximum for Mp3d, which is due to large number of cache misses to migratory blocks. The CB based system, by its access overlapping feature, provides optimization for migratory data to maximize the performance improvement. Hence the reason for the dramatic gains.

4 Conclusion

In this paper we have presented a programmer-centric approach to maintaining cache coherence in DSM systems. Our scheme obviates the need to maintain a directory, by ensuring that the memory is consistent at release synchronizations. The scheme uses a small cache, called Coherence Buffer (CB), with an associated controller, local to each processor to maintain the state of live shared variables between consecutive data sharing boundaries identified by release synchronizations (detected by annotations supplied by the compiler) in the program. Our scheme attempts to amortize the cost of global coherence by way of achieving early and local coherence. Through execution-driven simulations of 16 processor DSM configuration with MIPS R10000 type of processors, a speedup of 1.07 to 4.31 is obtained for a suite of SPLASH-2 benchmarks over an equivalent full-map directory protocol, under Release Consistent memory model. The performance improves because of taking coherence actions off the critical path, and overlapping them with other operations.

References

1. D. J. Lilja. Cache Coherence in Large-Scale Shared-Memory Multiprocessors: Issues and Comparisons. *ACM Computing Surveys*, 3(25):303–338, September 1993.
2. H. Sarojadevi, S.K. Nandy, and S. Balakrishnan. Coherence Buffer: An Architectural Support for Maintaining Early Cache Coherence at Data Sharing Boundaries. *Technical report, CAD Lab, IISc*, http://www.serc.iisc.ernet.in/~nandy, May 2002.
3. A.C. Lai and Babak Falsafi. Selective, Accurate, and Timely self-invalidation Using Last-Touch Prediction. In *Proceedings of the ISCA*, June 2000.
4. A. R. Lebeck and D. A. Wood. Dynamic self-invalidation: Reducing coherence overhead in Shared-memory multiprocessors. In *Proceedings of the ISCA*, pages 48–59, May 1995.
5. M. D. Hill, J. L. Larus, S. K. Reinhardt, and D. A. Wood. Cooperative Shared Memory: Software and Hardware for Scalable Multiprocessors. In *Proceedings of the ASPLOS*, pages 262–273, June 1992.
6. F. Dahlgren and P. Stenström. Using Write Caches to Improve Performance of Cache Coherence Protocols in Shared-Memory Multiprocessors. *Journal of Parallel and Distributed Computing*, pages 193–210, April 1995.
7. S.K.Nandy and Ranjani Narayan. An Incessantly Coherent Cache Scheme for Shared Memory Multithreaded Systems. Technical Report LCS, CSG-Memo 356, Massachusetts Institute of Technology, September 1994.
8. C. J. Hughes, V. S. Pai, P. Ranganathan, and S. V. Adve. Rsim: Simulating Shared-Memory Multiprocessors with ILP Processors. *IEEE Computer*, 35(2):40–49, February 2002.

CODACS Project: A Demand-Data Driven Reconfigurable Architecture

Lorenzo Verdoscia

Research Center on Parallel Computing and Supercomputers – CNR, Via Castellino, 111 80131 Napoli, Italy,
verdoscia.l@cps.na.cnr.it

Abstract. This paper presents CODACS (COnfigurable DAtaflow Computing System) architecture, a high performance reconfigurable computing system prototype with a highly scalable degree able to directly execute in hardware dataflow processes (dataflow graphs). The reconfigurable environment consists of a set of FPGA based *platform-processors* created by a set of identical Multi Purpose Functional Units (MPFUs) and a reconfigurable interconnect to allow a straightforward one-to-one mapping between dataflow actors and MPFUs. Since CODACS does not support the conventional processor cycle, the platform-processor computation is completely asynchronous according to the dataflow graph execution paradigm proposed in [8].

1 Introduction

The advent of in-circuit (re)programmable FPGAs(Field Programmable Gate Arrays) has enabled a new form of computing prompted by the (re)configurable computing paradigm [1] where a function calculation is computed configuring and interconnecting a number of cells. In particular, given its fine grain nature, the dataflow execution model is promising when applied to this platform because, as (re)configurable computing, it computes a dataflow graph configuring and interconnecting actors instead of carrying out in sequence a set of operations as a standard processor does.

However, despite the general scepticism regarding dataflow architectures due to its disappointing results, we believe they are still a valid proposal to increase performance and seriously attack, at least at a chip level, the von Neumann model.

There are at least four reasons that have motivated this proposal and consequently driven the design process [2]: the first is the demand to directly map in hardware dataflow graphs in dataflow mode; the second is the need to dispose of a straightforward data flow control and actor firing mechanism at a minimal hardware cost; the third is the requirement to reduce the continual LOAD and STORE operations and augment performance; the last is the possibility to adopt primitive functions of a functional language as assembler language for a processor.

B. Monien and R. Feldmann (Eds.): Euro-Par 2002, LNCS 2400, pp. 547–550.

2 CODACS Architecture

CODACS general architecture, shown in Fig. 1, is constituted by a set of identical nodes connected in a WK-Recursive topology [3] where each node is constituted by a *Smart Router Subsystem* (SRS), devoted to provide some kernel functions and all communication and routing activities, and a *Platform-Processor Subsystem* (PPS), devoted to execute dataflow graphs.

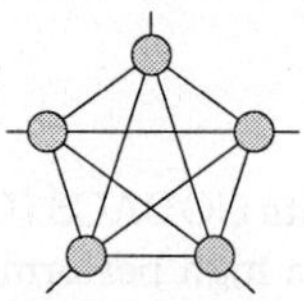

Fig. 1. CODACS Architecture connected as WK-recursive with $N_d = 5$ and $Level = 1$

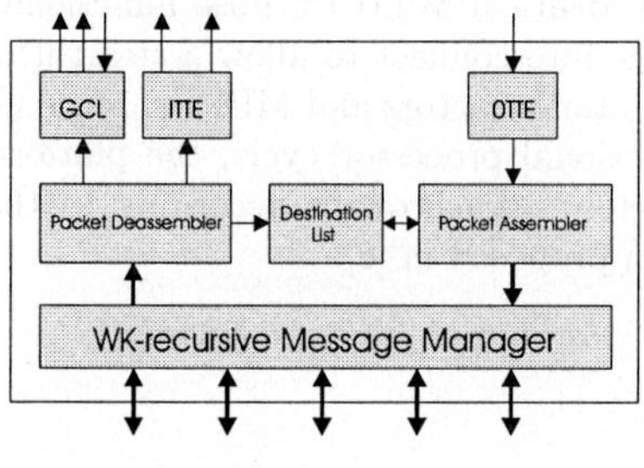

Fig. 2. Smart Router Subsystem Architecture

Smart Router Subsystem. When a message reaches a node, the SRS (Fig. 2) evaluates it. If that node is not the destination, the WK-recursive Message Manager (WKMM) routes the message through the appropriate output link according to the routing strategy described in [4]. If it is the destination node, the WKMM transfers the message to the Packet Disassembler (PD) for the processing. The PD unpacks it, evaluates its content, and transfers information to a) the Graph Configuration List (GCL), that contains the graph configuration table list assigned to the platform-processor; b) the Destination List (DL), that contains the list of the result destination node set (one set for each configuration); c) the Input Transfer Token Environment (ITTE), to transfer data tokens to the PPE.

In the ITTE, data token storage occurs in separate but associate buffers to transfer right and left tokens to MPFUs. When results coming from the PPE are ready inside the Output Token Transfer Environment (OTTE), the Packet Assembler (PA) scans the destination node list, associates nodes to results, prepares new messages (one for each receiver), and transfers them to the WKMM for delivering.

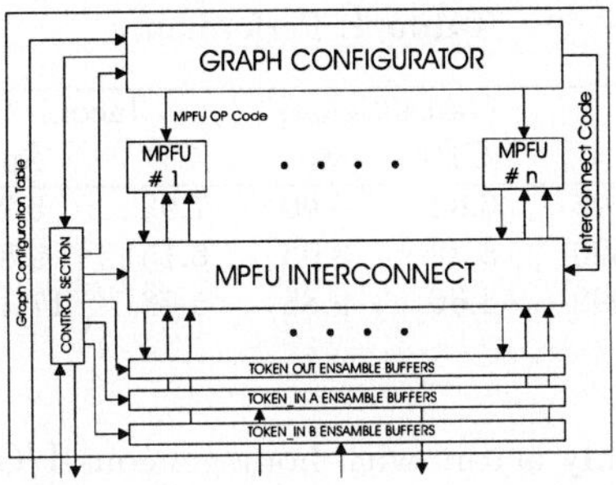

Fig. 3. Platform-Processor Subsystem Architecture

Platform-Processor Subsystem. This subsystem (Fig. 3) executes the dataflow graph assigned to that node. After receiving the graph configuration table from GCL, the Graph Configurator (GC) executes two operations: it sets the MPFU interconnect and assigns the operation code to each MPFU, thus carrying out the one-to-one correspondence (mapping) between graph nodes and computing units. Once the configuration phase terminates, it sends a signal to the control that enables the two Token_In buffers to start the computation. When a graph computation ends, results are stored in the Token_Out buffer and then transferred to the OTTE. If the same graph must process different input tokens (e.g. matrix inner product), the GC only checks for the input token availability. We point out that, thanks to the I/O Ensemble Buffers and Token Transfer Environments, the platform-processor and smart router environment become local to each subsystem allowing to overlap data load, message transfer, and computation activities.

3 Performance

To implement the proposed architecture, we have used ALTERA Quartus II [5] development software and a PCI board with 5 APEX20K15-C FPGA components. As a result [2], we obtained a platform-processor with 105 interconnected MPFUs that execute operations on 32 bit integer operands at time $t_{MPFU} = 1\ \mu$sec while the measured transfer time from/to the SRS $t_b = 8$ nsec. To evaluate CODACS, Jacobi and Gauss-Seidel iterative algorithms have been used because they face the same problem in a parallel and sequential mode. Table 1 shows the two performance indices CP (communication penalty) and Sp (speedup) for some values of n (number of equations). Due to fine grain dataflow operations, Jacobi performs better than Gauss-Seidel. However, most of the time is spent in communication.

4 Concluding Remarks

Principal features that distinguish this machine from similar ones ([6], [7]) are: the platform-processor executes dataflow graphs, including loops, without con-

Table 1. Performance

	Gauss-Seidel		Jacobi	
n	CP	Sp	CP	Sp
96	6.94	0.90	7.92	5.70
320	5.38	3.03	6.15	9.86
992	4.66	9.85	5.78	17.42

trol tokens but using only actors with homogeneous I/O conditions; MPFU assembly language is a high level programming language; graph execution and communication environments are separated to overlap data transfer and computation; no memory usage is required during a graph execution to reduce latency penalty; finally, it is characterized by a highly scalable general architecture. At the moment we are realizing a prototype employing 5 ALTERA APEX20K15-3C components.

Acknowledgments

This work was supported in part by Regione Campania under contract (POP-FESR Azione 5.4.2) "Sviluppo di Metodologie e Tecniche per la Progettazione e la Realizzazione dei Sistemi Avanzati di Supporto alle Decisioni" and by CNR-Agenzia2000 Program under Grant CNRC0014B3_008.

References

1. Gray, J.P., Kean, T.A.: Configurable Hardware: A New Paradigm for Computation. In *Proc. Decennial CalTech Conf. VLSI*, pages 277–293, Pasadena, CA, March 1989.
2. Verdoscia, L., Licciardo, G.: CODACS Project: The General Architecture and its Motivation. Technical report, CNR Research Center on Parallel Computing and Supercomputers, Via Castellino, 111 - 80131 Napoli - Italy, January 2002.
3. Chen, G.H., Du, D.R.: Topological Properties, Communication, and Computing on WK-Recursive Networks. *Networks*, 24:303–317, 1994.
4. Verdoscia, L., Vaccaro, R.: An Adaptive Routing Algorithm for WK-Recursive Topologies. *Computing*, 63(2):171–184, 1999.
5. ALTERA Corporation.: Quartus programmable logic development system and software. San Jose, CA, May 1999.
6. Singh, H., alii.: Morphosys: An Integrated Reconfigurable System for Data-Parallel and Computation Intensive Applications. *IEEE Trans. Computers*, 49(5):465–480, May 2000.
7. Murakawa, M., alii.: The GRD Chip: Genetic Reconfiguration of DSPs for Neural Network Processing. *IEEE Trans. on Computers*, 48(6):628–639, June 1999.
8. Verdoscia, L., Vaccaro, R.: A High-Level Dataflow System. *Computing*, 60(4):285–305, 1998.

Topic 9
Distributed Systems and Algorithms

Marios Mavronicolas and Andre Schiper

Topic Chairmen

This topic covers new exciting developments in the area of distributed systems and algorithms. It aims to address both theoretical and practical issues that arise in relation to the specification, design, implementation, verification and analysis of distributed systems and their algorithms. Loosely speaking, a distributed system is a collection of independent, autonomous computing elements that appears to its users as a single, coherent entity. Today, distributed systems are almost everywhere. Indeed, the immensely popular World Wide Web is arguably the biggest distributed system ever built. Even more so, the wide acceptance and use of Internet technologies and standards stresses the importance of distributed systems more than ever.

Particular areas of interest to the topic include, but are not limited to:

- techniques and formal methods for the design and analysis of distributed systems;
- architectures and structuring mechanisms for parallel and distributed systems;
- distributed operating systems and databases;
- resource sharing in distributed systems;
- openness and transparency issues in distributed systems;
- concurrency, performance and scalability in distributed systems;
- fault-tolerance in distributed systems;
- design and analysis of distributed algorithms;
- real-time distributed algorithms and systems;
- distributed algorithms in telecommunications;
- cryptography, security and privacy in distributed systems.

Out of twenty one submissions, seven papers were accepted and are presented in two sessions. The first session (Session 15) contains two regular papers and one short paper, whereas the second session (Session 16) contains one regular paper and three short papers. We next briefly describe the contributions of the accepted papers in these two sessions.

We start with Session 15. The paper by Datta *et al.*, *"A Self-stabilizing Token-Based k-out-of-l Exclusion Algorithm,"* presents the *first* self-stabilizing solution for the k-out-of-l mutual exclusion problem. In our opinion, the interest of the paper is not so much in the way the solution is obtained, but in the impressively precise and complete technical work. The paper by Ruiz *et al.*, *"An Algorithm for Ensuring Fairness and Liveness in Non-Deterministic Systems Based on Multiparty Interactions"* introduces the concept of k-fairness. The motivation for it is that the traditional notion of strong fairness is somehow limited, in the sense

B. Monien and R. Feldmann (Eds.): Euro-Par 2002, LNCS 2400, pp. 551–552.

that "eventual" selection for execution of an entity is too weak for a finite execution of the system; moreover, the new concept aims at preventing conspiracies (i.e., situations in which an unfortunate interleaving of atomic actions prevents an interaction from getting enabled in the first place). The last paper in the first session, namely the paper *"On Obtaining Global Information in a Peer-to-Peer Fully Distributed Environment"* by Jelasity and Preuss, is accepted as a short paper. The paper develops an interesting idea on how to extract global knowledge about a network out of the local knowledge of the individual nodes. More specifically, the paper addresses two specific problems whose solutions represent compilation of global knowledge: finding nodes in the network that fulfill certain criteria, and estimating the network size. The presented solutions are nice and elegant.

We continue with Session 16. The first paper (the only one regular paper in the session) is the paper *"A Fault-Tolerant Sequencer for Timed Asynchronous Systems"* by Baldoni *et al.* The paper pursues a primary-backup approach for providing a total order among processes in a distributed system. The nice idea behind the approach is to base total order on a fault-tolerant service allocating increasing integer numbers to client processes. The second paper by Gallard *et al.*, *"Dynamic Resource Management in a Cluster for Scalability and High-Availability,"* describes a distributed system that includes a distributed and dynamic service directory management; the system balances the directory management task across all active nodes in the cluster, while it can locate information about a specific service in an efficient manner. The session continues with the paper by Renault *et al.*, *"Progressive Introduction of Security in Remote-Write Communications with no Performance Sacrifice"*. This paper addresses a very interesting problem: how to make the remote-write operation secure while preserving performance. Different methods and their respective performance are compared in the paper. The session concludes with a paper on Java Applets, namely the paper by Suppi *et al.* *"Parasite: Distributed Processing Using Java Applets"*. The paper describes a novel infrastructure that enables distributed processing by transparently running Java Applets in the Web browsers of the users; this is, in our opinion, a remarkable idea. The paper describes the overall architecture of the distributed system, and the mechanisms employed to embed applets in Web pages as well.

Altogether, the contributions to Topic 09 at the Euro-Par in Paderborn overstress the wide variety of interesting, challenging and important issues in the area of distributed systems and algorithms. So, we are already looking forward to contributions to the topic to be made at the 2003 Euro-Par conference!

A Self-stabilizing Token-Based k-out-of-ℓ Exclusion Algorithm

Ajoy K. Datta[1], Rachid Hadid*[2], and Vincent Villain[2]

[1] Department of Computer Science, University of Nevada, Las Vegas
[2] LaRIA, Université de Picardie Jules Verne, France

Abstract. In this paper, we present the first self-stabilizing solution to the k out of ℓ exclusion problem [14] on a ring. The k out of ℓ exclusion problem is a generalization of the well-known mutual exclusion problem — there are ℓ units of the shared resources, any process can request some number k $(1 \leq k \leq \ell)$ of units of the shared resources, and no resource unit is allocated to more than one process at one time. The space requirement of the proposed algorithm is independent of ℓ for all processors except a special processor, called Root. The stabilization time of the algorithm is only $5n$, where n is the size of the ring.

Keywords: Fault-tolerance, k-out-of-ℓ exclusion, mutual exclusion, resource allocation, self-stabilization.

1 Introduction

Fault-tolerance is one of the most important requirements of modern distributed systems. Various types of faults are likely to occur at various parts of the system. The distributed systems go through the transient faults because they are exposed to constant change of their environment. The concept of self-stabilization [7] is the most general technique to design a system to tolerate arbitrary transient faults. A self-stabilizing system, regardless of the initial states of the processors and initial messages in the links, is guaranteed to converge to the intended behavior in finite time. In 1974, Dijkstra introduced the property of self-stabilization in distributed systems and applied it to algorithms for mutual exclusion.

The *ℓ-exclusion problem* is a generalization of the mutual exclusion problem–ℓ processors are now allowed to execute the critical section concurrently. This problem models the situation where there is a pool of ℓ units of a shared resource and each processor can request at most one unit. In the last few years, many self-stabilizing ℓ-exclusion algorithms have been proposed [2, 8–10, 18].

The *k-out-of-ℓ exclusion* approach allows every processor to request k $(1 \leq k \leq \ell)$ units of the shared resource concurrently, but, no unit is allocated to multiple processors at the same time [14]. One example of this type of resource sharing is the sharing of channel bandwidth: the bandwidth requirements vary among the requests multiplexing

* Supported in part by the Pôle de Modélisation de Picardie, France and the Fonds Social Européen.

B. Monien and R. Feldmann (Eds.): Euro-Par 2002, LNCS 2400, pp. 553–562.

on the channel. For example, the demand would be quite different for a video than an audio transmission request.

Algorithms for k-out-of-ℓ exclusion were given in [3, 12–15]. All these algorithms are permission-based: a processor can access the resource after receiving a permission from all the processors of the system [14, 15] or from the processors constituting the quorum it belongs to [12, 13].

Contributions. In this paper, we present the first self-stabilizing protocol for the k-out-of-ℓ exclusion problem. Our algorithm works on uni-directional rings and is *token-based*: a processor can enter its critical section, i.e., access the requested (k) units of the shared resource only upon receipt of k tokens. The space requirement of our algorithm is independent of ℓ for all processors except Root. The stabilization time of the protocol is only $5n$, where n is the size of the ring.

Outline of the Paper. In Section 2, we describe the model used in this paper, and present the specification of the problem solved. We propose a self-stabilizing k-out-of-ℓ exclusion protocol on rings in Section 3[1]. Finally, we make some concluding remarks in Section 4.

2 Preliminaries

2.1 The Model

The distributed system we consider in this paper is a uni-directional ring. It consists of a set of processors denoted by 0,1,..,n-1 communicating asynchronously by exchanging messages. Processors are anonymous. The subscripts 0,1,...,n-1 for the processors are used for the presentation only. We assume the existence of a distinguished processor (Processor 0), called Root. Each processor can distinguish its two neighbors: the left neighbor from which it can receive messages and the right neighbor it can send messages to. The left and right neighbors of Processor i are denoted by $i-1$ and $i+1$, respectively, where indices are taken modulo n. We assume that the message delivery time is finite but unbounded. We also consider a message to be in transit until it is processed by the receiving processor. Moreover, each link is assumed to be of bounded capacity, FIFO, and reliable (the messages are neither lost nor corrupted) during and after the stabilization phase. Our protocols are *semi-uniform* as defined in [6] — every processor with the same degree executes the same program, except one processor, Root. The messages are of the following form: $<$ *message-type*, *message-value* $>$. The *message-value* field is omitted if the message does not carry any value. Some messages contain more than one *message-value*. The program consists of a collection of actions. An action is of the form: <*guard*>$\longrightarrow$ < *statement*>. A *guard* is a boolean expression over the variables of the processor and/or an *input* message. A *statement* is a sequence of assignments and/or message sending. An action can be executed only if its guard evaluates to true. We assume that the actions are atomically executed, meaning that the evaluation of a guard and the execution of the corresponding statement of an action, if executed, are done in one atomic step. The atomic execution of an action of p is called a *step* of p. When several actions of a processor are simultaneously enabled, then only the first enabled

[1] Due to space limitations, the proof of correctness is omitted. See [5] for the proofs

action (as per the text of the protocol) is executed. The *state* of a processor is defined by the values of its variables. The *state* of a system is a vector of n+1 components where the first n components represent the state of n processors, and the last one refers to the multi-set of messages in transit in the links. We refer to the state of a processor and the system as a (*local*) *state* and *configuration*, respectively.

2.2 Self-stabilization

Definition 1 (Self-stabilization). *A protocol $\mathcal{P}$ is self-stabilizing for a specification $\mathcal{SP}$ (a predicate defined over the computations) if and only if every execution starting from an arbitrary configuration will eventually reach* (convergence) *a configuration from which it satisfies $\mathcal{SP}$ forever* (closure).

In practice, we associate to $\mathcal{P}$ a predicate $\mathcal{L_P}$ (called the *legitimacy predicate*) on the system configurations. $\mathcal{L_P}$ must satisfy the following property: Starting from a configuration α satisfying $\mathcal{L_P}$, $\mathcal{P}$ always behaves according to $\mathcal{SP}$, and any configuration reachable from α satisfies $\mathcal{L_P}$ (closure property). Moreover if any execution of $\mathcal{P}$ starting from an arbitrary configuration eventually reaches a configuration satisfying $\mathcal{L_P}$ (convergence property), we say that $\mathcal{P}$ stabilizes for $\mathcal{L_P}$ (hence for $\mathcal{SP}$). The worst delay to reach a configuration satisfying $\mathcal{L_P}$ is called the *stabilization time*.

2.3 The k-out-of-ℓ Exclusion Problem

In this section, we present the specification of the (k,ℓ)-exclusion problem. We will define the usual properties: *safety* and *fairness*. We also need to add another performance metric, called (k,ℓ)-*liveness*. An algorithm satisfying this property attempts to allow several processors to execute their critical section simultaneously. In order to formally define this property and get the proper meaning of the property, we assume that a processor can stay in the critical section forever. Note that we make this assumption only to define this property. Our algorithm does assume that the critical sections are finite. Informally, satisfying the (k,ℓ)-liveness means that even if some processors are executing their critical section for a long time, eventually some requesting processors can enter the critical section provided the safety and fairness properties are still preserved.

Definition 2. (k,ℓ)**-Exclusion Specification**

1. Safety: *Any resource unit can be used by at most one process at one time.*

2. Liveness: *(a)* Fairness: *Every request is eventually satisfied.*

(b) (k,ℓ)-liveness: *Let I be the set of processors executing their critical section forever, and every processor $i \in I$ using k_i units of the shared resource such that $\sum_{i\in I} k_i < \ell$. Let $\alpha = \ell - \sum_{i\in I} k_i$. Let J be the set of processors requesting the entry to their critical section such that every processor $j \in J$ needs $k_j \leq \alpha$ units of the resource. Then some of the processors in J will be eventually granted entry to the critical section provided they maintain the* safety *and* fairness *properties.*

Note that fairness and (k,ℓ)-liveness properties would not be related with each other if we did not include the fairness property in the (k,ℓ)-liveness property. On one hand, a classical mutual exclusion protocol can be a solution of the (k, ℓ)-exclusion problem which does not satisfy the (k, ℓ)-liveness property. On the other hand, it is easy to design a protocol that always allows a processor in J (as defined in (k, ℓ)-liveness property) to enter the critical section. However, if the set J remains non-empty forever, then a processor requesting more than α units (hence not in J) may never get a chance to enter the critical section (starvation).

In uni-directional rings, we can use a token-based algorithm to maintain an ordering among the requests by circulating the tokens in a consistent direction. Then this solution would guarantee both fairness and (k, ℓ)-liveness properties.

In the k-out-of exclusion problem, if the maximum number of units (denoted as K) any process can request to access the critical section is known, then the space requirement depends only on K. Obviously, $K \leq \ell$.

A k-out-of-ℓ exclusion algorithm is self-stabilizing if every computation starting from an arbitrary initial configuration, eventually satisfies the safety, fairness, and (k, ℓ)-*liveness* requirements.

2.4 Parametric Composition

The parametric composition of protocols P_1 and P_2 was first presented in [10]. This is a generalization of the *collateral composition* of [16] and *conditional composition* of [4]. It allows both protocols to read the variables written by the other protocol. This scheme also allows the protocols to use the predicates defined in the other protocol. Informally, P_1 can be seen as a tool used by P_2, where P_2 calls some "public" functions of P_1 (we use the term *function* here with a generic meaning: it can be the variables used in the collateral composition or the predicates as in the conditional composition), and P_1 can also use some functions of P_2 via the "parameters".

Definition 3 (Parametric composition). *Let P_1 be a protocol with a set of* parameters *and a* public *part. Let P_2 be a protocol such that P_2 uses P_1 as an "external protocol". P_2 allows P_1 to use some of its functions (function may return no result) by using the parameters defined in P_2. P_1 allows protocol P_2 to call some of its functions by using the public part defined in P_1. The* parametric composition *of P_1 and P_2, denoted as $P_1 \triangleright_P P_2$, is a protocol that has all the variables and all the actions of P_1 and P_2.*

The implementation scheme of P_1 and P_2 is given in Algorithm 1. Let $\mathcal{L}_1$ and $\mathcal{L}_2$ be predicates over the variables of P_1 and P_2, respectively. We now define a *fair* composition w.r.t. both protocols and define what it means for a parametrically composed algorithm to be self-stabilizing.

Definition 4 (Fair execution). *An execution e of $P_1 \triangleright_P P_2$ is fair w.r.t. P_i ($i \in \{1, 2\}$) if either e is finite, or e contains infinitely many steps of P_i, or contains an infinite suffix in which no step of P_i is enabled.*

Definition 5 (Fair composition). *$P_1 \triangleright_P P_2$ is fair w.r.t. P_i ($i \in \{1, 2\}$) if any execution of $P_1 \triangleright_P P_2$ is fair w.r.t. P_i.*

The following composition theorem and its corollary are obvious:

Algorithm 1 $P_1 \triangleright_P P_2$

Protocol $\mathbf{P_1}(\mathbf{F'}_1 : TF_1, \mathbf{F'}_2 : TF_2, ..., \mathbf{F'}_\alpha : TF_\alpha)$; **Public** $\mathbf{Pub}_1 : TP_1$ /* definition of Function $\mathbf{Pub}_1$ */ ... $\mathbf{Pub}_\beta : TP_\beta$ /* definition of Function $\mathbf{Pub}_\beta$ */ **begin** ... [] $< Guard > \longrightarrow < statement >$ /* Functions $\mathbf{F'}_i$ can be used in Guards and/or statements */ ... **end**	**Protocol** $\mathbf{P_2}$ **External Protocol** $\mathbf{P_1}(\mathbf{F}_1 : TF_1, \mathbf{F}_2 : TF_2, ..., \mathbf{F}_\alpha : TF_\alpha)$; **Parameters** $\mathbf{F}_1 : TF_1$ /* definition of Function $\mathbf{F}_1$ */ ... $\mathbf{F}_\alpha : TF_\alpha$ /* definition of Function $\mathbf{F}_\alpha$ */ **begin** ... [] $< Guard > \longrightarrow < statement >$ /* Functions $\mathbf{P_1}.\mathbf{Pub}_i$ can be used in Guards and/or statements */ ... **end**

Theorem 1. *If the composition $P_1 \triangleright_P P_2$ is fair w.r.t. P_1, is fair w.r.t. P_2 if P_1 is stabilized for $\mathcal{L}_1$, protocol P_1 stabilizes for $\mathcal{L}_1$ even if P_2 is not stabilized for $\mathcal{L}_2$, and protocol P_2 stabilizes for $\mathcal{L}_2$ if $\mathcal{L}_1$ is satisfied, then $P_1 \triangleright_P P_2$ stabilizes for $\mathcal{L}_1 \wedge \mathcal{L}_2$.*

Corollary 1. *Let $P_1 \triangleright_P P_2$ be a self-stabilizing protocol. If Protocol P_1 stabilizes in t_1 for $\mathcal{L}_1$ even if P_2 is not stabilized for $\mathcal{L}_2$ and Protocol P_2 stabilizes in t_2 for $\mathcal{L}_2$ after P_1 is stabilized for $\mathcal{L}_1$, then $P_1 \triangleright_P P_2$ stabilizes for $\mathcal{L}_1 \wedge \mathcal{L}_2$ in $t_1 + t_2$.*

3 Self-stabilizing k-out-of-ℓ Exclusion Protocol

The protocol presented in this section is token-based, meaning that a requesting processor receiving k ($1 \leq k \leq \ell$) tokens can enter the critical section. The protocol is based on a couple of basic ideas. First, we need a scheme to circulate ℓ tokens in the ring such that a processor cannot keep more than k tokens while it is in the critical section. Second, we use a method to make sure that any requesting processor eventually obtains the requested tokens. We use the parametric composition of two protocols: *Controller* (see Algorithm 2) and *ℓ-Token-Circulation* (see Algorithms 3 and 4), denoted as *Controller* $\triangleright_P$*ℓ-Token-Circulation*. We describe these two protocols next.

Algorithm 2 Controller.

For Root	**For Other Processors**
Controller(START, COUNT-ET) **Variables** MySeq: 0..MaxSeq	**Controller**(T-ENABLED) **Variables** MySeq: 0..MaxSeq

(A_{c1}) [] (**receive** $<CToken, Seq>$) $\wedge$ $(MySeq = Seq) \longrightarrow$
$MySeq := MySeq + 1$
START
send $<CToken, MySeq>$
	(A_{c4}) [] (**receive** $<CToken, Seq>$) $\longrightarrow$
	if $(MySeq \neq Seq)$ **then**
	$MySeq := Seq$
	for $t = 1$ **to** T-ENABLED **do**
	send $<Enabled\_Token>$
	send$<CToken, MySeq>$

(A_{c2}) [] **receive** $<Enabled\_Token> \longrightarrow$ COUNT-ET	(A_{c5}) [] **receive** $<Enabled\_Token> \longrightarrow$ **send** $<Enabled\_Token>$
(A_{c3}) [] **timeout** $\longrightarrow$ **send** $<CToken, MySeq>$	

Controller. The protocol **Controller** (presented in Algorithm 2) implements several useful functions in the process of designing the k-out-of-ℓ exclusion algorithm. The controller keeps track of the number of tokens in the system. If this number is less (more) than ℓ, it replenishes (resp., destroys) tokens to maintain the right number (ℓ) of tokens in the system. The main burden of the above tasks (of the controller) is taken by Root. Root maintains two special variables C_e and C_f to implement these tasks (in Algorithm 3). We should point out here that these two variables are maintained only at Root. The detailed use of these variables and the implementation of the controller are explained below.

Root periodically initiates a status checking process by sending a special token, called $CToken$ (Actions A_{c1} and A_{c3} of Algorithm of 3). (Note that we refer to the token used by the controller as $CToken$ to distinguish it from the tokens used by the k-out-of-ℓ exclusion algorithm.) The $CToken$ circulation scheme is similar to the ones in [1, 17]. Every time Root initiates a $CToken$, it uses a new sequence number ($MySeq$) in the $< CToken, Seq >$ message (Actions A_{c1} and A_{c3}). Other processors use one variable ($MySeq$) to store the old and new sequence numbers from the received $CToken$ messages (Actions A_{c4}). Now, we describe the maintenance of C_e and C_f at Root. Variable C_e records the number of "enabled tokens" in the system. Processors maintain two variables T_h and T_d in Algorithm 3. T_h indicates the number of tokens received (or originally held) by a processor. But, if a processor i is waiting to enter the critical section, i may be forced to "disable" some of these originally held active tokens. (We will describe this process in detail in the next paragraph.) T_d represents the number of disabled tokens. The disabled tokens cannot be used by a processor to enter the critical section until they are enabled later. The difference between T_h and T_d is what we call the "enabled tokens" in a processor. This is computed by Function T-ENABLED in Algorithm 3. On receiving a $CToken$ message from Root, a processor i computes the number of enabled tokens at i and then sends the same number of $<Enabled\_Token>$ message to its right number (Action A_{c4}). These $<Enabled\_Token>$ messages are forwarded by using Action A_{c5}. $<Enabled\_Token>$ messages eventually arrive at Root which then calculates the value of C_e (Action A_{c2}). Upon entering or exiting the critical section, processors send the extra enabled tokens (by using $<Token>$ message) to their right neighbor. As these $<Token>$ messages traverse the ring, the processors either use them (if needed) or forward to their right neighbor. The total number of these "free" enabled tokens are saved in C_f at Root. (See Algorithm 3 and 4 for details.)

Self-stabilizing ℓ-Token-Circulation. We briefly describe the interface between the ℓ-exclusion protocol and application program invoking the k-out-of-ℓ exclusion protocol. The interface comprises of three functions as described below:

1. Function STATE returns a value in $\{Request, In, Out\}$. The three values $Request$, In, and Out represent three modes of the application program "requesting to enter", "inside", and "outside" the critical section, respectively.

2. Function NEED returns the number of resource units (i.e., the tokens) requested by a processor.

3. Function ECS does not return a value. This function is invoked by the ℓ-exclusion protocol to send the permission to the application process to enter the critical section.

Algorithm 3 ℓ-Token-Circulation (Header).

```
For Root
ℓ-Exclusion(STATE() : {Request, In, Out},
        NEED(): 0..k (k ≤ K ≤ ℓ), ECS())
External Controller(START(),COUNT-ET())
Parameters
  Function START()
    if (Ce + Cf + MyTc + MyTa) > ℓ then
      send <Disable>
    else for t = 1 to (ℓ−(Ce+Cf+(MyTc+MyTa)) do
        send <Token>
      if (MyTc + MyTa > 0) then
        send <Allocate, MyTc + MyTa, MyOrder>
      send <Collect, 0>
      Ce := Th − Td;   Cf := 0
      MyTc := 0;   MyTa := 0
  end Function
  Function COUNT-ET()
    if Cf < ℓ then Cf := Cf + 1
  end Function
–
  Function LOCK(): Boolean
    Return(STATE() = Request ∧(0 < Th < NEED))
  end Function
Variables   Th, Td : 0..K (K ≤ ℓ)
            Ce, Cf: 0..ℓ
            MyTc: 0..MaxVc (K ≤ MaxVc ≤ ℓ)
            MyTa: 0..Min(2 × MaxVc, ℓ)
            MyOrder : Boolean

For Other Processors
ℓ-Exclusion(STATE() : {Request, In, Out},
        NEED(): 0..k (k ≤ K ≤ ℓ), ECS())
External Controller(T-ENABLED())
Parameters
  Function T-ENABLED(): Integer
    Return(Th − Td)
  end Function
–
  Function LOCK(): Boolean
    Return(STATE = Request ∧
        (0 < Th < NEED))
  end Function
Variables Th, Td : 0..K (K ≤ ℓ)
          MyTc: 0..MaxVc (K ≤ MaxVc ≤ ℓ)
          MyTa: 0..Min(2 × MaxVc, ℓ)
          MyOrder: Boolean
```

The basic objective of the algorithm in this section is to implement a circulation of ℓ tokens around the ring. A processor requesting k units of the resource can enter the critical section upon receipt of k tokens. The obvious approach to implement this would be the following: A requesting processor holds on to the tokens it receives until it gets the requested number (k) of tokens. When it receives k tokens, it enters the critical section. Upon completion of the critical section execution, it releases all the k tokens by sending them out to the next processor in the ring. Unfortunately, the above hold-and-wait approach is prone to deadlocks. Let α be the number of the (critical section entry) requesting processors in the system and β the total number of tokens requested by α processors. If $\beta \geq \ell + \alpha$, then ℓ tokens can be allocated in such a manner that every requesting processor is waiting for at least one token. So, the system has reached a deadlock configuration. We solve the deadlock problem by pre-emptying tokens. The method works in two phases as follows: 1. At least K tokens are disabled by pre-empting tokens from some processors. (Note that by definition, $k \leq K \leq \ell$.)

2. The disabled tokens are then used to satisfy the request of both the first waiting processor (w.r.t. the token circulation) with disabled tokens and the privileged[2] processor (say i). Processor i then enters the critical section. In order to ensure both fairness and (k, ℓ)-liveness, we construct a fair order in the ring (w.r.t. the token circulation) as follows: Every processor maintains a binary variable $MyOrder$ ($MyOrder \in \{true, false\}$). In Algorithms 3 and 4, two messages are used to implement the above two phases: <$Collect, T_c$> and <$Allocate, T_a, Order$> message. Root initiates both messages in

[2] The privileged processor is the first processor (w.r.t. the token circulation) whose $MyOrder$ is equal to that of Root. If all processors have their $MyOrder$ equal, then the privileged processor is Root.

Algorithm 4 ℓ-Token-Circulation (Actions).

For Root

(A_{l1}) [] STATE $\in \{Request, Out\} \wedge$
$(T_h + T_d > 0) \longrightarrow$
if STATE $= Out$ then
for $k = 1$ to $T_h - T_d$ do
send $<Token>$
$T_d := 0; T_h := 0$
else if $(T_h - T_d \geq$ NEED) then ECS

(A_{l2}) [] (receive $<Token>$) $\longrightarrow$
if $(C_f + C_e < \ell)$ then
$C_e := C_e + 1$
if (STATE $\in \{Out, In\}$) then
send $< Token >$
else if $T_h < NEED$ then $T_h := T_h + 1$
else $T_d := T_d - 1$

(A_{l3}) [] ((receive $<Allocate, T_a, Order>$)
$\wedge(MyOrder = Order)) \longrightarrow$
$MyT_a := T_a$
if (STATE $= Request$) then
if $(MyT_a \geq NEED - (T_h - T_d))$ then
$MyT_a := MyT_a - (NEED - (T_h - T_d))$
$T_h := NEED; T_d := 0$
$MyOrder := \overline{MyOrder}$
else if $(T_d \geq MyT_a)$ then
$T_d := T_d - MyT_a$
else $T_h := T_h + (MyT_a - T_d)$
$T_d := 0$
$MyT_a := 0$
else $MyOrder := \overline{MyOrder}$

(A_{l4}) [] (receive $<Allocate, T_a, Order>$) $\longrightarrow$
$MyT_a := T_a$
if $(T_d > 0 \wedge$ LOCK) then
if $(MyT_a \geq NEED - (T_h - T_d))$ then
$MyT_a := MyT_a - (NEED - (T_h - T_d))$
$T_h := NEED; T_d := 0$

(A_{l5}) [] (receive $<Collect, T_c>$)) $\longrightarrow$
$MyT_c := T_c$
if LOCK then
$MyT_c := Min(MyT_c + (T_h - T_d), MaxV_c)$
$T_d := T_d + (MyT_c - T_c)$

(A_{l6}) [] (receive $<Disable>$) $\longrightarrow$
$T_d := T_h$

For Other Processors

(A_{l7}) [] STATE $\in \{Request, Out\} \wedge$
$(T_h + T_d > 0) \longrightarrow$
if STATE $= Out$ then
for $k = 1$ to $T_h - T_d$ do
send $<Token>$
$T_d := 0; T_h := 0$
else if $(T_h - T_d \geq$ NEED) then ECS

(A_{l8}) [] (receive $<Token>$) $\longrightarrow$
if (STATE $\in \{Out, In\}$) then
send $< Token >$
else if $T_h < NEED$ then
$T_h := T_h + 1$
else $T_d := T_d - 1$

(A_{l9}) [] ((receive $<Allocate, T_a, Order>$)
$\wedge(MyOrder \neq Order)) \longrightarrow$
$MyT_a := T_a$
if (STATE $= Request$) then
if $(MyT_a \geq NEED - (T_h - T_d))$ then
$MyT_a := MyT_a - (NEED - (T_h - T_d))$
$T_h := NEED; T_d := 0$
$MyOrder := Order$
if $MyT_a > 0$ then
send $<Allocate, MyT_a, Order>$
else if $(T_d \geq MyT_a)$ then
$T_d := T_d - MyT_a$
else $T_h := T_h + (MyT_a - T_d)$
$T_d := 0$
$MyT_a := 0$
else $MyOrder := Order$
send $<Allocate, MyT_a, Order>$

(A_{l10}) [] (receive $<Allocate, T_a, Order>$) $\longrightarrow$
$MyT_a := T_a$
if $(T_d > 0 \wedge$ LOCK) then
if $(MyT_a \geq NEED - (T_h) - T_d)$ then
$MyT_a := MyT_a - (NEED - (T_h - T_d))$
$T_h := NEED; T_d := 0$
if $MyT_a > 0$ then
send $<Allocate, MyT_a>$
else send $<Allocate, MyT_a>$

(A_{l11}) [] (receive $<Collect, T_c>$)) $\longrightarrow$
$MyT_c := T_c$
if LOCK then
$MyT_c := Min(MyT_c + (T_h - T_d), MaxV_c)$
$T_d := T_d + (MyT_c - T_c)$
send $<Collect, MyT_c>$

(A_{l12}) [] (receive $<Disable>$) $\longrightarrow$
$T_d := T_h$; send $<Disable>$

Function START (Lines 05 and 06 of Algorithm 3). Root executes Function START before initiating a new $CToken$ message (see Algorithm 2).

The receipt of a $Collect$ message at a processor i has the following effect (see Actions A_{l5} and A_{l11} of Algorithms 4): If Processor i is waiting to enter the critical section (because it did not receive enough tokens yet) (verified by using Function LOCK), then the current enabled tokens at i are marked disabled and these tokens are added to the pool of collected tokens in the $Collect$ message. Finally, i forwards the $Collect$ message to its right neighbor. The field T_c in $<Collect, T_c>$ message represents the number of disabled tokens collected so far from the processors in the ring. Every processor maintains a variable MyT_c corresponding to the message field T_c.

When Root receives the $Collect$ message back (Action A_{l5}), it stores the total number of disabled tokens (collected from all the other processors) in its own variable MyT_c. When a processor i receives an $<Allocate, T_a, Order>$ message, (the field $Order$ corresponds to $MyOrder$ of Root), i does the following (see Actions A_{l3}, A_{l4}, A_{l9}, and A_{l10}): If i is waiting to enter the critical section (i.e., i is requesting and contains at least one disabled token) or i is privileged (i.e., i is requesting and $MyOrder_i = Order$), then it will use some (or all) tokens from the pool of available tokens in the message field T_a. This would allow i to enter the critical section by executing Action A_{l7} (Action A_{l1} for Root). If there are some available tokens, i.e., T_a is not zero, i will pass on those tokens to its right neighbor by sending an *Allocate* message. Thus, either Root receives an *Allocate* message containing some left-over tokens, or all the available tokens are consumed by other processors. It should be noted that *Allocate* message delivers its tokens (available in T_a) to a privileged processor i even if i's request cannot be granted (T_a is not enough) (see Actions A_{l3} and A_{l9}). But, if i is waiting, then *Allocate* message delivers its tokens to i only if its request can be granted (T_a is enough) (see Actions A_{l4} and A_{l10}). As discussed earlier, Root maintains two special counters: C_e and C_f. The sum of C_e, C_f, MyT_c, and MyT_a represents the total number of tokens in the ring at the end of the $CToken$ traversal. If this number is more than ℓ, then Root destroys (or disables) all the tokens by sending a special message $<Disable>$ (Lines 00-01 and Actions A_{l6} and A_{l12}). But, if Root sees that there are some missing tokens in the ring, it replenishes them to maintain the right number (ℓ) of tokens in the system (Lines 03-04).

Proof Outline. The movement of the $CToken$ and enabled tokens are independent of each other except that they are synchronized at the beginning of a new traversal of the $CToken$ (Action A_{c1} and Function START). So, we can claim that the composed *Controller* $\triangleright_P$ ℓ-$Token$-$Circulation$ is fair w.r.t. *Controller*. We can borrow the result of [1, 17] to claim that the $CToken$ stabilizes for the predicate "there exist only one $CToken$" in two $CToken$ traversal time, i.e., in $2n$. By the controller (Algorithm 2) (it maintains the right number ℓ of tokens in the system in at most three more $CToken$ traversal time) and the mechanism of pre-empting tokens, we can claim that deadlock cannot occur ($deadlock$-$freeness$). Moreover, it ensures the (k, ℓ)-liveness. By Algorithms 3 and 4 ($MyOrder$ construction), every processor i will be eventually privileged and i'request will eventually have higher priority than the rest of the requests in the system. Therefore, the composed *Controller* $\triangleright_P$ *ℓ-Token-Circulation* does not cause starvation of any processor. Then, our final result follows from Theorem 1 and Corollary 1: *Controller* $\triangleright_P$ *ℓ-Token-Circulation* stabilizes for the k-out-of-ℓ exclusion specification — safety, fairness, and (k, ℓ)-liveness — in at most five $CToken$ traversal time i.e., $5n$.

4 Conclusions

In this paper, we present the first self-stabilizing protocol for k-out-of-ℓ exclusion problem. We use a module called controller which can keep track of the the number of tokens in the system by maintaining only a counter variable only at Root. One nice characteristic of our algorithm is that its space requirement is independent of ℓ for all processors except Root. The stabilization time of the protocol is $5n$. Our protocol works on uni-directional rings. However, we can use a self-stabilizing tree construction protocol and the Euler tour of the tree (virtual ring) to extend the algorithm for a general network.

References

1. Afek, Y., Brown, G.M : Self-stabilization over unreliable communication media. Distributed Computing, Vol. 7 (1993) 27–34
2. Abraham, U., Dolev, S., Herman, T., Koll, I. : Self-Stabilizing ℓ-exclusion. In Proceedings of the third Workshop on Self-Stabilizing Systems, International Informatics Series 7, Carleton University Press (1997) 48–63
3. Baldoni, R. : An $O(N^{M/M+1})$ distributed algorithm for the k-out of-M resources allocation problem. In Proceedings of the 14th conference on Distributed Computing and System (1994) 81-85.
4. Datta, AK., Gurumurthy, S., Petit, F., Villain V. : Self-stabilizing network orientation algorithms in arbitrary rooted networks. In Proceedings of the 20th IEEE International Conference on Distributed Computing Systems (2000) 576–583
5. Datta, AK., Hadid, R., Villain V. : A Self-stabilizing Token-Based k-out-of-ℓ Exclusion Algorithm. Technical Report RR 2002-04, LaRIA, University of Picardie Jules Verne (2002).
6. Dolev, D., Gafni, E., Shavit, N. : Toward a non-atomic era: ℓ-exclusion as test case. In Proceeding of the 20th Annual ACM Symposium on Theory of Computing, Chicago (1988) 78–92
7. Dijkstra, EW. : Self stabilizing systems in spite of distributed control. Communications of the Association of the Computing Machinery, Vol. 17, No. 11 (1974) 643–644
8. Flatebo, M., Datta, AK., Schoone, AA. : Self-stabilizing multi-token rings. Distributed Computing, Vol. 8 (1994) 133-142
9. Hadid, R. : Space and time efficient self-stabilizing ℓ-exclusion in tree networks. In Journal of parallel and distributed computing. To appear.
10. Hadid, R., Villain, V : A new efficient tool for the design of Self-stabilization ℓ-exclusion algorithms: the controller, In Proceedings of the 5th International Workshop, WSS (2001) 136-151.
11. Lamport , L. : Time, clocks, and the ordering of events in a distributed system. Communications of ACM, Vol. 21 (1978) 145-159
12. Manabe, Y., Tajima, N. : (h, k)-Arbiter for h-out of-k mutual exclusion problem. In Proceedings of the 19th Conference on Distributed Computing Systems, (1999) 216-223.
13. Manabe, Y., Baldoni, R., Raynal, M., Aoyagi, S.: k-Arbiter: A safe and general scheme for h-out of-k mutual exclusion. Theoretical Computer Science, Vol. 193 (1998) 97-112
14. Raynal, M. : A distributed algorithm for the k-out of-m resources allocations problem. In Proceedings of the 1st conference on Computing and Informations, Lecture Notes in Computer Science, Vol. 497 (1991) 599-609
15. Raynal, M. : Synchronisation et état global dans les systèmes répartis. Eyrolles, collection EDF (1992)
16. Tel, G. : Introduction to distributed algorithms. Cambridge University Press (1994)
17. Varghese G. : Self-stabilizing by counter flushing. Technical Report, Washington University (1993)
18. Villain V. A Key Tool for Optimality in the State Model. DIMACS Workshop on Distributed Data and Structures, Proceedings in Informatics 6, Carleton Scientific, pages 133-148, 1999.

An Algorithm for Ensuring Fairness and Liveness in Non-deterministic Systems Based on Multiparty Interactions*

David Ruiz, Rafael Corchuelo, José A. Pérez, and Miguel Toro

Universidad de Sevilla, E. T. S. Ingenieros Informáticos,
Av. de la Reina Mercedes s/n, Sevilla, E–41012, Spain
druiz@lsi.us.es, http://tdg.lsi.us.es

Abstract. Strong fairness is a notion we can use to ensure that an element that is enabled infinitely often in a non–deterministic programme, will eventually be selected for execution so that it can progress. Unfortunately, "eventually" is too weak to induce the intuitive idea of liveliness and leads to anomalies that are not desirable, namely *fair finiteness* and *conspiracies*. In this paper, we focus on non–deterministic programmes based on multiparty interactions and we present a new criteria for selecting interactions called *strong k–fairness* that improves on other proposals in that it addresses both anomalies simultaneously, and k may be set a priori to control its goodness. We also show our notion is feasible, and present an algorithm for scheduling interactions in a strongly k–fair manner using a theoretical framework to support the multiparty interaction model. Our algorithm does not require to transform the source code to the processes that compose the system; furthermore, it can deal with both terminating and non–terminating processes.

1 Introduction

Fairness is an important liveliness concept that becomes essential when the execution of a programme is non–deterministic [8]. This may be a result of the inherently non–deterministic constructs that the language we used to code it offers, or a result of the interleaving of atomic actions in a concurrent and/or distributed environment.

Intuitively, an execution of a programme is fair iff every element under consideration that is enabled sufficiently often is executed sufficiently often, which prevents undesirable executions in which an enabled element is neglected forever. The elements under consideration may range from alternatives in a non–deterministic multi–choice command to high–level business rules, and combined with a precise definition of "*sufficiently often*" lead to a rich lattice of fairness notions that do not collapse, i.e., are not equivalent each other [3].

* This article was supported by the Spanish Interministerial Commission on Science and Technology under grant TIC2000-1106-C02-01.

B. Monien and R. Feldmann (Eds.): Euro-Par 2002, LNCS 2400, pp. 563–572.

There is not a prevailing definition, but many researchers agree in that so called *strong fairness* deserves special attention [8] because it may induce termination or eventual response to an event. Technically, an execution is said to be strongly fair iff every element that is enabled indefinitely often is executed infinitely often, i.e., it prevents elements that are enabled infinitely often, but not necessarily permanently, from being neglected forever.

In this paper, we focus on concurrent and/or distributed programmes that use the multiparty interaction model as the sole means for process[1] synchronisation and communication. This interaction model is used in several academic programming languages like Scripts [8], Raddle [7] or IP [9] and in commercial programming environments like Microsoft .NET Orchestration [4] too. In this paper, we focus on IP because it is intended to have a dual role: on the one hand, it is intended to be a distributed system specification language equipped with sound semantics that turn it into a language amenable to formal reasoning, a rather important property; on the other hand, it is intended to be an assembler language supporting more sophisticated high–level specification languages such as LOTOS, ESTELLE, SDL [10] or CAL [5]. Next, we report on those issues, present some approaches to address them and give the reader a bird's–eye view of the rest of the paper.

1.1 Known Issues

Figure 1 shows a solution to the well–known dining philosophers problem in IP. This classic multi-process synchronisation problem consists of five philosophers sitting at a table who do nothing but think and eat. There is a single fork between each philosopher, and they need to pick both forks up in order to eat. In addition, each philosopher should be able to eat as much as the rest, i.e., the whole process should be fair. This problem is the core of a large class of problems where a process needs to acquire a set of resources in mutual exclusion.

Get_i and Rel_i denote a number of three–party interactions that allow each philosopher P_i to get its corresponding forks F_i and $F_{i+1 \bmod N}$ in mutual exclusion with its neighbours ($i = 1, 2, \ldots, N$). For an interaction to become enabled, the set of processes that may eventually ready it, i.e., may eventually be willing to participate in the joint action it represents, need to be readying it simultaneously.

The only way to ensure that every philosopher that is hungry will eventually eat is by introducing a notion of fairness in the implementation of the language. However, strong fairness is not practical enough because of the following inherent problems:

Fair Finiteness: Every finite execution is strongly fair by definition. Figure 2.a shows a simple execution trace of an instantiation of the preceding programme in which $N = 5$. The notation $p.\chi$ means that process p readies the

[1] The term process refers to any autonomous, single–threaded computing artefact. It may be a process in an operating system, a thread, or even a hardware device.

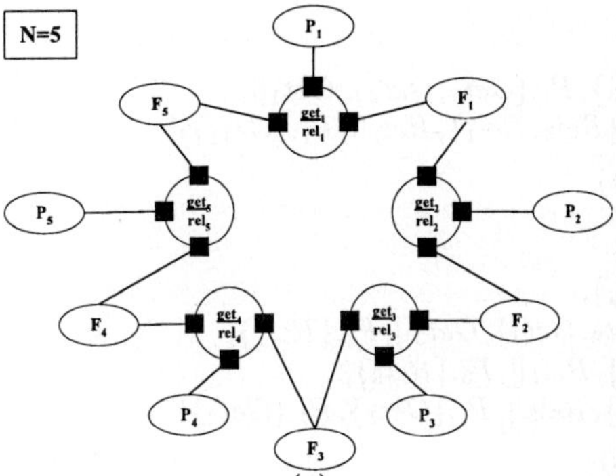

(a)

$\mathrm{S} :: [\|_{i=1}^{N} \mathrm{F}_i \| \|_{i=1}^{N} \mathrm{T}_i]$

P_i :: *[Get_i[] → *eat*; Rel_i[]; *think*]

F_i ::
*[Get_i[] → Rel_i[]
[] $\mathrm{Get}_{(i+1 \bmod N)}$[] → $\mathrm{Rel}_{(i+1 \bmod N)}$[]
]

(b)

Fig. 1. A solution to the dining philosophers problem in IP.

set of interactions χ. Notice that for any finite n, this execution is technically strongly fair, despite Get_2 being enabled n times but never selected. If $n = 10$, this execution may be considered fair from an intuitive point of view, but if $n = 1000$ it is not so intuitive to consider this behaviour fair.

Conspiracies: It does not take into account conspiracies in which an interaction never gets enabled because of an unfortunate interleaving of independent atomic actions. For instance, the execution shown in Figure 2.b is strongly fair for any $n \geq 0$, but notice that due to an unfortunate interleaving, interaction Get_2 is never readied by all of its participants at the same time and thus never gets enabled.

The above problems show that strong fairness (and other notions that rely on infiniteness and eventuality) fails to capture the intuitive idea of inducing liveliness. Although it may be the only way to proof termination or eventual response to an event during an infinite execution, "eventual" is usually too weak for practical purposes because any practical running programme must necessarily stop or be stopped a day.

1.2 Related Work

These issues motivated several authors to research stronger notions. Here we focus on two approaches called *strong finitary fairness* [1] and *hyperfairness* [2].

An execution is *strongly finitarily fair* iff there exists a natural number k (not known a priori) such that every interaction that is enabled infinitely often is executed at least every k steps. Although this notion introduces additional liveliness because it bounds the number of times an enabled interaction may be neglected, it has several drawbacks: (i) k is not known a priori, and thus it cannot be used to fine-tune a potential scheduler depending on the nature of the system it is scheduling; (ii) it does not prevent unfair finiteness; (iii) it does not prevent conspiracies; and, to the best of our knowledge, (iv) no general algorithm implementing it has been produced. (The authors do only present a

$$P_1.\{Get_1\}, P_2.\{Get_2\},$$
$$(P_5.\{Get_5, Get_1\}, P_2.\{Get_2, Get_3\}, P_1.\{Get_1, Get_2\}, Get_1[],$$
$$P_1.\{Rel_1\}, P_1.\{Rel_1, Rel_2\}, P_5.\{Rel_5, Rel_1\}, Rel_1[], P_1.\{Get_1\})^n$$
$$(a)$$

$$P_1.\{Get_1\}, P_2.\{Get_2\}P_3.\{Get_3\},$$
$$(P_5.\{Get_5, Get_1\}, P_3.\{Get_3, Get_4\},$$
$$P_1.\{Get_1, Get_2\}, Get_1[], P_2.\{Get_2, Get_3\}, Get_3[], P_1.\{Rel_1\},$$
$$P_1.\{Rel1, Rel_2\}, P_5.\{Rel_5, Rel_1\}, Rel_1[], P_3.\{Rel_3\},$$
$$P_2.\{Rel_2, Rel_3\}, P_3.\{Rel_3, Rel_4\}, Rel_3[], P_1.\{Get_1\}, P_3.\{Get_3\})^n$$
$$(b)$$

Fig. 2. Strong fairness anomalies.

transformational approach suitable to be used in the context of Büchi automata [12, 11].)

Hyperfairness also deserves attention because it alleviates the second problem. Technically, an execution is hyperfair iff it is finite or every interaction that may get enabled infinitely often, becomes enabled infinitely often. It is important to notice that this definition diverges from classical notions in that the latter imply eventual execution of an interaction if it gets enabled sufficiently often, whereas hyperfairness does only imply eventual enablement. Subsequent execution is under the criterion of an implied underlying classical fairness notion. Thus, this notion prevents conspiracies due to unfortunate interleaving of independent atomic actions but combined with finitary or strong fairness suffers from fair finiteness. To the best of our knowledge, no general algorithm able to implement hyperfairness has been produced. However, the authors presented a transformational approach by means of which we can transform an IP programme into an equivalent strongly hyperfair form, which implies modification of the source code and creation of explicit schedulers for each programme. This may be acceptable in the context of research languages, but it is not practical enough in real–world languages in which processes or components are available only in binary form and need to be scheduled without any knowledge of their internal details. Furthermore, it does not address the issue of fair finiteness.

1.3 Overview

In this paper, we present a new notion called *strong k–fairness* that solves fair finiteness and conspiracies. Intuitively, an execution is strongly k–fair iff no interaction is executed more than k times unless the set of interactions that share processes with it is stable, i.e., the processes that participate in them are waiting for interaction or finished, and it is the oldest in the group, i.e., the one that has not been executed for a longer period of time.

We present a theoretical interaction framework to formalize the multiparty interaction model. Furthermore, we present an algorithm that uses this framework for scheduling interactions in a strongly k–fair manner, and it is not dependent on the internal details of the processes that compose the system, i.e., it is not a transformational approach.

The succeeding sections are organised as follows: Section 2 presents our theoretical interaction framework; Section 3 presents a formal definition of strong k–fairness; Section 4 describes a scheduler we can use to implement this notion; finally, Section 5 reports on our main conclusions.

2 A Theoretical Framework to Support the Multiparty Interaction Model

Next, we present a formal definition of our abstract interaction framework.

Definition 1 (Static Characterisation of a System) *A system Σ is a 2–tuple (P_Σ, I_Σ) in which $P_\Sigma \neq \emptyset$ is a finite set of autonomous processes and $I_\Sigma \neq \emptyset$ is a finite set of interactions. We denote the set of processes that may eventually ready interaction x as $\mathbb{P}(x)$ (participants of interaction x). A configuration is a mathematical object that may be viewed as a snapshot of a system at run time. We denote them as $C, C', C_1, C_2 \ldots$*

An event is a happening that induces a system to transit from a configuration to another. In our model, we take into account the following kinds of events: $p.\iota$, which indicates that process p executes an atomic action that does only involve its local data; $p.\chi$, which indicates that process p is readying the interactions in set χ (notice that when $\chi = \emptyset$, process p arrives at a fixed point that we may interpret as its termination); and x, which indicates that interaction x has been selected and the processes participating in it can execute the corresponding joint action atomically.

Definition 2 (Dynamic Characterisation of a System) *An execution of a system Σ is a 3–tuple (C_0, α, β) in which C_0 denotes its initial configuration, $\alpha = [C_1, C_2, C_3, \ldots]$ is a maximal (finite or infinite) sequence of configurations through which it proceeds, and $\beta = [e_1, e_2, e_3, \ldots]$ is a maximal (finite or infinite) sequence of events responsible for the transition between every two consecutive configurations. Obviously $|\alpha| = |\beta|$. Finally, let $\lambda = (C_0, \alpha, \beta)$ be an execution of system Σ. We call α its configuration trace and denote it as λ_α, and β its event trace and denote it as λ_β.*

We denote the rule that captures the underlying semantics that control the transition between configurations as $\rightarrowtail_{\mathrm{L}}$. For instance, $C \xrightarrow{e}_{\mathrm{L}} C'$ indicates that the system may transit from configuration C to configuration C' on occurrence of event e. Thus, given an execution $\lambda = (C_0, [C_1, C_2, C_3, \ldots], [e_1, e_2, e_3, \ldots])$, we usually write it as[2]: $C_0 \xrightarrow{e_1}_{\mathrm{L}} C_1 \xrightarrow{e_2}_{\mathrm{L}} C_2 \xrightarrow{e_3}_{\mathrm{L}} \cdots$

[2] Notice that the exact formulation of $\rightarrowtail_{\mathrm{L}}$ depends completely on the language in which the system under consideration was written.

Definition 3 (Static Characterisation of a Process) *Process p is waiting a interaction set Υ at the i–th configuration in execution λ iff it has arrived at a point in its execution in which executing any $x \in \Upsilon$ is one of its possible continuations. Process p is finished at the i–th configuration in execution λ iff it can neither execute any local computation nor any interaction.*

$$\begin{aligned}\text{Waiting}(\lambda, p, \Upsilon, i) &\iff \exists k \in [1..i] \cdot \beta(k) = p.\chi \wedge \Upsilon \subseteq \chi \wedge \nexists j \in [k+1..i] \cdot \beta(j) = x \wedge x \in \chi \\ \text{Finished}(\lambda, p, i) &\iff \exists k \in [1..i] \cdot \beta(k) = p.\emptyset\end{aligned} \tag{1}$$

Definition 4 (Static Characterisation of an Interaction) *Interaction x is enabled at the i–th configuration in execution λ iff all of the processes in $\mathbb{P}(x)$ are readying x at that configuration. Interaction x is stable at the i–th configuration in execution λ iff it is either enabled or disabled at that configuration.*

$$\begin{aligned}\text{Enabled}(\lambda, x, i) &\iff \forall p \in \mathbb{P}(x) \cdot \text{Waiting}(\lambda, p, \{x\}, i) \\ \text{Stable}(\lambda, x, i) &\iff \forall p \in \mathbb{P}(x) \cdot \exists \Upsilon \subseteq I_\Sigma \cdot \text{Waiting}(\lambda, p, \Upsilon, i)\end{aligned} \tag{2}$$

Definition 5 (Dynamic Characterisation of an Interaction) *The set of interactions linked to interaction x at the i–th configuration in execution λ is the set of interactions such that there exists a process that is readying x and any of those interactions simultaneously. We define the execution set of interaction x at the i–th configuration in execution λ as the set of indices up to i that denote the configurations at which interaction x has been executed.*

$$\begin{aligned}\text{Linked}(\lambda, x, i) &= \{y \in I_\Sigma \cdot \exists p \in P_\Sigma \cdot \text{Waiting}(\lambda, p, \{x, y\}, i)\} \\ \text{ExeSet}(\lambda, x, i) &= \{k \leq i \cdot \beta(k) = x\}\end{aligned} \tag{3}$$

3 Strong k–Fairness

Intuitively, an execution is strongly k–fair iff no interaction is executed more than k times unless all of the interactions that are linked to it when it is executed are stable and it is the oldest amongst them.

Definition 6 (Strongly k–Fair Execution) *Let $\lambda = (C_0, \alpha, \beta)$ be an execution of a system, and k a non–null natural number. λ is strongly k–fair iff predicate $\mathcal{SKF}(\lambda, k)$ holds.*

$$\begin{aligned}\mathcal{SKF}(\lambda, k) \iff \forall x \in I_\Sigma, i \in \text{ExeSet}(\lambda, x, \infty) \cdot \mathit{Enabled}(\lambda, x, i) \wedge \\ (\text{LStable}(\lambda, x, i) \wedge \text{LOldest}(\lambda, x, i) \vee \\ \neg\, \text{LStable}(\lambda, x, i) \wedge \Delta(\lambda, x, i) \leq k)\end{aligned} \tag{4}$$

This definition relies on a number of auxiliary predicates and functions we have introduced for the sake of simplicity. LStable is a predicate we use to determine if an interaction and those that are linked to it are stable at a given configuration in an execution. Its formal definition follows:

$$\text{LStable}(\lambda, x, i) \iff \forall y \in \text{Linked}(\lambda, x, i) \cup \{x\} \cdot \text{Stable}(\lambda, y, i) \tag{5}$$

LOldest is a predicate we use to determine if an interaction is older than any of the interactions to which it is linked or, in the worst case, is the same age. Its definition follows:

$$\mathrm{LOldest}(\lambda, x, i) \iff \forall y \in \mathrm{Linked}(\lambda, x, i) \cdot \mathrm{Age}(\lambda, x, i) \geq \mathrm{Age}(\lambda, y, i) \tag{6}$$

The age of an interaction is the number of configurations that have elapsed since it was executed for the last time, or ∞ if it has never been executed so far.

$$\mathrm{Age}(\lambda, x, i) = \begin{cases} i - \max \mathrm{ExeSet}(\lambda, x, i) & \text{if } \mathrm{ExeSet}(\lambda, x, i) \neq \emptyset \\ \infty & \text{otherwise} \end{cases} \tag{7}$$

Δ is a function that maps an event trace, an interaction and an index into the number of times it has executed in the presence of a non–empty set of linked interactions that was not stable. Its definition follows:

$$\Delta(\lambda, x, i) = \sum_{\phi \leq k < i} (\lambda_\beta(k) = x \wedge \mathrm{Linked}(\lambda, x, k) \neq \emptyset \wedge \neg\, \mathrm{LStable}(\lambda, x, k)) \tag{8}$$

where $\sum$ denotes the counter quantifier ($\sum_{a \in A} P(a) \triangleq |\{a \in A \cdot P(a)\}|$), and ϕ is defined as follows (notice that we denote the maximum of an empty set as $\perp$):

$$\phi \triangleq \begin{cases} j & \text{if } j = \max\{k \in \mathrm{ExeSet}(\lambda, x, i) \cdot \mathrm{LStable}(\lambda, x, k)\} \wedge j \neq \perp \\ 1 & \text{otherwise} \end{cases} \tag{9}$$

4 A Strongly k–Fair Scheduler

Our algorithm is based on previous proposals which want to resolve another problems in the context of the multiparty interactions [6, 14, 13].

The idea behind our algorithm for scheduling interactions in a strongly k–fair manner consists of arranging the set of interactions into a queue τ so that the closer an interaction is to the rear, the less time has elapsed since it was executed for the last time. Furthermore, each interaction has an associated counter δ we use to count how many times it has been semi–enabled in presence of linked interactions.

To know the state in which a process or an interaction is, we use a map φ from the set of interactions into the set of processes that are readying it. We update it each time the selection module detects a transition $\xrightarrow{p.\chi}_{\mathrm{L}}$ occurs, or an interaction is executed. Our algorithm selects for execution interaction x as long as it is the first enabled one in queue τ, and the set of interactions linked to it is stable.

We describe the operational semantics of our strong k–fairness algorithm using transition rule $\longrightarrow_{\mathcal{SKF}}$ on configurations $D, D', D_1, \ldots$. These configurations are composed of the configuration C of the programme and the data structures we need to select interactions. Next, we define theses data structures and the functions that allows us to update it.

The extended configurations on which our algorithm works are of the form $(\tau, \varphi, \delta, \vartheta)$, where τ is an interaction queue, φ is a readiness map, δ is a semi–enablement map, and ϑ denotes the set of processes that are finished.

Definition 7 (Data Structures) *φ denotes a map from interactions into sets of processes. $\varphi(x)$ denotes the set of processes that are readying x. $\vartheta \subseteq P_\Sigma$ denotes the set of processes that are finished, i.e., can neither execute local computations nor ready any interaction. δ is a map from the set of interactions into the set of natural numbers. $\delta(x)$ denotes the number of times any interaction linked to x has been executed while x was semi–enabled. The higher $\delta(x)$ is, the higher the probability of conspiracy is. τ denotes a queue in which the set of interactions has been arranged so that the closer they are to the rear, the less time has elapsed since they were executed for the last time. As usual, we consider a queue of interactions is a map from a subset of natural numbers into the set of interactions.*

The initial extended configuration of our algorithm is of the form $(\tau_0, \varphi_0, \delta_0, \vartheta_0)$. It does not matter the order in which interactions are initially arranged into τ, but φ_0 must satisfy that $\forall x \in \operatorname{dom} \varphi_0 \cdot \varphi_0(x) = \emptyset$, δ_0 must satisfy that $\forall x \in \operatorname{dom} \delta_0 \cdot \delta_0(x) = 0$, and $\vartheta_0 = \emptyset$.

Definition 8 (Functions) *When process p readies a set of interactions χ, we use function* AddOffer(φ, p, χ) *to update map φ, and when interaction x is executed, we use function* RemoveOffer(φ, x). *By definition, process p finishes when it readies an empty set of interactions, and function* AddFinished(ϑ, p) *updates map ϑ. When interaction x is selected for execution, we use function Order to move it to the rear of τ, not necessarily to the last position. When interaction x is selected for execution, we use function* Update(φ, δ, x) *to create a new semi–enablement. Predicate* Stabilised$(\Upsilon, \varphi, \vartheta)$ *holds iff all of the processes that may eventually ready an interaction in Υ has readied it or are finished.* EnblDisj(τ, φ) *denotes the set of interactions that are enabled and it does not exists any preceding interaction in τ that is enabled and linked to them.*

$$
\begin{aligned}
\mathrm{AddOffer}(\varphi, p, \chi) =& \{x \mapsto \varphi(x) \cdot x \in \operatorname{dom} \varphi \wedge x \notin \chi\} \cup \\
& \{x \mapsto \varphi(x) \cup \{p\} \cdot x \in \operatorname{dom} \varphi \wedge x \in \chi\} \\
\mathrm{RemoveOffer}(\varphi, x) =& \{x \mapsto \varphi(x) \setminus \mathbb{P}(x) \cdot x \in \operatorname{dom} \varphi\} \\
\mathrm{AddFinished}(\vartheta, p) =& \begin{cases} \vartheta \cup \{p\} & \text{if } \chi \neq \emptyset \\ \vartheta & \text{if } \chi = \emptyset \end{cases} \\
\mathrm{Order}(\tau, \delta) =& \tau' \Leftrightarrow \operatorname{dom} \tau = \operatorname{dom} \tau' \wedge \operatorname{ran} \tau = \operatorname{ran} \tau' \wedge \\
& \forall x_1, x_2 \in \operatorname{ran} \tau \cdot (\tau'^{-1}(x_1) \leq \tau'^{-1}(x_2) \Rightarrow \delta(x_1) \geq \delta(x_2)) \\
\mathrm{Update}(\varphi, \delta, x) =& \delta \otimes \{x \mapsto 0\} \otimes \{y \mapsto \delta(y) + 1 \cdot y \in \mathcal{S} \setminus \{x\} \wedge \neg\, \mathrm{Stabilised}(\mathcal{S}, \varphi, \vartheta)\} \\
\mathrm{Stabilised}(\Upsilon, \varphi, \vartheta) \Leftrightarrow& \forall x \in \Upsilon \cdot \forall p \in \mathbb{P}(x) \cdot p \in \operatorname{ran} \varphi \vee p \in \vartheta \\
\mathrm{EnblDisj}(\tau, \varphi) =& \{x \in \operatorname{dom} \varphi \cdot \mathbb{P}(x) = \varphi(x) \wedge \nexists y \in \mathcal{S} \cdot \mathbb{P}(y) = \varphi(y) \wedge \tau^{-1}(y) < \tau^{-1}(x)\}
\end{aligned}
\tag{10}
$$

where $\mathcal{S} \triangleq \{z \in \operatorname{dom} \varphi \cdot \varphi(z) \cap \varphi(x) \neq \emptyset\}$.

Our algorithm is formally defined by means of the inference rules presented in Figure 3. Rule 11 is straightforward because it describes how the data structures are updated each time process p readies a set of interactions χ. Rule 12 describes which interaction must be selected so that the execution is strongly k–fair. The antecedent is complex, but the reasoning behind it is quite simple. Assume x is an enabled interaction and there is not a conflicting enabled interaction before x in queue τ, i.e., $x \in \text{EnblDisj}(\tau, \varphi)$; in this context x is selected for execution iff any of the following three conditions hold:

$$\frac{C \xrightarrow{p \cdot \chi}_{\text{L}} C' \wedge \\ \varphi' = \text{AddOffer}(\varphi, p, \chi) \wedge \vartheta' = \text{AddFinished}(\vartheta, p)}{(C, \tau, \varphi, \delta, \vartheta) \xrightarrow{p \cdot \chi}_{\mathcal{SKF}} (C', \tau, \varphi'\delta, \vartheta')} \tag{11}$$

$$\frac{\begin{array}{c} x \in \text{EnblDisj}(\tau, \varphi) \wedge \mathcal{S} = \{z \in \text{dom}\,\varphi \cdot (\varphi(z) \cap \varphi(x) \neq \emptyset)\} \wedge \\ \tau' = \text{Order}(\tau, \delta') \wedge \varphi' = \text{RemoveOffer}(\varphi, x) \wedge \delta' = \text{Update}(\varphi, \delta, x) \wedge \\ (\mathcal{S} = \{x\} \vee \text{Stabilised}(\mathcal{S}, \varphi, \vartheta) \vee (\mathcal{S} \neq \{x\} \wedge \max\limits_{y \in \mathcal{S} \setminus \{x\}} \delta(y) < k)) \end{array}}{(C, \tau, \varphi, \delta, \vartheta) \xrightarrow{x}_{\mathcal{SKF}} (C', \tau', \varphi', \delta', \vartheta) \wedge C \xrightarrow{x}_{\text{L}} C'} \tag{12}$$

Fig. 3. Algorithm for strongly k–fair scheduler.

1. It is not conflicting with other interactions ($\mathcal{S} = \{x\}$).
2. The interactions which it is conflicting with are stabilised, i.e., all of their participants are readying it or finished ($\text{Stabilised}(\mathcal{S}, \varphi, \vartheta)$).
3. The semi–enablement counter associated with each conflicting interaction is less than k ($\mathcal{S} \neq \{x\} \wedge \max\limits_{y \in \mathcal{S} \setminus \{x\}} \delta(y) < k$).

5 Conclusions and Future Work

In this paper, we have presented strong k–fairness in the context of concurrent and/or distributed programmes in which multiparty interaction models are the sole mean for process synchronisation and communication. An important contribution is the concept of semi–enablement that we use to forecast conspiracies and solve them. k can thus be viewed as a semi–enablement threshold that characterises the goodness of our notion because it makes our selection criterion more or less demanding. If k is minimum, the pace at which conflicting interactions are executed depends then on the participant that spends more time at doing local computations because, in this case, no interaction can be selected for execution unless the set of potentially conflicting interactions is consolidated. If k is very distant from its minimum, the algorithm introduces little delay because interactions may be scheduled as soon as they are enabled. The exact choice of k depends on the features of the programme under consideration and can only be tuned by means of experimentation.

In the future, we are going to research on a new algorithm in which k can be adapted at run–time. The idea is to record information about the execution of a programme so as to be able to forecast if an interaction is suffering from conspirations. Thus we might be able to introduce delays only when the forecast predicts it is necessary. Data mining techniques seem to be promising to forecast and will be paid much attention.

References

1. R. Alur and T. A. Henzinger. Finitary fairness. *ACM Transactions on Programming Languages and Systems*, 20(6):1171–1194, November 1998.
2. P.C. Attie, N. Francez, and O. Grumberg. Fairness and hyperfairness in multiparty interactions. *Distributed Computing*, 6(4):245–254, 1993.
3. E. Best. *Semantics of Sequential and Parallel Programs.* Prentice Hall, New York, 1996.
4. J. Conard. *Introducing .NET.* Wrox Press, Inc, 2001.
5. R. Corchuelo, J.A. Pérez, and M. Toro. A multiparty coordination aspect language. *ACM Sigplan*, 35(12):24–32, December 2000.
6. R. Corchuelo, D. Ruiz, M. Toro, and A. Ruiz. Implementing multiparty interactions on a network computer. In *Proceedings of the XXVth Euromicro Conference (Workshop on Network Computing)*, Milan, September 1999. IEEE Press.
7. M. Evangelist, V.Y. Shen, I.R. Forman, and M. Graf. Using Raddle to design distributed systems. In *Proceedings of the 10th International Conference on Software Engineering*, pages 102–115. IEEE Computer Society Press, April 1988.
8. N. Francez. *Fairness.* Springer–Verlag, 1986.
9. N. Francez and I. Forman. *Interacting processes: A multiparty approach to coordinated distributed programming.* Addison–Wesley, 1996.
10. D. Hogrefe. *Estelle, Lotos and SDL.* Springer–Verlag, Berlin, 1989.
11. J.R. Büchi. On a Decision Method in Restricted Second Order Arithmetic. In *Proceedings of the 1960 International Congress of Logic, Methodology and Philosophy of Science*, pages 1–12. Stanford University Press, 1960.
12. E. Olderog and K.R. Apt. Fairness in parallel programs: The transformational approach. *ACM Transactions on Programming Languages and Systems*, 10(3):420–455, July 1988.
13. J. A. Pérez, R. Corchuelo, D. Ruiz, and M. Toro. An order-based, distributed algorithm for implementing multiparty interactions. In *Fifth International Conference on Coordination Models and Languages COORDINATION 2002*, pages 250–257, York, UK, 2002. Springer–Verlag.
14. J.A. Pérez, R. Corchuelo, D. Ruiz, and M. Toro. An enablement detection algorithm for open multiparty interactions. In *ACM Symposium on Applied Computing SAC'02*, pages 378–384, Madrid, Spain, 2002. Springer–Verlag.

On Obtaining Global Information in a Peer-to-Peer Fully Distributed Environment*

Márk Jelasity[1] and Mike Preuß[2]

[1] Dept. of AI, Free Univ. of Amsterdam, jelasity@cs.vu.nl
and RGAI, Univ. of Szeged, Hungary
[2] Dept. of Computer Science, Univ. of Dortmund, mike.preuss@uni-dortmund.de

Abstract. Networking solutions which do not depend on central services and where the components posses only partial information are robust and scalable but obtaining global information like e.g. the size of the network raises serious problems, especially in the case of very large systems. We consider a specific type of fully distributed peer-to-peer (P2P) environment with many interesting existing and potential applications. We suggest solutions for estimating network size and detecting partitioning, and we give estimations for the time complexity of global search in this environment. Our methods rely only on locally available (but continuously refreshed) partial information.

1 Introduction

Peer-to-peer (P2P) systems are becoming more and more popular. The Internet offers an enormous amount of resources which cannot be fully exploited using traditional approaches. Systems that span many different institutions, companies and individuals can be much more effective for certain purposes such as information distribution (e.g. [3, 4]) or large scale computations (e.g. [2, 9]).

Systems exist that go to extremes in the sense of not using central services at all to achieve maximal scalability and minimal vulnerability to possible damages in components. Such an approach was chosen in e.g. [7] for broadcasting. We will focus on another architecture of this kind which we developed as part of the DREAM project [8] (described in more detail in Section 2). In a nutshell, the aim of the DREAM project is to create a complete environment for developing and running distributed evolutionary computation experiments on the Internet in a robust and scalable fashion. It can be thought of as a virtual machine or *distributed resource machine* (DRM) made up of computers anywhere on the Internet. The actual set of machines can (and generally will) constantly change and can grow immensely without any special intervention. Apart from security considerations, anyone having access to the Internet can connect to the DRM and can either run his/her own experiments or simply donate the spare capacity of his or her machine.

* This work is funded as part of the European Commission Information Society Technologies Programme (Future and Emerging Technologies). The authors have sole responsibility for this work, it does not represent the opinion of the European Community, and the European Community is not responsible for any use that may be made of the data appearing herein.

B. Monien and R. Feldmann (Eds.): Euro-Par 2002, LNCS 2400, pp. 573–577.

Although these fully distributed environments can grow to literally astronomical sizes [7] while automatically maintaining their own integrity they have a major drawback: exercising global control and obtaining global information becomes harder and harder as the size increases. Broadcasting or any global search becomes infeasible after a certain point.

This paper discusses methods for obtaining global information based only on *locally available partial information* in the nodes of our environment. These methods scale much better than e.g. broadcasting because their resource requirements are independent of the size of the network. The time complexity of global search based on (continuously refreshed) local information will be addressed in Section 3. In Section 4 a method for estimating the network size is presented. In Section 5 we suggest a way of detecting partitioning.

For the sake of completeness, let us mention that simulations with network sizes of up to 10000 nodes were performed to support our theoretical considerations here. Unfortunately, due to serious space limitations, we were forced to remove our simulation results to be able to keep most of our theoreical results which might be more appropriate to be published in a research note.

2 The Model

Focusing on the topic of this paper we discuss only a simplified version of our environment, in particular we ignore timestamp handling, and the mechanism of application execution. More information can be found in [5, 6].

The DRM is a network of DRM nodes. Let S denote that set of all nodes in the DRM, and let $n = |S|$. In the DRM every node is completely equivalent. Nodes must be able to know enough about the rest of the network in order to be able to remain connected to it. Spreading information over and about the network is based on epidemic protocols [1].

Every node $s \in S$ maintains an *incomplete* database containing descriptors of a set $D(s)$ of nodes ($|D(s)| = c$), where normally $n \gg c$. We call these nodes *the neighbours* of the node. The database is refreshed using a push-pull anti-entropy algorithm. Every node s chooses a node s' from $D(s)$ in every time-step. Then any differences between s and s' are resolved so that after the communication s and s' will both have the descriptors of the nodes from $D(s) \cup D(s')$. Besides this, s will receive a descriptor of s' and s' will also receive a descriptor of s. As mentioned before, the size of the database is limited in c. This limitation is implemented by removing randomly selected elements.

To connect a new node to the DRM one needs only one living address. The database of the new node is initialized with the entry containing the living address only, and the rest is taken care of by the epidemic algorithm described above. Removal of a node does not need any administration at all.

Fortunately, theoretical and empirical results show that limiting the size of the database does not affect the power of the epidemic algorithm, information spreads quickly and the connectivity (thus information flow) is not in danger [7, 5, 6]. For example a database size of 100 is enough to support a DRM of size 10^{33}.

3 Global Search

We would like to find nodes in the network which fulfill some criteria. Being able to do so is important in many situations. We might want to find a node that has lots of space or CPU capacity available, or nodes situated in a given geographical area, etc.

Our main purpose is to allow very large networks, in the order of $n = 10^5$ or more. In such networks any broadcasting approach is infeasible because every node must be able to do global search and for large networks too many broadcasts could be generated resulting in huge traffic. Collecting and storing information about the entire network is not the best solution either because we cannot assume large storage capacity in every node, and on the other hand the network changes constantly: nodes and running applications come and go.

In the following we will examine the limits and possibilities of using only the local database in a node to search the network. The idea is that we listen to the updates and when the appropriate node appears there, we return it. This might seem hopeless but theory and practice show that it is not necessarily the case. Note that this kind of search has practically no costs since we are using the database refreshment mechanism that is applied anyway. The only cost that increases with n is the waiting time.

Let $s^* \in S$ be the node we are looking for from node s ($s \neq s^*$). Let $D(s) = \emptyset$ at the start of the search. Let the set D_i denote the nodes in the database s is updated with during the ith database-exchange session according to the epidemic algorithm. Note that the elapsed time is not necessarily the same between the updates. In this section we assume that $D_1, D_2, \ldots$ are unbiased independent random samples of S.

Let the random variable ξ denote the index of the first update in which s^* can be found. In other words $s^* \in D_\xi$ and $\forall i < \xi : \; s^* \notin D_i$. From our assumption about the even distribution it follows that $P(s^* \in D_i) = c/n$ for $i = 1, 2, \ldots$ From the assumption of independence it follows that $P(\xi = i) = (1 - c/n)^{i-1}(c/n)^i$ thus ξ has a geometric distribution with the parameter c/n. This means that the expected value is $\mu_\xi = n/c$ and the variance is $\sigma_\xi^2 = n(n-c)/c^2$. Note that the optimal case in this framework is when we have $D_1 \cup \ldots \cup D_{\lceil n/c \rceil} = S$ when the information flow speed (the learning speed of s) is maximal. The expected value that belongs to this distribution is $\geq n/2c$. Compared to this the waiting time in the realistic situation is in the same order of magnitude which is rather surprising.

4 Estimating Network Size

Another promising possibility of exploiting the dynamics of the epidemic protocol is network size estimation. Since there is no central service we have no idea about the actual size of the network. It can be estimated however from the characteristics of information flow through the database of a server s. Intuitively, if there is much new information in the database of a peer then we expect to have a large network.

Let us examine a database exchange between s and s' (with databases D and D' respectively) during the normal functioning of our epidemic protocol. Let $d = |D' \backslash D|$, or in words the number of new elements in D'. If we assume that D and D' are independent unbiased samples from S then d has a binomial distribution $B((n - |D|)/n, |D'|)$ with the expected value

$$E(d) = |D'|(n - |D|)/n \tag{1}$$

(In this section we do not assume that $|D| = |D'| = c$, the results hold for the general case too.)

Of course we do not know the distribution of d because its parameters refer to the network size. However we can collect a sample for a fixed $|D|$ and $|D'|$ and we can approximate the expected value of d with the sample average $\overline{d}$. Using (1) and this approximation we can approximate n with the expression

$$n \approx \tilde{n} = \frac{|D'||D|}{|D'| - \overline{d}} \tag{2}$$

Since d has a binomial distribution, this approximation is optimal in the following Bayesian sense:

Proposition 1. *If ξ is a random variable from the binomial distribution $B(p, n)$ and $\{x_1, \ldots, x_k\}$ is an independently drawn sample of ξ then*

$$\arg\max_p P(x_1, \ldots, x_k | B(p, n)) = \frac{x_1 + \ldots + x_k}{kn}$$

Proof. After substituting the probability values, using the independence assumption and ignoring the binomial coefficients we get

$$\max_p P(x_1, \ldots, x_k | B(p, n)) = \max_p p^{x_1 + \ldots + x_k} (1 - p)^{kn - (x_1 + \ldots + x_k)}$$

Elementary calculus shows that the maximum of this polinom of p is at $p = (x_1 + \ldots + x_k)/(kn)$ which proves the proposition. □

5 Detecting Partitioning

During the operation of a DRM the underlying physical network may get partitioned. For a user it may be valuable to detect this partitioning because this could mean a serious degradation of his or her available computational resources. The approach we are presenting in this paper offers a potential solution for detecting such sudden changes. When the estimated network size decreases suddenly, it probably means that the node that percieves this change is part of a subnetwork that has just been separated from the original larger network.

6 Conclusions

In this paper techniques were presented that are able to provide global information in a distributed networking environment where no central services are available. The techniques are based on the dynamics of the epidemic protocol which is run in an environment where each node knows only a tiny bit about the whole network but where this knowledge is continuously updated by an epidemic protocol.

Possibilities of performing global search in this environment were analyzed. It was shown that the underlying epidemic algorithm pumps the complete system-state through every local node very quickly. It is notable that the design goals of our epidemic protocol did not include this requirement, it was an unexpected but useful side-effect. Depending on the waiting time available this makes global search feasible in many cases.

It was also suggested that the dynamics of information flow through a node can be exploited in many ways. One of these is estimating network size, another is predicting partitioning.

The possibilities were not fully exploited. Our goal was to give theoretical evidence which suggests that it is worth doing research in the direction of possible exploitations of information sources which are naturally present in certain types of distributed environments. We believe that techniques like the ones suggested in this work can many times offer a cheap yet effective alternative to implementing expensive additional protocols and services or introducing additional restrictions in the design.

Acknowledgments

The authors would like to thank the other members of the DREAM project for fruitful discussions, the early pioneers [8] as well as the rest of the DREAM staff, Maribel García Arenas, Emin Aydin, Pierre Collet and Daniele Denaro.

References

1. A. Demers, D. Greene, C. Hauser, W. Irish, J. Larson, S. Shenker, H. Sturgis, D. Swinehart, and D. Terry. Epidemic algorithms for replicated database management. In *Proceedings of the 6th Annual ACM Symposium on Principles of Distributed Computing (PODC'87)*, pages 1–12, Vancouver, Aug. 1987. ACM.
2. distributed.net. http://distributed.net/.
3. P. Druschel and A. Rowstron. Storage management and caching in PAST, a large-scale, persistent peer-to-peer storage utility. In *Proceedings of the 18th ACM Symposium on Operating Systems Principles (SOSP)*, Banff, Canada, 2001. ACM.
4. Gnutella. http://gnutella.wego.com/.
5. M. Jelasity, M. Preuß, and B. Paechter. A scalable and robust framework for distributed applications. Accepted for publication in the Proceedings of the 2002 Congress on Evolutionary Computation (CEC2002).
6. M. Jelasity, M. Preuß, M. van Steen, and B. Paechter. Maintaining connectivity in a scalable and robust distributed environment. Accepted for publication in the Proceedings of the 2nd IEEE International Symposium on Cluster Computing and the Grid (CCGrid2002), May 21-24, 2002, Berlin, Germany.
7. A.-M. Kermarrec, L. Massoulié, and A. J. Ganesh. Probablistic reliable dissemination in large-scale systems. Submitted for publication, available as http://research.microsoft.com/ camdis/PUBLIS/kermarrec.ps.
8. B. Paechter, T. Bäck, M. Schoenauer, M. Sebag, A. E. Eiben, J. J. Merelo, and T. C. Fogarty. A distributed resoucre evolutionary algorithm machine (DREAM). In *Proceedings of the Congress on Evolutionary Computation 2000 (CEC2000)*, pages 951–958. IEEE, IEEE Press, 2000.
9. SETI@home. http://setiathome.ssl.berkeley.edu/.

A Fault-Tolerant Sequencer for Timed Asynchronous Systems*

Roberto Baldoni, Carlo Marchetti, and Sara Tucci Piergiovanni

Dipartimento di Informatica e Sistemistica, Università di Roma "La Sapienza",
Via Salaria 113, 00198 Roma, Italia.
{baldoni,marchet,tucci}@dis.uniroma1.it

Abstract. In this paper we present the specification of a sequencer service that allows independent processes to get a sequence number that can be used to label successive operations (e.g. to allow a set of independent and concurrent processes to get a total order on their operations). Moreover, we provide an implementation of the sequencer service in a specific partially synchronous distributed system, namely the timed asynchronous model. As an example, if a sequencer is used by a software replication scheme then we get the advantage to deploy server replicas across an asynchronous distributed system such as the Internet.

1 Introduction

Distributed agreement among processes is one of the fundamental building blocks for the solution of many important problems in asynchronous distributed systems, e.g. mutual exclusion[9] and replication[10, 3, 8]. As an example, in the context of software replication replicas have to run a distributed agreement protocol in order to maintain replica consistency. In particular, in the case of active replication[10] the agreement problem reduces to the total order multicast problem and in the case of passive replication[3] to the view synchronous multicast problem[8]. In both cases these problems are not solvable in asynchronous distributed systems prone to process crash failures due to FLP impossibility result[7]. As a consequence, to solve these problems replicas have to be deployed over a partially synchronous distributed system i.e., an asynchronous distributed system which enjoys some timing assumption (e.g., known, or eventually known, bounds on message transfer delay and on relative process speeds). Practically, when working on a partially synchronous system, replication techniques can benefit of group communication primitives and services implemented by group toolkits such as total order multicast, view synchronous multicast, group membership, state transfer, etc.. These primitives employ agreement protocols to ensure replica consistency. Let us remark that the partial synchronous system assumption makes impossible the deployment of server replicas in real asynchronous distributed systems such as the Internet.

* Work partially supported by MIUR (DAQUINCIS) and by AleniaMarconiSystems.

B. Monien and R. Feldmann (Eds.): Euro-Par 2002, LNCS 2400, pp. 578–588.

In this paper we present the specification of a sequencer service. The specification ensures (i) that independent client processes obtain distinct sequence numbers for each distinct request submitted to the sequencer and (ii) that the sequence numbers assigned are consecutive, i.e. the sequence of assigned numbers does not present "holes".

A sequencer service can then be exploited, for instance, in the context of active replication. A sequencer implementation can be logically placed between clients and server replicas, and can be used to piggyback sequence numbers onto client requests, before sending them out to server replicas. In this way replicas can independently order client requests without the need of executing distributed agreement protocols. This actually makes possible the deployment of replicas over an asynchronous distributed system (interested readers can refer to [2] for additional details about this sequencer based replication technique) as there is no need of additional timing assumptions on the system model underlying the server replicas.

As second contribution, in this paper we provide a fault-tolerant implementation of the sequencer service. This implementation adopts a passive replication scheme in which the sequencer replicas run over a specific partially synchronous system, namely the timed asynchronous system model[6].

Coming back to the active replication example, the sequencer can then be seen as the component that embeds the partial synchrony neccssary to maintain consistency among a set of server replicas. Therefore server replicas do not need to run any distributed agreement protocol. This allows server replica deployment over a real asynchronous distributed system[2].

The remainder of this paper is organized as follows: Section 2 introduces the specification of the sequencer service. Section 3 presents the distributed system model. Section 4 details our implementation of the sequencer. Section 5 concludes the paper. Due to lack of space, a formal proof of the correctness of the implementation can be found in[1].

2 Specification of the Sequencer Service

A sequencer service receives requests from clients and assigns an integer positive sequence number, denoted *#seq*, to each *distinct* request. Each client request has a unique identifier, denoted *req_id*, which is a pair $\langle cl\_id, \#cl\_seq \rangle$ where *cl_id* is the client identifier and *#cl_seq* represents the sequence number of the requests issued by *cl_id*. As clients implement a simple retransmission mechanism to cope with possible sequencer implementation failures or network delays, the sequencer service maintains a state A composed by a set of assignments $\{a_1, \dots a_{k-1}, a_k\}$ where each assignment a corresponds to a pair $\langle req\_id, \#seq \rangle$, in which *a.req_id* is a request identifier and *a.#seq* is the sequence number returned by the sequencer service to client *a.req_id.cl_id*. A sequencer service has to satisfy the following properties:

P1. Assignment Validity. If $a \in A$ then there exists a client c that issued a request identified by *req_id* and *req_id* $=$ *a.req_id*.

P2. Response Validity. If a client c delivers a reply $\#seq$, then $\exists a = \langle req\_id, \#seq \rangle \in A$.
P3. Bijection. $\forall a_i, a_j \in A : a_i.\#seq \neq a_j.\#seq \Leftrightarrow a_i.req\_id \neq a_j.req\_id$
P4. Consecutiveness. $\forall a_i \in A : (a_i.\#seq \geq 1) \wedge (a_i.\#seq > 1 \Rightarrow \exists a_j : a_j.\#seq = a_i.\#seq - 1)$
P5. Termination. If a client c issues a request, then, unless the client crashes, it eventually delivers a reply.

P1 expresses that the state of the sequencer does not contain "spurious" assignments. P2 states that the client cannot deliver a sequence number that has not already been assigned by the sequencer. The predicate "P1 and P2" implies that each client, delivering a sequence number, has previously issued a request. P3 states that there is a one-to-one correspondence between the set of *req_id* and the set A. P4 says that the sequence, starting from one, of numbers assigned by the sequencer has not "holes". P5 states that the service is live.

3 System Model

We consider a distributed system in which processes communicate by message passing. Processes can be of two types: clients and replicas. The latter form a set $\{r_1, \ldots, r_n\}$ of processes implementing the fault-tolerant sequencer. A client c communicates with replicas using *reliable asynchronous channels.* Replicas run over a *timed asynchronous model*[6].

Client Processes. A client process sends a request to the sequencer service and then waits for a sequence number. A client copes with replica failures using a simple retransmission mechanism. A client may fail by crashing.

Communication between clients and replicas is *asynchronous* and *reliable.* This communication is modelled by the following primitives: **A-send**(m, p) to send an unicast message m to process p; and **A-deliver**(m, p) to deliver a message m sent by process p.

To label a generic event with a sequence number, a client invokes the GETSEQ() method. Such method blocks the client process until it receives an integer sequence number from a sequencer replica. In particular, the GETSEQ() method assigns to the ongoing request a unique request identifier $req\_id = \langle cl\_id, \#cl\_seq \rangle$, then (i) it sends the request to a replica and (ii) sets a local timeout. Then, a result is returned by GETSEQ() if the client receives a sequence number for the *req_id* request within the timeout expiration. Otherwise, another replica is selected (e.g. using a cyclic selection policy), and the request is sent again to the selected replica setting the relative timeout, until a reply is eventually delivered.

Replica Processes. Replicas have access to local hardware clock (which is not synchronized). Timeouts are defined for message transmission and scheduling delays. A performance failure occurs when an experienced delay is greater than the associated time-out. Replicas can also fail by crashing. A process is *timely* in a time interval $[s, t]$ iff during $[s, t]$ it neither crashes nor suffers a performance failure. For simplicity, a process that fails by crashing cannot recover. A message whose transmission delay is lesser than the associated time-out is *timely*. A subset of replicas form a *stable* partition in $[s,t]$ if any pair of replicas belonging to the subset is timely and each message exchanged between the pair in $[s,t]$ is timely. Timed asynchronous communications are achieved through a *datagram service* which filters out non-timely messages to the above layer. Replicas communicate among them through the following primitives: **TA-send**(m, r_i) to send an unicast message m to process r_i; **TA-broadcast**(m) to broadcast m to all replicas including the sender of m; **TA-deliver**(m, r_j) is the upcall initiated by the datagram service to deliver a *timely* message m sent by process r_j.

We assume replicas implement the leader election service specified in[5]. This service ensures that: (i) at every physical time there exists at most one *leader*, a *leader* is a replica in which the *Leader?()* boolean function returns *true*; (ii) the leader election protocol underlying the *Leader?()* boolean function takes at least 2δ for a leader change; (iii) when a majority of replicas forms a stable partition in a time interval $[t, t+\Delta t]$ $(\Delta t \gg 2\delta)$, then it exists a replica r_i belonging to that majority that becomes leader in $[t, t+\Delta t]$.

Note that the leader election service cannot guarantee that when a replica becomes leader it stays within the stable partition for the duration of its leadership (e.g. the leader could crash or send non-timely messages to other replicas).

In order to cope with *asynchronous interactions* between clients and replicas, to ensure the *liveness* of our sequencer protocol, we introduce the following assumption, i.e.:

eventual global stabilization: there exists a time[1] t and a set $\mathcal{S} \subseteq \{r_1, ... r_n\}$: $|\mathcal{S}| \geq \lceil \frac{n+1}{2} \rceil$ such that $\forall t' \geq t$, $\mathcal{S}$ is a *stable* partition.

The eventual global stabilization assumption implies (i) only a minority of replicas can crash[2] and (ii) there will eventually exist a leader replica $l_s \in \mathcal{S}$.

4 The Sequencer Protocol

In this section we present a fault-tolerant implementation of the sequencer service. A primary-backup replication scheme is adopted [3, 8]. In this scheme a particular replica, the *primary*, handles all the requests coming from clients. Other replicas are called *backups*. When a primary receives a client request, it processes the request, updates the backups and then sends the reply to the client. In our implementation the backup update lies on an update primitive (denoted WRITEMAJ()) that successfully returns if it *timely* updates at least a *majority* of replicas. This implies that inconsistencies can arise in some replica state.

If the primary fails, then the election of a new primary is needed. The primary election lies on: (i) the availability of the leader election service running among replicas (see Section 3). Leadership is a necessary condition to become primary

[1] Time t is not a priori known.

[2] Note that at any given time t' (with $t' < t$) any number of replicas can simultaneously suffer a performance failure.

and then to stay as the primary; (ii) a "reconciliation" procedure, namely "*computing sequencer_state*" procedure, (*css* in the rest of the paper) that allows a newly elected leader to remove possible inconsistencies from its state before becoming a primary. These inconsistencies, if kept in the primary state, could violate the properties defined in Section 2. Hence a newly elected leader, before becoming a primary, will read at least a majority of replica states, (this is done by a READMAJ() primitive during the *css* procedure). This allows a leader to have in its state all successfully updates done by previous primaries. Then the leader removes from its state all possible inconsistencies caused by unsuccessful primary updates.

4.1 Protocol Data Structures

A replica r_i endows: (1) a boolean variable called *primary*, which is set according to the role (either primary or backup) played by the replica at a given time; (2) an integer variable called *seq*, used to assign sequence numbers when r_i acts as a primary; (3) a *state* consisting of a pair $\langle TA, epoch\rangle$ where TA is a set $\{ta_1, ...ta_k\}$ of *tentative assignments* and *epoch* is an integer variable. *state.epoch* represents a value associated with the last primary seen by r_i. When r_i becomes primary, *state.epoch* is greater than any *epoch* value associated with previous primaries. *state.epoch* is set when a replica becomes primary and it does not change during all the time a replica is the primary. A *tentative assignment ta* is a triple $\langle req\_id, \#seq, \#epoch\rangle$ where *ta.#seq* is the sequence number assigned to the request *ta.req_id* and *ta.#epoch* is the epoch of the primary that executed *ta* [3].

The set *state.TA* is ordered by *TA.#seq* field and ties are broken using *TA.#epoch* field. We introduce *last*(*state.TA*) operation that returns the tentative assignment with greatest epoch number among those (if any) with greatest sequence number. If *state.TA* is empty, then *last*(*state.TA*) returns *null*.

4.2 Basic Primitives and Definitions

In this section we present the basic primitives used to update and to read replica states. Due to lack of space the pseudo-code of such primitives can be found in[1]. *UpdateMaj()*. Accepts as input parameter m, which can be either a tentative assignment *ta* or an epoch e, and returns as output parameter a boolean value b. Upon invocation, WRITEMAJ() executes **TA-broadcast**(m). Every replica r_i sends an acknowledgement upon the delivery of m. The method returns $\top$ (i.e., it *successfully* returns) if (i) the invoker receives at least a majority of *timely* acknowledgements, hence m is put into the replica state according to its type, and (ii) if the invoker is still the leader at the end of the invocation.

[3] Epoch numbers are handled by primaries to label their tentative assignments and by leaders to remove inconsistencies during the *css* procedure.

ReadMaj(). Does not take input parameters and returns as output parameter a pair $\langle b, maj\_state \rangle$ where b is a boolean value and maj_state is a state as defined in Section 4.1. Upon invocation, READMAJ() executes a **TA-broadcast**. Every replica r_i sends its state as reply. If READMAJ() receives at least a majority of *timely* replies, then it computes the union $maj\_state.TA$ of the tentative assignments contained in the just received states and sets $maj\_state.epoch$ to the maximum among the epochs contained in the just received states. If the invoker is still leader it returns $\langle \top, maj\_state \rangle$, otherwise $\langle \bot, - \rangle$.

Definitive Assignment: a tentative assignment ta is a *definitive* assignment *iff* exists a primary p such that p executed WRITEMAJ(ta)$= \top$.
Non-definitive Assignment: a tentative assignment which is not definitive.
Therefore, a definitive assignment is a tentative one. The viceversa is not necessarily true. Non-definitive assignments are actually inconsistencies due to unsuccessful WRITEMAJ() executions.

4.3 Protocol Description

Let us present in this section a preliminary explanation of the sequencer protocol and two introductory examples before showing the replica pseudo-code.
Primary Failure-Free Behaviour. A primary upon recciving a client request first checks if a sequence number was already assigned to the request, otherwise (i) it creates a new tentative assignment ta embedding the request identifier and a sequence number consecutive to the one associated with the last request, (ii) invokes WRITEMAJ(ta) to update the backups and (iii) if WRITEMAJ(ta) successfully returns, it sends back the sequence number to the client as ta is a definitive assignment.
Change of Primary. There are three events that cause a primary replica r_i to lose the primaryship: (i) r_i fails by crashing or (ii) WRITEMAJ(ta) returns $\bot$ (WRITEMAJ(ta) *could* have notified ta to less than a majority of replicas) or (iii) there is a leadership loss of r_i (i.e., the *Leader*?() value becomes false in r_i). If any of these events occurs, the protocol waits that a new leader is elected by the underlying leader election service. Then the *css* procedure is executed by the new leader before starting serving requests as primary.
The css *Procedure.* The first action performed by a newly elected leader r_i is to invoke READMAJ(). If READMAJ() returns $\langle \bot, - \rangle$ and r_i is always the leader, r_i will execute again READMAJ(). If r_i is no longer leader, the following leader will execute READMAJ() till this primitive will be successfully executed.

Once the union of the states of a majority of replicas, denoted maj_state, has been fetched by READMAJ(), the *css* procedure has three main goals. The first goal is to transform the tentative assignment $last(maj\_state.TA)$ in a definitive assignment *on behalf of a previous primary* that issued WRITEMAJ($last(maj\_state.TA)$). There is no way in fact for r_i to know if that WRITEMAJ() was successfully executed by the previous primary. The second goal is to remove from $maj\_state.TA$ all non-definitive assignments. Non-definitive assignments are filtered out using the epoch field of tentative assignments. More specifically, the implementation enforces bijection (Section 2) guaranteeing that when *there are multiple assignments with the same sequence number, the one with the greatest epoch number is a definitive assignment.* The third goal is to impose a primary epoch number e by using WRITEMAJ(). e is greater than the one returned by READMAJ() in *maj_state.epoch* and greater than all epoch numbers associated to previous primaries. If r_i successfully executed all previous points it starts serving requests as primary.

In the following we introduce two examples which point out how the previous actions removes inconsistencies from a primary state.

Example 1: Avoiding inconsistencies by redoing the last tentative assignment. The example is shown in Fig.1. Primary r_1 accepts a client request req_id_1, creates a tentative assignment $ta_1 = \langle req\_id_1, 1, 1\rangle$, performs WRITEMAJ($ta_1$)= $\top$ (i.e. ta_1 is a definitive assignment) and sends the result $\langle 1, req\_id_1\rangle$ to the client. Then r_1 receives a new request req_id_2, invokes WRITEMAJ($ta_2 = \langle req\_id_2, 2, 1\rangle$) and crashes during the invocation. Before crashing it updated only r_3. The next leader r_2 enters the *css* procedure: READMAJ() returns in *maj_state.TA* the union of r_2 and r_3 states (i.e., $\{ta_1, ta_2\}$) and in *maj_state.epoch* the epoch of the previous primary r_1 (i.e., 1). Therefore, as *last*(*maj_state.TA*) returns ta_2, r_2 executes WRITEMAJ(ta_2)= $\top$ on behalf of the previous primary (r_2 cannot know if ta_2 is definitive or not). Then r_2 executes WRITEMAJ(*maj_state.epoch* + 1) and it ends the *css* procedure. When r_2 receives the req_id_2, it finds ta_2 in its state then sends $\langle 2, req\_id_2\rangle$ to the client.

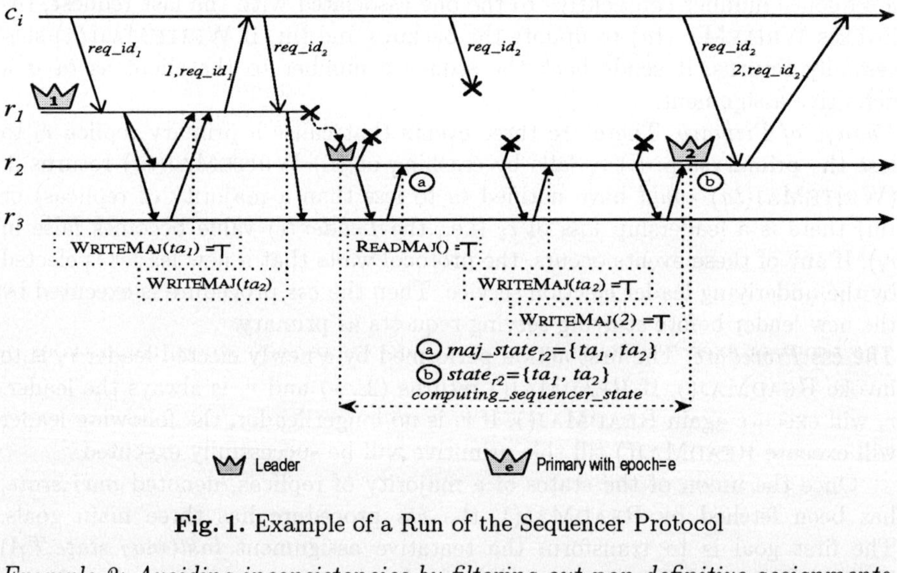

Fig. 1. Example of a Run of the Sequencer Protocol

Example 2: Avoiding inconsistencies by filtering out non-definitive assignments. The example is shown in Fig. 2. Primary r_1 successfully serves req_id_1. Then, upon the arrival of req_id_2, it invokes WRITEMAJ(), exhibits a performance failure and updates only replica r_3 (ta_2 is a non-definitive assignment). Then r_1 loses its primaryship and another leader (r_2) is elected. r_2 executes READMAJ() which

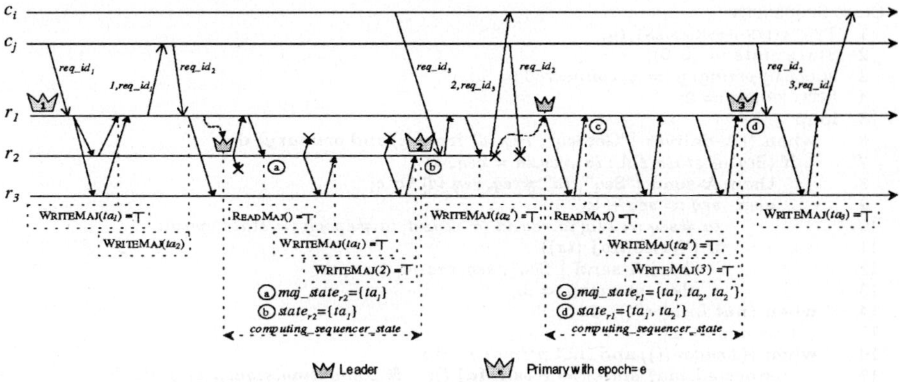

Fig. 2. Example of a Run of the Sequencer Protocol

returns in maj_state the union of r_1 and r_2 states (i.e., $\{ta_1\}$). Then r_2 executes WRITEMAJ$(ta_1)=\top$ and imposes its epoch. Upon the arrival of a new request req_id_3, primary r_2 successfully executes WRITEMAJ$(ta'_2 = \langle req_id_3, 2, 2\rangle)$ (i.e. ta'_2 is a definitive assignment) and sends back the result $\langle 2, req\_id_3\rangle$ to the client.

Note that r_1 and r_3 contain two distinct assignments (ta_2 and ta'_2) with same sequence number and different epoch numbers ($ta_2.\#epoch = 1$ and $ta'_2.\#epoch = 2$). However the $maj\_state.TA$ of a successive leader r_i (r_1 in Figure 2) includes the definitive assignment ta'_2 (as it contained in a majority of replicas). If ta_2 is also a member of $maj\_state.TA$, r_i is able to filter ta_2 out from $maj\_state.TA$ as $ta_2.\#epoch = 1 < ta'_2.\#epoch = 2$. After filtering, the state of the primary r_1 is composed only by definitive assignments. Note that without performing such filtering the bijection would result violated, as the state of a primary could contain two assignments with same sequence number.

Then, when r_1 receives the request req_id_2 it performs WRITEMAJ$(ta_3 = \langle req_id_2, 3, 3\rangle)$ and if it successfully returns, r_1 sends $\langle 3, req\_id_2\rangle$ to the client.

4.4 Behaviour of Each Replica

The protocol executed by r_i consists in an infinite loop where three types of events can occur (see Figure 3):

(1) receipt of a client request when r_i acts as a primary (line 6); (2) receipt of a "no leadership" notification from the leader election service (line 14); (3) receipt of a "leadership" notification from the leader election service when r_i is not primary (line 16).

Receipt of a client request req_id when r_i acts as a primary. r_i first checks if the client request has been already served (line 7). In the affirmative, r_i returns to the client the global sequence number previously assigned to the request (line 8). Otherwise, r_i (i) increases by 1 the *seq* variable (line 9) and (ii) creates a tentative assignment ta such that $ta.\#seq = seq$; $ta.req\_id = req\_id$; $ta.\#epoch = state.epoch$ (line 10). Then r_i executes WRITEMAJ(ta) (line 11). If it successfully

```
CLASS SEQUENCER
 1 TENTATIVE ASSIGNMENT ta;
 2 STATE state := (∅, 0);
 3 BOOLEAN primary := ⊥; connected := ⊥;
 4 INTEGER seq := 0;
 5 loop
 6   when ((A-deliver ["GetSeq", req_id] from c) and primary) do
 7     if (∃ta' ∈ state.TA : ta'.req_id = req_id)
 8       then A-send ["Seq", ta'.#seq, req_id] to c;
 9       else seq := seq + 1;
10            ta.#seq := seq; ta.req_id := req_id; ta.#epoch := state.epoch;
11            if (WriteMaj (ta))
12              then A-send ["Seq", seq, req_id] to c;
13              else primary := ⊥;
14   when (not Leader?()) do
15     primary := ⊥;
16   when ((Leader?()) and (not primary)) do
17     (connected, maj_state) := ReadMaj ();    % computing_sequencer_state %
18     if (connected)
19       then ta := last(maj_state.TA);
20            if (ta ≠ null)
21              then connected := WriteMaj (ta);
22                   if (connected)
23                     then for each ta_j, ta_ℓ ∈ maj_state.TA :
24                               (ta_j.#seq = ta_ℓ.#seq) and (ta_j.#epoch > ta_ℓ.#epoch)
25                          do maj_state.TA := maj_state.TA − {ta_ℓ};
26                        state.TA := maj_state.TA; seq := last(state.TA).#seq;
27            if (WriteMaj (maj_state.epoch + 1) and connected)
28              then primary := ⊤;
29 end loop
```

Fig. 3. The Sequencer Protocol Pseudo-code Executed by r_i

returns, *ta* becomes a definitive assignment and the result is sent to the client (line 12). Otherwise, the primary sets $primary = \perp$ (line 13) as WRITEMAJ(*ta*) failed and r_i stops serving client requests.

Receipt of a "leadership" notification when r_i is not primary. A *css* procedure (lines 17-28) is started by r_i to become primary. As described in the previous section, r_i has to successfully complete the following four actions to become primary: (1) r_i invokes READMAJ()(line 17). If the invocation is successful it timely returns a majority state in the *maj_state* variable[4]. (2) r_i extracts the last assignment *ta* from *maj_state.TA* (line 19) and invokes WRITEMAJ(*ta*) (line 21) to make definitive the last assignment of *maj_state.TA* (see the examples in the previous section). (3) r_i eliminates from *maj_state.TA* any assignment ta_ℓ such that it exists another assignment ta_j having the same sequence number of ta_ℓ but greater epoch number (lines 23-25). The presence of such a ta_j in *maj_state* implies that ta_ℓ is not definitive. This can be intuitively justified by noting that if an assignment ta_j performed by a primary p_k is definitive, no following primary will try to execute another assignment with the same sequence number. After the filtering, *state.TA* is set to *maj_state.TA* and *seq* to *last*(*state.TA*).#*seq* as this is the last executed definitive assignment (line 26).

[4] Due to the time taken by the the leader election protocol[5] (at least 2δ) to select a leader (see Section 3), it follows that any READMAJ() function starts after the arrival of all the *timely* messages broadcast through any previous WRITEMAJ().

(4) r_i invokes WRITEMAJ($maj\_state.epoch+1$) at line 27 to impose its primary epoch number (greater than any previous primary). Then, r_i becomes primary (line 28).

If any of the above actions is not successfully executed by r_i, it will not become primary. Note that if r_i is still leader after the unsuccessful execution of the *css* procedure, it starts to execute it again.

Receipt of a "no leadership" notification. r_i sets the *primary* variable to $\perp$ (line 15). Note that a notification of "no leadership" imposes READMAJ() and WRITEMAJ() to fail (i.e. to return $\perp$). Consequently if r_i was serving a request and executing statement 11, it sets *primary* to $\perp$ (line 13).

Note that the proposed implementation adopts an optimistic approach[4]: it allows internal inconsistencies among the sequencer replica states as it requires only a majority of replicas to be updated at the end of each definitive assignment. In other words the implementation sacrifices update atomicity to achieve better performances in failure-free runs. The price to pay is in the *css* phase carried out at each primary change.

It can be shown that the proposed protocol (along with the simple client invocation semantic described in Section 3) satisfies the sequencer specification given in Section 2. A detailed proof of correctness is given in[1].

5 Conclusions

In this paper we presented the specification of a sequencer service that allows thin, independent clients to get a unique and consecutive sequence number to label successive operations. We have then shown a fault-tolerant sequencer implementation based on a primary-backup replication scheme that adopts a specific partially synchronous model, namely the timed asynchronous model.

The proposed implementation adopts an optimistic approach to increase performances in failure-free runs with respect to (possible) implementations using standard group communication primitives, e.g. total order multicast. This follows because the proposed implementation only requires a majority of replicas to receive primary updates.

The practical interest of a fault-tolerant implementation of a sequencer service lies in the fact that it can be used to synchronize processes running over an asynchronous distributed system. For example, in the context of software replication, the sequencer actually embeds the partial synchrony necessary to solve the problem of maintaining server replica consistency despite process failures. This also allows to free server replicas from running over a partially synchronous system, i.e. to deploy server replicas over an asynchronous system.

References

1. R. Baldoni, C. Marchetti, and S. Tucci-Piergiovanni. Fault Tolerant Sequencer: Specification and an Implementation. Technical Report 27.01, Dipartimento di Informatica e Sistemistica, Università di Roma " La Sapienza", november 2001.

2. R. Baldoni, C. Marchetti, and S. Tucci-Piergiovanni. Active Replication in Asynchronous Three-Tier Distributed System. Technical Report 05-02, Dipartimento di Informatica e Sistemistica, Università di Roma " La Sapienza", february 2002.
3. N. Budhiraja, F.B. Schneider, S. Toueg, and K. Marzullo. The Primary-Backup Approach. In S. Mullender, editor, *Distributed Systems*, pages 199–216. Addison Wesley, 1993.
4. X. Défago, A. Schiper, and N. Sergent. Semi-passive replication. In *Proceedings of the 17th IEEE Symposium on Reliable Distributed Systems (SRDS)*, pages 43–50, West Lafayette, IN, USA, October 1998.
5. C. Fetzer and F. Cristian. A Highly Available Local Leader Election Service. *IEEE Transactions on Software Engineering*, 25(5):603–618, 1999.
6. C. Fetzer and F. Cristian. The Timed Asynchronous Distributed System Model. *IEEE Transactions on Parallel and Distributed Systems*, 10(6):642–657, 1999.
7. M. Fischer, N. Lynch, and M. Patterson. Impossibility of Distributed Consensus with One Faulty Process. *Journal of the ACM*, 32(2):374–382, April 1985.
8. R. Guerraoui and A. Schiper. Software-Based Replication for Fault Tolerance. *IEEE Computer - Special Issue on Fault Tolerance*, 30:68–74, April 1997.
9. M. Raynal. *Algorithms for Mutual Exclusion.* MIT Press, 1986.
10. F.B. Schneider. Replication Management Using State-Machine Approach. In S. Mullender, editor, *Distributed Systems*, pages 169–198. Addison Wesley, 1993.

Dynamic Resource Management in a Cluster for High-Availability

Pascal Gallard[1], Christine Morin[2], and Renaud Lottiaux[1]

[1] IRISA/INRIA – Paris Research group
[2] IRISA/Université de Rennes 1 – Paris Research group
{plgallar,cmorin,rlottiau}@irisa.fr

Abstract. In order to execute high performance applications on a cluster, it is highly desirable to provide distributed services that globally manage physical resources distributed over the cluster nodes. However, as a distributed service may use resources located on different nodes, it becomes sensitive to changes in the cluster configuration due to node addition, reboot or failure. In this paper, we propose a generic service performing dynamic resource management in a cluster in order to provide distributed services with high availability. This service has been implemented in the Gobelins cluster operating system. The dynamic resource management service we propose makes node addition and reboot nearly transparent to all distributed services of Gobelins and, as a consequence, fully transparent to applications. In the event of a node failure, applications using resources located on the failed node need to be restarted from a previously saved checkpoint but the availability of the cluster operating system is guaranteed, provided that its distributed services implement reconfiguration features.

1 Introduction

To efficiently execute high performance applications, cluster operating systems must offer some global resource management services such as a remote paging system[4], a system of cooperative file caches[7], a global scheduler[2] or a distributed shared memory[3, 1]. A cluster OS can be defined as a set of distributed services. Due to its distributed nature, the high-availability of such an operating system is not guaranteed when a node fails. Moreover, a node addition or shutdown should be done without stopping the cluster and its running applications.

In this paper, we propose a dynamic resource management service whose main goal is to hide any kind of change in the cluster configuration (node addition, eviction or failure) to the OS distributed services and to the applications, assuming process and page migration mechanisms are provided. A node failure should be also transparent for checkpointed applications. This work takes place in the framework of the design and implementation of Gobelins cluster OS. Gobelins is a single system image OS which aims at offering the vision of an SMP machine to programmers. Gobelins implements a set of distributed services for the global management of memory, processor and disk resources. Our generic

B. Monien and R. Feldmann (Eds.): Euro-Par 2002, LNCS 2400, pp. 589–592.

dynamic resource management service has been experimented with the global memory management service (a distributed shared memory)[6] of Gobelins for node addition and eviction.

In Section 2, we describe the proposed dynamic resource management service. Section 3 provides some details related to the service implementation and presents experimental results. Section 4 concludes.

2 Dynamic Resource Management

We call *configuration* the set of active nodes in the cluster. A node is considered to be *active* if it has not been detected failed by other active nodes and is not currently being added to or evicted from the cluster. A *configuration change* is due to a node addition, shutdown or failure. The cluster OS is said to be in the *stable state* when no configuration change is being processed. Otherwise, it is said to be in the *reconfiguration* state. Each of the distributed services that together form the cluster OS manages a collection of objects (for instance, a global memory management service manages a collection of memory pages). Objects may move between nodes at any time during the execution of an application on top of the cluster OS. A set of *metadata* is associated with each object. In particular the current location of an object is a metadata. In the model of distributed service we consider, each service implements a distributed directory with one entry per object, to store object metadata. On each node, the process responsible of the local directory entries is called a *manager*. In a given configuration, the manager of a particular directory entry is statically defined but when a configuration change occurs, the distribution of directory entries on the nodes belonging to the new configuration is updated.

The dynamic resource management service we have designed, called *adaptation layer*, is in charge of detecting configuration changes, updating the distribution of directory entries on cluster nodes in the event of a configuration change, triggering reconfiguration of distributed services when needed (for exemple after detection of a failure). Importantly, it is the adaptation layer which ensures that at any time, all cluster nodes have a consistent view of the current configuration. The adaptation layer is also used to locate directory managers[5]. At initialization time, each distributed service registers to the adaptation layer to benefit from its functions. The registration step allows distributed services to provide the adaptation layer service specific functions needed to perform the service reconfiguration. Note that the adaptation layer implements a single reconfiguration protocol to deal with any kind of configuration changes.

The adaptation layer is implemented by two processes on each node: the *locator* and the *supervisor*.

The locator process keeps track of directory managers for all distributed system services. It is activated each time an object is accessed by an application as the object metadata stored in the directory may be read to locate the considered object and/or updated depending on the operation performed on the object. The information used by the locator to locate managers is updated when the cluster

OS is in reconfiguration state. It does not change when the OS is in the stable state.

The supervisor process is responsible of the addition or the shutdown of the node on which it executes. It is the *supervisor* process that prepares its own node and notices the cluster. The set of *supervisors* in the cluster cooperate in order to maintain a consistent view of the cluster configuration. In this way, a node *supervisor* participates in the failure detection protocol. When a node failure happens (or is suspected) a consensus protocol, which is out of the scope of this paper, is executed. When a configuration change happens in the cluster, after a communication layer update, the *supervisor* triggers directory entries' migration. The functions registered by each service are used by the adapatation layer for the migration of directory entries.

3 Implementation in Gobelins and Evaluation

The dynamic resource management service described in the previous section has been implemented in Gobelins cluster OS and has been experimented with Gobelins global memory management service as an example of distributed service.

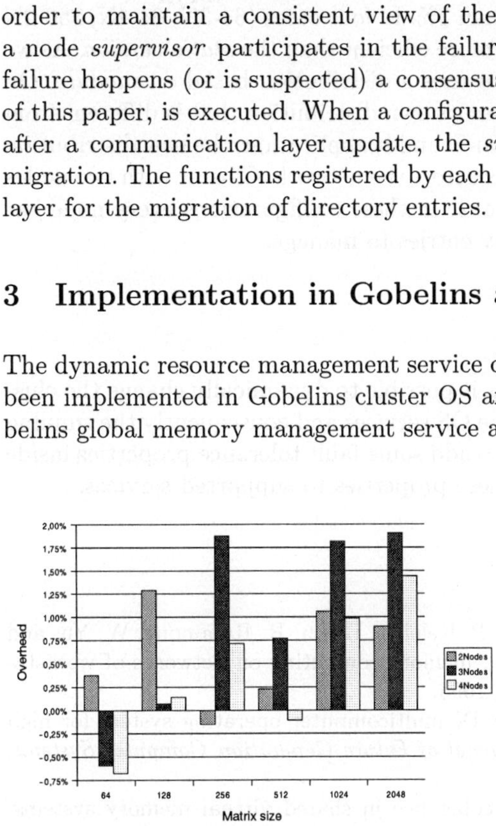

Fig. 1. Overhead evaluation

The cluster used for experimentation is made up of four *Pentium III* (500MHz, 512KB L2 cache) nodes with 512MB of memory. The nodes communicate with a Gigabit network. The Gobelins system used is an enhanced 2.2.13 Linux kernel. We consider here two of the Gobelins modules, the high performance communication system, and the global memory management system. We present in this paper an evaluation of the overhead due to the adaptation layer on the applications execution time. We have compared the execution time of the MGS application obtained with two different versions of Gobelins: the original one in which directory managers are located using a static modulo function (STAT) and a Gobelins version in which distributed services rely on the adaptation layer to locate directory managers (DYN).

The parallel application used in our tests is a Modified Gram-Schmidt (MGS) algorithm. The MGS algorithm produces from a set of vectors an orthonormal basis of the space generated by these vectors. The algorithm consists of an external loop running through columns producing a normalized vector and an inner loop performing for each normalized vector a scalar product with all the remaining ones. Time is measured on the external loop of the MGS program. Each test is repeated 10 times. During the tests, error checking mechanisms in the communication layer were disabled.

We made several sets of experiments with different matrix sizes (64, 128, 256, 512, 1024 and 2048) on different clusters (2, 3 and 4 nodes). The figure presents the measured overhead for MGS calculated as: $overhead = (\frac{DYN}{STAT} - 1) * 100$.

In all cases, the overhead is less than 2%. In four cases (64-3N, 64-4N, 256-2N and 512-4N), the *dynamic version* is more efficient than the *static version.* As we indicate previously, the *static version* uses a distribution based on *modulo.* On another side, the *dynamic version* uses its own distribution that is different from *modulo.* In the particular case of Gram-Schmidt application, the new distribution decreases the number of page requests across the network. Cluster with two nodes and four nodes are similar cases because in these configurations every node has exactly the same number of directory entries to manage.

4 Conclusion

The proposed adaptation layer makes it possible to dynamically change the cluster configuration without stopping the OS services and consequently the running applications. In the future, we want to add some fault tolerance properties inside the directories in order to provide these properties to supported services.

References

1. C. Amza, A. L. Cox, S. Dwarkadas, P. Keleher, H. Lu, R. Rajamony, W. Yu, and W. Zwaenepoel. Treadmarks: Shared memory computing on networks of workstations. *IEEE Computer*, 29(2):18–28, 1996.
2. A. Barak and 0. La'adan. The MOSIX multicomputer operating system for high performance cluster computing. *Journal of Future Generation Computer Systems*, 13(4-5):361–372, March 1998.
3. Kai Li and Paul Hudak. Memory coherence in shared virtual memory systems. *ACM Transactions on Computer Systems*, 7(4), November 1989.
4. Michael J. Feeley, William E. Morgan, Frederic H. Pighin, Anna R. Karlin, Henry M. Levy, and Chandramohan A. Thekkath. Implementing global memory management in a workstation cluster. In *Proc. of the 15th ACM Symposium on Operating Systems Principles*, pages 129–140, December 1995.
5. Pascal Gallard, Christine Morin, and Renaud Lottiaux. Dynamic resource management in a cluster for scalability and high-availability. Research Report 4347, INRIA, January 2002.
6. R.Lottiaux and C.Morin. Containers : A sound basis for a true single system image. In *Proceeding of IEEE International Symposium on Cluster Computing and the Grid*, pages 66–73, May 2001.
7. Thomas E. Anderson, Michael D. Dhalin, Jeanna M. Neefe, David A. Patterson, Drew S. Roselli, and Randolph Y. Wang. Serverless network file systems. *ACM Transactions on Computer Systems*, 14(1):41–79, February 1996.

Progressive Introduction of Security in Remote-Write Communications with no Performance Sacrifice

Éric Renault and Daniel Millot

Institut National des Télécommunications,
9, rue Charles Fourier, 91011 Évry Cedex, France,
{Eric.Renault,Daniel.Millot}@int-evry.fr

Abstract. In a framework where both security and performance are crucial, cluster users must be able to get both at the desired level. Remote-write communications bring high performance but expose physical addresses. In this paper, we present an approach which allows the user to secure remote-write while deciding the cost of that securization.

1 Introduction

Clusters of workstations have been a prominent architecture for some years now, and a large number of such platforms have been deployed. Efforts to interconnect those clusters into grids are on the way. However, there are still a lot of users looking for CPU power, while existing clusters are not busy all the time. When trying to give those users an opportunity to use idle periods on underloaded clusters, we have to achieve the highest performance from the available resources and a secure use of those resources. On the one hand, system and middleware overheads should be minimized so that users can manage the hardware in aggressive ways, for instance making the best out of the network interconnect when transferring data. On the other hand, allowing "foreign" users to access resources of a platform is a critical security issue, and provision has to be made in order to avoid misuses.

Therefore, it seems we pursue two opposite objectives: ensuring a high security level while minimizing overheads. In this paper, we show that we can meet both at the desired level, thanks to a secure use of the remote-write primitive. Section 2 first explains why remote-write is a good solution and then presents the GRWA architecture we propose. Section 3 focuses on the securization of remote-write in GRWA: different methods and their respective costs are presented. Finally, we conclude on the perspectives of this approach.

2 Security vs. Performance with Remote-Write?

A grid is an effort to make the best out of the CPU power available in a federation of computing resources, and could for instance allow users to run their

B. Monien and R. Feldmann (Eds.): Euro-Par 2002, LNCS 2400, pp. 593–597.

applications on remote clusters. The main objective of such a framework is performance. Although distributed programming traditionally relies on message passing libraries, such as MPI, it is not the most efficient since data movement and synchronization are intertwined in this approach. On the contrary, a remote-write primitive deals with data movement only, thereby leading to a programming model with better intrinsic performance. Furthermore, remote DMA capable NICs (for Network Interface Card) are available, making remote-write potentially efficient on such hardware. The remote-write protocol (where both local and remote physical addresses are requested) implements a zero-copy transfer which improves even more communications. Moreover, [1] showed that MPI can be efficiently implemented over a remote-write primitive for those who prefer the message-passing programming model.

Note that in a grid, foreign platforms can cooperate and users interact with distant sites, making security a big issue. Although remote-write is desirable, it is not recommended to deal with physical addresses without any protection, as accidental or intentional use of erroneous addresses could crash the kernel of the operating system. The paper shows how remote-write can be made secure while preserving performance.

The software architecture we propose, GRWA (for Global Remote-Write Architecture), is composed of independent modules which can be independently integrated in the kernel at three different levels, entailing different performance penalties: in user space, in kernel space with access through system calls or in kernel space with no system call (functionalities being accessed from other kernel modules). If setting the modules in user space is a way to provide the best performance, integrating all modules in the kernel of the operating system is the only way to ensure it is not possible for a user application to bypass the security set up by the administrator (like the protection of addresses for example).

Structuring the architecture in independent modules makes it possible for each module to use the implementation that best fits either the requests of the administrator of the machine or the underlying hardware. Moreover, providing our software architecture on another NIC, operating system or bus, just requires the corresponding module to be re-written, implementing the associated API.

3 Security and Performance in GRWA

In order to perform data transfers using the remote-write protocol, two kinds of addresses are given to the user: virtual addresses are used to manipulate data in the virtual address space and "structured" addresses are used by "normal" messages to specify memory locations for data transfer (unlike "short" messages where no address is involved). In some cases, virtual and structured addresses may be the same. Three successive steps might be used to protect information: organization of the information in structured addresses; integration of a fingerprint; an optional encryption of both information and fingerprint. When used, these methods may be tuned in order to provide a scale in which the better the performance, the lower the security.

In this section, we use upper cases to refer to the length of a data field (i.e. the number of bits necessary to store the information) and lower cases to indicate the value in the field. For an operating system where the memory is divided into pages, let 2^M be the memory size in bytes and 2^Q the size of a page in bytes. As page size is generally limited to a few kilo-bytes, let a "contiguous memory block" be a set of contiguous pages beginning at page number b and composed of $s+1$ pages (assuming that a contiguous memory block must be composed of at least one page). In this interval, let o be the offset inside the contiguous memory block which value ranges from 0 to $(s+1) \times 2^Q - 1$. Figure 1 shows the organization of a structured address. When the user specifies an address inside the contiguous memory block, the offset is the only part that may be modified.

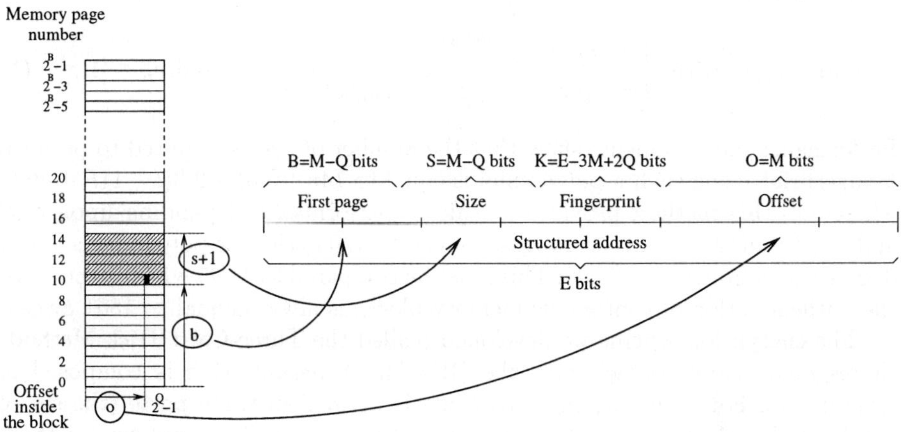

Fig. 1. Organization of structured addresses

Let $f(b,s)$ be the real fingerprint for the contiguous memory block. In order to get the address for the data transfer, the system compares the size of the contiguous memory block with the offset; then, it performs the real fingerprint using function f and checks that the fingerprint included in the structured address matches the real fingerprint.

The number of bits needed to store the position of the first page and the size of the contiguous memory block is $M-Q$; the number of bits for the offset inside the contiguous memory block is equal to M. By construction, $E = B+S+K+O$ and thus $K = E - 3M + 2Q$. The larger the fingerprint, the more reliable the structured address, so extensions may be provided to enlarge the fingerprint. Considering that the maximum size for a contiguous memory block depends upon the position of the first page of the block and as the larger the size of a block, the larger the maximum offset inside the block, a dynamic implementation can be derived from the static one described above. In this case, the size of the fingerprint is limited to $K = E - M - \lceil \ln_2(2^{M-Q} - b) \rceil - \lceil \ln_2(s+1) \rceil$. Performance measurements show that, for the static implementation, 23 cycles are required to create a structured address and 33 cycles are needed for checking; for the dynamic implementation, performance are respectively 103 and 117 cycles.

On our platform (233-MHz Pentium II), the time requested by MD5 [2] to authenticate two structured addresses (for the sender and the receiver) is more than twice as long as the one-way latency for a small message (13.4 μs vs. 5.4 μs). Moreover, as an encryption may be performed to hide information related to the contiguous memory block and its fingerprint, it is not necessary to use an extremely complex method. Therefore, we developed a method (called the Successive Square Method) based on the calculation of a polynomial (see (1)). In order to make sure fingerprints are statistically well distributed in $[0; 2^K[$, some constraints (on the parity of x and c_i and the maximum value for $R + r$) must be satisfied [3].

$$P(x) = v_l(x) \text{ with } \begin{cases} v_0(x) = c_0 x^{2^{R+r_0}} \\ v_{n+1}(x) = (v_n(x) + c_{n+1}) x^{2^{R+r_{n+1}}} \end{cases} \text{ and } r_n \in [1; r] \quad (1)$$

Performance measurements show that the number of cycles required to perform a fingerprint using such a polynomial is equal to $14r + 3.5l + 2.3(l-1)R + 46.6$. Therefore, this method provides a scale of performance depending upon l, R and r. The number of cycles ranges from 50 to 420 cycles for polynomials which degree varies from 0 to 320000. This must be compared to MD5 which latency for the authentication of contiguous memory blocks is always equal to 1561 cycles.

The encryption scheme we developed (called the Three-Card Trick Method) shares many caracteristics with the DES [4]. A permutation is composed of several cuts. For each cut, the set of bits is divided in three parts (the set of numbers of bits in each part is the key of the permutation) and two of them are swapped. There are three possibilities for the swapping. However, only those swapping adjacent sets of bits are used. Each cut performs a bijection of the set of bits on itself. Thus, in order to retrieve the original information, the same set of cuts must be performed in the reverse order. Moreover, it is easy to determine the minimum number of cuts needed to make sure the set of bits is well melted and the number of possibilities one must try to break down the encryption is very large even for a few cuts. Performance measurements show that 41 cycles are necessary to perform each cut.

Figure 2 compares the performance of the security methods discussed in this article. Performance was measured on a cluster composed of four 233-MHz Dual Pentium II linked in a daisy chain with 1-Gbit/s HSL links [5]. All these elements may be included when sending a message. Rectangles on the right show the one-way latency for both short and normal messages. As no security is required for short messages, the only extra latency that may be added is a system call. For normal messages, a large variety of solutions may be possible, from a highly unsecured version to a highly secure one which includes a dynamic organization of information, a medium-degree polynomial for the successive square method and a high number of cuts for the three-card trick method, all this located in the kernel of the operating system.

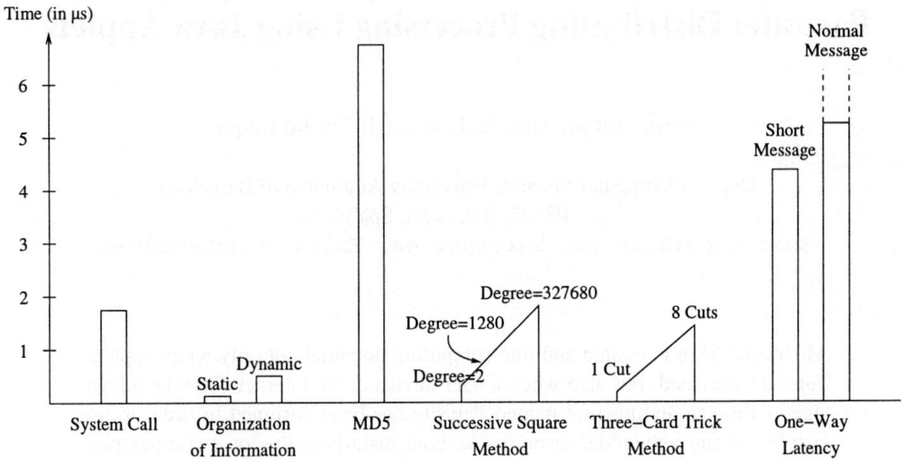

Fig. 2. Comparison of security elements overheads with one-way latency

4 Conclusion

In this article, we have presented the Global Remote-Write Architecture, which provides a set of programming interfaces whatever the underlying hardware and operating system, and the different ways security may be integrated to protect the use of addresses on both local and remote nodes. Performance measurements show that these methods are compatible with the use of a high performance network. At the moment, the architecture is available on the HSL network and an important effort is in progress to provide the same services on other high-speed networks, especially Myrinet-2000.

References

[1] O. Glück, A. Zerrouki, J.L. Desbarbieux, A. Fenyö, A. Greiner, F. Wajsbürt, C. Spasevski, F. Silva, and E. Dreyfus. Protocol and Performance Analysis of the MPC Parallel Computer. In *15th International Parallel & Distributed Processing Symposium*, page 52, San Francisco, USA, April 2001.

[2] R. Rivest. The MD5 Message-Digest Algorithm. Request for Comments 1321, April 1992.

[3] E. Renault. *Etude de l'impact de la sécurité sur les performances dans les grappes de PC.* Thèse de doctorat, Université de Versailles – Saint-Quentin-en-Yvelines, Décembre 2000.

[4] Federal Information Processing Standards Publication. Data Encryption Standard (DES), January 1988. FIPS PUB 46-2.

[5] F. Potter. *Conception et réalisation d'un réseau d'interconnexion à faible latence et haut débit pour machines multiprocesseurs.* Thèse de doctorat, Université Paris VI, Avril 1996.

Parasite: Distributing Processing Using Java Applets[1]

Remo Suppi, Marc Solsona, and Emilio Luque

Dept. of Computer Science, University Autonoma of Barcelona,
08193, Bellaterra, Spain
Remo.Suppi@uab.es, Marc@ptv.es, Emilio.Luque@uab.es

Abstract. There is wasted and idle computing potential not only when applications are executed, but also when a user navigates by Internet. To take advantage of this, an architecture named Parasite has been designed in order to use distributed and networked resources without disturbing the local computation. The project is based on developing software technologies and infrastructures to facilitate Web-based distributed computing. This paper outlines the most recent advances in the project, as well as discussing the developed architecture and an experimental framework in order to validate this infrastructure.

1 Introduction

In the last five years, a growing interest in distributed computation has been observed. Projects such as Seti@home and Distributed.net are examples of metacomputing popularity and extension [5,6]. These two projects are clear examples of the trends in using particular user equipment for distributed computing. Simply stated, metacomputing is a set of computers (whose geographical distribution is of no relevance) that are interconnected and that together act as a supercomputer. The metacomputing concept is a very generic definition that has undergone specialization through several proposals. [1-7]

Our proposal, referred to as *Parasite* (**Para**llel **Site**), results from the need for computing power and from the fact that it is possible to extract available resources with idle time and without disturbing the local user workload. This is the principle underpinning several metacomputing projects; however, our project introduces new ideas with respect to user intervention, available resources, net interconnection, the distributed programming paradigm or the resident software on each user's computer.

2 Our Proposal: Parasite (Parallel Site)

The main idea of Parasite is the utilization of personal computers as computing nodes and the interconnection network without carrying out modifications in the hardware

[1] This work has been supported by the CICYT under contract TIC98-0433 and TIC 2001-2592.

B. Monien and R. Feldmann (Eds.): Euro-Par 2002, LNCS 2400, pp. 598–602.

interconnection equipment and without the need to install software in the user's computers -UC- (machines that will integrate the Parasite distributed architecture). To this end, previously installed software in the UC connected to Internet is used: the Internet browsers (navigators). These applications, together with the possibility of executing Java Applets, open the possibility of creating computing nodes.

Our proposal is based on creating a hardware-software infrastructure that supports the embarrassingly parallel computation model and that the local user does not have to modify installed software (or install new software) in the local machine. This infrastructure must provide to the distributed applications programmer with the benefits of massive parallelism (using the CPU free time of the UC) and without the typically attendant costs (topology, tuning application-architecture, mapping & routing policies, communication protocols, etc.). Furthermore, all this must be transparent to the local user (only initial assent will be necessary). From the programmer's point of view (the user of the Parasite infrastructure), the distributed application code will be executed in the maximum number of available resources at each moment, without changes in the application code.

Figure 1 shows the Parasite architecture (clients & server), the information fluxes and Internet traffic in two different operation modes. The concept of Parasite is based on two operation forms: **collaborative** (for users who wish to grant their resources to the distributed computing process in any place of Internet) and **transparent** (the local user does not make an explicit web petition to the Parasite server; the Java Applet is sent transparently by the Parasite host to the user computer during the user navigation). The first continues the collaborative line of work set out by the projects referred to above [5-7]. The second form of work (the transparent form) is proposed for local network environments. The Parasite server (fig. 1) is who coordinates, distributes and collects the data between the UCs. The UCs can be working, according to the location, in collaborative mode (UC in any place of Internet) or transparent mode (UC in a private or corporative net). These UCs will execute a Java applet sent by the server and each applet will form part of the distributed application code.

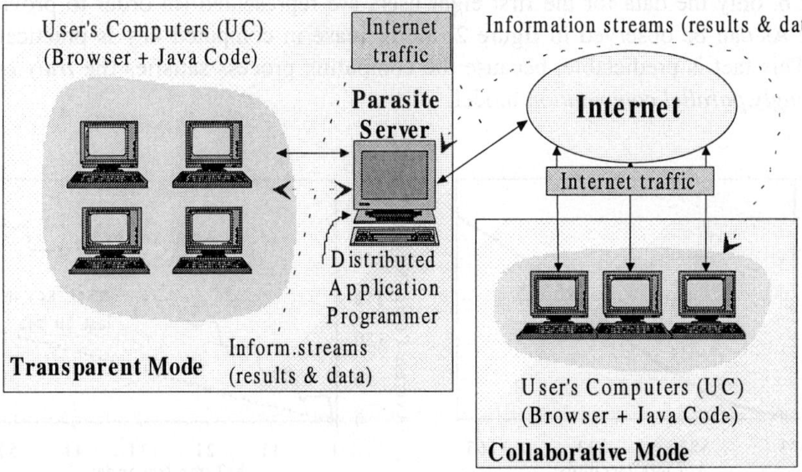

Fig. 1. Parasite architecture & working modes (collaborative & transparent)

The applet will be executed during the time that the user continues to use the navigator. It is therefore very important for the project objectives to analyze the users' navigation standards. This analysis will allow an estimation of the time (mean time) that the CPU remains free for the distributed computing (without affecting the user workload). With data obtained from [8,9], we can conclude that the average CPU time for distributed computing oscillates between 75% to 86% of the users' navigation time, according to user type and when considering the worse I/O case.

The Parasite architecture has been designed to sustain (but is not limited to) the "ideal" computation from the parallel computing point of view: a computation can be split into an independent number of tasks and each of these can be executed on a separate processor. This is known in the literature as *embarrassingly parallel computations* or *pleasantly parallel computations* [10]. There are a considerable number of applications appropriate for this model, such as the geometrical transformations of images, the Mandelbrot set, Monte Carlo methods, parallel random number generation, etc. The Parasite architecture also sustains variations of this model, such as the *nearly embarrassingly parallel computations,* where results need to be collected and processed, suggesting a master-worker organization.

3 Experimental Framework

In order to show the possibilities and performance of the Parasite architecture, real distributed computing experiments have been carried out. The program chosen for these experiments is based on the RSA laboratories proposal for 1997 to prove the robustness of the RC5 (RC5-32/12/8 –64 bits key-) encryption algorithm. [11,12]

Figure 2 shows the evolution of calculation (number of encrypted keys tested) versus the time without local workload whit Parasite working in collaborative mode. Figure 2.a shows the total number ($*10^6$) of key computed by the system. Figure 2.b shows average values: number of key ($*10^3$) per second and in the last 10 seconds. In figure 2.b, only the data for the first eight users are represented (in order to provide details). As can be observed in figure 2, the increase in computed key is practically linear. This fact is predictable, because the computing process satisfies the *truly embarrassingly parallel computation* model.

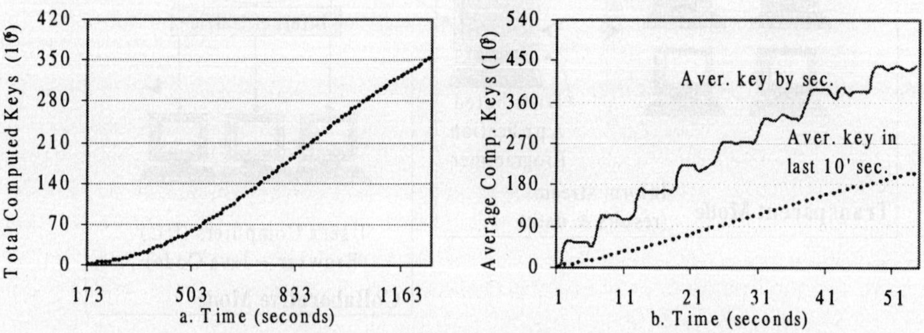

Fig. 2. Collaborative Mode: Evolution of Computed Keys.

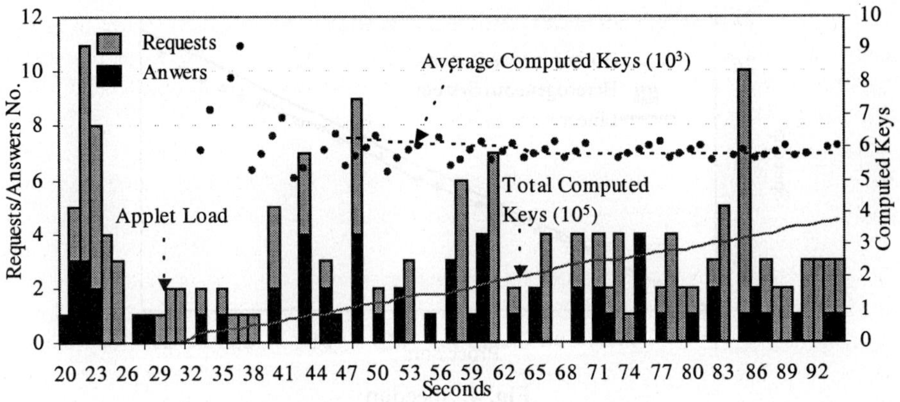

Fig. 3. Client behavior on google.com

In order to show client/server behavior in transparent mode, the www.google.com URL was selected. Figure 3 shows the navigator requests/answers number vs. the time and the behavior of the applet running in a representative node (computed keys/sec –dot line- and total keys –continuous line-). As can be observed in certain points of the computed keys/second (dots graph), there are some places where we do not find computed keys. This situation indicates a load increase in the local computer, and therefore the applet goes into a sleeping state. This situation generates a dispersion of the number of keys/sec, but if the tendency line (hyphens line) after the initial transient is observed, the system tends to be stabilized.

In order to compare system performance when the number of UCs is increased, a heterogeneous speedup has been defined. Figure 4 shows the speedup for a homogeneous system (continuous line) of 24 PC Pentium II 500 Mhz running in collaborative mode on a class C LAN. The dot line is the speedup for a heterogeneous system of Pentium III W9x, Pentium II Linux and Ultra 10 Sparc Solaris 2.x working in transparent mode in different LAN segments. As can be observed in figure 6, the results are excellent and the differences with respect to the linear speedup are due to the OS and net load. The results for the homogenous system and collaborative mode are better because the same LAN segment is used for the 24 machines and the server.

4 Conclusions

In a computer, there is wasted and idle computing potential not only when applications are executed, but also when a user navigates by Internet. To take advantage of this, an architecture named Parasite has been designed. This architecture allows jobs to be executed in the user computer without either affecting performance or modifying the user's work environment. The principal characteristic of the solution adopted is that the user does not have to install software in the local machine (only being required to give initial consent) and the Parasite system guarantees that it will not use more computing capacity than that which is idle. The system can work indistinctively in two modes: collaborative and transparent.

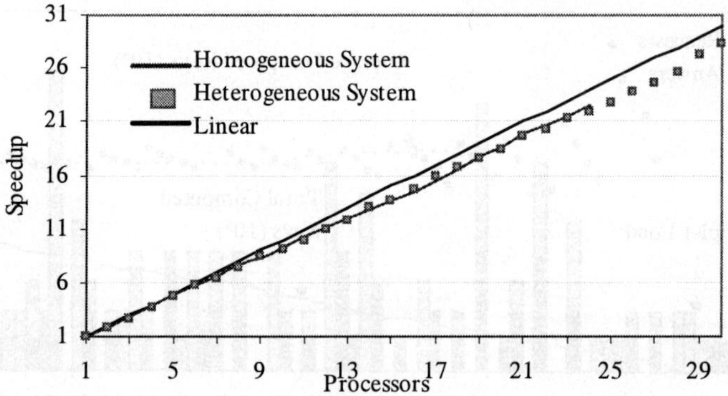

Fig. 4. Speedup

In order to show the capacities of the developed environment, a set of experiments based on the RSA laboratories proposal to prove the robustness of the RC5 encryption algorithm were undertaken. The conclusions for these experiments are that the environment is particularly suitable for applications based on the *(truly, nearly) embarrassingly parallel computations* model. The environment has been proven in homogenous-heterogeneous systems and the same or different LAN segments, the speedup obtained being close to the linear.

Future work will be guided towards: the need for a coordinated and hierarchical net of Parasite servers and the development of a set of applications based on *embarrassingly parallel computations* model in order to prove different granularity types and to determine their efficiency.

References

1. Anderson, T., Culler, D., Patterson, D. A case for NOW IEEE Micro (1995).
2. The Beowulf Project. (1998) http://www.beowulf.org
3. Litzkow, M., Livny, M., Mutka, W. Condor. A Hunter of Idle Workstations. Proc. 8th Int. Conf. Distributed Computing Systems. (1988) http://www.cs.wisc.edu/condor/
4. Foster, I., Kesselman, C. Globus: A Metacomputing Infrastructure Toolkit. International Journal of Supercomputer Applications. (1997). http://www.globus.org/
5. Search for Extraterrestrial Intelligence Project. (2002) http://setiathome.ssl.berkeley.edu/
6. Distributed.net Project. (2002) http://distributed.net
7. Neary, M., Phipps, A., Richman, S., Cappello, P. Javelin 2.0: Java-Based Parallel Computing on the Internet. EuroPar 2000. LNCS 1900 (2000).
8. Nielsen Net Ratings. (2001) http://www.nielsen-netratings.com
9. Sizing the Internet. Cyveillance Corporate. (2000) http://www.cyveillance.com
10. Wilkinson, B., Allen, M. Parallel Programming. Techniques and Applications using networked workstations and parallel computers. Prentice Hall. ISBN 0-13-671710-1. (1999)
11. RSA Data Security Secret-Key Challenge. (1997) http://www.rsa.com/rsalabs/97challenge
12. Ronald Rivest. RC5 Encryption Algorithm. Dr. Dobbs Journal. 226 (1995)

Topic 10
Parallel Programming: Models, Methods and Programming Languages

Kevin Hammond

Global Chair

1 Introduction

The greatest philosopher amongst us is as confined and hamstrung as the least significant thinker by the very language and notations in which his or her ideas can be expressed. By encapsulating complex concepts and ideas in simple words and phrases that we then reuse, we avoid the need to repeat the small and stumbling steps of our predecessors. So too is Computer Science advanced, allowing practitioners to benefit from the toil and wisdom of the pioneers through reusing models and abstraction.

This EuroPar workshop provides a forum for the presentation of the latest research results and practical experience in parallel programming models, methods and languages. Advances in algorithmic and programming models, design methods, languages, and interfaces are needed for construction of correct, portable parallel software with predictable performance on different parallel and distributed architectures.

2 The Research Papers

The 9 papers that have been selected for the workshop target various language paradigms and technologies: functional, object-oriented and skeletal approaches are all represented. A primary theme of the papers in this year's workshop is how technologies can cross over paradigm boundaries to find wider application. A second theme is exploitng abstraction mechanisms to reduce communication costs.

Two papers demonstrate cross-over from the functional community to conventional parallel systems. Firstly, Field, Kelly and Hansen show how the idea of *shared reduction variables* can be used to control synchronisation within SPMD programs. Shared variables can be introduced to eliminate explicit communications, thereby simplifying code structure. Furthermore, a *lazy evaluation* mechanism is used to fuse communications. The result is an improvement in performance over the original version due to the reduction in communication.

Secondly, Liniker, Beckman and Kelly propose to use *delayed evaluation*, to recapture context that has been lost through abstraction or compilation. In the initial stages of execution, evaluation is delayed and the system captures data flow information. When evaluation is subsequently forced through some demand, the data flow information can be used to construct optimised versions of the

B. Monien and R. Feldmann (Eds.): Euro-Par 2002, LNCS 2400, pp. 603–604.

software components as appropriate to the calling context. The approach has been tested experimentally in the context of four simple scientific applications using the BLAS linear algebra library.

Skeleton approaches promise to dramatically increase programming abstraction by packaging common patterns of parallelism in high level routines. There has, however, been a historical lack of support for skeletons in conventional languages such as C or C++. Kuchen's paper introduces a library of basic skeletons for such languages that supports required skeleton functionality including polymorphism, higher-order functions and partial applications at minimal cost in efficiency. The library is built on MPI and therefore portable and efficient.

Having the *right* skeletons available when required is equally important for effective program construction. Bischof and Gorlatch introduce a new skeleton construct, the *double-scan* primitive, a combination of two conventional scan operations: one left scan with one right counterpart. The work is applied to existing software components whose purpose is to solve a system of linear equations. The paper demonstrates both predictability of performance and absolute performance that is comparable to a hand-coded version of the problem.

Recent developments in FPGA technology provide the potential for cheap large-scale hardware parallelism. The paper by Hawkins and Abdallah shows how this potential can be exploited by using a high-level functional language as a behavioural specification that can be systematically transformed into Handel-C and thus to FPGA circuitry. The work is applied to a real-world problem: a JPEG decompression algorithm.

At a more abstract level, Pedicini and Quaglia introduce a new system for distributed execution of λ-terms, PELCR. Their approach used *Directed Virtual Reduction*, a parallel graph-rewriting technique, enhanced with a priority mechanism. Speedup is demonstrated for a standard λ-calculus benchmark, DDA.

Scalability and predictability are key concerns. Work by Sobral and Proença studies scalability issues for object-oriented systems. Their objective is to ensure scalability dynamically by automatically increasing task granularity and reducing communication through runtime coalescing of messages. The work has been evaluated empirically on a number of platforms using a farm-type application.

Finally, exception handling and I/O mechanisms that have been designed for sequential languages and systems can present difficulties for concurrency. One particular problem arises in the context of explicit asynchronous method invocation, where the caller may no longer be in a position to handle remotely induced exceptions at the point they are raised. The paper by Keen and Olsson addresses this issue, introducing new language constructs for forwarding remotely induced exceptions to appropriate handlers. The mechanism has been implemented in JR, an extended Java aimed at tightly coupled concurrent systems. Bougeé, Danjean and Namyst meanwhile consider how to improve responsiveness to I/O events in multithreaded reactive systems, by introducing a synchronous *detection server* to provide a managed service to such events. This approach is demonstrably superior to standard approaches based on polling.

Improving Reactivity to I/O Events in Multithreaded Environments Using a Uniform, *Scheduler-Centric* API

Luc Bougé[1], Vincent Danjean[2], and Raymond Namyst[2]

[1] PARIS Project, IRISA/ENS Cachan, Campus Beaulieu, F-35042 Rennes, France
[2] LIP, ENS Lyon, 46 allée d'Italie, F-69364 Lyon Cedex 07, France

Abstract. Reactivity to I/O events is a crucial factor for the performance of modern multithreaded distributed systems. In our *scheduler-centric* approach, an application detects I/O events by requesting a *service* from a *detection server*, through a simple, uniform API. We show that a good choice for this detection server is the thread scheduler. This approach simplifies application programming, significantly improves performance, and provides a much tighter control on reactivity.

1 Introduction

The widespread use of clusters of SMP workstations for parallel computing has lead many research teams to work on the design of portable multithreaded programming environments [1–3]. A major challenge in this domain is to reconcile *portability* with *efficiency*: parallel applications have to be portable across a wide variety of underlying hardware, while still being able to exploit much of its performance. Most noticeably, much effort has been focused on performing efficient communications in a portable way [4, 5], on top of high-speed networks [6, 7]. However, a major property has often been overlooked in the design of such distributed runtimes: the *reactivity* to communication events. We call *reactivity* of an application its ability to handle external, asynchronous events as soon as possible within the course of its regular behavior. The response time to a network event is indeed a critical parameter, because the observed *latency* of messages directly depends on it: if the application is not reactive enough, the observed external latency can be arbitrary larger the nominal, internal latency offered by the underlying communication library. For instance, all communication primitives including a back and forth interaction with a remote agent (e.g., to fetch some data) are extremely sensitive to the reactivity of the partner [8].

Berkeley's Active Messages Library [9] provides a good reactivity to network events. However, the communication system is highly dependent on the hardware, and it has only been implemented on specific message passing machines. Princeton's Virtual Memory Mapped Communication Library [10] can offer a good reactivity. However, once again, these mechanisms are highly hardware-dependent and need specific OS modifications or extensions. Our goal is not to propose yet another powerful I/O or communication library. Instead, we intend

B. Monien and R. Feldmann (Eds.): Euro-Par 2002, LNCS 2400, pp. 605–614.

to design a *generic* approach, allowing to use already *existing* I/O libraries in a multithreaded environment, so that we can *ensure a good reactivity*.

An application may use several strategies to detect I/O events. The most common approach is to use *active polling* which consists in checking the occurrence of some I/O events by repeatedly calling an appropriate function of the I/O subsystem. Such an elementary test is usually inexpensive, with an overhead of a few assembly instructions. However, repeating such a test millions of times may exhaust computing resources in a prohibitive way. Alternatively, the application can rely on *passive waiting*, using blocking system calls, signal handlers, etc. In this latter case, I/O events are signaled to the operating system by hardware interrupts generated by the I/O device, which makes the approach much more *reactive*. However, catching such an interrupt is usually rather costly, of the order of tenth of microseconds, disregarding the additional latency of rescheduling the application.

Usually, the choice of the I/O detection strategy is made within the application. This results in mixing *application-specific* algorithmic code with *system-dependent* reactivity management code. Moreover, this approach suffers from several severe drawbacks:

Determining the Available Methods. The operating system (i.e., the underlying I/O driver) may only offer a restricted set of methods. In some cases, only a single explicit polling primitive may be provided to the user. In other cases, handling interrupts may be the only way to check the completion of I/O operations. In this latter situation, the operating system may even provide no other choice but a single mechanism to handle interrupts. Moreover, complexity and portability requirements may often prevent the use of some mechanisms. For instance, raw asynchronous delivery of signals imposes hard reentrance constraints on the whole application code, if the consistency of *all* data structures accessed within signal handlers has to be guaranteed.

Selecting the Right One. When several methods are available at the OS level, selecting the most appropriate one depends on many factors. A key factor is the level at which the thread scheduler is implemented. Actually, there are many flavors of thread schedulers (user-level, kernel-level, hybrid) and each of them features its own characteristics as far as its interaction with the operating system is concerned. For instance, in the context of a pure user-level thread scheduler, operations such as blocking system calls, are usually prohibited, except if some sophisticated OS extensions (such as *Scheduler Activations* [11–13]) are available. Even hybrid schedulers, which essentially implement a user-level scheduler on top of a kernel-level one, suffer from this limitation.

Tuning for Performance. Most I/O subsystems (i.e., device drivers) natively provide a low-overhead polling mechanism. However, efficiently using such a mechanism is a difficult challenge in a multithreaded context [14, 15]. As for monothreaded applications, the polling frequency has a crucial impact on the overall application performance. If the I/O subsystem is not polled frequently enough, then the application reactivity may become severely altered. In contrast, a too aggressive polling policy leads to many unproductive polling operations, which wastes computing resources. Even if the optimal frequency can be predicted in advance, it may be difficult to instrument the application to effectively enforce it. Actually, those threads waiting for the completion of some I/O event, loop over a sequence of instructions: each iteration consists in a polling operation, followed by a `thread_yield` instruction in case the operation failed.

The contribution of this paper is to introduce a new approach to the problem of reacting to I/O events in multithreaded environments. We define it as *scheduler-centric*. In our view, the environment should provide the application with a *uniform* paradigm for reactivity management. The actual selection of the strategy, active polling and/or passive waiting, is then left to the scheduler. This allows to centralize all the reactivity-management mechanisms within the scheduler, thereby relieving the programmer from this difficult task. Moreover, this enables the scheduler to adjust its scheduling strategy with the reactivity level required by the applications, independently of the system load. Finally, it allows to aggregate multiple requests issued by concurrent applications to the same NIC, resulting in more efficient interactions. We demonstrate the feasibility of this new approach in the context of a the user-level thread scheduler of the PM^2 multithreaded, distributed environment. Significant performance gains are observed.

2 Our Proposition: a *Scheduler-Centric* Approach

We propose to *centralize* the management of I/O events at a single point within the scheduler, providing the application with a *uniform* mechanism to wait for the completion I/O events. Instead of making I/O completion detection an explicit part of the algorithmic code of the application, we view such an action as a *event detection service* requested by the application to an *external server*, namely the scheduler. The client thread is removed from the running list while waiting for the completion of the service. It is the task of the scheduler to determine the very best way of serving the request: polling, interrupt handling, etc., or any kind of dynamic, adaptive mix of them, and to return the control to the requesting thread.

2.1 Serving the I/O Event Detection Requests

We propose to let the thread scheduler serve the I/O event detection requests for several reasons. First, it is system-aware, in contrast with an application whose code has to be portable across various systems. Thus, the scheduler has full freedom to safely use all mechanisms provided by the OS, including the most sophisticated ones. For instance, a pure user-level thread scheduler "knows" that it is dangerous to invoke a system call that may potentially block the process, except when there is only one single active thread in the system. Furthermore, if some form of asynchronous mechanism is available, then the thread scheduler can provide signal-safe synchronization primitives to the threads waiting for I/O events, while providing regular and fast ones to other threads.

Second, the scheduler is probably the best place where *efficient* polling can be done. In particular, the specific frequency of polling for each requesting thread can be precisely mastered by the scheduler, as it can hold all relevant information, and an optimal decision can possibly be made at each context switch. Also, the scheduler can maintain for each request type a history of previous requests, so as to select the most efficient mechanism: a possible strategy should be to first actively poll and then switch to passively wait for a NIC interrupt after some time. Also, the scheduler can use the otherwise idle time to perform intensive polling if this has been given a high priority. Finally, the scheduler enjoys full freedom regarding the next thread to schedule: it can thus schedule a thread as soon as the corresponding I/O event has been detected.

Third, the scheduler appears thereby as the single *entry point* for event detection requests. This provides an interesting opportunity to *aggregate* the event detection requests issued by various threads. For instance, if several threads are waiting for a message on the same network interface, then there is no need in having all of them polling the interface: the scheduler can aggregate all the requests and poll the interface on their behalf; once an event has been detected, then it can lookup its internal requests tables to determine which thread is to be served. Observe that this aggregation ability is fully compatible with the other aspects listed above: one can well use a mixing of active polling and passive waiting in detecting common events for multiple I/O requests! Thus, our proposal generalizes the `MPI_testany()` functionality of MPI, to any kind of event detection request, using any kind of communication interface.

2.2 A Uniform API to Request Event Detection

We have designed the programming interface so as to insulate the application from the idiosyncrasies of the specific events under detection. The general idea is that the client application, most often a communication library, should register the specific callback functions to be used by the scheduler in serving its requests.

The application has first to register which kind of events it is intended to detect, and how, into the scheduler. This is done by filling the fields of a structure `params` with a number of parameters: callback functions to poll for the events and to group requests together, objective frequency for polling, etc. The `thread_IO_register` primitive returns a *handle* to be used for any subsequent request. Only requests issued with the same handle may be aggregated together.

```
thread_IO_t  thread_IO_register  (thread_IO_registration_t params);
```

Client threads are provided with a *single* primitive to wait for the occurrence of an I/O event. The `thread_IO_wait` primitive is a blocking one (for the caller thread). If needed, asynchronous I/O event detection can be achieved in multi-threaded environment by creating a new thread to handle the communication.

Argument arg will be transmitted to the previously registered, callback functions, so that these functions can get specific data about the particular request. The scheduler itself does not know anything about these functions. This primitive returns from the scheduler as soon as possible after an event is ready to be handled.

```
void  thread_IO_wait  (thread_IO_t IO_handle, void *arg);
```

For example, registering a *polling routine* and issuing an asynchronous receive for MPI would look like:

```
thread_IO_registration_t MPI_params;
thread_IO_t MPI_handle;
...
MPI_params.blocking_system_call = NULL;
MPI_params.group = &MPI_group();
MPI_params.poll = &MPI_poll();
MPI_params.frequency = 1;

MPI_handle=thread_IO_register(&MPI_params);
```

```
MPI_Request request;
MPI_IO_info_t MPI_IO_info;
....
MPI_Irecv(buf, size, ..., &request);

MPI_IO_info.request = request;
thread_IO_wait(MPI_handle,
      (void *) &MPI_IO_info);
```

3 Implementation Details

We implemented our generic mechanism within the "chameleon" thread scheduler of the PM^2 multithreaded environment [3] which can be customized to use any of the following scheduling flavors: *user-level* or *hybrid*. Our mechanism is virtually able to deal with a very large number of *scheduling flavors/device driver capabilities* combinations. We focus below on the most common situations.

3.1 Active Polling

A number of callback functions are needed for the scheduler to handle polling efficiently. They are passed to the scheduler through the params structure. If the I/O device interface allows it, then the function assigned to the group field should be able to aggregate all the requests for this device. Otherwise, a NULL pointer should be specified for this field. This function is called each time a new request is added or removed with respect to the given handle. The poll field holds the function which effectively does the polling job. This function should return -1 if no pending event exists, or the index of a ready request if there is any. Furthermore, a few other parameters have to be specified in the params structure including a frequency integer, which stores the number of time slices between each polling action. Thereby, various I/O devices can be polled with different frequencies, even though they are all accessed through the same interface. Figure 1 displays a skeleton of a callback poll function for the MPI communication interface, which actually generalizes the MPI_Testany primitive of MPI.

```
MPI_Request MPI_requests[MAX_MPI_REQUEST];
int MPI_count;

typedef struct {MPI_Request request; } MPI_IO_info_t;

void MPI_group(void) {
  MPI_IO_info_t *MPI_info;
  MPI_count=0;
  thread_IO_for_each_request(MPI_info) {/* Macro iterating on
                                          pending requests */
    MPI_requests[MPI_count++] = MPI_info->request;
  }
}

int MPI_poll(void) {
  int index, flag;
  MPI_Testany(MPI_count, MPI_requests, &index, &flag, ...);
  if (!flag) return -1;
  return index;
}
```

Fig. 1. Polling callback functions in the case of a MPI communication operation.

3.2 Passive Waiting

The end of a DMA transfer generates an interrupt. Most network interface cards are able to generate an interrupt for the processor when a event occurs, too. Because the processor handles interrupts in a special mode with kernel-level access, the application can not be directly notified by the hardware (network card, etc.) and some form of OS support is needed. Even when communication systems provide direct network card access at the user level (as specified in the VIA [16] standard for example), the card needs OS support to interrupt and notify a user process. Indeed, hardware interruption cannot be handled at user-level without losing all system protection and security. The simplest way to wait for an interrupt from the user space is thus to use blocking system calls. That is, the application issues a call to the OS, that suspends it until some interrupt occurs. When such blocking calls are provided by the I/O interface, it is straightforward to make them usable by the scheduler. The `blocking_system_call` field of the `params` structure should reference an intermediate application function, which effectively calls the blocking routine.

Note that I/O events may also be propagated to user space using Unix-like *signals*, as it is proposed by the POSIX *Asynchronous I/O* interface. When such a strategy is possible, our mechanism handles I/O signals by simply using the aforementioned polling routines to detect which thread is concerned when such a signal is caught. Threads waiting for I/O events are blocked using special *signal-safe* internal locks, without impacting the regular synchronization operations performed by the other parts of the application.

3.3 Scheduler Strategies

A main advantage of our approach consists in selecting the appropriate method to detect I/O events *independently* of the application code. Currently, this selection is done according two parameters: the flavor of the thread scheduler, and the range of methods registered by the application.

When the thread scheduler is entirely implemented at the user level, the active polling method is usually selected, unless some specific OS extensions (such as *Scheduler Activations* [11]) allow the user-level threads to perform blocking calls. Indeed, this latter method is then preferred because threads are guaranteed to be woken up very shortly after the detection of the interrupts. The same remark applies to the detection method based on signals, which is also preferred to active polling.

Two-level hybrid thread schedulers, which essentially run a user-level scheduler on top of a fixed pool of kernel threads, also prevent the direct use of blocking calls by application threads. Instead, we use a technique which uses specific kernel threads that are dedicated to I/O operations. When an application user thread is about to perform an I/O operation, our mechanism finds a new kernel thread on top of which the user thread executes the call. The remaining application threads will be left undisturbed, even if this thread gets blocked. Note that these specific kernel threads are idle most of the time, waiting for an I/O event, so little overhead will be incurred. Also, observe that the ability to aggregate event detection requests together has a very favorable impact: it decreases the number of kernel-level threads, and therefore alleviates the work of the OS.

Observe finally that all three methods (active polling, blocking calls and signals handling) are compatible with a kernel-level thread scheduler.

4 Experimental Evaluation

Most of the ideas of this paper have been implemented in our multithreaded distributed programming environment called PM^2 [3] (full distribution available at URL `http://www.pm2.org/`). First, we augmented our thread scheduler with our mechanism. It allows the applications to register any kind of event detected by system calls or active polling. (Support for asynchronous signals notification has not been implemented yet.) Then, we modified our communication library so that it uses the new features of the scheduler. At this time, MPI, TCP, UDP and BIP network protocols can be used with this new interface. Various platforms are supported, including Linux i386, Solaris SPARC, Solaris i386, Alpha, etc.

The aim of the following tests is to assess the impact of delegating polling to the scheduler, and of aggregating similar requests. They have been run with two nodes (bi-Pentium II, 450 MHz) over a 100 Mb/s Ethernet link. The PM^2 library provides us with both a user-level thread scheduler, and a hybrid two-level thread scheduler on top of Linux, so that it allows using blocking system calls. All durations have been measured with the help of the *Time-Stamp Counter* of x86 processors, allowing for very precise timing. All results have been obtained as the average over a large number of runs.

4.1 Constant Reactivity wrt. Number of Running Threads

A synthetic program launches a number of threads running some computation, whereas a single server thread waits for incoming messages and echoes them back as soon as it receives them. An external client application issues messages and records the time needed to receive back the echo. We list the time recorded by the client application with respect to the number of computing threads in the server program (Table 1).

Table 1. Reaction time for a I/O request wrt. the number of computing threads.

	# Computing threads				
Scheduler version	None	1	2	5	10
Naïve polling (ms)	0.13	5.01	10.02	25.01	50.01
Enhanced polling (ms)	0.13	4.84	4.83	4.84	4.84
Blocking system calls (ms)	0.451	0.453	0.452	0.457	0.453

With our original user-level thread library, with no scheduler support, the listening server thread tests for a network event each time it is scheduled (*naïve polling*). If no event has occurred, then it immediately yields control back. If n computing threads are running, a network event may be left undetected for up to n quanta of time. The quantum of the library is a classical 10 ms, so $10 \times n/2$ ms are needed to react on average, as shown on the first line of Table 1.

With the modified version of the thread library, the network thread delegates its polling to the user-level scheduler (*enhanced polling*). The scheduler can thus control the delay between each polling action, whatever the number of computing threads currently running. The response time to network requests is more or less constant. On average, it is half the time quantum, that is, 5 ms, as observed on the results.

Using blocking system calls provides better performance: we can observe a constant response time of 450 μs whatever the number of computing threads in the system. However, a two-level thread scheduler is needed to correctly handle such calls.

4.2 Constant Reactivity wrt. Number of Pending Requests

A single computing thread runs a computational task involving a lot of context switches, whereas a number of auxiliary service threads are waiting for messages on a TCP interface. All waiting service threads use a common handle, which uses the `select` primitive to detect events. An external client application generates a random series of messages. We report in Table 2 the time needed to achieve the computational task with respect to the number of auxiliary service threads.

This demonstrates that aggregating event detection requests within the scheduler significantly increases performance. Without aggregation, the execution time for the main task dramatically increases with the number of waiting threads.

Table 2. Completion time of a computational task wrt. the number of waiting service threads.

	# waiting service threads							
Scheduler version	1	2	3	4	5	6	7	8
Naïve polling (ms)	80.3	101.3	119.0	137.2	156.6	175.7	195.2	215.7
Enhanced polling (ms)	81.2	84.0	84.0	84.7	86.4	87.9	89.6	91.6

With aggregation, this time remains constant, although not completely, as the time to aggregate the requests depends in this case on the number of requests.

5 Conclusion and Future Work

We have proposed a generic *scheduler-centric* approach to solve the delicate problem of designing a portable interface to detect I/O events in multithreaded applications. Our approach is based on a uniform interface that provides a synchronous event detection routine to the applications. At initialization time, an application registers all the detection methods which are provided by the underlying I/O device (polling, blocking calls, signals). Then, the threads just call a unique *synchronous* function to wait for an I/O event. The choice of the appropriate detection method depends on various complex factors. It is entirely performed by the implementation in a transparent manner with respect to the calling thread.

We showed that the right place to implement such a mechanism is within the thread scheduler, because the behavior of the I/O event notification mechanisms strongly depends on the capabilities of the thread scheduler. Moreover, the scheduler has a complete control on synchronization and context-switch mechanisms, so that it can perform sophisticated operations (regular polling, signal-safe locks, etc.) much more efficiently than the application.

We have implemented our scheduler-centric approach within the PM^2 multithreaded environment and we have performed a number of experiments on both synthetic and real applications. In the case of an active polling strategy, for instance, the results show a clear improvement over a classical application-driven approach.

In the near future, we intend to investigate the use of adaptive strategies within the thread scheduler. In particular, we plan to extend the work of Bal *et al.* [14] in the context of hybrid thread schedulers.

References

1. Briat, J., Ginzburg, I., Pasin, M., Plateau, B.: Athapascan runtime: Efficiency for irregular problems. In: Proc. Euro-Par '97 Conf., Passau, Germany, Springer Verlag (1997) 590–599
2. Foster, I., Kesselman, C., Tuecke, S.: The Nexus approach to integrating multithreading and communication. Journal of Parallel and Distributed Computing **37** (1996) 70–82

3. Namyst, R., Méhaut, J.F.: PM2: Parallel multithreaded machine. a computing environment for distributed architectures. In: Parallel Computing (ParCo '95), Elsevier (1995) 279–285
4. Aumage, O., Bougé, L., Méhaut, J.F., Namyst, R.: Madeleine II: A portable and efficient communication library for high-performance cluster computing. Parallel Computing **28** (2002) 607–626
5. Prylli, L., Tourancheau, B.: BIP: a new protocol designed for high performance networking on Myrinet. In: Proc. 1st Workshop on Personal Computer based Networks Of Workstations (PC-NOW '98). Volume 1388 of Lect. Notes in Comp. Science., Springer-Verlag (1998) 472–485
6. Dolphin Interconnect: SISCI Documentation and Library. (1998) Available from `http://www.dolphinics.no/`.
7. Myricom: Myrinet Open Specifications and Documentation. (1998) Available from `http://www.myri.com/`.
8. Prylli, L., Tourancheau, B., Westrelin, R.: The design for a high performance MPI implementation on the Myrinet network. In: Proc. 6th European PVM/MPI Users' Group (EuroPVM/MPI '99). Volume 1697 of Lect. Notes in Comp. Science., Barcelona, Spain, Springer Verlag (1999) 223–230
9. von Eicken, T., Culler, D.E., Goldstein, S.C., Schauser, K.E.: Active messages: a mechanism for integrated communication and computation. Proc. 19th Intl. Symp. on Computer Architecture (ISCA '92) (1992) 256–266
10. Dubnicki, C., Iftode, L., Felten, E.W., Li, K.: Software support for virtual memory mapped communication. Proc. 10th Intl. Parallel Processing Symp. (IPPS '96) (1996) 372–381
11. Anderson, T., Bershad, B., Lazowska, E., Levy, H.: Scheduler activations: Efficient kernel support for the user-level managment of parallelism. In: Proc. 13th ACM Symposium on Operating Systems Principles (SOSP '91). (1991) 95–105
12. Danjean, V., Namyst, R., Russell, R.: Integrating kernel activations in a multithreaded runtime system on Linux. In: Proc. 4th Workshop on Runtime Systems for Parallel Programming (RTSPP '00). Volume 1800 of Lect. Notes in Comp. Science., Cancun, Mexico, Springer-Verlag (2000) 1160–1167
13. Danjean, V., Namyst, R., Russell, R.: Linux kernel activations to support multithreading. In: Proc. 18th IASTED International Conference on Applied Informatics (AI 2000), Innsbruck, Austria, IASTED (2000) 718–723
14. Langendoen, K., Romein, J., Bhoedjang, R., Bal, H.: Integrating polling, interrupts, and thread management. In: Proc. 6th Symp. on the Frontiers of Massively Parallel Computing (Frontiers '96), Annapolis, MD (1996) 13–22
15. Maquelin, O., Gao, G.R., Hum, H.H.J., Theobald, K.B., Tian, X.M.: Polling watchdog: Combining polling and interrupts for efficient message handling. In: Proc. 23rd Intl. Symp. on Computer Architecture (ISCA '96), Philadelphia (1996) 179–188
16. von Eicken, T., Vogels, W.: Evolution of the Virtual Interface Architecture. IEEE Computer **31** (1998) 61–68

An Overview of Systematic Development of Parallel Systems for Reconfigurable Hardware

John Hawkins and Ali E. Abdallah

Centre For Applied Formal Methods, South Bank University,
103, Borough Road, London, SE1 0AA, U.K.,
{John.Hawkins,A.Abdallah}@sbu.ac.uk

Abstract. The FPGA has provided us low cost yet extremely powerful reconfigurable hardware, which provides excellent scope for the implementation of parallel algorithms. We propose that despite having this enormous potential at our fingertips, we are somewhat lacking in techniques to properly exploit it. We propose a development strategy commencing with a clear, intuitive and provably correct specification in a functional language such as Haskell. We then take this specification, and, applying a set of formal transformation laws, refine it into a behavioural definition in Handel-C, exposing the implicit parallelism along the way. This definition can then be compiled onto an FPGA.

1 Introduction

Efficiency in implementations can be increased through the use of parallelism and hardware implementation. Unfortunately both of these introduce complexity into the development process. Complexity is a problem not only because it lengthens development times and requires additional expertise, but also because increased complexity will almost certainly increase the chance of errors in the implementation. The FPGA has provided huge benefits in the field of hardware development. Circuit design without reconfigurable hardware can be an exceedingly costly process, as each revision of the circuit implemented will come with a significant overhead in terms of both money and time. The FPGA allows a circuit to be implemented and re-implemented effortlessly and without cost. Furthermore, the Handel-C [6] language has been another great step forward in improving hardware development. This has allowed FPGA circuits to be specified in an imperative language, removing the requirement for an understanding of all the low level intricacies of circuit design.

However, there is still room for improvement in this design process. Parallelism in Handel-C is explicit, and so the responsibility for exploiting parallelism rests entirely with the programmer. Without a proper framework to guide the developer, it is likely the individual will resort to ad-hoc methods. Additionally, we feel that imperative languages are not a good basis for the specification of algorithms, as there is very little scope for manipulation and transformation. We propose that functional languages such as Haskell [4] provide a much better basis for specifying algorithms. We find that such languages can capture functionality

B. Monien and R. Feldmann (Eds.): Euro-Par 2002, LNCS 2400, pp. 615–619.

in a far more abstract way than an imperative language, and as such provide far greater scope for transformation and refinement. In this work we give an overview of a framework in which algorithms specified in a clear, intuitive functional style can be taken and refined into Handel-C programs, in part by composing together 'off the shelf' components that model common patterns of computation (higher order functions). This type of approach is often broadly referred to as Skeletons [5]. These programs can then be compiled into FPGA circuit designs. As part of this process, scope for parallelism implicit in the specification will be exposed.

2 Refining Functions to Handel-C

As already noted, functional languages such as Haskell provide an extremely good environment for clear specification of algorithms. Details of functional notation in general can be found in [4], which also includes more specific information relating to Haskell. Also, certain aspects and properties of the particular notation we use in this work are explored in [1, 2].

Handel-C [6] is a C style language, and fundamentally imperative. Execution progresses by assignment. Communication is effectively a special form of assignment. As previously noted, communication in Handel-C follows the style of CSP [7]. The same operators are used for sending and receiving messages on channels (`!` and `?`), and communication is synchronous - there must be a process willing to send and a process willing to receive on a given channel at the same time for the communication to take place. Parallelism in Handel-C can be declared with the `par` keyword.

Data refinement will form an important part of the development process, and will largely dictate the scope for, and type of, parallelism that will occur in our implementation. A list in our specification may correspond to two alternative types in our implementation. The **stream** communicates a list of items sequentially, as a sequence of messages on a single channel, followed by a signaling of the end of transmission (EOT). The **vector** communicates a list in parallel, with each item being communicated independently on a separate channel. Further communication possibilities arise from the combination of these primitives.

Let us consider an example of how a higher order function in the functional setting corresponds to a process in our implementation environment. Perhaps the most widely used higher order function is *map*. Functionally, we have:

$$map\ f\ [x_1, x_2, ..., x_n] \ = \ [f\ x_1, f\ x_2, ..., f\ x_n]$$

In stream terms we have the process `SMAP`, defined in Figure 1. This takes in a stream, and outputs a stream. It requires a process `p` as parameter, which should be a valid refinement of the function f in the specification. Alternatively, in vector terms we have the process `VMAP`, defined in Figure 2. This takes in a vector and outputs a vector. As before, it requires a process `p` as parameter, which should be a valid refinement of the function f in the specification.

```
macro proc SMAP (streamin, streamout, p) {
  Bool eot;
  eot = False;
  do {
    prialt {
      case streamin.eot ? eot: break;
      default: p(streamin,streamout) break;
    }
  } while (!eot);

  streamout.eot ! True;
}
```

Fig. 1. The process SMAP.

```
macro proc VMAP (size,vectorin, vectorout, p) {
  typeof (size) c;
  par (c=0;c<size;c++) {
    p(vectorin.elements[c],vectorout.elements[c]);
  }
}
```

Fig. 2. The process VMAP.

3 Closest Pair Example

Perhaps the best way to explain the development process is with a simple case study. Let us consider the closest pair problem. Given a set of distinct points *ps*, the task is to find the distance between the two closest points. Our intuition tells us a solution can be achieved by pairing each point with every other point in the set, calculating the distance between each of these pairs, then finding the minimum of all these distances. In essence we have:

$$closestpair = fold\ (\downarrow) \circ map\ dist \circ pairs$$

Here $\downarrow$ (pronounced min) is a binary minimum operator. The function *dist* takes a pair of co-ordinates and calculates the distance between them, and the function *pairs* takes a list of items, and returns a list where every item in the source list is paired with every other item. We can define *pairs* as follows:

$$pairs \ = \ fold\ (+\!\!+) \ \circ \ map\ mkpairs \ \circ \ tails_+$$

Here $tails_+$ takes in a list and returns all the non-empty final segments of that list. The function *mkpairs* takes in a list, and returns a list with the head of the source list paired with all following items. With some transformation, we can arrive at the following equivalent definition for *closestpairs*:

$$fold\ (\downarrow) \ \circ \ map\ (fold\ (\downarrow)) \ \circ map\ (map\ dist) \ \circ \ map\ mkpairs \ \circ \ tails_+$$

This definition is useful to us as the intermediate results are processed as a number of independent lists, and thus the scope for parallelism is greater. If we

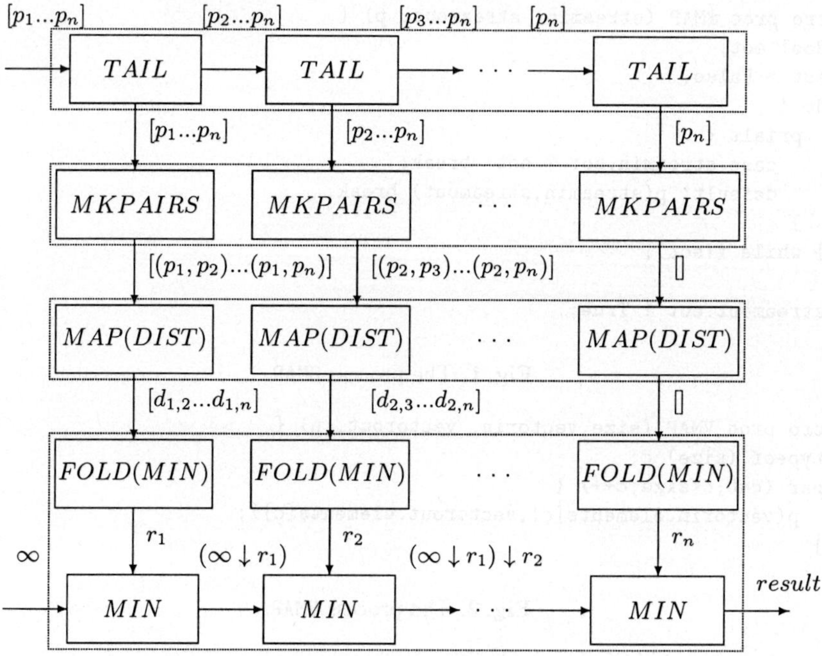

Fig. 3. The closest pair network.

then take this definition and refine it to make use of vectors of streams as the intermediate structure, we have:

$$vsfold\ (\downarrow) \circ vsmap\ dist \circ vmap\ smkpairs \circ vstails_{+}$$

Here *vsfold*, *vsmap* and *vmap* are taken from a library of functional refinements of higher order functions in terms of vectors, streams, and combinations thereof. Given a process refining *dist* and *smkpairs*, we can now construct the implementation from library components, corresponding directly to the above specification. The Handel-C definition is given in Figure 4, and the network is depicted in Figure 3. The dashed boxes in the diagram correspond to the four main stages in the definition. The original functional specification was quadratic time. Through processing each of the lists produced by $tails_{+}$ independently in parallel, with $O(n)$ processing elements, the parallel Handel-C implementation will run in linear time.

4 Conclusion

We have given a brief overview of a framework in which functional specifications can be implemented in parallel hardware. The development process starts with an intuitive functional specification. This forms an ideal basis for transformation such that we can manipulate it into a form that best suits our implementation

```
macro proc CLOSESTPAIR (n,streamin,channelout)
{
  VectorOfStreams (n,Coordinate,     vectora);
  VectorOfStreams (n,CoordinatePair, vectorb);
  VectorOfStreams (n,Distance,       vectorc);

  par
  {
    VSTAILSP (n,streamin,vectora);
    VMAP     (n,vectora,vectorb,MKPAIRS);
    VSMAP    (n,vectorb,vectorc,DIST);
    VSFOLD   (n,vectorc,channelout,MIN);
  }
}
```

Fig. 4. The CLOSESTPAIR process.

requirements. Data refinement can then be employed which will determine the scope for parallelism - both functional and data, in the implementation. Finally process refinement, with the help of a library of commonly used components, will allow us to construct our implementation. An extended example of this process, for a non-trivial problem, a JPEG decoder, can be found in [3].

References

1. A. E. Abdallah, Functional Process Modelling, in K Hammond and G. Michealson (eds), *Research Directions in Parallel Functional Programming*, (Springer Verlag, October 1999). pp339-360.
2. A. E. Abdallah and J. Hawkins, Calculational Design of Special Purpose Parallel Algorithms, in *Proceedings of 7th IEEE International Conference on Electronics, Circuits and Systems* (ICECS 2000), Lebanon, (IEEE, December 2000). pp261-267.
3. J. Hawkins and A. E. Abdallah, Synthesis of a Parallel Hardware JPEG Decoder from a Functional Specification, Technical Report, Centre For Applied Formal Methods, South Bank University, London, UK.
4. R. S. Bird *Introduction to Functional Programming Using Haskell*, (Prentice-Hall, 1998).
5. M. I. Cole, *Algorithmic Skeletons: Structured Management of Parallel Computation*, in Research Monographs in Parallel and Distributed Computing, (Pitman 1989).
6. *Handel-C Documentation*, Available from Celoxica (`http://www.celoxica.com/`).
7. C. A. R. Hoare, *Communicating Sequential Processes.* (Prentice-Hall, 1985).

A Skeleton Library

Herbert Kuchen

University of Münster, Department of Information Systems,
Leonardo Campus 3, D-48159 Münster, Germany,
kuchen@uni-muenster.de

Abstract. Today, parallel programming is dominated by message passing libraries such as MPI. Algorithmic skeletons intend to simplify parallel programming by increasing the expressive power. The idea is to offer typical parallel programming patterns as polymorphic higher-order functions which are efficiently implemented in parallel. The approach presented here integrates the main features of existing skeleton systems. Moreover, it does not come along with a new programming language or language extension, which parallel programmers may hesitate to learn, but it is offered in form of a library, which can easily be used by e.g. C and C++ programmers. A major technical difficulty is to simulate the main requirements for a skeleton implementation, namely higher-order functions, partial applications, and polymorphism as efficiently as possible in an imperative programming language. Experimental results based on a draft implementation of the suggested skeleton library show that this can be achieved without a significant performance penalty.

1 Introduction

Today, parallel programming of MIMD machines with distributed memory is typically based on message passing. Owing to the availability of standard message passing libraries such as MPI [1] [GLS99], the resulting software is platform independent and efficient. Typically, the SPMD (single program multiple data) style is applied, where all processors run the same code on different data. Conceptually, the programmer often has one or more distributed data structures in mind, which are manipulated in parallel. Unfortunately, the mentioned message passing approach does not support this view of the computation. The programmer rather has to split the conceptually global data structure into pieces, such that every processor receives one (or more) of them and cares about all computations which correspond to the locally available share of data. In the syntax of the final program, there is no indication that all these pieces belong together. The combined distributed data structure only exists in the programmer's mind. Thus, the programming level is much lower than the conceptual view of the programmer. This causes several disadvantages. First, the programmer often has to fight against low-level communication problems such as deadlocks and starvation which could be substantially reduced and often eliminated by using a more

[1] We assume some familiarity with MPI and C++.

B. Monien and R. Feldmann (Eds.): Euro-Par 2002, LNCS 2400, pp. 620–629.

expressive approach. Moreover, the local view of the computation makes global optimizations very difficult. One reason is that such optimizations require a cost model of the computation which is hard to provide for general message passing based computations.

Many approaches try to increase the level of parallel programming and to overcome the mentioned disadvantages. Few of them could gain significant acceptance by parallel programmers. It is impossible to mention all high-level approaches to parallel programming here. Let us just focus on a few particularly interesting ones.

Bulk synchronous parallel processing (BSP) [SHM97] is a restrictive model where a computation consists of a sequence of *supersteps*, i.e. independent parallel computations followed by a global communication and a barrier synchronization. BSP has been successfully applied to several data-parallel application problems, but owing to its restrictive model it cannot easily be used for irregularly structured problems.

An even higher programming level than BSP is provided by *algorithmic skeletons*, i.e. typical parallel programming patterns which are efficiently implemented on the available parallel machine and usually offered to the user as higher-order functions, which get the details of the specific application problem as argument functions. Thus, a parallel computation consists of a sequence of calls to such skeletons, possibly interleaved by some local computations. The computation is now seen from a global perspective. Several implementations of algorithmic skeletons are available. They differ in the kind of host language used and in the particular set of skeletons offered. Since higher-order functions are taken from functional languages, many approaches use such a language as host language [Da93,KPS94,Sk94]. In order to increase the efficiency, imperative languages such as C and C++ have been extended by skeletons, too [BK96,BK98,DPP97,FOT92].

Depending on the kind of parallelism used, skeletons can be classified into *task parallel* and *data parallel* ones. In the first case, a skeleton (dynamically) creates a system of communicating processes. Some examples are `pipe`, `farm` and `divide&conquer` [DPP97,Co89,Da93]. In the second case, a skeleton works on a distributed data structure, performing the same operations on some or all elements of this structure. Data parallel skeletons, such as `map`, `fold` or `rotate` are used in [BK96,BK98,Da93,Da95,DPP97,KPS94].

Although skeletons have many advantages, they are rarely used to solve practical application problems. One of the reasons is that there is not a common system of skeletons. Each research group has its own approach. The present paper is the result of a lively discussion within the skeleton community on a *standard set of skeletons*. By agreeing on some common set of skeletons, the acceptance of skeletons shall be increased. Moreover, this will facilitate the exchange of tools such as cost analyzers, optimizers, debuggers and so on, and it will boost the development of new tools.

The approach described in the sequel incorporates the main concepts suggested in the discussion and found in existing skeleton implementations. In particular, it provides task as well as data parallel skeletons, which can be combined based on the *two-tier model* taken from P^{3}L [DPP97]. In general, a computation consists of nested task parallel constructs where an atomic task parallel computation maybe sequential or data parallel. Purely data parallel and purely task parallel computations are special cases of this model.

Apart from the lack of standardization, another reason for the missing acceptance of algorithmic skeletons is the fact that they typically are provided in form of a new programming language. However, parallel programmers typically know and use Fortran, C, or C++, and they hesitate to learn new languages in order to try skeletons. Thus, an important aspect of the presented approach is that skeletons are provided in form of a library. Language bindings for the mentioned, frequently used languages will be provided. The C++ binding is particularly elegant, and the present paper will focus on this binding. The reason is that the three important features needed for skeletons, namely higher-order functions (i.e. functions having functions as arguments), partial applications (i.e. the possibility to apply a function to less arguments than it needs and to supply the missing arguments later), and polymorphism, can be implemented elegantly and efficiently in C++ using operator overloading and templates, respectively [St00]. Thus, the C++ binding does not cause the skeleton library to have a significant disadvantage compared to a corresponding language extension. For a C binding, the type system needs to be bypassed using questionable features like void pointers in order to simulate polymorphism (just as the C binding of MPI).The price is a loss of type safety.

The skeleton library can be implemented in various ways. The implementation considered in the present paper is based on MPI and inherits hence its platform independence.

This paper is organized as follows. In Section 2, we present the main concepts of the skeleton library. Section 3 contains experimental results. In Section 4 we conclude.

2 The Skeleton Library

2.1 Data Parallel Skeletons

Data parallelism is based on a *distributed data structure* (or several of them). This data structure is manipulated by operations (like `map` and `fold`, explained below) which process it as a whole and which happen to be implemented in parallel internally. These operations can be interleaved with sequential computations working on non-distributed data. In fact, the programmer views the computation as a sequence of parallel operations. Conceptually, this is almost as easy as sequential programming. Communication problems like deadlocks and starvation cannot occur. Currently, two distributed data structures are offered by the library, namely:

```
template <class E> class DistributedArray{...}
template <class E> class DistributedMatrix{...}
```

where E is the type of the elements of the distributed data structure. Other distributed data structures such as distributed lists may be added in the future. By instantiating the template parameter E, arbitrary element types can be generated. This shows one of the major features of distributed data structures and their operations. They are *polymorphic*. A distributed data structure is split into several partitions, each of which is assigned to one processor participating in the data parallel computation. Currently, only block partitioning is supported. Other schemes like cyclic partitioning may be added later.

Two classes of data parallel skeletons can be distinguished: computation skeletons and communication skeletons. *Computation skeletons* process the elements of a distributed data structure in parallel. Typical examples are the following methods in class `DistributedArray<E>`:

```
void mapIndexInPlace(E (*f)(int,E))
E fold(E (*f)(E,E))
```

`A.mapIndexInPlace(g)` applies a binary function `g` to each index position i and the corresponding array element A_i of a distributed array `A` and replaces A_i by $g(i,A_i)$. `A.fold(h)` combines all the elements of `A` successively by an associative binary function `h`. E.g. `A.fold(plus)` computes the sum of all elements of `A` (provided that `E plus(E,E)` adds two elements). The full list of computation skeletons including other variants of `map` and `fold` as well as different versions of `zip` and `scan` (parallel prefix) can be found in [Ku02a,Ku02b].

Communication consists of the exchange of the partitions of a distributed data structure between all processors participating in the data parallel computation. In order to avoid inefficiency, there is no implicit communication e.g. by accessing elements of remote partitions like in HPF [HPF93] or Pooma [Ka98]. Since there are no individual messages but only coordinated exchanges of partitions, deadlocks and starvation cannot occur. The most frequently used communication skeleton is

```
void permutePartition(int (*f)(int))
```

`A.permutePartition(f)` sends every partition $A_{[i]}$ (located at processor i) to processor $f(i)$. `f` needs to be bijective. This is checked at runtime. Some other communication skeletons correspond to MPI collective operations, e.g. `allToAll`, `broadcastPartition`, and `gather`. For instance `A.broadcastPartition(i)` replaces every partition of `A` by the one found at processor `i`.

Moreover, there are operations which allow to access attributes of the local partition of a distributed data structure, e.g. `get`, `getFirstCol`, and `getFirstRow` (see Fig. 1) fetch an element of the local partition and the index of the first locally available row and column, respectively. These operations are no skeletons but frequently used when implementing an argument function of a skeleton.

At first, skeletons like `fold` and `scan` might seem equivalent to the corresponding MPI collective operations `MPI_Reduce` and `MPI_Scan`. However, they are more powerful due to the fact that the argument functions of all skeletons can be *partial applications* rather than just C++ functions. A skeleton essentially defines some parallel algorithmic structure, where the details can be fixed

```
 1 inline int negate(const int a){return -a;}
 2
 3 template <class C, int n>
 4 C sprod(const DistributedMatrix<C,n,n>& A,
 5          const DistributedMatrix<C,n,n>& B, int i, int j, C Cij){
 6   C sum = Cij;
 7   for (int k=0; k<A.getLocalRows(); k++)
 8     sum += A.get(i,k+A.getFirstCol()) * B.get(k+B.getFirstRow(),j);
 9   return sum;}
10
11 template <class C, int n>
12 DistributedMatrix<C,n,n> matmult(DistributedMatrix<C,n,n> A,
13                                  DistributedMatrix<C,n,n> B){
14   A.rotateRows(negate);
15   B.rotateCols(negate);
16   DistributedMatrix<C,n,n> R(0,A.getBlocksInCol(),A.getBlocksInRow());
17   for (int i=0; i< A.getBlocksInRow(); i++){
18     R.mapIndexInPlace(curry(sprod<C,n>)(A)(B));
19     A.rotateRows(-1);
20     B.rotateCols(-1);}
21   return R;}
```

Fig. 1. Gentleman's algorithm with skeletons.

by appropriate argument functions. With partial applications as argument functions, these details can depend themselves on parameters, which are computed at runtime. Consider the code fragment in Fig. 1 taken from [Ku02b]. It is the core of Gentleman's algorithm for matrix multiplication (see e.g. [Qu94]). The idea is that two $n \times n$ matrices `A` and `B` are split into a matrix of $m \times m$ partitions (where $m = n/\sqrt{p}$ and p is the number of processors). Initially, the partitions of `A` and `B` are shifted cyclically in horizontal and vertical direction, respectively. More precisely, a partition in row i (column j) is shifted i (j) positions to the left (up) (lines 14,15). Then, the result matrix is initialized with zeros (line 16). The core of the algorithm is the repeated local matrix multiplication at each processor (line 18) followed by a cyclic shift of `A` and `B` by one position in horizontal and vertical direction, respectively. Note that the local multiplication at each processor (line 18) is achieved by partially applying the scalar product function `sprod` to `A` and `B`. The auxiliary function `curry` is used to transform the C++ function `sprod` in such a way that it can be partially applied, i.e. applied to less arguments than it actually needs. Note that `sprod` requires five arguments; two of them are provided by the partial application resulting in a function which needs three more arguments, and such a function is exactly what `mapIndexInPlace` expects as an argument. `R.mapIndexInPlace(curry(sprod<C,n>)(A)(B))` will apply `curry(sprod<C,n>)(A)(B)` to every row index i, column index j, and the corresponding element of $\mathtt{R}_{i,j}$ at position (i, j), i.e. it will provide the three missing arguments of `sprod`. Partial applications are frequently used in the example applications used in section 3. The "magic" curry function has been taken from the C++ template library Fact [St00].

2.2 Task Parallel Skeletons

Most parallel applications are data parallel, and they can be handled with data parallel skeletons alone. However, in some cases more structure is required. Consider for instance an image processing application where a picture is first improved by applying several filters, then edges are detected, and finally objects formed by these edges are identified possibly by comparing them with a data base of known objects. Here, the mentioned stages could be connected by a pipeline where each stage processes a sequence of pieces of the picture and delivers its results to the next stage. Each stage could internally use data parallelism resulting in a two tier model, where the computation is first structured by task parallel skeletons like the mentioned pipeline and where atomic task parallel computations can be data parallel.

Besides pipelines, the skeleton library offers farms and parallel composition. In a farm, a farmer process accepts a sequence of inputs and assigns each of them to one of several workers. Farms are convenient for applications which require some load balancing such as divide & conquer algorithms. The parallel composition works similar to the farm. However, each input is forwarded to every worker.

Each task parallel skeleton has the same property as an atomic process, namely it accepts a sequence of inputs and produces a sequence of outputs. This allows the task parallel skeletons to be arbitrarily nested. Task parallel skeletons like pipeline and farm are provided by many skeleton systems. The two-tier model and the concrete formulation of the pipeline and farm skeletons in our library have been taken from P^3L [DPP97].

In the example in Fig. 2, a pipeline of an initial atomic process, a farm of two atomic workers, and a final atomic process is constructed. In the C++ binding, there is a class for every task parallel skeleton. All these classes are subclasses of the abstract class `Process`. A task parallel application proceeds in two steps. First, a process topology is created by using the constructors of the mentioned class. This process topology reflects the actual nesting of skeletons. Then, this system of processes is started by applying method `start()` to the outermost skeleton. Internally, every atomic process will be assigned to a processor. For an implementation on top of SPMD, this means that every processor will dispatch depending on its rank to the code of its assigned process. When constructing an atomic process, the argument function of the constructor tells how each input is transformed into an output value. Again, such a function can be either a C++ function or a partial application. In Fig. 2, worker i multiplies all inputs by $i + 1$. The initial and final atomic processes are special, since they do not consume inputs and produce outputs, respectively.

2.3 Global vs. Local View

The matrix multiplication example (see Fig. 1) demonstrates that the programmer has a global view of the computation. Distributed data structures are manipulated as a whole. If you compare this to a corresponding program which

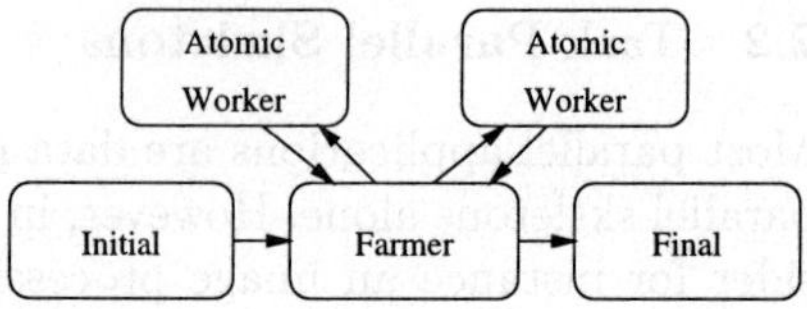


```
#include ''Skeleton.h"

static int current = 0;
static const int numworkers = 2;

int* init(){if (current++ < 99) return &current;
          else return NULL;}

int times(int x, int y){return x * y;}

void fin(int n){cout << ''result: '' << n << endl;}

int main(int argc, char **argv){
   InitSkeletons(argc,argv);

  // step 1: create a process topology (using C++ constructors)
   Initial<int>       p1(init);
   Process*           p2[numworkers];
   for (int i=0; i<numworkers; i++)
      p2[i] = new Atomic<int,int>(curry(times)(i+1),1);
   Farm<int,int>      p3(p2,numworkers);
   Final<int>         p4(fin);
   Pipe               p5(p1,p3,p4);

  // step 2: start the system of processes
   p5.start();

   TerminateSkeletons();}
```

Fig. 2. Task parallel example application.

uses MPI directly (see [Ku02b]), you will note that the latter is based on a local view of the computation. A typical piece of code is the following:

```
int A[m][m], B[m][m], R[m][m];
...
for (r=0; r<sqrtp; r++)
   for (k=0;k<m;k++)
      for (l=0;l<m;l++)
         for (q=0;q<m;q++)
            R[k][l] +=  A[k][q] * B[q][l];
```

It implements the multiplication of the local matrices. There are no global matrices here. A, B, R are just the locally available partitions of the global data structures, which only exist in the programmer's mind. This has several consequences. First, the index computations are different. In the above piece of code, all array indices start at 0, while the indices are usually global when using skeletons. For partition (i, j), they start at $(i * m, j * m)$ (where m is the number of rows and columns in each partition). If the actual computation depends on the global index, the global view is more convenient and typically more efficient. Otherwise the local view is preferable. Without skeletons, there is no choice: only the local view is possible. However for skeletons, we can provide the best of both worlds. We just have to add access operations, which support the local view or a combination of global and local view in order to provide the most convenient operations. In the matrix multiplication example, we could replace the code for the scalar product for instance by:

```
template <class C, int n>
C sprod(const DistributedMatrix<C,n,n>& A,
        const DistributedMatrix<C,n,n>& B, int i, int j, C Cij){
  C sum = Cij;
  for (int k=0; k<A.getLocalRows(); k++)
    sum += A.getGlobalLocal(i,k) * B.getLocalGlobal(k,j);
  return sum;}
```

Here, i and j are global indices, while k is a local one.

In general, the local view has the advantage that the computation may be more easily optimized to the special needs at each processor. Consider Gaussian elimination (see e.g. [Ku02a,Ku02b]). From a global perspective, one would have to apply the pivot operation to every element of the matrix. For the columns to the left of the pivot column this computation would be redundant, and it could be easily avoided based on a local perspective. For such situations, the skeleton library provides specialized versions of the corresponding skeletons, which do not refer to single elements of a distributed data structure but to every partition as a whole. This allows similar optimizations than the local view at the price of lesser elegance. In the above situation the skeleton `mapPartitionInPlace` could be used instead of `mapIndexInPlace` in order to process a partition as a whole.

3 Experimental Results

Since the skeleton library has been implemented on top of MPI, it will hardly be possible to outperform hand-written C++/MPI code. Thus, the question is whether the implementation of the skeletons causes some loss of performance and how large this will be. In order to investigate this, we have implemented a couple of example programs in both ways, with skeletons and using MPI directly. In particular, we have considered the following kernels of parallel applications:

Matrix Multiplication based on the algorithm of Gentleman [Qu94]:
this example uses distributed matrices and the skeletons `mapIndexInPlace`, `rotateRows`, and `rotateCols`.

All Pairs Shortest Paths based on matrix computations:
this uses essentially the same skeletons as matrix multiplication.

Gaussian Elimination: The matrix is split horizontally into partitions of several rows. Repeatedly, the pivot row is broadcasted to every processor and

Table 1. Runtimes for different benchmarks with and without skeletons.

		$p = 4$			$p = 16$		
example	n	skel.	MPI	quotient	skel.	MPI	quotient
matrix multiplication	256	0.459	0.413	1.11	0.131	0.124	1.06
	512	4.149	3.488	1.19	1.057	0.807	1.31
	1024	35.203	29.772	1.18	8.624	6.962	1.24
shortest paths	1024	393.769	197.979	1.99	93.825	44.761	2.10
Gaussian elimination	1024	13.816	9.574	1.44	7.401	4.045	1.83
FFT	2^{18}	2.127	1.295	1.64	0.636	0.403	1.58
samplesort	2^{18}	1.599	†	-	0.774	†	-

the pivot operation is executed everywhere. This example mainly uses `mapPartitionInPlace` and `broadcastPartition`.

FFT: This example is a variant of the FFT algorithm shown in [Qu94]. It uses `mapIndexInPlace` and `permutePartition`.

Samplesort: (see [Qu94]) This well-known sorting algorithm uses the skeletons `mapPartition`, `gather`, and `allToAll`.

It is important to note that our implementation of the skeleton library only uses `MPI_Send` and `MPI_Recv`, respectively, but refrains from collective operations, since we want to keep the implementation as portable as possible. In particular, we are able to port the library to any other message passing platform by changing only a few lines, even if this platform does not provide collective operations. Interestingly, this is the reason why the skeleton-based implementation of samplesort also works for large problem sizes, while the direct MPI implementation based on the rich set of collective operations happened to crash for medium problem sizes already (probably due to buffering problems). Table 1 shows the results for the above benchmarks on a Siemens hpcLine running RedHat Linux 7.2 [PC^2]. Columns 4, 5, 7, and 8 contain the runtimes (in seconds) with and without skeletons (i.e. using MPI directly), respectively. p is the number of processors, n is the problem size (#rows, #elements). Columns 5 and 8 show that the skeleton-based versions are between 1.1 and 2.1 times slower than their MPI-based counterparts. This is mainly caused by the overhead for parameter passing introduced by the higher-order functions. This overhead can be reduced by extensive inlining. Moreover, the mentioned global optimizations for skeletons have not yet been implemented. The scalability of the skeletons is similar to that of MPI.

4 Conclusions and Future Work

We have shown that it is possible to provide algorithmic skeletons in form of a library rather than within a new programming language. This will facilitate their use for typical parallel programmers. The library smoothly combines the main features of existing skeletons. In particular, it provides task parallel skeletons generating a system of communicating processes as well as data parallel skeletons working in parallel on a distributed data structure. Task and data parallelism are combined based on the two-tier model. The C++ binding of the skeleton library has been presented. Moreover, experimental results for some draft implementation based on MPI show that the higher programming level can be gained without a significant performance penalty. Communication problems like deadlocks and starvation are avoided since there are no individual messages but coordinated systems of messages for each skeleton.

As future work, tools like cost analyzers, optimizers, and debuggers will have to be developed for the standard skeleton library.

Acknowledgements. The presented skeleton library is the result of many stimulating discussions with Murray Cole, Sergei Gorlatch, Gaétan Hains, Quentin Miller, Susanna Pelagatti, Jörg Striegnitz and others. Their influence is gratefully acknowledged.

References

[BK96] G.H. Botorog, H. Kuchen: Efficient Parallel Programming with Algorithmic Skeletons, Proceedings of EuroPar '96, LNCS 1123, Springer, 1996.

[BK98] G. H. Botorog, H. Kuchen: Efficient High-Level Parallel Programming, Theoretical Computer Science 196, pp. 71-107, 1998.

[Co89] M. I. Cole: *Algorithmic Skeletons: Structured Management of Parallel Computation*, MIT Press, 1989.

[DPP97] M. Danelutto, F. Pasqualetti and S. Pelagatti Skeletons for Data Parallelism in p3l , EURO-PAR '97, Springer LNCS 1300, pp. 619-628, 1997.

[Da93] J. Darlington, A. J. Field, P. G. Harrison et al: Parallel Programming Using Skeleton Functions, in *Proceedings of PARLE '93*, LNCS 694, Springer, 1993.

[Da95] J. Darlington, Y. Guo, H. W. To, J. Yang: Functional Skeletons for Parallel Coordination, in *Proceedings of EURO-PAR '95*, LNCS 966, Springer, 1995.

[FOT92] I. Foster, R. Olson, S. Tuecke: Productive Parallel Programming: The PCN Approach, in *Scientific Programming*, Vol. 1, No. 1, 1992.

[GLS99] W. Gropp, E. Lusk, A. Skjellum: Using MPI, MIT Press, 1999.

[HPF93] High Performance Fortran Forum: High Performance Fortran Language Specification, Scientific Programming, Vol. 2(1), 1993.

[Ka98] S. Karmesin et al.: Array Design and Expression Evaluation in POOMA II. ISCOPE 1998, pp. 231-238.

[KPS94] H. Kuchen, R. Plasmeijer, H. Stoltze: Efficient Distributed Memory Implementation of a Data Parallel Functional Language, *PARLE*, LNCS 817, 1994.

[Ku02a] H. Kuchen: A Skeleton Library, Technical Report, Univ. Münster, 2002.

[Ku02b] H. Kuchen: The Skeleton Library Web Pages, http://danae.uni-muenster.de/lehre/kuchen/Skeletons/

[PC^2] PC^2: http://www.upb.de/pc2/services/systems/psc/index.html.

[Qu94] M. J. Quinn: *Parallel Computing: Theory and Practice*, McGraw Hill, 1994.

[SHM97] David B. Skillicorn, Jonathan M. D. Hill, W. F. McColl: Questions and answers about BSP, Scientific Programming 6(3): 249-274, 1997.

[Sk94] D. Skillicorn: Foundations of Parallel Programming, Cambridge U. Press, 1994.

[St00] J. Striegnitz: Making C++ Ready for Algorithmic Skeletons, Tech. Report IB-2000-08, http://www.fz-juelich.de/zam/docs/autoren/striegnitz.html

Optimising Shared Reduction Variables in MPI Programs

A.J. Field, P.H.J. Kelly, and T.L. Hansen

Department of Computing, Imperial College,
180 Queen's Gate, London SW7 2BZ, U.K.,
{ajf,phjk,tlh}@doc.ic.ac.uk

Abstract. CFL (Communication Fusion Library) is an experimental C++ library which supports shared reduction variables in MPI programs. It uses overloading to distinguish private variables from replicated, shared variables, and automatically introduces MPI communication to keep replicated data consistent. This paper concerns a simple but surprisingly effective technique which improves performance substantially: CFL operators are executed lazily in order to expose opportunities for run-time, context-dependent, optimisation such as message aggregation and operator fusion. We evaluate the idea using both toy benchmarks and a 'production' code for simulating plankton population dynamics in the upper ocean. The results demonstrate the library's software engineering benefits, and show that performance close to that of manually optimised code can be achieved automatically in many cases.

1 Introduction

In this paper we describe an experimental abstract data type for representing shared variables in SPMD-style MPI programs. The operators of the abstract data type have a simple and intuitive semantics and hide any required communication. Although there are some interesting issues in the design of the library, the main contribution of this paper is to show how lazy evaluation can expose run-time optimisations that may be difficult, or even impossible, to spot using conventional compile-time analysis. The paper makes the following contributions:

- We present a simple and remarkably useful prototype class library which simplifies certain kinds of SPMD MPI programs
- We discuss some of the design issues in such a library, in particular the interpretation of the associated operators
- We show how lazy evaluation of the communication needed to keep replicated variables consistent can lead to substantial performance advantages
- We evaluate the work using both toy examples and a large-scale application

This paper extends our brief earlier paper [2] in providing better motivation and further experimental results, as well as a more thorough description of the technique.

B. Monien and R. Feldmann (Eds.): Euro-Par 2002, LNCS 2400, pp. 630–639.

```
double s1, s2;

void sum( double& data ) {
  double s = 0.0;
  for ( j=jmin; j<=jmax; j++ ) {
    s += data[j];
  }
  MPI_Allreduce(&s,&s1,1,MPI_SUM,..);
}

void sumsq( double& data ) {
  double s = 0.0;
  for ( j=jmin; j<=jmax; j++ ) {
    s += data[j] * data[j];
  }
  MPI_Allreduce(&s,&s2,1,MPI_SUM,..);
}

for( i=0; i<N; i++ ) {
  sum(a[i]);
  sumsq(a[i]);
  var[i] = (s2 - s1*s1/N)/(N-1);
}
```

```
CFL_Double s1(0), s2(0);
/* Initial values assumed
   consistent across procs. */

void sum( double& data ) {
  s1 = 0.0;
  for ( j=jmin; j<=jmax; j++ ) {
    s1 += data[j];
  }
}

void sumsq( double& data ) {
  s2 = 0.0;
  for ( j=jmin; j<=jmax; j++ ) {
    s2 += data[j] * data[j];
  }
}

for( i=0; i<N; i++ ) {
  sum(a[i]);
  sumsq(a[i]);
  var[i] = (s2 - s1*s1/N)/(N-1);
}
```

Fig. 1. Variance calculation using MPI (left) and CFL (right).

2 The Idea

Fig. 1 shows a toy C++ application which computes the sample variance of N batches of M data items, stored in an $N \times M$ array. The data is replicated over P processors and each processor computes its contribution to the sum and sum-of-squares of each batch of data using appropriately defined methods. An MPI reduction operation sums these contributions. The main loop fills the variance array (`var`).

This program suffers two drawbacks. Firstly, the code is convoluted by the need to code the communication explicitly—an artefact of all MPI programs. Secondly, it misses an optimisation opportunity: the two reduction operations can be fused (i.e. resolved using a single communication to sum the contributions to `s1` and `s2` at the same time) since the evaluation of `sumsq` does not depend on that of `sum`. If the two reduction operations are brought out of the methods `sum` and `sumsq` and combined into a single reduction over a two-element vector in the outer loop a performance benefit of around 43% is achieved using four 300MHz UltraSparc processors of a Fujitsu AP3000 with `N=M=3000`. Further results for this benchmark are reported in Section 5.

Spotting this type of optimisation at compile time requires analysing across method boundaries. While perfectly feasible in this case, in general these operations may occur deep in the call graph, and may be conditionally executed, making static optimisation difficult. The alternative we explore in this paper is to attempt the optimisation at run-time, requiring no specialist compiler support.

We have developed a prototype library called CFL (Communication Fusion Library) designed to support shared reduction variables. The library can be freely mixed with standard MPI operations in a SPMD application. C++ operator overloading is used to simplify the API by using existing operators (e.g. `+`, `*`, `+=` etc.). Where an operation would normally require communication e.g. when a shared reduction variable is updated with the value of a variable local to each processor, the communication is handled automatically.

Fig. 1 shows how the CFL library can be used to model the shared quantities `s1` and `s2` in the previous example. This eliminates all the explicit communication, in the spirit of shared-memory programming. However, the main benefit comes from CFL's lazy evaluation: just prior to the assignment to `var[i]` no communication has yet taken place. The assignment forces both delayed communications and resolves them using a single reduction operation, akin to the manual optimisation outlined above. There are some overheads associated with the maintenance of these shared variables, so we would not expect to achieve the performance of the manually optimised code. Nonetheless, this very simple example, with scope for just two fusions per iteration, yields a performance improvement of around 37% when compared to the original code on the same platform. Again more detailed results are presented in Section 5.

In the remainder of this paper we present some relevant background to the work (Section 3), discuss the semantics of shared variables in the context of MPI programs (Section 4) and present some performance benchmarks for both contrived test programs and a production oceanography simulation (Section 5). The conclusions are presented in Section 6.

3 Related Work

The idea of delaying execution in order to expose optimisation opportunities has appeared before. POOMA [4] uses expression templates in C++ to support explicit construction and then evaluation of expressions involving arrays and communication. A delayed-evaluation self-optimising (DESO) numerical library for a distributed memory parallel computer is described in [6]. By delaying the evaluation of operations, the library is able to capture the data-flow graph of the computation. Knowing how each value is to be used, the library is able to calculate an optimised execution plan by propagating data placement constraints backwards through the DAG. This means that the library is able to calculate a very efficient initial distribution for the data across the processors, and hence fewer redistributions of the data will be necessary.

A related idea, which is exploited in BSP [3] and KeLP [8], is to organise communication in a global collective operation. This allows multiple small messages to be aggregated, and also provides the opportunity to schedule communication to use avoid network and buffer contention.

A shared-memory programming model can be supported on distributed-memory hardware using a page-based consistency protocol; sophisticated implementations such as TreadMarks [5] support some run-time adaptation, for example for pages with multiple writers. However, Treadmarks offers no special support for reductions.

4 Shared Variables in SPMD Programs

In a data-parallel SPMD program, a large data structure is distributed across each processor, and MPI is used to copy data to where it is needed. In contrast, we focus in this paper on the program's global state variables. In a distributed-memory implementation, each processor holds its own copy of each shared variable. When the variable is updated, communication is needed to ensure that each processor's copy is up to date.

In the context of this paper we focus exclusively on scalar double-precision floating-point variables. The semantics of arithmetic operations on private variables are very well-understood, but are not so straightforward for shared variables as an operation on a shared double will be executed on several processors.

The interesting case concerns the assignment of the result of an arithmetic expression to a variable. In what follows, `x`,`y` and `z` will refer to (global) shared variables (i.e. of type `CFL_Double`) and `a`, `b` to local variables private to each processor. Each processor maintains a local copy of each shared variable and the library must ensure that after each operation these copies are consistent.

If the target variable of an assigment is local, as in `a = x - b` then the assignment can be performed concurrently on each processor without (additional) communication. However, if the result is stored in a shared variable then the behaviour depends on the operator arguments. If both operator arguments are shared, as in `x = y * z` then again the assignment can be effected locally. However, if one of the arguments is local and the other shared, as in `x += a` or `x = y + a`, then our interpretation is that each processor contributes its own update to `x`, implying a global reduction operation, with the rule that `x -= a` is interpreted as `x += (-a)`. Because CFL is lazy, one or more of the shared variables on the right-hand side of an assignment may already contain a pending communication, either from an earlier assignment or an intermediate expression on the same right-hand side, as in `x = y + a - z`. Any new required communication is simply added to those currently pending.

Similar rules apply to the other operators `-`, `*`, `/` etc. and combined operations like `+=` have the same meaning as their expanded equivalents, e.g. `x += a` and `x = x + a`.

Assignment and Reduction. The way the assignment `v=e` is implemented now depends on the nature of `v` and `e`. It is tempting to think that any potential confusion can be overcome by using a different operator symbol when a global reduction is intended, for instance `x ++= a` instead of `x += a`. However the assignment `x = x + a` should have the same meaning so we would also need special versions of `+` (and the other operators) to cover all combinations of argument types. We thus choose to stick to the familiar symbols using the overloading, but propose the use of naming conventions to distinguish shared from local variables where any confusion may arise.

An attempt to assign a local variable to a shared variable either directly (e.g. `x = a`) or as a result of a calculation involving only local variables (e.g. `x = a - b`) is disallowed.

4.1 Delaying Communication

The parallel interpretation of some operator uses such as `x += a` means that at any point a shared variable may need to synchronise with the other processors. Because each processor sees the variable in the same state every processor will know that the variable needs synchronisation. Moreover, as operations are executed in the same order on all the processors (the SPMD model), shared variables will acquire the need for synchronisation in the same order on every processor. This means that, in order to delay communication, we need only maintain a list of all the variables that need to be synchronised, and in what way. When communication is forced (see below) these synchronisations are piggybacked onto a single message with an associated reduction operator. An alternative would be to initiate a non-blocking communication instead; although this might be appropriate for some hardware, little or no computation/communication overlap is possible in most current MPI implementations.

An assignment of a shared variable to a local variable constitutes a *force point.* At this point a communications manager marshalls all CFL variable updates into a single array and performs a single global reduction operation over that array. On completion, the resulting values are used to update all `CFL_Doubles` which were previously pending communication.

In principle, the synchronisation of a shared variable may be delayed until its global value is required (force point), but in practice the synchronisation may be forced earlier than this, e.g. when another shared variable synchronisation is forced before it. Forcing *any* delayed synchronisation will force *all* such synchronisations.

Limitations. In the prototype implementation of CFL only 'additive' operators (`+`, `-`, `+=`, `-=`) are handled lazily at present. This is sufficient for experimental evaluation of the basic idea. The other operators (and the copy constructor) are all supported but they force all pending communication. Implementing the remaining operators lazily requires a little more work to pack the data for communication and to construct the associated composite reduction operation, but is otherwise straightforward. This is left as future work.

5 Evaluation

Our performance results are from dedicated runs on three platforms: a Fujitsu AP3000 (80 nodes, each a 300MHz Sparc Ultra II processor with 128MB RAM, with Fujitsu's 200MB/s AP-Net network), a Quadrics/COMPAQ cluster (16 nodes, each a Compaq DS20 dual 667MHz Alpha with 1GB RAM), and a cluster of dual 400MHz Celeron PCs with 128MB RAM on 100Mb/s switched Ethernet. In each case there was one MPI process per node.

5.1 Toy Benchmark

Table 1 shows execution times for the toy benchmark of Fig. 1 for four problem sizes for the AP3000 platform using 4 processors. Here the data matrix is

Table 1. AP3000 execution times (in seconds) for Figure 1 for various problem sizes, with percentage speedup relative to the original, unoptimised code.

N	AP3000 (P=4) Execution time(s)		
	Original	Hand optimised	CFL
500	0.341	0.157 (53.9%)	0.192 (43.6%)
1000	0.748	0.347 (53.5%)	0.433 (42.0%)
1500	1.159	0.574 (50.5%)	0.724 (37.5%)
3000	2.544	1.463 (42.5%)	2.119 (16.7%)

assumed to be square, so the problem size defines both M and N. The figures in parentheses show the reduction in execution time, relative to the original unoptimised code. We see a significant performance improvement due to fusing communications, even though only two can be fused at any time. As expected, CFL's returns diminish, relative to hand-optimised code, as problem size increases because larger problems incur a smaller communication overhead, so the overhead of maintaining the shared variable state takes greater effect.

We would intuitively expect the performance of the CFL library to improve, relative to the hand-optimised code, for platforms with slower communication networks (measured by a combination of start-up cost and bandwidth) and vice versa. This is borne out by Table 2 which shows the performance of the same benchmark on our three reference platforms, using 4 processors in each case and with a problem size of 3000. Relative to the hand-optimised code, the CFL library performs extremely well on the PC cluster. However, on the COMPAQ platform, which has a very fast communication network, the overheads of supporting lazy evaluation outweigh the benefits of communication fusion.

The example of Fig. 1 enables exactly two reductions to be fused on each iteration of the loop. In some applications (see below, for example) it may be possible to do better. The variance example was therefore generalised artificially by introducing an extra loop that called the `sum` function (only) a given number of times, n, on each iteration of the outer (`i`) loop. The results were stored in an array and later summed (again arbitrary, but this has the effect of forcing communication in the CFL case). The objective was to measure the cost of performing repeated (explicit) MPI reduction operations relative to the cost of fusing them within CFL. The results for 4 processors on each platform with $N = 3000$ are shown in Fig. 2. The slope of the two curves (original MPI vs. CFL) in each case expose these relative costs and we can see why CFL wins out on both the AP3000 and PC cluster. Conversely, on the COMPAQ platform no amount of fusion opportunity can buy back the performance overheads inherent in the current CFL implementation.

5.2 Oceanography Simulation

We now present the results obtained when the CFL library was used to model shared variables in a large-scale simulation of plankton population dynamics in the upper ocean using the Lagrangian Ensemble method [7].

Table 2. Execution times for Figure 1 for various platforms.

Platform	Execution time(s) (N=3000)		
	Original	Hand optimised	CFL
AP3000	2.544	1.463 (42.5%)	2.119 (16.7%)
Cluster	7.154	3.670 (48.7%)	3.968 (44.5%)
COMPAQ	0.263	0.161 (38.9%)	0.418 (-58.8%)

The simulation is based on a one-dimensional water column which is stratified into 500 layers each 1m deep. The plankton are grouped into *particles* each of which represents a sub-population of identical individuals. The particles move by a combination of turbulence and sinking/swimming and interact with their local environment according to rules derived from laboratory observation.

The simulation is built by composing modules each of which models an aspect of the physics, biology or chemistry. The exact configuration may vary from one simulation to the next. To give a flavour for the structure of a typical code, the dominant (computationally speaking) component of a particular instance called "ZB" models phytoplankton by the sequential composition of four modules: `Move()`, `Energy()`, `Nutrients()` and `Evolve()` (motion, photosynthesis, nutrient uptake and birth/death). A similar structure exists for zooplankton. The model essentially involves calling these (and the many other) modules in the specified order once per time-step.

A vertical partitioning strategy is used to divide the plankton particles among the available processors, which cooperate through *environment variables* representing the chemical, physical and biological attributes of each layer. Each processor must see the same global environment at all times.

The various modules have been developed independently of the others, although they must fit into a common framework of global variables, management structures etc. Within these modules there are frequent updates to the shared variables of the framework and it is common for these to be assigned in one module and used in a later one. This relatively large distance between the producer and consumer provides good scope for message aggregation. However, manual optimisation will work only for that particular sequence of modules: adding a new module or changing the order of existing modules changes the data dependency. This is where the CFL library is particularly beneficial: it automatically fuses the maximum number of reduction operations (i.e. those that arise between force points).

We began with the original (parallel) version of ZB and then hand-optimised it by manually working out the data dependencies between the global shared quantities and identifying force points. The fusion was actually achieved by building a lazy version of `MPI_All_Reduce` [1]. This simplified the implementation significantly but introduced some overheads, very similar in fact to those in CFL. The MPI code was then rewritten using the CFL library, simply by marking the shared environment variables as `CFL_doubles`. The original code uses exclusively MPI reduction operations so the immediate effect of using CFL

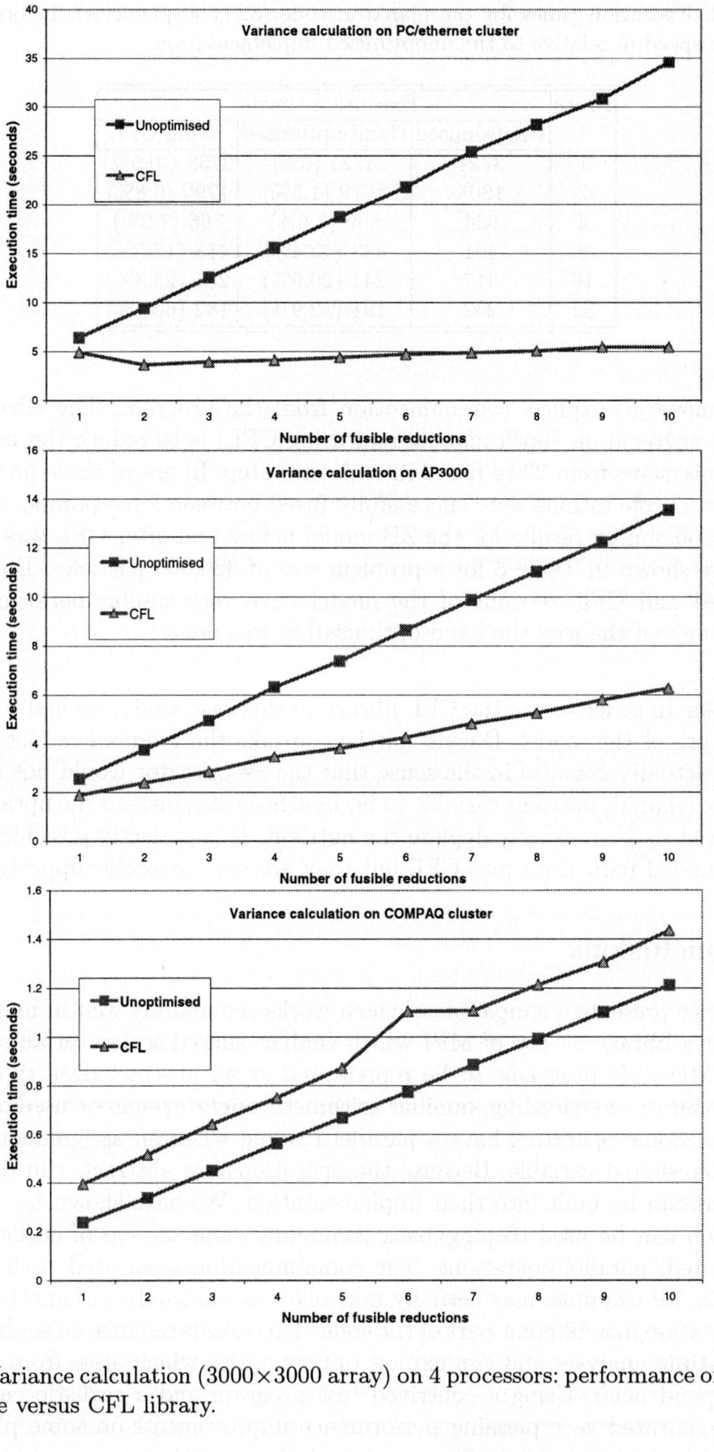

Fig. 2. Variance calculation (3000×3000 array) on 4 processors: performance of original MPI code versus CFL library.

Table 3. Execution times for the plankton code (320,000 particles). In brackets we show the speedup relative to the unoptimised implementation.

Procs	Execution time(s)		
	Unoptimised	Hand-optimised	Using CFL
1	3721	3721 (0%)	3738 (-0.5%)
2	1805	1779 (1.5%)	1790 (0.8%)
4	934	869 (7.5%)	866 (7.9%)
8	491	433 (13.4%)	418 (17.5%)
16	317	244 (29.9%)	257 (23.3%)
32	292	191 (52.9%)	182 (60.4%)

is to remove *all* explicit communication from the program. The effect of the message aggregation (both manual and using CFL) is to reduce the number of synchronisations from 27 to just 3 in each time step. In one of these no less than 11 reduction operations were successfully fused between force points.

AP3000 timing results for the ZB model before and after CFL was incorporated are shown in Table 3 for a problem size of 320,000 particles. Both hand-optimised and CFL versions of the model have very similar performance, as expected given the way the hand-optimisation was done.

Remarks In order to use the CFL library in this case study, we had to turn off one feature of the model. During nutrient uptake the required reduction operation is actually *bounded* in the sense that the -= operator would not normally allow the (shared) nutrient variable to become negative; instead the uptake would be reduced so as to exactly deplete the nutrient. It is perfectly possible to build such bounded reductions into CFL but they are not currently supported.

6 Conclusions

This paper presents a simple idea, which works remarkably well in many cases. We built a library on top of MPI which enables shared scalar variables in parallel SPMD-style programs to be represented as an abstract data type. Using C++'s operator overloading, familiar arithmetic operators can be used on shared variables. Some operators have a parallel reading when an assignment's target is another shared variable. Because the operations are abstract, run-time optimisations can be built into their implementation. We have shown how delayed evaluation can be used to piggyback communications on top of earlier, as yet unevaluated, parallel operations. The communication associated with a global reduction, for example, may actually take place as a side-effect of another reduction operation in a different part of the code. This avoids reliance on sophisticated compile-time analyses and can exploit opportunities which arise from dynamic data dependencies. Using a contrived test program and a realistic case study we demonstrated very pleasing performance improvements on some platforms. Unsurprisingly, greatest benefits occur on platforms with slower communication.

In essence, CFL is an application-specific cache coherence protocol, in the spirit of, among others, [9]. This hides consistency issues, and the associated communication, with obvious benefits in software engineering terms.

Could we achieve reduction fusion by executing standard MPI functions lazily? Not without compiler support, since the results of MPI operations are delivered to normal private data so we don't know when to force communication.

The prototype implementation is robust and has proven surprisingly useful. We now seek to it (e.g. to handle arrays as well as scalars) and to focus on internal optimisations to reduce management overheads.

Acknowledgements

This work was partially supported by the EPSRC-funded OSCAR project (GR/R21486). We are very grateful for helpful discussions with Susanna Pelagatti and Scott Baden, whose visits were also funded by the EPSRC (GR/N63154 and GR/N35571).

References

1. S. H. M. Al-Battran: *Simulation of Plankton Ecology Using the Fujitsu AP3000*, MSc thesis, Imperial College, September 1998
2. A J Field, T L Hansen and P H J Kelly: *Run-time fusion of MPI calls in a parallel C++ library.* Poster paper at LCPC2000, The 13th Intl. Workshop on Languages and Compilers for High-Performance Computing, Yorktown Heights, August 2000.
3. J.M.D. Hill, D.B. Skillicorn: *Lessons learned from implementing BSP.* J. Future Generation Computer Systems, Vol 13, No 4–5, pp. 327-335, March 1998.
4. S. Karmesin, J. Crotinger, J. Cummings, S. Haney, W.J. Humphrey, J. Reynders, S. Smith, T. Williams: *Array Design and Expression Evaluation in POOMA II.* ISCOPE'98 pp.231-238. Springer LNCS 1505 (1998).
5. P. Keleher, A. L. Cox, S. Dwarkadas, W. Zwaenepoel: *TreadMarks: Distributed Shared Memory on Standard Workstations and Operating Systems.* Proc. 1994 Winter Usenix Conference, pp. 115-131, January 1994
6. O. Beckmann, P. H. J. Kelly: *Efficient Interprocedural Data Placement Optimisation in a Parallel Library*, In LCR98, pp.123–138. Springer-Verlag LNCS 1511 (May 1998).
7. J. Woods and W. Barkmann, *Simulation Plankton Ecosystems using the Lagrangian Ensemble Method*, Philosophical Transactions of the Royal Society, B343, pp. 27-31.
8. S.J. Fink, S.B. Baden and S.R. Kohn, *Efficient Run-time Support for Irregular Block-Structured Applications*, J. Parallel and Distributed Programming, V.50, No.1, pp 61–82, 1998.
9. A.J. Bennett and P.H.J. Kelly, *Efficient shared-memory support for parallel graph reduction.* Future Generation Computer Systems, V.12 No.6 pp.481–503 (1997).

Double-Scan: Introducing and Implementing a New Data-Parallel Skeleton

Holger Bischof and Sergei Gorlatch

Technical University of Berlin, Germany

Abstract. We introduce a new reusable component for parallel programming, the *double-scan skeleton*. For this skeleton, we formulate and formally prove sufficient conditions under which the double-scan can be parallelized, and develop its efficient MPI implementation. The solution of a tridiagonal system of equations is considered as our case study. We describe how this application can be developed using the double-scan and report experimental results for both absolute performance and performance predictability of the skeleton-based solution.

1 Introduction

This work is motivated by the search for convenient, reusable, adaptable components for building parallel applications. We pursue the approach based on *skeletons* – typical algorithmic patterns of parallelism. The programmer composes an application using skeletons as high-level language constructs, whose implementations for different parallel machines are provided by a compiler or a library. Formally, skeletons are viewed as higher-order functions, customizable by means of application-specific functional parameters. Well-known examples of practical skeleton-based programming systems include P3L [9] and Skil [2] in the imperative setting, as well as Eden [3] and HDC [7] in the functional world.

This paper demonstrates by reference to an application case study – the solution of a tridiagonal system of linear equations – how a new data-parallel skeleton called *double-scan* is identified, implemented and added to an existing inventory of components.

The contributions and structure of the paper are as follows:

- We describe a basic skeleton repository containing well-known data-parallel skeletons, such as map, reduce, zip, and two scans – left and right (Section 2).
- We express the problem of solving a tridiagonal system of equations using the basic skeletons, demonstrating the need for a new skeleton (Section 3).
- We introduce the new double-scan skeleton – a composition of two scans, one of which has a non-associative base operator – and prove the sufficient condition under which it can be parallelized (Section 4).
- We demonstrate how the performance of programs using the double-scan skeleton can be predicted in advance and demonstrate both the absolute performance and the possibility of predicting the performance of our MPI implementation by experiments on a Cray T3E (Section 5).

We conclude the paper by discussing our results in the context of related work.

B. Monien and R. Feldmann (Eds.): Euro-Par 2002, LNCS 2400, pp. 640–647.

2 Basic Data-Parallel Skeletons

In this section, we present some basic data-parallel skeletons as higher-order functions defined on non-empty lists, function application being denoted by juxtaposition, i. e. $f\,x$ stands for $f(x)$:

- Map: Applying a unary function f to all elements of a list:

$$map\ f\ [x_1, \ldots, x_n] \;=\; [f\ x_1, \ldots, f\ x_n]$$

- Red: Combining the list elements using a binary associative operator $\oplus$:

$$red(\oplus)([x_1, \ldots, x_n]) \;=\; x_1 \oplus \cdots \oplus x_n$$

- Zip: Component-wise application of a binary operator $\odot$ to a pair of lists of equal length:

$$zip(\odot)([x_1, \ldots, x_n], [y_1, \ldots, y_n]) \;=\; [\,(x_1 \odot y_1), \ldots, (x_n \odot y_n)\,]$$

- Scan-left and scan-right: Computing prefix sums of a list by traversing the list from left to right (or vice versa) and applying a binary operator $\oplus$:

$$\begin{aligned} scanl(\oplus)([x_1, \ldots, x_n]) &= [\,x_1,\ (x_1 \oplus x_2), \ldots, (\cdots(x_1 \oplus x_2) \oplus x_3) \oplus \cdots \oplus x_n)\,] \\ scanr(\oplus)([x_1, \ldots, x_n]) &= [\,(x_1 \oplus \cdots \oplus (x_{n-2} \oplus (x_{n-1} \oplus x_n) \cdots),\ \ldots,\ x_n\,] \end{aligned}$$

We call these functions skeletons because each of them describes a whole class of functions, obtainable by substituting application-specific operators for parameters $\oplus$, $\odot$ and f.

Our skeletons have obvious data-parallel semantics: the asymptotic parallel complexity is constant for *map* and *zip* and logarithmic for *red* and both scans, if $\oplus$ is associative. If $\oplus$ is non-associative, then *red* and the scans are computed sequentially with linear time complexity.

3 Case Study: Tridiagonal System Solver

In this section, we consider an application example and try to parallelize it using the skeletons introduced in Section 2. Our case study is concerned with solving a tridiagonal system of linear equations, $A \cdot x = b$, where A is an $n \times n$ matrix representing coefficients, x a vector of unknowns and b the right-hand-side vector. The only values of matrix A unequal to 0 are on the main diagonal as well as above and below it (we call them the upper and lower diagonal, respectively), as demonstrated by equation (1).

$$\begin{pmatrix} a_{12} & a_{13} & & & \\ a_{21} & a_{22} & a_{23} & & \\ & \ddots & \ddots & \ddots & \\ & & a_{n-1,1} & a_{n-1,2} & a_{n-1,3} \\ & & & a_{n,1} & a_{n,2} \end{pmatrix} \cdot x = \begin{pmatrix} a_{14} \\ a_{24} \\ \vdots \\ a_{n-1,4} \\ a_{n,4} \end{pmatrix} \tag{1}$$

Our first step is to cast the tridiagonal system in the list notation since our skeletons work on lists. We choose a representation comprising a list of rows, each inner row i consisting of four values: the value $a_{i,1}$ that is part of the lower diagonal of matrix A, the value $a_{i,2}$ at the main diagonal, the value $a_{i,3}$ at the upper diagonal, and the value $a_{i,4}$ that is the i-th component of the right-hand-side vector b.

To make the first and last row consist of four values, too, we add to the matrix the fictitious zero elements a_{11} and $a_{n,3}$. The fictitious values obviously do not change the solution of the original tridiagonal system: imagine two additional unknowns, x_0 and x_{n+1}, and two additional columns, one on the left of the original matrix and the other on the right, whose values are equal to zero.

We can thus represent the whole tridiagonal system as the following list of rows, each of which is a quadruple:

$$\left[\,(a_{11}, a_{12}, a_{13}, a_{14}),\ (a_{21}, a_{22}, a_{23}, a_{24}),\ \ldots,\ (a_{n1}, a_{n2}, a_{n3}, a_{n4})\,\right]$$

A typical algorithm for solving a tridiagonal system is Gaussian elimination (see e. g. [8, 10]) which eliminates the lower and upper diagonal of the matrix as shown in Figure 1. Note that both the first and last column in the figure consist of fictitious zero elements.

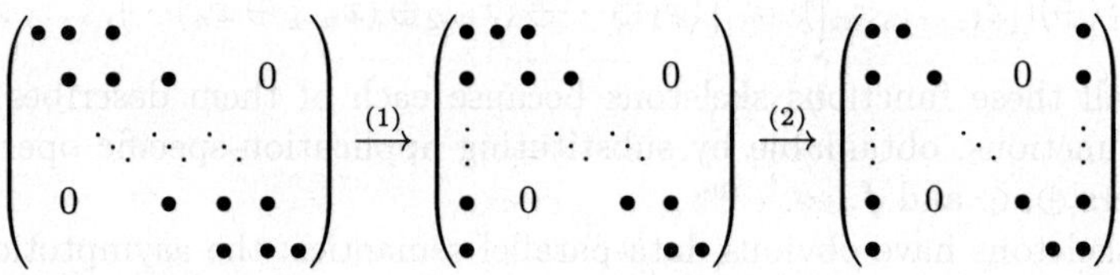

Fig. 1. The intuitive algorithm for solving a tridiagonal system of equations consists of two stages: (1) elimination of the lower diagonal; (2) elimination of the upper diagonal.

The two stages of the algorithm traverse the list of rows, applying operators denoted by ① and ②, which are informally defined as follows:

1. The first stage eliminates the lower diagonal by traversing matrix A from top to bottom according to the *scanl* pattern and applying operator ① on the rows pairwise:

$$(a_1, a_2, a_3, a_4) \text{ ① } (b_1, b_2, b_3, b_4) = \left(a_1,\ a_3 - \frac{b_2 a_2}{b_1},\ -\frac{b_3 a_2}{b_1},\ a_4 - \frac{b_4 a_2}{b_1}\right)$$

2. The second stage eliminates the upper diagonal of the matrix by a bottom-up traversal, i. e. using the pattern of *scanr* and applying operator ② on pairs of rows:

$$(a_1, a_2, a_3, a_4) \text{ ② } (b_1, b_2, b_3, b_4) = \left(a_1 - \frac{b_1 a_3}{b_2},\ a_2,\ -\frac{b_3 a_3}{b_2},\ a_4 - \frac{b_4 a_3}{b_2}\right) \quad (2)$$

Now we can specify the described Gaussian elimination algorithm as function *tds* (*tridiagonal system*), which performs in two stages:

$$tds \;=\; scanr(②) \circ scanl(①) \tag{3}$$

where $\circ$ denotes function composition from right to left, i. e. $(f \circ g)\ x = f(g(x))$.

If both customizing operators of the scan skeletons in (3), ① and ②, were associative, our task would now be completed: both scan skeletons have data-parallel semantics, and furthermore, they can be directly implemented using, for example, the MPI collective operation `MPI_Scan`. However, since operator ① is not associative, *scanl* in (3) prescribes strictly sequential execution.

An alternative representation of the algorithm eliminates first the upper and then the lower diagonal using two new row operators, ③ and ④:

$$(a_1, a_2, a_3, a_4)\ ③\ (b_1, b_2, b_3, b_4) = \left(a_1,\ a_2 - \frac{b_1 a_3}{b_2},\ -\frac{b_3 a_3}{b_2},\ a_4 - \frac{b_4 a_3}{b_2}\right)$$

$$(a_1, a_2, a_3, a_4)\ ④\ (b_1, b_2, b_3, b_4) = \left(a_1, -\frac{b_2 a_2}{b_1},\ a_3 - \frac{b_3 a_2}{b_1},\ a_4 - \frac{b_4 a_2}{b_1}\right)$$

This version of the algorithm can be specified as follows:

$$tds \;=\; scanl(④) \circ scanr(③) \tag{4}$$

Again, however operator ③ in (4) is not associative and thus the first step of algorithm (4), *scanr*(③) cannot be directly parallelized.

4 The Double-Scan Skeleton and Its Parallelization

This section deals with the algorithmic pattern we identified using the case study in Section 3 – a sequential composition of two scans. We call a composition of two scans, one of which is the "left" and the other the "right" scan, the *double-scan skeleton*. This skeleton has two functional parameters, which are the base operators of the constituting scans. The non-associativity of the first scan in the composition prevents parallelization of the double-scan.

To parallelize the double-scan skeleton, we relate it to a more complex algorithmic pattern than the basic skeletons introduced in Section 2. Let us consider a class of functions called *distributable homomorphisms* (DH), which was first introduced in [6]:

Definition 1. *A function h is a distributable homomorphism iff there exist binary operators $\oplus$ and $\otimes$, such that for arbitrary lists x and y of equal length, which is a power of two, it holds:*

$$h[a] = [a]\,, \qquad h(x +\!\!+\, y) = zip(\oplus)(h\ x, h\ y) +\!\!+\, zip(\otimes)(h\ x, h\ y) \tag{5}$$

For operators $\oplus$ and $\otimes$, we denote the corresponding DH by $h = (\oplus \updownarrow \otimes)$. Its computation schema is illustrated in Figure 2.

In [6], a generic parallel implementation of an arbitrary DH is developed, which works on a logical hypercube. For the sake of brevity, we give the MPI implementation in pseudocode:

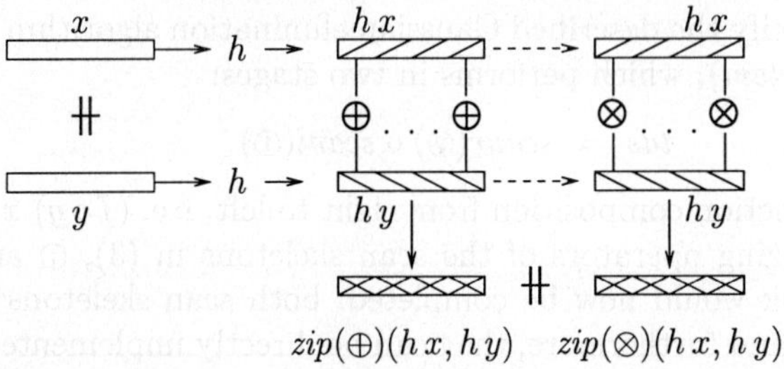

Fig. 2. Distributable homomorphism h on a concatenation of two lists, x and y: apply h to x and y, then combine the results elementwise with operators $\oplus$ and $\otimes$ and concatenate them (bottom of figure).

```
int MPI_DH(void *data, ..., MPI_User_function ⊕, ⊗, ...) {
  for (dim=1; dim<p; dim<<=1) {
    dest = my_rank^dim;          /* rank of hypercube neighbour */
    MPI_Sendrecv(data, ..., dest, temp, ..., dest, ...);
    if (my_rank<dest) data=data ⊕ temp; else data=temp ⊗ data; }}
```

Here, `MPI_Sendrecv` acts as a typical operational pattern of the hypercube behaviour: pairwise, bidirectional communication of the neighbouring nodes in dimension `dim`, followed by the application of either $\oplus$ or $\otimes$ in each of the nodes.

The novel result of this paper is the following sufficient condition under which the double-scan skeleton can be expressed using the DH pattern:

Theorem 1. *Let ①, ②, ③, ④ be binary operators, where ② and ④ are associative (①, ③ need not be associative). If the following equation holds:*

$$scanr(②) \circ scanl(①) \;=\; scanl(④) \circ scanr(③) \quad \stackrel{\text{def}}{=} \quad s \tag{6}$$

then s from (6) can be represented as follows:

$$s \;=\; map(\pi_1) \circ \oplus\!\updownarrow\!\otimes \circ \; map(triple) \tag{7}$$

with the operators $\oplus$ and $\otimes$ defined as follows:

$$\begin{pmatrix} a_1 \\ a_2 \\ a_3 \end{pmatrix} \oplus \begin{pmatrix} b_1 \\ b_2 \\ b_3 \end{pmatrix} = \begin{pmatrix} a_1 ② (a_3 ① b_2) \\ a_2 ② (a_3 ① b_2) \\ (a_3 ③ b_2) ④ b_3 \end{pmatrix} \quad \begin{pmatrix} a_1 \\ a_2 \\ a_3 \end{pmatrix} \otimes \begin{pmatrix} b_1 \\ b_2 \\ b_3 \end{pmatrix} = \begin{pmatrix} (a_3 ③ b_2) ④ b_1 \\ a_2 ② (a_3 ① b_2) \\ (a_3 ③ b_2) ④ b_3 \end{pmatrix} \tag{8}$$

Here, *triple* maps an element to a triple, $triple\, a = (a, a, a)$, and function π_1 extracts the first element of a triple, $\pi_1(a, b, c) = a$.

We do not present the theorem's proof here owing to the lack of space; it is contained in a technical report [1].

To apply the general result of Theorem 1 to our example of a tridiagonal system solver, it remains to be shown that operators ② in (3) and ④ in (4) are associative. For example, the associativity of ② defined in (2) is demonstrated below using the associativity of addition and multiplication and the distributivity of multiplication over addition:

$$\begin{pmatrix}a_1\\a_2\\a_3\\a_4\end{pmatrix} ② \left[\begin{pmatrix}b_1\\b_2\\b_3\\b_4\end{pmatrix} ② \begin{pmatrix}c_1\\c_2\\c_3\\c_4\end{pmatrix}\right] = \begin{pmatrix}a_1 - \frac{b_1 a_3}{b_2} + \frac{c_1 b_3 a_3}{c_2 b_2}\\ a_2\\ \frac{c_3 b_3 a_3}{c_2 b_2}\\ a_4 - \frac{b_4 a_3}{b_2} + \frac{c_4 b_3 a_3}{c_2 b_2}\end{pmatrix} = \left[\begin{pmatrix}a_1\\a_2\\a_3\\a_4\end{pmatrix} ② \begin{pmatrix}b_1\\b_2\\b_3\\b_4\end{pmatrix}\right] ② \begin{pmatrix}c_1\\c_2\\c_3\\c_4\end{pmatrix}$$

Analogously, the associativity of operator ④ can be proved. Now a parallel implementation of the DH skeleton can be used to compute the *tds* function.

5 Performance Prediction and Measurements

In this section, we look at the performance of programs using the double-scan skeleton from two perspectives: (1) whether performance for a particular application on a particular parallel machine can be predicted in advance, and (2) whether absolute performance is competitive with that of a hand-coded solution.

Programming with skeletons offers a major advantage in terms of performance prediction: performance has to be estimated only once for a skeleton. This generic estimate can then be tuned to a particular machine and application, rather than redoing the estimation procedure for each new machine and application. We demonstrate below possible tuning steps in performance prediction: we derive an estimate for the DH skeleton, and then tune it: first to a particular machine (Cray T3E), and then to a particular application (tridiagonal system solver).

Generic Performance Estimate. Let p be the number of processes and m the data-block size per process. Our implementation of the DH skeleton consists of $\log_2 p$ steps, each performing a bidirectional communication with data of length m and the computation with $\oplus$ or $\otimes$. Operators $\oplus$ and $\otimes$ are applied elementwise to the data, consisting of m elements. The resulting time estimate is:

$$t = \log_2 p \cdot (t_s + m \cdot (t_w + t_c)) \tag{9}$$

where t_s is the communication startup time, t_w the time needed to communicate one word, and t_c the maximum time for one computation with $\oplus$ or $\otimes$.

Estimate Tuned to Cray T3E. Variables t_s and t_w are machine-dependent. On a Cray T3E, t_s is approximately $16.4\,\mu s$. The value of t_w also depends on the size of the data type: $t_w = d \cdot t_B$, where t_B is the time needed to communicate one byte, and d is the byte count of the data type. Measurements show that for large array sizes the bidirectional bandwidth on our machine is approximately 300 MB/s, i. e. $t_B \approx 0.0033\,\mu s$. Inserting these values of t_s and t_B into (9) leads to the following runtime estimate for the double-scan skeleton on a Cray T3E:

$$t_{\text{Cray}} = \log_2 p \cdot (16.4\,\mu s + m \cdot (d \cdot 0.0033\,\mu s + t_c)) \tag{10}$$

Estimate Tuned to the Tridiagonal System Solver. The estimate for a particular machine can be further specialized for a particular application, i. e. an instance of the skeleton. Let us predict the runtime of the tridiagonal system solver on a Cray T3E. In computing *tds*, our data type is a triple of quadruples containing double values and has a size of $d = 96$ Bytes. Measurements show that one operator $\oplus$ or $\otimes$ in (8) for tridiagonal systems requires time $t_c \approx 2.44\,\mu s$ (measured by executing $\oplus$ in a loop over a large array). Inserting d and t_c into (10) results in:

$$t_{\text{Cray-tds}} = \log_2 p \cdot (16.4\,\mu s + m \cdot 2.76\,\mu s) \qquad (11)$$

The next specialization step is to substitute a particular problem size m. We compute with 2^{17} elements per process, where one element is a quadruple of float values in the sequential algorithm and a triple of quadruples of float values in the parallel algorithm; the *tds* function is computed elementwise on the input data. The resulting estimate for our problem size is:

$$T = \log_2 p \cdot 0.36\,s$$

Figure 3(a) compares the estimated and measured time for the tridiagonal system solver and demonstrates the high quality of our estimates.

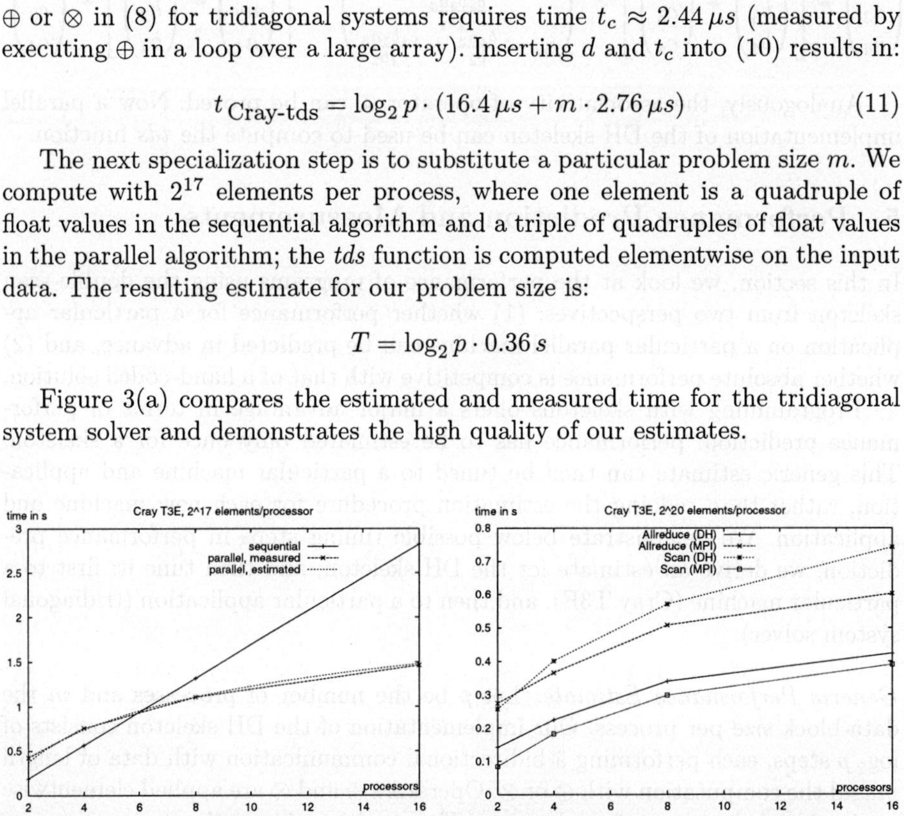

Runtimes of the sequential vs. parallel (estimated and measured) version of the tridiagonal system solver.

Comparison `MPI_DH` vs. `MPI_Scan` and `MPI_Allreduce` on Cray T3E with 2^{20}

Fig. 3. Measurements

Absolute Performance of DH Implementation. To show that our DH implementation is competitive with hand-coded implementations, we tested it under fairly hard conditions. Since allreduce and scan are both DHs ([6]), we use the `MPI_DH` function to compute scan and allreduce and compare it to the performance of two native collective MPI implementations – `MPI_Scan` and `MPI_Allreduce`. In other words, we compare the performance of particular instantiations of our generic DH implementation with specially developed, hand-coded native MPI implementations, optimized for a particular machine.

Figure 3(b) shows the results for a Cray T3E with 2^{20} elements per process using elementwise integer addition as the base operator. Note that the overall number of elements grows proportionally to the number of processes. The results of measurements show that our DH implementation, despite its simplicity, exhibits very competitive performance.

6 Related Work and Conclusion

The desire to be able to name and reuse "programming patterns", i. e. to capture them in the form of parameterizable abstractions, has been a driving force in the evolution of high-level programming languages in general. In the sequential setting, design patterns [5] and components [11] are recent examples of this. In parallel programming, where algorithmic aspects have traditionally been of special importance, the approach using algorithmic skeletons [4] has emerged. Related work on skeletons is manifold and was partially cited in the introduction.

In this paper, we systematically developed a parallel implementation for solving a tridiagonal system of equations using a novel generic program component (the double-scan skeleton), which is reusable for other classes of applications. The use of the skeleton also allowed a systematic performance prediction, the results of which have been confirmed in machine experiments.

References

1. H. Bischof, S. Gorlatch, and E. Kitzelmann. The double-scan skeleton and its parallelization. Technical Report 2002/06, Technische Universität Berlin, 2002.
2. G. Botorog and H. Kuchen. Efficient parallel programming with algorithmic skeletons. In L. Bougé et al., editors, *Euro-Par'96: Parallel Processing*, Lecture Notes in Computer Science 1123, pages 718–731. Springer-Verlag, 1996.
3. S. Breitinger, R. Loogen, Y. Ortega-Mallén, and R. Peña. The Eden coordination model for distributed memory systems. In *High-Level Parallel Programming Models and Supportive Environments (HIPS)*. IEEE Press, 1997.
4. M. I. Cole. *Algorithmic Skeletons: A Structured Approach to the Management of Parallel Computation.* PhD thesis, University of Edinburgh, 1988.
5. E. Gamma, R. Helm, R. Johnson, and J. Vlissides. *Design patterns: elemets of reusable object-oriented software.* Addison Wesley, 1995.
6. S. Gorlatch. Systematic efficient parallelization of scan and other list homomorphisms. In L. Bougé, P. Fraigniaud, A. Mignotte, and Y. Robert, editors, *Euro-Par'96: Parallel Processing, Vol. II*, Lecture Notes in Computer Science 1124, pages 401–408. Springer-Verlag, 1996.
7. C. A. Herrmann and C. Lengauer. HDC: A higher-order language for divide-and-conquer. *Parallel Processing Letters*, 10(2–3):239–250, 2000.
8. F. T. Leighton. *Introduction to Parallel Algorithms and Architectures: Arrays, Trees, Hypercubes.* Morgan Kaufmann Publ., 1992.
9. S. Pelagatti. *Structured development of parallel programs.* Taylor&Francis, 1998.
10. M. J. Quinn. *Parallel Computing.* McGraw-Hill, Inc., 1994.
11. C. Szyperski. *Component software: beyond object-oriented programming.* Addison Wesley, 1998.

Scheduling vs Communication in PELCR

Marco Pedicini[1] and Francesco Quaglia[2]

[1] IAC, Consiglio Nazionale delle Ricerche, Roma, Italy
[2] DIS, Università "La Sapienza", Roma, Italy

Abstract. PELCR is an environment for λ-terms reduction on parallel/distributed computing systems. The computation performed in this environment is a distributed graph rewriting and a major optimization to achieve efficient execution consists of a message aggregation technique exhibiting the potential for strong reduction of the communication overhead. In this paper we discuss the interaction between the effectiveness of aggregation and the schedule sequence of rewriting operations. Then we present a Priority Based (BP) scheduling algorithm well suited for the specific aggregation technique. Results on a classical benchmark λ-term demonstrate that PB allows PELCR to achieve up to 88% of the ideal speedup while executing on a shared memory parallel architecture.

1 Introduction

PELCR (Parallel Environment for Lambda-Calculus Reduction) is a recent software [7] for efficient optimal reduction of λ-terms on parallel/distributed computing systems. The development of this software is based on results in the field of functional programming, which have shown how the reduction of λ-terms can be mapped onto a particular graph rewriting technique known as Directed Virtual Reduction (DVR) [1–5]. In DVR, each computational step corresponds to a transition from a graph G to a graph G' obtained through the *composition* of two labeled edges, say e' and e'', insisting on a node v. Such a composition has the following effects: (i) a new node v' is created with two exiting labeled edges that point to the source nodes of e' and e'' respectively, (ii) the labels of e' and e'' are modified to reflect that the two edges must not be composed anymore.

PELCR allows edge compositions to be performed concurrently by supporting the graph distribution among multiple machines. As respect to this point, the nodes dynamically originated by DVR steps are distributed according to a load balancing mechanism able to prevent overload on any machine.

An additional optimization embedded by PELCR deals with communication, implemented in the form of message exchange based on the MPI layer. More precisely, PELCR adopts a message aggregation technique that collects application messages destined to the same machine (each of those messages notifies the existence of a new edge in the graph), and delivers them using a single MPI message. The advantage of aggregation is in the reduction of the network path setup time, due to the reduction in the amount of MPI messages. On the other hand, aggregation delays the delivery of application messages since they are not sent as

B. Monien and R. Feldmann (Eds.): Euro-Par 2002, LNCS 2400, pp. 648–655.

soon as they are produced, but only after the construction of the aggregate, with possible performance loss anytime the destination machine for the aggregate is idle and is waiting for new load (i.e. for new edges to be composed) to come in. Therefore, the higher the arrival rate of application messages into a given aggregate, the more valuable the aggregate since adequately small send delay for the aggregate allows in any case strong reduction of the communication overhead due to network path setup. The frequency of application messages arrival into a given aggregate, namely the goodness of the aggregate, depends on the order of DVR steps occurring on the machine. Currently those steps are scheduled in FCFS (First Come First Served) order, which means that edge compositions are performed in the same order according to which the machine becomes aware of new edges in the graph. In this paper we extend PELCR functionalities by embedding a Priority Based (PB) scheduling mechanism which aims at increasing the amount of valuable aggregates of the execution. This is done by associating with each edge a scheduling priority computed on the basis of the amount of application messages destined to remote machines which are expected to be produced due to DVR steps involving that edge. Higher priority is assigned in case of higher expected number of application messages, which should increase the frequency of application messages insertion into the aggregates. We report an experimental evaluation whose results point out how the effectiveness of the aggregation process can be increased thanks to PB. In particular, using a classical benchmark λ-term, we show that PB allows PELCR to achieve up to 88% of the ideal speedup due to parallelization, with an increase of up to 10% as compared to the case of FCFS schedule.

The remainder of this paper is structured as follows. In Section 2 we report a short, technical description of PELCR's software engine, which will allow us to easily describe PB in Section 3. Performance data are reported in Section 4.

2 PELCR Engine

The PELCR engine running on the i-th machine involved in the λ-term reduction keeps track of information related to locally hosted nodes of the graph through a list, namely $nodes_i$. The element of the list associated with a node v is denoted as $nodes_i(v)$ and has a compound structure. A relevant field of the structure, called $nodes_i(v).composed$, is a list containing the edges incident on v that have already been composed with each other according to DVR. A buffer $incoming_i$ is used to store received application messages. Each of those messages carries information related to a new edge that must be added to the graph, i.e. it must insist on a locally hosted node, and must be composed with edges, if any, insisting on the same node. Note that $incoming_i$ contains also those application messages that the engine on the i-th machine sends to itself due to notification of new edges produced by locally executed DVR steps. Denoting with m an application message, with e_m the edge carried by the message m, and with $e.target$ (resp. $e.source$) the target (resp. source) node of an edge e, the main loop executed by the engine can be schematized as shown in Figure 1.

```
1  while not end_computation do
2    ⟨collect all incoming application messages and store them in incoming_i⟩;
3    while not empty(incoming_i) do
4      ⟨extract a message m from incoming_i⟩;
5      if e_m.target ∈ nodes_i   'node already in the local list'
6      then
7        for each edge e ∈ nodes_i(e_m.target).composed do
8          ⟨compose e_m with e⟩;
9          ⟨select the destination machine for hosting the node originated by the composition⟩;
10         ⟨send the edges produced by the composition to
             the machines hosting e_m.source and e.source, respectively⟩;
11       endfor
12     else ⟨add e_m.target to nodes_i⟩;   'add new node to the local list'
13     ⟨add e_m to nodes_i(e_m.target).composed⟩;
14   endwhile;
15   ⟨send out all the non-empty aggregates⟩;
16 endwhile
```

Fig. 1. PELCR Engine's Main Loop

It is interesting to note that PELCR adopts message exchange only for the notification of new edges, while it does not adopt message exchange for notification of new nodes in the graph. Any new node destined to the i-th machine is added to $nodes_i$ as soon as the i-th machine becomes aware of the existence of at least one edge that should insist on that node (see line 12). This technique allows to reduce the amount of notification messages at the application level thus tackling the communication overhead problem in a way orthogonal to what done by aggregation. Also, when a new edge e must be added to the graph, access to the structure $nodes_i(e.target)$ is implemented efficiently through a hashing mechanism that allows fast retrieve of the memory address of the structure itself.

To support aggregation, the communication module on the i-th machine collects application messages destined to the j-th machine into an aggregation buffer $out\_buff_{i,j}$ and periodically sends the aggregate content. To determine when an aggregate must be sent, the module keeps an age estimate for each aggregation buffer $out\_buff_{i,j}$ by periodically incrementing a local counter $c_{i,j}$. The value of $c_{i,j}$ is initialized to zero and is set to zero each time the application messages aggregated in the buffer are sent via an MPI message. At the end of the composition phase associated with the extraction of an edge from $incoming_i$, $c_{i,j}$ is increased by one if the corresponding buffer stores at least one application message to be sent. Therefore, one tick of the age counter is equal to the average time for DVR steps associated with an edge extraction from $incoming_i$. Also, the counter value represents the age of the oldest message stored in the aggregation buffer. Messages aggregated into $out\_buff_{i,j}$ are sent when $c_{i,j}$ reaches a well suited maximum value $max_{i,j}$ which is dynamically determined by PELCR in an application specific manner. Actually $max_{i,j}$ represents a kind of time-out for the send delay of application messages aggregated into $out\_buff_{i,j}$.

Note that any non-empty aggregate is sent either when the corresponding time-out expires, or after the completion of the internal cycle in the engine main loop (see line 15). In other words, the application forces the communication module to send out any non-empty aggregate as soon as the extraction phase of edges from $incoming_i$ (and the corresponding compositions) terminates. This is done since delaying the sent of those aggregates until the time-out expiration is expected to not achieve benefits since the application remains busy due to the receipt phase of application messages (see line 2), with no possibility to produce new messages to be inserted into the aggregates for a while. Once inserted the first message into an aggregation buffer, say $out\_buff_{i,j}$, the higher the frequency of message aggregation into that buffer within the expiration of the time-out $max_{i,j}$ (or before the send operation forced by the application), the more valuable the aggregate. Currently, the extraction of application messages in line 4 is performed in FCFS order, i.e. the messages are extracted according to their insertion order in the buffer $incoming_i$. As a consequence there is no attempt to determine a schedule sequence of DVR steps which tends to increase the frequency of application messages insertion into the aggregation buffers. This is exactly the objective of the PB solution we present in the next section.

3 Priority Based Scheduling

Let us consider a snapshot of the content of the buffer $incoming_i$ before the engine starts the execution of the internal **while** cycle in line 3 of its main loop. In other words, the snapshot refers to the buffer content just after the collection phase of application messages to be inserted into the buffer itself. We use the notation M to identify the set of application messages contained into the buffer at the snapshot time. M might be an empty set. We study the problem of determining a well suited schedule of message extractions from $incoming_i$, to be adopted while extracting messages during the execution of the **while** internal cycle. Actually, well suited means a schedule sequence that amplifies the benefits from aggregation performed by the communication module. We use the notation $S = (M, \stackrel{sc}{\rightarrow})$ to indicate such a schedule sequence, where M has the meaning defined above and $\stackrel{sc}{\rightarrow}$ is a total ordering relation determining the order of message extractions from $incoming_i$. In other words, given two messages m and m' in M, $m \stackrel{sc}{\rightarrow} m'$ means that m must be extracted before m' in the cycle.

Each time an edge e_m carried by the application message m is extracted from the buffer $incoming_i$ and composed with an edge e already in the graph and insisting on the node $e_m.target$, the composition typically produces a "non-null" result, which means that a new node and two new edges are originated. However, "null" result might arise in some circumstances depending on the labels of the edges to be composed. In this case, the only effect of composition is to modify the labels of the edges e_m and e in order to keep track that they must not be composed anymore. This is done in practice by adding e_m to the list of already composed edges associated with node $e_m.target$ (see line 13 of the engine main loop). Let us denote as $EDGES(e_m)$ the amount of new edges that would be added to the graph due to the extraction of m from the buffer $incoming_i$, and due to the composition of e_m with other edges already insisting on the node $e_m.target$, computed on the basis of the current state of the portion of the graph locally maintained at the time the schedule sequence must be determined. The value of $EDGES(e_m)$ depends on both (i) the amount of edges currently insisting on $e_m.target$ at the time the schedule sequence must be determined and (ii) the labels of those edges (since composition with e_m might produce null result, with no new edge to be added to the graph). Note that $EDGES(e_m)$ might be different from the real amount of new edges that will be really added to the graph due to the extraction and composition of e_m since the state of the graph at the time of the extraction might be different from the state of the graph at the time the schedule sequence is determined. In particular, changes in the state might be caused by message extractions preceding the extraction of m in the schedule sequence adopted while executing the internal **while** cycle. $EDGES(e_m)$ can be rewritten as $EDGES(e_m) = LOCAL(e_m) + REMOTE(e_m)$, where the two functions $LOCAL(e_m)$ and $REMOTE(e_m)$ denote the amount of new edges that will insist on locally hosted and remotely hosted nodes of the graph, respectively. Note that whether new edges will insist on local or on remote nodes depends on which machines host the source node of e_m and the source nodes of the edges insisting on $e_m.target$ at the time the schedule sequence is determined.

The idea underlying PB is to define a schedule sequence of message extractions from $incoming_i$, to be adopted in the internal **while** cycle, which originates a sequence of DVR steps aiming at maximizing the frequencies of application message insertion into the aggregation buffers. For each buffer $out\_buff_{i,j}$, the frequency refers to the interval between the instant of the first message insertion into $out\_buff_{i,j}$ and the send time of the aggregate, i.e. the lifetime of the aggregate maintained into that buffer. We recall that finding an optimal schedule is not viable in practice since the outcome of DVR steps (and thus the effects of DVR steps on aggregation) depends on the current state of the graph at the time the steps are performed. Such a state is unpredictable at the time the schedule sequence must be determined. For this reason we have taken the practical approach to define PB as a heuristic that finds the best suited schedule sequence on the basis of the current graph state at the time of the determination of the schedule sequence, i.e. before starting the execution of the internal **while** cycle.

Let us define a precedence relation among messages in the set M.

Definition 1. *Message $m \in M$ precedes message $m' \in M$, denoted $m \prec m'$ if, and only if, $REMOTE(e_m) > REMOTE(e'_m)$.*

In other words, the relation $\prec$ defines a partial order $\widehat{M} = (M, \prec)$ among all the messages belonging to M. Introducing messages into the schedule sequence according to any linear extension of the partial order defined by the relation $\prec$ has the effect to maximize the predictable number of application messages to be inserted into the aggregation buffers per tick of the $c_{i,j}$ counters. This, in its turn, allows the maximization of the average message insertion frequency, computed overall those buffers, predictable at the time the schedule is defined. (We use the term predictable just because those quantities refer to the current state of the graph at the time the schedule must be defined.) Therefore PB is a heuristic that gives extraction priorities to messages according to the message ordering associated with the linear extension. The algorithm for PB is as follows:

$\langle$**find a linear extension** EXT **of** $\widehat{M} = (M, \prec)\rangle$;
$\langle S = (M, \overset{sc}{\rightarrow})$ **is such that, given** $m \in M$ **and** $m' \in M$, $m \overset{sc}{\rightarrow} m' \Rightarrow m$ **precedes** m' **in** $EXT\rangle$;

3.1 Implementation Issues

Approximating the Function $REMOTE$. The function $REMOTE(e_m)$ measures the amount of application messages that would be produced by the composition of the edge e_m on the basis of the current graph state evaluated at the time of the determination of the schedule sequence of message extractions from $incoming_i$. PB needs to compute the value of this function to determine the relation $\prec$ (see Definition 1) required to construct the partial order $\widehat{M}$. Computing this function could be a time consuming operation since it requires: (i) label check for all the edges insisting on the node $e_m.target$ in order to determine the non-null compositions associated with the edge e_m and (ii) identification of the edges insisting on $e_m.target$ having sources hosted by remote machines. These operations require $O(n)$ time if n is the amount of edges currently in the list $nodes_i(e_m.target).composed$ (i.e. the edges currently insisting on $e_m.target$). To reduce the time complexity we have decided to approximate the function $REMOTE(e_m)$ with its upper bound value, namely with the maximum number of remote application messages that would been produced by the composition of the edge e_m (considering the current state of the graph) in case all the compositions themselves were non-null. The upper bound, that we denote as $UB(REMOTE(e_m))$ can be computed in $O(1)$ time by simply maintaining two counters associated with the structure $nodes_i(e_m.target)$. One keeps track of the total amount of egdes currently in the list $nodes_i(e_m.target).composed$. The other keeps track of the amount of edges in the list $nodes_i(e_m.target).composed$ having sources hosted by remote machines. The counters are updated when inserting an edge into the list ([1]).

On the Use of Priority Queues. Although the complexity of the evaluation of the function $REMOTE$ can be reduced to $O(1)$ through the upper bound approximation presented in the previous paragraph, the cost of PB remains at least $O(n \times log_2 n)$ where n is the amount of messages in the buffer $incoming_i$ at the time the schedule sequence must be determined (i.e. $n = |M|$). This cost is due to the construction of the linear extension EXT of the partial order $\widehat{M}$. To bound the cost due to the schedule determination, we have taken the design choice to implement an approximation of PB, which is based on the use of multiple priority queues for buffering the incoming application messages. We call this approximation Multiple Queues Priority Based (MQPB) scheduling.

To implement MQPB, incoming application messages are received and buffered into a set of N distinct buffers, namely $incoming_i^0, incoming_i^1, \ldots, incoming_i^{N-1}$. The buffers are associated with different priorities that decrease with the increase in the buffer index. When an application message m is received, the function $UB(REMOTE(e_m))$ is evaluated (this takes $O(1)$ time). Then the priority level, i.e. the index of the destination buffer for the message, is computed as:

$$\text{index} = \lfloor (N-1) \times (1 - \frac{UB(REMOTE(e_m))}{MAX\_UB} \rfloor \tag{1}$$

[1] In PELCR each edge explicitly carries the information on the location of its source.

where MAX_UB is the relative maximum of the UB functions observed during the current determination of the schedule sequence ([2]). While executing the internal **while** cycle in line 3 of the engine main loop, the messages are extracted starting from the buffer with the highest priority and then going to the one with the lowest priority. Messages into the same buffer are extracted according to their order of insertion into the buffer itself. This type of extraction is actually an approximation of the sequence determined by PB since it does not guarantee that given two messages m and m', with $UB(REMOTE(m)) > UB(REMOTE(m'))$, m precedes m' in the schedule sequence. However, it guarantees that messages associated with larger values of $UB(REMOTE)$ are likely to be extracted before. The operational advantage of MQPB is that the assignment of a message to a given buffer, which determines the position of the message in the extraction sequence, can be computed in $O(1)$ time upon the receipt of the message in line 2 of the engine structure.

4 Performance Data

We report in this section some performance data in order to evaluate the capability of MQPB in increasing the effectiveness of message aggregation. As hardware platform we have used a Sun Ultra Enterprise 4000 machine, which is a shared memory multiprocessor equipped with 4 CPUs SuperSparc 167 MHz. The experiments have been performed using a classical benchmark for parallel executions, namely DD4 [7], corresponding to the λ-term $(\delta)(\delta)\underline{4}$ where $\delta = \lambda x(x)x$ represents the self application. The normal form of this term represents the Church's integer $(4^4)^{4^4}$, whose graph representation manipulated by PELCR has a number of nodes which is in the order of hundreds of thousands.

We report two series of plots (see Figure 2). The first one relates to speedup as a function of the number of used CPUs (from 2 to 4). Speedup results are plotted for the case of parallel execution with both FCFS schedule and MQPB schedule. The speedup is computed against the execution time on a single CPU when FCFS schedule of message extractions is used (FCFS schedule is not penalizing in case of execution on a single CPU since aggregation is activated only in case of execution on multiple CPUs). Also, for the case of MQPB we report results for three different values of the number of priority levels, say 3, 5 and 10. The second series of plots report the average number of application messages (ANAM) delivered through a single MPI message. This parameter is an indicator of the effectiveness of message aggregation. For all the series, each reported value is the average of ten runs. The data show two major tendencies. First, MQPB achieves higher values of ANAM. This reduces the communication overhead, especially for the case of a larger number of CPUs, where application message exchange occurs more frequently. As a consequence, we get an acceleration of

[2] Actually, using for MAX_UB the relative maximum during the current schedule sequence determination, instead of the relative maximum overall the schedule determinations performed since the beginning of the execution, allows PELCR to better react to dynamics in the graph rewriting.

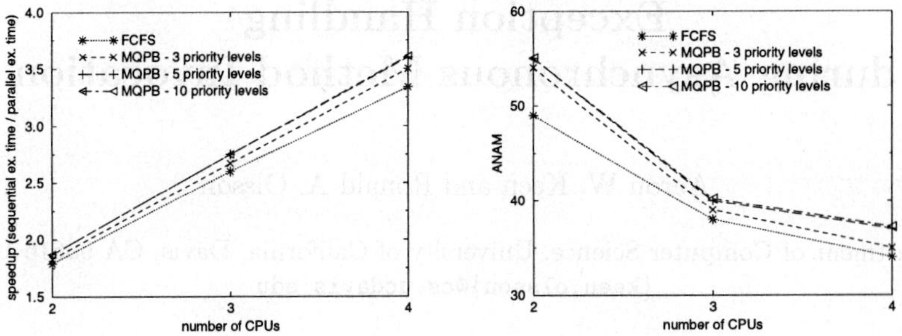

Fig. 2. Speedup and ANAM vs the number of CPUs

the parallel execution (as compared to that achievable with FCFS schedule) that for the case of 4 CPUs is in the order of 10%. The second tendency is related to the behavior of MQPB vs the number of priority levels. Specifically, moving it from 3 to 5 allows improvements in the aggregation process (see the values of ANAM in Figure 2). Instead, moving it from 5 to 10 does not produce additional advantages, i.e. the behavior of MQPB stabilizes when the number of priority levels gets over a given threshold. As a final observation, the speedup with MQPB is between 85% and 88% of the ideal one. This is an indication that MQPB actually brings PELCR near an ideal parallel execution.

References

1. V. Danos, M. Pedicini, and L. Regnier. Directed virtual reductions. In M. B. D. van Dalen, editor, *Computer Science Logic, 10th Int. Workshop, CSL '96*, volume 1258 of *Lecture Notes in Computer Science*, pages 76–88. EACSL, Springer Verlag, 1997.
2. V. Danos and L. Regnier. Local and asynchronous beta-reduction (an analysis of Girard's EX-formula). In *Logic in Computer Science*, pages 296–306. IEEE Computer Society Press, 1993. Proceedings of the Eight Annual Symp. on Logic in Computer Science, Montreal, 1993.
3. J.-Y. Girard. Geometry of interaction 1: Interpretation of system F. In R. Ferro, C. Bonotto, S. Valentini, and A. Zanardo, editors, *Logic Colloquium '88*, pages 221–260. North-Holland, 1989.
4. G. Gonthier, M. Abadi, and J.-J. Lévy. The geometry of optimal lambda reduction. In *Proc. of 17th Annual ACM Symp. on Principles of Programming Languages*, pages 15–26. Albuquerque, New Mexico, January 1992. ACM Press.
5. J. Sousa Pinto. Parallel implementation models for the lambda-calculus using the geometry of interaction. In *5th International Conference on Typed Lambda Calculi and Applications*, pages 385–399. Krakow, Poland, May 2001. LNCS 2044.
6. J. Lamping. An algorithm for optimal lambda calculus reduction. In *Proc. of 17th Annual ACM Symp. on Principles of Programming Languages*, pages 16–30, San Francisco, California, January 1990. ACM Press.
7. M. Pedicini, and F. Quaglia. A Parallel implementation for optimal lambda-calculus reduction. In *Proc. of 2nd ACM Int. Conference on Principles and Practice of Declarative Programming*, pages 2–13, Montreal, Canada, September 2000. ACM Press.

Exception Handling during Asynchronous Method Invocation

Aaron W. Keen and Ronald A. Olsson

Department of Computer Science, University of California, Davis, CA 95616 USA,
{keen,olsson}@cs.ucdavis.edu

Abstract. Exception handling mechanisms provided by sequential programming languages rely upon the call stack for the propagation of exceptions. Unfortunately, this is inadequate for handling exceptions thrown from asynchronously invoked methods. For instance, the invoking method may no longer be executing when the asynchronously invoked method throws an exception. We address this problem by requiring the specification of handlers for exceptions thrown from asynchronously invoked methods. Our solution also supports handling exceptions thrown after an early reply from a method and handling exceptions after forwarding the responsibility to reply. The JR programming language supports the exception handling model discussed in this paper.

1 Introduction

Asynchronous method invocation facilitates the dynamic creation of concurrently executing threads and the communication between such threads. When a method is asynchronously invoked, the invoking thread continues to execute while another thread executes the body of the invoked method. Such concurrent execution can benefit both the design of and the performance of a program. The use of asynchronous method invocation, however, complicates the use of exceptions and, in particular, the handling of thrown exceptions.

Exception handling mechanisms for sequential programming languages are well-understood. Goodenough [2] presents a general discussion of the issues that exception handling mechanisms need to address and the different semantics (e.g., terminate, retry, or resume) provided by such mechanisms. Yemini and Berry [9] propose an additional model (replacement) that allows an erroneous subexpression (one that raises an exception) to be *replaced* by the handler's result.

The exception handling mechanisms provided by sequential programming languages rely upon the call stack for the propagation of exceptions. A thrown exception is propagated (either implicitly or explicitly) up the call stack until an appropriate handler is found. In Figure 1, method `baz` throws an exception. The exception is propagated through method `bar` and into method `foo`, where it is finally handled.

Unfortunately, the reliance of such mechanisms on the call stack is inadequate for handling exceptions thrown from an asynchronously invoked method. Figure 2 depicts the same program as before, but with the method `bar` invoked

B. Monien and R. Feldmann (Eds.): Euro-Par 2002, LNCS 2400, pp. 656–660.

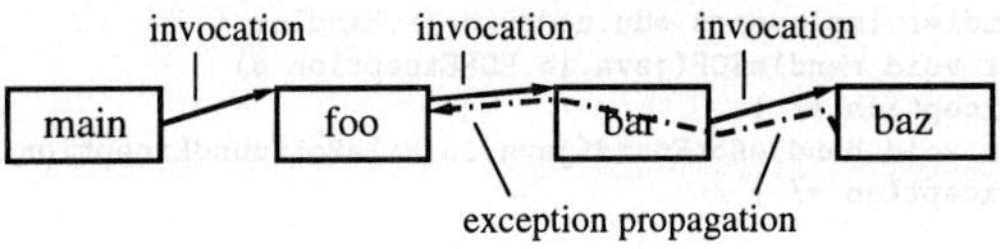

Fig. 1. Exception propagated through call stack.

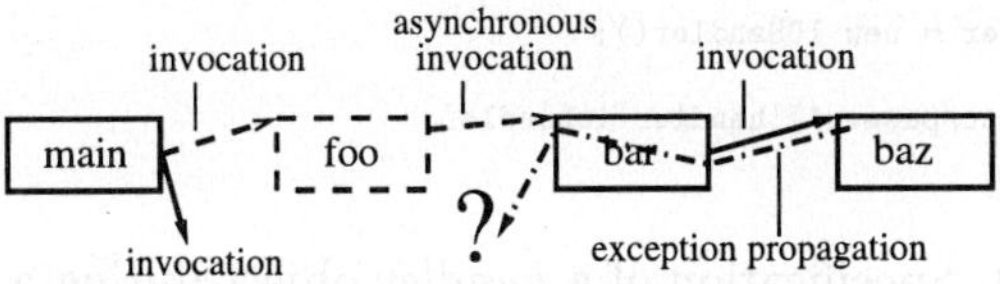

Fig. 2. Exception propagated from method invoked asynchronously.

asynchronously. Again, an exception is thrown from method `baz` and propagated to method `bar`. If method `bar` does not have an appropriate exception handler, then the exception must be further propagated. But, since method `bar` was invoked asynchronously, the preceding call stack is not accessible. In fact, the method that invoked `bar` (i.e., `foo`) may no longer be executing. As such, the call stack cannot be used to propagate exceptions from methods invoked asynchronously.

We address this problem by requiring the specification of handlers for exceptions thrown from methods invoked asynchronously. The exception handling support presented in this paper bears some resemblance to that provided in both ABCL/1 [3] and Arche [4], which are discussed in greater detail in Section 4. Our solution, however, differs in the support for static checks for handling exceptions, handling exceptions after an early reply from a method, and handling exceptions after forwarding within a method. Both early reply and forwarding are useful in distributed programming [7, 1]. Early reply allows the invoked method to transmit return values to its invoker, yet the invoked method continues to exist and executes concurrently with its invoker.[1] Forwarding passes the responsibility for replying to the invoker from the invoked method to another method. The exception handling support discussed in this paper has been designed and implemented as part of the JR programming language [6, 5].

2 Exceptions during Asynchronous Invocation

The JR programming language extends Java with support for, among other features, asynchronous method invocation via the `send` statement. To facilitate the handling of exceptions thrown from an asynchronously invoked method, we require the specification of a *handler* object as part of a `send`. Any exceptions propagated out of the invoked method will be directed to the *handler* object. As such, in accordance with the verbose nature of Java's exception handling

[1] Further discussion of support for reply is discussed in [5].

```
public class IOHandler implements edu.ucdavis.jr.Handler {
    public handler void handleEOF(java.io.EOFException e)
    { /* handle exception */ }
    public handler void handleNotFound(java.io.FileNotFoundException e)
    { /* handle exception */ }
}
```

Fig. 3. Class definition for a simple *handler* object.

```
IOHandler iohandler = new IOHandler();
...
send readFile("/etc/passwd") handler iohandler;
...
```

Fig. 4. Specification of a *handler* object during a send.

model, JR requires that the specified *handler* object is capable of handling any exception potentially thrown from the invoked method.

To be used as a *handler*, an object must implement the `Handler` interface and define a number of *handler* methods. A method is defined as a *handler* through the use of the `handler` modifier (much like the `public` modifier). A *handler* method takes only a single argument: a reference to an exception object. Each *handler* method specifies the exact exception type that it can handle. When an exception is delivered to a *handler* object, it is handled by the *handler* method of the appropriate type. If two *handler* methods may handle the exception, then the method handling the most specific type is selected. *Handler* methods cannot propagate exceptions out of their method body.

An example definition of a *handler* object's class is given in Figure 3. In this example, *handler* objects of type `IOHandler` can handle end-of-file and file-not-found exceptions. An exception of type `java.io.EOFException` directed to such a *handler* object will be handled by the `handleEOF` method.

A `send` statement must specify, using a `handler` clause, the *handler* object that is to be used to handle any exceptions propagated from the asynchronously invoked method. An example of the specification of a handler object is given in Figure 4. The JR compiler statically checks that the specified *handler* object can handle each of the potentially thrown exceptions.

3 Exceptions after Forwarding

A modification of the standard synchronous invocation semantics is forwarding. An invoked method may *forward* to another method the responsibility of replying. For example, in Figure 5, method `foo` does some calculations and then forwards responsibility to method `bar`, after which the two methods execute concurrently.

A `forward` statement must specify a *handler* object to handle any exceptions thrown by the forwarding method after executing the `forward` statement. Any exceptions thrown prior to executing a `forward` statement are propagated according to the manner in which the method was invoked. Any exceptions thrown

```
int baz(String filename) throws java.io.EOFException {
    int retval = foo(filename);
    // retval actually comes from bar because of foo's forward
    ...
}
int foo(String filename) throws java.io.EOFException {
    ...
    IOHandler iohandler = new IOHandler();
    forward bar(filename) handler iohandler;  // forward invocation
    ...  // continue executing
}
int bar(String filename) throws java.io.EOFException {
    ... // potentially throw java.io.EOFException
}
```

Fig. 5. Forwarding.

by the forwarding method after executing a `forward` statement are directed to the *handler* object. The method to which responsibility is forwarded inherits the handler of the forwarding method: the call stack link if invoked synchronously and a handler object if invoked asynchronously.

4 Discussion

Handler objects are implemented as *Serializable* Java objects. The *handler* methods are implemented as normal methods, but are gathered during compilation to generate a dispatch method (named by the *Handler* interface). The dispatch method is necessary as a single, well-named entry point into each *handler* object. All exceptions are directed to the dispatch method, which routes each exception to the appropriate *handler* method. A *handler* object is sent, as an additional parameter, to an invoked method, and used only when an exception is raised. As such, the *handler* object will exist during the duration of the method execution.

A previous approach [8] to handling exceptions thrown from asynchronously invoked methods allows for the specification of an exception handler when an exception is actually raised. Unfortunately, specifying the handler at the point an exception is raised introduces some limitations. For example, it might be desirable for a method that can be invoked both synchronously and asynchronously to propagate exceptions up the call stack or to a handler, as appropriate . Such a distinction would require support for a method to determine how it was invoked.

As mentioned previously, the solution proposed in this paper bears some resemblance to the solutions for ABCL/1 [3] and Arche [4]. ABCL/1 allows synchronous, asynchronous, and future-based method invocation. Each invocation may specify a “complaint” destination. Any exception raised during the execution of the method will be directed to the “complaint” object, if specified. Otherwise, the exception is propagated to the invoker through the call stack. ABCL/1, due to the nature of the language, does not perform static checks on the “complaint” destination to ensure that it can handle the thrown exceptions.

The exception handling support in Arche is similar to that provided by ABCL/1. A set of handler objects can be specified as part of an asynchronous invocation. Exceptions thrown from the invoked method are directed to each of the specified handler objects. Arche also statically checks that each of the handler objects can actually handle the potentially thrown exceptions.

5 Conclusion

This paper presented the design and implementation of an exception model that supports handling exceptions thrown from an asynchronously invoked method, handling exceptions thrown after an early reply from a method, and handling exceptions after forwarding. The JR programming language supports *handler* objects and the presented exception model.

References

1. G.R. Andrews. *Concurrent Programming: Principles and Practice.* Benjamin/Cummings Publishing Company, Inc., Redwood City, CA, 1991.
2. J. B. Goodenough. Exception handling issues and a proposed notation. *Communications of the ACM*, 18(12):683–696, 1975.
3. Y. Ichisugi and A. Yonezawa. Exception handling and real time features in an object-oriented concurrent language. In *Proceedings of the UK/Japan Workshop on Concurrency: Theory, Language, and Architecture*, pages 604–615, 1990.
4. V. Issarny. An exception handling mechanism for parallel object-oriented programming: toward reusable, robust distributed software. *Journal of Object-Oriented Programming*, 6(6):29–40, 1993.
5. A. W. Keen. *Integrating Concurrency Constructs with Object-Oriented Programming Languages: A Case Study.* PhD dissertation, University of California, Davis, Department of Computer Science, June 2002.
6. A. W. Keen, T. Ge, J. T. Maris, and R. A. Olsson. JR: Flexible distributed programming in an extended Java. In *Proceedings of the 21st IEEE International Conference on Distributed Computing Systems (ICDCS 2001)*, pages 575–584, April 2001.
7. B. Liskov, M. Herlihy, and L. Gilbert. Limitations of remote procedure call and static process structure for distributed computing. In *Proceedings of 13th ACM Symposium on Principles of Programming Languages*, St. Petersburg, FL, January 1986.
8. A. Szalas and D. Szczepanska. Exception handling in parallel computations. *ACM SIGPLAN Notices*, 20(10):95–104, October 1985.
9. S. Yemini and D. M. Berry. A modular verifiable exception handling mechanism. *ACM Transactions on Programming Languages and Systems*, 7(2):214–243, 1985.

Designing Scalable Object Oriented Parallel Applications

João Luís Sobral and Alberto José Proença

Departamento de Informática - Universidade do Minho
4710 - 057 Braga – Portugal
{jls, aproenca}@di.uminho.pt

Abstract. The SCOOPP (Scalable Object Oriented Parallel Programming) system efficiently adapts, at run-time, an object oriented parallel application to any distributed memory system. It extracts as much parallelism as possible at compile time, and it removes excess of parallel tasks and messages through run-time packing. These object and call aggregation techniques are briefly presented. A design methodology was developed for three main types of scalable applications: pipeline, divide & conquer and farming. This paper reviews how the method can help programmers to design portable and efficient parallel applications. It details its application to a farming case study (image threshold) with measured performance data, and compares with programmer's tuned versions in a Pentium cluster.

1 Introduction

The development of portable parallel applications that *efficiently* run on several platforms imposes a platform-tuning overhead. This paper addresses issues that may reduce this overhead, namely to up/down scale an object oriented parallel application, including the automatically tuning of the application for each platform.

Applications may require dynamic granularity control to get an acceptable parallel execution performance in time-shared platforms, namely when the parallel tasks are dynamically created and whose behaviour cannot be accurately estimated at compile-time. Few systems provide dynamic granularity control [1][2][3][4], based on fork/join parallelism constructs, ignoring the fork construct and executing tasks sequentially (parallelism serialisation).

The SCOOPP system [5] is a hybrid compile and run-time system, that extracts parallelism, supports explicit parallelism and dynamically serialises parallel tasks and packs communication in excess at run-time. A design methodology was developed for three main types of scalable applications: object pipelines [6], static object trees (e.g., farming) and dynamic object trees (e.g., divide & conquer). Several case studies have been tested on various platforms: a 7 node Pentium cluster, a 16 node PowerPC based Parsytec PowerXplorer and a 56 node Transputer based MC-3.

This paper shows evaluation results that were experimentally obtained by executing a farm type application and a comparison with a programmer's optimised

* This work was partially supported by grant PRAXIS 2/2.1/TIT/1557/95

B. Monien and R. Feldmann (Eds.): Euro-Par 2002, LNCS 2400, pp. 661–665.

version, on a 7 node Pentium III based cluster, under Linux with a threaded PVM on TCP/IP. The cluster nodes are inter-connected through a 1,2 Gbit Myrinet switch.

Section 2 presents an overview of the SCOOPP system. Section 3 introduces a farm type application, the design methodology applied for scalability, and presents performance results. Section 4 closes the paper with suggestions for future work.

2 SCOOPP System Overview

SCOOPP system scales parallel applications to any distributed memory systems, in two steps: at compile-time, the compiler and/or the programmer specifies a large number of fine-grained parallel tasks; at run-time, parallel tasks are packed into larger grains - according to the application/platform behaviour and based on security and performance issues – and communications are packed into larger messages.

The SCOOPP programming model is based on an OO paradigm supporting both active and passive objects. Active objects (parallel objects in SCOOPP) specify explicit parallelism: they model parallel tasks, they may be placed at remote processing nodes and they communicate through either asynchronous or synchronous method calls. Passive objects take advantage of existing code; they are placed in the context of the parallel object that created them, and only copies can be moved between parallel objects; method calls on these objects are always synchronous.

SCOOPP extracts parallelism by transforming selected passive objects into parallel ones [7], and at run-time it removes parallelism overheads by transforming (packing) parallel objects in passive ones and by aggregating method calls [8].

These run-time optimisations are implemented through:

- method call aggregation: (delay and) combine a series of asynchronous method calls into a single aggregate call message; this reduces message overheads and per-message latency;
- object agglomeration: when a new object is created, create it locally so that its subsequent (asynchronous parallel) method invocations are actually executed synchronously and serially.

The decision to pack objects and method calls considers several factors: the latency of a remote "null-method" call (λ), the inter-node communication bandwidth, the average overhead of the method parameters passing (ν) and the average local method execution time on each node type (ε). More details of the run-time granularity control and how decision factors are estimated can be found in [6].

3 Design and Performance Evaluation of Farming Applications

This section presents and analyses a farming parallel algorithm, based on master and slaves. The master executes the sequential part of the work, e.g., it divides the work into several tasks, sends them to the slaves and joins the received processed results.

In SCOOPP, two parallel objects classes implement the farming applications: master and the slave classes. Both classes have other methods, which promote code reuse, since the master and slave classes are generic.

The design of a scalable farming application to efficiently run on several target platforms, must adequately address three main issues: the number of slaves to specify, the master hierarchy (if any), and the task granularity.

A high number of slaves helps to scale the application to a larger system, since the number of slaves limits the number of nodes that the application can use. However, if the slave/node is high, performance may suffer due to the slave management time. Specifying a number of slaves equal to the number of nodes limits dynamic changes on the number nodes used and when more than one master is used, it may be easier to use a number of slaves proportional to the number of masters.

Using just one master may limit the application performance, since the tasks and slaves management is centralised. A high number of slaves should be followed by a decentralisation of management work, by using a master hierarchy. However, using several masters introduces overheads due to the coordination among masters.

The specification of the task size depends on the number of slaves and on the target platform. The work division should provide a number of tasks higher than the number of slaves to provide enough work for all slaves. A high number of tasks helps the load distribution, but also introduces higher overheads, due to the additional work to join and split work, and each task may be too small for a platform.

The SCOOPP methodology can help to achieve an adequate solution to these issues, showing the feasibility to develop parallel applications, that are portable and scalable on several target platforms, without requiring source code changes. The SCOOPP methodology suggests the programmer to specify a high number of slaves and masters (e.g., parallel objects) and a high number of parallels tasks (e.g., method invocations). The SCOOPP run-time system is responsible to pack excess of masters and slaves, and to aggregate method invocations, reducing the impact of the overhead due to excess of parallelism and communication.

3.1 Packing Policy for Farming Applications

Packing policies defines “when” and “how much” to pack. These are grouped according to the structure of the application: object pipelines, static object trees and dynamic object trees. This section focuses on packing policies for farming.

When communication overhead becomes higher than the task processing time, method calls should be packed. This occurs in SCOOPP when the overhead of a remote method call - given by the sum of the average latency of a remote "null-method" call (λ) and the overhead of the method parameters passing (ν) is higher than the average method execution time, ε, e.g., $(\lambda+\nu)>\varepsilon$. This is the turnover point to pack, where the communication overhead is considered too high.

The communication grain-size, G_m, is computed from the λ, ν and ε and defines “how many” method calls to pack into a single message. Sending a message that packs G_m method calls introduces a time overhead of $(\lambda+G_m\nu)$ and the time to execute this pack is $G_m\varepsilon$. Packing should ensure that $(\lambda+G_m\nu)<G_m\varepsilon$, e.g., that the overhead to send a pack of several remote method calls is lower than the time to locally execute the pack of method calls, e.g., $G_m=\lambda/(\varepsilon-\nu)$. This point defines the minimum number of method calls to pack into a message. The packing policy is based on a dynamic

criterion, which initially packs $\lambda/(\varepsilon\text{-}\nu)$ method calls on each message, to spread the work rapidly, and progressively increases the number of method calls on each pack, proportionally to the number of packs sent to each node (μ), to further reduce the overheads, e.g., $Gm=\lambda(1+\mu)/(\varepsilon\text{-}\nu)$.

Object farming places some limitations on object packing. When only one master is provided, the excess of parallel objects (slaves) cannot be effectively removed, since to remove parallelism they should be packed with the master, which concentrates most of the slaves on the master's node. Note that packing slaves together do not remove parallelism overheads, since no method calls are performed between slaves. However, this limitation can be overcome by using a hierarchy of masters, which allows the packing of slaves with intermediate masters.

3.2 A Farming Case Study

An image processing case study, using a farming structure, was selected to show the impact of SCOOPP: a dynamic threshold with a predefined window size. The application was developed on a C++ SCOOPP prototype [7][8]. Performance results were obtained using an image of 512x512 (windows of 11x11) and 16 slaves (~2/node), and a 2-level master hierarchy.

In this application the master splits the image into frames and sends the frames to the slaves (call a method on the slaves). Each slave processes the frames and sends them back to the master (call a method on the master). On this case study, frames are square-shaped and there is some frame overlap to reduce intermediate communication during frame processing.

To select the frame size, a programmer may opt for a higher number of frames, if she/he aims scalability for large distributed systems. To guarantee an efficient execution, the programmer has to tune the application for each target platform. Fig.1 shows the several experiments data she/he had to collect (using 4096 frames and manually packing several frames per message), just for two platforms: a 4- and a 7-node cluster. Using SCOOPP, a single point is automatically obtained for any platform, where performance is very close to the optimum value.

Table 1 shows how the grain-size of the selected frame may affect the overall performance, without any further tuning (worst case). This table clearly suggests that

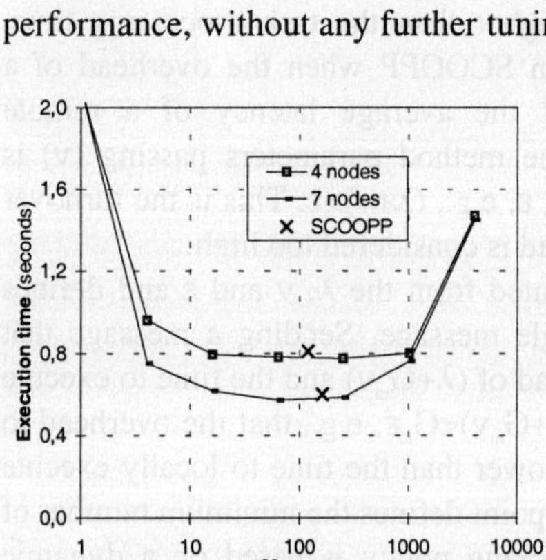

Fig. 1. Execution times using a single master

Table 1. Execution times with and without SCOOPP and decision factors (7-nodes).

Frames	Programmer	SCOOPP	Gm	ε	λ+ν
1	2,11	2,10	1	672k	54k
4	0,57	0,56	1	218k	14k
16	0,45	0,45	1	60k	4129
64	0,39	0,39	1	15k	1415
256	0,36	0,37	2	4k	673
1024	0,49	0,40	6	997	455
4096	**1,45**	**0,56**	**29**	**260**	**385**
16384	5,20	1,07	95	66	329

Conclusion

The success of parallel computation in the past did not had the expected results mainly due to (i) the lack of adequate tools to support automatic mapping of the applications into distinct target platforms, without compromising efficiency, and (ii) the portability costs due to excessive overhead to tune the application for each target platform. Current trends (time-shared clusters and the Grid) place additional challenges, namely on dynamic tuning. SCOOPP attempts to overcome these limitations by providing dynamic and efficient scalability of object oriented parallel applications across several target platforms, without requiring any code modification. The presented results show the effectiveness of the SCOOPP methodology when applied to farming applications; it dynamically increases grain-sizes, improving execution times and showing that this methodology successively identifies and removes most parallelism overheads.

Current experimental results gathered data related to both the platform and the application behaviour; the latter is dynamically updated, while the former was, so far, statically gathered before execution. Research work is also being carried out to support a dynamic decision mechanism based on stochastic approaches [9].

References

1. Mohr, E., Kranz, A., Halstead, R.: Lazy Task Creation: A Technique for Increasing the Granularity of Parallel Programs, IEEE Trans. on Par.& Dist. Proc., Vol. 2(3), July (1991)
2. Goldstien, S., Schauser, K., Culler, D: Lazy Threads: Implementing a Fast Parallel Call, Journal of Parallel and Distributed Computing, Vol. 37(1), August (1996)
3. Karamcheti, V., Plevyak, J., Chien, A.: Runtime Mechanisms for Efficient Dynamic Multithreading, Journal of Parallel and Distributed Computing, Vol. 37(1), August (1996)
4. Taura, K., Yonezawa, A.: Fine-Grained Multithreading with Minimal Compiler Support, Proc. ACM SIGPLAN CPLDI'97, Las Vegas, July (1997)
5. Sobral, J., Proença, A.: Dynamic Grain-Size Adaptation on Object-Oriented Parallel Programming - The SCOOPP Approach, Proc. 2nd IPPS/SPDP, Puerto Rico, April (1999)
6. Sobral, J., Proença, A.: A SCOOPP Evaluation on Packing Parallel Objects in Run-time, VecPar'2000, Porto, Portugal, June (2000)
7. Sobral, J., Proença, A.: ParC++: A Simple Extension of C++ to Parallel Systems, Proc. of the 6th Euromicro Work. on Par. & Dist. App. (PDP'98), Madrid, Spain, January (1998)
8. Sobral, J., Proença, A.: A Run-time System for Dynamic Grain Packing, Proceedings of the 5th Int. EuroPar Conference (Euro-Par'99), Toulouse, France, September (1999)
9. Santos, L., Proença, A.: A Bayesian RunTime Load Manager on a Shared Cluster, Scheduling and Load Balancing on Clusters (SLAB'2001), special session in IEEE Int. Symp. on Cluster Computing and the Grid (CCGrid'2001), Brisbane, Australia, May, 2001

Delayed Evaluation, Self-optimising Software Components as a Programming Model

Peter Liniker, Olav Beckmann, and Paul H.J. Kelly

Department of Computing, Imperial College,
180 Queen's Gate, London SW7 2BZ, United Kingdom,
{pl198,ob3,phjk}@doc.ic.ac.uk

Abstract. We argue that delayed-evaluation, self-optimising scientific software components, which dynamically change their behaviour according to their calling context at runtime offer a possible way of bridging the apparent conflict between the quality of scientific software and its performance. Rather than equipping scientific software components with a *performance interface* which allows the caller to supply the context information that is lost when building abstract software components, we propose to recapture this lost context information at runtime. This paper is accompanied by a public release of a parallel linear algebra library with both C and C++ language interfaces which implements this proposal. We demonstrate the usability of this library by showing that it can be used to supply linear algebra component functionality to an existing external software package. We give preliminary performance figures and discuss avenues for future work.

1 Component-Based Application Construction

There is often an apparent conflict between the quality of scientific software and its performance. High quality scientific software has to be easy to re-use, easy to re-engineer, easy to maintain and easy to port to new platforms, as well as suited to the kind of thorough testing that is required for instilling confidence in application users. Modern software engineering achieves these aims by using abstraction: We should only have to code one version of each operation [16], independently of the context in which it is called or the storage representation of participating data. The problem with this kind of abstract, component-based[1] software is that abstraction very often blocks optimisation: the fact that we engineer software components in isolation means that we have no context information available for performing certain types of optimisation.

Performance Interfaces. One common solution to this problem is to equip software components with a *performance interface* that allows a calling program to tune not only those parameters that affect the semantics of a component, but also those that affect performance. One example for this might be PBLAS [5]: The P_DGEMV parallel matrix-vector product routine takes 19 parameters, 3 of which are themselves arrays of 9 integers. This compares with 11 parameters for the equivalent sequential routine from BLAS-2 [4].

[1] In this paper we use the term *component* to refer to separately deployable units of software reuse, including *e.g.* subroutines from libraries like the BLAS [4].

B. Monien and R. Feldmann (Eds.): Euro-Par 2002, LNCS 2400, pp. 666–673.

The additional parameters in PBLAS are used to select parallel data placement. Thus, when a calling program contains a series of PBLAS routines, these parameters can be used to choose a set of data placements that minimise the need for redistributions between calls. Assuming that the application programmer knows what the optimal data layout is, the performance interface solution is of course "optimal". However, calling routines with such large numbers of parameters is very tedious and highly likely to induce programming errors. Furthermore, selecting optimal data placements is often an NP-hard problem [14], so expecting application programmers to make the right choice without access to suitable optimisation algorithms is unrealistic.

1.1 Background: Related Work

Code-Generating Systems. Several systems have been described that automatically adapt numerical routines to new computer architectures: PHiPAC [3] uses parameterised code generators and search scripts that find optimal parameters for a given architecture to generate matrix multiply routines that are competitive with vendor libraries. ATLAS (automatically tuned linear algebra software) [19] uses code generators to automatically adapt the performance-critical BLAS library [4] to new architectures.

Telescoping Languages. The telescoping languages work [13] is in some aspects similar to code-generating systems discussed above; however, the aim is not to optimise individual routines to exploit machine architectures, but rather to optimise library routines according to the context in which they are called. The strategy is to exhaustively analyse a library off-line, generating specialised instances of library routines for different calling contexts. This is combined with a language processor that recognises library calls in user programs and selects optimised implementations according to context. This work is currently still very much in-progress.

Template Meta-programming. Generic Programming techniques in C++ have been used for example in MTL [16]: each algorithm is implemented only once as an abstract template, independently of the underlying representation of the data being accessed. Optimisation in this framework is achieved by using C++ effectively as a two-level language, with the template mechanism being used for partial evaluation and code generation [18]. However, as pointed out by Quinlan *et al.* [15], a serial C++ compiler cannot find scalable parallel optimisations. A further possible problem with this technique is that templates make heavy demands of C++ compilers which on at least some high-performance architectures are much less developed than C or Fortran compilers.

Incorporating Application Semantics into Compilation. Many library-based programming systems effectively provide programmers with a semantically rich meta-language. However, this meta-language is generally not understood by compilers, which means that both syntactic checking and optimisation of the meta-language are impossible. MAGIK [9] is a system that allows programmers to incorporate application-specific semantics into the compilation process. This can be used for example in specialising remote procedure calls or in enforcing rules such that application programs should check the return code of system calls. A related system, ROSE [15], is a tool for generating library-specific optimising source-to-source preprocessors. ROSE is demonstrated through an optimising pre-processor for the P++ parallel array class library.

1.2 Delayed Evaluation, Self-optimising (DESO) Libraries

Our approach is to use delayed evaluation of software components in order to re-capture lost context information from within the component library at runtime. While execution is being delayed, we can build up a DAG (directed acyclic graph) representing the *data flow* of the computation to be performed [1]. Evaluation is eventually forced, either because we have to output result data, or because the control-flow of the program becomes data dependent (in conditional expressions).[2] Once execution is forced, we can construct an optimised execution plan at runtime, automatically and transparently changing the behaviour of components according to calling context.

We have implemented a library of delayed evaluation, self-optimising routines from the widely used set of BLAS kernels. The library performs cross-component data placement optimisation at runtime, aiming to minimise the cost of data redistributions between library calls. Our library has both a C language interface, which is virtually identical to the recently proposed C bindings for BLAS [4], and a C++ interface. The C++ interface uses operator overloading to facilitate high-level, generic coding of algorithms. This paper is accompanied by a public release of this library [7].

Contributions of This Paper. We have previously described the basic idea behind this library [2,1]. The distinct contributions of this paper are as follows:

1. We demonstrate the *usability* of our approach by showing how a number of common iterative numerical solvers can be implemented in a high-level, intuitive manner using this approach.
2. We show that the C++ interface, which we have not previously described, implements the API required for instantiating the algorithm templates in the IML++ package by Dongarra *et al.* [8].
3. We give performance figures for four iterative solver algorithms from the IML++ package, which show that fairly good parallel performance can be obtained by simply using our library together with an existing generic algorithm.
4. We discuss the techniques used in implementing the C++ interface to our library.

2 Usability and Software Quality

One of the main requirements for a high-level parallel programming model is that it should be easy for application programmers to implement scientific algorithms in parallel. IML++ by Dongarra *et al.* [8] provides generic C++ algorithms for solving linear systems using a variety of iterative methods. Figure 1 (left) shows the generic IML++ code for the preconditioned biconjugate gradient algorithm. Note that this C++ code is almost as high-level as pseudocode, the only likely difference being various type

[2] We show examples of both kinds of force points in Section 2.

```
template < class Matrix, class Vector,
           class Preconditioner, class Real >
int BiCG( const Matrix &A, Vector &x,
          const Vector &b, const Precond &M,
          int &max_iter, Real &tol) {
  Vector rho_1(1), rho_2(1), alpha(1), beta(1);
  Vector z( x.size() ), ztilde( x.size() );
  Vector p( x.size() ), ptilde( x.size() );
  Vector q( x.size() ), qtilde( x.size() );
  Vector r( x.size() ); r = b - A * x;
  Vector rtilde( x.size() ); rtilde = r;
  Real resid, normb; normb = norm( b );
  // Omitted check whether already converged

  for( int i = 1; i <= max_iter; i++ ) {
    z = M.solve(r);
    ztilde = M.trans_solve(rtilde);
    rho_1(0) = dot(z, rtilde);
    // Omitted check for breakdown
    if (i == 1) {
      p = z;
      ptilde = ztilde;
    }
    else {
      beta(0) = rho_1(0) / rho_2(0);
      p = z + beta(0) * p;
      ptilde = ztilde + beta(0) * ptilde;
    }
    q = A * p;
    qtilde = A.trans_mult(ptilde);
    alpha(0) = rho_1(0) / dot(ptilde, q);
    x += alpha(0) * p;
    r -= alpha(0) * q;
    rtilde -= alpha(0) * qtilde;

    // DESO++: Need to force evaluation of x
    deso::evaluate( x );

    rho_2(0) = rho_1(0);
    if ((resid = norm(r) / normb) < tol) {
      tol = resid; max_iter = i; return 0;
    }
  }

  tol = resid; return 1;
}
```

```
#include <ParDeso.h++>
#include "include/bicg.h" // IML++ BiCG template

int main( int argc, char * argv[] ) {
  int SZ, max_iter;
  int result = -1;        // CG return code

  deso::initialise(&argc, &argv);

  SZ         = atoi(*(++argv));
  max_iter   = atoi(*(++argv));

  // Create and read in matrix
  Matrix<double> A( SZ, SZ );
  deso::fileRead( A, "filename_A" );

  // Create rhs and solution vectors
  Vector<double> b( A.xsize() );
  Vector<double> x( A.ysize() );
  deso::fileRead( b, "filename_b" );
  deso::fileRead( x, "filename_x" );

  // Create identity preconditioner
  DiagPreconditioner<double> I( SZ, "I");

  // Convergence tolerance
  double tol = (50.0 * DBL_EPSILON);
  Scalar<double> err( tol );

  deso::startTimer();

  result = BiCG( A, x, b, I, max_iter, err );

  tol = deso::returnValue( err );

  deso::stopTimer();

  if( deso::isController() ) {
    printf( "\nFinal␣tolerance:␣%.10f.\n", tol );
  }

  deso::printTime(SZ);

  deso::finalise();

  return (result == 1 ? 0 : -1);
}
```

Fig. 1. IML++ Preconditioned BiConjugate Gradient template function (left), together with calling program (right).

declarations. We believe that the API defined by IML++ satisfies the requirements of being easy-to-use, high-level and abstract. Since the C++ `BiCG` function is a templated (generic) function, it has to be instantiated with a `Matrix`, `Vector`, `Precond` and `Real` class in order to be called. The template function implicitly defines the API these classes need to implement, such as overloaded operators for vector-matrix computations.

DESO++, the C++ interface for our delayed evaluation, self-optimising linear algebra library, provides parallel matrix, vector, scalar and preconditioner types that implement the API required for instantiating IML++ template algorithms. Figure 1 (right) shows an executable parallel program which is obtained by instantiating the `BiCG` template. This demonstrates:

- A parallel BiCG solver can be implemented simply by creating DESO++ objects for initial matrices and vectors, choosing a DESO++ preconditioner and then calling the IML++ template.
- Note that each operator in the BiCG template will call a delayed evaluation parallel function, building up a DAG representing the computation to be performed. Execution is *forced* either transparently on conditionals, such as the convergence test in

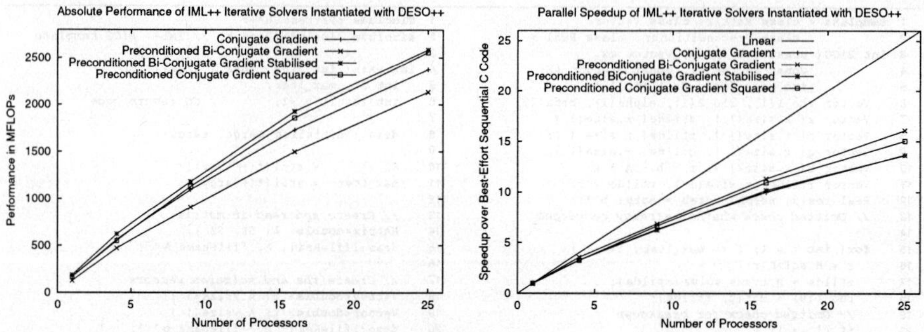

Fig. 2. Performance of four different parallel iterative algorithms implemented using IML++ templates with DESO++. The platform is a (heterogeneous) cluster of AMD Athlon processors with 1.0 or 1.4 GHz clockspeed, 256 KB L2 cache, 256 or 512 MB RAM, running Linux 2.4.17 and connected via a switched 100Mbit/s ethernet LAN. The problem size in each case is a dense matrix of size 7200 × 7200. The left graph shows absolute performance in MFLOP/s, the right graph shows speedup over a handwritten sequential C-language version of the same algorithm.

line 40, or explicitly by using the `deso::evaluate` function. The latter can be seen in line 37. The reason why we have to manually force evaluation of the solution vector `x` here is because the control flow of the program never directly depends on `x`. Alternatively, we could wait until function exit when `x` would normally be written to disk, which would also force evaluation.

IML++ was written with the aim of being usable with a diverse range of vector and matrix classes. Since the code we instantiated required virtually no changes, we believe that our parallel library should be suitable for transparently parallelising a range of existing applications that currently rely on sequential vector and matrix classes written in C++ to implement an API similar to IML++.

3 Performance

We have implemented four different iterative solvers in parallel in the manner shown in Section 2: Conjugate Gradient, preconditioned Bi-Conjugate Gradient, preconditioned Bi-Conjugate Gradient Stabilised and Conjugate Gradient Squared. The performance we obtain is shown in Figure 2.

– Note that all these algorithms have $O(N^2)$ computation complexity on $O(N^2)$ data, which means that there is only limited scope for getting good sequential performance because of memory re-use.
– The measurements we show in Figure 2 are obtained without performing data placement optimisation at runtime [1]. We believe that the performance can be improved by optimising data placement to eliminate unnecessary communication.
– Even without data placement optimisation, a speedup of about 15 on a 25-processor commodity cluster platform is encouraging, given how easy it was to obtain.

4 C++ Interface

In this section we discuss some of the design decisions and C++ programming techniques that were used in implementing the DESO++ interface. The DESO++ interface is built fully on top of the C interface, *i.e.* it calls the functions and uses the datatypes from our C language library API. In the C language interface, the results of delayed operations are represented by handles (which ultimately are integer indices into the data structure storing the DAG for the computation being performed). The application programmer has to force evaluation of such handles explicitly before being able to access the data. In C++, we can do better by using operator overloading: For example, the *force* that happens on the conditional statements in line 40 of Figure 1 is entirely transparent.

Reference-Counting Smart Pointers. The following example illustrates a potential problem that could arise due to our use of delayed evaluation:

```
Vector &fun ( const Vector &x, const Scalar &beta ) {
  Vector a;
  a = beta * x;
  return (x + a);
}
```

In our system, this function would return a handle for a delayed expression, to be evaluated when the return value of the function is eventually forced. The problem is that on function exit, a would normally be destructed, leaving the return value of the function having an indirect reference to an invalid handle. We resolve this issue by using reference-counting smart pointers, via an extra level of indirection, for accessing delayed handles.

Expression Templates. We use expression templates similar to those in Blitz++ [17] and POOMA II [12] for parsing array expressions such as `r = b - A * x`. Construction of such expressions is fully in-lined. Execution of the assignment operator = triggers the actual construction of the DAG of delayed operations representing the expression.

Careful Separation of Copy Constructors and Assignment Operators. Non-basic types such as our handles for the results of delayed operations trigger copy constructors in C++ even for the purpose of parameter passing. We initially defined copy constructors as making delayed calls to the BLAS copying routine `_copy`. This resulted in vast numbers of superfluous data copies. We therefore took the design decision to define copy constructors as making aliases, whilst the assignment operator actually copies data.

Traits. The traits technique [18] allows programmers to write functions that operate on and return *types*. This technique is very useful when implementing generic functions, in particular generic operators such as `*`. We could envisage writing a generic interface for `*` as follows:

```
template< typename T1, typename T2 >
inline Return_Type operator* ( const T1 &m1, const T2 &m2 ) {
  // ...
}
```

What should Return_Type be? Traits allow us to define a function that gives the correct type:

```
template< typename T1, typename T2 >
class _promote_product {
  // General case: type of product is type of first operand.
  typedef T1 Value_Type;
};

template< typename T2 >
class _promote_product< Scalar<double>, T2 > {
  // But Scalar * any T2 is always T2
  typedef T2 Value_Type;
};

template< >
class _promote_product< Vector<double>, Vector<double> > {
  // Special case for dot product: Vector * Vector = Scalar
  typedef Scalar<double> Value_Type;
};
```

The return type for * would then be `_promote_product<T1, T2>::Value_Type`. Note that this example has been very much simplified in order to illustrate the programming technique used.

5 Conclusion

We have described delayed evaluation, self-optimising software components as a possible way of bridging the apparent conflict between the quality of scientific software and its performance. We have presented a library which implements this proposal and have shown that this can be used to write parallel numerical algorithms in a very high-level intuitive manner as well as to transparently parallelise some existing sequential codes.

Skeletons without a Language. It is interesting to consider how our work compares with the Skeletons approach to parallel programming [6, 10]. Typically, skeletons provide a language for expressing the composition of computational components. The benefit of this is that we have very precise high-level structural information about application programs available for the purpose of optimisation. This information can be hard to capture automatically when using compilers for common imperative languages. In our approach, the information which is provided through high-level constructs in skeleton programs is instead captured at runtime by using delayed evaluation.

Acknowledgements

This work was supported by the United Kingdom EPSRC-funded OSCAR project (GR/R21486). We are very grateful for helpful discussions with Susanna Pelagatti and Scott Baden, whose visits to Imperial College were also funded by the EPSRC (GR/N63154 and GR/N35571).

References

1. O. Beckmann. *Interprocedural Optimisation of Regular Parallel Computations at Runtime*. PhD thesis, Imperial College of Science, Technology and Medicine, University of London, Jan. 2001.

2. O. Beckmann and P. H. J. Kelly. Runtime interprocedural data placement optimisation for lazy parallel libraries (extended abstract). In *Proceedings of Euro-Par '97*, number 1300 in LNCS, pages 306–309. Springer Verlag, Aug. 1997.
3. J. Bilmes, K. Asanovic, C.-W. Chin, and J. Demmel. Optimizing matrix multiply using PhiPAC: A portable, high performance, ANSI C coding methodology. In ICS '97 [11], pages 340–347.
4. BLAST Forum. Basic linear algebra subprograms technical BLAST forum standard, Aug. 2001. Available via `www.netlib.org/blas/blas-forum`.
5. J. Choi, J. J. Dongarra, S. Ostrouchov, A. Petitet, D. Walker, and R. C. Whaley. LAPACK working note 100: a proposal for a set of parallel basic linear algebra subprograms. Technical Report CS–95–292, Computer Science Department, University of Tennessee, Knoxville, July 1995.
6. J. Darlington, A. J. Field, P. G. Harrison, P. H. J. Kelly, D. W. N. Sharp, Q. Wu, and R. L. While. Parallel programming using skeleton functions. In *PARLE '93: Parallel Architectures and Languages Europe*, number 694 in LNCS. Springer-Verlag, 1993.
7. Release of DESO library. `http://www.doc.ic.ac.uk/~ob3/deso`.
8. J. Dongarra, A. Lumsdaine, R. Pozo, and K. A. Remington. LAPACK working note 102: IML++ v. 1.2: Iterative methods library reference guide. Technical Report UT-CS-95-303, Department of Computer Science, University of Tennessee, Aug. 1995.
9. D. R. Engler. Incorporating application semantics and control into compilation. In *DSL '97: Proceedings of the Conference on Domain-Specific Languages*, pages 103–118. USENIX, Oct. 15–17 1997.
10. N. Furmento, A. Mayer, S. McGough, S. Newhouse, T. Field, and J. Darlington. Optimisation of component-based applications within a grid environment. In *Supercomputing 2001*, 2001.
11. *Proceedings of the 11th International Conference on Supercomputing (ICS-97)*, New York, July7–11 1997. ACM Press.
12. S. Karmesin, J. Crotinger, J. Cummings, S. Haney, W. J. Humphrey, J. Reynders, S. Smith, and T. Williams. Array design and expression evaluation in POOMA II. In *ISCOPE'98: Proceedings of the 2nd International Scientific Computing in Object-Oriented Parallel Environments*, number 231–238 in LNCS, page 223 ff. Springer-Verlag, 1998.
13. K. Kennedy. Telescoping languages: A compiler strategy for implementation of high-level domain-specific programming systems. In *IPDPS '00: Proceedings of the 14th International Conference on Parallel and Distributed Processing Symposium*, pages 297–306. IEEE, May 1–5 2000.
14. M. E. Mace. *Memory Storage Patterns in Parallel Processing*. Kluwer Academic Press, 1987.
15. D. J. Quinlan, M. Schordan, B. Philip, and M. Kowarschik. Compile-time support for the optimization of user-defined object-oriented abstractions. In *POOSC '00: Parallel/High-Performance Object-Oriented Scientific Computing*, Oct. 2001.
16. J. G. Siek and A. Lumsdaine. The matrix template library: A generic programming approach to high performance numerical linear algebra. In *ISCOPE '98: International Symposium on Computing in Object-Oriented Parallel Environments*, number 1505 in LNCS, pages 59–71, 1998.
17. T. L. Veldhuizen. Arrays in Blitz++. In *ISCOPE'98: Proceedings of the 2nd International Scientific Computing in Object-Oriented Parallel Environments*, number 1505 in LNCS, page 223 ff. Springer-Verlag, 1998.
18. T. L. Veldhuizen. C++ templates as partial evaluation. In *PEPM '99: Partial Evaluation and Semantic-Based Program Manipulation*, pages 13–18, 1999.
19. R. C. Whaley, A. Petitet, and J. J. Dongarra. Automated empirical optimizations of software and the ATLAS project. *Parallel Computing*, 27(1–2):3–35, Jan. 2001.

Topic 11
Numerical Algorithms

Iain S. Duff, Wolfgang Borchers, Luc Giraud, and Henk A. van der Vorst

Topic Chairpersons

Following the traditions of previous Euro-Par Conferences, Euro-Par 2002 includes the topic "Numerical Algorithms". This topic has been scheduled this year for two afternoon sessions on Thursday, 29 August.

Current research and its applications in nearly all areas of natural sciences involves, with increasing importance, mathematical modelling and numerical simulation. Indeed, such a basis for computation also has been extended to other disciplines like engineering and economics. At the same time, there is a continuously increasing number of requests to solve problems with growing complexity. It is thus nearly superfluous to say that the efficient design of parallel numerical algorithms becomes more and more essential. We have addressed this development by proposing two sessions this year and have accepted over half of all submitted papers. The support of the Euro-Par Organizing Committee is gratefully acknowledged here.

As for previous Euro-Par Conferences, we have accepted only papers in which fundamental numerical algorithms are addressed and which are therefore of a wide range of interest, including parallel numerical algebra with new developments for the Fast Fourier Transform, QR-decomposition, and parallel methods for solving numerically large scale ordinary differential equations. Closely interlinked to these topics we have included recent results on modern solution procedures for partial differential equations, treatment of large sparse linear systems, multigrid and domain decomposition methods.

According to this selection, we have divided the presentations into two sessions. One is mainly devoted to methods of parallel linear algebra and the second to parallel methods for PDEs.

The first paper in the first session for this topic, by R.D. da Cunha, D. Becker and J.C. Patterson presents a new rank-revealing QR-factorization algorithm. The algorithm is integrated into an algorithm due to Golub and van Loan to determine the rank. Coded in F90 with MPI and running on several parallel systems, the authors obtained good parallel performance of the resulting algorithm which, as is well known, has so many important applications. Having in mind applications to control theory, the next paper by J.M. Badia, P. Benner, R. Mayo and E.S. Quintana-Orti presents a new parallel method for solving Lyapunov's equation, extending existing methods for the corresponding small-scale problem with dense matrices to the large scale case with sparse matrices. This paper is then followed by a new parallel blocking procedure for a very large scale FFT given by D. Takahashi, T. Boku and M. Sato who investigated the challenging problem of a parallel version. Based on a block nine-step idea, the authors were able to reduce the number of global communications. The paper by S.H.M. Bui-

B. Monien and R. Feldmann (Eds.): Euro-Par 2002, LNCS 2400, pp. 675–676.

jssen and S. Turek discusses the loss of efficiency of the multigrid method when only block smoothers are allowed, as is the case for complex applications on parallel systems. The analysis is carried out for a parallel finite-element program which solves the Navier-Stokes equations and which has been applied successfully to a variety of industrial problems. As an alternative approach to the parallel multigrid method, the second session starts with a contribution by I.G. Graham, A. Spence and E. Vainikko who use a domain decomposition method combined with a block-preconditioned GMRES version for parallelizing a Navier-Stokes solver. The preconditioner, based on a recent proposal of Kay, Loghin and Wathen, has been generalized here to the case of block-systems. In addition, the authors discuss the stability of the flow problem by applying their linear solver. The resulting generalized eigenvalue problem (large scale with indefinite matrices) is then iteratively solved by the abovementioned variant of GMRES. These last two contributions on PDEs are supplemented by a paper on a parallel 3D-multifrontal mesh generator written by D. Bouattoura, J.-P. Boufflet, P. Breitkopf, A. Rassineux and P. Villon. The method presented by these authors is based on existing corresponding sequential programs for subdomains and is thus attractive for both the parallel multigrid and the domain decomposition approach. The method is constructed for very large scale problems on complex domains. Finally, in the paper of M. Koch, T. Rauber and G. Rünger a parallel implementation is given for solving numerically large scale ordinary differential equations. The authors constructed a parallel version of an embedded Runge-Kutta method and discussed the improvements in efficiency of different variants of their program.

We hope and we believe again that the sessions contain a highly interesting mix of parallel numerical algorithms. We would like to take this opportunity of thanking all contributing authors as well as all reviewers for their work. We owe special thanks to Jan Hungershöfer, Rainer Feldmann and Bernard Bauer from the staff in Paderborn for their patience and assistance.

New Parallel (Rank-Revealing) QR Factorization Algorithms

Rudnei Dias da Cunha, Dulcenéia Becker, and James Carlton Patterson

Programa de Pós-Graduação em Matemática Aplicada, Instituto de Matemática, Universidade Federal do Rio Grande do Sul, 91509-900 Porto Alegre – RS, Brazil
{rudnei,dubecker,pattersn}@mat.ufrgs.br

Abstract. We present a new algorithm to compute the QR factorization of a matrix $A_{m \times n}$ intended for use when $m \gg n$. The algorithm uses a reduction strategy to perform the factorization which in turn allows a good degree of parallelism. It is then integrated into a parallel implementation of the QR factorization with column pivoting algorithm due to Golub and Van Loan, which allows the determination of the rank of A. The algorithms were coded in FORTRAN 90 using the MPI library. Results are presented for several different problem sizes on an IBM 9076 SP/2 parallel computer.

1 Introduction

The QR factorization of a matrix $A_{m \times n}$ is required in some situations, especially when solving a (block) linear least-squares problem,

$$A_{m \times n} X_{n \times k} = B_{m \times k}, \quad m \geq n,\ k \geq 1. \tag{1}$$

This is also required inside some solvers for block systems of linear equations, like the Block GMRES method by Saad [12].

The QR factorization of A leads to a orthogonal matrix $Q_{m \times m}$ and an upper triangular matrix $R_{m \times n}$ such that

$$QR = Q \begin{pmatrix} R_1 \\ 0 \end{pmatrix} = A, \tag{2}$$

where R_1 is $n \times n$ and X can easily be obtained solving the (block) triangular system

$$R_1 X = Q^T B. \tag{3}$$

To compute Q, one usually uses either Householder reflections or Givens's rotations, or both (see [6, pp. 193-220]); Q may not be stored explicitly, since its action over a matrix or vector may be obtained via a sequence of rank-1 updates or applying Givens's rotations to that matrix/vector.

However, if $r = rank(A) < n$, we can avoid doing unnecessary work if we can determine which columns of A are multiples of others. This can be accomplished via the QR factorization with column pivoting, from which one obtains

$$QR = Q \begin{pmatrix} R_{11} & R_{12} \\ 0 & R_{22} \end{pmatrix} = AP, \quad \| R_{22} \| \approx 0 \tag{4}$$

B. Monien and R. Feldmann (Eds.): Euro-Par 2002, LNCS 2400, pp. 677–686.

where P is a permutation matrix, Q has orthonormal columns, R is upper triangular and R_{11} is of order r, the numeric rank of A.

For large problems, one may need to use a parallel algorithm to perform either of these factorizations. Parallel algorithms to compute (2) are presented in [10] (which requires $O(n^2)$ processors), [4] (where a recursive strategy to obtain Q and R is presented) and [13]. The latter presents a number of different parallel algorithms, ranging from the modified Gram-Schmidt to hybrid algorithms using Householder reflections and Givens's rotations.

The algorithms available to compute the QR factorization with column pivoting (see, for instance, [5], [2], [6, pp. 233-236] and [1]) use the non-pivoting QR factorization as an inner-step. Our main aim in this paper is to present a new algorithm to compute the non-pivoting QR factorization of A and its use inside a parallel implementation of the QR factorization with column pivoting – a *rank-revealing* QR algorithm – presented in [6, pp. 233-236], for matrices where the number of columns is very small when compared to the number of rows. The parallel QR factorization algorithm presented here share some similarities with the row merging strategy proposed by [9] and multifrontal QR factorizations [11]. The novelty in our algorithm is to introduce the merging of the triangular factors in a tree-like reduction scheme.

Our motivation lies in the need to solve large, sparse block linear systems of equations using the Block GMRES method [12] in parallel, using message-passing constructs; an initial implementation of this method used Bischof's (sequential) rank-revealing QR package [1]. However this approach did not provide an adequate performance, leading us to develop the parallel algorithms presented here.

1.1 Parallel QR Factorization

In some applications, one will need the full Q matrix, whereas in others – like the solution of system (3) – just a few columns of Q are necessary. Our algorithm, which will be presented now, can cope with both situations, though for some larger problems the full Q matrix will not be stored explicitly.

The parallel QR algorithm (to which we will refer as PQR) we have developed shares some concepts with those presented in [13]. In that paper, a number of parallel QR algorithms are presented, their main differences being in the way data is partitioned (rows or columns); the most efficient one uses a row partitioning and is made up of two parts, one completely parallel, and the other involving communication between the processors, as if they were interconnected in a ring. Our algorithm differs from this exactly in the latter part. We have also been inspired by the recursive algorithm presented in [4].

The PQR algorithm is written using the *Single-Program, Multiple-Data, SPMD* model. We partition the m rows of the matrix to be factored (of size $m \times n$) among p processors, each having at most $\hat{m} = \lfloor m/p \rfloor + 1$ rows. For a processor q, we call $[i_q, f_q]$ the range of rows of A stored on it ($\hat{m}_q = f_q - i_q + 1$).

A restriction we place on the algorithm is that $\hat{m} \geq n$. This restriction is made since we assume that $m \gg n$ and n is, typically, $2 \leq n \leq 10$; therefore, the triangular system (3) is very small and if one intended to solve it in parallel, it would yield a very poor performance (see, for instance, [7], [8] and [3]). The matrix R_1 may be made available in all processors at the end of PQR; this would allow for an easy solution of (3) by all processors involved.

Before applying the PQR algorithm we copy the factoring matrix into a matrix R; the algorithm then proceeds in two parts. At first, every processor performs a *local* Householder factorization, obtaining as a result a triangular block of order n; the Householder reflectors are stored in the strictly lower triangular portion of R. This part of the algorithm is perfectly parallel and has a *flop* count of $(2\hat{m}n+\hat{m}+n)\gamma$, where γ is the time to perform either a multiplication or addition.

Now one must annihilate all the triangular blocks but the first; these are stored in processors $1, 2, \ldots, p-1$. The key idea here is to note that those blocks *do not* need to be annihilated in an orderly fashion, say $p-1$ with $p-2$; then $p-2$ with $p-3$, etc. In fact, it is possible to perform this elimination phase using pairs of blocks, as if one were traversing a binary tree upwards; all the annihilations at the same level in that tree can be made in parallel, independently. This is commonly called a *reduction* operation in parallel computing terminology.

In this latter phase we use Givens's rotations to update the topmost triangular block of a given pair of processors, each processor exchanging with the other its triangular block, and applying the rotations locally. The sine/co-sine values are stored using a single value τ as given in [6, pp. 202]; between each pair of processors, there will be $(n^2+n)/2$ τ elements computed; they are stored in an array, to be used later for computing Q. The total operation cost for this phase is $\lceil \log_2 p \rceil (T_c((n^2+n)/2)+3(n^2+n)\gamma$, where $T_c(w)=\alpha+w\beta$ is the time taken to transfer w words between two processors, α being the latency and β the rate of transfer. Figure 1 gives an example of the workings of the PQR algorithm for a matrix with 8 rows and 2 columns, partitioned across 4 processors.

The pairs of processors used in the second part of the PQR algorithm are given by simply grouping together, initially, a processor and its neighbour, e.g. $(0,1)$, $(2,3)$, …, $(p-2, p-1)$; then in the next recursion level we group the leftmost processor of a pair, $(0,2)$, $(4,6)$, …, and proceed in this fashion until the last pair is $(0, \lfloor p/2 \rfloor)$, with the result that the triangular block stored in processor 0 contains the actual triangular matrix R_1 in the QR factorization. If

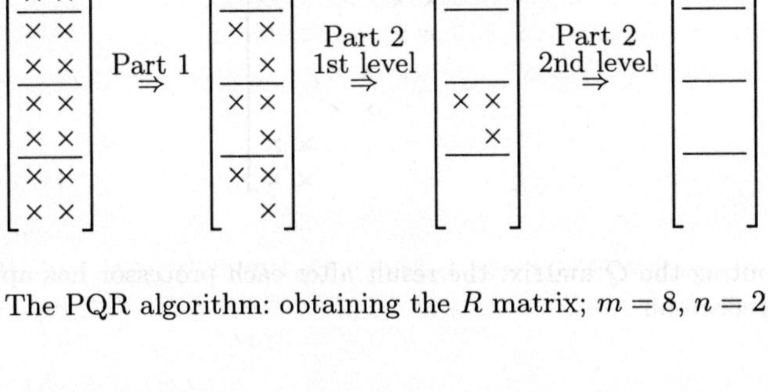

Fig. 1. The PQR algorithm: obtaining the R matrix; $m=8$, $n=2$, $p=4$.

the number of active processors in each level of recursion is not even, we simply add another pair including it and another suitable processor. We note that other groupings could be used to take advantage of a given topology of a network of processors, minimizing the length of the path needed for the messages to traverse.

1.2 Obtaining Q

The orthogonal matrix Q (or some of its first columns) is obtained after A is triangularized. Again, Q is partitioned by rows, in the same way as A was partitioned. The computation of Q is done in two phases, using backward accumulation. The Householder reflectors, which are stored in each processor in the strictly lower portion of R, are applied to Q. As they were computed for each block $\hat{m} \times n$ of R, the overall effect of this part is that Q ends up with a very definite block structure, as shown in Figure 2. Note the relationship between the sizes of the blocks and the number of rows of A stored in each processor (see the end of the last section).

The Givens's rotations are now to be applied to Q, again pairing two processors as was done for triangularizing A. For a pair (s,t), $s < t$ (storing rows $[i_s, f_s]$ and $[i_t, f_t]$ of A), this will lead to $f_t - i_t + 1$ rows being added, *below* the Q block on s, to the first m columns of that block (therefore, due to the partitioning of Q, these new elements will be stored in t); and, conversely, $f_s - i_s + 1$ rows will be added, *above* the Q block on t, to the first m columns of that block (but these will be stored in s). Of course, those elements of Q already stored in a processor will be modified when applying the rotations. Figure 3-(a) shows Q as a result of the pairs $(0,1)$ and $(2,3)$ having applied their Givens's rotations.

Following this scheme, consider the case when we move upwards one level on the recursion tree and have, now, to operate on a pair (s,u), having previously applied the rotations to pairs (s,t) and (u,v). Then, on processor s, the number of rows created will be $f_v - i_u + 1$ on the first m columns of the Q block respective to s; and there will be $f_t - i_s + 1$ new rows on the first m columns of the block corresponding to u on Q.

$$\begin{bmatrix} \times & \times & & & & & & \\ \times & \times & & & & & & \\ & & \times & \times & & & & \\ & & \times & \times & & & & \\ & & & & \times & \times & & \\ & & & & \times & \times & & \\ & & & & & & \times & \times \\ & & & & & & \times & \times \end{bmatrix}$$

Fig. 2. Computing the Q matrix: the result after each processor has applied its own Householder reflectors.

$$
(a)\;\begin{array}{r} i_s \rightarrow \\ f_s \rightarrow \\ i_t \rightarrow \\ f_t \rightarrow \\ i_u \rightarrow \\ f_u \rightarrow \\ i_v \rightarrow \\ f_v \rightarrow \end{array}
\begin{bmatrix}
\times & \times & \diamond & \diamond & & & & \\
\times & \times & \diamond & \diamond & & & & \\
\diamond & \diamond & \times & \times & & & & \\
\diamond & \diamond & \times & \times & & & & \\
& & & & \times & \times & \diamond & \diamond \\
& & & & \times & \times & \diamond & \diamond \\
& & & & \diamond & \diamond & \times & \times \\
& & & & \diamond & \diamond & \times & \times
\end{bmatrix},
\quad (b)\;
\begin{bmatrix}
\times & \times & \diamond & \diamond & + & + & & \\
\times & \times & \diamond & \diamond & + & + & & \\
\diamond & \diamond & \times & \times & + & + & & \\
\diamond & \diamond & \times & \times & + & + & & \\
+ & + & & & \times & \times & \diamond & \diamond \\
+ & + & & & \times & \times & \diamond & \diamond \\
+ & + & & & \diamond & \diamond & \times & \times \\
+ & + & & & \diamond & \diamond & \times & \times
\end{bmatrix}
$$

Fig. 3. Computing the Q matrix: (a) Q after the pairs of processors $(s,t) = (0,1)$ and $(u,v) = (2,3)$ have applied Givens's rotations to their columns of Q; the symbol $\diamond$ indicates those elements of Q created as a result. (b) Q after the pair of processors $(s,u) = (0,2)$ have applied Givens's rotations to their columns of Q, where those elements of Q created during the process are indicated by $+$.

When we reach the root of the recursion tree, we will have the pair $(0, \lfloor p/2 \rfloor)$. Therefore, if we proceed with the above procedure, the first m columns of Q will be filled-in completely, as in Figure 3-(b), where we show Q after the pair $(0,2)$ applies their rotations to Q. The total cost of computing Q, simultaneously in each processor, is $(4\,(\hat{m}^2 n - \hat{m}n^2 + n^3/3) + 3\hat{m}(n^2+n))\gamma$.

Now, if $H^{(q)} = H_1^{(q)} H_2^{(q)} \ldots H_m^{(q)}$ is the matrix resulting from applying all necessary Householder transformations to triangularize a block of A stored in processor q; and $G = G_1 G_2 \ldots G_h$, where $h = \lceil \log_2 p \rceil$ is the number of levels of recursion in the tree, then $Q = H^{(1)} H^{(2)} \ldots H^{(p)} G$.

The PQR algorithm is outlined below, according to the previous description. Note that step 1 is completely parallel, without any communication whatsoever between the processors; steps 2 to 5 entail communication between the processors, with messages of *constant size* $(n^2+n)/2$ words (single- or double-precision). Step 6 does not require communication as the Givens's rotations are applied with the sine/co-sine pairs already computed in steps 2 to 5.

ALGORITHM PQR:

1. *Every processor triangularizes its block* $A_{\hat{m}\times n}$ *using Householder reflections;*
2. *Compute pairs of processors* (s,t)*;*
3. *for all pairs* (s,t) *along the reduction tree do*
4. *processor s receive triangular block from t (of size n);*
5. *processor s retriangularize its own block using the block from t (as if the latter had been annihilated);*

   endfor
6. *Compute Q as in §1.2.*

1.3 The PQR Algorithm and the Least-Squares Problem

If one wishes to solve (1), then the solution to (3) must be obtained. Since B is $m \times k$, we need only the first m columns of Q to compute $Q^T B$, whose rows are partitioned across the processors; this product can be easily computed then.

We note that while in our explanation we have shown the full matrix Q, the actual implementation of the algorithm stores only the number of columns of Q (in each block of rows per processor) needed by the application, plus a small number of auxiliary arrays. Of course, if one needs the full Q matrix, it may not fit in the memory available; in this case, the above accumulation of Q must be modified to compute the action of the individual transformation matrices into a vector or a matrix.

2 The Rank-Revealing *QR* Algorithm

The PQR algorithm is used inside a parallel version of the RRQR algorithm by Golub and Van Loan [6, pp. 233-236], called PRRQR, which is also written using the SPMD model. This RRQR algorithm is easily parallelizable, requiring just parallel inner-products; we do not require a parallel swapping procedure for the columns, since the matrix to be factored is partitioned by rows (therefore, all processors have the same columns and any column swappings are made locally in every processor simultaneously).

A change made to that algorithm is that the QR factorization – performed by PQR – is done only once. After PQR is completed, the matrix R_1, which ends up in processor 0, is broadcast to all other processors. Afterwards, if a column swapping is performed by the PRRQR algorithm, then only the upper $n \times n$ block needs to be re-triangularized (and Q must be updated); this re-triangularization is done using Givens's rotations. However, now all processors hold a copy of R_1 and their own blocks of rows of Q, allowing the *local* use of Givens's rotations to re-triangularize R_1 and update of Q. We note that just a few of the rows of R_1 will need to be updated and our implementation performs only the necessary work, keeping track of the rows changed. The algorithm is outlined below; communication occurs only in steps 1 (inside PQR) and 2 (a broadcast of a message $(n^2 + n)/2$ words long).

ALGORITHM PRRQR:

1. *Every processor calls PQR;*
2. *The root processor, 0, broadcasts R_1 to the other processors;*
3. *In every processor, independently:*
4. *for all columns of R_1*
5. *if near-dependency is detected, perform:*
6. *column swapping in R_1;*
7. *retriangularize its own copy of R_1;*
8. *update its block of rows of Q.*

3 Experiments

The PQR and RRQR algorithms were implemented using FORTRAN90 and MPI, and there are versions in real and complex arithmetic, each of which in single- and double-precision.

We have tested our double-precision implementations of both algorithms on a number of random matrices of different sizes. The tests were carried out on an IBM 9076 SP/2 parallel computer housed at the Brazilian *"Laboratório Nacional de Computação Científica, LNCC"* (National Scientific Computing Laboratory). The computer was not entirely dedicated to our testings and therefore there was some fluctuation in the traffic across its network.

Tables 1 and 2 show the running times (in seconds) and the respective speed-ups ($S_p = T_1/T_p$) of PQR and PRRQR for some large matrices, with $m \gg n$.

Table 1. Timings (in seconds) and speed-ups for PQR; the column labelled "LAPACK" shows the timings obtained with solving the same problem using LAPACK's DGEQPF/DORGQR routines.

m	n	LAPACK	$p=1$	$p=2$	$p=4$	$p=8$	$p=16$
20000	5	0.07	0.16	0.15	0.07	0.05	0.05
				1.07	2.29	3.20	3.20
	10	0.22	0.52	0.32	0.19	0.12	0.09
				1.63	2.74	4.33	5.78
40000	5	0.15	0.31	0.30	0.12	0.09	0.07
				1.03	2.58	3.44	4.43
	10	0.47	1.00	0.65	0.38	0.23	0.23
				1.54	2.63	4.35	4.35
80000	5	0.28	0.63	0.59	0.24	0.17	0.10
				1.07	2.63	3.71	6.30
	10	0.94	1.93	1.25	0.75	0.46	0.26
				1.54	2.57	4.20	7.42
200000	5	0.72	1.63	0.97	0.57	0.34	0.24
				1.68	2.86	4.79	6.79
	10	2.35	4.77	2.93	1.76	1.01	0.70
				1.63	2.71	4.72	6.81
400000	5	1.44	3.00	1.78	1.07	0.62	0.45
				1.69	2.80	4.84	6.67
	10	4.69	9.50	5.57	3.43	2.01	1.35
				1.71	2.77	4.73	7.04
800000	5	2.88	6.56	3.86	2.30	1.33	0.85
				1.70	2.85	4.93	7.72
	10	9.38	19.06	11.38	6.71	4.07	2.71
				1.67	2.84	4.68	7.03

Table 2. Timings (in seconds) and speed-ups for PRRQR.

m	n	$p=1$	$p=2$	$p=4$	$p=8$	$p=16$
20000	5	0.21	0.12	0.08	0.06	0.05
			1.75	2.63	3.50	4.20
	10	0.73	0.43	0.25	0.16	0.10
			1.70	2.92	4.56	7.30
40000	5	0.40	0.24	0.13	0.09	0.07
			1.67	3.08	4.44	5.71
	10	1.55	0.93	0.52	0.29	0.17
			1.67	2.98	5.34	9.12
80000	5	0.75	0.47	0.26	0.25	0.16
			1.60	2.88	3.00	4.69
	10	2.83	1.83	1.04	0.59	0.33
			1.55	2.72	4.80	8.58
200000	5	2.02	1.09	0.61	0.37	0.27
			1.85	3.31	5.46	7.48
	10	7.65	4.36	2.39	1.29	0.83
			1.75	3.20	5.93	9.22
400000	5	3.86	2.08	1.19	0.70	0.50
			1.86	3.24	5.51	7.72
	10	13.89	7.55	4.76	2.25	1.63
			1.84	2.92	6.16	8.52
800000	5	7.76	4.15	2.35	1.40	0.94
			1.87	3.30	5.54	8.22
	10	28.39	14.79	7.56	4.55	3.30
			1.92	3.76	6.24	8.60

Here T_1 is the time taken by a sequential implementation of either the QR or RRQR algorithm where the QR factorization is obtained via Householder reflections. On Table 1 we also show the timings obtained when solving the same problem with the LAPACK DGEQPF and DORGQR routines, on a single processor.

The entries on the tables show that up to four processors, the speed-up decreases for fixed m and increasing n; afterwards, it increases. When there are few processors available, the amount of work done in each processor during the second phase, to triangularize A and compute Q as a result, is considerable and this imposes a reduction in the speed-up for a larger n; also, increasing n leads to larger triangular blocks being exchanged between the processors.

However, as p increases, with fixed m and n, less rows are stored in each processor and the binary-tree reduction mechanism used in PQR allows for a greater level of parallelism, since more and more data exchanges are made si-

Table 3. Ratios T_H/T_p, T_G/T_p and speed-ups for PQR, with $m = 4000$.

	$n = 5$			$n = 10$		
p	T_H/T_p	T_G/T_p	S_p	T_H/T_p	T_G/T_p	S_p
2	0.98	0.02	1.94	0.96	0.04	1.93
4	0.92	0.08	3.74	0.86	0.14	3.49
8	0.80	0.20	6.64	0.70	0.30	5.90
16	0.63	0.37	10.85	0.49	0.51	8.47

multaneously and less work is done in each processor. The data confirms that both algorithms scale very well for a large number of processors, with PRRQR having a little better performance since the updates of Q and R_1 during any retriangularization caused by a column swapping is made locally, without communication between the processors.

Table 3 shows the ratios between the time taken for the first phase (Householder triangularization, perfectly parallel) and the total time (T_H/T_p); and also for the second phase (Givens's triangularization, with communication between the processors) and the total time (T_G/T_p), for two problems of size $m = 4000$ and $n = 5$, $n = 10$, using the PQR algorithm. It is interesting to note that the first phase is responsible for nearly all the execution time and it is the main reason for the good performance of the algorithm, as it does not involve any communication between the processors; also note that that the ratio for the first phase decreases with increasing p. However, as mentioned above, increasing n means that the second phase will have a larger contribution on the execution. As the time for this phase grows by a factor $C \log_2 p$, for some value of p both ratios will be approximately equal; the speed-up attained after this value of p will be less than $p/2$.

We should mention that we have tested our implementations for some situations where even if A has full rank, a block allocated to a processor may be defficient in rank (for instance, two or more of its columns may be linearly dependent; this has also been noted in [13]). The results obtained, with regards to those test matrices, have shown that our algorithms can cope with this problem.

4 Final Remarks

We have presented a new parallel algorithm, based on a reduction strategy, for computing the QR and rank-revealing QR factorizations of large matrices $A_{m \times n}$, $m \gg n$.

The performance obtained with our actual implementations in FORTRAN90 and MPI has been shown to present a good scalability when increasing the number of processors.

We would like to note that using PQR and RRQR in our parallel implementations of the Block GMRES method has shown to provide similar performance as that presented here.

Acknowledgements

The authors would like to thank LNCC for allowing our using its computing facilities. We also thank the referees for their very useful comments which have helped improve this paper.

References

1. C.H. Bischof and G. Quintana-Ortí. Computing rank-revealing QR factorizations of dense matrices. *ACM Transactions on Mathematical Software*, 24(2):226–253, June 1998.
2. T.F. Chan. Rank revealing QR factorizations. *Linear Algebra and its Applications*, 88/89:67–82, 1987.
3. R.D. da Cunha and T.R. Hopkins. The parallel solution of triangular systems of linear equations. In Durand, M. and El Dabaghi, F., editor, *High-Performance Computing II – Proceedings of the Second Symposium on High Performance Computing*, pages 245–256, Amsterdam, October 1991. North-Holland. Also as Report No. 86, Computing Laboratory, University of Kent at Canterbury, U.K.
4. E. Elmroth and F.G. Gustavson. Applying recursion to serial and parallel QR factorization leads to better performance. *IBM Journal of Research and Development*, 44(4):605–624, July 2000.
5. L.V. Foster. Rank and null space calculations using matrix decomposition without column interchanges. *Linear Algebra and its Applications*, 74:47–71, 1986.
6. G.H. Golub and C.F. Van Loan. *Matrix Computations.* Johns Hopkins University Press, Baltimore, 2nd edition, 1989.
7. M.T. Heath and C.H. Romine. Parallel solution of triangular systems on distributed-memory multiprocessors. *SIAM Journal of Scientific and Statistical Computing*, 9:558–588, 1988.
8. G. Li and T.F. Coleman. A new method for solving triangular systems on distributed-memory message-passing multiprocessors. *SIAM Journal of Scientific and Statistical Computing*, 10:382–396, 1989.
9. J.W.H. Liu. On general row merging schemes for sparse Givens transformations. *SIAM Journal of Scientific and Statistical Computing*, 7:1190–1211, 1986.
10. F.T. Luk. A rotation method for computing the QR-decomposition. *SIAM Journal on Scientific and Statistical Computing*, 7(2):452–459, 1986.
11. D.J. Pierce and J.G. Lewis. Sparse multifrontal rank revealing QR factorization. *SIAM Journal on Matrix Analysis and Applications*, 18(1):159–180, 1997.
12. Youcef Saad. *Iterative methods for sparse linear systems.* PWS Publishing Company, Boston, 1995.
13. R.B. Sidje. Alternatives for parallel Krylov subspace basis computation. *Numerical Linear Algebra with Applications*, 4(4):305–331, 1997.

Solving Large Sparse Lyapunov Equations on Parallel Computers*

José M. Badía[1], Peter Benner[2], Rafael Mayo[1], and Enrique S. Quintana-Ortí[1]

[1] Depto. de Ingeniería y Ciencia de Computadores, Universidad Jaume I, 12080–Castellón, Spain,
{badia,mayo,quintana}@icc.uji.es,
Tel.: +34-964-728257, Fax: +34-964-728486

[2] Institut für Mathematik, Technische Universität Berlin, D-10623 Berlin, Germany,
benner@math.tu-berlin.de,
Tel.: +30-314-28035, Fax: +30-314-79706

Abstract. This paper describes the parallelization of the low-rank ADI iteration for the solution of large-scale, sparse Lyapunov equations. The only relevant operations involved in the method are matrix-vector products and the solution of linear systems. Experimental results on a cluster, using the SuperLU library, show the performance of this approach.

Key words: Lyapunov equations, ADI iteration, low-rank approximation, sparse linear systems.

1 Introduction

The *Lyapunov equation*

$$AX + XA^T + BB^T = 0, \tag{1}$$

where $A, X \in \mathbb{R}^{n\times n}$, $B \in \mathbb{R}^{n\times m}$, and X is the sought-after solution, plays a fundamental role in control theory; see, e.g., [1]. In this paper, we assume that the spectrum of A lies in the open left half plane. Under this assumption, there exists a unique solution, $X = X^T \geq 0$ [8], which can be factorized as $X = SS^T$; here, S is known as the "Cholesky" factor of the solution. Usually $m \ll n$ and often, S is of low (column) rank [9].

Numerical methods for the solution of small-scale, dense Lyapunov equations are introduced in [2, 5]. These methods require the application of the QR algorithm, resulting in a parallelism poorly attractive [6]. The use of the matrix sign function [11] allows the solution of moderate-scale dense Lyapunov equations (n up to a few thousands) on parallel computers [3]. However, none of these algorithms are appropriate for large-scale sparse Lyapunov equations.

In this paper, we follow an approach based on the LR-ADI iteration [10], which provides a low-rank approximation of the Cholesky factor of the solution. Many large-scale applications, e.g. in model reduction and optimal control,

* Supported by the Fundació Caixa-Castelló/Bancaixa PI-1B2001-14.

B. Monien and R. Feldmann (Eds.): Euro-Par 2002, LNCS 2400, pp. 687–690.

employ this approximation rather than the solution. The LR-ADI solver only requires matrix operations such as the matrix-vector product and the solution of linear systems. The parallelization of these matrix operations on distributed-memory computers has been largely studied in the literature and several libraries are currently available like, e.g., SuperLU and MUMPS.

The rest of the paper is structured as follows. In Section 2 we briefly review the LR-ADI solver for large sparse Lyapunov equations. Details on the implementation and the parallelization are given in Section 3. Finally, experimental results and concluding remarks follow, respectively, in Sections 4 and 5.

2 LR-ADI Solver for Large Sparse Lyapunov Equations

The *cyclic low-rank alternating direction implicit* (LR-ADI) Lyapunov solver [10] benefits from the usual low-rank property of matrix B to provide low-rank approximations to the Cholesky factor of X. This iterative algorithm also includes a heuristic to determine a set of "shift" parameters to accelerate the convergence.

Specifically, given an "l–cyclic" set of complex shift parameters $\{p_1, p_2, \ldots\}$, $p_k = a_k + b_k i$, $p_k = p_{k+l}$, the LR-ADI iteration can be formulated as follows:

$$\begin{aligned} V_0 &= (A + p_1 I_n)^{-1} B, & S_0 &= \sqrt{-2\, a_1}\, V_0, \\ V_{k+1} &= V_k - \delta_k (A + p_{k+1} I_n)^{-1} V_k, & \delta_k &= p_{k+1} + \overline{p_k}, \\ S_{k+1} &= [S_k\ ,\ \gamma_k V_{k+1}]\,, & \gamma_k &= \sqrt{a_{k+1}/a_k}, \end{aligned} \tag{2}$$

where I_n denotes the identity matrix of order n. On convergence, after $\tilde{k}$ iterations, a low-rank matrix $\tilde{S}$ of order $n \times \tilde{k}m$ is computed such that $\tilde{S}\tilde{S}^T$ approximates X. Matrix $\tilde{S}$ can be employed in many situations where the Cholesky factor S is required. Further details on the LR-ADI iteration are given in [10].

The only matrix operation required in the LR-ADI iteration is the solution of linear systems with a sparse coefficient matrix shifted by a certain (complex) scalar.

The performance of the iteration strongly depends on the selection of the shift parameters; see [10]. A heuristic procedure is proposed in [10] which employs approximations of certain eigenvalues of A. The procedure is based on an Arnoldi iteration with A and A^{-1}, and therefore only requires matrix-vector products and the solution of linear systems, respectively.

3 Implementation and Parallelization

The LR-ADI iteration (2) involves the same coefficient matrix in iterations k and $k+l$. Thus, when more than l iterations are required and sufficient storage space is available, a large computational cost is saved by computing factorizations of $(A+p_i I_n)$, $i = 1, 2, \ldots, l$, in advance and using these factors in the solution of the subsequent linear systems. This approach benefits from the use of direct solvers; besides, as all matrices have the same sparsity structure, the preliminary analysis phase, common in direct sparse solvers, needs to be performed only once.

In practice, the LR-ADI iteration produces a sequence of matrices V_k of decreasing norm [9]. A practical criterion for detecting the convergence of the iteration is therefore to stop when $\|V_k\|$ is small.

Although A is real, the iteration may require complex arithmetic in case any of the shift parameters is complex. For symmetric matrices all the shifts should be chosen to be real, using a Lanczos-based procedure, and the iteration only employs real arithmetic.

The parallelization of the solver requires parallel routines for the computation of the matrix-vector product, the solution of linear systems involving sparse (complex) matrices, and some other minor operations. Here we employ the SuperLU library [4] for the solution of linear systems. In the current version of SuperLU both the coefficient and the right-hand side matrices have to be completely stored in all processes (processors). We take advantage of this replication to easily implement an efficient parallel matrix-vector product. Future extensions of our parallel solvers will benefit from employing parallel kernels from other libraries like MUMPS, MCSPARSE, ScaLAPACK, etc.

4 Numerical Experiments

All the experiments presented in this section were performed on a cluster of Intel Pentium-II processors (300MHz, 128 MBytes of RAM), connected with a *Fast Ethernet* switch, using IEEE double-precision floating-point arithmetic. The BLAS routine `DGEMV` for the matrix-vector product achieved around 46 Mflops (millions of flops/sec.) on this architecture.

We report experimental results for the solution of large-scale Lyapunov equations (1) where A is tridiagonal stable matrix with random entries and B is a random vector ($m = 1$).

Figure 1 reports the execution time of the first LR-ADI iteration (2) using n_p processors. This iteration requires the solution of a complex linear system of order n with a single right-hand side, and several other minor computations (e.g., convergence criterion). A profile of the parallel algorithm showed that, when 4 processors were used on a problem of order n=150,000, the iteration spent roughly 75% in the LU factorization (including negligible times for equilibration, row and column permutations and symbolic factorization; almost half of this time was employed to distribute the matrices). The solution of the linear system from the LU factors used 24% of the time (including one iterative refinement step); the remaining time was spent in minor computations.

The figure also shows the convergence rate, measured as $\|V_k\|_1$, when l different shifts are used in the LR-ADI iteration and n=150,000. Actually, $k_p + k_m$ shifts are computed: $k_p = 2l$ are obtained using the Arnoldi iteration on A and $k_m = 2l$ with the Arnoldi iteration on A^{-1}. A selection procedure determines then the "best" l shifts to use in the iteration.

Solving a Lyapunov equation of order n=150,000 on 4 processors, with $k_p = k_m = 2l = 12$, required about 35 minutes; 39% of this time was spent in the computation of the shifts and the remaining 61% in the LR-ADI iteration.

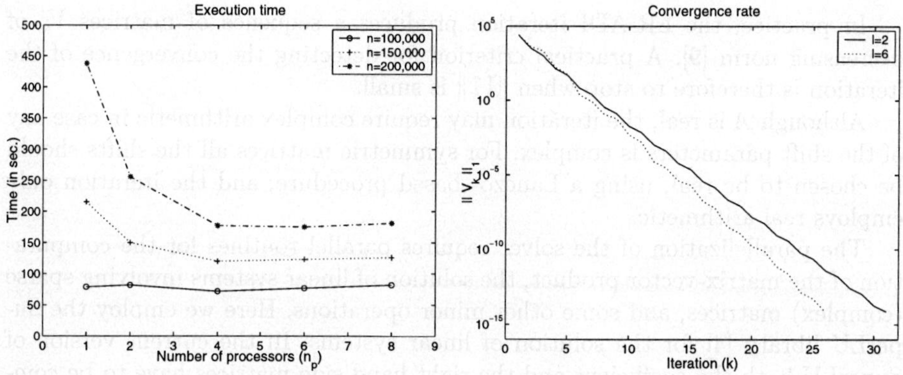

Fig. 1. Execution time of a single LR-ADI iteration (left) and convergence rate (right).

5 Concluding Remarks

We have described the parallelization of a numerical solver, based on the LR-ADI iteration, for large-scale, sparse Lyapunov equations on a cluster. The algorithm benefits from the availability and efficiency of parallel kernels for the matrix-vector product and the solution of sparse linear systems using direct methods.

References

1. B.D.O. Anderson and J.B. Moore. *Optimal Control – Linear Quadratic Methods.* Prentice-Hall, Englewood Cliffs, NJ, 1990.
2. R.H. Bartels and G.W. Stewart. Solution of the matrix equation $AX + XB = C$: Algorithm 432. *Comm. ACM*, 15:820–826, 1972.
3. P. Benner, J.M. Claver, and E.S. Quintana-Ortí. Parallel distributed solvers for large stable generalized Lyapunov equations. *Parallel Proc. Lett.*, 9:147–158, 1999.
4. J.W. Demmel, J.R. Gilbert, and X.S. Li. *SuperLU User's Guide*, 1999. Available from `http://www.nersc.gov/~xiaoye/SuperLU`.
5. S.J. Hammarling. Numerical solution of the stable, non-negative definite Lyapunov equation. *IMA J. Numer. Anal.*, 2:303–323, 1982.
6. G. Henry and R. van de Geijn. Parallelizing the QR algorithm for the unsymmetric algebraic eigenvalue problem: myths and reality. *SIAM J. Sci. Comput.*, 17:870–883, 1997.
7. A.S. Hodel, B. Tenison, and K.R. Poolla. Numerical solution of the Lyapunov equation by approximate power iteration. *Linear Algebra Appl.*, 236:205–230, 1996.
8. P. Lancaster and M. Tismenetsky. *The Theory of Matrices.* Academic Press, Orlando, 2nd edition, 1985.
9. T. Penzl. Eigenvalue decay bounds for solutions of Lyapunov equations: the symmetric case. *Sys. Control Lett.*, 40(2):139–144, 2000.
10. T. Penzl. A cyclic low rank Smith method for large sparse Lyapunov equations. *SIAM J. Sci. Comput.*, 21(4):1401–1418, 2000.
11. J.D. Roberts. Linear model reduction and solution of the algebraic Riccati equation by use of the sign function. *Internat. J. Control*, 32:677–687, 1980.

A Blocking Algorithm for Parallel 1-D FFT on Clusters of PCs

Daisuke Takahashi, Taisuke Boku, and Mitsuhisa Sato

Center for Computational Physics, Institute of Information Sciences and Electronics, University of Tsukuba, 1-1-1 Tennodai, Tsukuba-shi, Ibaraki 305-8577, Japan,
{daisuke,taisuke,msato}@is.tsukuba.ac.jp

Abstract. In this paper, we propose a blocking algorithm for a parallel one-dimensional fast Fourier transform (FFT) on clusters of PCs. Our proposed parallel FFT algorithm is based on the six-step FFT algorithm. The six-step FFT algorithm can be altered into a block nine-step FFT algorithm to reduce the number of cache misses. The block nine-step FFT algorithm improves performance by utilizing the cache memory effectively. We use the block nine-step FFT algorithm to design the parallel one-dimensional FFT algorithm. In our proposed parallel FFT algorithm, since we use cyclic distribution, all-to-all communication is required only once. Moreover, the input data and output data are both can be given in natural order. We successfully achieved performance of over 1.3 GFLOPS on an 8-node dual Pentium III 1 GHz PC SMP cluster.

1 Introduction

The fast Fourier transform (FFT) [1] is an algorithm widely used today in science and engineering. Parallel FFT algorithms on distributed-memory parallel computers have been well studied [2–6].

Many FFT algorithms work well when data sets fit into a cache. When a problem size exceeds the cache size, however, the performance of these FFT algorithms decreases dramatically. The key issue of the design for large FFTs is to minimize the number of cache misses.

In this paper, we propose a blocking algorithm for a parallel one-dimensional FFT algorithm on clusters of PCs.

Our proposed parallel one-dimensional FFT algorithm is based on the six-step FFT algorithm [7, 8]. The six-step FFT algorithm requires two multicolumn FFTs and three data transpositions. The three transpose steps typically are the chief bottlenecks in cache-based processors.

Some previously presented six-step FFT algorithms [8, 9] separate the multicolumn FFTs from the transpositions.

Taking the opposite approach, we combine the multicolumn FFTs and transpositions to reduce the number of cache misses, and we modify the six-step FFT algorithm to reuse data in the cache memory [10]. We call this a block six-step FFT algorithm. The block six-step FFT algorithm can be altered into a block

B. Monien and R. Feldmann (Eds.): Euro-Par 2002, LNCS 2400, pp. 691–700.

nine-step FFT algorithm to reduce the number of cache misses. We use the block nine-step FFT algorithm to design the parallel one-dimensional FFT algorithm.

We have implemented the block nine-step FFT-based parallel one-dimensional FFT algorithm on an 8-node dual Pentium III 1 GHz PC SMP cluster, and we report the performance in this paper.

The rest of the paper is organized as follows. Section 2 describes the six-step FFT algorithm. In section 3, we propose a nine-step FFT algorithm. Section 4, we propose a block nine-step FFT algorithm used for problems that exceed the cache size. Section 5, we propose a parallel FFT algorithm based on the block nine-step FFT. Section 6 describes the in-cache FFT algorithm used for problems that fit into a data cache. Section 7 gives performance results. In section 8, we present some concluding remarks.

2 The Six-Step FFT

The discrete Fourier transform (DFT) is given by

$$y_k = \sum_{j=0}^{n-1} x_j \omega_n^{jk}, \quad 0 \le k \le n-1, \tag{1}$$

where $\omega_n = e^{-2\pi i/n}$ and $i = \sqrt{-1}$.

If n has factors n_1 and n_2 ($n = n_1 \times n_2$), then the indices j and k can be expressed as:

$$j = j_1 + j_2 n_1, \quad k = k_2 + k_1 n_2. \tag{2}$$

We can define x and y as two-dimensional arrays (in Fortran notation):

$$x_j = x(j_1, j_2), \quad 0 \le j_1 \le n_1 - 1, \quad 0 \le j_2 \le n_2 - 1, \tag{3}$$

$$y_k = y(k_2, k_1), \quad 0 \le k_1 \le n_1 - 1, \quad 0 \le k_2 \le n_2 - 1. \tag{4}$$

Substituting the indices j and k in equation (1) with those in equation (2), and using the relation of $n = n_1 \times n_2$, we can derive the following equation:

$$y(k_2, k_1) = \sum_{j_1=0}^{n_1-1} \sum_{j_2=0}^{n_2-1} x(j_1, j_2) \omega_{n_2}^{j_2 k_2} \omega_{n_1 n_2}^{j_1 k_2} \omega_{n_1}^{j_1 k_1}. \tag{5}$$

This derivation leads to the following six-step FFT algorithm [7, 8]:

Step 1: Transpose

$$x_1(j_2, j_1) = x(j_1, j_2).$$

Step 2: n_1 individual n_2-point multicolumn FFTs

$$x_2(k_2, j_1) = \sum_{j_2=0}^{n_2-1} x_1(j_2, j_1) \omega_{n_2}^{j_2 k_2}.$$

Step 3: Twiddle-factor multiplication

$$x_3(k_2, j_1) = x_2(k_2, j_1)\omega_{n_1 n_2}^{j_1 k_2}.$$

Step 4: Transpose

$$x_4(j_1, k_2) = x_3(k_2, j_1).$$

Step 5: n_2 individual n_1-point multicolumn FFTs

$$x_5(k_1, k_2) = \sum_{j_1=0}^{n_1-1} x_4(j_1, k_2)\omega_{n_1}^{j_1 k_1}.$$

Step 6: Transpose

$$y(k_2, k_1) = x_5(k_1, k_2).$$

The distinctive features of the six-step FFT algorithm can be summarized as:

- Two multicolumn FFTs are performed, one in step 2 and the other in step 5. Each column FFT is small enough to fit into the data cache.
- The six-step FFT algorithm has three transpose steps, which typically are the chief bottlenecks in cache-based processors.

In order to reduce the number of cache misses, the block six-step FFT algorithm has been proposed [10].

3 A Nine-Step FFT Algorithm

We can extend the six-step FFT algorithm in another way into a three-dimensional formulation. If n has factors n_1, n_2 and n_3 ($n = n_1 n_2 n_3$), then the indices j and k can be expressed as:

$$\begin{aligned} j &= j_1 + j_2 n_1 + j_3 n_1 n_2, \\ k &= k_3 + k_2 n_3 + k_1 n_2 n_3. \end{aligned} \tag{6}$$

We can define x and y as three-dimensional arrays (in Fortran notation), e.g.,

$$\begin{aligned} x_j = x(j_1, j_2, j_3), &\quad 0 \le j_1 \le n_1 - 1, \\ 0 \le j_2 \le n_2 - 1, &\quad 0 \le j_3 \le n_3 - 1, \end{aligned} \tag{7}$$

$$\begin{aligned} y_k = y(k_3, k_2, k_1), &\quad 0 \le k_1 \le n_1 - 1, \\ 0 \le k_2 \le n_2 - 1, &\quad 0 \le k_3 \le n_3 - 1. \end{aligned} \tag{8}$$

Substituting the indices j and k in equation (1) by those in equation (6) and using the relation of $n = n_1 n_2 n_3$, we can derive the following equation:

$$\begin{aligned} y(k_3, k_2, k_1) = \sum_{j_1=0}^{n_1-1} \sum_{j_2=0}^{n_2-1} \sum_{j_3=0}^{n_3-1} & x(j_1, j_2, j_3)\omega_{n_3}^{j_3 k_3} \\ & \omega_{n_2 n_3}^{j_2 k_3} \omega_{n_2}^{j_2 k_2} \omega_{n}^{j_1 k_3} \omega_{n_1 n_2}^{j_1 k_2} \omega_{n_1}^{j_1 k_1}. \end{aligned} \tag{9}$$

This derivation leads to the following nine-step FFT:

Step 1: Transpose

$$x_1(j_3, j_1, j_2) = x(j_1, j_2, j_3).$$

Step 2: $n_1 n_2$ individual n_3-point multicolumn FFTs

$$x_2(k_3, j_1, j_2) = \sum_{j_3=0}^{n_3-1} x_1(j_3, j_1, j_2)\omega_{n_3}^{j_3 k_3}.$$

Step 3: Twiddle-factor multiplication

$$x_3(k_3, j_1, j_2) = x_2(k_3, j_1, j_2)\omega_{n_2 n_3}^{j_2 k_3}.$$

Step 4: Transpose

$$x_4(j_2, j_1, k_3) = x_3(k_3, j_1, j_2).$$

Step 5: $n_1 n_3$ individual n_2-point multicolumn FFTs

$$x_5(k_2, j_1, k_3) = \sum_{j_2=0}^{n_2-1} x_4(j_2, j_1, k_3)\omega_{n_2}^{j_2 k_2}.$$

Step 6: Twiddle-factor multiplication

$$x_6(k_2, j_1, k_3) = x_5(k_2, j_1, k_3)\omega_{n}^{j_1 k_3}\omega_{n_1 n_2}^{j_1 k_2}.$$

Step 7: Transpose

$$x_7(j_1, k_2, k_3) = x_6(k_2, j_1, k_3).$$

Step 8: $n_2 n_3$ individual n_1-point multicolumn FFTs

$$x_8(k_1, k_2, k_3) = \sum_{j_1=0}^{n_1-1} x_7(j_1, k_2, k_3)\omega_{n_1}^{j_1 k_1}.$$

Step 9: Transpose

$$y(k_3, k_2, k_1) = x_8(k_1, k_2, k_3).$$

The distinctive features of the nine-step FFT algorithm can be summarized as:

– Three multicolumn FFTs are performed in steps 2, 5 and 8. The locality of the memory reference in the multicolumn FFT is high. Therefore, the nine-step FFT is suitable for cache-based processors because of the high performance which can be obtained with high hit rates in the cache memory.
– The matrix transposition takes place four times.

For extremely large FFTs, we should switch to a four-dimensional formulation and higher approaches.

4 A Block Nine-Step FFT Algorithm

We combine the multicolumn FFTs and transpositions to reduce the number of cache misses, and we modify the nine-step FFT algorithm to reuse data in the

```
COMPLEX*16 X(N1,N2,N3),Y(N3,N2,N1)
COMPLEX*16 U2(N3,N2),U3(N1,N2,N3)
COMPLEX*16 YWORK(N2+NP,NB),ZWORK(N3+NP,NB)
DO J=1,N2
  DO II=1,N1,NB
    DO KK=1,N3,NB
      DO I=II,II+NB-1
        DO K=KK,KK+NB-1
          ZWORK(K,I-II+1)=X(I,J,K)
        END DO
      END DO
    END DO
    DO I=1,NB
      CALL IN_CACHE_FFT(ZWORK(1,I),N3)
    END DO
    DO K=1,N3
      DO I=II,II+NB-1
        X(I,J,K)=ZWORK(K,I-II+1)*U2(K,J)
      END DO
    END DO
  END DO
END DO
DO K=1,N3
  DO II=1,N1,NB
    DO JJ=1,N2,NB
      DO I=II,II+NB-1
        DO J=JJ,JJ+NB-1
          YWORK(J,I-II+1)=X(I,J,K)
        END DO
        END DO
      END DO
      DO I=1,NB
        CALL IN_CACHE_FFT(YWORK(1,I),N2)
      END DO
      DO J=1,N2
        DO I=II,II+NB-1
          X(I,J,K)=YWORK(J,I-II+1)*U3(I,J,K)
        END DO
      END DO
    END DO
    DO J=1,N2
      CALL IN_CACHE_FFT(X(1,J,K),N1)
    END DO
  END DO
  DO II=1,N1,NB
    DO JJ=1,N2,NB
      DO KK=1,N3,NB
        DO I=II,II+NB-1
          DO J=JJ,JJ+NB-1
            DO K=KK,KK+NB-1
              Y(K,J,I)=X(I,J,K)
            END DO
          END DO
        END DO
      END DO
    END DO
  END DO
END DO
```

Fig. 1. A Block Nine-Step FFT Algorithm

cache memory. As in the nine-step FFT above, it is assumed in the following that $n = n_1 n_2 n_3$ and that n_b is the block size. We assume that each processor has a multi-level cache memory. A block nine-step FFT algorithm can be stated as follows.

1. Consider the data in main memory as an $n_1 \times n_2 \times n_3$ complex array X. Fetch and transpose the data n_b rows at a time into an $n_3 \times n_b$ work array ZWORK. The $n_3 \times n_b$ work array ZWORK fits into the L2 cache.
2. For each n_b columns, perform n_b individual n_3-point multicolumn FFTs on the $n_3 \times n_b$ array ZWORK in the L2 cache. Each column FFT fits into the L1 data cache.
3. Multiply the resulting data in each of the $n_3 \times n_b$ complex matrices by the twiddle factors U2. Then transpose each of the resulting $n_3 \times n_b$ matrices, and return the resulting n_b rows to the same locations in the main memory from which they were fetched.
4. Fetch and transpose the data n_b rows at a time into an $n_2 \times n_b$ work array YWORK.
5. Perform n_b individual n_2-point multicolumn FFTs on the $n_2 \times n_b$ work array YWORK in the L2 cache. Each column FFT fits into the L1 data cache.
6. Multiply the resulting data in each of the $n_2 \times n_b$ complex matrices by the twiddle factors U3. Then transpose each of the resulting $n_2 \times n_b$ matrices, and return the resulting n_b rows to the same locations in the main memory from which they were fetched.

7. Perform $n_3 n_2$ individual n_1-point multicolumn FFTs on the $n_1 \times n_2 \times n_3$ array X. Each column FFT fits into the L1 data cache.
8. Transpose and store the resulting data on an $n_3 \times n_2 \times n_1$ complex matrix.

We note that this algorithm is a *three-pass* algorithm. Fig. 1 gives the pseudo-code for this block nine-step FFT algorithm. Here the twiddle factors $\omega_{n_2 n_3}^{j_2 k_3}$ and $\omega_n^{j_1 k_3} \omega_{n_1 n_2}^{j_1 k_2}$ are stored in arrays U2 and U3, respectively. The arrays YWORK and ZWORK are the work arrays. The parameters NB and NP are the blocking parameter and the padding parameter, respectively. If an out-of-place algorithm (e.g. Stockham autosort algorithm [11]) is used for the individual FFTs, the additional scratch requirement for performing the individual FFTs in steps 2, 5 and 7 is $O(n^{1/3})$ at most.

If we do not require an ordered transform, the $n_3 \times n_2 \times n_1$ complex matrix (the array Y in Fig. 1) and the transposition in step 8 can be omitted.

5 Parallel FFT Algorithm Based on the Block Nine-Step FFT

We can adopt the idea of the block nine-step FFT as described in section 4.

Let N have factors N_1, N_2 and N_3 ($N = N_1 \times N_2 \times N_3$). The original one-dimensional array $x(N)$ can be defined as a three-dimensional array $x(N_1, N_2, N_3)$ (in Fortran notation). On a distributed-memory parallel computer which has P nodes, the array $x(N_1, N_2, N_3)$ is distributed along the first dimension N_1. If N_1 is divisible by P, each node has distributed data of size N/P. We introduce the notation $\hat{N}_r \equiv N_r/P$ and we denote the corresponding index as $\hat{J}_r$ which indicates that the data along J_r are distributed across all P nodes. Here, we use the subscript r to indicate that this index belongs to dimension r. The distributed array is represented as $\hat{x}(\hat{N}_1, N_2, N_3)$. At node m, the local index $\hat{J}_r(m)$ corresponds to the global index as the *cyclic* distribution:

$$J_r = \hat{J}_r(m) \times P + m, \quad 0 \le m \le P-1, \quad 1 \le r \le 3. \tag{10}$$

To illustrate the all-to-all communication it is convenient to decompose N_i into two dimensions $\tilde{N}_i$ and P_i, where $\tilde{N}_i \equiv N_i/P_i$. Although P_i is the same as P, we are using the subscript i to indicate that this index belongs to dimension i.

Starting with the initial data $\hat{x}(\hat{N}_1, N_2, N_3)$, the nine-step FFT-based parallel FFT can be performed according to the following steps:

Step 1: Transpose

$$\hat{x}_1(J_3, \hat{J}_1, J_2) = \hat{x}(\hat{J}_1, J_2, J_3).$$

Step 2: $(N_1/P) \cdot N_2$ individual N_3-point multicolumn FFTs

$$\hat{x}_2(K_3, \hat{J}_1, J_2) = \sum_{J_3=0}^{N_3-1} \hat{x}_1(J_3, \hat{J}_1, J_2) \omega_{N_3}^{J_3 K_3}.$$

Step 3: Twiddle-factor multiplication and rearrangement

$$\hat{x_3}(\hat{J_1}, J_2, \tilde{K_3}, P_3) = \tilde{x_2}(P_3, \tilde{K_3}, \hat{J_1}, J_2)\omega_{N_2 N_3}^{J_2 K_3}$$
$$\equiv \hat{x_2}(K_3, \hat{J_1}, J_2)\omega_{N_2 N_3}^{J_2 K_3}.$$

Step 4: All-to-all communication

$$\hat{x_4}(\tilde{J_1}, J_2, \hat{K_3}, P_1) = \hat{x_3}(\hat{J_1}, J_2, \tilde{K_3}, P_3).$$

Step 5: Rearrangement

$$\hat{x_5}(J_2, \tilde{J_1}, \hat{K_3}, P_1) = \hat{x_4}(\tilde{J_1}, J_2, \hat{K_3}, P_1).$$

Step 6: $N_1 \cdot (N_3/P)$ individual N_2-point multicolumn FFTs

$$\hat{x_6}(K_2, \tilde{J_1}, \hat{K_3}, P_1) = \sum_{J_2=0}^{N_2-1} \hat{x_5}(J_2, \tilde{J_1}, \hat{K_3}, P_1)\omega_{N_2}^{J_2 K_2}.$$

Step 7: Twiddle-factor multiplication and rearrangement

$$\hat{x_7}(J_1, K_2, \hat{K_3}) \equiv \hat{x_7}(P_1, \tilde{J_1}, K_2, \hat{K_3})$$
$$= \hat{x_6}(K_2, \tilde{J_1}, \hat{K_3}, P_1)\omega_N^{J_1(\hat{K_3}+K_2 N_3)}.$$

Step 8: $N_2 \cdot (N_3/P)$ individual N_1-point multicolumn FFTs

$$\hat{x_8}(K_1, K_2, \hat{K_3}) = \sum_{J_1=0}^{N_1-1} \hat{x_7}(J_1, K_2, \hat{K_3})\omega_{N_1}^{J_1 K_1}.$$

Step 9: Transpose

$$\hat{y}(\hat{K_3}, K_2, K_1) = \hat{x_8}(K_1, K_2, \hat{K_3}).$$

The distinctive features of the nine-step FFT-based parallel FFT algorithm can be summarized as:

- $N^{2/3}/P$ individual $N^{1/3}$-point multicolumn FFTs are performed in steps 2, 6 and 8 for the case of $N_1 = N_2 = N_3 = N^{1/3}$.
- The all-to-all communication occurs just once. Moreover, the input data x and the output data y are both can be given *natural order*.

If both of N_1 and N_3 are divisible by P, the workload on each node is also uniform.

6 In-Cache FFT Algorithm

We use the radix-2, 4 and 8 Stockham autosort algorithm for in-cache FFTs.

Table 1 shows the number of operations required for radix-2, 4 and 8 FFT kernels on Pentium III processor. The higher radices are more efficient in terms of both memory and floating-point operations. A high ratio of floating-point instructions to memory operations is particularly important in a cache-based processor. In view of the high ratio of floating-point instructions to memory operations, the radix-8 FFT is more advantageous than the radix-4 FFT. A power-of-two FFT (except for 2-point FFT) can be performed by a combination of radix-8 and radix-4 steps containing at most two radix-4 steps. That is, the power-of-two FFTs can be performed as a length $n = 2^p = 4^q 8^r$ ($p \geq 2$, $0 \leq q \leq 2$, $r \geq 0$).

Table 1. Real inner-loop operations for radix-2, 4 and 8 FFT kernels based on the Stockham autosort algorithm on Pentium III processor

	Radix-2	Radix-4	Radix-8
Loads and stores	8	16	32
Multiplications	4	12	32
Additions	6	22	66
Total floating-point operations ($n \log_2 n$)	5.000	4.250	4.083
Floating-point instructions	10	34	98
Floating-point/memory ratio	1.250	2.125	3.063

7 Performance Results

To evaluate the block nine-step FFT-based parallel FFT, we compared its performance against that of the block nine-step FFT-based parallel FFT and that of the FFT library of the FFTW (version 2.1.3) [12] which is known as one of the fastest FFT libraries for many processors. We averaged the elapsed times obtained from 10 executions of complex forward FFTs. These parallel FFTs were performed on double-precision complex data and the table for twiddle factors was prepared in advance.

An 8-node dual Pentium III 1 GHz PC SMP cluster (i840 chipset, 1 GB RDRAM main memory per node, Linux 2.2.16) was used. The nodes on the PC SMP cluster are interconnected through a 1000Base-SX Gigabit Ethernet switch.

MPICH-SCore [13] was used as a communication library. We used an intra-node MPI library for the PC SMP cluster.

For the block nine-step FFT-based parallel FFT, the compiler used was g77 version 2.95.2. For the FFTW, the compiler used was gcc version 2.95.2.

Table 2 compares the block nine-step FFT-based parallel FFT and the FFTW in terms of their run times and MFLOPS. The first column of a table indicates the number of processors. The second column gives the problem size. The next four columns contain the average elapsed time in seconds and the average execution performance in MFLOPS. The MFLOPS value is based on $5N \log_2 N$ for a transform of size $N = 2^m$.

Table 3 shows the results of the all-to-all communication timings on the dual Pentium III PC SMP cluster. The first column of a table indicates the number of processors. The second column gives the problem size. The next two columns contain the average elapsed time in seconds and the average bandwidth in MB/sec.

For $N = 2^{26}$ and $P = 8{\times}2$, the block nine-step FFT-based parallel FFT runs about 2.75 times faster than the FFTW, as shown in Table 2.

In Tables 2 and 3, we can clearly see that all-to-all communication overhead dominates the execution time. Although the FFTW requires three all-to-all communication steps, the block nine-step FFT-based parallel FFT requires only one all-to-all communication step. Moreover, the performance of the block nine-step

Table 2. Performance of parallel one-dimensional FFTs on dual Pentium III PC SMP cluster

P (Nodes × CPUs)	N	Block Nine-Step FFT Time	Block Nine-Step FFT MFLOPS	FFTW Time	FFTW MFLOPS
1×1	2^{23}	5.40606	178.45	10.50152	91.86
1×2	2^{23}	3.33968	288.86	7.49437	128.72
2×1	2^{24}	7.46566	269.67	16.55127	121.64
2×2	2^{24}	4.99214	403.29	11.96556	168.26
4×1	2^{25}	8.22695	509.82	17.79209	235.74
4×2	2^{25}	6.01907	696.84	15.44108	271.63
8×1	2^{26}	8.68712	1004.26	19.33295	451.26
8×2	2^{26}	6.58020	1325.82	18.06414	482.95

Table 3. All-to-all communication performance on dual Pentium III PC SMP cluster

P (Nodes × CPUs)	N	Time	MB/sec
1×2	2^{23}	0.46537	72.10
2×1	2^{24}	2.18825	30.67
2×2	2^{24}	2.00209	25.14
4×1	2^{25}	2.48046	40.58
4×2	2^{25}	2.60625	22.53
8×1	2^{26}	3.01393	38.97
8×2	2^{26}	3.46417	18.16

FFT-based parallel FFT remains at a high level even for the larger problem size, owing to cache blocking. This is the reason why the block nine-step FFT-based parallel FFT is most advantageous with the dual Pentium III PC SMP cluster.

These results clearly indicate that the block nine-step FFT-based parallel FFT is superior to the FFTW.

We note that on an 8-node dual Pentium III 1 GHz PC SMP cluster, over 1.3 GFLOPS was realized with size $N = 2^{26}$ in the block nine-step FFT-based parallel FFT as in Table 2.

8 Conclusion

In this paper, we proposed a blocking algorithm for parallel one-dimensional FFT on clusters of PCs. We reduced the number of cache misses for the six-step FFT algorithm. In our proposed parallel FFT algorithm, since we use cyclic distribution, all-to-all communication is required only once. Moreover, the input data and output data are both can be given in natural order.

Our block nine-step FFT-based parallel one-dimensional FFT algorithm has resulted in high-performance one-dimensional parallel FFT transforms suitable for clusters of PCs. The block nine-step FFT-based parallel FFT is most advantageous with processors that have a considerable gap between the speed of the cache memory and that of the main memory.

We successfully achieved performance of over 1.3 GFLOPS on an 8-node dual Pentium III 1 GHz PC SMP cluster. The performance results demonstrate that the block nine-step FFT-based parallel FFT has low communication cost, and utilize cache memory effectively.

Implementation of the extended split-radix FFT algorithm [?], which has a lower operation count than radix-8 FFT on clusters of PCs is one of the important problems for the future.

The proposed block nine-step FFT-based parallel one-dimensional FFT routines can be obtained in the "*FFTE*" package at `http://www.ffte.jp`.

References

1. Cooley, J.W., Tukey, J.W.: An algorithm for the machine calculation of complex Fourier series. Math. Comput. **19** (1965) 297–301
2. Swarztrauber, P.N.: Multiprocessor FFTs. Parallel Computing **5** (1987) 197–210
3. Agarwal, R.C., Gustavson, F.G., Zubair, M.: A high performance parallel algorithm for 1-D FFT. In: Proc. Supercomputing '94. (1994) 34–40
4. Hegland, M.: A self-sorting in-place fast Fourier transform algorithm suitable for vector and parallel processing. Numerische Mathematik **68** (1994) 507–547
5. Edelman, A., McCorquodale, P., Toledo, S.: The future fast Fourier transform? SIAM J. Sci. Comput. **20** (1999) 1094–1114
6. Mirković, D., Johnsson, S, L.: Automatic performance tuning in the UHFFT library. In: Proc. 2001 International Conference on Computational Science (ICCS 2001). Volume 2073 of Lecture Notes in Computer Science., Springer-Verlag (2001) 71–80
7. Bailey, D.H.: FFTs in external or hierarchical memory. The Journal of Supercomputing **4** (1990) 23–35
8. Van Loan, C.: Computational Frameworks for the Fast Fourier Transform. SIAM Press, Philadelphia, PA (1992)
9. Wadleigh, K.R.: High performance FFT algorithms for cache-coherent multiprocessors. The International Journal of High Performance Computing Applications **13** (1999) 163–171
10. Takahashi, D.: A blocking algorithm for FFT on cache-based processors. In: Proc. 9th International Conference on High Performance Computing and Networking Europe (HPCN Europe 2001). Volume 2110 of Lecture Notes in Computer Science., Springer-Verlag (2001) 551–554
11. Swarztrauber, P.N.: FFT algorithms for vector computers. Parallel Computing **1** (1984) 45–63
12. Frigo, M., Johnson, S.G.: The fastest Fourier transform in the west. Technical Report MIT-LCS-TR-728, MIT Lab for Computer Science (1997)
13. Sumimoto, S., Tezuka, H., Hori, A., Harada, H., Takahashi, T., Ishikawa, Y.: High performance communication using a commodity network for cluster systems. In: Proc. Ninth International Symposium on High Performance Distributed Computing (HPDC-9). (2000) 139–146
14. Takahashi, D.: An extended split-radix FFT algorithm. IEEE Signal Processing Letters **8** (2001) 145–147

Sources of Parallel Inefficiency for Incompressible CFD Simulations

Sven H.M. Buijssen[1,2] and Stefan Turek[2]

[1] University of Heidelberg, Institute of Applied Mathematics, Interdisciplinary Center for Scientific Computing (IWR), INF 294, 69120 Heidelberg, Germany

[2] University of Dortmund, Institute for Applied Mathematics and Numerics, Vogelpothsweg 87, 44227 Dortmund, Germany

Abstract. Parallel multigrid methods are very prominent tools for solving huge systems of (non-)linear equations arising from the discretisation of PDEs, as for instance in *Computational Fluid Dynamics* (CFD). The superiority of multigrid methods in regard of numerical complexity mainly stands and falls with the smoothing algorithms ('smoother') used. Since the inherent highly recursive character of many global smoothers (SOR, ILU) often impedes a direct parallelisation, the application of block smoothers is an alternative. However, duc to the weakened recursive character, the resulting parallel efficiency may decrease in comparison to the sequential performance, due to a weaker total numerical efficiency. Within this paper, we show the consequences of such a strategy for the resulting total efficiency if incorporated into a parallel CFD solver for 3D incompressible flow. Moreover, we compare this parallel version with the related optimised sequential code in FEATFLOW and we analyse the numerical losses of parallel efficiency due to communication costs, numerical efficiency and finally the choice of programming language (C++ vs. F77). Altogether, we obtain quite surprising, but more realistic estimates for the total efficiency of such a parallel CFD tool in comparison to the related 'optimal' sequential version.

1 Numerical and Algorithmic Approach

A parallel 3D code for the solution of the incompressible nonstationary Navier-Stokes equations

$$\mathbf{u}_t - \nu \Delta \mathbf{u} + (\mathbf{u} \cdot \nabla)\mathbf{u} + \nabla p = \mathbf{f}, \quad \nabla \cdot \mathbf{u} = 0 \tag{1}$$

has been developed. This code is an adaptation of the existing sequential FEATFLOW solver (see `www.featflow.de`). For a detailed description of the numerical methods applied see [1, 3]. Here we restrict ourselves to a very brief summary of the mathematical background. Equation (1) is discretised separately in space and time. First, it is discretised in time by one of the usual second order methods known from the treatment of ordinary differential equations (Fractional-Step-θ-scheme, Crank-Nicolson-scheme). Space discretisation is performed by applying

B. Monien and R. Feldmann (Eds.): Euro-Par 2002, LNCS 2400, pp. 701–704.

a special finite element approach using the non-conforming $\tilde{Q}_1/Q_0$ spaces (non-parametric version). The convective term is stabilised by applying an upwind scheme (weighted Samarskij upwind). Adaptive time stepping for this implicit approach is realised by estimating the local truncation error. Consequently, solutions at different time steps are compared. Within each time step the coupled problem is split into scalar subproblems using the Discrete Projection method. We obtain definite problems in $\mathbf{u}$ (Burgers equations) as well as in p (Pressure-Poisson problems). Then we treat the nonlinear problems in $\mathbf{u}$ by a fixed point defect correction method, the linearised nonsymmetric subproblems are solved with multigrid. For the ill-conditioned linear problems in p a preconditioned conjugate gradient method is applied. As preconditioner, multiplicative as well as additive multigrid (using Jacobi/SOR/ILU smoothers) has been implemented.

In order to parallelise the multigrid method the coarse mesh is split into parallel blocks by a graph-oriented partitioning tool (Metis, PARTY). Subsequently, each block is uniformly refined. Consistency with the sequential algorithm (MV application, grid transfer) is guaranteed through local communication between at most two parallel blocks (this is possible because of the face-oriented $\tilde{Q}_1/Q_0$ ansatz). The inherent recursive character of global smoothers impedes a direct parallelisation. Therefore, the global smoothing is replaced by smoothing within each parallel block only (block smoothers). To minimise the communication overhead for solving the coarse grid problem, it is treated on a single processor with an optimised sequential algorithm. The costs is two global communications (setting up the right side and propagation of the solution vector).

The code has been tested [1] for many configurations including standard benchmarks like Lid-Driven-Cavity and "DFG-Benchmark" [2] as well as some problems with industrial background: computation of drag values on model car surfaces (automotive industry), simulation of molten steel being poured into a mould (steel industry). Hexahedral meshes with aspect ratios up to 500 and problems with 100 million degrees of freedom in space and up to several thousand time steps have been successfully handled. Examples are presented in the talk.

2 Comparison of Sequential vs. Parallel Implementation

As mentioned earlier, the difference between sequential and parallel implementation is the way how smoothing operations are performed. If invoked on more than one node the parallel smoothers work on subdomains that differ from the one the sequential algorithm uses such that, besides communication costs, the number of multigrid sweeps increases with increasing number of parallel processes. The second major difference is the programming language. The sequential implementation has been done in F77, the parallel in C++. These two aspects have to be taken into account when comparing run times. Table 1 shows for a given problem the run times of both implementations. One notices that there is a significant loss associated with stepping from the sequential CFD solver to a parallel one and from F77 to C++. On the very same architecture 4 nodes are necessary to match the sequential F77 run time. If using a parallel supercomputer whose

Table 1. Run time comparison for nonstationary DFG-Benchmark 3D-2Z [2] using sequential (F77) and parallel implementation (C++).

implementation	architecture	d.o.f.		cpu time
		space	time	[min]
sequential	Alpha ES40	5,375,872	1,455	2086
parallel (1 node)	Alpha ES40	5,375,872	1,446	6921
parallel (4 nodes)	Alpha ES40	5,375,872	1,466	2151
parallel (4 nodes)	Cray T3E-1200	5,375,872	1,466	4620
parallel (32 nodes)	Cray T3E-1200	5,375,872	1,522	831
parallel (70 nodes)	Cray T3E-1200	5,375,872	1,502	615
parallel (130 nodes)	Cray T3E-1200	5,375,872	1,514	688

single nodes have approximately half the performance of a workstation node, the sequential run time can of course be beaten, but only using brute compute power. Several reasons play a role: the parallel implementation uses a more abstract programming language and is less close to hardware, the optimisation skills of compilers probably differ. The main question, however, is: How good/bad is the total efficiency of the parallel implementation? What causes the run time losses observed? Which influence has the different (blockwise) smoothing?

3 Examination of Parallel Efficiency

In order to explain the behaviour of the parallel implementation seen in Table 1 we primarily focused on its scalability. For a mesh with small aspect ratios (AR ≈ 3) a medium and a big size problem were studied for different platforms and numbers of processes. Figure 1 shows the measured parallel efficiencies. It can be observed that parallel efficiency is not too bad if the problem size is sufficiently large (Figure 1 (b)) and if the parallel infrastructure is well-designed (as on a

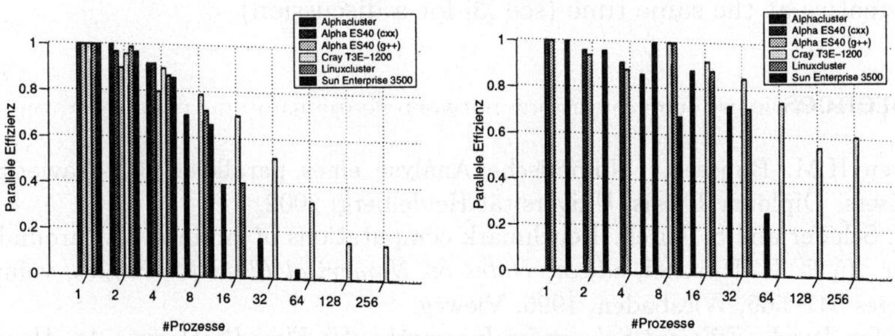

(a) 1.3 million d.o.f. in space (b) 10.6 million d.o.f. in space

Fig. 1. Parallel efficiency for a problem with 1.3 resp. 10.6 million d.o.f. in space on a mesh with small aspect ratios. (Platforms: Alphacluster ALiCE Wuppertal, Alpha ES40 with cxx/g++ compiler, Cray T3E-1200 Jülich, Linuxcluster PIII 650 MHz Heidelberg, Sun Enterprise 3500 Dortmund)

Cray T3E). The performance on multi-processor workstations and clusters is worse. This is clearly due to increasing communication loss on these platforms. But besides communication loss (which can be expected) there is another effect playing an important role as soon as the number of parallel processes increases: the number of iterations needed to solve the Pressure-Poisson problems increases by a factor of 3 if stepping from 1 to 256 processes.[1] This means the amount of time the program spends solving the Pressure Poisson problems increases from 9 to 33 percent. Without this effect, parallel efficiency for the big problem using 256 processes on a Cray T3E-1200 would be 0.72 instead of nearly 0.60.

If a mesh with worse aspect ratios (AR ≈ 20) is probed, the problem becomes much more obvious. Using such an anisotropic mesh and comparing two runs with 1 and 64 processes, respectively, the mean number of iterations needed to solve the (elliptic) Pressure Poisson problems increases by a factor of 4-5: More than half of the run time is now spent solving these subproblems. Consequently, parallel efficiency regresses even more. Additionally, solving Pressure Poisson problems needs more communication than any other part of the program. Thus, the increasing mean number of iterations gives us additional communication loss.

4 Conclusions

The detailed examinations in [1] show that our "standard" parallel version of an optimised sequential 3D-CFD solver has (at least) three sources of parallel inefficiency: Besides the obvious overhead due to inter-processor communication, the change from F77 to C++ compilers is a factor of 3.[2] However, the biggest loss is due to the weakened numerical efficiency since only blockwise smoothers can be applied. Consequently, the number of multigrid cycles strongly depends on the anisotropic details in the computational mesh and the number of parallel processors. As a conclusion, for many realistic configurations, more than 10 processors are needed to beat the optimised sequential version in FEATFLOW. Thus, new and improved numerical and algorithmic techniques have to be developed to exploit the potential of recent parallel supercomputers and of modern Mathematics at the same time (see [3] for a discussion).

References

[1] Sven H.M. Buijssen. Numerische Analyse eines parallelen 3-D-Navier-Stokes-Lösers. Diploma Thesis, Universität Heidelberg, 2002.

[2] M. Schäfer and S. Turek. Benchmark computations of laminar flow around cylinder. In E. H. Hirschel, editor, *Notes on Numerical Fluid Mechanics*, volume 52, pages 547–566, Wiesbaden, 1996. Vieweg.

[3] Stefan Turek. *Efficient Solvers for Incompressible Flow Problems - An Algorithmic and Computational Approach.* Springer-Verlag, Berlin Heidelberg, 1999.

[1] Replacing ILU smoother by SOR results in less degeneration, but longer run time. Jacobi smoothing is at no time competitive, it represents an upper run time bound.

[2] PCs and Sun systems were tested, too; similar behaviour was observed.

Parallel Iterative Methods for Navier-Stokes Equations and Application to Stability Assessment*

Ivan G. Graham, Alastair Spence, and Eero Vainikko

Department of Mathematical Sciences, University of Bath, Bath BA2 7AY, U.K.,
I.G.Graham@bath.ac.uk, A.Spence@bath.ac.uk, E.Vainikko@bath.ac.uk,
http://www.maths.bath.ac.uk/~parsoft

Abstract. We describe the construction of parallel iterative solvers for finite element approximations of the Navier-Stokes equations on unstructured grids using domain decomposition methods. The iterative method used is FGMRES, preconditioned by a parallel adaptation of a recent block preconditioner proposed by Kay, Loghin and Wathen. The parallelisation is achieved by adapting the technology of our domain decomposition solver DOUG (previously used for scalar problems) to block-systems. An application of the resultant linear solver to the stability assessment of flows is briefly indicated.

1 Introduction

When steady solutions of complex physical problems are computed numerically it is often crucial to compute their *stability* in order to, for example, check that the computed solution is "physical", or carry out a sensitivity analysis, or help understand complex nonlinear phenomena near a bifurcation point. An important example is in the computation of fluid flows governed by the steady-state Navier-Stokes equations. Suppose that a velocity field $\mathbf{w}$ has been computed for some particular parameter values. To assess its stability it is necessary to solve the PDE eigenvalue problem:

$$\left.\begin{aligned} -\epsilon\Delta\mathbf{u} + \mathbf{w}\cdot\nabla\mathbf{u} + \mathbf{u}\cdot\nabla\mathbf{w} + \nabla p = \lambda\mathbf{u} \\ \nabla\cdot\mathbf{u} = 0\,, \end{aligned}\right\} \tag{1}$$

for some eigenvalue $\lambda \in \mathbb{C}$ and nontrivial eigenfunction $(\mathbf{u}, p)$, satisfying suitable homogeneous boundary conditions. Here the parameter ϵ is the viscosity, which is inversely proportional to the Reynolds number *Re*. The eigenfunction $(\mathbf{u}, p)$ consists of the velocity $\mathbf{u}$, a vector field in 2D (resp. 3D) and the pressure p (a scalar field) both defined on a 2D (resp. 3D) computational (usually finite) domain. (See, for example [8].) For ease of exposition we shall restrict here to the 2D case.

* Supported by UK Engineering and Physical Sciences Research Council Grant GR/M59075

B. Monien and R. Feldmann (Eds.): Euro-Par 2002, LNCS 2400, pp. 705–714.

The computation of the reference velocity $\mathbf{w}$ is, in itself, a large task, especially for 3D problems, or for 2D problems with multiple scales. The eigenvalue problem is generally accepted to be at least as hard as this, and in many cases considerably harder. Thus it is of great interest to find reliable and efficient algorithms for this task.

Restricting for convenience to 2D applications, discretisation of (1) with mixed finite elements yields the algebraic generalised eigenvalue problem:

$$A\mathbf{x} = \lambda M\mathbf{x} \tag{2}$$

where $\mathbf{x} = (\mathbf{U}_1^T, \mathbf{U}_2^T, \mathbf{P}^T)^T$ is a vector of n degrees of freedom approximating $(u_1, u_2, p)^T$, and the matrices A and M take the form:

$$A = \begin{bmatrix} F_{11} & F_{12} & B_1^T \\ F_{21} & F_{22} & B_2^T \\ B_1 & B_2 & 0 \end{bmatrix}, \; M = \begin{bmatrix} M_u & 0 & 0 \\ 0 & M_u & 0 \\ 0 & 0 & 0 \end{bmatrix}. \tag{3}$$

Here, for example, F_{11} is the finite element stiffness matrix corresponding to the operator $u_1 \to -\epsilon\Delta u_1 + \mathbf{w}\cdot\nabla u_1 + (\partial w_1/\partial x)u_1$ and M_u is the mass matrix corresponding to the identity operator operating on a typical velocity component. Overall, A is unsymmetric, M is positive semi-definite, both are sparse and in real applications n will be very large ($10^5 - 10^9$ is common). Full details of this type of discretisation can be found, for example, in [8].

In stability analysis only a small number of "dangerous" eigenvalues are required, often those (possibly complex) eigenvalues nearest the imaginary axis. In this context it is usually necessary to perform "shift-invert" iterations, which require repeated solution of systems of the form

$$(A - \sigma M)\mathbf{y} = M\mathbf{x}, \tag{4}$$

for some *shift* σ (which may be near a spectral point) and for various right-hand sides $\mathbf{x}$. Examples of algorithms which require solves of the form (4) include classical inverse iteration, its subspace variant, or Arnoldi's method, based on Krylov spaces.

If σ is close to an eigenvalue of (2) then the condition number of $A-\sigma M$ will be large. Consequently one might think that if a Krylov-based iterative method were used to solve (4) convergence would be slow. In fact for the eigenvalue problem there are likely to be only a small number of isolated eigenvalues near zero and it is well known that in such cases the convergence of Krylov solvers is hardly affected. An alternative approach is the Jacobi-Davidson method (see [16]) which seeks to remove the near singularity in the linear solve. We have not chosen this approah.

Previous algorithms developed at Bath [1] were able to find the eigenvalues of (2) nearest the imaginary axis on the assumption that systems like (4) could be solved exactly. These algorithms form a basis of the stability assessment module in the commercial finite element code ENTWIFE [7]. However, requiring exact solutions of (4) is impracticable for large problems.

In large applications, systems (4) have to be solved iteratively. For the rest of this paper we concentrate on describing a parallel solver for (4) in the special case when A and M arise from Navier-Stokes equations, as in (3). At the end of the paper we describe some sample eigenvalue calculations using Arnoldi's method [14], each iteration of which requires a solve of the form (4).

2 Iterative Method

Our iterative method for (4) is the Flexible Generalised Minimal Residual algorithm (FGMRES), which requires the repeated computation of matrix-vector multiplication with $A-\sigma M$ and converges with a rate determined by the distribution of the spectrum of $A-\sigma M$. The rate of convergence can be accelerated by introducing a (right) preconditioner C and solving, instead of (4), the system

$$(A-\sigma M)C\tilde{\mathbf{y}} = M\mathbf{x}\ , \tag{5}$$

after which $\mathbf{y}$ is retrieved by calculating $\mathbf{y} = C\tilde{\mathbf{y}}$. For best performance, C should firstly be a good approximate inverse for $A-\sigma M$. When solving (5) we require repeated matrix-vector multiplication with $(A-\sigma M)C$ which means secondly that this multiplication operation also needs to be sufficiently fast.

Our experience suggests that even though the shift σ might be near an eigenvalue λ of (2), a good preconditioner for $A-\sigma M$ in the PDE context can be found by using a good preconditioner for A (essentially since A represents a higher order operator than M). Thus from now on we concentrate on describing our preconditioner for A.

Note that we choose to discuss the case of right preconditioning here because it fits in with the standard formulation of FGMRES [15].

3 Preconditioner for A

3.1 Block preconditioners for A

An ideal (right) preconditioner for A suggested by Elman and Silvester [6] is of the form

$$P = \begin{bmatrix} F_{11} & F_{12} & B_1^T \\ F_{21} & F_{22} & B_2^T \\ 0 & 0 & -X \end{bmatrix},$$

where

$$X = \mathbf{B}\mathbf{F}^{-1}\mathbf{B}^T,\ \ \mathbf{F} = \begin{bmatrix} F_{11} & F_{12} \\ F_{21} & F_{22} \end{bmatrix},\ \ \mathbf{B} = [B_1\ B_2]\,.$$

This lead to a preconditioned matrix AP^{-1} with a minimal polynomial of degree two, thus guaranteeing convergence in two iterations. This "perfect" preconditioner requires assembling and inverting the Schur complement X and inverting the 2×2 block matrix $\mathbf{F}$ and is not a practical choice. Elman [5] suggested the more practical approximation for X^{-1}:

$$X_B^{-1} = (\mathbf{B}\mathbf{B}^T)^{-1}(\mathbf{B}\mathbf{F}\mathbf{B}^T)(\mathbf{B}\mathbf{B}^T)^{-1}. \tag{6}$$

To motivate the approximation X_B^{-1} note that since $\mathbf{B}$ is rectangular there is no simple formula for X in terms of a product of inverses of its factors. On the other hand $\mathbf{B}\mathbf{B}^T$ acts on the pressure space. X_B^{-1} is a plausible approximation of X^{-1} in that it combines (i) the action of F and (ii) the inverse of $\mathbf{B}\mathbf{B}^T$.

However, numerical experiments with X_B^{-1} indicate that this choice degrades considerably as $h \to 0$ and as the Reynolds number Re increases. Kay, Loghin and Wathen [12, 13] considered the Oseen problem (equivalent to deleting the term $\mathbf{u} \cdot \nabla \mathbf{w}$ in (1)) and for this they proposed instead

$$X_F^{-1} = M_p^{-1} F_p A_p^{-1}, \tag{7}$$

where F_p and A_p are discrete versions of $-\epsilon\Delta + \mathbf{w} \cdot \nabla$ and Δ respectively and M_p is the pressure mass matrix. With this choice the preconditioner is observed to be independent on h and has only a tiny dependence on Re. But the drawback is that (7) requires rediscretisation of the convection-diffusion operator $-\epsilon\Delta + \mathbf{w} \cdot \nabla$ and Laplacian Δ on the pressure space. Our aim is to develop a preconditioner corresponding to (7) which does not need rediscretisation of any of the operators and uses only submatrices of A. We restrict here to the case of the $Q_2 - Q_1$ finite element discretisation (see, for example [8]) and use a suitable restriction of F_{11} to the pressure space instead of explicitly forming F_p. Since $Q_1 \subset Q_2$ we can define a linear interpolation operator Π from pressure to velocity freedoms by simple injection and we use the following approximation of X:

$$X_\Pi^{-1} = \Pi^T F_{11} \Pi (\mathbf{B}\mathbf{B}^T)^{-1}. \tag{8}$$

This is another plausible approximation of X^{-1} based on the same discussion as that following equation (6). Action of the whole block preconditioning step $(\mathbf{u}, p)^T = P^{-1}(\mathbf{w}, r)^T$ is achieved with the following algorithm:

(i) Solve $(\mathbf{B}\mathbf{B}^T)s = r$,
(ii) apply $p = -\Pi^T F_{11} \Pi s$,
(iii) apply $\mathbf{v} = \mathbf{w} - \mathbf{B}^T p$,
(iv) solve $\mathbf{F}\mathbf{u} = \mathbf{v}$.

Note that we assemble the matrix $\mathbf{B}\mathbf{B}^T$ by a sparse matrix multiplication algorithm and the product remains sparse. In our implementation we replace $\mathbf{F}$ in (iv) with a block diagonal approximation $\mathrm{diag}(F_{11}, F_{22})$.

This "block" preconditioner calls for the fast inversion of F_{ii} and $\mathbf{B}\mathbf{B}^T$. These are essentially discrete scalar PDEs and for this we use the DOUG technology [4].

3.2 The DOUG Package

The general parallel iterative solver DOUG (Domain Decomposition on Unstructured Grids, see [4, 10]) for linear symmetric and unsymmetric systems in 2D or

3D has been developed at Bath University since 1996. The current version of the code can be used for the efficient solution of scalar elliptic finite element problems defined in terms of element stiffness matrices on parallel computers. (See, for example [2].) The next version of the code will handle also systems with assembled matrices and block systems (like in the case of linearised Navier Stokes equations discussed here). Krylov subspace methods like CG, MINRES, BiCGSTAB and (F)GMRES are available with additive Schwarz type preconditioners with overlapping subdomains.

For parallel scalability in the presence of unstructured grids a nonmatching coarse grid technique is used. The coarse grid (if required) is computed automatically and adaptively so that it reflects the qualitative features of the fine grid.

DOUG is written in Fortran77, the parallelisation is based on MPI (Message Passing Interface, see for example [9]) and is written using a master-slave setup. The master node handles input and output and is responsible for the coarse problem solutions. The slaves are responsible for the subdomain solves. With N slaves the computational domain is split into N subdomains using the mesh partitioning software METIS [11]. To minimise computation time and to achieve efficient parallelisation, DOUG chooses automatically a suitable size of actual subdomains. For this purpose, the slaves may perform further mesh partitioning to achieve the desired subdomain size. Subdomain solves are performed exactly using the direct multifrontal solver UMFPACK [3]. The LU-factors are computed once at the beginning of computations and stored for each subproblem so that, during the course of the iterative solution, only the corresponding back-substitution phase needs to be called.

For efficient parallel implementation, computations are desired to overlap with communication. With this in mind, DOUG reorders the freedoms suitably for matrix-vector operations so that freedoms at overlaps always come first in the freedom arrays. In the block preconditioner case, freedoms are additionally grouped together according to their block-rank. Thus, our parallel matrix-vector multiplication computations start with overlapping freedoms, after which simultaneously overlapped communication and multiplication operations on inner freedoms are performed. For a more detailed overview of the parallel implementation please refer to [10].

3.3 Test Results: A Typical Solve with A

The test problem of our choice is the problem of flow past a cylinder. A typical discretisation is given in Fig. 1. This mesh has 1728 elements. When the Navier Stokes equations are approximated using $Q_2 - Q_1$ elements on this mesh, the resulting system has 15286 freedoms. The applied velocity is $(1,0)^T$ on the top, bottom and left hand end of the domain, $(0,0)^T$ on the cylinder and $(\mathbf{u},p)^T$ satisfies the outflow conditions $\epsilon\frac{\partial u_1}{\partial x} + p = 0 = \frac{\partial u_2}{\partial x}$ at the right hand end. We compare the number of FGMRES iterations for solving (4) with $\sigma = 0$ with preconditioner

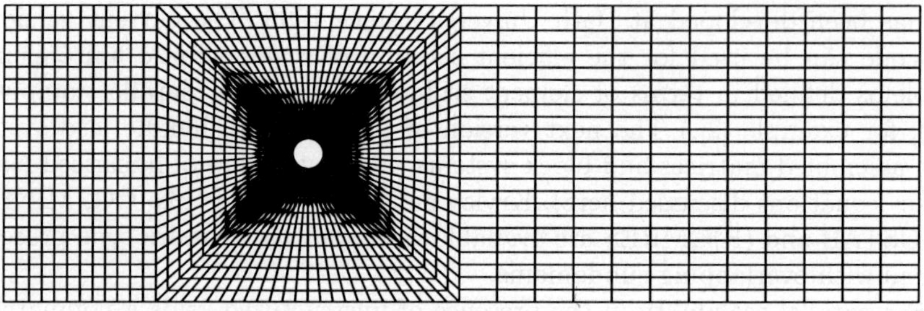

Fig. 1. Typical discretisation grid of the flow past a cylinder.

$$P = \begin{bmatrix} F_{11} & 0 & B_1^T \\ 0 & F_{22} & B_2^T \\ 0 & 0 & -X \end{bmatrix}$$

with X^{-1} approximated using either (6) or (8). We employ the stopping criteria that the initial residual norm should be reduced by six orders of magnitude.

In the implementation of each solve with P it is required to implement (subproblem) solves with F_{ii} $i = 1, 2$ and $\mathbf{B}\mathbf{B}^T$ (the latter is done twice in the case of $X^{-1} \approx X_B^{-1}$). We have tested our preconditioner with exact and inexact subproblem solves. In the exact case all the solves with matrices $\mathbf{B}\mathbf{B}^T$, F_{11} and F_{22} are done directly. In the case of inexact subsolvers we apply K iterations of an inner FGMRES solver with two level additive Schwarz preconditioner. $K = 1$ corresponds to simply replacing the exact matrix to be inverted with one application of its additive Schwarz approximate inverse. In general, the larger the value of K the more precisely the subproblems are solved inside the preconditioner.

An example of the mesh partitioned into subdomains is given in Fig. 2 and an example of an adaptively refined coarse grid is given in Fig. 3. The results with $Re = 1.0$ and $Re = 25.0$ are presented in Table 1 and in Table 2 respectively. Note that the iteration counts in the tables reflect also the exact number of matrix vector operations in FGMRES (since we use a zero initial guess). Generally, the method with X_{Π}^{-1} converges better. Particularly, with exact subsolves we observe

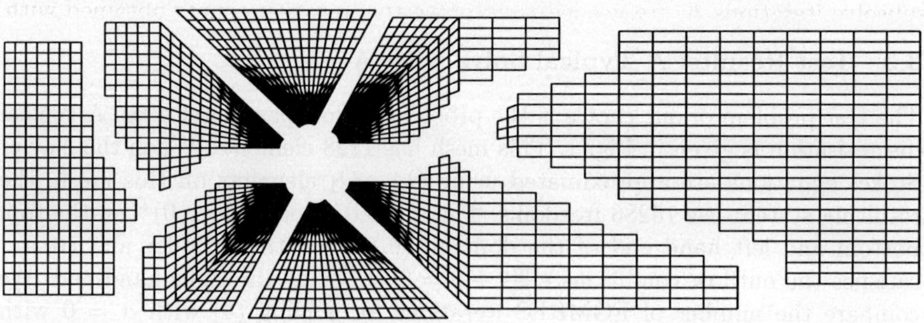

Fig. 2. Flow past a cylinder: the grid partitioned into eight subdomains.

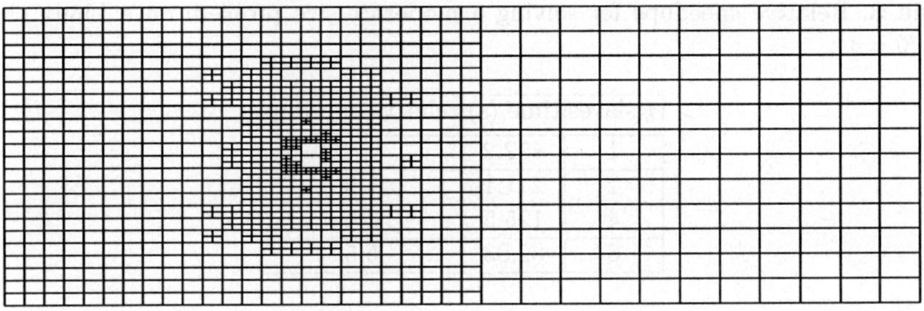

Fig. 3. Flow past a cylinder: an adaptively refined coarse grid.

Table 1. Number of iterations for exact and inexact subproblem solvers with $Re = 1.0$.

$Re = 1.0$	X_B^{-1}						X_{Π}^{-1}				
#dof \ K	exact	16	8	4	2	1	exact	8	4	2	1
6734	54	55	60	130	170	133	50	50	53	66	114
27294	87	104	350	¿800	¿800	234	63	65	78	259	171
61678	116	188	791	-	-	311	73	78	173	781	205
109886	139	352	-	-	-	382	81	91	281	¿800	250

Table 2. Number of iterations for exact and inexact solvers with $Re = 25.0$.

$Re = 25.0$	X_B^{-1}						X_{Π}^{-1}				
#dof \ K	exact	16	8	4	2	1	exact	8	4	2	1
6734	123	126	136	170	338	244	144	150	191	273	295
27294	194	235	529	¿800	-	335	133	146	174	275	384
61678	258	346	¿800	-	-	475	132	162	173	261	407
109886	309	621	¿800	-	-	593	127	153	183	272	459

that the preconditioner with X_B^{-1} has some dependence on the meshsize but for X_{Π}^{-1} with $Re = 1.0$ such dependence is very small and with $Re = 25.0$ does not appear at all. Similar observations appear in [12, 13] for preconditioner (7). Similarly, the preconditioner with X_B^{-1} is much more demanding on the accuracy of the subsystem solves – with the number of degrees of freedom increasing, more subsolve iterations K are needed to improve the iteration counts obtained with $K = 1$. Too few inner solves can damage the convergence. (For the smallest problem $K \geq 4$ is needed to improve on the result with $K = 1$ and at least $K \simeq 16$ inner iterations are needed for the largest problem.) For X_{Π}^{-1} with $Re = 1.0$ the same behaviour is observed but is less significant. Finally, in the case of the inexact subsolves, X_{Π}^{-1} with $Re = 25.0$ gives essentially no increase in iteration counts with mesh refinement for each $K > 1$. This is a particularly good feature for stability calculations as in many problems computations over a large range of Re values are required.

We have tested the parallel efficiency of our code in the case of solving a system with 61678 freedoms. We have measured the total time for the solution of the whole problem (including initialisation and data distribution phases.) The

Table 3. Relative speedups for solving a moderate size problem on a Linux cluster (for $K = 1$).

#slaves	time (s)	relative speedup
1	452.2	-
2	241.1	1.9
4	125.5	3.6
8	82.03	5.5

same problem is solved in the case of one up to eight slaves. Since the DOUG code optimises the choice of subdivisions even for one processor, we can think of the time for one processor as "best" in this computation. The timings for a nine-node 1.3 GHz AMD Athlon Linux cluster are presented in Table 3. We conclude that the algorithm has a good parallel efficiency. Better efficiency may be obtained for finer discretisations.

4 Test Results: The Eigenvalue Problem

To compute the dangerous eigenvalues of the Navier-Stokes system (1), in the stability analysis we use PARPACK – the parallel version of the ARPACK package – which is based on the Implicitly Restarted Arnoldi Method [14] and is suitable for large scale eigenvalue problems. The software provides the option to compute a few (k) eigenvalues of a general $n \times n$ matrix with user specified features. For example, one can ask that eigenvalues of largest real part or largest magnitude be computed. The storage requirements of the algorithm are of the order of $n * k$ locations. From (4) it follows that for our purposes the matrix under consideration is $(A - \sigma M)^{-1}M$, and we call PARPACK in the shift-invert mode. (P)ARPACK uses a reverse communication interface which in this case requires calling one of the following three operations on each outer iteration step (depending on a PARPACK subroutine output parameter value):

a) $\mathbf{y} = (A - \sigma M)^{-1}M\mathbf{x}$,
b) $\mathbf{y} = (A - \sigma M)^{-1}\mathbf{x}$,
c) $\mathbf{y} = M\mathbf{x}$.

After one of these calls control is looped back to the PARPACK subroutine. When convergence is achieved another PARPACK subroutine is called to extract the corresponding computed eigenvalues.

We used the PARPACK computational mode 'LI' – as we know from some preliminary investigations that the desired eigenvalues have large imaginary part. In our case $k = 16$ was chosen which was enough for the eigenvalue nearest to the imaginary axis (using the zero shift $\sigma = 0$) to be found. The eigenvalue tolerance in PARPACK was fixed at 10^{-6}. The paths of a few dangerous eigenvalues as Re increases are depicted in Figs. 4 and 5 for two different discretisations. The figures illustrate the necessity to solve the eigenvalue problem with a large number of degrees of freedom. A Hopf bifurcation for this system occurs at $Re < 24$. An

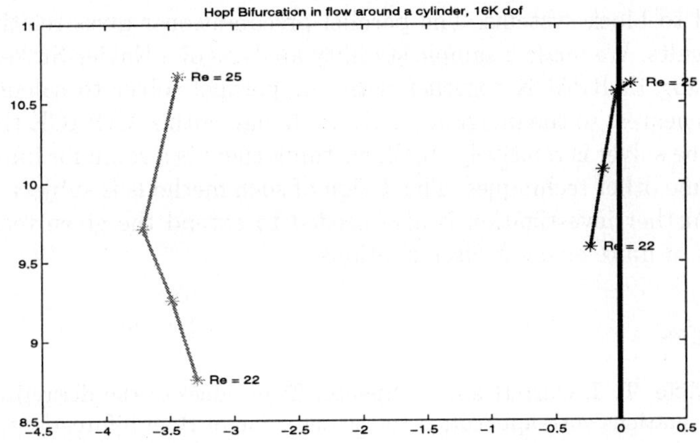

Fig. 4. The paths of a few eigenvalues as Re increases, 16K dof.

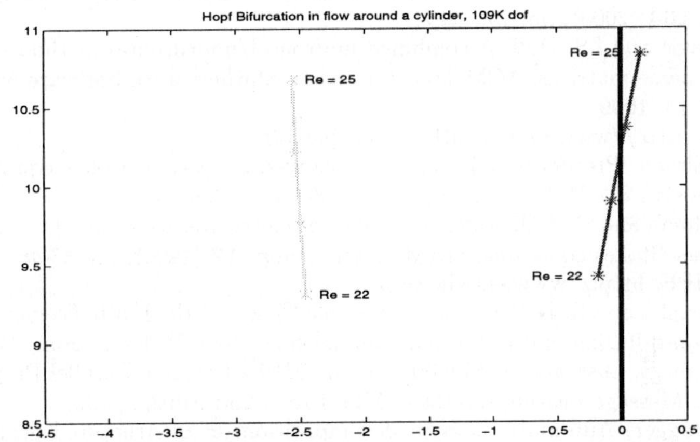

Fig. 5. The paths of a few eigenvalues as Re increases, 109K dof.

incorrect approximation would result if one only examined the eigenvalues of the small system (Fig. 4).

The number of calls to the linear solver requested by PARPACK was ≈ 130 for each Re.

5 Conclusions

In this paper we have discussed some possibilities of building up parallel preconditioners for the linearised Navier-Stokes problem. We investigated two types of preconditioners based on Schur complement techniques. The preconditioners are built up using only existing submatrices of the original problem stiffness matrix and are relatively cheap to construct. For parallelisation the Domain Decomposition technique is used and for the implementation the DOUG package is extended to block systems. The parallel preconditioner gives relatively good speedup results. We made a sample stability analysis of a Navier-Stokes problem problem using PARPACK together with our parallel solver to calculate a few eigenvalues nearest to the imaginary axis. Although with PARPACK the number of calls to the solver is relatively small, we think there is a room for improvement by using some other techniques. The design of such methods is subject to further research. Further investigation is also needed to extend the given technique to other types of finite element discretisations.

References

1. K. A. Cliffe, T. J. Garrett and A. Spence, Eigenvalues of the discretized Navier-Stokes equations with application to the detection of Hopf bifurcations. Advances in Computational Mathematics **1**, (1993), pp. 337-356.
2. K. A. Cliffe, I.G. Graham, R. Scheichl and L. Stals, Parallel computation of flow in heterogeneous media modelled by mixed finite elements, Journal of Computational Physics **164** (2000), pp. 258 -282.
3. T.A. Davis and I.S. Duff, A combined unifrontal/multifrontal method for unsymmetric sparse matrices. ACM Transactions on Mathematical Software, vol. **25**, no. 1, pp. 1-19, 1999.
4. DOUG: http://www.maths.bath.ac.uk/~parsoft.
5. H. C. Elman, Preconditioning for the steady-state Navier-Stokes equations with low viscosity, SIAM J. Sci. Comp. **20** (1999), pp. 1299-1316.
6. H. C. Elman and D. J. Silvester, Fast nonsymmetric iterations and preconditioning for Navier-Stokes equations, SIAM J. Sci. Comp. **17** (1996), pp. 33-46.
7. ENTWIFE: http://www.entwife.com.
8. P. M. Gresho and R. L. Sani, Incompressible Flow and the Finite Element Method. Advection-diffusion and isothermal laminer flow. John Wiley & Sons, 1998.
9. W. Gropp, E. Lusk and A. Skjellum, Using MPI - Portable Parallel Programming with the Message Passing Interface. MIT Press, 2nd Edition, 1999.
10. M. J. Hagger, Automatic domain decomposition on unstructured grids DOUG, Advances in Computational Mathematics **9** (1998), pp. 281 - 310.
11. G. Karypis and V. Kumar, METIS: Unstructured graph partitioning and sparse matrix ordering system, Department of Computer Science, University of Minnesota Report, Minneapolis, Aug. 1995.
12. D. Kay and D. Loghin, A Green's function preconditioner for the steady-state Navier-Stokes equations. Oxford Numerical Analysis Group Research Report NA-99/06, 1999.
13. D. Kay, D. Loghin and A. Wathen. A preconditioner for the steady-state Navier-Stokes equations. SIAM J. Sci. Computing, to appear.
14. R. B. Lehoucq, D. C. Sorensen and C. Yang, ARPACK user's guide: Solution of large scale eigenvalue problems with implicitly restarted Arnoldi methods. http://www.caam.rice.edu/software/ARPACK.
15. Y. Saad. A flexible inner-outer preconditioned GMRES algorithm. SIAM J. Sci. Comput., **14** (1993), pp. 461-469.
16. G. Sleijpen and H. van der Vorst. Jacobi-Davidson Methods (Section 4.7 pp 88 - 105) In Z. Bai, J. Demmel, J. Dongarra, A. Ruhe and H. van der Vorst, editors. Templates for the Solution of Algebraic Eigenvalue Problems: A Practical Guide. SIAM, Philadelphia, 2000.

A Modular Design for a Parallel Multifrontal Mesh Generator

Jean-Paul Boufflet[1], Piotr Breitkopf[2], Alain Rassineux[2], and Pierre Villon[2]

[1] UMR 6599 HeuDiaSyc, Université de Technologie de Compiègne, 60205 Compiègne Cedex, France,
{Jean-Paul.Boufflet}@utc.fr

[2] UMR 6066 Roberval, Université de Technologie de Compiègne, 60205 Compiègne Cedex, France,
{Alain.Rassineux,Pierre.Villon,Piotr.Breitkopf}@utc.fr

Abstract. The proposed approach consists in extending an existing sequential mesh generator in order to design a parallel mesh generator. A sequential three-dimensional mesh generator builds the internal mesh by applying advancing front techniques from a surface mesh defining the volume of the initial domain. We geometrically decompose the domain by recursively splitting it into subdomains using cutting planes. We run a sequential mesh generator code on different processors, each of them working on a single subdomain. The interface mesh compatibility on the two sides of two subdomains makes it possible to merge the results. We present in this paper our modular approach, an example of a interface mesh generation, and we specify the cutting plane decomposition method.

1 Introduction

The goal of this paper is to present a parallel algorithm of volume mesh generation in 3D. The data is the surface triangular mesh. Two strategies are possible for the generation of internal tetrahedral:

1. parallelize a mesh generation code;
2. decompose the data.

The first stategy is subject to current investigation [1]. In the current work we develop an original technique for the second strategy – the domain decomposition of the surface mesh prior to the volume mesh generation. The obvious benefit of this approach is the re-use of an existing sequential volume mesh generator [5] in parallel over the subdomains. Several issues have however to be addressed. The most important is that splitting of a closed envelope gives open subdomains. We have therefore to generate an interface surface mesh in order to close the subdomains. The second issue is the quality of the resulting volume mesh obtained after merging the subdomains. The special load balancing criteria have also to be defined. The chosen approach may be resumed as follows:

B. Monien and R. Feldmann (Eds.): Euro-Par 2002, LNCS 2400, pp. 715–723.

1. we define a “cutting plane” for partitioning the initial surface;
2. we generate the interface surface mesh;
3. we run the sequential volume mesh generator over the subdomains;
4. we merge the subdomain meshes in order to obtain the global volume mesh;
5. we apply the mesh optimization techniques in order to meet the mesh quality criteria in vicinity of the interface.

The actual approach should not be confused with the established mesh partitioning techniques [7–9] developed for the computational meshes. In the primary stage of our algorithm the volume mesh does not yet exist and the only information is the triangulation of the envelope. We could obviously use the standard domain decomposition code in order to split the initial data. Two problems would however arise:

– the interface surface mesh has to be generated anyway;
– we have no guarantee that the interface could be easily meshes with surface elements.

Therefore, we propose to simplify the interface geometry by using an explicit dividing surface - in the actual work, a cutting plane. The benefit of this strategy is the possibility of re-use of a standard surface (plane) mesh generator [6] for the interface. The drawback is the impact of the cutting plane on the form of neighbouring volume elements. This issue is treated by usual techniques of tetrahedral mesh optimization [6].

The current work may be related to the technique of Recursive Inertial Bisection (RIB) [2,3]. The fundammental differences between the two techniques is in the definition of the cutting plane by the Moving Least Square (MLS) approximation [10] and in the strategy of updating the cutting plane position and direction. The MLS approximation permits to get a smooth evolution of the cutting plane based on the local information. In fact, only the points “close enough” to the current cutting plane are taken into account and their influence decreases with the distance. This approach permits to treat more complex and multiply connected geometries corresponding to industrial parts. Rather than the domain decomposition, this paper concerns the envelope splitting and interface generation for a standard volume mesh generator. The usual domain decomposition techniques [7–9] may be used at the initialisation stage.

2 The Modular Design Strategy

The overall process is presented in figure 1. From left to right clockwise, there is the data flow through the modules.

The first module decomposes a geometric domain into two subdomains by computing a cutting plane as we detail in section 4. The second module generates a triangular mesh of the interface.

At this stage, we have generated two subdomains having a complete surface mesh by building a compatible interface mesh on the two sides of the cutting

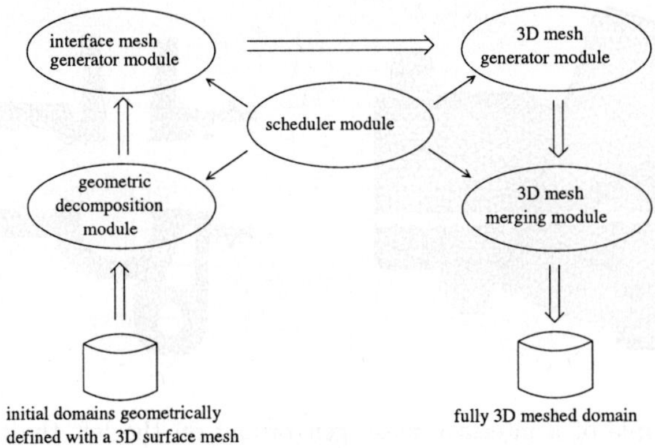

Fig. 1. The modules of the parallel multifrontal mesh generator

plane for the two subdomains that ensures merging the results. Two processors using the sequential mesh generator can generate the internal meshes of the two subdomains in parallel. The sequential mesh generator is the third module. The fourth module merges the meshes to obtain the full mesh.

One can recursively generate the subdomains by applying the first and the second module, and then compute each sub-mesh using the sequential mesh generator. In this case the merging module has to be accordingly invoked. Consequently a scheduling module has to pilot the process.

3 The Interface Mesh Generation

At this stage the goal is to generate the surface mesh in order to close the two open subdomains obtained by applying the cutting plane Π to the domain D. We perfom as follows:

1. we define the interface surface nodes that are close to Π;
2. we project this interface nodes to Π, that defines the geometry needed for the surface mesh generator;
3. we generate a surface mesh using this geometry with a standard 2D mesh generator [6];
4. we fit this surface mesh to the coordinates of the interface nodes.

The step 1 can be achieved by assigning first the triangles of the envelope to three sets:

- $\mathcal{S}_1$: the triangular finite element on one side of Π;
- $\mathcal{S}_2$: the triangular finite element on the other side of Π;
- $\mathcal{S}_3$: the triangular finite element intersected by Π.

Then, using the geometric informations, we assign each triangular finite elements of $\mathcal{S}_3$ either to $\mathcal{S}_1$ or $\mathcal{S}_2$. We therefore obtain the nodes defining the

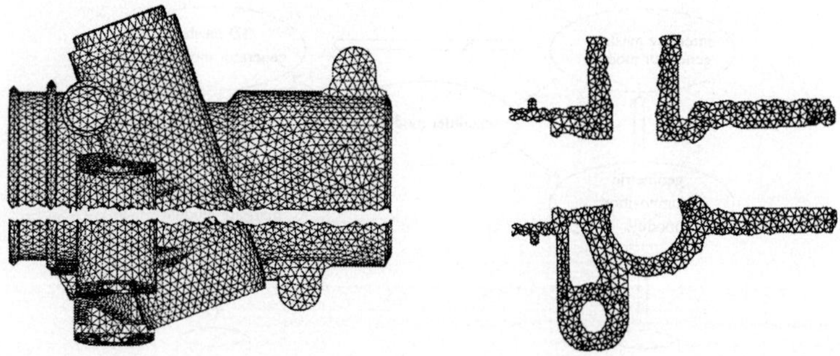

Fig. 2. An example of a interface mesh generation: on the left the two subdomains obtained by applying a cutting plane, on the right the interface mesh generated at the boundary

boundary between the two subdomains. Then, we project the nodes on Π (step 2). These projections define a two-dimensionnal geometry and a two-dimensional mesher can be applied on this classical problem [6]. At this stage we obtain a set of plane two dimensionnal meshes composed of new triangular finite elements. Finally, by applying geometrical transformations, we adapt the new triangles near the interface nodes in order to fit with the three-dimensional shape of the boundary. Consequently the interface mesh generated is not plane and can be composed of several part if there are holes.

By applying this process we build a unique interface mesh that geometricaly fits. By supplementing $\mathcal{S}_1$ and $\mathcal{S}_2$ with this interface mesh we finally obtain two subdomains with two complete envelopes.

On the left of figure 2 we show an example of two subdomains obtained by applying the process. One can notice the "cutted eggshell" shape aspect on the boundary: we do not split the triangular finite elements of the initial envelope. On this exemple the cutting plane Π applied to the complex geometry of the initial domain leads to several areas with dense parts and holes. However, we show on the right of figure 2 that the interface mesh is correctly generated at the boundary between the subdomains.

Whatever the results of the three-dimensional meshing applied on these two new complete envelopes can be, the unique interface mesh generated for the two subdomains ensures the perfect merging of the two results.

4 Geometric Decomposition

The decomposition method that we present in this section exploits only the geometric information of the nodes of the surface mesh. We choose to partition a geometric domain by splitting it into two parts balancing the number of nodes on both sides of the cutting plane. We have to choose the cutting plane defined by its direction and its position with regard to the domain we want to partition in order to satisfy the quality criteria of the submeshes.

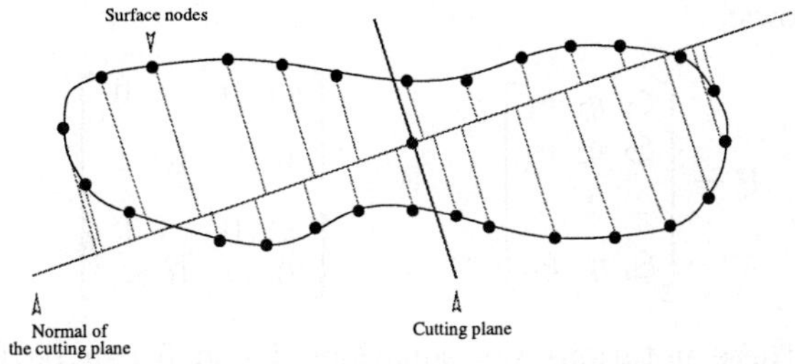

Fig. 3. An example of a domain D with the cutting plane principle

Let D be a domain having a surface mesh (its envelope) such that its volume is defined by the n nodes of the surface (cf. figure 3). Each node X_i is defined by its Cartesian coordinates x_i, y_i and z_i (for $i = 1, \ldots, n$). The purpose is to determine the mean plane that ensures a balanced partition of the domain. First, we calculate the centre of gravity $\bar{X}$:

$$\bar{X} = (\bar{x}, \bar{y}, \bar{z}) = (\frac{1}{n}\sum_i x_i, \frac{1}{n}\sum_i y_i, \frac{1}{n}\sum_i z_i) \quad (1)$$

Then, we compute the nodal coordinates centered around $\bar{X}$. For each node X_i:

$$(\xi_i, \eta_i, \nu_i) = (x_i - \bar{x}, y_i - \bar{y}, z_i - \bar{z}) \qquad for \;\; i = 1, \ldots, n \quad (2)$$

We now determine the plane passing through the point $\bar{X}$ and as near as possible to the nodes X_i. The equation of this plane is *a priori*:

$$a(x - \bar{x}) + b(y - \bar{y}) + c(z - \bar{z}) = 0 \quad (3)$$

We have to compute the triplet (a, b, c) minimizing the following cost:

$$J(a, b, c) = \frac{1}{2}\sum_i w_i\,(a\,\xi_i + b\,\eta_i + c\,\nu_i)^2 \quad (4)$$

with respect to the constraint:

$$a^2 + b^2 + c^2 = 1 \quad (5)$$

where the weights are $w_i = w_{ref}(dist(h(X_i), \bar{X})/r)$. The term $h(X_i)$ is the projection of each node X_i to the normal of the cutting plane passing through the point $\bar{X}$ (cf. figure 3). The quantity r is a reference radius that permits the selection of the nodes contributing to the calculus of the weights w_i. The function w_{ref} is the reference attenuation function. For instance, one can choose $w_{ref}(d) = 0.5\,(1 + \cos(\pi\, d))$ for $d \in [0, 1]$, and 0 otherwise. This function permits to assign higher weights to the nodes near $\bar{X}$ for building the plane.

In order to compute the plane we set $\alpha^T = (a, b, c)$ and we built the matrices P and W as follows:

$$P = \begin{bmatrix} \xi_1 & \eta_1 & \nu_1 \\ \xi_2 & \eta_2 & \nu_2 \\ \vdots & \vdots & \vdots \\ \xi_n & \eta_n & \nu_n \end{bmatrix} \qquad W = \begin{bmatrix} w_1 & 0 & \cdots & 0 \\ 0 & \ddots & 0 & \vdots \\ \vdots & 0 & \ddots & 0 \\ 0 & \cdots & 0 & w_n \end{bmatrix}$$

Considering these notations the equations 4 and 5 can be formulated as: $\min_\alpha J(\alpha)$ under the constraint $\alpha^T\alpha = 1$ and where $J(\alpha) = \frac{1}{2}\,\alpha^T P^T W P\,\alpha$. The expression of the Lagrangian of this problem is:

$$\mathcal{L}(\alpha, \lambda) = \frac{1}{2}\,\alpha^T P^T W P\,\alpha - \lambda\,\alpha^T\alpha \tag{6}$$

and the associated optimality system is:

$$P^T W P\,\alpha - \lambda\,\alpha = 0 \tag{7}$$

$$\alpha^T\alpha = 1 \tag{8}$$

The equation 7 is equivalent to $(\lambda I - P^T W P)\alpha = 0$. That is to say (λ, α) are respectively eigenvalues and eigenvectors of the matrix $P^T W P$. This matrix is positive definite symmetric, therefore there are 3 real eigenvalues λ_1, λ_2, λ_3 such that $0 < \lambda_1 \leq \lambda_2 \leq \lambda_3$ and 3 associated orthonormal eigenvectors. Consequently the solution of the system of equation (7-8) is either (λ_1, α_1) or (λ_2, α_2) or (λ_3, α_3). These three solutions represent the *extrema* of the cost function $J(\alpha)$. In order to find the solution that minimizes, we evaluate $J(\alpha_j)$ for $(j = 1, 2, 3)$.

It follows:

$$J(\alpha_j) = \frac{1}{2}\,\alpha_j^T P^T W P\,\alpha_j = \frac{1}{2}\,\alpha_j^T \lambda_j\,\alpha_j = \frac{1}{2}\,\lambda_j\,\alpha_j^T\alpha_j = \frac{1}{2}\,\lambda_j \tag{9}$$

To summarize, the vector $\alpha = (a, b, c)$ minimizing $J(\alpha)$ corresponds to the normalized eigenvector associated with the smallest eigenvalue of the matrix $P^T W P$. From a practical point of view, we use the inverse iterate power method to compute an approximation of the eigenvector associated with the smallest modulus eigenvalue.

5 The Partitioning Algorithm

The outline of the algorithm consists in bringing up the cutting plane Π from an initial position until we obtain a decomposition of the domain into two equal parts subdomains.

The cutting plane is characterized by the point $\bar{Y}$ and a normal direction α (cf. former section). The initialization step includes the load of the coordinates of the nodes of the surface mesh of the domain (the envelope) from data files and the determination of the initial cutting plane. This initial cutting plane passes through the centre of gravity $\bar{X}$ (cf. equation 1) and the associated normal vector is the normalized eigenvector associated with the smallest eigenvalue of the matrix $P^T P$. At this initialization step we take $W = I$.

The main loop that updates the cutting plane is as follows:

- select the nodes X_i that are the nearest from the intersection of the cutting plane with the envelope;
- compute their centre of gravity $\bar{Y}$;
- compute the projections Y_i of all the nodes X_i to the straight line Δ in normal direction α from the cutting plane;
- sort the points Y_i in an increasing order of distance from $\bar{Y}$;
- compute the weights $w_i = w_{ref}(dist(\bar{Y} - Y_i)/r)$;
- build the matrix $P^T W P$;
- compute the new normal vector α, (normalized eigenvector associated with the smallest eigenvalue of $P^T W P$);
- actualize the cutting plane.

The actualization of the cutting plane is performed as follows: the new cutting plane passes through the point X_k and its new normal vector is the new normal vector α. The point X_k has the point Y_k as its projection to the straight line Δ such that Y_k is the nearest point from $\bar{Y}$ decreasing the balancing criterion. We define the balancing criterion as $C(\Pi) = \frac{abs(n_1 - n_2)}{n_1 + n_2}$ where n_1 is the number of nodes of the surface mesh (the envelope) located at the side with direction $+\alpha$ from the cutting plane (while n_2 is the number of nodes located at the other side with direction $-\alpha$ from the cutting plane).

The main loop of the algorithm stops when the criterion $C(\Pi)$ is lower or equal than a chosen threshold. The parameter r we defined for computing the weights w_i can evolve during the process in order to escape from local minima or to avoid cycling.

We tested our algorithm on three examples having complex geometric shapes. On the right part of figure 4, we use two level of grey to identify the subdomains. On the left part, for each mesh, we report the graph of the percentage of unbalance over the two subdomains as a function of the number of iterations of the main loop.

We observe a rapid convergence to a cutting plane that balances the number of surface nodes over the two subdomains. We also notice that the stopping criterion we proposed for the algorithm depends on the detection of the first local minimum encountered before stabilizing. We reach a relative gap roughly of 2 % between the number of nodes by spending a computing time between one and two minutes on a SUN Solaris workstation, that is satisfying with respect to the time spent by the three-dimensional mesh generator on the domain.

The cutting plane Π may geometrically split the surface mesh into more than two subdomains. The figure 5 show an example of such situation. We observe on the left part of figure 5, more precisely top left near the cutting plane, that a small part of the surface mesh has been assigned to the dark grey subdomain. There is no connection with the other part of the dark grey subdomain. However, the two separate dark grey parts belong to the same side of Π. We associate a graph

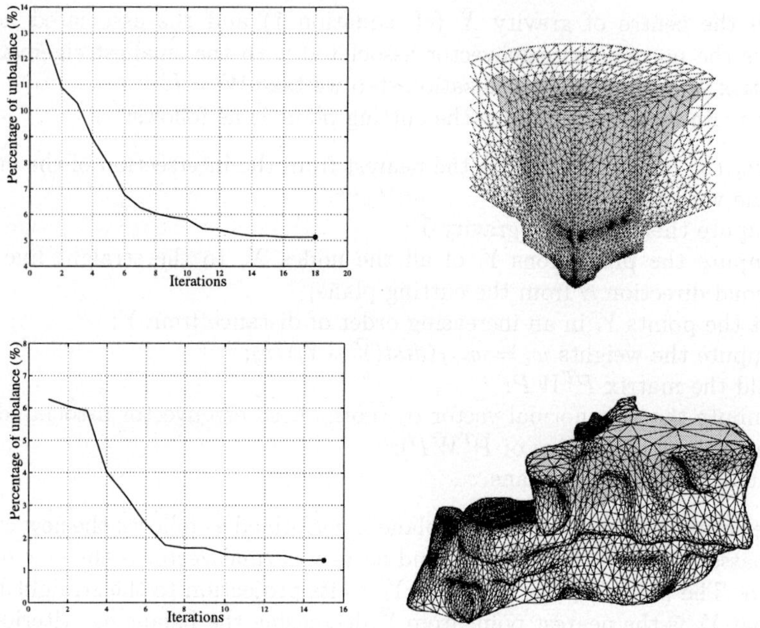

Fig. 4. Two examples of geometric partitioning (light grey and grey parts)

$G(D)$ with the surface mesh of D. The post-processing consists in detecting first all the connected components issued from Π in $G(D)$. Considering the exemple of figure 5 we obtain three connected components. Next, we merge the smallest connected component with its biggest neighbour. The right part of figure 5 shows the repaired mesh. By assigning correctly the parts we finally obtain two connected components. This can be done by applying basic graph algorithms on the parts of the surface mesh issued from the cutting plane.

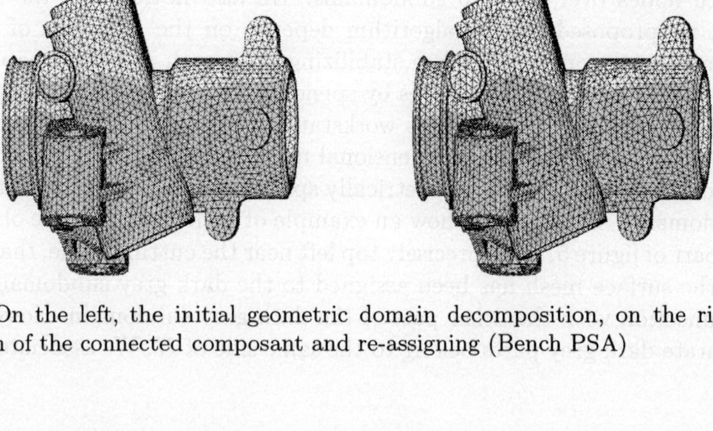

Fig. 5. On the left, the initial geometric domain decomposition, on the right after detection of the connected composant and re-assigning (Bench PSA)

6 Conclusion

The three main modules needed to design the parallel multifrontal mesh generator are: the sequential multifrontal mesh generator, the geometric decomposition and the interface mesh generator. The preliminary results have to be confirmed on other benchmarks and have to be compared with the meshes computed using the sequential mesh generator alone. The behavior of our approach has to be studied concerning the load balancing in order to design the scheduling module.

References

1. Chen M-B, Chuang T-R, Wu J-J.: Experience in parallelizing mesh generation code with High Performance Fortran. In 9th SIAM Conference on Parallel Processing for Scientific Computing. San Antonio, Texas, USA, SIAM Press. March (1999).
2. Hendrickson B., Devine K.: Dynamic Load Balancing in Computational Mechanics. Comp. Meth. Applied Mechanics & Engineering. 184(2-4):485-500, (2000).
3. Simon H. D. Partioning of unstructured problems for parallel processing. Computing Systems in Engineering, 2(2/3):135-148, 1991.
4. Bouattoura D., Boufflet J.P., Rassineux A., Villon P. : Mailleur Parallèle Multifrontal. Proceedings of Quatrième Colloque National en Calcul des Structures, (1999) 297-302
5. Rassineux A.: 3D Mesh Adaptation - Optimization of Tetrahedral Meshes by Advancing front Technique. Compt. Methd. Appli. Mech. Eng., Vol. 141. (1997)
6. Frey P. J., George P-L.: Maillages, applications aux éléments finis. HERMES Science Publication (1999).
7. Karypis G. and Kumar V. METIS : A Software Package for Partitioning Unstructured Graphs, Partitioning Meshes, and Computing Fill-Reducing Orderings of Sparse Matrices. Tech. Report University of Minnesota, Department of Computer Science, sept, (1998).
8. Pellegrini F. SCOTCH 3.0 User's guide. Tech. Report LaBRI, URA CNRS 1304, Université Bordeaux I,(1996).
9. Hendrickson B. and Leland R. The *Chaco* user's guide, version 2.0. Tech. Report Sandia National Laboratories, Albuquerque SAND95–2344, jul, (1995).
10. Lancaster, P., Salkauskas, K., Surfaces Generated by Moving Least Squares Methods, Math. of Comp. Vol, 155, pp. 141-158, (1981).

Pipelining for Locality Improvement in RK Methods

Matthias Korch[1], Thomas Rauber[1], and Gudula Rünger[2]

[1] Universität Halle-Wittenberg, Institut für Informatik,
{korch,rauber}@informatik.uni-halle.de

[2] Technische Universität Chemnitz, Fakultät für Informatik,
ruenger@informatik.tu-chemnitz.de

Abstract. We consider embedded Runge-Kutta (RK) methods for the solution of ordinary differential equations (ODEs) arising from space discretizations of partial differential equations and study their efficient implementation on modern microprocessors with memory hierarchies. For those systems of ODEs, we present a block oriented pipelining approach with diagonal sweeps over the stage and approximation vector computations of RK methods. Comparisons with other efficient implementations show that this pipelining technique improves the locality behavior considerably. Runtime experiments are performed with the DOPRI5 method.

1 Introduction

Time-dependent partial differential equations (PDEs) with initial conditions can be solved by discretizing the spatial domain using the method of lines. This leads to an initial value problem (IVP) for a system of ODEs in the time domain of the form

$$\mathbf{y}'(x) = \mathbf{f}(x, \mathbf{y}(x)) \text{ with } \mathbf{y}(x_0) = \mathbf{y}_0 \tag{1}$$

where $\mathbf{y} : \mathbb{R} \to \mathbb{R}^n$ is the unknown solution, $\mathbf{y}_0$ is the initial vector at start time x_0, $n \geq 1$ is the system size, and $\mathbf{f} : \mathbb{R} \times \mathbb{R}^n \to \mathbb{R}^n$ is the right hand side function describing the structure of the ODE system.

RK methods with embedded solutions are one of the most popular one-step methods for the numerical integration of non-stiff IVPs of the form (1). At each time step, these methods compute a discrete approximation vector $\eta_{\kappa+1} \in \mathbb{R}^n$ for the solution function $\mathbf{y}(x_{\kappa+1})$ at position $x_{\kappa+1}$ using the previous approximation vector η_κ. We consider an s-stage RK method that uses s stage vectors $\mathbf{v}_1, \ldots, \mathbf{v}_s \in \mathbb{R}^n$ with

$$\mathbf{v}_l = \mathbf{f}(x_\kappa + c_l h_\kappa, \eta_\kappa + h_\kappa \sum_{i=1}^{l-1} a_{li} \mathbf{v}_i) \ , \qquad l = 1, \ldots, s \ , \tag{2}$$

to compute two approximation vectors of different order according to:

$$\eta_{\kappa+1} = \eta_\kappa + h_\kappa \cdot \sum_{l=1}^{s} b_l \mathbf{v}_l \ , \qquad \hat{\eta}_{\kappa+1} = \eta_\kappa + h_\kappa \cdot \sum_{l=1}^{s} \hat{b}_l \mathbf{v}_l \ . \tag{3}$$

B. Monien and R. Feldmann (Eds.): Euro-Par 2002, LNCS 2400, pp. 724–733.

$\hat{\eta}_{\kappa+1}$ is an additional vector for error control. The s–dimensional vectors $b = (b_1, \ldots, b_s)$, $\hat{b} = (\hat{b}_1, \ldots, \hat{b}_s)$, $c = (c_1, \ldots, c_s)$ and the $s \times s$ matrix $A = (a_{li})$ specify the particular RK method. For non-stiff ODE systems of the form (1), explicit RK methods with an error control and stepsize selection mechanism are robust and efficient [6] and guarantee that the obtained discrete approximation of $\mathbf{y}$ is consistent with a predefined error tolerance [6, 4].

In this article, we investigate the efficient implementation of RK methods on recent microprocessors. Modern microprocessors exhibit a complex architecture with multiple functional units and a storage hierarchy with registers, two or three caches of different size and associativity, and a main memory. Memory hierarchies provide improved average memory access times due to the locality of reference principle. As a consequence, spatial and temporal locality of a program have a large influence on the execution time.

Because of their large impact on the performance, optimizations to increase the locality of memory references have been applied to many methods from numerical linear algebra including factorization methods like LU, QR and Cholesky [3] and iterative methods like 2D Jacobi [5] and multi-grid methods [8]. Based on BLAS, there are efforts like PHiPAC [2] and ATLAS [9] to provide efficient implementations of BLAS routines. Approaches for dense linear algebra algorithms or grid based methods are given in [3, 5]. Locality optimizations for general purpose RK methods solving non-stiff ODEs with step-size control are investigated in [7]. In this paper, we consider ODE systems resulting from discretized PDEs which have a specific structure with a low coupling density induced by the coupling of the original PDE system and we investigate how this structure can be exploited to increase the locality of memory references. In particular, we use the specific access structure of $\mathbf{f}$ for a block-oriented pipelined computation of stage and approximation vectors. This is the basis for a reordering of the computation such that the working space of the algorithm is significantly decreased, which leads to a considerably better locality behavior. We show that these new computation schemes lead to significant reductions of the execution time on modern microarchitectures like the Pentium III or the UltraSPARC III processors.

2 Data Access Structure and Pipelining

For an embedded RK method applied to an ODE system of type (1), several equivalent program variants realizing different execution orders have been derived in [7] and the impact on the resulting execution time has been investigated for recent processors with memory hierarchies. The use of an arbitrary right hand side function $\mathbf{f}$ of (1) implies the conservative assumption that every component of $\mathbf{f}$ depends on all components of its argument vector and this limits the possibilities of code arrangements.

In this paper, we consider ODE systems (1) with a more specific right hand side function $\mathbf{f}$ depending on only a few components of the argument vector. Those ODE systems arise when applying the method of lines to a time-dependent PDE. As a typical example, we use a non-stiff ODE system resulting from a 2D

```
for (i = 0; i < s; i++) {
    for (j = 0; j < n; j++) w_i[j] = η[j];
    for (l = 0; l < i; l++)
        for (j = 0; j < n; j++) w_i[j] += h a_il v_l[j];
    for (j = 0; j < n; j++) v_i[j] = f_j(x + c_i h, w_i); }
for (j = 0; j < n; j++) { t[j] = 0.0; u[j] = 0.0; }
for (i = 0; i < s; i++)
    for (j = 0; j < n; j++) { t[j] += b̃_i v_i[j]; u[j] += b_i v_i[j]; }
for (j = 0; j < n; j++) { η[j] += h u[j]; ε[j] = h t[j]; }
```

Fig. 1. Basic program variant ($\tilde{\mathbf{b}} = \mathbf{b} - \hat{\mathbf{b}}$).

Brusselator equation, a reaction-diffusion equation for two chemical substances. A standard five-point-star discretization of the spatial derivatives on a uniform $N \times N$ grid with mesh size $1/(N-1)$ leads to an ODE system of dimension $n = 2N^2$ for the discretized solution $\{U_{ij}\}_{i,j=1,\dots,N}$ and $\{V_{ij}\}_{i,j=1,\dots,N}$, see [6]. We start our investigations with an implementation variant from [7] for which pipelining is possible if the access structure of $\mathbf{f}$ fulfills some requirement, see also Fig. 1.

Basic Program Variant: The RK method (2) and (3) is implemented in a straightforward way with a set of stage vectors $\mathbf{v}_1, \dots, \mathbf{v}_s$ and a separate set of argument vectors $\mathbf{w}_1, \dots, \mathbf{w}_s$ for function $\mathbf{f}$ with $\mathbf{v}_i = \mathbf{f}(x_\kappa + c_i h_\kappa, \mathbf{w}_i)$, $i = 1, \dots, s$, so that the loop structure has a minimal number of data dependencies which enables many loop restructurings. In the program code, the stages are computed successively. At each stage $i = 1, \dots, s$ a nested loop is executed that computes the elements of the argument vectors $\mathbf{w}_i$ by calculating the weighted sum of the stage vectors $\mathbf{v}_j$, $j = 1, \dots, i-1$. The working space of stage i of this implementation consists of $i \cdot n$ stage vector elements, n argument vector elements, and n elements of the approximation vector.

For comparison, we include a specialized program variant from [7].

Specialized Program Variant: After applying the transformation $\mathbf{v}_i = \mathbf{f}(x_\kappa + c_i h_\kappa, \mathbf{w}_i)$, $i = 1, \dots, s$, see [7], the loop structure is re-arranged such that only one scalar value is needed to store all stage vector components temporarily. Moreover, a specific RK method with fixed coefficients and fixed number of stages (DOPRI5) is coded with further optimizations like the unrolling of loops over stages.

Storage Schemes. We consider two different linearizations of the grid points $\{U_{ij}\}_{i,j=1,\dots,N}$ and $\{V_{ij}\}_{i,j=1,\dots,N}$. The **row-oriented** organization

$$U_{11}, U_{12}, \dots U_{NN}, V_{11}, V_{12}, \dots, V_{NN} \tag{4}$$

results in function components f_l accessing argument components $l-N, l-1, l, l+1, l+N$, and $l+N^2$ $(l-N^2)$, if available, for $l = 1, \dots, N^2$ $(l = N^2+1, \dots, 2N^2)$. This is a typical access structure for grid-based computations. A specific disadvantage concerning the locality of memory references is that for the computation of each component f_l a component of the argument vector in distance N^2 is accessed. A **mixed row-oriented** organization

$$U_{11}, V_{11}, U_{12}, V_{12}, \ldots, U_{ij}, V_{ij}, \ldots U_{NN}, V_{NN} \tag{5}$$

stores corresponding components of U and V next to each other and results in function component f_l accessing argument components $l-2N, l-2, l, l+1, l+2, l+2N$ (if available) for $l = 1, 3, \ldots, 2N^2 - 1$ and $l - 2N, l - 2, l - 1, l, l + 2, l + 2N$ (if available) for $l = 2, 4, \ldots, N^2$. For this access structure the most distant components of the argument vector to be accessed for the computation of one component of $\mathbf{f}$ have a distance equal to $2N$.

Pipelining. For the basic computation scheme with storage scheme (5), a pipelined computation based on a division of the stage, the argument and the approximation vectors into N blocks of size $2N$ can be exploited. The computation of an arbitrary block $J \in \{1, \ldots, N\}$ of $\eta_{\kappa+1}$ and $\hat{\eta}_{\kappa+1}$ requires the corresponding block J of $\mathbf{v}_s$, which itself depends on block J of $\mathbf{w}_s$ and, if available, the neighboring blocks $J - 1$ and $J + 1$ of $\mathbf{w}_s$ because of the access pattern of $\mathbf{f}$. The computation of the blocks $J - 1$, J, $J + 1$ of $\mathbf{w}_s$ requires the corresponding blocks of $\mathbf{v}_{s-1}$. But these blocks cannot be computed before the computation of the blocks $J - 2$ to $J + 2$ of $\mathbf{w}_{s-1}$ is finished. Altogether, each block J of $\eta_{\kappa+1}$ and $\hat{\eta}_{\kappa+1}$ depends on at most $\sum_{i=1}^{s}(2i + 1) = s(s + 1) + s = s(s + 2)$ blocks of $\mathbf{w}_1, \ldots, \mathbf{w}_s$ of size $2N$ and $\sum_{i=1}^{s}(2i - 1) = s(s + 1) - s = s^2$ blocks of $\mathbf{v}_1, \ldots, \mathbf{v}_s$ of size $2N$, see Fig. 2 (left).

This dependence structure can be exploited in a pipelined computation order for the blocks of the stage vectors $\mathbf{v}_1, \ldots, \mathbf{v}_s$ and the argument vectors $\mathbf{w}_1, \ldots, \mathbf{w}_s$ in the following way: the computation is started by computing the first $s + 1$ blocks of argument vector $\mathbf{w}_1$. Since the computation of component $(\mathbf{v}_1)_l$ requires the evaluation of $f_l(x_\kappa, \mathbf{w}_1)$ and since $\mathbf{f}$ has the specific access

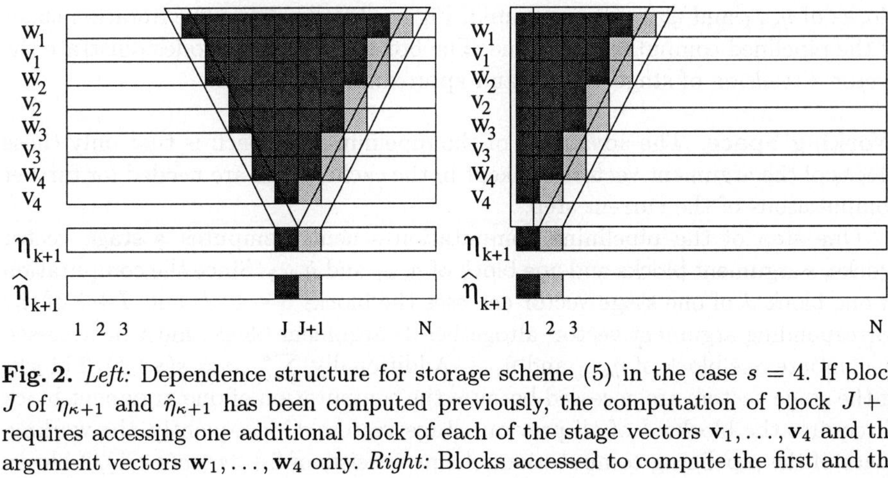

Fig. 2. *Left:* Dependence structure for storage scheme (5) in the case $s = 4$. If block J of $\eta_{\kappa+1}$ and $\hat{\eta}_{\kappa+1}$ has been computed previously, the computation of block $J + 1$ requires accessing one additional block of each of the stage vectors $\mathbf{v}_1, \ldots, \mathbf{v}_4$ and the argument vectors $\mathbf{w}_1, \ldots, \mathbf{w}_4$ only. *Right:* Blocks accessed to compute the first and the second blocks of $\eta_{\kappa+1}$ and $\hat{\eta}_{\kappa+1}$.

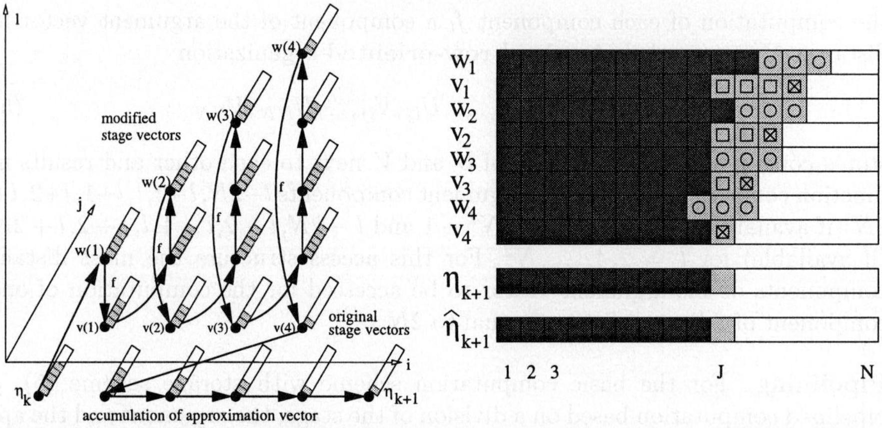

Fig. 3. *Left:* Illustration of pipelined computation for $s = 4$. The dimension of the vectors is shown to demonstrate the pipelined computation from Fig. 2. Filled boxes denote blocks of (intermediate) result vectors. The first block of $\eta_{\kappa+1}$ depends on all filled blocks shown in the figure. The filling structure of the vectors $\mathbf{w}_1, \ldots, \mathbf{w}_s$ shows the triangular structure given in Fig. 2 (right). *Right:* Illustration of the working space of one pipelining step. Argument blocks marked by a circle are accessed during the function evaluation executed to compute the stage vector blocks tagged by a cross. Stage vector blocks used to compute blocks of argument and approximation vectors are marked by a square.

structure described above, the computation of s blocks of $\mathbf{v}_1$ is enabled, which again enables the computation of s blocks of $\mathbf{w}_2$ and so on. Finally one block of $\mathbf{v}_s$ is computed and used to compute the first block of $\eta_{\kappa+1}$ and $\hat{\eta}_{\kappa+1}$. The next block of $\eta_{\kappa+1}$ and $\hat{\eta}_{\kappa+1}$ can be determined by computing only one additional block of $\mathbf{w}_1$ which enables the computation of one additional block of $\mathbf{v}_1, \ldots, \mathbf{v}_s$ and $\mathbf{w}_2, \ldots, \mathbf{w}_s$, see Fig. 2 (right). This computation is repeated until the last blocks of $\eta_{\kappa+1}$ and $\hat{\eta}_{\kappa+1}$ are computed. Figure 3 (left) shows the iteration space of the pipelined computation scheme. The boxes attached to nodes illustrate the vector dimension of stage vectors and approximation vectors.

Working Space. The advantage of the pipelining approach is that only those blocks of the argument vectors are kept in the cache which are needed for further computations of the current step.

One step of the pipelining computation scheme computes s stage vector blocks, s argument blocks and one block of $\eta_{\kappa+1}$ and $\hat{\eta}_{\kappa+1}$. Since the computation of one block J of one stage vector accesses the blocks $J-1$, J, and $J+1$ of the corresponding argument vector, altogether $3s$ argument blocks must be accessed to compute one block of $\eta_{\kappa+1}$ and $\hat{\eta}_{\kappa+1}$. Additionally, $\sum_{i=1}^{s} i = s(s+1)/2$ blocks of the stage vectors are accessed because the computation of one argument block J requires the blocks J of all previous stage vectors. Consequently, the working space of the pipelining computation scheme consists of $2 + 3s + s(s+1)/2$ blocks of size $2N$, see Fig. 3 (right). For the DOPRI5 method with $s = 7$ stages, at most 51 blocks would have to be kept in cache to minimize the number of cache misses. This is usually a small part of the N blocks of size $2N$ that each stage vector contains. Taking $\eta_{\kappa+1}$ and $\hat{\eta}_{\kappa+1}$ into consideration, the proportion of the total number of blocks that have to be held in cache is

$$\frac{(2+3s+s(s+1)/2)}{(2s+2)N} = O\left(\frac{s}{4N}\right)$$

with usually $s \ll N$.

Implementation. The pipelining approach has been implemented in C in the following two program variants:

Basic Pipelining: The main body (Fig. 4 (a)) of the implementation consists of three phases: initialization of the pipeline, diagonal sweep over the argument vectors, and finalization of the pipeline. We introduce the following three macros (Fig. 4 (b)): STAGE0(A) is used to compute one block of vector $\mathbf{w}_0$. Starting at offset A, STAGE(A, m) computes one block of the stage vector $\mathbf{v}_{m-1}$ and one block of the argument vector $\mathbf{w}_m$. The macro FINAL(A) evaluates the function values of one block of the last argument vector to obtain the corresponding stage vector block and finally computes one block of $\eta_{\kappa+1}$ and one block of the local error estimate $\epsilon_{\kappa+1} = \eta_{\kappa+1} - \hat{\eta}_{\kappa+1}$.

Specialized Pipelining: The second implementation is a pipelined version of the specialized implementation, which is optimized for a fixed number of $s = 7$ stages and exploits locality in the solution of Brusselator-like systems.

```
k = s · 2N;

for (j = 0; j < k; j += 2N)
    STAGE0(j);

for (i = 1, l = k − 2N; i < s; i++, l -= 2N)
    for (j = 0; j < l; j += 2N)
        STAGE(j, i);

for (j = k, l = k + 2N; j < n, j += l)
{
    STAGE0(j);
    for (i = 1, j -= 2N; i < s; i++, j -= 2N)
        STAGE(j, i);
    FINAL(j);
}

for (l = 1, k = n − 2N; l < s; l++)
{
    for (i = l, j = k; i < s; i++, j -= 2N)
        STAGE(j, i);
    FINAL(j);
}
FINAL(k);
```

(a) Body

```
STAGE0(A):
for (p = A; p < A + 2N; p++)
    w_0[p] = η[p];

STAGE(A, m):
for (p = A; p < A + 2N; p++)
    w_m[p] = η[p];
for (p = A; p < A + 2N; p++)
    v_{m−1}[p] = f_p(x + c_{m−1}h, w_{m−1});
for (r = 0; r < m; r++)
    for (p = A; p < A + 2N; p++)
        w_m[p] += h a_{mr} v_r[p];

FINAL(A):
for (p = A; p < A + 2N; p++) {
    v_{s−1}[p] = f_p(x + c_{s−1}h, w_{s−1});
    t[p] = 0.0; u[p] = 0.0; }
for (r = 0; r < s; r++)
    for (p = A; p < A + 2N; p++) {
        t[p] += b̃_r v_r[p];
        u[p] += b_r v_r[p]; }
for (p = A; p < A + 2N; p++) {
    η[p] += h u[p];
    ε[p] = h t[p]; }
```

(b) Macros

Fig. 4. Basic pipelining.

3 Runtime Experiments

In this section, we investigate the performance enhancements achieved by the pipelining approach and compare the results with the general implementations for the two different storage schemes (4) and (5) of the Brusselator function. Different target platforms with varying memory hierarchies have been investigated:

1. UltraSPARC III at 750 MHz, 64 KB L1 data cache (4-way associative), 32 KB L1 instruction cache (4-way associative), 8 MB L2 cache (2-way associative),
2. UltraSPARC II at 450 Mhz, 16 KB L1 data cache (1-way associative), 16 KB L1 instruction cache (2-way associative), 4 MB L2 cache (1-way associative),
3. Pentium III at 600 MHz, 16 KB L1 data cache (4-way associative), 16 KB L1 instruction cache (4-way associative), 256 KB L2 cache (8-way associative),
4. MIPS R5000 at 300 MHz, 32 KB L1 data cache (2-way associative), 32 KB L1 instruction cache (2-way associative), 1 MB L2 cache.

For the comparison, we present measurements for the basic version with storage scheme (4) (*basic row*) and storage scheme (5) (*basic mixed row*), for the specialized version with storage scheme (4) (*specialized row*), and with storage scheme (5) (*specialized mixed row*), for the pipelined version (*pipelined*), and for the pipelined specialized version (*pipelined specialized*). As RK method we use the DOPRI5 method with $s = 7$ stages. Figure 5 shows the execution times for one time step for the Brusselator equation on the target systems introduced above. Figure 6 shows the number of instructions executed and the L1 and L2 cache misses measured on the UltraSPARC III, and Fig. 7 shows those measurements for the Pentium III system. The data in Figs. 6 and 7 are obtained using the PCL library [1].

Comparison of Storage Schemes. Except for the Pentium III system, on all machines the mixed row-oriented storage scheme is significantly faster than the pure row-oriented scheme. The best results have been obtained on the MIPS processor. On this processor the use of the mixed row-oriented storage scheme leads to 10.13 % faster execution times for the basic implementation and 12.77 % for the specialized implementation when the size of the system is $n = 294\,912$. The execution times on the Pentium III system are very similar to each other for both storage schemes.

Figure 6 shows that the numbers of cache misses on the UltraSPARC III are very similar for both storage schemes. There are only slight improvements for the L2 and L1 data cache misses. The number of L1 instruction cache misses for the basic implementation with the mixed row-oriented storage scheme is even noticeably higher than the number measured with the original storage scheme. Thus, the improvements of the execution times achieved with the mixed row-oriented storage scheme on this machine seem to be caused by the lower number of instructions executed. The difference in the numbers of instructions executed for both storage schemes is caused by the different code of the function **f** and, as a consequence, the different number of machine instructions the two implementations of **f** are compiled to.

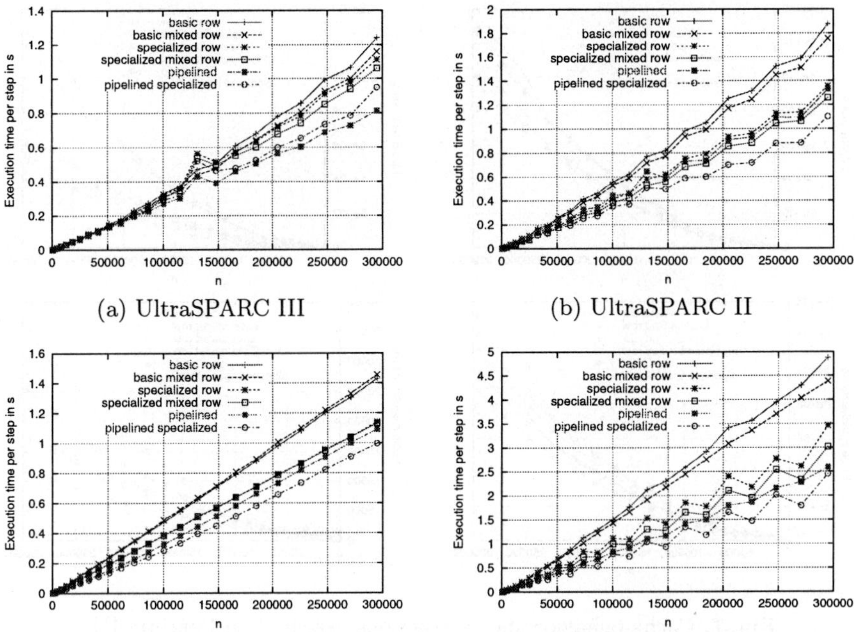

(a) UltraSPARC III (b) UltraSPARC II

(c) Pentium III (d) MIPS R5000

Fig. 5. Execution times of the RK implementations.

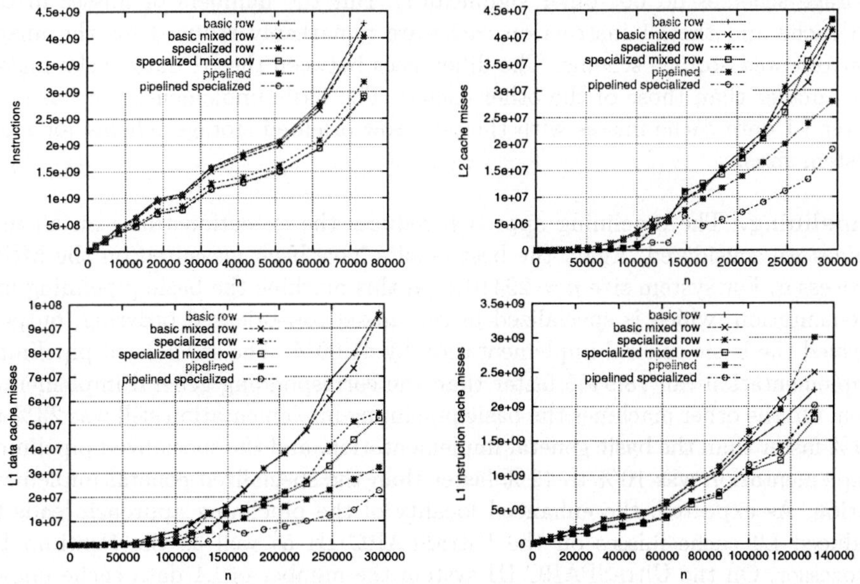

Fig. 6. Cache behavior and instructions executed on UltraSPARC III.

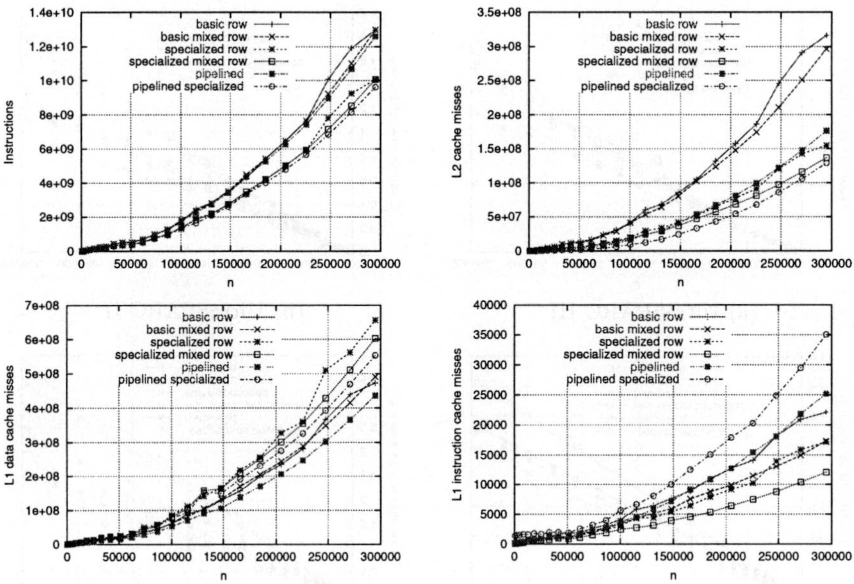

Fig. 7. Cache behavior and instructions executed on Pentium III.

On the Pentium III system the numbers of instructions executed for both storage schemes do not differ significantly. But the numbers of misses in the L2 cache and the L1 instruction cache are remarkably reduced for the mixed row-oriented storage scheme. The differences between the L1 data cache misses are smaller than those of the other caches. The basic implementation has even fewer L1 data cache misses with the pure row-oriented storage scheme for most system sizes.

Pipelining. The pipelining approach reduces the execution times on all machines we considered. Again the best results have been measured on the MIPS processor. For system size $n = 294\,912$, on this machine the basic pipelining implementation, which is specialized in the mixed row-oriented ordering, outperformed the basic general implementation by 41.00 %. The specialized pipelining implementation ran 18.94 % faster than the corresponding general implementation. On the other machines the basic pipelining implementation still was 23 % to 29 % faster than the basic general implementation, and the specialized pipelining implementation was 10 % to 12 % faster than the specialized general implementation. As expected, the enhanced locality of the pipelining approach leads to reduced L2 cache misses on the UltraSPARC III as well as the Pentium III processor. On the UltraSPARC III system the number of L1 data cache misses is also decreased. The number of L1 instruction cache misses does not change significantly on the UltraSPARC III but is increased on the Pentium III machine. Similarly, the number of instructions executed has hardly changed on the UltraSPARC III but is slightly smaller on the Pentium.

4 Conclusions

Runtime experiments have shown that for general RK implementations the mixed row-oriented storage scheme outperforms the row-oriented scheme on most of the processors considered. These results are due to higher locality caused by the smaller distance of the components accessed in one evaluation of the right hand side function **f**. Because of the increase in locality obtained by the pipelining computation scheme, we have measured reductions in execution time between 10 % and 41 %.

References

1. R. Berrendorf and B. Mohr. *PCL - The Performance Counter Library, Version 2.0.* Research Centre Jülich, September 2000.
2. J. Bilmes, K. Asanovic, C.-W. Chin, and J. Demmel. Optimizing matrix multiply using PHiPAC: a portable, high-performance, ANSI C coding methodology. In *11th ACM Int. Conf. on Supercomputing*, 1997.
3. J. Choi, J. J. Dongarra, L. S. Ostrouchov, A. P. Petitet, D. W. Walker, and R. C. Whaley. Design and implementation of the ScaLAPACK LU, QR and Cholesky factorization routines. *Scientific Programming*, 5:173–184, 1996.
4. Wayne H. Enright, Desmond J. Higham, Brynjulf Owren, and Philip W. Sharp. A survey of the explicit Runge-Kutta method. Technical Report 94-291, University of Toronto, Department of Computer Science, 1995.
5. K. S. Gatlin and L. Carter. Architecture-cognizant divide and conquer algorithms. In *Proc. of Supercomputing'99 Conference*, 1999.
6. E. Hairer, S. P. Nørsett, and G. Wanner. *Solving Ordinary Differential Equations I: Nonstiff Problems.* Springer–Verlag, Berlin, 1993.
7. Thomas Rauber and Gudula Rünger. Optimizing locality for ODE solvers. In *Proceedings of the 15th ACM International Conference on Supercomputing*, pages 123–132. ACM Press, 2001.
8. C. Weiß, W. Karl, M. Kowarschik, and U. Rüde. Memory characteristics of iterative methods. In *Proceedings of the ACM/IEEE SC99 Conference*, Portland, Oregon, November 1999.
9. R. C. Whaley and J. J. Dongarra. Automatically tuned linear algebra software. Technical report, University of Tennessee, 1999.

Topic 12
Routing and Communication in Interconnection Networks

Michele Flammini, Bruce Maggs, Jop Sibeyn, and Berthold Vöcking

Topics Chairpersons

This topic is devoted to communication in parallel computers, networks of workstations, and the Internet. All aspects of communication, including the design and packaging of interconnection networks, routing and communication algorithms, and the communication costs of parallel algorithms have been invited. Contributed papers should present significant, original advances in the theory and/or practice of communication in parallel computers and large interconnection networks. The call for papers focussed on the following aspects:

- interconnection networks
- routing algorithms
- communication costs of parallel algorithms
- fault-tolerant communication
- collective communication support
- high-speed SAN/LAN's for cluster computing
- network adapters for cluster computing
- IP network measurement and analysis
- high-speed multimedia and I/O traffic support
- synchronization support for parallel computation
- graph embedding

Altogether 24 papers have been submitted to this topic. We selected eleven (46%) of the submitted papers for publication, four full papers (17%) and seven short papers (29%). We would like to thank all authors for their submissions. Our special thanks go to the 22 external referees who kindly participated in the selection process.

B. Monien and R. Feldmann (Eds.): Euro-Par 2002, LNCS 2400, p. 735.

On Multicasting with Minimum Costs for the Internet Topology*

Young-Cheol Bang[1] and Hyunseung Choo[2]

[1] Department of Computer Engineering, Korea Polytechnic University, Kyunggi-Do, Korea,
ybang@kpu.ac.kr

[2] School of Electrical and Computer Engineering, Sungkyunkwan University, Suwon 440-746, Korea,
choo@yurim.skku.ac.kr

Abstract. We have developed and evaluated a novel heuristic algorithm for the construction of a multicast tree with minimizing tree costs. Our algorithm works on directed asymmetric networks and is shown here to have a perform gain in terms of tree costs for Internet like networks over existing algorithms. The time complexity of our algorithm is $O(D \times m)$ for a m-arc network with D number of members in the multicast group and is comparable to well-known algorithms for multicast tree construction. We have performed empirical evaluation that compares our algorithms with the others on large networks.

1 Introduction

Multicasting in a network is the process of sending the same information from a source node to a set of destination nodes called *multicasting group*. New communication services involving multicast and multimedia applications (*High-Bandwidth* applications) are becoming widespread. These applications need very large reserved bandwidth to satisfy their quality-of-service (QoS) requirements. Limitations in network resources make it critical to design multicast routing paths that use the resources, such as bandwidth, optimally.

The general problem of multicasting is well-studied in the areas of computer networks and algorithmic network theory. Depending on the specific cost or criterion, the multicast problem could be of varying levels of complexity. The Steiner tree, studied extensively in network theory, deals with minimizing the "cost" of a tree that connects a source to a subset of nodes of a network [11, 4]. The Steiner tree has a natural analog of the general multicast-tree in computer networks [5, 10, 6, 13]. The computation of Steiner tree is NP-complete [3], and two interesting polynomial-time algorithms called KMB and TM are proposed in [7] and [9], respectively. Also, an interesting polynomial-time algorithm for a

* This work was supported in part by Brain Korea 21 project and grant No. 2000-2-30300-004-3 from the Basic Research Program of Korea Science and Engineering Foundation. Dr. H. Choo is the corresponding author.

B. Monien and R. Feldmann (Eds.): Euro-Par 2002, LNCS 2400, pp. 736–744.

fixed parameter has been proposed in [8], wherein an overview of several other approximation algorithms are also provided; Distributed algorithms based on Steiner heuristics are provided in [2].

We consider a computer network represented by a graph $G = (V, E)$ with n nodes and m arcs or links, where V and E are a set of nodes and a set of arcs (links), respectively. Each link $e(i, j) \in E$ is associated with *cost* $C(e) \geq 0$. Consider a *simple directed path* P from i_0 to i_k given by $(i_0, i_1), (i_1, i_2), \ldots, (i_{k-1}, i_k)$, where $(i_j, i_{j+1}) \in E$, for $j = 0, 1, \ldots, (k-1)$, and all $i_0, i_1, \ldots, i_k$ are distinct. Subsequently, a simple directed path is referred to simply as a path. The *path-cost* of P is given by $C(P) = \sum_{j=0}^{k-1} C(e_j)$, where $e_j = (i_j, i_{j+1})$. The *tree-cost* of tree T is given by $C(T) = \sum_{e} C(e)$, for all $e \in T$. Let s be a node in the network, called the source node, and $D = d_1, d_2, ..., d_k$, where $k \leq n-1$ be the set of destination nodes. Our objective is to minimize the $C(T)$ for networks with asymmetric links.

To evaluate the performance of the multicast tree construction algorithms, network models constructed by random graph generators are used. Oftentimes, algorithms provide a wide range of results depending upon the random network generation model. Currently, the Waxman model [10] is the most frequently used one. Recently, it was shown in [12] that it is important to consider networks that accurately reflect the Internet topology for evaluation studies. To this end, a random network generator called the *Transit-Stub* model was proposed in [12]. Our proposed algorithm was empirically compared with others on networks that are generated using the Transit-Stub model.

The algorithm (TM) due to Takahashi and Matsuyama [9] is a shortest-path-based algorithm and was further studied and generalized by Ramanathan [8]. The algorithm (KMB) by Kou, Markowsky, and Berman [7] is a minimum spanning tree based algorithm. Ramanathan [8], in his comparison between parameterized TM and KMB, has shown that TM outperforms KMB in terms of the cost of the tree constructed. Moreover, unlike the KMB algorithm, TM works on asymmetric directed networks. Our proposed algorithm like the TM is a shortest path based multicast tree construction algorithm. We show in this paper that our algorithm produces multicast trees with lower tree costs in comparison with the TM algorithm for Internet like networks.

Our algorithm consists of the following two summarized steps which we will elaborate in the next section.

1. Using the cost $C(e)$ associated with each link e, determine all the minimum cost paths (will be referred as also the shortest path) from s to each $d_i \in D$ in the given network. Each minimum cost path from s to $d_i \in D$ forms a directed path. Merge all the minimum cost paths from s to each of the d_i's to form a directed acyclic graph G'. Note that each link in G' is a link in the original network and has the same associated cost.
2. Find a multicast tree given s and $MC = \{d_1, d_2, ..., d_k\}$ on the directed acyclic graph G'.

We will show in the next section that using a simple modification to the Dijkstra's shortest path algorithm we can indeed construct G'. In order to find the multicast tree on the graph G' we propose a new heuristic called the *most shared link* (MSL) and show by empirical evaluation that our algorithm is superior to both TM and KMB for networks that follow the Internet topology. Given a directed acyclic graph, no polynomial algorithm for finding the multicast tree with minimum cost is known to exist [11].

Our contribution in this research is three-fold. First, we propose a new heuristic for multicast tree construction in directed asymmetric networks. Second, the sizes of input random networks are large and they reflect the Internet topology. We use the Transit-Stub network generation model proposed by [12]. Third, we have compared our heuristic with the two well-known hueristics, the TM and KMB and determined that the cost of the tree constructed by our algorithm is smaller than those constructed by TM and KMB. The time-complexity of our algorithm is the same as that of the TM algorithm.

The rest of the paper is organized as follows. In section 2, we present details of our algorithm. We discuss the network model and the results of our simulation in section 3. Conclusions are presented in section 4.

2 The Proposed Algorithm

We will first present the modified Dijkstra's shortest path algorithm to compute the network G' discussed in the introduction.

Let us define the predecessor data structure as an array *Pred*$[j]$ for $j \in V$, where each element of the array points to a linked list. If *Pred*$[j] \rightarrow p \rightarrow q$, then there are two paths from s to j with one of them passing through the link (p, j) and the other through the link (q, j). Thus the data structure *Pred* can store all shortest paths for every pair of nodes (s, d), where s is the source and d is a member of the multicast group (MC). The following algorithm is a modification of the Dijkstra's algorithm that is used to determine *Pred*.

Algorithm Compute_Pred ($s \in V$)
// Assume that the minimum costs for paths (s, i) from s to all i with $i \in V$ is
//precomputed.
// Let minimum cost for any (s, i) be $min[i]$.
cost$[v] = \infty$ for each node $v \in V$
cost$[s] = 0$
Pred$[i]$ = NULL for $i \in V$
$S \leftarrow \phi$
$Q \leftarrow V[G]$
while $Q \neq \phi$ **do**
    $u \leftarrow$ a node with minimum of cost$[u]$ from Q
    $S = S + \{u\}$
    for each vertex $v \in Adj[u]$ **do**
        if cost$[v] > cost[u] + C(u, v)$ **then**

cost$[v]$ $=$ $cost[u] + C(u,v)$
if cost$[v]$ $= min[v]$ **then**
$Pred[v] \rightarrow$ node = u

The time-complexity of the above algorithm is $O(m+n \log n)$ using Fibonacci heaps implementation of the Dijkstra's algorithm [1]. Using the *Pred* information we can construct the network G' in time $O(m)$. Note that G' is a directed acyclic network.

In order to determine the multicast tree on G' using our most shared links approach, we have to first label the arcs in G'. An arc (u, v) in G' has a set of labels, L_{uv}. A node label (or identifier) $d \in MC$ is in L_{uv} if and only if there is path from node u to d using the arc (u, v). That is, the label on an arc indicates the set of destinations that can be reached by using that arc. The label for each arc in G' can be determined by backtracking from each d to s after temporarily reversing the arcs in G' in $O(m)$ time. Next we need the following definition.

Definition 1. *If $L_{uv} \subseteq L_{uw}$ then an arc (u, v) is redundant, otherwise non-redundant.*

The heuristic to determine the multicast tree T from G' consists of the following steps. Note that we use a clever implementation of the following steps to reduce the time complexity of the algorithm.

1. Perform a breadth first search starting from the node s (the root of tree T).
2. For each node u encountered during the breadth first search perform the following two steps.
   a. Remove arcs (u, v) that are redundant.
   b. Modify the labels L_{uv} of non-redundant arcs as follows: Let $L' = L_{uv} \cap L_{uw}$ with $|L_{uv}| \leq |L_{uw}|$ and $v < w$. Assume that $L' \neq L_{uv}$ and $L' \neq L_{uw}$. Now label $L_{uv} = L_{uv} \setminus L'$.
3. Prune the breadth first search tree to remove arcs that do not reach any $d \in MC$. This can be accomplished by performing a depth first search.

Steps 1 and 2 are implemented using clever data structuring in the following algorithm.

Algorithm RRL(G', MC)
Input : G' the acyclic directed network and MC the multicast group with $D = |MC|$.
Output : The multicast tree T.
/* Let CV be a current Boolean vector of size D. Let FP be a Boolean vector of size D for each node in G'. Let A be a Boolean vector of size D for each arc in G'. Let $N(u, v)$ be the number of destinations that can be reached by using the arc (u, v). Note that $N(u, v) \leq D$. Let L_{uv} be the label on arc (u, v). $Q = \emptyset$ initally. */
Assign L_{uv} for each (u, v) in G' by backtracking from each $d \in MC$.
for $i = 1$ to D **do**
$CV[i] \leftarrow$ false

4. $FP_s[i] \leftarrow$ true
5. $Q \leftarrow Q \cup \{s\}$
6. **while** Q not empty **do**
7. remove p from Q
8. **for** each arc (p, v) **do**
9. **for** $i = 1$ to D **do**
10. $A_{pv}[i] \leftarrow$ false
11. **for** each $x \in L_{pv}$ **do**
12. $A_{pv}[x] \leftarrow$ true
13. $A_{pv} \leftarrow FP_p \wedge A_{pv}$
14. compute $N(p, v)$ based on true value in A_{pv}
15. sort the arcs (p, v) based on $N(p, v)$ in descending order using the bucket sort. Let $(p, v_1), (p, v_2), \ldots, (p, v_k)$ be the sorted arcs.
16. $CV \leftarrow A_{pv_1}$
17. **for** $i = 2$ to k **do**
18. **for** $j = 1$ to D **do**
19. **if** $A_{pv_i}[j]$ == true **then**
20. **if** $CV[j]$ == false **then**
21. $CV[j] \leftarrow$ true
22. **else**
23. $A_{pv_i}[j] \leftarrow$ false
24. $N(p, v_i) \leftarrow N(p, v_i)$ - 1
25. remove all arcs (p, v_i) with $N(p, v_i) = 0$
26. place all neighbor v_i of p with $N(p, v_i) > 0$ in Q
27. $FP_{v_i} \leftarrow A_{pv_i}$

The time-complexity of the algorithm RRL is evaluated as follows. Step 1 can be completed in $O(n+m)$ time using depth first search. Steps 2-4 take $O(D)$ time which is at most $O(n)$. The while statement in step 6 is executed at most $O(n)$ time. Steps 8-13 is executed at most $O(m \times D)$ during the entire execution of the algorithm. Steps 14-15 can be completed in $O(n \times D)$ during the breadth first search as we visit nodes level by level. Steps 17-24 can be executed in $O(m \times D)$ during the entire execution of the algorithm. Hence the total time complexity of the above algorithm is $O(m \times D)$. The total time complexity of our entire algorithm including the construction of G' is $O(m \times D)$. The time complexity of KMB is $O(D \times n^2)$ [7] and TM has a time-complexity of $O(D \times n^2)$ [9].

Figure 1(a) shows a directed asymmetric network. Assuming that the source node is labeled 0 and the $MC = \{6, 7\}$, the trees constructed by algorithms KMB, TM, and our algorithm are shown in Figures 1(b), 1(c), and 1(d), respectively. The cost of the trees generated by KMB, TM and our algorithm are 6, 6, and 4, respectively. The subnetwork G' for the network in Figure 1(a) with $s = 0$ and $MC = \{6, 7\}$ is shown in Figure 2(a). The text on each arc is the arc label. Figure 2(b) is the multicast tree T for G'.

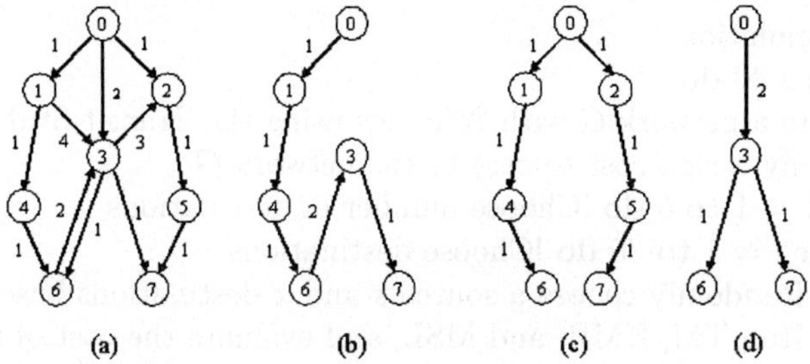

Fig. 1. Given a network (a), a multicast tree based on KMB is shown in (b), a multicast tree based on TM is shown in (c), and a multicast tree based on MSL is shown in (d), where a source is 0, and $MC = \{6, 7\}$.

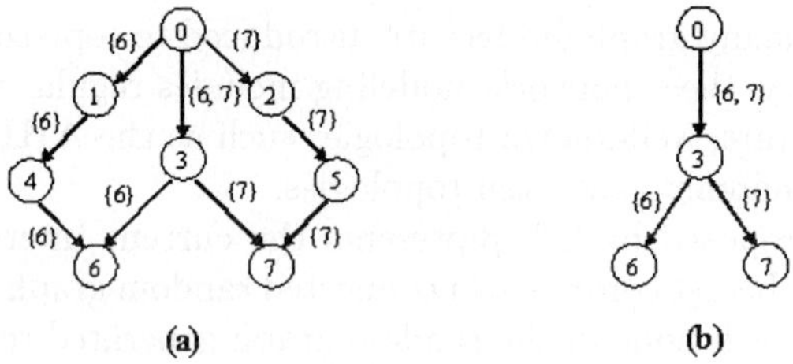

Fig. 2. Subnetwork by merging all minimum cost paths from source to each destination is shown in (a) and tree derived from (a) is shown in (b).

3 Performance Evaluation

To compare performances of algorithms introduced, we used extensive simulations. We implemented the algorithms in language C under the Linux environment. As mentioned earlier we used the Transit-Stub model proposed by Zegura et. al [12] to generate our random networks. Note that the Waxman model [10] is a popular graph generation model, but generates graphs that do not reflect the Internet topology [12].

The weights on the arc represent the cost of using the arc. In the Transit-Stub model, the cost of the arcs in the backbone network is less compared with other arcs in the network. Semantically, it is advisable to use the backbone to route traffic between inter-domain nodes since the backbone has a higher bandwidth in comparison with other arcs.

We considered networks with number of nodes equal to 117, 204, 315, 420, and 500. We generated 30 different networks for each size given above. The random networks used in our experiments are directed and connected, where each node in graphs has the average degree 4. Without loses of generality, a source and a set of destinations are uniformly selected for each multicast session. The number of destinations chosen by our simulation was in range from 10–300 depending upon the size of the graph.

The overall simulation experiment is organized as in the following algorithm.

Algorithm Simulation

```
for i = 1 to 30 do
    Generate a network G with N nodes using the Transit-Stub model.
    Randomly assign cost to arcs to the network G
        for k = 1 to 6 do !Choose number of destinations
            for l = 1 to 30 do !Choose destinations
                Randomly choose a source s and x destinations based on N
                Run TM, KMB, and MSL, and evaluate the cost of the tree
```

3.1 Topology Model

In the literature, many graph models are introduced to represent actual network topologies. Broadly, these network modeling includes regular topologies such as trees, rings, and stars, well-known topologies such as the **ARPA**net or **NSF**net backbone, and randomly generated topologies.

This model proposed by [12] represents the current Internet very well (for details, read [12]). It first constructs a connected random graph using any random graph model and each node in the random graph generated represents an entire transit domain. Then, to create the backbone topology of one transit domain (150 *km* x 150 *km* in our simulation), each node, which represents an entire transit domain, is replaced by a newly generated random network. For each node in the backbone of each transit domain, a number of random networks, called stub domains which are connected to that node, are generated. The size of each stub domain is 50 *km* x 50 *km*, in our simulation. After that, additional links between any pair of nodes, where one is from a transit domain and the other is from a stub domain, or both nodes are from two different stub domains, are generated.

3.2 Simulation Results

To evaluate performances of MSL, we compare with previous hueristics KMB [7] and TM [9]. Since algorithm proposed in [8] has the best performance when it is equivalent to TM, we omit the algorithm proposed in [8] in our evaluation. Our simulation indicate that the relative performances of two heuristics KMB [7] and TM [9] remained almost same for all different sizes of the multicast group. However, the simulation revealed that the performance of our algorithm MSL is relatively better than those of both KMB and TM. We will present results for networks with 117 nodes, 204 nodes, 315 nodes, 420 nodes, and 500 nodes.

Since it is impractical to find the optimal solution for large graphs, we used the normalized surcharge δ_H of algorithm [8] with respect to MSL defined as follows:

$$\delta_H = \frac{C(T_H) - C(T_{MSL})}{C(T_{MSL})} \tag{1}$$

In the above equation $C(T_H)$ is the cost of tree based on algorithm H and $C(T_{MSL})$ is the cost of tree based on algorithm MSL. To depict relative performances by plots, δ_H is multiplied by 100 to express as a percentage.

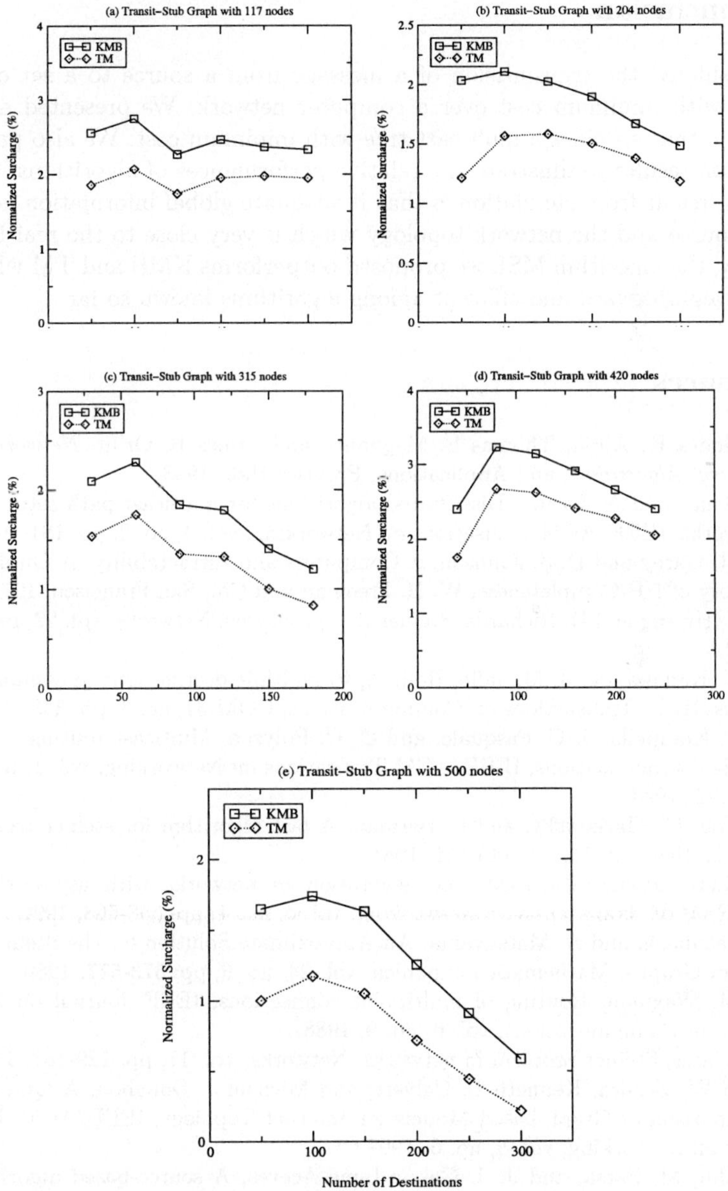

Fig. 3. Normalized Surcharges versus the number of destinations for network with 117 nodes, 204 nodes, 315 nodes, 420 nodes, and 500 nodes, with respect to MSL .

As indicated in Figure 3, it can be easily noticed the MSL always outperforms KMB. Notice that the relative performance of MSL is highest when the number of destinations are about 20% of number of nodes, and it becomes decreasing after then.

4 Conclusion

We considered the transmission of a message from a source to a set of destinations with minimum cost over a computer network. We presented a simple algorithm that specifies a multicast tree with minimum cost. We also presented simulation results to illustrate the relative performances of algorithms. One interesting result from simulation is that if adequate global information is known at the source and the network topology which is very close to the real Internet topology, the algorithm MSL we proposed outperforms KMB and TM which are most straightforward and efficient among algorithms known so far.

References

1. Ravindra K. Ajuja, Thomas L. Magnanti, and James B. Orlin, *Network Flows: Theory, Algorithms,* and Applications, Prentice-Hall, 1993.
2. F. Bauer and A. Varma, Distributed algorithms for multicast path setup in data networks, IEEE/ACM Transations on Networking, vol. 4, no. 2, pp. 181-191, 1996.
3. M. R. Garey and D. S. Johnson, it Computers and Intractability: A Guide to the Theory of NP-Completeness, W. H. Freeman and Co., San Francisco, 1979.
4. F. K. Hwang and D. Richards, Steiner tree probkems, Networks, vol. 22, pp. 55-89, 1992.
5. B. K. Kadaba and J. M. Jaffe, Routing to multiple destinations in computer networks, IEEE Transactions on Communications, COM-31, no. 3, pp. 343-351, 1983.
6. V. P. Kompella, J. C. Pasquale, and G. C. Polyzoa, Multicast routing fo rmultimedia communications, IEEE/ACM Transations on Networking, vol. 1, no. 3, pp. 286-292, 1993.
7. L. Kou, G. Markowsky, and L. Berman, A fast algorithm for steiner trees, Acta Informatica, vol. 15, pp. 145-151, 1981.
8. S. Ramanathan, Multicast tree generation in networks with asymetric links, IEEE/ACM Transations on Networking, vol. 4, no. 4, pp. 558-568, 1996.
9. H. Takahashi and A. Matsuyama, An Approximate Solution for the Steiner Problem in Graphs, Mathematica Japonica, vol. 24, no. 6, pp. 573-577, 1980.
10. B. M. Waxman, Routing of multipoint connections, IEEE Journal on Selected Areas in Communications, vol. 6, no. 9, 1988.
11. P. Winter, Steiner problem in networks, Networks, vol. 17, pp. 129-167, 1987.
12. Ellen W. Zegura, Kenneth L. Calvert, and Michael J. Donahoo, A Quantitative Comparison of Graph-based Models for Internet Topology, IEEE/ACM Transactions on networking, vol. 5, no. 6, 1997.
13. Q. Zhu, M. Parsa, and J. J. Garcia-Luna-Aceves, A source-based algorithm for delay constrained minimum-cost multicasting, Proceedings of INFOCOM, pp. 377-385, 1995.

Stepwise Optimizations of UDP/IP on a Gigabit Network*

Hyun-Wook Jin[1], Chuck Yoo[1], and Sung-Kyun Park[2]

[1] Department of Computer Science and Engineering, Korea University
Seoul, 136-701 Korea,
{hwjin,hxy}@os.korea.ac.kr
[2] SK Telecom, Seoul, 110-110 Korea

Abstract. This paper describes stepwise optimizations of UDP/IP on Myrinet. We eliminate internal overheads such as fragmentation, checksum computation, data copy, and multiple DMA initializations. In addition, we reduce the NIC overhead with a faster NIC. Stepwise optimizations clearly illustrate how much bandwidth is wasted by each overhead and what factors should be considered for designing a gigabit network protocol. As a result, we show that UDP/IP can achieve 926Mbps on 32bit 33MHz PCI platform.

1 Introduction

The physical network bandwidth is dramatically increasing in recent years. Accordingly, in order to fully utilize the bandwidth, many research groups have been trying to develop a user-level communication protocol [2, 8] or optimize a traditional protocol such as TCP/IP and UDP/IP [3, 9]. Existing works, however, are focusing on the removal of only one specific overhead, such as data copy overhead.

This paper aims to stepwisely eliminate several critical overheads of UDP/IP one by one on Myrinet [1]. This paper also demonstrates that the performance of UDP/IP is improved by reducing the overhead of Network Interface Card (NIC). As a result, we can show that the optimized UDP/IP fully utilizes the bandwidth of Myrinet. Our stepwise optimizations contribute to the understanding of how much bandwidth is encroached by each overhead and what factors cause bottlenecks on a gigabit network.

2 Optimizations of UDP/IP

The details of stepwise optimizations are described in the following subsections. Optimizations are performed using a pair of workstations equipped with an Intel Pentium III 450MHz processor and an M2F-PCI32C (LANai-4) Myrinet NIC.

* This research was supported by University Software Research Center Supporting Project from Korea Ministry Information and Communication.

B. Monien and R. Feldmann (Eds.): Euro-Par 2002, LNCS 2400, pp. 745–748.

Workstations are directly connected via a Myrinet cable. The kernel is Linux (kernel version 2.2), and we adopt Myrinet Software (version 3.22c) [5] for the device driver and the firmware of the NIC. In Subsection 2.4, we altered the Myrinet NIC and Myrinet Software into an M2L-PCI64B (LANai-9) Myrinet NIC and GM (version 1.4) [6], respectively. We measure the throughput using *ttcp*.

2.1 Large Maximum Transmission Unit (MTU)

A small MTU size leads to a fragmentation that results in a per-packet overhead. Actually, we observe that the (de)fragmentation overhead is roughly $31\mu s$ on the sender and $9\mu s$ on the receiver. In addition, the packet header per fragment wastes the network bandwidth. Moreover, the defragmentation routine of the Linux kernel version 2.2 performs the copy operation that moves all received fragments into a large buffer in order to merge fragments. This copy operation induces about $10\mu s$ overhead per 1KB. As a result, the throughput reaches its peak at the MTU size of 4KB while it remains slightly below the peak maximum with a larger packet size than MTU as the line labeled *UDP/IP* in Figure 1.

A notable characteristic of gigabit networks is to provide us with a large MTU size (e.g., jumbo frame of Gigabit Ethernet). In the case of Myrinet, the MTU size really has no limit. We enlarge the Myrinet MTU size large enough to support the user data up to 32KB. Figure 1 shows that the throughput with large MTU size increases continuously even for data sizes larger than 4KB.

2.2 Checksum Offloading and Zero-Copy

Per-byte overheads, such as checksum computation and data copy overheads, increase as the packet size increases. UDP performs the checksum computation

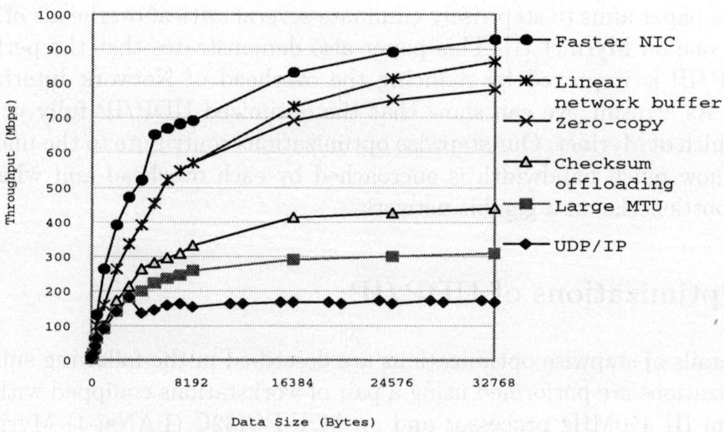

Fig. 1. Throughput gains by removing overheads one by one

for the whole network data and also moves it between the user and kernel buffers via a copy operation. These per-byte overheads are principal factors in a network bottleneck because the packet size tends to become large in order to achieve a better network utilization.

There are some works to improve the performance of checksum computation, which achieve a significantly improved rate [4, 7]. In spite of those improvements, the checksum computation still burns the host processor.

In order to offload the checksum computation from the host processor, we adopt a hardware checksum computation. Many gigabit network NICs include the function of the checksum computation, which is generally integrated with the DMA engine. Figure 1 shows that UDP/IP without the checksum computation achieves 44% higher throughput than the UDP/IP of the improved version as described in the previous subsection.

Moreover, to eliminate the copy operation, we implement the mechanism that moves a data directly between the user buffer and NIC without copying it to/from the kernel buffer. An interesting result to be noted is that the throughput of UDP/IP without both checksum computation and copy operation is much higher than that without only the checksum computation as shown in Figure 1. The reason why the removal of the copy operation (after the checksum computation) results in a higher improvement is due to the cache effect. Because the received data is not cached, any operation to touch the data in the first time takes longer. When the checksum computation is followed by the copy operation, the checksum computation is done with the uncached data. If the checksum computation is removed in UDP/IP, the copy operation has to be done with the uncached data. So removing both of the checksum computation and the copy operation eliminate the overhead of touching uncached data.

2.3 Linear Network Buffer

The performance of DMA between the host and NIC memories is affected by the linearity of the network buffer. If the network buffer is not contiguous, NIC needs a scatter or gather to move a chunk of data. A scatter or gather is a list of vectors, each of which indicates the location and length of one linear memory segment in the overall receiving or sending request. Therefore, several vectors in a scatter or gather induce multiple DMA initializations as much as the number of vectors. In the case of traditional protocols, the kernel allocates a physically linear memory area for the network buffer. On the other hand, the zero-copy UDP/IP of the previous subsection uses the memory area in user space as a network buffer, which is not physically linear. Therefore, the network buffer crossing the page boundary leads to multiple DMA initializations.

To reduce the DMA initialization overhead, we add a system call that allocates a physically linear memory area and also maps this area into the user space allowing the application to use it as a network buffer. Figure 1 shows the physically linear buffer improves the throughput up to 10% compared with the last improvement described in the previous subsection.

2.4 Faster NIC

Another important component of an end system is NIC because of its relatively high overhead when compared with the host overhead reduced by previous subsections.

In order to reduce the NIC overhead, we adopt another Myrinet NIC equipped with a faster processor and an effective firmware. We measure the throughput of the improved UDP/IP on GM with M2L-PCI64B (LANai-9) Myrinet NIC. The GM provides a highly optimized firmware for Myrinet NIC, and the LANai-9 RISC operates at 132MHz that is 4 times faster than LANai-4. Figure 1 shows the measurement results. The maximum throughput of the present optimized UDP/IP is 926Mbps that is a remarkable performance with a traditional protocol on a 32bit 33MHz PCI platform.

3 Conclusions

We eliminate fragmentation, checksum computation, data copy, and multiple DMA initializations from UDP/IP on Myrinet in a stepwise manner. An interesting result is that we can significantly improve the performance by removing both checksum computation and data copy. In addition, this paper shows how the low overhead NIC influences the throughput. The optimized UDP/IP achieves 926Mbps on a 32bit 33MHz PCI platform.

References

1. Boden, N. J., Cohen, D., Felderman, R. E., Kulawik, A. E., Seitz, C. L., Seizovic, J. N., and Su, W. -K.: Myrinet – A Gigabit-per-Second Local-Area Network. IEEE-Micro, Vol. 15, No. 1, pp. 29-36, February 1995.
2. Dunning, D., Regnier, G., McAlpine, G., Cameron, D., Shubert, B., Berry, A. M., Gronke, E., and Dodd, C.: The Virtual Interface Architecture. IEEE Micro, Vol. 8, pp. 66-76, March-April 1998.
3. Gallatin, A., Chase, J., and Yocum, K.: Trapeze/IP: TCP/IP at Near-Gigabit Speeds. In Proceedings of 1999 USENIX Technical Conference, June 1999.
4. Kay, J. and Pasquale, J.: Measurement, Analysis, and Improvement of UDP/IP Throughput for the DECstation 5000. In Proceedings of 1993 USENIX Winter Conference, pp. 249-258, 1993.
5. Myricom Inc.: Myrinet User's Guide. http://www.myri.com, 1996.
6. Myricom Inc.: The GM Message Passing System. http://www.myri.com, January 1998.
7. Partridge, C. and Pink, S.: A Faster UDP. IEEE/ACM Transactions on Networking, Vol. 1, No. 4, pp. 429-440, August 1993.
8. Prylli, L. and Tourancheau, B.: BIP: a new protocol designed for high performance networking on myrinet. In Proceedings of IPPS/SPDP98, 1998.
9. Yoo, C., Jin, H. -W., and Kwon, S. -C.: Asynchronous UDP. IEICE Transactions on Communications, Vol.E84-B, No.12, pp. 3243-3251, December 2001.

Stabilizing Inter-domain Routing in the Internet*

Yu Chen[1], Ajoy K. Datta[1], and Sébastien Tixeuil[2,**]

[1] Department of Computer Science, University of Nevada Las Vegas
[2] LRI-CNRS UMR 8623, Université Paris Sud, France

Abstract. This paper reports the first self-stabilizing Border Gateway Protocol (BGP). BGP is an inter-domain routing protocol. Self-stabilization is a technique to tolerate arbitrary transient faults. The purpose of self-stabilizing BGP is to solve the routing instability problem. The routing instability in the Internet can occur due to errors in configuring the routing data structures, transient physical and data link problems, software bugs, and memory corruption. This instability can increase the network latency, slow down the convergence of the routing data structures, and can also cause the partitioning of networks. The self-stabilizing BGP presented here provides a way to detect and automatically recover from this kind of faults.

Keywords: Border Gateway Protocol, routing, routing instability, self-stabilization.

1 Introduction

Self-stabilization. The concept of self-stabilization was first introduced by Edsger W. Dijkstra in 1974 [2]. It is now considered to be the most general technique to design a system to tolerate arbitrary transient faults. A self-stabilizing system guarantees that starting from an arbitrary state, the system converges to a legal configuration in a finite number of steps, and remains in a legal state until another fault occurs (see also [3]).

Routing Instability. Routing instability in the Internet has a number of origins including route configuration errors, transient physical and data link problems, software bugs, and sometimes memory corruption. Moreover, in a recent work of Varghese and Jayaram (see [12]), it is proven that the crash failures of routers can lead every other router in the network to an arbitrary state, and that if links can reorder and loose messages, then any incorrect global state is reachable. Therefore, it is very important to have a self-stabilizing [2] routing protocol which can recover from any arbitrary faults without any external intervention.

The Border Gateway Protocol (BGP) is an inter-domain routing protocol [10]. It is used to exchange network reachability information among autonomous systems (ASs). The BGP gained a lot of popularity in the last few years due to the significant growth in the number of ISPs.

* The extended version of this article is available in [1].

** This author was supported in part by the french project STAR.

B. Monien and R. Feldmann (Eds.): Euro-Par 2002, LNCS 2400, pp. 749–752.

Related Work. Some experimental studies [9] were done to investigate the Internet stability and the origins of failure in the Internet backbones, but no solution was given. In 1997 [9], the Internet routing instability was investigated based on the data collected from BGP routing messages generated by border routers at five Internet public exchange points during a nine month period. Overall, the study showed that the Internet continued to exhibit high levels of routing instability despite the increased emphasis on aggregation (combining smaller IP prefixes into a single route announcement) and the deployment of route dampening technology (refusing to believe the updates that exceed certain parameters of instability). A Simple Path Vector Protocol (SPVP) was presented in [6–8] to eliminate the possible route oscillations in BGP. But, none of the above protocols is self-stabilizing.

The correct routing information in an autonomous system helps BGP achieve a stable routing among the autonomous systems. Numerous intra-domain routing schemes were developed, such as OSPF and RIP.

Contributions. None of the previous work on BGP is self-stabilizing. In this paper, we present the first self-stabilized Border Gateway Protocol. The proposed protocols also dynamically allocate/deallocate storage for the routing information as the network size changes. Our algorithm requires $O(IDiam)$ to stabilize, where $IDiam$ is the maximum diameter of an autonomous system. If we assume that the degree of each router and the capacity of each data link is significantly smaller than the number of autonomous systems, then the memory used at each router is $O(NBR \log NBR)$, where NBR is the number of autonomous systems.

Specification of Inter-domain Routing Problem. We define $NEXT\_BR$ as the IP address of the border router that should be used as the next closest neighbor towards a specific destination.

Specification 1 *We consider a computation e of the Inter-domain routing problem to satisfy the specification $\mathcal{SP}_{\mathcal{IDR}}$ when the routing tables* (i) *do not contain any information about the unreachable nodes in the system and* (ii) *contain the IP address of $NEXT\_BR$ from which a specific destination can be reached.*

2 Stabilizing BGP

We now present a self-stabilizing BGP algorithm, called Algorithm $SBGP$. The algorithm starts in an arbitrary state without any initialization. A node in Algorithm $SBGP$ periodically sends its own routing table to its neighboring peers. Therefore, if a node loses all its routing information due to memory corruption, it would be able to reconstruct the table by receiving the complete routing tables from its neighboring peers. In the original algorithm, the nodes send only the update messages. So, in case of memory corruption, the system would crash because a node might never regain the complete routing information. A part of our algorithm that eliminates the possibility of the routing oscillation is an adaptation of the $SPVP$ algorithm proposed in [6–8].

The abstract version of the $SBGP$ algorithm is presented as Algorithm 2.1. (Detail version of the algorithm is omitted due to space limitations.) A router p may have two different types of neighbors: internal and external. An internal neighbor q may not be directly connected to p, but is in the same autonomous system as p. We call q an $Internal\_Peer$ of p. An external neighbor q' is directly connected to p, but resides in a different autonomous system than p. We refer to q' as an $External\_Peer$ of p. When an $SBGP$ router wants to send a message to its internal peers or external peers, it checks its $Interal\_Peers$ or $External\_Peers$ sets to make sure that the corresponding peer router is active. We now explain how each $SBGP$ router p maintains the sets $Internal\_Peers_p$ and $External\_Peers_p$. From the underlying OSPF protocol, p gets the set $Internal\_Peers_p$. From the local topology maintenance algorithm (e.g. of [4]), p obtains the set of its neighbors, from which it may withdraw those that do not run the $SBGP$ algorithm, and hence obtains the set $External\_Peers_p$.

Algorithm $SBGP$ is message reactive (apart from the timeout mechanism needed as shown by [5] and that is not presented in Algorithm 2.1) and aware of three kinds of messages. Update and acknowledgment messages (Lines 1.01 – 1.12 of Algorithm 2.1) are used to implement a self-stabilizing *cycle of broadcasting* by using the counter flushing mechanism of [11]. Each $SBGP$ router sends update messages to keep the routing tables up-to-date, and expects acknowledgment from all external peers. A counter is added to each message so that eventually, messages that do not belong to the current cycle of broadcasting are removed from the network. The third kind of message is the IP packet (Lines 1.15 – 1.20 of Algorithm 2.1) that may be used within an AS to carry update or acknowledgment messages among routers that do not run Algorithm $SBGP$.

3 Conclusions

In this paper, we presented a self-stabilizing BGP algorithm, Algorithm $SBGP$, which is based upon a practical (Internet) protocol. The algorithm $SBGP$ takes $O(IDiam)$ to stabilize after the underlying local topology maintenance protocol is stabilized, where $IDiam$ is the diameter of an autonomous system that has the largest diameter. Our solution makes use of the counter flushing scheme [11] to ensure the reliable delivery of the control messages, and of the Update algorithm of [4] to maintain local topology information.

Since BGP-4 became the standard inter-domain routing protocol, a lot of new features, such as AS confederations, route flap damping, communities, etc, have been developed. Those extensions make the BGP a more scalable and robust routing protocol. A future research topic would be to design a self-stabilizing BGP with all the new features. Also, in our implementation of BGP, we avoided the interaction between BGP and the underlying IGP. Future work can investigate the alternatives of carrying the transit traffic, such as propagation of BGP information via IGP.

Algorithm 2.1 (*SBGP*) Abstract Version of Stabilizing BGP

```
upon receipt of  an update message from neighbor q
    if my counter value and the received counter value are different then
        Save the message in the waiting queue;
        Send an acknowledgment back to q;
        if the previous broadcasting cycle is done then
            Update Adj_RIB_In_qp; Decision Process;
        endif
    else
        Send an acknowledgment back to q;
    endif
upon receipt of  an acknowledgment message from external peer q
    Record the acknowledgment;
upon receipt of  an IP packet from internal peer q
    if I am the destination then
        Decapsulate the IP packet;
        Run the update part or the acknowledgment part of the algorithm;
    else
        Forward the IP packet to the best
        neighbor towards the destination using OSPF routing scheme;
    endif
```

References

1. Y Chen, A K Datta, and S Tixeuil. Stabilizing inter-domain routing in the internet. Technical Report 1302, LRI-CNRS UMR 8623, 2002.
2. EW Dijkstra. Self stabilizing systems in spite of distributed control. *Communications of the Association of the Computing Machinery*, 17:643–644, 1974.
3. S Dolev. *Self-stabilization*. The MIT Press, 2000.
4. S Dolev and T Herman. Superstabilizing protocols for dynamic distributed systems. *Chicago Journal of Theoritical Computer Science*, 3(4), 1997.
5. MG Gouda and N Multari. Stabilizing communication protocols. *IEEE Transactions on Computers*, 40:448–458, 1991.
6. T Griffin, FB Shepherd, and G Wilfong. Policy disputes in path-vector protocols. In *Proceedings of ACM SIGCOMM 1999*, 1999.
7. T Griffin and G Wilfong. An analysis of bgp convergent properties. In *Proceedings of ACM SIGCOMM 1999*, 1999.
8. T Griffin and G Wilfong. A safe path vector protocol. In *Proceedings of IEEE INFOCOM 2000, Book 2*, pages 490–499, 2000.
9. C Labovitz, GR Malan, and F Jahanian. Internet routing instability. In *Proceedings of ACM SIGCOMM 1997*, 1997.
10. Y Rekhter and T Li. A border gateway protocol 4(bgp-4). Technical report, Network Working Group, 1995.
11. G Varghese. Self-stabilization by counter flushing. In *PODC94 Proceedings of the Thirteenth Annual ACM Symposium on Principles of Distributed Computing*, pages 244–253, 1994.
12. G Varghese and M Jayaram. The fault span of crash failures. *Journal of the ACM*, 47(2):244–293, March 2000.

Performance Analysis of Code Coupling on Long Distance High Bandwidth Network*

Yvon Jégou

IRISA / INRIA, Campus de Beaulieu, 35042 Rennes Cedex, France,
Yvon.Jegou@irisa.fr

Abstract. Coupling numerical simulations running on distant nation-wide sites becomes realistic through the introduction of high bandwidth networks. The code coupling library we experiment in this paper relies on the presence of a software DSM for locating the data in the memories of PC clusters. In this paper, we show on some experiments that a strict share of the network bandwidth between the computation nodes improves the performance even in the case where the charge of transferring the data is balanced between the computation nodes.

1 Introduction

With the interconnection of parallel computers through wide-area networks, new applications in the numerical simulation domain become realistic. Code coupling allows different simulations to collaborate and to dynamically exchange data. When the parallel codes run on distributed memory architectures such as PC clusters, the data involved in the exchanges are scattered in the memories and must be located by the coupling interface. Moreover, in general, each individual computation node of a cluster has a limited communication capacity and all the nodes must contribute to the transfers when large volumes of data must be transferred. Component models such as CORBA [5] allow to connect codes running on distinct platforms, but their application domain is currently limited to shared memory architectures. Extensions to these models such as the Parallel CORBA Objects [6] have been proposed and consider the distribution of the object elements on the memories. Simulation codes running on parallel architectures can also be coupled using specialized coupling libraries such as PAWS [1] or COCOLIB [2]. These libraries consider various kinds of data representations and distributions in the parallel codes but their implementation do not specifically target high performance networks. The Mome distributed shared memory (DSM) has been developed as a base for building run-time systems for HPF and OpenMP [3]. The Mome coupling library [4] takes profit of this linear shared address space when accessing the coupled data and leaves more freedom for the organization of the transfers on the computation nodes than in COCOLIB or PAWS. The current implementation of this library targets high performance interconnection networks and high transfer rates between clusters of workstations or PCs. If the number of nodes involved in the computations is large enough, the network interfaces compete for network bandwidth during the transfers. The experimental results

* Part of this work has been founded by the French government under the framework of the RNRT VTHD program

B. Monien and R. Feldmann (Eds.): Euro-Par 2002, LNCS 2400, pp. 753–756.

of the next section show that, through a fair share of the available bandwidth between these network interfaces, it is possible to exploit the whole throughput of wide area high performance IP networks.

2 Performance Analysis

2.1 Experimentation Platform

The Mome coupling library was experimented on two PC clusters connected through a high bandwidth interconnection network, the French experimental VTHD network <http://www.vthd.org>. The computation nodes of each cluster are connected to an Ethernet switch using 100 Mbits/s FastEthernet links. These switches are connected to the VTHD network by 1 Gbits/s fiber optic links. The throughput of the VTHD backbone is 2.5 Gbits/s. The clusters are located in Rennes and Sophia-Antipolis in France, separated by about 1000 kilometers. The machines used for these experiments are dual 1Ghz processors PCs running Linux 2.2.18.

The coupling library was experimented using a finite difference numerical simulation code on the server cluster located in Rennes and a visualization application running on the client cluster in Sophia-Antipolis. For the purpose of these experimentations, a 2400×2400 mesh computed by the server (46 Mbytes) is transfered after each iteration of the server to the visualization client. The transfer and the computation of the next iteration overlap.

2.2 Preliminary Tests

The same number of computation nodes is used on both side for the preliminary test. In this simple case, the coupling library organizes a simple point-to-point communication scheme between the two sites using TCP/IP sockets. The transfer load is well-balanced on the server and on the client nodes. If more than eleven nodes are in use on both sides, then server nodes can potentially saturate the connection to the VTHD network which is limited to 1 Gbits/s. Each new transfer generates a traffic burst which must be regulated by the TCP/IP congestion protocol. Using this configuration, with twelve nodes on both sides, the total throughput between the two sites is limited to 800 Mbits/s.

2.3 Send and Receive TCP Buffer Size

The TCP/socket interface allows some control, for instance through the specification of the send and receive TCP buffer size. Table 1 shows the total throughput between twelve servers and twelve clients, depending on the TCP send and receive buffer sizes. For this test, each point-to-point connection is implemented by two sockets.

In this table, the best performance values can be seen on the row corresponding to a send buffer length 48Kb and on the column corresponding to receive buffer length 96Kb. In the case where the two buffer lengths are higher than these optimal values (bottom right of the table), the throughput drops below 800 Mbits/s, with variations during the simulation. TCP/IP statistics show packet losses in this region. Below the optimal values, the throughput increases linearly with the buffer length. In the remaining tests, the receive buffer length was fixed to the maximum value 256K bytes.

Table 1. Throughput (Mbits/s) / length of TCP buffers

TCP send buffer (Kb)	Length of TCP receive buffer (Kb)							
	16	32	48	64	80	96	112	128
16	165.5	336.7	337.3					337.3
32	165.3	367.2	538.5	655.4	664.8	665.0	665.0	665.4
48		365.0	535.1	719.1	903.8	928.2	926.8	924.6
64			536.4	719.1	896.5	939.4	870.0	817.9
80				720.6	876.2	926.8	842.5	846.9
96					890.1	929.2	845.0	817.9
112						929.0	845.9	748.1
128							863.2	794.4

Table 2. Throughput (Mbits/s) versus TCP send buffer length and number of channels

TCP send buffer length (Kb)	Number of channels		
	1	2	4
16	169.8	338.9	672.7
32	334.5	666.2	807.8
48	489.1	936.5	811.8
64	668.6	840.6	769.3
80	816.8	803.1	779.5
96	922.3	787.8	820.9
112	930.7	762.9	790.2
128	773.6	815.5	762.8
160	795.9	780.1	846.1
192	809.6	765.8	813.9
224	856.9	806.0	798.9
256	842.8	820.9	820.0

Table 3. Retransmitted packets versus TCP send buffer length and number of channels

TCP send buffer length (Kb)	Number of channels		
	1	2	4
16	0	0	0
32	0	0	407
48	0	0	1013
64	0	93	1466
80	0	258	1188
96	0	258	1300
112	0	343	1357
128	46	378	1392
160	112	416	1100
192	196	477	1162
224	169	324	1328
256	129	320	1288

2.4 Send Buffer Length versus Number of Sockets

Table 2 shows the total throughput depending on the TCP send buffer length and on the number of sockets associated to each point-to-point connection. For a fixed number of sockets, the throughput linearly increases with the buffer length until 930 Mbits/s and then drops to 800 Mbits/s. In the four channels case, this limit is reached using a 24 Kbytes TCP send buffer length (this result does not appear in the table). This comportment can be explained from the observation of the number of packet retransmissions in table 3. Below the optimal buffer length, no packet is lost during the transfers. Above the optimal send buffer length, some packets are lost and the TCP layer seems to take the control, trying to adjust the throughput to the available bandwidth.

2.5 Balancing Transfer Duration

The previous experimentations considered a simple use of the coupling library: identical number of computation nodes on both sides, identical data distributions. This simple case results in a perfect balance of the block sizes. This perfect balance is not always possible, for instance, when the number of computation nodes differ on the server and the client sides. Running our codes with different numbers of processors results in rather poor performance, 750 Mbits/s with twelve server nodes and fourteen client nodes. All sockets do not transfer the same volume of data. The network is overfilled at the beginning of the transfer and under-used at the end.

To overcome this limitation, another send buffer length computation has been implemented in the library. Using this new algorithm, the send buffer size associated to some socket is derived from the number of bytes sent on this socket in order to reach the same transfer duration on all sockets. Using this new buffer length computation, the total throughput is measured above 900 Mbits/s.

3 Conclusion and Future Work

In this paper, we analyzed the behavior of code coupling through a large bandwidth, long distance network. The first experimentations show that it is possible to really exploit all the bandwidth even from off-the-shelf PC cluster. But these experimentations also show a rather large performance loss (between 10 and 20%) as soon as the requested throughput exceeds the capacity of the network.

Finding correct TCP buffer sizes and number of sockets greatly improves performance but is a difficult task. The optimal values depend on the network throughput and latency, and on the application needs. It should be possible to automatically evaluate correct values for these parameters using a simple test.

References

1. Peter H. Beckman, Patricia K. Fasel, and William F. Humphrey. Efficient Coupling of Parallel Applications Using PAWS. In *High Performance Distributed Computing (HPDC)'7*, Chicago, IL, July 1998.
2. Erik Brakkee, Klaus Wolf, Dac Phuoc Ho, and Anton Schuller. The COupling COmmunications LIBrary. In *Proceedings of the fifth Euromicro Workshop on Parallel and Distributed Processing*, pages 155–162. IEEE Computer Society Press, January 1997.
3. Y. Jégou. Controlling distributed shared memory consistency from high level programming languages. In *5th International Workshop on High-Level Parallel Programming Models and Supportive Environments*, mai 2000.
4. Yvon Jégou. Coupling DSM-Based Parallel Applications. In *Proceedings of the First IEEE/ACM International Symposium on Cluster Computing and the Grid*, pages 82–89, May 2001.
5. Object Management Group. *Common ObjectRequest Broker: Architecture and Specification*, October 1999.
6. Christophe René and Thierry Priol. MPI Code Encapsulation Using Parallel CORBA Object. In *Proceedings of the Eighth IEEE International Symposium on High Performance Computing*, pages 3–10, August 1999.

Adaptive Path-Based Multicast on Wormhole-Routed Hypercubes

Chien-Min Wang[1], Yomin Hou[1], and Lih-Hsing Hsu[2]

[1]Institute of Information Science
Academia Sinica, Taipei 115, Taiwan, ROC
{cmwang, ymhou}@iis.sinica.edu.tw
[2]Department of Computer and Information Science
National Chiao-Tung University, Hsinchu, Taiwan, ROC
lhhsu@cis.nctu.edu.tw

Abstract. In this paper, we consider path-based multicast on wormhole-routed hypercubes. A minimum set of routing restrictions is used as the base routing algorithm. To correctly perform multicast, we present the natural list to sort the destination nodes. It can be proved to be deadlock-free for both one-port and multi-port systems. Furthermore, it creates only one worm for each multicast. Between each pair of nodes in a multicast path with distance k, on the average, there are at least $(k+1)!/2^k$ adaptive shortest paths, which is superior to previous works. We also propose a heuristic algorithm to reduce the path length. The simulation result shows that its sustainable throughput is much better than related works. In addition, unicast and broadcast can be treated as degenerated cases and use the same routing algorithm. Therefore, it offers a comprehensive routing solution for communication on hypercubes.

1 Introduction

Multicast is a collective communication in which the same message is delivered from a source node to an arbitrary number of destination nodes on a parallel computer. Both *unicast*, which involves a single destination, and *broadcast*, which involves all nodes in the network, are special cases of multicast. Multicast has several uses in large-scale multiprocessors, including direct use in various parallel algorithms, implementation of data parallel programming operations, such as replication and barrier synchronization [15], and support of shared-data invalidation and updating in systems using a distributed shared-memory paradigm [5].

Wormhole routing has been widely adopted recently due to its effectiveness in inter-processor communication [1], [11]. It divides each message into small pieces called flits and transmits flits in a pipeline fashion. To support multicast on wormhole-routed networks, different approaches had been considered in previous research. Some of them used the *unicast-based* strategy [9], [14]. In this strategy, if there are k destination nodes, then k unicasts will be generated to send the message. This strategy raises a problem that the increased traffic load of those unicasts may hinder the system performance. In order to minimize the network traffic, some research

B. Monien and R. Feldmann (Eds.): Euro-Par 2002, LNCS 2400, pp. 757–766.

tem performance. In order to minimize the network traffic, some research considered the *path-based* multicast. In this approach, a *multicast path* consists of a set of consecutive channels, starting from the source node and traversing each destination in the set. *Hamiltonian path* approach had been used to prevent deadlock from the channel waiting cycle. However, there are also possible deadlocks due to dependencies on consumption channels in a path-based wormhole-routed network. Boppana *et al.* showed how such a deadlock might happen and proposed the *column-path* routing to eliminate this problem for 2D meshes [2]. For *k*-ary *n*-cubes, Panda *et al.* proposed a framework of a *base-routing-conformed-path* (BRCP) model [12], [13]. In these algorithms, there may be several worms required for a multicast, and a multi-port model is necessary to perform these worms concurrently.

For multicast on hypercubes, Lin proposed the UD-path method without the need of multiple worms [6]. It is an extension of the deterministic path-based method [7]. Based on a node labeling method, the author defined *up path* (*U-path*), *down path* (*D-path*), and *up-down path* (*UD-path*). For a unicast, all the UD-paths can be used. For a multicast, the destination nodes are carefully ordered so that there is at least one multicast path, which is an UD-path and can be used to send the message. However, from a node to the next destination node, there may be only U-paths or D-paths that can be used. Therefore, the routing algorithms for sending messages between two nodes are different for unicast and multicast. Moreover, it required at least two input ports (consumption channels) to prevent deadlock.

In this paper, we proposed an adaptive multicast routing algorithm for wormhole-routed hypercubes. For each multicast, the proposed approach creates only one worm and provides adaptive shortest paths between each pair of nodes in the multicast. All the multicasts, including unicasts and broadcasts, follow the same routing rules. Therefore, it offers a comprehensive routing solution for communication on hypercubes. It is proved that the proposed approach is deadlock-free and can be applied on either one-port or multi-port systems. We also propose a heuristic algorithm to reduce the path length. The potential adaptivity of the proposed algorithm is superior to that of the UD-path algorithm. The simulation results clearly show significant performance improvement provided by the proposed approach.

2 Background

In an *n*-dimensional hypercube, each node x corresponds to an *n*-bit binary number, where $x(i)$ denotes the *i*th bit of x, $0 \le i \le n-1$. If there are k different bits between binary strings of two nodes x and y, then k is said to be the *distance* of these two nodes. Two nodes are connected with a pair of channels, one for each direction, if and only if their distance is 1. A channel from node x to x', $x(k) \neq x'(k)$, is said to be at dimension k and denoted by $c^{k}_{x,x'}$. In case $x(k)=0$, the channel is called *positive* and denoted by $c^{k+}_{x,x'}$, otherwise it is *negative* and denoted by $c^{k-}_{x,x'}$. Communications are handled by *routers* as shown in Fig. 1. The *external channels* connect a router to its neighboring routers, and the *internal channels* connect to its local processor/memory. The internal input channels are also known as *consumption channels*. The *port model* refers to the number of internal channels in each node [10]. If each node possesses more than one pair of internal input/output channels, it is called a *multi-port* system. Otherwise, it is called a *one-port* system. An *all-port* system is a special case of a multi-port system, in which every external channel corresponds to a distinct internal channel. Let $c_i^{IN(j)}$ and $c_i^{OUT(j)}$ be the *j*th internal input and output channel of node *i*, respectively. The number *j* may be omitted for the one-port model.

Lin *et al.* have developed an approach to hardware-supported multicast, called *path-based routing* [7]. In this approach, a *multicast path* for a source and a set of destinations consists of a set of consecutive channels, starting from the source node and traversing each destination in the set. The message sent by the source node may be replicated at intermediate destination nodes and forwarded along the multicast path to the next destination node. A multicast can be denoted as $(s, \{ d_0, d_1, \ldots, d_{r-1} \})$, where *s* is the source node and $\{d_0, d_1, \ldots, d_{r-1}\}$ is the set of destination nodes. A *multicast list* is an ordered list, which indicates the order of the destination nodes in the multicast path. We shall use $<s: d'_0, d'_1, \ldots, d'_{r-1}>$ to represent a multicast list, where $(d'_0, d'_1, \ldots, d'_{r-1})$ is a permutation of $(d_0, d_1, \ldots, d_{r-1})$. Fig.2 shows a possible multicast path for the multicast list <1: 3, 6, 4> on a 3D hypercube.

Routing algorithms can be represented as a set of routing restrictions. The routing restrictions specify which external input channels can forward messages to certain external output channels, called *legal channels*. As the header flit(s) arrives a router, the router determines which legal channel is used. The base routing algorithm used in this paper is based on the MIN routing algorithm [4]. However, it was proposed for unicast and might not be suitable for multicast. Hence, the routing restrictions are modified so that messages can be forwarded from $c^l_{x,y}$ to $c^m_{y,z}$ on the router of node *y* if and only if one or both of the following conditions are true: (1) $m < l$, or (2) c^m is positive. In our algorithm, a router determines which channels are required according to the destination node of a message. Among the channels that are both required and legal, the one at the highest dimension is chosen if it is negative. Otherwise, any one can be chosen. The only difference between our algorithm and the MIN algorithm is that, in the former, messages cannot be forwarded from a lower dimensional channel to a negative higher dimensional channel; while, in the latter, messages cannot be forwarded from a higher dimensional channel to a negative lower dimensional channel. Therefore, our algorithm is also deadlock-free for unicast and provides a minimum set of deadlock-free routing restrictions as the MIN algorithm [4].

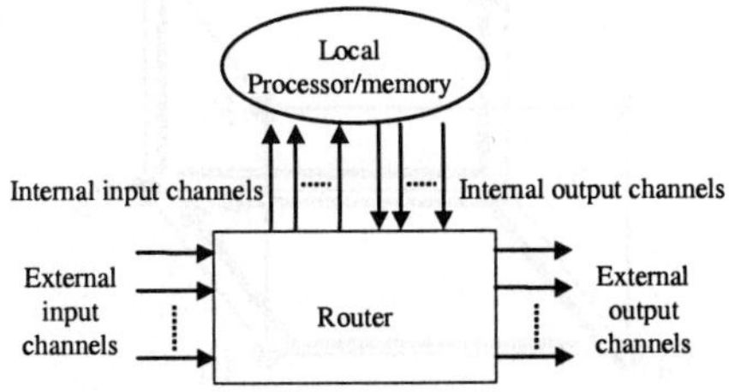

Fig. 1. The architecture for each node.

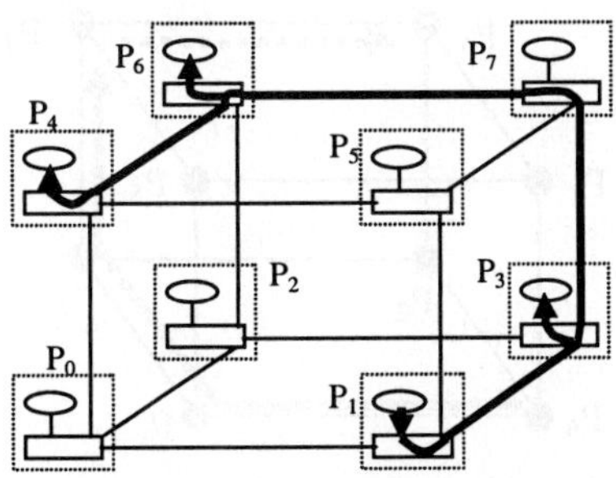

Fig. 2. An example multicast path.

3 The Multicast Routing Strategy

The most fundamental requirement for an efficient path-based multicast routing algorithm is to avoid deadlock. Next, it is expected that the degenerate cases such as unicast and broadcast also use the same algorithm, thereby offering a comprehensive routing solution. Finally, it is desired to minimize the time required to deliver the messages. However, due to channel contention, it is hard to accurately predict the message delivery time in an analytical way. Therefore, in this paper, we aim to provide more paths between source/destination nodes and to minimize the path length of a multicast. In particular, a unicast message routed according to the algorithm should always follow a shortest path. Upon developing an efficient path-based multicast routing algorithm, two issues must be considered: (1) How to route a multicast message? (2) How to order the destination nodes, i.e., how to decide the multicast list for a multicast? Our answer to the first question is the base routing algorithm presented in the previous section. In this section, we shall present the natural list method to construct a multicast list for a multicast so that no deadlock may be incurred. We will also give a heuristic algorithm for the natural list in order to reduce the path length.

3.1 The Natural List

For a multicast with r destination nodes, there are $r!$ possible multicast lists. However, only a fraction of them can be performed correctly under the base routing algorithm without violating its routing restrictions. Fig. 3 shows such an example based on e-cube routing [11], which traverses required channels according to their dimensions in increasing order. For the multicast list <0: 7, 6>, the message is first sent from P_0 to P_7 through the path $(c_{0,1}, c_{1,3}, c_{3,7})$. Then, it tries to forward the message from P_7 to P_6 through channel $c_{7,6}$. Since $c_{3,7}$ is at dimension 2 and $c_{7,6}$ is at dimension 0, the message cannot be forwarded from $c_{3,7}$ to $c_{7,6}$ under e-cube routing. A multicast list is said to be *legal* if there exists at least a legal channel to forward the message at any router in its path before reaching its last destination node; otherwise it is *illegal*.

It's not a trivial job to find a legal multicast list for any base routing algorithm. Fortunately, for the base routing algorithm in the previous section, we can construct a legal multicast list as follows. Consider the multicast $(s, \{d_0, d_1, \ldots, d_{r-1}\})$. The set of

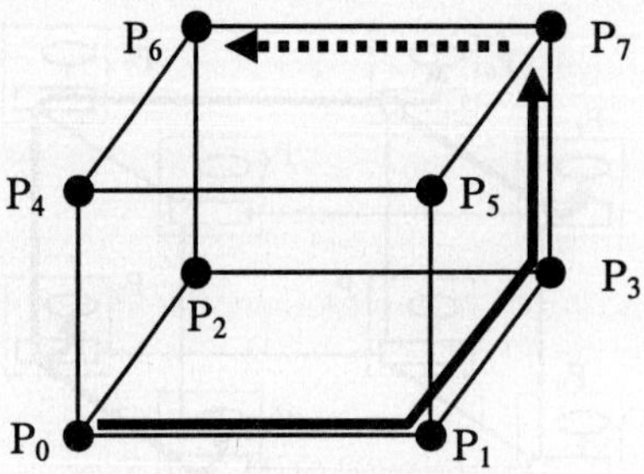

Fig. 3. An illegal multicast list.

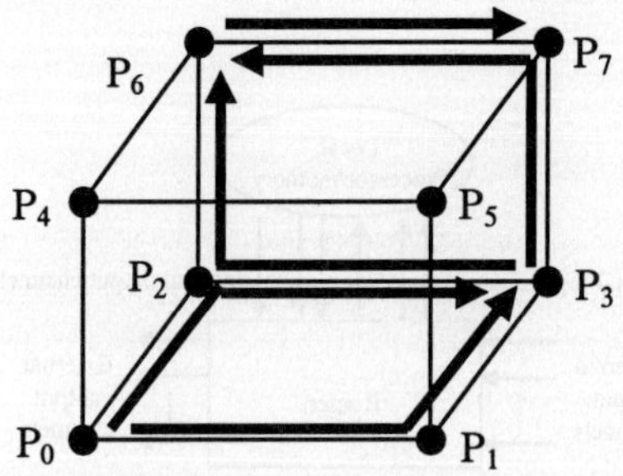

Fig. 4. Paths for an example natural list.

destination nodes can be sorted increasingly such that we can obtain a permutation $(d'_0, d'_1, \ldots, d'_{r-1})$ of $\{d_0, d_1, \ldots, d_{r-1}\}$, where $d'_i<d'_{i+1}$ for $0\leq i<r-1$. Then, a multicast list $<s: d'_0, d'_1, d'_2, \ldots, d'_{r-1}>$ can be constructed for $(s, \{d_0, d_1, \ldots, d_{r-1}\})$. Such a multicast list is called the *natural list.* Before showing the natural list is legal under the routing restrictions, let's see the example illustrated in Fig.4. The multicast (0, {3, 6, 7}) is performed by the natural list under the base routing algorithm. Between the source node P_0 and the first destination node P_3, there are two available paths, $(c_{0,1}, c_{1,3})$ and $(c_{0,2}, c_{2,3})$. If $(c_{0,1}, c_{1,3})$ is selected, then there are two available paths, $(c_{3,2}, c_{2,6})$ and $(c_{3,7}, c_{7,6})$, for forwarding the message from P_3 to P_6. On the other hand, if $(c_{0,2}, c_{2,3})$ is selected, then there is only one available path, $(c_{3,7}, c_{7,6})$. The reason is that $c_{2,3}$ and $c_{3,2}$ are both at dimension 0, and $c_{3,2}$ is negative. Therefore, it is not allowed to forward the message from $c_{2,3}$ to $c_{3,2}$. The last channel in the multicast path is $c_{6,7}$. Since $c_{6,7}$ is a positive channel, it is always legal for forwarding the message from P_6 to P_7. The following theorems show that the natural list is legal and deadlock-free under the base routing algorithm.

Theorem 1. The natural list is legal under the base routing algorithm.

Proof: Let $<s: d'_0, d'_1, \ldots, d'_{r-1}>$ be the natural list for multicast $(s, \{d_0, d_1, \ldots, d_{r-1}\})$. First, the multicast message can be sent from s to d'_0 as a unicast. Accordingly, there exists at least a legal channel to forward the message from s to d'_0 at any router in this path. Next, consider any (d'_i, d'_{i+1}), $0\leq i<r-1$, in the natural list. Let the highest dimension among the required channels from d'_i to d'_{i+1} be denoted by h and its corresponding channel be denoted by c^h. Because $d'_i<d'_{i+1}$, c^h must be positive. When the multicast message arrives d'_i, no matter which channel it comes from, c^h can be selected as the first channel to forward the message from d'_i to d'_{i+1}. After the first channel is selected, the base routing algorithm can be used continually so that the message can be sent to d'_{i+1} correctly. Hence, there exists at least a legal channel to forward the message from d'_i to d'_{i+1} at any router in this path. Iteratively applying the above reasoning for $0\leq i<r-1$, we can conclude there exists at least a legal channel to forward the message in the multicast path. Therefore, the natural list is legal under the base routing algorithm. ■

Theorem 2: The natural list is deadlock-free under the base routing algorithm.

3.2 The Heuristic List

In the above subsection, we have shown the natural list approach. It is depth contention-free and convenient for implementation. However, sometimes the path length of a natural list for a multicast is longer than necessary. For example, the path length of the natural list <0: 3, 6, 7> in Fig. 4 is 5. Suppose the message is forwarded from P_3 through channels $c_{3,7}$ and $c_{7,6}$ to P_6. Obviously, the message has already passed P_7. Hence, there is another multicast list <0: 3, 7, 6> whose path length is only 4 as shown in Fig. 5. Because a shorter path requires fewer system resources, the system

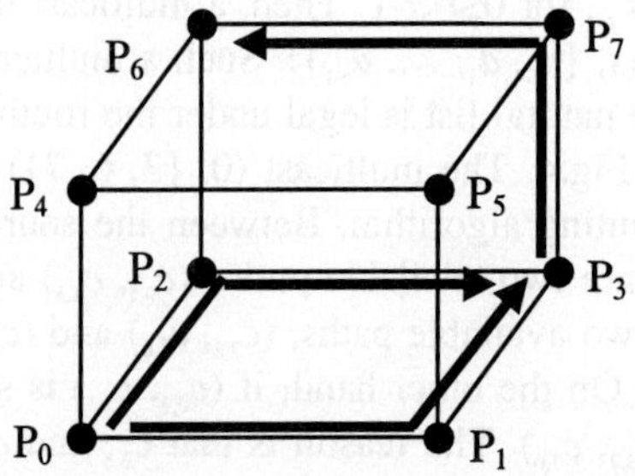

Fig. 5. Paths for an example heuristic list.

```
Heuristic_list(<s:d'0,…,d'r-1>){
  NewList=<s:d'0>;
  for(i=1; i<r; i++){
    1. Use Theorem 3 to determine
    all the legal positions to
    insert d'i to NewList.
    2. Choose a legal position to
    minimize the path length.
    3. Insert d'i to the chosen
    position in NewList.
  }
  return(NewList);
}
```

Fig. 6. The proposed heuristic algorithm.

performance can thereby be improved. In the following paragraphs, we will give a heuristic algorithm for the natural list in order to reduce the path length.

Let $<s: d'_0, d'_1, \ldots, d'_{r-1}>$ be the natural list of $(s, \{d_0, d_1, \ldots, d_{r-1}\})$. The heuristic algorithm is given in Fig. 6. The basic concept is to insert destination nodes one-by-one into a legal multicast list. At each step, we determine where to insert the next node so that the new multicast list is still legal and the path length can be minimized. The time require for step i is $O(i)$, Hence, the complexity of the heuristic algorithm is $O(r^2)$ if there are r destination nodes. We shall use the *heuristic list* to denote the new list generated by the heuristic algorithm. According to the base routing algorithm, it can be derived directly that the heuristic list is legal and deadlock-free.

Theorem 3: Given a legal multicast list $<s=d_0: d_1, d_2, \ldots, d_i, \ldots, d_k>$, $k<r-1$, if all the following conditions are true, then $<s=d_0: d_1, d_2, \ldots, d_{i-1}, d_{k+1}, d_i, \ldots, d_k>$ is also legal. (1) $d_{i-1}<d_{k+1}$ or $d_{i-1}=s$. (2) (the lowest required dimension from d_{i-1} to d_{k+1}) > (the highest required dimension from d_{k+1} to d_i). (3) If $i \neq k$ and $d_i>d_{i+1}$, (the lowest required dimension from d_{k+1} to d_i) > (the highest required dimension from d_i to d_{i+1}).

Theorem 4: The heuristic list is legal and deadlock-free under the base routing algorithm.

4 Performance Analysis

To show the advantage of the proposed approach, we analyze its potential adaptivity and compare the result with that of the UD-path method. Simulations are also conducted to show the performance improvement of the proposed approach.

4.1 Adaptivity Analysis

The potential adaptivity is measured by the average numbers of paths from the source node to the first destination node and from an intermediate destination node to the next destination node, respectively. The former is also equivalent to the average number of paths for unicast. Suppose that the distance in a unicast (s, d) is k. We may assume that (s, d) is a pair of antipodal nodes in a k-dimensional hypercube. Let the set of all combinations be denoted as S_k. It can be divided into two subsets S_k^+ and S_k^- by the sign of the highest dimensional channel c^{k-1}, i.e., $S_k^+ = \{(s, d) | (s, d) \in S_k$ and c^{k-1} is positive$\}$, and $S_k^- = \{(s, d) | (s, d) \in S_k$ and c^{k-1} is negative$\}$. Correspondingly, we can define all legal paths for these three sets as P_k, P_k^+, and P_k^-. Let t_k, t_k^+ and t_k^- be the cardinality of P_k, P_k^+, and P_k^-, respectively. Obviously, we have $t_k = t_k^+ + t_k^-$. For each path in P_k^+, since c^{k-1} is positive, it is a legal channel and can be selected at any time. Therefore, we can derive that $t_k^+ = k \times t_{k-1}$. For each path in P_k^-, since c^{k-1} is negative, it must be the first channel to be passed. Hence, $t_k^- = t_{k-1}$. Therefore, we can derive the average number, e_k, of paths from s to d with distance k as follows.

$$t_k = t_k^+ + t_k^- = k \times t_{k-1} + t_{k-1} = (k+1) \times t_{k-1} \text{ and } t_1 = 2. \Rightarrow t_k = (k+1)! \quad (1)$$

$$e_k = (k+1)! / 2^k \quad (2)$$

Next, suppose that $<s: d'_0, d'_1, \ldots, d'_{r-1}>$ is the natural list for $(s, \{d_0, d_1, \ldots, d_{r-1}\})$. We shall measure the average number of paths from a destination node d'_i to the next destination node d'_{i+1}. The number of paths from d'_i to d'_{i+1} depends on not only their addresses but also the external input channel c^e of d'_i, which the message is forwarded through to arrive d'_i. Similar to the above discussion, we may assume that (d'_i, d'_{i+1}) is a pair of antipodal nodes in a k-dimensional hypercube. Note that c^{k-1} must be positive because $d'_i < d'_{i+1}$ in the natural list. Hence, without considering the impact of c^e, we can derive the upper bound on the average number, e'_k, of paths from d'_i to d'_{i+1} with distance k as follows.

$$e'_k \leq k \times t_{k-1} / 2^{k-1} = k \times k! / 2^{k-1} \quad (3)$$

The impact of c^e will reduce the number of legal paths. Suppose $P_k^{\mathrm{P}} = \{p | p \in P_k$ and the first channel of p is positive$\}$ and $P_k^{\mathrm{N}} = \{p | p \in P_k$ and the first channel of p is negative$\}$. Let t_k^{P} and t_k^{N} be the cardinality of P_k^{P} and P_k^{N}, respectively. No matter which channel c^e is, we can derive

$$t_k^{\mathrm{N}} = t_{k-1} + (k-1) \times t_{k-1}^{\mathrm{N}} = (k+1)! / 2 \text{ and } t_k^{\mathrm{P}} = t_{k-1}^{\mathrm{N}} + k \times t_{k-1}^{\mathrm{P}} = (k+1)! / 2 \quad (4)$$

$$e'_k \geq (k \times t_{k-1}^{\mathrm{P}} + t_{k-1}^{\mathrm{N}}) / 2^{k-1} = (k+1)! / 2^k \quad (5)$$

This result is compared with that of the UD-path method and shown in Table 1. The first row gives the distances between two nodes. The average numbers of paths for the UD-path method is given in the second row. In the third and fourth rows, the lower bound and the upper bound of the proposed approach are presented. It can be

Table 1. Average number of paths between two nodes.

Distance	1	2	3	4	5	6	7	8	9	10
UD-Path	1.00	1.00	1.50	3.00	7.50	22.50	78.75	315.00	1417.50	7087.50
Natural List (Lower Bound)	1.00	1.50	3.00	7.50	22.50	78.75	315.00	1417.50	7087.50	38981.25
Natural List (Upper Bound)	1.00	2.00	4.50	12.00	37.50	135.00	551.25	2520.00	12757.50	70875.00

observed from the table that the average number of paths for the proposed approach is at least $(k+1)/2$ times than that of the UD-path method. It is of great improvement on the potential adaptivity. The potential adaptivity for the heuristic list is much more difficult to analyze. Therefore, we implement it on a simulation system to see the network behavior in the following subsection.

4.2 Performance Simulation

To investigate the performance improvement of the proposed approach, some experiments are made by simulating the network behavior of a 7-dimensional hypercube for the UD method, the natural list, and the heuristic list. For the UD method, there are two algorithms to construct the multicast list. One is greedy and the other is optimal for the path length under the UD routing algorithm. The simulation is conducted by using a tool based on MultiSim [8]. In the simulation, all the channels are assumed to have the same bandwidth, 1 flit/cycle, i.e., each flit requires one cycle to be transmitted through a channel, either internal or external channel. The fist-come-first-served policy is used for all the resources in the network.

All the processors generate messages at time intervals given by a negative exponential distribution random variable. The destination nodes are randomly chosen. The measures of interest are average message latency and average sustainable network throughput. The message latency is the number of cycles spent by a message in traveling from its source processor to its last destination, taking the queuing delay into account. The average network throughput indicates the average number of flits delivered per cycle per processor. It is sustainable if the number of messages queued at their source processors is small and bounded. For a given system, the average message latency, in general, grows as the throughput increases. At low throughput, the network latency is contributed mainly by the message length and the distance to travel because there is little queuing delay involved. As the throughput increases, more channel contention and longer queuing delay happened, give rise to a higher message latency.

Fig. 7 shows the simulation result for multicast. The number of destination nodes in a multicast is 16. It can be observed that the message latency for the four strategies is similar for low throughput. As the throughput increases, the message latency for the natural list increases much slower than that of the UD method. The main reason is due to its higher potential adaptivity. The heuristic list performs even better due to its shorter path length. In Fig. 8, we considered a mixture of unicast and multicast traffic as in the experiments of [2]. In all the communications, 90% are unicasts and 10% are multicasts. The number of destination nodes of a multicast is a uniform discrete random variable with values between 2 to 15. This traffic pattern could be representative of communication in cache-coherent shared memory multiprocessors; the majority of traffic is due to remote fetches of cache blocks and the rest is a mixture of traffic for invalidations of shared cache blocks and synchronizations. Figs. 8(a) and 8(b) show the result for message size = 20 flits and 200 flits, respectively. It can be observed that in both cases, the performance of the proposed approach is better than that of the

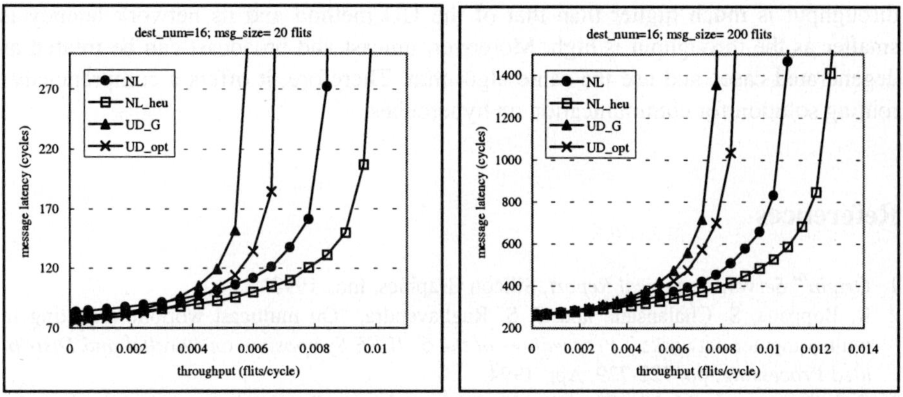

Fig. 7. Simulation result for multicast. (a) Message size = 20 flits. (b) Message size = 200 flits.

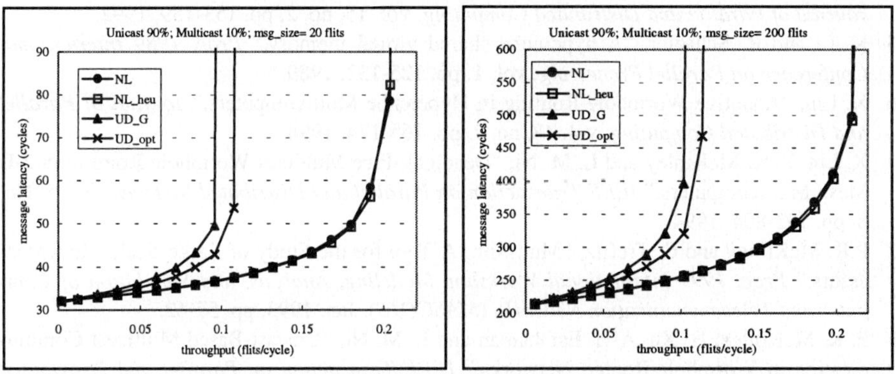

Fig. 8. Simulation results for the mixture of unicast and multicast traffic. (a) Message size = 20 flits. (b) Message size = 200 flits.

UD method. The sustainable throughput for the UD method is about 0.1 flits/cycle, while for the proposed algorithms, it can reach about 0.2 flits/cycle.

5 Conclusion

In this paper, we consider path-based multicast on wormhole-routed hypercubes. A set of routing restrictions is given to prevent deadlock for unicast. To perform multicast correctly, we present the natural list to sort the destination nodes. It can be proved to be deadlock-free for both one-port and multi-port systems. Furthermore, it creates only one worm for each multicast. Between each pair of nodes with distance k in a multicast list, the average number of paths, e'_k, is bounded by $(k+1)!/2^k \leq e'_k \leq k \times k!/2^{k-1}$. We also propose a heuristic algorithm to reduce the path length. The applicability of the proposed approach is demonstrated by simulation. Its sustainable throughput is much higher than that of the UD method and its network latency is smaller as the throughput is high. Moreover, unicast and broadcast can be treated as degenerated cases and use the same algorithm. Therefore, it offers a comprehensive routing solution for communication on hypercubes.

References

1 *Origin™ Servers Technical Report*, Silicon Graphics, Inc., 1998.
2 R. Boppana, S. Chalansani, and C. S. Raghavendra, "On multicast wormhole routing in multicomputer networks," *Proceedings of the 6th IEEE Symposium on Parallel and Distributed Processing*, pp. 722-729, Apr. 1994.
3 C. J. Glass and L. M. Ni, "The Turn Model for Adaptive Routing," *Journal of the Association for Computing Machinery*, vol. 41, no. 5, pp. 874-902, Sep. 1994.
4 Q. Li, "Minimum Deadlock-Free Message Routing Restrictions in Binary Hypercubes," *Journal of Parallel and Distributed Computing*, vol. 15, no. 2, pp. 153-159, 1992.
5 K. Li and R. Schaefer, "A hypercube shared virtual memory," *Proc. 1989 International Conference on Parallel Processing*, vol. 1, pp. 125-132, 1989.
6 X. Lin, "Adaptive Wormhole Routing in Hypercube Multicomputers," *Journal of Parallel and Distributed Computing*, vol. 48, no. 2, pp. 165-174, 1998.
7 X. Lin, P. K. McKinley and L. M. Ni, "Deadlock-Free Multicast Wormhole Routing in 2-D Mesh Multicomputers," *IEEE Transaction on Parallel and Distributed Systems*, Vol. 5, No. 8, pp. 793-804, 1994.
8 P.K. McKinley and C. Trefftz, "MultiSim: A Tool for the Study of Large-Scale Multiprocessors," *Proc. 1993 International Workshop Modeling, Analysis, and Simulation of Computer and Telecommunication Networks* (*MASCOTS*), Jan. 1993, pp. 57-62.
9 P. K. McKinley, H. Xu, A-H. Esfahanian and L. M. Ni, "Unicast-Based Multicast Communication in Wormhole-Routed Networks," *IEEE Transaction on Parallel and Distributed Systems*, Vol. 5, No. 12, pp. 1252-1265, Dec. 1994.
10 P. K. McKinley, Y. Tsai, and D. F. Robinson, "Collective Communication in Wormhole-Routed Massively Parallel Computers," *IEEE Computer*, Vol. 28, No. 12, pp. 39-50, Dec. 1995.
11 L. M. Ni and L. M. McKinley, "A Survey of Wormhole Routing Techniques in Direct Networks," *IEEE Computer*, vol. 26, pp. 62-76, Feb. 1993.
12 D. K. Panda, S. Singal, and P. Prabhakaran, "Multidestination Message Passing Mechanism Conforming to Base Wormhole Routing Scheme," *Parallel Computer Routing and Communication: First International Workshop, PCRCW'94*, Lecture Notes in Computer Science 853, Springer-Verlag, pp. 131-145, 1994.
13 D. K. Panda, S. Singal, and R. Kesavan, "Multidestination Message Passing in Wormhole *k*-ary *n*-cube Networks with Base Routing Conformed Paths," *IEEE Transaction on Parallel and Distributed Systems*, vol. 10, no. 1, pp. 76-96, Jan. 1999.
14 David F. Robinson, Philip K. McKinley and Betty H.C. Cheng, "Optimal Multicast Communication in Wormhole-Routed Torus Networks," *IEEE Transaction on Parallel and Distributed Systems*, Vol. 5, No. 12, Dec. 1994, pp. 1252-1265.
15 H. Xu, P. K. Mckinley, and L. M. Ni, "Efficient implementation of barrier synchronization in wormhole-routed hypercube multicomputers," *Journal of Parallel and Distributed Computing*, vol. 16, no. 2, pp. 172-184, 1992.

A Mixed Deflection and Convergence Routing Algorithm: Design and Performance *

D. Barth[1], P. Berthomé[2], T. Czarchoski[3],
J.M. Fourneau[1], C. Laforest[4], and S. Vial[4]

[1] PRiSM, Université de Versailles Saint-Quentin, 78000 Versailles, France
[2] LRI, CNRS UMR 8623, Université de Paris Sud Orsay, 91405 Orsay, France
[3] IITIS-PAN, Ul. Balticka, Gliwice, Poland
[4] LaMI, CNRS UMR 8042, Université d'Evry, 91000 Evry, France

1 Introduction

All-optical packets networks represent a challenging and attractive technology to provide a large bandwidth for future networks. The motivations for an all-optical network and a description of the ROM project may be found in [5]. With current optical technology, optical switches do not have large buffers or even buffers at all. Delay loops allow some computation time for the routing algorithms but they are not designed to store a large amount of packets. Therefore routing algorithms are quite different from the algorithms designed for store and forward networks based on electronic buffering. In this paper, we study packet routing strategies without intermediate storage of data packets (hereafter simply called packets) [8], such as deflection routing [3, 10] and Eulerian routing [2, 6, 1]. These routing strategies do not allow packet loss. Thus the performance guarantee in terms of packets loss is just the physical loss rate which is very low for optical fibers. However these strategies keep the packets inside the network and reduce the bandwidth. The usable bandwidth (i.e., the goodput) of the network and the routing protocol is therefore a major measure of interest. In this paper, we focus on two performance criteria: the goodput of the network, and the ending guarantee. First let us define them more precisely.

Definition 1 (Goodput). *It is defined as the average number of packets which leave the network at their destination per time slot as the network operates in synchronized mode even if it is not physically synchronous [5]. The goodput is dependent of the traffic characteristics, the network topology and the routing algorithm.*

Definition 2 (Ending guarantee). *There exists a finite value $\mathcal{G}_e$ such that any emitted packet in the network will reach its destination within a maximal number of steps $\mathcal{G}_e$.*

This guarantee implies that there is no livelock in the network. The ending guarantee is a bound on the transport delay. It does not take into account the

* This work has been supported by the French RNRT project *ROM*

B. Monien and R. Feldmann (Eds.): Euro-Par 2002, LNCS 2400, pp. 767–774.

waiting time before entering the optical part of the network. In shortest-path Deflection Routing, switches attempt to forward packets along a shortest path to their destinations. Each link can send a finite number of packets per time-slot (the link capacity); here it is equal to one. No packet is queued. At each slot, incoming packets have to be sent to their next switch along the path. If the number of packets which require a link is larger than the link capacity, only some of them will receive the link they ask for and the other ones have to be misdirected or deflected. Thus, deflected packets travel on longer paths to their destination. These routing algorithms are known to clearly avoid deadlocks (packets in the network do not move) but livelocks could occur (packets move but never reach their destination). Simulations show that the goodput is quite good but the number of deflections may be not negligible, especially if the traffic is unbalanced. The mean number of deflection is not that large but the standard deviation is quite important and a large number of packets are heavily deflected. We have observed several packets with more than 1000 deflections during a simulation of a 10×10 2D-mesh. The packets with a significant number of deflections are a real problem because they are considered as lost due to large delay and they increase significantly the loss rates at heavy load with unbalanced traffic. Thus, the lack of ending guarantee is a real problem. To obtain a finite ending guarantee, a solution is to use a *convergence routing* technique [7, 8]. In such a routing, packets are routed along a global sense of direction, which gives an ending guarantee. As proposed in various works [4, 7], such a global sense of direction can be created by using some decompositions of the target directed graphs (or of a covering sub-digraph of it) into circuits [11]. In [4], Feige gives such a technique, based on an Eulerian circuit in a sub-digraph, ensuring an ending guarantee equal to $O(n^{3/2})$ for any graph with a minimal number of edges. Here, we use an Eulerian routing, i.e., in which packets follow an Eulerian circuit in the network, a routing technique we described and studied in [2, 6, 1]. Consider an Eulerian circuit $\mathcal{C}$ in a digraph G, where G represents the network. Thus, such an Eulerian circuit can be seen as a circuit of $|A(G)|$ (the set of arcs of G) arcs, where a vertex v of G has $d_G(v)$ (*degree* of v in G) occurrences. Each emitted packet follows $\mathcal{C}$ and, at each step, has priority on the next arc on this circuit. Hence, a packet emitted by a node u for a final destination v will eventually reach it. Then, $d_{\mathcal{C}}(u, v)$, the maximal number of arcs on $\mathcal{C}$ between one occurrence of the source vertex u and the first following occurrence in $\mathcal{C}$ of the destination vertex v, is the major parameter of this routing strategy. It represents the longest delay for packet transport from vertex u to vertex v. Using $\mathcal{C}$, any packet emitted in G reaches its destination in at most $stretchW_{\mathcal{C}}$ steps, where

$$stretchW_{\mathcal{C}} = \max_{\substack{u,v \in V(G) \\ u \neq v}} d_{\mathcal{C}}(u, v).$$

$V(G)$ represents the set of vertices of G. Thus, the ending guarantee obtained by using an Eulerian routing on $\mathcal{C}$ in G is $\mathcal{G}_e = stretchW_{\mathcal{C}}$. However, this ending guarantee has a cost for networks with many nodes: a significant reduction of the goodput. For various Eulerian directed cycles, we have found by simulation goodput from 3 up to 6 packets per slot for a 10×10 2D-mesh [1]. For similar networks and traffic, deflection routing algorithms provide a goodput of more than 32 packets per slot [1]. This is due to the large average transport delay experienced with Eulerian routing. Thus, we design and studied the performance of a new routing algorithm which combines the ending guarantee of the Eulerian routing with the goodput efficiency of deflection routing. The remaining of the paper is organized as follows. In Section 2, we describe our new routing algorithm based on deflection and Eulerian routing and its ending guarantee. Section 3 is devoted to the minimization of ending guarantee for a 2D mesh network.

2 Design of the Routing Algorithm

As shown in the previous section, each of the routing strategy has some desirable performances or guarantee we would like to obtain but they do not have all of them. So we design a new routing strategy which combines both basic routing strategies in order to preserve the best characteristics of each individual ones. The main ideas of the algorithm follow:

- Use two modes for routing: a deflection mode and an Eulerian mode.
- Packets in deflection mode follow a shortest path deflection routing. Packets in Eulerian mode follow the Eulerian circuit associated to the network by the routing algorithm.
- At each time slot, in each router, the routing algorithm has two phases: first route the packets in Eulerian mode and then route the packets in deflection mode.
- A packet entering the network is in deflection mode.
- Packets carry a counter for the deflections they have already experienced.
- The routing algorithm is parameterized by a *deflection threshold* S.
- Once a message number of deflections has reached this *deflection threshold*, the message switches to the Eulerian routing mode.
- Once a message is in Eulerian mode, it stays in this mode until it reaches its destination.

As the packets in Eulerian mode are routed in a first phase, they have a higher priority to access the output links. And by definition, two Eulerian packets in a same node at a same step never ask for a same output link. It is worth remarking that any message can be switched at any moment to the Eulerian routing mode without any additional delay. This is simply due to the fact that all the edges of the network belong by definition to the Eulerian circuit. Thus, a message is always on a portion of the Eulerian circuit, and uses the next arc in the Eulerian circuit. The ending guarantee of the Eulerian routing strategy ensures that the message will reach its destination, whatever the load of the network is. It is worthy to remark that the complexity of our algorithm has the same order of magnitude as the Deflection routing algorithm. The header of the message has to be increased in order to code the number of deflections a message can afford before using the Eulerian routing strategy. Thus, when this number is S, the message uses the Eulerian strategy. This modification of the header of a packet is compatible with optical technology, where all the headers are globally regenerated at each router. We refer to [5] for technical details of optical technologies of the project we consider. Even if the question of the bounded transport delay is quite natural, it seems that only one reference already exists on the extension of deflection routing with an ending guarantee. In [3], a routing mechanism based on deflection and priorities is presented for a hypercube. But the proof of the ending guarantee depends of the network topology and the bound is not stated in a closed form. Unlike this restricted case, our approach is much more general for the network topology (an Eulerian directed graph) and we prove the value of the bound for transport delay in terms of the parameters of the graph and the algorithm. We now prove the ending guarantee of the mixed routing.

Theorem 1. *Let $G = (V, E)$ a symmetric digraph and $\mathcal{C}$ an Eulerian circuit of G. Using the mixed routing strategy with threshold $S > 0$, any message reaches its destination in at most $D(G, S, \mathcal{C})$ steps, with:*

$$D(G, S, \mathcal{C}) = diam_G + 2S + stretchW_{\mathcal{C}} - 3.$$

where $diam_G$ is the diameter (maximal distance) of G. When $S = 0$, we have $D(G, 0, \mathcal{C}) = stretchW_{\mathcal{C}}$.

Proof. Whenever there is a deflection, a message does not move closer to its destination, and might even move one step farther away. So a message might move closer to its destination $diam_G - 1$ times, and might move one step farther away $2(S - 1)$, and finally might have to follow an Eulerian path. □

We can note immediately that once the network is known, we have two ways to shorten the ending guarantee: decrease the threshold S or decrease the value of $stretchW_{\mathcal{C}}$ by a clever selection of the Eulerian circuit which covers the network.

3 Application on the 2D-Mesh

In the following we consider the square *2D-mesh* $N \times N$ network which seems to be an sufficiently realistic network topology. We also assume that the link capacity is only 1. Of course, real networks will have a link capacity much larger [5]. This restriction is made to allow the comparison of routing algorithms where the deflection part is an optimal algorithm in terms of number of deflected packets. Thus the comparison takes only into account the relative performance and properties of the routing algorithms and is not influenced by greedy algorithms used to choose the deflected packets. Note that this model is also useful for the analysis of networks where wavelength converters are not allowed. As the diameter of the $N \times N$ mesh is $2N - 2$, we derive from Theorem 1 a new formula for the ending guarantee:

$$D(G, S, \mathcal{C}) = 2N + 2S + stretchW_{\mathcal{C}} - 5.$$

This network allows us to derive theoretical results on the Eulerian circuits. It is also used in the simulator we designed to study the performance of the algorithms or exhibit sample-paths with unexpected behaviors.

3.1 Simulator Design

We have designed the simulator with QNAP II modelling tool [9]. Objects represented in QNAP are queues and customers. Every switch is modelled by two queues: one for the packets present in the optical part of the network and the other one to store the packets which wait to enter the optical network. The algorithm used to choose the deflected packets minimizes the number of deflection for each switch and for each time unit. The source, destination, hop count, time and statistics are carried by the customers. As the destination of the packet is carried by the customer, we can model different types of traffic. The simulator reports statistics about the queue utilization and the customers sample-paths: mean delay to enter the network, switch utilization, distribution of the number of deflections for all the packets, distribution of the transport delay, number of packets in Eulerian mode, and of course, averages and other statistics from these distributions. The confidence intervals for the means are computed but they are not depicted in the figures as they are generally smaller than 1 percent. For the sake of concision, we restrict ourselves here to the presentation of results for a very simple traffic, even if the simulator design allows much more general traffic description. We assume Poisson arrivals of packets and an uniform traffic for source and destination among the nodes of the networks.

3.2 Eulerian Circuits in the 2D-Mesh

It has been proved in [2] that the stretch of an Eulerian circuit can be as bad as the number of the arcs minus 3 (i.e., $m - 3$) and cannot be less than the number of arcs of the corresponding graph divided by the minimum degree of the graph (more precisely, $\frac{m}{\delta} - 1$). Therefore, for a $N \times N$ 2D-mesh, the stretch lies between $2N(N-1) - 1$ and $4N(N-1) - 3$. In the following, we study 2 Eulerian circuits, namely *the Antenna* (see [6] for the description) and *the Double Snake* (due to space limitations we do not describe it). We can show that $stretchW_C = 4N(N-1) - 2N + 1$ for the Antenna. For the double snake, we can prove that $stretchW_{C'} = 2N(N-1) + 3$, for odd N. This corresponds to the best general circuit we have found for the squared grid. When N is even, a similar construction leads to a stretch value of $2N(N-1) + 2N - 3$. Remark that these values are very close to the lower bounds.

3.3 The Threshold Problem

Another solution to shorten the maximum transport delay is to decrease the threshold S used to change the routing of a packet from deflection mode to Eulerian mode. It is definitely not a good solution to use a small value for S. First, if the threshold is very low, the mixed routing has the same goodput as the Eulerian routing. For instance a 11×11 mesh with $S = 2$ has a goodput of only 8 packets per slot. Furthermore for larger value of S we can observe a phase transition phenomenon (see for instance Fig. 1 where a sample-path with $S = 8$ is shown for a 11×11 mesh). The arrivals follow a Poisson process with rate 20.

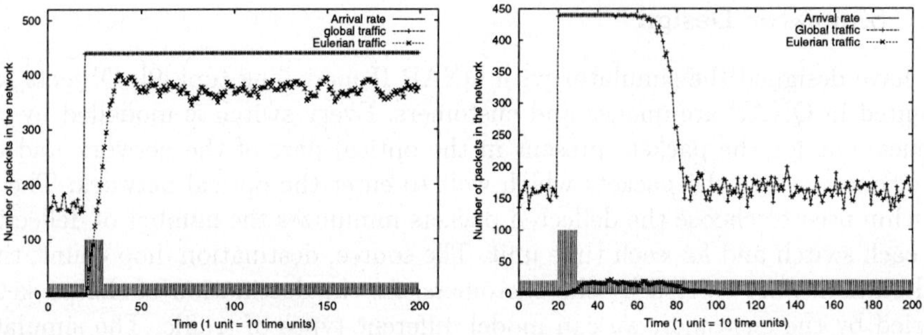

Fig. 1. Antenna, Left) phase transition sample-path with $S = 8$, Right) Effect of a burst of packets when $S = 20$

But, during a small period (i.e., between instant 200 and 250), the intensity is changed to model a burst (the arrival rate is now 100). Due to this modification of the arrival process, we consider three parts in the sample path: before the modification, during the modification and after the return to the initial rate. Before slot 200, the total number of packets in the network is quite low (i.e., between 150 and 200). Indeed the arrival rate in this first part implies a very small utilization of the network. At this load, there is only a few deflections and no packets in Eulerian mode can be observed by simulation within up to 10^7 time units. At time 200, due to the modification of the arrival process, the number of packets jumps up to almost 440. The network is saturated. When the arrival process goes back to 20 the network stays saturated. During this third part of the sample-path, the saturation and the low goodput are not due to the arrival process. The real reason of these low performances is the number of packets in Eulerian mode. Indeed the two basic modes of routing have very different goodputs and very different mean transport delay. Remember that for a 11×11 mesh the Eulerian routing has a goodput less than 6 packets par time unit while the deflection performance is at least 32 packets. Remark that an arrival rate of 20 is between these two values. The deflection mode is stable but if all the packets are in Eulerian mode, the queue at the entrance of the network will grow up to infinity almost surely. The sudden increase of the arrival rate increases the number of packets inside the optical part of the network. At heavy load, a significant number of packets experienced a large number of deflections. If the threshold S is low, this large number of packets will switch to the Eulerian mode of routing. But the average transport delay is much larger for Eulerian mode. Thus packets in Eulerian mode stay longer (in general) and as they have priority for the routing algorithms, they imply more deflections and again more packets in Eulerian routing. This domino effect (Eulerian packets will create new Eulerian packets) leads to a network where almost all the packets are in Eulerian mode and the goodput of the network is now very low. As the arrival rate is larger than the goodput of the Eulerian routing, the network will stay in this mode almost surely. This is depicted in the left part of Fig. 1. How to avoid such a undesirable behavior? First, it must be clear that this phase transition is always possible but we can decrease its probability. And we can increase the probability of the reverse transition. The basic parameters are the threshold S and the value of $stretchW_C$. If we decrease $stretchW_C$, we expect that the mean transport delay will decrease. Thus, the goodput of the Eulerian mode will increase and the probability of a domino effect will decrease. And if we increase S, the probability of switching from deflection mode to Eulerian mode will decrease. This is illustrated in the right part of Fig. 1 where we have considered the same arrival process with the same burst and the same network as in lefat part of Fig. 1. But now the threshold value is 20. The population inside the network increases, so does the number of packets in Eulerian mode. But after a delay, the number of packets in Eulerian mode will decrease and quickly the total number of packets will go back to the initial range (140,200). Finally, we assume a large value for the threshold S. We consider that S will be equal to the diameter of the network. Simulations show that this value is sufficient, even if is not a formal proof. For instance, in the left part of Fig. 2, we have considered a sample-path for a 11×11 mesh where the initial point of the simulation is a network with 200 packets in Eulerian mode. Despite this very bad initial point (whose probability is clearly very low) and an arrival rate of 25 new packets per slot, the number of packets in Eulerian mode moves to an equilibrium region between 20 and 40. Thus, if we want to minimize the transport delay guarantee, we have to decrease the stretch but the threshold S must be kept large enough to avoid the phase transition depicted in Fig. 1.

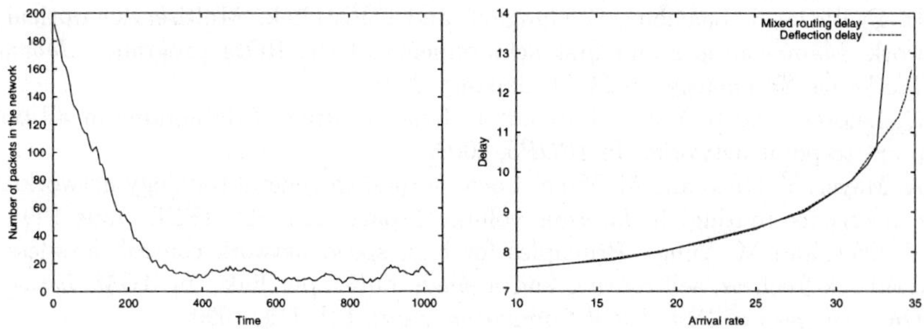

Fig. 2. $S = 20$, Antenna, 11×11 mesh, Left) Sample path of the number of packets from an initial catastrophic state in Eulerian mode, Right) Delay vs arrival rate for Mixed routing and Deflection routing

4 Conclusion

In this paper, we have shown that a mixed routing strategy can be used to obtain a ending guarantee with a small cost: the goodput is only weakly reduced. The algorithmic cost is almost free if the Deflection part of the algorithm do not have a linear complexity. The goodput is much more efficient than in the case of the Eulerian routing. Such properties are in fact independent from the topology of the network. The two important things that have to be well chosen are the Eulerian circuit and the way to determine the threshold. Some works are in progress to study mixed strategies with new convergence routing with better bounds or average transport delays with optimization in the algorithms due to the small number of packets in Eulerian mode.

Acknowledgements

Authors thank the referee who suggested an improvement of proof of theorem 1.

References

1. D. Barth, P. Berthomé, A. Borrero, J.M. Fourneau, C. Laforest, F. Quessette, and S. Vial. Performance comparisons of Eulerian routing and deflection routing in a 2d-mesh all optical network. In *ESM'2001*, 2001.
2. D. Barth, P. Berthomé, and J. Cohen. The Eulerian stretch of a digraph and the ending guarantee of a convergence routing. Technical Report A00-R-400, LORIA, BP 239 - F54506 Vandoeuvre-lès-Nancy, France, 2000.
3. J. C. Brassil and R. L. Cruz. Bounds on maximum delay in networks with deflection routing. *IEEE Trans. on Parallel and Distributed Systems*, 6(7):724–732, 1995.
4. U. Feige. Observations on hot potato routing. In *ISTCS: 3rd Israeli Symposium on the Theory of Computing and Systems*, 1995.
5. P. Gravey, S. Gosselin, C. Guillemot, D. Chiaroni, N. Le Sauze, A. Jourdan, E. Dotaro, D. Barth, P. Berthomé, C. Laforest, S. Vial, T. Atmaca, G. Hébuterne, H. El Biaze, R. Laalaoua, E. Gangloff, and I. Kotuliak. Multiservice optical network: Main concepts and first achievements of the ROM program. *Journal of Ligthwave Technology*, 19:23–31, January 2001.
6. C. Laforest and S. Vial. Short cut Eulerian routing of datagrams in all optical point-to-point networks. In *IPDPS*, 2002.
7. A. Mayer, Y. Ofek, and M. Yung. Local fairness in general-topology networks with convergence routing. In *Infocom*, volume 2, pages 891–899. IEEE, June 1995.
8. Y. Ofek and M. Yung. Principles for high speed network control: loss-less and deadlock-freeness, self-routing and a single buffer per link. In *ACM Symposium On Principles of Distributed Computing*, pages 161–175, 1990.
9. D. Potier. New user's introduction to QNAP2. Technical Report RT40, INRIA, Roquencourt FRANCE, 1984.
10. A. Schuster. *Optical Interconnections and Parallel Processing: The Interface*, chapter Bounds and analysis techniques for greedy hot-potato routing, pages 284–354. Kluwer Academic Publishers, 1997.
11. B. Yener, S. Matsoukas, and Y. Ofek. Iterative approach to optimizing convergence routing priorities. *IEEE/ACM Trans. on Networking*, 5(4):530–542, August 1997.

Evaluation of Routing Algorithms for InfiniBand Networks*

M.E. Gómez, J. Flich, A. Robles, P. López, and J. Duato

Department of Computer Science, Universidad Politécnica de Valencia,
P.O.B. 22012, 46071 – Valencia, Spain,
{megomez,jflich,arobles,plopez,jduato}@gap.upv.es

Abstract. Storage Area Networks (SANs) provide the scalability required by the IT servers. InfiniBand (IBA) interconnect is very likely to become the de facto standard for SANs as well as for NOWs. The routing algorithm is a key design issue in irregular networks. Moreover, as several virtual lanes can be used and different network issues can be considered, the performance of the routing algorithms may be affected. In this paper we evaluate three existing routing algorithms (up*/down*, DFS, and smart-routing) suitable for being applied to IBA. Evaluation has been performed by simulation under different synthetic traffic patterns and I/O traces. Simulation results show that the smart-routing algorithm achieves the highest performance.

1 Introduction

In IBA [2] networks, switches can be arranged freely in order to provide wiring flexibility and incremental expansion capability. The irregularity in the topology makes the routing quite complicated. Several routing algorithms for irregular topologies have been proposed. The up*/down*, smart-routing and DFS routings[1] are suitable for IBA networks due to the fact that they can be implemented in a deterministic way. The three routing algorithms have been already evaluated in [1] and for wormhole-switched Myrinet networks by using different synthetic traffic patterns that might not be representative of SANs. However, in this paper, we also use I/O traces for IBA interconnects that use virtual cut-through switching.

In a SAN environment, the use of a particular routing algorithm together with the distribution of storage devices may significantly affect the overall system performance. Moreover, IBA allows the use of several virtual lanes (VL). In a SAN environment, it could be thought that some disks can be addressed through a particular VL. In this paper, we will also evaluate how the disk distribution affects the performance of the routing algorithms and the use of different VLs.

* This work was supported by the Spanish CICYT under Grant TIC2000-1151-C07 and by Generalitat Valenciana under Grant GV00-131-14.

[1] These routing algorithms can be implemented on IBA networks by the strategies proposed in [3, 4].

B. Monien and R. Feldmann (Eds.): Euro-Par 2002, LNCS 2400, pp. 775–780.

The paper is organized as follows. In the next section, the main simulator considerations are described. In section 3 the simulation results are discussed. Finally, some conclusions are drawn in Section 4.

2 Simulation Model

We have developed a detailed simulator that allows us to model the network at the register transfer level following the IBA specifications [2]. We will use with a non-multiplexed crossbar on each switch with a simple crossbar arbiter based on FIFO request queues per output crossbar port. The routing time at each switch will be set to 100 ns. This time includes the time to access the routing tables, the crossbar arbiter time, and the time to set up the crossbar connections. The link injection rate will be fixed to the 1X configuration [2].

We have used different message destination distributions. In the uniform distribution, the destination of a message is chosen randomly. In the hot-spot distribution, a percentage of traffic is sent to one host. In the distribution with several hot-spot hosts, 10% of traffic is sent to them. When using synthetic traffic, we will use short packets with a payload of 32 bytes, and long packets with a payload of 256 bytes. Buffer size (input and output) will be fixed to 1 KB. We will analyze irregular networks of 8, 16, 32 and 64 switches randomly generated. We will assume that every switch in the network has 8 ports, using 4 ports to connect to other switches and leaving 4 ports to connect to hosts (servers and storage devices).

The I/O traces were provided by Hewlett-Packard Labs. They include all the I/O activity generated from 1/14/1999 to 2/28/1999 at the disk interface of the *cello* system. A detailed description of similar traces of 1992, collected in the same system, can we found in [5]. We will use packets with a payload equal to the size specified in the trace for the I/O accesses, but if the access is larger than 512 bytes, we will split it into packets with a payload of 512 bytes at most. Buffer size (input and output) will be fixed to 8KB. The disks will be attached to twenty-three ports. The rest of switch ports will be connected to hosts. When using I/O traces, three different evaluations[2] will be performed. Firstly, we will use only one virtual lane (VL). The disks will be randomly distributed over the network. Secondly, we will use different disk distributions. In particular we will distribute the disks: (1) randomly; (2) concentrated (disks are grouped in 6 switches selected randomly); and (3) uniformly (only one disk will be attached to a particular switch). And finally, we use different VLs. When using different VLs we need a different SL for each VL. We will refer to this assignment as SL/VL. All the traffic injected into a particular VL remains in the the same VL until delivered.

[2] Regarding performance of routing algorithms, latency is the elapsed time between the generation of a packet at the source host until it is delivered at the destination node, accepted traffic is the amount of information delivered by the network per time unit.

3 Performance Evaluation

3.1 Results for Synthetic Traffic

Figures 1.a and 1.b show the behavior of the three routing algorithms for uniform distribution of packet for different network sizes. SMART routing is not shown for the 64-switch network due to its high computation time. As it was expected, SMART achieves the highest network throughput for all the evaluated topologies. In particular, for 32 switches SMART increases network throughput by factors of 2.29 and 1.11 with respect to UD and DFS, respectively. The higher network throughput achieved by SMART and DFS routings is due to their better traffic balance, as can be seen in Figure 1.c.

Table 1 shows minimum, maximum, and average increases of network throughput when comparing UD, DFS, and SMART using 10 random topologies for each network size. We observe that SMART always increases network throughput with respect to UD and DFS. We also observe that, as network grows, DFS also increases its improvement over UD. For 64-switch networks DFS improves over UD by a factor of 2.66, on average. For the hot-spot traffic pattern, on average, DFS and SMART decrease their throughput (with respect to UD) when changing

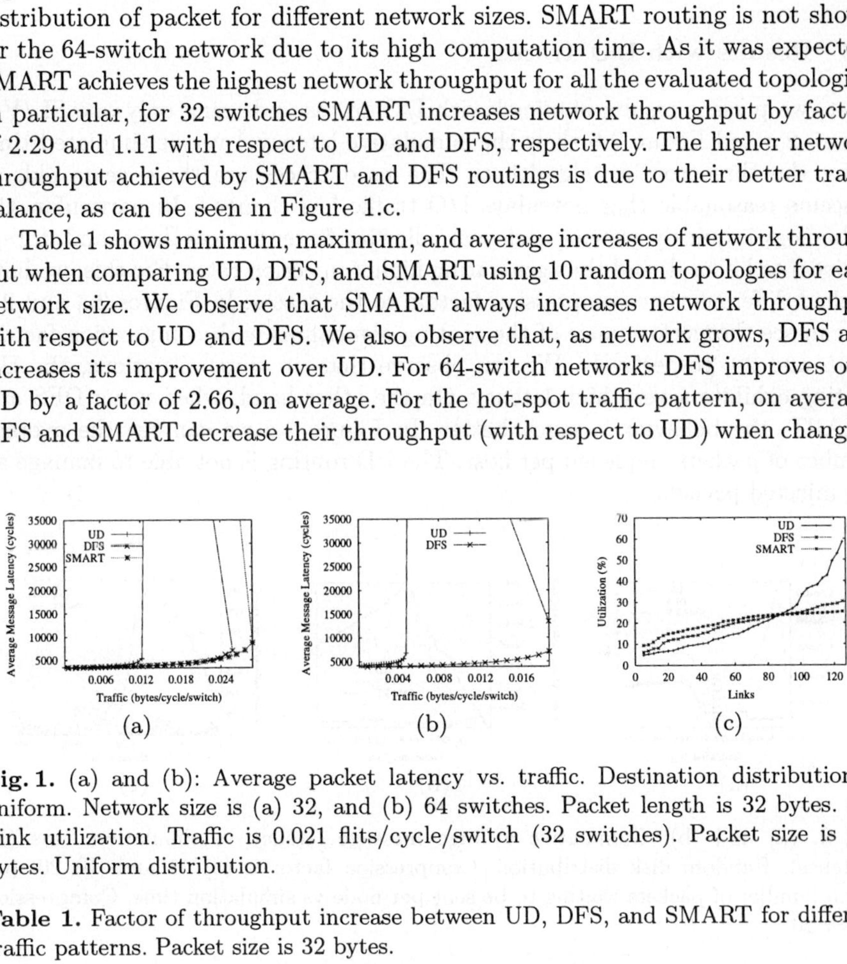

Fig. 1. (a) and (b): Average packet latency vs. traffic. Destination distribution is uniform. Network size is (a) 32, and (b) 64 switches. Packet length is 32 bytes. (c): Link utilization. Traffic is 0.021 flits/cycle/switch (32 switches). Packet size is 256 bytes. Uniform distribution.

Table 1. Factor of throughput increase between UD, DFS, and SMART for different traffic patterns. Packet size is 32 bytes.

				SMART vs UD			SMART vs DFS			DFS vs UD		
Sw.	Distr	Number HS	Percentage	Min	Max	Avg	Min	Max	Avg	Min	Max	Avg
16	Unif.	-	-	1.3	2.12	1.72	1.00	1.32	1.09	1.13	1.95	1.52
32	Unif.	-	-	1.67	3.29	2.46	1.07	1.75	1.23	1.23	2.63	2.03
64	Unif.	-	-	N/A	N/A	N/A	N/A	N/A	N/A	2.11	3.73	2.66
32	HS	1	5%	1.24	1.92	1.51	0.98	1.33	1.09	1.23	1.98	1.44
32	HS	1	10%	0.94	1.26	1.11	0.97	1.05	1.00	0.97	1.22	1.11
32	HS	1	20%	0.97	1.08	1.03	0.97	1.01	1.00	0.99	1.10	1.04
32	HS	2	10%	1.23	2.38	1.68	0.94	1.21	1.04	1.08	2.24	1.60
32	HS	4	10%	1.53	2.73	2.07	0.87	1.50	1.13	1.08	2.5	1.86
32	HS	8	10%	1.86	3.14	2.35	1.04	1.70	1.23	1.09	2.67	1.94

from a 5% hot-spot (1.51 and 1.44) to a 20% hot-spot (1.03 and 1.04). Finally, in Table 1 we show the same results for several hot-spot hosts. As the number of hot-spots increases in the network, the traffic is better balanced and therefore we can take advantage of a better designed routing algorithm (SMART and DFS).

3.2 Results with I/O Traces

First, we present results obtained with I/O traces and using only one SL/VL. Figure 2.a and Figure 2.b show the cumulative latency [3] versus simulated time using the three routing algorithms. The used traces are three years old, so it seems reasonable that nowadays I/O traffic has changed. In particular, the technology is quickly growing each year, allowing faster devices (hosts and storage devices) to be used, and thus generating higher injection rates. For this, we have applied different time compression factors to the traces. In Figures 2.a and 2.b we can see the performance of the routing algorithms with compression factors of 15 and 20, respectively. We can observe that, in these situations, the UD routing exhibits a very high latency. On the other hand, when using DFS and SMART, the behavior is much better. In Figure 2.c we can see the average number of packets enqueued per host. The UD routing is not able to manage all the injected packets.

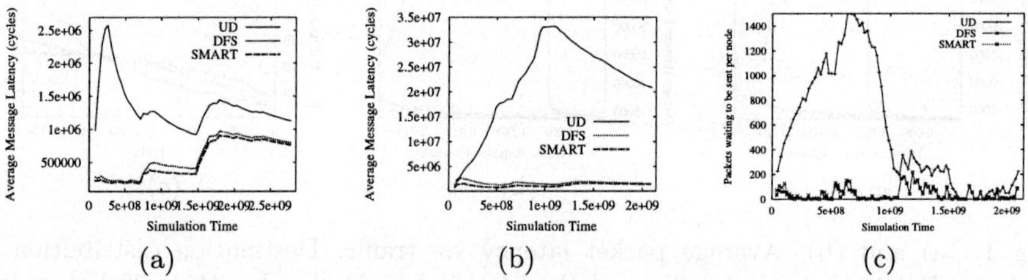

Fig. 2. (a) and (b): Cumulative average message latency vs simulation time (32 switches). Random disk distribution. Compression factor is (a) 15 and (b) 20. (c): Mean number of packets waiting to be sent per node vs simulation time. Compression factor 20.

Now, we analyze how the disk distribution over the network affects the different routing schemes. Figure 3 compares the three disk distributions for every scheme. UD is the most sensitive routing to disk distributions. For example, in Figure 3.a we can observe that, at the beginning of the simulation, concentrating the disks in some switches is better than randomly distributing them, whereas later the random distribution of disks has a much better behavior. The other routings (DFS and SMART) are much more robust to the disk distribution.

[3] The cumulative latency is obtained by adding the latency of all the messages (from the beginning of the simulation) and dividing by the number of messages received at each simulation time.

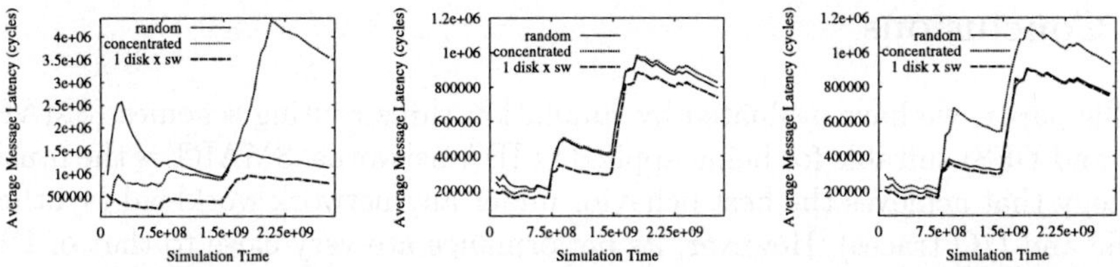

(a) (b) (c)

Fig. 3. Cumulative average message latency from generation vs simulation time using different disk distributions. 1 SL/VL (32 switches) for different disk distribution. Compression factor is 15. Routing scheme is (a) UD, (b) DFS, and (c) SMART.

For all the schemes, the best option is distributing the disks among switches, having one disk per switch. By doing this, the workload is better balanced in the network. Moreover, SMART obtains similar results for randomly distributed disks and one disk per switch distribution. Hence, SMART is the least sensitive routing algorithm to changes in the disk distribution.

Finally, we analyze how the use of SL/VLs affects the performance of the routing schemes. Figure 4 shows the behavior of the three routing schemes using different numbers of SL/VL. Disks are assigned randomly to SL/VLs. bAs we can see, the UD routing (Figure 4.a) benefits from using an additional SL/VL (2 SL/VL). Latency is noticeably reduced. The peak latency is reduced from 2.5 million of cycles to 1 million of cycles. Using two additional SL/VLs (4 SL/VL) helps even more. When using 8 SL/VL additional improvements are not achieved. The other routing algorithms (DFS and SMART) obtain low improvements on performance when using additional SL/VLs. The congestions caused by the UD routing is reduced when using different SL/VLs. However, the DFS and SMART routings can handle traffic in an efficient way. As a conclusion, we can see that even by using a large number of network resources (8 SL/VLs) the UD is not able to obtain the good network performance achieved by the other routings (DFS and SMART) with only one SL/VL.

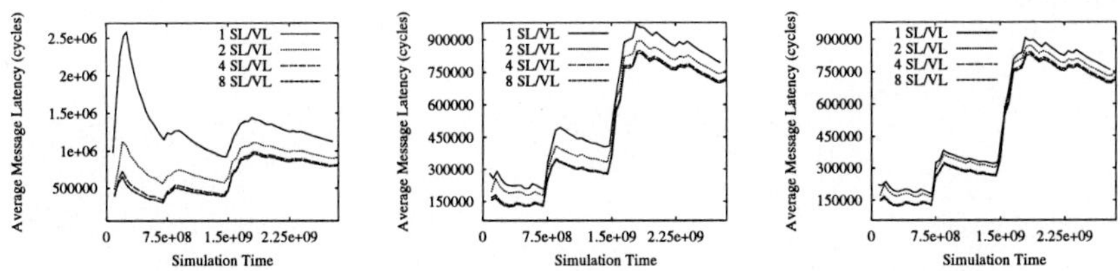

(a) (b) (c)

Fig. 4. Cumulative average message latency from generation vs. time using different numbers of SL/VL (32 switches). Compression factor is 15. Routing scheme is (a) UD, (b) DFS, and (c) SMART.

4 Conclusions

In this paper, we have evaluated by simulation three routing schemes (SMART, UD, and DFS) suitable for being applied to IBA networks. SMART is the routing strategy that achieves the best behavior under any network workload (synthetic traffic and I/O traces). However, its performance are very close to that of DFS. This behavior is mainly due to the better traffic balance exhibited by SMART and DFS. When analyzing the behavior under I/O traces, it is observed that UD has not enough capacity to manage the traffic generated by the trace. This causes an increase in the number of packets stored in queues and, in turn, a significant increase in the packet latency. However, SMART and DFS have no problem to follow near the injected traffic. Moreover, these routing algorithms exhibit a greater robustness than UD, facing eventual changes in the disk distribution. On the other hand, unlike SMART and DFS, UD takes advantage of using additional SLs/VLs in order to reduce the head-of-line blocking effect in the input buffers. Despite it, with only one SL/VL, SMART and DFS continue to outperform UD, even when the latter strategy uses 8 SLs/VLs.

References

1. J. Flich, P. Lopez, M.P. Malumbres, J. Duato, and T. Rokicki, "Combining In-Transit Buffers with Optimized Routing Schemes to Boost the Performance of Networks with Source Routing," *Proc. of Int. Symp. on High Performance Computing*, Oct. 2000.
2. InfiniBand^{TM} Trade Association, *InfiniBand^{TM} architecture. Specification Volumen 1. Release 1.0.a.* Available at http://www.infinibandta.com.
3. J.C. Sancho, A. Robles, and J. Duato, Effective Strategy to Compute Forwarding Tables for InfiniBand Networks, in *Proc. of 2001 International Conference on Parallel Processing (ICPP'01)*, Sept. 2001.
4. P. López, J. Flich, and J. Duato, Deadlock-free Routing in InfiniBand^{TM} through Destination Renaming, in *Proc. of 2001 International Conference on Parallel Processing (ICPP'01)*, Sept. 2001.
5. C. Ruemmler, J. Wilkes, *Unix Disk Access Patterns*, Winter Usenix Conference, Jan 1993.

Congestion Control Based on Transmission Times*

E. Baydal, P. López, and J. Duato

Dept. of Computer Engineering, Universidad Politécnica de Valencia,
Camino de Vera s/n, 46071 – Valencia, Spain,
elvira,plopez,jduato@disca.upv.es

Abstract. Congestion leads to a severe performance degradation in multiprocessor interconnection networks. Therefore, the use of techniques that prevent network saturation are of crucial importance to avoid high execution times.
In this paper, we propose a new mechanism that uses only local information to avoid network saturation in wormhole networks. In order to detect congestion, each network node computes the quotient between the real transmission time of messages and its minimum theoretical value. If this ratio is greater than a threshold, the physical channel used by the message is considered congested. Depending on the number of congested channels, the available bandwidth to inject messages is reduced.
The main contributions of the new mechanism are three: i) it can detect congestion in a remote way, but without transmitting control information through the network; ii) it tries to dynamically adjust the effective injection bandwidth available at each node; and iii) it is starvation-free.
Evaluation results show that the proposed mechanism avoids network performance degradation for different network loads and topologies. Indeed, the mechanism does not introduce any penalty for low and medium network loads, where no congestion control mechanism is required.

1 Introduction

Massively parallel computers provide the performance that most scientific and commercial applications require. Their interconnection networks offer the low latencies and high bandwidth that is needed for different kinds of traffic. Usually, wormhole switching with virtual channels and adaptive routing is used [6]. However, multiprocessor interconnection networks may suffer from severe saturation problems with high traffic loads, which may prevent reaching the wished performance.

This problem can be stated as follows. With low and medium network loads, the accepted traffic rate is the same as the injection rate and latency slightly increases due to contention. When traffic injection rate exceeds certain level (the network saturation point), accepted traffic falls and message latency increases considerably. Notice that both latency and accepted traffic are dependent variables on injected traffic. When this situation is reached, we say that the interconnection network is congested. Performance degradation due to congestion is specially important when messages stay in the network in case of contention, which is the case of wormhole.

* This work was supported by the Spanish CICYT under Grants TIC2000-1151-C07 and 1FD97-2129

B. Monien and R. Feldmann (Eds.): Euro-Par 2002, LNCS 2400, pp. 781–790.

Message throttling has been the most frequently used method to avoid network congestion. Several mechanisms have been already proposed, they have important drawbacks that will be analyzed in Section 2.

In this paper, we propose and evaluate a new mechanism to prevent network congestion that tries to overcome those drawbacks. The mechanism is based on locally estimating network traffic by using message transmission times and applying message throttling when congestion is presumed. The rest of the paper is organized as follows. In order to design an efficient mechanism, in Section 2 we describe what is expected from a good congestion control mechanism, criticizing previous approaches. Section 3 describes the new proposal. Performance evaluation results are presented in Section 4. Finally, some conclusions are drawn.

2 Features of a Good Congestion Control Mechanism

In order to design an efficient congestion control mechanism, in this section, we will describe the desirable features of such a mechanism.

First, the mechanism should be robust. As saturation point depends on network load and topology, a given mechanism may not always work properly. However, many of the previously proposed mechanisms have been analyzed for only one network size [1], [16] and for the uniform distribution of message destinations [4], [8], [7], [14] or do not achieve good results for different traffic patterns [15].

Second, the mechanism should not penalize network behavior when the network is not saturated, which is the most frequent situation [13]. However, some of the previous proposals increase message latency before the saturation point [4], [15], [7].

Finally, the new mechanism should not complicate network design. Some of the proposed mechanisms increase network complexity by adding new signals [14], [8] or even a sideband network [16], [15]. Others need to send extra information through the network [8].

3 Congestion Detection Based on Measuring Packets Transmission Time

This section describes the new mechanism proposed in this paper.

First, let us analyze the effect of traffic rate on message latency. With low traffic rate, messages flow smoothly through the network, only suffering the routing and switching delays at each traversed node. Indeed, physical channels will be seldom multiplexed into virtual channels.

As traffic increases, the probability that the message find busy channels also increases, thereby, multiplexing physical channels and slowing down message advance speed. Indeed, once one of the channels used by the message is multiplexed, the whole message advances at a lower speed. In this situation, message latency may increase until v times, v being the maximum number of virtual channels per physical channel.

If traffic continues to increase, the contention for the use of channels gets worse. Therefore, all the possible virtual output channels for reaching a given destination may

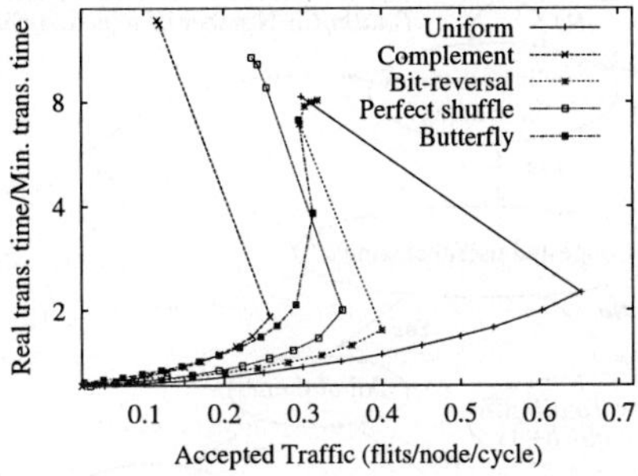

Fig. 1. Quotient between actual and minimum message transmission time versus accepted traffic. 8-ary 3-cube (512 nodes). TFAR routing with 3 virtual channels per physical channel and deadlock recovery. 16-flit messages.

be busy. In this situation, the message header stops, preventing also the advance of the remaining flits of the message. Channel buffers may allow data flits to continue flowing for some time, until they become full. In this case, message latency may considerably increase because of the combined effects of header contention and physical channel multiplexing.

Therefore, with low network load, message latency will be close to the minimum theoretical value, and it will increase as network load does. The mechanism proposed in this paper uses this idea in order to detect network congestion. In particular, the proposed method computes, for each virtual channel, the elapsed time between the arrival of a message header to the node and the transmission of its tail flit. This is the actual transmission time. The minimum value (what we call the theoretical transmission time) is the product of message size and the time required to transfer one flit across the link. Although the absolute value of actual transmission time depends on message destination distribution and message length, the quotient between this value and its minimum theoretical time does not strongly depend on network load, as Figure 1 shows. Indeed, the obtained values when the network is near to the saturation point are quite similar (around 2.0) for all the destination distributions considered. On the other hand, to make the mechanism independent on message length, this ratio is computed periodically, every time the time required to transfer a given number of flits (for instance 8 flits) has elapsed.

This ratio may be used to detect network congestion. When it is greater than a given threshold $u1$, the message using the virtual channel is having problems to advance, then we mark the physical channel containing the virtual channel as congested during some interval $u2$.

The information about physical channel status -congested or not- is used when new messages have to be injected into the network. Only the status of useful channels[1] will be considered. In particular, if any of the useful channels to route the new message is congested, message throttling will be applied. On the other hand, the proposed mecha-

[1] A physical channel is useful to route a message if it is a minimal route between the current node and the destination node

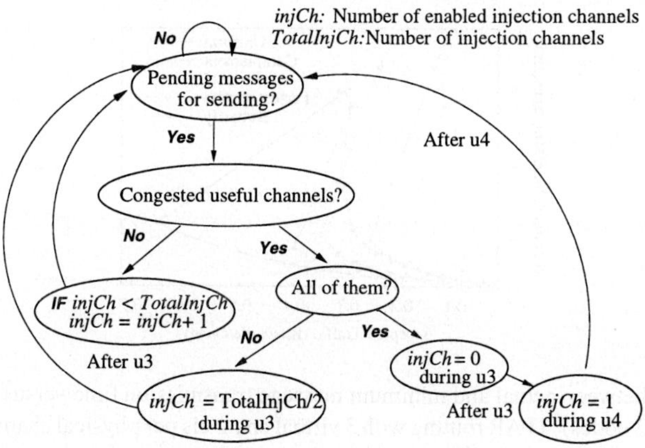

Fig. 2. Operation of the congestion control mechanism based on message transmission time.

nism tries to consider the saturation level of the network, in order to apply congestion control measures in a step-by-step fashion. In particular, different restrictions are applied depending on the number of congested useful channels, reducing accordingly the number of injection channels[2]. If some of the useful physical channels are congested (but not all of them), we reduce the number of injection channels by half (e.g. from 4 to 2 injection channels) during an interval time $u3$. If all the useful channels are congested, the measures have to be more restrictive. In this case, newly generated messages cannot be injected during $u3$ cycles. After that the performed actions depend on the level of detected congestion. In the former case (soft congestion), the status of useful physical channels is analyzed again, proceeding in the same way as we described. In the latter case (serious congestion), we will enable one injection channel during an interval $u4$, regardless of the network status, in order to prevent starvation. After $u4$, if there are pending messages in the injection queue, we will analyze again the status of the useful physical channels. Finally, if congestion is not longer detected, the number of injection channels is progressively increased. Figure 2 shows the behavior of the mechanism.

The proposed mechanism has the advantage of being able to detect the congestion in a remote way, but by using only local information. Moreover, all the nodes that are along the path followed by the message that finds congestion will detect the problem. The sender node of the message may do not detect the problem if the message tail has already left the node when the header flit reach the congested area. However, if the congestion situation is persistent, the next sent messages will every time increase their transmission times in nearer areas. Hence, the sender node will also detect the problem.

4 Evaluation

In this section, we will evaluate by simulation the behavior of the proposed congestion control mechanism. The evaluation methodology used is based on the one proposed in

[2] This is equivalent to reduce the bandwidth associated to message injection at each node

[5]. The most important performance measures are latency (time required to deliver a message, including the time spent at the source queue) and throughput (maximum traffic accepted by the network). Accepted traffic is the flit reception rate. Latency is measured in clock cycles, and traffic in flits per node per cycle.

4.1 Network Model

Message injection rate is the same for all nodes. Each node generates messages independently, according to an exponential distribution. Destinations are chosen according to the *Uniform, Butterfly, Complement, Bit-reversal,* and *Perfect-shuffle* traffic patterns, which illustrate different features. Uniform distribution is the most frequently used in the analysis of interconnection networks. The other patterns take into account the permutations that are usually performed in parallel numerical algorithms [9]. For message length, 16-flit and 64-flit messages are considered.

The simulator models the network at the flit level. Each node has a router, a crossbar switch and several physical channels. Routing time, transmission time across the crossbar and across a channel are all assumed to be equal to one clock cycle. Each node has four injection/ejection channels. Although most commercial multiprocessors have only one injection/ejection channel, previous works [8], [6], [2] have highlighted that the bandwidth available at the network interface may be the bottleneck to achieve a high network throughput.

Concerning deadlock handling, we use software-based deadlock recovery [12] and a True Fully Adaptive routing algorithm (TFAR) [13, 12] with 3 and 4 virtual channels per physical channel. This routing algorithm allows the use of any virtual channel of those physical channels that forwards a message closer to its destination. In order to detect network deadlocks, we use the mechanism proposed in [10] with a deadlock detection threshold equal to 32 cycles.

We have evaluated the performance of the proposed congestion control mechanism on a bidirectional k-ary n-cube. In particular, we have used the following network sizes: 256 nodes (n=2, k=16), 512 nodes (n=3, k=8), and 4096 nodes (n=3, k=16).

4.2 Performance Comparison

In this section, we will analyze the behavior of the mechanism proposed in section 3, which will be referred to as **Trans-t** (*Transmission time*). For comparison purposes, we will also evaluate the behavior of the mechanism proposed in [16], which will be referred to as **Self-Tuned**. In this mechanism, nodes detect network congestion by using global information about the total number of full buffers in the network. If this number surpasses a threshold, all nodes apply message throttling. The use of global information requires to broadcast data among all the network nodes. A way of transmitting this control information is to use a sideband network with a far from negligible bandwidth [16]. To make a fair comparison, as the mechanism proposed in this paper does not need to exchange control messages, the bandwidth provided by the sideband network should be considered as additional available bandwidth in the main interconnection network. However, in the results that we present we do not consider this fact. If this additional bandwidth were considered, the differences, not only in throughput but also in latency, between **Self-Tuned** and the new mechanism would be greater than the ones shown. Moreover, results without any congestion control mechanism (**No-Lim**) are also shown.

First of all, the **Trans-t** mechanism has to be tuned. We have found that the most critical threshold is $u1$ (the threshold used for comparing the quotient of transmission times). As once congestion is detected the injection bandwidth is dynamical adjusted, the other thresholds have a high tolerance variations. The other thresholds tolerate bigger variations. In the case of $u2$ and $u4$, if they are very high, congestion control measures may be applied more time than needed. Therefore, if the network is not longer congested, message latency increases. On the other hand, if they are too low, the status of physical channel is unnecessarily checked many times. Concerning $u3$, if it is too low, it does not apply enough injection limitation restrictions to reduce congestion and the mechanism does not work. After several experiments, the thresholds $u2$, $u3$, and $u4$ have been established in 8, 16 and 8 clock cycles, respectively.

With respect to threshold $u1$, we used a value equal to the number of virtual channels per physical channel as the starting point. The explanation is simple. When congestion appears for the first time, many physical channels will be completely multiplexed in virtual channels. Therefore, the message transmission time across each virtual channel will be roughly equal to the theoretical minimum value multiplied by the multiplexing degree. However, we must consider also that physical channels may not be fully multiplexed and the contention experimented by message header. Hence, the optimal value for this threshold can slightly change. The results show that, for a given network topology, the same value of threshold $u1$ is well suited for different message destinations and message sizes. The design factors that impact threshold $u1$ are the topology radix (k), and the number of virtual channels per physical channel. For a given k-ary n-cube, the optimal threshold increases with the number of virtual channels. On the contrary, when k increases, the optimal threshold decreases. The justification is simple. The higher the number of virtual channels per physical channel, the lower the network congestion, thereby injection policy can be more relaxed. On the contrary, increasing k (the number of nodes per dimension) without increasing the number of dimensions, there is a higher number of paths which shared links among them, which exacerbates congestion. Hence, a more strict injection policy is required (a lower threshold should be used).

Figure 3 shows the average message latency versus traffic for different $u1$ threshold values for the uniform and perfect-shuffle traffic patterns for a 8-ary 3-cube (512 nodes). As it can be seen, the lowest threshold values lead to apply more injection limitation than necessary. As a consequence, message latency is increased due to the fact that messages are waiting at the source nodes. On the other hand, the highest threshold value allows a more relaxed injection policy and trends saturating the network. In this case, a good $u1$ threshold value is 4. Table 1 shows the optimal thresholds found for other topologies and number of virtual channels.

Once tuned the new mechanism, we can compare it with other proposals. Figures 4 through 7 show some of the results obtained, for different message destination distributions, network and message sizes.

In all the cases, simulations finish after receiving 500,000 messages, but only the last 300,000 ones are considered to calculate average latencies. As we can see, the new mechanism (**Trans-t**) avoids the performance degradation in all the cases. Indeed, it always

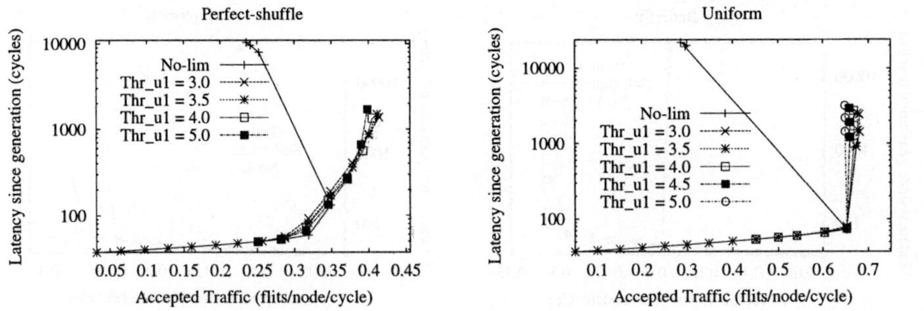

Fig. 3. Average message latency vs. accepted traffic for different $u1$ threshold values for the **Trans-t** mechanism. 8-ary 3-cube (512 nodes). 16-flit messages. 3 virtual channels per physical channel.

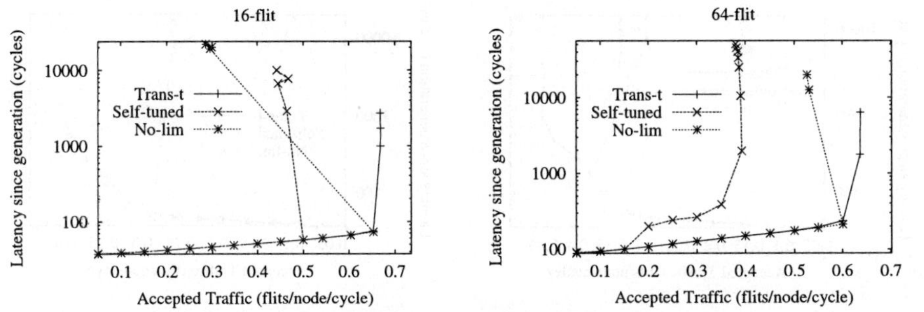

Fig. 4. Average message latency vs. accepted traffic. Uniform distribution of message destinations. 8-ary 3-cube (512 nodes). 3 virtual channels per physical channel. $u1 = 4$.

Table 1. Optimal $u1$ threshold values for different topologies and number of virtual channels per physical channel.

Nodes	Nr. of vc	$u1$	Nr. of vc	$u1$
512 ($8 \times 8 \times 8$)	3	4	4	5
256 (16×16)	3	2.5	4	3.75
4096 ($16 \times 16 \times 16$)	3	2.25	4	3.5
1024 (32×32)	3	2	4	3.25

improves network performance by increasing the throughput achieved when no congestion control mechanism is used. On the other hand, although the **Self-Tuned** mechanism helps in alleviating network congestion, it strongly reduces network throughput and increases network latency with low and medium loads.

We have also used a bursty load that alternates periods of high message injection rate, with periods of low traffic. In this case, we inject a given number of messages into the network and simulation goes on until all messages arrive to their destinations. Figure 8 shows the results for a 2-ary 16-cube (256 nodes) with a uniform distribution

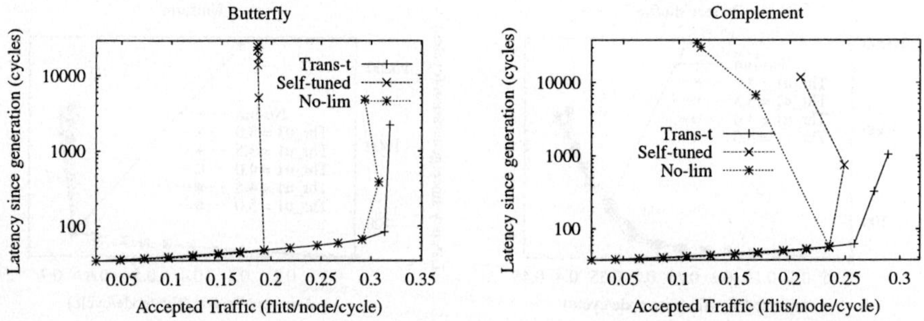

Fig. 5. Average message latency vs. accepted traffic. 16-flit messages. 8-ary 3-cube (512 nodes). 3 virtual channels per physical channel. $u1 = 4$.

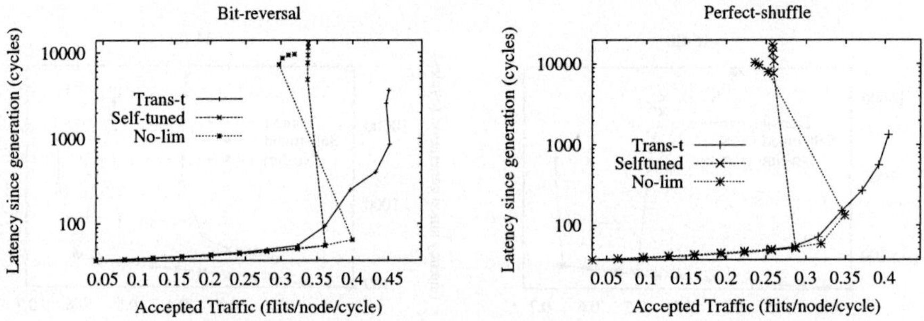

Fig. 6. Average message latency vs. accepted traffic. 16-flit messages. 8-ary 3-cube (512 nodes). 3 virtual channels per physical channel. $u1 = 4$.

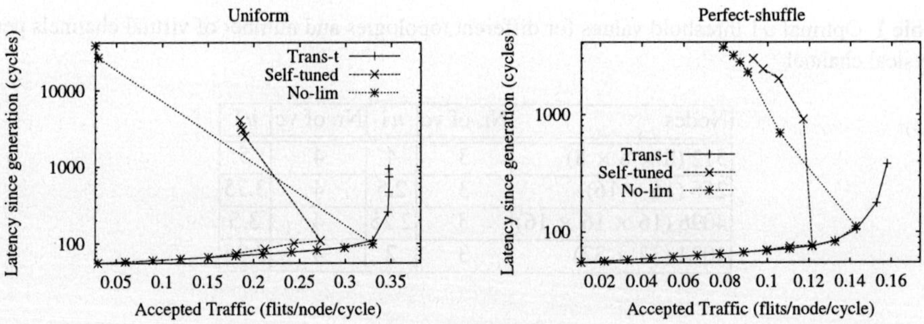

Fig. 7. Average message latency vs. accepted traffic. 16-flit messages. 16-ary 3-cube (4096 nodes). 3 virtual channels per physical channel. $u1 = 2.25$.

of message destinations, 3 virtual channels per physical channel, 400,000 messages generated at a rate of 0.34 flits/node/cycle (high load period) and 200,000 messages at 0.23 flits/node/cycle. These loads are applied alternatively twice. As we can see, with the **Trans-t** mechanism the network accepts the injected bursty traffic without problems.

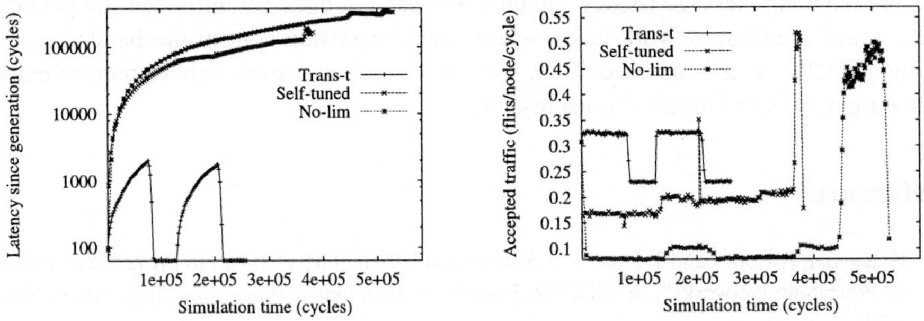

Fig. 8. Performance with variable injection rates. Uniform distribution of message destinations. 16-flits messages. 16-ary 2-cube (256 nodes), 3 virtual channels per physical channel. $u1 = 2.5$.

On the contrary, when no congestion control mechanism is applied, as soon as the first burst is applied into the network, congestion appears. As a consequence, latency strongly increases and accepted traffic falls down. Lately, after some time injecting at the low rate, network traffic starts recovering but the arrival of a new traffic burst prevents it. Congestion only disappears in the last period of time, when no new messages are generated. Concerning the **Self-Tuned** mechanism, we can see that it excessively limits the injection rate, significantly reducing the highest value of accepted traffic and increasing the time required to deliver all the injected messages. This time is another performance measure that is strongly affected by the presence of network congestion. As Figure 8 shows, **Trans-t** delivers the required number of messages in half the time than **No-Lim**, while **Self-Tuned** achieves an intermediate value between both of them.

5 Conclusions

In this paper, we propose a new mechanism (**Trans-t**) based on message throttling to avoid network congestion. This mechanism estimates network traffic by using only local information. In particular, the relationship between the actual and the minimum theoretical transmission time of messages sent across channels is used. The transmission time of a message is the elapsed time between the transfer of its header and tails. If this quotient exceeds a threshold for a given channel, the mechanism assumes that there is congestion in that direction. Although the threshold has to be empirically tuned, it does neither strongly depend on message destination distribution nor on message size, although it depends on the topology radix k (number of nodes per dimension) and the number of virtual channels per physical channel. This is not a problem, as these design parameters are fixed once the machine is built. The information about channels status -congested or not- is considered every time a node tries to inject a new message into the network, adjusting the injection bandwidth depending on the number of congested physical channels that are useful to route the message.

The mechanism has been evaluated for different network loads and topologies. The evaluation results show that the mechanism is able to avoid performance degradation in all the analyzed conditions, outperforming recent proposals, increasing network throughput and reducing message latency. On the other hand, it does not introduce any penalty for low and medium network loads, when none congestion control mechanism is required. Finally, as it is based only on local information, it does neither require extra signaling nor control message transmission.

References

1. E. Baydal, P. López and J. Duato, "A Simple and Efficient Mechanism to Prevent Saturation in Wormhole Networks", in *14th. Int. Parallel & Distributed Processing Symposium*, May 2000.
2. E. Baydal, P. López and J. Duato, "Avoiding network congestion with local information", *4th. Int. Symposium High Performance Computing*, May 2002.
3. T. Callahan and S.C . Goldstein, "NIFDY: A Low Overhead, High Throughput Network interface", *Proc. of the 22th Int. Symposium on Computer Architecture*, June 1995.
4. W. J. Dally and H. Aoki, "Deadlock-Free Adaptive Routing in Multicomputer Networks Using Virtual Channels", *IEEE Trans. on Parallel and Distributed Systems, vol. 4, no. 4, pp. 466–475*, April 1993.
5. J. Duato, "A new theory of deadlock-free adaptive routing in wormhole networks", *IEEE Trans. on Parallel and Distributed Systems*, vol. 4, no. 12, pp. 1320–1331, December 1993.
6. J. Duato, S. Yalamanchili and L.M. Ni, *Interconnection Networks: An Engineering Approach*, IEEE Computer Society Press, 1997.
7. C. Hyatt and D. P. Agrawal, "Congestion Control in the Wormhole-Routed Torus With Clustering and Delayed Deflection" *Workshop on Parallel Computing, Routing, and Communication*, June 1997, Atlanta, GA.
8. J. H. Kim, Z. Liu and A. A. Chien, "Compressionless routing: A framework for Adaptive and Fault-Tolerant Routing", in *IEEE Trans. on Parallel and Distributed Systems*, Vol. 8, No. 3, 1997.
9. F. T. Leighton, "Introduction to Parallel Algorithms and Architectures: Arrays, Trees, Hypercubes", *San Mateo, CA, USA, Morgan Kaufmann Publishers*, 1992.
10. P. López, J.M. Martínez and J. Duato, "A Very Efficient Distributed Deadlock Detection Mechanism for Wormhole Networks", Proc. of High Performance Computer Architecture Workshop, February 1998.
11. P. López, J.M. Martínez and J. Duato, "DRIL: Dynamically Reduced Message Injection Limitation Mechanism for Wormhole Networks", *1998 Int. Conference Parallel Processing*, August 1998.
12. J.M. Martínez, P. López and J. Duato, "A Cost–Effective Approach to Deadlock Handling in Wormhole Networks", *IEEE Trans. on Parallel and Distributed Processing*, pp. 719–729, July 2001.
13. T.M. Pinkston and S. Warnakulasuriya, "On Deadlocks in Interconnection Networks", *24th Int. Symposium on Computer Architecture*, June 1997.
14. A. Smai and L. Thorelli, "Global Reactive Congestion Control in Multicomputer Networks", *5th Int. Conference on High Performance Computing*, 1998.
15. M. Thottetodi, A.R. Lebeck and S.S. Mukherjee, "Self-Tuned Congestion Control for Multiprocessor Networks", *Technical Report CS-2000-15*, Duke University, November 2000.
16. M. Thottetodi, A.R. Lebeck and S.S. Mukherjee, "Self-Tuned Congestion Control for Multiprocessor Networks", *Proc. of High Performance Computer Architecture Workshop*, February 2001.

A Dual-LAN Topology with the Dual-Path Ethernet Module*

Jihoon Park, Jonggyu Park, Ilsuk Han, and Hagbae Kim

Department of Electrical and Electronic Engineering, Yonsei University,
134 Shinchon-Dong, Seodamun-Ku, Seoul 120-749, Korea,
hbkim@yonsei.ac.kr, http://dipl.yonsei.ac.kr

Abstract. A Dual-Path Ethernet Module (DPEM) is developed to improve Local Area Network (LAN)'s performance and High Availability (HA). Since a DPEM simply locates at the front end of any network device as a transparent add-on, it does not require sophisticated server reconfiguration. Our evaluation results show that the developed scheme is more efficient than the conventional LAN structures in various aspects.

1 Introduction

There are many efforts to guarantee High-Performance and H/A on the LAN environment [1, 2]. In this paper, we suggest the Data-link layer LAN path dualizing method, which can improve the availability and transmission bandwidth. Specifically, a Dual-Path Ethernet Module (DPEM) is developed to construct an effective dual-LAN structure. Since DPEM is thoroughly transparent to neighborhood devices, it can be applied regardless of network topology, device and operating system.

2 Implementation and Deployment of a DPEM

2.1 Internal Operating Mechanism

A DPEM is an independent hardware device with some software modules built in the linux-based kernel to reduce the functional overheads of the application-level processes.

The DPEM adopting Dual-LAN transmits one packet stream into two separate paths. A packet from a host is transmitted into the DPEM via port 0. After some internal transactions, it is separated to exit through port 1 or 2. The internal packet modification is accomplished by passing though two software modules as described in the following.

Gateway Management Unit (GMU) modifies the gateway MAC address of the outgoing packet. Generally, a normal host can designate only one gateway address [3]. Without modifying the host configuration, a normal packet is transmitted from a NIC to port 0. Next, the GMU separates and forwards the packet

* This work is supported by the MOCIE project of Korea

B. Monien and R. Feldmann (Eds.): Euro-Par 2002, LNCS 2400, pp. 791–794.

to two gateways on the basis of the results of the path conditions that were acquired by periodic path monitoring.

Local Management Unit (LMU) controls a packet between a host and the LAN, or in the internal loop of the LAN. It basically consists of two sub-modules as follows.

Packet Distributor (PD) transfers a modified packet by the GMU to the designated router. After going through the PD, the packet stream is actually separated by the toggling operation of the PD.

Packet Filter (PF) examines the packet's destination address. The internal loop cannot destroy the outward packets until the Time To Live (TTL) is expired. To remove these useless packets, the PD examines the packet transmission from port 1 to port 2.

The block diagram of functional modules in the DPEM is captured in Fig. 1.

2.2 Deployment of the DPEM

The DPEM's transparent add-on type structure does not require sophisticated server reconfiguration. As showed in Fig. 2, only a simple attachment to the host is enough to build the Dual-LAN. It does not modify the current network topology, but needs only the addition of a DPEM in front of the host machine.

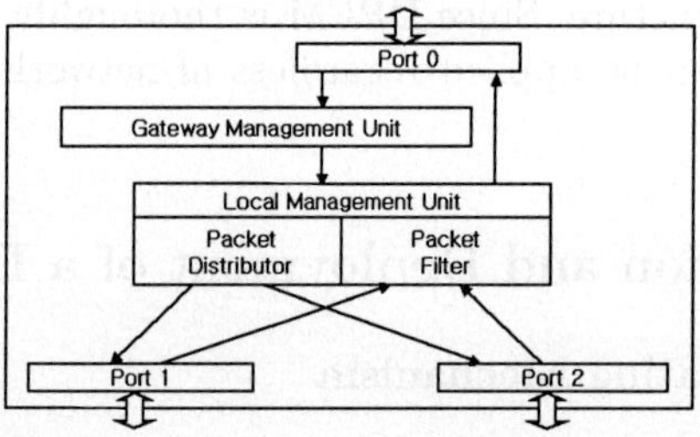

Fig. 1. Functional modules and operations of the DPEM

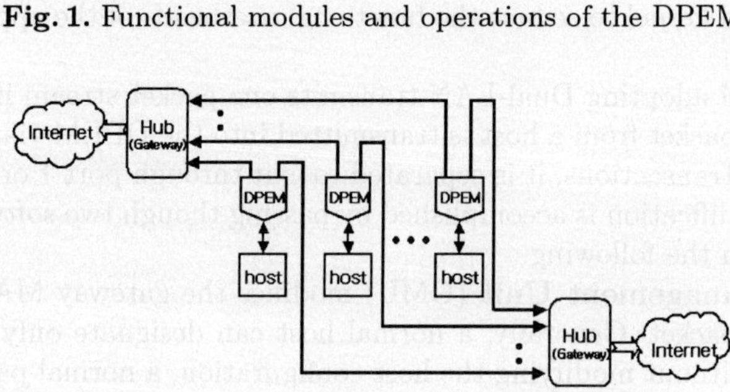

Fig. 2. Dual-LAN structure with DPEMs

3 Performance Evaluation of the DPEM with the Dual-LAN

3.1 Test Conditions

To evaluate the DPEM's performance in general circumstances, we select the test conditions with three hosts as depicted in Fig. 3.

One machine is a file server and the other is a client (as a data requester). Another machine, that locates between the file server and the client measures the data-transmission rates as a controller to analyze the characteristics of the network. Each router also works as a gateway.

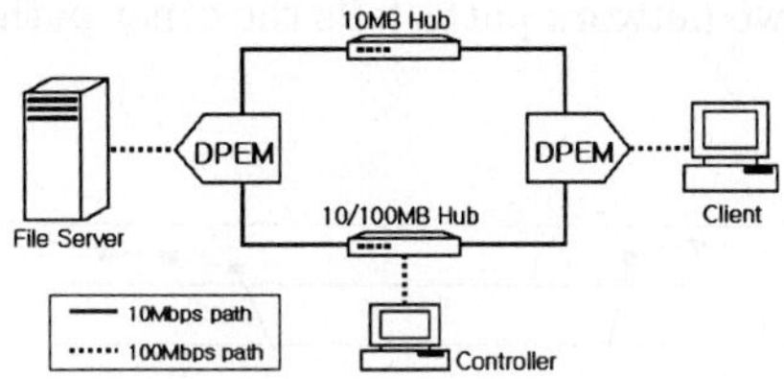

Fig. 3. DPEM test conditions

3.2 Performance Comparison

This test is for evaluating the traffic efficiency and bandwidth limitation. The client increases the data request rate by additional 2Mbps from 8Mbps until reaching the bandwidth limit, maximally 40Mbps. This evaluation method is applied to both the Single and the Dual-LAN.

The throughput of a Single-LAN path is limited below 10 Mbps. It follows the physical specification limit of a normal UTP cable. However, the throughput of the dual LAN paths is shown to have significantly improved performance achieving higher speed than numerical summation of two 10 Mbps paths. Fig. 4 compares the total throughputs of the Single-LAN with that of the Dual-LAN.

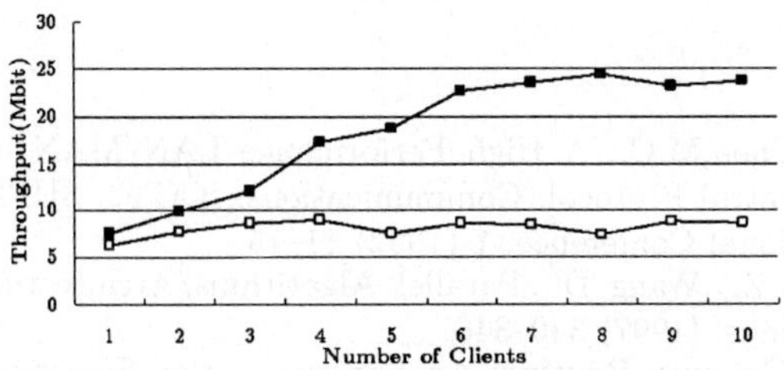

Fig. 4. Throughputs of the Single and the Dual-LAN

3.3 Failure and Recovery Process of the Dual-LAN

The goal of this test is to characterize the recovery operation upon failure occurrence. The fault detection interval is set for 10 seconds, where any status change of the network is reflected on the forwarding entry every 10 seconds.

The test begins with a normal status. After 10 seconds, a failure occurs to drive one path to be disabled. In this case, since the DPEM cannot detect the failure by the next aging time, the throughput becomes lower than 10 Mbps. However, eventually after 20 seconds, the DPEM detects the failure of one path, and then the GMU changes the entry to transmit the data through a healthy path. After 30 seconds both paths become available, but the DPEM cannot detect yet. After 40 seconds the Dual-LAN regains its stable condition. The throughput variation resulted from above procedures is shown in Fig. 5. It verifies that though either of two network paths fails the other path can still continuously deliver the packets.

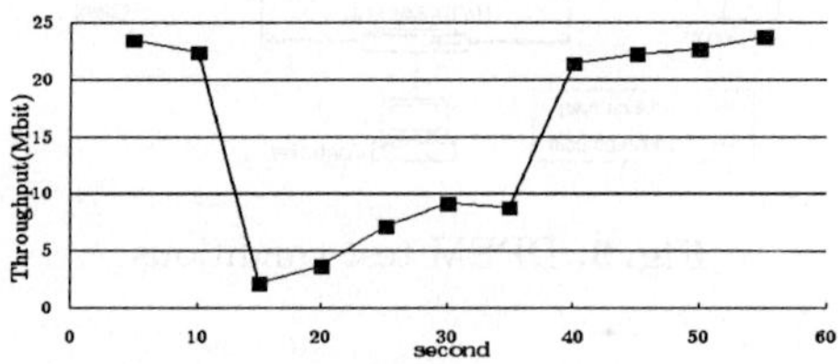

Fig. 5. Throughput variation in the presence of a failure

4 Conclusion

We suggest the data-link layer Dual-LAN building method, which can significantly improve the availability, transmission bandwidth, and security, respectively. A Dual-Path Ethernet Module (DPEM) is developed to construct an effective Dual-LAN structure. The DPEM locates at the front end of any network-connected machine, which can be a server, a client, or a router. Since the DPEM is thoroughly transparent to neighborhood devices, it can be applied regardless of network topologies, devices, and operating systems.

References

[1] Yuang,M.C., Chen,M.C.: A High Performance LAN/MAN using a Distributed Dual Mode Control Protocol. Communications. ICC'92. SUPERCOMM/ICC'92. IEEE International Conference. **1** (1992) 11–15

[2] Wang H., Yin Z., Wang D.: Parallel Algorithms/Architecture Synthesis. 1997. IEEE Proceedings. (1997) 340–346.

[3] Strazisar. V.: Gateway Routing: An Implementation Specification. IEN 30. Bolt Beranek and Newman. (1979)

A Fast Barrier Synchronization Protocol for Broadcast Networks Based on a Dynamic Access Control*

Satoshi Fujita and Shigeaki Tagashira

Department of Information Engineering, Graduate School of Engineering,
Hiroshima University, Higashi-Hiroshima, 739-8527, Japan

Abstract. In this paper, we propose a fast barrier synchronization protocol for broadcast networks. A key point of the protocol is to reduce the performance degradation by introducing a dynamic control of shared bus accesses. The proposed method is implemented on actual LAN and is compared with several protocols implemented in GAMMA system.

1 Introduction

Let $S = \{0, 1, \ldots, N-1\}$ be a set of processes being executed on different hosts, where process 0 is referred to as the **master** and the other processes are referred to as **workers**. In this paper, we consider the **barrier synchronization problem** [1,3] among processes in S on a network connected by a shared bus. Note that a barrier synchronization in the above environment requires at least N messages, since every process in S must transmit a message at least once.

A natural idea for realizing a barrier synchronization efficiently in terms of the number of transmitted messages is to "count" the number of messages in a centralized manner. More concretely, we can count it by using a counter prepared on the master process in the following manner: 1) Upon arriving at a synchronization point, each worker sends an enter_barrier message to the master through the bus. It then waits for a reply from the master, and quits the barrier after receiving an exit_barrier message. 2) The master counts the number of received enter_barrier messages, and broadcasts an exit_barrier message after receiving $N-1$ enter_barrier messages if its execution also arrives at the synchronization point. In this paper, we propose a fast barrier synchronization protocol for broadcast networks. A key point of the protocol is to reduce the performance degradation due to naive access to the shared bus by introducing a dynamic control mechanism. The proposed protocol is implemented on actual LAN, and is compared with several protocols implemented in GAMMA system [2].

This paper is organized as follows. Section 2 describes related work. Section 3 proposes a new protocol, whose performance is compared with GAMMA protocols in Section 4. Section 5 concludes the paper with future problems.

* This research was partially supported by the Ministry of Education, Culture, Sports, Science and Technology of Japan (# 13680417).

B. Monien and R. Feldmann (Eds.): Euro-Par 2002, LNCS 2400, pp. 795–798.

2 Related Work

In GAMMA system developed in Genoa University, the following three barrier synchronization protocols have been implemented [2]: 1) *concurrent* protocol under which all workers independently try to transmit their enter_barrier message; 2) *serialized* protocol under which workers $N-1, N-2, \ldots, 1$ transmit their enter_barrier message sequentially in this order; and 3) a *hybrid* protocol under which c chains of serialized protocol proceeds in a concurrent manner.

In [2], several experiments are conducted to evaluate the efficiency of the protocols. Result of the experiments can be summarized as follows: 1) in the concurrent protocol, the synchronization time increases *exponentially* as N increases; 2) in the serialized protocol, the synchronization time increases *linearly* as N increases; and 3) in the hybrid protocol, the case $c = 2$ exhibits the best performance among others; it is pointed out that the concurrent access by two chains is well interleaved on the shared bus used in their experiments.

3 Proposed Method

The proposed protocol is designed to achieve the following two goals: 1) to bound the number of candidates for accessing the shared bus by some constant α during the execution of the protocol; and 2) to determine the order of bus accesses in such a way that there exist at least α candidates at any time, where the set of *candidates* consists of workers that have not succeeded to transmit an enter_barrier message through the bus (note that it can contain workers that have not yet arrived at the synchronization point). By selecting α to be sufficiently large, we could avoid a situation in which no processes try to transmit a message through the bus, and by selecting α to be sufficiently small, we can reduce the frequency of conflicts on the shared bus.

A formal description of the protocol is given as follows. For simplicity, let us assume $\alpha = 2$ and N satisfies $N = 2^m + 1$ for some integer $m \geq 1$. (Extension to general cases can be done in a straightforward fashion.) The protocol is executed in m phases. Let p and s be local variables representing current phase and step numbers, respectively, that are initialized to zero at the beginning of the protocol. Given a pair of phase and step numbers (p, s), function Next returns a pair of subsequent phase and step numbers, determined as follows:

$$\mathsf{Next}(p, s) \stackrel{\mathrm{def}}{=} \left(p + \left\lceil \frac{s+1}{2^{m-1-p}} \right\rceil, s+1 \bmod 2^{m-1-p} \right).$$

In the protocol, each process keeps the latest value of (p, s). In addition, the master keeps a set U of workers who have *not* succeeded to transmit an enter_barrier message to the master. Let $S(p, s)$ denote a subset of S, defined as follows:

$$S(p, s) \stackrel{\mathrm{def}}{=} \begin{cases} S & \text{if } p = m \\ \{i \in S : i \equiv s+1 \pmod{2^{m-1-p}}\} & \text{if } p \neq m. \end{cases}$$

Suppose that $(p, s, U) = (p_0, s_0, U_0)$ holds at some time instance, where variables p and s are locally maintained by each process and variable U is maintained only by the master process. At that time, worker i is given a permission to broadcast an enter_barrier message through the bus if $i \in S(p_0, s_0) \cap U_0$. Suppose that worker i succeeds to broadcast a message. A message broadcast by worker i is attached with a triple (p_0, s_0, i), in order to inform the current step and phase numbers to the other processes. After receiving the message, each process j (including i itself) updates its local variables as $(p, s) := \mathsf{Next}(p_0, s_0)$. In addition, the master removes i from U, and broadcasts $(s, p, 0)$ while $U \cap S(p, s) = \emptyset$ and it has not received $N - 1$ enter_barrier messages from the workers (note that this operation is necessary to skip vacant steps in which no workers try to broadcast a message). It is worth noting here that the protocol allows a situation in which several workers in $S(p_0, s_0) \cap U_0$ succeed to broadcast a message in a step; in fact, we cannot avoid such a situation, since shared variables (p, s) that are locally maintained by each process, are not locked in the protocol.

4 Experiments

To evaluate the goodness of the proposed method, we conducted several experiments on a LAN consisting of 9 identical hosts with the following parameters: CPU: Celeron 500MHz, Memory: 256MBytes, NIC: Intel EtherExpress Pro/100, HUB: Fast Ethernet Shared HUB (100Base-TX), and OS: Linux Kernel 2.4.9 with Gamma-01-11-26. Two parameters used in the protocols are fixed as $\alpha = 2$ and $c = 2$. In the following, we compare the hybrid protocol with our proposed protocol since the other two protocols are less scalable than the hybrid one.

First, we compared the performance of the two protocols by using the following simple banchmark program: after receiving the declaration message from the master, each worker selects a random value r from $\{1, 2, \ldots, Max\}$ for some positive integer Max, repeats an empty loop for r times, and after completing the loop, it enters a barrier. Figure 1 illustrates the result, where (a) is the synchronization time and (b) is the response time to the last worker. Each point in the figure is an average of three units of experiments. As is shown in the figure, two curves for the response time to the last worker crosses around $Max = 17000$ (see Figure 1 (b)), whereas it always takes longer time than GAMMA in terms of the synchronization time (see Figure 1 (a)). This phenomenon is probably due to the high maintenance cost for keeping shared variables (p, s), that should be realized by a broadcasting/snooping mechanism on LAN[1].

Second, we compared the performance of the schemes under a worst case situation for GAMMA, in which a waiting loop is inserted only to two first workers in the chains (note that our method does not have such a worst case situation). The waiting time is randomly selected, as before. Figure 2 illustrates the result. Under such a worst case setting, the cross point moves from $Max =$

[1] GAMMA uses a simple point-to-point message passing (from process $i+1$ to process i), which could bound the communication overhead by a very small value.

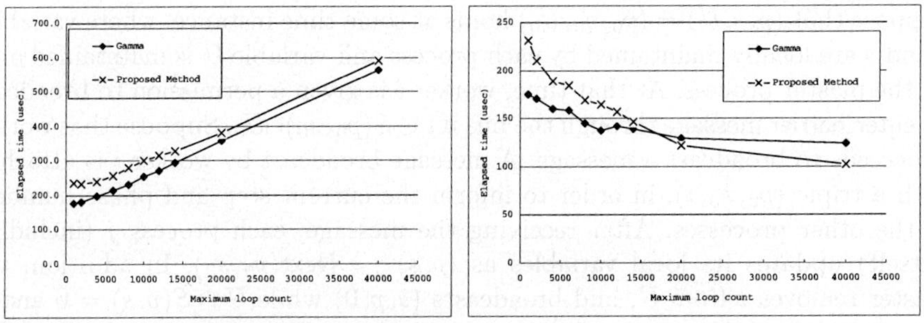

(a) (b)

Fig. 1. Results for random waiting times.

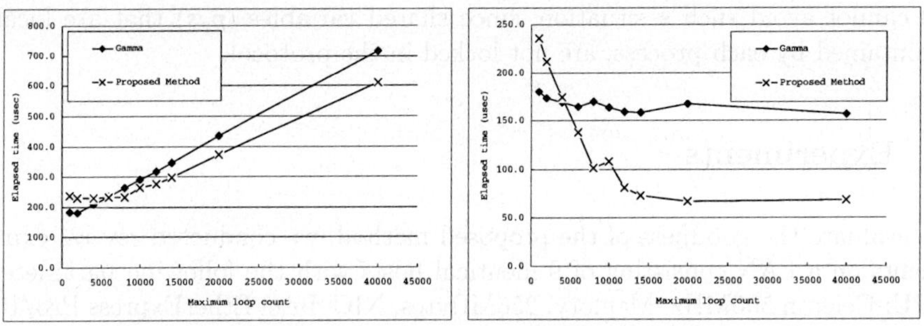

(a) (b)

Fig. 2. Results for worst case waiting times.

17000 to $Max = 5000$ (see Figure 2 (b)), and we could observe such a cross point even for the synchronization time (see Figure 2 (a)).

5 Concluding Remarks

Extension to wireless networks is an interesting direction for future research. It should also be important to extend the scheme to general networks by using a multicast to relevant processes instead of broadcasting.

References

1. J. Anderson. "Simulation and analysis of barrier synchronization methods," Technical Report, HPPC-95-04, University of Minnesota, High-Performance Parallel Computing Research Group (1995).
2. G. Chiola and G. Ciaccio. "Fast Barrier Synchronization on Shared Fast Ethernet," in Proc. CANPC'98, LNCS 1362 (1998).
3. D. Johnson, D. Lilja, J. Riedl, J. Anderson. "Low-cost, high-performance barrier synchronization on networks of workstations," JPDC 40(1): 131-137 Jan. 1997.

The Hierarchical Factor Algorithm for All-to-All Communication

Peter Sanders[1,*] and Jesper Larsson Träff[2]

[1] Max-Planck-Institut für Informatik,
Stuhlsatzenhausweg 85, 66123 Saarbrücken, Germany,
sanders@mpi-sb.mpg.de, http://www.mpi-sb.mpg.de/~sanders/
[2] C&C Research Laboratories, NEC Europe Ltd.,
Rathausallee 10, 53757 Sankt Augustin, Germany,
traff@ccrl-nece.de

Abstract. We present an algorithm for *regular, personalized all-to-all communication*, in which every processor has an individual message to deliver to every other processor. Our machine model is a cluster of processing nodes where each node, possibly consisting of several processors, can participate in only one communication operation with another node at a time. The nodes may have different numbers of processors. This general model is important for the implementation of all-to-all communication in libraries such as MPI where collective communication may take place over arbitrary subsets of processors. The algorithm is optimal up to an additive term that is small if the total number of processors is large compared to the maximal number of processors in a node.

1 Introduction

A successful approach to parallel programming is to write a sequential program executing on all processors and delegate interprocessor communication and coordination to a communication library such as MPI [9]. With this approach, many parallel computations can be expressed in terms of a small number of *collective communication operations*, where "collective" means that a subset of processors is cooperating in a nontrivial way. One such frequently used collective communication operation is *regular, personalized all-to-all message exchange*: Each of p processors has to transmit a personalized message to itself and each of $p-1$ other processors, i. e., for every pair of processor indices i and j a message m_{ij} has to be sent from processor i to processor j. In *regular* all-to-all exchange, all messages are assumed to have the same length. Examples of subroutines using all-to-all communication are matrix transposition and FFT.

This paper presents an algorithm for regular all-to-all communication on clusters of processing nodes where each node may consist of several processors. We assume that only a single processor from each node can be involved in internode communication at a time. Prime examples of such hierarchical systems are

* Partially supported by the Future and Emerging Technologies programme of the EU under contract number IST-1999-14186 (ALCOM-FT).

B. Monien and R. Feldmann (Eds.): Euro-Par 2002, LNCS 2400, pp. 799–803.

clusters of SMP nodes, where processor groups of 2–16 processors communicate via a shared memory, and where some medium to large number of nodes are interconnected via a commodity interconnection network. For example, the Earth Simulatorand the NEC SX-6 supercomputerhave up to 8 processors per node; the IBM SP POWER3 allows up to 16 processors per node.The difficult case is when nodes have differing numbers of processors participating in the all-to-all exchange. This situation must be handled efficiently in a high-quality communications library because it arises naturally if a job is assigned only part of the machine, or if the exchange is only among a subset of the processors in a job.

We use a simple machine model that allows an efficient implementation portable over a large spectrum of platforms. The nodes are assumed to be fully connected. Communication is *single ported* in the sense that at most one processor per node can communicate with a processor on another node at a time. The single-ported assumption is valid for current interconnection technologies like Myrinet, Giganet, the Scalable Coherent Interface (SCI), or for the crossbar switch used on the NEC machines and the Earth Simulator.

Our algorithm for all-to-all communication extends a well-known algorithm for non-hierarchical systems based on *factoring* the complete graph into matchings. The new algorithm is optimal with respect to the time a processor spends waiting or transmitting data up to an additive term that is bounded by the time needed for data exchange inside a node. This time is comparatively small if the total number of processors is large compared to the maximum number of processors in a node. Our algorithm runs in phases, in each phase getting rid of nodes with the minimum number of processors among the surviving nodes. The main issue is to balance the communication volume of nodes with many processors over the phases so that the number of communication steps is minimized.

All-to-all communication has been studied intensively, and we mention only a sample of the known results. Most work focuses on non-hierarchical systems with specific interconnection networks [7, 2, 10]. Trade-offs between communication volume and number of communication start-ups were studied in [1, 2], which achieve algorithms that are faster for small messages. Collective communication on hierarchical systems has recently received some attention [8, 5, 4]. Huse [4] reports experiments with a regular all-to-all algorithm which ensures that only one processor per node is involved in inter-node communication at a time. Algorithmic details and properties are not stated.

2 The Non-hierarchical Factor Algorithm

The basis for our algorithm is a well-known algorithm for the single-ported, non-hierarchical case [2, 6, 10]. This algorithm exploits the existence of a 1-*factorization* of the complete graph [3]. Our formulation of the all-to-all communication problem requires the inclusion of self-loops in the graph, whereas the usual construction has no self-loops. We give the construction and the proof here. It is perhaps interesting to note that self-loops *simplify* the construction.

Lemma 1. *Let G be the complete graph with p vertices including self-loops. G is* 1-factorizable, *i. e., $G = (V, E)$ can be decomposed into p subgraphs $G^i = (V, E_i), i = 0, \ldots, p-1$ in which each vertex has degree 1 (*1-factors*).*

Proof. Let $V = \{0, \ldots, p-1\}$. The ith factor $G^i = (V, E_i)$, is constructed as follows. For $u \in V$ define $v^i(u) = (i - u) \bmod p$. Define $E_i = \{(u, v^i(u)) | u \in V\}$. Since $v^i(v^i(u)) = (i - ((i-u) \bmod p)) \bmod p = u$ all vertices have degree exactly one. Furthermore, any edge $(u, v) \in E$ will find itself in some factor, namely in factor $G^{(u+v) \bmod p}$. In particular, the self-loop (u, u) will find itself in $G^{2u \bmod p}$. ∎

The non-hierarchical *factor algorithm* is the basis for our hierarchical algorithm explained in the next section. It requires p communication rounds for any number of p processors. In the ith round, all processors u and v that are neighbors in G^i are *paired* and exchange their messages m_{uv} and m_{vu}.

3 All-to-All Communication on Hierarchical Systems

We now generalize the factor algorithm to clustered, hierarchical systems. Let N be the number of processor nodes, and let G denote the N-node complete graph with self-loops. Let G_A denote the subgraph of G induced by a subset of nodes A, and G_A^i the ith 1-factor of G_A. We use U and V to denote processor nodes of the system, and u and v for individual processors. By size(U) we denote the number of processors in node U, and by $l(u)$ the *local index* of processor u within its node, $0 \leq l(u) < \text{size}(U)$ for $u \in U$. To specify what messages should be exchanged when two nodes U and V are paired we impose the node ordering

$$U \preceq V \text{ if } \text{size}(U) < \text{size}(V), \text{or } \text{size}(U) = \text{size}(V) \wedge U \leq V$$

where $U \leq V$ relates to an arbitrary total ordering of the nodes. The algorithm is shown in Fig. 1 using this notation.

The outermost loop iterates over a number of phases, each of which considers a 1-factorization of the set of *active nodes* A that have not yet exchanged all their messages. The second loop iterates over the 1-factors G_A^i of G_A. The parallel loop considers all node pairs (U, V) that are neighbors in the given 1-factor G_A^i. The node ordering $U \preceq V$ is used to conveniently describe the message exchange between processors on node U and processors on nodes V necessary for reestablishing the invariant for the outermost loop after 'done' has been increased to 'current'. When $U = V$, the bidirectional exchange is replaced by a unidirectional send because otherwise, intra-node messages would be transmitted twice. Sending m_{uu} from u to u means copying m_{uu} from source buffer to destination buffer of u.

Theorem 1. *Algorithm Hierarchical-AllToAll performs a personalized all-to-all exchange in a number of steps equal to the maximal number of messages that the processors in a node have to send.*

Algorithm Hierarchical-AllToAll:

```
A ← {0,...,N−1}                                            // set of active nodes
done ← 0
while A ≠ ∅ do                                             // phase
    loop invariant: ∀(U,V) ∈ G : ∀u ∈ U, v ∈ V :
        (U ≼ V ∧ 0 ≤ l(u) < done) ⇒ m_uv and m_vu have been delivered
    current ← min{size(U) | U ∈ A}
    for i = 0,...,|A| − 1 do                               // round
        for all (U,V) ∈ G^i_A where U ≼ V pardo
            for each u ∈ U, done ≤ l(u) < current do
                for each v ∈ V do                          // step
                    if U = V then send m_uv from u to v
                    else exchange m_uv and m_vu between u and v
    done ← current
    A ← A \ {U | size(U) = done}
```

Fig. 1. The hierarchical factor algorithm.

Proof (Outline). Regarding correctness, let $0 = S_0 < S_1 < \ldots < S_k$ be the sequence of different node sizes. The algorithm performs k phases. In phase i nodes U with $\text{size}(U) \geq S_i$ are active. In particular, the outer loop terminates. Furthermore, at the end of the algorithm done $= \max_{U \in \{0,\ldots,N-1\}} \text{size}(U)$. The loop invariant implies that all messages have been exchanged.

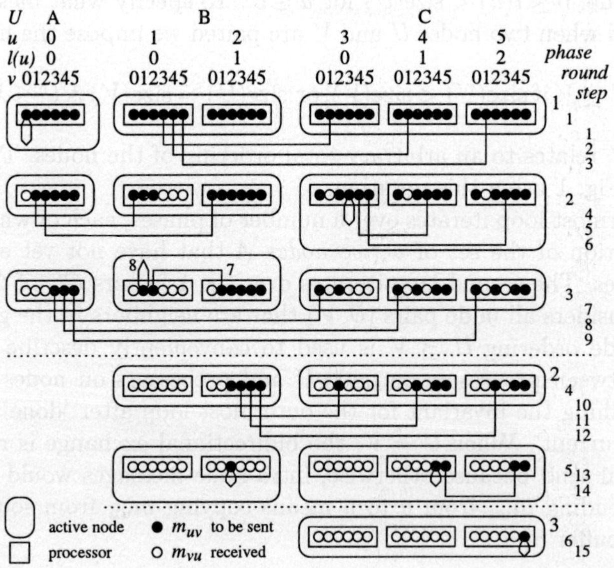

Fig. 2. Example execution of algorithm Hierarchical-AllToAll for three nodes with size 1, 2, and 3 respectively. The algorithms goes through 3 phases, and 3, 2 and 1 rounds respectively are required, for a total of 15 steps. Step 6 of phase 1 in which no inter-node communication takes place, can easily be moved to the end of the computation.

The bound on the number of steps follows since all nodes with the maximum number of processors are participating in a communication in every step and because no message is sent twice. ∎

The reason why the algorithm is not optimal in all cases is that a node paired with itself communicates only unidirectionally in each step. If at the same step two other nodes with maximum number of processors are paired, they communicate bidirectionally and hence take longer to complete a round. This is not optimal since at least in some cases there are schedules which avoid such situations. However, there are only few such inefficient steps: Consider a node U with maximal number $n = \text{size}(U)$ of processors. Our algorithm performs pn steps. At most n^2 of these steps — a fraction of n/p — can be inefficient for node U. Hence, the inefficient steps are few compared to the efficient steps for $p \gg n$.

Although the algorithm was formulated for single-ported communication using 1-factorizations, generalizations to multi-ported communications are possible. The same basic scheme applies if the complete graph is decomposed into graphs with degrees at most k or into permutations (directed cycles). Decomposition into permutations is particularly interesting since several all-to-all algorithms for non-fully connected networks are known that are based on this approach [7, 2, 10].

References

1. J. Bruck, C.-T. Ho, S. Kipnis, E. Upfal, and D. Weathersby. Efficient algorithms for all-to-all communications in multiport message-passing systems. *IEEE Transactions on Parallel and Distributed Systems*, 8(11):1143–1156, 1997.
2. S. E. Hambrusch, F. Hameed, and A. A. Khokar. Communication operations on coarse-grained mesh architectures. *Parallel Computing*, 21:731–751, 1995.
3. F. Harary. *Graph Theory*. Addison-Wesley, 1967.
4. L. P. Huse. MPI optimization for SMP based clusters interconnected with SCI. In *7th European PVM/MPI User's Group Meeting*, volume 1908 of *Lecture Notes in Computer Science*, pages 56–63, 2000.
5. N. T. Karonis, B. R. de Supinski, I. Foster, W. Gropp, E. Lusk, and J. Bresnahan. Exploiting hierarchy in parallel computer networks to optimize collective operation performance. In *Proceedings of International Parallel and Distributed Processing Symposium (IPDPS'2000)*, pages 377–384, 2000.
6. P. Sanders and R. Solis-Oba. How helpers hasten h-relations. *Journal of Algorithms*, 41:86–98, 2001.
7. D. S. Scott. Efficient all-to-all communication patterns in hypercube and mesh topologies. In *Sixth Distributed Memory Computing Conference Proceedings*, pages 398–403, 1991.
8. S. Sistare, R. vandeVaart, and E. Loh. Optimization of MPI collectives on clusters of large-scale SMPs. In *Supercomputing*, 1999. `http://www.supercomp.org/sc99/proceedings/techpap.htm#mpi`.
9. M. Snir, S. Otto, S. Huss-Lederman, D. Walker, and J. Dongarra. *MPI – The Complete Reference*, volume 1, The MPI Core. MIT Press, second edition, 1998.
10. Y. Yang and J. Wang. Optimal all-to-all personalized exchange in self-routable multistage networks. *IEEE Transactions on Parallel and Distributed Systems*, 11(3):261–274, 2000.

Topic 13
Architectures and Algorithms for Multimedia Applications

Andreas Uhl

Salzburg University, Department of Scientific Computing, Jakob Haringer-Str. 2, A-5020 Salzburg, Austria,
uhl@cosy.sbg.ac.at, http://www.cosy.sbg.ac.at/sc/

In the recent years multimedia technology has emerged as a key technology, mainly because of its ability to represent information in disparate forms as a bit-stream. This enables everything from text to video and sound to be stored, processed, and delivered in digital form. A great part of the current research community effort has emphasized the delivery of the data as an important issue of multimedia technology. However, the creation, processing, and management of multimedia forms are the issues most likely to dominate the scientific interest in the long run. The aim to deal with information coming from video, text, and sound will result in a data explosion. This requirement to store, process, and manage large data sets naturally leads to the consideration of programmable parallel processing systems as strong candidates in supporting and enabling multimedia technology. Therefore, this fact taken together with the inherent data parallelism in these data types makes multimedia computing a natural application area for parallel and distributed processing. In addition to this, the concepts developed for parallel and distributed algorithms are quite useful for the implementation of distributed multimedia systems and applications. Thus, the adaptation of these methods for distributed multimedia systems is an interesting topic to be studied.

These facts are also reflected by a number of conferences and workshops exclusively devoted to parallel and distributed multimedia processing. The "Workshop on Parallel and Distributed Computing in Image Processing, Video Processing, and Multimedia (PDIVM)" is an annual workshop co-organized by the author of this introduction in the framework of the "International Parallel and Distributed Processing Symposium (IPDPS)" (see http://www.cosy.sbg.ac.at/~uhl/pdivm.html). "Parallel and Distributed Methods for Image Processing I – IV" is an annual conference organized in the context of SPIE's annual meeting (published so far as SPIE proceedings no. 3166, 3452, 3817, and 4118). Also, several special sessions at various conferences have been devoted to these or similar topics. For example, the "2002 International Conference on Parallel and Distributed Processing Techniques and Applications (PDPTA'2002)" featured special sessions on "Parallel and Distributed Image Processing (PDIP 2002)" and "Parallel and Distributed Multimedia Processing & Retrieval (PDMPR 2002)". The EuroPar Topic 13 "Architectures and Algorithms for Multimedia Applications" stands out from comparable conference special sessions due to its high scientific quality ensured by a rigorous review process. Out of 12 submitted pa-

B. Monien and R. Feldmann (Eds.): Euro-Par 2002, LNCS 2400, pp. 805–806.

pers, 4 have been accepted as regular papers (33%), 3 as short papers (24%), and 5 contributions have been rejected (41%). In total, 46 reviews have been received for the 12 papers (3.83 reviews per paper on average !).

The accepted papers reflect very well the diversity of research themes covered by the titel of the topic. Four out of seven accepted papers represent the classical parallel computing field whereas the remaining three papers cover distributed multimedia. Dedicated hardware (extensions) suited for multimedia processing are discussed in "Novel Predication Scheme for a SIMD System-On-Chip" and "Performance Scalability of Multimedia Instruction Set Extensions", whereas "MorphoSys: A Coarse Grain Reconfigurable Architecture For Multimedia Applications" focuses on multimedia processing with reconfigurable hardware which is an important trend in this area. Mapping of a recent multimedia standard (i.e. H.26L) to a general purpose parallel architecture is investigated in "A Parallel Implementation of H.26L Video Encoder". Distributed multimedia systems are represented by the papers "Deterministic Scheduling of CBR and VBR Media Flows on Parallel Media Servers" and "Double P-Tree: A Distributed Architecture for Large-Scale Video-on-Demand". Finally, a distributed multimedia application is described in "Message Passing in XML-based Language for Creating Multimedia Presentations".

The diversity of papers shows the liveliness of this research area. We feel that this topic is an important contribution to EuroPar and to the field of parallel computing in general. Last but not least, we wish to thank the topic vice-chairs Suchendra M. Bhandarkar and Michael V. Bove as well as the local chair Reinhard Lüling for organising the reviews and the referees for their valuable suggestions and comments.

Deterministic Scheduling of CBR and VBR Media Flows on Parallel Media Servers

Costas Mourlas

Department of Computer Science, University of Cyprus,
75 Kallipoleos str., CY-1678 Nicosia, Cyprus,
mourlas@ucy.ac.cy

Abstract. We study a new design strategy for the implementation of Parallel Media Servers with a predictable behavior. This strategy makes the timing properties and the quality of presentation of a set of media streams predictable. The proposed real-time scheduling approach exploits the performance of parallel environments and seems a promising method that brings the advantages of parallel computation in media servers. The proposed mechanism provides deterministic service for both Constant Bit Rate (CBR) and Variable Bit Rate (VBR) streams. A prototype implementation of the proposed parallel media server illustrates the concepts of server allocation and scheduling of continuous media streams.

1 Introduction

Continuous media servers differ enough from traditional storage servers since they store and manipulate continuous media data (video and audio) which consist of *streams* of media quanta (video frames and audio samples) that must be presented using the same timing sequence with which they were captured. This implies that special scheduling and resource allocation strategies must be provided by the CM (Continuous Media) server such that the required CM data will be available for the time they are needed. Hence, media servers need to ensure that the retrieval and storage of such CM streams proceed at their pre-specified real-time rates.

During the last few years, much interesting work has been focused on the design of parallel on-demand media servers [2, 9]. A great part of the current research community effort has been also focused on new data layout schemes [7], striping mechanisms, admission control and disk scheduling [9] for storage device arrays of parallel [5] and clustered multimedia servers [4]. The problem of providing deterministic service for VBR streams in a single system has been studied in [10] where the notion of the *demand trace* of every VBR stream has been introduced.

A very interesting approach described in [6] and [8] provides accurate scheduling of video streams with the restrictive assumption that all the streams have been encoded using a unique base stream rate R. This paper presents an extension of that scheduling strategy and supports both CBR and VBR media streams

B. Monien and R. Feldmann (Eds.): Euro-Par 2002, LNCS 2400, pp. 807–815.

(s_is) video and audio encoded using different playback rates (R_is), which is an interesting feature not supported in the original version of the algorithm. We focus mainly on resource management of the parallel server in order to provide on-demand support for a large number of concurrent continuous media objects in a *predictable* manner.

2 The Architecture of the Parallel Media Server

In order to design a general purpose architecture which can be adapted to the current user requirements, a scalable parallel multimedia server shall be designed. We will use the traditional model for a parallel media server previously described in [11, 3]. In that architecture there exist three kinds of nodes: storage nodes, delivery nodes and one control node. The three kinds of nodes are explained in greater detail below:

Storage Nodes are responsible for storing video and audio clips, retrieving requested data blocks and sending them to delivery nodes within a time limit. In addition, partitioned video blocks are wide striped among storage nodes in a round-robin fashion to balance the workload.

Delivery Nodes are responsible for serving stream requests that have been previously accepted for service. Their main function is to request the striped data from the storage nodes through the internal interconnection network, re-sequence the packets received if necessary and then send the packets over the wide area network to the clients.

The Control Node receives all incoming requests for media objects. It has knowledge of which storage node stores the first data block of the object and the workload of the delivery and storage nodes. In a typical request-response scenario, the control node receives a request for a media object. If the resource requirements of the request are consistent with the system load at that time, then the request is accepted. A delivery node to serve the stream is chosen by the control node and the delivery node then takes over the authority of serving the stream. To that end, it retrieves the stream fragments from the storage nodes and transmits them at the required rate to the client.

The logical storage and delivery nodes can be mapped to different as well as to the same physical node. The model where a node can be both a storage node and a delivery node is called "flat" architecture and it is more suitable to be implemented on a cluster of workstations interconnected by high-speed links. In this paper, we are focused on "flat" architectures.

3 The Proposed Scheduling Algorithm

As mentioned earlier, the data is compressed and striped across all storage nodes in a round-robin fashion. Although data blocks are wide striped, without properly scheduling of data retrievals, resource conflicts may be occurred such as *port contention* where two storage nodes are transmitting to a single delivery node at the same time. Another resource conflict that may also happen is *disk contention* where more than one request retrieve blocks from the same storage node at the same time instance.

The work described in this paper, is concentrated on the special case of conflict-free scheduling that provides deterministic guarantees for both CBR and VBR stream requests. It is well understood that providing deterministic service for CBR streams is easier due to the fact that a CBR stream requests the same amount of data in every interval. For presentational purposes, the initial version of our scheduling algorithm that schedules CBR requests is presented first. In a following subsection we extend our scheduling algorithm to include VBR streams taking into account the fluctuations in the bit rates of multiple requests that may overload throughput capacity of the storage nodes.

3.1 Deterministic Guarantees for CBR Streams Encoded at Different Playback Rates

We will describe how requests for media streams can be modeled as a set of periodic tasks and we give a formal evaluation of some components such as the period and the data retrieval section of each task. Time is divided into time rounds (or cycles) where the length of every time round T_i equals to T which is a constant value. R_i is the required playback rate for stream s_i that has been pre-determined during the compression phase of that stream. Note that, for a CBR stream s_i, the value R_i is a constant during the length of the stream. Our aim in the design of a parallel media server is to supply the stream with enough data to ensure that the playback processes do not starve.

Therefore, every stream s_i is represented by a periodic task τ_i where in every period (i.e. in every time round) T needs to retrieve $F_i = T * R_i$ amount of data to guarantee that the stream s_i will meet its real-time requirements. The above equation determines the stripe fragment size F_i of stream s_i which is different in general for every stream according to its playback rate R_i. Every media stream s_i is striped across all nodes in a round-robin fashion where the stripe fragment size of s_i equals to F_i. The average time to retrieve F_i bytes from the storage node and transmit them to a delivery node is given by equation

$$t_s^i = t_{avg\_seek} + t_{avg\_rot} + t_{r\_F_i} + t_{nw\_Fi} \tag{1}$$

where t_{avg_seek} and t_{avg_rot} are the average seek and rotational latencies for the disks being used, $t_{r\_F_i}$ is the disk data transfer time for F_i bytes and t_{nw_Fi} is the internal network latency to transport F_i bytes from a storage node to a delivery node. Thus, t_s^i is the length of the data retrieval section of the periodic task τ_i.

Since the stripe fragments of a continuous media are consecutively distributed in all N storage nodes, if a task τ_i at time round m retrieves data from node k, it will retrieve data from node $(k+l)$ *mod* N at time round $(m+l)$. A complete schedule is represented by a schedule table consisting of N consecutive time rounds. Let u_i be the set of tasks allocated to the delivery node i for service. We define as the utilization factor U_i of a delivery node i, the sum given by the formula:

$$U_i = \sum_{\tau_j \in u_i} \frac{t_s^j}{T}, \qquad 0 \leq i \leq N-1 \tag{2}$$

The value U_i of a delivery node i changes only when a new request is allocated to the delivery node i by the control node, or when an existing request completed its execution and quits. U_i represents the load of the delivery node i and its value can never be greater than one. Note that, the utilization factor of a storage node varies from one time round to the other. More precisely, the load of one storage node in the round T_i moves to the next storage node in round T_{i+1} and returns in round T_{i+N}.

Our current work is concentrated on the special case of scheduling called conflict-free scheduling [11], which is an extension of the work presented in [6] and [8]. In the next subsection, we extend further our scheduling strategy to accommodate VBR stream requests.

A conflict-free schedule is a schedule that in every time instance the following scenario will never occur: two media streams request data from the same storage server or two storage servers transmit data to the same delivery node. In order to construct such a schedule we implement every round (or cycle) in such a way that only one storage node transmits data to one delivery node. Note that, a starting sequence which designates the transmission order between storage nodes and delivery nodes needs to be assigned at the first basic time round T_0. Different but equivalent basic time rounds exist each one with a different starting sequence and any of that rounds can be selected as the basic round T_0. Since the blocks of the media streams are consecutively distributed in all N storage nodes, when delivery node 0 schedules a request that retrieves a block from storage node 1 the same request retrieves blocks from storage nodes 2,3,...,N-1,0 in the next N-1 rounds (see Figure 1).

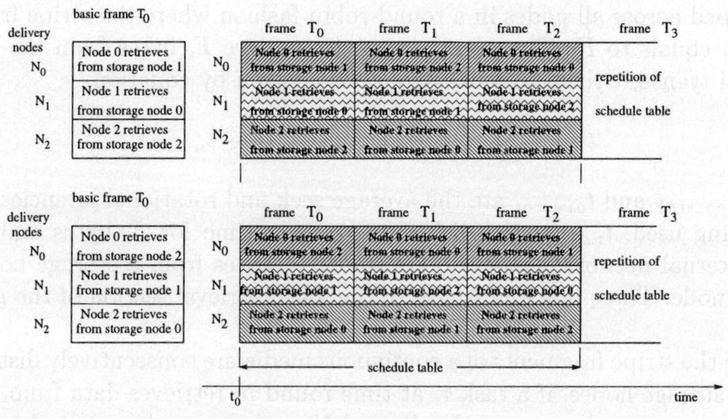

Fig. 1. Equivalent schedule tables starting from a different basic time round T_0.

The proposed algorithm schedules the stream requests as follows: When a new request r_k for a media object arrives where its starting block is stored in node j the first step is to test schedulability of the new request. The control node checks the loads of the delivery nodes and finds the node i, $(0 \leq i \leq N-1)$ with the minimum load U_i. Then, it checks if the condition $U_i + \frac{t_s^k}{T} \leq 1$ is satisfied for that node. If the condition is satisfied then the node i is declared as the delivery node of the stream s_k and it will serve together with the previous streams the new one during its lifetime. The new request r_k starts receiving data when delivery node i is connected with storage node j for first time after the receipt of request r_k and the schedule table is updated accordingly. Notice the possibility to delay the beginning of service for a request till delivery node is connected for the first time after the receipt of the request with the corresponding storage node. In case that the above condition cannot be satisfied the request is rejected or postponed for later service. An important property of the proposed algorithm is that when the control node finds a delivery node i to serve the request r_k and the condition $U_i + \frac{t_s^k}{T} \leq 1$ can be satisfied, it is guaranteed immediately that the new load $U_i' = U_i + \frac{t_s^k}{T}$ can be accommodated also by the storage nodes.

The proposed conflict-free scheduling algorithm can be illustrated by a single example (see Figure 2), where a parallel media server is used with a "flat" architecture that supports 3 nodes (N=3). Noticc that request r_4 is delayed for T time units before it is served.

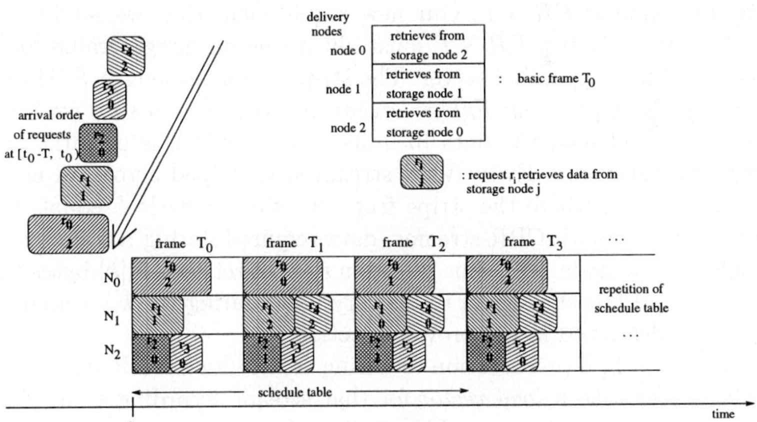

Fig. 2. The basic round T_0, the arrival order of the requests and the complete schedule of the example

3.2 Deterministic Guarantees for VBR Streams

The problem of providing deterministic guarantees for VBR streams is harder due to the following two reasons:

1. the load of a stream on the storage units varies from one round to the other, and
2. scheduling the first block of a VBR stream does not mean that the rest blocks of the stream can be scheduled.

One approach for the solution of the problem is to compute the peak rate of the stream and reserve enough bandwidth on storage nodes to satisfy the peak requirements of the stream. This pessimistic approach results to underutilization of resources since the peak demand is observed only for short durations compared to the whole duration of the stream.

In the our approach, every CBR stream s_i is represented by a periodic task τ_i where in every time round T needs to retrieve a constant data length block F_i determined by the equation $F_i = T * R_i$. R_i is the required playback rate for stream s_i which is a constant value for every different CBR stream. Using VBR streams, the bit rate R_i is variable which means that the stripe fragment length for VBR streams is also variable and thus the workload for the storage devices changes from one round to the other.

Our approach considers the variations in loads and provides guarantees for VBR streams as follows: Time is divided as described above into time rounds of equal size T. We introduce the parameter FR_i which denotes the frame rate of real-time playback (frames per second) for a video stream s_i (or samples per second for audio stream) determined when the media was captured. Thus, the number of frames included in every stripe fragment of the media stream s_i is given by the number $FR_i * T$. Our new requirement that we set here is that the result of the product $FR_i * T$ must always be an integer value for all the streams stored in our parallel server. The stripe fragment length $F_i[k]$ on round k is given by the expression: $F_i[k] = sizeof(FR_i * T\ frames\ on\ round\ k)$. We therefore store and read the data in units of $F_i[k]$ which is of variable length on every different round. Every VBR stream s_i is striped across all nodes in a round-robin fashion where the stripe fragment size for cycle k equals to $F_i[k]$. Notice that just as with CBR streams, data required during a round is located on a single storage node. The time duration $t_s^i[k]$ to retrieve $F_i[k]$ bytes from the storage node and transmit them to delivery node during the k_{th} round is given by equation 1 described in the previous section.

In our approach, a presentation requiring service for a VBR stream supplies the media server with a *load vector* for that stream according to its demands. More precisely, the load vector $LV_i[k]$, $0 \leq k < dur(s_i)$ describes the time required for a storage node to retrieve and transmit $F_i[k]$ data units to a delivery node on each round k. The term $dur(s_i)$ defines the length of the stream s_i in rounds. The load vector LV_i for stream s_i can be stored on the Control Node in a form of a special file. We can easily conclude that compared to the size of the video or audio file, the size of the load vector file is not significant.

The Control Node keeps track also of the utilization of every delivery node i of the server in the form of a utilization vector $U_i[]$. The utilization vector $U_i[]$ for delivery node i stores the actual utilization of every delivery node in each round over sufficient period of time. We define as the utilization $U_i[k]$ of a delivery node i during the k_{th} round, the sum given by the formula:

$$U_i[k] = \sum_{\tau_j \in u_i} \frac{t_s^j[k]}{T}, \quad 0 \leq i \leq N-1 \tag{3}$$

The value $U_i[k]$ of a delivery node i changes not only when a new request is allocated to the delivery node i by the control node, or when an existing request completed its execution and quits. It changes also from one time round to the other due to the variable bit rate of the streams. $U_i[k]$ represents the load of the delivery node i on round k and its value can never be greater than one. The utilization vector $U_i[]$ of a delivery node i stores the current as well as the future utilization values for that delivery node. Before a stream is accepted by the control node, its load vector is combined with the utilization vector of every delivery node to check if there exists a delivery node where its new load never exceeds its capacity (i.e. its utilization is always lower than or equal to one during the length of the candidate stream). Let $U_i[j]$ denote the utilization of delivery node i in round j when a new request for stream s_r arrived at the server. Let the starting block of the stream s_r be on a storage node where delivery node i will be connected to that node after k time rounds ($0 \leq k \leq N-1$, where N is the total number of storage nodes). Then, the stream can be admitted if there exists a delivery node i that all the following m conditions:

$$U_i[j+k+m] \ + \ \frac{t_s^r[m]}{T} \leq 1, \quad 0 \leq m < dur(s_r) \quad \textit{where} \tag{4}$$

$$U_i[j+k+m] = \sum_{\tau_j \in u_i} \frac{t_s^j[j+k+m]}{T}$$

can be satisfied. The term k denotes the startup latency for that stream. In case that multiple delivery nodes satisfy the above conditions, the delivery node with the minimum value of k can be selected to provide the minimum startup latency. If the request is accepted then the utilization vector of the selected delivery node is updated accordingly. Let $dur(s_r)$ be the length of the stream s_r measured in time rounds. The control node requires at most $N * dur(s_r)$ additions to determine if a stream can be accepted or not for service, where N is the number of delivery nodes of the parallel server.

4 Experimental Results

A first prototype of the proposed parallel server has been implemented using a network of three Unix workstations, running Solaris 8 Operating System. It is a flat cluster architecture where the three nodes are connected by a 100 Mbps Ethernet LAN. Each node is equipped with 128 MB RAM and one local SCSI disk. CBR media streams are partitioned as media blocks with variable length according to the playback rate of each stream. We used in our experiments CBR streams encoded in different playback rates with values between 1.17 Mbps and

1.95 Mbps. The selected length of every time round T_i equals to 0.5 sec. The calculated stripe fragment size F_i of every CBR stream s_i is between 75 and 125 Kbytes. Here, the length of a media block represents half of the second of playback time. For these experiments, we used three VBR MPEG-encoded video streams. Each one contained 40,000 samples at a frame rate of 24 frames per second (about half an hour playing time). The stripe fragment of every VBR stream contains 12 frames and it is varied from 42 KB to 280 KB. The load vector $LV_i[]$ of every VBR stream contains 3334 entries where each entry describes the time required for a storage node to retrieve and transmit the corresponding stripe fragment on a specific round. The stripe fragments of both CBR and VBR streams are wide striped into the three processing nodes in a round-robin fashion.

The prototype has been encoded in C using MPI[1] calls that implement the communication between the three processing nodes. We initially tested the performance of a single node media server using media streams with the previously described parameters. The maximum value of the throughput observed was 26 concurrent streams. In a subsequent step we tested the performance of the three-node cluster architecture. The computation overhead for the evaluation of conditions (4) was not significant in our experiments (less than 1 ms). The overall throughput of the parallel server had a minimum value of 42 streams and a maximum value of 46 streams for the same set of requests. That variation of throughput is due to the fact that the internal network latency was not constant (values between 10 and 15 ms) in our computing environment. Reducing the network latency improves the performance of the server reaching higher values of stream throughput.

5 Conclusion

This paper is focused on the resource management problem for parallel media servers. A new conflict-free scheduling scheme was presented that provides on-demand support for a large number of concurrent continuous media objects in a predictable manner. The proposed algorithm supports both CBR and VBR encoded media streams video and audio at different playback rates. Using that algorithm, we are able to achieve optimal scheduling in distributed memory parallel architectures. More precisely, our real-time scheduling algorithm guarantees the QoS level of every accepted stream, efficiently utilizes server resources, reduces the required buffer size and increases system throughput. A prototype implementation version of the proposed parallel media server was presented, which demonstrates and confirms the practicability and the efficiency of the method in real-life applications.

In our future work we will try to support different classes of quality for client connections through scalability. The scheduling subsystem of the server will be able at runtime to allow appropriate media frames to be dropped in a predictable manner without fully suspending service to any one user.

References

1. W. Gropp, E. Lusk, N. Doss, and A. Skjellum. A high-performance, portable implementation of the MPI message passing interface standard. *Parallel Computing*, 22(6):789–828, September 1996.
2. D. Jadav, C. Srinilta, and A. Choudhary. Batching and dynamic allocation techniques for increasing the stream capacity of an on-demand media server. *Parallel Computing - Special Issue on parallel processing and multimedia*, 23(12):1727–1742, December 1997.
3. D. Jadav, C. Srinilta, A. Choudhary, and P. Bruce Berra. An evaluation of design tradeoffs in a high performance media-on-demand server. *Multimedia Systems*, 5(1):53–68, 1997.
4. A. Khaleel and A. Reddy. Evaluation of data and request distribution policies in clustered servers. In *Proceedings of the High Performance Computing*, 1999.
5. Jack Y.B. Lee. Parallel video servers: A tutorial. *IEEE Multimedia*, 5(2):20–28, 1998.
6. Chow-Sing Lin, W. Shu, and M.Y. Wu. Performance study of synchronization schemes on parallel cbr video servers. In *Proceedings of the Seventh ACM International Multimedia Conference*, November 1999.
7. V. Rottmann P. Berenbrink and R. Luling. A comparison of data layout schemes for multimedia servers. In *European Conference on Multimedia Applications, Services, and Techniques (ECMAST'96)*, pages 345–364, 1996.
8. A. L. Narasimha Reddy. Scheduling and data distribution in a multiprocessor video server. In *Proceedings of the 2nd IEEE Int'l Conference on Multimedia Computing and Systems*, pages 256–263, 1995.
9. V. Rottmann, P. Berenbrink, and R.I. Lüling. A simple distributed scheduling policy for parallel interactive continuous media servers. *Parallel Computing - Special Issue on parallel processing and multimedia*, 23(12):1757–1776, December 1997.
10. Ravi Wijayaratne and A. L. Narasimha Reddy. Providing QOS guarantees for disk I/O. *Multimedia Systems*, 8(1):57–68, 2000.
11. M.Y. Wu and W. Shu. Scheduling for large-scale parallel video servers. In *Proceedings of the Sixth Symposium on the Frontiers of Massively Parallel Computation*, pages 126–133, October 1996.

Double P-Tree: A Distributed Architecture for Large-Scale Video-on-Demand

Fernando Cores, Ana Ripoll, and Emilio Luque[1]

Computer Science Department - University Autonoma of Barcelona – Spain
{Fernando.Cores, Ana.Ripoll, Emilio.Luque}@uab.es

Abstract. In order to ensure a more widespread implementation of video-on-demand (VoD) services, it is essential that the design of cost-effective large-scale VoD (LVoD) architectures be able to support hundreds of thousands of concurrent users. The main keys for the designing of such architectures are high streaming capacity, low costs, scalability, fault tolerance, load balance, low complexity and resource sharing among user requests. To achieve these objectives, we propose a distributed architecture, called double P-Tree, which is based on a tree topology of independent local networks with proxies. The proxy functionality has been modified in such a way that it works at the same time as cache for the most-watched videos, and as a distributed mirror for the remaining videos. In this way, we manage to distribute main server functionality (as a repository of all system videos, server of proxy-misses and system manager) among all local proxies. The evaluation of this new architecture, through an analytical model, shows that double P-Tree architecture is a good approach for the building of scalable and fault-tolerant LVoD systems. Experimental results show that this architecture achieves a good tradeoff between effective bandwidth and storage requirements.

1 Introduction

Video-on-demand (VoD) refers to video services in which a user is able to request any video content at any time from a server. This technology is important for many multimedia applications such as distance learning, digital libraries, videoconferences, Internet, TV broadcasting, and video-on-demand systems.

The service of a video request involves a high volume of information with real-time requisites, very strict quality of service (QoS) levels and great disk bandwidths. These requirements imply that a multimedia server can support only a limited number of users depending on its capacity. Moreover, VoD system capacity is also restricted by available network bandwidth, which requires a massive investment in infrastructure.

Therefore, in order to provide VoD services to accommodate hundreds of thousands of concurrent users, the design of large-scale VoD architectures is required. In addition to high streaming capacity, the main keys for these LVoD systems are:

- Low-cost. An LVoD system may require a high investment; therefore, it is imperative to reduce server, network and storage costs.

[1] This work was supported by the MCyT-Spain under contract TIC 2001-2592 and partially supported by the Generalitat de Catalunya- Grup de Recerca Consolidat 2001SGR-00218.

B. Monien and R. Feldmann (Eds.): Euro-Par 2002, LNCS 2400, pp. 816–825.

- Scalability. It is essential that the system can adjust initial system-size to user requirements, but maintaining the possibility of easy expansion to accommodate more users or new services.
- Low complexity. LVoD architecture components should not be too complex as this can make system implementation, design and management more difficult.
- Fault-tolerance. VoD systems have to continue giving service to users even if one or more architecture components crash.
- Load distribution is important due to the nonuniform distribution of user requests, which leads to load imbalances among servers and poor utilization of the overall system resources.
- Resource sharing. Nowadays, resource sharing (broadcast, multicast, etc.) is the key for the design and implementation of cost-effective VoD systems.

In order to attain an LVoD system and to accomplish the previous requirements, we proposed a full-distributed architecture based on a tree topology of independent networks with proxies. This architecture, called Double P-Tree, modifies the typical proxy functionality in such a way that it works at the same time as cache for the most-watched videos, and as a distributed mirror for the remaining system videos. To enhance topology connectivity and to improve architecture efficiency and fault tolerance, we group several local networks (brother nets) from the same level.

This paper is organized in the following way. In section 2, we will first undertake an overview of the solutions proposed in the literature for the construction of LVoD systems. Following this, in section 3, we will describe the Double P-Tree architecture. In section 4, we will describe an analytical model for the architecture and in section 5 we show its evaluation. Finally, in the last section, we will indicate the main conclusions to be drawn from our discussion, and will suggest future lines of research.

2 Related Work

In recent years, research into VoD systems has mainly been focused on policies that attempt to improve available bandwidth efficiency. These techniques basically aim at increasing the number of users that can be served with a limited bandwidth. Such approaches are grouped into two broad policy groups: broadcasting (pyramid [14] and skyscraper [7]) and multicasting techniques (batching [4], patching [8], merging [5] and chaining [12]). Nevertheless, all of these techniques aim to improve the performance of available bandwidth within the system, but do not increase it.

Currently, there are several alternatives that can be used to implement LVoD systems:

- Centralized systems are the simplest approach to an LVoD system, but require high costs, very complex servers and are not scalable [1].
- Independent servers [2][13]. In these systems, users are grouped into network segments known as local networks; in such a way that system bandwidth is able to be that accumulated by each one of the individual nets. The key to the success of these systems is based on placing VoD servers close to clients' nets so that these users do not need to access the main server, and thereby create a system of hierarchical servers. These systems have an unlimited scalability (adding new servers) but with high storage costs, load imbalance and poor resource sharing.

- Proxy-based systems [2][6]. Previous architecture involves considerable storage cost; as a result, certain proposals have opted for reducing the size of local servers in such a way that they do not store all system videos, but rather, only those with a higher access frequency. These servers are managed as main-server caches and are called proxies, just as their Internet counterparts. The main problem with these systems is that requests that cannot be served locally end up in the main server, which becomes both a bottleneck and a growth-limiting factor.

In spite of these architectures being able to obtain a high streaming capacity, they do not successfully achieve all LVoD requirements, which limit their implementation.

Current research on scalability and distributed systems are focused on server design and implementation [10][11], more than in the entire system. Therefore, they do not consider net bandwidth costs and scalability as main goal. Recently studies about distributed VoD architectures are oriented in the system management [9] or how to map the media assets onto hierarchical architectures to improve QoS in [15].

3 Designing a Fully Distributed VoD System

In this section we present the different elements that integrate our architecture. First, we show the basic tree topology and the new proxy functionality; we then extend the tree topology, obtaining our final proposed architecture and its implementation.

3.1 Basic Topology (P-Tree) with Mirroring

Our first approach consists of an expansion of the proxy architecture, currently restricted to a single level, using a tree topology that provides the system with unlimited scaling capacity, as well as greater flexibility when deciding on its size and form of growth. This topology, called Proxy-Tree (or P-Tree) presented in [3], consists of a series of levels, in accordance with the number of local networks and the order of the tree (binary, tertiary, etc.). Every hierarchy level is made up of a series of local networks with its local-proxy and clients that forms the following tree level. Subsequent networks are successively hung from each one of the previous levels of local networks until reaching the final level.

To decentralize the architecture we remove the main server, distributing is functionality (as a repository of all system videos, server of proxy-misses and system manager) among all local proxies. To do this, we modify proxy functionality, dividing the proxy storage into two parts: one of these will continue being managed as a cache, storing the most requested videos; and the remainder will be used for making a distributed mirror of system videos.

However, this distributed architecture, with mirroring, does not achieve as much performance as similar systems such as one-level proxies, because it has a smaller proxy-hit probability and a bigger average request-service distance. This smaller proxy-hit probability is due to our having to distribute proxy storage into two different schemes (caching and mirroring), reducing the total popularity of proxy videos and affecting proxy efficiency.

On the other hand, total net traffic also has an important influence on VoD system performance. In proxy systems, net traffic depends on request-service distance. For example: In one-level proxy when a local-proxy cannot serve a request, this only has to cross one network (level) to reach the main server, and its miss penalty is restricted to twice the bandwidth required for a video stream. With the mirroring scheme, the proxy-miss service-distance is affected by distributed-mirror capacity, which depends on the number of proxies situated at the same distance from the local network. If distance-1 distributed mirror does not have enough capacity to store all system videos, then some requests have to cross two or more levels to be attended, having a penalty of 3 or more times the bandwidth required for a stream.

Moreover, both factors are dependent: increasing proxy-storage dedicated to caching implies a better proxy-hit probability but increases the average-distance needed to serve the proxy-misses from distributed mirroring. On the other hand, if we use more storage for mirroring, we reduce average mirror-service distance, but increase proxy-misses.

The only way to enhance distributed-mirror capacity, without modifying caching performance or proxy capacity, is by augmenting the number of adjacent proxies from every local network.

3.2 Improving Connectivity (Double P-Tree Architecture)

One way to increase connectivity is to use a bigger tree order, such as a tertiary-tree. However, we have realized that bigger tree-order topologies are not a good solution because they have a broader last-level that is proportionally far larger than in the binary tree topology. These last-level networks have poor connectivity, as they have no child networks, and this increases service-distance. Moreover, their effect on system efficiency is worse because the last level is the largest of the tree topologies.

A better solution is to increase topology connectivity by grouping several local networks (brothers nets) of the same level. In this way, we can increase the number of adjacent local networks without changing the topology size or last level width.

Using the concept of brother networks, we have designed a new topology shown in Fig 1a. This topology is called Double P-Tree due to brother networks forming a second tree within the original proxy tree topology.

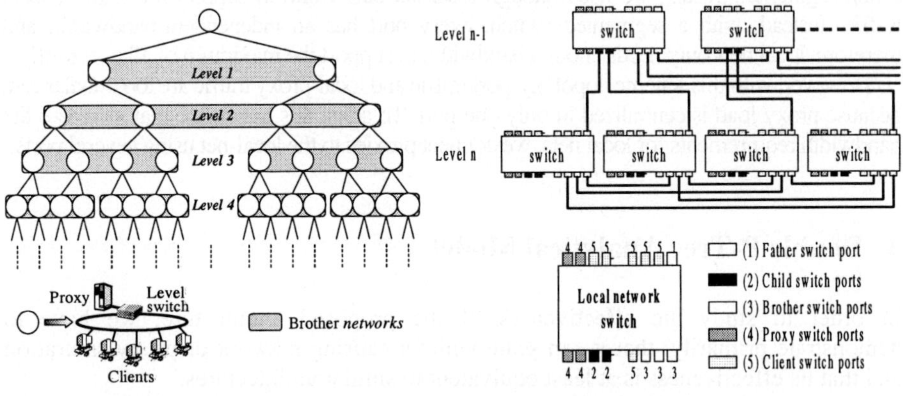

(a) Topology. (b) Implementation.

Fig. 1. Double P-Tree architecture.

The main advantage of this topology is that it considerably reduces the service-distance of user requests. The most evident enhancement is the reduction of distance-2 traffic. Upgrading topology connectivity, the number of distance-1 proxies is augmented, and increases distributed-mirror capacity and efficiency. In this way, it is more feasible that the distributed mirror is able to store all the system videos by using only distance-1 proxies, and therefore, all requests would be satisfied without the need to access upper levels. Double P-Tree architecture not only increases the efficiency of the mirroring, it also improves caching. Since the distributed mirror is larger, this scheme does not need as much proxy-storage as previously required. This unneeded proxy-capacity can then be assigned for caching.

This topology also has a better fault-tolerance. In a simple binary-tree topology, a network crash can create a sub-tree that is totally isolated from the rest of the system, which can cause request reneging if this sub-tree has insufficient proxies to store a full copy of the system videos. This is more unlikely to occur with the new topology.

3.3 Double P-Tree Architecture Implementation

Double P-Tree topology implementation is performed in local-network switches. In Fig 1b, we show the reserved ports to build the topology and how they are interconnected. Every local-net has a port (father-port) to connect the local net with the upper topology-level, several ports (child-ports), depending on tree order, to connect all the local children nets (above level), some ports (brother-ports) to connect the net with its brothers in the same level and finally one or several ports for the local-proxy. The remaining switch-ports are used to connect clients, usually using several switches that form a tree hierarchy.

Distributed architectures usually have a performance penalty compared with centralized approaches. This efficiency reduction is the sacrifice required in order to obtain a scalable and fault-tolerant system. An important characteristic in distributed systems with respects to centralized ones is that load is distributed between different sources. Meanwhile in a centralized approach, only one source supports all load.

In order to take advantage of this feature, so as to reduce the penalty imposed on distribute architectures, we propose the use of segmented switches in local nets. Using a non-segmented switch, we need enough local-net bandwidth to support the total of local traffic. Instead, with a segmented switch, every port has an independent-bandwidth, and therefore, local nets only needs enough bandwidth to support the maximum of all port traffic.

However, with this scheme, topology-port traffic and local-proxy traffic are too unbalanced, because proxy load is centralized in only one port. To solve this unbalance that increases the bandwidth requirements for local nets, we connect proxies to the local-net using several ports.

4 Double P-Tree Analytical Model

In order to study the effectiveness of the proposed architecture, we have to demonstrate primarily: that it can scale without causing network or proxy saturation and that its effectiveness is at least equivalent to similar architectures.

The system scalability is estimated evaluating the growth of traffic generated by the system itself and the evolution of bandwidth requirements in architecture components when the system grows. For our study, we are going to use the number of

Table 1. Notation used in the analytical model.

Symbol	Definition	Symbol	Definition
B_T	Total system bandwidth	T_p	Proxy switch-ports
B_e	Effective system bandwidth	C	Proxy cache size (number of videos)
B_p	Main net bandwidth	M	Proxy mirror size (number of videos)
B_c	Local net bandwidth	P_{mc}	Proxy's cache miss probability
B_u	Local net user bandwidth	P_{hc}	Proxy's cache hit probability
L	Number of topology levels	P_{mm}	Proxy's cache+mirror miss probability
N	Number of local networks	P_{mc}	Proxy's cache+mirror hit probability
O	Tree-order	Lf_i	Load received by net *i* from father net
B	Number of brother-nets	Lc_{ib}	Load received by net *i* from brother net *b*
V	System videos	Ls_{is}	Load received by net *i* from son net *s*

independent-streams supported by the system (effective bandwidth) as a main performance-metric. Other measurements such as total system bandwidth, local network bandwidths, server/proxies streaming capacity and storage requirements will provide us with an idea of the system's limitation with respect to the number of users that it can admit, its costs, and its grade of scalability.

All these parameters are evaluated using an analytical model of architecture behavior. Also, to contrast Double P-Tree system with similar approaches, we will study one-level proxies architecture performance. shows the notation used during this analysis. In Table 2 we show the analytical model for one-level proxy-based architectures (which can also be used to model centralized and independent severs systems). Table 3 shows the expressions used for Double P-tree analytical model.

In the following models, we assume a unicast policy, i.e., each user is assigned to its own dedicated stream. This assumption is valid since our study is directed at evaluating the system streaming-capacity (effective bandwidth) and its scalability. These results will be independent of whether bandwidth management policies can later be used to increase the efficiency and number of clients managed by the system.

The traffic managed by topology ports (*Lf, Lc, Ls*), in expression (6) of table 3, is the sum of outgoing (local proxy-misses served from other proxies), incoming (proxy misses from remainder of system served by the local proxy) and passing traffic (proxy misses from the remainder of the system served by other proxies). Total outgoing traffic consists of the proxy-misses, and the total incoming traffic to the network can be calculated in the same way as in (7). However, its distribution among different ports and the passing traffic have to be evaluated by using an analytical simulator.

Table 2. One-level proxy-based architecture analytical model

Measure	Expression	Nº.
Total bandwidth	$B_T = B_p + B_c \cdot n$	**(1)**
Main network required bandwidth	$B_p = B_u \cdot n \cdot p_{mc}$	**(2)**
Local networks required bandwidth	$B_c = B_u \cdot Max\left(\frac{1-p_{mc}}{T_p}, p_{mc}\right)$	**(3)**

Table 3. Double P-Tree Architecture analytical model.

Measure	Expression	Nº.
Effective system bandwidth	$B_e = B_u \cdot n$	(7)
Total system bandwidth	$B_T = \sum_{i=0}^{L} o^i \cdot Bc_i$	(8)
Bandwidth required by local net *I*	$Bc_i = Max(\frac{Lp_i}{Tp}, Lf_i, Lc_i1, \cdots, Lc_io, Ls_i1, \cdots, Ls_ib)$	(9)
Load managed by local proxy	$Lp_i = [Bu \cdot (1 - p_{mm}(i,0))] + \left[Bu \cdot \sum_{d=1}^{L*2} Ne(i,d) \cdot \frac{p_{mm}(i,d)}{r_{\ln}(i,d)} \right]$	(10)

5 Double P-Tree Evaluation

In order to evaluate Double P-Tree architecture and to compare its performance with other systems we assume a total system bandwidth of 127.000 Mbps (every 127 local net has 1.000 Mbps of bandwidth), that proxies/servers are connected using 4 switch-ports in all architectures (without taking topology ports into account), a proxy-capacity for 20% of system videos and in Double P-Tree a binary topology with 7 levels.

5.1 Scalability

Fig 2, illustrates the bandwidth requirements for most critical scalability elements in distributed architectures when the system size grows. In one-level proxies (Fig. 2a), the most critical element is main network bandwidth and in the Double P-Tree, it is the bandwidth required by local networks and proxies (Fig. 2b).

With the results in Fig. 2a, we confirm our previous statements about the limited scalability of one-level architectures. In these architectures, system growth is obtained at the expense of increasing bandwidth requirements for the main network independently of proxy capacity (charts 2 and 3). Consequently, its scalability is limited by its centralized components (main network and server).

In the Double P-Tree, Fig. 2b, we notice that even when exponentially increasing the system capacity (chart 2), the maximum bandwidth required for local networks and its proxies (charts 3 and 4) is stable and small (400 and 1200 Mbps respectively). These results allow us to conclude that Double P-Tree architecture has an unlimited scalability, even when using small architecture components.

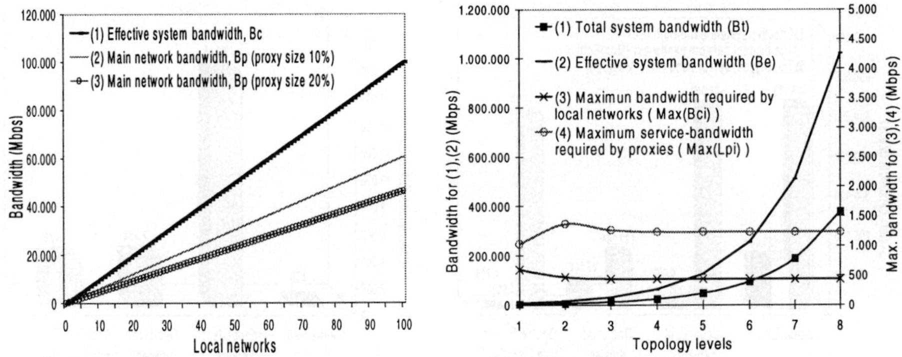

(a) One-level proxy-based system. (b) Double P-Tree system.

Fig. 2. Scalability of distributed architectures.

5.2 Proxy-Storage Requirements

In Fig. 3, following our comparison between proxy architectures, we study the proxy-storage requirements for both systems.

It is important to emphasize that, by using the same analytical model, in Fig. 3a we can evaluate the 3 main approaches for LVoD architectures: centralized systems (when proxy capacity is 0%), independent server systems (when proxy capacity is 100%) and one-level proxy systems in the remaining cases.

Comparing both proxy-based architectures, we can see that double P-Tree effective bandwidth (Fig. 3b) has a bigger growth gradient, achieving its maximum peak with a proxy capacity of only 25%. Meanwhile in a one-level system, the same peak is only achieved with a proxy capacity of 100% (the system then becomes an independent server architecture). In the same way, the balance-point between system and effective bandwidth is achieved with a proxy capacity of 8% as against the 17% (more than double) in the one-level proxy approach. These results also show that without the need for a main server, Double P-Tree architecture can operate by using proxies with a capacity of only 1% of system videos.

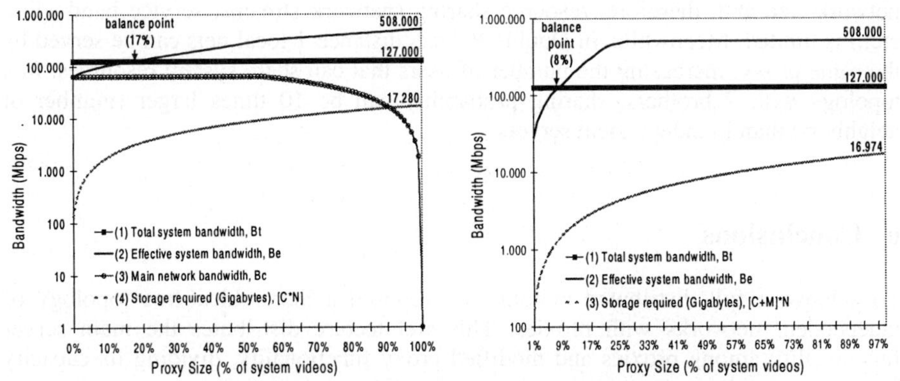

(a) One-level proxy-based system. (b) Double P-Tree system.

Fig. 3. Storage requirements in proxy-based architectures.

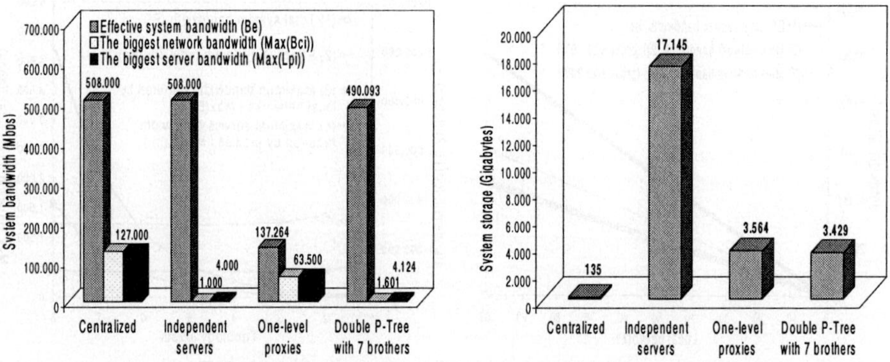

(a) LVoD systems bandwidth requirements. (b) LVoD systems storage requirements

Fig. 4. Performance and requirements in Large-scale VoD architectures.

5.3 Efficiency

We now compare Double P-Tree with the main current approaches for scalable LVoD systems. The centralized approach is only used as a reference, due to its null scalability and high costs, which are not consistent with our design goals.

In Fig. 46 a, we observe that Double P-Tree improves one-level effective bandwidth more than 350%. Moreover, the biggest Double P-Tree network and server are 15 times smaller (4000 against 63.500), requiring similar proxy-storage (Fig. 46 b) of around 3.500 Gigabytes.

Additionally, by now comparing results with the independent servers, it can be seen that this architecture obtains 4% more effective bandwidth than in our approach (508.000 against 490.000 Mbps), but uses 5 times more storage (17.200 against 3.500 Mbps), as shown in Fig. 46 b. We believe that this perfomance gap would be resolved and even overcome by our approach when the effect of multicast techniques (the key to VoD system performance) is taken into account. Our optimism is due to the fact that multicast efficiency broadly depends on the number of users accessing the same server. In independent server systems, this number is limited by local-network size and, therefore, resource sharing (network streams, service bandwdith, etc..) is limited. Meanwhile, in Double P-Tree, distance-1 local nets can be served by the same proxy, increasing the number of users that can share system resources. In a topology with 7 brothers, sharing probability can be 10 times larger (number of neighbors) than in independent servers.

6 Conclusions

To achieve a full distributed system we proposed a hierarchical-tree topology of independent networks with proxies. This architecture distributes the main server functionality among proxies and modified proxy functionality, dividing its capacity between two schemes: caching and mirroring. The caching scheme stores the most requested movies, whilst mirroring is used for making a distributed mirror. Moreover, to improve topology connectivity and system performance, we have modified the basic-tree topology by adding the concept of brother networks to local group nets.

The double P-Tree guarantees an unlimited and low-cost growth, high fault tolerance, a better load-balance (due to larger local-network connectivity), and a large streaming capacity independent of the technology available, and without requiring high storage requirements or complex and costly servers/networks.

The results show that new architecture outperforms streaming capacity (effective bandwidth) by more than 350%, compared with similar architectures (one-level proxies). In comparison with independent server systems, the effectiveness is similar but has 5 times less storage and allows for better resource sharing between users.

Our future research will focus on the study of proxy management policies that allow proxy size reduction. In addition, we intend to study the performance and changes required by policies such as prefix caching, chaining, patching and other classical policies for their subsequent incorporation into our architecture.

References

1. S. A. Barnett and G. J. Anido, "*A cost comparison of distributed and centralized approaches to video-on-demand*," IEEE Journal on Selected Areas in Communications, vol. 14, pp. 1173-1183, August 1996.
2. S.-H. G. Chan and F. Tobagi, "Distributed Servers Architecture for Networked Video Services", in IEEE/ACM Transactions on Networking, Vol. 9, No. 2 April 2001.
3. F. Cores, A. Ripoll, E. Luque, "A fully scalable and distributed architecture for video-on-demand", in Procedings of PROMS'01, Twente, Holland, Oct 2001
4. A. Dan, D. Sitaram, and P. Shahabuddin, "Dynamic batching policies for an on-demand video server," Multimedia Systems 4, pp. 112-121, June 1996.
5. D. L. Eager, M. K. Vernon, and J. Zahorjan, "Minimizing bandwidth requirements for on-demand data delivery", Proc. MIS'99, pp. 80-87, Indian Wells, CA, Oct. 1999.
6. H. Fabmi, M. Latif, S. Sedigh-Ali, A. Ghafoor, P. Liu, L.H. Hsu, "Proxy servers for scalable interactive video support" , IEEE Computer , Vol. 34 Iss. 9, pp. 54 -60 Sept. 2001.
7. K. A. Hua and S. Sheu, "Skyscraper broadcasting: a new broadcasting scheme for metropolitan VoD systems," in SIGCOMM 97, pp. 89--100, ACM, Sept. 1997.
8. K. A. Hua, Ying Cai and S. Sheu, Patching: A multicast tecnique for true video-on-demand services, ACM Multmedia'98, pages 191-200.
9. F. Kon, R. Campbell, and K. Nahrstedt. "Using Dynamic Configuration to Manage a Scalable Multimedia Distributed System." Computer Communication Journal, Special Issue on QoS-Sensitive Distributed Network Systems and Applications, 2000.
10. A. Krikelis, "Scalable Multimedia Servers", IEEE Concurrency, pp. 8-10, Oct-Dec, 1998.
11. D.N Serpanos, A. Bouloutas, A, "Centralized versus distributed multimedia servers", IEEE Transactions on Circuits and Systems for Video Technology, Vol. 10 Issue: 8, pp. 1438-1449, Dec. 2000.
12. S. Sheu, K. A. Hua, and W. Tavanapong "Chaining: a generalized batching technique for video-on-demand systems", In Proc. IEEE Int'l Conf. On Multimedia Compunting and Systems (ICMCS)'97, Ottawa, Canada, June 1997, pp. 110-117.
13. F. A. Tobagi, "Distance learning with digital video," IEEE Multimedia Magazine, pp. 90-94, Spring 1995.
14. S. Viswanathan and T. Imielinski, "Metropolitan area video-on-demand service using pyramid broadcasting," Multimedia Systems 4, pp. 197--208, Aug. 1996.
15. Xiaobo Zhou; R. Luling, Li Xie, "Solving a Media Mapping Problem in a Hierarchical Server Network with Parallel Simulated Annealing", Procs. 2000 International Conference on Parallel Processing, pp. 115 -124, 2000.

Message Passing in XML-Based Language for Creating Multimedia Presentations

Stanisław Polak[1], Renata Słota[1], and Jacek Kitowski[1,2]

[1] Institute of Computer Science, AGH, al. Mickiewicza 30, 30-059 Cracow, Poland, phone: (+48 12) 6173 964, fax: (+48 12) 6339 406,
{polak,rena,kito}@uci.agh.edu.pl, http://www.icsr.agh.edu.pl/

[2] ACC CYFRONET, ul. Nawojki 11, 30-950 Cracow, Poland

Abstract. The paper describes a new language and supporting tools for creation and display of multimedia presentations, specially for the Internet usage. The language implements the message passing paradigm to set up communication between elements of presentations. Model of the language and its flexibility are discussed.

1 Introduction

There are a few languages for creation Internet-enabled multimedia presentations. Simple presentations are implemented in HTML [1]. Tags like `<img>`, `<embed>` or `<bgsound>` are used for inclusion of multimedia data (e.g. graphics, movies, sounds) to the WWW pages. Only some WWW browsers recognize these tags correctly, furthermore the processes of presenting those data cannot be synchronized.

Complicated multimedia presentations can be built using Java [2] and Java Media Framework (JMF) [3]. Processes of presenting media data can be synchronized but knowledge of Java programming language is required.

Another possibility is to build a presentation with Synchronized Multimedia Integrated Language (SMIL) [4]. This language is based on eXtensible Markup Language (XML) [5]. SMIL enables one to create complicated multimedia presentations, including audio or video sequences, animations, graphics and texts. Media elements can be placed in any position on the screen, displayed according to user's computer capabilities and coordinated in three ways: sequential, parallel and exclusive. The sequence of presentation is determined by tags, therefore modification of the sequence requires changes in the SMIL document. In SMIL, semantically-related XML elements, attributes, and attribute values are grouped in modules. In the reported approach we propose a different method of synchronization of element presentation to have more flexibility in controlling the sequence of the presentation. The method of synchronization is based on events and messages generated on-line according to user's needs. Instead of modules we propose libraries which extend language capabilities without any additional modifications of the interpreter code.

In this paper first we shortly describe the MUltimedia Lecture description Language (MULL) [6, 7] and two tools related to the language. Next, we discuss

B. Monien and R. Feldmann (Eds.): Euro-Par 2002, LNCS 2400, pp. 826–829.

the flexibility of the language that results from the message passing paradigm used for media synchronization as well as a few libraries for extending scope of the language. Conclusions summarize the paper.

2 Multimedia Lecture Description Language and MULLtools System

The set of MULL elements consists of objects, messages, buffers, events and libraries. The object is an abstract representation of reality. 'Audio', 'Video', 'Picture', 'Answer', 'Timer', 'Text', and 'Page' objects are defined. Their properties describe various object features, for example the name and the position on the screen. The object contains buffers and it can execute a predefined basic operation (e.g., "show", "hide", "start", "stop" and "display"). The objects communicate to others through Message Manager by message passing. The messages are placed in buffers, which are associated with events. Every object can send any kind of the message, while it can interpret particular messages only. There are four message types: operational, setting up, information and library. The set of operations for the objects and the set of messages can be extended by the libraries *olibraries* and *mlibraries* respectively. MULL is implemented in XML. Objects and libraries are represented by tags. Buffers and properties establish a set of tags attributes. The message is a value of the attribute representing the buffer.

Two programs – Creator and Browser, called MULLtools – are associated with the MULL. The Creator is used for design of lecture appearance, which is defined by the resulting MULL file. For example, the content of the file describes presentation modes of the individual elements, connection between the elements, etc. Since the MULL file is a plain text file, a similar result could be obtained using any text editor for the price of more programming effort and knowledge of the MULL language. To show the lecture, the Browser is executed. The MULLtools system is written in Java 1.1.

3 Flexibility Obtained from the Message Passing

Capabilities of the MULL language depend on the messages because the messages define the sequence of the presentation. Message passing is used to set up communication between parallel events, i.e., to synchronize, to initiate and to activate them. Flexibility obtained from the message passing is correlated with different kinds of messages but especially with: a) possibility of changing messages in buffers during presentation and b) possibility of extending a set of messages by specialized libraries. These two possibilities are shortly described below.

The messages are placed in buffers. Every buffer has an attribute equivalent in the MULL document. Changing values of the attributes results in automatic change of the content of the buffers, that influences the sequence of the presentation. So the sequence can be determined by the author of the presentation in the MULL document. But if the buffers contain the setting up messages, the sequence of the presentation can be modified on-line, according to various circumstances, e.g, student's actions.

The MULL contains a basic set of messages, which can be extended with the libraries. Three examples of the libraries are described below – for sending a few messages at the same time, for exchanging messages among different MULL presentations and for creating simple conditions respectively.

The "Complex Message" Library. A few basic messages are placed in one buffer, therefore, the "complex message" is regarded as a single message. The *mlibrary* recognizes this kind of message and extracts the basic messages to be send in the order of their appearance in the "complex message". This library is useful if one has defined a sequence of events in advance.

The "Communication Message" Library. A single WWW page can contain several named applets, which communicate to others. Because the MULL Browser is an applet, so it can send messages to other MULL Browsers placed on the same WWW page thanks to the "communication message" library. Therefore two or more presentations could control each other.

This library provides direct and indirect modes of message sending. In the direct mode the messages sent by the objects placed at the sender side (i.e. by the first Browser instance) are received by the objects localized both at the sender and at the receiver sides (by the second Browser instance). In the indirect mode, the object localized at the sender side can send only selected messages to the objects placed at the receiver side. The message transmitted to the receiver is encapsulated in the "communication message".

The "Condition Message" Library. The purpose of the library is to create a simple conditional "instructions". Two types of the "condition message" are introduced: ordinary and special. Using the special "condition message", modifications of values of the variables previously defined are permitted. The value of these messages is tested by the ordinary "condition message", and according to the result, a corresponding message is sent.

One example of the library usage is to display a picture under condition that a process of reproducing a movie has been finished and at the same time a mouse cursor has been placed within a specified active area of an another picture.

4 Conclusions

The result of our work is the MULL language and the tools for composing and for presenting multimedia lectures. The tools consist of the MULL Creator and the MULL Browser. The language is useful for displaying text and pictures during playing audio/video file, building simple multiple-choice tests, synchronous starting, stopping and playing audio/video files, synchronous displaying pictures, creating simple animations, building a course which depends on user preferences, actions, and state-of-the-art, etc.

The message passing is responsible for controlling the sequence of presentation and the appearance of its individual elements. The messages are placed in buffers and the content of them can change according to current circumstances, like user's actions. If the existing capabilities are not sufficient, the user can take advantage of the libraries. Own libraries can be developed as well. Capabilities of the Browser, the Creator and the language automatically follow the developed libraries – every library extends the set of available messages.

The Browser has been designed to share message libraries with other language interpreters, however these languages must be based on the presented paradigm. For example, we can create a new language similar to X3D [8] for which we can use the same libraries without any recompilation of the new language interpreter.

Acknowledgments

The work has been supported by Scientific and Technological Cooperation Joint Project between Poland and Austria: KBN-OEADD grant No. 3633/R01/R02 and by AGH grant.

References

1. World Wide Web Consortium: HyperText Markup Language. (2002) `http://www.w3.org/MarkUp/`
2. Sun Microsystems: The Java Tutorial. (2002) `http://java.sun.com/docs/books/tutorial/`
3. deCarmo, L.: Unlocking the secrets of the Java Media Framework. Java Developer's Journal, Feb.-Apr. (2001) `http://www.sys-con.com/java/`
4. World Wide Web Consortium: Synchronized Multimedia Integration Language (SMIL 2.0) Specification – W3C Recommendation. (2001) `http://www.w3.org/TR/smil20/`
5. Cagle, C., Gibbons, D., Hunter, D., Ozu, N., Pinnock, J.: Beginning XML. Amazon (2000).
6. Polak, S., Słota, R., Kitowski, J.: Creation and Presentation of Multimedia Courses on the Internet. In Domingo, M.E., Cebollada, J.C.G., Salvador, C.P., eds.: Proceedings of Six Annual Scientific Conference on Web Technology, New Media, Communications and Telematics Theory, Methods, Tools and Applications, SCS, Delft (The Netherlands) (2001) 222–226.
7. Polak, S., Słota, R., Kitowski, J., Otfinowski, J.: Xml-based tools for multimedia course preparartion. Archiwum Informatyki Teoretycznej i Stosowanej, Vol. 13, No. 1 (2001) 3–21.
8. The Web3D Consortium: X3D Specification. (2001) `http://www.web3d.org/TaskGroups/x3d/specification/`

A Parallel Implementation of H.26L Video Encoder*

J.C. Fernández[1] and M.P. Malumbres[2]

[1] Dept. de Ingeniería y Ciencia de los Computadores, Universidad Jaume I, 12071-Castellón (Spain), Phone: +34-964-728265, Fax: +34-964-728435, jfernand@icc.uji.es

[2] Dept. de Informática de Sistemas y Computadores, Universidad Politécnica de Valencia, 46071-Valencia (Spain), Phone: +34-96-3879703, Fax: +34-96-3877579, mperez@gap.upv.es

Abstract. Over the last decade a lot of research and development efforts have gone into designing competitive video coding standards for several kinds of applications. Some video encoders like MPEG-4 and H.26L, exhibit a high computational cost, far from real-time encoding, with medium to high quality video sequences. So, for these kinds of video coding standards, it is very difficult to find software solutions able to code video in real-time. In this paper, we design a parallel version of the ITU-T H.26L video coding standard, showing different implementation approaches and evaluate their performance.

1 Introduction

The storage, processing and delivery of multimedia data in their raw form is very expensive; for example, a standard 35mm photograph could require about 18 MBytes of storage and one second of NTSC-quality colour video requires almost 23 MBytes of storage. To make widespread use of digital imagery practical, some form of data compression must be used. Digital images can be compressed by eliminating redundant information.

In the last few years, many video compression algorithms have been proposed. As a result, several image and video compression standards have been approved [H.26X, MPEG-X, JPEG2000] and many hardware/software solutions are now available. Furthermore, clusters of workstations (COWs) are currently being considered as a cost-effective alternative for small-scale parallel computing.

With respect to video parallel programming, several MPEG parallel [1], [4] and [5] have been developed. Also, several H.263 parallel implementations have been developed on multiprocessors [6] and COWs [2].

In this paper, a parallel ITU-T H.26L video encoder is proposed because the computational cost of the H.26L encoder is extremely high. We propose two parallel versions: the first divides the overall video sequence among the working nodes (GOP-level parallelism). The second divides one frame among the working

* This work is supported by the CICYT Project TIC2000-1151-C07-06.

B. Monien and R. Feldmann (Eds.): Euro-Par 2002, LNCS 2400, pp. 830–833.

nodes (Frame-level parallelism). Both approaches were evaluated on a Myrinet-based COWs.

In section (2) we give a brief description of the H.26L encoder. In section (3) the two parallel versions are explained. Section (4) shows some evaluation results and, finally, in section (5) some conclusions are drawn.

2 The H.26L Video Encoder

H.26L [3] is the current project of the ITU-T Video Coding Experts Group (VCEG), a group officially chartered as ITU-T Study Group 16 Question 6. The primary goals of the H.26L project are: Improved coding efficiency, improved Network Adaptation and simple syntax specification.

Different from the previous MPEG and ITU-T standards, some new techniques, such as spatial prediction in intra coding, motion compensation with adaptive block size, 4x4 integer DCT, UVLC (Universal Variable Length Coding), CABAC (Context-based Adaptive Binary Arithmetic Coding) and loop filter are adopted by H.26L.

3 Parallel Algorithms

In this section, we present two versions of an H.26L parallel video encoder: GOP (Group Of Pictures) division and frame division. The difference between them consists of the degree of parallelism employed.

3.1 GOP Division

In this version each processor computes a GOP of the video sequence. Each GOP begins with an I-Frame, the rest being P-Frames and, optionally, B-Frames. So, if the first picture is an I-Frame, that does not depend on previous pictures, then a GOP is defined as a closed group of pictures that can be decoded independently. Let us consider the following values, n_frames is the number of frames in the video sequence, n_frames_gop is the number of frames in one GOP, n_gops (the number of GOPs) is given by n_frames/n_frames_gop and p is the number of processors. The number of not assigned GOPs is $n\_gops\_not = mod(n\_gops, p)$ and the number of assigned GOPs to each processor is $n\_gops\_as = (n\_gops - n\_gops\_not)/p$. The total number of GOPs assigned to the processor P_k, $k = 0 : p-1$, is $n\_gops\_p = n\_gops\_as + 1$ if $k < n\_gops\_not$, or $n\_gops\_p = n\_gops\_as$ if $k \geq n\_gops\_not$.

To determine which frames will be assigned to each processor, two parameters have been defined, if_k, the initial frame, and ff_k the final frame belonging to processor P_k. Then P_k calculates the frames $if_k, if_{k+1}, \cdots, ff_k$. The values of these parameters are the following: $if_k = k * (n\_gops\_p * n\_frames\_gop)$ if $k < n\_gops\_not$ or $if_k = n\_frames - ((p-k) * (n\_gops\_p * n\_frames\_gop))$ if $k \geq n\_gops\_not$, and $ff_k = ((k+1) * (n\_gops\_p * n\_frames\_gop)) - 1$ if $k < n\_gops\_not$ or $ff_k = (n\_frames - ((p-k-1)*(n\_gops\_p*n\_frames\_gop))) - 1$ if $k \geq n\_gops\_not$.

3.2 Frame Division

In this case the task unit is the slice, a group of consecutive macroblocks in a frame. Each frame is divided in slices, and those are assigned to working nodes. Since most test sequences use only one slice per row of macroblocks (the slice width is the same as the frame width), each frame usually contains a small number of slices. With QCIF formats a slice will contain 11 macroblocks, so the total number of slices is limited to 9. Let us consider the following values, n_frames is the number of frames in the overall sequence, n_m_t is the number of macroblocks in one frame, n_m_s is the number of macroblocks in one slice, $n\_s = n\_m\_t/n\_m\_s$ is the number of slices and p is the number of processors. The number of not assigned slices is $n\_s\_n = mod(n\_s, p)$ and the number of assigned slices to each processor is $n\_s\_as = (n\_s - n\_s\_n)/p$. The total number of slices assigned to the processor P_k, $k = 0 : p-1$, is given by $n\_s\_p = n\_s\_as+1$ if $k < n\_s\_n$ or $n\_s\_p = n\_s\_as$ if $k \geq n\_s\_n$.

To determine the macroblocks assigned to each processor, two parameters have been defined, im_k, the initial macroblock, and fm_k the final macroblock of processor P_k. Then P_k calculates the macroblocks $im_k, im_{k+1}, \cdots, fm_k$. The values of these parameters are given by $im_k = k * (n\_s\_p * n\_m\_s)$ if $k < n\_s\_n$ or $im_k = n\_m\_t - ((p-k) * (n\_s\_p * n\_m\_s))$ if $k \geq n\_s\_n$, and $fm_k = ((k+1) * (n\_s\_p * n\_m\_s)) - 1$ if $k < n\_s\_n$ or $fm_k = (n\_m\_t - ((p-k-1) * (n\_s\_p * n\_m\_s))) - 1$ if $k \geq n\_s\_n$.

4 Experimental Results

The results have been obtained using a Beowulf cluster of 32 nodes interconnected with a Myrinet switch. Each node consists of an Intel Pentium-II processor at 300MHz with 128 MBytes RAM. Communication routines in MPI and C language are used. We have run experiments with 2, 3, 4, 9, 16 and 24 nodes. We have used the public available sources of H.26L TML 8.4 as the starting point for this study. During the experiments a carphone QCIF (176x144) video sequence, of 315 frames was used. The sequence of frame types used in all tests is the following $IBBPBBPI \cdots$. In the GOP division method $n\_frames\_gop = 7$ and $n\_gops = 45$. Nine slices of 11 macroblocks have been considered. In the frame division method $n\_m\_t = 99$, $n\_m\_s = 11$ and $n\_s = 11$. The deblocking filter optional mode was not used.

Table 1 shows the experimental results in seconds for carphone video sequence.

In general the behaviour of frame division, in terms of encoding time, is worse than GOP division, because the former requires more communications than the latter. The best efficiencies are obtained with GOP division, which achieves a good load balance. In the frame division there is one communication per frame, but in the GOP division there is only one communication during the total encoding time.

Table 1. Experimental results using the carphone video sequence.

GOP division				Frame division			
p	T_p	Speed-up	Efficiency	p	T_p	Speed-up	Efficiency
1	9365.362			1	9365.362		
2	4714.796	1.986	99.3%	2	5160.32	1.81	90.74%
3	3233.557	2.896	96.54%	3	3369.01	2.77	92.66%
4	2468.485	3.794	94.84%	4	3359.01	2.78	69.68%
9	1099.195	8.52	94.66%	9	1324.061	7.07	78.59%
16	620.031	15.10	94.40%				
24	425.395	22.01	91.73%				

5 Conclusions

We have presented a preliminary study of two parallel implementations of a H.26L video encoder. The first method distributes the whole video sequence between system nodes, dividing the original sequence in Groups Of Pictures (GOPs). The second method proposed, divides each video frame among the nodes. The division was made at slice level, requiring a barrier synchronization between working nodes in order to exchange the necessary information to properly code the next frame. The best results are obtained with the GOP division method, because it requires only one communication step. As future work, we are planning to make several optimizations to the frame division method in order to hide its communication overhead.

References

1. Akramullah, S.M., Ahmad, I., Liou, M.L.: Parallelization of MPEG-2 Video Encoder for Parallel and Distributed Computing System. Midwest Symposium on Circuits and Systems, vol. 2. Rio Janeiro, Brasil (1995) 834-837
2. Akramullah, S.M., Ahmad, I., Liou, M.L.: A Software Based H.263 Video Encoder Using a Cluster of Workstations. Optical Science Engineering and Instrumentation Symposium, vol. 3166, San Diego- California (1997)
3. ITU-T: H.26L Test Model Long Project (TML-8), July (2001)
4. Olivares, T., Quiles, F., Cuenca, P., Orozco-Barbosa, L., Ahmad, I.: Study of Data Distribution Techniques for the Implementation of an MPEG-2 Video Encoder. Proc. of IASTED Int. Conference Parallel and Distributed Computing Systems, MIT Boston (1999)
5. Yu, Y., Anastassiou, D.: Software Implementation of MPEG2 Video Encoding using Socket Programming in LAN. Proceedings of the SPIE conference of Digital Video Compression on Personal Computer: Algorithm and Technology 7-8. San José-California, Feb. (1994) 229-240
6. Zheng, L., Cosmas, J., Itagaki, T.: Real Time H.263 Video Encoder Using Mercury Multiprocessor Workstation.

A Novel Predication Scheme for a SIMD System-on-Chip

Alexander Paar[1], Manuel L. Anido[2], and Nader Bagherzadeh[3]

[1] Universität Karlsruhe, Fakultät für Informatik,
76128 Karlsruhe, Germany
AlexPaar@ieee.org
[2] Federal University of Rio de Janeiro, NCE, Brazil
mlois@nce.ufrj.br
[3] University of California, Irvine, CA 92697
nader@ece.uci.edu
http://www.eng.uci.edu/comp.arch

Abstract. This paper presents a novel predication scheme that was applied to a SIMD system-on-chip. This approach was devised by improving and combining the unrestricted predication model and the guarded execution model. It is shown that significant execution autonomy is added to the SIMD processing elements and that the code size is reduced considerably. Finally, the implemented predication scheme is compared with predication schemes of general purpose processors, and it is shown that it enables more efficient if-conversion compilations than previous architectures.[1]

1 Introduction

Many regular data-parallel problems have benefited from the tremendous computational power provided by Single Instruction Multiple Data (SIMD) machines. However, there is also a wide variety of irregular data-parallel problems that are much more difficult to map on SIMD architectures. A major problem has been the difficulty of those architectures to deal efficiently with parallel *if-then-else* clauses.

Reflecting the importance of exploring data parallelism to improve performance, some recent microprocessors have incorporated SIMD execution units and it has been shown that significant improvements can be achieved [1, 2]. This has been driven by the need to accelerate multimedia and digital signal processing applications. Other applications like computer image generation, volume rendering and scientific visualization in real-time have also employed SIMD architectures [3]. The next section will give a brief overview of the general SIMD computation model and a particular instance of such architecture. It will further introduce the principles of

[1] The work that led to this paper was conducted at the Department of Electrical & Computer Engineering at University of California, Irvine, where the two first named authors had appointments as visiting researchers. Dr. M. Anido acknowledges the financial support by CNPq. This work was supported by DARPA (DoD) under contract F-33615-97-C-1126 and the National Science Foundation under grant CCR-0083080.

B. Monien and R. Feldmann (Eds.): Euro-Par 2002, LNCS 2400, pp. 834–843.

predicated execution before the main part of this paper describes how both techniques were combined to achieve improvements in performance and programmability over existing comparable architectures.

2 Previous Work

2.1 SIMD Architectures and MorphoSys Reconfigurable System-on-Chip

SIMD lets one instruction operate at the same time on multiple data items. What normally requires a repeated succession of instructions can now be performed in one instruction. For that purpose standard SIMD architectures incorporate an array of processing elements that are centrally controlled by a general-purpose processor. Though there have been efforts to add execution autonomy to those array elements, such as allowing different subsets of PEs to have masked/complemented execution of one instruction based on a predicate flag [4, 5], none of these approaches enables a processing element to take a local decision about executing entire sections within if-then-else clauses nor do previous approaches provide the capability of nested predication.

MorphoSys [6] is a coarse grain, integrated reconfigurable system-on-chip targeted at data-parallel applications. It incorporates a reconfigurable array of processing elements, a RISC processor core, an efficient memory interface unit, and an 8x8 array of SIMD processing elements (PEs). MorphoSys processing elements will be referred to as reconfigurable cells (RC). A reconfigurable cell comprises of a 32-bit context register that contains the SIMD instruction (*context word*, CW) for the current cycle, two input data multiplexers, and a combinational network that includes an ALU, a multiplier and a shifter. Data is stored in a set of RC registers and an output register for inter cell data exchange (see Fig. 1, 2).

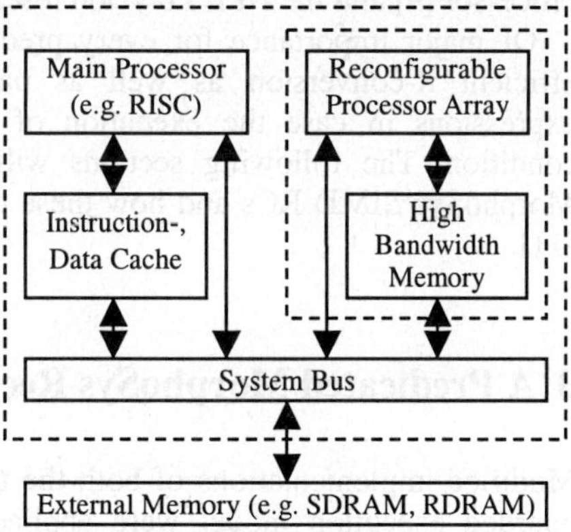

Fig. 1. MorphoSys architecture

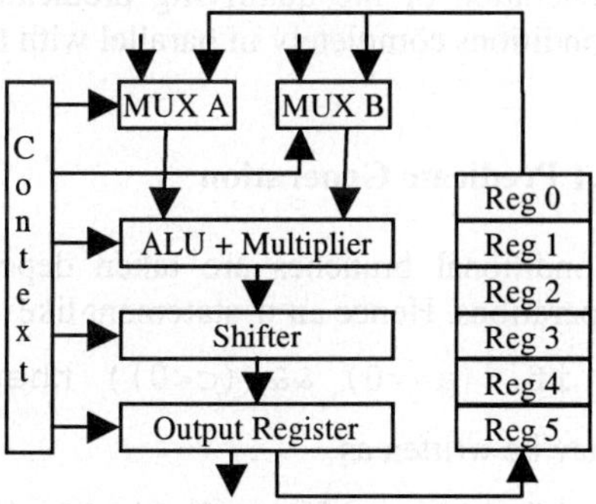

Fig. 2. MorphoSys reconfigurable cell

2.2 Predication

Predication refers to the conditional execution of an instruction based on the Boolean value of a *qualifying predicate*. If the value of the predicate is true, the instruction is allowed to execute normally (commit), otherwise the instruction is nullified, preventing it from modifying the processor state. Predication is able to remove branch instructions from a program code. This is why it can be implemented even in an SIMD processing element that does not have a program counter.

The basic compiler transformation to exploit predicated execution is known as *if-conversion*. If-conversion replaces conditional branches in the code with comparison instructions that define one or more predicates. Instructions that are control dependent on the branch are then converted to predicated instructions, utilizing the appropriate predicate value. In this manner, control dependencies are converted to data dependencies.

There are two basic predication models. In the *unrestricted predication model* [7], all instructions can be predicated. The *guarded execution model* [8] approach is to introduce a special instruction, which controls the conditional execution of following non-predicated instructions. Using these two basic models further refinements can be made via *conditional move instructions* and *instruction nullification*. Some of the better known processor architectures that incorporate predicated ISAs are the Alpha processor [9] and the HPL PlayDoh research architecture [10].

Of major importance for every predication scheme are instructions that support efficient if-conversion as well as parallel computation of high fan-in logical expressions in case the execution of an instruction depends on more than one condition. The following sections will describe how predication was applied to MorphoSys SIMD RCs and how these techniques compare to the IA-64 architecture [11].

3 A Predicated MorphoSys Reconfigurable Cell

Modified implementations of both the unrestricted predication model as well as the guarded execution model were applied to the SIMD reconfigurable cells of the MorphoSys system-on-chip. A main design objective was to perform both the generation of the qualifying predicate as well as the evaluation of up to 4 basic conditions completely in parallel with the actual arithmetic/logical execution.

3.1 Predicate Generation

Conditional branches are taken depending on the outcome of several arithmetic operations. Hence an if-statement like

```
if ((a==0) && (c<0)) then .. end if;
```

may be written as

```
if ((Zero flag Op1) && (Sign flag Op2)) then .. end if;
```

Assuming the referred ALU flags are implicitly generated by several instructions they need to be stored until the if-statement that depends on these flags is reached. In our design these stored ALU flags are called *basic predicates*.

The MorphoSys RC ALU was extended with a flag generation unit (Fig. 3) that computes eight different flags: $Sign_{16}$, $Sign_{32}$, $Zero_{32}$, $Carry_{16}$, $Carry_{32}$, $Overflow_{16}$, $Underflow_{16}$, and $Parity_{32}$. The index determines the specified bit range. These flags are written to a dedicated ALU flag register whenever an instruction commits.

An RC context word for a predicated instruction carries the information as to which ALU flag (or its negation) of the current flag register is to be stored into the *basic predicate register file*. Since the Boolean terms in the above if-statement have already been reduced to ALU flags, they may now be replaced by the Boolean value of a 1-bit basic predicate:

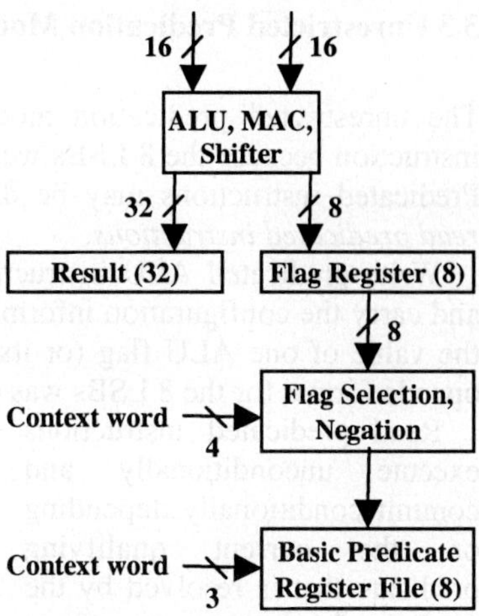

Fig. 3. Predicate generation

```
if (bp1 && bp2) then .. end if;
```

3.2 Predicate Evaluation

We refer to predicate evaluation as the process of generating the qualifying predicate that determines whether an instruction is to be committed or not. A qualifying predicate represents the 1-bit result of a Boolean equation given by an if-statement. According to our design objective an if-statement may incorporate up to four basic conditions that may be combined in an arbitrary way. A look-up table based approach was chosen to implement the evaluation stage.

Four multiplexers supply a look-up table with the 1-bit value of four basic predicates. A Boolean equation with four input variables can be described by a 16 bit wide truth table. The current implementation of the look-up table holds 16 different equations that could represent 16 different if-statements (Fig. 4). The *equation selection* input selects one evaluation equation. Predicates A, B, C, and D are assigned values of up to four different basic predicates accordingly to the *predicate selection* configuration. Their values are concatenated and determine the index in the currently selected evaluation equation at which the value of the resolving qualifying predicate is found.

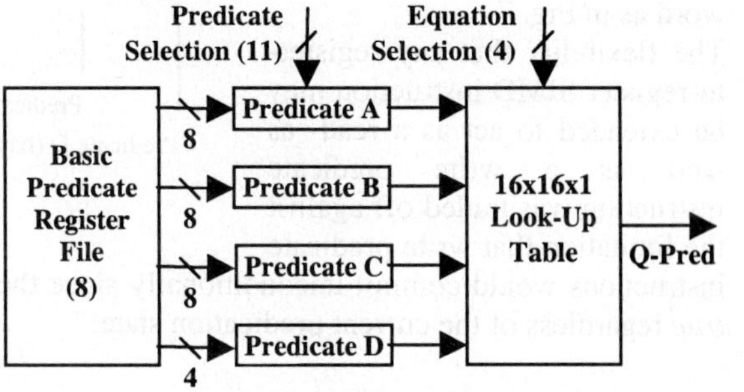

Fig. 4. Predicate Evaluation

3.3 Unrestricted Predication Model

The unrestricted predication model was applied to every register-to-register RC instruction because the 8 LSBs were available to contain the predication information. Predicated instructions may be distinguished as *write predicated instructions* and *read predicated instructions*.

Write predicated ALU instructions are executed and committed unconditionally and carry the configuration information for the predicate generation units. They store the value of one ALU flag (or its negation) into a specified predicate register. The opcode format for the 8 LSBs was extended as in Fig. 5.

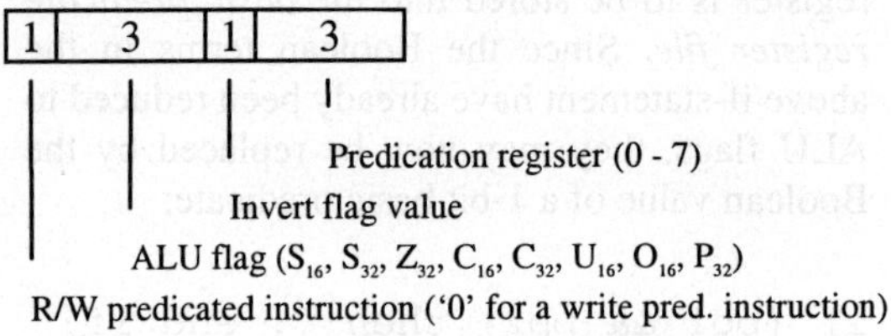

Fig. 5. WP instruction

Read predicated instructions execute unconditionally and commit conditionally depending on the current qualifying predicate that is resolved by the predicate evaluation stage from the current content of the predicate register file and the configuration of the evaluation stage (Sec. 3.2). The opcode format for read predicated instructions was extended as in Fig. 6.

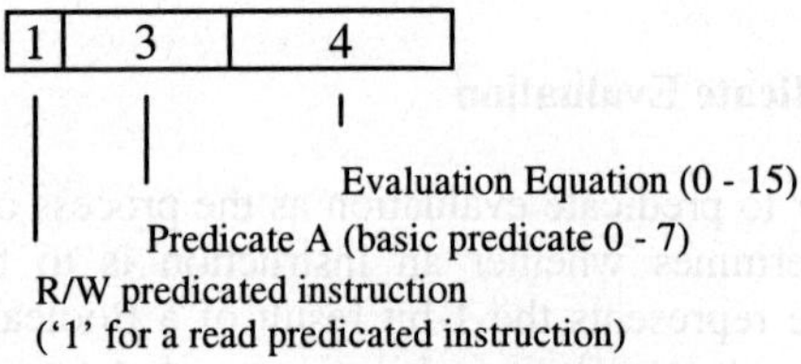

Fig. 6. RP instruction

A read predicate instruction can implicitly specify only one input basic predicate. The other three multiplexers of the evaluation stage are configured by an explicit PREDCONF context word as in Fig. 7.

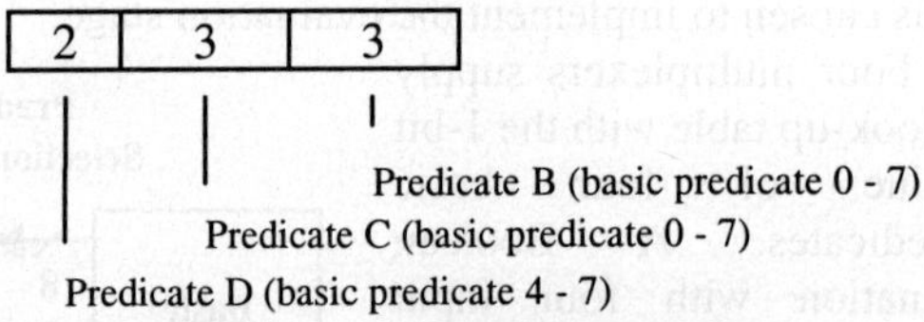

Fig. 7. PREDCONF

The flexibility that any register-to-register SIMD instruction may be extended to act as a read- as well as a write predicate instruction was traded off against the limitation that write predicate instructions would commit unconditionally since they set the qualifying predicate to *true* regardless of the current predication state.

3.4 Guarded Execution Model

To overcome the above restriction a guarded execution model was incorporated into the unrestricted predication model. It generates a further qualifying predicate that is referred to as *guard flag*. According to this predication scheme two new SIMD instructions were introduced:

GUARD: This command is a read predicate instruction (Sec. 3.3) that conditionally sets/resets the guard flag depending on the current qualifying predicate. Beyond this, it carries *label* information (in the present implementation only 1 bit) that is written to a label register upon commit.

WAKEUP: This command also carries label information and compares it with the current content of the label register. If these values match, it resets the guard flag.

3.5 Resulting Predication Model

The resulting predication model is depicted in Fig. 8. The *final qualifying predicate* that enables an SIMD instruction to commit is a combination of the *qualifying predicate* of the unrestricted predication model and the *guard flag* generated by the guarded execution model. Such a scheme allows *nested predication*. The current implementation supports up to three levels as depicted in Fig. 9.

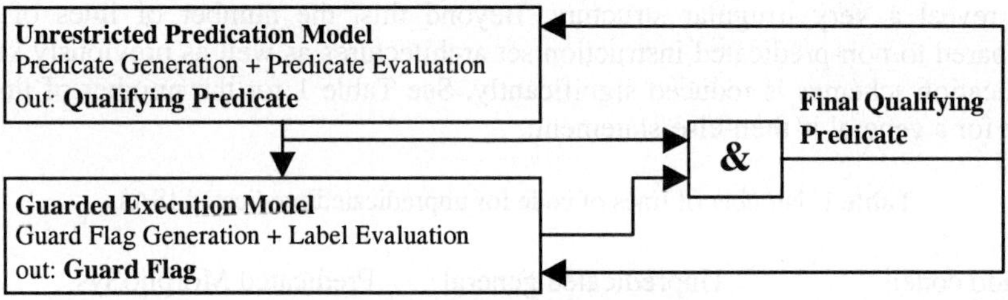

Fig. 8. Final Predication Model

```
Guard (Label 0)
   ...
   Guard (Label 1)
      ...
      Read Predicated
      Instruction
      ...
   WakeUp (Label 1)
   ...
WakeUp (Label 0)
```

Fig. 9. Nested predication

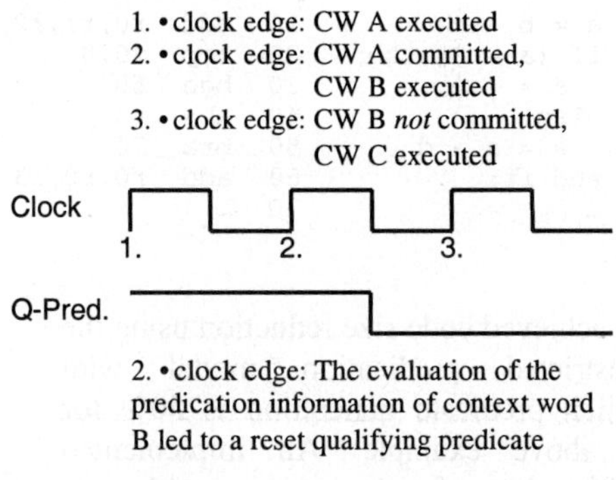

Fig. 10. Pipelined predication

3.6 Pipelined Predication

Two major design objectives for adding predication to a MorphoSys RC were that these features did not have a significant impact on the maximum clock frequency of the entire system and they should require as few additional clock cycles as possible. Both requirements were met by a *pipelined predication* approach.

Since the evaluation of the current predication state and the generation of the final qualifying predicate can wait until the commit time of the succeeding SIMD instruction, it was possible to defer both to the next clock cycle. Fig. 10 shows a context word (CW) timing sequence for a read predicate instruction that is preceded by a write predicate instruction. It becomes clear that the generation of the final qualifying predicate could take up to one clock period and still be in time.

4 Evaluation and Comparison

4.1 Evaluation

The benefits of adding predication to an SIMD processing element are much more remarkable than for a predicated general purpose processor. A predicated SIMD processing element offers a high level of execution autonomy and thus a significantly enhanced programmability. The application scope can be widened to input streams that reveal a very irregular structure. Beyond this, the number of lines of code compared to non-predicated instruction set architectures as well as previously known predication schemes is reduced significantly. See Table 1 for the number of lines of code for a general if-then-else-statement:

Table 1. Number of lines of code for unpredicated/predicated ISAs

Pseudo code:	Unpredicated general purpose processor:	Predicated MorphoSys SIMD RC:
`10 a = b + c`	`10 add r0,r1,r2`	`10 add r0,r1,r2,z,p1`
`20 if (a < 0) then`	`20 cmp r0,0`	`20 (qp) abs r0,pcnf`
`30 a = |a|`	`30 bge 60`	`30 (qp) add r0,r0,r3,pcnf`
`40 else`	`40 abs r0`	`40 ...`
`50 a = a + d`	`50 bra 70`	
`60 end if;`	`60 add r0,r0,r3`	
`70 ...`	`70 ...`	

The achieved code size reduction using the unrestricted predication model with implicit predicate generation is 50% for the above example. An implemented 32x32 bit fixed point arithmetic multiplication using 16-bit integer arithmetic resulted in a code size reduction by 56% compared to unpredicated x86 assembly language code and a 35% code size reduction compared to a pseudo code predication model with explicit predicate calculation. A further optimization by 3% is achieved by code reordering so that previous configurations of the predicate

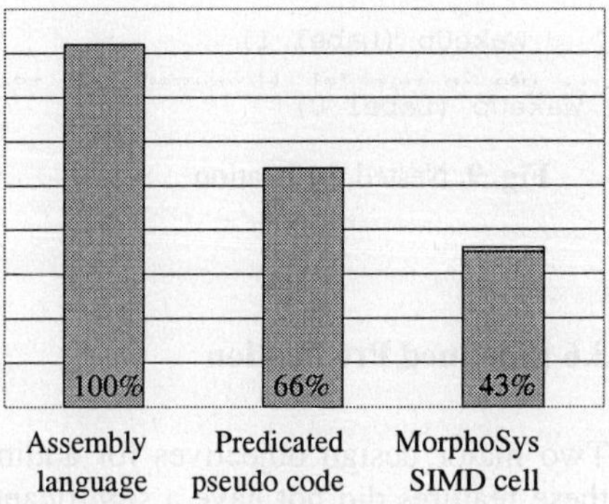

Fig. 11. Number of lines of code

evaluation stage (Sec. 3.2) can be reused for different if-conversions. The measured values are shown in Fig. 11. These numbers are the results of some very particular test cases and exemplify a magnitude of speedup that could be achieved.

Due to the SIMD concept of the MorphoSys architecture, the execution of a sequence of context words requires not only less clock cycles but is executed with different input data on up to 64 RCs at the same time.

4.2 MorphoSys Predication versus the IA-64 Model

The MorphoSys predication model distinguishes between read- and write-predicated instructions. Write predicated instructions implicitly generate ALU flags and store one of those flags into a basic predicate register where this information is available for further evaluation. The information about how these basic predicates are combined to create the qualifying predicate is again contained in a read predicated instruction itself. Both evaluation and generation take place in parallel to the actual arithmetic calculations.

In the IA-64 architecture only dedicated compare instructions write to a predicate register file. These predicates then serve directly as a qualifying predicate for succeeding instructions (though their content can be transferred to/from a general register). The evaluation of if-converted statements that depend on more than one comparison result is achieved by parallel compare instructions that update the target predicate register only for a particular comparison result. This allows multiple simultaneous OR-type or multiple simultaneous AND-type compares to target the same predicate register.

An if-converted statement like

```
R1 = R1 - R2;
R2 = R3 - R4;
R5 = R1 + R3;
if (((R1<0) && (R3==R4)) || (R5>0)) then
   R1 = R2 + R5;
end if;
```

that depends on three comparison results, could be assembled to the IA-64 / MorphoSys architecture as in Table 2. Two simultaneous IA-64 comparisons could execute in parallel. However, this requires a multiple issue processor. The third compare instruction must be deferred to the next clock cycle. Both examples contain an initializing command that configures the predication units. This setup may not be necessary in certain cases. Disregarding individual speedups that may be achieved by pipelined/SIMD execution, for the above case MorphoSys predication is 30% faster than IA-64 code. Beyond this, up to four basic comparison results can be combined in an *arbitrary* way. The IA-64 compare instruction is limited to six basic types and to perform four comparisons at the same time, all of them must be of the same basic type, always presuming a four issue implementation of the IA-64 architecture.

Table 2. Number of lines of code IA-64 / MorphoSys

IA-64 architecture:	MorphoSys architecture:
`10 (p0) sub r1 = r1,r2`	`10 pcf mask`
`20 (p0) sub r2 = r3,r4`	`20 sub r1,r1,r2,s,p0`
`30 (p0) add r5 = r1,r3`	`30 sub r2,r3,r4,z,p1`
`40 (p0) mov pr = r6,mask`	`40 add r5,r1,r3,s,p2`
`50 (p0) cmp.gt.and p1,p2 = 0,r1`	`50 (qp) add r1,r2,r5,pcnf`
`(p0) cmp.eq.and p1,p2 = r3,r4`	
`60 (p0) cmp.lt.or p1,p2 = 0,r5`	
`70 (p1) add r1 = r2,r5`	

5 Implementation Results

A dedicated implementation of a MorphoSys reconfigurable cell was developed to examine the above predication approaches. A combined behavioral/structural VHDL design synthesized with a 0.13 •m target library and constraints for a clock frequency of 200 MHz resulted in the following numbers. An unpredicated RC with a 16x16 multiplier takes 0.026 mm^2 and 0.015 mm^2, respectively. Combinational cells' area for the predication units is 0.009 mm^2. The lookup table requires an area of 0.0095 mm^2. Predication related units take about 30% of the die size. Despite the fact that this relative overhead would be smaller for an RC with more non-predication related features such as a MAC unit or additional RAM, a future objective is to optimize the implementation of the predicate evaluation unit. A lookup table with 16 entries was chosen to keep enough equations for more than one kernel to avoid external memory access in case there are two or more kernels executing alternately. We emphasize that embedding the nested guarded predication model into an existing unrestricted predication model increased the cell area overhead by less than 1%.

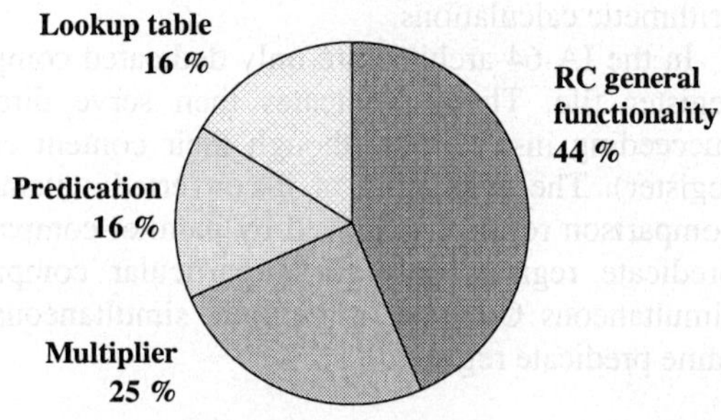

Fig. 12. Cell area percentages

6 Summary and Conclusions

A novel predication scheme was developed and applied to an existing high performance reconfigurable SIMD system-on-chip. Thus, this paper addressed two fundamental issues of SIMD architectures. Firstly, it was shown that the code size could be reduced significantly and the effect of conditional branches could be minimized, which is essential to cope well with parallel if-then-else statements on

SIMD machines. Secondly, the execution autonomy of SIMD processing elements could be increased considerably because a data dependant decision could be taken locally on each PE, even for nested if-then-else clauses. Increased execution autonomy could unload tasks from a central control unit, consequently increasing the SIMD array efficiency and widening the application scope.

Finally, it was shown that the implemented predication scheme did not increase the number of clock cycles. We believe that general-purpose microprocessors could also benefit from this approach.

References

1. Patwardhan, B.: "Introduction to the Streaming SIMD Extensions in the Pentium III, Parts I- III", Dr. Dobb's Journal, February, (2002)
2. Fomithchev, M.: "AMD 3Dnow", Dr. Dobb's Journal, Vol. 25, No. 8, pp.40-42, August, (2000)
3. Meiβner, M., Grimm, S., Straβer, W., Packer, J. and Latimer, D,: "Parallel Volume Rendering on a Single-Chip SIMD Architercture", IEEE 2001 Symposium on Parallel and Large-Data Visuallization and Graphics, San Diego, CA, October, (2001)
4. Blevins, D.W., Davis, E.W., Heaton, R.A. and Reif, J.H.: "BLITZEN: A Highly Integrated Massively Parallel Machine," J. Parallel and Distributed Computing, vol. 8, no. 2, pp. 150-160, Feb. (1990)
5. Nickolls, J. and Reush, J.: "Autonomous SIMD Flexibility in MP-1 and MP-2", SPAA '93 – 5th Annual ACM Symp. On Parallel Algorithms and Architectures, pp. 98-99, Velen, Germany, June, (1993)
6. Singh, H., Lee, M.-H., Lu, G., Kurdahi, F., Bagherzadeh, N., Filho, E.: "MorphoSys: An Integrated Reconfigurable System for Data-Parallel Computation-Intensive Applications", IEEE Transactions on Computers (2000)
7. Mahlke, S.A., Hank, R.E., McCormick, J.E., August, D.I., Hwu, W.W.: "A Comparison of Full and Partial Predicated Execution Support for ILP Processors", 22nd International Symposium on Computer Architecture, pages 138-149, (June 1995)
8. Pnevmatikatos, D.N., Sohi, G.S.: "Guarded Execution and Branch Prediction in Dynamic ILP Processors", 21st International Symposium on Computer Architecture, pages 120-129, (June 1994)
9. "Alpha Architecture Handbook", Compaq Computer Corporation (1998)
10. Kathail, V., Schlansker, M.S., Rau, B.R.: "HPL-PD Architecture Specification: Version 1.1", Compiler and Architecture Research, HP Laboratories, Palo Alto (2000)
11. "IA-64 Application Instruction Set Architecture Guide, Revision 1.0", Intel Corporation / Hewlett-Packard Company (1999)

MorphoSys: A Coarse Grain Reconfigurable Architecture for Multimedia Applications

Hooman Parizi, Afshin Niktash, Nader Bagherzadeh, and Fadi Kurdahi

Department of Electrical and Computer Engineering
University of California, Irvine 92697
{hparizi, aniktash, nader, kurdahi}@ece.uci.edu
http://www.eng.uci.edu/morphosys

Abstract. MorphoSys is a reconfigurable architecture for computation intensive applications. It combines both coarse grain and fine grain reconfiguration techniques to optimize hardware, based on the application domain. M2, the current implementation, is developed as an IP core. It is synthesized based on the TSMC 0.13 micron technology. Experimental results show that for multimedia applications MorphoSys has a performance comparable to ASICs with the added benefit of being able to be reconfigured for different applications in one clock cycle.

1 Introduction

Reconfigurable systems are an intermediate approach between the Application Specific Integrated Circuits (ASICs) and general purpose processors. They have wider applicability than ASICs while their performance is comparable to them. On the other hand, multimedia applications comprise of several subtasks with different characteristics. This feature in addition to a large set of input and output data, lead to an uneconomical solution in ASIC, and low performance solution on general purpose architectures. Reconfigurable systems are considered as an alternative approach for developing architectures for multimedia and DSP applications, Raw [1], PipeRench [2], Garp [3] and MorphoSys[4] are ongoing research projects in this area.

In this paper M2, a new implementation of MorphoSys is introduced. Comparison to M1, the previous implementation, M2 consists of high performance functional units, and optimized memory architecture. M2 implementation goes through a fully automated methodology starting from an IP core.

In Sections 2 basic structure of MorphoSys are described. Section 3 describes M2 reconfigurable cell with emphasis on the new features. Section 4 discuses the parallel structure of MorphoSys. Section 5 evaluates some multimedia applications on the M2.

B. Monien and R. Feldmann (Eds.): Euro-Par 2002, LNCS 2400, pp. 844–848.

2 MorphoSys Architecture

Figure 1 shows the basic building blocks and their connections in MorphoSys. RC-Array is the reconfigurable part of the system. It consists of an 8 by 8 array of reconfigurable cells (RCs). The configuration data is stored in the context memory. During the execution, the context word is loaded from the context memory to the context registers of the reconfigurable cell. Frame buffer is an embedded data memory in the MorphoSys. It gets the data from the external memory and feed the RC-Array with the appropriate data. All the data movements between the MorphoSys memory elements and the external memory are handled by the DMA controller. TinyRisc [5] is a general purpose 32-bit RISC processor. It controls the sequence of operations in MorphoSys as well as executing non-data parallel operations.

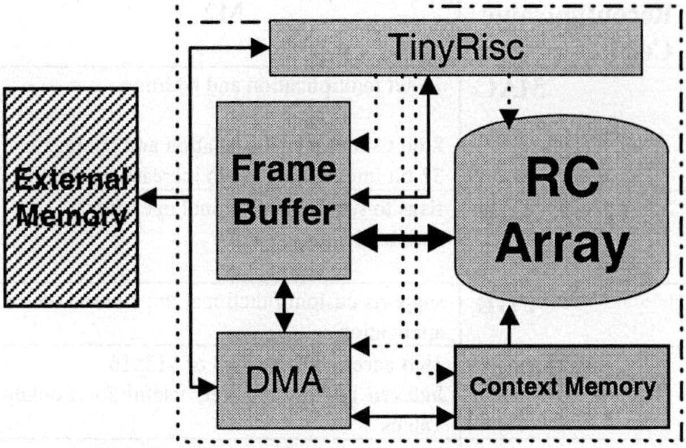

Figure 1- MorphoSys Architecture

3 Reconfigurable Cell Architecture

Reconfigurable Cells (RCs) are the main processing units in MorphoSys. Each RC consists of four types of basic elements. Functional units for arithmetic and logic operations, Memory element to feed the functional units and store their results, input and output modules to connect cells together to form the RC-Array architecture and a fine grain reconfigurable logic block. Figure 2 shows the block diagram of each RC.

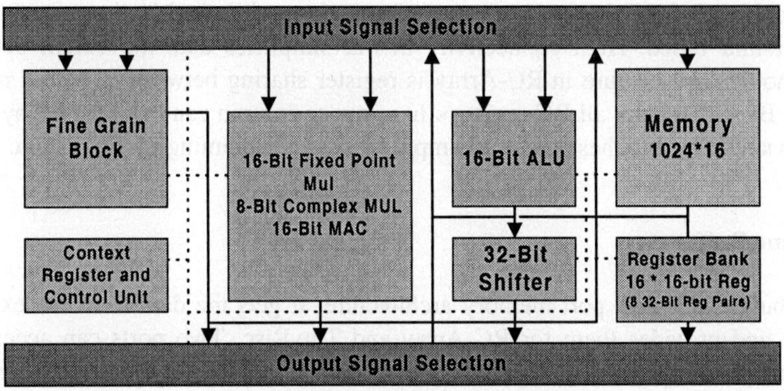

Figure 2 - Reconfigurable Cell Block Diagram

Table 1- Reconfigurable Cell features in M2 versus M1 implementation

Reconfigurable Cell	M2	M1
MAC	16-bit multiplication and addition 8-bit Complex multiplication and conjugate 32-bit internal register to increase precision	16x12-bit multiplication and addition NOT supported NOT supported
ALU	flags to support conditional operations variable length shifter	NOT supported Shifter with fixed number of shifts
FGB	supports custom functional units based on the application	NOT supported
Memory	1KB accessible as 1Kx8 or 512x16 Indexed address register, useful for Lookup Tables	NOT supported NOT supported
Register File	Fourteen 16-bit Registers Auto-increment and auto decrement registers	Four 16-bit Registers NOT supported
Context Reg	32-bit configuration register	32-bit configuration register

Table 1 summaries the RC features in M2 and our previous implementation M1. More details on RC functional units are in [6].

4 MorphoSys Parallel Architecture

MorphoSys parallel architecture is based on the connection of reconfigurable cells and the way that they are connected to the memory. In this section these structures will be discussed.

4.1 Reconfigurable Cell Array

RC-Array forms the parallel architecture of MorphoSys. In M2 all RCs in a single row or column of RC-Array are connected together, while in M1 the connection is only pyramid based. High connectivity in M2 simplifies data movement between RCs. Another new feature in RC-Array is register sharing between RCs in a row or column. By this feature all RC registers in a row or column can be accessed by other RCs in a single cycle. These features simplifies the programming of the system.

4.2 Frame Buffer

Frame buffer is a dual port memory architecture. It gets the data from the external memory and provides them for RC-Array and TinyRisc. Two ports can access the frame buffer simultaneously, so reading or writing can be done by the RC-Array or TinyRisc and at the same time DMA is transferring data between the frame buffer and external memory. Figure 3 shows the frame buffer interface in the system. Frame buffer interface to RC-Array is through a reconfigurable bus. To exchange data between the frame buffer and the RC Array first the bus configuration should be established by loading the appropriate data to the frame buffer configuration tables and then the read or write operation can be done. The flexibility of the frame buffer makes it simple to access data, reorder them and feed the processing elements as fast as possible.

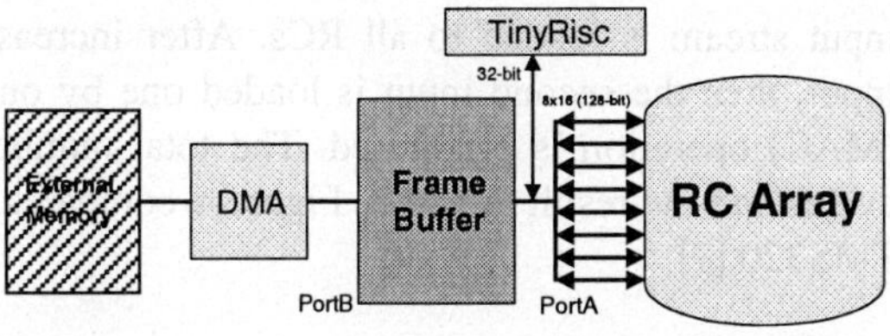

Figure 3. Frame Buffer Interface in M2

5 Algorithm Mapping and Performance Analysis

A number of kernels have been studied for the purpose of benchmarking and performance analysis of M2.

DCT. The forward and inverse DCT is used in MPEG encoders and decoders for transformation of image pixels to the frequency domain and back to the spatial domain. Chen's algorithm for an eight-point 1-D DCT is considered for mapping. M2 requires 21 cycles to complete 2-D DCT (or IDCT) on 8x8 block of pixel data. 2 extra cycles are required for loading/storing data from/into Frame Buffer. The cycle count for this benchmark on TMS320C64 is 154 and on C62 is 208.

FFT. We used the similar approach to DCT mapping for implementation of 64 point complex 2-D FFT. The computational cycles of 2-D FFT in this case is 42 cycles. This is much less than 276 cycles on C64 and 835 cycles on C62.

Correlation. For 64 point correlation of 16-bit real inputs, one can use one RC/point for first input stream. At each iteration the first

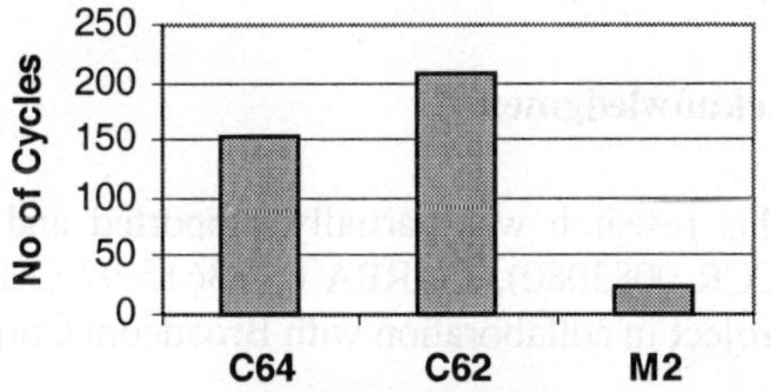

Figure 4a- DCT Performance Comparision

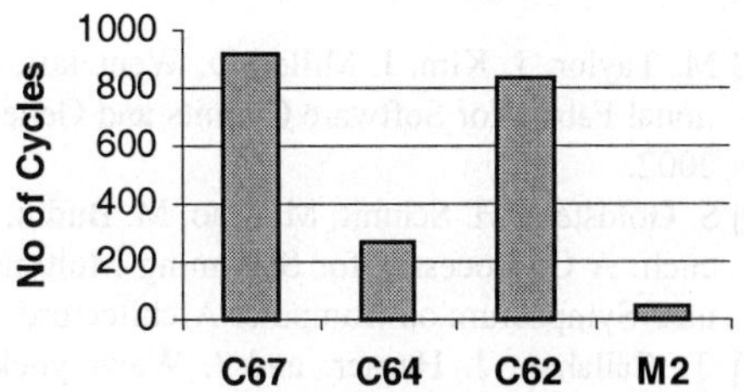

Figure 4b- FFT Performance Comparison

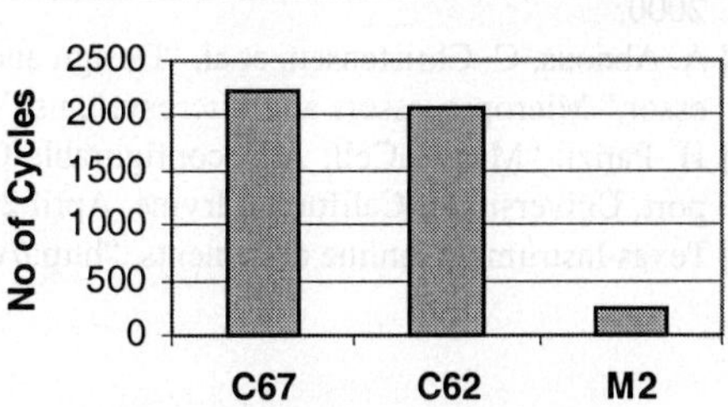

Figure 4c- Correlation Performance Comparison

input stream is loaded to all RCs. After increasing the address pointer of the first input, then the second input is loaded one by on to the whole RC and multiply add (MAC) operation is performed. The total number of cycles in this approach is 256 cycles and the result is 32 bit. Figure 4 compares the result of these benchmarks with TMS320 [7].

Conclusion

In this paper M2, a new implementation of MorphoSys has been introduced. M2 follows the basic concepts of MorphoSys, but it is optimized for computation intensive applications. It is a coarse grain reconfigurable architecture, with a set of fine grain blocks. Its memory architecture is highly optimized to overcome the high demands for data movement and shuffling in multimedia applications. Experimental results show that Morphoys architecture has a performance comparable to multimedia processors and ASICs.

Acknowledgments

This research was partially supported and funded by National Science Foundation (CCR-0083080), DARPA (F-33615-97-C-1126), and University of California CoRe Project in collaboration with Broadcom Corp (99-10061).

References

[1] M. Taylor, J. Kim, J. Miller, D. Wentzlaff, et al, "The Raw Microprocessor: A Computational Fabric for Software Circuits and General Purpose Programs," IEEE Micro, Mar/Apr 2002.

[2] S. Goldstein, H. Schmit, M. Mao, M. Budiu, S. Cadambi, R. Taylor and R. Laufer, "PipeRench: A Coprocessor for Streaming Multimedia acceleration," Proceeding of the 26th Annual Symposium on Computer Architecture, pages 28-39, May 1999.

[3] T. Callahan, J. Hauser, and J. Wawrzynek, "The Garp Architecture and C Compiler," *IEEE Computer*, April 2000.

[4] H. Singh, et al, "MorphoSys: An Integrated Reconfigurable System for Data-Parallel and Computing-Intensive Applications", IEEE Transaction on Computers, Vol. 49, No. 5, May 2000.

[5] A. Abnous, C. Christensen, et al, "Design and Implementation of the TinyRISC Microprocessor," Microprocessors and Microsystems, Vol. 16, No. 4, pp. 187-194, 1992.

[6] H. Parizi, "MorphoCell: A Reconfigurable Cell Targets DSP applications," Technical Report, University of California, Irvine, April 2001.

[7] Texas Instrument online documents, "http://www.ti.com/"

Performance Scalability of Multimedia Instruction Set Extensions

D. Cheresiz[1], B. Juurlink[2], S. Vassiliadis[2], and H. Wijshoff[1]

[1] LIACS, Leiden University, The Netherlands
[2] Computer Engineering Laboratory, Electrical Engineering Department, Delft University of Technology, The Netherlands

Abstract. Current media ISA extensions such as Sun's VIS consist of SIMD-like instructions that operate on short vector registers. In order to exploit more parallelism in a superscalar processor provided with such instructions, the issue width has to be increased. In the Complex Streamed Instruction (CSI) set exploiting more parallelism does not involve issuing more instructions. In this paper we study how the performance of superscalar processors extended with CSI or VIS scales with the amount of parallel execution hardware. Results show that the performance of the CSI-enhanced processor scales very well. For example, increasing the datapath width of the CSI execution unit from 16 to 32 bytes improves the kernel-level performance by a factor of 1.56 on average. The VIS-enhanced machine is unable to utilize large amounts of parallel execution hardware efficiently. Due to the huge number of instructions that need to be executed, the decode-issue logic constitutes a bottleneck.

1 Introduction

In the last years multimedia applications started to play an important role in the design of computing systems because they provide highly appealing and valuable services to the consumers, such as video-conferencing, digital content creation, speech recognition, virtual reality and many others. Applications like JPEG image compression/decompression, MPEG video and MP3 players are now common on most desktop systems. Multimedia applications impose high requirements on the performance of computing systems because they need to process huge amounts of data under stringent real-time constraints.

To meet these requirements, most processor vendors have extended their instruction set architectures (ISAs) with new instructions which allow key multimedia algorithms to be implemented efficiently. Examples of such extensions are Intel's *MMX* and *SSE* [12,14], Motorola's *AltiVec* [4] and Sun's *Visual Instruction Set (VIS)* [15]. All of them are, essentially, load-store vector architectures with short vector registers. These vector registers (also referred as multimedia registers) are usually 64 bits wide and contain vectors consisting either of eight 8-bit, four 16-bit, or two 32-bit elements. Multimedia instructions exploit SIMD parallelism by concurrently operating on all vector elements. More recent extensions (SSE, AltiVec) increase the register size to 128 bits allowing to process up

B. Monien and R. Feldmann (Eds.): Euro-Par 2002, LNCS 2400, pp. 849–859.

to 16 8-bit values in parallel. Multimedia extensions have proven to provide significant performance benefits [13] by exploiting the data-level parallelism present in multimedia codes. However, these extensions have several features which are likely to hinder further performance improvements.

First, the fixed size of the multimedia registers limits the amount of parallelism that can be exploited by a single instruction to at most 8 (VIS, MMX) or 16 (SSE, AltiVec) parallel operations, while more parallelism is present in multimedia applications. In many important multimedia kernels, the same operation has to be performed on data streams containing tens to hundreds of elements. If existing ISA extensions are employed, such long data streams have to be split into sections which fit in the multimedia registers to process one section at a time. This results in a high number of instructions that need to be executed. Second, implementations of multimedia kernels with short-vector SIMD extensions require a significant amount of overhead for converting between different packed data types and for data alignment, increasing the instruction count even further. According to [13], for VIS up to 41% of the total instruction count constitutes overhead.

With the number of instructions being fixed, more instructions have to be fetched, decoded and executed in each cycle in order to increase performance. It is generally accepted, however, that increasing the issue width requires a substantial amount of hardware [5] and negatively affects the cycle time [11]. One way to achieve higher performance without increasing the issue width is to reduce the instruction traffic by increasing the size of the multimedia registers, but this approach implies a change of the ISA and, therefore, requires recompilation or even rewriting existing codes. Moreover, increasing the register size beyond 256 bits is not likely to bring much benefit, because many multimedia kernels process small 2-dimensional sub-matrices and only a limited number of elements, typically 8 or 16, are stored consecutively.

In order to overcome the limitations of short-vector SIMD extensions mentioned above, the *Complex Streamed Instruction* (CSI) set was proposed and has proven to provide significant speedups [8]. A single CSI instruction can process 2-dimensional data streams of arbitrary length, performing sectioning, aligning and conversion between different data formats internally in hardware.

In this paper we study how the performance of a superscalar processor enhanced with a short-vector SIMD extension (VIS) or with CSI scales with the amount of parallel execution hardware. This study is motivated by current trends in microprocessor technology, which exhibits growing levels of integration and transistor budgets. For example, in a contemporary 0.15-micron CMOS technology, a 32-bit adder requires less than 0.05 mm^2 of chip area [9], allowing designers to put dozens of such units on a single chip. The challenge is to feed these units with instructions and data. We explore the design space by varying key parameters of the processor, such as the issue width and instruction window size, in order to identify the performance bottlenecks.

This paper is organized as follows. In Section 2 we give a brief description of the CSI architecture and some related proposals. Section 3 describes the simulated benchmarks, the modeled processors and presents the experimental results. Conclusions and areas of future research are presented in Section 4.

2 Background

2.1 The CSI Architecture

In this section we briefly sketch the CSI multimedia ISA extension. Further details on the CSI architecture and a possible implementation can be found in [8]. CSI is a memory-to-memory architecture for two-dimensional streams. Usually, a CSI instruction fetches two input streams from memory, performs simple arithmetic operations on corresponding elements, and stores the resulting output stream back to memory. The streams follow a matrix access pattern, with a fixed *horizontal stride* (distance between consecutive elements of the same row) and a fixed *vertical stride* (distance between rows). The stream length is not architecturally fixed. The programmer visible state of CSI consists of several *sets of stream control registers (SCR-sets)*. Each SCR-set consists of a several registers that hold the base address, the horizontal and vertical strides, the number of elements, etc., and completely specifies one stream.

The key advantages of the CSI architecture can be summarized as follows:

- **High parallelism**
  CSI increases the amount of parallelism that can be exploited by a single instruction. This is achieved by having no restriction on the stream length and by supporting 2-dimensional streams.
- **Reduced conversion overhead**
  CSI eliminates the overhead instructions required for converting between different packed data types by performing it internally in hardware and by overlapping it with useful computations.
- **Decoupling of data access and execution**
  Because the data streams in CSI are fully described by the SCR-sets, address generation, data access and execution can be decoupled. The CSI execution unit can generate addresses and send them to the memory hierarchy well before the data is used, thereby reducing the number of stalls due to late data arrival.
- **Compatibility**
  The same CSI code can be run without recompilation on a new implementation with a wider SIMD datapath and fully utilize it. This is possible because the number of elements that are actually processed in parallel is not part of the architecture.

2.2 Related Work

CSI instructions process two-dimensional data streams of arbitrary length stored in memory. Several early vector architectures such as the TI ASC and the Star-100 [6] were memory-to-memory as well. These architectures, however, suffered from a long startup time, which was mainly caused by execution of the overhead instructions needed for setting up the vector parameters and by long memory latency. The CSI implementation described in [8] does not suffer from these limitations for the following reasons. First, since CSI is implemented next to a fast superscalar core, the overhead instructions needed to set up the stream control registers take very little time. Second, the stream data is accessed through the L1 data cache. Because the L1 access time is short and the hit rates of the considered benchmarks are high, the time between the request for stream data and its arrival is short.

A related proposal that also exploits the 2-dimensional data-level parallelism available in multimedia applications is the Matrix Oriented Multimedia (MOM) ISA extension [2]. MOM instructions can be seen as vector versions of current SIMD media extensions. Two key features distinguish MOM from CSI. First, MOM is a register-to-register architecture (which implies sectioning when the streams do not fit into the MOM registers). Second, MOM requires overhead instructions for explicit data conversion.

Another related proposal is the *Imagine* processor [9], which has a load/store architecture for one-dimensional streams of data records. The SIMD processing hardware of Imagine consists of 8 arithmetic clusters. Each cluster is a VLIW engine with 6 functional units. Data records are distributed one per cluster and are processed in a SIMD fashion by executing the same sequence of VLIW instructions at each cluster. Contrary to CSI, which is an ISA extension and implemented in a CPU, Imagine is a stand-alone multimedia coprocessor.

3 Experimental Evaluation

The main goal of the experiments is to evaluate how the performance of a superscalar processor extended with VIS or CSI scales with the amount of SIMD execution hardware and to identify the performance bottlenecks. The benchmark set consists of MPEG-2 and JPEG codecs (*mpeg2encode, mpeg2decode, cjpeg, djpeg*) from the *Mediabench* suite [10] and of several image processing kernels (*add8, scale8, blend8, convolve3x3*) taken from the VIS Software Development Kit (VSDK) [16].

3.1 Tools and Simulation Methodology

In order to evaluate the performance of the VIS-enhanced and CSI-enhanced processors, we modified the `sim-outorder` simulator of the SimpleScalar toolset (version 3.0) [1]. This is a cycle-accurate execution-driven simulator of an out-of-order superscalar processor with a 5-stage pipeline based on the *Register Update Unit (RUU)*. A corrected version of the SimpleScalar memory model based on the SDRAM specifications given in [3] was used.

VIS- and CSI-executables of each *Mediabench* benchmark were obtained by manually rewriting the most time-consuming kernels in assembly. These are the *idct* and *Add_Block* routines in *mpeg2decode*, *dist1* in *mpeg2encode*, *idct* and *ycc_rgb_convert* in *djpeg*, and *rgb_ycc_convert* and *h2v2_downsample* in *cjpeg*. VIS- and CSI-executables of the VSDK kernels were obtained by completely rewriting them in assembly. Whenever available, we used the vendor-provided assembly codes for VIS.

We remark that out-of-order execution of CSI instructions is not allowed. Out-of-order execution of a CSI instruction with a potentially conflicting memory reference is also not allowed.

3.2 Modeled Processors

A wide range of superscalar processors was simulated by varying the issue width from 4 to 16 instructions per cycle and the instruction window size (i.e, the number of entries in the RUU unit) from 32 to 512. Table 1 summarizes the basic parameters of the processors and lists the functional unit latencies. Processors were made capable of processing VIS or CSI instructions by adding the VIS functional units or the CSI unit. The CSI unit is interfaced to L1 cache as described in [8].

Each time the issue width is doubled, the number of integer and floating-point units is scaled accordingly. The number of cache ports is fixed at 2, however, because multi-ported caches are expensive and hard to design (for example, currently there are no processors with a 4-ported cache). Note, that the CSI-enhanced processors do not require more ports to cache then the VIS-enhanced ones. The CSI-enhanced processors use two cache ports, sharing loads and stores generated by the CSI unit between them. The VIS- and the CSI-enhanced CPUs

Table 1. Processor parameters.

Clock frequency	666 MHz
Issue width	4/8/16
Instruction window size	16-512
Load-store queue size	8-128
Branch Prediction	
Bimodal predictor size	2K
Branch target buffer size	2K
Return-address stack size	8
Functional unit type and number	
Integer ALU	4/8/16
Integer MULT	1/2/4
Cache ports	2
Floating-point ALU	4/8/16
Floating-point MULT	1/2/4
VIS adder	2/4/8
VIS multiplier	2/4/8

FU latency/recovery (cycles)	
Integer ALU	1/1
Integer MUL	
multiply	3/1
divide	20/19
Cache port	1/1
FP ALU	2/2
FP MUL	
FP multiply	4/1
FP divide	12/12
sqrt	24/24
VIS adder	1/1
VIS multiplier	
multiply and pdist	3/1
other	1/1

Table 2. Memory parameters.

Instruction cache	ideal
Data caches	
L1 line size	32 bytes
L1 associativity	direct-mapped
L1 size	32 KB
L1 hit time	1 cycle
L2 line size	128 bytes
L2 associativity	2-way
L2 size	128 KB
L2 replacement	LRU
L2 hit time	6 cycles

Main memory	
type	SDRAM
row access time	2 bus cycles
row activate time	2 bus cycles
precharge time	2 bus cycles
bus frequency	166 MHz
bus width	64 bits

were simulated with different amounts of SIMD-processing hardware by varying the number of the VIS functional units and the width of the CSI unit's datapath, respectively.

The parameters of the cache and memory subsystems are summarized in Table 2. The memory latencies are expressed in memory clock cycles. A memory cycle is 6 ns, corresponding to a clock frequency of 166 MHz, which is typical for contemporary DRAM chips. The system bus (between the L2 cache and the memory controller) is also clocked at 166 MHz as in current PCs [7]. The ratio of CPU clock frequency to memory clock frequency was set to 4, corresponding to a CPU clock rate of 666 MHz.

3.3 Experimental Results

In this section we study the impact of the RUU size on the performance of the CSI- and VIS-enhanced processors. Then we analyze the performance behavior of these processors with respect to the number of SIMD functional units. After this, the performance bottleneck of the VIS-enhanced processors is identified. Finally, we present the speedups attained by CSI relative to VIS.

Increasing the Window Size. First, we study the influence of the instruction window (RUU) size on the performance. Because the RUU is a rather costly resource, it is important to minimize its size. The results (which are not presented here due to space limitations) show that for the studied kernels increasing the window size has a positive effect on the performance of a VIS-enhanced processor but does not influence the performance of a CSI-enhanced one. This is not surprising for the following reason. The kernels usually operate on long data streams, performing the same operation independently on all stream elements. In VIS, long data streams are split into sections which fit into VIS registers and a separate instruction is generated for each section. The instructions are independent which means that the VIS translates the parallelism present in the kernels into instruction-level parallelism (ILP). Larger instruction windows allow a VIS-enhanced superscalar CPU to expose and utilize larger amounts of ILP. On the other hand, in CSI the parallelism is exposed by a single CSI instruction which processes the whole stream. Therefore, a CSI-enhanced CPU does not need large instruction windows in order to exploit it. These results mean that the CSI-enhanced CPUs can be implemented with smaller (and cheaper) instruction windows then the VIS-enhanced ones.

The simulation results also show that, for each issue width, increasing the RUU size beyond a certain limit yields diminishing returns. The speedup of the 4-way VIS-enhanced processor saturates when the RUU consists of 64 entries, the 8-way machine when the RUU size is 128, and the 16-way processor reaches its peak performance when the RUU size is 256. Therefore, we select these instruction window sizes both for VIS- and CSI-enhanced processors for all the experiments presented in the rest of the paper.

Increasing the Amount of Parallel Execution Hardware. The next question we consider is: how does the performance of the superscalar processors equipped with VIS or CSI scale with the amount of parallel execution hardware? This issue is important because of current trends in microprocessor technology. Larger scales of integration and, hence, growing transistor budgets allow to put dozens of functional units on a chip. The challenge for the designer is to keep these units utilized. The amount of SIMD processing hardware is characterized by the number of bytes which can be processed in parallel. For VIS this is determined by the number of VIS adders and multipliers. For example, to operate on 16 bytes in parallel, the machines are configured with 2 VIS adders and 2 VIS multipliers. For CSI, the amount of parallelism is determined by the width of the datapath of the CSI execution unit. So, in order to process 32 or 64 bytes in parallel the number of VIS units of the VIS-enhanced processor is doubled or quadrupled, whereas for CSI, the width of its datapath is increased appropriately.

As the baseline processors, we considered the 4-, 8-, and 16-way superscalar CPUs with instruction windows of 64, 128, and 256 entries, respectively. Each baseline processor was equipped with sufficient SIMD hardware so that it can process 16, 32, or 64 bytes in parallel. Let a VIS-extended processor with an issue width of x, a window size of y, and a SIMD processing width of z bytes be denoted by $VIS(x, y, z)$. A similar notation will be used for CSI-enhanced CPUs. Figure 1(a) depicts the speedups attained by all simulated VIS-enhanced processors relative to $VIS(4, 64, 16)$. The speedups achieved by all simulated CSI-enhanced CPUs relative to $CSI(4, 64, 16)$ are plotted in Figure 1(b). All VIS and CSI kernels exhibited similar behavior and, therefore, we only present the performance figures of four selected kernels. Figure 1(a) shows that, for a fixed issue width, increasing the number of the VIS functional units does not yield any benefit. Contrary to VIS, CSI is able to utilize additional SIMD execution hardware (see Figure 1(b)), and its performance scales almost linearly with the amount of SIMD resources. It can also be observed that the performance of the

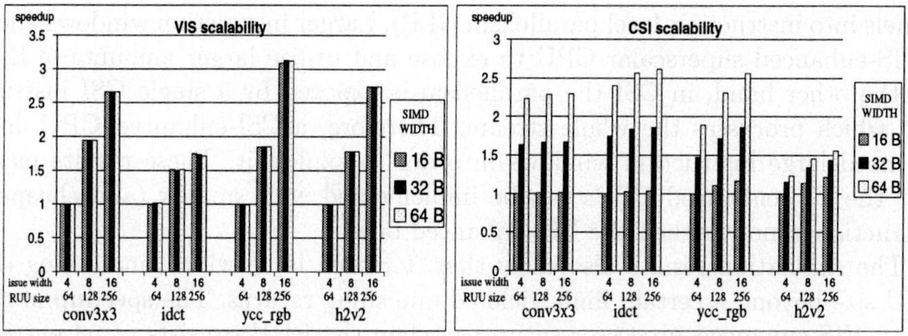

(a) (b)

Fig. 1. Scalability of the CSI and VIS performance w.r.t. available SIMD hardware.

CSI-extended processors is insensitive to the issue width. This is expected since CSI does not exploit instruction-level parallelism and, therefore, does not need to issue a lot of instructions in parallel.

Figure 2 presents the speedups attained by CSI-enhanced processors with respect to VIS-extended processors equipped with the same amount of SIMD execution hardware, i.e., the figure shows the speedups of $CSI(x, y, z)$ over $VIS(x, y, z)$ for various values of x, y, and z. It shows that CSI clearly outperforms VIS, especially for the 4-way and 8-way configurations. We remark that these processors are less costly and can probably achieve a higher clock frequency than a 16-way one.

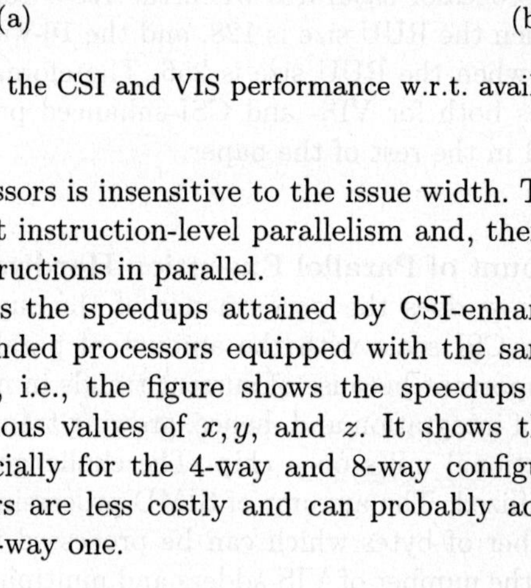

Fig. 2. Speedup of CSI over VIS for several kernels.

Identifying the Bottleneck. It is important to identify where the performance bottleneck of the VIS-enhanced machines lies. Figure 1(b) shows that the VIS-enhanced machines perform better when the issue width is increased. This suggests that the issue width is the limiting factor for VIS. To investigate this, we study the IPCs attained by the VIS-enhanced CPUs. For each issue width, Figure 3 depicts the attained IPCs normalized to the corresponding ideal IPCs (e.g., for a 4-way machine, the ideal IPC is 4). It shows that, for most kernels,

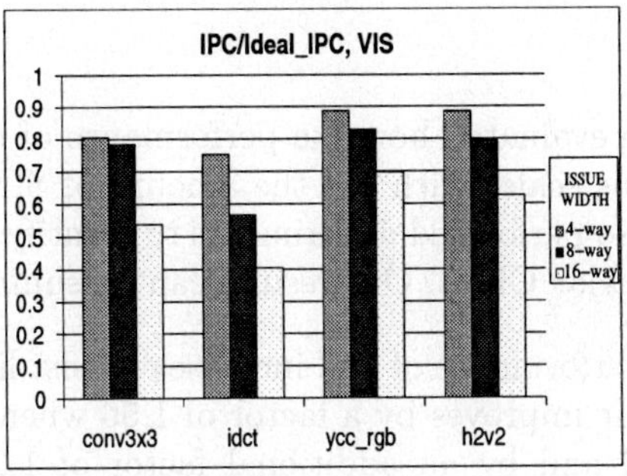

Fig. 3. Ratio of achieved IPC to ideal IPC for several VIS kernels and issue widths.

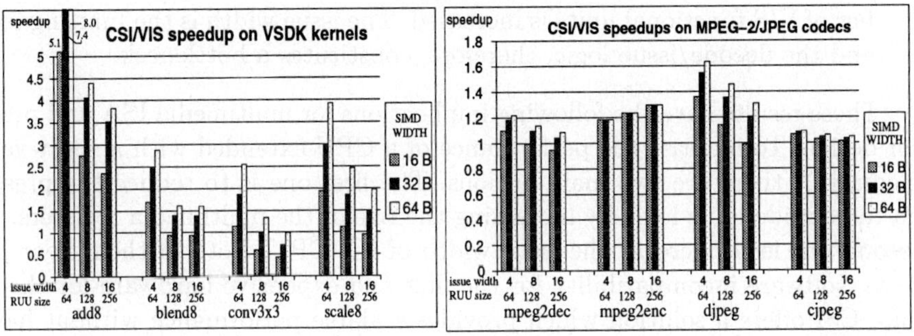

(a) (b)

Fig. 4. Speedup of CSI over VIS on VSDK kernels and MPEG/JPEG codecs.

the performance achieved by the 4-way and 8-way superscalar VIS-enhanced processors is close to ideal. These CPUs achieve IPCs which are within 78% to 90% of the ideal IPC for all kernels except *idct*. This means that they cannot be accelerated by more than 11-28% (1/0.9 – 1/0.78) and, therefore, even when they achieve the ideal IPC their performance will not approach that of the corresponding CSI-enhanced CPUs. This shows that the issue width indeed limits the performance of the VIS-enhanced CPUs. The IPCs attained by the 16-way machine are far from the ideal and are within 55% to 70% of the ideal IPC, for all kernels except *idct*, meaning that the accelerations of 42-81% are possible. However, to achieve such increases in IPC a very accurate branch predictor is needed.

In conclusion, Figure 4 depicts the VSDK kernel and application speedups achieved by CSI over VIS. Of course, the speedups achieved by CSI on MPEG-2/JPEG codecs are smaller than those achieved on VSDK kernels due to Amdahl's law. Still rather impressive application-level speedups are achieved by CSI, especially for smaller (and more realistic) issue widths of 4 and 8. For example, when the issue width is 4 and 32 bytes can be processed in parallel, CSI achieves speedups ranging from to 1.08 (*cjpeg*) to 1.54 (*djpeg*) with an average of 1.24.

4 Conclusions

In this paper we have evaluated how the performance of the CSI and VIS multimedia ISA extensions scales with the the amount of SIMD parallel execution hardware. We have also performed experiments to identify the bottlenecks of the VIS-enhanced superscalar CPUs. Out results can be summarized as follows:

- The kernel-level performance of CSI increases almost linearly with the width of CSI datapath. It improves by a factor of 1.56 when the width of the CSI datapath doubled and by an additional factor of 1.27 when the width is quadrupled. Furthermore, the performance of the CSI-enhanced processor is not sensitive to the issue width.
- The VIS-enhanced CPUs do not perform substantially better when the number of VIS functional units is increased. The issue width is the limiting factor and the decode/issue logic, therefore, constitutes a bottleneck.

These results have the following implications for multimedia ISA and processor design. To increase the performance of a CPU extended with a short-vector media ISA there are two main options. The first one is to reduce the pressure on the decode-issue logic by increasing the size of the multimedia registers. The second one is to increase the issue width of the CPU. Both of them bear huge costs: software incompatibility for the first and expensive hardware for the second. CSI offers a solution which provides scalable performance without having to increase the issue width as well as software compatibility.

We plan to further investigate the CSI paradigm in several directions. For example, since the interface of the CSI execution unit with the memory subsystem is not fixed, a designer has freedom in its implementation. The CSI execution unit can be interfaced to the L1 cache, L2 cache or even main memory. We plan to evaluate such implementations.

References

1. D. Burger and T.M. Austin. The SimpleScalar Tool Set, Version 2.0. Technical Report 1342, Univ. of Wisconsin-Madison, Comp. Sci. Dept., 1997.
2. Jesus Corbal, Mateo Valero, and Roger Espasa. Exploiting a New Level of DLP in Multimedia Applications. In *MICRO 32*, 1999.
3. M. Gries. The Impact of Recent DRAM Architectures on Embedded Systems Performance. In *EUROMICRO 26*, 2000.
4. L. Gwennap. AltiVec Vectorizes PowerPC. *Microprocessor Report*, 12(6), 1998.
5. J.L. Hennessy and D.A. Patterson. *Computer Architecture - A Quantitative Approach.* Morgan Kaufmann, second edition, 1996.
6. Kai Hwang and Faye A. Briggs. *Computer Architecture and Parallel Processing.* McGraw-Hill, second edition, 1984.
7. PC SDRAM Specification,Rev 1.7. Intel Corp., November 1999.
8. B. Juurlink, D. Tcheressiz, S. Vassiliadis, and H. Wijshoff. Implementation and Evaluation of the Complex Streamed Instruction Set. In *Int. Conf. on Parallel Architectures and Compilation Techniques (PACT)*, 2001.

9. B. Khailany, W.J. Dally, U.J. Kapasi, P. Mattson, J. Namkoong, J.D. Owens, B. Towles, A. Chang, and S. Rixner. Imagine: Media Processing With Streams. *IEEE Micro*, 21(2):35–47, 2001.
10. C. Lee, M. Potkonjak, and W.H. Mangione-Smith. MediaBench: A Tool for Evaluating and Synthesizing Multimedia and Communication Systems. In *MICRO 30*, 1997.
11. S. Palacharla, N.P. Jouppi, and J.E. Smith. Complexity-Effective Superscalar Processors. In *ISCA'97*, 1997.
12. Alex Peleg, Sam Wilkie, and Uri Weiser. Intel MMX for Multimedia PCs. *Communications of the ACM*, 40(1):24–38, January 1997.
13. P. Ranganathan, S. Adve, and N.P. Jouppi. Performance of Image and Video Processing with General-Purpose Processors and Media ISA Extensions. In *ISCA 26*, pages 124–135, 1999.
14. Shreekant Thakkar and Tom Huff. The Internet Streaming SIMD Extensions. *Intel Technology Journal*, May 1999.
15. Marc Tremblay, J. Michael O'Conner, Venkatesh Narayanan, and Lian He. VIS Speeds New Media Processing. *IEEE Micro*, 16(4):10–20, August 1996.
16. VIS Software Developer's Kit.
Available at `http://www.sun.com/processors/oem/vis/`.

Topic 14
Meta- and Grid-Computing

Michel Cosnard and Andre Merzky

Topic Chairpersons

Since the origins of computer networks connecting spatially distributed resources, attempts have been made to create an infrastructure for distributing applications across regional, national and administrative boundaries. New paradigms have been introduced to support the programming and deployment of such computational networks, generally as extensions to existing parallel computing models (such as SPMD) and as distributed computing techniques based on distributed object technologies. The primary objective in many such systems is to achieve uniformity in programming and use, whilst supporting heterogeneity and transparency in architectures, operating systems and environments.

During the last couple of years, these effort have been culminating in the emerging field of Grid Computing. Grid Computing tries to turn Computational Networks into Computational Grids, that means to add an ubiquitous and consistent infrastructure to the distributed resources. This infrastructure provides generalized operations like caching, authentication, resource discovery, resource scheduling etc, and hence enables uniformity in programming and use.

This new emerging field of Grid Computing still finds many critizism, from "new name for old ideas", over "solution looking for a problem" to "yet another hype wthout any contents". But fact is that the Grid paradigm seems currently to be the only one available showing good chances for keeping its promises.

The strength of Grid Computing paradigm lies in the simplicity of its idea and in its consequent focus on standardization, for architecture, protocols and interfaces, and in its attempt to enclose all the known facets of classical distributed computing and meta computing. The Global Grid Forum (GGF) plays a crucial role here and provides substantial momentum to this field.

The papers in the Europar topic "Meta- and Grid Computing" illuminate various details of that huge area, and reflect the wide range of problems the community is facing and solving in this context today.

The paper by Vijay Dialani, Simon Miles, Luc Moreau, David De Roure, and Michael Luck, an infrastructure to implement fault tolerance services in Grid environments. As Web services are a major focus right now in the Grid community, and life cycle management of Grid Services get a real important issue in this context, the authors add valuable input to that topic.

The paper by P.H.J Kelly, S. Pelegatti and M. Rossitzer adds an interesting perspective to the well researched problem of idle CPU cycle utilization in a network of workstations, by focusing on response time and on minimal user interferenz instead of throughput.

The paper by Darin Nikolow, Renata Slota, Mariusz Dziewierz and Jacek Kitowski presents two approaches for access time estimations for tertiary storage

B. Monien and R. Feldmann (Eds.): Euro-Par 2002, LNCS 2400, pp. 861–862.

systems. These approaches fill a gap present in many Grid environments focusing on large scale data management problems.

The paper by Martin Alt, Holger Bischof and Sergei Gorlach adresses the widely neglected problem of algorithm design for Grid computing environments. It is based on high level components called skeletons, and focuses on Java applications.

The paper by Eddy Caron, Frédéric Desprez, Frédéric Lombard, Jean-Marc Nicod, Laurent Philippe, Martin Quinson and Frédéric Suter describes an distributed interactive engineering toolbox (DIET) which consists of a hierarchical set of components to build network enabled servers applications.

The paper by Jaroslaw Pytlinski, Lukasz Skorwider, Piotr Bala Miroslaw Nazaruk and Konrad Wawruch presents a case study for deploying an commercial application with the Unicore Grid environment.

The paper by J. Santon, S. Newhouse and J. Darlington describes the design and implementation of a component for 'collaborative scientific visualisation'. The component is part of the ICENI Grid middleware project. Some of its capability is explored by means of an astrophysical application simulating coronal mass ejections.

Instant-Access Cycle-Stealing for Parallel Applications Requiring Interactive Response

P.H.J. Kelly, S. Pelagatti*, and M. Rossiter**

Department of Computing, Imperial College, London SW7 2BZ, UK,
{p.kelly,s.pelagatti}@doc.ic.ac.uk

Abstract. In this paper we study the use of idle cycles in a network of desktop workstations under unfavourable conditions: we aim to use idle cycles to improve the responsiveness of interactive applications through parallelism. Unlike much prior work in the area, our focus is on response time, not throughput, and short jobs - of the order of a few seconds. We therefore assume a high level of primary activity by the desktop workstations' users, and aim to keep interference with their work within reasonable limits. We present a fault-tolerant, low-administration service for identifying idle machines, which can usually assign a group of processors to a task in less than 200ms. Unusually, the system has no job queue: each job is started immediately with the resources which are predicted to be available. Using tracc-driven simulation we study allocation policy for a stream of parallel jobs. Results show that even under heavy load it is possible to accommodate multiple concurrent guest jobs and obtain good speedup with very small disruption of host applications.

Keywords: parallel computing, cycle stealing, performance prediction

1 Introduction

This paper concerns the feasibility of on-the-fly recruitment of idle workstations to enable parallel execution of short computationally-intensive phases of an interactive application, as commonly arise in a computer-aided design environment. In such applications, when the user is constructing the design, little processing power is required, however when the user selects 'Generate Photo-realistic Image', the computation required increases dramatically. Ideally, the user would not want to wait long for the image to be produced, possibly grabbing spare processing time from unused workstations. Our objective is to exploit the fact that (as we quantify below) even when a machine is actually being used interactively (the "host" job), there are often periods of inactivity lasting several seconds or more. We focus on the challenging goal of using these brief periods of idleness to execute short "guest" jobs in parallel in order to enhance response time. In addition to presenting a simple and effective software tool, we explore the potential for achieving this objective. We have chosen an extremely difficult environment - a heavily-used student laboratory of 32 Linux PCs; see Figure 1. We show that a typical (albeit rather

---

* On leave from Dipartimento di Informatica, Corso Italia, 40, 56125 Pisa, Italy.
** Now with Telcordia, Inc, New Jersey, USA

B. Monien and R. Feldmann (Eds.): Euro-Par 2002, LNCS 2400, pp. 863–872.

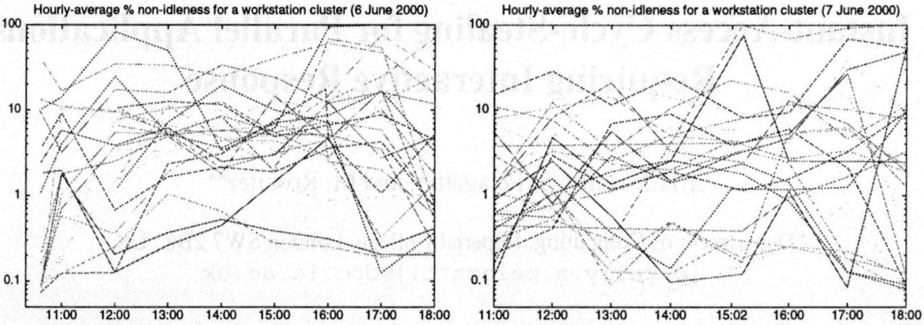

Fig. 1. This graph shows (on a log scale), the hourly-averaged percentage utilisation (see Section 3.1) of our 32 Linux PCs over two typical days. Although not always 100% busy, the machines are essentially in continuous use.

simple) parallel task can reliably achieve a speedup of 3 or more (reducing runtime to ca.14 seconds), while interfering with only 6-7% of host user seconds. Furthermore, we evaluate a simple allocation policy which handles intermittent arrival of such tasks.

Cycle Stealing on Networks of Desktop Workstations. The idea of making use of this wasted processing power is attractive and exploiting idle workstations has been a popular research area. Studies have shown that in typical networks of workstations (NOWs), most machines are idle most of the time [2, 1]. Batch systems like Condor [9] have been in use for years to utilize idle workstations for running independent sequential jobs. There have also been studies on the possibility of using idle workstations for parallel processing on coarse grain parallel jobs. Arpaci *et al.* [2] study the availability traces of a 60-workstation pool using a job arrival trace for a 32 node CM-5 partition. They find that the pool is able to sustain the 32-node parallel workload in addition to the sequential load imposed by interactive tasks. Similarly, Acharya *et al.* [1] show that for three non-dedicated pools of workstations it was possible to achieve a performance equal to that of a dedicated parallel machine between one third and three quarters the size of the pool. The results were achieved on relatively coarse grain *adaptive* parallel applications which could dynamically reconfigure to cope with changes in the pool of idle workstations available.

Instant-Access Cycle-Stealing for Interactive Response. In contrast to this earlier work, we focus on interactive applications with intermittent bursts of computation load. This requires optimizing the average response time for individual guest jobs and not the global system throughput. Second, our computation bursts are quite short (10-20 seconds if executed in parallel). This rules out the possibility of expensive process migrations during computation and makes crucial the ability to foresee idle times accurately. It is impractical to ship code and data to distant specialized nodes as happens in grid-oriented metacomputing environments [4, 7, 12]. Finally, since the guest jobs arise from interactive applications, we have to exploit idle workstations during busy day hours and we are not interested in patterns of idleness during nights or weekends.

To our knowledge, this is the first attempt to investigate idle workstation harvesting in this particular setting. There are two, linked challenges: (1) Can we achieve a useful speedup? Parallel programs (especially short-running ones) rely on all processors making progress. If just one of the participating machines is poorly-chosen, the entire parallel task will be delayed. (2) Is the interference with the host machines' other user(s) excessive?

Contributions. The main contributions of this paper are: (1) We present a low-overhead distributed recruitment service, which automatically identifies the available workstations on a local-area network. By autonomously electing a leader, the service requires minimal administration and handles failures gracefully. (2) We analyse traces of workstation utilisation, in order to quantify the idle time available on a network of workstations, its predictability, and the potential for using idle time for parallel processing. (3) Using a simulation driven by these traces, we evaluate the guest job performance achievable, and the amount of interference to host jobs. (4) We investigate scheduling policies to deal with a workload consisting of multiple users generating occasional computationally-intensive guest jobs.

The paper is organized as follows. Section 2 gives an overview of the architecture of the system proposed, Section 3 reports on the experimental results obtained, Section 4 discusses some related work and Section 5 concludes.

2 System Overview

The system is organized as a network of daemon processes, one for each workstation. Daemons monitor local load, provide job startup services and cooperate to predict future load and to schedule incoming guest jobs. As the guest jobs are fairly short (10-20 seconds), we restrict our attention to clients and servers on a single LAN running under the same administration/domain.

The mpidled Monitor Process. Allocation is orchestrated by a *leader* daemon which acts as a central server. The leader is elected using the distributed protocol by Garcia-Molina [8]. The protocol ensures automatic substitution of a leader in case of suspected failure. When a client wants to spawn a new guest job it makes a recruitment request to the leader which, after querying the daemon processes, returns a list of machines predicted to be idle for the near future. Then, the client can contact the daemon on each machine to inform it of the program to be executed. Each daemon process is responsible for monitoring the system status and computing a load prediction (Section—3.1). The leader, which may be any one of the daemons, is responsible for allocating resources (Section—3.2).

The mpidle Application and API. A client can initiate a request for resources using a command line utility (`mpidle`) which produces a list of idle workstations, as a parameter of an MPI job. Alternatively, a lower-overhead API is provided for direct invocation from within client applications.

3 Experimental Evaluation

Overview. Section 3.1 quantifies the amount of idle time likely to be found in a typical LAN environment during the day. Section 3.2 discusses and evaluates our load prediction strategy. Section 3.3 evaluates the time spent in finding a suitable workstation pool to execute new guest jobs (*recruitment overhead*). Section 3.4 presents the simulation results under various scheduling strategies.

3.1 Idle Workstation Recruitment

We consider a workstation *idle* if it is not executing user processes and has a significant amount of spare CPU time. More precisely we define a workstation as *idle* if, over a one second period, less than 10% of CPU time is spent executing user processes, and at least 90% of CPU time could be devoted to a new process.

Experimental Environment. To measure idleness patterns using our recruitment policy we carried out observations of load traces collected over two weeks on a pool of 32 very similar non-dedicated workstations (300MHz and 350MHz Pentium II, 128MB, Redhat Linux 6.1) located at Imperial College London. This is a uniform pool of publicly available machines used fairly intensively by undergraduate computer science students for course assignments, software development projects, web browsing and email. Traces were collected during the busy daytime hours, weekdays 9am to 6pm.

Pattern of Workstation Utilisation. Of all the one-second samples, 86% were idle. Idle periods occur very frequently. Figure 2.*left* shows the distribution of time between idle periods – 55% of intervals are 1s or less. Figure 2.*right* shows the distribution of length of idle periods over all workstations. 50% of idle periods last for at least 3.3s. One quarter of all idle periods last for longer than 10s: idle workstations often remain idle long enough to perform another useful task. (Note the small inflection in the plot at 60s, indicating that there are occasionally 'periodic' processes running on the workstations that cut-short idle periods that would have otherwise exceeded 60s.). To evaluate scope for parallel guest jobs, we studied the patterns of idleness across *groups* of workstations. Figure 3.*left* shows the probability of having a group of workstations of a given size at any given time. A group of 15 idle workstations is very nearly guaranteed to be available at any time, and a group of 22 is available with a rather high probability. The stability of such groups is shown in Figure 3.*right*. A group of 15 idle workstations is unlikely to remain idle for very long - there is only a 15% chance of them lasting for more than 5 seconds. Smaller groups are normally more robust (Figure 4.*left*).

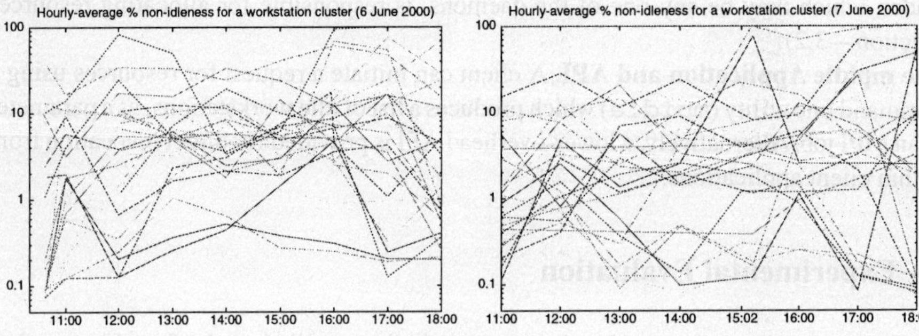

Fig. 2. Distribution of time between idle periods (left) and distribution of length of idle periods (right).

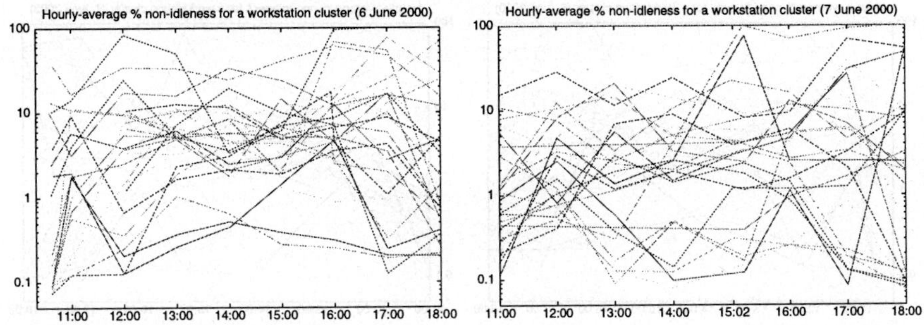

Fig. 3. Likelihood of having x idle workstations at the same time (left) and expected lifetime of a group of 15 idle workstations (right).

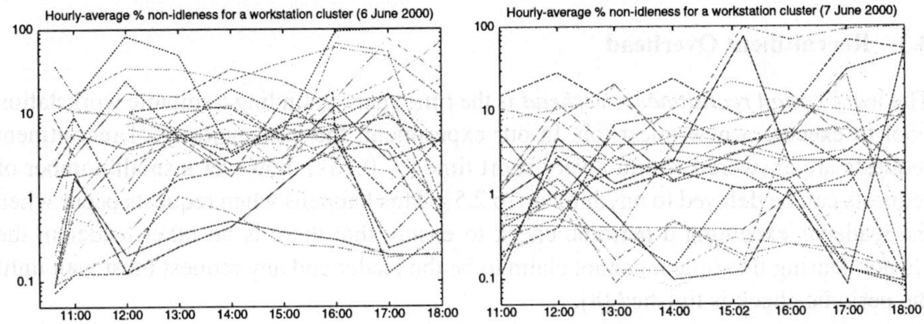

Fig. 4. Likely length of idle periods for different group sizes (left) and effect of window size on mean absolute error (right).

3.2 Predicting Short Term Workstation Load

Each daemon process monitors its CPU load once every second. When an availability request is received from the leader a load prediction is computed and returned. Load is predicted using a windowed mean of recent load measurements to predict the load over the next few seconds. Previous studies [5, 14] have shown that accurate short-term load prediction is possible and that good predictions can be made simply by taking the mean of recent load measurements. However, the load metric considered in [5, 14] (UNIX 'Load Average' - the average length of the run-queue) is different from the metric being considered here (CPU activity) and so we evaluated the accuracy of their prediction scheme with our metric. The windowed-mean prediction scheme was applied to our load traces, and the prediction errors were computed. We found that the error obtained using a window of 5 measurements is usually very small - 35% of predictions correctly forecast the average load over the following 10*s*. We also studied the relationship between the window size and the length of the period for which the prediction is needed. Figure 4.*right* shows the effects of window length on the mean absolute error for a particular desired prediction length (in this case 10*s*). Figure 5.*left* shows optimal window sizes for different forecast length periods.

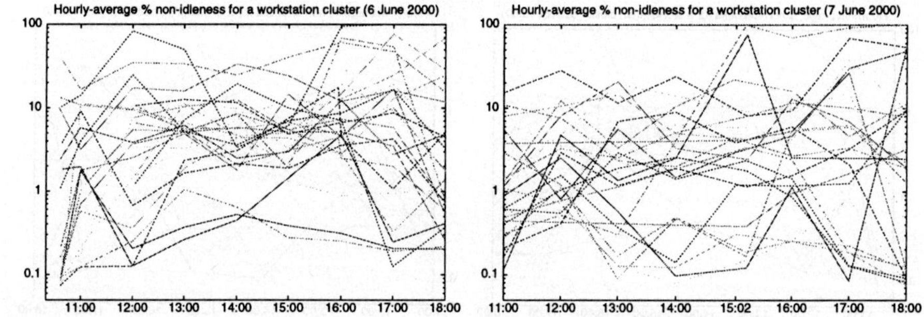

Fig. 5. Optimum window size for forecast length (left) and speedup on a dedicated workstation network (right).

3.3 Recruitment Overhead

The *workstation recruitment overhead* is the time spent in finding a suitable workstation pool to execute a new guest job. In our experiment, the vast majority of recruitment requests are answered within a very short time ($\leq 0.15s$), however a small number of requests can be delayed to anything up to $2.5s$. This happens when requests occur when the leader is executing a periodic check to ensure that there is no other leader in the cluster. During this time it cannot claim to be the leader and any request must wait until the periodic check is finished [8].

3.4 Evaluating Scheduling Strategies

Trace-Driven Simulation. To ensure reproducibility of results and allow for closer insight of the system behavior, we constructed a simulation using the load traces discussed in Section 3.1, varying various parameters. We tested the system with a sample rendering application which takes $42s$ on a single workstation. Figure 5 shows its speedup behavior when executed on a dedicated cluster of the workstations. The simulation uses the application's speedup curve to predict the expected completion time of each task on the resources available. It also accounts for the delay incurred (to all participating processors) when a guest process contends with a host process for CPU time. The contention which occurs is determined from the load traces, which record the number of running processes during each second so that a process's CPU time share can be computed.
The Simulated Usage Regime. To exercise the resource allocation mechanism, we simulate a fairly intensive situation in which clients request execution of rendering jobs at random intervals. The rendering jobs are all of the same size ($42s$ on one processor). Requests arrive with an exponential distribution, with a mean inter-arrival time of $20s$.
Scheduling Strategies. We experimented with three different scheduling strategies: **random, no reserve** and **x-reserve**. The results are shown in Table 1. For each scheduling strategy we measured the following:

- *Jobs Refused* the proportion of submitted guest jobs for which there were no available participants;

Table 1. System behavior results for different scheduling strategies.

Scheduling strategy	**rand**	**no res**	**10-res**	**20-res**	**30-res**	**40-res**	**50-res**	**60-res**
Jobs Refused	nil	16.3%	3.3%	1.5%	2.0%	0.5%	0.2%	0.1
Idle Seconds Used	25.0%	21.6%	23.3%	23.2%	22.6%	22.1%	21.5%	21.0%
Mean Group Size	17.0	17.2	15.24	13.6	12.0	10.3	8.7	7.1
Mean Speedup	3.68	3.58	4.58	4.88	4.96	4.82	4.46	3.92
Seconds Disrupted	44.4%	5.28%	6.3%	6.5%	5.9%	6.0%	5.9%	6.2%

- *Idle Seconds Used* the proportion of idle seconds in the day that were put to good use by the system;
- *Mean Group Size* the mean size of the group of workstations allocated to incoming guest jobs;
- *Mean Speedup* the mean speedup for guest jobs including those for which no workstations were available (i.e. those that were forced to execute sequentially);
- *Seconds Disrupted* the proportion of busy seconds that were disrupted by the execution of guest jobs, i.e. how often did a misprediction lead to disruption of ordinary workstation applications.

The Random Policy. We show the performance of a random allocation policy as a control experiment. A constant number of workstations is recruited for every job and this set is chosen at random among all the workstations regardless to their load. We used a constant group size of 17, which is near to the mean under the no-reserve policy.

The "No-Reserve" Policy. The no-reserve policy allocates all the idle workstations available to each recruitment request. Should a second request arrive shortly afterwards, no idle workstations will be left.

- This led to a slightly worse speedup than random allocation (3.58).
- However, a large proportion (84%) of recruitment requests were satisfied.
- 20% idle seconds were exploited, out of the average 25% of seconds belonging to periods of at least 10 seconds. This could be improved, especially since jobs were refused.
- The proportion of seconds that were disrupted by inappropriate allocation of jobs was low (5.3%), although not low enough for the system to be considered completely non-intrusive.

The x-Reserve Policies. The x-reserve strategies try to save $x\%$ of the resources available at any given time for (near) future requests in order to have a better distribution of the group sizes and to lower the percentages of guest jobs refused. With no-reserve, a large proportion of jobs were executed on small numbers of workstations or were forced to be executed serially because no workstation was available. Table 1 shows the results obtained with x-reserve strategies keeping a different proportion x of reserve at each allocation (the no-reserve strategy is the same as the 0-reserve strategy). The results are as follows:

- By choosing the right reserve percentage we can achieve an average speedup of up to 4.96.

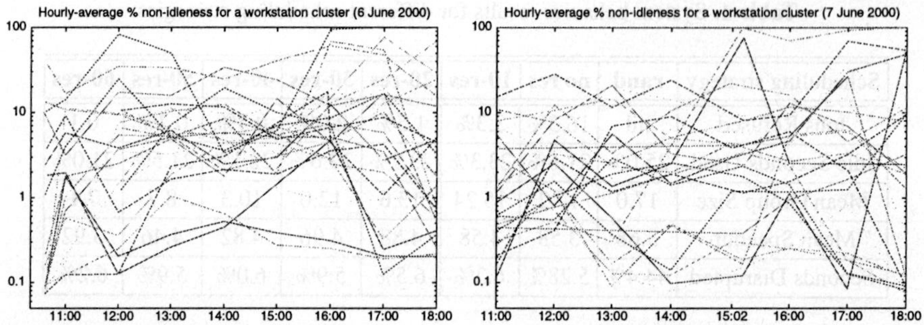

Fig. 6. The effect of x-reserve scheduling policies on distribution of group sizes.

- Furthermore, this increase in speedup is achieved without significantly increasing the proportion of the seconds disrupted.
- The speedup falls when too many workstations are kept in reserve as the average group size drops.
- Keeping reserves reduces the percentage of jobs refused, reducing the variance of the speedup experienced by different guest jobs.

The effect of the reserve percentage on the distribution of group sizes is illustrated in Figure 6. As expected, for small reserves, groups are either very large or very small, while for larger reserves the group sizes are close to the mean.

4 Related Work

With Condor [9], the aim is to speed up independent sequential guest jobs using idle workstations in a LAN. Usually, the jobs require a large amount of computation (hours more than seconds) and a network of monitor daemons is used to collect information on the current load of machines on the net. Disruption of host jobs is minimised by migrating the guest job as soon as the host jobs need a workstation. Linger-Longer [11] works in the same scenario but allows a guest job to remain on a host machine when it ceases to be idle. To avoid disruption it employs a set of Linux kernel extensions which use a new guest process priority to prevent guest processes from stealing time from host processes and a new page replacement policy which limits the slow down caused by guest pages in the virtual page system. With this new scheme, the authors claim a much effective usage of workstations, allowing gains up to 60% in the total compute time with respect to Condor. Although the techniques used in these systems can be used in our setting, the focus of our work is on *interactive* parallel guest jobs posing a quite different set of challenges. As mentioned in the Introduction, the use of idle workstations to execute a batch queue of parallel jobs has been studied by Acharya *et al.* and Arpaci *et al.* [1, 2]. With the longer-running jobs they study, processes can be migrated from machine to machine during execution. Furthermore, their objective was to minimize the execution time of a whole batch, which can mean very long execution time for single jobs in order to achieve better global resource arrangement.

Some of the problems addressed in our research, such as workstation load prediction and load sensitive guest job scheduling have been addressed recently in the broader framework of WAN scale metacomputing systems [7, 4, 12]. This setting is much more complex than ours and requires network load prediction to be addressed. Moreover, the higher overhead due to non local job scheduling is more suitable for coarser grain guest jobs than the ones addressed in this study.

Finally, scheduling parallel computations on batch parallel systems has attracted considerable attention [3, 6, 13, 10]. The usual metric to be optimized here is global batch throughput. However, Subholk *at al.* [13] proposes strategies to minimize response time for individual applications. They take both communication load and computation load into account and select a pool of workstation and communication links to be used. Our research addresses LAN environments in which only computational load is relevant for node selection. The strategy proposed by Subholk *et al.* for our specific problem corresponds to our no-reserve policy. As we discussed in Section 3 this strategy penalizes future jobs and leads to smaller average speedup figure with respect to x-reserve. Although more experiments are needed, we believe that, in our setting, a strategy aiming to optimize the average speedup experienced by competing guest jobs leads to better resource usage and more reliable behavior than optimizing the response time of a guest job in isolation.

5 Conclusions and Directions for Further Research

We have provided evidence that interactive performance of applications with intermittent computational demands can be substantially enhanced through opportunistic parallel execution on other instantaneously-idle workstations on the same LAN. Some interference with host tasks is incurred, but the effect is small. When guest job requests arrive frequently, much better performance is achieved by holding back some of the available resource on each allocation.

While there is enormous scope for further work, this paper has demonstrated "mpidled" to be a simple yet surprisingly effective tool. The software is in regular use at Imperial College and a public release is planned.

Further research is needed: (1) How would our results change with different levels of host load? We have taken a fairly extreme situation of essentially continuous utilisation - many realistic environments would give better results. Our simple policy of holding back some resources for future requests appears fairly stable, but we would like to characterise how the policy should be adjusted as task arrival rate and host load are varied. Some kind of adaptive scheme looks attractive. (2) Our definition of "idle" is somewhat arbitrary (Section 3.1). We need to evaluate how lowering the idleness threshold would reduce interference, and reduce speedup. In our environment, external users (and Windows users) often connect to our Linux systems remotely, so some level of interference to desktop responsiveness is already tolerated. Other organisations have a different culture. (3) We used a rather simple parallel application to exercise the system. Although our rendering application has less-than ideal speedup, it is relatively loosely-synchronised. We have been using mpidled to run a tightly-synchronised CFD solver and have positive practical experience but have not yet been able to quantify the resulting performance. (4)

Realistic applications often (like the CFD solver) have large input and output files. This is easily addressed by using the local filesystem on the machines allocated by mpidled - but the next interactive use of the application (which uses the results from the previous run) is likely to be allocated a differing set of machines. We plan to explore strategies for achieving parallel file access while retaining the necessary scheduling flexibility.

subsection*Acknowledgements

This work was partially supported by an EPSRC Visiting Fellowship (GR/N63154).

References

1. A. Acharya, G. Edjlali, and J. Saltz. The utility of exploiting idle workstations for parallel computing. In *Proc. 1997 ACM SIGMETRICS Intl. Conf. on Measurement & Modeling of Computer Systems*, pages 225–236.
2. R. H. Arpaci, A. C. Dusseau, A. M. Vahdat, L. T. Liu, T. A. Anderson, and D. A. Patterson. The interaction of parallel and sequential workloads on a network of workstations. In *Proc. 1995 ACM SIGMETRICS Intl. Conf. on Measurement & Modeling of Computer Systems*, pages 267–278, Ottawa, May 1995.
3. M. J. Atallah, C. L. Black, D. C. Marinescu, H. J. Siegel, and T. L. Casavant. Models and algorithms for co-scheduling compute-intensive tasks on networks of workstations. *J. Parallel & Distr. Computing*, 16:319–327, 1992.
4. H. Casanova and J. Dongarra. Netsolve: A network server for solving computational science problems. *Intl. J. Supercomputer Appl. & HPC*, 11(3):212–223, 1997.
5. P. Dinda and D. O'Hallaron. An evaluation of linear models for host load prediction. In *Proc. 8th IEEE HPDC-8*, Redondo Beach, CA, Aug. 1999.
6. A. C. Dusseau, R. H. Arpaci, and D. E. Culler. Effective distributed scheduling of parallel workloads. In *Proc. 1996 ACM SIGMETRICS Intl. Conf. on Measurement & Modeling of Computer Systems*, Philadelphia, PA, 1996.
7. I. Foster and C. Kesselman. Globus: A metacomputing infrastructure toolkit. *Intl. J. Supercomputer Applications*, 11(2):115–128, 1997.
8. H. Garcia-Molina. Elections in a distributed computing system. *IEEE Trans. Comp.*, C-31(1):47–59, Jan. 1982.
9. M. Litzkow, M. Livny, and M. Mutka. Condor - a hunter of idle workstations. In *Proc. 8th Intl. Conf. of Distributed Computing Systems*, pages 104–111, San Jose, California, June 1988. IEEE-CS Press.
10. F. Petrini and W. Feng. Buffered coscheduling: a new methodology for multitasking parallel jobs on distributed systems. In *Proc. IPDPS 2000*, Cancun, MX, May 2000.
11. K. D. Ryu and J. K. Hollingsworth. Linger-Longer : Fine-grain cycle stealing for networks of workstations. In *Proc. Supercomputing'98*, Orlando, Nov. 1998.
12. S. Sekiguchi, M. Sato, H. Nakada, S. Matsuoka, and U. Nagashima. Ninf: Network based information library for globally high performance computing. In *Proc. Parallel Object-Oriented Methods & Applications (POOMA)*, Santa Fe, New Mexico, Feb. 1996.
13. J. Subholk, P. Lieu, and B. Lowekamp. Automatic node selection for high performance applications on the networks. In *Proc. 7th ACM SIGPLAN PPoPP'99*.
14. R. Wolski, N. Spring, and J. Hayes. Predicting the CPU availability of time-shared Unix systems. In *Proc. 8th IEEE HPDC 8*, Redondo Beach, CA, Aug. 1999.

Access Time Estimation for Tertiary Storage Systems

Darin Nikolow[1], Renata Słota[1], Mariusz Dziewierz[1], and Jacek Kitowski[1,2]

[1] Institute of Computer Science, AGH, al. Mickiewicza 30, 30-059, Cracow, Poland
[2] Academic Computer Centre CYFRONET AGH, Cracow, Poland,
{darin, rena, kito}@uci.agh.edu.pl
phone: (+48 12) 6173964, fax: (+48 12) 6338054

Abstract. We propose two approaches for estimating the Tertiary Storage System access time: open approach and gray-box approach. In the first case the source code of the storage system is available, so changes in the code are made by adding event reporting functions. In the second approach - the essential system information is accessible via its native tools only. In this paper we describe an implementation of the open approach for access time estimation. The second approach we have shortly described as our future work.

1 Introduction

As the requirements for storage capacity grow exponentially each year many applications of different types (e.g. archiving, backup, scientific, DBMS, and multimedia) make use of Tertiary Storage Systems (TSS). The access time of a file stored on the TSS can vary a lot: from few seconds to hours. This depends mainly on the system load at the time of issuing a request, but other parameters like location of data on the storage medium, transfer rates and seek times of the drives are also important. In some cases a priori knowledge of the access time is essential, e.g., in the case of a Grid data replication system [1, 2]. This will allow more efficient usage of storage resources and will decrease the overall latency times. In addition, the user's satisfaction increases when the service time of his or her request is predicted (e.g. user waiting to watch a selected video sequence requested from a near-on-demand video server; administrator recovering from backup).

Estimating the access time of a request for a given TSS state by using a single analytical function is not applicable here because of the algorithmic nature of the request processing. The Queuing Systems Theory could be used to compute analytically the average access time of requests for a certain TSS state. The goal of this study is to develop a method for accurate estimating the latency time of a given request to the TSS. Since the analytically computed mean values are not sufficient in many cases, the event simulation approach to estimate the latency for a real system is adopted.

B. Monien and R. Feldmann (Eds.): Euro-Par 2002, LNCS 2400, pp. 873–880.

We propose two approaches for estimating the TSS access time:

- **Open TSS Approach**, in which the source code of the TSS is available, so event reporting functions can be introduced, and
- **Gray-Box TSS Approach**, in which the essential system information is accessible via its native tools only.

In this paper we describe an implementation of the open TSS approach for access time estimation of own developed FiFra (File Fragmentation) TSS. The second approach is shortly presented as our future work.

The rest of the paper is organized as follows. The next section presents some related works. The third section describes briefly FiFra TSS. The fourth one represents design and implementation details as well as experimental results for the open TSS approach. Future work is presented in the following section and the last one concludes the paper.

2 Related Work

Rodney Van Meter in [3] proposes Storage Latency Estimation Descriptors (SLEDs) as a method of supplying to the client a predictive information about the I/O performance of the underlying storage systems. By using SLEDs the application can be more efficient by rescheduling its I/O calls in such a way that the less expensive (for instance cached) I/O calls are invoked first. Rodney Van Meter and Minxi Gao in [4] implement SLEDs for the Linux operating system and show significant performance improvement of the applications modified to take advantages of SLEDs. In their implementation the I/O performance estimation is based on latency and bandwidth measurements done during the boot process for each storage device attached to the system.

Shen et. al in [5,6] present a multi-storage architecture and a performance prediction method to increase I/O efficiency of scientific applications. They also developed a run time library on top of Storage Resource Broker (SRB) [7] for optimizing tertiary storage access. Their prediction algorithm is based on time measurements of basic SRB file operations (open, seek, read, close, etc.) and assumes that the file is in the disk cache. Therefore, they do not concentrate on prediction of the staging time for data located on tertiary storage.

3 Description of the FiFra TSS

3.1 Background

Since files stored on TSS get bigger the problem of efficient access to fragments of them arises. This is essential for systems which require that the latency is kept below a certain limit, like continuous media stream servers for instance. The problem of efficient access to fragments of files stored on TSS is one of the subjects covered in the task which we will carry out within the CrossGrid project [2].

A TSS capable to access video sequences called Video Tertiary Storage System (VTSS) [8] was developed during our previous work. Next, based on VTSS, a more general version of the system (allowing access to plain file fragments) was developed. We called it FiFra (File Fragmentation) TSS.

The access to the data stored in the FiFra TSS is sequential. Examples of applications with sequential data access are: restoring data from backup copy, CD-image archiving, anonymous ftp server, serving software depots, applications using multimedia data (video, audio, image), logging systems, etc. Some of these applications will probably need to access fragments of the files, e.g., playing interesting parts of a video file, extracting log records for a certain period of time or continue an interrupted file transfer.

3.2 Architecture of the FiFra TSS

The architecture of FiFra TSS is shown in Fig. 1. The system consists of two main daemons: the Repository Daemon (REPD) and the Tertiary File Manager Daemon (TFMD). Since the FiFra TSS is based on the VTSS described in [8] only brief descriptions of REPD and TFMD are given.

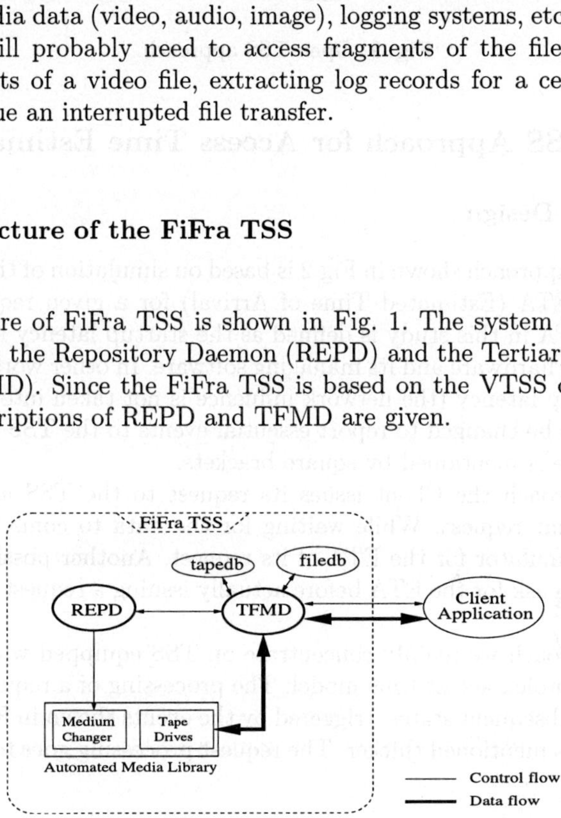

Fig. 1. Architecture of the FiFra TSS.

REPD keeps repository information in its internal data structures. When a mount request is received REPD issues appropriate SCSI commands to the robot arm of the library. TFMD manages information about files and media and transfers the files from the removable media devices to the client. In the case of tapes the files should be stored with the hardware tape drive compression turned off if we want to enable a direct file fragment retrieval from tape.

Since client requests (like write a new file or read a file fragment) can compete for access to the storage resources they are served according to the FIFO strategy. The most often used request is expected to be the read fragment request.

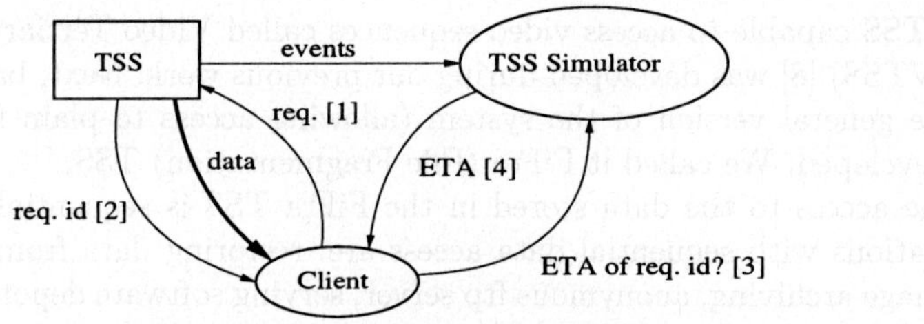

Fig. 2. Open TSS approach.

4 Open TSS Approach for Access Time Estimation

4.1 General Design

The open TSS approach shown in Fig.2 is based on simulation of the TSS in order to obtain the ETA (Estimated Time of Arrival) for a given request processed by the TSS. ETA in this study is defined as the startup latency imposed by the tertiary storage hardware and its managing software. In other words it represents the local startup latency (the network influence is not taken into account). The real TSS has to be changed to report essential events to the TSS simulator. The calling sequence is mentioned by square brackets.

In this approach the Client issues its request to the TSS and receives an identifier for that request. While waiting for the data to come the Client can ask the TSS Simulator for the ETA of its request. Another possibility (just for prediction) is to ask for the ETA before actually issuing a request (not presented in the figure).

In this approach we mainly concentrate on TSS equipped with DLT drives, due to their complex access time model. The processing of a request by the TSS goes through subsequent states, triggered by the events shown in Fig.3. The most probable path is mentioned thicker. The request processing goes to *Waiting* state

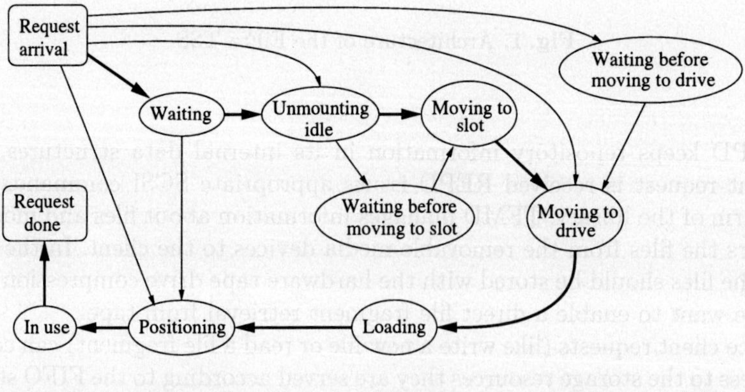

Fig. 3. State transition diagram of a request processed by the TSS.

if there are no resources (drive or tape) available to proceed further. The state *Unmounting idle* means preparing an idle tape for ejecting. The state *Moving to slot* indicates that the idle tape is being moved to an empty slot. If the robot arm is busy serving another request when *Unmounting idle* is finished then the *Waiting before move to slot* state will be visited. The next state *Moving to drive* points that the needed tape for the current request is being moved to an empty drive. *Loading* represents loading the tape into the drive. The state *Positioning* indicates that the tape is being positioned and the state *In use* means that the tape is being read or written. When the transfer is finished the tape becomes idle and the request is done.

Shortcuts of the described path are possible for certain cases: for instance if the needed tape is already mounted then the request processing starts from the *Positioning* state.

The simulation algorithm is based on the state transition diagram shown before. Based on previous measurements the time of completion for each state (except the *Waiting* state) is estimated. It can be fixed or dependent on such parameters like position of the tape, size of fragment, transfer rate. The completion time of the *Waiting* state is estimated by simulating processing of the previous requests, which have possibly occupied resources for which the current request is waiting. The overall estimation is done by summing up the completion times for the states passed through during the processing.

4.2 Implementation

For the FiFra TSS implementation of the open approach an additional daemon, called SIMUD (Simulator Daemon) is introduced. It communicates with REPD, TFMD and the client application. At its initializing phase SIMUD requests information from REPD about the current state of the TSS and starts waiting for events. REPD has been modified by adding a new command `sendstat`, which is used by SIMUD to retrieve the initial TSS state information. Event reporting functions have been added to the source code of the REPD and TFMD daemons. The reporting is done via dispatching an appropriate command to SIMUD. This command can be `NEWREQ` reporting a new request, `STATCH` reporting state change of request, `DELREQ` reporting that a request has been finished. The client can ask SIMUD to simulate ETA of a given request by using `SIMETA` command.

4.3 Preliminary Results

Comparisons between the latency times for the real and simulated systems are presented below. Measurements were done for a sequence of 100 requests, generated every 60 or 100 seconds, according to the zipf distribution. The simulated TSS was configured with two DLT tape drives.

The results presented in Fig.4 show that the TSS is overloaded, because the latency time of the subsequent requests tends to be higher. The characteristics for the real and the estimated startup latency are similar - the fluctuations occur for the same request number. In Fig.5 where the interval between requests

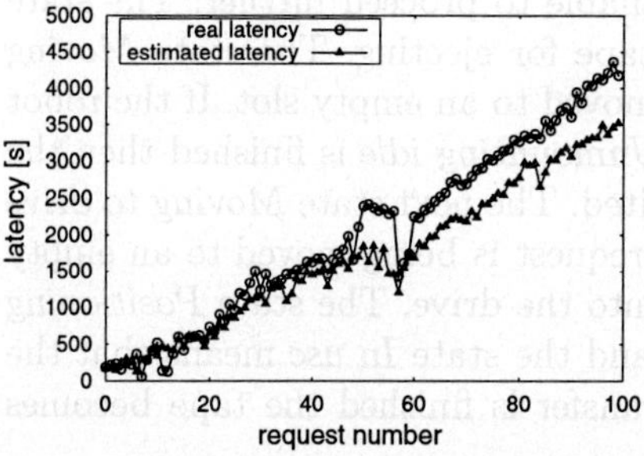

Fig. 4. Comparison between the real and estimated startup latency (with 60 seconds interval between requests).

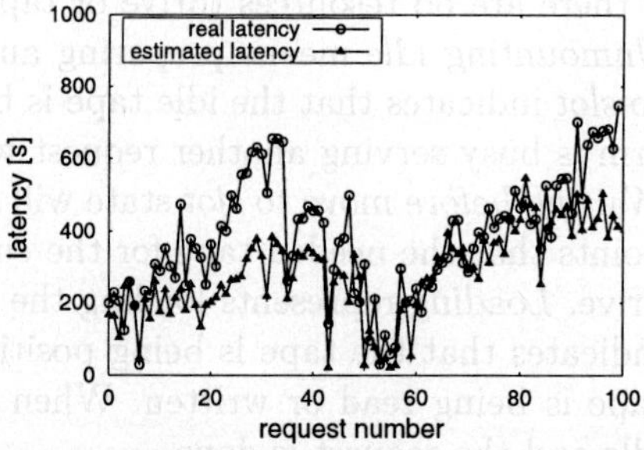

Fig. 5. Comparison between the real and estimated startup latency (with 100 seconds interval between requests).

is higher the system can handle the coming requests better (with no increasing tendency). In this case the characteristics are also similar but the relative error is higher.

In Fig.6 and Fig.7 the histograms of the relative error of the estimated latency are presented. For the overloaded system (Fig.6) about 80% of the estimations are obtained with the error below 20%. For the latter case (see Fig.7) 80% of the estimations have error below 45%. The relative error in this case is higher due to smaller latency absolute values, because for only few requests in the queue the positioning time error can vary a lot. This feature follows from the ideal DLT tape positioning model implemented in this study. The difference is higher especially when the block to position to is far from the beginning of the tape. When there are more requests this error is smaller because of averaging of positioning times. More accurate results could be obtained by using the low cost access time model for serpentine tape drives proposed in [9] which we plan to use.

5 Future Works

The future work concentrates on the estimation of the access time for tertiary storage systems in which no changes of the source code are allowed. Our first target is the UniTree HSM system with the gray-box approach implemented. Estimation of the access time will be based on knowledge about UniTree HSM system operations and on various data gathered from the available utilities delivered by the vendor.

In Fig.8 the mentioned approach is presented. The core of the system is the TSS Simulator similar to that presented in Section 4. Again, the square brackets represent the calling sequence. TSS Monitor will collect the necessary data about the UniTree state. Request Monitor & Proxy catches client requests in order to

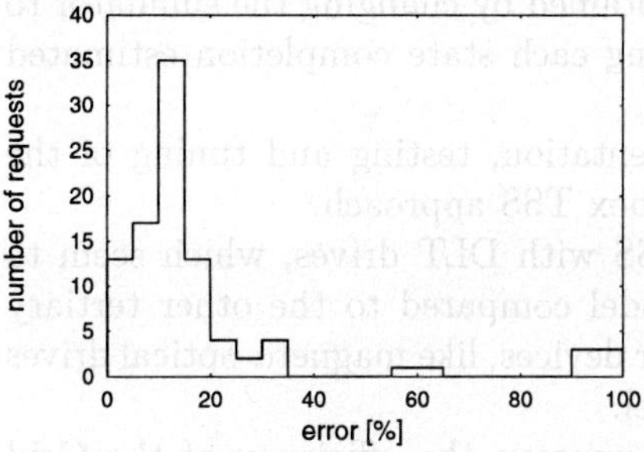

Fig. 6. Histogram of relative error of the estimated latency (with 60 seconds interval between requests).

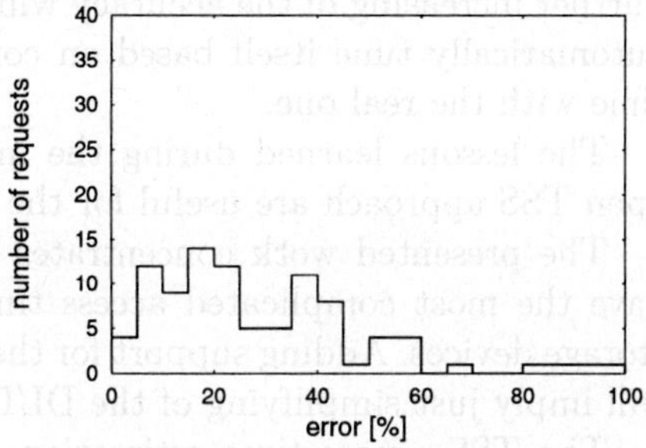

Fig. 7. Histogram of relative error of the estimated latency (with 100 seconds interval between requests).

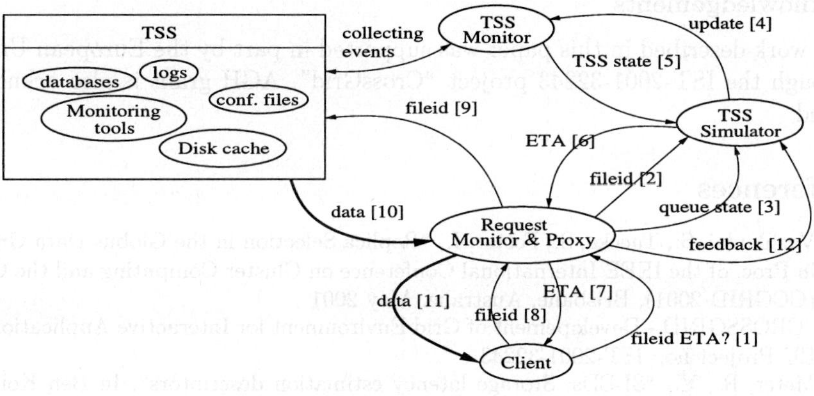

Fig. 8. Gray-box TSS approach.

get more detailed information (queue order, file identifiers, time statistics) about the requests being processed by the UniTree.

We also plan to check (and change eventually) the DLT access time model for the cases when the hardware compression is on (which is the usual case).

6 Conclusions

In this paper we have focused on the problem of estimating the access time of data stored on the tertiary storage. We implemented the access time estimation method for a TSS (developed at our site) using the open system approach. The system has been tested and preliminary results obtained. The results show that the estimation errors are lower when the TSS is overloaded. A further improving of the system is necessary since our goal is to estimate the latency time more accurately even if there is only one or few requests in the queue. This will be done by implementing better access time model for the DLT tape drives.

Further increasing of the accuracy will be obtained by changing the simulator to automatically tune itself based on comparing each state completion estimated time with the real one.

The lessons learned during the implementation, testing and tuning of the open TSS approach are useful for the gray-box TSS approach.

The presented work concentrates on TSS with DLT drives, which seem to have the most complicated access time model compared to the other tertiary storage devices. Adding support for the other devices, like magneto-optical drives will imply just simplifying of the DLT model.

The TSS access time estimation could improve the efficiency of the Grid replica selection and migration services. The TSS Simulator could be used to supply information about the TSS state to the Grid monitoring services for optimizing data access for the Grid applications.

Acknowledgements

The work described in this paper was supported in part by the European Union through the IST-2001-32243 project "CrossGrid". AGH grant is also acknowledged.

References

1. Vazhkudai, S., Tuecke, S., Foster, I., "Replica Selection in the Globus Data Grid", in Proc. of the IEEE International Conference on Cluster Computing and the Grid (CCGRID 2001), Brisbane, Australia, May 2001.
2. "CROSSGRID - Developement of Grid Environment for Interactive Applications", EU Project no.: IST-2001-32243.
3. Meter, R., V., "SLEDs: Storage latency estimation descriptors", In Ben Kobler, editor, in Proc. 6th NASA Goddard Conference on Mass Storage Syst. and Tech. in Coop. with 15th IEEE Symp. on Mass Storage Syst., pp. 249-260, March 1998.
4. Meter, R., V., Gao, M., "Latency Management in Storage Systems", in Proc. of the 4th Symp. on Operating Syst. Design and Implementation (OSDI'00), October 2000.
5. Shen, X., Choudhary, A., "A Distributed Multi Storage Resource Architecture and I/O Performance Prediction for Scientific Computing" in Proc. 9th IEEE Symp. on High Performance Distributed Computing, pp.21-30, IEEE Computer Society Press, 2000.
6. Shen, X., Liao, W., Choudhary, A., "Remote I/O Optimization and Evaluation for Tertiary Storage Systems through Storage Resource Broker", in IASTED Applied Informatics, Innsbruck, Austria, February, 2001.
7. Baru, C., Moore, R., Rajasekar, A., Wan, M., "The SDSC Storage Resource Broker", in Proc. CASCON'98 Conference, Toronto, Canada, Dec. 1998.
8. Nikolow, D., Słota, R., Kitowski, J., Nyczyk, P., Otfinowski, J., "Tertiary Storage System for Index-based Retrieving of Video Sequences", Lecture Notes in Computer Science, **2110**, pp.435-444, Springer, 2001.
9. Sandstå, O., Midstraum, R., "Low-Cost Access Time Model for Serpentine Tape Drives", in Proc. of 16th IEEE Symposium on Mass Storage Systems the 7th NASA Goddard Conference on Mass Storage Systems and Technologies, San Diego, California, USA, March 1999, pp. 116-127.

BioGRID – Uniform Platform for Biomolecular Applications

Jarosław Pytliński[1], Łukasz Skorwider[1], Piotr Bała[1], Mirosław Nazaruk[2], and Konrad Wawruch[2]

[1] Faculty of Mathematics and Computer Science, N. Copernicus University, Chopina 12/18, 87-100 Toruń, Poland, {pyciu,luksoft,bala}@mat.uni.torun.pl

[2] ICM Warsaw University, Pawińskiego 5a, 02-106 Warsaw, Poland, {mirnaz,kwaw}@icm.edu.pl

Abstract. In this paper we describe developed grid tools for biomolecular applications. We have used UNICORE infrastructure as framework for development dedicated user interface to the Gaussian98 and Amber 6.0. The user interface is integrated with the UNICORE client based on plugin mechanism which provides general grid functionality such as single login, job submission and control mechanism.

1 Introduction

Research in the area of molecular biology and quantum chemistry requires computer resources usually not available at the user workstation. Recent advances in computer technology, especially grid tools makes them good candidate for development of user interfaces to computing programs and resources [1]. Computational grids enable sharing a wide variety of geographically distributed resources and allow selection and aggregation of distributed resources across multiple organizations for solving large scale computational and data intensive problems in science. Molecular biology and quantum chemistry traditionally provides large number of important, nontrivial and computationally intensive problems.

User point of view is most important principle in the development of the UNICORE [2] software. UNICORE is uniform interface to the computer resources which allows user to prepare, submit and control application specific jobs and file transfers.

The user can find WebMo [3] which is web based submition systems for quantum chemistry codes such as Gaussian, Gamess and Mopac but this system is limited to the local batch systems and has no grid capabilities. The web submission to the geographically distributed systems is possible within BioCore [5] which is web interface to the molecular dynamics code NAMD. Currently this system is limited to the particular MD code and single visualization package (VMD). Similar functionality for Gamess provides NPACI Portal [4].

B. Monien and R. Feldmann (Eds.): Euro-Par 2002, LNCS 2400, pp. 881–884.

2 The UNICORE Architecture

The UNICORE architecture is based, as other grid middleware, on the three tier model. It consists of user, server and target system tier. The user tier consists of the graphical user interface – UNICORE client – written as Java application. It offers the functions to prepare and control jobs and to set up and maintain the user's security environment.

The UNICORE client generates from the user's input an Abstract Job Object (AJO) which is sent to the other components of the UNICORE infrastructure. The AJO compromises the UNICORE protocol between user interface and Network Job Supervisor (NJS) together with the abstract job specification generated from user input. The Gateway is the first part of the UNICORE Server tier. It takes care of user authentication, secure communication between client and server. Gateway provides client with information on resources available at site. It also talks to the Network Job Supervisor at the site to send jobs and data, status requests and control commands for further processing and to receive data to make it available to the user. The security is based on the Secure Socket Layer (SSL) protocol and the X509 certificates.

The graphical user interface offers functions to maintain security, to prepare UNICORE jobs, submit and monitor them. Basic elements are: (i) script task, to submit job scripts; (ii) transfer task to specify data transfer between different jobs groups; (iii) job groups, to build subjobs for other systems. All elements can be add and edited with the graphical interface in the UNICORE client. User can define target system for each job group as well as dependences between job groups and file transfers. Each job group must have defined resources it requires such as number of processors or nodes, memory, CPU time and disk space. Target system chosen by the user is checked if it can fulfill this requirements. This tasks is performed during process of preparing job, before it is submitted.

Within UNICORE infrastructure the user program can be run as script submitted to the target system in the way analogous to the batch execution mode. The advantage is single login and transparent file transfer but submitted job is similar to the one used with the traditional queuing system. The main advantage of the UNICORE is easy development of sofisticated user interfaces to the application programs. Such interface - plugin - can be easily integrated with the UNICORE client taking advantage of all functionality already present.

2.1 Gaussian Plugin for UNICORE Client

The UNICORE script job still requires knowledge of the input files for the biomolecular applications. In most cases it is performed with standard text editor and requires significant experience from the user.

We have used UNICORE as framework for development dedicated user interface to the Gaussian98 [6]. Plugin is written in Java and is loaded to the UNICORE client during start, or on request. Once it is available in the client, in addition to the standard futures such as preparation of the script job, user gets access to the menu which allows for preparation application specific job. User can set up type of *ab initio* calculations and add options specific to the particular type of simulations. Plugin allows only for options which do not conflict with the chosen simulation type.

In the separate window user adds atoms coordinates required for the calculations. Currently the most popular Cartesian coordinates can be used together with Z-matrix format, specific to the *ab initio* calculations. User can input coordinates in dedicated window or load from the file. The most popular formats are supported. Read coordinates can be easily modify by the user.

Once all parameters are set user can generate valid Gaussian98 input with pressing proper button in the plugin. The text input in the application specific format is presented in the client window. Because input file is generated automatically, the user gets advantage of the correct syntax and proper combination of the options. Any further modifications to the input can be simple performed by the change of the chosen parameter at any time during input preparation.

An advanced user can edit generated input and introduce changes he wants based on his own experience with the Gaussian98. At any time he can return to the automatically generated input by pressing generate input button. The job file can be saved and used in the future. Gaussian plugin is also able to import valid Gaussian98 input prepared by the user, either by hand using any text editor either by an external application.

Once input is ready user can specify target system and resources required for jobs using standard UNICORE client facilities. For example user can check job status, monitor execution and retrieve output to the local workstation. All these functions can be performed from the UNICORE client with the single login during client startup. User can monitor job status and retrieve output to any computer connected to the network with the UNICORE client installed, in particular other than one used for job submition.

2.2 Amber Plugin

The experience with the development of plugin for quantum chemical application was used for development of the user interface to the Amber 6.0, one of the most popular molecular dynamics code. developed user interface allows for preparing input for MD simulations in various modes: constant energy, constant temperature or constant pressure. In each mode user is able to specify most important parameters of the simulations such as time step, total simulation time, initial time and others.

Because MD job requires input files with initial coordinates, the topology, parameters and possibly others. User has to specify names of proper files using dedicated window. This files are included in job build by plugin and will be automatically transferred to the target system. In the same way user defines names of the files which will be created and should be transferred back to the user workstation. Target system together with required resources are specified by the user using UNICORE client which is also used for job submission and control.

3 Conclusions

We have used UNICORE as the main grid middleware for the development of the BioGRID - computational grid for molecular biology and quantum chemistry. Compare to other grid tools and applications this solution is easy to install, both at the server and client side and provides user with simple and intuitive interface. The UNICORE client provides general mechanisms for user authentication and job preparation, submition and control. It is also used as framework for development application specific interfaces. In particular plugins for Gaussian98 and Amber 6.0 have been developed. Build interface allows user to prepare and run typical biomolecular applications in the grid environment. All components were tested on the computational grid based on the EUROGRID infrastructure.

Acknowledgements

This work is supported by European Commission under IST grant 20247. The software was used using EUROGRID facilities at ICM Warsaw University (Poland), Forschungcentrum Jülich (Germany), University of Manchester – CSAR (UK), IDRIS (France) and University of Bergen – Parallab (Norway).

References

1. C. Kesselman I. Foster, editor. *The Grid: Blueprint for a Future Computing Infrastructure.* Morgan Kaufman Publishers, USA, 1999.
2. Unicore. Pallas. http://www.unicore.org.
3. WebMo. http://www.webmo.net.
4. Gamess NPACI Portal. http://gridport.npaci.edu/GAMESS.
5. M. Bhandarkar, G. Budescu, W. F. Humphrey, J. A. Izaguirre, S. Izrailev, L. V. Kalé, D. Kosztin, F. Molnar, J. C. Phillips, and K. Schulten. Biocore: A collaboratory for structural biology. In A. G. Bruzzone, A. Uchrmacher, and E. H. Page, editors, *Proceedings of the SCS International Conference on Web-Based Modeling and Simulation.* San Francisco, California, 1999.
6. M. J. Frisch, G. W. Trucks, H. B. Schlegel, G. E. Scuseria, M. A. Robb, J. R. Cheeseman, V. G. Zakrzewski, J. A. Montgomery, Jr., R. E. Stratmann, J. C. Burant, S. Dapprich, J. M. Millam, A. D. Daniels, K. N. Kudin, M. C. Strain, O. Farkas, J. Tomasi, V. Barone, M. Cossi, R. Cammi, B. Mennucci, C. Pomelli, C. Adamo, S. Clifford, J. Ochterski, G. A. Petersson, P. Y. Ayala, Q. Cui, K. Morokuma, P. Salvador, J. J. Dannenberg, D. K. Malick, A. D. Rabuck, K. Raghavachari, J. B. Foresman, J. Cioslowski, J. V. Ortiz, A. G. Baboul, B. B. Stefanov, G. Liu, A. Liashenko, P. Piskorz, I. Komaromi, R. Gomperts, R. L. Martin, D. J. Fox, T. Keith, M. A. Al-Laham, C. Y. Peng, A. Nanayakkara, M. Challacombe, P. M. W. Gill, B. Johnson, W. Chen, M. W. Wong, J. L. Andres, C. Gonzalez, M. Head-Gordon, E. S. Replogle, and J. A. Pople. Gaussian 98. 2001.
7. P. Kollman. *AMBER (Assisted Model Building with Energy Refinement).* University of California, San Francisco, USA, 2001.

Implementing a Scientific Visualisation Capability within a Grid Enabled Component Framework

J. Stanton, S. Newhouse, and J. Darlington

London e-Science Centre, 180 Queen's Gate, London SW7 2BZ, UK,
lesc@ic.ac.uk, internet: http://www.lesc.ic.ac.uk*

Abstract. Collaborative scientific visualisation and computational steering are key enabling technologies within many e-science applications. Grid infrastructures provide an ideal environment for providing such a capability, and one approach is to use encapsulation within a Grid enabled component framework. This research note discusses the initial development of a collaborative visualisation component within the ICENI Grid middleware infrastructure.

1 Introduction

The delivery of collaborative scientific visualisation and computational steering of large data sets has been a goal for many years but has in practice been difficult to achieve.

As computational power has increased and high speed networks and sophisticated visualisation tools have become commonplace, so has risen the demand for infrastructures to support real-time collaborative visualisation and computational steering. To satisfy this demand the emerging Computational Grids, federations of distributed computational, storage and networking resources, must provide mechanisms to support collaborative visualisation environments that can be readily applied to scientific applications.

This research note describes the integration of such a capability within the Imperial College e-Science Networked Infrastructure (ICENI), an integrated Grid middleware for component based applications [1].

2 Deployment of Component Based Grid Applications

Effective Grid infrastructures provide a range of underlying services necessary for the efficient deployment and execution of applications. These are reviewed in detail in [2], but we can summarise a mimimum subset for the transparent use of Grid resources as follows:

* Research supported by the OST Core e-Science Programme on equipment provided by the HEFCE/JREI grant GR/R04034/01 and EPSRC equipment grant 2001

B. Monien and R. Feldmann (Eds.): Euro-Par 2002, LNCS 2400, pp. 885–888.

– registration services to provide the pool of high performance computational resources within the virtual Grid communities;
– security services to define access to these resources;
– analytic services to provide a basis for performance based deployment of applications onto these resources;
– monitoring services to support process re-deployment according to changes in network performance and the application's computational requirements.

Within the ICENI component based Grid infrastructure, the components which are composed to define applications are additionally annotated with performance data which is then used to determine the optimum deployment on available resources. Components which have large or diverse computational demands (such as for numeric computation and for visualisation) may therefore be instantiated by the Grid middleware on separate resources, or co-located if transport costs would negate any benefits that this would provide.

3 Implementing Collaborative Visualisation and Steering within ICENI

3.1 Design Architecture

Previous work had developed a prototype system for collaborative visualisation and steering within a distributed environment [3]. This has been extended by encapsulating its functionality within a co-ordinating server component that can be deployed by the ICENI grid middleware.

Users wishing to interface with an application do so by connecting to its corresponding server instance using a local visualisation client program. As each client connects the the server, a rendering process is deployed onto Grid resources to process the visualisation pipeline and transport the resulting image to the user's workstation using Chromium [4].

A data pull model is used, with each rendering process requesting new data as required to refresh its image. The server collects these requests and passes them to the data generating application. One copy of each data set requested is then returned to the server at the end of the following computation cycle, for distribution by the server to the rendering processes of the requesting clients.

Conceptually therefore the server component provides the inferface between the data generating application, the client processes and the rendering processes. In practice the transport mechanisms used for data exchange will depend on the deployment of the processes and may range from direct access for co-located components to GridFTP for high volume transfer between remote components.

The communication model for an example session involving two clients and deployment of all processes on remote resources is shown in Figure 1.

3.2 Demonstrator Application

A scientific application simulating coronal mass ejections [5] was selected to demonstrate the functional capabilities of the initial implementation. Originally

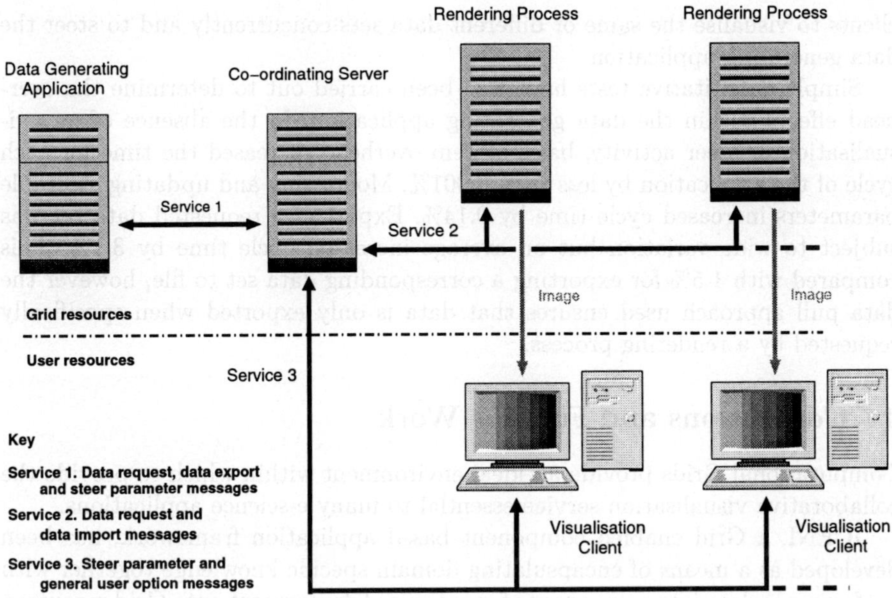

Fig. 1. Process and Communication Model for Collaborative Visualisation within the ICENI Framework

structured as a sequential Fortran program capable of resolutions necessary for 2-D batch mode analysis on a high performance UNIX workstation, it has recently become practical to execute the code on high-end commodity PC's. Results are output to file for subsequent visualisation with a typical run of 600 cycles capable of generating a data file of 160 MB.

A simple restructuring of this code was carried out to facilitate the insertion of the library routines necessary to set up the control and data paths to the co-ordinating server. A rendering process capable of generating a sequence of contoured colour maps from a corresponding sequence of arbitrary two or three dimensional scalar arrays was defined using the VTK library [6].

Three scalar arrays were identified for export, consisting of magnetic field data which would generate images representing the motion of magnetic flux tubes within a magnetised plasma. A single steerable parameter was identified defining the computational time step, which would provide a means to balance the accuracy of the simulation with its rate of development (and stability).

3.3 Operational Results

Proof of concept trials have been carried out using resources within the London e-Science Centre [7]. The data generating application was executed on a 24 processor Sun E6800 machine, while the co-ordinating server and rendering processes executed on a 20 processor Linux cluster. The visualisation clients were established on desktop PC's. These tests demonstrated the capability of multiple clients to visualise the same or different data sets concurrently and to steer the data generating application.

Simple quantitative tests have also been carried out to determine the overhead effects within the data generating application. In the absence of any visualisation or steer activity, basic system overhead increased the time for each cycle of the application by less than 0.001%. Monitoring and updating steerable parameters increased cycle time by 0.14%. Export of a requested data set was subject to wide variation but on average increased cycle time by 3.1%. This compared with 1.5% for exporting a corresponding data set to file, however the data pull approach used ensures that data is only exported when specifically requested by a rendering process.

4 Conclusions and Further Work

Computational Grids provide an ideal environment within which to provide the collaborative visualisation service essential to many e-science applications.

ICENI, a Grid enabled component based application framework, has been developed as a means of encapsulating domain specific knowledge together with performance data into components for optimised deployment onto Grid resources. The functional requirements of a collaborative visualisation and steering capability can similarly be componentised for integration within a user's application.

An initial implementation of such a component has been developed and has demonstrated the basic viability of this approach with a simple sequential application. This implementation has used a simple token based mechanism to control read access to data and read/write access to the steerable parameters. Key areas for further work are therefore to integrate the access control mechanisms with the policy based mechanisms of the ICENI framework, and to provide support for interaction models appropriate to more complex applications.

References

1. Furmento, N., Mayer, A., McGough, S., Newhouse, S., Darlington, J.: Optimisation of Component-based Applications within a Grid Environment. SuperComputing 2001, `http://www.icpc.doc.ic.ac.uk/components/papers/index.html`
2. Foster, I., Kesselmann, C. (Editors): The Grid: Blueprint for a New Computing Infrastructure. Morgan Kauffmann, 1998
3. Stanton, J.: Enabling Computational Steering through Visualization in a Distributed Environment. Master's Thesis, Imperial College, Department of Computing, `http://www.doc.ic.ac.uk/~jks100/msc`
4. Sourceforge: Project: The Chromium Project, `http://sourceforge.net/projects/chromium/`
5. Cargill, P., Chen, J., Spicer, D., Zalasak, S.: Magnetohydrodynamic Simulations of the Motion of Magnetic Flux Tubes through a Magnetised Plasma, Journal of Geophysical research, Vol 101, March 1996
6. Kitware Inc.: The Visualisation Toolkit, `http://public.kitware.com/VTK/`
7. London e-Science Centre: Resources, `http://www.lesc.ic.ac.uk/resources/index.html`

Transparent Fault Tolerance for Web Services Based Architectures

Vijay Dialani, Simon Miles, Luc Moreau, David De Roure, and Michael Luck

Department of Electronics and Computer Science, University of Southampton, Southampton SO17 1BJ, UK,
{vkd00r,sm,L.Moreau,dder,mml}@ecs.soton.ac.uk

Abstract. Service-based architectures enable the development of new classes of Grid and distributed applications. One of the main capabilities provided by such systems is the dynamic and flexible integration of services, according to which services are allowed to be a part of more than one distributed system and simultaneously serve different applications. This increased flexibility in system composition makes it difficult to address classical distributed system issues such as fault-tolerance. While it is relatively easy to make an individual service fault-tolerant, improving fault-tolerance of services collaborating in multiple application scenarios is a challenging task. In this paper, we look at the issue of developing fault-tolerant service-based distributed systems, and propose an infrastructure to implement fault tolerance capabilities transparent to services.

1 Introduction

The *Grid problem* is defined as flexible, secure, coordinated resource sharing, among dynamic collections of individuals, institutions and resources [10]. Grid Computing and eBusiness share a large number of requirements, such as interoperability, platform independence, dynamic discovery, etc. In the eBusiness community, Web Services have emerged as a set of open standards, defined by the World Wide Web consortium, and ubiquitously supported by IT suppliers and users. They rely on the syntactic framework XML, the transport layer SOAP [3], the XML-based language WSDL [2] to describe services, and the service directory UDDI [1].

The benefit of open standards has recently been acknowledged by the Grid Community, as illustrated by three projects embracing Web Services in various ways. Geodise (`www.geodise.org`) is a Grid project for engineering optimisation, which makes Grid services such as Condor available as Web Services [7]. myGrid (`www.mygrid.org.uk`) is a Grid middleware project in a biological setting, which addresses the integration of Web Services with agent technologies [16]. More recently, the Open Grid Service Architecture (OGSA) [9] extends Web Services with support for the dynamic creation of transient Grid Services.

Grid computing is characterised by applications that may be long-lived and involve a very large number of computing resources. Hence, applications need to be designed with fault tolerance in order to be robust. As a result, the Grid

B. Monien and R. Feldmann (Eds.): Euro-Par 2002, LNCS 2400, pp. 889–898.

community, and more generally, the distributed computing community have devised multiple algorithms for fault tolerance. However, the Web community has not focused on this aspect, and therefore, there is no standard way to develop fault-tolerant Web Services.

It is this specific problem that we address in this paper. Our approach may be summarised as follows: implementors of a Web Service have to implement an interface (e.g. checkpoint and rollback); the architecture dynamically extends the service interface (published as a WSDL document) with methods for fault tolerance; applications making use of different Web Services have to declare their inter-dependencies, which are used by a fault-manager to control fault recovery; an extension of the SOAP communication layer is able to log and replay messages.

The specific contributions of this paper are: *(i)* The design of an architecture for fault-tolerance of Web Services which supports multiple algorithms for fault-tolerance. *(ii)* The specification of the interfaces between the different architecture components. *(iii)* An overview of our implementation. The paper is organised as follows. In Section 2, we summarise some of the techniques for fault tolerance which we support in our architecture, while in Section 3, we present the Web Services stack. We describe our architecture in Section 4, and its implementation in Section 5 and we conclude the paper in Section 6.

2 Fault Tolerance Background

Before expanding on our design, we present a brief introduction to fault tolerance for distributed systems. This is intended as an aid to understanding terms used later in the paper, and not as an extensive survey.

Fault tolerance is the ability of an application to continue valid operation after the application, or part of it, fails in some way. Such failure may be due to, for example, a processor crashing. In order for an application suffering a failure to continue, the state of processes, and the data they use, must be returned to a *previous consistent state*. For example, object data may return (*rollback*) to previous values if the current values are lost, and processes may return to a state in which a message is re-sent, if the previous attempt apparently failed.

In order to return to a previous consistent state, an application must record a replica of its previous state. The entire state of a process can be copied using a *checkpointing* mechanism, or only the incremental changes to the process state using a *logging* mechanism. Both methods can be used to rollback to a previous valid state [8].

Fault tolerance becomes considerably more difficult in distributed applications, made up of several processes that communicate by passing messages between themselves. One process may fail without the other processes being aware of the failure. This can lead to the state of the application as a whole (the global state) being inconsistent. An application is in a *globally consistent state* if whenever the receipt operation of a message has been recorded in the state of some process, then the send operation of that message must have been recorded also

[15]. It is the aim of a fault tolerance mechanism for distributed applications to keep an application to a consistent global state, or return to the last known consistent state (also known as *maximal state*) in case of failure.

Fault tolerance mechanisms should have transparency, low overhead, portability and scalability [19]. Transparency implies that there exists a mechanism such that applications implemented using it can largely ignore processes failing or recovering, as this will all be dealt with by the mechanism. It is important that transparency exists so that both the developers' burden is eased and the fault tolerance mechanism can be replaced by another without the rest of the application requiring modification. The requirement for low overhead, portability and scalability can lead to a choice in the fault tolerance mechanisms to apply. Fault tolerance can also be achieved by using fault tolerance Object Replication techniques, e.g. [13].

Checkpointing can also be performed in a variety of ways in distributed applications. *Consistent* or *synchronous* checkpointing involves all processes being forced to *globally synchronise* before the state of all the processes is recorded [12]. Global synchronisation means that all processes are in a state in which they have processed all received messages and are blocked from sending any messages [12]. As blocking processes may reduce the speed of the application, consistent checkpointing may not always be the preferred means of fault tolerance. On failure, all processes using consistent checkpointing rollback to the last global checkpoint. A *quasi-synchronous* approach is suggested by Manivannan and Singh [15], where, rather than requiring global synchronisation, processes force each other to checkpoint at almost the same time through sending messages.

An alternative to consistent checkpointing is *independent* or *asynchronous* checkpointing. In this case, each process records its own state without attempting to coordinate with other processes, so potentially avoiding the overhead of global synchronisation. However, communicating processes may depend on each other, i.e. require that they are in certain states. The rollback of one process, on failure, may require that other processes also rollback to previous checkpoints. It is then possible that these rollbacks will require the original process to rollback even futher to attempt to reach a consistent global state. This repetition of rollbacks can lead to a *domino effect* where each process must rollback many times to reach a consistent global state, losing a lot of processing that has occurred without failure [20].

If the state of one process becomes invalid by the rollback of another, we consider there to be a *dependency* of the former process on the latter. When a rollback (or checkpoint) should take place on multiple processes, a fault tolerance mechanism should ensure that it does not create extra dependencies. In order to achieve this, the mechanism can employ a *two-phase commit*, in which processes are first put into a blocking state to prevent messages being sent and new dependencies forming, and then later rolled back (or requested to checkpoint) at an appropriate moment.

Independent checkpointing mechanisms deal with process dependency in two ways. *Pessimistic* independent checkpointing [19], requires that each process logs the changes since the last checkpoint after sending or receiving any message. As dependencies between processes are only due to messages passed between them, this ensures that rollback of one process to the previous checkpoint will not affect dependent processes. *Optimistic* independent checkpointing requires that dependencies are explicitly recorded somewhere in the system, so that on rollback of a process, dependent processes will be informed appropriately and possibly also rolled back [6, 11, 20]. The pessimistic approach places more restrictions on a process' autonomy in checkpointing and may require more checkpointing than optimistic approaches. Optimistic mechanisms will have more overhead in rollback, on the other hand. However, it should be noted that no single mechanism is universally applicable. The suitability of algorithms differ for each *application type*, namely that there exists a different set of algorithms for batch processing, shared memory and MPI based applications. In this paper we restrict ourselves to discussions on fault tolerance requirement of Web Services architecture.

3 Web Services

The World Wide Web is more and more used for application to application communication. The programmatic interfaces made available are referred to as Web Services.[http://www.w3c.org/2002/ws/]. To ensure interoperability between different architectures, the Web Services architecture describes standards for definition, discovery, binding and communication between services. A service provides a set of application functionality through a bound and advertised interface. This architecture provides an abstraction over the implementation of services.

Service discovery mechanisms such as *Universal Description, Discovery and Integration* (UDDI), aid in discovering services, statically or dynamically bound. To facilitate binding, services describe their behaviour by using a description language, such as WSDL[17]. However, there exists no explicit information about its lifetime and instance creation, management policy differs across implementations.

The Web Services Stack: A number of Web Services implementations exist, each with a proprietary Web Services stack. Such stacks vary in the way that they gel or interact with legacy systems and proprietary technologies. A generalized conceptual Web Service stack is represented in figure 1.

The network layer, messaging layer and the service description layer have been standardized to ensure interoperability. SOAP is supported as the de-facto XML-messaging protocol for most of the Web Service implementations. Detailed discussions on SOAP protocol and WSDL are described in [4], [2] respectively. The "vertical layers" describe attributes of the framework and must be addressed at each level. At present, security, management and QoS are the widely accepted system attributes.

Error Handling in Web Services: Different layers in the conceptual stack employ different types of error handling. At the description layer, WSDL pro-

Fig. 1. A generalized Conceptual Web Services Stack

vides a mechanism by the way of `<wsdl:fault>` for applications to specify the error characteristics. This is similar to the way we define exceptions raised by the methods in a Java interface. Similarly the underlying SOAP messaging layer, provides a `<soap:fault>` for applications to communicate the error information. The error mechanisms of SOAP and WSDL help support errors raised by an application, but no mechanism exists for handling framework failures and system errors.

Service Lifetime Management: Web Services differ from usual message-based distributed systems. SOAP omits features often found in messaging systems and distributed object systems [4], such as: *(i)* distributed garbage collection; *(ii)* boxcarring or batching of messages; *(iii)* objects-by-reference (which requires distributed garbage collection); *(iv)* activation (which requires objects-by-reference). As Web Services do not support explicit activation or deactivation of services, it becomes difficult to have any lifetime management control. Most common implementations [5] use time-based expiry mechanism for controlling the underlying resources.

Fault Tolerance for Web Services: In service-based infrastructures, a single process may be part of multiple applications. Therefore, rollback cannot be initiated using the standard fault tolerance mechanisms mentioned earlier. We propose fault-tolerance as one of the vertical layers of the Web Services stack. In our earlier discussion, we described the need of fault-tolerance for service-based infrastructure. In the rest of the paper, we discuss the special requirements of each layer.

4 Architecture

In figure 2, we present an overview of the fault tolerant system, as applied to the Web Services architecture. The top of the figure shows various components of the fault tolerance infrastructure that are specific to an application instance and are location independent. The lower half represents modifications to the existing

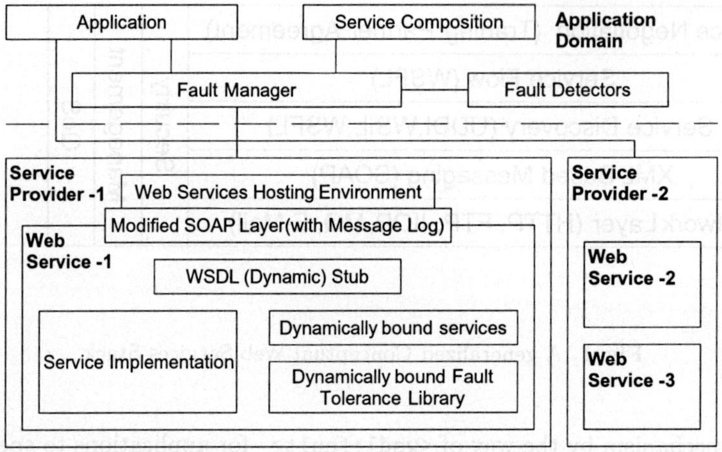

Fig. 2. Architecture for Fault Tolerant Web Services

hosting environment for services. The modifications in the latter case can be classified into a set of changes to the messaging layer (refer to next section for a detailed description) and a set of interfaces supported by individual services. Henceforth, we refer to the upper half as the *application layer* and to the lower half as the *service layer*. In general, the overall framework provides the capability to:

1. Detect a fault or failure,
2. Estimate the damage caused and decide on the strategy for recovery,
3. Repair a fault, and
4. Restore the application state.

The framework differs from traditional frameworks, such as CORBA [18] as it employs a two-pronged strategy to recover from a fault, namely the *local recovery mechanism* and the *global recovery mechanism*. The context of a local recovery is restricted to recovery of an individual service instance, while global recovery applies to the entire application. The local recovery mechanism tries to revive the service instance with minimal or no intervention by the global recovery mechanism. A local recovery mechanism escalates the failure notification to the global recovery mechanism in case of its failure to recover the fault locally. The architecture imitates an hour glass model to restrict the dependency between the two layers to a minimal set of interfaces, for co-ordination between the two layers.

The application layer assumes that an application instance aggregates a set of service instances to provide the overall capability for the application. The concept of service aggregation, also known as service composition, is central to the definition of a Virtual Organisation (VO)[10].

However, our definition of service composition is not restricted to VOs and can be extended to service composition expressed by the way of workflow specification, e.g. WSFL [14] or X-LANG [21] . The composition of a service may be created statically at design time or can be created dynamically by using negotiation techniques, enactment description of WSFL, or any other technique. A detailed discussion on negotiations and composition of services is outside the scope of this paper. The application layer assumes that there exists a description of service composition that it can refer to for obtaining a list of collaborating services. The application layer can be initialized by the application instance or by enactment of a composition. The instantiation data can be held within the composition definition or it can be provided explicitly during creation. The application layer implements a set of key components, namely:

1. *Application*: It uses the services in a composition to provide the overall capability. An application can directly interact with the global fault manager(refer to the definition below) or allow the application framework to interact on its behalf.
2. *Global Fault Manager*: A coordinator that interacts with the applications or framework and the underlying services to implement a fault tolerant system. A fault manager is responsible for monitoring, fault diagnosis and checkpoint and rollback co-ordination; it may be central or distributed.
3. *Service*: An entity that is bound by its interface definition, usually a WSDL description, and executes as an independent process or within the process of Web Services Hosting Environment [5].
4. *Fault detector*: A fault detector detects a change in the perceived ideal environment and uses software interrupts to the fault manager to notify of any failure. In addition to providing the context for the fault, it may also provide behavioral override, allowing applications to extend the fault notification mechanism.

A global fault manager interacts with a set of services specified in the service composition. Each of the underlying services needs to support a set of interfaces to enable communication between the local and global fault managers. A local fault manager coordinates independent checkpointing and rollback of an individual service; it monitors the service and supports the fault detector interface for creating fault notifications. The local fault manager interacts with the messaging layer to initiate a blocking or non-blocking recovery, with or without the replay of messages. The global fault manager relies on a set of fault detectors to send a fault notifications. An application can register a custom list of detectors in addition to those supported by the individual services.

Our modified SOAP layer provides the message logging, message replay and a capability to acknowledge either the receipt or the processing of a message. It provides interfaces for interaction with the global fault manager and local fault managers. Modifications to the layer allow the application framework to maintain a log of messages and also to selectively suspend the communication between the services. They also enable the framework to isolate a service instance from the rest of the system during a local recovery. In addition, the ability to suspend communication helps rollback by providing the ability to isolate the affected set of services. Modifications to the SOAP messaging layer enable us to support both fault tolerance by message-based checkpointing and rollback, and fault tolerance by object replication. In the following section, we describe interactions between the various components in the framework for implementing message based checkpointing and rollback. Later in our discussion, we describe how the framework could support fault tolerance by object replication.

5 Implementation

IBM WSTK-3.0, Apache SOAP, IBM Web-Sphere Application Server, IBM Web Hosting Environment were used to implement our proposed framework. Our implementation provides a modified SOAP layer, different libraries to initialise the application framework, and a set of plug-ins for various application types. The Application framework allows an application to specify a service composition; we support both design-time and run-time compositions of services. In its current implementation, the framework assumes service compositions to be static and immutable; however, the framework can be modified to allow dynamic compositions, to complement UDDI and WSIL support for dynamic discovery and binding of services. The Application layer can be implemented to be a part of the application instances execution space or be a Web Service by itself. In either case, the application framework creates and initializes a global fault manager. The global fault manager accepts fault tolerance mechanism specific parameters and the application type as immutable parameters; it uses service composition to discover and establish contact with the local fault managers; it performs a two-phase commit checkpoint operation to coordinate the checkpointing activity across the service instances.

The local fault manager is implemented as a set of libraries that can be bound dynamically to the service code. The local fault manager interacts with the modified SOAP layer to control the flow of messages during the recovery as the mechanism used by it may or may not support non-blocking checkpoint and/or rollback. In case of a failure, the local fault manager categorizes the fault, and then tries to recover the fault. In certain cases, it may be possible to recover the service locally and rollback to the current state by replaying messages. In case a full recovery is not possible the local fault manager tries recovering to a maximal state and escalates the fault notification to the global fault manager.

On notification, the global fault manager initiates a roll back by notifying the affected services. The dependency set for recovery can be provided by the application. Additionally, the fault detectors can provide a dependency set for the current fault: the provision is specifically useful in case of compositions that use different protocols or support different end-points. For example, service composition may consist of a set of Intranet and Internet services; services within an Intranet may use IIOP for inter-service communication and connect to the Internet using a SOAP layer. The rollback is also implemented as a two-phase commit operation. The framework ensures loose coupling, by supporting different fault mechanisms for local and global fault managers. In addition, to checkpointing and rollback mechanism for fault tolerance, Object Replication can also be used to improve fault tolerance of applications. One of the possible ways is to enable the Web Services hosting environment to create a set of redundant services and define a mechanism for active or passive replication of services, client redirection. However, a detailed discussion on replication based fault-tolerance is beyond the scope of present discussion.

6 Conclusion and Future Work

We have successfully conceptualised and implemented a fault tolerant architecture for Web Services, without affecting interoperability of existing services. The framework demonstrates a method of effectively decoupling the local and global fault recovery mechanisms. It provides a capability for monitoring the individual service instances as well the service hosts. The algorithm independence and support for different application types allow us to provide fault tolerant capabilities to Web Services that internally employ different programming models. Dynamic varying composition of services is an issue that needs to be addressed. However, much depends upon the composition schemes that will evolve from research in Web Services.

Acknowledgement

This research is funded in part by EPSRC myGrid project (reference GR/R67743/01) and EPSRC combichem project (reference GR/R67729/01)

References

[1] Uddi standards. http://www.uddi.org.
[2] W3c wsdl spec. http://www.w3c.org/TR/wsdl.
[3] Xml protocol working group. http://www.w3c.org/2000/xp/Group/.
[4] W3c soap standards, 2001.
[5] Web services hosting technology. http://www.alphaworks.ibm.com/tech/wsht, December 2001.
[6] B. Bhargava and S. Lian. Independent checkpointing and concurrent rollback for recovery in distributed systems–an optimistic approach. In *Proceedings of the 7th IEEE Symposium on Reliable Distributed Systems*, pages 3–12, 1988.
[7] S. J. Cox, M. J. Fairman, G. Xue, J. L. Wason, and A. J. Keane. The Grid: Computational and Data Resource Sharing in Engineering Optimisation and Design Search. In *IEEE Proceedings of the 2001 ICPP Workshops*, pages 207–212, Valencia, Spain, September 2001.
[8] E. N. Elnohazy, D. B. Johnson, and Y.M. Wang. A survery of rollback-recovery protocols in message-passing systems.
[9] Ian Foster, Carl Kesselman, Jeffrey M. Nick, and Steven Tuecke. The Physiology of the Grid — An Open Grid Services Architecture for Distributed Systems Integration. Technical report, Argonne National Laboratory, 2002.
[10] Ian Foster, Carl Kesselman, and Steve Tuecke. The anatomy of the grid. enabling scalable virtual organizations. *International Jounral of Supercomputer Applications*, 2001.

[11] David B. Johnson and Willy Zwaenepoel. Recovery in distributed systems using optimistic message logging and checkpointing. In *Proc. 7th Annual ACM Symp. on Principles of Distributed Computing*, pages 171–181, Toronto (Canada), 1988.

[12] M. Frans Kaashoek, Raymond Michiels, Henri E. Bal, and Andrew S. Tanenbaum. Transparent fault-tolerance in parallel orca programs. In *Proceedings of the Symposium on Experiences with Distributed and Multiprocessor Systems III*, pages 297–312, 1992.

[13] Sean Landis and Silvano Maffeis. Building reliable distributed systems with CORBA. *Theory and Practice of Object Systems*, 3(1):31–43, 1997.

[14] Prof. Dr. Frank Leymann. Web services flow language. http://www-4.ibm.com/software/solutions/webservices/pdf/WSFL.pdf, May 2001. Member IBM Academy of Technology, IBM Software Group.

[15] Manivannan and Singhal. Comprehensive low-overhead process recovery based on quasi-synchronous checkpointing.

[16] Luc Moreau. Agents for the Grid: A Comparison for Web Services (Part 1: the transport layer). In *IEEE International Symposium on Cluster Computing and the Grid*, Berlin, Germany, May 2002.

[17] Judith M. Myerson. Web services architectures. http://www.webservicesarchitect.com/content/articles/myerson01.asp, January 2002.

[18] OMG, http://www.omg.org/docs/formal/01-12-63.pdf. *Fault Tolerant CORBA*, December 2001. Version 2.6.

[19] D. J. Scales and M. S. Lam. Transparent fault tolerance for parallel applications on networks of workstations. In *Proceedings of the USENIX 1996 Annual Technical Conference*, pages 329–341, San Diego, CA, USA, 1996.

[20] Robert E. Strom and Shaula Yemini. Optimistic recovery in distributed systems. *ACM Transactions on Computer Systems*, 3(3):204–226, 1985.

[21] Satish Thatte. 'xlang'- web services for business process design. http://www.gotdotnet.com/team/xml_wsspecs/xlang-c/default.htm, 2001.

Algorithm Design and Performance Prediction in a Java-Based Grid System with Skeletons

Martin Alt, Holger Bischof, and Sergei Gorlatch

Technical University of Berlin, Germany

Abstract. We address the challenging problem of algorithm design for the Grid by providing the application user with a set of high-level, parameterized components called *skeletons*. We describe a Java-based Grid programming system in which algorithms are composed of skeletons. The advantage of our approach is that skeletons are reusable for different applications and that skeletons' implementations can be tuned to particular machines of the Grid with quite well-predictable performance.

1 Introduction

One of the main challenges in application programming for the Grid is the phase of algorithm design and, in particular, performance prediction early on in the design process: it is difficult to choose the right algorithmic structure and perform architecture-specific optimizations of an application, because the type of machine the program will actually be executed on is not known in advance.

We propose providing Grid application programmers with a set of high-level algorithmic patterns, called *skeletons*. Skeletons are used as program components, customizable for particular applications. Computational servers of the Grid provide possibly different, architecture-dependent implementations of the skeletons, which can be tuned for execution on particular Grid servers.

The advantage of our approach is that applications can be conveniently expressed using reusable algorithmic skeletons, for which reliable performance estimates on a particular Grid server are available. This facilitates systematic rather than *ad hoc* design decisions on both the algorithmic structure of an application and the assignment of application parts to servers.

In this paper, we describe an experimental Java-based programming system with skeletons for a Grid environment, with focus on the critical problem of performance prediction in the course of algorithm design. The particular contributions and the structure of the paper are as follows:

– An architecture of the Grid programming system is proposed, in which the user chooses a suitable server for each skeleton invocation in his application program (Section 2).
– A Java+RMI experimental implementation is presented, with Java bytecodes used as application-specific parameters of skeletons (Section 3).
– A simple performance model for remote execution of skeletons on the Grid servers is proposed and tested using system measurements. (Section 4).

We conclude the paper by discussing our findings in the context of related work.

B. Monien and R. Feldmann (Eds.): Euro-Par 2002, LNCS 2400, pp. 899–906.

2 System Architecture and Skeletons

The idea of Grid programming with skeletons is to separate two phases of programming – algorithm design and implementation. The user composes his program using predefined algorithmic patterns (skeletons), which appear as function calls with application-specific parameters. The actual organization of parallelism is left to the skeleton implementation, which is provided on the server side and is geared to a particular architecture of a Grid server, e. g. distributed- or shared-memory, multithreaded, etc.

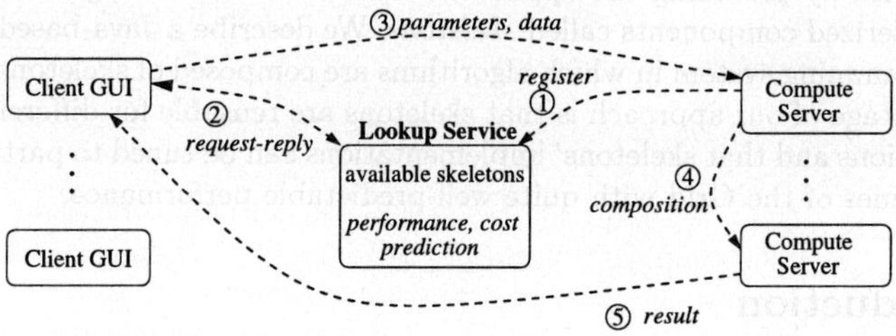

Fig. 1. System architecture and interaction of its parts

We propose the following system architecture, consisting of three kinds of components: user machines (*clients*), target machines (*servers*) and the central entity, called *lookup service* (see Fig. 1).

Each server provides a set of skeletons that can be invoked from the clients. Invoking skeletons remotely involves the following steps, also shown in Figure 1:

① Registration: Each server registers the skeletons it provides with the lookup service to make them accessible to clients. Together with each skeleton, a performance estimation function is registered, as explained below.

② Service request-reply: A client queries the lookup service for a skeleton it needs for an application and is given a list of servers implementing the skeleton. The skeletons that will actually be used are selected (using heuristics or tool-driven by the user).

③ Skeleton invocation: During the program execution, skeletons are invoked remotely with application-specific parameters.

④ Composition: If the application consists of a composition of skeletons, they may all be executed either on the same server or, alternatively, in a pipelined manner across several servers.

⑤ Skeleton completion: When the compute server has completed the invoked skeleton, the result is sent back to the client.

Skeleton Examples. In this paper, we confine our attention to so-called *data-parallel* skeletons whose parallelism stems from partitioning the data among processors and performing computations simultaneously on different data chunks.

For the sake of simplicity, our running example in the paper is an elementary data-parallel skeleton called *reduction*: function $reduce(\oplus)$ computes the "sum" of all elements in a data structure using the associative customizing operator $\oplus$. For example, for a list of three elements $[a, b, c]$, the result is $a \oplus b \oplus c$.

Formally, *reduce* is a *higher-order function* whose argument is the customizing operator $\oplus$. Implementations of *reduce* on particular servers are parameterized with $\oplus$ as well; they expect the operator to be provided during invocation. Note that the customizing operator $\oplus$ itself may be time-consuming: e. g. below we consider reduction on a list of matrices, where the operator denotes matrix multiplication. Another, quite similar example is the scan skeleton, $scan(\oplus)$, which computes the prefix sums using an associative operator $\oplus$, i. e. applying $scan(\oplus)$ to list $[a, b, c]$ would result in $[a, a \oplus b, a \oplus b \oplus c]$. Skeletons can also express more complex algorithmic patterns. For example, divide-and-conquer recursion can be expressed as a skeleton $DC(d, e, c)$, whose parameters are a divide function d (mapping a list to two sublists), function e to apply to lists of length one and a conquer function c (to combine two partial results into one).

3 System Implementation

The system sketched in Figure 1 was implemented in Java, using RMI for communication. Java has several advantages for our purposes. First of all, Java bytecodes are portable across a broad range of machines. The skeleton's customizing functional parameters can therefore be used on any of the server machines without rewriting or recompilation. Moreover, Java and RMI provide simple mechanisms for invoking a method (skeleton) remotely on the server.

The interaction between the system components – client, compute server and lookup server – is realized by implementing a set of remote interfaces known to all components. Figure 2 shows a simplified UML class diagram for the most important classes and interfaces of our implementation. Dashed lines with a solid triangle arrowhead connect interfaces and their implementing classes, while open arrowheads denote the "uses" relationship.

Compute Servers provide data-parallel skeletons that are remotely invoked by the clients. For each skeleton, a corresponding interface can be implemented on several servers. For example, for the reduction skeleton, interface `Reduce` is implemented by `ReduceImpl` in Figure 2; for the scan skeleton, interface `Scan` is implemented, etc. Skeleton implementations are adaptable to the server's machine type: e.g. they may be multithreaded Java programs for UMA multiprocessors, MPI programs for clusters, etc. Customizing operators are obtained from the client as classes implementing appropriate remote interfaces, e. g. operators for the reduction skeleton implement the `BinOp` interface. The necessary code shipping is handled transparently by RMI. The system is easily extensible: to add a new skeleton, an appropriate interface must be specified and copied to the codebase, along with any other necessary interfaces (e. g. operators with three parameters). The interfaces can then be implemented on the server and registered with the lookup service in the usual manner.

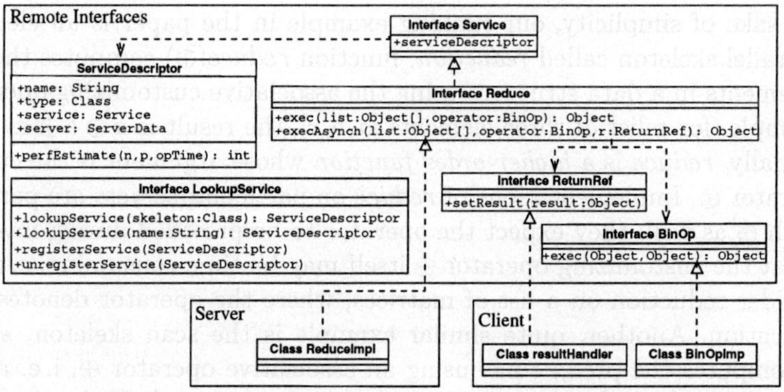

Fig. 2. Simplified class diagram of the implementation

Lookup Service administers the list of available skeletons and is running on its own server. Its hostname and port are known to all servers and clients. Each entry in the list consists of an object of class `ServiceDescriptor`, containing the skeleton's name and the implementing servers, a remote reference to the skeleton's implementation on the server side and a performance-estimation function (cf. Section 4).

Clients and servers interact with the lookup service by calling methods of the `LookupService` interface shown in the class diagram: `registerService` is used by the servers to register their skeletons, and `lookupService` is used by the clients to query for a particular skeleton.

Clients run a simple GUI for program development and server selection. On startup, the lookup service is contacted to obtain a list of all available skeletons. The user can then specify the structure of the program graphically: selected skeletons are displayed as nodes of a directed graph, with an edge between two skeletons if one provides input data for the other (composition). A special skeleton "local" is used to represent local (client-sided) computations in the graph.

From the graphical representation, a partial Java program is generated. It contains all skeleton calls and class definitions for classes that must be implemented by the user (e.g. customizing operators for skeletons). For each pair of server and skeleton, a performance estimate for that pair is computed, as explained in more detail in Section 4. Using this prediction, the user assigns a server to each particular skeleton invocation.

Skeleton Invocation is implemented using Java's RMI mechanism. This has the advantage that all parameter marshalling and unmarshalling as well as code shipping are handled transparently by the RMI system. One drawback, however, is the absence of asynchronous method invocation in RMI. We remedy this situation by providing a second implementation for each skeleton, implementing an asynchronous invocation. The `executeAsynch` method of a skeleton's interface immediately returns a remote reference to an object of class `rObject` which resides on the server side. To obtain the skeleton's result, the client invokes the `rObject`'s `getResult()` method, which blocks until the results are available. The transmission of data is again handled by RMI.

Skeleton composition is also handled using `rObjects`. For each skeleton, another `execute` method is provided, which receives parameters of type `rObject`. Thus, it is possible to express composition by simply writing

```
result=skeleton2.execute(skeleton1.executeAsynch(...));.
```

As `executeAsynch` only returns a remote reference, the results are not sent from the server back to the client, and from there on to the next server; instead only remote references are passed on. The second server can then obtain the data directly via the `getResult()` method.

4 Skeleton Performance Prediction

Intuitively, a client should delegate the skeleton execution to a particular Grid server if the skeleton is expected to execute faster on the server than locally or on another server. Invoking a skeleton remotely involves the time costs of sending arguments to and receiving results from the server. Thus, the decision about where to execute a skeleton is influenced by two main factors: performance gain and communication costs.

To decide whether to compute a skeleton remotely – and, if so, then on which server – it is necessary to predict both communication costs and performance gain. Thus, each server in our system provides a function describing the performance of each skeleton implemented by it. A client obtains this function t_{skel} from the lookup service for every server on which the skeleton *skel* is available. The total time T for remote execution can be computed as follows:

$$T = 2t_s + (n + o + r)t_w + t_{skel}(n, p, t_\oplus) \tag{1}$$

n being the size of the parameters, o the size of the operator's bytecode (which must be sent to the server as well) and r the size of the result. The number of processors is p, and $t_\oplus$ is the time taken to execute the customizing operator $\oplus$.

Let us consider the reduce skeleton example. Our experimental parallel implementation of reduction partitions the list into p sublists, with one thread computing the reduction sequentially on each sublist. The partial results are then reduced sequentially in one thread. The time taken to execute the reduce skeleton using this algorithm is

$$t_{red}(n, p, t_\oplus) = (\lceil n/p \rceil - 1)t_\oplus + (p - 1)t_\oplus \tag{2}$$

This algorithm can obviously be improved by organizing the reduction of partial results in a tree-like manner.

We measured the execution time for the reduce skeleton's implementation mentioned above, with 20×20 matrices as list elements and matrix multiplication as the operator. For all measurements, a SUN Ultra 5 Workstation with an UltraSparc-IIi processor running at 360 MHz ("client") and a SunFire

6800 shared-memory SMP system with 16 UltraSparc-III processors at 750 MHz ("server") were used. They are connected via a WAN, the client being at the University of Erlangen and the server at the Technical University of Berlin, with a distance of about 500 km between them.

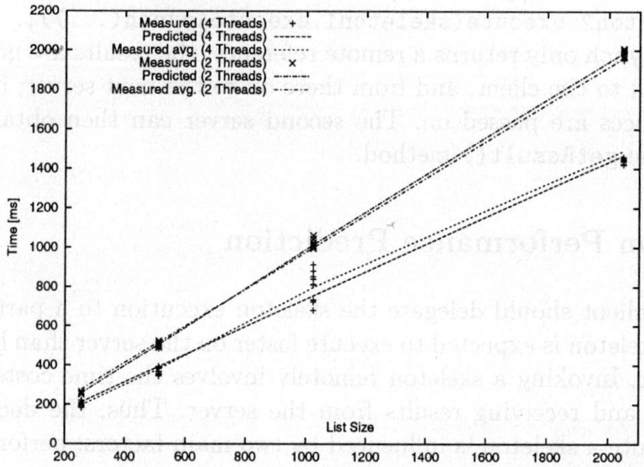

	t_{red}	t'_{red}	$t_\oplus$	t_s	t_w	T_1	T_2
Time [ms]	312.18	320.4	1.21	13.371	0.406	755	797

Fig. 3. Predicted and measured execution time for reduce skeleton

The table in Figure 3 contains the measured time $t_\oplus$ for executing the customizing operator (matrix multiplication) on the server. The time t_{red} for reducing a list of 1024 matrices was predicted, using the prediction function for the reduce skeleton given by equation (2) above. The obtained value is quite close to the measured time (t'_{red} in the table), though the latter is slightly higher owing to the overhead for synchronizing threads. The table also contains values t_s and t_w for sending messages via the WAN from Erlangen to Berlin (note that t_w is the time taken to transmit one matrix). The values were measured by sending messages of varying size from the client to the server, without invoking any skeletons. For the total time T for executing the reduce skeleton remotely, two values are given: T_1 is obtained using the predicted value for t_{red} together with equation (1); T_2 is the actual value measured when invoking the reduce skeleton remotely (the average over 10 invocations). The measured value is approximately 5% larger than the predicted one.

In the graph shown in Figure 3, the measured values for lists of sizes 256, 512, 1024 and 2048 are compared to the predicted values. For each list size, the values were measured ten times using the same setting as described above, with two and four threads. The measured values vary considerably (up to 20% for list size 1024 and 4 threads) owing to load changes on the server and varying network traffic. Thus, the predicted value differs up to 18% from the measured one. Most measured values are, however, much closer to the predicted ones, with differences of only 2 or 3%. Comparing the predicted values to the average values over all ten measurements, the differences are less than 7%.

4.1 Operator Performance Prediction

To achieve realistic time estimates for skeleton execution, it is important to predict accurately both the runtime of the customizing functional arguments (which are Java bytecodes) and the Grid network parameters. While many tools for predicting network performance are available, e. g. the Network Weather Service [6], very little is known about predicting the performance of Java bytecodes.

The simplest way to predict the runtime of an operator would be to send the operator's bytecode to the server, along with a sample set of operands, execute it there and obtain the execution time. Although very accurate time values can be expected, this method consumes a considerable amount of both network and computational resources on the server side, which is a significant drawback. We therefore investigate two approaches that do not involve computations on the server side or communication between client and server: (1) bytecode analysis, and (2) bytecode benchmarking.

Performance Prediction through Bytecode Analysis. To estimate the operator's runtime, we can execute it in a special JVM on the client side, counting how often each instruction is invoked. The obtained numbers for each instruction are then multiplied by a time value for that instruction. Our experiments have shown, that this approach, however, poses the problem of finding valid time values for single instruction runtimes. One way to obtain values that give promising results is reported in [1].

Performance Prediction through Benchmarking. An alternative way of estimating the performance of the operator's Java bytecode is to measure the achievable speedup a priori, by running a benchmark on the server and comparing the result for the benchmark execution on the server side to the runtime of the same benchmark on the client. Then, to obtain an estimate for the execution time of an operator on the server, the time for execution on the client is taken and multiplied by the measured speedup for the benchmark. To increase the accuracy of such performance predictions, several benchmarks containing different instruction mixes (arithmetic instructions, comparison operations, etc.) could be executed on the server. The client then picks the benchmark that appears suitable for the operator in question and uses this to compute a performance estimate as described above.

5 Related Work and Conclusions

Initial research on Grid computing focused, quite naturally, on developing the enabling infrastructure, systems like Globus, Legion and Condor being the prominent examples presented in the seminal "Grid-book" [3]. The next wave of interest focused on particular classes of applications and supporting tools for them, with such important projects as Netsolve [2], GridPP [5] and Cactus.

We feel that algorithmic and programming methodology aspects have been partly neglected at this early stage of Grid research and are therefore not yet properly understood. The main difficulty is the unpredictable nature of the Grid resources, resulting in difficult-to-predict behaviour of algorithms and programs. Initial experience has shown that entirely new approaches to software development and programming are required for the Grid [4].

Our work attempts to overcome the difficulties of algorithm design for Grids by using higher-order, parameterized programming constructs called skeletons. The advantage of skeletons is their high level of abstraction combined with an efficient implementation, tuned to a particular node of the Grid. Even complex applications composed of skeletons have a simple structure, and at the same time each skeleton can exemplify quite a complicated parallel or multithreaded structure in its implementation.

In the experimental programming system described in this paper, we focused on the problem of mapping the parts of an application to particular Grid machines and on the especially challenging problem of performance predictability. The initial results reported here are quite encouraging: they show that even in a heterogeneous, geographically distributed Grid environment the use of skeletons leads to more structured, predictable applications.

The grid system presented in our paper is still in an early stage and lacking many important services, such as resource allocation and authentication services. These issues can be addressed by using a more sophisticated infrastructure, such as provided by the Globus toolkit ([3]).

References

1. M. Alt, H. Bischof, and S. Gorlatch. Program development for computational grids using skeletons and performance prediction. In *Third Int. Workshop on Constructive Methods for Parallel Programming (CMPP 2002)*, Technical Report. Technische Universität Berlin, 2002. To appear.
2. H. Casanova and J. Dongarra. NetSolve: A network-enabled server for solving computational science problems. *Int. J. of Supercomputing Applications and High Performance Computing*, 3(11):212–223, 1997.
3. I. Foster and C. Kesselmann, editors. *The Grid: Blueprint for a New Computing Infrastructure.* Morgan Kaufmann, 1998.
4. K. Kennedy et al. Toward a framework for preparing and executing adaptive grid programs. In *Proceedings of NSF Next Generation Systems Program Workshop (International Parallel and Distributed Processing Symposium 2002)*, Fort Lauderdale, April 2002.
5. R. Perrot. Testbeds for the GridPP. First US-UK Workshop on Grid Computing.
6. R. Wolski, N. Spring, and J. Hayes. The network weather service: A distributed resource performance forecasting service for metacomputing. *Journal of Future Generation Computing Systems*, 15(5-6):757–768, October 1999.

A Scalable Approach to Network Enabled Servers*

Eddy Caron[1], Frédéric Desprez[1], Frédéric Lombard[2], Jean-Marc Nicod[2], Laurent Philippe[2], Martin Quinson[1], and Frédéric Suter[1]

[1] LIP, ENS Lyon. INRIA Rhône – Alpes, France,
Frederic.Desprez@ens-lyon.fr

[2] LIFC, Faculté des sciences et techniques, Besançon, France

Abstract. This paper presents the architecture of DIET (Distributed Interactive Engineering Toolbox), a hierarchical set of components to build Network Enabled Server applications in a Grid environment. This environment is built on top of different tools which are able to locate an appropriate server depending of the client's request, the data location and the dynamic performance characteristics of the system.

1 Introduction

Huge problems can now be computed over the Internet thanks to Grid Computing Environments. Because most of current applications are numerical, the use of high-performance libraries is mandatory. The integration of such libraries in high level applications using languages like Fortran or C is far from easy. Moreover the computational power and memory needs of such applications obviously may not be available on every workstation. Thus the RPC paradigm seems to be a good candidate to build Problem Solving Environments (PSE) for numerical applications on the Grid. Several tools following this approach exist, like NetSolve [1] or Ninf [5]. They are commonly called Network Enabled Server (NES) environments [4] and usually have five different components: *Clients* that submit problems they have to solve to*Servers*, a *Database* that contains information about software and hardware resources, a *Scheduler* that chooses an appropriate server depending on the problem sent and the information contained in the database, and finally *Monitors* that acquire information about the status of the computational resources.

But the environments previously cited have a centralized scheduler which can become a bottleneck when many clients try to access several servers. Moreover as networks are highly hierarchical, the location of the scheduler has a great impact on the performance of the overall architecture. This paper presents the architecture of DIET (Distributed Interactive Engineering Toolbox), a hierarchical set of components to build NES applications [2].

* This work was supported in part by the ACI GRID and the RNTL of the french department of research.

B. Monien and R. Feldmann (Eds.): Euro-Par 2002, LNCS 2400, pp. 907–910.

2 DIET Architecture and Related Tools

The aim of a NES environment such as DIET is to provide a transparent access to a pool of computational servers. DIET focuses on offering such a service at a very large scale. A **client** that has a problem to solve should be able to obtain a reference to the server that is best suited for him. A problem can be submitted from a web page, a PSE such as Scilab (a Matlab-like tool), or from a compiled program. DIET is designed to take into account the data location when scheduling jobs. Data are kept as long as possible on (or near to) the computational servers in order to minimize transfer times.

DIET is built upon ***Computational Resource Daemons*** **(CRD)** and ***Server Daemons*** **(SeD)**. A CRD is a set of hardware and software components that can perform sequential or parallel computations on data sent by a client (or an other server). For instance a CRD can be the entry point of a parallel computer. It usually provides a set of libraries and is managed by an SeD which encapsulates a computational server. The information stored on an SeD is the list of the data available on a server, the list of problems that can be solved on it, and every information concerning its load (available memory, number of available resources, ...). An SeD declares the problems it can solve to DIET and provides an interface to clients for submitting their requests. A SeD can give performance prediction for a given problem using FAST [6], an application-centric performance forecasting tool based on NWS [7]. The scheduler is scattered across a hierarchy of ***Local Agents*** **(LA)** and ***Master Agents*** **(MA)**. An MA receives computation requests from clients. These requests are generic descriptions of problems to be solved. The SLiM (Scientific Libraries Metaserver) module is used to find all the implementation of these generic problem. Using SLiM, an MA collects computation services from the SeDs and chooses the best one. The reference of this server is returned to the client. An LA aims at transmitting requests and information between MAs and SeDs. The information stored on an LA is the list of requests and, for each of its sub-trees, the number of servers that can solve a given problem and information about data distributed in this sub-tree. Depending on the underlying network architecture, a hierarchy of LAs may be deployed between an MA and its SeDs.

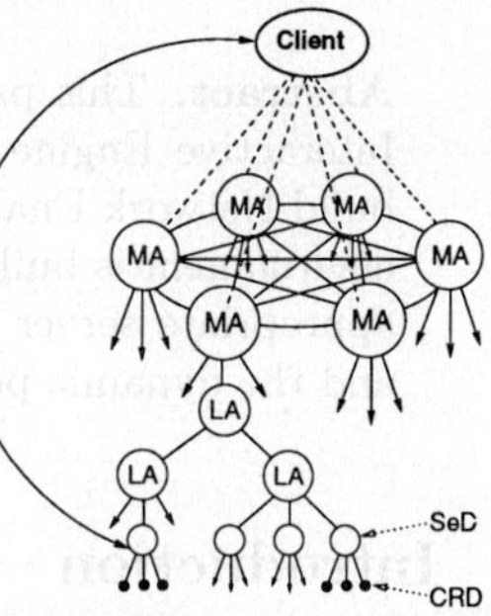

Fig. 1. Diet architecture.

Our first DIET prototype is based upon *OmniORB*, a free Corba implementation which provides good communication performance. Corba systems provide a remote method invocation facility with a high level of transparency. This transparency should not dramatically affect the performance, communication layers being well optimized in most Corba implementations [3]. Moreover, the time to select a server using corba should be short with regard to the computation time.

3 Experimental Result

Our first experiments made with the DIET prototype involve one MA and a hierarchy of LAs. Our goal is to validate our architecture on basic examples and demonstrate that the hierarchical approach of DIET decreases the request processing time. For each experience, we build a DIET tree and run 10000 clients sequentially. We assume the problem submitted by the clients is known by every SeDs. Thus, all SeDs are contacted at each request. The measured value was the average submission time (i.e., the latency time). Experiments have been made on a local switched Fast Ethernet network dedicated to the experiment. All DIET components are running on a different computer.

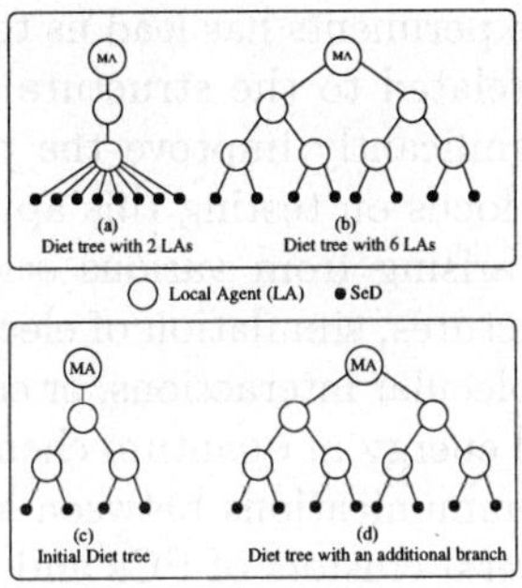

Fig. 2. Comparison between two DIET trees (top). Adding a branch to a DIET tree (bottom).

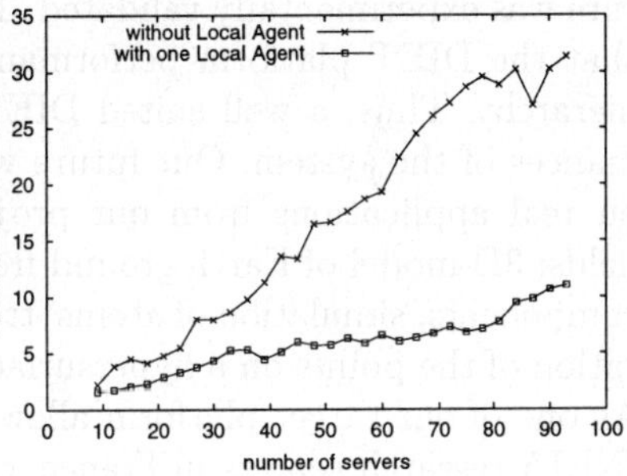

Fig. 3. Time to process a request.

Comparing Two Architectures of Similar Depth: In Figure 2 (top) we compare two DIET architectures that have the same depth and the same number of SeDs. The average request processing time is 52.2ms on the first architecture (a) while 33.5ms for the second one (b). Administrators have to carefully build their LAs hierarchy to improve the performance.

Adding a Branch to a DIET Tree: Figure 2 (bottom) shows how it is possible to have twice as many SeDs just by adding one son to the MA. This son leads a hierarchy similar to the existing branch, which consists in a binary tree. The mean submission time is 32.3ms for the first architecture (c) and 33.5ms for the second one (d). The increase of 1.2ms is nearly three times less than when we add servers directly in an existing branch. This experiment shows how requests are processed in parallel when we add servers on an independent branch. The strategy to add new servers has to be carefully examined.

Request Broadcast Evaluation: In this case, LAs are used to perform an efficient request broadcast from the MA to the SeDs. The solw network link between client and SeDs has a bandwidth of 2Mb/s shared with other links. The local network that supports the SeDs (and the LA in the second case) is a switched Fast Ethernet network. The MA and the client are running on the same workstation on the other end of the slow link. For each number of SeDs, 50 clients are run sequentially and the average client execution time has been considered as the submission time. With one LA, the request submission time appears to be 3 (with 42 SeDs) up to 4 (with 72 SeDs) times less than without LA. Furthermore the slope is 4 times greater whith no LA deployed. This result corroborates our idea that bottlenecks are an important issue in NES environments. Moreover, LAs deployment on the network is crucial for the efficiency of DIET.

4 Conclusion and Future Work

In this paper we presented our view of a scalable NES system. We propose a hierarchical approach using existing software like NWS or Corba. Our architecture was experimentally validated. The different experiments has lead us to state that the DIET platform performance is closely related to the structure of the hierarchy. Thus, a well suited DIET tree can significantly improve the performances of the system. Our future work will first focus on testing this approach on real applications from our project partners arising from various scientific fields: 3D-model of Earth ground from satellite pictures, simulation of electronic components, simulation of atoms' trajectory in molecular interactions, or computation of the points on a hypersurface of potential energy in quantum chemistry. As one of our target platform allows 2.5 Gb/s communications between several INRIA research centers in France, connecting several clusters of PCs and parallel machines, we think that tightly coupled applications written in a RPC mode could benefit of such an approach. Thus the powerful computational resources needed for such application can be utilized that could not otherwise be obtained.

References

1. D. Arnold, S. Agrawal, S. Blackford, J. Dongarra, M. Miller, K. Sagi, Z. Shi, and S. Vadhiyar. Users' Guide to NetSolve V1.4. U Tennessee TR. CS-01-467, 2001.
2. E. Caron, P. Combes, S. Contassot-Vivier, F. Desprez, F. Lombard, J.-M. Nicod, M. Quinson, and F. Suter. A Scalable Approach to Network Enabled Servers. Technical Report RR2002-21, Lab. de l'Inf. du Parallélisme (LIP), May 2002.
3. A. Denis, C. Pérez, and T. Priol. Towards high performance CORBA and MPI middlewares for grid computing. In Craig A. Lee, editor, *Proc. of the 2nd International Workshop on Grid Computing*, number 2242 in LNCS, pages 14–25, Nov. 2001.
4. S. Matsuoka, H. Nakada, M. Sato, , and S. Sekiguchi. Design Issues of Network Enabled Server Systems for the Grid, 2000. Grid Forum, APM WG whitepaper.
5. H. Nakada, M. Sato, and S. Sekiguchi. Design and Implementations of Ninf: towards a Global Computing Infrastructure. *FGCS*, 15(5-6):649–658, 1999.
6. M. Quinson. Dynamic Performance Forecasting for Network-Enabled Servers in a Metacomputing Environment. In *Proc. of the Int. Work. on Performance Modeling, Evaluation, and Optimization of Par. and Dist. Systems (PMEO-PDS'02)*, 2002.
7. R. Wolski, N. T. Spring, and J. Hayes. The Network Weather Service: A Distributed Resource Performance Forecasting Service for Metacomputing. *FGCS*, 15(5–6):757–768, Oct. 1999.

Topic 15
Discrete Optimization

Rainer Feldmann and Catherine Roucairol

Topic Chairpersons

Discrete optimization problems arise in various applications such as airline crew scheduling and rostering, vehicle routing, frequency assignment, communication network design, etc.

The advent of parallel computation has raised many expectations in this field: faster resolution of known problem instances, the possibility to solve larger instances of difficult problems, the opportunity to address new problems, etc. But parallel computation also challenges researchers to rethink their models and algorithms, and imposes a number of specific issues related, in particular, to efficient data structures, communication and information sharing mechanisms, workload distribution, performance measures, etc. Thus, the aim of this theme is to present recent developments related to the main areas in the edge of Parallel Computing and Operations Research.

In contrast to most of the other topics, the topic *Discrete Optimization* was new at this years Euro-Par conference. From a total of 7 submissions one was accepted as a regular paper, while two submissions were accepted as research notes.

The papers will cover different fields: In the first, the authors present a new parallel algorithm for a graph coloring problem arising in matrix partitioning problems in the field of numerical optimzation. The second paper deals with a parallel implementation of GRASP, a heuristic local search algorithm. In a third paper the authors describe MALLBA, a library containing exact, heuristic and hybrid resolution methods for combinatorial optimization problems.

The papers altogether show that the effort required to harness the potential power of parallel computers should not be underestimated, and that the use of parallel computers leads to substantial savings and even makes it possible to modelize and solve larger problem instances than before.

Finally, we would like to thank all the contributors and referees for the excellent work they have performed in helping to make the workshop possible.

B. Monien and R. Feldmann (Eds.): Euro-Par 2002, LNCS 2400, p. 911.

Parallel Distance-k Coloring Algorithms for Numerical Optimization

Assefaw Hadish Gebremedhin[1], Fredrik Manne[1], and Alex Pothen[2,*]

[1] Department of Informatics, University of Bergen, N-5020 Bergen, Norway,
{assefaw,fredrikm}@ii.uib.no

[2] Computer Science Department, Old Dominion University, Norfolk, VA 23529 USA,
CSRI, Sandia National Labs, Albuquerque NM 87185 USA,
ICASE, NASA Langley Research Center, Hampton, VA 23681-2199 USA,
pothen@cs.odu.edu

Abstract. Matrix partitioning problems that arise in the efficient estimation of sparse Jacobians and Hessians can be modeled using variants of graph coloring problems. In a previous work [6], we argue that *distance-2* and *distance-*$\frac{3}{2}$ graph coloring are robust and flexible formulations of the respective matrix estimation problems. The problem size in large-scale optimization contexts makes the matrix estimation phase an expensive part of the entire computation both in terms of execution time and memory space. Hence, there is a need for both shared- and distributed-memory parallel algorithms for the stated graph coloring problems. In the current work, we present the first practical shared address space parallel algorithms for these problems. The main idea in our algorithms is to randomly *partition* the vertex set equally among the available processors, let each processor *speculatively* color its vertices using information about already colored vertices, detect eventual conflicts in parallel, and finally re-color conflicting vertices sequentially. *Randomization* is also used in the coloring phases to further reduce conflicts. Our PRAM-analysis shows that the algorithms should give almost linear speedup for sparse graphs that are large relative to the number of processors. Experimental results from our OpenMP implementations on a Cray Origin2000 using various large graphs show that the algorithms indeed yield reasonable speedup for modest numbers of processors.

1 Introduction

Numerical optimization algorithms that rely on derivative information often need to compute the Jacobian or Hessian matrix. Since this is an expensive part of the computation, efficient methods for estimating these matrices via finite differences (FD) or automatic differentiation (AD) are needed. It is known that the problem of minimizing the number of function evaluations (or AD passes)

* This author's research was supported by NSF grant DMS-9807172, DOE ASCI level-2 subcontract B347882 from Lawrence Livermore National Lab; and by DOE SCI-DAC grant DE-FC02-01ER25476.

B. Monien and R. Feldmann (Eds.): Euro-Par 2002, LNCS 2400, pp. 912–921.

required in the computation of these matrices can be formulated as variants of graph coloring problems [1–3, 8, 11]. The particular coloring problem differs with the optimization context: whether the Jacobian or the Hessian matrix is to be computed; whether a direct or a substitution method is employed; and whether only columns, or only rows, or both columns and rows are to be used to evaluate the matrix elements. In addition, the type of coloring problem depends on the kind of graph used to represent the underlying matrix. In [6], we provide an integrated review of previous works in this area and identify the *distance-2* (D2) graph coloring problem as a unifying, generic, and robust formulation. The D2-coloring problem has also noteworthy applications in other fields such as channel assignment [10] and facility location problems [12].

Large-scale PDE-constrained optimization problems can be solved only with the memory and time resources available on parallel computers. In these problems, the variables defined on a computational mesh are already distributed on the processors, and hence parallel coloring algorithms are needed for computing, for instance, the Jacobian. It turns out that the problems of efficiently computing the Jacobian and Hessian can be formulated as the D2- and D$\frac{3}{2}$-coloring problems, respectively. The latter coloring problem is a relaxed variant of the former, and will be described in Section 2.2. In these formulations, the bipartite graph associated with the rows and columns of the matrix is used for the Jacobian; in the case of the Hessian matrix, the adjacency graph corresponding to the symmetric matrix is used.

In this paper, we present several new deterministic as well as probabilistic parallel algorithms for the D2- and D$\frac{3}{2}$-coloring problems. Our algorithms are practical and effective, well suited for shared address space programming, and have been implemented in C using OpenMP primitives. We report results from experiments conducted on a Cray Origin2000 using large graphs that arise in finite element methods and in eigenvalue computations.

In the sequel, we introduce the graph problems in Section 2, present the algorithms in Section 3, discuss our experimental results in Section 4, and conclude the paper in Section 5.

2 Background

2.1 Matrix Partition Problems

An essential component of the efficient estimation of a sparse Jacobian or Hessian using FD or AD is the problem of finding a suitable *partition* of the columns and/or the rows of the matrix. The particular partition chosen defines a system of equations from which the matrix entries are determined. A method that utilizes a diagonal system is called a *direct* method, and one that uses a triangular system is called a *substitution* method. A direct method is more restrictive but the computation of matrix entries is straightforward and numerically stable. A substitution method, on the other hand, is less restrictive but it may be subject to approximation difficulties and numerical instability. Moreover, direct methods offer more parallelism than substitution methods. In this paper, we focus on direct methods that use column partitioning.

A partition of the columns of an unsymmetric matrix A is said to be *consistent* with the direct determination of A if whenever a_{ij} is a non-zero element of A then the group containing column j has no other column with a non-zero in row i [1]. Similarly, a partition of the columns of a symmetric matrix A is called *symmetrically consistent* with the direct determination of A if whenever a_{ij} is a non-zero element of A then *either* (i) the group containing column j has no other column with a non-zero in row i, or (ii) the group containing column i has no other column with a non-zero in row j [2]. From a given (symmetrically) consistent partition $\{C_1, C_2, \ldots, C_\rho\}$ of the columns of A, the nonzero entries can be determined with ρ function evaluations (matrix-vector products).

Thus we have the following two problems of interest. Given the sparsity structure of an unsymmetric $m \times n$ matrix A, find a consistent partition of the columns of A with the fewest number of groups. We refer to this problem as UNSYMCOLPART. Similarly, the problem that asks for a symmetrically consistent partition of a symmetric A is referred to as SYMCOLPART.

2.2 Graph Problems

In a graph, two distinct vertices are said to be *distance-k neighbors* if the shortest path connecting them consists of at most k edges. A *distance-k ρ-coloring* (or (k, ρ)-coloring for short) of a graph $G = (V, E)$ is a mapping $\phi : V \rightarrow \{1, 2, \ldots, \rho\}$ such that $\phi(u) \neq \phi(v)$ whenever u and v are distance-k neighbors. We call a mapping $\phi : V \rightarrow \{1, 2, \ldots, \rho\}$ a $(\frac{3}{2}, \rho)$-coloring of a graph $G = (V, E)$ if ϕ is a $(1, \rho)$-coloring of G and every path containing three edges uses at least three colors. Notice that a $(\frac{3}{2}, \rho)$-coloring is a restricted $(1, \rho)$-coloring, and a relaxed $(2, \rho)$-coloring, and hence the name. For instance, consider a path u, v, w, x in a graph. The assignment $2, 1, 2, 3$ to the respective vertices is a valid D1- and D$\frac{3}{2}$-coloring, but not a valid D2-coloring. The *distance-k graph coloring problem* asks for a (k, ρ)-coloring of a graph with the least possible value of ρ.

Let A be an $m \times n$ rectangular matrix with rows $r_1, r_2, \ldots, r_m$ and columns $a_1, a_2, \ldots, a_n$. We define the *bipartite graph* of A as $G_b(A) = (V_1, V_2, E)$ where $V_1 = \{r_1, r_2, \ldots, r_m\}$, $V_2 = \{a_1, a_2, \ldots, a_n\}$, and $(r_i, a_j) \in E$ whenever $a_{ij} \neq 0$, for $1 \leq i \leq m$, $1 \leq j \leq n$. When A is an $n \times n$ symmetric matrix with non-zero diagonal elements, the *adjacency graph* of A is defined to be $G_a(A) = (V, E)$ where $V = \{a_1, a_2, \ldots, a_n\}$ and $(a_i, a_j) \in E$ whenever a_{ij}, $i \neq j$, is a non-zero element of A. Note that the non-zero diagonal elements of A are not explicitly represented by edges in $G_a(A)$. In our work, we rely on the bipartite and adjacency graph representations. However, in the literature, an unsymmetric matrix A is often represented by its *column intersection graph* $G_c(A)$. In this representation, the columns of A constitute the vertex set, and an edge (a_i, a_j) exists whenever the columns a_i and a_j have non-zero entries at the same row position (i.e., a_i and a_j are not structurally orthogonal). As argued in [6], the bipartite graph representation is more flexible and robust than the 'compressed' column intersection graph.

Coleman and Moré [1] showed that problem UNSYMCOLPART is equivalent to the D1-coloring problem when the matrix is represented by its column intersection graph. We have shown [6] that the same problem is equivalent to a partial D2-coloring when a bipartite graph is used and discussed the relative merits of the two approaches. The word 'partial' reflects the fact that only the vertices corresponding to the columns need to be colored.

McCormick [11] showed that the approximation of a Hessian using a direct method is equivalent to a D2-coloring on the adjacency graph of the matrix. One drawback of McCormick's formulation is that it does not exploit symmetry. Later Coleman and Moré [2] addressed this issue and showed that the resulting problem (SYMCOLPART) is equivalent to the D$\frac{3}{2}$-coloring problem.

3 Parallel Coloring Algorithms

The distance-k coloring problem is NP-hard for any fixed integer $k \geq 1$ [11]. A proof-sketch showing that D$\frac{3}{2}$-coloring is NP-hard is given in [2]. Furthermore, Lexicographically First $\Delta + 1$ Coloring (LFC), the polynomial variant of D1-coloring in which the vertices are given in a predetermined order and the question at each step is to assign the vertex the smallest color not used by any of its neighbors, is P-complete [7]. The practical implication of this is that designing efficient fine-grained parallel algorithm for LFC is hard.

In practice, greedy sequential D1-coloring heuristics are found to be quite effective [1]. In a recent work [5], we have shown effective methods of parallelizing such greedy algorithms in a coarse-grained setting. Here, we extend this work to develop parallel algorithms for the D2- and D$\frac{3}{2}$-coloring problems.

Jones and Plassmann [9] describe a parallel distributed memory D1-coloring algorithm that uses randomization to assign priorities to the vertices, and then colors the vertices in the order determined by the priorities. There is no speculative coloring in their algorithm. It was reported that the algorithm slows down as the number of processors is increased. Finocchi et al.[4] suggest a parallel D1-coloring algorithm organized in several rounds; in each round, currently uncolored vertices are assigned a tentative pseudo-color without consulting their neighbors mapped to other processors; in a conflict resolution step, a maximal independent set of vertices in each color class is assigned these colors as final; the remainder of the vertices are uncolored, and the algorithm moves into the next round. However, they do not give any implementation and we believe that this algorithm incurs too many rounds, each with its synchronization and communication steps, for it to be practical on large graphs.

Our algorithm (in its generic form) may be viewed as a compromise between these algorithms, where we permit speculative coloring, but limit the number of synchronization steps to two in the whole algorithm. However, our current algorithms rely on the shared address space programming model; we will adapt our algorithms to distributed memory programming models in future work. We are unaware of any previous work on parallel algorithms for D2- and D$\frac{3}{2}$-coloring problems.

3.1 Generic Greedy Parallel Coloring Algorithm (GGPCA)

The steps of our generic parallel coloring algorithm can be summarized as shown below; refer to [5] for a detailed discussion of the D1-coloring case. Let $G = (V, E)$ be the input graph and p be the number of processors.

Phase 0: Partition
  Randomly partition V into p equal blocks $V_1 \ldots V_p$. Processor P_i is responsible for coloring the vertices in block V_i.
Phase 1: Pseudo-color
  for $i = 1$ **to** p **do in parallel**
    for each $u \in V_i$ **do**
      assign the smallest available color to u, paying attention to already colored vertices (both local and non-local).
Phase 2: Detect conflicts
  for $i = 1$ **to** p **do in parallel**
    for each $u \in V_i$ **do**
      check whether the color of u is valid. If the colors of u and v are the same for some $(u, v) \in E$, then
        $L_i = L_i \cup min\{u, v\}$
Phase 3: Resolve conflicts
  Color the vertices in the conflict list $L = \cup L_i$ sequentially.

In Phase 0, the vertices are randomly partitioned into p equal blocks each of which is assigned to some processor. In Phase 1, when two adjacent vertices are on different processors, the two processors could color both simultaneously, possibly assign them the same value, and cause a conflict. The purpose of Phase 2 is to detect and store any such conflict vertices which are subsequently re-colored sequentially in Phase 3.

3.2 Simple Distance-2 (SD2) Coloring Algorithm

The meaning of 'available' color in GGPCA depends on the required coloring. In the case of a D2-coloring, a vertex is assigned the smallest color not used by any of its D2-neighbors. Let Δ denote the maximum degree in the graph. It can easily be verified that, in a D2-coloring, a vertex can always be assigned one of the colors from the set $\{1, 2, \ldots, \Delta^2 + 1\}$. Moreover, since the D1-neighbors of a vertex are D2-neighbors with each other, the D2-chromatic number (the least number of colors required in a D2-coloring) is at least $\Delta + 1$. Thus, the greedy approach is an $O(\Delta)$-approximation algorithm. We refer to the variant of GGPCA that applies to D2-coloring as Algorithm SD2.

Note that the sequential time complexity of greedy D2-coloring is $O(\Delta^2|V|)$. The following results show that the number of conflicts discovered in Phase 2 of Algorithm SD2 is often small for sparse graphs, making the algorithm scalable when the number of processors $p = O(\sqrt{\frac{|V|^2}{\Delta|E|}})$. The proofs, which are essentially similar to those of the D1-coloring case given in [5], have been omitted for space considerations. Let $\overline{\delta} = 2|E|/|V|$ denote the average vertex degree in G.

Lemma 1. *The expected number of conflicts created at the end of Phase 1 of Algorithm SD2 is at most* $\approx \frac{\Delta\bar{\delta}(p-1)}{2}$.

Theorem 2. *On a CREW PRAM, Algorithm SD2 distance-2 colors the input graph consistently in expected time* $O(\Delta^2(\frac{|V|}{p}+\Delta\bar{\delta}p))$ *using at most* Δ^2+1 *colors.*

Corollary 3. *When* $p = O(\sqrt{\frac{|V|^2}{\Delta|E|}})$, *the expected runtime of Algorithm SD2 is* $O(\frac{\Delta^2|V|}{p})$.

The number of conflicts predicted by Lemma 1 is an over estimate. The analysis assumes that whenever two D2-adjacent vertices are colored simultaneously, they are assigned the same color, thereby resulting in a conflict. However, a more involved probabilistic analysis that takes the distribution of colors used into account may provide a tighter bound. Besides, the actual number of conflicts in an implementation could be significantly reduced by choosing a random color from the allowable set, instead of the smallest one as given in Phase 1 of GGPCA.

3.3 Improved Distance-2 (ID2) Coloring Algorithm

The number of colors used in Algorithm SD2 can be reduced using a 'two-round-coloring' strategy. The underlying idea in the D1-coloring case is due to Culberson and was used in the parallel algorithm of [5]. Here we extend the result to the D2-coloring case.

Lemma 4. *Let* ϕ *be a distance-2 coloring of a graph* G *using* χ *colors, and* π *a permutation of the vertices such that if* $\phi(v_{\pi(i)}) = \phi(v_{\pi(l)}) = c$, *then* $\phi(v_{\pi(j)}) = c$ *for* $i < j < l$. *Applying the greedy sequential distance-2 algorithm to* G *where the vertices have been ordered by* π *will produce a coloring* ϕ' *using* χ *or fewer colors.*

The idea is that if the greedy coloring algorithm is re-applied on a graph, with the vertices belonging to the same color class (in the original coloring) listed consecutively, then the new coloring obtained is better or at least as good as the original. One ordering (among many) that satisfies this condition, with a good potential for reducing the number of colors used, is to list the vertices consecutively for each color class in the reverse order of the introduction of the color classes. Based on Lemma 4, we modify Algorithm SD2 and introduce an additional parallel coloring phase between Phases 1 and 2. Algorithm ID2 below outlines the resulting 4-phase algorithm.

Phases 0 and 1. Same as Ph. 0 and 1 of GGPCA. Let s be the number of colors.
Phase 2. **for** $k = s$ **downto** 1 **do**
  Partition ColorClass(k) into p equal blocks $V_1', \ldots, V_p'$
  for $i = 1$ **to** p **do in parallel**
    for each $u \in V_i'$ **do**
      assign the smallest available color to vertex u.
Phases 3 and 4. Same as Phases 2 and 3 of GGPCA, respectively.

In Phase 2, most of the vertices in a color class are at a distance greater than two edges from each other, and the exceptions arise from the conflict vertices colored incorrectly in Phase 1. Since the number of such conflict vertices from Phase 1 is low, the number of conflict vertices at the end of the re-coloring phase will be even lower. Phases 3 and 4 are included to detect and resolve any eventual conflicts not resolved in Phase 2.

3.4 Simple Distance-$\frac{3}{2}$ (SD$\frac{3}{2}$) Coloring Algorithm

Recall that a D$\frac{3}{2}$-coloring is a *relaxed* D2-coloring (see the example in Section 2.2). One way of relaxing the requirement for D2-coloring in GGPCA so as to obtain a valid D$\frac{3}{2}$-coloring is to let two vertices at a distance of *exactly* two edges from each other share a color as long as the vertex in between them is already colored with a (color of) lower value. We refer to the variant of GGPCA that employs this technique to achieve a distance-3/2 coloring as Algorithm SD$\frac{3}{2}$.

Note that both Algorithms ID2 and SD$\frac{3}{2}$ take asymptotically the same time as Algorithm SD2.

3.5 Randomization

The potential scalability of GGPCA depends on the number of conflicts discovered in Phase 2, since these are resolved sequentially in Phase 3. For dense graphs the number of conflicts could be large enough to destroy the scalability of the algorithm. To overcome this problem, we use randomization as a means for reducing the number of conflicts. In the randomized variants of Algorithms SD2, SD$\frac{3}{2}$, and ID2, a vertex is assigned the next available color with probability q, where $0 < q \leq 1$. The first attempt is made with the smallest available color, i.e., the color is chosen with probability q. An attempt is said to be successful if the vertex is assigned a color. If an attempt is not successful, then the next available color is tried with probability q, and so on, until the vertex gets a color. Algorithms SD2, SD$\frac{3}{2}$, and ID2, can be seen as the deterministic variants where $q = 1$. We refer to the randomized versions of the respective algorithms as RSD2, RSD$\frac{3}{2}$, and RID2.

Let u and v be two vertices with the same (infinite) set of allowable colors. If u and v are colored concurrently, it can be shown that the probability that u and v get the same color is $q/(2-q)$. This shows how randomization leads to a reduction in the number of conflicts; the lower the value of q, the lower chance for a conflict to arise. It should however be noted that a 'low' value of q may result in an increase in the number of colors used. Choosing the right value for q thus becomes a design issue.

4 Experimental Results and Discussion

Our test bed consists of graphs that arise from finite element methods and from eigenvalue computations [5]. Due to space limitations, we report on results obtained using two representative graphs, one from each field. Table 1 provides the test graphs' structural information. We point out that similar tendencies were observed in experiments using the other graphs from our test bed.

Table 1. Test graphs: the last 3 columns list the max., min., and average degree

Problem	$\|V\|$	$\|E\|$	Δ	δ	$\bar{\delta}$
fe144	144,649	1,074,393	26	4	14
ev02	19,845	3,353,890	749	78	338

Table 2 provides coloring and timing information of the different algorithms. Since the deterministic algorithms gave acceptable results on the relatively sparse graph fe144, the randomized variants were not run on this graph. Table 2 shows that Algorithm SD2 uses many fewer colors than the bound $\Delta^2 + 1$ and that Algorithm ID2 reduces the number of colors by up to 10% compared to SD2. The advantage of exploiting symmetry in Problem SYMCOLPART can be seen by comparing the number of colors used in SD2 and SD$\frac{3}{2}$; the latter can be as much as 37% fewer than the former. In general, the number of colors used in our deterministic algorithms increases only slightly with increasing p. For the randomized algorithms, the number of colors in some cases even decreases as p increases, but we think this is a random phenomenon.

Table 2. Results for fe144 (upper) and ev02 (lower) half. χ(alg x) and t(alg x) give the number of colors and the time in sec., respectively, used by alg x

p	χ(sd2)	χ(id2)	χ(sd$\frac{3}{2}$)	χ(rsd2)	χ(rid2)	χ(rsd$\frac{3}{2}$)	t(sd2)	t(id2)	t(sd$\frac{3}{2}$)	t(rsd2)	t(rid2)	t(rsd$\frac{3}{2}$)
1	41	37	35	-	-	-	6.8	9.6	8.1	-	-	-
12	43	38	36	-	-	-	1.1	1.6	1	-	-	-
1	4260	4016	2697	5564	4024	3142	255	456	247	257	380	257
12	5168	4455	3049	5536	4104	3081	214	112	45	31	106	29

Fig. 1 shows how the different phases of the algorithms scale as more processors are employed. For graph fe144, the time used in resolving conflicts sequentially is negligible compared to the overall time. Moreover, it can be seen that Phases 1 and 2 of Algorithms SD2 and SD$\frac{3}{2}$ scale rather well on this graph as the number of processors is increased. However, Phase 2 of Algorithm ID2 does not scale as well. This is due to the existence of many color classes with few vertices which entails extra synchronization and communication overhead in the parallel re-coloring phase.

The picture for the more dense graph ev02 is different: the time elapsed in Phase 3 of Algorithm SD2 is significant and increases as the number of processors is increased. The situation is somewhat better for SD3/2 and ID2. Note that ev02 is about 100 times denser than fe144 (where density $= \frac{|E|}{|V|^2}$), and the results in

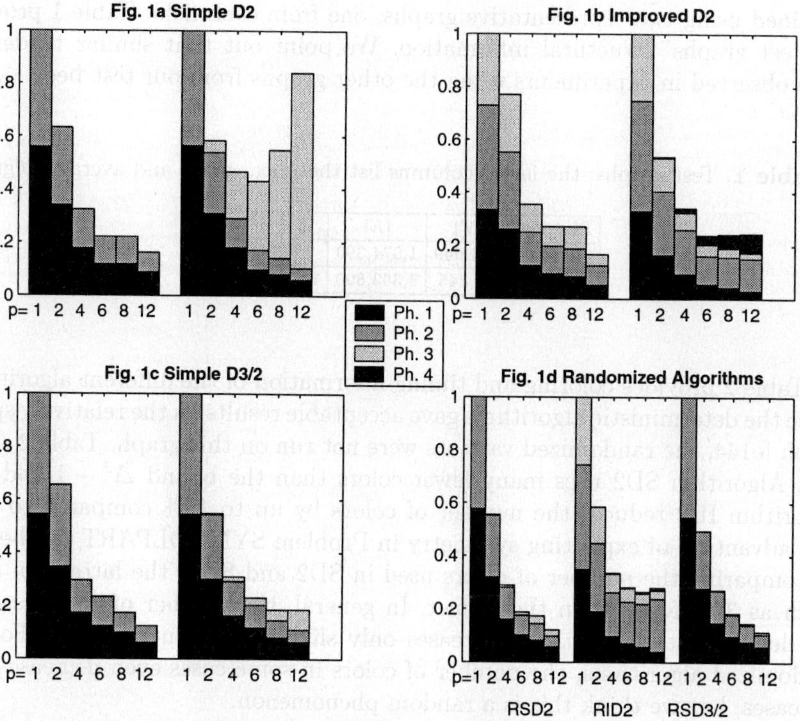

Fig. 1. a-c: Relative performance of deterministic algorithms on graphs fe144 (left-half on each sub-figure) and ev02 (right-half). Fig. 1d: Relative performance of randomized algorithms on ev02.

Table 2 and Fig. 1 agree well with the results in Corollary 3: when the average degree is high, we lose scalability.

Fig. 1d shows how using probabilistic algorithms solves the problem of high number of conflicts for ev02. The improvement in scalability comes at the expense of increased number of colors used (see Table 2). We have experimented using different values for q and found good results when $q = 1/20$ for RSD2, and $q = 1/6$ for RID2 and RSD$\frac{3}{2}$.

It should be noted that the speedups observed in Fig. 1 are all relative to the respective parallel algorithm run with $p = 1$. A comparison against a sequential version (with no conflict detecting and resolving phase) would yield less speedup. In particular, relative to a sequential version, the ideal speedup obtained by Algorithms SD2 and ID2 is roughly $\frac{1}{2}p$ and $\frac{2}{3}p$, respectively.

Our algorithms in general did not scale well beyond around 16 processors. We believe this is due to, among other things, the relatively high cost associated with non-local physical memory accesses. It would be interesting to see how this affects the behavior of the algorithms on different parallel platforms.

5 Conclusion

We have presented several simple and effective parallel approximation algorithms as well as results from OpenMP-implementations for the D2- and D$\frac{3}{2}$-coloring problems. The number of colors produced by the algorithms in the case where $p = 1$ is generally good as it is typically off from the lower bound $\Delta+1$ of SD2 by a factor much less than the approximation ratio Δ. As more processors are employed, the algorithms provide reasonable speedup while maintaining the quality of the solution. In general, our deterministic algorithms seem to be suitable for sparse graphs and the probabilistic variants for more dense graphs. We believe the functionality provided by our algorithms is useful for many large-scale optimization codes, where parallel speedups while desirable, are not paramount, as long as running times for coloring are low relative to the other steps in the optimization computations. The three sources of parallelism in our algorithms – partitioning, speculation, and randomization – can be exploited in developing distributed parallel algorithms, but the algorithms would most likely differ significantly from the shared memory variants presented here.

References

1. T. F. Coleman and J. J. Moré. Estimation of sparse Jacobian matrices and graph coloring problems. *SIAM J. Numer. Anal.*, 20(1):187–209, February 1983.
2. T. F. Coleman and J. J. Moré. Estimation of sparse Hessian matrices and graph coloring problems. *Math. Program.*, 28:243–270, 1984.
3. T. F. Coleman and A. Verma. The efficient computation of sparse Jacobian matrices using automatic differentiation. *SIAM J. Sci. Comput.*, 19(4):1210–1233, July 1998.
4. I. Finocchi, A. Panconesi, and R. Silvestri. Experimental analysis of simple, distributed vertex coloring algorithms. In *Proceedings of the Thirteenth ACM-SIAM Symposium on Discrete Algorithms (SODA 02)*, San Francisco, CA, 2002.
5. A. H. Gebremedhin and F. Manne. Scalable parallel graph coloring algorithms. *Concurrency: Pract. Exper.*, 12:1131–1146, 2000.
6. A. H. Gebremedhin, F. Manne, and A. Pothen. Graph coloring in optimization revisited. Technical Report 226, University of Bergen, Dept. of Informatics, Norway, January 2002. Available at: http://www.ii.uib.no/publikasjoner/texrap/.
7. R. Greenlaw, H. J. Hoover, and W. L. Ruzzo. *Limits to Parallel Computation: P-Completeness Theory*. Oxford University Press, New York, 1995.
8. A.K.M. S. Hossain and T. Steihaug. Computing a sparse Jacobian matrix by rows and columns. *Optimization Methods and Software*, 10:33–48, 1998.
9. M. T. Jones and P. E. Plassmann. A parallel graph coloring heuristic. *SIAM J. Sci. Comput.*, 14(3):654–669, May 1993.
10. S. O. Krumke, M. V. Marathe, and S. S. Ravi. Approximation algorithms for channel assignment in radio networks. In *DIAL M for Mobility*, Dallas, Texas, 1998.
11. S. T. McCormick. Optimal approximation of sparse Hessians and its equivalence to a graph coloring problem. *Math. Program.*, 26:153–171, 1983.
12. V. V. Vazirani. *Approximation Algorithms*. Springer, 2001. Chapter 5.

A Parallel GRASP Heuristic for the 2-Path Network Design Problem

Celso C. Ribeiro[1] and Isabel Rosseti[1]

Department of Computer Science, Catholic University of Rio de Janeiro,
Rua Marquês de São Vicente 225, Rio de Janeiro, RJ 22453-900, Brazil
{celso,rosseti}@inf.puc-rio.br

Abstract. We propose a parallel GRASP heuristic with path-relinking for the 2-path network design problem. A parallel strategy for its implementation is described. Computational results illustrating the effectiveness of the new heuristic are reported. The parallel implementation obtains linear speedups on a cluster with 32 machines.

1 Introduction

Let $G = (V, E)$ be a connected undirected graph, where V is the set of nodes and E is the set of edges. A k-path between nodes $s, t \in V$ is a sequence of at most k edges connecting them. Given a non-negative weight function $w : E \rightarrow R_+$ associated with the edges of G and a set D of pairs of origin-destination nodes, the *2-path network design problem* (2PNDP) consists in finding a minimum weighted subset of edges $E' \subseteq E$ containing a 2-path between every origin-destination pair. Applications can be found in the design of communication networks, in which paths with few edges are sought to enforce high reliability and small delays.

Dahl and Johannessen [3] proved that the decision version of 2PNDP is NP-complete. They proposed a greedy heuristic for 2PNDP, based on the linear relaxation of its integer programming formulation. They also gave an exact cutting plane algorithm and presented computational results for randomly generated problems with up to 120 nodes, 7140 edges, and 60 origin-destination pairs.

In the next section, we propose a GRASP with path-relinking heuristic for 2PNDP. An independent-thread parallelization strategy is given in Section 3. Computational results illustrating the effectiveness of the new heuristic and the speedups achieved by its parallel implementation on a cluster with 32 machines are reported in Section 4. Final remarks are presented in the last section.

2 A GRASP Heuristic

GRASP is a multistart process, in which each iteration consists of two phases: construction and local search [4]. The best solution found after `MaxIterations` iterations is returned. The reader is referred to Resende and Ribeiro [10] for implementation strategies, variants, and extensions. In the following, we customize a GRASP for the 2-path network design problem. We describe the construction and local search procedures, as well as a path-relinking intensification strategy.

B. Monien and R. Feldmann (Eds.): Euro-Par 2002, LNCS 2400, pp. 922–926.

2.1 Construction Phase

The construction of a new solution begins by the initialization of modified edge weights with the original edge weights. Each iteration of the construction phase starts by the random selection of an origin-destination pair still in D. A shortest 2-path between the extremities of this pair is computed, using the modified edge weights. The weights of the edges in this 2-path are set to zero until the end of the construction procedure, the origin-destination pair is removed from D, and a new iteration resumes. The construction phase stops when 2-paths have been computed for all origin-destination pairs.

2.2 Local Search Phase

The local search phase seeks to improve each solution built in the construction phase. Each solution may be viewed as a set of 2-paths, one for each origin-destination pair in D. To introduce some diversity by driving different applications of the local search to different local optima, the origin-destination pairs are investigated at each GRASP iteration in a circular order defined by a different random permutation of their original indices.

Each 2-path in the current solution is tentatively eliminated. The weights of the edges used by other 2-paths are temporarily set to zero, while those which are not used by other 2-paths in the current solution are restored to their original values. A new shortest 2-path between the extremities of the origin-destination pair under investigation is computed, using the modified weights. If the new 2-path improves the current solution, then the latter is modified; otherwise the previous 2-path is restored. The search stops if the current solution was not improved after a sequence of $|D|$ iterations along which all 2-paths have been investigated. Otherwise, the next 2-path in the current solution is investigated for substitution and a new iteration resumes.

2.3 Path-Relinking

Successful applications of path-relinking [5] used as an intensification strategy within a GRASP procedure are reported in [1, 7, 11]. Implementation strategies are investigated in detail by Resende and Ribeiro [10].

In this context, path-relinking is applied to pairs (z, y) of solutions, where z is an initial solution randomly chosen from a pool with a limited number of `MaxElite` previously found elite solutions and y is the guiding solution obtained by local search. The pool is originally empty. Each locally optimal solution is considered as a candidate to be inserted into the pool if it is different from every other solution currently in the pool. If the pool already has `MaxElite` solutions and the candidate is better than the worst of them, then the former replaces the latter. If the pool is not full, the candidate is simply inserted.

The algorithm starts by determining all origin-destination pairs whose associated 2-paths are different in the two solutions z and y. These computations amount to determining a set of moves $\Delta(z, y)$ which should be applied to the initial solution to reach the guiding one. Each move is characterized by a pair of 2-paths, one to be inserted and the other to be eliminated from the current solution. The best solution $\bar{y}$ along the path to be constructed from z to y is initialized with the initial solution z. At each path-relinking iteration the best yet unselected move is applied to the current solution, the incumbent $\bar{y}$ is updated, and the selected move is removed from $\Delta(z, y)$, until the guiding solution is reached. The incumbent $\bar{y}$ is returned as the best solution found by path-relinking and inserted into the pool if it satisfies the membership conditions. The pseudo-code with the complete description of procedure GRASP+PR_2PNDP is given in Figure 1, with $f(x)$ denoting the weight of a solution x.

```
procedure GRASP+PR_2PNDP;
   Set f* ← ∞ and Pool ← ∅;
   for k = 1,...,MaxIterations do;
      Construct a randomized solution x (construction phase);
      Find y by applying local search to x (local search phase);
      if y satisfies the membership conditions then insert y into Pool;
      Randomly select a solution z ∈ Pool (z ≠ y) with uniform probability;
      Compute the set Δ(z, y) of moves;
      Let ȳ be the best solution found by applying path-relinking to (z, y);
      if ȳ satisfies the membership conditions then insert ȳ into Pool;
      if f(ȳ) < f* then do; x* ← ȳ; f* ← f(ȳ); end;
   end;
   return x*;
end GRASP+PR_2PNDP;
```

Fig. 1. Pseudo-code of the sequential GRASP with path-relinking heuristic

3 Parallel Implementation

Strategies for the parallel implementation of metaheuristics were reviewed by Cung et al. [2]. Parallel implementations of metaheuristics are much more robust than their sequential versions. Typical parallelizations of GRASP correspond to multiple-walk independent-thread strategies, based on the distribution of the iterations over the processors [8, 9]. The iterations may be evenly distributed over the processors or according with their demands, to improve load balancing.

The processors perform MaxIterations/p iterations each, where p is the number of processors and MaxIterations the total number of iterations. Each processor has a copy of algorithm GRASP+PR_2PNDP, a copy of the problem data, and its own pool of elite solutions. One of the processors acts as the master, reading and distributing the problem data, generating the seeds which will be used by the pseudorandom number generators at each processor, distributing the iterations, and collecting the best solution found by each processor.

4 Computational Results

The parallel GRASP heuristic was implemented in C, using version egcs-2.91.66 of the gcc compiler and the LAM 6.3.2 implementation of MPI. The computational experiments were performed on a cluster of 32 Pentium II-400 processors, each one with 32 Mbytes of RAM and running under version 2.2.14-5.0 of the Red Hat implementation (release 6.2) of the operating system Linux. Since neither the instances nor the codes used in [3] were available for a straightforward comparison with the parallel GRASP algorithm, we randomly generated 100 new test instances with 70 nodes and 35 origin-destination pairs, with the same parameters used in [3]. Each instance was run five times, with different seeds.

We compared the results of the parallel GRASP algorithm with those obtained by the greedy heuristic in [3], using two samples of solution values and the statistical test t for unpaired observations [6]. The main statistics are summarized in Table 1. These results made it possible to conclude with 40% of confidence that GRASP finds better solutions than the other heuristic. The average value of the solutions obtained by GRASP was 2.2% smaller than that of the solutions obtained by the other heuristic. The dominance of the former over the latter is even stronger when harder or larger instances are considered. The parallel GRASP was applied to problems with up to 400 nodes, 79800 edges, and 4000 origin-destination pairs, while the other heuristic solved problems with no more than 120 nodes, 7140 edges, and 60 origin-destination pairs.

Table 1. Statistics for GRASP (sample A) and the greedy heuristic (sample B)

	Parallel GRASP (sample A)	Greedy [3] (sample B)
Size	$n_A = 100$	$n_B = 30$
Mean	$\mu_A = 443.73$	$\mu_B = 453.67$
Standard deviation	$S_A = 40.64$	$S_B = 61.56$

We give in Table 2 the average elapsed times in seconds over four runs of the parallel GRASP heuristic for a fixed number of 3200 iterations on the same instance of 2PNDP with 400 nodes, using 1, 2, 4, 8, 16, and 32 processors. Almost-linear speedups are observed in all cases. We also report in this table the solution values obtained in the first run for each number of processors. As expected from an independent-thread parallelization strategy, solution quality slightly deteriorates (approximately 4%) when the number of processors increases from 1 to 32. Preliminary results obtained with a cooperative strategy show that the latter is able to overcome this drawback and preserves the linear speedups.

5 Concluding Remarks

We proposed in this work a parallel GRASP heuristic with path-relinking for the 2-path network design problem. Statistical tests on instances with 70 nodes

Table 2. Elapsed times and solution values for the parallel GRASP heuristic

Processors	Elapsed time (s)	Speedup	Solution value
1	27599.71	1.00	2968
2	13898.29	1.99	2976
4	6987.33	3.95	2977
8	3495.03	7.90	3012
16	1760.68	15.68	3049
32	885.72	31.16	3093

showed that the GRASP heuristic compares favourably with another heuristic reported in the literature. This dominance is even stronger when harder or larger instances are considered. Linear speedups for up to 32 processors are observed for the parallel implementation on a cluster. A cooperative parallelization strategy using a centralized pool and detailed results obtained on a broader set of larger instances will be reported elsewhere.

References

1. S.A. Canuto, M.G.C. Resende, and C.C. Ribeiro, "Local search with perturbations for the prize-collecting Steiner tree problem in graphs", *Networks* 38 (2001), 50–58.
2. V.D. Cung, S.L. Martins, C.C. Ribeiro, and C. Roucairol, "Strategies for the parallel implementation of metaheuristics", in *Essays and Surveys in Metaheuristics* (C.C. Ribeiro and P. Hansen, eds.), pages 263–308, Kluwer, 2001.
3. G. Dahl and B. Johannessen, "The 2-path network design problem", submitted for publication, 2000.
4. T.A. Feo and M.C.G. Resende, "Greedy randomized adaptive search procedures", *Journal of Global Optimization* 6 (1995), 109–133.
5. F. Glover, "Tabu search and adaptive memory programing – Advances, applications and challenges", in *Interfaces in Computer Science and Operations Research* (R.S. Barr, R.V. Helgason, and J.L. Kennington, eds.), pages 1–75, Kluwer, 1996.
6. R. Jain, *The art of Computer Systems Performance Analysis: Techniques for Experimental Desgin, Measurement, Simulation, and Modeling*, Wiley, 1991.
7. M. Laguna and R. Martí, "GRASP and path relinking for 2-layer straight line crossing minimization", *INFORMS Journal on Computing* 11 (1999), 44–52.
8. S.L. Martins, M.G.C. Resende, C.C. Ribeiro, and P. Pardalos, "A parallel GRASP for the Steiner tree problem in graphs using a hybrid local search strategy", *Journal of Global Optimization* 17 (2000), 267–283.
9. S.L. Martins, C.C. Ribeiro, and M.C. Souza, "A parallel GRASP for the Steiner problem in graphs", *Lecture Notes in Computer Science* 1457 (1998), 285–297.
10. M.G.C. Resende and C.C. Ribeiro, "GRASP", to appear in *State-of-the-Art Handbook of Metaheuristics* (F. Glover and G. Kochenberger, eds.), Kluwer.
11. C.C. Ribeiro, E. Uchoa, and R.F. Werneck, "A hybrid GRASP with perturbations for the Steiner problem in graphs", to appear in *INFORMS Journal on Computing.*

MALLBA: A Library of Skeletons for Combinatorial Optimisation*

E. Alba[3], F. Almeida[2], M. Blesa[1], J. Cabeza[2], C. Cotta[3], M. Díaz[3], I. Dorta[2], J. Gabarró[1], C. León[2], J. Luna[2], L. Moreno[2], C. Pablos[2], J. Petit[1], A. Rojas[2], and F. Xhafa[1]

[1] LSI – UPC. Campus Nord C6. 08034 Barcelona (Spain)
[2] EIOC – ULL. Edificio Física/Matemáticas. 38271 La Laguna (Spain)
[3] LCC – UMA. E.T.S.I. Informática. Campus de Teatinos. 29071 Málaga (Spain)

Abstract. The MALLBA project tackles the resolution of combinatorial optimization problems using algorithmic skeletons implemented in C++. MALLBA offers three families of generic resolution methods: exact, heuristic and hybrid. Moreover, for each resolution method, MALLBA provides three different implementations: sequential, parallel for local area networks, and parallel for wide area networks (currently under development). This paper explains the architecture of the MALLBA library, presents some of its skeletons, and offers several computational results to show the viability of the approach.

1 Introduction

Combinatorial optimization problems arise in various fields such as control theory, operations research, biology, and computer science. Several tools offering parallel implementations for generic optimization techniques such as Simulated Annealing, Branch and Bound or Genetic Algorithms have been proposed in the past (see, e.g. [4,7–9]). Some existing frameworks, such as *Local++*, its successor *EasyLocal++* [3], *Bob++* [2], and the IBM COIN open source project [6] provide sequential and parallel generic implementations for several exact, heuristic and hybrid methods, but lack features to integrate them.

The MALLBA project is an effort to develop an integrated library of skeletons for combinatorial optimization (including exact, heuristic and hybrid methods) dealing with parallelism in a user-friendly and, at the same time, efficient manner. Its three target environments are sequential computers, LANs of workstations and WANs. The main features of MALLBA are: integration of all the skeletons under the same design principles, facility to switch from sequential to parallel optimization engines, and cooperation among engines to provide more

* http://www.lsi.upc.es/~mallba. Work partially supported by: Spanish CICYT TIC-1999-0754 (MALLBA), EU IST program IST-2001-33116 (FLAGS), Future and Emerging Technologies of EU contract IST-1999-14186 (ALCOM-FT) and Canary Goverment Project PI/2000-60. C. León partially supported by TRACS program at EPCC. M. Blesa partially supported by Catalan 2001FI-00659 pre-doctoral grant.

B. Monien and R. Feldmann (Eds.): Euro-Par 2002, LNCS 2400, pp. 927–932.

powerful hybrid skeletons, ready to use on commodity machines. Clusters of PCs under Linux are currently supported and the software architecture is flexible and extensible (new skeletons can easily be added, alternative communication layers can be used, etc.).

In MALLBA, each resolution method is encapsulated into a skeleton. At present, the following skeletons are available: Divide and Conquer (DC), Branch and Bound (BnB), Dynamic Programming (DP), Hill Climbing, Metropolis, Simulated Annealing (SA), Tabu Search (TS), Genetic Algorithms (GA) and Memetic Algorithms. Moreover hybrid techniques have been implemented combining the previous skeletons, e.g., GA+TS, GA+SA, BnB+SA.

2 The MALLBA Architecture

MALLBA skeletons are based on the separation of two concepts: the concrete problem to be solved and the general resolution method to be used. While the particular features related to the problem must be given by the user, the main method and the knowledge to parallelize the execution of the resolution method is implemented in the skeleton. The users do not need to deal neither with the algorithmic part of the method nor with any parallelism issues.

Skeletons are implemented by a set of *required* and *provided* C++ classes which represent object abstractions of the entities participating in the resolution method. The *provided* classes implement internal aspects of the skeleton in a problem-independent way. The *required* classes specify information and behavior related to the problem. This conceptual separation allows us to define required classes with a fixed interface but without any implementation, so that provided classes can use required classes in a generic way. Fig. 1 depicts this architecture.

More specifically, each skeleton includes the `Problem` and `Solution` required classes, that encapsulate the problem-dependent entities needed by the resolution method. The `Problem` class abstracts the features of the problem that are relevant to the selected optimization method. The `Solution` class abstracts the features of the feasible solutions that are relevant to the selected resolution method. Depending on the skeleton, other auxiliary classes may be required. On the other hand, each skeleton offers two provided classes: `Solver` and `Setup`. The former abstracts the selected resolution method. The later contains the setup parameters needed to perform and tune the execution. The `Solver` class provides methods to run the skeleton and also to consult or change its state dynamically. The only information the solver needs is an instance of the problem to solve and the setup parameters. In order to enable an skeleton to have different solver engines, the `Solver` class defines a unique interface and provides several subclasses that provide different sequential and parallel implementations (see Fig. 1).

3 Parallel Implementations

The skeletons of the MALLBA library are currently implemented for two target environments: sequential and LAN. The user is able to use different paralleliza-

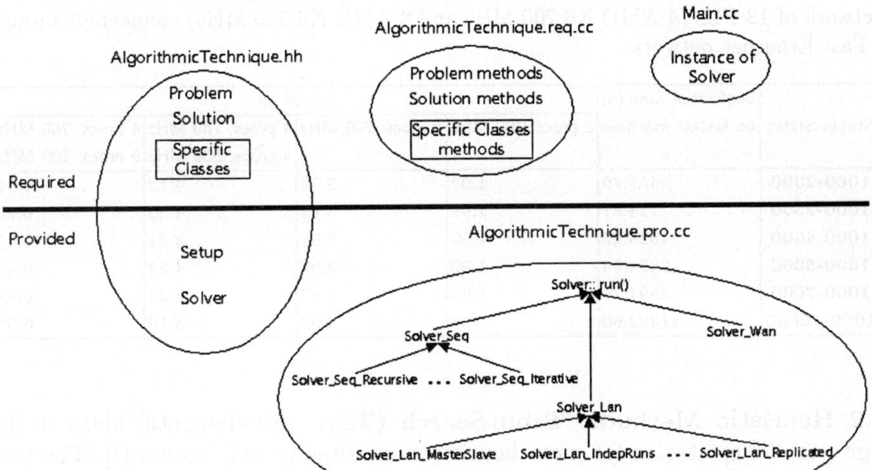

Fig. 1. Architecture of a MALLBA skeleton

tion engines just by easily extending the sequential instantiations. These different implementations can be obtained by creating separate subclasses of the `Solver` abstract class (see Fig. 1). At present, we are using our own middleware layer `NetStream` implemented on top of MPI to ease communications.

3.1 Exact Methods. The MALLBA library follows Ibaraki's discrete Dynamic Programming (DP) approach for Multistage Problems to represent DP problems and the general parallelization scheme described in [5]. The parallelization performs a cyclic mapping of a pipeline on a ring topology. Divide and Conquer (DC) and Branch and Bound (BnB) have been parallelized using a master-slave strategy. While a queue of tasks suffices for the BnB, the DC requires of a hierarchy of queues. This structure gives support to the required synchronizations, since the corresponding combination phase has to occur after all the children have been solved. Factors to take into account are the relationship between the number of available processors and the number of generated subproblems, the depth of the generated subproblems and the communication, and computation capabilities of the hosting computer. The user can choose among several strategies provided by the skeletons. Table 1 shows the speedup obtained from an instantiation of the Dynamic Programming skeleton for the Resource Allocation Problem. The parallel engine shows a good scalability until four processors. Between four and eight processors performance decreases due to the slower machines introduced, but it remains increasing when introducing more processors.

Table 1. Results for the Resource Allocation Problem using the DP skeleton, over a network of 13 PCs (4 AMD K6 700 MHz and 9 AMD K6 500 MHz) connected through a Fast Ethernet network.

Stages-States	Sequential time (s) on fastest machine	Speed-up			
		2 procs. 700 MHz	4 procs. 700 MHz	4 procs. 700 MHz 4 procs. 500 MHz	4 procs. 700 MHz 9 procs. 500 MHz
1000-2000	457.79	1.97	3.92	4.12	6.01
1000-2500	714.87	1.98	3.94	4.30	6.02
1000-4000	1828.22	1.99	3.96	4.31	6.41
1000-5000	2854.04	1.99	3.97	4.24	6.42
1000-7000	5594.74	1.99	3.97	4.22	6.41
1000-10000	11422.60	1.98	3.97	4.18	6.38

3.2 Heuristic Methods: Tabu Search (TS). Fundamental ideas to design parallel strategies for meta-heuristics are already well-known [1]. The parallel implementations for TS in the MALLBA library include: Independent Runs (IR), Independent Runs with Search Strategies (IRSS), Master-Slave (MS) and Master-Slave with Neighborhood Partition (MSNP).

We give in Table 2 some computational results obtained from the IRSS and the MSNP for the 0-1 Multidimensional Knapsack problem, for instances from the standard OR-Library. In the IRSS the communication time is almost irrelevant while in the MSNP there is a considerable communication time. This explains the fact that, for some instances, the solution found by IRSS is better than the one found by the MSNP.

3.3 Hybrid Methods: Genetic Annealing. The software architecture of MALLBA has allowed us a fast prototyping of several parallel hybrid skeletons: (GA) a parallel Genetic Algorithm with collaboration among sub-algorithms, (SA1) a parallel Simulated Annealing without collaboration among elementary SA's, (SA2) a parallel SA with collaboration among elementary SA's, and (GASA1) a hybrid skeleton where parallel GA's are applying SA as a local search

Table 2. Results from IRSS and MSNP over a network of 9 AMD K6-2 450 MHz conected through a Fast Ethernet network. Maximum execution time fixed to 900s. An instance name like OR5x250-00 is an instance of 5 constraints and 250 variables. Averages calculated over 100 executions.

instance	best cost known	avg. deviation % from the best known					
		IR with Strategies			MS with Neighb. Part.		
		2 proc.	4 proc.	8 proc.	2 proc.	4 proc.	8 proc.
OR5x250-00	59312	0.028	0.015	0.020	0.051	0.024	0.020
OR5x250-29	154662	0.005	0.006	0.003	0.012	0.007	0.006
OR10x250-00	59187	0.045	0.047	0.045	0.079	0.064	0.064
OR10x250-29	149704	0.012	0.010	0.004	0.012	0.012	0.009
OR30x250-00	56693	0.023	0.022	0.017	0.041	0.028	0.030
OR30x250-29	149572	0.009	0.010	0.003	0.016	0.009	0.007

operator inside their sequential main loop. Finally, we implemented two other hybrid skeletons where parallel GA's are run under different populations of solutions and then parallel SA's are applied, each one selecting some strings drawn from the final GA's population to improve them (by tournament -GASA2- or randomly -GASA3-). The construction of these hybrid algorithms has been extremely easy by using the MALLBA architecture.

We have solved the Maximum Cut problem: a high edge-density 20-vertex instance (`cut20-0.9`) and a 100-vertex instance (`cut100`) (see Table 3). In the hybrid parallel skeletons, SA is used at each reproduction event with probability 0.1 for 100 iterations. In the parallel-cooperation modes (all but SA1), the algorithms are arranged in ring topology (i.e., not master-slave), asynchronously migrating individuals each 200 iterations. Regarding the 100-vertex graph, the best result is provided by SA1 -4 sec.-, a parallel SA without cooperation (78% success in finding the optimum). The cooperative parallel version (SA2) is faster yet less effective (just 20% success); the reason for this result strives in that there exists a large number of local optima that SA2 repeatedly visits, while SA1 (without collaboration) focuses the search faster to an optimum in every parallel sub-algorithm, thus avoiding this "oscillation" effect and visiting a larger number of different search regions. The model GASA1 achieves an intermediate 36%-success value for `cut100`, but it is much more computationally expensive.

Table 3. Results for Maximum Cut. The experiments are done using 6 PCs (Pentium III, 700 MHz, 128 Mb) connected through a Fast Ethernet network. The "iterations" and "time" columns refer to the mean values of successful runs (those in which an optimal solution is found).

	Cut20-0.9			Cut100		
Algorithm	iterations	time(s)	successful runs	iterations	time (s)	successful runs
SA1	1742	0.11	92%	4048	3.96	78%
SA2	856	0.08	100%	2660	2.66	20%
GA	36	0.36	86%	399	78.66	6%
GASA1	5	0.50	100%	54	104.54	36%
GASA2	56	0.59	100%	178	43.84	4%
GASA3	71	0.73	94%	171	33.65	8%

4 Concluding Remarks and Future Work

We have sketched the architecture of the MALLBA library. Also, we have given some computational results obtained over a cluster of heterogeneous PCs using Linux. Our results indicate that: skeletons can be instantiated for a large number of problems, sequential instantiations provided by users are ready to use in parallel, parallel implementations are scalable, general heuristic skeletons can provide solutions whose quality is comparable to *ad hoc* implementations for concrete problems, and the architecture supports easy construction of powerful hybrid algorithms. Our future work will focus on new skeletons for WAN.

References

1. T. Crainic, M. Toulouse, and M. Gendreau. Towards a taxonomy of parallel tabu search heuristics. *INFORMS Journal on Computing*, 9(1):61–72, 1997.
2. B. L. Cun. Bob++ library illustrated by VRP. In *European Operational Research Conference (EURO'2001)*, page 157, Rotterdam, 2001.
3. L. Di Gaspero and A. Schaerf. EasyLocal++: an object-oriented framework for the flexible design of local search algorithms and metaheuristics. In *4th Metaheuristics International Conference (MIC'2001)*, pages 287–292, 2001.
4. J. Eckstein, C. A. Phillips, and W. E. Hart. Pico: An object-oriented framework for parallel branch and bound. Technical report, RUTCOR, 2000.
5. D. González, F. Almeida, J. Roda, and C. Rodríguez. From the theory to the tools: Parallel dynamic progr. *Concurrency: Practice and Experience*, (12):21–34, 2000.
6. IBM. COIN: Common Optimization INterface for operations research, 2000. http://oss.software.ibm.com/developerworks/opensource/coin/index.html.
7. K. Klohs. Parallel simulated annealing library. http://www.uni-paderborn.de/fachbereich/AG/monien/SOFTWARE/PARSA/, 1998.
8. D. Levine. PGAPack, parallel genetic algorithm library. http://www.mcs.anl.gov/pgapack.html, 1996.
9. S. Tschöke and T. Polzer. Portable parallel branch-and-bound library, 1997. http://www.uni-paderborn.de/cs/ag-monien/SOFTWARE/PPBB/introduction.html.

Topic 16
Mobile Computing, Mobile Networks

Friedhelm Meyer auf der Heide, Mohan Kumar,
Sotiris Nikoletseas, and Paul Spirakis

Topic Chairpersons

The development of small and powerful computing devices and wireless, mobile communication systems offer a great variety of new applications. The design and analysis of efficient and robust mobile networks impose new challenges, which necessitate a complementary use of technology and algorithms. The aim of this topic is to bring together computer scientists and engineers in the areas of wireless mobile networks, mobile computing and parallel and distributed computing.

19 papers were submitted to Topic 16. 1 paper was withdrawn. All papers were reviewed by at least three referees, while the vast majority of them were reviewed by four referees. To support the reviewing process, a total of 23 reviews from 18 external referees specializing in the areas under consideration were collected. The Topic Committee would like to sincerely thank all those who contributed papers and the colleagues who helped with the reviewing process. Also, we would like to thank the EUROPAR 2002 Organizing Committee for valuable help.

As a result of this thorough reviewing process, 7 papers out of the 19 submissions were accepted: 1 as a distinguished paper, 3 as regular papers and 3 as short papers.

The paper titled "Distributed Maintenance of Resource Efficient Wireless Network Topologies", by M. Gruenewald, T. Lukovszki, C. Schindelhauer and K. Volbert, was accepted as a distinguished paper. This work investigates the energy, congestion and dynamic performance properties of efficient wireless network topologies and proposes a new one which is best with respect to dynamic behavior.

The paper "Weak Communication in Radio Networks", by T. Jurdzinski, M. Kutylowski and Jan Zatopianski, compares the computational power of weak and strong models of radio networks and presents an efficient simulation of the strong model by the weak one. The paper was selected as a regular one.

The work titled "A Source Route Discovery Optimization Scheme in Mobile Ad hoc Networks", by A. Boukerche, proposes, implements and evaluates an optimization technique for routing in ad-hoc mobile networks that uses GPS information and significantly increases the efficiency of the network load. This paper was accepted as a regular paper.

The regular paper titled "A Local Decision Algorithm for Maximum Lifetime in Ad Hoc Networks", by A.Clemantis, D. D'Agostino and V. Gianuzzi, proposes and experimentally evaluates a new routing algorithm that allows local selection of the next routing hop to optimize the ad hoc network's lifetime.

B. Monien and R. Feldmann (Eds.): Euro-Par 2002, LNCS 2400, pp. 933–934.

The short paper "An Efficient Time-based Checkpointing Protocol for Mobile Computing Systems over Wide Area Networks", by C-Y. Lin, S-C. Wang and S-Y. Kuo, proposes an efficient and non-blocking coordinated checkpointing method.

The short paper titled "Coordination of Mobile Intermediaries Acting on behalf of Mobile Users", by N. Zaini and L. Moreau, deals with building distributed applications across mobile devices and fixed infrastructure, by introducing a protocol for coordinating mobile Intermediaries, which are called Shadows.

Finally, the short paper by S-H. Hwang and K-J. Han, titled "A Discriminative Collision Resolution Algorithm for Wireless MAC Protocol", proposes a wireless Medium Access Control method to support the Quality of Service requirements of real-time applications.

Distributed Maintenance of Resource Efficient Wireless Network Topologies
(Extended Abstract)

Matthias Grünewald*, Tamás Lukovszki**,
Christian Schindelhauer**, and Klaus Volbert**

Heinz Nixdorf Institute, Paderborn University

Abstract. Multiple hop routing in mobile ad hoc networks can minimize energy consumption and increase data throughput. Yet, the problem of radio interferences remains. However if the routes are restricted to a basic network based on local neighborhoods, these interferences can be reduced such that standard routing algorithms can be applied.

We compare different network topologies for these basic networks with respect to degree, spanner-properties, radio interferences, energy, and congestion, i.e. the **Yao-graph** (aka. Θ-graph) and some also known related models, which will be called the **SymmY-graph** (aka. YS-graph), the **SparsY-graph** (aka.YY-graph) and the **BoundY**-graph. Further, we present a promising network topology called the **HL-graph** (based on **H**ierarchical **L**ayers).

Further, we compare the ability of these topologies to handle dynamic changes of the network when radio stations appear and disappear. For this we measure the number of involved radio stations and present distributed algorithms for repairing the network structure.

1 Motivation

Our research aims at the implementation of a mobile ad hoc network based on distributed robust communication protocols. Besides the traditional use of omni-directional transmitters, we want to investigate the effect of space multiplexing techniques and variable transmission powers on the efficiency and capacity of ad hoc networks. Therefore our radios can send and receive radio signals independently in k sectors of angle θ using one frequency. Furthermore, our radio stations can regulate its transmission power for each transmitted signal. To show that this approach is also suitable in practical situations, we are currently developing a communication module for the mini robot Khepera [11, 8] that can transmit and receive in eight sectors using infrared light with variable transmission distances up to one meter, see Fig. 1. A colony of Khepera robots will be

* Dept. of Electrical Engineering and Information Technology, System & Circuit Technology, gruenewa@hni.uni-paderborn.de. Partially supported by the DFG-SFB 376.

** Dept. of Mathematics and Computer Science, tamas@hni.upb.de, {schindel, kvolbert}@upb.de. Partially supported by the DFG-SFB 376 and by the Future and Emerging Technologies programme of the EU, contract nr. IST-1999-14186 (ALCOM-FT).

B. Monien and R. Feldmann (Eds.): Euro-Par 2002, LNCS 2400, pp. 935–946.

equipped with this modules to establish ad hoc networks and to evaluate our research results under realistic conditions.

We assume that most of the time the network is stable and performs a point-to-point communication protocol according to an adequately chosen routing protocol. In [10] it is shown that the quality of the routing depends on the choice of the underlying network that we call *basic network*. In this paper we investigate how such networks can be maintained when stations enter and leave the network.

Little is known about the efficient design of topology-preserving dynamic algorithms. Many approaches consider a model where a central algorithm controls the network structure, using the exact coordinates in $\mathbb{R}^2$ of the radio stations (e.g., [3, 14]). In contrast to this model we want to investigate a distributed network model where the only information available is given by incoming radio signals and which sector it is received, which gives a rough estimation of the direction to the sender.

The dynamics we are investigating is that a single radio station enters or leaves the system, while the rest of the system is stable. We claim that a node entering a network knows this situation, e.g. because it is switched on or it eavesdrops on existing communication from the network. A node leaving the system is equivalent to a complete node failure. This means that it is not necessary that the leaving node informs the network. Such dynamic changes are the most frequent changes of a radio network besides the motion of radio stations.

In our view its very unlikely that all mobile radio station would start (or leave) at the same time. And even if this is enforced one can easily add a probabilistic strategy that prevents this situation. Then the establishment of the complete network turns out to be a series of single stations entering an existing network. This approach makes sense, since nobody expects that a radio connection to the network is instantly established and we will see that there exist network structures where entering and leaving will only need some logarithmic communication rounds.

In this paper, we do not address the problem of moving radio stations. However, if the movement is not too fast, the moving node can reestablish the correct network by triggering a `leave` and an `enter`-operation. Furthermore, we hope that the basic routines developed for this switching dynamics provide basic techniques for more sophisticated maintenance techniques of mobile ad hoc networks.

2 Model

Our investigations concentrate on the implementation of distributed algorithms for mobile ad hoc networks with radio stations with specific hardware features. However, some network topologies (like the HL-graph) can be used in a much more general hardware model.

2.1 Communication Model

In this paper we assume that if a station enters the system it will send out control messages to stop normal packet routing for the (hopefully short) time needed to update the network structure. All packets are stored on the radio stations and delivered when the network structure has been restored. In contrast to this reactive approach, one can also take advantage of synchronized clocks if available. If a periodically time period is reserved that is known to all nodes (including new ones), the maintenance of the network can be done in this special maintenance period. Thus, no control packets are necessary to stop the packet routing mode and collisions caused by the control packets can be prevented.

In our communication model, we assume that a radio station w, also called node, is able to detect three types of incoming signals: *No signal* indicates that no radio signal is transmitted at all or that all radio stations r in distance d send with transmission distance $d' < d$. The *interference signal* indicates that at least two radio stations u and v send in this time step t with transmission distance $d(u,t) > ||u,w||_2$ and $d(v,t) > ||v,w||_2$, where $||u,w||_2$ denotes the Euclidean distance. A *clear signal* is received by w if one radio signal with appropriated transmission power to cancel out weaker incoming signals is reaching u's antenna. Then it can read the transmitted information $m \in \{0,1\}^p$ of some length p. A communication round is the time necessary to send one packet of length p, where p is large enough to carry some elementary information like the sending station, the addressed stations (if specified), the transmission distance, and some control information.

We assume that there is a timing schedule adapted to the basic network topology that allows the stations in a static time period, i.e. no nodes enter or leave, to transmit and acknowledge packets over the network routes with only small number of interfering packets. During such a phase we can neglect the interfering impact of acknowledgment signals.

However when a connection is established the sending and answering signal have the same small length, because only control information needs to be transmitted. Then the impact of answering signals is the same as those of sending signals. Therefore, we consider two types of interferences: The *uni-directional interferences* in the routing mode and the *bi-directional interferences* when connections are established or network changes are compensated.

2.2 Hardware Model

Every node can choose the transmitting power according to s discrete choices $p_1, \ldots, p_s$. The energy to send over distance d is given by $\text{pow}(d) := d^c$ for some constant $c \geq 2$ (constant factors are omitted for simplicity). This defines the transmitting distances $d_i = (p_i)^{1/c}$ for all $i \in [s]$, where $[s] := \{1, \ldots, s\}$.

Every node u has k sending and receiving devices, which are located such that they can communicate in parallel within each of k disjoint sectors with angle $\theta = \frac{2\pi}{k}$. Every node u has been rotated by a angle α_u, which is unknown to u. Note that the radio stations have different offset angles α_u. If u sends a signal in the ith sector it actually sends into a direction described by the interval $R = [\alpha_u + i\theta, \alpha_u + (i+1)\theta)$ and can be received by node v in sector j if $R \cap [\alpha_v + j\theta, \alpha_v + (j+1)\theta) \neq \emptyset$. Of course v receives u only if in addition u sends this signal with transmission distance $d_i \geq ||u,v||_2$.

Furthermore, we allow that radio stations can measure distances only by sending messages with varying transmission power. Then the receiving party can only decide whether the signal arrives or not. This restricts transmission distances to the set $S = \{d_1, \ldots, d_s\}$.

Define $D : \mathbb{R} \to \{\varnothing, 1, \ldots, s\}$ as the minimum discrete choice of transmission power to send over a given distance by $D(x) := \min\{i \mid d_i \geq x\}$ if $x \leq d_s$ and $D(x) := \varnothing$ if $x > d_s$.

Define $\sphericalangle(u, v) := \left\lfloor \frac{[\measuredangle(v-u)-\alpha_u]}{\theta} \right\rfloor$ mod k as the number of u's sector containing the edge (u, v), where $\measuredangle(x)$ denotes the angle of a vector x in $\mathbb{R}^2$.

2.3 Location of Nodes

One of the most delimiting properties is that radio stations do not know their locations. The following restriction prevent the vertex set from taking abnormal positions.

Definition 1. *A vertex set V is in* **general position**, *if there are no vertices $u, v, w \in V$ with $v \neq w$ and $||u, v||_2 = ||u, w||_2$. The (Euclidean) distance between two nodes $u, v \in V$ is given by $||u, v||_2$. We call a vertex set* **normal**, *if for a fixed polynomial $p(n)$ we have $\frac{\max_{u,v \in V} ||u,v||_2}{\min_{u,v \in V} ||u,v||_2} \leq p(n)$. We call the locations of radio stations* **nice**, *if for all $u, v, w \in V$ we have $D(u, v) \neq \varnothing$ and $v = w \iff \sphericalangle(u, v) = \sphericalangle(u, w) \wedge D(u, v) = D(u, w)$.*

3 Basic Network Topologies

3.1 Yao-Graphs and Variants

The underlying hardware model allows to communicate in k disjoint sectors in parallel. Therefore a straight-forward approach is to choose as a communication partner the nearest neighbor in a sector. This leads to the following definition.

Definition 2. *The* **Yao-graph** *(aka. Θ-graph) is defined by the following set of directed edges: $E := \{(u, v) \mid \forall w \neq u \ : \ \sphericalangle(u, v) = \sphericalangle(u, w) \ \Rightarrow \ D(u, v) \leq D(u, w)\}$.*

Throughout this paper we assume vertex sets to be nicely located, hence every node has at most one neighbor in a sector. The out-degree is therefore bounded by k. However, a node can be the nearest node of many nodes. To overcome this problem of high in-degree resulting in time-consuming interference resolution schedules, we present three Yao-graph based topologies.

The **symmetric Yao Graph** is a straight-forward solution of the high in-degree problem. An edge (u, v) is only introduced if u is the nearest neighbor of v and vice versa.

Definition 3. *Let G_θ be the Yao-graph of a vertex set V. Then, the* **Symmetric Yao graph (SymmY)** *is defined by the edge set $E := \{(u, v) \in E(G_\theta) \mid (v, u) \in E(G_\theta)\}$.*

Although such a graph reduces interferences to a minimum (because in every sector only at most one neighbor appears) very long detours may appear, which make such a graph incapable of bearing short routes and allowing routing without bottlenecks.

Following the approach of [13] we consider also a graph topology which allows at most two neighbors in a sector and call this graph **sparsified Yao-graph**, which is a Yao-graph where, when the in-degree of a sector exceeds one, only the incoming shortest edge will be chosen.

Definition 4. *For a given vertex set V the edge set of the* **Sparsified Yao graph** *(***SparsY-graph***) is defined by* $E := \{(u,v) \in E(G_\theta) \mid \forall w \in V : ((w,v) \in E(G_\theta)$ *and* $\sphericalangle(v,w) = \sphericalangle(v,u)) \Rightarrow D(v,w) > D(v,u)\}$*, where G_θ denotes the Yao-graph of V.*

It is an open problem whether all SparsY-graphs are c-spanners, i.e. the shortest path between vertices in the network is at most c-times longer than the Euclidean distance.

To construct a c-spanner with constant degree Arya et al. [2] introduced the following transformation. Like in [9] we apply this technique to the Yao-graph and call the resulting graph a **Bounded Degree Yao graph (BoundY graph)**.

For this, let $G = (V, E)$ be a c'-spanner with bounded out-degree. Let $N(v) = \{w \in V : wv \in E\}$ the set of in-neighbors of $v \in V$. For each $v \in V$, the star defined by the edges $\{wv \in N(v)\}$ will be replaced by a so-called v-single sink c''-spanner, $c'' = c/c'$, $T(v)$, which has a bounded in- and out-degree, i.e. $G^* = (V, E^*)$, where $E^* = \bigcup_{v \in V, uw \in E(T(v))} uw$.

A graph with a vertex set U is called a *v-single sink c''-spanner (a (v, c'')-SSS)*, if from each vertex $w \in U$ there is a c''-spanner path to the vertex v. Such a (v, c'')-SSS for U can be constructed as follows.

Let $\alpha = 2 \arcsin \frac{c''-1}{2c''}$. We divide the plane around v into sectors of an angular diameter at most α. For each sector C, let U_C be the set of all vertices of $U \setminus \{v\}$ contained in C. If a subset U_C contains more than $|U|/2$ vertices, then we partition it arbitrarily into two subsets $U_{C,1}$ $U_{C,2}$, each of size at most $|U|/2$ For each subset U_C, let $w_c \in U_C$ be the vertex wich is closest to v. We add the edge $w_C v$ and then we recursively construct a (w_C, c'')-SSS for each subset U_C. This recursion ends after $\log |U|$ steps, since we halve (at least) the number of vertices at each level of the recursion. In this way we obtain a directed tree $T(v)$ with root v which is a (v, c'')-SSS for $N(v) \cup \{v\}$. Since each vertex v had a bounded out-degree in G, and therefore it can be contained in a constant number of in-neighborhoods $N(u)$, $u \in V$, its degree in G^* will be also bounded. This completes the construction of the BoundY-graph.

The above recursive construction allows the distributed construction of the BoundY graph given the Yao graph. Furthermore, for compass routing it provides suitable rerouting information: If a message wants to use an edge uv in the Yao-Graph, then it will use the tree-path from u to v in $T(v) \subset G^*$, which has at most $O(\log n)$ hops.

3.2 The Hierarchical Layer Graph

Adopting ideas from clustering [6,7] and generalizing an approach of [1] we present a graph consisting of w layers $L_0, L_1, \ldots, L_w$. The union of all this graphs gives the **Hierarchical Layer graph (HL graph)**. The lowest layers L_0 contains all vertices V. The vertex set of a higher layer is a subset of the vertex set of a lower layer until in the highest layer there is only one vertex, i.e. $V = V(L_0) \supseteq V(L_1) \supseteq \cdots \supseteq V(L_w) = \{v_0\}$.

The crucial property of these layers is that in each layer L_i vertices obey a minimum distance: $\forall u, v \in V(L_i) : ||u,v||_2 \geq r_i$. Furthermore, all nodes in the next-lower layer must be covered by this distance: $\forall u \in V(L_i)\ \exists v \in V(L_{i+1}) : ||u,v||_2 \leq r_{i+1}$. Our construction uses parameters $\alpha \geq \beta > 1$, where for some $r_0 < \min_{u,v \in V} ||u,v||_2$ we use radii $r_i := \beta^i \cdot r_0$ and we define in layer L_i the edge set $E(L_i)$ by $E(L_i) := \{(u,v) \mid u, v \in V(L_i) \wedge ||u,v||_2 \leq \alpha \cdot r_i\}$.

Clearly, for a normal vertex set we have a maximum number of $w = O(\log n)$ layers. For HL-graphs we need not assume nice or normal locations, as long as our hardware models supports the following transmission distances: $\{r_i, \alpha r_i \mid i \in \{0,\dots,w\}\} \subseteq S = \{d_1,\dots,d_s\}$, $d_0 \le \min_{u,v\in V} ||u,v||_2$ and $d_w \ge \max_{u,v\in V} ||u,v||_2$.

4 Elementary Graph Properties

We can show the following inclusions. Note that $A \not\#\, B$ denotes $A \not\subseteq B$ and $B \not\subseteq A$.

Lemma 1. *Let V be a nice vertex set. Then,* SymmY$(V) \subseteq$ SparsY$(V) \subseteq$ BoundY(V) *and* SparsY$(V) \subseteq$ Yao(V). *For some V it holds that* BoundY$(V) \not\#$ Yao(V).

Lemma 2. *For normal and nice vertex sets V consisting of n nodes we observe:*

Topology	*Yao*	*SymmY*	*SparsY*	*BoundY*	*HL*
in-degree	$n-1$	k	k	$(k+1)^2$	$O(\log n)$
out-degree	k	k	k	k	$O(\log n)$
degree	$n-1$	k	$2k$	$k+(k+1)^2$	$O(\log n)$

Definition 5. *A graph $G = (V, E)$ is a* **c-spanner**, *if for all $u, v \in V$ there exists a (directed) path p from u to v with $||p||_2 \le c \cdot ||u,v||_2$. G is a* **weak c-spanner**, *if for all $u, v \in V$ there exists a path p from u to v which is covered by a disk of radius $c \cdot ||u,v||_2$ centered at u. G is a (c, d)-***power spanner**, *if for all $u, v \in V$ there is a path $p = (u = u_1, u_2, \dots, u_m = v)$ from u to v in G such that $\sum_{i=1}^{m-1}(||u_i, u_{i+1}||_2)^d \le c \min_{(u=v_1,v_2,\dots,v_w=v)} \sum_{i=1}^{w-1}(||v_i, v_{i+1}||_2)^d$. If for all $d > 1$ there exists a constant c such that G is a (c, d)-power spanner we call G a* **power spanner**.

On the positive side the following results are known.

Lemma 3. *Let $V \subset \mathbb{R}^2$.*
- *[12] For $k > 6$ the Yao graph is a c-spanner with $c = 1/(1 - 2\sin\frac{\theta}{2})$.*
- *[5] For $k \ge 6$ and $c = \max\left((1 + 48\sin^4(\theta/2))^{1/2}, \sqrt{5} - \cos\theta\right)$ the Yao-graph is a weak c-spanner.*
- *[4] For $k = 4$, the Yao-graph is a weak c-spanner with $c = \sqrt{3+\sqrt{5}}$.*
- *[2] For $k > 6$ the BoundY-graph is a c-spanner for a constant c.*
- *[13] For $k > 6$ the SparsY-graph is a power spanner*

It is an open problem whether SparsY-graphs are c-spanners, but we show that they are weak spanners (and the proof of this theorem can be used give a proof of the power spanner property without assuming that the angle k is depending on V as done in [13]). In theorem 3, we prove the open problem stated in [13].

Theorem 1. *For $k > 6$ the SparsY-graph is a weak c-spanner where $c = \frac{1}{1-2\sin\frac{\theta}{2}}$.*

Proof. Let $G = (V, E)$ be the SparsY-graph and $G_Y = (V, E_Y)$ be the underlying Yao-graph. Starting from two vertices u, v we will show how to find a directed path from u to v in the SparsY-graph that is inside a disk with center at u of radius $||u,v||_2/(1 - 2\sin\frac{\theta}{2})$. For a sector i, define the Yao-neighbor v of a vertex u as the (unique) vertex v with $(u, v) \in E_Y$. Then we know:

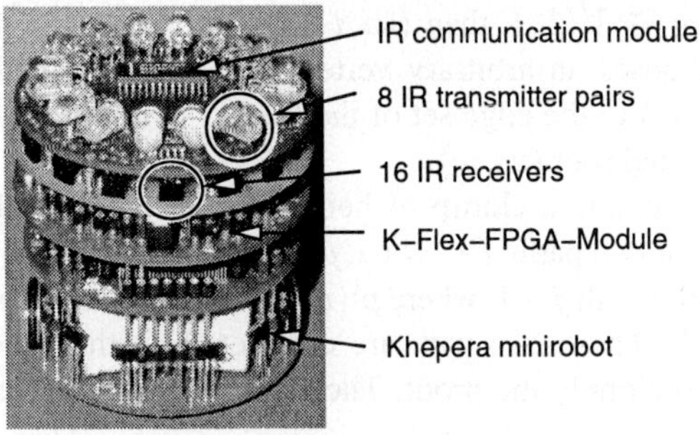

Fig. 1. The mini-robot Khepera equipped with an infrared communication module designed for sector-based variable-power communication.

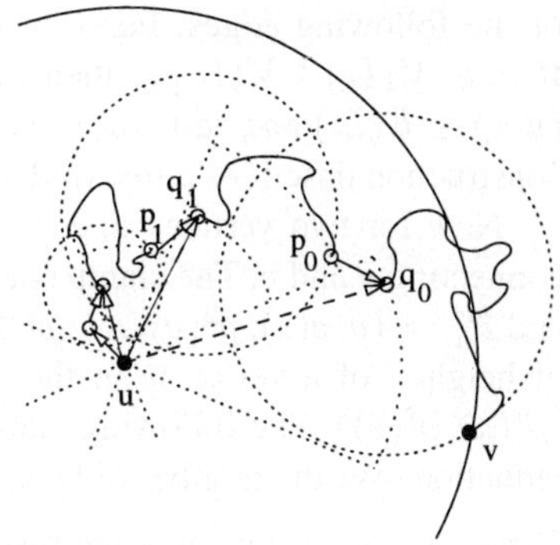

Fig. 2. Weak spanner property of the SparsY-graph

- If a node u has no directed edge in a sector i, then either the sector is empty (i.e. no edge in the Yao-graph), or there is a Yao-neighbor v (i.e., $(u, v) \in E_Y$) incident to an edge $(w, v) \in E$, where w is in another sector of u. Furthermore, $||u, w||_2 < ||u, v||_2$, because $\theta < \pi/3$ and $||v, w||_2 < ||u, v||_2$.
- Every node u has at least one neighbor v, i.e. $\exists v \in V \, : \, (u, v) \in E$.

Now, we recursively construct the path $P(u, v)$ using some of the Yao-neighbors of u (see Fig. 2). If $(u, v) \in E$ then $P(u, v) = ((u, v))$, if $u = v$ then $P(u, v) = ()$. If in sector $i = \sphericalangle(u, v)$ the Yao-neighbor, called q_0, is not directly connected to u. Then, we know that there exists an edge $(p_0, q_0) \in E$, where p_0 is in a sector $i_1 \neq i_0$ of u and $||p_0, u||_2 < ||q_0, u||_2$. Furthermore we have that $||q_0, u||_2 \leq ||u, v||_2$. Then, we repeat this consideration for the sector i_1 and replace v by p_1. This iteration ends when a Yao-neighbor q_m or p_m is directly connected to u, i.e. $(u, q_m) \in E$ or $(u, p_m) \in E$. Because every node has at least one neighbor in E this process terminates.

Now we recursively define the path $P(u, v)$ from u to v that terminates at node q_m (for p_m the path can be defined analogous: replace (u, q_m) by $(u, p_m) \circ (p_m, q_m)$) by

$$P(u, v) = (u, q_m) \circ P(q_m, p_{m-1}) \circ (p_{m-1}, q_{m-1}) \circ \ldots \circ P(q_1, p_0) \circ (p_0, q_0) \circ P(q_0, v).$$

Note that all nodes p_i, q_i are inside the disk with center u and radius $||u, v||_2$. Furthermore, we have $||q_i, p_{i-1}||_2 < ||u, v||_2$. In the next recursion vertices of the path may lie outside of this disk. However it is straight-forward that the maximum disk amplification of a recursion step is $2 \sin \frac{\theta}{2} \, ||u, v||_2$. That means, that the maximum amplification of the disk with center u and radius $||u, v||_2$ can be at most $||u, v||_2 + 2 \sin \frac{\theta}{2} \, ||u, v||_2$ in each recursion step. Let r be the depth of the recursion, then by $\sum_{i=0}^{r} (2 \sin \frac{\theta}{2})^i \, ||u, v||_2 \leq ||u, v||_2 / (1 - 2 \sin \frac{\theta}{2})$ it follows, that $P(u, v)$ is inside the disk with center u of radius $||u, v||_2 / (1 - 2 \sin \frac{\theta}{2})$ and so we get $c = 1/(1 - 2 \sin \frac{\theta}{2})$.

Theorem 2. *If $\alpha > 2\frac{\beta}{\beta - 1}$ the HL-graph is a c-spanner for* $c = \max\left\{\beta \frac{\alpha(\beta-1)+2\beta}{\alpha(\beta-1)-2\beta}, \frac{\alpha}{\beta}\right\}$.

Proof. Define a directed tree T on the vertex set $V_0 \times \{0, \ldots, w\}$ as follows. The leaves of T are all pairs $V_0 \times \{0\}$. If $u \in V(L_i)$, then (u, i) is a vertex of T. T consists of the following edges: For $i > 0$ if $u \in V(L_i)$, then $((u,i-1),(u,i)) \in E(T)$. If $u \in V(L_i) \setminus V(L_{i+1})$ then one chooses an arbitrary vertex $v \in V(L_{i+1})$ with $(u,v) \in E(L_i)$ and add $((u,i),(v,i+1))$ to the edge set of the tree T. Note that this construction describes a tree of depth w and root (v_0,w).

Now for two vertices $u,v \in V$ we define a **clamp** of height j, which is a path connecting u and v. The clamp consists of two paths $P_u^j := (u,p(u),p^2(u),\ldots,p^j(u))$ and $P_v^j := (v,p(v),p^2(v),\ldots,p^j(v))$ of length $j-1$, where $p^i(w)$ denotes the ancestor of height i of a vertex w in the tree T. These two path are connected by the edge $(p^j(u),p^j(v))$. The following claims now imply the proof. They can be proved by an induction over the number of layers.

Claim. If for vertices u,v the distance is bounded by $||u,v||_2 \leq d_j$, then a clamp of height j is contained in the HL-graph, where $d_0 = \alpha r_0$ and $d_{j+1} := (\alpha\beta-\alpha-2\beta)r_j+d_j$.

Claim. A clamp C of height j has maximum length ℓ_j, where $\ell_0 := \alpha r_0$ and $\ell_{j+1} := (\alpha\beta-\alpha+2\beta)r_j+\ell_j$.

Theorem 3. *The SymmY-graph is* **neither** *a weak c-spanner for any constant $c \in \mathbb{R}$,* **nor** *a (c,d)-power spanner for any $d > 1$.*

Proof. We show an example for n points in the plane, such that the SymmY-graph of that points is not a weak c-spanner for any c. Let ℓ_1 and ℓ_2 be two vertical lines of unit distance from each other, such that ℓ_2 is right to ℓ_1. Rotate ℓ_1 clockwise around its intersection point with the x-axis by a very small angle δ_c, and rotate ℓ_2 counterclockwise around its intersection point with the x-axis by an angle δ_c. We denote the rotated lines by ℓ_1' and ℓ_2'. Consider the vertex sets $U = \{u_1,\ldots,u_m\}$ and $V = \{v_1,\ldots,v_m\}$, $m = n/2$, placed on ℓ_1' and ℓ_2', respectively, as follows. Assume that for each point $u \in U$, the half-line, halving the ith sector of u is horizontal and directed in positive x-direction, and for $v \in V$, the half-line, halving the i'th sector of v is horizontal and directed in negative x-direction. The vertex u_1 is placed on the intersection point of ℓ_1 and the x-axis. We place v_1 on ℓ_2' such that v_1 is in the ith sector of u_1 and it is very close to the upper boundary of the ith sector of u_1. The vertex u_2 is placed on ℓ_1' in the i'th sector of v_1 close to the upper boundary of that sector. The vertex v_2 is placed on ℓ_2' in the ith sector of u_2 close to the upper boundary of that sector, etc... Then the SymmY-graph does not contain any edge (u,v) such that $u \in U\setminus\{u_m\}$ and $v \in V\setminus\{v_m\}$. The nearest neighbor of u_1 in sector i is v_1, while v_1 has u_1 and u_2 also in sector i', where u_2 is nearer, etc... Only the last link u_m,v_m will be established. Therefore, even if there is a path from u_1 to v_1 in the SymmY-graph, its length is at least $||u_1,u_m||_2+||u_m,v_m||_2+||v_m,v_1||_2$. For any given c we can choose δ_c appropriately small, in order to get $||u_1,u_m||_2, ||v_m,v_1||_2 \geq c/2$. This proves the claim.

Theorem 4. *For $k \geq 6$ and for general vertex sets the SymmY-graph is connected.*

Proof. We prove this by an induction over the distance of vertices $||u,v||_2$. First, note that the closest pair of vertices form an edge of the SymmY-graph. Now observe for two vertices u,v: Either there is an edge from u to v or there is a vertex w with $\sphericalangle(u,v) = \sphericalangle(u,w)$ and $||u,w||_2 < ||u,v||_2$ (or symmetrically $\sphericalangle(v,u) = \sphericalangle(v,w)$ and $||v,w||_2 < ||u,v||_2$). Because $\theta \leq \pi/3$ we have $||v,w||_2 < ||u,v||_2$. By induction there is a path from u to w and a path from w to v. Therefore a path from u to v exists.

5 Network Properties

In [10] we investigate the basic network parameters **interference number**, **energy**, and **congestion**. In this paper we extend the definition of interference number to directed communication. The reason is that we allow two communication modes. In the packet routing mode acknowledgment signals are very short and we can neglect its impact on the interferences. When control messages have to be exchanged sending and answering signals are both short, then we have to consider all combination of interferences. Therefore we distinguish the following of interferences.

Definition 6. *The edge* (r, s) *has a* **uni-directional interference** *caused by* (u, v), *denoted by* $(r, s) \in \mathrm{UInt}(u, v)$, *if* $\sphericalangle(s, r) = \sphericalangle(s, u)$ *and* $\sphericalangle(u, s) = \sphericalangle(u, v)$ *and* $D(u, v) \geq D(u, s)$. *The edge* (r, s) **bi-directionally interferes** *with* (u, v), *denoted by* $(r, s) \in \mathrm{BInt}(u, v)$, *if* $(r, s) \in \mathrm{UInt}(u, v)$ *or* $(s, r) \in \mathrm{UInt}(u, v)$ *or* $(r, s) \in \mathrm{UInt}(v, u)$ *or* $(s, r) \in \mathrm{UInt}(v, u)$. *The (bi-directional)* **interference number** *of a basic network* G *is defined by* $\max_{e \in E}\{1 + |\mathrm{BInt}(e)|\}$, *where* $\mathrm{BInt}(e)$ *denotes the set of edges that interfere with e if packets are simultaneously transmitted.*

Analogously, we define the **uni-directional interference number** *of a graph, by replacing* $\mathrm{BInt}(e)$ *by* $\mathrm{UInt}(e)$.

Note that both types of interferences are asymmetric, i.e. $u \in \mathrm{BInt}(v) \not\Leftrightarrow v \in \mathrm{BInt}(u)$ and analogously for UInt. This stems from the fact that we use adjustable transmission distances.

A routing protocol can be described by a set of paths $\mathcal{P}$, called path system, that optimizes network parameters. We assume that the path system is chosen according to a demand $w : V \times V \to \mathbb{N}$ representing the point-to-point communication traffic within the network. Since the locations of the vertex sets are nice for every combination of vertices there is at least a path p from u to v in the path system if $w(u, v) > 0$.

Definition 7. *The load* $\ell(e)$ *of an edge e is the number of packets that are using this edge. The interfering load of an edge is* $\ell(e) + \sum_{e' \in \mathrm{UInt}(e)} \ell(e')$. *The edge with the maximum interfering load defines the* **congestion of a path system**. *The* **energy of a path system** $\mathcal{P}$ *is given by* $\sum_{P \in \mathcal{P}} \sum_{e \in P} \mathrm{pow}(e)$, *where* $\mathrm{pow}(e) = (||e||_2)^2$. *This is* $\sum_e \ell(e)(||e||_2)^2$.

It turns out that energy and congestion are connected to power spanners and weak spanners. The link between these geometric properties and the networking features is described by the following theorem:

Lemma 4. *(i) If the basic network is a* (c, d) *power-spanner, then it allows a path system that approximates the optimal energy path system by a constant factor of* c.
(ii) Every c *spanner is* (c^d, d) *power-spanner.*
(iii) If for a normal vertex set the basic network is a weak c*-spanner* G *with uni-directional interference number* q *then there is a path system in* G *that approximates the optimal path system minimizing the congestions by a factor of* $O(q \log n)$.

Combining this Lemma with the basic graph properties investigated in section 3 we obtain:

Theorem 5. *For a nicely located vertex set V the following table describes the features of our graph topologies.*

Topology	*uni-directional interference nr.*	*Spanner*	*Energy approx. factor*	*Congestion approx. factor*
Yao-graph	$n-1$	*yes*	$O(1)$	—
SymmY-graph	1 *(bi-direct.!)*	*no, but connected*	—	—
SparsY-graph	1	*weak and power spanner*	$O(1)$	$O(\log n)$
BoundY-graph	$\Theta(n)$	*yes*	$O(1)$	—
HL-graph	$O(\log n)$	*yes*	$O(1)$	$O(\log^2 n)$

6 Maintaining the Network

The standard mode of an ad hoc network is the packet routing mode. In the lucky case of SymmY-graphs there are no interferences between messages and acknowledgments of different edges. For the SparsY-graph packets sent along the direction of the edges cannot interfere with other packets on different edges. However, acknowledgment signals of such edges can interfere. Since in the normal transportation mode data packets are long compared to the short acknowledgments, we neglect this interaction.

In all other graphs we have to resolve (uni-directional) interference. There are two strategies:

Non-interfering Deterministic Schedule. In general it is an $\mathcal{NP}$-hard problem to compute a schedule that resolves all interferences within optimal time.

However, in the HL-graph in each layer the bi-directional interference number is a constant. Hence, it is easy to define a deterministic schedule that ensures each edge a time frame of $\frac{1}{c \log n}$, which in the worst case slows down communication only by this logarithmic factor.

For the Yao graph the (uni-) directional interferences are given by the in-degree. Hence a straight-forward strategy is to assign each of these incoming senders a time frame of same size. Unlike as for the HL-graph this schedule is far from being optimal, since it does not reflect the actual load on the edges.

The main advantage of such a non-interfering schedule is that collisions immediately indicate that dynamic changes have occurred.

Interfering Probabilistic Schedule. Following the ideas presented in [1] every edge e of the basic network is activated with some independent probability $p(e) \leq \frac{1}{2}$, where for all edges e it holds $p(e) + \sum_{e' \in \mathrm{UInt}(e)} p(e') \leq 1$.

Then, there is a constant probability of at least $\frac{1}{4}$ that a packet is transferred without being interfered by another packet.

The detection of dynamic network changes may need more time than in non-interfering schedules. Here, since with probability of at least $\frac{1}{4}$ every receiver does not get an input signal, it suffices to repeat the dynamic change signal for some $O(\log n)$ rounds. Then all nodes are informed with probability $1 - 1/p(n)$ (for some polynomial $p(n)$).

The only information necessary to maintain such a probabilistic schedule is the local number of uni-directional interferences, or an approximation of that number. In the case of the BoundY-graph this number is not given by a graph property as in the other topologies. Therefore, a node has to inform all m interfering nodes, that they interfere and how many of them interfere. A straight-forward approach shows that this takes time $O(m)$. However we state a general approach that computes and transmits an appropriate approximation of that number in time $O(\log m)$.

We investigate two elementary dynamic operations necessary to maintain dynamic wireless networks:

Enter: While the network is distributing some packets, one radio station wants to enter the network. It will send a special signal causing a special interference signature that will cause all radio station in some specified distance to stop the point-to-point communication mode and switch to a special enter node.

Then, this part of the network devotes its communication to insert the new node into the network topology. After this, it will resume to the normal transportation mode.

Leave: A single station stops sending and receiving. At some time a neighbored node notices this failure and signals it to other nodes of the network. These nodes halt routing packets and rebuild the network.

The two important resources in these update processes are **time** and **number of involved processors**. If these parameters are minimized, then the impact of the network disturbance can be kept to a minimum.

Theorem 6. *For a normal and nicely located vertex set V the $\Theta(|V|)$ edges need to be changed if an enter/leave operation happens in a Yao-, SymmY-, SparsY-, or BoundY-graph. For the HL-graph this number is bounded by $O(\log |V|)$.*

Clearly, this worst case behavior is not the typical situation. Therefore we introduce the number of involved vertices m as an additional parameter into the analysis of the time behavior of the enter/leave algorithms.

Theorem 7. *For a normal and nicely located vertex set V and m edges are involved an enter/leave operation can be performed in the Yao based graphs in time $O(m \log s)$. For the HL-graph the time is bounded by $O(\log |V| + \log s)$.*

7 Conclusions

The following table summarizes the results concerning communication and dynamic performance of the five graph topologies. It turns out that the best dynamic behavior can be achieved by the HL-graph. From the Yao graph variants the SparsY graph outperforms the HL-graph on the approximation factor of congestion. In this overview the SymmY-graph gives the worst impression. Nevertheless, it guarantees that no signals interfere at all. Therefore for a small number of radio stations or average locations it may outperform all the other graph types.

Topology	Congestion approx. factor	Energy approx. factor	time for `enter` & `leave`	`enter/leave` involved nodes
Yao-graph	—	$O(1)$	$O(n \log s)$	$\Theta(n)$
SymmY-graph	—	—	$O(n \log s)$	$\Theta(n)$
SparsY-graph	$O(\log n)$	$O(1)$	$O(n \log s)$	$\Theta(n)$
BoundY-graph	—	$O(1)$	$O(n \log s)$	$\Theta(n)$
HL-graph	$O(\log^2 n)$	$O(1)$	$O(\log n + \log s)$	$O(\log n)$

References

1. M. Adler and Ch. Scheideler. Efficient Communication Strategies for Ad-Hoc Wireless Networks (extended Abstract). In *Proc. SPAA'98*, pages 259–268, 1998.
2. S. Arya, G. Das, D. M. Mount, J. S. Salowe, and M. H. M. Smid. Euclidean Spanners: Short, Thin, and Lanky. In *Proc. STOC*, pages 489–498, 1995.
3. B. Chen, K. Jamieson, H. Balakrishnan, and R. Morris. Span: An Energy-Efficient Coordination Algorithm for Topology Maintenance in Ad Hoc Wireless Networks. In *Proc. MobiCom*, 2001.
4. M. Fischer, T. Lukovszki, and M. Ziegler. Geometric searching in walkthrough animations with weak spanners in real time. In *Proc. ESA*, pages 163–174, 1998.
5. M. Fischer, F. Meyer auf der Heide, and W.-B. Strothmann. Dynamic data structures for realtime management of large geometric scenes. In *Proc. ESA*, pages 157–170, 1997.
6. J. Gao, L. J. Guibas, J. Hershberger, L. Zhang, and A. Zhu. Discrete mobile centers. In *Proc. Symposium on Computational Geometry*, pages 188–196, Medford, MA, USA, 2001.
7. J. Gao, L. J. Guibas, J. Hershberger, L. Zhang, and A. Zhu. Geometric spanner for routing in mobile networks. In *Proc. Symposium on Mobile Ad Hoc Networking and Computing*, pages 45–55, 2001.
8. K-Team S.A. *Khepera miniuature mobile robot*, 2000. http://www.k-team.com/robots/khepera.
9. T. Lukovszki. New results on fault tolerant geometric spanners. In *Sixth Workshop on Algorithms and Data Structures*, pages 193–204, 1999.
10. F. Meyer auf der Heide, Ch. Schindelhauer, K. Volbert, and M. Grünewald. Congestion, Energy and Delay in Radio Networks. To appear at SPAA, 2002.
11. F. Mondada, E. Franzi, and A. Guignard. The development of khepera. In *Proc. of the 1st International Khepera Workshop*, pages 7–13, Paderborn, Germany, December 10.-11. 1999.
12. J. Ruppert and R. Seidel. Approximating the *d*-dimensional complete Euclidean graph. In *3rd Canadian Conference on Computational Geometry (CCCG '91)*, pages 207–210, 1991.
13. Y. Wang and X.-Y. Li. Distributed Spanner with Bounded Degree for Wireless Ad Hoc Networks. In *Parallel and Distributed Computing Issues in Wireless networks and Mobile Computing*, 2002.
14. Y. Xu, J. Heidemann, and D. Estrin. Geography-informed Energy Conservation for Ad Hoc Routing. In *Proc. Mobicom*, 2001.

A Local Decision Algorithm for Maximum Lifetime in ad Hoc Networks

Andrea Clematis[1], Daniele D'Agostino[1], and Vittoria Gianuzzi[2]

[1] IMATI -CNR Via De Marini 6,
16149 Genova, Italy
{clematis, dago}@ima.ge.cnr.it
[2] DISI, Università di Genova, Via Dodecaneso 35
16146 Genova, Italy
gianuzzi@disi.unige.it

Abstract. Mobile hosts of ad-hoc networks operate on battery, hence optimization of system lifetime, intended as maximization of the time until the first host drains-out its battery, is an important issue. Some routing algorithms have already been proposed, that require the knowledge of the future behavior of the system, and/or complex routing information. We propose a novel routing algorithm that allows each host to locally select the next routing hop, having only immediate neighbor information, to optimize the system lifetime. Simulation results of runs performed in different scenarios are finally shown.

1 Introduction

An ad hoc network is a collection of wireless hosts forming a temporary network without the aid of any centralized administration. It can be mobile, with more or less frequent topology changes over time, or static, such as sensor based monitoring network. Hosts communicate establishing multi-hop paths by means of a route discovery algorithm, usually based on the flooding mechanism.

The routing is maintained in a route cache by each wireless host, until the destination is reachable. If not, the route discovery algorithm is again started.

A host can be no more reachable either in a mobile or in a static network: some host of the route could move, falling outside the transmission range of the sender, or the signal propagation conditions could change.

Different route discovery algorithms have been proposed in literature, depending on the knowledge of the network (e.g. host positioning and link congestion), on the selected metric and on the possibility of adjusting the transmission energy level. A short selection follows.

In [1] the shortest-path routing in hops number is selected, considering fixed transmission energy, in [2] the route discovery is optimized, having information about the positioning and the mobility of the host. The aim of these algorithms is to minimize the energy consumed in routing a data packed. The same problem has been addressed

B. Monien and R. Feldmann (Eds.): Euro-Par 2002, LNCS 2400, pp. 947–956.

addressed in [3] and [4], but considering adjustable transmission power. In this case, an approximation of the MINIMUM PATH ENERGY GRAPH is considered.

Minimum energy paths finding is not the unique metric considered: for example in [5] the route selection metric is the host routing load. The path is reconstructed when intermediate hosts of the route have their interface queue overloaded. In [6] the objective is to maximize the system lifetime, defined as the time until the first host battery drain-out, when the rate at which each information is generated at every host is known, and transmission energy can be adjusted.

In [7] a similar approach is proposed, considering as metric the remaining battery capacity of each node: nodes with low capacity have some "reluctance" to forward packets, and such a reluctance is taken in account in the cost function definition for the routing algorithm. No previous knowledge of the information generation rate is required. However, it is necessary to evaluate alternative routes for each destination and to know the minimum value of the battery capacities, dynamically changing, of the hosts belonging to each route.

In this paper we refer to battery operating devices, with adjustable transmission energy, arranged either in static or in mobile networks with few topology changes over time, where it is possible to reason as the network is static for most of the time. Our objective is to study how to prolong the system life time, as defined in [6], but without a-priory knowledge of the information rate generation, using a limited amount of information at each node, and without calculating (or recalculating) alternative paths for each pair of source-destination.

The starting point has been the analysis of the influence of the path selection on the path cost. Selecting an alternative path with respect to the better one could have the advantage of resulting in a more uniform energy consumption, and probably also in a better load balancing, but could lead to a higher global energy consumption, with negative effects on overall routing. In fact, the goal of optimizing different metrics is not always reachable.

Then, instead of considering how each node could split incoming traffic among different paths, we define a novel decision algorithm that allows each node to locally select the next relay node, having only the knowledge of the remaining battery capacity of the neighbor hosts falling in its radio coverage range.

This algorithm, called ME+LS, has been at the moment studied and simulated only considering the power consumption for the routing.

It can be applied as "additional selection level" starting from an already existing route discovery algorithm that minimizes the path energy, such as the one presented in [4]. Some additional information must be transmitted and recorded during the route discovery and the data transmission.

2 The Network Model

In wireless network minimization of power consumption is an important requirement

Radio transmission between two devices at distance d requires power consumption proportional to d^n : usually n, the path-loss exponent of outdoor radio propagation, assumes a value out of 2, 3 or 4. We will consider only transmission power, the most important parameter to determine the energetic balance, ignoring receiving power (always constant, independently from the position of the sending device) and computational power consumption, since usually communication costs are more expensive than the others.

Given a network *N* of wireless devices, let us consider the complete weighted graph *G=(V, E)* where *V* is the set of nodes, each one corresponding to a device, and *E* is the set of bi-directional edges *(u, v)*, for each *u, v* ∈ *V*. The weight *w* of an edge *e*, that is *w(e)*, is the transmission cost of a single data packet over that link, that is:

$$w(e) = c \mid (u, v) \mid^{n}. \tag{1}$$

for some *n* and constant *c*. Supposing W_{max} be the maximum power value at which a node can transmit, we delete from *G* all the edges having weight greater than W_{max}. The graph *G* so obtained is called *Reachability Graph* (*RG*), the edges of which represent a transmission link between the edge end points. Considering the *RG* of a network, a *route* between two nodes n_1 and n_k is a paths $P = (n_1, n_2, \ldots n_k)$ over *RG*, the cost of which is defined as:

$$C(P) = \Sigma_i \ w((n_{i-1}, n_i)). \tag{2}$$

for every $n_i \in P$. The cost of a route is also the total power consumption required to transmit a data packet from the host n_1 to the host n_k.

Finally, the Minimum Energy (*ME*) route between two nodes is the path linking the two nodes having the minimum cost. Actually, two hosts apparently linked in the *RG* could not to communicate because of obstacles that prevent the communication. In this case, additional edges can be removed from *RG*, without deteriorating our results.

The sub-graph of *RG* including all and only the edges belonging to some Minimum Energy path is called *Minimum Energy Reachability Graph (MERG)* for the network *N*. *ME* routes guarantee the minimization of the total power consumption, since they allow the transmission of a message with the lower power cost.

3 Pros and Cons of Minimum Energy Reachability Graph

Distributed algorithms that build approximations of *MERG* for a network, generally use local decisions that however guarantee some global optimization. They start with every node sending a beacon at the maximum power, or growing it transmission power, until it finds neighbor hosts satisfying a predetermined property and ensuring the connectivity of the network.

*MERG*s have also another advantageous property: the low cost of the edges (then the low transmission power) and the small node degree ensure a minimal interference and allow a high throughput.

However, they also have a critical disadvantage: due to the limited node degree, and consequently, the low number of edges and paths of the *MERG*, some host maybe part of an increasing number of routes. This fact leads to a quick consumption of its battery capacity. Power consumption thus depends from such the degree of a host and from the energy that it must consume to forward a packet to the next relay host. The hosts that behave like collectors of route branches and like bridges to far hosts are candidates to faster drain-out their batteries.

This situation is likely to occur when the message traffic is mainly in the form of many-to-one, such as in sensor networks where information flows to a Base Station, or in a mobile environment where communications are directed to a coordinator. In these cases, the resulting network topology is a spanning tree that leads to highly non-uniform energy utilization and concentrates routes in few nodes.

The problem of maximizing the network lifetime is equivalent to solve a linear programming problem to find the maximum flow, under the flow conservation condition, and supposing to know the set of origin nodes and the set of destination nodes for each commodity, and the rate at which information is generated at every node belonging to some commodity. Building in a distributed way the solution of this problem could be heavy in energy consumption, if the network is not static, and it is not useful when the dynamic of the information generation is not known a priori.

4 The ME + Link Selection Algorithm

Our algorithm may inherit one already existing *MERG* approximation and routing discovery algorithm and extends it by adding a Link Selection strategy during transmission operation to prevent, at some extent, early energy consumption in critical nodes. Link Selection requires that some values be piggybacked with messages and aks during the transmission. It is useful to distinguish two different phases: a set-up phase (routing discovery), and the message transmission phase.

4.1 Set up for Link Selection

Let us start considering a generic existing algorithm that finds an approximation of the *MERG* during the route discovery phase (hereafter called the basic algorithm).

During this phase, a host **a** collects information about the hosts within its transmission radius. With respect to **a** there are two kinds of hosts: those that are its *neighbors* in the approximated *MERG*, composing the set *H(a)*, and those that are *reachable* (that is, which are neighbors of **a** in the *RG*) but that are not neighbors in the *MERG*, composing the set *R(a)*.

The ME+LS algorithm requires each host to transmit few additional information together with the control messages needed for the basic algorithm:

1. Each node must know not only the first hop host of the route to a destination, but also the second one,
2. Each node transmits (with the maximum energy possible with respect to the basic algorithm) its *R* set.

The meaning of this additional information is represented in Fig. 1. The circle represents the maximum communication radius of host **a** allowed by the basic algorithm, solid lines represent the edges belonging to the approximated *MERG*, dashed lines represent edges in *RG* not in *MERG* and dotted lines represent possible multi hops paths to the destination hosts **A**, **B** and **C**. Host **a** knows that **d**, **f** and **g** are the second relay hosts on the path to, respectively, **A**, **B** and **C** (from point 1 above). Moreover, **a** also knows that **f** is reachable from **e** (from point 2 above).

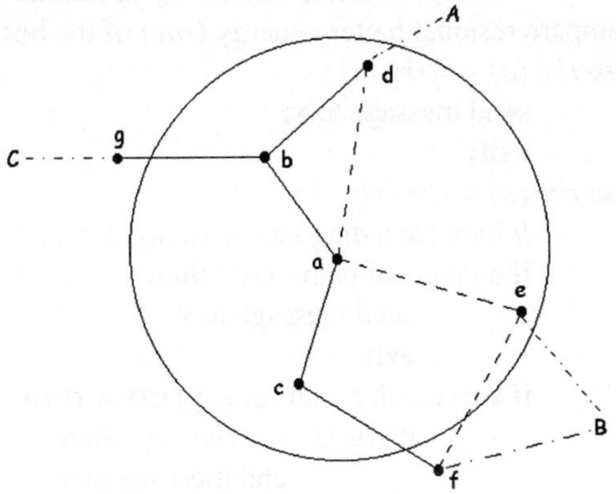

Fig. 1. Environmental knowledge of a host **a**

As an example, to route a message from **a** to the destination **A**, the host **d** is the next hop host after **b**, and **d** can be reached directly from **a**, even if with a greater energy consumption. Considering destination **B**, **f** is the next hop host after **c** and, even if not directly reachable from **a**, it can be reached by **a** through **e**. However, a message routed through **e** to reach the destination **B**, could not to follow the same path as it was routed through **f**, since a different path could be selected by **e**.

Path (**a**, **d**) is called *diagonal alternative path* from **a** to **d**, while path (**a**, **e**, **f**) is called *triangular alternative path* from **a** to **f**.

4.2 Transmission Phase with Link Selection

The Link Selection algorithm acts when the devices are transmitting and routes are effective.

Link Selection algorithm requires that when a host forwards a message or an ack, it also piggybacks its remaining battery level. Considering some host **a**, it can receive this additional information from hosts in *R(a)*, provided that the transmitting radius used for this communication is sufficiently large. Link Selection is a symmetric algorithm acting as follows during the message forwarding:

```
Algorithm Transmission with Link Selection:
// for host a
begin
    when a message arrive at host a to be routed to destination A:
            Look in the routing table for destination A, finding a path having
                    host x as first hop and host y as second hop;
            Compare residual battery energy (rbe) of the hosts:
            case rbe (a) <= rbe (x) :
                    send message to x;
                    exit;
            case rbe (a) > rbe (x):
                    // look for a diagonal or triangular path connecting a to y
                    if a diagonal path exists then
                            send message to y,
                            exit;
                    if a triangular path (a, z, y) exists then
                            if rbe (z) <= rbe (x),  then
                                    send message to x;
                                    exit;
                            if rbe (z) > rbe (x) then
                                    send message to z;
                                    exit;
                    send message to x  // no alternative path found
end
```

Referring to Fig. 1, if a message has to be forwarded to destination **A**, a diagonal alternative path can be considered; if the destination is **B** a triangular alternative path would have been taken in account, through host **e**. No alternative paths exist to destination **C**. If battery consumption for some host in *R* is not known, an approximated value can be considered as for example a weighted average, to be updated after the receipt of the ack for a message routed through it.

Roughly, the ME+LS algorithm tries to limit the energy consumption of the ME path hosts, using alternative paths.

It is worthwhile to put in evidence that the presented algorithm has two nice properties:

1. It exhibits an adaptive behavior so that, if the communication pattern changes (e.g. in case of temporary clusters), the selected paths are modified in order to better use the energy of nodes with the largest availability;
2. It is based on a local decision criteria thus requiring for each node to maintain only a limited amount of information, more precisely a table with the first 2 hosts for each ME path.

As it will be discussed in the following Section, simulation results show that the algorithm provides, on the average, good results but with some variance. It is possible to look for improvements considering at least the following aspects:

- If we have information about relative node positions then we may refine the criteria to select between diagonal and triangular alternative paths;
- A threshold difference between the residual battery energy of two hosts could be considered, instead to evaluate only the simple difference;
- Additional information could be transmitted by the acks, for example the minimum residual battery energy of the path.

5 Simulation Results

In this simulation, we considered only the transmission phase, considering static networks, since the power consumption during message exchange is prevailing in network having few topological changes over time. For these reasons, we do use any particular *MERG* approximated evaluation and routing discovery algorithm, but instead we base simulation on the actual Minimum Energy Path graph, building it by means of the Bellman Ford algorithm. Again to not depend on any particular set-up algorithm, the alternative links considered in the *ME+LS* algorithm are Delaunay edges that approximate the possible environmental knowledge of each host.

*MERG*s have been generated considering 50, 100 and 150 nodes placed randomly within a 1500*m* x *1500m* area, with radio propagation range for each host of 450*m*. in order to simulate different densities. For each different number of hosts, 300 runs have been conducted, and finally the data collected have been averaged.

To investigate the impact of our algorithm, we also considered two completely different communication scenarios.

- *Scenario A*: many-to-one communication to a unique sink host, like in sensor networks where devices send information to a Base Station or in mobile networks where hosts send messages to a fixed installation.
- *Scenario B*: a completely random communication scheme, where hosts communicate with each other. Such an environment is not realistic, but it is useful to test the algorithm in an extreme situation.

Finally, with respect to formula (1) , we performed simulations using two values of n, that is n=*2* and n=*4*. Performances obtained with the ME+*LS* algorithm have thus been compared with the results obtained using only minimum energy paths, and also using the minimum hop path, with the maximum transmission energy for each host. All the hosts had the same initial energy, but different for each value of n.

As an example, let us consider a single trial, with 100 hosts randomly distributed, sending information to a sink host and outdoor radio propagation n=2. Fig. 2. (a) shows the minimum energy paths built over the *MERG*, while Fig. 2. (b) shows the Delaunay graph. Fig. 3 shows the *ME+LS* paths graph, obtained after the simulation.

The final energy distributions have been evaluated.

Nodes labeled with letter **a** terminated the simulation with less than 1/6 of the initial battery energy, with letter **b** if the final energy is less than 1/3 of the initial, with letter **c** less than 1/2 and letter **d** less than 2/3.

Transmitting over the *MERG,* the host labeled with **a** drains-out the battery after that 1299 messages have been received from the Base Station, while using the *ME+LS* algorithm 2488 messages are received.

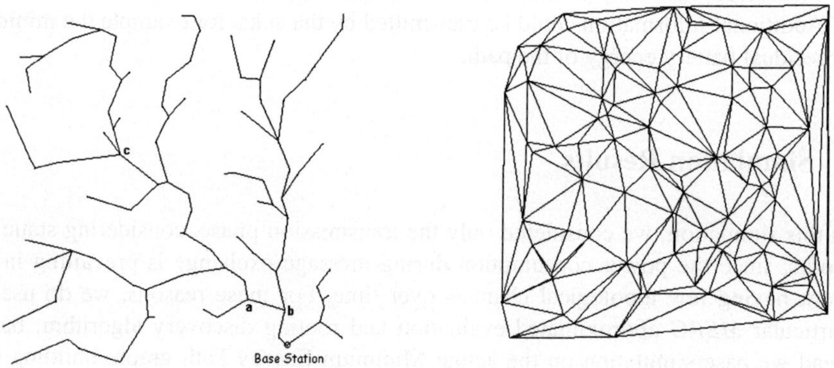

Fig. 2. (a) Minimum energy path graph (on the left) and (b) Delaunay graph (on the right)

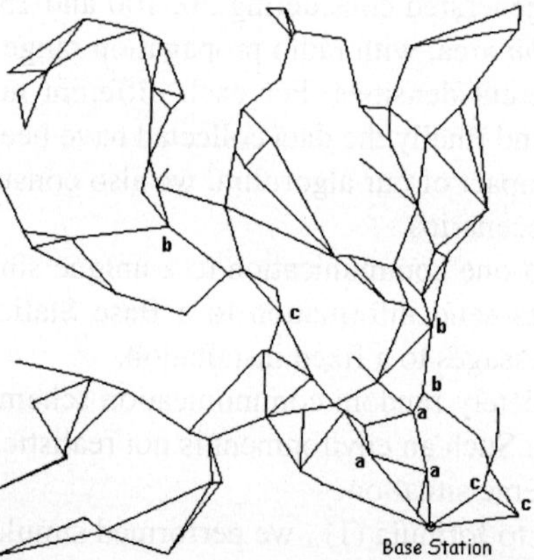

Fig. 3. Paths followed by packets using the MERG+LS algorithm

Using *n=4,* the Base Station receives 1081 messages in the first case, 1873 in the second case. In this last trial, on a total of 21601 hops, 2208 of them have been made using diagonal alternatives and 2340 with triangular alternatives.

Remark that diagonal hops allow bypassing hosts with low battery energy, but require higher energy consumption from sending hosts. Triangular hops sometimes allow jumping on a different branch, leading to more uniform energy consumption. Looking at Fig. 3, the branch on the right, composed by 7 hosts, routes to the Base Station part of the data that otherwise would be flowed over the branches on the left.

In Tables 1 and 2 the results of the simulations are shown as follows: in *ME* columns the results obtained routing data through *ME* paths (that is, the average number of messages sent before the first host drain-out its battery), in *MH* columns the results obtained using the minimum hop routing algorithm, in *LS* columns the results obtained running the *MERG+LS* algorithm, and finally, in columns % the percentage difference between column ME and column LS.

Table 1 shows results of the simulation of scenario A, considering on the left $n=2$, on the right $n=4$. The same for Table 2, but with respect to scenario B.

Table 1. Scenario A: columns on the left show results obtained with a path-loss exponent of outdoor radio propagation n=2, with n=4 on the right

Nodes	ME	MH	LS	%	ME	MH	LS	%
50	1050	677	1390	32	679	88	839	23
100	1849	508	2470	33	2140	111	2621	22
150	3056	530	4124	35	5614	113	6976	24

Table 2. Scenario B: columns on the left show results obtained with a path-loss exponent of outdoor radio propagation n=2, with n=4 on the right

Nodes	ME	MH	LS	%	ME	MH	LS	%
50	3042	999	3645	20	835	102	879	5
100	7370	1386	9183	24	3896	153	3983	2
150	12208	1596	15337	25	9798	176	9887	1

The three different distributions show similar results: in fact, the average outdegrees of the ME and ME+LS routing graphs of Scenario A are respectively: 4.20 and 4.93 for 50 nodes, 4.41 and 5.49 for 100 nodes, 4.46 and 5.66 for 150 nodes. That is, the different density leads to a little increase of the number of outcoming edges per node, and, consequently, to a related performance improvement.

A final remark can be made: the routing performed using only the ME graph is not sufficient to guarantee the maximization of the system lifetime. In fact, in Scenario A the minimum energy path graph is a spanning tree, thus the hosts behaving as collector of branches risk to run down very quickly. In this case *ME+LS* algorithm performs well, since it is possible to bypass overloaded hosts, increasing the power consumption of non-critical hosts.

However, in Scenario B, with random paths, the energy is consumed in a more uniform way. In this case alternative routings, consuming more energy, are not always convenient, at least when *n* is greater than 2, because the selection of a non minimum weight alternative path has a very high impact on the total energy consumption.

This consideration confirms the fact that the two goals, minimize the total transmission power and maximize the system lifetime, cannot be reached simultaneously when the characteristics of the network are not known in advance.

However, the *ME+LS* algorithm significantly improves the system lifetime when some kind of organization is present in the network, for example when some hosts are more important than the others, being fixed installations or cluster coordinators. Moreover, one promising approach in dealing with the obstacle above stated, may be to have a host choose an alternative path only if some threshold in the remaining battery energy difference has been overcome.

Given the encouraging results, our future work will be to perform simulations of the complete algorithm also considering the threshold correction.

6 Conclusion

The route selection mechanisms based on minimum energy paths do not guarantee the maximum system lifetime and can lead to network congestion, since the routing load is likely to be concentrated on certain nodes. The *MERG+LS* helps to prolong the system lifetime and, selecting alternative paths, to better balance the load.

It could also give higher stability to a route: for example a device could temporally be unable to relay data packets due to changed signal propagation conditions, or signal interference. The possibility of selecting different local routing before to run the route discovery algorithm, at least for some time, increases the longevity of the routes. In this manner, the routes are likely to be long-lived and hence there is no need to restart frequently, resulting in higher attainable throughput.

Finally, *ME+LS* algorithm benefits of the fact that it uses only immediate neighbor information in selecting the alternative routing, thus requiring limited amount of memory.

References

1. Johnson D. B.: Routing in Ad Hoc Networks of Mobile Hosts. Proc. IEEE Work. on Mobile Computing Systems and Applications (1994).
2. Ko Y-B., Vaidya N.H.: Location-Aided Routing (LAR) in Mobile Ad Hoc Networks. Proc. ACM/IEEE Conf. Mobile Computing and Networking (1998) 66-75.
3. V. Rodoplu, T.H. Meng: Minimum energy mobile wireless networks. IEEE J. Selected Areas in Communications, 17(8) (1999) 1333-1244.
4. Li L., Halpern J.Y.: Minimum energy mobile wireless networks revisited. Proc. IEEE Int. Conference on Communications (ICC), Helsinki (2001).
5. Lee S-J., Gerla M.: Dynamic Load-Aware Routing in Ad hoc Networks. IEEE Int. Conference on Communications (ICC), Helsinki (2001).
6. Chang J-H, Tassiulas L.: Energy Conserving Routing in Wireless Ad-hoc Networks. Proc. INFOCOM (2000) 22-31.
7. Toh C.K.: Maximum Battery Life Routing to Support Ubiquitous Mobile Computing in Wireless Ad Hoc Networks. IEEE Communications Magazine (2001).
8. Wattenhofer R. et al.: Distributed topology Control for Power efficient Operation in Multihop Wireless Ad Hoc Networks. Proc. INFOCOM (2001) 1388-1397.

A Performance Study of Distance Source Routing Based Protocols for Mobile and Wireless ad Hoc Networks[1]

Azzedine Boukerche, Joseph Linus, and Agarwal Saurabha

Department of Computer Sciences, University of North Texas, USA
{boukerche, linus, saurabha}@cs.unt.edu

Abstract. In this paper, we focus upon the distance source routing (*DSR*) protocol. First, we describe an optimization scheme, which we refer to as *GDSR*, a reactive protocol that makes use of DSR scheme and the Global positioning system (GPS). As opposed to the DSR protocol GDSR consists of propagating the route request messages only to the nodes that are further away from the query source. Next, we consider a randomized version of both DSR and GDSR algorithms, which we refer to as *RDSR* and *GDSR* respectively. We discuss the algorithms, their implementation and present the experimental results we have obtained to study their performance. Our results clearly indicate that the GDSR protocol outperforms the DSR protocol by significantly decreasing the number of route query packets thereby increasing the efficiency of the network load. Our simulations experiments show that GPS screening and randomization paradigm are very effective in improving the performance of DSR under various performance conditions. Our results also indicate that a probabilistic congestion control scheme based on local tuning of protocol parameters is feasible, and that such a mechanism can be effective in reducing the amount of traffic routed through a node which is temporarily congested.

1 Introduction

Ad hoc networks [2] are useful for providing communication support where no fixed infrastructure exists or the deployment of a fixed infrastructure is not economically profitable, and movement of communicating parties is allowed. In this paper, we describe an adaptive routing protocol for mobile ad hoc networks, which we refer to as GDSR, based upon the Dynamic Source Routing protocol (DSR) [2,7]. GDSR uses the position of the neighboring mobile nodes and takes advantage of the Global Positioning System (GPS) [3,6] to reduce the number of route query packets that are generated to find a route from source node to the destination. DSR broadcasts the route requests to all its neighbors, which in turn broadcast it to all their neighbors causing a route request flood. By selecting and limiting the number of nodes to which a query is forwarded, our scheme reduces the network traffic thereby increasing the bandwidth utilization and the efficiency of the network. We also consider a

[1] This work was supported by Texas Advanced Research Program (ARP/ATP), and UNT Research grants. Contact person: boukerche@cs.unt.edu

B. Monien and R. Feldmann (Eds.): Euro-Par 2002, LNCS 2400, pp. 957–964.

randomized version of both DSR and GDSR for better congestion control within the network.

2 Dynamic Source Routing Protocol (DSR)

The Dynamic Source Routing Protocol [2,7] is an on-demand (reactive) routing protocol. It is source-routed, i.e. the entire path form the source to destination is stored in the header of the data packet. A node maintains route cache containing the source routes that it is aware of. The node update entries in the route cache as and when it learns about new routes. The two major phases of DSR protocol are *route discovery* and *route maintenance*.

In the route discovery phase, a node S wishing to send a packet to node D obtains the route to node D. The route discovery is initiated "on demand" i.e. only when the nodes need to send data to other nodes. Route table is not maintained for all destinations. Suppose a source node S wants to send a data packet to destination node D. First *S* will check its cache to see if it contains a route to *D*. If the route is present in the cache it will construct a data packet with the entire route stored in its header and send it to the next hop as listed in the cached route. If the route is not found in the cache then it constructs a route request message containing unique request identifier, source identifier, destination identifier and a list that records all the nodes through which the packet has been forwarded. The list is empty in the beginning. The source will then broadcast this packet to all its neighbors, which in turn will forward it to their neighbors till the destination node is found. Whenever an intermediate node receives a route request packet from another node, it first checks to see the destination address. If the destination address is not same as its own address then it will append its own address in the list and simply broadcast the packet to all its neighbors. When the packet reaches the destination, the destination node constructs a route reply message. To send the route reply message, destination node will first look into its route cache to find the source address. If the route to the source exists in its cache, it will send the route reply message using the cached route else it will start a new route discovery for the source. But in the route request packet for the source it will piggyback the route reply to avoid the route request cycles

The second phase of DSR deals with the route maintenance. The network topology keeps changing because of the random movement of the node. Hence the routes stored in cache become invalid. It is quite possible that an intermediate node is unable to deliver the data packet according to the source route stored in packet because the node is moved or the link is broken. In such situation, the intermediate node that detects the broken route or broken link sends a route error message to the originator of the data packet. The originator invalidates that route entry in its cache and looks for an alternative path to the destination in its cache. If the path is found it will use that path else it will initiate the route discovery again. The route maintenance phase is responsible for maintaining up to date routes using route error messages.

3 GDSR: A DSR Route Discovery Optimization Scheme Using GPS

In this section we describe our optimization technique GDSR that uses the node location provided by Global Positioning Systems (GPS) to decrease the number of route requests packets as compared to DSR. GPS is a satellite navigation system designed to provide instantaneous position, velocity and time information almost anywhere on the globe at any time. The details of GPS can be found in [3,6] and other related sites.

In DSR protocol, route discovery of the destination node is an expensive operation in terms of bandwidth utilization. If the nodes can find the geographical position of all other nodes, then the route request query from the source node can be directed in a certain direction avoiding its unnecessary broadcast to all the neighboring nodes. The use of GPS information allows the selection of neighbor nodes to which a node should forward the request query for determining the route, thereby decreasing the network-wide broadcasts. One such technique is based on the GPS screening angle [3] where the nodes takes the forwarding decision based on angle between the previous node, itself and the next node. Let us suppose that source node S wants to send some message to destination node D. So it will find the route to D by constructing a route request packet that has unique request ID, source ID, its position, velocity, a route record and destination position and velocity. The source S will then forward this request packet to all of its neighbors i.e. to nodes B, C and E. When node B receives the request packet from node S, it will not broadcast it to all its neighbor nodes. Instead it will first calculate the GPS screening angle. The angle between the processing node (B), the previous node (the node through which current node received the packet i.e. S) and the next node (the node for which the forwarding decision is to be taken like H, A for B) is called as the GPS screening angle.

Each node in the network is assumed to know the location of its immediate neighbor. The processing node B calculates the screening angle between S and all its neighbor nodes one by one. Only if the screening angle between the previous node, itself and the next neighbor node is greater than some value x the packet is forwarded to that node. If the value of x is zero the processing node will forward the packet to all its neighbors just like in DSR protocol. As the value of the cut-off angle x is increased, the number of nodes to which the route request is forwarded will decrease. In our approach when selecting the neighbors based on screening angle, there is a possibility that the request packet never reaches the destination i.e. the route to the destination is not discovered. When the route request is unable to reach a destination, the source times out waiting for the route reply and it restarts the route discovery for the same destination.

4 RDSR and RGDSR: Randomized Versions of DSR and GDSR Routing Protocols

In the original DSR protocol, whenever a node wants to send a data packet or establish communication with any other node in the network, it initiates a Route Discovery process. In the route discovery process, the source node sends route request query messages to all its neighboring nodes, and each of those nodes sends it to all of its neighbors. This process continues until the message reaches the destination node and the whole route is discovered. Whenever the route request query arrives at a certain node, if that node is not the destination node, then it simply adds its IP address on the list. This list is contained in the header of the route request message. Thus address of every node the request message has gone through is appended on the list and when the node eventually reaches the destination, the whole route is available. At this point, the destination node checks whether it has a route to the source so that it can send the request message back to the source as a confirmation of route discovery. If the destination node doesn't have a route to the source node, it may generate a route discovery for the source or the third option is it may just reverse the route so that it can eventually reach the source node. But this kind of route reversal is not always possible because sometimes communication path between two nodes is not bi-directional but only unidirectional because of several reasons.

The other point of concern here is, since the network is ad hoc, its continuously changing its configuration and topology. Nodes are free to move and hence they may drastically change positions with respect to each other. In such kind of uncertain and unpredictable situations many existing routes might not be valid routes any more and more and more route discoveries may be initiated. It may also so happen then during route discovery certain nodes might become unreachable and the route discovery might have to be started all over again. In the wake of all these things, the network may always be flooded with a lot of messages. Needless to say that every message or packet that is sent over the wireless network uses network resources, and the most important resource here is the bandwidth. This kind of overuse of the bandwidth may give rise to network congestion that may render the data communication inefficient and ineffective.

In this paper, we are using the methodology of randomly selecting the route request query messages for sending them to the neighboring nodes. The randomization we have achieved works in following two steps:

1. *Whenever a node receives a route request query message, it first decides whether it will forward the message or not. This decision is made based on a randomly generated distribution.*
2. *If the node decides to forward the route request message, it does not forward the message to all its neighbors. Based on a randomly generated distribution, neighbors are selected for receiving the route request message.*

Thus, the route request messages won't be sent to all the neighbors of the node but we would adopt a probabilistic approach. What we are doing here is, we are having a certain probability of sending or blocking a route request to the neighboring node. If the randomly generated number falls in the probability range we send the request message, otherwise we just drop the message. If the node decides to send the message, then again on a probabilistic basis we decide which neighboring nodes are actually going to receive the message. By doing this we will reduce the message overhead, which as we just explained, is a major concern. This kind of randomization is being utilized in both Dynamic Source Routing (DSR) and DSR with Global Positioning System (GDSR). Thus we consider randomized versions of both these Routing protocols and we will call them RDSR and RGDSR. It's quite obvious that there could be some kind of performance degradation in the network because of randomly dropping the request messages. This performance degradation might manifest itself in the form of extended delay in discovering the route. If the price we pay for the reduction of request messages is not substantial then it's worth sacrificing the performance to that extent. That is the basic idea behind this project.

5 Simulation Model

We implemented GDSR using network simulator (ns-2 version 2.1b7a) provided by CMU [8]. The traffic and mobility patterns that we describe here are same as in [1] and are used by other performance comparison papers as well [1,2,3,5]. Traffic sources are CBR (continuous bit rate). The source-destination pairs are spread out randomly over the network. Node mobility is based on *random waypoint model* in a rectangular field. Each node starts its journey from a random location to a random destination with a randomly chosen speed uniformly distributed between 0-20m/s. Once the destination is reached, another destination is targeted after a pause. Varying the pause time changes the frequency of node movement

Figure 1 shows the percentage throughput of GDSR vs. GPS screening angles for different network sizes. As we can see from the figure, the throughput decreases when the network size is increased. The throughput is below 80% with network of 100 nodes. This is same for DSR as well. Note that the GPS angle of 0 degrees corresponds to DSR and the throughput of DSR is almost same as that of GDSR with screening angle from 0 to 135 degrees. For network sizes as 100, 80 and 50 nodes, when the GPS angle is 180 degrees, we notice a sharp drop in the throughput, which implies loss of data packets. Data packets may be dropped because the route to the destination cannot be found by forwarding the requests to few neighbor nodes that are at an angle (absolute value of angle is considered) of 180 degrees from the current node. Thus route discovery failure leads to decrease in throughput. For small size networks like 20 nodes, the throughput remains unaffected even with an angle of 180 degrees possibly because the source destination pairs are too close to apply the GPS angle technique. The destinations are one or two hops away and routes are discovered quickly.

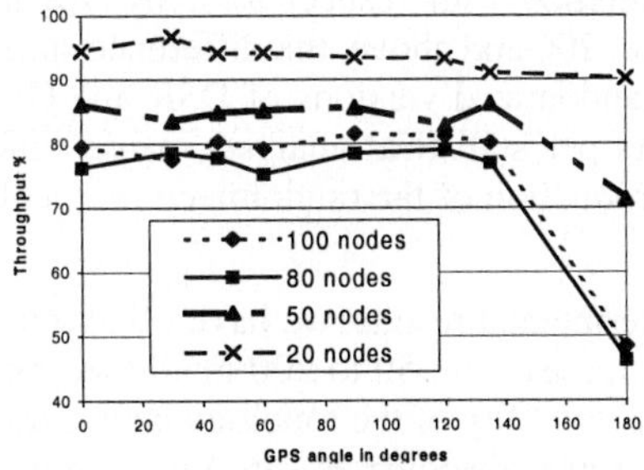

Fig. 1. Throughput vs. Screening angle

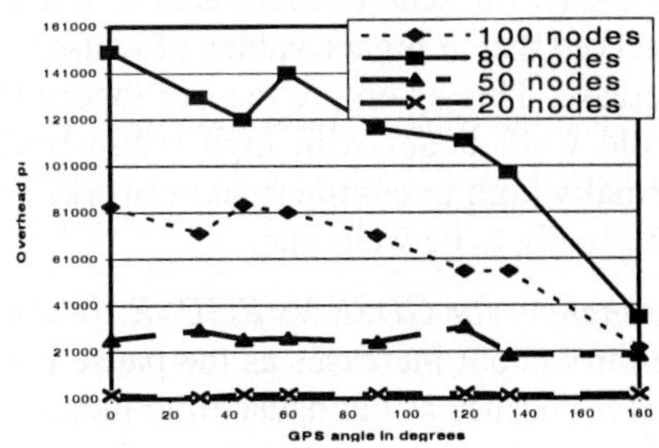

Fig. 2. Routing packets vs. screening angle

Figure 2 depicts the total number of route request packets that are generated during the entire simulation time for network of size 100, 80, 50 and 20 nodes. As the network size increases, the number of request packet also increases causing increase in the traffic. But with increase in GPS angle from 0 to 180 degrees, the number of route packets decreases and as we can see from the figure, DSR (angle 0) generates 150110 packets compared to 35596 packets for GDSR (angle 180) for network of size 100. For network of 20 nodes, the number of packets is below 2000. It appears from the graph that the number of overhead packets does not change for 20 nodes network but it is not true. There is a minor change that is not visible due to the scaling of the y-axis. There is a sharp drop (100 and 80 nodes network) when angle changes from 135 to 180 similar to what we have in throughput.

The delay also shows the same characteristics as throughput and overhead packets i.e. a sharp increase after angle 135 degrees. The source tries for the route to destination many times before it declares a route failure. For this entire duration the data packet is buffered. The delay for the buffered packet increases with the number of retries for the route to its destination.

6 DSR and GDSR Protocols and Their Randomized Versions: A Comparative Study

Delay for GDSR – RGDSR: our experiments indicate that the delay introduced while using RGDSR is slightly more than GDSR. The reason for this is, we have introduced randomization in case of RGDSR that results in randomly dropping the route request query packets. Because of this randomization the time required for the route discovery goes up. We notice that the difference in delay is not substantial at higher pause time. With pause time values of more than 150 seconds the difference becomes quite negligible. Hence, we can conclude that RGDSR performs almost as good as GDSR at higher pause times with respect to delay. There is a definite increase in the delay as we increase the number of nodes. This happens for GDSR and RGDSR. Also, the delay in both these protocols is considerably more than DSR, which proves that the introduction of GPS gives rise to a substantial rise in delay that is almost 250%.

Delay for DSR – RDSR: Our results indicate that delay introduced due to the randomization process was evident for different size of the network models. Although the delay for RDSR maintains a fixed difference with respect to DSR, we have observed that at higher values of pause time i.e. 200 and above, the difference starts to decrease. This again proves our theory that randomized versions of DSR and GDSR would work better with high pause times. As per statistical analysis, pause time is normally high in commercial networks so introduction of the randomized protocols is most likely to be welcome.

Throughput for GDSR Vs RGDSR: In our experimental results, we have observed that the throughput increases as the pause time increases from 50 to 200 in all the graphs. This is because as the pause time increases the volatility of the topology comes down. This results in a reduced need for new route discoveries. In our simulation experiments, a comparison between GDSR and RGDSR showed a slightly lower throughput for the RGDSR. This is because of the fact that randomization results in packet dropping. This results in a lesser efficiency due to failed route discovery attempts. But the loss in throughput in the RGDSR was not very high. It may also be noted that RGDSR holds its ground even with an increase in number of nodes. Our results indicate that the difference in throughput for the 100 nodes simulation at pause time 200 sec was only about 2 to 3 %.

Throughput for DSR Vs RDSR: The results we obtained indicate that for the 50 nodes and 75 nodes the throughput for DSR is higher at a pause time of 0 seconds but the RDSR shows a higher throughput very close to DSR at higher pause times due to the reduced need for route discoveries. As we can see, our results indicate that R-DSR performs well with respect to throughput in most scenarios. Our results shows that even for a 100-node scenario RDSR performs well at high pause time with a throughput lower by only 1%-3% as compared to DSR. The reduction in throughput is within the tolerance limits considering the fact that randomization leads to a loss in throughput. But the gains appear in the form of a drastic reduction in routing overhead. This is evident in the discussions that follow.

Overhead GDSR Vs RGDSR: Our results indicate, as expected, that the overhead packets are lesser in number with higher pause times. The effect of randomization starts to show up only at higher pause times closer to what is really expected in commercial or other real implementations of the protocol. In this set of graphs it can be seen that RDSR has a definitely lower overhead for all the scenarios at pause times of over 150 seconds. The only case we have obtained that shows poor performance by RDSR where when the overhead for RGDSR stays above GDSR until the pause time is as high as 150. At this point it catches up with GDSR and in fact performs really well at a pause time of 200 seconds. Simulations with a bigger scope actually consider pause times ranging from 0 to 2000.So Thus, RGDSR performs better at higher pause times over 200 seconds.

Overhead DSR Vs RDSR: Our experimental results indicate number of overhead packets for DSR and RDSR declines as the pause time increases, which is a general trend and is expected. But interestingly the overhead nosedives in the case of RDSR beyond a pause time of 50sec.The reduction in overhead is substantial for RDSR in the 75 nodes and 100 nodes simulations in general and in particular at 200sec pause time. This difference is about 50,000 packets for 75 nodes and 100 nodes .It may also be noted that the performance of RDSR is very consistent even under very strenuous environments as in a 100 node scenario .It should also be noted that the volatility is very high in all the scenarios as the pause time is below 200 seconds. This shows that RDSR would perform really well in a real world scenario.

7 Conclusion and Future Work

In this paper, we have focused upon reactive routing protocols in ad hoc networks. We have presented an efficient routing algorithm, which we refer to as GDSR, based upon dynamic source routing protocol (DSR) and a GPS query optimization scheme. We also presented a Randomized version of both DSR and GDSR protocols. The objective is to reduce the number of the route queries within the DSR protocol and improve its efficiency. Our simulation results indicate clearly that GPS screening angle the and randomization paradigm have a profound impact on reducing the number of route queries, and thereby, GDSR and its randomized versions can be used in a large network where there is high mobility of nodes and strict limitation on bandwidth usage. Furthermore, our experiments show that randomized version of GDSR and DSR outperforms both DSR and GDSR.

References

[1] Boukerche, A. "Simulation Based Comparative Study of Ad Hoc Routing Protocols", 34th Annual Simulation Symposium, 2001, pp. 85-92.

[2] Perkins. C.E. "Mobile Ad Hoc networks" (Addison Wesley,2001)

[3] Boukerche A., and S. Roger, "GPS Query Optimization in Mobile and Wireless Ad Hoc Networking", 6th IEEE Symp. on Computers and Comm., pp.198-203, July 2001.

[4] Ko, Y. and Vaidya, N. "Location aided Routing (LAR) in Mobile Ad hoc Networks", Mobicom 98.

[5] Broch, J., Maltz, D. A., Johnson, D.B.,Hu, Y, Jetcheva, J. "A Performance comparison of Multi-hop wireless Ad hoc Network Routing Protocols", Mobicom 2000.

[6] GPSoverview:http://www.colorado.edu/geography/gcraft/notes/gps/gps_f.html/

[7] Johnson, D.B., Maltz, D. A. "The Dynamic Source Routing Protocol for Mobile AdHoc Networks" IETF, http://www.ietf.org/internet-drafts/draft-ietf-manet-dsr-03.txt

[8] CMU Network Simulator, http://www.isi.edu/nsnam/ns/ns-documentation .html

Weak Communication in Radio Networks*

Tomasz Jurdziński[1,2], Mirosław Kutyłowski[3], and Jan Zatopiański[2]

[1] Institute of Computer Science, Technical University of Chemnitz, Germany
[2] Institute of Computer Science, Wrocław University, Poland
[3] Institute of Mathematics, Wrocław University of Technology
and Dept. of Math. and Computer Science, A. Mickiewicz University, Poznań, Poland

Abstract. Quite often algorithms designed for no-collision-detection radio networks use a hidden form of collision detection: it is assumed that a station can simultaneously send and listen. Then, if it cannot hear its own message, then apparently a collision has occurred. IEEE Standard 802.11 says that a station can either send or listen to a radio channel, but not both. So we consider a *weak* radio network model with no collision detection where a station can either send or receive signals. Otherwise we talk about *strong model.*
We show that power of weak and strong radio networks differ substantially in deterministic case. On the other hand, we present an efficient simulation of strong by weak ones, with randomized preprocessing of $O(n)$ steps and $O(\log\log n)$ energy cost.

1 Introduction

Radio networks became today an attractive alternative for wired ones. Luck of a fixed infrastructure and mobility allow to create self-adjusting network of independent devices that is useful for application areas such as disaster-relief, law-enforcement, and so on. The main advantages are dynamic topology and simple communication protocol. However, colliding messages create many challenging problems.

We are interested in networks of *stations* communicating via radio channels. The stations are hand-held bulk produced devices running on batteries. They have no ID's or serial numbers, some of the stations are switched off, but somehow the network has to organize itself, despite the lack of a central control. It is often overlooked by designers of algorithms for radio networks that many current technologies do not provide capability of listening while a station is sending (IEEE Standard 802.11). On the other hand, simultaneous sending and listening for collisions is a feature used in many algorithms. Our goal is to check how does it influence problem complexity on radio networks.

Model. A radio network (RN, for short) consists of several processing units called *stations*. A RN is synchronized by a global clock (for instance based on GPS). Communication is possible in time slots, called here *steps*. We assume that there is a single communication channel available to all stations (so we consider *single-channel, single-hop* RN's). If during a step stations may either send (broadcast) a message or listen to

* the first author was supported by DFG grant GO 493/1-1, the second author by KBN grant 7 T11C 032 21, and the third author by KBN grant 7 T11C 032 20.

B. Monien and R. Feldmann (Eds.): Euro-Par 2002, LNCS 2400, pp. 965–972.

the channel, then we talk about *weak* RN. If both operations can be performed by a station, then we have to do with the *strong* RN. (In the literature, by a RN the authors usually mean the strong RN.) If exactly one station sends, then all stations that listen at this moment receive the message sent. We assume in this paper that if at least two stations send, then a *collision* occurs and the stations that listen do not receive messages and even cannot recognize that messages have been sent (*no collision-detection RN*). If a station is the only station that sends a message during step i, then we say that the station *succeeds* and that step i is *successful*.

The stations of a RN fall into two categories. Some of them are switched off and do not participate in the protocol. The other stations are *active* and we assume that they stay active during the whole protocol. The set of all active stations is called *active set*. During a step an active station might be either *awake* or *asleep*. The first case occurs when a station sends a message or listens to the communication channel.

Complexity Measures. There are two main complexity measures for RN algorithms: *time complexity* and *energy cost* [12]. Time complexity is the number of steps executed. Energy cost is the maximum over all stations of the number of steps in which the station is awake. Low energy cost is a desirable feature: if a station which is awake for a longer time, it may fail to operate due to batteries exhaustion [3]. So a station which is awake much longer than the other stations is likely to fail and break down the protocol.

For further discussion on the RN model and its complexity measures, see e.g. [2, 8–12, 14, 15].

Fundamental Tasks. Due to the lack of ID numbers, uncertainty about the number of active stations, switching off the stations, and so on, certain problems become hard for RN's, while they are much easier in the wired networks. Some of such basic tasks are the following:

Leader Election: active stations have to choose a single station, called *leader*, among themselves. That is, at the end of the protocol a single station has status *leader* and all other active stations have status *non-leader*.

Initialization: consecutive ID numbers are to be assigned to active stations with no ID's.

Renumeration: consecutive ID numbers are to be assigned to active stations having some unique ID's, but in the range bigger than the number of stations.

2 Previous Results

Due to the role of energy cost, there is a lot of interest for algorithms that are time and energy efficient. However, research has been focused on the strong model.

For leader election, the first energy efficient algorithm has been designed in [11]. The authors present a randomized algorithm that for n stations elects a leader (n must be known to the stations) in time $O(\log f)$ and energy $O(\log\log f + \frac{\log f}{\log n})$ with probability $1-1/f$ for any $f \geq 1$. Moreover, they get algorithms that elect a leader within $O(\log n)$ energy cost and $O(\log^2 n)$ time with probability $1 - \frac{1}{n}$ if n is unknown. [6] presents an algorithm which achieves $O(\log n)$ time and $O(\log^* n)$ energy cost with probability

$1 - \frac{1}{n}$ when n is known to the stations. If n is unknown, a general algorithm that approximates the value of n within a constant multiplicative factor can be applied [5]. It uses time $O(\log^{2+\varepsilon} n)$ and energy cost $O((\log\log n)^{\varepsilon})$ with probability $1 - \frac{1}{n}$.

It is impossible to elect a leader deterministically when stations are indistinguishable, due to symmetry. So we apply deterministic algorithms only when the stations have unique ID's, say in the range $[1..n]$ (but any subset of these ID's may correspond to active stations). A simple leader election deterministic algorithm for this case is given in [11, 12]. Its energy cost is $\log n + O(1)$, run time is $n + O(1)$. A recursive procedure that works in time $O(n)$ and energy $O(\log^{\epsilon} n)$ is given in [6].

An energy efficient solution to initialization problem is proposed in [12]: with probability at least $1 - \frac{1}{n}$ its energy cost is $O(\log\log n)$ and execution time is $O(n)$. However, the number of stations n must be known beforehand. The algorithm can be easily generalized to the case when an approximation of n within a constant multiplicative factor is known. Hence using size-approximation algorithm from [5], the same time and energy efficiency can be obtained for an unknown n.

3 New Results

The main result of this paper is a randomized protocol for simulating algorithms for strong RN's by weak RN's.

Theorem 1. *An algorithm for strong RN with run time T and energy cost E can be simulated by a randomized weak RN with run time $O(T)$ and energy cost $O(\max(E, T/n))$. A preprocessing is required with time $O(n)$ and energy cost $O(\log\log n)$.*

Before we construct the simulation algorithm (presented in Sections 5-7), we show in Section 4 that the situation is much different in deterministic setting:

Theorem 2. *Leader election has energy cost $\Omega(\log n)$ on deterministic weak RN's.*

This result is interesting since on the other hand there is a solution for the strong model that works in time $O(n)$ and energy $O(\log^{\epsilon} n)$ [6]. For a more practical solution with energy cost $O(\sqrt{\log n})$ see [7].

4 Gap between Weak and Strong Models

In this section we prove Theorem 2. Let the stations have distinct identifiers in the range $\{1, \ldots, n\}$, but the number of active stations might be arbitrary up to n. We construct sets $A_0, A_1, \ldots$ such that for every i, if $I \subseteq A_i$ is the active set, $j \in I$ and $|A_i| > 1$, then the station j cannot decide if it is the leader after step i. Moreover, we impose some other properties that guarantee that $|A_i| > 0$ and the equality $A_i = \{j\}$ implies that station j is awake at least $\Omega(\log n)$ times up to step i (if the active set is equal to A_i).

Let $A_0 = \{1, \ldots, n\}$. For $i > 0$ let S_i (R_i resp.) be a set of stations that send a message (listen, resp.) during step i, if the active set is A_{i-1}. For definition of A_i we use an auxiliary "weight" $w_i(j)$ of station j at step i. It fulfills the following invariant: $\sum_{j \in A_i} w_i(j) = n$ for $i \in \mathbb{N}$. We start with $w_0(j) = 1$ for $i = 1, \ldots, n$. We put $A_i = A_{i-1} \backslash R_i$ if $\sum_{j \in S_i} w_i(j) > \sum_{j \in R_i} w_i(j)$ and $A_i = A_{i-1} \backslash S_i$ otherwise. If $A_i = A_{i-1} \backslash R_i$, then

- $w_i(j) = w_{i-1}(j)$ for every $j \in A_i \backslash S_i$
- $w_i(j) = w_{i-1}(j) + v_j$ for every $j \in S_i$, where $v_j = \frac{w_{i-1}(j)}{\sum_{l \in S_i} w_{i-1}(l)} \cdot \sum_{l \in R_i} w_{i-1}(l)$.

If $A_i = A_{i-1} \backslash S_i$, then the weights are defined similarly. Obviously, the weights defined in such a way satisfy the invariant. Note that above rules guarantee that $w_i(j) \leq 2w_{i-1}(j)$ for every i, j. Moreover, $w_i(j) > w_{i-1}(j)$ if and only if j is awake at the step i (for active sets contained in A_i).

Proposition 1. *Let $j \in A_i$. Then station j does not hear any message up to step i for any active set $I \subseteq A_i$ such that $j \in I$.*

Proof. The proof is by induction on i. The case $i = 0$ is obvious. Assume that the property holds for $i-1$. If $A_i = A_{i-1} \backslash R_i$, then all awake stations from A_i are sending and nobody listens. So no information is reaching anybody. If $A_i = A_{i-1} \backslash S_i$, then all awake stations are listening, but nobody sends a message at step i, so no awake station of A_i gets any information. □ Proposition 1

Proposition 2. *If the active set is equal to $I \subseteq A_i$ and $j \in I$, then $w_i(j) \leq 2^l$, where l is the number of steps among steps 1 through i during which station j was awake.*

Proof. We show it by induction with respect to i. The case $i = 0$ is obvious. For $i > 0$, it is enough to observe that the new factor added to the weight is smaller than the "old" weight and the weight may increase only when a station is awake. □ Proposition 2

By Proposition 1, if $|A_i| > 1$ and $j \in A_i$, then the station j is unable to distinguish after i steps between active sets $\{j\}$ and A_i. Let l be the leader elected for active set A_i and $j \in A_i \backslash \{l\}$. Then station j cannot decide if it is the leader after the step i (it becomes the leader if the active set is equal to $\{j\}$ and it is not the leader if the active set is A_i). We conclude that if $|A_i| > 1$, then the algorithm cannot terminate after step i. Since $|A_i| > 0$ for every $i > 0$, the algorithm can terminate after step i for the active set A_i if and only if $|A_i| = 1$. Let j be the only element of such A_i. Then $w_i(j) = n$ and by Proposition 2, station j is awake at least $\log n$ times during steps $1, \ldots, i$.

5 Algorithmic Tricks for Weak RN's

Although energy cost of deterministic leader election is different for weak and strong RN's, in the randomized case the best algorithms can be run on weak RN's [6], using tricks presented below. Moreover, a linear approximation of the number of active stations may be obtained efficiently by randomized algorithm, using a method designed for the strong model. This indicates that randomization may really help. In this section we make some observations that help to design "strong algorithms" in weak RN's.

Previous RN algorithms use very often the following *basic trick* ([4, 12, 11, 15]): given n active stations, each of them sends with probability $1/n$. Then with probability $n \cdot \frac{1}{n} \cdot \left(1 - \frac{1}{n}\right)^{n-1} \approx \frac{1}{e}$ exactly one station sends. In the strong model, the station that succeeds may recognize it. This experiment can be used for randomized leader election in the strong model: we apply the experiment $\ln n$ times and use Basic Algorithm for leader election over those stations that have been successful during $\ln n$ experiments (with ID's equal to numbers of successive steps).

Pairing Trick in the Weak Model. The method described above cannot be implemented directly in the weak model. However, this problem can be eluded as follows:

1. each station chooses uniformly at random to be either *sender* or *receiver*,
2. each sender transmits a message with probability $\frac{2}{n}$,
3. each receiver listens with probability $\frac{2}{n}$ and if a message is received, it responds with the same message,
4. each sender that have sent a message at step 2 listens. If it receives its own message now, then it was the only station sending at step 2.

In this procedure it may happen that a station sending a message is successful, but it does not receive a confirmation since more than one receiver is sending it. Nevertheless, we show that the probability that one station sends a message and receives a confirmation is bounded from below by a constant close to $\frac{1}{e^2}$. First observe that the probability that there are more than $n/2 + n^{2/3}$ senders is at most $e^{n^{1/3}/6}$ by Chernoff bound [13]. If there are $\frac{n}{2} + d$ senders, $d < n^{2/3}$, then probability of getting a confirmation equals

$$\left((\tfrac{n}{2} + d) \cdot \tfrac{2}{n} \cdot (1 - \tfrac{2}{n})^{n/2-1+d}\right) \cdot \left((\tfrac{n}{2} - d) \cdot \tfrac{2}{n} \cdot (1 - \tfrac{2}{n})^{n/2-1-d}\right)$$
$$\approx \tfrac{1}{e^2} \cdot \left(1 - \tfrac{d^2}{(n/2)^2}\right) > \tfrac{1}{e^2} \cdot (1 - \tfrac{4}{n^{2/3}})$$

Partial Initialization. Assume that among n stations we have already a fraction n/d stations that have assigned unique ID numbers in the range $1..n/d$ - we call them the *first group*. Then we are able to run an algorithm for the strong model on the remaining $n - n/d$ stations, called the *second group*, in the following way: step i of the algorithm is simulated by the steps $2i - 1$ and $2i$. At step $2i - 1$ the processors of the second group execute step i of the algorithm. Additionally, the station $1 + i \bmod \frac{n}{d}$ from the first group listens. At step $2i$ it sends the message received (if any). In turn, all stations of the second group that have been sending during step $2i - 1$ are listening. This enables them to recognize a collision, if it occurred at step $2i - 1$.

Observe that for simulating n steps of an algorithm for strong model each of the stations of the first group is awake for $2d$ steps. If d is small, for instance $\log\log n$, then this energy cost might be smaller than the energy cost occurring for the stations of the second group, and therefore acceptable.

6 Initialization

The best known algorithm for initialization [12] works in the strong model only: its key point is that the stations assign themselves random ID's in the range $1, \ldots, m$ for some $m \in \mathbb{N}$ and verify their uniqueness in such a way that stations with ID equal to j send at step j for $j = 1, \ldots, m$ and check for collisions. For the weak model one may try to re-use the pairing trick, i.e. split stations at random into senders and receivers, then each sender sends a message and each receiver listens in the step equal to picked number and they confirm uniqueness as in the pairing trick. However, it does not work well if there are few stations choosing ID's from a big set – but just this situation leads to double logarithmic energy cost in [12].

Let us outline our initialization algorithm. It consists of three main phases:

1. Assign temporary unique ID's from the set $\{1, \ldots, c_1 n\}$ to a subset U of stations such that $|U| \geq c_2 n$, for some constants c_1 and c_2.
2. Run "renumeration" procedure on the elements of U. This procedure assigns distinct consecutive ID's, starting with 1, to a large subset of U, say I.
3. Using the partial initialization trick run the initialization algorithm from [12] on the stations that do not belong to I (with I as the *first group*).

Phase 1. Let us describe in detail the first phase:

(A) Split the set of stations into subsets S_0 and S_1. Each station chooses bit b uniformly at random. The stations that set $b = i$ belong to the set S_i.
(B) Each station picks an integer from the set $\{1, \ldots, 3n\}$ uniformly at random. Let S_i^j be the set of elements of S_i which choose number j.
(C) `for` $j \leftarrow 1$ `to` n `do`
the stations from S_1^j and S_0^j verify if $|S_1^j| = |S_0^j| = 1$: first all elements of S_1^j send and all elements of S_0^j listen; in the next step all elements of S_1^j listen and each element of S_0^j sends the message just heard; if the message comes through, then $|S_1^j| = |S_0^j| = 1$ and the only element of S_1^j sets ID$\leftarrow j$.

Lemma 1. *With probability at least $1 - \frac{1}{n^2}$ the number of stations with assigned ID's is bigger than cn, for a constant c independent of n.*

Proof. As we have already seen $n/2 - n^{2/3} \leq |S_b| \leq n/2 + n^{2/3}$ for $b = 0, 1$ with probability $1 - e^{-n^{1/3}/6}$. Thus assume that $|S_i| = c_i n$, for $i = 0, 1$, where c_0, c_1 are very close to $1/2$, and $c_0 + c_1 = 1$. Let us consider the number of sets S_b^j such that $|S_b^j| = 1$. We call the stations from such S_b^j's successful. Imagine that we pick numbers for stations sequentially. If the jth station of S_b picks a number chosen already by the other station, then we say that a collision occurs. Let p_j be the probability that a collision occurs in the jth step. It depends very much on the previous choices, but every time $p_j \leq \frac{j-1}{3n}$. Let X be the number of collisions. We can bound (from above) the probability that $X > \alpha$ (for any α) by the probability that $Y = \sum_{i=1}^{c_b n} Y_i > \alpha$, where Y_i are independent random "0-1" variables such that $\mathbf{P}[Y_j = 1] = \frac{j}{3n}$. Observe that $\mathbf{E}[Y] = \sum_{j=1}^{c_b n} \frac{j}{3n} = \frac{c_b n(c_b n - 1)}{2 \cdot 3n}$. Together with Chernoff Bound [13] this yields

$$\mathbf{P}\left[X > \frac{(c_b n)^2}{3n}\right] \leq \mathbf{P}\left[Y > \frac{(c_b n)^2}{3n}\right] \leq e^{-\frac{c_b n(c_b n - 1)}{2 \cdot 3 \cdot 3n}} = O(n^{-4})$$

for n large enough. A collision may prevent at most two stations from being successful. Thus the number of stations that are not successful is at most $2X$. So the number of successful stations in S_b is at least $c_b n - 2X \geq c_b - \frac{c_b^2 n^2}{3n} = n(c_b - \frac{c_b^2}{3}) \geq c_b' n$ with high probability, where c_b' is a constant very close to $\frac{5}{12}$.

Now let us assume that the set of successful stations in S_b (for $b = 0, 1$) consists of $d_b n$ elements, where $d_b \geq \frac{1}{3}$. Fix $d_0 n$ numbers associated to successful stations from S_0 (each subset of $\{1, \ldots, 3n\}$ of size $d_0 n$ has equal probability). Now imagine that we assign numbers to successful stations from S_1 sequentially. The probability that the $(j+1)$st station is "paired" with a successful station from S_0 is not smaller than $\frac{d_0 n-j}{3n-j} \geq \frac{d_0 n-j}{3n}$. The expected number of "successful pairs" is not smaller than $\sum_{j=1}^{d} \frac{d_0 n-j}{3n} \geq \frac{1}{3n} \cdot \frac{dn(dn-1)}{2} \geq d'n$, where $d = \min(d_0, d_1)$ and d' is a constant. Finally, by Chernoff Bound [13], the number of "successful pairs" is bigger than $d'n/2$ with high probability, assuming that the sizes of S_0 and S_1 are close each other and the number of successful stations in them are bigger than $n/3$. But this assumptions are satisfied with high probability, too. □

Renumeration. Now, we recall the following result from [6]: For no-collision detection RN there is a deterministic leader election algorithm with $\log^* n$ energy cost and time $O(n)$ that works for an arbitrary set of stations with distinct identifiers in the range $1..n$ such that the number of active stations is at least $c \cdot n$ for any fixed constant $c > 0$. In fact this algorithm works in the weak model. Moreover, it consists of $O(\log^* n)$ phases and the following properties are satisfied in the phase i:

– At least $c_i x_i$ active stations called masters participate in the phase.
– A set of at least $\frac{c}{2} \cdot s_i$ active stations called slaves is assigned to each master. The slaves of the master are numbered by consecutive numbers $1, 2, \ldots$
– Only one master remains after the last phase.

We have $x_1 = n, s_1 = 2/c, c_1 = c$ and $c_{i+1} = c_i/2, x_{i+1} = x_i/2^{s_i}, s_{i+1} = c_{i+1} \cdot s_i \cdot 2^{s_i}$ for $i > 1$. So, the number of masters and their slaves at phase i is not smaller than $\frac{c}{2} \cdot s_i c_i x_i$. Observe that $c_i = \frac{c}{2^{i-1}}$, $\frac{c}{2} \cdot c_1 x_1 s_1 = cn$, and $c_{i+1} x_{i+1} s_{i+1} = \frac{c_i}{2} \cdot \frac{x_i}{2^{s_i}} \cdot \frac{c_i}{2} s_i 2^{s_i} = \frac{c}{2^{i+1}} c_i x_i s_i$. Thus $c_{i+1} x_{i+1} s_{i+1} = (c_1 x_1 s_1) \cdot \frac{c}{2^3} \cdot \frac{c}{2^4} \cdot \ldots \cdot \frac{c}{2^{i+1}} = cn \cdot \frac{c^{i-1}}{2^{(i+1)i/2}} \geq (c')^{i^2} n$ for a constant c'. So, for $i = \Theta(\log^* n)$, there are at least $(c')^{\Theta((\log^* n)^2)} n = \Omega\left(\frac{n}{\log\log n}\right)$ stations that participate in phase i, for every n large enough. It means that the algorithm initializes $\Omega\left(\frac{n}{\log\log n}\right)$ stations, i.e. labels them uniquely with $1, 2, \ldots$.

By Lemma 1, at least a linear fraction of stations are assigned in Phase 1 unique ID's from the set $1, \ldots, 3n$. In Phase 2 we run the algorithm just described and we get a set of stations I labeled uniquely by numbers $1, 2, \ldots, |I|$ such that $|I| = \Omega(\frac{n}{\log\log n})$.

Phase 3. Paper [12] presents an algorithm for the strong RN that initializes n stations in time $O(n)$ with energy cost $O(\log\log n)$ and probability at least $1 - 1/n$. It can be easily applied to the case when the approximation of the number of stations up to a constant multiplicative factor is given. Let I be the set of stations that obtained consecutive ID's during the renumeration, $|I| = \Omega(\frac{n}{\log\log n})$ with high probability. In Phase 3 we simulate the initialization algorithm from [12] using partial initialization trick (the elements of I with ID's $1, \ldots, \min(|I|, n/2)$ play a role of the *first group* and all other elements belong to the *second group*).

Recall that the initialization algorithm [12] works in time $O(n)$ and has energy cost $O(\log\log n)$. So one can easily check that our algorithm for the weak model works in time $O(n)$ and energy $O(\log\log n)$, too.

7 General Simulation

An algorithm for the strong RN can be simulated on a weak RN as follows: first we run initialization procedure as described in Section 6. Then we run the algorithm based on the partial initialization trick from Section 5: the stations $1, \ldots, n/2$ are responsible for simulating the stations of the strong RN, station i simulating stations $2i - 1$ and $2i$, the stations $n/2 + 1, \ldots, n$ are used as the first group for the trick described in Section 5. Simulation of a step j of the strong RN looks as follows: a station simulating stations $2i - 1$ and $2i$ of the strong RN sends the bit 0 when both these stations send, does not send anything if neither $2i - 1$ nor $2i$ send and send the bit 1 followed by the message sent by the station $2i$ or $2i-1$ otherwise. Station $n/2+j$ listens at this step. If it receives the bit 1 followed by a message then it sends the message received at the next step, and these stations among $1, \ldots, n/2$ listen, which represent the awake stations of the strong RN.

References

1. Y. Azar, A.Z. Broder, A.R. Karlin, U. Upfal, *Balanced Allocations*, SIAM Journal on Computing : 180-200 (1999).
2. J.L. Bordim, J. Cui, T. Hayashi, K. Nakano, S. Olariu, *Energy efficient initialization protocols for ad-hoc radio networks*, ISAAC'99, LNCS 1741, Springer-Verlag, 215–224.
3. W.C. Fifer, F.J. Bruno, *Low Cost Packet Radio*, Proc. of the IEEE, 75 (1987), 33–42.
4. L. Gąsieniec, A. Pelc, D. Peleg, *The wakeup problem in synchronous broadcast systems*, ACM PODC'2000, 113–121.
5. T. Jurdziński, M. Kutyłowski, J. Zatopiański, *Energy-Efficient Size Approximation for Radio Networks with no Collision Detection*, Computing and Combinatorics, COCOON'2002.
6. T. Jurdziński, M. Kutyłowski, J. Zatopiański, *Efficient Algorithms for Leader Election in Radio Networks*, ACM PODC'2002.
7. Jurdziński, T., Kutyłowski, M., Zatopiański, J.: Weak Communication in Radio Networks, Tech. Rep. CSR-02-04, Techn. Universität Chemnitz, Fakultät für Informatik, 2002.
8. E. Kushilevitz, Y. Mansour, *Computation in Noisy Radio Networks*, ACM-SIAM SODA '98, 236–243.
9. K. Nakano, S. Olariu, *Randomized $O(\log\log n)$-round leader election protocols in radio networks*, ISAAC'98, LNCS 1533, Springer-Verlag, 209-218.
10. K. Nakano, S. Olariu, *Randomized leader election protocols for ad-hoc networks*, SIROCCO'2000, Carleton Scientific, 253-267.
11. K. Nakano, S. Olariu, *Randomized Leader Election Protocols in Radio Networks with No Collision Detection.* ISAAC'2000, LNCS 1969, Springer-Verlag, 362–373.
12. K. Nakano, S. Olariu, *Energy Efficient Initialization Protocols for Radio Networks with no Collision Detection*, ICPP'2000, IEEE, 263–270.
13. R. Motwani, P. Raghavan, *Randomized Algorithms*, Cambridge University Press, 1995.
14. S. Rajasekaran, J. H. Reif, J. D. P. Rolim, (Eds.), *Handbook on Randomized Computing*, Kluwer Academic Publishers, 2001.
15. D.E. Willard, *Log-logarithmic selection resolution protocols in multiple access channel*, SIAM Journal on Computing 15 (1986) , 468-477.
16. IEEE Standard for Information Technology - LAN/MAN: Wireless LAN Medium Access Control (MAC), *http://grouper.ieee.org/groups/802/11/main.html.*

Coordination of Mobile Intermediaries Acting on Behalf of Mobile Users

Norliza Zaini[1] and Luc Moreau*

Department of Electronics and Computer Science, University of Southampton, Southampton SO17 1BJ, UK
{nmz00r,L.Moreau}@ecs.soton.ac.uk

Abstract. We introduce the notion of a mobile intermediary, called Shadow, which is a mobile agent located in the network infrastructure, interacting with complex applications on behalf of mobile users. Due to intermittent connectivity, multiple Shadows may simultaneously coexist. In this paper, we introduce a protocol capable of coordinating Shadows and we present an abstraction layer, hiding away communication and coordination details, which offers a substrate to build distributed applications across mobile devices and fixed infrastructure.

1 Introduction

The context of this paper is the "ubiquitous computing environment" [7] where embedded devices and artifacts abound in buildings and homes, and have the ability to sense and interact with devices carried by people in their vicinity. Mobile devices' networking capabilities offer opportunities for a new range of services, such as customised access to news updates or exchange of information with other mobile users discovered dynamically. However, communications between mobile devices and the infrastructure have some limitations, in the form of intermittent connectivity and low bandwidth. Furthermore, processing power and memory capacity of compact mobile devices remain relatively small. As a result, such an environment would prevent the *large scale* deployment of *advanced* services to mobile users, as they tend to be communication and computation intensive.

We believe that applications can be offloaded to the fixed infrastructure, and act semi-autonomously on behalf of the user. Such an approach does not rely on permanent connectivity with mobile devices, which can save device's resources and take advantage of the available resources on the wired network [3]. Here, we introduce an *intermediary process* in the fixed infrastructure, whose responsibility is to spawn applications in reaction to user's requests and to store and forward messages between devices and applications, according to the available connectivity. Our vision is that of a *mobile* intermediary, which is a mobile agent [2], acting as a *Shadow* of the mobile user, migrating to the user's vicinity when

* This research is funded in part by QinetiQ and EPSRC Magnitude project (reference GR/N35816).

B. Monien and R. Feldmann (Eds.): Euro-Par 2002, LNCS 2400, pp. 973–977.

prevailing conditions permit it. This benefits from a number of advantages: *(i)* Shadow and mobile device can communicate using specialised protocols, possibly dynamically chosen according to the current location or to a negotiation between parties; *(ii)* newly created applications would run in the user's vicinity, making use of the local infrastructure; *(iii)* local services on a local network could be accessed; *(iv)* Shadows and applications can communicate reliably using transparent routing of messages to mobile agents [4, 5].

When a user moves to a new location, their mobile device will request the Shadow to migrate to a new location. However, this may fail when the local network is not connected with the user's previous location. To support services in the current vicinity, we opted for a solution where new Shadows can be created dynamically. As result, a user may be associated with multiple Shadows that need to be coordinated. The purpose of this paper is to describe the interactions between mobile devices, Shadows, applications and fixed infrastructure. Our specific contributions are: *(i)* An architecture supporting multiple Shadows; *(ii)* A coordination protocol between mobile devices and Shadows; *(iii)* An abstraction layer, encapsulating migration and coordination offering a substrate to program applications directly between mobile devices and fixed infrastructure.

In the next section, we overview the architecture. Then, we present the algorithms to be implemented by all its components. In Section 4, we discuss related work, and describe work in progress on the implementation.

2 Architecture Overview

Our proposed architecture is composed of three major components, namely a mobile device, a Shadow and a Shadow Manager, which we describe with the assumptions we make concerning their communication capabilities. A mobile device has the ability to connect to a network in its vicinity and we assume that it is allocated an address, which can be used by networked entities to communicate with it. Shadow Managers and Shadows are agents that need to run on agent platform which is a runtime environment that is able to perform the tasks of supporting the agents' creation, execution, localization, migration, communication and security control. A Shadow Manager acts as a local daemon in a local network, first contact point of a mobile device with the local network. It is responsible for starting or migrating Shadows on behalf of devices.

A Shadow is a mobile agent, acting as an intermediary between a mobile device and infrastructure applications. Being able to migrate allows it to move "closer" to the mobile device, and to communicate with it using the address allocated by the local network. Shadow functions include: *(i)* to create applications on behalf of the mobile device; *(ii)* to send messages to the applications on behalf of the mobile device; *(iii)* to store and forward messages for the mobile device; *(iv)* to migrate to a location closer to the mobile device, whenever the mobile device changes its location, network connectivity permitting.

Our architecture may be summarised as follows. When connected to a network, a mobile device makes contact with a Shadow Manager, and requests its

Shadows to migrate to the manager's location. In the simplest case, there exists a single Shadow. If successful, the Shadow can start interacting locally with the mobile device after its migration. The Shadow spawns new applications as requested by the device and forwards messages to and from them; in essence, the Shadow acts as a router of messages to the applications. Communications between Shadow and applications are robust to the migration of Shadows, based on a transparent routing algorithm [4, 5]. If migration failed for all Shadows, a new Shadow is spawned locally, and the device keeps a log of all created Shadows. When several Shadows are requested to migrate to a specific destination, the first Shadow to reach the location is assigned to be the "main Shadow"; the others coordinate with it to offload information about applications they were routing messages to.

In the following section, we describe the algorithm of each component. Our goal is to define an abstraction layer, which hides the details of communication and coordination between mobile devices, Shadows and applications. On top of this abstraction layer, we will be able to construct applications involving mobile devices: in the mobile device, a programming API will be provided to communicate transparently with fixed infrastructure applications, while applications will be given the possibility to interact transparently with mobile devices; the abstraction layer takes care of all necessary routing and coordination.

3 The Algorithm

In this section, we describe the algorithm coordinating the interactions between mobile devices, Shadows, Shadow Managers and applications.

Mobile Device. When connected to a network, a mobile device sends a "MigrateRequest" message to a discovered Shadow Manager requesting its Shadows to migrate "closer" to its current location. Then, if it receives a "ShadowInformation" message, it sets the sender as the main Shadow by sending an "MSAssignment" message. To each subsequent "ShadowInformation" message received, the sender is notified about the current main Shadow by using an "MSInformation" message. The application layer on a mobile device may request an application to be created or a message to be sent to a particular application on the fixed infrastructure. This request would be forwarded to the main Shadow.

A mobile device may receive "TerminationMessage" from the main Shadow which informs about a Shadow that has recently terminated. On every attempt to send message to a Shadow, a failure handler is provided which adds any message failed to be sent to a queue of outgoing messages. In parallel to other activities, messages from the queue of outgoing messages are sent to the respective receivers. On failure the messages are added back to the queue.

Shadow Manager. When a Shadow Manager is started, it advertises its presence through e.g. Jini or LDAP, and then waits for messages. A Shadow Manager may receive a request from a mobile device to migrate Shadows; the Shadow

Manager then sends a "MigrateRequest" message to all Shadows requesting them to migrate to the platform on which it is operating. If no Shadow was able to migrate, it starts a new Shadow to which an "MDInformation" message which contains information on the requesting mobile device is sent.

Shadow. In a Shadow, there is a hook for intelligent decision making about migration, in which the output of this decision making process is obtained by the "callback" canMigrate(), which returns true if the application layer decides to migrate. A Shadow may create application on behalf of mobile device. Each application has an identifier and an address. The application identifier to application address mappings are placed in a list (LAM). Messages failed to be sent to the mobile device are added to the list of outgoing messages (LOM), while messages failed to be sent to applications are added to a list of incoming messages (LIM). In parallel to other activities, messages from these two lists are attempted to be sent to the respective receivers.

On creation, a Shadow waits for an "MDInformation" message, which contains information about a mobile device. Then the Shadow sends a "ShadowInformation" message to the mobile device. If it receives an "MSAssignment" message, it is assigned to be the main Shadow and responsible for sending "LocationInformation" messages to all other active Shadows of the device. The message indicates current location of the mobile device. If a Shadow receives an "MSInformation" or a "LocationInformation", another Shadow is acting as the main Shadow and it has to handover its function to the main Shadow by transferring its LAM, LOM and LIM. Then the Shadow informs all applications it is interacting with, that the main Shadow is the new intermediary to communicate with the mobile device. When this is done, the Shadow is ready for termination; before terminating itself, it sends a "TerminationMessage" to the main Shadow.

A Shadow may receive a "MigrateRequest" from a Shadow Manager. If canMigrate() returns true, it migrates to the platform on which the Shadow Manager is running. On arrival at the new platform, it sends a "ShadowInformation" message to the mobile device. Otherwise, the Shadow stays on the same platform and may receive a "LocationInformation" from the main Shadow. As usual, on receiving this message, a Shadow has to hand over its function to the main Shadow before terminating itself.

A main Shadow is expected to receive LAM, LIM and LOM from other Shadows. When received, these lists are extended to the Shadow's local lists. The main Shadow may also receive a "TerminationMessage" which requires it to relay this message to the mobile device indicating the termination of a Shadow. As for messages coming from the mobile device, a main Shadow may receive requests to create application or send message to an application on the fixed infrastructure. An application is started according to the type and identifier included in the "CreateApplication" request, while its identifier to address mapping is added to LAM. A "SendMessage" request includes identifier of an application to which a message should be forwarded. Before sending the message, the application address is extracted from LAM. If the Shadow failed to create application or send a message, a failure notification is returned to the mobile device.

4 Discussion and Related Work

This paper has focused on the coordination of multiple Shadows, and on the communication between devices and Shadows. We have not given details on how applications can communicate with Shadows: this issue has been studied extensively in previous papers, for instance by using a message routing algorithm for mobile agents [4]. In our approach, we wish to promote the flexibility of the system, by allowing multiple Shadows to be created, according to the prevailing network conditions, and by allowing Shadows to make intelligent decisions as whether to migrate. The handover of function of a Shadow to the main Shadow shortcuts chains of forwarding pointers when the Shadow does not migrate.

There are other projects applying mobile agent technology to support applications for mobile users. In TACOMA on PDAs [1] and MobiAgent [6], there exists a stationary proxy communicating with mobile devices and mobile agents. This offers less flexibility than ours as it requires a connectivity to exist to the stationary proxy regardless of the distance.

As far as implementation of this system is concerned, we are currently prototyping the coordination algorithm. The transparent routing of messages to mobile agents is already available in the Southampton Framework for Agent Research (SoFAR). In this coordination algorithm, we have not yet considered robustness to failures: in complement to [5], we would like to introduce some redundancy to become tolerant to failures of intermediary nodes. We are planning to develop two applications on this abstraction layer: an application to share documents with mobile users in virtual meeting rooms, and a virtual briefing room, where documents from multiple (mobile) sources are collated and presented to the user.

References

1. Kjetil Jacobsen and Dag Johansen. Mobile software on mobile hardware – experiences with tacoma on pdas. Technical Report 97-32, Department of Computer Science,University of Troms , Norway, 1997.
2. Danny B. Lange and Mitsuru Ishima. *Programming and Deploying Java Mobile Agents with Aglets.* Addison-Wesley, 1998.
3. Patrik Mihailescu and Walter Binder. A Mobile Agent Framework for M-Commerce. *Computer Science 2001*, GI/OCG annual Convention:2:959–967. .
4. Luc Moreau. Distributed Directory Service and Message Router for Mobile Agents. *Science of Computer Programming*, 39(2–3):249–272, 2001.
5. Luc Moreau. A Fault-Tolerant Directory Service for Mobile Agents based on Forwarding Pointers. In *The 17th ACM Symposium on Applied Computing (SAC'2002) — Track on Agents, Interactions, Mobility and Systems*, Madrid, March 2002.
6. Mahmoud Q.H. MobiAgent – An Agent-based Approach to Wireless Information Systems. In *Proceedings of the 3rd International Bi-Conference Workshop on Agent-Oriented Information Systems (AOIS-2001)*, Montreal, 2001.
7. Mark Weiser. Some Computer Science Problems in Ubiquitous Computing. *Communications of the ACM*, 36(7):74–84, July 1993.

An Efficient Time-Based Checkpointing Protocol for Mobile Computing Systems over Wide Area Networks

Chi-Yi Lin, Szu-Chi Wang, and Sy-Yen Kuo

Department of Electrical Engineering
National Taiwan University
Taipei, Taiwan
sykuo@cc.ee.ntu.edu.tw

Abstract. In this paper, an efficient time-based coordinated checkpointing protocol for mobile computing systems is proposed. The main difference from traditional time-based protocols is that our protocol tries to reduce the number of checkpoints per checkpointing process, at the expense of only a small number of coordinating messages. Additionally, the protocol improves the mechanism of timer synchronization by taking advantage of the reliable timers in the mobile support stations, so that it is well adapted to mobile computing systems over wide area networks.

1 Introduction

Checkpointing and rollback-recovery techniques for parallel and distributed systems can also be used in mobile computing systems [1-7]. A common goal of checkpointing protocols in a mobile environment is to avoid extra coordinating messages and unnecessary checkpoints. Prakash and Singhal [2] proposed a nonblocking protocol that requires only a minimum number of processes to take checkpoints. However, Cao and Singhal [4] proved that such a min-process nonblocking checkpointing algorithm does not exist. Later, they introduced the concept of *mutable checkpoints* [6] in their nonblocking algorithm, which forces only a minimum number of processes to take their checkpoints on the stable storage.

Time-based protocols [3, 8] use *time* to indirectly coordinate the creation of checkpoints, so that explicit coordinating messages can be avoided. However, time-based protocols require *every* process to save its checkpoint on the stable storage. Moreover, since timers are never perfectly synchronized, the consistency between the checkpoints can be a problem. In [8] they solved the problem by using a blocking protocol. In [3], however, processes are nonblocking because the inconsistency can be resolved by the information piggybacked in each message. Timer synchronization can also be done using the piggybacked information, but the scheme is less useful when the transmission delay between two mobile hosts becomes relatively large.

In this paper, we first improve the mechanism of timer synchronization in [3]. By taking advantage of the reliable timers in the mobile support stations, our mechanism adapts well to wide area networks. Based on the improved scheme, we then propose a time-based protocol that no unnecessary checkpoints are taken.

B. Monien and R. Feldmann (Eds.): Euro-Par 2002, LNCS 2400, pp. 978–982.

2 System Model and Background

A mobile computing application is executed by a set of *N* processes running on several mobile hosts (*MH*s). Processes communicate with each other by sending messages. These messages are received and then forwarded to the destination host by mobile support stations (*MSS*s), which are interconnected by a fixed network. In the system, every process takes a checkpoint periodically. Each checkpoint is associated with a monotonically increasing *checkpoint number*. The time interval between the kth and $k+1^{th}$ checkpoints is called the kth *checkpoint interval*.

Every *MH* and *MSS* contains a system clock, with typical clock drift rate ρ in the order of 10^{-5} or 10^{-6}. The system clocks of *MSS*s can be synchronized using Internet time synchronization services such as *Network Time Protocol*, which makes the maximum deviation of all the clocks within tens of milliseconds. However, in a wide area network environment, *MSS*s may belong to different organizations. So, we may use the clock synchronization protocol to sync the *logical* clocks of *MSS*s. The clocks of *MH*s can be synchronized likewise, but we use synchronized *timers* instead. The advantages of using timers to coordinate the creation of checkpoints are that the checkpointing protocol does not have to rely on synchronized system clocks of the participating hosts, and no explicit synchronization is needed.

Before a mobile computing application starts, a predefined *checkpoint period T* is set on the timers. When the local timer expires, the process saves its system state as a checkpoint. If all the timers expire at exactly the same time, the set of *N* checkpoints taken at the same instant forms a globally consistent checkpoint. Since timers are not perfectly synchronized, the checkpoints may not be consistent because of *orphan* messages. An orphan message *m* represents an inconsistent system state with the event *receive*(*m*) included in the system state while the event *send*(*m*) not in the state. Orphan messages may lead to *domino effect*, which causes unbounded, cascading rollback propagation. As a result, we need to ensure that no orphan message exists in a global checkpoint, so that the recovery can be free from domino effect.

3 Improved Timer Synchronization

In this section we show the mechanism of improved timer synchronization based on [3] by Neves and Fuchs. The mechanism in [3] uses piggybacked timer information from the *sender* to adjust the timer at the receiver. To achieve more accurate and reliable timer synchronization, we can utilize the timers in *MSS*s as an absolute reference because timers in the fixed hosts are more reliable than those in *MH*s. In our design, the local *MSS* of the receiver is responsible for piggybacking its own "*time to next checkpoint*" (represented as *timeToCkp*) in every message destined to the receiver, because the *MSS* is the *closest* fixed host to the receiver.

In the system every *MH* and *MSS* maintains a checkpoint number. The checkpoint number is incremented whenever the local timer expires. In the following we use cn_S, cn_D, and cn_{MSS} to represent the checkpoint number of the sender, the receiver, and the local *MSS* of the receiver, respectively. On the route from the sender to the receiver, the sender first piggybacks its own checkpoint number cn_S in each message, and then the local *MSS* of the receiver piggybacks its *timeToCkp* (represented as *m.timeToCkp*) and cn_{MSS} in the message. The relationship between cn_D, cn_{MSS}, and cn_S determines how the timer is adjusted, as described in the following cases.

I. Checkpoint numbers of the sender and the receiver are the same ($cn_S = cn_D$)
   (1) $cn_{MSS} = cn_S = cn_D$: The receiver resets its *timeToCkp* to *m.timeToCkp*.
   (2) $cn_{MSS} > cn_S = cn_D$ (Fig. 1(a)): The timer of MH_D is *late* compared to that of MSS_2. So as soon as message *m* is processed, MH_D takes a checkpoint with checkpoint number cn_{MSS}, and then resets its *timeToCkp* to *m.timeToCkp*.
   (3) $cn_{MSS} < cn_S = cn_D$: The timers of MH_S and MH_D are both *early* compared to that of MSS_2, so MH_D resets its *timeToCkp* to "*T + m.timeToCkp*".

II. Checkpoint number of the sender is smaller than that of the receiver ($cn_S < cn_D$)
   (1) $cn_S < cn_{MSS} = cn_D$: Since MH_D and its local *MSS* are within the same checkpoint period, MH_D just resets its *timeToCkp* to *m.timeToCkp*.
   (2) $cn_S = cn_{MSS} < cn_D$: $cn_{MSS} < cn_D$ means that the timer of MH_D expires too early, so MH_D resets its *timeToCkp* to "*T + m.timeToCkp*".

III. Checkpoint number of the sender is larger than that of the receiver ($cn_S > cn_D$)
   (1) $cn_S > cn_{MSS} = cn_D$ (Fig. 1(b)): Before MH_D can process *m*, it has to take a checkpoint with checkpoint number cn_S in order not to make *m* an orphan message. Then MH_D resets its *timeToCkp* to "*T + m.timeToCkp*".
   (2) $cn_S = cn_{MSS} > cn_D$: As in the previous case, MH_D has to take a checkpoint before processing *m*. Since the timer of MH_D is late compared to that of MSS_2 ($cn_{MSS} > cn_D$), MH_D then resets its *timeToCkp* to *m.timeToCkp*.

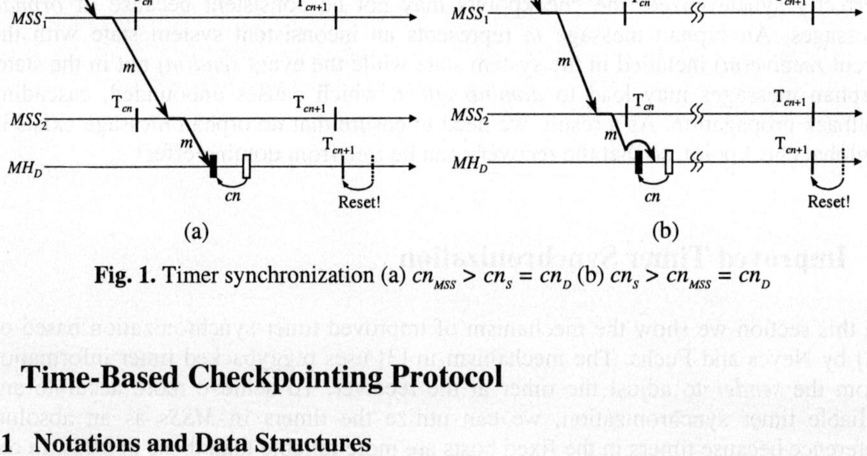

Fig. 1. Timer synchronization (a) $cn_{MSS} > cn_S = cn_D$ (b) $cn_S > cn_{MSS} = cn_D$

4 Time-Based Checkpointing Protocol

4.1 Notations and Data Structures

- *SoftCkpt*^{cn} / *HardCkpt*^{cn} / *PermCkpt*^{cn}: The cn^{th} *soft/hard/permanent* checkpoint of a process, which is saved in the main memory of an *MH*/local disk of an *MH*/stable storage of a local *MSS*.
- $Recv_i$: An array of *N* bits of process P_i. In the beginning of every checkpoint interval, $Recv_i[j]$ is initialized to 0 for $j = 1$ to *N*, except $Recv_i[i]$ always equals 1. When P_i receives a message from P_j, $Recv_i[j]$ is set to 1.

- *Sent*$_i$: A Boolean variable of process P_i. In the beginning of every checkpoint interval, $Sent_i = 0$. When P_i has sent a message in the current checkpoint interval, $Sent_i$ is set to 1.
- *Last*$_i$: A record (*Recv*, *Sent*) maintained by process P_i, in which *Recv*/*Sent* is the $Recv_i/Sent_i$ of P_i of the preceding checkpoint interval.
- *Cell*$_k$: The wireless cell served by MSS_k.

4.2 Checkpointing Protocol

- **Checkpoint Initiation.** A process P_i takes a *soft* checkpoint (say $SoftCkpt^{cn}$) when the local timer expires. When the soft checkpoint is taken successfully, P_i sends its $Recv_i$ to the local *MSS*. Then P_i saves its ($Recv_i$, $Sent_i$) in $Last_i$, resets $Recv_i$ and $Sent_i$, and resumes its computation. Here we assume one of the processes will play the role of the checkpoint initiator during each checkpointing process. Before a checkpointing process begins, the initiator P_j sends a checkpoint request *only* to its local *MSS* (denoted by MSS_{init}). Note that $Sent_j$ must be 1 after P_j sends out the checkpoint request. During the checkpointing process, MSS_{init} is then responsible for collecting and calculating the dependency relationship between the processes.
- **Determining the Dependency Relationship.** After the local timer of MSS_{init} expires, MSS_{init} broadcasts a *Recv_request* message to all *MSS*s. Upon receiving *Recv_request*, each *MSS* forwards the dependency vectors *Recv*s for every process in its cell to MSS_{init}. Then MSS_{init} is able to construct an $N \times N$ dependency matrix D with one row per process. Here we adopt the algorithm in [4] that by matrix multiplications, all the processes on which the initiator transitively depends can be calculated. After finishing the calculation, MSS_{init} acquires a dependency vector D_{init}, in which $D_{init}[i] = 1$ represents that the initiator transitively depends on P_i.
- **Discarding Unnecessary Soft Checkpoints.** $SoftCkpt^{cn}$ can be discarded for those processes that have no dependency with the initiator during the cn-1th checkpoint interval. To do that, MSS_{init} obtains a set $S\_Discard^{cn}$ from D_{init}, which consists of any process P_i such that $D_{init}[i] = 0$, and sends a notification $DISCARD^{cn}$ to the processes in $S\_Discard^{cn}$. When a process P_j receives $DISCARD^{cn}$, it deletes its $SoftCkpt^{cn}$, renumber $HardCkpt^{cn\text{-}1}$ as $HardCkpt^{cn}$, and then sets $Recv_j = (Last_j.Recv \vee Recv_j)$, $Sent_j = (Last_j.Sent \vee Sent_j)$. For those processes that do not receive $DISCARD^{cn}$ during the cn^{th} checkpoint interval, they save $SoftCkpt^{cn}$ as $HardCkpt^{cn}$ before $SoftCkpt^{cn+1}$ can be taken. During the cn+1th checkpoint interval, a process sends its $HardCkpt^{cn}$ to the local *MSS*, which is then saved as $PermCkpt^{cn}$.
- **Handling Disconnections and Handoffs.** When a mobile host MH_i within its cn^{th} checkpoint interval is about to disconnect with the local *MSS* (say MSS_p), the processes on MH_i are required to take $SoftCkpt^{cn+1}$, and then along with vectors (*Recv*, *Sent*) are sent to MSS_p before MH_i is granted to disconnect. When the timer of MSS_p expires, MSS_p provides MSS_{init} with the saved *Recv* vectors of the disconnected processes to calculate the dependency relationship. When MH_i switches from $Cell_p$ to $Cell_q$, MSS_p sends the permanent checkpoints of the processes in MH_i to MSS_q. For a successful handoff, only those checkpoints need to be transferred to the new *MSS*.

4.3 Performance Analysis

- **Blocking Time**: Our protocol is nonblocking so that the blocking time is 0.
- **Number of New Hard Checkpoints**: Since only those processes that are depended by the initiator generate new hard checkpoints, the number of new hard checkpoints is minimum (represented as N_{min}). That is, only N_{min} out of N checkpoints need to be sent to the fixed hosts.
- **Number of Coordinating Messages**: Here we only consider coordinating messages transmitted in the wireless network because the bottleneck of communication is at the wireless part. The number of coordinating messages is N *Recv* vectors and N-N_{min} *DISCARD* notifications, so it is $2N$-N_{min} messages totally.

5 Conclusions

In this paper we have proposed a time-based checkpointing protocol for mobile computing systems over wide area networks. Our protocol alleviates the need for every process to take a hard checkpoint during a checkpointing process, at the expense of only a small number of coordinating messages. These coordinating messages not only help in reducing the number of hard checkpoints, but also ensure regular timer synchronization for every mobile host. The accuracy of timer synchronization in *MH*s is improved by making the reliable timers in *MSS*s as the absolute reference. In this way, as long as the accuracy of the timers in *MSS*s is ensured, the timers in *MH*s can be well synchronized even when *MH*s are spread across a wide area network. The analysis shows that the overhead of our protocol is relatively small.

References

1. A. Acharya and B. R. Badrinath, "Checkpointing Distributed Applications on Mobile Computers," *Proc. of Int'l Conf. on Parallel and Distributed Information Systems*, pp. 73-80, Sep 1994.
2. R. Prakash and M. Singhal, "Low-Cost Checkpointing and Failure Recovery in Mobile Computing Systems," *IEEE Trans. on Parallel and Distributed Systems*, Vol. 7(10), pp. 1035-1048, Oct 1996.
3. N. Neves and W. K. Fuchs, "Adaptive Recovery for Mobile Environments," *Comm. of the ACM*, pp. 68-74, Jan 1997.
4. G. Cao and M. Singhal, "On the Impossibility of Min-Process Non-Blocking Checkpointing and An Efficient Checkpointing Algorithm for Mobile Computing Systems," *Proc. of the 27th Int'l Conf. on Parallel Processing*, pp. 37-44, Aug 1998.
5. Y. Morita, H. Higaki, "Hybrid Checkpoint Protocol for Supporting Mobile-to-Mobile Communication," *Proc. of the IEEE Int'l Conf. on Information Networking*, pp. 529-536, Jan 2001.
6. G. Cao and M. Singhal, "Mutable Checkpoints: A New Checkpointing Approach for Mobile Computing," *IEEE Trans. on Parallel and Distributed Systems*, Vol. 12(2), pp. 157-172, Feb 2001.
7. T. Park, N. Woo, and H. Y. Yeom, "An Efficient Recovery Scheme for Mobile Computing Environments," *IEEE Int'l Conf. on Parallel and Distributed Systems*, pp. 53-60, June 2001.
8. N. Neves and W. K. Fuchs, "Coordinated Checkpointing Without Direct Coordination," *Proc. of the IEEE Int'l Computer Performance & Dependability Symp.*, pp. 23-31, Sep 1998.

Discriminative Collision Resolution Algorithm for Wireless MAC Protocol

Sung-Ho Hwang and Ki-Jun Han

Department of Computer Engineering, Kyungpook National University, Daegu, Korea
sungho@comeng.ce.knu.ac.kr, kjhan@knu.ac.kr

Abstract. This paper proposes a discriminative collision resolution algorithm for the wireless medium access control protocols to support the quality of service requirements of real-time applications. Our algorithm deals with access requests in different ways depending on their delay requirements. In our algorithm, a collision resolution period is used to quickly resolve collisions for the delay sensitive traffic in order to support their delay requirements. Performance analysis shows that our algorithm may successfully meet the delay requirement of real time applications by reducing access delays and collisions.

1 Introduction

Most wireless MAC protocols are based on a demand-assignment scheme, which employs a frame structure consisting with a contention and a reservation period. PRMA/DA and MASCARA are examples of the wireless MAC protocols [1,2]. These protocols employ the Slotted ALOHA (S-ALOHA) to resolve the contention of multiple access requests. With S-ALOHA, initial and retry access requests use the same contention window without considering their priorities. Therefore, all access requests have the same opportunity regardless of whether the requests are delay sensitive or not. As a result, the access delay requirement cannot be guaranteed, a result undesirable for delay sensitive multimedia traffic [3,4,5]. As traffic load increases, this problem becomes more serious and the system throughput drastically drops.

The aim of this paper is to propose an algorithm for wireless MAC protocols to support the QoS requirements of real-time applications. Proposed algorithm deals with access requests in different ways depending on their delay requirements. Thereby, it could successfully support the delay requirement of real time applications by reducing access delays and collisions

2 Discriminative Collision Resolution Algorithm (DCRA)

Proposed algorithm, DCRA, is based on traffic discrimination depending on its delay QoS requirement when it accesses the contention period and resolves the collisions. To do this, traffic types are classified into delay-sensitive and delay-insensitive traffic

B. Monien and R. Feldmann (Eds.): Euro-Par 2002, LNCS 2400, pp. 983–987.

on its delay requirement. Fig. 1 shows the frame structure used for our new algorithm, which is divided by two sub-periods; the Contention Period (CP) and the Reservation Period (RP). Furthermore, CP is consisted with three sub-periods; the Collision Resolution Period (CRP), the Urgent Period (UP), and the Normal Period (NP).

First, each Mobile Terminal (MT) sends its request to the access point (AP) through UP when it has delay-sensitive traffic in its queue; otherwise, the MT sends a request through NP. These request messages will be contended with other request messages from other MTs in UP or NP. The AP collects the information on the requests made in CP and determines which terminals are successful as well as how many time slots are collided on the *i*-th frame.

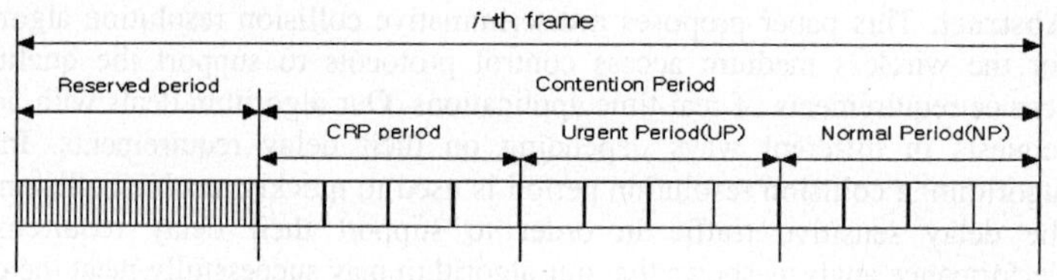

Fig. 1. Frame structure for DCRA

This collected information is then broadcast to MTs through the downlink. If the request was successful at a MT, its traffic can be sent through the reservation period, which is allocated by the AP. On the other hand, if MT knows that its request was collided, it has to do retry. The backoff time that the collided MT has to wait for the next retransmission is computed in different ways with ordinary algorithms depending on where the collision has occurred in UP or NP. If collision was happened in NP, it executes the standard S-ALOHA algorithm. On the other hand, if collision was in UP on the *i*-th frame, AP allocates CRP for the (*i*+1)-th frame. Similarly, if the collision is occurred in the allocated CRP on the *i*-th frame, AP allocates another CRP for the (*i*+1)-th frame. It should be noted that the CRP is used only for resolving collisions that occur in UP or CRP. Note that AP does not allocate a CRP when there was no collision in NP. The length for CRP allocated for the (*i*+1)-th frame, denoted by T_{CRP}, is determined by (1a) when the number of collided time slots in UP and CRP on the *i*-th frame is given by N_{COL}. The collided MT determines the slot position within CRP to access for retransmission by picking up a random value using (1b).

$$T_{CRP} = MIN[2^{N_{COL}}, MAX_{CRP}] \tag{1a}$$

$$T_{bo} = (T_{CRP} * rand()) * T_s \tag{1b}$$

where MAX_{CRP} is the maximum allowing size of CRP or predefined backoff window (T_{window}) to access the contention period.

Therefore, DCRA supports a faster collision resolution for delay-sensitive traffic to win the next contention for retransmission than ordinary algorithm. Fig. 2 summarizes the DCRA algorithm described to this point.

```
DCRA (N_COL, T_s )
{
    T_CRP = 2^N_COL    ;
    while(Resolve the collision != TRUE) {
        if ((IsTrafficDelaySensitive == TRUE) && ( T_CRP ≤ T_window ))
            T_bo = (T_CRP * rand())*T_s   ;
        else
            T_bo = (T_window * rand())*T_s ;
    }
}
```

Fig. 2. The DCRA

3 Performance Evaluation

It is assumed that both new arrivals and retransmissions due to collisions form an on-off process with mean arrival rates of $\lambda_{ds} = G_{ds} / T(packets/\sec)$ and $\lambda_{di} = G_{di} / T(packets/\sec)$ for delay-sensitive and delay-insensitive traffic, respectively. Denote T be the packet transmission time and G_{ds} and G_{di} be the offered loads of delay-sensitive traffic and delay-insensitive traffic, respectively, in T interval.

The probability that packet transmission is successful, denoted by P_0, is given by $P_0 = e^{-G}$, since success means that there was no collision in the interval. And, the collision probability is given by $(1-P_0)$. Throughputs of delay-sensitive traffic and delay-insensitive traffic, denoted by S_{ds} and S_{di}, respectively, can be expressed by

$$S_{ds} = G_{ds} P_0 \tag{2a}$$

$$S_{di} = G_{di} P_0 \tag{2b}$$

since these represent the number of successful transmissions in interval T. Now, we should consider the total transfer delay experienced by a packet in the network including the slot synchronization time. It means the waiting time after arrival until the beginning of the time slot, the delay due to retransmissions, and the packet transmission time. Since the arrival process for new packets is assumed to be Poisson, all arrival times during a slot are equally likely. Thus, the slot synchronization time is given by $T/2$. When the average number of retransmissions is given by H, the average transfer delay for each retransmission cycle is determined to be

$$T_{di} = \left(1 + r + \frac{(2^H + 1)}{2}\right) T \tag{3a}$$

for delay-insensitive traffic and

$$T_{ds} = \left(1 + r + \frac{2^{(1-P_0) N_{ds}}}{2}\right) T \tag{3b}$$

for delay-sensitive traffic, where N_{ds} represents the number of delay-sensitive requests on UP and r represents the waiting time for the allocation information broadcast through the downlink from BS.

After simple calculus, the value of H is determined by

$$H = \sum_{i=1}^{\infty} i(1-P_0)(1-P_0)^{i-1}P_0 = \frac{(1-P_0)}{P_0} \tag{4}$$

Now, the average transfer delay can be obtained by

$$T_{di} = T + (T/2) + \left(\frac{1-P_0}{P_0}\right)\left(1 + r + \frac{(2^H+1)}{2}\right)T \tag{5a}$$

for delay-insensitive traffic and

$$T_{ds} = T + (T/2) + \left(\frac{1-P_0}{P_0}\right)\left(1 + r + \frac{2^{(1-P_0)N_{ds}}}{2}\right)T \tag{5b}$$

for delay-sensitive traffic.

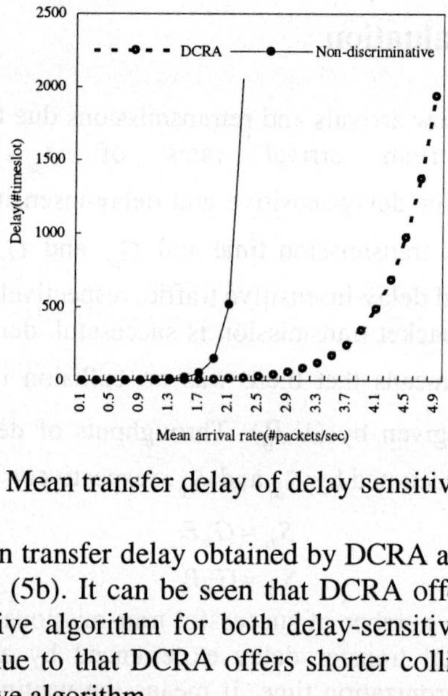

Fig. 3. Mean transfer delay of delay sensitive traffic

Fig. 3 depicts the mean transfer delay obtained by DCRA and a non-discriminative algorithm using (5a) and (5b). It can be seen that DCRA offers a lower mean delay than the non-discriminative algorithm for both delay-sensitive and delay-insensitive traffics. This is mainly due to that DCRA offers shorter collision resolution periods than the non-discriminative algorithm.

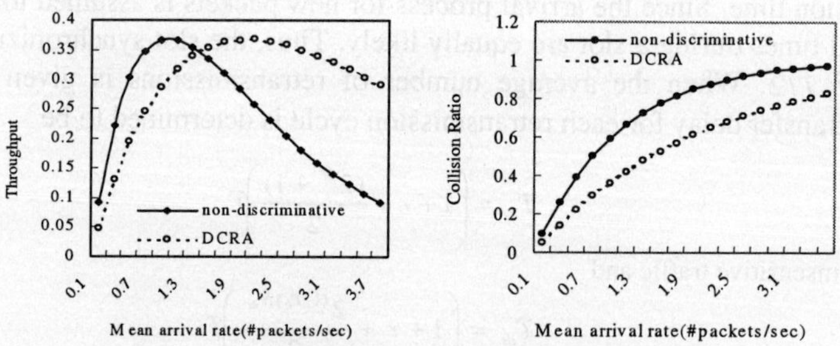

Fig. 4. Throughput and collision probability

Fig. 4 shows that DCRA provides a lower collision probability and a higher throughput than the non-discriminative algorithm. It is the reason why that DCRA deals with packets discriminatively depending on their delay sensitivity requirements. As a result, there will be no collision by different types of data in our DCRA.

4 Conclusions

This paper proposed a discriminative collision resolution algorithm, named DCRA, which considers traffic delay sensitivity for wireless MAC protocols to support the QoS requirements of real-time applications. Proposed algorithm deals with access requests in different ways depending on their delay requirements. Performance analysis and simulation results showed that DCRA offers better performance in terms of transfer delay, collision probability, and throughput than a non-discriminative algorithm. DCRA could be applicable to wireless networks, including cellular based networks and wireless LAN operating with a centralized MAC such as the IEEE802.16 HyperLAN.

References

1. J Sanchez, R Martinez and M. W. Marcellin, "A survey of MAC protocols Proposed for Wireless ATM", *IEEE Network* , Vol 11, issue 6, pp. 52-62, 1997.
2. O. Kubbar and H.T. Mouftah, "Multiple Access Control Protocols for Wireless ATM: Problem Definition and Design Objectives." *IEEE Communications Magazine*, Vol. 35, Issue 11, pp. 93-99, Nov. 1997.
3. O. Kubbar and H.T. Mouftah, "An Aloha-Based Channel Access Scheme Investigation for Broadband Wireless Networks," *Proc. IEEE International Symposium on Computers and Communications*, pp. 203 –208, 1999.
4. M. Natkaniec and A. R. Pach, "An Analysis of the Backoff Mechanism used in IEEE 802.11 Networks," *2000. Proc. ISCC 2000.* 5_{th} *IEEE Symposium of Computers and Communications*, pp. 444 –449, 2000.
5. N. Passas et al., "Quality-of-Service-Oriented Medium Access Control for Wireless ATM Networks," *IEEE Communications Magazine*, pp 42-50, Nov. 1997.
6. D. Raychaudhuri et al., "WATMnet: A Prototype Wireless ATM System for Multimedia Personal Communication," *IEEE Journal on Selected Areas in Communications*, Vol. 15, No. 1, pp.83-95, Jan. 1997.
7. A. Acampora, "Wireless ATM: A Perspective on Issues and Prospects," *IEEE Personal Communications*, pp 8-17, Aug. 1996.

Author Index

Lecture Notes in Computer Science

For information about Vols. 1–2346
please contact your bookseller or Springer-Verlag

Vol. 2347: P. De Bra, P. Brusilovsky, R. Conejo (Eds.), Adaptive Hypermedia and Adaptive Web-Based Systems. Proceedings, 2002. XV, 615 pages. 2002.

Vol. 2348: A. Banks Pidduck, J. Mylopoulos, C.C. Woo, M. Tamer Ozsu (Eds.), Advanced Information Systems Engineering. Proceedings, 2002. XIV, 799 pages. 2002.

Vol. 2349: J. Kontio, R. Conradi (Eds.), Software Quality – ECSQ 2002. Proceedings, 2002. XIV, 363 pages. 2002.

Vol. 2350: A. Heyden, G. Sparr, M. Nielsen, P. Johansen (Eds.), Computer Vision – ECCV 2002. Proceedings, Part I. XXVIII, 817 pages. 2002.

Vol. 2351: A. Heyden, G. Sparr, M. Nielsen, P. Johansen (Eds.), Computer Vision – ECCV 2002. Proceedings, Part II. XXVIII, 903 pages. 2002.

Vol. 2352: A. Heyden, G. Sparr, M. Nielsen, P. Johansen (Eds.), Computer Vision – ECCV 2002. Proceedings, Part III. XXVIII, 919 pages. 2002.

Vol. 2353: A. Heyden, G. Sparr, M. Nielsen, P. Johansen (Eds.), Computer Vision – ECCV 2002. Proceedings, Part IV. XXVIII, 841 pages. 2002.

Vol. 2355: M. Matsui (Ed.), Fast Software Encryption. Proceedings, 2001. VIII, 169 pages. 2001.

Vol. 2356: R. Kohavi, B.M. Masand, M. Spiliopoulou, J. Srivastava (Eds.), WEBKDD 2002 – Mining Log Data Across All Customers Touch Points. Proceedings, 2001. XI, 167 pages. 2002. (Subseries LNAI).

Vol. 2358: T. Hendtlass, M. Ali (Eds.), Developments in Applied Artificial Intelligence. Proceedings, 2002 XIII, 833 pages. 2002. (Subseries LNAI).

Vol. 2359: M. Tistarelli, J. Bigun, A.K. Jain (Eds.), Biometric Authentication. Proceedings, 2002. X, 197 pages. 2002.

Vol. 2360: J. Esparza, C. Lakos (Eds.), Application and Theory of Petri Nets 2002. Proceedings, 2002. X, 445 pages. 2002.

Vol. 2361: J. Blieberger, A. Strohmeier (Eds.), Reliable Software Technologies – Ada-Europe 2002. Proceedings, 2002 XIII, 367 pages. 2002.

Vol. 2362: M. Tanabe, P. van den Besselaar, T. Ishida (Eds.), Digital Cities II. Proceedings, 2001. XI, 399 pages. 2002.

Vol. 2363: S.A. Cerri, G. Gouardères, F. Paraguaçu (Eds.), Intelligent Tutoring Systems. Proceedings, 2002. XXVIII, 1016 pages. 2002.

Vol. 2364: F. Roli, J. Kittler (Eds.), Multiple Classifier Systems. Proceedings, 2002. XI, 337 pages. 2002.

Vol. 2365: J. Daemen, V. Rijmen (Eds.), Fast Software Encryption. Proceedings, 2002. XI, 277 pages. 2002.

Vol. 2366: M.-S. Hacid, Z.W. Raś, D.A. Zighed, Y. Kodratoff (Eds.), Foundations of Intelligent Systems. Proceedings, 2002. XII, 614 pages. 2002. (Subseries LNAI).

Vol. 2367: J. Fagerholm, J. Haataja, J. Järvinen, M. Lyly. P. Råback, V. Savolainen (Eds.), Applied Parallel Computing. Proceedings, 2002. XIV, 612 pages. 2002.

Vol. 2368: M. Penttonen, E. Meineche Schmidt (Eds.), Algorithm Theory – SWAT 2002. Proceedings, 2002. XIV, 450 pages. 2002.

Vol. 2369: C. Fieker, D.R. Kohel (Eds.), Algorithmic Number Theory. Proceedings, 2002. IX, 517 pages. 2002.

Vol. 2370: J. Bishop (Ed.), Component Deployment. Proceedings, 2002. XII, 269 pages. 2002.

Vol. 2371: S. Koenig, R.C. Holte (Eds.), Abstraction, Reformulation, and Approximation. Proceedings, 2002. XI, 349 pages. 2002. (Subseries LNAI).

Vol. 2372: A. Pettorossi (Ed.), Logic Based Program Synthesis and Transformation. Proceedings, 2001. VIII, 267 pages. 2002.

Vol. 2373: A. Apostolico, M. Takeda (Eds.), Combinatorial Pattern Matching. Proceedings, 2002. VIII, 289 pages. 2002.

Vol. 2374: B. Magnusson (Ed.), ECOOP 2002 – Object-Oriented Programming. XI, 637 pages. 2002.

Vol. 2375: J. Kivinen, R.H. Sloan (Eds.), Computational Learning Theory. Proceedings, 2002. XI, 397 pages. 2002. (Subseries LNAI).

Vol. 2377: A. Birk, S. Coradeschi, T. Satoshi (Eds.), RoboCup 2001: Robot Soccer World Cup V. XIX, 763 pages. 2002. (Subseries LNAI).

Vol. 2378: S. Tison (Ed.), Rewriting Techniques and Applications. Proceedings, 2002. XI, 387 pages. 2002.

Vol. 2379: G.J. Chastek (Ed.), Software Product Lines. Proceedings, 2002. X, 399 pages. 2002.

Vol. 2380: P. Widmayer, F. Triguero, R. Morales, M. Hennessy, S. Eidenbenz, R. Conejo (Eds.), Automata, Languages and Programming. Proceedings, 2002. XXI, 1069 pages. 2002.

Vol. 2381: U. Egly, C.G. Fermüller (Eds.), Automated Reasoning with Analytic Tableaux and Related Methods. Proceedings, 2002. X, 341 pages. 2002 .(Subseries LNAI).

Vol. 2382: A. Halevy, A. Gal (Eds.), Next Generation Information Technologies and Systems. Proceedings, 2002. VIII, 169 pages. 2002.

Vol. 2383: M.S. Lew, N. Sebe, J.P. Eakins (Eds.), Image and Video Retrieval. Proceedings, 2002. XII, 388 pages. 2002.

Vol. 2384: L. Batten, J. Seberry (Eds.), Information Security and Privacy. Proceedings, 2002. XII, 514 pages. 2002.

Vol. 2385: J. Calmet, B. Benhamou, O. Caprotti, L. Henocque, V. Sorge (Eds.), Artificial Intelligence, Automated Reasoning, and Symbolic Computation. Proceedings, 2002. XI, 343 pages. 2002. (Subseries LNAI).

Vol. 2386: E.A. Boiten, B. Möller (Eds.), Mathematics of Program Construction. Proceedings, 2002. X, 263 pages. 2002.

Vol. 2387: O.H. Ibarra, L. Zhang (Eds.), Computing and Combinatorics. Proceedings, 2002. XIII, 606 pages. 2002.

Vol. 2388: S.-W. Lee, A. Verri (Eds.), Pattern Recognition with Support Vector Machines. Proceedings, 2002. XI, 420 pages. 2002.

Vol. 2389: E. Ranchhod, N.J. Mamede (Eds.), Advances in Natural Language Processing. Proceedings, 2002. XII, 275 pages. 2002. (Subseries LNAI).

Vol. 2390: D. Blostein, Y.-B. Kwon (Eds.), Graphics Recognition. Proceedings, 2001. XI, 367 pages. 2002.

Vol. 2391: L.-H. Eriksson, P.A. Lindsay (Eds.), FME 2002: Formal Methods – Getting IT Right. Proceedings, 2002. XI, 625 pages. 2002.

Vol. 2392: A. Voronkov (Ed.), Automated Deduction – CADE-18. Proceedings, 2002. XII, 534 pages. 2002. (Subseries LNAI).

Vol. 2393: U. Priss, D. Corbett, G. Angelova (Eds.), Conceptual Structures: Integration and Interfaces. Proceedings, 2002. XI, 397 pages. 2002. (Subseries LNAI).

Vol. 2395: G. Barthe, P. Dybjer, L. Pinto, J. Saraiva (Eds.), Applied Semantics. IX, 537 pages. 2002.

Vol. 2396: T. Caelli, A. Amin, R.P.W. Duin, M. Kamel, D. de Ridder (Eds.), Structural, Syntactic, and Statistical Pattern Recognition. Proceedings, 2002. XVI, 863 pages. 2002.

Vol. 2398: K. Miesenberger, J. Klaus, W. Zagler (Eds.), Computers Helping People with Special Needs. Proceedings, 2002. XXII, 794 pages. 2002.

Vol. 2399: H. Hermanns, R. Segala (Eds.), Process Algebra and Probabilistic Methods. Proceedings, 2002. X, 215 pages. 2002.

Vol. 2400: B. Monien, R. Feldmann (Eds.), Euro-Par 2002 – Parallel Processing. Proceedings, 2002. XXIX, 993 pages. 2002.

Vol. 2401: P.J. Stuckey (Ed.), Logic Programming. Proceedings, 2002. XI, 486 pages. 2002.

Vol. 2402: W. Chang (Ed.), Advanced Internet Services and Applications. Proceedings, 2002. XI, 307 pages. 2002.

Vol. 2403: Mark d'Inverno, M. Luck, M. Fisher, C. Preist (Eds.), Foundations and Applications of Multi-Agent Systems. Proceedings, 1996-2000. X, 261 pages. 2002. (Subseries LNAI).

Vol. 2404: E. Brinksma, K.G. Larsen (Eds.), Computer Aided Verification. Proceedings, 2002. XIII, 626 pages. 2002.

Vol. 2405: B. Eaglestone, S. North, A. Poulovassilis (Eds.), Advances in Databases. Proceedings, 2002. XII, 199 pages. 2002.

Vol. 2406: C. Peters, M. Braschler, J. Gonzalo, M. Kluck (Eds.), Evaluation of Cross-Language Information Retrieval Systems. Proceedings, 2001. X, 601 pages. 2002.

Vol. 2407: A.C. Kakas, F. Sadri (Eds.), Computational Logic: Logic Programming and Beyond. Part I. XII, 678 pages. 2002. (Subseries LNAI).

Vol. 2408: A.C. Kakas, F. Sadri (Eds.), Computational Logic: Logic Programming and Beyond. Part II. XII, 628 pages. 2002. (Subseries LNAI).

Vol. 2409: D.M. Mount, C. Stein (Eds.), Algorithm Engineering and Experiments. Proceedings, 2002. VIII, 207 pages. 2002.

Vol. 2410: V.A. Carreño, C.A. Muñoz, S. Tahar (Eds.), Theorem Proving in Higher Order Logics. Proceedings, 2002. X, 349 pages. 2002.

Vol. 2412: H. Yin, N. Allinson, R. Freeman, J. Keane, S. Hubbard (Eds.), Intelligent Data Engineering and Automated Learning – IDEAL 2002. Proceedings, 2002. XV, 597 pages. 2002.

Vol. 2413: K. Kuwabara, J. Lee (Eds.), Intelligent Agents and Multi-Agent Systems. Proceedings, 2002. X, 221 pages. 2002. (Subseries LNAI).

Vol. 2414: F. Mattern, M. Naghshineh (Eds.), Pervasive Computing. Proceedings, 2002. XI, 298 pages. 2002.

Vol. 2415: J.R. Dorronsoro (Ed.), Artificial Neural Networks – ICANN 2002. Proceedings, 2002. XXVIII, 1382 pages. 2002.

Vol. 2417: M. Ishizuka, A. Sattar (Eds.), PRICAI 2002: Trends in Artificial Intelligence. Proceedings, 2002. XX, 623 pages. 2002. (Subseries LNAI).

Vol. 2418: D. Wells, L. Williams (Eds.), Extreme Programming and Agile Methods – XP/Agile Universe 2002. Proceedings, 2002. XII, 292 pages. 2002.

Vol. 2419: X. Meng, J. Su, Y. Wang (Eds.), Advances in Web-Age Information Management. Proceedings, 2002. XV, 446 pages. 2002.

Vol. 2420: K. Diks, W. Rytter (Eds.), Mathematical Foundations of Computer Science 2002. Proceedings, 2002. XII, 652 pages. 2002.

Vol. 2421: L. Brim, P. Jančar, M. Křetínský, A. Kučera (Eds.), CONCUR 2002 – Concurrency Theory. Proceedings, 2002. XII, 611 pages. 2002.

Vol. 2423: D. Lopresti, J. Hu, R. Kashi (Eds.), Document Analysis Systems V. Proceedings, 2002. XIII, 570 pages. 2002.

Vol. 2426: J.-M. Bruel, Z. Bellahsene (Eds.), Advances in Object-Oriented Information Systems.Procedings, 2002. IX, 314 pages. 2002.

Vol. 2430: T. Elomaa, H. Mannila, H. Toivonen (Eds.), Machine Learning: ECML 2002. Proceedings, 2002. XIII, 532 pages. 2002. (Subseries LNAI).

Vol. 2431: T. Elomaa, H. Mannila, H. Toivonen (Eds.), Principles of Data Mining and Knowledge Discovery. Proceedings, 2002. XIV, 514 pages. 2002. (Subseries LNAI).

Vol. 2436: J. Fong, C.T. Cheung, H.V. Leong, Q. Li (Eds.), Advances in Web-Based Learning. Proceedings, 2002. XIII, 434 pages. 2002.

Vol. 2440: J.M. Haake, J.A. Pino (Eds.), Groupware – CRIWG 2002. Proceedings, 2002. XII, 285 pages. 2002.

Vol. 2442: M. Yung (Ed.), Advances in Cryptology – CRYPTO 2002. Proceedings, 2002. XIV, 627 pages. 2002.

Vol. 2443: D. Scott (Ed.), Artificial Intelligence: Methodology, Systems, and Applications. Proceedings, 2002. X, 279 pages. 2002. (Subseries LNAI).

Vol. 2444: A. Buchmann, F. Casati, L. Fiege, M.-C. Hsu, M.-C. Shan (Eds.), Technologies for E-Services. Proceedings, 2002. X, 171 pages. 2002.